McGraw-Hill
Encyclopedia of
Ocean and
Atmospheric
Sciences

McGraw-Hill Encyclopedia of Ocean and Atmospheric Sciences

Sybil P. Parker

Editor in Chief

McGraw-Hill Book Company

New York	Mexico
St. Louis	Montreal
San Francisco	New Delhi
	Panama
Auckland	Paris
Bogotá	São Paulo
Hamburg	Singapore
Johannesburg	Sydney
London	Tokyo
Madrid	Toronto

Library of Congress Cataloging in Publication Data:

McGraw-Hill encyclopedia of ocean and atmospheric sciences.

   Much of the material published previously in the McGraw-Hill
encyclopedia of science and technology, 4th ed., 1977.
   Includes index.
   1. Oceanography—Dictionaries. 2. Atmosphere—Dictionaries.
I. Parker, Sybil P.   II. McGraw-Hill Book Company.   III.
McGraw-Hill encyclopedia of science and technology.   IV. Title:
Encyclopedia of ocean and atmospheric sciences.
GC9.M32   551.4'6'003   79-18644

ISBN 0-07-045267-9

Contents

Preface

The oceans and phenomena of the atmosphere have held a fascination for humankind throughout history. Early ocean explorations were primarily geographical, for the purpose of describing navigable routes, and information gathered about the oceans was only incidental. Even more mysterious than the uncharted oceans were the changing atmospheric conditions which affect the environment and constitute weather.

Comprehensive investigations of the oceans and atmosphere began during the 19th century and gradually evolved into the modern sciences of oceanography and meteorology. While the subjects of these disciplines are unique, the fluid nature of water and air and their dynamic relationships underlie the need for considering them together as a single system. In fact, the interaction between oceans and the atmosphere is so complex that cause and effect are impossible to separate.

Meteorologists and oceanographers are concerned with the chemical and physical properties and phenomena of the atmosphere and the oceans, respectively. By observing characteristics of the atmosphere, such as temperature, precipitation, and winds, weather information is obtained which has application in the operational problems of industry and agriculture. Oceanographic investigations, on the other hand, are many-sided, relating to geological, biological, chemical, physical, and meteorological aspects of the oceans. The information provides a means for increasing the benefits as well as for avoiding the harmful effects of the oceans.

The oceans and the atmosphere are directly and indirectly of great importance. In addition to sustaining life on Earth, the atmosphere controls the environment, constitutes weather, shields harmful solar radiations, and is a source of energy in the form of wind power. Many valuable food and mineral resources are extracted from the oceans. Sea water contains deuterium, a potentially important source of power, and offers the promise of providing potable water.

This Encyclopedia is an interdisciplinary treatment of the ocean and atmospheric sciences. The depth and scope of coverage provide both theoretical and practical information on subjects such as weather forecasting, mining and farming the seas, atmospheric pollution, satellite programs, industrial meteorology, climate modification, and deep-sea diving.

The 236 articles are alphabetically arranged. They were selected and reviewed by the Board of Consulting Editors and were written by a group of international authorities. Some of the articles have been written especially for this Encyclopedia; others were taken from the *McGraw-Hill Encyclopedia of Science and Technology*. There are almost 500 illustrations to supplement the text. Other useful features include cross-references and the analytical index.

SYBIL P. PARKER
Editor in Chief

McGraw-Hill
Encyclopedia of
Ocean and
Atmospheric
Sciences

Aeronautical meteorology

That branch of meteorological study which deals with atmospheric effects on the operation of aircraft—heavier than air and lighter than air—rockets, missiles, and projectiles. Eight major effects are considered in this article. *See* NAVAL METEOROLOGY.

Low visibility at terminals. Fog, snow, and rain prevent landings and takeoffs when horizontal visibility at the surface falls below regulatory minimum values for manned aircraft. The minimum is generally 800 meters (m) in the United States for airplanes equipped to operate under standard instrument conditions. Runway visual range (RVR), a new operational concept of visibility, is rapidly supplanting the use of conventional meteorological visibility. It is the distance at which high-intensity lights may be observed down the instrument runway in the direction of landing or takeoff. The RVR minimum for landing has been reduced to 600 m at certain airports and is expected to be cut down to about 300 m after adequate electronic and visual aids have been installed at major airports.

Since RVR is usually appreciably greater than meteorological visibility, particularly with fog, this has brought aviation closer to the hitherto elusive goal of all-weather flying. It also calls for more exacting terminal forecasting requirements. Some of the methods that have been developed to meet these new requirements include mesoscale and microscale synoptic analyses, electronic computer calculations, and radar observations of fog trends.

Turbulence. Atmospheric turbulence is defined as any deviation from the normal, steady, horizontal flow of air and occurs in two ways: sharp-edged gusts in the horizontal or vertical and large-scale up- and downdrafts which cause aircraft to sink or rise more gradually and for longer periods. Extreme turbulence may render manned aircraft uncontrollable or even cause structural set or failure. The major forms of turbulence are (1) mechanical, produced by surface friction and flow of air over uneven ground surfaces; (2) thermal, produced by rising currents set off by heating of air in contact with the surface; (3) thunderstorm, which may be extreme (Fig. 1); (4) mountain wave (lee wave), an organized regime of disturbed airflow in

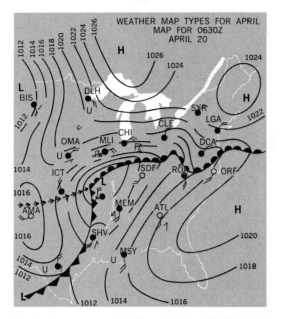

Fig. 1. Severe warm-front thunderstorm situation over the eastern United States. Open or solid circles with three initials represent reporting stations, such as OMA for Omaha and CHI for Chicago. (*United Air Lines*)

the lee of mountains or hills, especially when stable air is forced over them; and (5) wind shear, occuring in the horizontal or vertical along airmass boundary surfaces, temperature inversions (including the tropopause), and the jet stream.

[HENRY T. HARRISON]

Clear-air turbulence (CAT). The cases of turbulence mentioned under (4) and (5) above frequently are encountered in the absence of clouds. The pilot therefore has no forewarning of their occurrence. The unexpected jolts which an aircraft may receive in CAT add to the danger of this phenomenon, especially with respect to passenger safety and comfort. Forecasting of CAT has only been moderately successful in the past. Since clear-air turbulence is a small-scale phenomenon—patches of turbulence frequently measure only a few kilometers in diameter and a few hundred to a few thousand feet in thickness—the present radiosonde network is insufficient in locating or predicting the precise locations of CAT zones. Hope rests with the development of remote sensors that might detect zones of wind shear and temperature inversions ahead of the aircraft, so that evasive actions could be taken or warning signs could be flashed on to alert passengers. *See* CLEAR-AIR TURBULENCE.

The intensity of CAT which is experienced very much depends on the type and speed of the aircraft flying through the turbulent zone. Generally, high-speed aircraft suffer more than low-speed ones. CAT may be expected to exist even at flight levels of future supersonic transport aircraft. It will have to be considered in the design of such aircraft and in its possible effects on engine performance.   [ELMAR R. REITER]

Upper winds and temperature. Since aircraft are part of the air current in which they are embedded, they experience aiding or retarding effects

determined by wind direction in relation to the course being flown. Wind direction and speed vary only moderately from day to day and from winter to summer in certain parts of the world, but fluctuations at middle and high latitudes in the troposphere and lower stratosphere are commonly extreme. Flight schedules therefore must be based on climatological samplings which anticipate that a certain percentage (70–80%) of flights will make their schedule but that the remainder will suffer some degree of wind delay.

Careful planning of long-range flights is completed with the aid of prognostic upper-air charts. These make it possible to select flight tracks and cruising altitudes where more favorable winds and temperature will result in a lower elapsed flight time in spite of added ground distance. This planned procedure, known as minimal time track (least time track, pressure pattern) flying, is capable of effecting considerable savings in operation. The role of the aeronautical meteorologist is to provide accurate forecasts of the wind and temperature field, in space and time, through the operational ranges of each aircraft involved. For civil jet-powered aircraft, the optimum flight plan must always represent a compromise between wind, temperature, and turbulence conditions. The fastest track may be costly in fuel or turbulent for passengers.

Jet stream. A meandering, shifting current of relatively swift windflow is embedded in the general westerly circulation at upper levels. Sometimes girdling the globe at middle and subtropical latitudes where the strongest jets are found, this band of strong winds—generally 300–500 kilometers (km) in width—has great operational significance for aircraft flying at cruising levels of 6–15 km. It is the cause of most serious flight-schedule delays and may result in unscheduled fuel stops.

Average speed in the core of a well-developed jet near the tropopause at middle latitudes in winter is close to 100 mph, but speeds as high as 175 mph are fairly common and extremes have been measured at close to 300 mph. The jet stream chal-

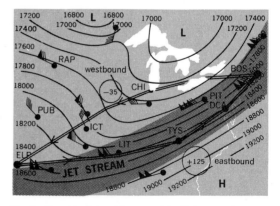

Fig. 2. Use of minimal time track in flight planning over the eastern United States. The solid circles represent weather-reporting places, landing places, or both, such as CHI for Chicago and BOS for Boston. The plus or minus figures shown in the large circles indicate the relation of air-to-ground speed (mph) to be expected in the direction of flight. (*United Air Lines*)

lenges the forecaster and the flight planner to utilize the tailwind to the utmost downwind and to avoid a retarding headwind as much as practicable upwind (Fig. 2). *See* JET STREAM.

Tropopause. This boundary between troposphere and stratosphere is generally defined in terms of points in the vertical air-mass soundings where the temperature lapse rate becomes less than 2°C/km. It is sometimes defined as a discontinuity in the thermal wind shear. Thought by some to occur in multiple layers, or overlapping "tropical" and "polar" leaves, the tropopause, in the current synoptic trend, is analyzed as a single unbroken surface extending from pole to Equator.

The tropopause's significance to aviation arises from its frequent association with mild forms of clear-air turbulence and change of vertical temperature gradient with altitude, which has considerable effect on turbine-engine performance. Transcontinental flights at 10–12 km may penetrate the layer several times during one flight. Since turbine-engine performance varies with each deviation from standard atmosphere conditions (ICAO standard tropopause is −56.5°C at 11 km), each penetration must be evaluated for engine performance and efficiency. The tropopause also marks the vertical limit of clouds and storms; however, thunderstorms are known to punch through the tropopause surface to heights of 23 km at times, and lenticular clouds of strong mountain waves, strongly developed over mountains, have been observed at similar heights. *See* TROPOPAUSE.

Lightning strikes. Such strikes or static discharges, when experienced inside an aircraft, cause a blinding flash and usually a muffled explosive sound audible above the roar of the engines. Structural damage to the craft is commonly limited to small molten spots in the outer skin at the point of entry or exit, fusing of antenna wires, and small

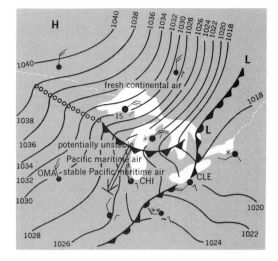

Fig. 4. Heavy icing situation over the Great Lakes area in the United States. The solid circles are used to indicate weather stations, such as CHI for Chicago or CLE for Cleveland. (*United Air Lines*)

punctures in such surfaces as radome, nose, or tail. Atmospheric conditions favorable for strikes follow a consistent pattern: solid clouds or enough clouds for the aircraft to be intermittently on instruments, active precipitation of an icy character, and ambient air temperature close to 0°C. St. Elmo's fire, radio static, and choppy air often precede the strike. Evasive action by the pilot usually consists of one or more of the following: air speed reduction, change of course as indicated by radar echoes, or change of altitude. Pilots unable to avoid typical strike situations turn up cockpit lights and avert their eyes from strike flashes to decrease the risk of being blinded. *See* LIGHTNING.

Icing. Flight through a subcooled cloud, freezing rain, or wet snow commonly results in this condition (Figs. 3 and 4). Once one of the major hazards of flying, icing has been reduced to a minor weather factor by effective antiicing and deicing devices for the protection of critical components of aircraft. Automatic protection is also provided for high-speed jet aircraft by the compressional heating effect upon airfoils. Icy or deeply snow-covered runways offer operational problems for aviation, but these factors are being reduced in importance through modern snow-removal methods at major airports. [HENRY T. HARRISON]

Agricultural meteorology

The study and application of meteorology and climatology (a branch of meteorology) to the specific problems of agriculture. Agriculture is the production of food and fiber in all its forms—crops for human and animal consumption, pasture and range for animal grazing, crops (including trees) as raw materials for manufactured products. Hence, agricultural meteorology deals with farming, ranching, and forestry as well as with the transportation of substances required for production—water for irrigation, fertilizer, and agricultural chemicals—to the producer and transportation of the products to markets.

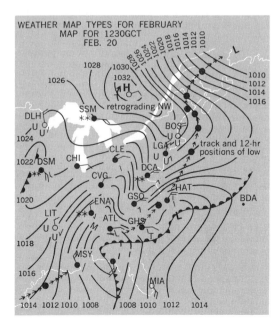

Fig. 3. Heavy snowstorm situation along United States Atlantic coast. Open or solid circles mark weather stations, such as HAT for Hatteras. (*United Air Lines*)

Problems. Agricultural meteorologists deal with a variety of problems in cooperation with many other types of specialists. Examples of these follow.

Predicting weather effects. Ultimate crop yields are influenced by weather factors. Statistical regression models treat yield as a function of such factors as preplanting soil moisture supply, date of plant seeding or emergence, and temperature and rainfall occurrences as the season progresses. Other models of a more deterministic nature predict dry matter accumulation as the difference between photosynthesis and respiration. The latter processes require knowledge of the solar radiation, carbon dioxide concentration, tissue temperature, plant size, physiological condition, nutritional status of the crop, and many other complex factors of the environment.

Weather effects influence the demand for water in crop fields, both irrigated and unirrigated, and in pastures, rangeland, and forests. Predictions of these effects require knowledge of the processes of turbulent transport of heat and water vapor in the lowest layers of the atmosphere. *See* CROP MICROMETEOROLOGY; EVAPOTRANSPIRATION.

Water use efficiency. Cultural practices (such as row spacings, population densities, plant architectures, and land shaping) must be defined and tested to best increase water use efficiency—defined as photosynthetic production of dry matter/water consumption or photosynthesis/evapotranspiration.

Weather-related diseases and pests. To predict the outbreak of weather-related crop and animal diseases or insect pests, the ways in which organism life cycles are affected by weather must be learned. When these relations are understood, the climatologist can determine the probabilities of an outbreak during any part of the growing season. Agricultural meteorologists can prepare special advisories on the basis of the actual weather conditions occurring and the forecast of coming weather. This information guides the grower in preparing control measures for the periods of especial danger and provides for additional alerts when specific pest control actions will be needed. Climatologists, weather forecasters, and agricultural meteorologists work closely with pathologists and entomologists to refine these techniques.

Growing season. It is important to predict the probable length of the growing season for regions in which historical weather records are available. This prediction aids in planning planting and harvest operations and helps in the choice of new crops for introduction to an area. *See* CLIMATOLOGY.

Frost. Agrometeorologists are involved in predicting the timing and severity of frosts and their probable impacts on crops, especially high-value fruits, vegetables, and ornamentals. Climatologists use historical weather records to specify the degree of frost risk for a general area. Micrometeorologists refine these specifications by study of the topography of the specific orchard, the patterns of cold air drainage into low spots, and the likely soil temperature differences due to slope and aspect since the "lay of the land" determines exposure to the Sun's rays during daytime. Forecasters can predict the severity of frost on a given night from knowledge of general weather conditions, particularly sky cover (clouds trap much of the Earth's thermal radiation to space) and windspeed (in calm conditions thermal inversions develop and cold air piles up near the surface). Engineers and agricultural meteorologists work on techniques for frost protection.

Agricultural meteorologists are involved in developing means of frost protection. Coverings are used to protect low-growing crops against frost. Space heating and thermal radiation devices are used for warming orchard trees during a frost night. Sprinkling devices are used to keep the frost-threatened vegetation ice-covered as insulation against temperatures lower than freezing and to liberate heat of fusion to warm the air slightly. Agricultural meteorologists have also developed foams to cover small plants and insulate them against freezing temperatures. Foliar sprays of aluminum powder have been used to lower the rate of thermal emission from citrus trees so that they cool more slowly during a frost.

Wind. The problem lies in predicting the usual or predominant speed and direction of winds for locating houses, barns, or other farm buildings. Odor problems may be minimized by properly locating the farmstead in the lee of natural obstructions to the wind or planted windbreaks. Energy costs for heating and cooling can also be reduced by proper placement of buildings.

The impact of windiness on animal comfort and on the growth of crops is another area of study. Windbreaks of trees (shelterbelts) or of construction materials can be designed to provide effective shelter for animals against extreme chilling winds in winter or hot, desiccating winds in summer. Agricultural meteorologists assist in the design of windbreaks for this purpose.

Crops generally grow better in the lee of windbreaks, especially in the arid, semiarid, and subhumid regions of the world. The more uniform spread of snow in winter results in a favorable water supply when crops are grown in spring. During the growing season a more favorable microclimate—generally warmer and more humid—results in less frequent and less severe moisture stress in the sheltered plants. Agricultural meteorologists conduct studies to determine optimum design of tree windbreaks and of ways to use annual or perennial tall crops (such as corn, elephant grass, wheat grass, and sunflowers) to shelter shorter crops.

Severe weather conditions. Agrometeorologists work with animal husbandry specialists, animal physiologists, and engineers to develop protective techniques against, and prepare advisory alerts on, severe weather conditions which can endanger animals on the open range or in exposed feedlots. Growers need warnings, as far in advance as possible, of blizzards which are likely to cut cattle off from supplies or feed or which may cause animals to wander into low-lying areas where they may become buried by snow. Forecasting periods of very hot weather is also important since animals on feedlots suffer greatly when very high temperatures occur in conjunction with high humidity. Under such conditions animals are unable to dissipate their body heat effectively. Emergency rations can be fed during these periods to reduce body respiration rates, and emergency cooling can be effected with sprinklers. Shade can also be provided to minimize radiation load on the animals.

Turbulence and remote sensing. Some other areas of research and applications in agrometeorology are in turbulence and remote sensing. The Earth's surfaces (terrestrial and marine) exchange radiation and mass with the atmosphere and absorb momentum of the wind. Understanding the mechanisms of these exchanges is the function of micrometeorology. The exchanges of mass (CO_2, water vapor, pollen, dust, pollutants, and so on) is effected through the processes of turbulent transport. Agrometeorologists are necessarily interested in the development and testing of theoretical explanations of the mechanisms of turbulent transport and in its applications, which include predicting the rates at which heat and water vapor will be removed from or delivered to the surface, and the rates at which CO_2 is delivered to the plant while it is photosynthesizing, and is removed at night when only respiration is occurring. Effectiveness of pollination and of the applications of dusts and sprays are also determined by the intensity of turbulence. See MICROMETEOROLOGY.

The space age has been a boon to agricultural meteorologists in providing techniques for remotely sensing the condition of the land and the vegetation growing on it. Multispectral scanners on a number of National Oceanic and Atmospheric Administration (NOAA) and National Aeronautics and Space Administration (NASA) satellites and on aircraft provide data on the extent, density, and vigor of vegetation. The flux density of radiation in the visible-wave-band range can indicate whether tissues are healthy or necrotic. The near-infrared waveband provides information on whether or not leaves are hydrated and turgid. The thermal wavebands provide information on temperature of the emitting surfaces. Plants short of water may be considerably warmer than others not so stressed. The microwave and radar bands can be used to determine wetness of the surface soil. Agricultural meteorologists have been especially active in conducting "ground truth" studies, which aid in the interpretation of remotely sensed information.

Major agencies. The State Agricultural Experiment Station network and the Science and Education Administration of the U.S. Department of Agriculture (USDA) are active in agricultural meteorology research. Applications of the research and the provision of advice to farmers, ranchers, agricultural industries, and the general public are functions of the Cooperative Extension Services in each state. The NOAA provides special forecasts of interest to agriculturalists through the National Weather Service networks and also special weather advisories and other advisory services in limited areas of the country served by Environmental Science Service Centers. The World Food Outlook and Situation Board, operated by USDA, and the Center for Climatic and Environmental Assessment of NOAA attempt to monitor the impacts of weather on food production worldwide. A world weather and crop update is published in the *Weekly Weather and Crop Bulletin*, a joint publica-

tion of NOAA and USDA. In some projects—for example, the recently completed Large Area Crop Inventory Experiment (LACIE) to predict wheat production in countries around the world—several government agencies cooperate. In the LACIE project and continuing efforts to evaluate worldwide production of other crops, NASA, NOAA, and USDA are involved.

The Commission for Agricultural Meteorology (CAgM) of the World Meteorological Organization (WMO) carries out an international exchange of information on all facets of agricultural meteorology. Standardization of instrumentation and procedures and compilation of information on specific problems (such as drought) are carried out by CAgM. The organization also engages in training agrometeorologists for the developing countries and assists in the creation of agrometeorological services which function within ministries of agriculture, development, water resources, and such. A recent program with this objective is AGRHYMET, aimed at providing agrometeorological and operational hydrology services to the countries of Sahelian Africa.

[NORMAN J. ROSENBERG]

Bibliography: G. Campbell, *An Introduction to Environmental Physics*, 1977; R. Geiger, *The Climate near the Ground*, 4th ed., 1965; R. Lee, *Forest Microclimatology*, 1978; J. L. Monteith, *Principles of Environmental Physics*, 1973; J. L. Monteith, *Vegetation and the Atmosphere*, vol. 2; *Case Studies*, 1976; R. E. Munn, *Descriptive Micrometeorology*, 1966; N. J. Rosenberg, *Microclimate: The Biological Environment*, 1974; O. G. Sutton, *Micrometeorology*, 1953.

Air

A predominantly mechanical mixture of a variety of individual gases enveloping the terrestrial globe to form the Earth's atmosphere. In this sense air is one of the three basic components, air, water, and land (atmosphere, hydrosphere, and lithosphere), that interblend to form the life zone at the face of the Earth.

Some aspects of terrestrial air are well known, but others are in various stages of investigation or remain little understood. For outlines of composition, chemical attributes, and structural and physical characteristics of the Earth's atmosphere *see* AIR PRESSURE; AIR TEMPERATURE; ATMOSPHERE; ATMOSPHERIC CHEMISTRY; ATMOSPHERIC POLLUTION; ATMOSPHERIC WATER VAPOR; and for the science of the Earth's atmosphere, especially with regard to weather and climate, *see* METEOROLOGY.

Radio, radar, rockets, satellites, and growing interest in interplanetary space and interplanetary travel are stimulating investigation of the upper atmosphere and the transition zone to outer space. *See* IONOSPHERE.

[CHARLES V. CRITTENDEN]

Air mass

A term applied in meteorology to an extensive body of the atmosphere which approximates horizontal homogeneity in its weather characteristics. An air mass may be followed on the weather map as an entity in its day-to-day movement in the general circulation of the atmosphere. The expres-

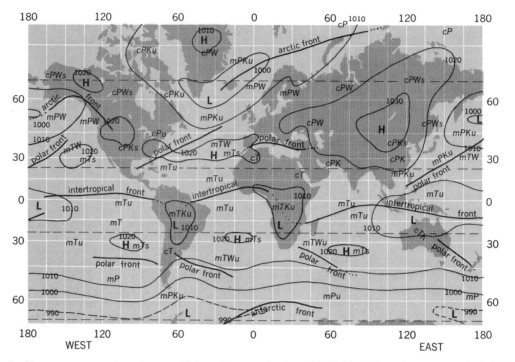

Fig. 1. Air-mass source regions, January. High- and low-atmospheric-pressure centers are designated H and L within average pressure lines numbered in millibars (such as 1010). Major frontal zones are labeled along heavy lines. (*From H. C. Willett and F. Sanders, Descriptive Meteorology, 2d ed., Academic Press, 1959*)

sions air mass analysis and frontal analysis are applied to the analysis of weather maps in terms of the prevailing air masses and of the zones of transition and interaction (fronts) which separate them. The relative horizontal homogeneity of an air mass stands in contrast to the sharp horizontal changes in a frontal zone. The horizontal extent of important air masses is reckoned in millions of square miles. In the vertical dimension an air mass extends at most to the top of the troposphere, and frequently is restricted to the lower half or less of the troposphere. The frontal zones between air masses usually slope in such a manner that the colder air mass underlies the warmer as a wedge. In the vertical direction the properties of an air mass, specifically its content of heat and moisture, may vary between a high degree of stratification and one of homogeneity produced by vertical mixing. *See* FRONT; METEOROLOGY; WEATHER MAP.

Development of concept. Practical application of the concept to the air mass and frontal analysis of daily weather maps for prognostic purposes was a product of World War I. A contribution of the Norwegian school of meteorology headed by V. Bjerknes, this development originated in the substitution of close scrutiny of weather map data from a dense local network of observing stations for the usual far-flung internation network. The advantage of air-mass analysis for practical forecasting became so evident that during the three decades following World War I the technique was applied in more or less modified form by nearly every progressive weather service in the world. However, the rapid increase of observational weather data from higher levels of the atmosphere during and since World War II has resulted in a progressive tendency to drop the careful application of air-mass analysis techniques in favor of those involving the kinematical or dynamic analysis of upper-level air-flow patterns. *See* UPPER SYNOPTIC AIR WAVES.

Origin. The occurrence of air masses as they appear on the daily weather maps depends upon two facts, the existence of air-mass source regions, and the large-scale character of the branches or elements of exchange of the general circulation. Air-mass source regions consist of extensive areas of the Earth's surface which are sufficiently uniform so that the overlying atmosphere acquires similar characteristics throughout the region; that is, it approximates horizontal homogeneity. The designation of an area of the Earth's surface as a source region assumes that the overlying atmosphere in that area normally remains there long enough to approximate thermodynamic equilibrium with respect to the underlying surface, or in other words, to acquire the weather characteristics that typify that particular source region.

The large-scale character of the elements by which the general circulation is accomplished is observed on the daily weather maps in the major atmospheric currents of polar or tropical origin, whose southward or northward progress can be traced from day to day. These major currents, together with the associated polar and tropical air masses, are the means by which surplus tropical heat is effectively transported to polar latitudes. *See* ATMOSPHERIC GENERAL CIRCULATION.

Weather significance. The thermodynamic properties of air mass determine not only the general character of the weather in the extensive area covered by the air mass but also, to some ex-

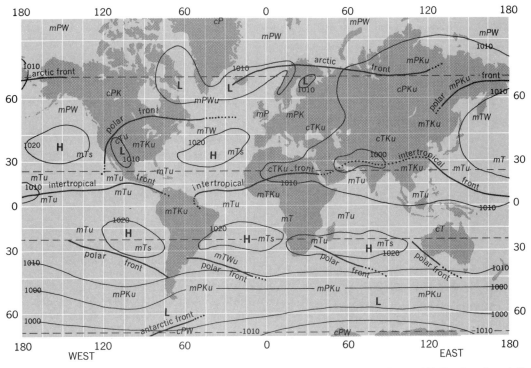

Fig. 2. Air-mass source regions, July. The symbols which are used in this figure are the same as those for

Fig. 1. (*From H. C. Willett and F. Sanders, Descriptive Meteorology, 2d ed., Academic Press, 1959*)

tent, the severity of the weather activity in the frontal zone of interaction between air masses. Those properties which determine the primary weather characteristics of an air mass are defined by the vertical distribution of the two elements, water vapor and heat (temperature). On the vertical distribution of water vapor depend the presence or absence of condensation forms and, if present, the elevation and thickness of fog or cloud layers. On the vertical distribution of temperature depend the relative warmth or coldness of the air mass and, more importantly, the vertical gradient of temperature, known as the lapse rate. The lapse rate determines the stability or instability of the air mass for thermal convection and, consequently, the stratiform or convective cellular structure of the cloud forms and precipitation. The most unstable moist air mass, in which the vertical lapse rate may approach 1°C/100 m, is characterized by severe turbulence and heavy showers or thundershowers. In the most stable air mass there is observed an actual increase (inversion) of temperature with increase of height at low elevations. With this condition there is little turbulence, and if the air is moist there is fog or low stratus cloudiness and possible drizzle, but if the air is dry there will be low dust or industrial smoke haze. *See* TEMPERATURE INVERSION.

Classification. A wide variety of systems of classification and designation of air masses was developed by different weather services around the world. The usefulness of a system with type designators to be applied in the analysis of weather maps is directly proportional to its effectiveness in accurately expressing the thermodynamic properties which determine the weather characteristics

of the air mass. These properties are imparted to the air masses primarily by the particular source region of its origin, and secondarily by the modifying influences to which it is subjected after leaving the source region. Consequently, most systems of air-mass classification are based on a designation of the character of the source region and the subsequent modifying influences to which the air mass is exposed. Probably the most effective and widely applied system of classification is a modification of the original Norwegian system that is based on the following four designations.

Polar versus tropical origin. All primary air-mass source regions lie in polar (P in Figs. 1 and 2) or in tropical (T) latitudes. In middle latitudes there occur the modification and interaction of air masses initially of polar or tropical origin. This difference of origin establishes the air mass as cold or warm in character.

Maritime versus continental origin. To be homogeneous, an air-mass source region must be exclusively maritime or exclusively continental in character. On this difference depends the presence or absence of the moisture necessary for extensive condensation forms. However, a long trajectory over open sea transforms a continental to a maritime air mass, just as a long land trajectory, particularly across major mountain barriers, transforms a maritime to a continental air mass. On Figs. 1 and 2, m and c are used with P and T (mP, cP, mT, and cT) to indicate maritime and continental character, respectively.

Heating versus cooling by ground. This influence determines whether the air mass is vertically unstable or stable in its lower strata. In a moist air mass it makes the difference between convective cumulus clouds with good visibility on the one hand and fog or low stratus clouds on the other. Symbols W (warm) and K (cold) are used on maps—thus, mPK or mPW.

Convergence versus divergence. Horizontal convergence at low levels is associated with lifting and horizontal divergence at low levels with sinking. Which condition prevails is dependent in a complex manner upon the large-scale flow pattern of the air mass. Horizontal convergence produces vertical instability of the air mass in its upper strata (u on maps), and horizontal divergence produces vertical stability (s on maps). On this difference depends the possibility or impossibility of occurrence of heavy air-mass showers or thundershowers or of heavy frontal precipitation. Examples of the designation of these tendencies and the intermediate conditions for maritime polar air masses are $mPWs$, mPW, $mPWu$, mPs, mPu, $mPKs$, mPK, and $mPKu$.

[HURD C. WILLETT]

Bibliography: W. L. Donn, *Meteorology*, 4th ed., 1975; H. C. Willett and F. Sanders, *Descriptive Meteorology*, 2d ed., 1959.

Air pressure

The force per unit area that the air exerts on any surface in contact with it, arising from the collisions of the air molecules with the surface. It is equal and opposite to the pressure of the surface against the air, which for atmospheric air in normal motion approximately balances the weight of the atmosphere above. It is the same in all directions and is the force that balances the weight of the column of mercury in the Torricellian barometer, commonly used for its measurement. *See* BAROMETER.

Units. In 1975 the American Meteorological Society adopted the International System of Units (SI). In this system the basic unit of pressure is the pascal (Pa), which is defined as equal to 1 newton per square meter. Since this is a very small unit, air pressures are normally measured in kilopascals; 1 kilopascal (kPa) equals 1000 pascals. Before the adoption of SI, the units of pressure most frequently used in meteorology were those based on the bar, which is defined as equal to 10^5 pascals or 10^6 dynes/cm²; 1 bar equals 1000 millibars (mb) equals 100 centibars (cb). Such units are still permitted, but tolerance for their use is likely to be transitory.

Also widely used in practice are units based on the height of the mercury barometer under standard conditions, expressed commonly in millimeters or in inches. The standard atmosphere (760 mmHg) is also used as a unit, mainly in engineering where large pressures are encountered. The following equivalents show the conversions between the commonly used units of pressure, where $(\text{mmHg})_n$ and $(\text{in. Hg})_n$ denote the millimeter and inch of mercury, respectively, under standard (normal) conditions, and where $(\text{kg})_n$ and $(\text{lb})_n$ denote the weight of a standard kilogram and pound mass, respectively, under standard gravity.

$$1\,\text{kPa} = 1000\,\text{Pa} = 10^4\,\text{dynes/cm}^2 = 10\,\text{mb}$$
$$= 7.50062\,(\text{mmHg})_n = 0.295300\,(\text{in. Hg})_n$$

$$1\,\text{mb} = 100\,\text{Pa} = 0.1\,\text{kPa} = 1000\,\text{dynes/cm}^2$$
$$= 0.750062\,(\text{mmHg})_n = 0.0295300\,(\text{in. Hg})_n$$
$$1\,\text{atm} = 101.3250\,\text{kPa} = 1013.250\,\text{mb}$$
$$= 760\,(\text{mmHg})_n = 29.9213\,(\text{in. Hg})_n$$
$$= 14.6959\,(\text{lb})_n/\text{in.}^2 = 1.03323\,(\text{kg})_n/\text{cm}^2$$
$$1\,(\text{mmHg})_n = 1\,\text{torr} = 0.1333224\,\text{kPa}$$
$$= 1.333224\,\text{mb} = 0.03937008\,(\text{in. Hg})_n$$
$$1\,(\text{in. Hg})_n = 3.38639\,\text{kPa} = 33.8639\,\text{mb}$$
$$= 25.4\,(\text{mmHg})_n$$

Variation with height. Because of the almost exact balancing of the weight of the overlying atmosphere by the air pressure, the latter must decrease with height, according to the hydrostatic equation, Eq. (1), where P is air pressure, ρ is air

$$dP = -g\rho\,dZ \qquad (1)$$

density, g is acceleration of gravity, Z is altitude above mean sea level, dZ is infinitesimal vertical thickness of horizontal air layers, dP is pressure change which corresponds to altitude change dZ, P_1 is pressure at altitude Z_1, and P_2 is pressure at altitude Z_2. The expressions on the right-hand side of Eq. (2) represent the weight of the column of air between the two levels Z_1 and Z_2.

$$P_1 - P_2 = \int_{Z_1}^{Z_2} \rho g\,dZ \qquad (2)$$

In the special case in which Z_2 refers to a level above the atmosphere where the air pressure is nil, one has $P_2 = 0$, and Eq. (2) yields an expression for air pressure P_1 at a given altitude Z_1 for an atmosphere in hydrostatic equilibrium.

By substituting in Eq. (1) the expression for air density based on the well-known perfect gas law and by integrating, one obtains the hypsometric equation for dry air under the assumption of hydrostatic equilibrium, Eq. (3), valid below about 90

$$\log_e\left(\frac{P_1}{P_2}\right) = \frac{10^{-3}M}{R}\int_{Z_1}^{Z_2}\frac{g}{T}\,dZ \qquad (3a)$$

$$Z_2 - Z_1 = \frac{10^3 R}{M}\int_{P_2}^{P_1}\frac{T}{g}\frac{dP}{P} \qquad (3b)$$

km, where g is the gravitational acceleration in m/s²; M is the gram-molecular weight, 28.97 for dry air; R is the gas constant for 1 mole of ideal gas, or 8.31470 joules/(mole)(K); T is the air temperature in K; and the altitude Z is expressed in meters.

Equation (3) may be used for the real moist atmosphere if the effect of the small amount of water vapor on the density of the air is allowed for by replacing T by T_v, the virtual temperature given by Eq. (4), in which e is partial pressure of water va-

$$T_v = T\left[1 - \left(1 - \frac{M_w}{M}\right)\frac{e}{P}\right]^{-1} \qquad (4)$$

por in the air, M_w is gram-molecular weight of water vapor (18.0160 g/mole), and $(1 - M_w/M) = 0.37803$.

Equation (3a) is used in practice to calculate the vertical distribution of pressure with height above sea level. The temperature distribution in a standard atmosphere, based on mean values in middle latitudes, has been defined by international agreement. The use of the standard atmosphere permits the evaluation of the integrals of

Eqs. (3a) and (3b) to give a definite relation between pressure and height. This relation is used in all altimeters which are basically barometers of the aneroid type. The difference between the height estimated from the pressure and the actual height is often considerable; but since the same standard relationship is used in all altimeters, the difference is the same for all altimeters at the same location, and so causes no difficulty in determining the relative position of aircraft. Mountains, however, have a fixed height, and accidents have been caused by the difference between the actual and standard atmospheres.

Horizontal and time variations. In addition to the large variation with height discussed in the previous paragraph, atmospheric pressure varies in the horizontal and with time. The variations of air pressure at sea level, estimated in the case of observations over land by correcting for the height of the ground surface, are routinely plotted on a map and analyzed, resulting in the familiar "weather map" representation with its isobars showing highs and lows. The movement of the main features of the sea-level pressure distribution, typically from west to east, produces characteristic fluctuations of the pressure at a fixed point, varying by a few percent within a few days. Smaller-scale variations of sea-level pressure, too small to appear on the ordinary weather map, are also present. These are associated with various forms of atmospheric motion, such as small-scale wave motion and turbulence. Relatively large variations are found in and near thunderstorms, the most intense being the low-pressure region in a tornado. The pressure drop within a tornado can be a large fraction of an atmosphere, and is the principal cause of the explosion of buildings over which a tornado passes. *See* ISOBAR; WEATHER MAP.

It is a general rule that in middle latitudes at localities below 1000 m (3280 ft) in height above sea level, the air pressure on the continents tends to be slightly higher in winter than in spring, summer, and autumn; whereas at considerably greater heights on the continents and on the ocean surface, the reverse is true.

Various maps of climatic averages indicate certain regions where systems of high and low pressure predominate. Over the oceans there tend to be areas or bands of relatively high pressure, most marked during the summer, in zones centered near latitude 30°N and 30°S. The Asiatic landmass is dominated by a great high-pressure system in winter and a low-pressure system in summer. Deep low-pressure areas prevail during the winter over the Aleutian, the Icelandic-Greenland, and Antarctic regions. These and other centers of action produce offshoots which may travel for great distances before dissipating.

Thus during the winter, spring, and autumn in middle latitudes over the land areas, it is fairly common to experience the passage of a cycle of low- and high-pressure systems in alternating fashion over a period of about 6–9 days in the average, but sometimes in as little as 3–4 days, covering a pressure amplitude which ranges on the average from roughly 15–25 mb (1.5–2.5 kPa) less than normal in the low-pressure center to roughly 15–20 mb (1.5–2.0 kPa) more than normal in the high-pressure center. During the summer in middle latitudes the period of the pressure changes is generally greater, and the amplitudes are less than in the cooler seasons (see table).

Within the tropics where there are comparatively few passages of major high- and low-pressure systems during a season, the most notable feature revealed by the recording barometer (barograph) is the characteristic diurnal pressure variation. In this daily cycle of pressure at the ground there are, as a rule though with some exceptions, two maxima, at approximately 10 A.M. and 10 P.M., and two minima, at approximately 4 A.M. and 4 P.M., local time.

The total range of the diurnal pressure variation is a function of latitude as indicated by the following approximate averages (latitude N and range in millibars): 0°, 0.3 kPa; 30°, 0.25 kPa; 35°, 0.17 kPa; 45°, 0.12 kPa; 50°, 0.09 kPa; 60°, 0.04 kPa. These results are based on the statistical analysis of thousands of barograph records for many land stations. Local peculiarities appear in the diurnal variation because of the influences of physiographic features and climatic factors. Mountains, valleys, oceans, elevations, ground cover, temperature variation, and season exert local influences; while current atmospheric conditions also affect it, such as amount of cloudiness, precipitation, and sunshine. Mountainous regions in western United States may have only a single maximum at about 8–10 A.M. and a single minimum at about 5–7 P.M., local time, but with a larger range than elsewhere at the same latitudes, especially during the

Mean atmospheric pressure and temperature in middle latitudes, for specified heights above sea level*

Altitude above sea level			
Standard geopotential meters, m'	m at latitude 45°32'40"	Air pressure, kPa	Assumed temperature, K
0	0	1.01325×10^3	288.15
11,000	11,019	2.2632×10^2	216.65
20,000	20,063	5.4747×10^1	216.65
32,000	32,162	8.6798×10^0	228.65
47,000	47,350	1.1090×10^0	270.65
52,000	52,429	5.8997×10^{-1}	270.65
61,000	61,591	1.8209×10^{-1}	252.65
79,000	79,994	1.0376×10^{-2}	180.65
88,743	90,000	1.6437×10^{-3}	180.65†

*Approximate annual mean values based on radiosonde observations at Northern Hemisphere stations between latitudes 40 and 49°N for heights below 32,000 m and on observations made from rockets and instruments released from rockets. Some density data derived from searchlight observations were considered. Values shown above 32,000 m were calculated largely on the basis of observed distribution of air density with altitude. In correlating columns 1 and 2, G is 9.80665 m²/s² per standard geopotential meter (m'). Data on first three lines are used in calibration of aircraft altimeters.

†Above 90,000 m there occurs an increase of temperature with altitude and a variation of composition of the air with height, resulting in a gradual decrease in molecular weight of air with altitude.

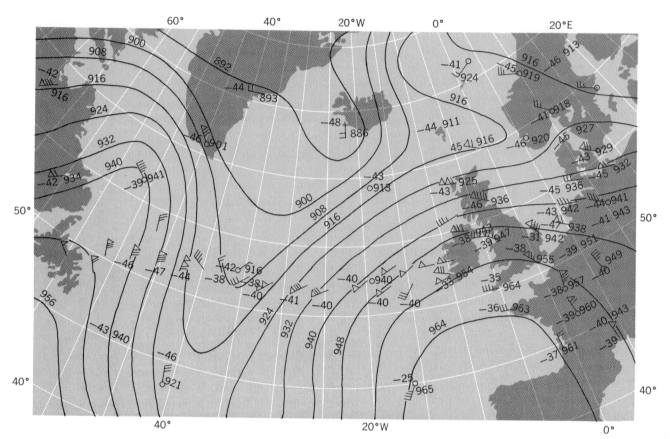

Contours of 300-mb (30-kPa) surface, in tens of meters, with temperature in °C, and measured winds at the same level, on June 16, 1960. Winds are plotted with arrow pointing in direction of the wind, with each bar of the tail representing 10 m/s. Triangle represents 50 m/s.

warmer months (for instance, about 4 mb difference between the daily maximum and minimum).

At higher levels in the atmosphere the variations of pressure are closely related to the variations of temperature, according to Eq. (3a). Because of the lower temperatures in higher latitudes in the lower 10 km, the pressures at higher levels tend to decrease toward the poles. The figure shows a typical pattern at approximately 10 km above sea level. As is customary in representing pressure patterns at upper levels, the variation of the height of a surface of constant pressure, in this case 300 mb (30 kPa), is shown, rather than the variation of pressure over a horizontal surface.

Besides the latitudinal variation, the figure also shows the characteristic wave pattern in the pressure field, and the midlatitude maximum in the wind field known as the jet stream, with its "waves in the westerlies." In the stratosphere the temperature variations are such as to reduce the pressure variations at higher levels, up to about 80 km, except that in winter at high latitudes there are relatively large variations above 10 km. At altitudes above 80 km the relative variability of the pressure increases again. Although the pressure and density at these very high levels are small, they are important for rocket and satellite flights, so that their variability at high altitudes is likewise important.

Relations to wind and weather. The practical importance of air pressure lies in its relation to the wind and weather. It is because of these relationships that pressure is a basic parameter in weather forecasting, as is evident from its appearance on the ordinary weather map.

Horizontal variations of pressure imply a pressure force on the air, just as the vertical pressure variation implies a vertical force that supports the weight of the air, according to Eq. (1). This force, if unopposed, accelerates the air, causing the wind to blow from high to low pressure. The sea breeze is an example of such a wind. However, if the pressure variations are on a large scale and are changing relatively slowly with time, the rotation of the Earth gives rise to geostrophic or gradient balance such that the wind blows along the isobars. This situation occurs when the pressure variations are due to the slow-moving lows and highs that appear on the ordinary weather map, and to the upper air waves shown in the figure, in which the relationship is well illustrated. *See* GEOSTROPHIC WIND.

The wind near the ground, in the lowest few hundred meters of the atmosphere, is retarded by friction with the surface to a degree that depends on the smoothness or roughness of the surface. This upsets the balance mentioned in the previous paragraph, so that the wind blows somewhat across the isobars from high to low pressure.

The large-scale variations of pressure at sea level shown on a weather map are associated with characteristic patterns of vertical motion of the air, which in turn affect the weather. Descent of air in a high heats the air and dries it by adiabatic compression, giving clear skies, while the ascent of air in a low cools it and causes it to condense and produce cloudy and rainy weather. These processes at low levels, accompanied by others at higher levels, usually combine to justify the clear-cloudy-rainy marking on the household barometer.

[RAYMOND J. DELAND]

Bibliography: H. R. Byers, *General Meteorology*, 1974; R. G. Fleagle and J. A. Businger, *An Introduction to Atmospheric Physics*, 1963; A. Miller, *Meteorology*, 1976; O. G. Sutton, *Challenge of the Atmosphere*, 1961; J. Van Mieghem, *Atmospheric Energetics*, 1973.

Air temperature

The temperature of the atmosphere represents the average kinetic energy of the molecular motion in a small region, defined in terms of a standard or calibrated thermometer in thermal equilibrium with the air.

Measurement. Many different types of thermometer are used for the measurement of air temperature, the most common depending on the expansion of mercury with temperature, the variation of electrical resistance with temperature, or the thermoelectric effect (thermocouple). The electrical methods are especially useful for the automatic recording of temperature. The basic problems of ensuring that the temperature of the thermometer be as close as possible to that of the air are the same for all methods of measurement. For the atmosphere, probably the most serious difficulty is the heating or cooling of the thermometer by radiation to and from other bodies at different temperatures, the most obvious being the Sun. The representativeness of temperature measurements, meaning the degree to which they provide information about the temperature of the air over a region much larger than the thermometer, is also an important practical requirement. The well-known standard meteorological measurement of the air temperature in a louvered shelter, about 2 m above a natural ground surface, with a mercury-in-glass thermometer that averages the temperature over a period of about 1 min because of its thermal inertia, is designed to satisfy the above requirements. *See* METEOROLOGICAL INSTRUMENTATION.

Causes of variation. The temperature of a given small mass of air varies with time because of heat added or subtracted from it, and also because of work done during changes of volume, according to Eqs. (1) and (2). Here h represents heat added, P

$$\frac{dT}{dt} = \frac{1}{C_v}\left(\frac{dh}{dt} - P\frac{d\alpha}{dt}\right) \tag{1}$$

$$\frac{dT}{dt} = \frac{1}{C_p}\left(\frac{dh}{dt} + \frac{1}{\rho}\frac{dP}{dt}\right) \tag{2}$$

the pressure, ρ the density and α its reciprocal, and C_v and C_p the specific heat at constant volume and constant pressure, respectively. The heat added or subtracted may be due to many different physical processes, of which the most important are absorption and emission of radiation, heat

conduction, and changes of phase of water involving latent heat of condensation and freezing. In the upper atmosphere, above about 20 km, photochemical changes are also important; for example, those that occur when ultraviolet radiation dissociates oxygen molecules to atomic oxygen, which then recombines with molecular oxygen to form ozone. Because of the variation of air pressure with height, rising and sinking of air causes expansion and contraction and thus temperature changes due to the work of expansion, Eqs. (1) and (2).

A spectacular example of temperature rise due to sinking of air from higher levels is the chinook, a warm wind that sometimes blows down the eastern slope of the Rocky Mountains in winter. The slow temperature changes associated with the changes of season are mainly due to a combination of radiational heat exchange and conduction to and from the ground surface, whose temperature itself changes in response to the varying radiational exchange with the Sun and the atmosphere. On a shorter time scale the diurnal variation of temperature throughout the day is caused by the same processes. *See* AIR PRESSURE; ATMOSPHERIC GENERAL CIRCULATION; TERRESTRIAL ATMOSPHERIC HEAT BALANCE.

The rate at which the temperature changes at a particular point, that is, as measured by a fixed thermometer, depends on the movement of air as well as the physical processes discussed above. The large changes of air temperature from day to day are mainly due to the horizontal movement of air, bringing relatively cold or warm air masses to a particular point, as the large-scale pressure-wind systems move across the weather map. *See* AIR MASS.

Temperature near the surface. Temperatures are read at one or more fixed times daily, and the day's extremes are obtained from special maximum and minimum thermometers, or from the trace (thermogram) of a continuously recording instrument (thermograph). The average of these two extremes, technically the midrange, is considered in the United States to be the day's average temperature. The true daily mean, obtained from a thermogram, is closely approximated by the mean of 24 hourly readings, but may differ from the midrange by 1 or 2°F, on the average. In many countries temperatures are read daily at three or four fixed times, chosen so that their weighted mean closely approximates the true daily mean. These observational differences and variations in exposures complicate comparison of temperatures from different countries and any study of possible climatic changes.

Averages of daily maximum and minimum temperature for a single month for many years give mean daily maximum and minimum temperatures for that month. The average of these values is the mean monthly temperature, while their difference is the mean daily range for that month. Monthly means, averaged through the year, give the mean annual temperature; the mean annual range is the difference between the hottest and coldest mean monthly values. The hottest and coldest temperatures in a month are the monthly extremes; their averages over a period of years give the mean monthly maximum and minimum (used extensively in Canada), while the absolute extremes for the month (or year) are the hottest and coldest temperatures ever observed. The interdiurnal range or variability for a month is the average of the successive differences, regardless of sign, in daily temperatures.

Over the oceans the mean daily, interdiurnal, and annual ranges are slight, because water absorbs the insolation and distributes the heat through a thick layer. In tropical regions the interdiurnal and annual ranges over the land are small also, because the annual variation in insolation is relatively small. The daily range also is small in humid tropical regions, but may be large (up to 40°F) in deserts. Interdiurnal and annual ranges increase generally with latitude, and also with distance from the ocean; the mean annual range defines continentality. The daily range depends on aridity, altitude, and noon Sun elevation.

Extreme temperatures arouse much popular interest and often are cited uncritically, despite their possible instrumental, exposure, and observational errors of many kinds. The often given absolute maximum temperatures of 134°F for the United States in Death Valley, Calif. (July 10, 1913), and 136°F for the world in Azizia, Tripoli (Sept. 13, 1922) are both questionable; in the subsequent years, Death Valley's hottest reading has been only 127°F, and the Azizia reading was reported by an expedition, not a regular weather station. Lowest temperatures in the Northern Hemisphere are −90°F at Verkhoyansk (−67.6°C on Feb. 5 and 7, 1892) and Oimekon (−67.7°C on Feb. 6, 1933), Siberia; −87°F at Northice, Greenland (Jan. 9, 1954); −81°F at Snag, Yukon Territory, Canada (Feb. 3, 1947); −70°F at Rogers Pass, Mont. (the current United States record, on Jan. 20, 1954). The first winter at Vostok, 78°27′S, 106°52′E, encountered a minimum temperature of −125°F on Aug. 25, 1958, and the third winter a minimum of −127°F on Aug. 24, 1960, much lower than the lowest at the United States station at the South Pole.

Vertical variation. The average vertical variation of temperature in the atmosphere is shown in the figure. The atmosphere is seen to consist of layers, each of which has a characteristic variation of temperature with height. The decrease of temperature with height in the lowest layer, the troposphere, is basically due to the presence of a heat source resulting from the solar radiation absorbed at the Earth's surface, giving an excess of heat that is carried away from the surface mainly by convection currents and lost to space by reradiation. Heating a compressible fluid such as air from below results in a decrease in temperature with height because rising masses of air cool as they expand, according to Eq. (2). This is the main reason for the decrease of temperature with height in the troposphere. *See* ATMOSPHERE.

In the mesosphere and higher layers the exchange of heat energy between layers of air by emission and absorption of infrared radiation is the most important factor determining the distribution of temperature with height. *See* INSOLATION; RADIATION.

[RAYMOND J. DELAND]

Bibliography: H. R. Byers, *General Meteorology*, 1974; R. G. Fleagle and J. A. Businger, *An Introduction to Atmospheric Physics*, 1963; A. Miller, *Meteorology*, 1976; O. G. Sutton, *Challenge of the Atmosphere*, 1961.

Air-velocity measurement

The measurement of the rate of displacement of air or gas at a specific location, for example, wind speed, air velocity in the test section of a wind tunnel, airspeed of an aircraft, or air velocity produced by a fan or blower in air conditioning. Velocity indicates the magnitude and direction of the flow rate; the latter is usually measured separately. To obtain the total flow velocity as required in pipes or conduits, integration over the cross sec-

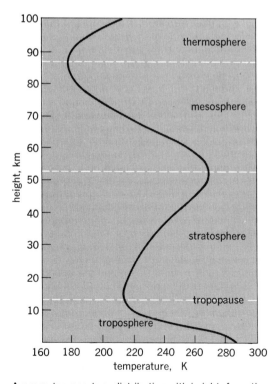

Average temperature distribution with height, from the International Reference Atmosphere. Note that the nomenclature for stratosphere and mesosphere varies, the upper part of the stratosphere being considered part of the mesosphere by some authorities. (*After R. G. Fleagle and J. A. Businger, An Introduction to Atmospheric Physics, Academic Press, 1963*)

Methods of measuring air velocity

Method	Instruments
Pressure on an element in the airstream	Pitot-static tube Venturi tube Bridled pressure plate
Speed or number of revolutions of a rotating element in the airstream	Cup anemometer Vane anemometer
Effect of velocity on a heated element in the airstream	Hot-wire anemometer Kata thermometer

tion is required, or another method of measurement is used.

Three primary methods are used in measuring air velocity. These methods and the instruments which make use of each are listed in the table. For discussion of the three anemometers *see* ANEMOMETER.

Pitot-static tube. This, coupled with temperature-measuring devices, is widely used in wind tunnels and in aircraft to measure airspeed. For all velocity ranges (subsonic or supersonic), measurements of total (pitot) or stagnation pressure and temperature, p_0 and T_0, respectively, and static pressure p are required. For any gas the relation between these and velocity v is shown in Eq. (1), where T_0 is the absolute temperature, M the

$$v = \sqrt{\left(\frac{2k}{k-1}\right)\left(\frac{RT_0}{M}\right)\left[1-\left(\frac{p}{p_0}\right)^{(k-1)/k}\right]} \quad (1)$$

molecular weight, R the universal gas constant (8.314×10^7 ergs/g-mole K for cgs units), and k the ratio of specific heats. For air $k = 1.4$ and $M = 29$ g/g-mole, Eq. (2) holds, where v is in meters/second and T_0 in kelvins.

$$v = 4.480 \, T_0^{1/2} \, [1 - (p/p_0)^{0.286}]^{1/2} \quad (2)$$

The methods of measurement of p_0 and p are different for subsonic and supersonic flow. For subsonic flow a typical configuration is shown in Fig. 1. The airstream enters the open end of the tube and decelerates until its velocity is zero; the pressure measured by a tube connected to this region is p_0. Small holes are drilled in the walled-off section, or static tube, about 10 tube diameters from the open end. The air passes externally over these holes; a tube open to the walled-off section measures p.

For supersonic flow the configuration is basically the same. Provided the holes in the walled-off section are set far enough back (determined by calibration), the static pressure may be determined in the same way. The shock wave around the instrument does not permit direct measurement of p_0, however (Fig. 2); instead the total pressure behind the shock p_{0s} is monitored. A unique but complicated relation exists between p, p_{0s}, and p_0, thus determining the pressure terms in the above relation.

The pressure gage on aircraft is calibrated in units of indicated airspeed and requires an additional correction for air density (or temperature).

Venturi tube. This has limited application in measuring gas velocity in industry. A reference pressure is required, either static or pitot. The tube is open at both ends and reduces in diameter sharply but smoothly from the end that heads into the airstream. The original diameter is recovered

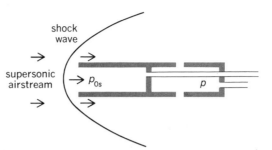

Fig. 2. Supersonic configuration of pitot-static tube.

smoothly at the rear end. The rear section is three or more times longer than the front. Suction is developed at the constriction and is measured by a pressure gage connected to it by tubing. The gage is also connected to a static or pitot tube. The differential pressure measured is a function of the velocity and also of the air density.

Bridled pressure plate. This is useful in measuring air velocities in gusts, since it has a greater speed of response than do rotating elements. The velocity pressure on the plate exposed to the wind is balanced by the force of a spring. The deflection is measured by an inductance-type transducer. The signal from the transducer is amplified and transmitted to a recorder. The indication is dependent upon the air density. Simple mechanical designs are also used to measure low air velocities.

Kata thermometer. This is used to measure low velocities in air-circulation problems. An alcohol thermometer with a large bulb is heated above 100°F, and its time to cool from 100 to 95°F, or some other interval above the ambient temperature, is measured. This time interval is a measure of the air current at the location.

Shielded thermocouple. This instrument is used for obtaining the total temperature inside a pitot tube using a thermocouple rather than a pressure-sensing device. The thermocouple must be shielded to minimize losses due to radiation and conduction. Calibration takes account of probe configurations, wall conductivity, and gas stream conditions. The same instrument may be used for subsonic and supersonic flows.

[HERBERT FOX]

Bibliography: H. W. Liepmann and A. Roshko, *Elements of Gasdynamics*, 1957; E. H. J. Pallett, *Aircraft Instrument Manual*, 1964; L. Prandtl and O. S. Tietjens, *Applied Hydro and Aeromechanics*, 1957.

Airglow

Weak, nonthermal emissions from the atmosphere above 60 km, other than aurora. Subdivisions are dayglow, twilight glow, and nightglow. The energy of the nightglow comes from reactions of atomic oxygen between 70 and 100 km, and from ionic recombination around 300 km. The lower layer radiates the very intense vibration-rotation bands of the OH radical, various bands of O_2, the forbidden green "auroral" line of atomic oxygen, and the yellow D lines of sodium. The upper layer produces the forbidden red doublet of atomic oxygen and a fainter oxygen green line. Many additional emissions are found in the dayglow, as well as large enhancements of the upper nightglow layer

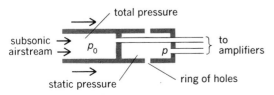

Fig. 1. Subsonic configuration of pitot-static tube.

and the sodium lines. A forbidden line of atomic nitrogen appears in the green. Infrared bands of O_2 are strong in the region 30–120 km. Band systems of N_2, N_2^+, and NO appear in the ultraviolet, and strong resonance lines of atomic oxygen, nitrogen, and hydrogen in the far-ultraviolet. Many of these emissions are the direct result of sunlight, scattered by the appropriate species; others are produced by electrons ejected by solar photons, or by enhanced nightglow processes. The twilight period, besides showing the transition between day and night, allows observation of some dayglow emissions from the ground, with much less interference from the overwhelming foreground light of the day sky. *See* ALKALI EMISSIONS.

[DONALD M. HUNTEN]

Bibliography: J. W. Chamberlain, *Physics of the Aurora and Airglow*, 1961; D. M. Hunten, Spectroscopic studies of the twilight airglow, *Space Sci. Rev.*, 6:493, 1967; B. M. McCormac and A. Omholt, *Atmospheric Emissions*, 1969; J. F. Noxon, Day airglow, *Space Sci. Rev.*, 8:92, 1968.

Albedo

That fraction of the total light incident on a reflecting surface, especially a celestial body, which is reflected back in all directions by the surface. A perfect diffuse reflector has, by definition, unit albedo; however, all actual surfaces have albedoes less than unity. Albedoes observed astronomically may be compared with those of known substances measured in the laboratory, and such a comparison can give valuable clues as to the nature of the celestial body's surface. The Moon, for example, reflects only about 7% of the total sunlight it receives and must therefore have a fairly dark surface.

A perfect diffuse reflector obeys Lambert's cosine law that the luminance (surface brightness) of the reflecting surface is proportional to the cosine of the angle of incidence and is independent of the angle of reflection.

The geometrical albedo is the ratio of the mean luminance of a planet at normal incidence and reflection to the mean luminance of a perfect diffusor of the same size at the same distance from the Sun. If r_0 is the mean distance of the planet from the Sun in astronomical units, s_0 its apparent semidiameter at mean opposition, and i_0/I_0 the ratio of the apparent brightness i_0 of the planet at mean opposition to that of the Sun I_0 at unit distance, the geometrical albedo is $p = i_0 r_0^2/I_0 \sin^2 s_0$. This can also be expressed in the logarithmic stellar magnitude scale, $\log p = 0.4(M_0 - m_1) - 2 \log \sin s_1$, where M_0 is the apparent stellar magnitude of the Sun at unit distance, m_1 that of the planet, and s_1 the planet's apparent semidiameter at unit distance.

The physical or spherical albedo includes the effects of the departures from Lambert's law represented by the phase law giving the relative brightness $\phi(\theta)$ of the planet, reduced to unit distance, as a function of the planet's phase angle θ (see illustration). For a discussion of planetary phases *see* PLANET.

If $\phi(\theta)$ is normalized to $\phi(\theta) = 1$ for $\theta = 0°$, the phase integral is given by the equation below. For

$$q = 2 \int_0^\pi \phi(\theta) \sin \theta \, d\theta$$

an ideally smooth sphere scattering according to Lambert's law, $q = 1.50$. The spherical albedo is then $A = pq$ and the diffuse reflection coefficient $r = 1.5p = 1.5A/q$ (see table).

Determinations of the monochromatic albedo at several wavelengths give the spectral reflectivity curve of a planet, which can be compared with similar curves determined in the laboratory for terrestrial substances.

The term albedo is also used in nuclear reactor terminology. It refers to the reflection factor which a surface, for example, paraffin, has for neutrons. If N neutrons pass through air and strike a slab of paraffin with the result that, at the end of the process, n neutrons are in the air and $N - n$ neutrons are in the paraffin, then n/N is the albedo.

[GERARD DE VAUCOULEURS]

Alkali emissions

The alkali metals lithium, potassium, and especially sodium can be observed in twilight by the scattering of their resonance lines. Analysis has revealed that the free atoms are concentrated in a layer 90–100 km above the Earth's surface. The existence of free atoms can be explained by reactions of the type $NaO + O \rightarrow Na + O_2$. Thus, the alkali layer coincides with the region where oxygen is dissociated. At present, the most likely source of the material is vaporized meteoric material, but for sodium and perhaps potassium a con-

Visual albedoes of planets and Moon*

	Mercury	Venus	Earth	Moon	Mars
p	0.09	0.55	0.35	0.11	0.14
q	0.65	1.26	1.15	0.66	1.10
$A = pq$	0.06	0.70	0.40	0.07	0.15

Jupiter	Saturn	Uranus	Neptune	Pluto
0.37	0.42	0.41	0.50	?
1.10	1.0:	1.1:	1.1:	?
0.41	0.42	0.45	0.55	?

*The colons indicate uncertain estimates.

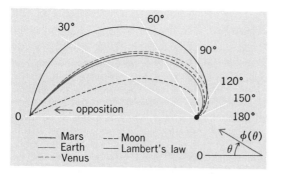

Polar scattering diagrams of Mars, Venus, Earth, and Moon with curve for Lambert's law. Relative intensity $\phi(\theta)$ is plotted as a function of phase angle θ.

tribution may also be made by marine salt particles which have been lifted to mesospheric heights. *See* MESOSPHERE.

Sodium emission. The sodium abundance (integrated density) has a marked seasonal variation which is also a function of latitude, reversing in the Southern Hemisphere. The variations observed at 52°N, 32°N, and 44°S are shown schematically in the figure. These curves cover only 1 or 2 years and may not be entirely representative, as substantial year-to-year variations exist. The data show a large winter maximum at high latitudes; at lower latitudes it is much less obvious. A study by D. M. Hunten and W. L. Godson provides evidence that source material is being raised at times of stratospheric warmings over the polar cap in winter. A superposed-epoch analysis shows a clear peak in sodium abundance at times of warming events. It is likely that the material is raised in the form of small dust or aerosol particles. The less marked winter maximum at lower latitudes is thus explained by remoteness from the source.

Sodium dayglow emission has been observed both from the ground and from rockets. Surprisingly, the sodium abundance is enhanced by a factor of 2 to 4 above the twilight value, but the height of the layer is unchanged. The sodium distribution is often observed to be remarkably sharp, with scale height much less than that of the atmosphere. These observations remain unexplained; they may be related to the evaporation of small meteoric particles, either during entry into the atmosphere or after slowing down to a final settling velocity.

Sodium and other alkali atoms are readily ionized by sunlight, and it is not clear that a rapid means of recombination exists. The resulting electrons may play an important role in the D and lower E regions, especially at night and in some forms of the sporadic E ionization. Similarly, a large degree of ionization would require large corrections to the observed metal abundances, since the ions do not scatter the same wavelengths as the neutral atoms. These suggestions must be considered as only tentative until more is known about the recombination. In particular, the dust particles should act as very efficient centers for recombination, and ionization might well be important only above 100 km, where the dust concentration is very small.

Lithium emission. Natural lithium emission can usually be observed by very sensitive equipment, but many artificial enhancements have been measured as well. The two largest of these were probably the result of thermonuclear explosions about 50 km above Johnston Island in 1958 and 1962. The Soviet 60-megaton explosion of 1961 produced a much smaller effect, and this is the only low-altitude explosion that has injected observable amounts. Two explosions outside the atmosphere were also slightly effective.

In all the above cases, the observations have been made thousands of kilometers from the source, and the effects can be assumed to have been worldwide. Less violent injections of lithium have also been made from time to time by sounding rockets, and some of these have been detected

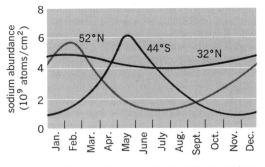

Annual variation of sodium abundance at three stations, two northern and one southern.

from nearby stations. A release at Churchill, Manitoba, in March, 1965, was observed at Saskatoon, Saskatchewan, by W. A. Gault and H. N. Rundle and a little later in Alaska. Releases above White Sands, N.Mex., in June, 1966, were observed visually at Tucson, Ariz., and affected the twilight photometer there for about a week.

In the face of all this activity, there is some doubt that natural lithium has ever been observed at all. However, there is a small, steady level of about 10^8 atoms cm^{-2}, above which there appear to be sporadic enhancements. These observations have been made at Saskatoon by H. M. Sullivan and Hunten and more recently by W. A. Gault and H. N. Rundle. At Tucson, similar equipment almost never records any detectable emission, and it appears that the amounts must be less at low latitudes.

The behaviors of lithium and sodium are strikingly different, and it seems likely that the lithium enhancements are related to meteoric influx; perhaps certain "dust showers" are responsible. Or it may be that lithium atoms condense almost irreversibly on the dust particles, whereas the sodium is easily reevaporated.

Potassium emission. Readily observed with sensitive equipment, potassium emission seems to have about the same intensity at high and low latitudes. The seasonal variation at Saskatoon is very small, according to Gault and Rundle, in contrast to the sodium. In winter, the Na/K ratio is about 100, and attempts have been made to deduce the source of the material from the ratio. But subsequent information that the ratio is variable does not encourage this attempt, and incomplete evaporation of dust would be expected to fractionate the elements.

One may conclude that the best tracer for atmospheric circulation is artificial lithium, although too many injections could raise the background so much as to spoil the method; this has already happened to a large extent with the natural lithium. The most likely source of material is vaporized meteoric material, although some salt particles may be raised by the violent meteorology of the polar winter. The variety of behavior exhibited by the three elements is astonishing and as yet largely unexplained. *See* ATMOSPHERE; ATMOSPHERIC CHEMISTRY.

[DONALD M. HUNTEN]

Bibliography: J. Heicklen (ed.), *Atmospheric Chemistry*, 1976; D. M. Hunten, Metallic emissions in twilight, *Science*, 145:26, 1964; D. M. Hunten, Spectroscopic studies of the twilight airglow, *Space Sci. Rev.*, 6:493, 1967; M. J. McEwan and L. F. Phillips, *The Chemistry of the Atmosphere*, 1975.

Anemometer

A device which measures the magnitude of air velocity. Anemometers are commonly known as the devices which measure wind magnitude, but they are also used to measure the rate of flow of air or of other gases in other applications, for example, in wind tunnels and on aircraft. The most common types are the cup, vane, and hot-wire anemometers. *See* AIR-VELOCITY MEASUREMENT.

Cup anemometer. This is widely used to measure horizontal wind speed, independent of direction. The use of three hemispherical cups mounted on a vertical shaft is standard in the United States. Commonly the cups rotate a worm gear which operates an electric contact every mile of wind, or fraction thereof. These contacts are recorded on a chart driven at constant speed. From the number of contacts in a selected time interval, the wind speed is deduced. Since the number of cup revolutions per mile varies with wind speed, particularly at low speeds, a correction must be made.

Wind speed practically independent of air density is directly obtained if the cups rotate an electric generator (Fig. 1). The generator must operate on a minimum of torque to obtain valid indication at low speeds. Its output, often amplified, drives an electric indicator calibrated in wind speed units.

Vane anemometer. This portable instrument is used to measure low wind speeds and airspeeds in large ducts. It consists of a number of vanes radiating from a common shaft and set to rotate when facing the wind (Fig. 2); a guard ring surrounds the vanes. The vanes operate a counter to indicate the number of rotations, which when timed with a stopwatch serve to determine the speed. The parts are made of lightweight materials; friction is kept low to obtain reliable measurements.

Hot-wire anemometer. This is an important device used principally in research on air turbulence and boundary layers. There are two types, known as constant-voltage and constant-temperature. The cooling of an electrically heated fine wire placed in a gas stream which alters wire resistance depends on the fluid velocity.

With the constant-voltage type (Fig. 3), as soon as the gas starts flowing, the hot wire cools off, the voltage across a Wheatstone bridge connected to the wire is kept constant, and the calibrated galvanometer shows a reading related to the velocity. This is useful only for very low velocities but is extremely sensitive – velocities down to 0.2 in./sec are typical.

The constant-temperature, or constant-resistance, type (Fig. 4) is more useful. The resistance in

ANEMOMETER

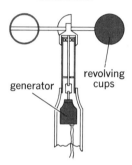

Fig. 1. Cutaway diagram of revolving-cup electric anemometer. (*From D. M. Considine and S. D. Ross, eds., Process Instruments and Controls Handbook, McGraw-Hill, 2d ed., 1974*)

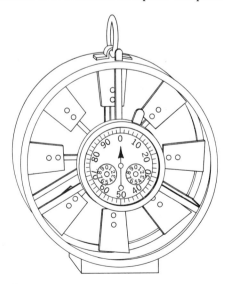

Fig. 2. Portable revolving-vane anemometer. (*From D. M. Considine and S. D. Ross, eds., Process Instruments and Controls Handbook, McGraw-Hill, 2d ed., 1974*)

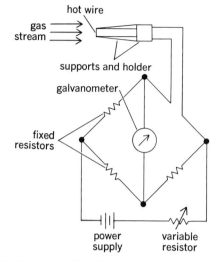

Fig. 3. Constant-voltage hot-wire anemometer.

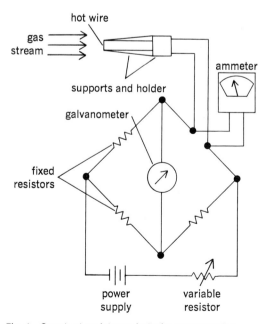

Fig. 4. Constant-resistance hot-wire anemometer.

the battery across the Wheatstone bridge is increased to such a value that the wire which was cooled by the stream is again brought to its original temperature. The current in the hot wire, read by a calibrated ammeter, gives a measure of the air velocity.

Another important application is the measurement of turbulent fluctuations. These fluctuations show up as variations in wire resistance and are useful for boundary-layer studies.

Typical materials for such hot wires are platinum and tungsten with diameters ranging from 10^{-5} to 10^{-3} in.   [HERBERT FOX]

Bibliography: D. M. Considine (ed.), *Process Instruments and Controls Handbook*, 1957; D. M. Considine and S. D. Ross (eds.), *Handbook of Applied Instrumentation*, 1964; H. W. Liepmann and A. Roshko, *Elements of Gasdynamics*, 1957.

Antarctic Ocean

The three oceans of the Earth, in order of decreasing area are the Pacific, the Atlantic, and the Indian. In the Northern Hemisphere and most of the Southern Hemisphere, the oceans are connected only by narrow, shallow passages which prohibit significant interactions. Such isolation would in time lead to each of the oceans having differing characteristics were it not for a major interconnecting artery farther south. This passage exists between the southern shores of the Australian, South American, and African continents and the coastline of Antarctica. It is a broad, deep, circumpolar ocean belt and is essential in equalizing the characteristics of the three oceans and in maintaining the deep and bottom layers of these oceans as an aerobic, cold environment. Because of the importance of this ocean area and its obvious homogeneity in climate (due to its subpolar geographic position), it is convenient to name it the Antarctic Ocean. This region is not officially recognized as a separate ocean, mainly because of its lack of a northern boundary in the classical sense, that is, a complete (or nearly so) land mass. However, from studying its sea-water characteristics, it becomes possible to set a northern oceanographic limit as the Antarctic Convergence or Subtropical Convergence (Fig. 1).

Oceanographic investigations. There has been much exploration of the Antarctic Ocean since the 16th century. At first curiosity, exploration for rich, unclaimed lands, and the need for new ship routes were the impetuses. Associated with the first two was the search for Terra Australis Incognita, the proposed great southern continent. Sightings of many of the larger islands in the southwestern Pacific Ocean were first considered to be this long-sought land, but circumnavigation proved that these were islands and, with the exception of a few of the smaller ones, not economically worth exploiting. In the 1770s, under the command of Capt. James Cook, the first scientific voyages were carried out in the Antarctic Ocean and added much to the knowledge of the Southern Hemisphere's geography and climate. From the size of the icebergs observed and a latitudinal variation of the pack ice field, Cook surmised a frozen continent not symmetrical with the geographic South

Pole. The 19th century witnessed further explorations of the Antarctic Ocean and the delineation of the coastlines of Antarctica. This series of explorations was sparked mainly by the growing seal and whaling industries.

Further scientific studies in the 20th century have been carried out by the German ship *Deutschland* and the circumpolar expeditions of the English vessels *Discovery II* and *William Scoresby*. The United States has contributed much through the U.S. Navy Deep Freeze operations and the 1962–1972 circumpolar study of the Antarctic Ocean by the National Science Foundation–sponsored ship USNS *Eltanin*.

Currents. The major flow is the Antarctic Circumpolar Current, or West Wind Drift (Fig. 1). Along the Antarctic coast is the westward-flowing East Wind Drift. The strongest currents are in the vicinity of the polar front zone and restricted passages such as the Drake Passage and deep breaks in the meridionally oriented submarine ridge systems.

Though the magnitude of the current is small, there is little attenuation of flow with depth, which results in the great volume transport of the Antarctic Circumpolar Current. It is estimated that over 200×10^6 m³/s of water goes through the Drake Passage, about three times the flow of the Gulf Stream. The nonzonal character of the flow is associated with irregularities in the bottom topography (Fig. 2). On approaching a ridge, the current accelerates and turns northward. It becomes more diffuse and turns southward when approaching a basin.

The total meridional transport across a latitude from the sea surface to the sea floor must be zero, but between 1000 and 4000 m there is a net southward flow. This is balanced by a northward flow above 1000 m and below 4000 m (Fig. 3). The total southward transport between 1000 and 4000 m is

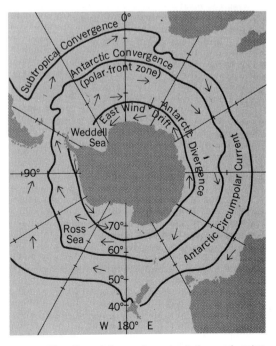

Fig. 1. Direction of the surface circulation and major surface boundaries of the Antarctic Ocean.

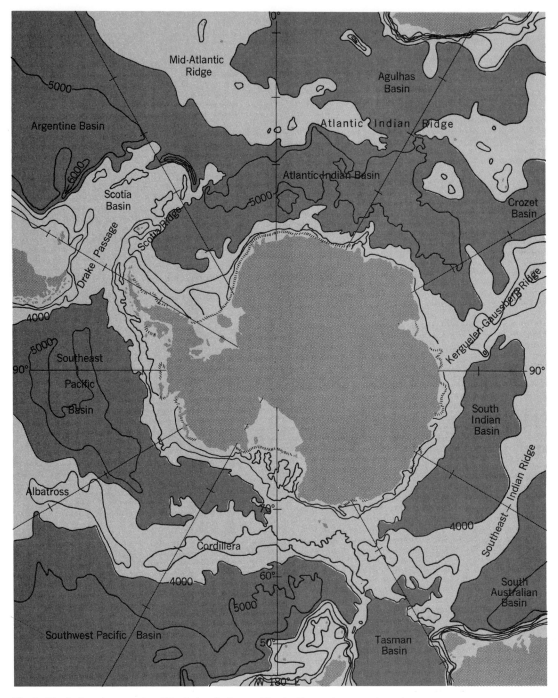

Fig. 2. The bottom topography of the Antarctic Ocean. The depths as shown are in meters.

about one-half of the zonal transport. The greater area through which this southward transport passes indicates that the mean southward velocity is much less than the zonal velocity. The meridional circulation is of great importance since the southward flow brings into the Antarctic Ocean relatively warm water depleted of much of its oxygen but high in nutrients. The warm water eventually reaches the near-surface layers, where through sea-air interaction it is cooled and renewed in its oxygen content. At the same time it enriches the euphotic zone with much-needed nutrients for planktonic life. The cooling is accounted for by a

heat transfer from sea to air. This transfer significantly warms the subantarctic atmosphere. The interaction with the atmosphere causes large thermohaline (temperature and salinity) alterations of the surface water. *See* OCEAN CURRENTS.

Water masses. The sea water comprising the Antarctic Ocean can be conveniently divided into several water masses. The criteria used for this division are based on the temperature and salinity (T/S relation) of the water. Each water mass will fall in a different region of the diagram (Fig. 4). Figure 3 shows, in schematic form, the positions of these water masses and the average meridional

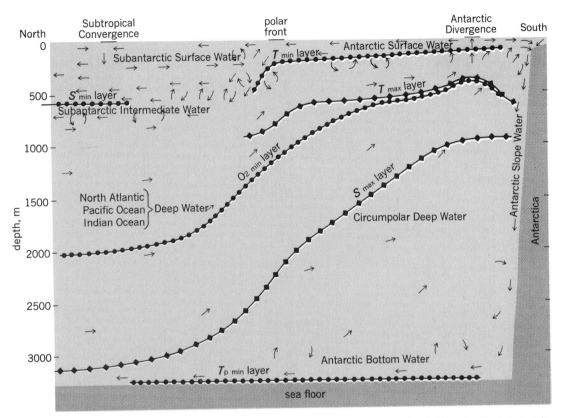

Fig. 3. Generalized diagram of the meridional flow in the Antarctic Ocean. The symbol T_p represents the potential temperature, that is, the temperature a water sample would attain if it were raised adiabatically to the surface. (*After A. L. Gordon, in V. Bushnell, ed., Antarctic Map Folio Series, number 6, American Geographical Society Publication, 1967*)

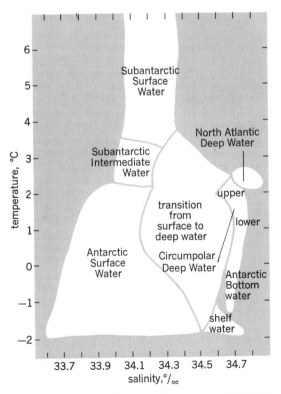

Fig. 4. Generalized diagram of the several water masses of the Antarctic Ocean; division of the water masses is based on the temperature-salinity relationship.

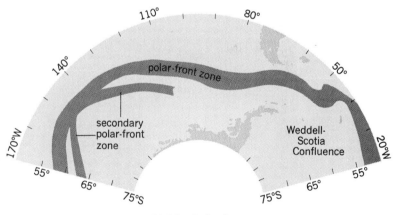

Fig. 5. Polar front zone and Weddell-Scotia Confluence.

circulation of the Antarctic Ocean. Within each of the water masses is an associated "core layer." This is a layer which contains the most undiluted water characterizing the water mass. It is observed as an extreme (maximum or minimum) in temperature, salinity, or dissolved-oxygen concentration.

As observed in the T/S diagram, the coldest, freshest water is the Antarctic Surface Water. The cooling results from the loss of heat to the atmosphere, and the low salinity is due to the excess of precipitation over evaporation. The heat and salt needed to maintain this water at a constant temperature and salinity over a period of years are sup-

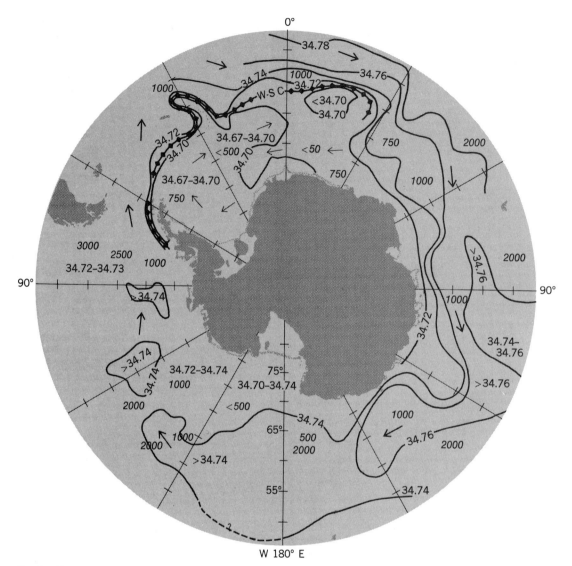

Fig. 6. Circumpolar maximum-salinity layer. Numbers in italics represent depth of maximum salinity. Direction of circulation at maximum-salinity layer is shown. W-S C represents the Weddell-Scotia Confluence.

plied by the upwelling of the southward-flowing water discussed above. This water mass is called the Circumpolar Deep Water.

The Antarctic Surface Water is bounded to the north by thick, homogeneous Subantarctic Surface Water, which is warmer and slightly more saline than the Antarctic Surface Water. A polar front zone separates these water masses (Fig. 1).

The polar front zone is characterized by a large surface temperature gradient. It is also called the Antarctic Convergence, because there is occasionally evidence of a convergence of the two surface waters. During convergence, the mixture of both surface waters sinks and flows northward as Subantarctic Intermediate Water. However, later work suggests that this does not always occur; at times, signs of divergences are found or signs of neither convergence nor divergence but more a simple meeting of two water masses. The position and thermal character of the polar front zone seem to be controlled by wind and bottom topography.

Work on board the *Eltanin* in the Scotia Sea and Pacific sector of the Antarctic Ocean has shown

that the Antarctic Convergence moves within a zone of 3–4° latitude in width (Fig. 5). In some locations a weaker zone (secondary polar front zone) exists to the south of the primary zone. The Weddell-Scotia Confluence (Fig. 5) is the line of separation of the very cold water flowing eastward, which is derived from the east of the Antarctic Peninsula (Weddell Sea), and the warmer water passing through the Drake Passage into the Scotia Sea. The Weddell-Scotia Confluence is exceptionally well defined and stationary in deep water. Within the surface layer it has much spatial variation, resulting in large changes in the surface characteristics of the water in the southern Scotia Sea.

The Circumpolar Deep Water can be divided into two subdivisions: the upper and lower. The lower part, characterized by a salinity maximum (Fig. 6), makes up the bulk of the deep water. The upper part is identified by its contrast with the overlying surface water, which leads to a temperature maximum and oxygen minimum (50–60% full saturation). The heat and salt losses from the Circumpolar Deep Water to Antarctic Surface Water

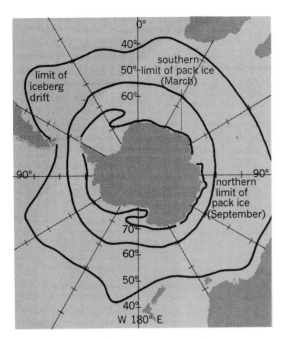

Fig. 7. Various ice limits in the Antarctic Ocean.

are renewed by the influx of the North Atlantic Deep Water in the Atlantic and western section of the Indian sectors of the Antarctic Ocean. In the region from the Scotia Sea to 70°E, the North Atlantic Deep Water is integrated into the Circumpolar Deep Water. From 70°E to the Drake Passage, the Circumpolar Deep Water interacts with water of lower salinity along its northern boundary, and an axis of salinity maximum occurs (Fig. 6). This axis is found below and slightly south of the polar front zone. A low-salinity, clockwise gyre exists in the Weddell Sea. The lower salinity suggests recirculation many times before substantial renewal occurs with the deep water to the north and east.

The Antarctic Bottom Water is a cold, highly oxygenated (70–80% of saturation) water, primarily a result of processes occurring during the winter season, when the cold salty slope, or shelf, water forms. The shelf water eventually mixes with the Circumpolar Deep Water and sinks to the bottom as Antarctic Bottom Water. The shelf water is produced by pack ice formation. The pack ice, having a lower salinity than sea water, concentrates the salt in the remaining water. This shelf water, which is at freezing point (about −1.9°C), increases in salinity until it sinks. The bottom water is made up of shelf water and deep water in roughly a 1:1 ratio. Most of the bottom water forms in the Weddell Sea, though minor amounts of a saltier bottom water form in the Ross Sea and vicinity and perhaps along the coasts in the Indian Ocean sector of the Antarctic Ocean. The Antarctic Bottom Water flows far to the north, and its cold oxygenated water affects large areas of the world's oceans. *See* SEA WATER.

Sea ice. Naturally the extreme cold of the polar regions causes an extensive ice field to form over the southern regions of the Antarctic Ocean. The extent of the ice is seasonal (Fig. 7) in that during the October-to-March period the area decreases, and it increases during the remaining months. The

seasonal difference in the volume of sea ice is estimated as 2.3×10^{19} g. Satellite photographs reveal that the sea ice field is not uniform, but has many large ice-free regions, called polynyas. The sea ice plays an important role in the heat balance since it reflects much more solar radiation (and therefore heat) back into space than would be the case for a water surface. The polynyas would therefore be of special interest in radiation and heat balance studies. In addition to the ice formed at sea, the ice calving at the coast of Antarctica introduces icebergs into the ocean at a rate of approximately 1×10^{18} g/year (Fig. 7). *See* ICEBERG; SEA ICE; TERRESTRIAL ATMOSPHERIC HEAT BALANCE.

Sediments. The composition of the sediments are influenced by two factors: the high biological productivity in the nutrient-rich euphotic zone, and the input of glacial debris carried seaward by icebergs. The siliceous ooze (diatoms and radiolaria) is found roughly between the polar front zone and the limit of the pack ice. To the north, the siliceous ooze is replaced by calcareous ooze and red clay in deeper basins. Within the pack ice field, sediments are of glacial marine type.

The differences in concentration of diatom and radiolaria species with depth in the ooze indicate past minor climatic changes of the Antarctic region. A major transition in sediment types is found to have occurred 1,600,000 years ago when the sediments were red clay, indicating low productivity. This suggests that the present Antarctic circulation has existed for at least 1,600,000 years. *See* MARINE SEDIMENTS. [ARNOLD L. GORDON]

Bibliography: G. E. R. Deacon, The southern ocean, in M. N. Hill (ed.), *The Sea*, vol. 2, 1963; A. L. Gordon, Structure of Antarctic waters between 20°W and 170°W, in V. Bushnell (ed.), *Antarctic Map Folio Series*, no. 6, American Geographical Society, 1967; J. D. Hays, Quaternary sediments of the Antarctic Ocean, *Progr. Oceanogr.*, 4:117–131, 1967; R. Priestley et al. (eds.), *Antarctic Research*, 1964; J. L. Reid (ed.), *Antarctic Oceanology One*, Antarctic Research Service, vol. 15, 1971.

Arctic Ocean

The north polar ocean lying between North America and Asia, extending over about 10^6 km². It is nearly completely covered by 2–3 m of ice in winter, and in summer it becomes substantially open only at its peripheries. Its extent has been variably defined, but it is oceanographically appropriate to consider it bounded on the south by a line running from northern Greenland through Smith, Jones, and Lancaster sounds, along northwestern Baffin Island to the Canadian mainland, thence to the Alaskan coast, across Bering Strait, along the Siberian coast to Novaya Zemlya, across to Franz Josef Land and Spitsbergen, and over to northern Greenland (Fig. 1). This definition omits the Barents, Norwegian, and Greenland seas and Baffin Bay, which have a pronounced North Atlantic character.

Bottom features. The central polar basin, somewhat triangular in shape, is surrounded by continental shelves which are interrupted only by the deep passage running through Fram Strait. The shelf from Greenland to Barrow is only about 100

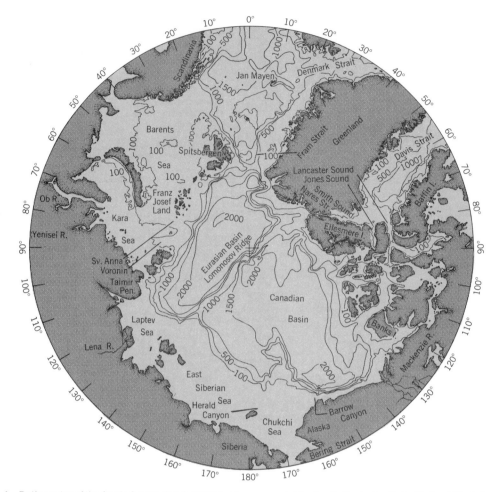

Fig. 1. Bathymetry of the Arctic Ocean; depth in fathoms (1 fathom = 1.83 m). (*From J. E. Sater, Oceanography: Submarine topography and water masses, in J. E. Sater,* coord., *The Arctic Basin, Arctic Institute of North America, 1969*)

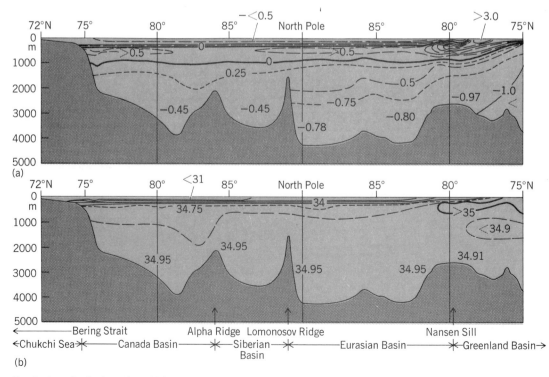

Fig. 2. Longitudinal section of (*a*) temperature (°C) and (*b*) salinity (°/∞) across the Arctic Ocean. (*From L. K. Coachman and K. Aagaard, Physical oceanography of arctic and subarctic seas, in Y. Herman, ed., Marine Geology and Oceanography of the Arctic Seas, 1974*)

Fig. 3. Surface (5 m) salinity (°/∘∘) of the Arctic Ocean. (*From Arctic Institute of North America, Oceanography, in* *Naval Arctic Manual ATP-17(A), pt. 1: Environment, De-* *partment of National Defense, Montreal, 1967*)

km wide, but in the Chukchi, East Siberian, Laptev, and Kara seas the shelf width is typically about 800 km. A number of submarine canyons indent the continental shelf, the largest being Svataya Anna Canyon in the northern Kara Sea, over 500 km long. *See* CONTINENTAL SHELF AND SLOPE; SUBMARINE CANYON.

The central basin is divided by the Lomonosov Ridge (sill depth about 1400 m) into the Canadian Basin (greatest depth 4000 m) and the considerably smaller Eurasian Basin (greatest depth 5100 m). The Eurasian Basin is further divided by the Nansen Ridge, apparently a continuation of the Mid-Atlantic Ridge, and the Canadian Basin is subdivided by the Alpha Ridge, both of which run parallel to the Lomonosov Ridge. *See* MARINE GEOLOGY.

Upper waters. The upper 200 m of the Arctic Ocean, referred to as Surface Water or Arctic Water, is characterized by a significant density stratification produced by the strong increase in salinity downward from the surface (Fig. 2). This density stratification is of considerable importance, for it prevents a deep-reaching convection from

developing within the Arctic Ocean and also prevents the heat of the underlying warm Atlantic Water from reaching the surface. The relatively low salinity at the surface (Fig. 3) is maintained against the upward diffusion of salt by the addition of fresh water, principally through river outflow. Except for the Mackenzie, all the major rivers enter the Arctic Ocean on the Siberian side, and over 40% of the river outflow is to the Kara Sea alone. The total fresh-water discharge to the Arctic Ocean is close to 100,000 m³/s, and about the same amount exits through Fram Strait in the form of low-salinity ice carried by the East Greenland Current.

The upper 30–50 m of Surface Water tends to be relatively uniform vertically in temperature and salinity. Except for areas which become ice-free in summer (such as the Chukchi Sea), the water will be near the freezing point. Away from the immediate vicinity of river mouths, the salinity is 27–34°/∘∘, with a general increase from Bering Strait to Fram Strait (Fig. 3); the salinity in winter is somewhat higher than in summer. Below 30–50 m there are two different subsurface layers. In the Eurasian Basin the salinity increases rapidly with depth to

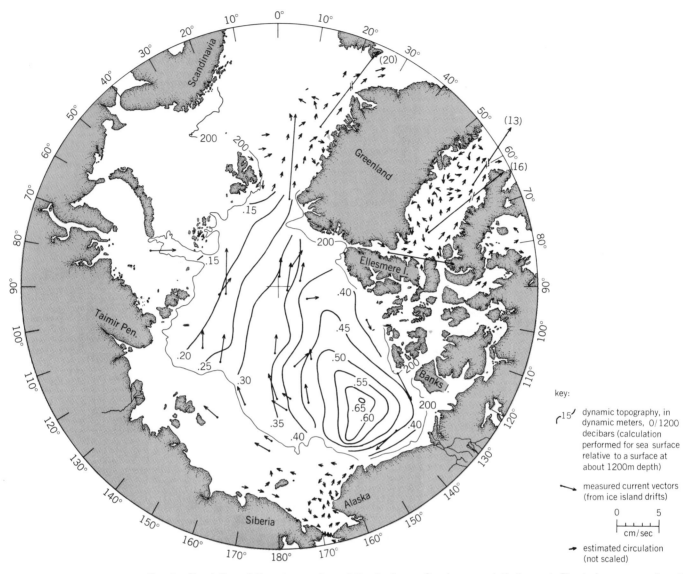

Fig. 4. Circulation of the upper waters of the Arctic Ocean. Mean currents are expected to lie parallel to lines of equal dynamic topography, with speed inversely proportional to spacing between the lines. (*From L. K. Coachman and K. Aagaard, Physical oceanography of arctic and subarctic seas, in Y. Herman, ed., Marine Geology and Oceanography of the Arctic Seas, 1974*)

key:

dynamic topography, in dynamic meters, 0/1200 decibars (calculation performed for sea surface relative to a surface at about 1200m depth)

measured current vectors (from ice island drifts)

0   5
cm/sec

estimated circulation (not scaled)

$34.9-35.0°/_{\infty}$ at 200 m, while the temperature remains below $-1.5°C$ to 100 m and then begins increasing, reaching 0°C at about 200 m. This cold, saline water appears to be formed in the submarine canyons and near the outer edge of the continental shelf bordering the Eurasian Basin. In the Canadian Basin the salinity increases less rapidly with depth, and there is a small relative maximum in the temperature between 50–100 m (but always less than 0°C), followed by a minimum of about $-1.5°C$ at 150 m, before increasing to 0°C at 200–300 m. The subsurface temperature maximum is maintained by relatively warm water coming from the Bering Sea in summer and mixing with shelf water, and the minimum is formed by water coming from the Bering Sea in winter. *See* SEA ICE.

Currents in the upper waters (Fig. 4) tend to be relatively slow (10 cm/s or less), and they are similar in both speed and direction to the ice motion. The Transpolar Drift Stream is directed from

north of the Laptev and East Siberian seas across the length of the Eurasian Basin, exiting through western Fram Strait as the ice-laden East Greenland Current. Initially the stream is very slow (2–3 cm/s), but past the pole it becomes more concentrated, increasing to 8–10 cm/s or more as it reaches Fram Strait. In the Canadian Basin the mean currents form a large clockwise gyre, again with predominantly low speeds (2 cm/s), although the flow is intensified north of Alaska to 5–10 cm/s. The overall circulation in the upper waters has its ultimate cause in the prevailing wind pattern over the Arctic Ocean.

As in other oceans, the current can at any instant vary greatly from the mean condition. The most spectacular example observed in the Arctic Ocean occurs on an occasional basis in the Canadian Basin, consisting of a high-speed current core (speeds in excess of 50 cm/s have been recorded) centered near 150 m and contained completely

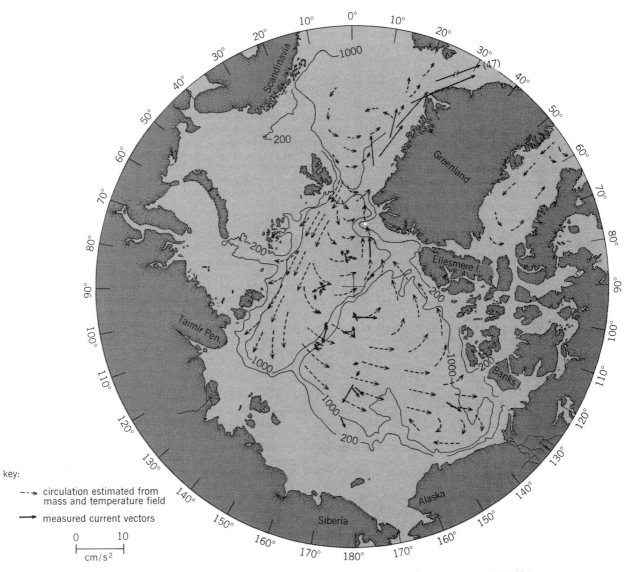

key:

- - → circulation estimated from
     mass and temperature field

⟶ measured current vectors

0   10
|———|———|
cm/s²

Fig. 5. Circulation of the Atlantic Water in the Arctic Ocean. (*From J. A. Galt, Current measurements in the Canadian Basin of the Arctic Ocean, summer, 1965, Univ. Wash. Dep. Oceanogr. Tech. Rep. no. 184, 1967*)

below the surface. These fast currents represent small circular current rings, or eddies, about 10 km across, which are being carried slowly around within the ocean, and those observed so far probably originated in the vicinity of the Chukchi Sea. Other measured variable current conditions can be associated with storms or changes in the wind, and with tides. While semidiurnal tides generally predominate, there are also clear diurnal effects. *See* OCEAN CURRENTS; SEA WATER; TIDE.

Intermediate and deep waters. Below the Surface Water, the temperature increases to a maximum, which over most of the region is about 0.5°C and lies between 300 and 500 m. The salinity is nearly uniform (about 34.95°/₀₀), and since at low temperatures the density of sea water depends almost solely on salinity (disregarding pressure effects), there is virtually no density stratification beneath the upper waters. Significant deviations from the stated temperature occur only in the southern Eurasian Basin closest to Spitsbergen, for it is there that the warm and saline water (called

Atlantic Water) which maintains the temperature maximum throughout the Arctic Ocean first enters. This water has its origin in the North Atlantic and flows northward in the Norwegian and West Spitsbergen currents, passing through Fram Strait. Once into the Arctic Ocean it sinks because of its high salinity and moves eastward along the Eurasian continental slope. The initially high temperature and salinity (over 3°C and 35°/₀₀ in Fram Strait) are then quickly reduced. The general circulation of the Atlantic Water is shown in Fig. 5. The speeds appear to be quite low, typically less than 5 cm/s.

Beneath the Atlantic Water lies cold, nearly uniform Bottom Water (Fig. 2). These two water masses together constitute over 90% of the volume of the Arctic Ocean. The Bottom Water is formed in the Greenland Sea, is mixed, and flows north through Fram Strait. Within the Arctic Ocean it is nearly isothermal within each of the two major basins, but beneath 1400–1500 m a sharp temperature difference of nearly 0.5°C occurs across the Lomonosov Ridge (Fig. 6). The 1400–1500 m level

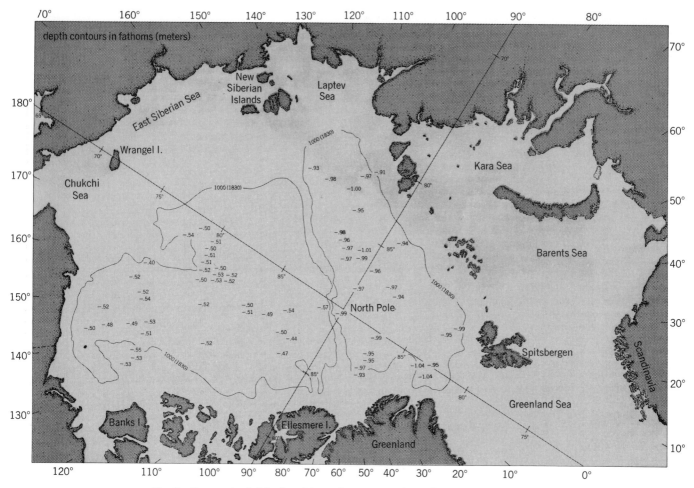

Fig. 6. Deep potential (reduced to surface pressure) temperatures in the Arctic Ocean. (*From L. K. Coachman, Physical oceanography of the Arctic Ocean, 1965, in J. E. Sater, coord., Arctic Drifting Stations, Arctic Institute of North America, 1968*)

Water, salt, and heat budgets for the Arctic Ocean (annual mean)

Source	Volume transport, 10^6 m³/s	Heat transport (relative to −0.1°C), 10^9 kcal/s	Mean temperature, °C	Salt transport, 10^3 metric tons/s	Mean salinity, ‰
Bering Strait					
Water	1.5	0.9	0.5	48.6	32.4
Ice	negligible	−0.4		negligible	
Canadian Archipelago	−2.1	1.3	−0.7	−71.8	34.2
East Greenland Current					
Polar Water	−1.8 ⎫ −7.1	2.0	−1.2	−61.2 ⎫ −246.2	34.0
Atlantic Water	−5.3 ⎭	−3.2	0.5	−185.0 ⎭	34.9
Ice	−0.1	8.0		−0.3	3.0
West Spitsbergen Current	7.1	16.3	2.2	248.9	35.0
Spitsbergen – Franz Josef Land	−0.1	−0.3	2.7	−3.5	34.9
Franz Josef Land – Novaya Zemlya (including Karskiye Vorota Strait)	0.7	0.7	0.9	24.3	34.7
Runoff	0.1	0.5	5.0	0	
Total inflow	9.4		1.8	321.8	34.6
Total outflow	−9.4		−0.1	−321.8	34.6
Total advective heat gain		29.7			
Total advective heat loss		−3.9			
Net exchange	0.0	25.8		0.0	

probably represents the effective sill depth for the ridge, below which a free exchange of water is prevented. The few current measurements that have been made in the Deep Water conform to those in the Atlantic Water.

Mass and heat exchanges. The mean oceanic water, salt, and heat budgets for the Arctic Ocean are shown in the table. The oceanic exchange is dominated by the flows through Fram Strait, with about 95% of the net heat exchange being accomplished by the transport of sensible heat by the West Spitsbergen Current and of ice by the East Greenland Current. It is quite possible that the heat transport could vary from year to year by 50%.

[KNUT AAGAARD]

Bibliography: K. Aagaard and L. K. Coachman, Arctic oceanography, *Oceans*, 6:22–31, 1973; Y. Herman (ed.), *Marine Geology and Oceanography of the Arctic Seas*, 1974; *Climates of the Polar Region*, World Survey of Climatology Series, vol. 14, 1971; J. E. Slater (coord.), *Arctic Drifting Stations*, Arctic Institute of North America, 1968.

Atlantic Ocean

The large body of sea water separating the continents of North and South America in the west from Europe and Africa in the east and extending southward from the Arctic Ocean to the continent of Antarctica.

The Atlantic is the second largest ocean water body and in area covers nearly one-fifth of the Earth's surface. It receives the weathering products from a continental drainage area approximately four times larger than that draining into either the Pacific or Indian Ocean. The two major divisions, North and South Atlantic oceans, have the Equator as the common boundary. The North Atlantic, because of projecting land areas and island arcs, has numerous subdivisions. These include three large mediterranean-type seas, the Mediterranean Sea, the Gulf of Mexico plus Caribbean Sea, and the Arctic Ocean; two small mediterranean-type seas, the Baltic Sea and Hudson Bay; and four marginal seas, the North Sea, English Channel, Irish Sea, and Gulf of St. Lawrence. Parts of the Atlantic are given special names but lack precise boundaries, such as the Bahama Sea, Irminger Sea, Labrador Sea, Sargasso Sea, and Gulf of Guinea. *See* OCEANS AND SEAS.

Oceanographic research. The first data concerning surface currents in the Atlantic were recorded about 1650. Research along systematic lines may be attributed to the efforts of M. F. Maury in 1853. Maury suggested obtaining regular observations of hydrographic and meteorologic data on merchant ships and the collection of this data in central offices. The present-day monthly charts (pilot charts) are based on this material. Deep-sea research in the Atlantic began with the first deep-sea soundings of J. C. Ross in 1839 and the biological investigations of L. Agassiz and W. Thomsen in about 1860. Many research ships have worked in the Atlantic since then. In recent years oceanographic research in the Atlantic has been concentrated in special institutions of the bordering countries. *See* OCEANOGRAPHY.

Bottom topography. The first depth chart of the North Atlantic, compiled by M. F. Maury in 1854, was based on a few deep-sea soundings. Since 1860 work on the transoceanic telegraph cables has accelerated progress in this field, and since 1922 use of the echo sounder has made it possible to obtain a great number of depth soundings. The mean depth of the Atlantic Ocean is 3868 m, and its volume is 318,000,000 km³. *See* ECHO SOUNDER; MARINE GEOLOGY.

Broad shelves with depths less than 200 m are found in the region of the North Sea and the British Isles, on the Grand Banks of Newfoundland, and off the coasts of northeastern South America and Patagonia (Fig. 1). The Mid-Atlantic Ridge, which extends from the Arctic Ocean to 55°S, is less than 3000 m beneath the surface and is characterized by a pronounced relief. It separates the east and west Atlantic troughs, both of which have relatively uniform relief. The east and west troughs are connected in the vicinity of the Equator by the Romanche Deep, the only deep submarine passage through the Mid-Atlantic Ridge, with a depth of 7728 m. This deep, although relatively narrow, is important for the distribution of bottom waters.

Three marked east-west ridges—the Greenland-Scotland Ridge in the North Atlantic and the Walvis and Rio Grande Ridges in the South Atlantic—and several less conspicuous east-west rises separate the two Atlantic troughs into a series of basins including the West European, Canary, and Angola in the eastern Atlantic and the North American, Brazilian, and Argentine basins in the western Atlantic. Only isolated seamounts (such as the Great Meteor, Altair, and Atlantis seamounts) rise from the floor of the deep basins. Greatest depths occur in the narrow trenches along island arcs: 9219 m in the Puerto Rico Trench and 8264 m in the South Sandwich Trench. *See* SEAMOUNT AND GUYOT.

Islands. The islands of Jan Mayen, Iceland, St. Paul, Ascension, St. Helena, Tristan da Cunha, Gough, and Bouvet are parts of the Mid-Atlantic Ridge and are of purely volcanic origin. Other islands of volcanic origin but which lie outside the Mid-Atlantic Ridge are the Faeroes, Madeira, Fernando Poo, Príncipe, São Tomé, Annobón, Fernando Noronha, Trinidad, and the South Sandwich Islands. The Azores, Canary Islands, Cape Verde Islands, and Lesser Antilles are of predominantly volcanic origin. The Bermudas are the northernmost coral reefs. They rise from an old submarine volcanic cone. All the other islands in the Atlantic Ocean are continental in character, such as Spitsbergen, Bear Islands, the British Isles, Greater Antilles, Falkland Islands, and South Georgia.

Bottom sediments. About 73% of the Atlantic Ocean floor, including the adjacent seas, is covered with pelagic sediments; the remainder is covered with hemipelagic or littoral deposits. Of the pelagic sediments, about 47% is calcareous Globigerina ooze, 5% is siliceous diatom ooze, and 18% is red clay. Diatom ooze is predominant south of the Antarctic Convergence, between 50 and 60°S. Red clay is predominant in the great depths of the Argentine, Brazilian, and North American basins. Littoral deposits are found on the shelves and slopes of continents. Turbidity currents have carried some of the littoral deposits far into the deep-sea basins. The zone of littoral sediments fringing polar and subpolar lands is especially extensive. Transported glacial materials of varying size may be found at any depth lying within the seaward range of drifting icebergs. *See* MARINE SEDIMENTS; TURBIDITY CURRENT.

Climate. The primary circulation of surface winds over the Atlantic Ocean is characterized by a zonal distribution pattern oriented in an east-

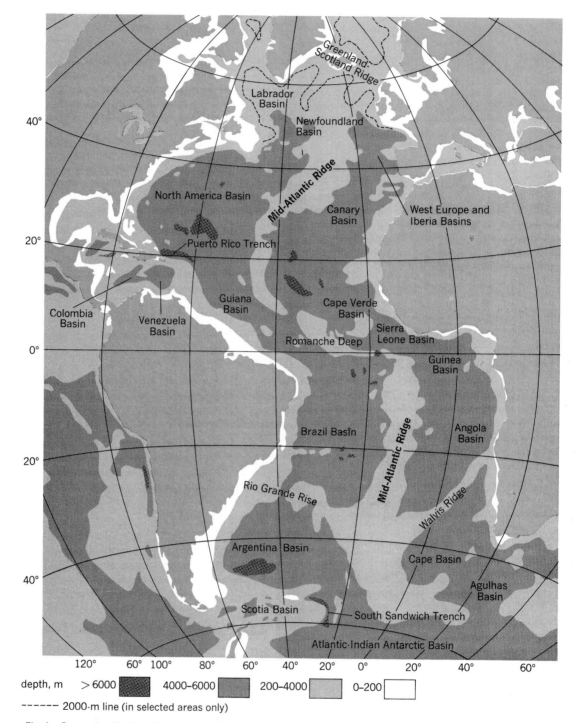

depth, m   >6000   4000–6000   200–4000   0–200
------ 2000-m line (in selected areas only)

Fig. 1. Generalized bottom topography of the Atlantic Ocean, showing its major features.

west direction. For a description of surface zonal winds see WIND.

The greatest storm frequency, more than 30% in winter, is in the zone of the prevailing westerlies. Winds of Beaufort force 8 or greater occur. Tropical cyclones, or hurricanes, occur only in the western North Atlantic in the late summer. Their mean annual frequency is 8; the lowest number recorded was 2; the greatest number was 21, in 1933. See HURRICANE.

Air temperatures are observed to follow a similar zonal pattern of distribution. They are lower in the South Atlantic than in the North Atlantic, and lower in the tropics and subtropics over the eastern Atlantic than they are in the same latitudes over the western Atlantic. Maximum precipitation occurs in the doldrum zone (2000 mm/year). Precipitation also is relatively great in the zone of westerlies but is low in the trade-wind zones. The occurrence of fog is greater where the water tem-

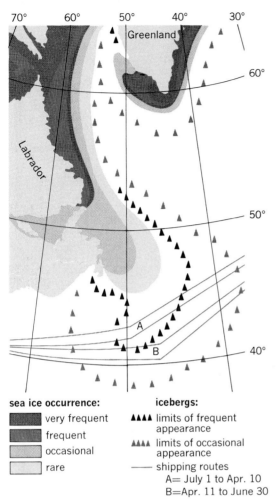

Fig. 2. Mean ice conditions as they occur in May in the northwestern Atlantic Ocean.

sea ice occurrence:
- ■ very frequent
- ■ frequent
- ■ occasional
- □ rare

icebergs:
- ▲▲▲▲ limits of frequent appearance
- ▲▲▲▲ limits of occasional appearance
- —— shipping routes
  A= July 1 to Apr. 10
  B=Apr. 11 to June 30

perature is particularly low compared to the air temperature. Thus the Grand Banks have more than 40% fog frequency in summer because of the cold water of the Labrador Current and the coastal areas of southwest Africa have more than 20% fog frequency throughout the year because of cold upwelling water. *See* CLIMATOLOGY.

Surface temperature and salinity. Surface temperatures are generally 1°C warmer than the overlying air temperatures. Deviations from this pattern of distribution are caused by horizontal water transport in strong currents and by vertical transport in regions of upwelling. Surface salinity is low (less than 35 °/oo) in the doldrums, where there is heavy precipitation. On the equatorial sides of the horse latitudes, the salinity is greater than 37 °/oo. In the northern westerlies the salinity is about 35 °/oo, and in the southern westerlies it is about 34 °/oo. Salinity values below 30 °/oo occur where currents, such as the East Greenland, West Greenland, and Labrador, transport melting ice.

Ice conditions. Sea ice is formed in the northernmost and southernmost parts of the Atlantic Ocean. From these areas drift ice moves equatorward into neighboring regions where it becomes a hazard to sea traffic and limits fishing. Many ice-

bergs drift southward into the sea lanes of the North Atlantic. Most of these have their origin in the valley glaciers of western Greenland. As the glaciers empty into Disco Bay, the icebergs break off and are carried southward by the Labrador Current (Fig. 2). Icebergs generally drift south of the Grand Banks, and some are known to have drifted southeast of Bermuda. During the period of greatest frequency (between March and July), their paths are observed and reported by the International Ice Patrol. In the South Atlantic large, tabular icebergs separate from the Antarctic ice shelf and drift northward. One of the largest, seen in 1953, was 140 km long, 40 km wide, and rose 30 m out of the sea. *See* ICEBERG; SEA ICE.

Surface currents. Surface currents in the Atlantic Ocean flow in much the same direction as the prevailing surface winds. Deflections from these directions are caused by the bottom topography and the latitude or increased effect of Coriolis forces. The fairly constant flow of the North and South Equatorial currents is sustained largely by the trade winds. As a result, warm water is piled

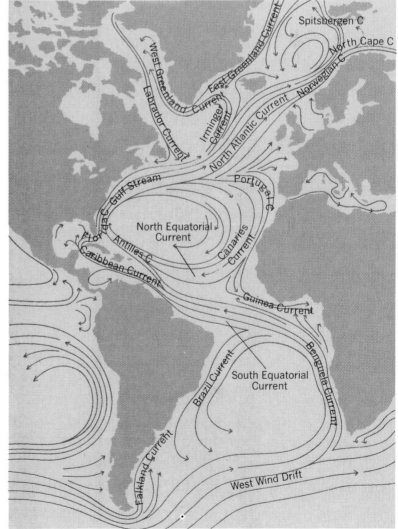

Fig. 3. Currents of the Atlantic Ocean. (*Adapted from J. Bartholomew, Advanced Atlas of Modern Geography, McGraw-Hill, 3d ed., 1957*)

Waters of various areas ranked according to decreasing density

Area	Temperature, °C	Salinity, °/oo
West Norwegian Sea	<−1	34.92
Weddell Sea	<−1	34.65
Labrador and Irminger seas	3	35.00
Antarctic Convergence	6	34.20

up along the poleward borders of these currents and on the western sides of the Atlantic Ocean. *See* OCEAN CURRENTS.

North Atlantic. The greater part of the water transported by the North Equatorial Current enters the American mediterranean as the Caribbean Current and leaves it as the Florida Current, which together with the Antilles Current forms the Gulf Stream (Fig. 3). South of the Grand Banks the Gulf Stream divides into several branches. Of these, the strongest current is the North Atlantic Current. This current flows across the Mid-Atlantic Ridge and forms several offshoots of relatively warm, saline waters that continue flowing in a northeasterly direction. One of these, the Irminger Current, reaches Iceland. Another flows across the Greenland-Scotland Ridge between the Faeroes and the Shetland Islands and thence along the Norwegian coast as the Norwegian Current. Part of this flow continues into the Barents Sea as the North Cape Current, and part of it reaches the Arctic Ocean as the West Spitsbergen Current. As the waters enter the Arctic Ocean, they flow beneath the salt-starved surface waters as a relatively warm, saline undercurrent, which may still be detected at the New Siberian Islands. The transport of cold water southward in the East Greenland and Labrador currents compensates for the transport of relatively warmer water northward into the Arctic Ocean. Other branches of the North Atlantic Current flow southward as the Portugal and Canaries currents. These currents join the North Equatorial Current to complete the circuit in the North Atlantic. *See* ARCTIC OCEAN; CARIBBEAN SEA; GULF OF MEXICO; GULF STREAM.

South Atlantic. The currents in the South Atlantic are, in many respects, the counterparts of those of the North Atlantic, for example, the Brazilian Current and the Gulf Stream, the Benguela and Canaries currents, and the Falkland and the Labrador currents. A circumpolar current, the West Wind Drift, is present, and at about 50°S there is a pronounced converging movement, the so-called Antarctic Convergence. The Equatorial Countercurrent, which flows in an easterly direction between the North and South Equatorial currents, is clearly defined as the Guinea Current along the Gold Coast of Africa. *See* ANTARCTIC OCEAN.

Deep circulation. The surface water in certain areas takes on a particularly high density in winter under the influence of climatic conditions (see the table). These water masses sink to a depth where the surrounding waters have a corresponding density and then spread out at that level. At the same time they are constantly mixing with the surrounding waters. In this way a multistoried stratification arises with some of the characteristics outlined below:

1. Subarctic Bottom Water. This very cold water flows pulsatingly over the Greenland-Scotland Ridge but seldom penetrates farther south than 50°N.
2. Antarctic Bottom Water. This very cold water may be traced as far north as 40°N, where it is recognizable in the North American Basin. It also enters the east Atlantic trough through the Romanche Deep.
3. North Atlantic Deep Water. This originates in the Labrador and Irminger seas, is dispersed at a depth of 2000–3000 m, and flows southward to mix with the salty water of the Mediterranean Sea, which flows out from the Strait of Gibraltar. It can be traced to the Antarctic where it rises.
4. Subantarctic Intermediate Water. This water sinks at the Antarctic Convergence to depths of 700–800 m, where it spreads northward as far as 20°N.

The warm-water sphere overlies the Subantarctic Intermediate Water between the oceanic polar fronts. Temperatures in these waters are greater than 8°C. Definite currents are found only at the four main levels of the deep circulation and on the west side of the Atlantic Ocean. They have speeds up to 10 cm/sec, or 0.2 knot. The renewal of water at great depths is made possible by these currents. Compared with that of the Indian and Pacific oceans, the deep circulation in the Atlantic Ocean is very vigorous, and the deeper water is therefore rich in oxygen. The abundance of nutrients permits a greater rate of organic production where the nutrient-rich waters nearly reach the surface, as in the Antarctic waters (Fig. 4). *See* SEA WATER; SEA-WATER FERTILITY.

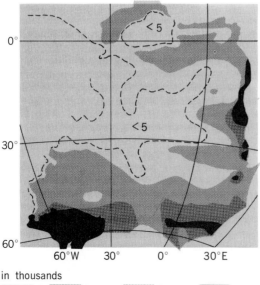

in thousands

---- 5   ▓ 10–50   ▒ 50–100   ■ >100

Fig. 4. Number of plankton organisms per liter in the upper 50-m layer in the South Atlantic. (*After E. Hentschel and H. Wattenberg, from G. Dietrich and K. Kalle, General Oceanography, 1963*)

Tides. The semidiurnal tidal form predominates in the Atlantic Ocean. The semidiurnal tidal wave is distinguished by an amphidromic point at 55°S, a node between southern Brazil and the Gold Coast, another node between the Lesser Antilles and West Africa, and an amphidromic point at 52°N. The mean tidal range is about 1 m in the open ocean, but it decreases to 16 cm in the nodes off Rio Grande do Sul in southern Brazil and to 9 cm off Puerto Rico. Tidal ranges increase beyond broad shelves under favorable physical conditions. They are 9.74 m in the Bay of Bahia Grande in Patagonia (51°S), 10.58 m in the Bay of St. Malo in the English Channel, and 11.47 in the Bristol Channel. The highest range for spring tides in any of the world's seas is 14.14 m in the Bay of Fundy in the Gulf of Maine. The tides of the mediterranean and marginal seas are cooscillations of the tides of the Atlantic Ocean. *See* TIDE.

Fishing. About 36.5% of the catch comes from the Atlantic Ocean. The shelf waters account for most of the Atlantic catch as follows: 20.0%, northwestern Atlantic; 49.2%, northeastern Atlantic; 5.3%, Mediterranean; 6.3% west-central Atlantic; 6.8%, east-central Atlantic; 2.9%, southwestern Atlantic, and 9.7%, southeastern Atlantic. *See* MARINE RESOURCES.

Whaling is concentrated in the Antarctic waters of the Atlantic, Pacific, and Indian oceans.

[GUNTER O. DIETRICH]

Bibliography: A. Defant, *Physical Oceanography*, 2 vols., 1961; G. Dietrich and K. Kalle, *General Oceanography*, 1963; G. Schott, *Geographie des Atlantischen Ozeans*, 1942; H. Stommel, *The Gulf Stream: A Physical and Dynamical Description*, 1964, reprint 1977.

Atmosphere

The gaseous envelope surrounding a celestial body. The terrestrial atmosphere, by its composition, control of temperature, and shielding effect from harmful wavelengths of solar radiation, makes possible life as known on Earth. The atmosphere, which is retained on the Earth by gravitational attraction and in a large measure rotates with it, is a system whose chemical and physical properties and fields of motion constitute the subject matter of meteorology. The changing atmospheric conditions which affect man's environment, particularly temperature, wind, humidity, cloudiness, and precipitation, constitute weather, and the synthesis of these conditions over a period determines the climate at any place. *See* CLIMATOLOGY; METEOROLOGY; WEATHER.

The average atmospheric pressure at the Earth's surface is about 1013 mb and the density 1.2 kg m^{-3}, and these vary by only a few percent over the globe. They both decrease rapidly and roughly exponentially with height, and at several earth radii the density can be said to have fallen to that of interplanetary space.

Composition. The atmosphere is thought to have developed as the result of chemical and photochemical processes combined with differential escape rates from the Earth's gravitational field. Chemical abundances in the atmosphere, therefore, are not directly related to cosmic abundances; in particular, the atmosphere is highly oxidized and contains very little hydrogen. *See* ATMOSPHERE, EVOLUTION OF.

The atmosphere, apart from its highly variable water-vapor content in the troposphere, its liquid droplets and solid matter in suspension, and its variable ozone content in the stratosphere, is well mixed and constant in composition up to about 100 km. This region is termed the homosphere. At higher levels where there is little mixing, diffusive separation tends to take place, with the lighter elements becoming progressively more dominant with height. Moreover, in this region oxygen and the minor constituents, such as carbon dioxide and water vapor, are dissociated by solar ultraviolet radiation. At about 300 km atomic oxygen probably becomes the most important constituent, until about 800 km where helium and hydrogen in turn predominate. This region of highly variable composition is termed the heterosphere. Ionization of the various constituents also due to absorption of ultraviolet solar radiation becomes a major factor above about 60 km, the base of the ionosphere, which is of major importance to radio communications. At levels above 600–800 km collisions between atmospheric particles become so infrequent that some traveling outward may escape from the atmosphere. This region is termed the exosphere. *See* ATMOSPHERIC CHEMISTRY; ATMOSPHERIC POLLUTION.

Table 1 gives a summary of the composition of the atmosphere. The water vapor is a very variable constituent having a mass mixing ratio of 10^{-3} to 10^{-2} gram per gram of dry air (gg^{-1}) in the lower troposphere and about 2.10^{-6} gg^{-1} in the stratosphere. The ozone layer in the stratosphere has its maximum concentration of about 10^{-5} gram per gram of air at 30–35 km, and is formed by molecular-atomic collisions following dissociation of molecular oxygen by solar radiation wavelengths

Table 1. Composition of the atmosphere*

Molecule	Fraction by volume near surface	Vertical distribution
Major constituents		
N_2	7.8084×10^{-1}	Mixed in homosphere; photochemical dissociation high in thermosphere
O_2	2.0946×10^{-1}	Mixed in homosphere; photochemically dissociated in thermosphere, with some dissociation in mesosphere and stratosphere
A	9.34×10^{-3}	Mixed in homosphere with diffusive separation increasing above
Important radiative constituents		
CO_2	3.1×10^{-4}	Mixed in homosphere; photochemical dissociation in thermosphere
H_2O	highly variable	Forms clouds in troposphere; little in stratosphere; photochemical dissociation above mesosphere
O_3	variable	Small amounts, 10^{-8}, in troposphere; important layer, 10^{-6}–10^{-5}, in stratosphere; dissociated above
Other constituents		
Ne	1.82×10^{-5}	Mixed in homosphere with diffusive separation increasing above
He	5.24×10^{-6}	
Kr	1.14×10^{-6}	
CH_4	1.5×10^{-6}	Mixed in troposphere; dissociated in upper stratosphere and above
H_2	5×10^{-7}	Mixed in homosphere; product of H_2O photochemical reactions in lower thermosphere, and dissociated above
NO	$\sim 10^{-8}$	Photochemically produced on stratosphere and mesosphere

*Other gases, for example, CO, N_2O, and NO_2, and many by-products of atmospheric pollution also exist in small amounts.

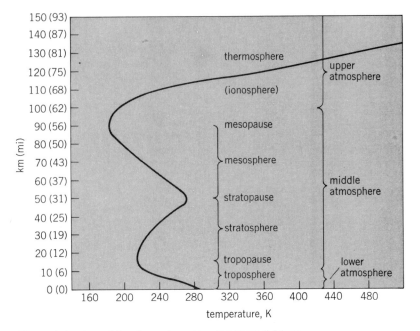

Thermal structure of the atmosphere, showing major divisions.

below about 2400 A. The minor constituents water vapor, carbon dioxide, and ozone play a vital role in the atmosphere's primary biological, heating, and shielding functions.

Thermal structure and circulation. A convenient division of the atmosphere is by spherical shells, "spheres," each characterized by the way its temperature varies in the vertical, and with tops denoted by "pauses," as in the figure. Table 2 lists characteristic values for temperature, pressure, density, and mean molecular weight at a selection of levels within these shells.

Troposphere. The average temperature in the troposphere decreases with height from the surface (~300 K in low latitudes and 260 K in high latitudes) to its upper level, the tropopause, which is around 16 km (temperature 180 K) in the tropics and 10 km (230 K) near the poles. The troposphere includes the layer in which man lives and is the seat of all the important weather phenomena af-

fecting the environment. Its thermal structure is primarily due to the heating of the Earth's surface by solar radiation, followed by upward heat transfer by turbulent mixing and convection. Heat is also transferred polewards by atmospheric motions from the more strongly heated equatorial regions. *See* INSOLATION; TERRESTRIAL RADIATION.

The processes involved include evaporation of water from the surface and condensation with release of latent heat, leading to clouds and precipitation. The general circulation of the troposphere includes wind systems on all scales—prevailing winds, monsoons, long waves, anticyclones and depressions, fronts, hurricanes, thunderstorms, and shower clouds—each system being associated with characteristic weather patterns. The application of physical and dynamical principles to prognosis of atmospheric motions and weather is one of the main tasks of modern meteorology. *See* CLOUD PHYSICS; WEATHER FORECASTING AND PREDICTION.

Stratosphere. The stratosphere extends from about 10–16 km to about 50 km. Its thermal structure is mainly determined by its radiation balance, in contrast to the troposphere, where convective turbulent exchange is of major importance. It is generally very stable with low humidity and no weather in the popular sense. The only clouds in this region are the "mother-of-pearl" clouds seen infrequently at 20–30 km in high latitudes in winter. The ozone layer strongly absorbs solar radiation of 2000–3000 A in the mesosphere and upper stratosphere, which results in the high temperature (250–290 K) region around the stratopause. The main energy source driving the circulation at these levels is thought to be the excess of absorbed energy over infrared energy emitted by the atmosphere (mainly by carbon dioxide and ozone) in the summer hemisphere, compared with the deficit or sink in the winter hemisphere. This produces an atmospheric circulation system in the upper stratosphere and mesosphere which is separate from that of the lower stratosphere, whose circulation is broadly driven by that of the troposphere below. However, in winter large-scale wave motions also propagate upward into the upper stratosphere and mesosphere and have important effects

Table 2. Atmospheric structure, a selection of mean midlatitude values*

Height, km	Pressure, mb	Temperature, K	Density, kg m^{-3}	Mean molecular weight	Layer
0	1.01×10^3	288	1.23×10^0	28.96	Troposphere
5	5.40×10^2	256	7.36×10^{-1}	28.96	
10	2.65×10^2	223	4.14×10^{-1}	28.96	
20	5.53×10^1	217	8.89×10^{-2}	28.96	Stratosphere
40	2.87×10^0	250	4.00×10^{-3}	28.96	
60	2.20×10^{-1}	247	3.10×10^{-4}	28.96	Mesosphere
80	1.05×10^{-2}	199	1.85×10^{-5}	28.96	
100	3.20×10^{-4}	195	5.60×10^{-7}	28.40	Thermosphere
150	4.54×10^{-6}	634	2.08×10^{-9}	24.10	
200	8.47×10^{-7}	855	2.54×10^{-10}	21.30	
300	8.77×10^{-8}	976	1.92×10^{-11}	17.73	
400	1.45×10^{-8}	996	2.80×10^{-12}	15.98	
500	3.02×10^{-9}	999	5.22×10^{-13}	14.33	
600	8.21×10^{-10}	1000	1.14×10^{-13}	11.51	

*Based on United States Standard Atmosphere, 1976.

on the circulation there. *See* MESOPHERE; STRATOSPHERE.

Mesosphere. There is a decrease of temperature with height throughout the mesosphere from 55 km to about 80 km, the mesopause, where there is a temperature minimum. In summer its value is about 150 K at high latitudes where occasionally noctilucent clouds are formed, while in winter it has a higher value, around 220 K. This distribution is probably mainly due to dynamic effects, as there is comparatively little direct absorption of solar ultraviolet radiation at this level. The lowest ionized region, the D layer, with $10-10^3$ electrons cm^{-3}, is from about 60 to 90 km.

Thermosphere. The thermosphere is a very high temperature region and extends from 80 km to the outer edge of the atmosphere. It receives its energy by the direct absorption of solar radiation below about 2000 A. The response to change in solar radiation, for example, from night to day, with solar activity and with sunspot cycle, is very marked, and this region is under direct solar control. The atmospheric motions seem to be mainly thermally induced solar tides, but the dynamics of this region are not well understood. The physical phenomena also are complicated and include excitation, dissociation, and ionization of the constituents and effects, such as the aurora following corpuscular radiation from the Sun. *See* AIRGLOW; AURORA.

The increasing importance of ionization with height also means that electrical and magnetic forces have important effects on the atmospheric motions and these, in turn, produce measurable geomagnetic effects at the Earth's surface. The principal ionized layers or wedges are the E layer at 90–160 km, the F1 layer at 160–200 km, and the F2 layer above 200 km with 10^4-10^6 electrons cm^{-3}. The effects of viscosity increasing with height also become manifest, and above 100–110 km, the turbopause, atmospheric flow no longer appears to be turbulent, and diffusive separation of the atmospheric constituents becomes increasingly important with height.

[R. J. MURGATROYD]

Bibliography: R. A. Craig, *The Upper Atmosphere: Meteorology and Physics*, 1965; R. M. Goody and J. C. G. Walker, *Atmospheres*, 1972; F. K. Hare, *The Restless Atmosphere*, 1966; E. N. Lorenz, *The Nature and Theory of the General Circulation of the Atmosphere*, World Meteorol. Organ. Tech. Publ. no. 115, 1967; E. Palmén and C. W. Newton, *Atmospheric Circulation Systems*, 1969; J. A. Ratcliffe (ed.), *Physics of the Upper Atmosphere*, 1960; K. Rawer (ed.), *Winds and Turbulence in Stratosphere, Mesosphere and Ionosphere*, 1968; A. N. Strahler, *The Earth Sciences*, 1963; J. M. Wallace and P. V. Hobbs, *Atmospheric Science: An Introductory Survey*, 1978.

Atmosphere, evolution of

Variation with time of the chemical composition and total weight of the Earth's atmosphere. The atmosphere is a most tenuous envelope; its mass is less than one-millionth that of the solid Earth; its density even at sea level is less than one-thousandth that of rocks, and virtually all of the at-

Table 1. Nonvariable constituents of atmospheric air*

Constituent	Content
N_2	78.084%
O_2	20.946%
CO_2	0.033%
A	0.934%
Ne	18.18×10^{-6}
He^4	5.24×10^{-6}
He^3	6.55×10^{-12}
Kr	1.14×10^{-6}
Xe	0.087×10^{-6}
H_2	0.5×10^{-6}
N_2O	0.5×10^{-6}

*The classification of H_2 and N_2O as nonvariable constituents is uncertain.

mosphere is below a height only one-hundredth of an earth radius above the surface of the Earth. But the atmosphere is taken so much for granted that one tends to be surprised at the thought that it has a history, that its chemical composition and total weight have varied through time. Indeed, it would be odd to find that the atmosphere had not changed during the long years of the Earth's existence, and that its weight and composition had not responded in some way to the complicated series of events that have left their mark on the Earth's crust.

The study of these changes is difficult, because there seems to be no way of obtaining reliable samples of the atmosphere in the past. But the quest is not altogether hopeless. The origin of life, the continued existence of animals since at least 600,000,000 years ago, the marks which interaction of the atmosphere with surface rocks have left on ancient sediments, and the nature of the input of volcanic gases through geologic time all give clues about the chemical evolution of the atmosphere. At present one cannot pretend to solve all the puzzles of the history of the Earth's atmosphere with these clues, but the broad pattern is emerging and a greater understanding is well within reach.

It may be wise to define the problem rather precisely before looking at the clues. Today the atmosphere contains a small number of major components and a very large number of minor compo-

Table 2. Variable constituents of dry air

Constituent	Content
O_3	0 to 0.07 ppm (summer)
	0 to 0.02 ppm (winter)
SO_2	0 to 1 ppm
NO_2	0 to 0.02 ppm
CH_4	0 to 2 ppm
CH_2O	Uncertain
I_2	0 to 10^{-10} g cm^{-3}
NaCl	Order of 10^{-10} g cm^{-3}
NH_3	0 to trace
CO	0 to trace
	(0.8 cm atm)

*Also from I_2 evaporation from the oceans following photooxidation of I^- in the ocean surface.

nents. Each component exerts a pressure, which is essentially constant for some components at sea level and which is variable in time and space for other components. The most important constituents of the atmosphere are shown in Tables 1 and 2. One could consider the problem of the chemical evolution of the Earth's atmosphere solved when an accurate plot can be made of the variation with time of the pressure of these and other components that may have existed in the atmosphere in the past. This is obviously a large undertaking, but it is made somewhat less so by the mutual exclusion of certain gases as major components of the atmosphere. At present, for instance, gases like ammonia, NH_3, methane, CH_4, and hydrogen can exist in the atmosphere as trace components only, because they are unstable in the presence of the large quantities of oxygen. Conversely, oxygen could not have been a major component of an atmosphere in which ammonia, methane, and hydrogen were abundant. *See* ATMOSPHERIC CHEMISTRY.

Origin of free oxygen. Free oxygen is somewhat of an anomaly on the Earth. Rocks more than a few feet below the Earth's surface are out of equilibrium with free oxygen and are oxidized in contact with the atmosphere. This is seen most easily in the development of red hydrous ferric oxide minerals in soil zones above many rock types in temperate and tropical areas and in the development of the extensive red to reddish-brown sediments of the western United States.

The origin of oxygen in the Earth's atmosphere has been a source of continuing controversy for many decades. Of the two theories that have been dominant, the first proposes that atmospheric oxygen has been produced through geologic time by the continuing effect of photosynthesis, during which carbon is effectively separated from oxygen in carbon dioxide. In a very rough manner this reaction can be written as shown in Eq. (1).

$$6CO_2 + 6H_2O \xrightarrow{\text{Photosynthesis}} \underset{\text{Carbohydrate}}{C_6H_{12}O_6} + 6O_2 \quad (1)$$

This reaction runs in the opposite direction during the decay of plant material and the breathing of animals. It can be shown that nearly all of the oxygen produced by photosynthesis during a given period of time is lost by plant decay, but that the small amount not lost could account for the present rather large quantity of atmospheric oxygen.

The second theory proposes an alternate way of producing free oxygen. Ultraviolet light from the Sun decomposes water molecules in the upper atmosphere. Most of these recombine, but there is a finite possibility that a given hydrogen atom will manage to escape from the Earth's atmosphere before recombination has taken place. Oxygen atoms, being 16 times as heavy as hydrogen atoms, escape very much more slowly or not at all. The decomposition of water vapor followed by hydrogen escape is therefore a distinctly plausible manner of generating free atmospheric oxygen. The escape rate of hydrogen from the atmosphere is almost certainly rapid. The critical factor in determining escape rates is the temperature in the upper atmosphere, and this is now well known from rocket measurements. But a strongly limiting factor for oxygen production by this mechanism is the formation of ozone, O_3, as a by-product of the photodissociation of water. Ozone absorbs ultraviolet light very readily and tends to form a screen preventing it from reaching the lower levels of the atmosphere, where water vapor is abundant. M. McElroy pointed out that most of the hydrogen which does manage to reach the upper atmosphere today is a constituent not of water but of biologically produced methane, and that the rate of hydrogen escape from the Earth's atmosphere is controlled largely by the rate of upward transport of methane rather than of water. Nevertheless, the hydrogen loss rate is currently a small fraction of the oxygen production rate by photosynthesis. The oxygen content of the atmosphere today is therefore only slightly influenced by hydrogen loss from the upper atmosphere; rather, it depends almost exclusively on the operation of a feedback system which links the rate of oxygen production during

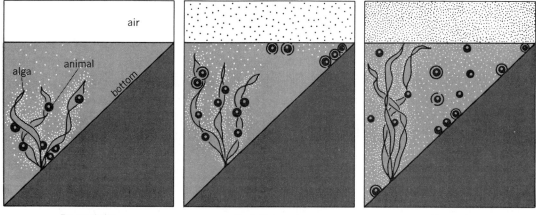

| Precambrian | Early Paleozoic | Present |

Fig. 1. Relationship between atmospheric oxygen and animal evolution. Precambrian (stage 1): Atmosphere is essentially devoid of free oxygen; animals are living in "oxygen oases" in complete respiratory dependence on host plants. Early Paleozic (stage 2): Oxygen pressure has increased to level at which animals may leave plants but flock toward air-water interface. Present (stage 3): Atmosphere and water are highly oxygenated; animals are widely distributed. (*A. G. Fischer, Proc. Nat. Acad. Sci., 53:1205–1213, 1965*)

photosynthesis to the rate of oxygen use by weathering and the decay of organic matter. *See* ATMOSPHERIC OZONE.

Biologic evidence. The proposition that the present high concentration of oxygen in the atmosphere is largely due to oxygen production during photosynthesis implies that oxygen was much less abundant prior to the existence of photosynthesis. This view is in harmony with the requirement that free oxygen was absent from the atmosphere during the development of life.

It is very likely that life evolved early in Earth history. The oldest known unmetamorphosed sedimentary rocks, those found in the Barberton Mountain area of South Africa, are about 3.3×10^9 years old and contain microscopic bits of carbon which, according to M. Muir, could well be of biologic origin. Limestones of similar age in South Africa contain stromatolitic structures, which are almost certainly the work of calcareous algae. There is little doubt, then, that life began more than 3×10^9 years ago. It is not yet clear when organisms developed which were able to produce free oxygen, but evidence from the nature of sedimentary rocks suggests that this event occurred well before 2×10^9 years ago. Biologic evidence for the rise of oxygen to levels approaching those of the present day is still scant until the close of the Precambrian era, some 600×10^6 years ago, despite the discovery of superb microfossils by J. W Schopf and P. E. Cloud in earlier rocks.

D. C. Rhoads pointed out that the sequence of animals which developed during the latest part of the Precambrian era and the beginning of the Phanerozoic era parallels rather strikingly the zoning of animals in contemporary settings of progressively greater oxygen content. This suggests, but hardly proves, that evolutionary events were related to changes in the levels of atmospheric oxygen 600×10^6 years ago. A. G. Fischer suggested that, prior to the general rise in the level of atmospheric oxygen, animals were restricted to "oxygen oases" in the vicinity of photosynthetic organisms (Fig. 1), and that they were able to populate the oceans as a whole only after the level of oxygen in the atmosphere had risen above some threshold value.

L. V. Berkner and L. C. Marshall proposed that the invasion of the land by plants and animals about 400×10^6 years ago occurred when the oxygen pressure had risen to about one-tenth of its present value. At this oxygen pressure the intensity of ultraviolet radiation at the Earth's surface would have been so low that it no longer presented a health hazard. Although this is no more than an interesting speculation, the persistence of animals requiring oxygen in more or less large quantities indicates that oxygen levels in the atmosphere have probably never been lower than one-tenth of the present value during the past 400×10^6 years. There may well have been times when the oxygen pressure was greater than at present.

Evidence from sediments. Today oxidation of rocks at the Earth-atmosphere interface is widespread. If oxygen had been essentially absent from the atmosphere in times past, one might expect to see relatively less oxidation in the minerals of an-

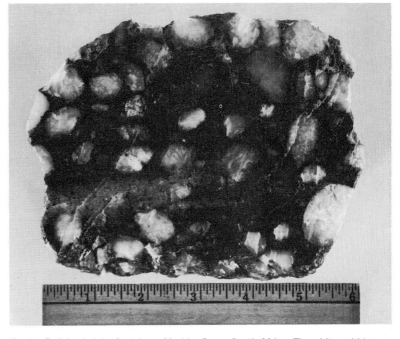

Fig. 2. Polished slab of gold ore, Modder Deep, South Africa. The white pebbles are quartz, SiO_2, the major metallic constituent is pyrite, FeS_2. Most of the gold ore also contains uraninite, UO_2, which apparently is detrital.

cient sediments, and to see the formation of new minerals in these sediments which are less oxidized than their modern counterparts. Iron, manganese, uranium, and sulfur are among the elements which today respond most readily to oxidation at the Earth-atmosphere interface. Uraninite, UO_2, for instance, reacts rapidly with atmospheric oxygen to form a variety of higher oxides and hydrous oxides; concerted search for uraninite in black sands during World War II was quite unsuccessful. And yet, uraninite, which has apparently survived weathering and transport, occurs in the ores of the Dominion Reef series and the Witwatersrand series of South Africa (Figs. 2 and 3), as well as at Blind River, Canada, and at Serra de Jacobina in Brazil.

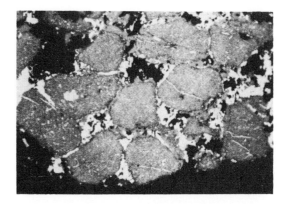

Fig. 3. Photomicrograph ($\times$ 280) of a polished section of uranium ore, West Rand Cons Mine, South Africa. Gray grains are uraninite, UO_2, and probably detrital. White material in cracks and between grains of uraninite is brannerite, an oxide of uranium, titanium, and calcium. (*Ramdohr, Abhandl. Deut. Akad. Wiss. Berlin, 1958*)

In these areas the sediments are more than 1.8×10^9 years old. The origin of the uraninite as a residue of weathering has been proposed by geologists who have closely studied these ore deposits, and has been corroborated by work on the age of the uraninite grains. These appear to be considerably older than the time of accumulation of the sediments. On the other hand, C. F. Davidson maintained that the uraninite is not a weathering residue but was introduced into the sediments after they were deposited. If these grains are indeed detrital, then the rate of oxidation of uraninite must have been much slower prior to 1.8×10^9 years ago than it is today. A careful study by D. Grandstaff of the rate of oxidation of uraninite has shown that the oxygen pressure was probably less than 4×10^{-3} atm (1 atm = 101.325 Pa) to permit the survival of uraninite during weathering and transport.

Macgregor's observation of the abnormal abundance of detrital pyrite in ancient sediments from South Africa is certainly also in accord with this conclusion. But at the same time large amounts of calcite, dolomite, biogenically precipitated pyrite, and iron ore, much of it apparently in the form of hematite or a hydrated ferric oxide, were being deposited in North America. The precipitation of calcite and dolomite demands that at least some of the atmospheric carbon was present as CO_2. Although hematite and its hydrated analogs are stable at extremely low oxygen pressures, the total amount of oxygen which must have been used up in oxidizing the predominantly ferrous iron of most igneous rocks to ferric iron was large. It therefore seems likely that the atmosphere between 2 and 3×10^9 years ago was sufficiently less oxidizing than at present to prevent the oxidation or uraninite, at least under some circumstances, yet was sufficiently oxidizing to permit the formation of widespread accumulations of ferric oxide and hydrous ferric oxide.

Evidence from gaseous emissions. Fortunately, a third line of evidence is available to tell something about the oxidation state of the atmosphere in a nonbiotic state. W. W. Rubey showed that many of the volatile constituents of the atmosphere and oceans cannot have been derived from the weathering of igneous rocks. He has developed a very plausible argument for the concept that these "excess volatiles" have boiled out of the interior of the Earth during the course of its long history. Volcanoes are eloquent evidence for such boiling out today, and it seems likely that a portion of the discharge of at least some hot springs has a deep source. In a nonbiotic state the chemistry of the atmosphere would be controlled largely by the chemistry of such emanations and by their interaction with surface rocks (Fig. 4). Thus, if one knew something about the oxidation of volcanic gases in the past, one might be able to predict, at least within broad limits, the oxidation state of an atmosphere unaffected by biologic processes.

At present, volcanic gases consist mainly of water, carbon dioxide, sulfur dioxide, hydrogen, carbon monoxide, and nitrogen. Table 3 contains an average of representative analyses of Hawaiian volcanic gases. Free oxygen is almost completely absent: The oxygen pressure in these gases as they emerge is about 10^{-7} atm. If a gas is defined as being neutral from an oxidation-reduction point of view when it contains neither free hydrogen nor free oxygen, then these gases are slightly on the reduced side since they do contain a small amount of free hydrogen and CO. Today, these reduced gases react rapidly with atmospheric oxygen. In the absence of atmospheric oxygen a variety of reactions might take place. All of these would tend to produce an essentially neutral atmosphere which would be similar to the present atmosphere with the difference that oxygen would be virtually absent.

The oxidation state of volcanic gases today is controlled in large part by the oxidation state of the associated lavas. This in turn is reflected in the ratio of ferrous iron, Fe^{+2}, to ferric iron, Fe^{+3}, in lavas, a ratio which tends to be preserved on cooling. Studies by S. Steinthorssen of this ratio in basalts have shown that it has probably not changed significantly with time during at least the past 2×10^9 years. One can be fairly certain, then, that the oxidation state of volcanic gases has not varied greatly during the second half of Earth history. There are good theoretical reasons for believing that this same state of affairs prevailed at least back to 3×10^9 years ago, but it is quite possible that shortly after its birth the Earth vented volcanic gases which were more highly re-

Fig. 4.   Eruption of Mount Vesuvius in 1943. The gases emitted by volcanoes throughout geologic time have played a major role in the chemical evolution of the atmosphere of the Earth.

Table 3. Composition of typical Hawaiian volcanic gases*

Gas	Vol. %
H_2O	79.3
CO_2	11.6
SO_2	6.5
N_2	1.3
H_2	0.6
CO	0.4
Cl_2	0.05
Ar	0.04

*After Eaton and Murata, 1960.

ducing. The cause for this state of affairs would have been the presence of metallic iron in the upper part of the Earth's mantle. Today this iron is largely concentrated in the Earth's core, far below the part of the mantle where lavas are generated.

Rise of oxygen pressure. It may be well to summarize the evidence regarding the history of the pressure of oxygen in the Earth's atmosphere before considering the pressure of the other constituents through geologic time. The largest area of ignorance surrounds the events prior to the accumulation of the earliest known sediments, some 3.3×10^9 years ago. It is possible that the atmosphere was quite reducing as a consequence of the injection of highly reduced volcanic gases into the atmosphere during the first few hundred years of Earth history.

Shortly after 3×10^9 years ago, the atmosphere contained less than 4×10^{-3} atm, but more than 0 atm, oxygen. That is indicated on the one hand by the presence of detrital uraninite and on the other hand by the presence of hematite, calcite, and dolomite in these early sediments. Photosynthesis was already under way, but presumably at a rate which was insufficient to support more than a small oxygen pressure in the atmosphere. This state of affairs probably prevailed until about 1.8×10^9 years ago. The first widespread red bed sequences were deposited about that time, and no uranium deposits of the Blind River type younger than 1.8×10^9 years have been discovered. Both observations are consistent with an increase in atmospheric oxygen at that time.

Another rise may well have taken place at the end of the Precambrian and may have been important in the development of animal life. Finally, the colonization of the land became possible about 400×10^6 years ago after a further rise in the oxygen pressure and a further drop in the intensity of ultraviolet radiation at sea level. Since that time the oxygen pressure has, presumably, continued to rise; whether a plateau has been reached or whether there have been fluctuations in the oxygen pressure during the last 400×10^6 years is uncertain. It is most unlikely, however, that large fluctuations have occurred recently. Figure 5 is an effort at illustrating the curve of oxygen pressure versus time; the assigned uncertainties are reasonably generous, but future work may well prove that they are not generous enough.

Rare gases other than helium. Of the atmospheric constituents other than oxygen, the rare gases, nitrogen, and carbon dioxide are probably the most interesting. H. Brown and H. E. Suess pointed out that the Earth is quite depleted in rare gases in comparison with the Sun. Even xenon, an element which certainly cannot escape from the Earth's atmosphere today at anything but a geologically insignificant rate, appears to be only about one-millionth as abundant in comparison with silicon as in the Sun. Brown concluded that the Earth was essentially devoid of an atmosphere after it had reached its present size and gravitational field. It is possible but unlikely that an original gaseous envelope containing the missing rare gases was swept away by strong magnetic

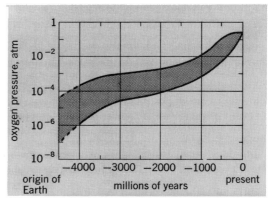

Fig. 5. The pressure of oxygen in the atmosphere during geologic time. Available evidence suggests that there has been a nearly continuous increase in the oxygen pressure since the Earth was formed.

fields which penetrated throughout the solar system, or during the violent Hayashi phase early in the history of the Sun. Whichever view turns out to be correct, the Earth seems to have been left nearly devoid of an atmospheric blanket at some time not long after the end of the accumulation phase.

A corollary of this conclusion is that the present atmosphere has been acquired since that time. Rubey's argument for the degassing of the Earth's interior has been given above, and this is strongly confirmed by the rather anomalous abundance and isotopic composition of atmospheric argon. This gas is much more abundant than the other rare gases, and consists almost entirely of the isotope of mass 40, which should, in the normal course of nuclear events, not be particularly more abundant than the argon isotopes of mass 36 and mass 38. The anomaly is readily explained by the degassing of argon-40 from the Earth after its production from the radioactive decay of potassium-40. Although the potassium content of the Earth is not very well known, there is little doubt that the available potassium could account for the abundance of atmospheric argon-40. It seems likely that degassing was most intense during the early history of the Earth. N. C. Craig's discovery of excess He[3] in the Pacific Ocean indicates that degassing of primary material is still continuing. The pressure of neon, argon, krypton, and xenon have probably been building up gradually in the atmosphere.

Other gases. The abundance of helium in the atmosphere is anomalously low. The quantity of helium which has been generated in the solid Earth by the decay of uranium and thorium series nuclides and has entered the atmosphere with other gases is much larger than the quantity of helium now present in the atmosphere. The inference that helium has escaped into interplanetary space is amply corroborated by analyses of the mechanics of the escape of helium from the Earth's atmosphere.

Nitrogen is a good deal more reactive than the rare gases. It is cycled through living organisms at a geologically rapid rate but seems to be accumulating mainly in the atmosphere. From a thermodynamic point of view nitrate should today be quite abundant in the oceans. Its concentration is, however, kept to very low levels by living organisms for whom nitrate is an important and frequently limiting nutrient.

Ammonia is thermodynamically unstable in the presence of free oxygen, and it is easy to show that a major fraction of the nitrogen now in the atmosphere would be converted to ammonia only in the presence of a hydrogen pressure in excess of 10^{-3} atm. Such a hydrogen pressure could have existed only very early in the Earth's history.

However, the value of the hydrogen pressure, and even the very existence of a highly reducing early atmosphere, is presently very uncertain. R. Brett pointed out that the relatively large nickel content of olivine in the upper mantle is inconsistent with the presence of large amounts of metallic iron in the upper mantle early in Earth history. If this is correct, volcanic gases may never have been much more reducing than at present. Interestingly, McElroy suggested that a sufficient concentration of H_2 could have been built up in the early atmosphere from the injection of volcanic gases no more reducing than those of the present day to permit the production of the organic compounds which were necessary for the development of life.

Methane could have been a major constituent of an early reducing atmosphere. It is likely that its pressure was already less than 10^{-3} atm during the deposition of the sediments of the Issua area in West Greenland more than 3.7×10^9 years ago. Methane must have been a very minor component ever since atmospheric oxygen became more than a trace constituent of the atmosphere. The CH_4 concentration since then has almost certainly been controlled by the balance between the rate of biologic production and the rate of its photochemical decomposition.

It now seems likely that the CO_2 pressure has been hovering near the value at which hydrous magnesium aluminum silicate minerals are in equilibrium with kaolinite, quartz, calcite, and dolomite, as shown in Eq. (2). All of these minerals

$$Mg_5Al_2Si_3O_{10}(OH)_8 + 5CaCO_3 + 5CO_2 \rightleftharpoons$$
$$\text{Chlorite} \qquad \text{Calcite}$$

$$5CaMg(CO_3)_2 + Al_2Si_2O_5(OH)_4$$
$$\text{Dolomite} \qquad \text{Kaolinite}$$

$$+ SiO_2 + 2H_2O \qquad (2)$$
$$\text{Quartz}$$

are abundant in sediments and sedimentary rocks, and it is likely that such equilibria have helped to maintain the CO_2 pressure near its present value for much of geologic time. But it also seems likely that the CO_2 pressure in the atmosphere was somewhat greater than today, perhaps by as much as a factor of 5, before the invasion of the land by plants near the end of the Silurian, some 400×10^6 years ago.

The CO_2 content of the atmosphere has increased by about 10% since the turn of the century because of fossil fuel burning. If all of the current reserves of coal, oil, and gas were suddenly burned, the CO_2 content of the atmosphere would increase by about a factor of 10. Most projections suggest that the peak in CO_2 pressure will occur about the year 2100 at a level about five times the present CO_2 pressure. After that date the CO_2 pressure will decrease in response to the removal of CO_2 from the atmosphere, largely into the oceans, and CO_2 pressures close to the present values should again prevail by the year 3000.

Today, carbon dioxide is second in abundance only to water in gases issuing from volcanoes. Yet the ratio of CO_2 to water in the atmosphere and oceans is minuscule. CO_2 has clearly been scavenged very thoroughly from the atmosphere-hydrosphere system. Its resting place is easily discovered in the elemental carbon in sedimentary rocks and in the carbonate rocks which have been deposited throughout the entire range of Earth history that is accessible through the study of sedimentary rocks (Fig. 6). The two dominant carbonate minerals, calcite, $CaCO_3$, and dolomite, $CaMg(CO_3)_2$, are the major components of limestones and dolomites. The removal of CO_2 from the atmosphere, its reaction with surface rocks, and its ultimate burial are processes which are chemically, mineralogically, and biologically complex.

Three stages of evolution. The discussion seems to lead quite naturally to the threefold division in Table 4 of the history of the Earth's atmosphere. During the first stage, very shortly after the accretion of the Earth, the atmosphere may have been quite reducing. Reduced gases issuing from volcanoes probably would have given rise to an atmosphere consisting of methane with minor quantities of hydrogen, nitrogen, and ammonia. After metallic iron was removed from the upper mantle, the oxidation state of volcanic gases probably approached its present value, methane was replaced by carbon dioxide, and ammonia was converted to nitrogen. In the atmosphere during this (its second) stage, nitrogen was its dominant component, and carbon dioxide and argon were its most important minor constituents.

The third stage opened when the rate of oxygen

Fig. 6. Sample of Tertiary limestone from Florida. Nearly all carbon dioxide injected into the atmosphere has been removed in limestones and dolomites. During the past 600×10^9 years much removal has taken place through the action of organisms with carbonate shells. Prior to that time, removal by inorganic precipitation and by stromatolites was probably dominant.

Table 4. Summary of data on the probable chemical composition of the atmosphere during stages 1, 2, and 3

Components	Stage 1	Stage 2	Stage 3
Major components: $P > 10^{-2}$ atm	CH_4 H_2 (?)	N_2	N_2 O_2
Minor components: $10^{-2} > P > 10^{-4}$ atm	H_2 (?) H_2O N_2 H_2S NH_3 Ar	H_2O CO_2 A O_2 (?)	Ar H_2O CO_2
Trace components: $10^{-4} > P > 10^{-6}$ atm	He	Ne He CH_4 NH_3 (?) SO_2 (?) H_2S (?)	Ne He CH_4 Kr

production by photosynthesis became sufficiently great so that oxygen became more than a trace component of the atmosphere. The transition from stage 2 to stage 3 may well have occurred about 1.8×10^9 years ago. Since the opening of stage 3, the oxygen pressure has climbed to its present value by a path which was probably simple but which could have been complex in detail. During this period the nitrogen, neon, argon, krypton, and xenon pressures have also gradually climbed to their present value, while the helium and CO_2 pressures have remained reasonably constant, the former suspended between the rate of input and the rate of escape, the latter controlled in large part by reactions which involve carbonate and silicate minerals.

This sequence seems a reasonable synthesis of the rather diverse sets of clues which have turned up in atmospheric research. But in conclusion, even if the analysis of the plot turns out to be correct in its main features, it would be too much to ask for an absence of new surprises as the search continues for a more complete understanding of the history of the atmosphere.

[HEINRICH D. HOLLAND]

Bibliography: P. J. Brancazio and A. G. W. Cameron (eds.), *The Origin and Evolution of Atmospheres and Oceans*, 1964; R. M. Garrels and F. T. Mackenzie, *Evolution of Sedimentary Rocks*, 1971; P. M. Hurley (ed.), *Advances of Earth Science*, 1966; K. B. Krauskopf, *Introduction to Geochemistry*, 1967; K. A. Kvenvolden (ed.), *Geochemistry and the Origin of Life*, 1974; B. Mason, *Principles of Geochemistry*, 3d ed., 1966; S. I. Rasool (ed.), *Chemistry of the Lower Atmosphere*, 1973; J. C. Walker, *Evolution of the Atmosphere*, 1977.

Atmospheric chemistry

A subdivision of atmospheric science concerned with the chemistry and physics of atmospheric constituents, including studies of their sources, circulation, and sinks and their perturbations caused by anthropogenic activity. *See* ATMOSPHERIC POLLUTION.

Known gaseous constituents have mixing ra-

tios with air by volume (or equivalently by number), f, ranging from 0.78 for N_2 to 6×10^{-20} for Rn. Known particulate constituents (solid or liquid) have mixing ratios with air by mass, χ, ranging from about 10^{-3} for liquid water in raining clouds to about 10^{-16} for large hydrated ions in otherwise clear air. These constituents are involved in cyclic processes of varying complexity which, in addition to the atmosphere, may involve the hydrosphere, biosphere, lithosphere, and even the deep interior of the Earth. The subject of atmospheric chemistry has assumed considerable importance in recent years because a number of the natural chemical cycles in the atmosphere may be particularly sensitive to perturbation by the industrial and related activities of humans.

Atmospheric composition. A summary of the important gaseous constituents of tropospheric air is given in the table. The predominance of N_2 and O_2 and the presence of the inert gases ^{40}Ar, Ne, ^4He, Kr, and Xe are considered to be the result of a very-long-term evolutionary sequence in the atmosphere. These seven gases have extremely long atmospheric lifetimes, the shortest being 10^6 years for ^4He, which escapes from the top of the atmosphere. In contrast, all the other gases listed in the table participate in relatively rapid chemical cycles and have atmospheric residence times of a few decades or less. *See* ATMOSPHERE, EVOLUTION OF.

Particles in the atmosphere range in size from about 10^{-3} to more than $10^2 \mu$m in radius. The term "aerosol" is usually reserved for particulate material other than water or ice. A summary of tropospheric aerosol size ranges and compositions is given in Fig. 1. Concentrations of the very smallest aerosol particles in the atmosphere are limited by coagulation to form larger particles, and the con-

Composition of tropospheric air

Gas	Volume mixing ratio
Nitrogen, N_2	0.781 (in dry air)
Oxygen, O_2	0.209 (in dry air)
Argon, ^{40}Ar	9.34×10^{-3} (in dry air)
Water vapor, H_2O	Up to 4×10^{-2}
Carbon dioxide, CO_2	2 to 4×10^{-4}
Neon, Ne	1.82×10^{-5}
Helium, ^4He	5.24×10^{-6}
Methane, CH_4	1 to 2×10^{-6}
Krypton, Kr	1.14×10^{-6}
Hydrogen, H_2	4 to 10×10^{-7}
Nitrous oxide, N_2O	2 to 6×10^{-7}
Carbon monoxide, CO	1 to 20×10^{-8}
Xenon, Xe	8.7×10^{-8}
Ozone, O_3	Up to 5×10^{-8}
Nitrogen dioxide, NO_2	Up to 3×10^{-9}
Nitric oxide, NO	Up to 3×10^{-9}
Sulfur dioxide, SO_2	Up to 2×10^{-8}
Hydrogen sulfide, H_2S	2 to 20×10^{-9}
Ammonia, NH_3	Up to 2×10^{-8}
Formaldehyde, CH_2O	Up to 1×10^{-8}
Nitric acid, HNO_3	Up to 1×10^{-9}
Methyl chloride, CH_3Cl	Up to 3×10^{-9}
Hydrochloric acid, HCl	Up to 1.5×10^{-9}
Freon-11, $CFCl_3$	About 8×10^{-11}
Freon-12, CF_2Cl_2	About 10^{-10}
Carbon tetrachloride, CCl_4	About 10^{-10}

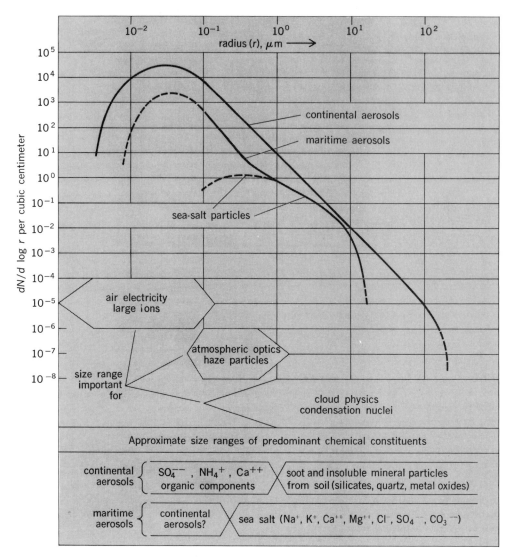

Fig. 1. Chart of the average size distributions and the predominant chemical constituents of some natural aerosols. The size ranges which are important for the various fields of meteorology are shown.

centrations of the larger aerosols are restricted by sedimentation, the rate of which increases as the square of the aerosol radius. In addition, the size distributions of water droplets and ice crystals are affected by evaporation, condensation, and coalescence processes. Some dry aerosols are water-soluble and can also grow by condensation. Aerosols are important in the atmosphere as nuclei for the condensation of water droplets and ice crystals, as absorbers and scatterers of radiation, and as participants in various chemical cycles.

Above the tropopause, the composition of the atmosphere begins to change, primarily because of the decomposition of molecules by ultraviolet radiation and the subsequent chemistry. For example, decomposition of O_2 produces an O_3 layer in the stratosphere. Although the peak value of f for O_3 in this layer is only about 10^{-5}, this amount of ozone is sufficient to shield the Earth's surface from biologically lethal ultraviolet radiation. About 100 km above the surface, ultraviolet dissociation of O_2 is so intense that the predominant atmos-

pheric constituents become N_2 and O. There is also a layer of aerosols in the lower stratosphere composed primarily of sulfuric acid and dust particles. The sulfuric acid is probably produced by oxidation and hydration of sulfur gases. This same process, but strongly amplified, is the probable source of the much thicker clouds of sulfuric acid recently identified on the planet Venus. *See* ATMOSPHERIC OZONE.

A number of radioactive nuclides are formed naturally in the atmosphere by decay of Rn and by cosmic radiation. Radon, which is produced by decay of U and Th in the crust, enters the atmosphere, where it in turn decays to produce a number of radioactive heavy metals. These metal atoms become attached to aerosol particles and sediment out. Cosmic rays striking N_2, O_2, and Ar principally in the stratosphere give rise to a number of radioactive isotopes, including ^{14}C, [7]Be, [10]Be, and [3]H. The incorporation of ^{14}C into organic matter, where it decays with a half-life of about 5600 years, forms the basis of the radiocarbon dat-

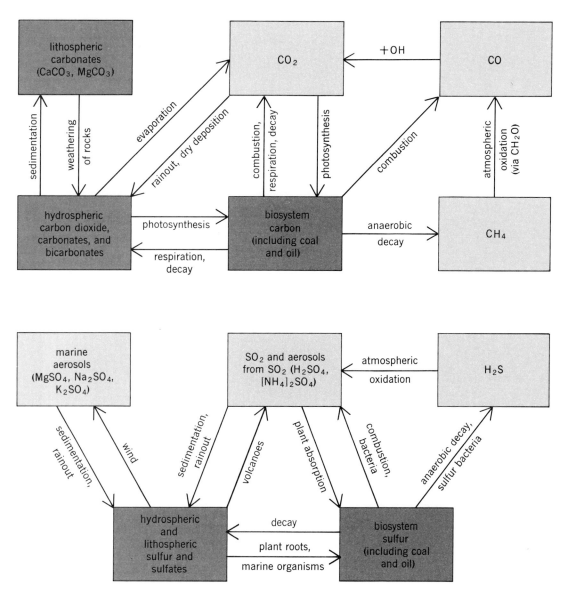

Fig. 2. The carbon and sulfur cycles.

ing method. In addition to naturally occurring radioactivity, large quantities of radioactive material have been injected into the atmosphere as a result of nuclear bomb tests. The most dangerous isotope is ^{90}Sr, which can be incorporated into human bones, where it radioactively decays with a half-life of about 28 years. Both natural and anthropogenic radioisotopes have been used as tracers for the study of tropospheric and stratospheric motions.

Atmospheric chemical models. In order both to adequately understand the present chemistry of the atmosphere and to predict the effects of anthropogenic perturbations on this chemistry, it has been necessary to construct quantitative chemical models of the atmosphere. In general, gas and particle mixing ratios show considerable variation with space and time. This variability can be quantitatively analyzed using the continuity equation for the particular species. In terms of χ, time t, wind velocity v, average particle sedimentation

velocity W_p, and air density ρ, this equation is conveniently written as shown below, where an over-

$$\frac{\partial \bar{\chi}}{\partial t} \approx -\bar{\mathbf{v}} \cdot \nabla \bar{\chi} - W_p \frac{\partial \bar{\chi}}{\partial z} - \frac{1}{\bar{\rho}} \nabla \cdot (\bar{\rho}\overline{\chi'\mathbf{v}'}) + \frac{d\bar{\chi}}{dt}$$

bar denotes an average over a time scale that is long compared to that associated with turbulence, and a prime denotes an instantaneous fluctuation from this average value. The first term on the right-hand side of the equation describes the changes in $\bar{\chi}$ due to the mean atmospheric circulation; the second term gives the $\bar{\chi}$ alternation due to sedimentation which is, of course, zero for gases; the third term denotes fluctuations in $\bar{\chi}$ due to turbulence or eddies, and in this term the eddy flux $\bar{\rho}\overline{\chi'\mathbf{v}'}$ is often roughly approximated by $-K\bar{\rho}\nabla\bar{\chi}$, where K is a three-dimensional matrix of eddy diffusion coefficients; and the last term describes changes in $\bar{\chi}$ caused by chemical production or destruction which may involve simple condensa-

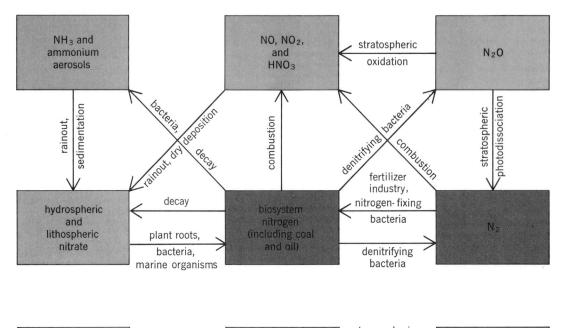

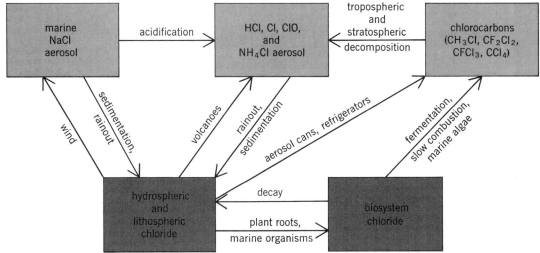

Fig. 3. The nitrogen and chlorine cycles.

tion or evaporation or more complex chemical reactions.

From the equation above, it is seen that the variability of gas or aerosol concentrations is generally due to a combination of transport and true production or destruction. If a constituent has a destruction time T_0, then $d\bar{\chi}/dt = -\bar{\chi}/T_0$, and the equation then implies that the larger the T_0 value the less variability one expects to see in the atmosphere. A constituent for which $\bar{\chi}$ is completely independent of space and time is said to be well mixed. For example, the very-long-lived gases N_2, O_2, ^{40}Ar, Ne, ^4He, Kr, and Xe are essentially well mixed in the lower atmosphere.

Atmospheric chemical models involve simultaneous solution of the above equation for each atmospheric constituent involved in a particular chemical cycle. Often, simplified versions of this equation can be utilized, for example, when chemical lifetimes are much shorter than typical atmospheric transport times, or vice versa. Models for

the ozone layer have now progressed from historical one-dimensional models which neglected transport to sophisticated three-dimensional models which, in addition to solving the above equation, including transport, solve the equations of motion to obtain **v** as a function of position and time.

Chemical cycles. In studying the chemical cycles of atmospheric gases, it is important to consider both the overall budgets on a global scale and the kinetics of the elementary chemical reactions on a local scale. The study of chemical cycles is in its infancy. Some of the minor details in the cycles outlined below may be subject to change, but this present lack of definition does not detract from their importance.

Carbon cycle. The atmospheric cycle which is of primary significance to life on Earth is that of carbon, which is illustrated in Fig. 2. The CO_2 content of the oceans is about 60 times that of the atmosphere and is controlled by the temperature and acidity of sea water. Release of CO_2 into the atmosphere over tropical oceans and uptake by polar oceans result in a CO_2 residence time of about 5 years. On the other hand, the cycle of CO_2 through the biosphere has a turnover time of a few decades. The amount of CO_2 buried as carbonate in limestone, marble, chalk, dolomite, and related deposits is about 600 times that in the ocean-atmosphere system. If all this CO_2 were released to the atmosphere, a massive CO_2 atmosphere similar to that on the planet Venus would result. Because CO is a poisonous gas, its production from automobile engines in urban areas must be closely monitored. On a global scale, the principal sources of CO are combustion of oil and coal and oxidation of the CH_4 produced naturally during anaerobic decay.

Sulfur cycle. The important aspects of the sulfur cycle are also illustrated in Fig. 2. Sulfur dioxide is produced in the atmosphere by combustion of high-sulfur fuels, by plants and bacteria, and by oxidation of H_2S introduced by anaerobic decay. On a global scale, the SO_2 budget is perturbed significantly by the anthropogenic source, and this perturbation is even more pronounced in urban localities. Oxidation of SO_2 produces sulfuric acid, which is a particularly noxious pollutant; thus regulation of high-sulfur fuel combustion, at least on the local scale, is now required.

Nitrogen cycle. The nitrogen cycle is shown in Fig. 3. The nitrogen oxides (NO, NO_2) are important because they are presently the major compounds governing ozone concentrations in the stratosphere. The main natural source of stratospheric NO and NO_2 is decomposition of the N_2O produced from soil nitrate ions (NO_3^-) by denitrifying bacteria. There has been considerable concern about injection of nitrogen oxides directly into the stratosphere by supersonic aircraft. Projected fleet levels for the year 2000 suggest a future anthropogenic source for these oxides comparable to their natural source. This anthropogenic perturbation would cause a decrease of about 10% in total atmospheric ozone and would significantly increase the ultraviolet radiation dosage at the Earth's surface.

Chlorine cycle. The main natural source for atmospheric chlorine is acidification of chloride aerosols producing HCl near the surface. The chlorine cycle (Fig. 3) has received considerable attention because any significant concentrations of Cl and ClO in the stratosphere will lead to depletion of ozone in a manner similar to that caused by NO and NO_2. Because the HCl produced at the ground is severely depleted by rain-out, it does not give rise to significant stratospheric chlorine concentrations. However, the chlorocarbons CCl_4, $CFCl_3$, CF_2Cl_2, and CH_3Cl are relatively insoluble and are not rained out. They appear to decompose in the stratosphere, releasing chlorine; and although the influence of this chlorine on the present stratospheric ozone budget is small, an unchecked buildup of these compounds could lead to significant ozone depletion. The compounds CF_2Cl_2 and $CFCl_3$ are manufactured for use as propellants in aerosol cans, for use in refrigerators and air conditioners, and for manufacture of plastic foams. Methyl chloride is produced naturally by microbial

fermentation and by combustion of vegetation, and CCl_4 is probably derived from both natural and industrial sources.

[RONALD G. PRINN]

Bibliography: S. S. Butcher and R. J. Charlson, *An Introduction to Air Chemistry*, 1972; J. Heicklen, *Atmospheric Chemistry*, 1976; C. E. Junge, *Air Chemistry and Radioactivity*, 1963; M. J. McEwan and F. F. Phillips, *The Chemistry of the Atmosphere*, 1975; L. T. Pryde, *Chemistry of the Air Environment*, 1973; E. Robinson and R. E. Robbins, in W. Strauss (ed.), *Air Pollution Control*, pt. 2, 1972.

Atmospheric electricity

The electrical processes constantly taking place in the lower atmosphere. This activity is of two kinds, the intense local electrification accompanying storms, and the much weaker fair-weather electrical activity over the entire globe, which is produced by the many electrified storms continuously in progress over the Earth.

The mechanisms by which storms generate electric charge are presently unknown, and the role of atmospheric electricity in meteorology has not been determined. Some scientists believe that electrical processes may be of importance in precipitation formation and in severe tornadoes.

Disturbed-weather phenomena. Almost all precipitation-producing storms throughout the year are accompanied by energetic electrical activity. The most intense of these are the thunderstorms, in which the electrification attains values sufficient to produce lightning. Electrical measurements show that most other storms, even though they do not give lightning, are also quite strongly electrified.

Thunderstorms begin as little fair-weather clouds that usually form and disappear without producing rain or electrical effects. When the air is sufficiently thermally unstable, a few of these clouds undergo a rapid and dramatic change. Suddenly, for no obvious reason, one will begin a very rapid growth, increasing in height as fast as 15 m/s. When this happens, other significant changes occur. Often in only a few minutes, strong electric fields and rain appear; shortly after, if the cloud is sufficiently vigorous, lightning appears.

The usual height of a thunderstorm is about 10 km; however, they can be as low as 4 km or as high as 20 km. A common feature of these storms is their strong updrafts and downdrafts, which often have speeds in excess of 30 m/s.

The electric fields are most intense within the cloud, reaching values as high as 5000 V/cm. Above the top of the cloud, values in excess of 1000 V/cm have been measured. Over land surfaces beneath the storm, the fields seldom exceed 100 V/cm, but over water surfaces, which lack points to produce point discharge, they can be as high as 500 V/cm. Although the distribution of the electric fields in and about the thunderstorm is complex and variable, most storms approximate a tipped dipole with positive charge above and negative charge below. A few storms appear to have just the reverse polarity.

The origin and nature of the charged regions responsible for the electric fields of thunderstorms are not understood. A variety of explanations has been proposed, most of which are based on the idea that electrification is caused by the falling of precipitation particles electrified within the cloud by such processes as ion capture, contact electrification, freezing, and drop breakup. Alternatively it is suggested that the transport of charged cloud particles by updrafts and downdrafts produces the regions of charge in the cloud. In the absence of reliable data on charge-carrying particles and their motions within the cloud, there is no general agreement on the relative contributions of the various mechanisms that have been postulated.

The electric fields of thunderstorms cause three currents to flow, each of a few amperes: lightning, point discharge from the ground beneath, and conduction in the surrounding air. Because the external field and conductivity are greatest over the top of the cloud, most of the conduction current flows to the ionosphere, the upper, highly conductive layer of the atmosphere.

Fair-weather field. Fair-weather measurements, irrespective of place and time, show the invariable presence of a weak negative electric field caused by the estimated several thousand electrified storms continually in progress. Together these storms cause a 2000-A current from the earth to the ionosphere that raises the ionosphere to a positive potential of about 300,000 V with respect to the earth. This potential difference is sufficient to cause a return flow of positive charge to the earth by conduction through the intervening lower atmosphere equal and opposite to the thunderstorm supply current. The fair-weather field is simply the voltage drop produced by the flow of this current through the atmosphere. Because the electrical resistance of the atmosphere decreases with altitude, the field is greatest near the Earth's surface and gradually decreases with altitude until it vanishes at the ionosphere.

Because the ionosphere is a good electrical conductor, it is often assumed that at any time the ionosphere is everywhere at the same potential. Observations show, however, that geomagnetic activity produces horizontal electric fields that can, over large distances, give rise to potential differences of tens of kilovolts. Influences of solar activity on the ionospheric potential and on the electrical conductivity of the upper atmosphere have been suggested as possible links between solar activity and the Earth's weather.

Human activities affect atmospheric electricity. Measurements indicate that in the Northern Hemisphere the electrical conductivity of the lower atmosphere has been decreasing, probably as the result of manufactured aerosols. Local temporary increases in conductivity have been noted as the result of nuclear tests. Calculations indicate that, if continued, the release into the atmosphere of krypton-85 gas, a radioactive by-product of nuclear power plants, will in 50 years have decreased the resistance between the Earth and the ionosphere by 15%.

The fair-weather field at the Earth's surface is observed to fluctuate somewhat with space and time, largely as a result of local variations in atmospheric conductivity. However, in undisturbed locations far at sea or over the polar regions, the field is observed to have a diurnal cycle independent of position or local time and quite similar to the diurnal variation of thunderstorm activity over the globe. No importance is presently attached to fair-weather atmospheric electricity except that according to some theories it is responsible for the initiation of the thunderstorm electrification process. *See* CLOUD PHYSICS; LIGHTNING; SFERICS; STORM DETECTION; THUNDER; THUNDERSTORM; TORNADO.   [BERNARD VONNEGUT]

Bibliography: J. A. Chalmers, *Atmospheric Electricity*, 2d ed., 1967; H. Israel, *Atmospheric Electricity*, vol. 1: *Fundamentals, Conductivity, Ions*, 1971, vol. 2: *Fields, Charges, Currents*, 1973; V. P. Kolokolov et al. (eds.), *Studies in Atmospheric Electricity*, 1974.

Atmospheric general circulation

The statistical description of atmospheric motions over the Earth, their role in transporting energy, and the transformations among different forms of energy. Through their influence on the pressure distributions that drive the winds, spatial variations of heating and cooling generate air circulations, but these are continually dissipated by friction. While large day-to-day and seasonal changes occur, the mean circulation during a given season tends to be much the same from year to year. Thus, in the long run and for the global atmosphere as a whole, the generation of motions nearly balances the dissipation. The same is true of the long-term balance between solar radiation absorbed and infrared radiation emitted by the Earth-atmosphere system, as evidenced by its relatively constant temperature. Both air and ocean currents, which are mainly driven by the winds, transport heat. Hence the atmospheric and oceanic general circulations form cooperative systems. *See* OCEAN CURRENTS; OCEAN-ATMOSPHERE RELATIONS.

Owing to the more direct incidence of solar radiation in low latitudes and to reflection from clouds, snow, and ice, which are more extensive at high latitudes, the solar radiation absorbed by the Earth-atmosphere system is about three times as great in the equatorial belt as at the poles, on the annual average. Infrared emission is, however, only about 20% greater at low than at high latitudes. Thus in low latitudes (between about 35°N and 35°S) the Earth-atmosphere system is, on the average, heated and in higher latitudes cooled by radiation. The Earth's surface receives more radiative heat than it emits, whereas the reverse is true for the atmosphere. Therefore, heat must be transferred generally poleward and upward through processes other than radiation. At the Earth-atmosphere interface, this transfer occurs in the form of turbulent flux of sensible heat and through evapotranspiration (flux of latent heat). In the atmosphere the latent heat is released in connection with condensation of water vapor. *See* CLIMATOLOGY; TERRESTRIAL ATMOSPHERIC HEAT BALANCE.

Considering the atmosphere alone, the heat gain by condensation and the heat transfer from the Earth's surface exceed the net radiative heat loss in low latitudes. The reverse is true in higher latitudes. The meridional transfer of energy, necessary to balance these heat gains and losses, is accomplished by air currents. These take the form of organized circulations, whose dominant features are notably different in the tropical belt (roughly the half of the Earth between latitudes 30°N and 30°S) and in extratropical latitudes. It is convenient to discuss these circulations in terms of the zonal mean circulation (air motions averaged at all longitudes around the Earth) and the eddies (waves, cyclones, and so forth, representing deviations from the mean circulation). *See* METEOROLOGY; STORM; TROPICAL METEOROLOGY.

Mean circulation. In a mean meridional cross section the zonal (west-east) wind component is almost everywhere dominant and in quasigeostrophic balance with the mean meridional pressure gradient. The pressure gradient changes with height in accordance with the distribution of air density, which at a given pressure is inverse to temperature. Hence the distribution of zonal wind is related to that of temperature, as expressed by Eq. (1), where u denotes zonal wind component; z,

$$\frac{\partial \bar{u}}{\partial z} \approx -\frac{g}{2(\Omega \sin \phi + \bar{u}a^{-1}\tan \phi)\bar{T}}\frac{\overline{\partial T}}{\partial y} \quad (1)$$

height above sea level; g, acceleration of gravity; Ω, angular velocity of the Earth; a, Earth radius; φ, latitude; T, Kelvin temperature; and y, distance northward. Overbars denote values averaged over longitude and time. *See* GEOSTROPHIC WIND.

Only in the lowest kilometer or so, where surface friction disturbs the geostrophic balance, and in the vicinity of the Equator is the mean meridional (south-north) component comparable to the zonal wind. Because of the nature of the atmosphere as a shallow layer, the mean vertical wind component is weak. Whereas the magnitude of the mean zonal wind varies between 0 and 45 m/s, and the mean meridional wind varies up to 3 m/s, the mean vertical wind nowhere exceeds 1 cm/s. The vertical component cannot be observed directly, but can be calculated from the distribution of horizontal motions.

In the troposphere and lower stratosphere, the zonal circulation is similar in winter and summer, with easterlies in low latitudes and westerlies in higher latitudes, except in small regions of low-level easterlies around the poles (Fig. 1). The strongest west winds, about 40 m/s, are in the winter hemispheres observed near latitudes 30° at about 12 km. In summer, the west-wind maxima are weaker and located farther poleward. *See* STRATOSPHERE; TROPOSPHERE.

In the troposphere, the zonal wind increases upward according to Eq. (1) and with a general poleward decrease in temperature. In the lower stratosphere over most of the globe, this temperature gradient is reversed, and the wind decreases with height. Above about 20 km, separate wind systems exist, with prevailing easterly winds in summer and westerlies in winter that attain speeds up to 60–80 m/s near 60 km height. *See* JET STREAM.

The much weaker meridional circulation consists of six separate cells. Their general locations

and nomenclatures are shown in Fig. 1, along with the approximate circulations in terms of mass flux. For each cell, only central streamlines are shown, but these represent flows that are several kilometers deep in the horizontal branches, while each vertical branch represents gentle ascending or descending motions over latitude belts some thousands of kilometers wide. The tropical Hadley cells, best developed in the winter hemisphere, are mainly responsible for maintaining the westerly winds that are strongest near the poleward bounds of their upper branches, in subtropical latitudes.

Balance requirements. Equation (1) and certain principles related to the generation of air motions prescribe that the wind distribution over the globe must be consistent with the thermal structure of the atmosphere. Sources and sinks (physical processes such as friction and radiation that act to increase or decrease the momentum or heat content of the air) continually tend to change the existing distributions. The sources and sinks are largely a function of latitude and elevation, such that meridional and vertical fluxes of heat energy and momentum are required. Two methods by which these fluxes are calculated from different kinds of observations give results that offer a check upon one another.

In the first method estimates are made of the rate of change of any property X per unit area of the Earth's surface due to sources and sinks. To maintain an unchanged condition, their integrated value, over the area north of a latitude φ, must equal the northward flux F_ϕ of the property across the latitude circle. This requirement is expressed by Eq. (2), in which t denotes time.

$$F_\phi = -2\pi a^2 \int_\phi^{90°N} \frac{d\overline{X}}{dt}\cos\phi\,d\phi \quad (2)$$

The second method is to compute the fluxes directly from aerological observations (made by balloon-borne radiosondes). If x denotes a given property per unit mass of air, the flux is given by Eq. (3), where v is the meridional wind component

$$F_\phi = \frac{2\pi a \cos\phi}{g}\int_0^{p_0}(\bar{x}\bar{v} + \overline{x'v'})dp \quad (3)$$

(positive northward). The integration, with pressure p as a vertical coordinate (related to height z as in Fig. 1), is extended from the bottom ($p = p_0$) to the top of the atmosphere ($p = 0$). Here $\bar{x}$ and $\bar{v}$ denote values averaged over time and longitude, and x' and v' are deviations from these mean values at a given pressure surface. A corresponding expression can be written for the vertical fluxes.

The first term of the integrand may be called the circulation flux, and the second term, the eddy flux. The circulation flux arises from a correlation, in the vertical, between the meridional wind component $\bar{v}$ and the atmospheric variable $\bar{x}$. Eddy flux results from correlations between the fluctuations x' and v' at given pressure levels. The eddy fluxes generally dominate in the meridional transfer of properties, except in the central parts of the tropical Hadley cells.

Eddy fluxes can be divided into two groups: large-scale disturbances (such as cyclones) and small-scale eddies associated either with mechani-

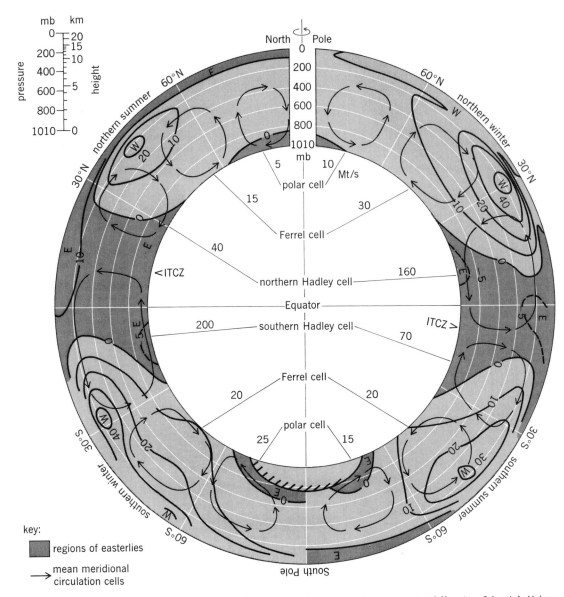

Fig. 1. Pattern of mean zonal (west-east) wind speed averaged over all longitudes, as a function of latitude, height, and season. Height, greatly exaggerated relative to Earth radius, is shown on a linear pressure scale with geometrical equivalent given at upper left. Mean zonal wind (westerly positive, easterly negative, shaded regions) has the same value along any one line and is shown in meters per second (1 m/s ≈ 2 knots). Values are for the geostrophic wind; in the lowest kilometer or so, the actual zonal wind is somewhat weaker. Mean meridional circulation cells in each season are named and intensities are given in terms of mass flow in megatons per second (Mt/s). ITCZ denotes mean latitude of intertropical convergence zone in each season.

key:
■ regions of easterlies
→ mean meridional circulation cells

cal turbulence or with thermal convection (especially cumulus clouds). Only the large-scale disturbances play a significant role in the meridional eddy fluxes of properties. Concerning the vertical flux, however, the entire spectrum of eddies has to be considered. Generally, mechanical turbulence tends to transport heat and zonal momentum downward but moisture upward, whereas convective turbulence transports heat and moisture upward and momentum downward. Especially in the tropics, convective turbulence is very active in this respect. In the equatorial belt (ITCZ in Fig. 1), practically all of the ascending motion in the Hadley cells actually takes place in the warm updrafts of convective clouds, wherein intense rising currents occupy a small part of the total region.

In extratropical regions, where convective phenomena are less prominent especially in winter, the large-scale eddies, associated with polar-front cyclones, dominate in transporting heat and moisture upward as well as poleward. *See* FRONT; THUNDERSTORM.

Angular momentum balance. In an absolute framework, the motion of the atmosphere plus the eastward speed of a point on the Earth's surface represent the total motion relative to the Earth's axis. The angular momentum of a unit mass of air is given by Eq. (4). Considering the mean value in a

$$M = (u + \Omega a \cos \phi) a \cos \phi \qquad (4)$$

zonal ring around a latitude circle, this quantity is conserved unless the ring is subjected to a torque.

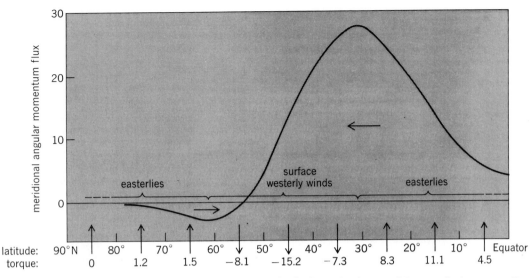

Fig. 2. Average annual meridional flux (positive values northward) of angular momentum in the Northern Hemisphere, with corresponding surface torques in belts of latitude. Units are 10^{18} kg m²s⁻². Maximum flux is 50% larger in winter and 75% smaller in summer than the value shown; seasonal variations are much smaller in the Southern Hemisphere.

The surface easterlies in low latitudes and westerlies in middle latitudes (Fig. 1) exert the principal torques upon the atmosphere, due to frictional drags that are opposite to the direction of the surface winds. The frictional stress τ_x can be estimated from the surface wind velocity by an empirical formula. The torque per unit area, $\tau_x a \cos \phi$, can then be inserted in Eq. (2) to give estimates of the meridional flux of angular momentum.

Since the torques would tend to diminish the westerlies and easterlies, and this is not observed to occur over the long run, it follows that angular momentum has to be transferred from the belts of surface easterlies to the zones of westerlies. Calculations using Eq. (3), with x replaced by M from Eq. (4), show that this meridional flux occurs mainly in the upper troposphere. Hence angular momentum has to be brought upward from the surface layer in low latitudes, transferred poleward, primarily in higher levels, and ultimately brought down to the Earth in the belt of westerlies. Figure 2 shows the annual average meridional flux of angular momentum, and the surface torque in each belt.

Heat energy balance. The energy balance may be calculated by substituting the quantities summarized in Table 1 into Eq. (2). The listed heat sources comprise R_a, net (absorbed minus emitted) radiation of the atmosphere; R_e, net radiation at the Earth's surface; Q_s, flux of sensible heat from the surface to the atmosphere; LE, flux of latent heat from the surface (E denoting rate of evapotranspiration and L, heat of vaporization); and LP, release of latent heat in the atmosphere as estimated from the observed rate of precipitation P.

Alternatively, direct calculations of the fluxes of quantities in Table 2 can be made from aerological observations, by use of Eq. (3). The atmospheric energy comprises the sum $(c_p T + gz)$, the two quantities being interchangeable during vertical air movement; decompression of rising air dimin-

ishes its sensible heat content by an amount equal to the increase of potential energy associated with change of elevation. Almost everywhere $(c_p T + gz)$ increases upward, and Lq generally but not always decreases upward.

Various components of the energy balance are summarized, by 30° belts of latitude, in Fig. 3. Of the net radiation absorbed at the Earth's surface over the whole globe, 81% is expended in evaporation. Correspondingly, 81% of the net radiative loss by the atmosphere is compensated by release of latent heat when water vapor condenses and falls out as rain, snow, or hail, and 19% by transfer of sensible heat from the Earth. In the tropical belt 30°N–30°S, the Earth-atmosphere system gains heat by radiation; the excess is exported to higher latitudes as atmospheric heat and latent heat, and by ocean currents. Considering the tropical- and temperate-latitude belts of the two hemispheres,

Table 1. Sources of atmospheric properties

Different sources	$\overline{dX}/dt$
Atmospheric heat	$\overline{R}_a + \overline{Q}_s + L\overline{P}$
Latent heat	$L(\overline{E} - \overline{P})$
Heat of Earth's surface	$\overline{R}_e - \overline{Q}_s - L\overline{E}$
Heat of atmosphere and Earth	$\overline{R}_a + \overline{R}_e$

Table 2. Atmospheric properties used for flux computations

Property	x (per unit mass)*
Atmospheric energy	
Sensible heat	$c_p T$
Potential energy	gz
Total atmospheric energy	$c_p T + gz$
Latent heat	Lq

*Here c_p is specific heat of air; q is specific humidity, or mass of water vapor per unit mass of air.

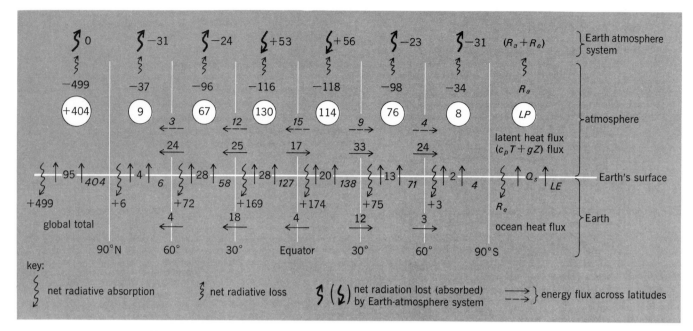

Fig. 3. Annual heat balance for the Earth as a whole and for each 30° latitude belt. Units are 10^{14} W. All values are keyed to the column at right. Italic numbers apply to water vapor (latent heat) flux.

significant differences in the apportionments of evaporation and sensible heat transfer from Earth to atmosphere, and of the meridional transports of energy in the various forms, arise from the greater dominance of continents in the Northern Hemisphere. On the annual average, water-vapor-laden trade winds (lower branches of Hadley cells in Fig. 1) converge at about 5°N, where the greatest precipitation is observed. Minima of precipitation occur in the belts near 30°N, where mean descending motions occur. Secondary maxima of precipitation, in latitudes 40–50°, are associated with frequent extratropical cyclones. See HYDROMETEOROLOGY.

The frequency and intensity of cyclones, and the contrasts between their cold and warm air masses, are much greater in winter than in summer in the Northern Hemisphere; these variations are much less pronounced in the oceanic Southern Hemisphere. Thus there is a fourfold greater poleward transport of sensible heat in middle latitudes of the Northern Hemisphere in winter than in summer, contrasted with only about a 30% seasonal variation in the Southern Hemisphere. In the tropics, large seasonal changes in the intensities of the Hadley circulation, together with a migration of the rain belt of the intertropical convergence zone (Fig. 1), are most pronounced in the monsoon regions of Asia-Australia and Africa. See MONSOON.

Between late spring and early autumn a given hemisphere receives solar radiation far in excess of the amount it loses by infrared radiation. The excess heat is stored mainly in the oceans during the warm seasons and given up to the atmosphere as sensible and latent heat during the cooler seasons. Thus the oceans serve as an energy reservoir that tempers the seasonal changes of atmospheric temperature over them and over neighboring land areas invaded by marine air masses. See MARINE INFLUENCE ON WEATHER AND CLIMATE.

[C. W. NEWTON]

Bibliography: B. L. Dzendzeevski and P. P. Pogosyan, *Atmospheric Circulation*, 1973; J. R. Holton, *An Introduction to Dynamic Meteorology*, 1972; S. Petterssen, *Introduction to Meteorology*, 3d ed., 1969; W. D. Sellers, *Physical Climatology*, 1965.

Atmospheric ozone

Ozone is found in trace quantities throughout the atmosphere, the largest concentrations being located in a layer in the lower stratosphere between the altitudes of 15 and 30 km (Fig. 1). This ozone results almost entirely from the dissociation of molecular oxygen by solar ultraviolet radiation in the upper atmosphere.

Although present in only trace quantities, this atmospheric ozone plays a critical role for the biosphere by absorbing the ultraviolet radiation with wavelength λ between 2400 and 3200 A (240 and 320 nm), which would otherwise be transmitted to the Earth's surface. This radiation is lethal to simple unicellular organisms (algae, bacteria, protozoa) and to the surface cells of higher plants and animals. It also damages the genetic material of cells (DNA) and is responsible for sunburn in human skin. Also, incidence of skin cancer has been statistically correlated with observed surface intensities of ultraviolet wavelengths between 2900 and 3200 A (290 and 320 nm), which are not totally absorbed by the ozone layer.

Ozone also plays an important role in heating the upper atmosphere by absorbing solar ultraviolet and visible radiation ($\lambda < 7100$ A or 710 nm) and thermal infrared radiation ($\lambda \simeq 9.6$ μm). As a consequence, the temperature increases steadily from about 220 K at the tropopause (8–16 km altitude) to about 280 K at the stratopause (50 km altitude). This ozone heating provides the major energy source for driving the circulation of the upper stratosphere and mesosphere. See ATMOSPHERE.

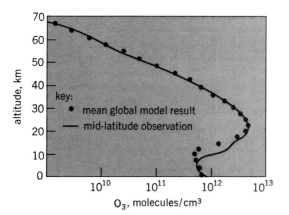

Fig. 1. The global average ozone concentrations predicted in the MIT Stratospheric Circulation Model are compared with mid-latitude rocket observations of ozone. (*From D. Cunnold et al., A three-dimensional dynamical-chemical model of atmospheric ozone, J. Atmos. Sci., 32:170–194, 1975*)

Chemistry of stratospheric ozone. Above about 30 km, oxygen is dissociated during the daytime by ultraviolet photons, $h\nu$, as shown in reaction (1).

$$O_2 + h\nu \rightarrow O + O \qquad (1)$$

$$\lambda < 2400 \text{ A or } 240 \text{ nm}$$

The oxygen atoms produced then form ozone by reaction (2), where M is any other molecule. Ozone

$$O + O_2 + M \rightarrow O_3 + M \qquad (2)$$

has a short lifetime during the day because of photodissociation as shown in reaction (3). However,

$$O_3 + h\nu \rightarrow O_2 + O \qquad (3)$$

$$\lambda < 7100 \text{ A or } 710 \text{ nm}$$

except above 90 km where O_2 begins to become a minor component of the atmosphere, reaction (3) does not lead to a net destruction of ozone. Instead the O is almost exclusively converted back to O_3 by reaction (2). If the odd oxygen concentration is defined as the sum of the O_3 and O concentrations, then odd oxygen is produced by reaction (1) and, in the earliest theories of the ozone layer, removed by reaction (4). One can see that reactions (2) and (3)

$$O + O_3 \rightarrow O_2 + O_2 \qquad (4)$$

do not affect the odd oxygen concentrations but merely define the ratio of O to O_3. Because the rate of reaction (2) decreases with altitude while that for reaction (3) increases, most of the odd oxygen below 60 km is in the form of O_3 while above 60 km it is in the form of O.

Research studies have disclosed that reaction (4) is responsible for only about 18% of the odd oxygen removal rate. The bulk of the removal (about 70%) is caused by the trace gases nitric oxide, NO, and nitrogen dioxide, NO_2, which serve to catalyze reaction (4) by way of reactions (5) and (6).

$$NO + O_3 \rightarrow NO_2 + O_2 \qquad (5)$$

$$NO_2 + O \rightarrow NO + O_2 \qquad (6)$$

This catalytic destruction cycle is partially short-

circuited in the daytime because reaction (5) can be followed by photodissociation, reaction (7),

$$NO_2 + h\nu \rightarrow NO + O \qquad \lambda < 3950 \text{ A or } 395 \text{ nm} \qquad (7)$$

which regenerates odd oxygen. The gases NO and NO_2 constitute only about 3 parts per billion (ppb) of the air in the ozone layer. They are produced naturally in this layer by the decomposition of atmospheric nitrous oxide, N_2O, with excited oxygen atoms, $O(^1D)$, derived from ozone, as shown in reactions (8) and (9). The NO_2 is removed by reac-

$$O_3 + h\nu \rightarrow O_2 + O(^1D) \qquad \lambda < 3100 \text{ A or } 310 \text{ nm} \qquad (8)$$

$$O(^1D) + N_2O \rightarrow NO + NO \qquad (9)$$

tion with hydroxyl radicals, OH, to form hydrogen nitrate, HNO_3, by reaction (10), and reformed by photodissociation of the HNO_3, in reaction (11). If

$$OH + NO_2 + M \rightarrow HNO_3 + M \qquad (10)$$

$$HNO_3 + h\nu \rightarrow OH + NO_2 \qquad (11)$$

$$\lambda < 3450 \text{ A or } 345 \text{ nm}$$

the odd nitrogen family is defined to consist of the species NO, NO_2, and HNO_3, about half of the stratospheric odd nitrogen is present as HNO_3, and downward transport followed by rainout of this water-soluble HNO_3 is probably the main stratospheric odd nitrogen removal mechanism.

Another catalytic cycle which is responsible for about 11% of the total odd oxygen removal rate involves H atoms and OH and HO_2 (hydroperoxyl) radicals. These species are produced in the upper atmosphere by dissociation of water vapor (H_2O), methane (CH_4), and hydrogen peroxide (H_2O_2), and removed by various reactions which re-form H_2O and H_2O_2. A summary of the three main chemical destruction mechanisms for odd oxygen is given in

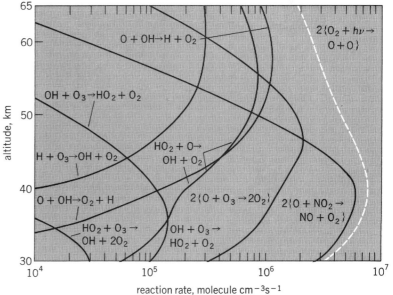

Fig. 2. The rates of reactions important in odd oxygen loss are given by the solid lines as a function of altitude. The total odd oxygen production rate which essentially equals the total oxygen loss rate is given by the broken line. (*From M. McElroy et al., Atmospheric ozone: Possible impact of stratospheric aviation, J. Atmos. Sci., 31:287–303, 1974*)

Fig. 2. Measurements of O, O$_3$, NO, NO$_2$, N$_2$O, HNO$_3$, OH, CH$_4$, and H$_2$O in the stratosphere have largely confirmed the chemistry of the ozone layer outlined above. In addition to these mechanisms, about 1% of the odd oxygen is removed by downward transport of ozone into the troposphere, where it is destroyed at or near the ground.

Circulation of stratospheric ozone. In a purely chemical model of stratospheric ozone with no transport, the maximum ozone concentrations would be obtained at altitudes and latitudes where reaction (1) is fastest; that is, in equatorial regions above 30 km. However, the observed maximum ozone concentrations occur at about 18 km altitude in polar regions. A substantial poleward and downward transport of ozone must therefore occur in the lower stratosphere. This poleward transport is illustrated in Fig. 3, where it is seen that the column abundances of ozone are about 50% greater at the poles than at the Equator, with a maximum in polar ozone in the spring.

Most of the energy for driving the lower stratospheric circulation comes from planetary-scale waves which are generated in tropospheric air by ocean-land temperature contrasts and topographic forcing, and then travel upward. As they ascend, these long waves are damped by emission and

absorption of thermal radiation, and reflected downward when the stratosphere has easterly winds (as in summer) or strong westerly winds (as in the winter upper stratosphere). One would therefore expect the strongest circulation in the fall, spring, and lower winter stratospheres, where there are usually weak westerlies. This period of maximum activity in the lower stratospheric forcing produces the observed spring maximum in ozone. Computations of ozone circulation and chemistry using the Massachusetts Institute of Technology (MIT) Three-Dimensional Stratospheric Circulation Model (MITSCM) have succeeded in simulating these observed seasonal ozone variations (see Fig. 3). At a particular locality, daily variations of a few percent in total ozone are also observed, and these are probably associated with short-term variations in stratospheric motions.

Although the horizontal winds in the stratosphere can be quite large, the increase in temperature with altitude in this region makes it much more stable to vertical motions than the troposphere. Consequently, once destructive species such as NO and NO$_2$ are introduced into the ozone layer, it takes roughly 3 years before they can be removed again by mixing down to the ground. This long stratospheric residence time, combined with the catalytic aspect of the ozone destruction caused by these and similar species, makes the ozone layer particularly sensitive to perturbations in their concentrations. *See* ATMOSPHERIC GENERAL CIRCULATION.

Perturbations to stratospheric ozone. Supersonic aircraft designed to fly in the lower stratosphere (such as the Anglo-French Concorde, Russian Tupolev-144, and American Boeing 2707) produce NO and NO$_2$ in their engines by thermal decomposition of air (N$_2$ and O$_2$). A fleet of about 500 of the now-canceled American aircraft was expected to be flying by the year 2000. Such a fleet would inject NO and NO$_2$ into the stratosphere at a rate some three times greater than that from the present N$_2$O source and would result in a significant perturbation to the atmospheric nitrogen cycle. Computations using the MITSCM for this fleet flying at 20 km altitude and 45°N latitude imply ozone reductions of 8 and 16% in the Southern and Northern Hemispheres, respectively. These reductions would cause increases in the sunburn rate for Caucasian skin of roughly 16 and 32%, respectively. Similar ozone reductions would result from fleets of several thousand of the smaller, lower-flying Concordes. International restrictions on fleet sizes for stratospheric aircraft appear necessary.

There has also been concern that industrial production of the chlorofluoromethanes CFCl$_3$ (Freon 11) and CF$_2$Cl$_2$ (Freon 12) may be significantly altering the natural chlorine cycle. These very inert species are principally used as refrigerants and aerosol-can propellants. Once they are released into the atmosphere, their only presently recognized removal mechanism involves photodissociation in the stratosphere, as in reactions (12) and (13). The chlorine atoms released in these and subsequent reactions can catalytically destroy ozone by conversion to chlorine oxide radicals, ClO.

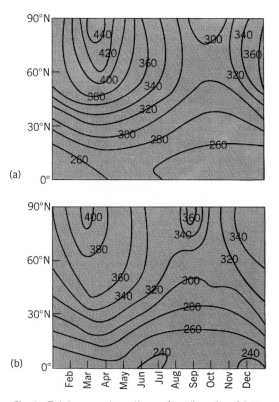

Fig. 3. Total ozone above the surface (in units of 2.7 × 10^{16} molecules per square centimeter) in the Northern Hemisphere as a function of season and latitude: (a) as determined from observations, and (b) as predicted by the MIT Stratospheric Circulation Model. (*From F. Alyea, D. Cunnold, and R. Prinn, Stratospheric ozone destruction by aircraft-induced nitrogen oxides, Science, 188:117–121; copyright 1975 by the American Association for the Advancement of Science*)

$$CFCl_3 + h\nu \rightarrow CFCl_2 + Cl \qquad (12)$$
$$\lambda < 2260\ A\ or\ 226\ nm$$

$$CF_2Cl_2 + h\nu \rightarrow CF_2Cl + Cl \qquad (13)$$
$$\lambda < 2140\ A\ or\ 214\ nm$$

Reactions (14)–(18) play roles similar to reactions

$$Cl + O_3 \rightarrow ClO + O_2 \qquad (14)$$
$$ClO + O \rightarrow Cl + O_2 \qquad (15)$$
$$ClO + h\nu \rightarrow Cl + O \qquad (16)$$
$$\lambda < 3035\ A\ or\ 303.5\ nm$$
$$Cl + CH_4 \rightarrow HCl + CH_3 \qquad (17)$$
$$OH + HCl \rightarrow Cl + H_2O \qquad (18)$$

(5)–(7), (10), and (11). Computations in one-dimensional atmospheric chemical models imply that the chlorofluoromethanes are presently responsible for at most a 1% reduction in stratospheric ozone. However, when injection of $CFCl_3$ and CF_2Cl_2 into the atmosphere is continued at current rates, the models predict about a 10% decrease in ozone by the year 2050. If reactions (12) and (13) are indeed the only atmospheric removal mechanisms for $CFCl_3$ and CF_2Cl_2, then their atmospheric lifetimes are about 45 and 70 years, respectively. Therefore, even if chlorofluoromethane production were halted, the effects would still be felt for several decades thereafter. The chlorine chemistry is not as fully understood as the nitrogen oxide chemistry, but recent stratospheric observations of $CFCl_3$, CF_2Cl_2, and HCl provide qualitative support for the proposed chemical cycles.

A number of other potential ozone-destroying processes with anthropogenic origins have also been identified. Atmospheric tests of thermonuclear weapons inject NO and NO_2, produced by thermal decomposition of air, into the stratosphere. Increased use of nitrate (NO_3^-) fertilizers may increase the rate of production of N_2O from soil nitrate by denitrifying bacteria, resulting in increased stratospheric N_2O concentrations. Solid-fuel rockets using ammonium perchlorate (NH_4ClO_4) as the oxidizer (Minuteman missiles, NASA space shuttle booster) inject small amounts of hydrogen chloride (HCl) into the stratosphere. Expanded production of methyl bromide (CH_3Br), an agricultural fumigant, may result in significant stratospheric bromine concentrations. Catalytic odd oxygen destruction by bromine is similar to but faster than that by chlorine. The global measurements of ozone which are required to ascertain the reality of all these possible long-term ozone depletions have become possible from satellites. Continuous ozone monitoring of this type is clearly mandatory for the future. *See* ATMOSPHERIC CHEMISTRY; ATMOSPHERIC POLLUTION.

[RONALD G. PRINN]

Bibliography: F. Alyea, D. Cunnold, and R. Prinn, Stratospheric ozone destruction by aircraft-induced nitrogen oxides, *Science*, 188:117–121, 1975; P. Crutzen, A review of upper atmospheric photochemistry, *Can. J. Chem.*, 52:1569–1581, 1974; D. Cunnold et al., A three-dimensional dynamical-chemical model of atmospheric ozone, *J. Atmos. Sci.*, 32:170–194, 1975; H. Johnston, global ozone balance in the natural stratosphere, *Rev. Geophys. Space Phys.*, 13:637–649, 1975; A. K. Khrgian, in P. Greenberg (ed.), *The Physics of Atmospheric Ozone*, 1976; M. McElroy et al., Atmospheric ozone: Possible impact of stratospheric aviation, *J. Atmos. Sci.*, 31:287–303, 1974; F. Rowland and M. Molina, Chlorofluoromethanes in the environment, *Rev. Geophys. Space Phys.*, 13:1–35, 1975.

Atmospheric pollution

Alteration of the atmosphere by the introduction of natural and artificial particulate contaminants. Most artificial impurities are injected into the atmosphere at or near the Earth's surface. The atmosphere cleanses itself of these quickly, for the most part. This occurs because in the troposphere, that part of the atmosphere nearest to the Earth, temperature decreases rapidly with increasing altitude (Fig. 1), resulting in rapid vertical mixing: the rainfall sometimes associated with these conditions also assists in removing the impurities. Exceptions, such as the occasional temperature inversion layer over the Los Angeles Basin, may have notably unpleasant results. *See* ATMOSPHERE; TROPOSPHERE.

In the stratosphere, that is, above the altitude of

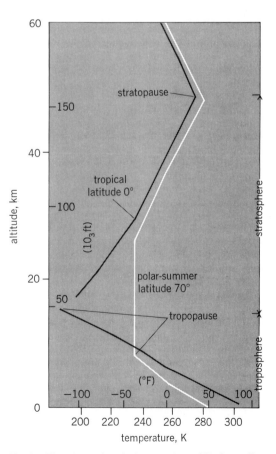

Fig. 1. The atmosphere's temperature-altitude profile. (*Adapted from R. E. Newell, Radioactive Contamination of the Upper Atmosphere, in Progress in Nuclear Energy, ser. 12: Health Physics, vol. 2, p. 538, Pergamon Press, 1969*)

the temperature minimum (tropopause), either temperature is constant or it increases with altitude, a condition that characterizes the entire stratosphere as a permanent inversion layer. As a result, vertical mixing in the stratosphere (and hence, self-cleansing) occurs much more slowly than that in the troposphere. Contaminants introduced at a particular altitude remain near that altitude for periods as long as several years. Herein lies the source of concern: the turbulent troposphere cleanses itself quickly, but the relatively stagnant stratosphere does not. Nonetheless, the pollutants injected into the troposphere and stratosphere have impact on humans and the habitable environment. *See* STRATOSPHERE; TROPOPAUSE.

All airborne particulate matter, liquid and solid, and contaminant gases exist in the atmosphere in variable amounts. Typical natural contaminants are salt particles from the oceans or dust and gases from active volcanoes; typical artificial contaminants are waste smokes and gases formed by industrial, municipal, household, and automotive processes, and aircraft and rocket combustion processes. Another postulated important source of artificial contaminants is certain fluorocarbon compounds (gases) used widely as refrigerants, as propellants for aerosol products, and for other applications. Pollens, spores, rusts, and smuts are natural aerosols augmented artificially by humans' land-use practices. *See* ATMOSPHERIC CHEMISTRY; SMOG.

Sources and types. Sources may be characterized in a number of ways. A frequent classification is in terms of stationary and moving sources. Examples of stationary sources are power plants, incinerators, industrial operations, and space heating. Examples of moving sources are motor vehicles, ships, aircraft, and rockets. Another classification describes sources as point (a single stack), line (a line of stacks), or area (a city).

Different types of pollution are conveniently specified in various ways: gaseous, such as carbon monoxide, or particulate, such as smoke, pesticides, and aerosol sprays; inorganic, such as hydrogen fluoride, or organic, such as mercaptans; oxidizing substances, such as ozone, or reducing substances, such as oxides of sulfur and oxides of nitrogen; radioactive substances, such as iodine-131, or inert substances, such as pollen or fly ash; or thermal pollution, such as the heat produced by nuclear power plants.

Air contaminants are produced in many ways and come from many sources. It is difficult to identify all the various producers. For example, it is estimated that in the United States 60% of the air pollution comes from motor vehicles and 14% from plants generating electricity. Industry produces about 17% and space heating and incineration the remaining 9%. Other sources, such as pesticides and earth-moving and agricultural practices, lead to vastly increased atmospheric burdens of fine soil particles, and of pollens, pores, rusts, and smuts; the latter are referred to as aeroallergens because many of them induce allergic responses in sensitive persons.

The annual emission over the United States of many contaminants is very great (Fig. 2). As mentioned, motor vehicles contribute about 60% of total pollution: nearly all the carbon monoxide, two-thirds of the hydrocarbons, one-half of the nitrogen oxides, and much smaller fractions in other categories.

Pollution in the stratosphere. Sources of contaminants in the stratosphere are effluents from high-altitude aircraft such as the supersonic transport (SST), powerful nuclear explosions, and volcanic eruptions. There are also natural and artificial sources of gases which diffuse from the troposphere into the stratosphere.

Table 1 lists the natural burden of gases and particles injected into the stratosphere by high-flying aircraft, assuming the consumption of 2×10^{11} kg of fuel during a period of one year. It is to be noted that the percentage increase over the natural burden is substantial for NO, NO_2, HNO_3, and SO_2. Since the concentrations are substantial, they may adversely affect humans' living environment.

Other manufactured pollutants which diffuse from the troposphere into the stratosphere are the halogenated hydrocarbons, specifically dichloromethane (CF_2Cl_2) and trichlorofluoromethane ($CFCl_3$) gases which are used as propellants in many of the so-called aerosol spray cans for deodorants, pesticides, and such, and as refrigerants. These gases have an average residence time (residence time is the time required for a substance to reduce its concentration by $1/e$, approximately 1/3) of 1000 years or more. $CFCl_3$ is one of a family of halogenated hydrocarbons (also known as fluorocarbons) which are widely used. For example, in 1973 world production of these compounds was 1.7×10^9 lb (770×10^6 kg), which represented an 11% growth over the 1972 production. Almost all halogenated hydrocarbons are ultimately released to the atmosphere.

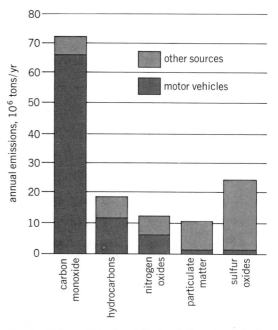

Fig. 2. Motor vehicles' contribution to five atmospheric contaminants in the United States.

Table 1. The natural stratospheric background of several atmospheric gases from 13 to 24 km compared to engine emissions

Gas	Mass mixing ratio	Natural burden, kg		Increase in mass due to aircraft emission, %‡
		IDA*	Penndorf†	
CO_2	480 E-6	500 E-12§	480 E-12	0.1
H_2	2.7 E-6	2 E-12	2.7 E-12	9
CH_4	0.55 E-6	1 E-12	0.55 E-12	0.02
CO	0.05–0.1 E-6	30 E-9	50–100 E-9	0.6–1.2
NO	0.5 E-9	1 E-9	0.52 E-9	100
NO_2	1.6 E-9	3 E-9	1.8 E-9	100
HNO_3	4 E-9	<10 E-9	3.6 E-9	85
SO_2	1 E-9	4 E-9 (?)	1.4 E-9	10–40
Aerosol ($\alpha > 0.1\ \mu m$)	2 E-9	0.3 E-9	2 E-9	10

*Estimation by R. Oliver, Institute for Defense Analyses, 1974.
†Estimation by R. Penndorf, *CIAP Atmospheric Monitoring and Experiments, The Program and Results*, DOT-TST-75-106, pp. 4–7, 1975.
‡One-year fuel consumption by stratospheric aircraft of 2×10^{11} kg.
§Read 500 E-12 as 5×10^{12}.

The nitrous oxide (N_2O) and the halogenated hydrocarbons reach the upper regions of the stratosphere, where they are photodissociated by the Sun's radiation to produce nitric oxide (NO) and chlorine (Cl). The NO and Cl react with the ozone, as in Eqs. (1)–(6). The ozone is destroyed

$$N_2O + O \rightarrow 2NO \qquad (1)$$

$$NO + O_3 \rightarrow NO_2 + O_2 \qquad (2)$$

$$NO_2 + O \rightarrow NO + O_2 \qquad (3)$$

$$CF_2Cl_3 + h\nu \ (\text{solar energy}) \rightarrow CFCl_2 + Cl \qquad (4)$$

$$Cl + O_3 \rightarrow ClO + O_2 \qquad (5)$$

$$ClO + O \rightarrow Cl + O_2 \qquad (6)$$

by NO and Cl respectively, whereas NO and Cl are conserved. Stratospheric ozone is valuable as a filter for solar ultraviolet (uv) radiation. A decrease of its concentration results in an increase in the amount of uv impinging on the surface of the Earth. As an illustration based on theoretical considerations, an injection 2×10^9 kg/yr of $NO_x (NO_x = NO + NO_2)$ at 17 km (see Fig. 1) will result in a reduction of the total amount of ozone by about 3%. This represents an increase of vertically incident uv on the Earth of 6%. An increase of uv could adversely affect humans and plants. *See* ATMOSPHERIC OZONE.

SO_2-aerosol-climate relations. Aircraft engine effluents contain SO_2, as shown in Table 1, and could add a considerable amount of aerosol particles at the end of this century for the predicted aircraft fleet sizes. Table 1 indicates aircraft effluents could increase the natural background by 10–40%—seemingly large, yet small compared to volcanic injections, for which estimates range up to 10,000%.

Why are particles so important? They scatter and absorb (in specific wavelength regions) solar radiation, and thereby influence the radiative budget of the Earth-atmospheric system, and finally perhaps the climate on the ground. The particles formed from aircraft effluents may increase the optical thickness of the layer, the upwelling and downwelling infrared radiation, and the albedo of the Earth. The average global albedo of the Earth-atmospheric system has been measured as 28% with probably some short-term variation of unknown but small magnitude. It has been calculated that for an additional mass of 0.1 $\mu g/m^3$ particles over a 10-km layer from 15 to 25 km (equivalent to 5.1×10^8 kg for the whole Earth, or about 20% of the "natural" background concentration), the albedo increases by about 0.05% (from 28 to 28.05%) at low latitudes all year and at high latitudes in summer, but by about 0.1–0.15% from September to February in high latitudes. If the added mass is larger than 0.1 $\mu g/m^3$, the albedo increases proportionally to the cited numbers. For the optical thickness of the stratosphere, a value of 0.02 is generally assumed. While the present subsonic flights increase this value by a very small amount (10^{-4}), a large fleet of high-flying aircraft (Table 1) could increase it by about 10%.

The chemistry of the natural stratospheric aerosols is dominated by sulfate, presumably of volcanic origin. These naturally occurring aerosols are concentrated in thin layers at altitudes between 15 and 25 km. R. Cadle has described the chemical composition of stratospheric particles at an altitude of 20 km during the period 1969–1973 as consisting of 48%, by mass, of sulfate; 24% of stony elements (such as silicon, aluminum, calcium, and magnesium); and 20% of chlorine; other constituents make up the remainder. The amount of sulfate introduced into the stratosphere as a result of a large fleet of SSTs operating at about 18–20 km may, in a worst-case estimate, equal the total amount occurring naturally. The influence of dynamic motions of the stratosphere on the distribution of aerosols is indicated by Fig. 2, which illustrates high correlation of the aerosol and ozone-rich layers in the 15-km region. Moreover, the water vapor mixing ratio also increases in layers at about 15–20 km. Since there is no known chemical link between the production of aerosols and that of water vapor and ozone, the observation illustrated by Fig. 3 may be a dynamic rather than a chemical effect, the implication being that the dynamic effects are far more important in determining the relative profiles of these constituents than any chemical effect at this altitude.

The perturbation of the lower stratosphere by the engine effluents of a large fleet of vehicles may strongly increase its optical thickness to visual-band solar radiation, which has a natural value of about 0.02. An increase in optical thickness results in a reduction of solar radiation, the principal source of atmospheric heating, by about the same fraction. This effect can be likened to a reduction in solar constant by about the same amount at the subsolar point, and about double that value when the solar zenith angle is 60° or greater, as it may be at high latitudes. The sensitivity of the troposphere to changes in the solar constant has been studied by M. MacCracken. In the modeling for which the computed precipitation is illustrated by Fig. 4, a 10% reduction of solar constant leads to a reduction of average temperatures, from 3°C at the Equator to 10°C average from latitude 40° to the

pole. The winds are substantially weakened, and total precipitation reduced, mainly in the summer. Snowfall increases, and the winter snow line moves lower in latitude by about 5°.

An increase in the solar constant of 13% results in an annual mean temperature increase of 2–5°C at all latitudes. Although the overall precipitation increases, as illustrated in Fig. 4, the relative humidity decreases slightly, and thus cloudiness decreases. The total snowfall decreases, and the snow line moves higher in latitude by 10°. Extensive melting of the polar ice caps also begins. The effects of doubling of the stratospheric optical thickness, for example, are an order of magnitude smaller than the changes of precipitation and snowfall depicted in Fig. 4.

Sinks. A sink is defined as a process by which gases or particles are removed from a given volume of atmosphere. It could be chemical, homogenous or heterogenous (gas-solid reactions), or dynamical, such as dispersion (transport), diffusion, gravity, or precipitation.

In the stratosphere, three important contaminants are NO, Cl, and SO_2. The sink for NO_2 is a chemical reaction by which NO_2 is combined by a complex method with (OH)—a derivative of water (H_2O) present in the stratosphere in parts per million (ppm)—to form nitric acid (HNO_3).

The SO_2 reacts chemically with oxygen (O), water, and its derivatives (OH, HO_2) to form H_2SO_4. Then, water molecules are absorbed by H_2SO_4 to form $nH_2O \cdot (H_2SO_4)$ cluster or aerosol (where n, number, is equal to about 2). Aggregates of these polymolecules, growing larger with each successive collision, eventually become aerosols; when the diameter is greater than 0.1 μm, they act as scatterers of the visual band of sunlight.

The chlorine (Cl) chemically reacts with oxygen and water to form Cl, ClO, and HCl. In the stratosphere, these reactions involving small concentrations have not been measured adequately.

In the troposphere, the predominant sinks are dispersion, transport, and precipitation. The chemical reactions in the troposphere are less active in general than in the stratosphere.

Dispersion. Dispersion of pollution is dependent on atmospheric conditions. Winds transport and diffuse contaminants; rain may wash them to the surface; and under cloudless skies, solar radiation may induce important photochemical reactions.

Wind direction, speed, and turbulence influence atmospheric pollution. Wind direction determines the area into which the pollution is carried. Dilution of contaminants from a source is directly proportional, other factors being constant, to wind speed, which also determines the intensity of mechanical turbulence produced as the wind flows over and around surface objects, such as trees and buildings.

Eddy diffusion by wind turbulence is the primary mixing agency in the troposphere; molecular diffusion is negligible in comparison. In addition to mechanical turbulence, there is thermal turbulence which occurs in an unstable layer of air. Thermal turbulence and associated intense mixing develop in an unsaturated layer in which the temperature decreases with height at a rate greater than 1°C/100 m, the dry adiabatic rate of

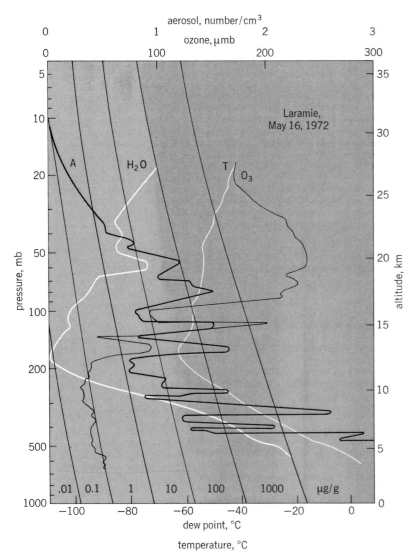

Fig. 3. A simultaneous measurement of the vertical distribution of ozone (O_3), water vapor (H_2O), and temperature (T). The aerosol (A) sounding was made about 5 hr before the ozone–water vapor soundings. The smooth curves are lines of constant water vapor mixing ratio. (*From T. J. Pepin, J. M. Rosen, and D. H. Hoffman, The University of Wyoming Global Monitoring Program, in Proceedings of the AIAA/AMS International Conference on the Environmental Impact of Aerospace Operations in the High Atmosphere, AIAA Pap. no. 73-521, 1973*)

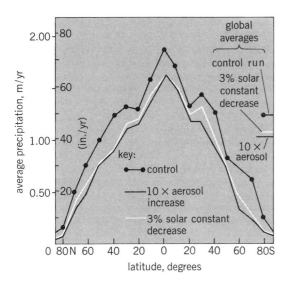

Fig. 4. Latitudinal distribution of total precipitation. (*From M. C. MacCracken, in Report of Findings: Effects of Stratospheric Pollution by Aircraft, DOT-TST-75-50, 1974*)

cooling. When the temperature decreases at a lower rate, the air is stable, and turbulence and mixing, now primarily mechanical, are less intense. If the temperature increases with height—its normal behavior in the stratosphere, creating a condition known as an inversion—the air is stable, and horizontal turbulence and mixing are still appreciable, but vertical turbulence and mixing are almost completely suppressed. *See* TEMPERATURE INVERSION.

Inversion. Precipitation, fog, and solar radiation exert secondary meteorological influences. Falling raindrops may collect particles with radii greater than 1 μm or may entrain gases and smaller particles and carry them to the ground. Gas reactions with aerosols also occur; neutralizing cations in fog droplets or traces of ammonia (NH_3) in the air act as catalysts to accelerate reaction rates leading to rapid oxidation of sulfur dioxide (SO_2) in fog droplets. For highly polluted city air, it is estimated that in the presence of NH_3, the oxidation of the SO_2 to ammonium sulfate, $(NH_4)_2SO_4$, is completed in 1 hr for fog droplets 10 μm in radius. Photochemical oxidation of hydrocarbons in sunlight is frequent. Most hydrocarbons do not have appropriate absorption bands for a direct photochemical reaction; nitrogen dioxide (NO_2), when present, acts as an oxidation catalyst by absorbing solar radiation strongly and subsequently transferring the light energy to the hydrocarbon and thereby oxidizing it.

Natural ventilation in the atmosphere is best when the winds are strong and turbulent so that mixing is good, and when the volume in which mixing occurs is large so that dilution of pollution is rapid. As cities have grown in size, air pollution has become more widespread. It has become necessary to think of whole urban complexes as large area sources of pollution. The rate of natural ventilation of an urban area is dependent on two quantities: the wind speed and the mixing volume over the city. Active mixing upward is often limited by a stable layer, perhaps even a very stable

inversion layer, aloft. The upward extent of this region of active mixing, known as the mixing height, determines the magnitude of the mixing volume of the city.

The number of air changes per unit time in this mixing volume specifies the rate of natural ventilation of the urban area. The problems of air pollution become highly complex, however, because the mixing height is rarely constant for long. Some of the factors causing it to vary are described below.

At night when the sky is clear and the wind light, Earth's surface loses heat by long-wave radiation to space. As a result, the ground cools and a surface radiation inversion is formed. The inversion inhibits mixing, so that pollution accumulates. Solar heating of the ground causes a reversal of the lapse rate, which may exceed the dry adiabatic rate of cooling and enhances active mixing in the unstable layer.

The mixing may bring pollution from aloft, causing a temporary peak in the surface concentrations, a process known as an inversion breakup fumigation. By midafternoon, the height of mixing is a maximum for the day, and surface concentrations tend to be low as the natural ventilation improves. In the evening, the lapse rate becomes stable, and accumulation of contaminants may begin again.

Subsidence inversion. The accumulation of pollution for longer periods of time is especially likely to occur if a persistent inversion aloft exists. Such an inversion aloft is the subsidence inversion formed by the sinking and vertical convergence of air in an anticyclone, illustrated in Fig. 5. A layer of air at high levels descends, diverging horizontally and hence converging in the vertical, and warms at the dry adiabatic rate of heating of 1°C/100 m. Figure 5a shows how a low-level inversion may result from this process, while Fig. 5b depicts how the mixing height H is limited in vertical extent by the subsidence inversion aloft, so that pollution accumulates within and just above the city. It is the presence of such a subsidence inversion aloft associated with the Pacific subtropical anticyclone which is the primary cause of Los Angeles and other California smogs; these are made even worse by local mountain and valley sides which prevent horizontal dispersion.

Fog. The worst pollution occurs when, in addition to subsidence inversions accompanying slowly moving or stationary anticyclones, fog also develops. All the major air pollution disasters, such as those listed in a later section, took place when fog persisted during protracted stagnant anticyclonic conditions. The reasons for the adverse influence of fog are shown in Fig. 6. When there is no fog (Fig. 6a), solar radiation heats the ground, which in turn causes a lapse rate equal to, or greater than, the dry adiababic rate of cooling, with good mixing and hence a substantial mixing height H. On the other hand, with a fog layer (Fig. 6b), up to 70% of the solar radiation incident at the top of the fog is reflected to space, with relatively little left to heat the fog and ground below. With the cloudless skies characteristic of anticyclonic weather, there is a continuous loss of heat to outer space from the upper surface of the fog bank,

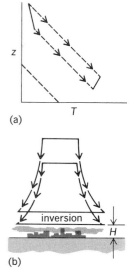

Fig. 5. Subsidence inversion. (*a*) Solid lines show temperature T and height z before and after dry adiabatic descent of air; dashed lines represent dry adiabatic rate of heating. (*b*) Inversion limits mixing height H over ci

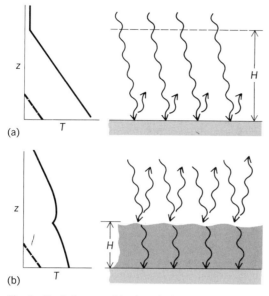

Fig. 6. The influence of fog in reducing mixing height H. (a) Without fog. (b) With fog. The dashed lines represent the dry adiabatic rate of cooling. Arrows represent solar and reflected radiation.

which acts radiatively as an elevated ground surface. More heat is lost to space than is gained from the Sun, and an inversion develops above the fog and persists night and day until the anticyclone dissipates or moves away. If the air is polluted, the fog particles may become acids and salts in solution; the saturation vapor pressure over such particles may decrease to 90 or 95% of the pure water value, so the smog becomes even more persistent

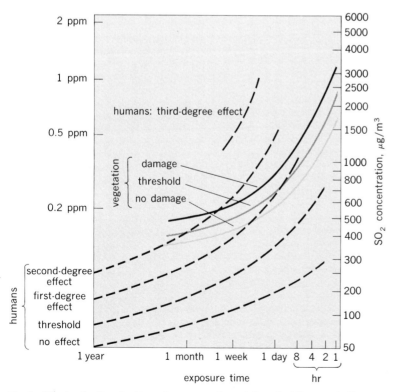

Fig. 7. Effects of sulfur dioxide on humans and vegetation. (*L. J. Brasser et al.*)

than if it remained as a pure water fog. Disastrous concentrations of contaminants may accumulate during prolonged foggy conditions of this kind.

Warm fronts. Another significant inversion aloft is associated with a slowly moving warm frontal surface. Consider two cities, one lying to the southwest and the other lying to the northeast of a warm front extending from the southeast to the northwest (and moving in a northeast direction), as illustrated in Fig. 7. City B lies in the cool air, with the warm frontal above it. In the cool air ahead of the warm front, the pollution from City B is trapped below the warm frontal inversion and may travel for many miles with large surface concentrations. On the other hand, the prevailing southwest winds in the warm sector will carry pollution from City A up and above the warm frontal inversion, which effectively prevents its diffusion downward to the surface. This situation brings out an important point: an inversion layer may be advantageous, not disadvantageous, if it inhibits diffusion down to the ground.

Stack dispersion. Dispersion from an elevated point source, such as a stack, is conveniently expressed by Eq. (7), where χ is ground level con-

$$\chi = \frac{Q}{\pi\sigma_y\sigma_z\bar{u}} \exp\left[-\frac{1}{2}\left(\frac{y^2}{\sigma_y^2}+\frac{h^2}{\sigma_z^2}\right)\right] \quad (7)$$

centration of contaminant in mass per unit volume; Q is the source strength in mass per unit time; $\bar{u}$ is mean wind speed; y is the horizontal direction perpendicular to the mean wind $\bar{u}$; σ_y and σ_z are diffusion coefficients expressed in length units in the y and z directions, respectively, the z direction being vertical; and h is the height of the source above ground.

This diffusion equation should be used only under the simplest conditions, for example, in flat uniform terrain and well away from hills, slopes, valleys, and shorelines. Table 2 lists various meteorological categories, and Table 3 gives values of the diffusion coefficients appropriate for each category. It should be noted that the values to be used depend on distance from the source at which concentrations are to be calculated. A variety of other forms is available for more complex conditions of terrain and meteorology.

Natural cleansing processes. Pollution is removed from the atmosphere in such ways as washout, rain-out, gravitational settling, and turbulent impaction. Washout is the process by which contaminants are washed out of the atmosphere by raindrops as they fall through the contaminants; in rain-out the contaminants unite with cloud droplets, which may later grow into precipitation.

Gravitational settling is significant mainly for large particles, those having a diameter greater than 20 μm. Agglomeration of finer particles may result in larger ones which settle out by gravitation. Fine particles may also impact on surfaces by centrifugal action in very small turbulent eddies. Gases may be converted to particulates, as by photochemical action of sunlight in Los Angeles, Denver, and Mexico City. These particulates may then be removed by settling or impaction.

The rate of natural cleansing may be slower than

Table 2. Meteorological categories*

| Surface wind speed, m/s | Daytime insolation | | | Thin overcast or $\geq$ 4/8 cloudiness† | $\leq$ 3/8 cloudiness |
	Strong	Moderate	Slight		
<2	A	A-B	B		
2	A-B	B	C	E	F
4	B	B-C	C	D	E
6	C	C-D	D	D	D
<6	C	D	D	D	D

*A = Extremely unstable conditions. D = Neutral conditions (applicable to heavy overcast, day or night).
B = Moderately unstable conditions. E = Slightly stable conditions.
C = Slightly unstable conditions. F = Moderately stable conditions.
†The degree of cloudiness is defined as that fraction of the sky above the local apparent horizon which is covered by clouds.

the rate of injection of pollutants into the atmosphere, in which case pollution may increase on a global scale. There is evidence that the concentration of atmospheric carbon dioxide has been increasing slowly since the beginning of the century because of combustion of fossil fuels. The tropospheric burden of very small particles and of Freon gas may also be increasing.

Effects of stratospheric pollution. Pollution of the stratosphere with nitrogen oxide causes reduction of stratospheric ozone. Ozone reduction in the stratosphere has been linked to biological effects such as skin cancer in two steps: (1) reduced ozone in the stratosphere causes an increase in uv radiation reaching the Earth's surface, and (2) increased uv radiation enhances the normal biological effects of natural uv radiation.

Biological damage. In step 1, the relation between the reduction of stratospheric ozone and the increase in solar uv flux effective in causing sunburn and, presumably, also skin cancer is readily calculable. The factors of interest here are the uv wave band of 290–320 nanometers (nm), the middle latitudes (where the summer sun is nearly overhead and where the worst cases of skin cancer occur), and small decreases in ozone. For these factors of interest, the percent increase in solar ultraviolet flux is about twice the percent decrease in ozone.

Step 2, from uv radiation to the enhancement of skin cancer incidence, involves the assumption, supported by some scientific evidence but not proven by experiments on humans, that skin cancer in humans is induced by exposure to uv radiation of the same wavelength (290–320 nm) that causes erythema (sunburn), and that the relative effectiveness of the various wavelengths for car-

cinogenesis is the same as that for sunburn. Estimates of biological damage to humans from exposure to uv radiation are based on inferences from the statistics of epidemiological surveys of humans and of laboratory experiments with animals. Nonmelanomic skin cancer, which is almost never fatal if given proper care, occurs primarily on sun-exposed areas of the skin, especially the face and hands. It is relatively common—about 250 cases per 100,000 persons occur in fair-skinned Caucasians in the United States. The incidence of nonmelanomic skin cancer is correlated with latitude—and, therefore, with sunlight, including uv flux, since average sunlight varies with latitude.

Climatic changes. Pollution of the stratosphere may involve also a climate chain of cause-and-effect relations by which aircraft engine effluents, notably sulfur dioxide (SO_2), and to a lesser degree water vapor (H_2O) and nitrogen oxides (NO_x), affect climatic variables such as temperature, wind, and rainfall.

If enough particles larger than 0.1 μm in diameter were added to the stratosphere, they could alter the radiative heat transfer of the Earth-Sun system, and thereby influence climate. Particles of this size are produced by several constituents of engine emissions, in particular those of SO_2. When considering the large numbers of aircraft postulated for future operations in the stratosphere, the amount of particles developed from the SO_2 engine emissions are potentially serious, unless the fuels used have a sulfur content smaller than that of today's fuels.

The sequence in the climate cause-and-effect chain is proposed as follows. Stratospheric SO_2, after first being oxidized, interacts with the abundant water vapor exhaust from jet engines to pro-

Table 3. Values of diffusion coefficients

| Distance from source, m | Diffusion coefficient, m², in various meteorological categories | | | | | |
	A	B	C	D	E	F
$10^2, \sigma_y$	22	16	12	8	6	4
10^3	210	150	105	75	52	36
10^4	1700	1300	900	600	420	360
10^5	11,000	8500	6300	4100	2800	2000
$10^2, \sigma_z$	14	11	7.6	4.8	3.6	2.2
10^3	500	120	70	32	24	14
10^4	—	—	420	140	90	46
10^5	—	—	2100	440	170	92

duce solid sulfuric acid particles that build up to sizes greater than 1 μm. These particles disperse within the stratosphere, principally within the hemisphere in which they are injected, where they remain for periods as long as 3 years, depending on their altitude.

The effects of SO_2 emissions are summarized in Table 4, where the stratosphere's opacity to sunlight is represented by optical thickness. The natural optical thickness of the stratosphere is about 0.02, which means that sunlight and heat are reduced by about 2% while passing through the stratosphere.

The present subsonic fleet of about 1700 aircraft operates at times in the low stratosphere, burning a total of about 2×10^{10} kg (2×10^7 tons) of fuel per year and producing an estimated increase in the stratosphere's optical thickness of about 0.0001 or 0.5% of the natural value.

As a matter of history, the variability of temperature due to natural causes is a substantial fraction of 1°C, even over a few decades. The year-to-year variation is several tenths of a degree. In the 1890–1940 period, a warming of ½°C occurred. During the 1940–1960 period, a general cooling of ⅕°C took place.

Effects of tropospheric pollution. A number of the many effects of tropospheric pollution are described briefly below.

Humans. The effects of many pollutants on human health under most ordinary circumstances of living, rural or urban, are difficult to specify with confidence. In the United States, a number of animal experiments under controlled conditions and epidemiological studies have been made, but the results are difficult to interpret in terms of human health. A study on the lower East Side of New York City indicated that, in children less than 8 years old, the occurrence of respiratory symptoms was associated with the levels of particulate matter and of carbon monoxide in the atmosphere. With heavy smokers, however, eye irritation and headache were directly related to increasing concentrations of carbon monoxide.

Air pollution is suspected as a causative agent in the occurrence of chronic bronchitis, emphysema, and lung cancer, but the evidence is not clear cut. On the other hand, from mid-August to late September the potent aeroallergen, ragweed pollen, is a substantial cause of allergic rhinitis and bronchi-

al asthma for over 10,000,000 persons living east of the Rockies. In Los Angeles County the effects of smog on health are becoming better understood. For example, people living in less smoggy areas of the county survive heart attacks more readily than others: In 1958 in high-pollution areas the mortality rate per 100 hospital admissions for heart attacks averaged 27.3 in comparison with 19.1 for low-smog areas. Other studies show a small but significant relation between motor vehicle accidents and oxidant levels. Medical authorities in Los Angeles are becoming concerned about the long-term influences of various pollutants, including photochemical smog, despite the lack of comprehensive knowledge of the nature of such effects.

In Great Britain there is a similar lack of precise knowledge. There are indications that emphysema, bronchitis, and other respiratory diseases are not caused primarily by increased atmospheric pollution, but because more people are living longer. Despite a substantial reduction in the concentrations of atmospheric particulates since the Clean Air Act was passed in 1956, the respiratory disease rate continues to rise. These facts do not prove that air pollution is not a factor, but that its influence may be synergistic and therefore difficult to identify precisely. For example, it is known that the combined effect of sulfur oxides and particulates is substantially greater than the sum of the two separate effects, and many other such synergistic effects doubtless occur.

In the Netherlands special efforts have been made to relate SO_2, both concentration values and exposure times, to effects on humans and on vegetation. The results of these studies, based on investigations in England, the United States, West Germany, Italy, and the Netherlands, are illustrated in Fig. 7. Influences on humans are shown in the lower family of curves: A first-degree effect is a small increase in functional disturbances, symptoms, illnesses, diseases, and deaths; a second-degree effect is a more prevalent or more pronounced effect of the same kind; and a third-degree effect is a substantial increase in the number of deaths. It should be emphasized that the exposures were, in general, to SO_2 in dusty and sooty atmospheres.

Under extreme circumstances when stagnant atmospheric conditions with persistent low wind

Table 4. Estimated increase in stratospheric optical thickness per 100 aircraft

Subsonic aircraft type*	Fuel burned, kg/yr†	Altitude, km (10³ ft)	Maximum SO₂ EI‡ without controls, g/kg fuel	Percent change in stratospheric optical thickness in Northern Hemisphere	
				Without controls	With future EI controls achieving only 1/20 of emission of present-day aircraft
707/DC-8	1×10^9	11 (36)	1	0.023	0.0012
DC-10/L-1011	1.5×10^9	11 (36)	1	0.032	0.0016
747	2×10^9	11 (36)	1	0.044	0.0022
747-SP	2×10^9	13.5 (44)	1	0.10	0.0050

*The present subsonic fleet consists of 1217 707/DC-8s, 232 DC-10/L-1011s, and 232 747s flying at a mean altitude of 11 km (36,000 ft), and is estimated to cause an increase in stratospheric optical thickness of 0.5%.

†Subsonics are assumed to operate at high altitude, 5.4 hr per day, 365 days per year.

‡EI is emission index, which is defined as grams of pollutants per kilogram of fuel used.

and fog exist, major disasters involving many deaths occurred, as in and around London in 1873, 1880, 1891, 1948, 1952, 1956, and 1962. Similar diasters occurred in the Meuse Valley of Belgium in 1930 and at Donora, Pa., in 1948.

Atmospheric pollution has a substantial influence on the social aspects of human life and activity. For example, the distribution of urban populations is being increasingly affected by such pollution, and recreational patterns are similarly influenced. The atmospheric burden of pollution is thus becoming more and more important as a determinant in social decision making.

Animals. Studies of the response of laboratory animals to specified concentrations of pollutants have been conducted for many years, but the interpretation of the results in terms of corresponding human response is most difficult. Assessment of the effects of certain contaminants on livestock is relatively straightforward, however. Thus contamination of forage by airborne fluorides and arsenicals from certain industrial operations has led to the loss of large numbers of cattle in the areas adjacent to such chemical industries.

Plants. Damage to vegetation by air pollution is of many kinds. Sulfur dioxide may damage such field crops as alfalfa, and trees such as pines, especially during the growing season; some general relations are presented in Fig. 7. Both hydrogen fluoride and nitrogen dioxide in high concentrations have been shown to be harmful to citrus trees and ornamental plants which are of economic importance in central Florida. Ozone and ethylene are other contaminants which cause damage to certain kinds of vegetation.

Materials. Corrosion of materials by atmospheric pollution is a major problem. Damage occurs to ferrous metals; to nonferrous metals, such as aluminum, copper, silver, nickel, and zinc; to building materials; and to paint, leather, paper, textiles, dyes, rubber, and ceramics.

Weather. Tropospheric pollution may affect weather in a number of ways. Heavy precipitation at Laporte, Ind., is attributed to a substantial source of air pollution there, and similar but less pronounced effects have been observed elsewhere. Industrial smoke reduces visibility and also ultraviolet radiation from the Sun, and polluted fogs are more dense and more persistent than natural fogs occurring under similar conditions. Possible major effects of air pollution on Earth's climate have been mentioned earlier. *See* CLIMATE MODIFICATION.

Costs. It is extremely difficult to estimate accurately the economic costs of air pollution. Dollar values are not readily established for losses due to illness caused by exposure to air pollution. At the present time a reasonable estimate appears to be that air pollution costs the United States about $10,000,000,000 a year. For Great Britain the corresponding figure — probably a conservative one — is £400,000,000 a year.

Controls. Four main methods of air-pollution control are indicated below.

Prevention. This method was originally applied mainly to reduce pollution from combustion processes. Improved equipment design and smoke-less fuels have reduced pollution both from industrial and motor vehicle sources.

Collection. Collection of contaminants at the source has been one of the important methods of control. Many types of collectors have been employed successfully, such as settling chambers, cyclone units employing centrifugal action, bag filters, liquid scrubbers, gas-solid adsorbers, ultrasonic agglomerators, and electrostatic precipitators. The optimum choice for a given industrial process depends on many factors. A major problem is disposal of the collected materials. Sometimes they can be used in by-product manufacture on a profitable or a break-even basis.

Containment. This method is useful for pollutants whose noxious characteristics may decrease with time, such as radioactive contaminants from nuclear power plants. For contaminants with a short half-life, containment may allow the radioactivity to decay to a level which permits their release to the atmosphere. Containment, with destruction or conversion of the offending substances, often malodorous or toxic, is used in certain chemical, oil refining, and metallurgical processes and in liquid scrubbing.

Dispersion. Atmospheric dispersion as a control method has a number of advantages, especially for industrial processes which can be varied to take advantage of the periods when dispersion conditions are so good that contaminants may be distributed very widely in such small concentrations that they inconvenience no one. Some coal-burning electrical power stations are building high stacks, up to 1000 ft (300 m), to lift the SO_2-bearing stack gases well above the ground. Some plants store low-sulfur anthracite coal for use when atmospheric dispersion is poor.

Laws. Many laws designed to limit air pollution have been enacted. Major steps forward were taken by the Netherlands in 1952, by Great Britian in 1956, by Germany in 1959 and 1962, by France in 1961, by Norway in 1962, by the United States in 1963 and 1967, and by Belgium in 1964.

Efforts to control air pollution by legal means commenced many years ago in Great Britain. In 1906 the Alkali Act consolidated and extended previous similar acts, the first of which was passed in 1863. This calls for the annual registration of scheduled industrial processes, and requires that the escape of contaminants to the atmosphere from scheduled processes must be prevented by the "best practicable means." The Alkali Act functions by interpretation and not by statutory requirement, the Alkali Inspector being the sole judge of the "best practicable means." The Clean Air Act of 1956 provided more effective ways of limiting air pollution by domestic smoke, industrial particulates, gases and fumes from the processes registrable under the Alkali Act, and smoke from diesel engines. This legislative program has had considerable success in alleviating air-pollution problems in Great Britain.

In the United States, air-pollution control had been considered to be a matter of local concern only. By 1963 only one-third of the states had air-pollution control programs, most of which were relatively ineffective. Only in California were local programs, at the city and county level, supported adequately. The Clean Air Act of 1963 brought the Federal government into a regulatory position of increased scope by granting the Secretary of Health, Education, and Welfare specific abatement powers under certain circumstances. It also established a Federal program of financial assistance to local control agencies and recommended more vigorous action to combat pollution by motor vehicle exhausts and by smoke from incinerators.

The Air Quality Act of 1967 brought the Federal government into a more substantial regulatory role. One of its important effects has been to change the emphasis in legislation from standards based on emissions from sources, such as stacks, to standards based on concentrations of contaminants in the ambient air which result from such emissions. The Air Quality Act of 1967 consists of three main portions, listed below.

Title I: Air-Pollution Prevention and Control. The first section of the Air Quality Act amends the Clean Air Act to encourage cooperative activities by states and local governments for the prevention and control of air pollution and the enactment of uniform state and local laws; to establish new and more effective programs of research, investigation, training, and related activities; to give special emphasis to research related to fuel and vehicles; to make grants to agencies to support their programs; to provide strong financial support for interstate air-quality agencies and commissions; to define atmospheric areas and to assist in establishing air quality control regions, criteria, and control techniques; to provide for abatement of pollution of the air in any state or states which endangers the health or welfare of any persons and to establish the necessary procedures; to establish the President's Air Quality Advisory Board and Advisory Committees; and to provide for control of pollution from Federal facilities.

Title II: National Emission Standards Act. This section is concerned mainly with pollution from motor vehicles which accounts for some 60% of the total for the United States. The act covers such matters related to motor vehicle emissions as the following: establishment of effective emission standards and of procedures to ensure compliance by means of prohibitions, injunction procedures, penalties, and programs of certification of new motor vehicles or engines and registration of fuel additives. It also calls for a comprehensive report on the need for, and the effect of, national emission standards for stationary sources.

Title III. The final section is general and covers matters such as comprehensive economic cost studies, definitions, reports, and appropriations.

There is no doubt that this far-reaching legislative program, stimulating new approaches at the local, state, and Federal levels, will play a major role in controlling air pollution within the United States. The other industrial nations of the world are preparing to meet their growing air-pollution problems by initiatives appropriate to their own particular circumstances.   [A. J. GROBECKER; S. C. CORONITI; E. WENDELL HEWSON]

Bibliography: L. J. Brasser et al., *Sulphur Dioxide: To What Level Is It Acceptable?*, Research Institute of Public Health Engineering, Delft, Netherlands, Rep. no. G300, 1967; J. H. Chang and H. Johnston, *Proceedings of the 3d CIAP Conference*, DOT-TSC-OST-74-15, pp. 323–329, 1974; R. E. Dickinson, in *Proceedings of the AIAA/AMS International Conference on the Environmental Impact of Aerospace Operations in the High Atmosphere*, AIAA Pap. no. 73-527, 1973; Federal Task Force on Inadvertent Modification of the Stratosphere, *IMOS Report*, prepared for the Federal Council for Science and Technology, 1975; J. Friend, R. Liefer, and M. Tichon, *Atmos. Sci.*, 30:465–479, 1973; D. Garvin and R. F. Hampson, *Proceedings of the AIAA/AMS International Conference on the Environmental Impact of Aerospace Operations in the High Atmosphere*, AIAA Pap. no. 73-500, 1973; A. J. Grobecker, S. C. Coroniti, and R. H. Cannon, *Report of Findings: The Effects of Stratospheric Pollution by Aircraft*, U.S. Department of Transportation, DOT-TST-75-50, 1974; P. A. Leighton, Geographical aspect of air pollution, *Geogr. Rev.*, 56:151, 1966; M. C. MacCracken, *Tests of Ice Age Theories Using a Zonal Atmospheric Model*, UCRL-72803, Lawrence Livermore Laboratory, 1970; A. R. Meetham, *Atmospheric Pollution*, 1964; M. J. Molina and F. S. Rowland, *Geophys. Rev.*, pp. 810–812, 1974; National Academy of Sciences, *Environmental Impact of Stratospheric Flight*, pp. 128–129, 1975; R. S. Scorer, *Pollution in the Air: Problems, Policies and Priorities*, 1973; A. Stern, *Air Pollution*, vols. 2 and 5, 1977; R. S. Stolarski and R. J. Cicerone, *Can. J. Chem.*, 52:1610–1615, 1974; S. C. Wofsy and M. B. McElroy, *Can. J. Chem.*, 52:1582–1591, 1974.

Atmospheric tides

In analogy to the ocean tides, the worldwide atmospheric oscillations with the same periods are called atmospheric tides. However, their excitation is due not only to gravitational action of the Moon and the Sun, but also to the daily solar heating of the atmosphere. It has long been recognized that the 12-hourly solar oscillation S_2 must be caused by this daily heating since it is about 20 times larger than the corresponding lunar oscillation L_2, even though the tidal force of the Moon is 2.2 times that of the Sun. But if this is true, the 24-hourly solar oscillation S_1 should be larger than S_2 because, in the daily temperature wave, the 24-hr term is larger than the 12-hr term, the latter being largely due to the fact that the temperature follows a sine curve only during the daylight hours, whereas it decreases continuously during the night. Actually, S_2, as observed in the surface barometric pressure variation, is twice as large as S_1.

The 12-hourly solar oscillation. Theoretical analysis based on the thermo- and hydrodynamic equations governing the atmosphere and on the knowledge of its vertical structure has shown that the effects of periodic heating by direct atmospheric absorption of solar radiation can account for the observed amplitudes of S_2 without appreciable resonance magnification. The main contribution (constituting about three-quarters of the cause) to this thermal forcing function stems from the absorption of solar radiation by ozone in the strato-

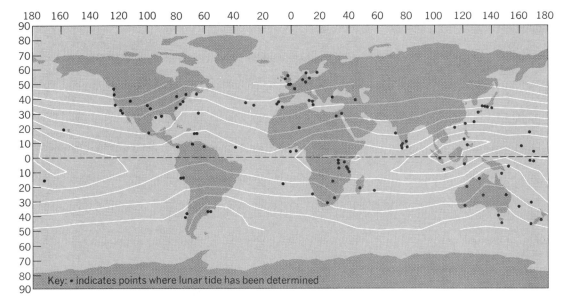

Lines of equal mean amplitude of the lunar barometric tide (contour interval 0.01 mb). Amplitude maximum over Indonesia is 0.08 mb. Degrees of latitude and longitude are indicated at the left and top. (*From B. Haurwitz and Ann D. Cowley, The lunar barometric tide, Rev. Pure Appl. Geophys., PAGEOPH, 77(6):122–150, 1969*)

sphere and mesosphere, and the remainder is due mainly to water vapor absorption, largely in the troposphere. The latitudinal distribution of the thermal forcing function is very similar to that of the dominant mode of S_2. This similarity accounts for the observed regular global distributions of S_2. Moreover, the theory shows that the vertical wavelength of the dominant mode of S_2 is long, well over 100 km. Thus, the contributions of the different atmospheric layers are essentially additive and produce a strong S_2. *See* MESOSPHERE; STRATOSPHERE; TROPOSPHERE.

The 24-hourly solar oscillation. In the case of the diurnal oscillation S_1, the situation is quite different. Although the latitude dependence of the thermal forcing function is smooth, the latitude dependence of the different modes of S_1 is quite complicated. Therefore, at least three or four modes are needed to represent the forcing function. Since some of these modes do not propagate vertically, they are appreciable only in the layer of excitation; or, since they have short vertical wavelengths, destructive interference between different atmospheric layers can occur. Thus, the excitation mechanism of S_1 is not very effective. Nevertheless, S_1 is not negligible compared to S_2. Considering only the largest modes, one may say that at the Equator the observed amplitude of the semidiurnal surface pressure oscillation is 1.16 mb (1 mb = 100 Pa) and that of S_1 is 0.59 mb, and both decrease poleward approximately as $(\cos \varphi)^3$, where φ is the latitude; this approximation is much poorer for S_1 than for S_2, as is to be expected since the observed S_1 is due to the interaction of a number of nodes. The maximum of S_2 occurs on the average at about 10 A.M. and 10 P.M. local mean solar time, that of S_1 at about 5 A.M. local mean solar time; but as in the case of the amplitudes, there are large deviations from these values at individual stations, especially for S_1. Besides S_1 and S_2, two additional thermally forced oscillations are observed with periods of 8 and 6 hr. Since their forcing functions are much weaker, they are considerably smaller than S_1 and S_2.

Periodic wind changes. These pressure oscillations are much smaller than the irregular pressure variations accompanying weather disturbances in the low atmosphere, and the corresponding periodic wind changes have amplitudes of the order 10–20 cm/s. But since the kinetic energy of these wave motions remains nearly constant as they propagate upward, the amplitudes should increase in inverse ratio to the square root of the air density. Observations have confirmed this increase of the periodic winds with elevation. Up to 30-km winds are measured by meteorological balloons, winds from 30 to 60 km by a small number of stations using meteorological rockets. Between 80 and 120 km, the drifts of ionized meteor trails are observed by ground-based Doppler radar giving the air motion. The patterns emerging from these still scanty data confirm in general the expected increase of the periodic winds with the elevation, with amplitudes of the order of 10 m/s and more at the meteor levels. Often these amplitudes are larger than the mean winds. Thus, in contrast to the weather in the low atmosphere, the meteorology of these high levels is strongly modulated by periodic winds. *See* WIND.

Lunar atmospheric tide. The lunar atmospheric tide is entirely due to the tidal force of the Moon. Its main constituent M_2 produces a surface pressure oscillation $L_2(p_0)$, which is now known for more than 100 stations despite the difficulty inherent in determining such a small variation from the barometric pressures, which have much larger unperiodic changes. At the Equator, the amplitude is about 0.06 mb, and it decreases at the approximate rate $(\cos \varphi)^3$ with the latitude φ. The pressure maximums occur about ½ hr after the Moon's transit through the meridian. But like $S_2(p_0)$, although to a much lesser degree than $S_1(p_0)$, the times of maximum pressure and the amplitudes vary considerably from place to place. As an example, the illustration shows the actual amplitude distribution with its deviations from the zonal symmetry, which would be expected since M_2 is proportional to $(\cos \varphi)^2$. In the case of S_1 and S_2, these irregularities are produced to a large extent by the irregular distribution of the heating function. But in the case of L_2, with its gravitational excitation, the deviations must be caused by the unequal response of the atmosphere to M_2, and by the Earth and ocean tides, which raise and lower the solid and liquid surface of the Earth sufficiently to influence L_2. The theory accounts readily for the gross features of the observed lunar semidiurnal tide, but attempts to explain the details of the global distribution have only partly succeeded. See EARTH TIDES.

Among other still unexplained features of L_2 is a pronounced annual variation: In both hemispheres simultaneously the daily pressure maximums are more than half an hour later during the Northern winter than during the rest of the year; and at many stations the amplitude is smallest at this time. Since the lunar tidal potential does not show such an annual variation, it must be due to atmospheric changes, possibly to the seasonal variation of the large-scale wind systems. Another puzzling phenomenon is that the maximums of $L_2(p_0)$ and, during some seasons, also those of $S_2(p_0)$ and $S_1(p_0)$, occur earlier in the Southern than in the Northern Hemisphere.

The other, smaller terms in the lunar tidal potential generate also atmospheric oscillations. But only those due to the terms O_1 and N_2 have so far been identified in the pressure data for a few stations.

Like S_1 and S_2, L_2 has its characteristic wind system. At the surface the wind amplitudes are very small, of the order of 1 cm/s. They must increase with height approximately in inverse proportion to the square root of the density, but they have not yet been directly determined in the high atmosphere because sufficient data are lacking. However, these high-level motions cause lunar-tidal variations in the geomagnetic parameters recorded at the Earth's surface. See ATMOSPHERE; TIDE.

[BERNHARD HAURWITZ]

Atmospheric water vapor

Water in the form of vapor diffused in the atmosphere. Water vapor is a relatively small and variable constituent of the atmosphere, but because water may exist in the liquid, solid, and vapor states within the usual range of temperatures which occur in the atmosphere, water vapor plays an important role in the world environment. Water vapor is usually less than 4% of the atmospheric gas, but this small amount helps to determine the temperature distribution over the Earth and is the source of the world's fresh water.

The primary source of atmospheric water vapor is the oceans, which cover the major part of the Earth. Evaporation from lakes, rivers, ice, snow, and soil, and transpiration by plants also contribute large amounts of water vapor to the atmo-

sphere, although these sources are small in comparison with evaporation from the oceans which is estimated to be about 492,000 km³ of water annually.

Heat of vaporization. The heat of vaporization of water at 21°C is 2445 joules/gram (584 cal/g or 1047 Btu/lb). Because of this heat requirement, water vapor in air masses transports large quantities of heat from warmer to cooler regions of the world. When an air mass containing vapor is cooled, the vapor condenses to form clouds of small water droplets. Under some conditions the released heat may be important in increasing the buoyancy of the air, which causes the air to rise and cool, leading to more condensation.

Evaporation cools a water surface because the escaping water carries with it the heat of vaporization. As the water cools, its vapor pressure decreases and the evaporation rate drops. Unless additional heat is provided, evaporation will cease. In a deep water body the cooled surface water may sink and be replaced by warmer water from below.

The transformation from ice to vapor at 0°C (called sublimation) requires a heat exchange of 2834 J/g (677 cal/g), which is the sum of the heat of vaporization at 0°C of 2500 J/g (597 cal/g) and the heat of fusion of 335 J/g (80 cal/g). Conversely, the condensation of a gram of vapor on a snow surface releases 2500 J (597 cal), which is sufficient heat to melt 7.5 g of snow.

Evapotranspiration. Evaporation from the continents returns the greater portion of the precipitation directly to the atmosphere. Transpiration by plants plays an important role in this process. Evaporation from bare soil becomes very small after the surface layer has been dried, because of the slow movement of water upward in the soil. Plants whose roots may extend several meters into the soil are important in removing moisture deep in the soil. Water taken up by the roots moves through the plant system and, except for the small amount incorporated in the plant, is released to the atmosphere through the leaves. See EVAPOTRANSPIRATION.

Measurement of water vapor. The vapor content of the atmosphere (humidity) is expressed quantitatively in several ways. The actual vapor content in g/kg of moist air is called specific humidity. The temperature at which moisture begins to condense is known as the dew-point temperature and is extensively used in meteorology as a measure of humidity. The partial pressure exerted by the water vapor in the atmosphere independently of the other gases is called the vapor pressure. The amount of water in a vertical column of the atmosphere from the ground to the outer limits of the atmosphere is the precipitable water, a unit often used in studies of rainfall. The relative humidity is probably the most widely used measure of humidity. It is the ratio of the actual moisture content of the air at some location to that which would be present if the space were saturated with vapor at prevailing temperature and pressure. Since the saturation value is a function of temperature, relative humidity varies with temperature even though the specific humidity remains constant. Hence, relative humidity is not a very useful measure of humidity for scientific purposes. It is, however, an index to human comfort.

Water vapor content of the atmosphere can be measured with various devices. A psychrometer consists of two thermometers, one with its bulb covered with wet muslin. When the device is ventilated, the temperature of the wet bulb is depressed below that of the dry bulb by evaporation of water from the muslin. The humidity is a function of the two temperatures. The dew-point temperature can be measured by measuring the temperature of a polished metal surface at the moment when condensation begins to form on it. Devices which indicate relative humidity by measuring the change in length of human hair or chemically treated paper are also available. Such devices are cheap but not very reliable. By measuring the attenuation of light at a wavelength where there is high absorption by water vapor, it is possible to calculate the mass of water vapor in the light path. With the Sun as a light source, the precipitable water can be determined in this fashion.

Estimating evaporation. Evaporation can result in serious losses of water from reservoirs. Hence, methods of estimating evaporation from existing or proposed reservoirs have been developed. J. Dalton in 1802 first pointed out that evaporation is proportional to the difference in vapor pressure between the water surface and the air above it. Hence, evaporation is possible only when the dew-point temperature of the air is less than the temperature of the water surface. The greater this temperature difference, the greater is the evaporation rate. In absolutely still air a vapor blanket forms at the water surface, and evaporation decreases rapidly to rates limited by the diffusion of vapor from the blanket. With wind the vapor is carried away and replaced by dryer air, and evaporation continues.

Another basis for estimating evaporation is the energy balance method. If the total radiation input to the water body less the heat loss by long-wave radiation and convection can be determined, the excess of heat input over heat outgo divided by the latent heat of vaporization indicates the volume of evaporation.

Evaporimeters are pans of water from which evaporation can be measured. Because of heat transfers through the sides and bottoms of such pans, evaporation from pans is always higher than from lakes. The National Weather Service class A pan is 1.2 m (4 ft) in diameter and 25 cm (10 in.) deep. It is supported a few inches above the ground on a timber grid. Evaporation from the class A pan must be reduced about 30% to estimate lake evaporation.

Salt, ice, and snow factors. Addition of salts to water reduces the vapor pressure at the water surface. Thus evaporation per unit area from the oceans and salt lakes is slightly less than from a fresh-water body. Roughly, evaporation is reduced about 1% for each percent of dissolved salts, so that evaporation from the oceans is about 3.5% less than from an equivalent fresh-water surface. Evaporation from ice and snow is also less than from water at the same temperature, because the vapor pressure over ice is somewhat less than over

water. Evaporation from snow and ice is also low, because vapor pressure (and vapor-pressure differences) are very low at temperatures below freezing.

Based largely on evaporimeter data, it can be said that evaporation varies from 5 to 7 m (15 to 20 ft) of water per year in tropical deserts to near zero in polar regions.   [RAY K. LINSLEY]

Bibliography: T. A. Blair and R. C. Fite, *Weather Elements*, 5th ed., 1965; R. K. Linsley, M. A. Kohler, and J. L. H. Paulhus, *Hydrology for Engineers*, 2d ed., 1975; R. C. Ward, *Principles of Hydrology*, 2d ed., 1975.

Atoll

An annular coral reef, with or without small islets, that surrounds a lagoon without projecting land area.

Physiography. Most atolls are isolated reefs rising from the deep sea, and vary considerably in size. Small rings, usually without islets, may be less than a mile in diameter, but many atolls have a diameter of about 20 mi and bear numerous islets. The largest atoll, Kwajalein in the Marshall Islands, has an irregular shape and covers 840 mi^2.

The reefs of the atoll ring are flat, pavementlike areas, large parts of which, particularly along the seaward margin, may be exposed at times of low tide (Fig. 1). The reefs vary in width from narrow ribbons to broad bulging areas more than a mile across. The most prominent feature may be a low, cuestalike, wave-resistant ridge whose steeper side faces the sea. Usually brownish or pinkish in color, this ridge is composed mainly of calcareous algae. To seaward of the ridge the gentle submarine slope is cut by a series of regularly spaced grooves. These may be slightly sinuous but in general they extend seaward at right angles to the reef front. They are separated by flattened spurs or buttresses that appear to have been developed mainly by coral and algal growth. The grooves and buttresses form the toothed edge that is so conspicuous when an atoll is viewed from the air. Beyond the toothed edge may be a shallow terrace, then a steep (35°) submarine slope that flattens progressively at deeper levels until it touches the ocean floor. Some grooves of the toothed margin extend through the marginal algal ridge as surge channels, ending on the reef flat in a series of "blow holes" due to constriction by algal growth. These structures taken together form a most effective baffle that robs the incoming waves of much of their destructive power, and at the same time bring a constant supply of refreshing sea water with oxygen, food, and nutrient salts to wide expanses of reef.

Inside the margin of the reef may be several zones that are rich in corals or other reef builders, but large parts of the reef flat are composed of cemented reef rock, bare or thinly veneered with loose sand or coarser reef debris. When the reef flats of Eniwetok Atoll were drilled, a solid plate of hard limestone was found below that extends to depths of 10–15 ft. Growth of organisms and deposition of sediments on the reef are controlled by the prevailing winds and currents and other ecologic factors. The sediments, including cemented

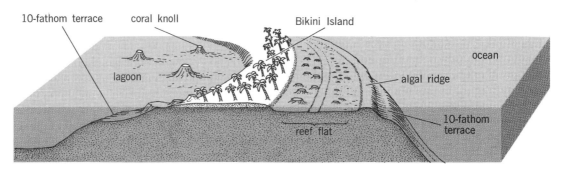

Fig. 1. Diagrammatic section showing features of ring of Bikini Atoll.

rock, that make up the reef surface and its islets are composed entirely of carbonates secreted by shallow-water organisms.

The reef rings of most atolls are broken, usually to leeward, by one or more passes. Some of these are as deep as the lagoon and give ready access to that body of water. In general, lagoon depths are proportional to size. A lagoon with a diameter of 20 mi may have a depth of 300 ft. Hundreds of coral knolls rise from the lagoon floor. The broad crests of these structures, covered by a growth of coral, lie at all depths, the highest approaching low-tide level.

On many atolls islets are spaced at irregular intervals along the reef. Most of these are small and rarely rise as much as 20 ft above the reef flat. They are composed of reef detritus (coral heads and sand) piled up by waves and winds. The sediments are unconsolidated except for layers of beach rock (cemented conglomerate and sandstone) at intertidal levels. Islets located on reefs that are subjected periodically to violent storms may exhibit one or more ramparts built of blocks of reef rock, coral heads, and finer material. These structures apparently can be built, moved, or destroyed by the waves of a single major storm. Inland from the beaches and ramparts, sand and gravel flats occur and large areas are covered with

coconut palms, screw pines, and thickets of bay cedar and other plants. Low sand dunes may occur on the borders of islands inside sand beaches.

Many atolls exhibit terraces around their margins and inside their lagoons. These structures were probably developed during the Pleistocene glacial epochs when, at intervals, sea level stood appreciably lower than now. The most recent change in sea level may have been a negative shift of 6 ft, but convincing evidence for such a change has not been preserved on most atolls. Numerous investigations on reef islands other than atolls are now under way to determine—with the aid of age measurements such as the carbon-14 and uranium natural series techniques—if there was a eustatic shift of this magnitude in fairly recent times. If such a shift did occur, it would offer a plausible explanation for the origin of atoll islets: When the sea level dropped, the reefs were eroded by waves and the debris concentrated to form the existing islets.

Distribution. Existing atolls, like other types of coral reefs, require strong light and warm waters and are limited in the existing seas to tropical and near-tropical latitudes. A large percentage of the world's atolls are contained in an area known as the former Darwin Rise that covers much of the central and southwestern Pacific. Atolls are also numerous in parts of the Indian Ocean and a number are found, mostly on continental shelves, in the Caribbean area. *See* REEF.

Structures properly described as atolls existed in shallow Paleozoic seas. These ancient atolls were built by a variety of organisms; the types of corals and algae that are primarily responsible for today's reefs were not then in existence. Many Paleozoic atolls became reservoirs of petroleum, and are being studied and mapped in detail by deep drilling.

Origin and development. The many atolls that rise from the deep sea have probably been built on submerged volcanoes. The reef caps of two such atolls (Eniwetok and Midway) have been drilled and the existence of a foundation of volcanic flows confirmed. In both instances it was found that the volcanoes had been at least partly truncated by erosion prior to the initiation of reef growth. Many of the existing atolls probably originated as near-surface reefs in Tertiary time. The oldest limestones beneath Eniwetok are late Eocene, and Midway's oldest limestones are early Miocene.

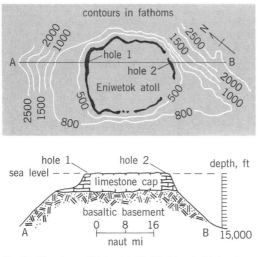

Fig. 2. Structure of Eniwetok Atoll in Marshall Islands.

The bases of other atolls may prove to be appreciably younger.

Wherever extended three-dimensional studies have been carried out on atolls (Funafuti in the Ellice Islands, Kita-Daito-Jima, Bikini, and Eniwetok in the Marshall Islands, and Midway in Hawaii), they have proved the atolls to be complex structures; surface similarities may mask striking differences in underground structure. Wherever tested by the drill, the thicknesses of reef caps of atolls greatly exceed the depth limits of the reef builders, and these caps appear to have been developed during long periods of subsidence. *See* SEAMOUNT AND GUYOT.

The structure of Eniwetok Atoll in the Marshall Islands, as revealed by drilling, is shown in Fig. 2. The limestone cap, nearly a mile in thickness, rests on the summit of a basaltic volcano that rises 2 mi above the ocean floor. In hole 1 the lower limestones are late Eocene and formed as forereef deposits on an outer slope. Correlative limestones in hole 2 were formed in shallow water. The 4000-ft section of limestone that forms the cap of Eniwetok records an overall subsidence of that amount over a period exceeding 40,000,000 years. Subsidence of this magnitude even exceeds the amounts so brilliantly postulated by Charles Darwin. The sinking has not been steady or continuous, but was interrupted by periods of emergence. The emergent stages are recorded in zones of recrystallized and partially dolomitized sediments. Such zones are termed solution unconformities and were formed when the top of the atoll stood above the sea for appreciable time. During such periods land snails lived on the island and land plants left a record of spores and pollen; Eniwetok and perhaps many other atolls functioned as stepping stones in the distribution of shallow-water marine life in the Pacific. *See* OCEANIC ISLANDS.

[HARRY S. LADD]

Bibliography: R. Berner, Dolomitization of the mid-Pacific atolls, *Science*, 147:1297–1299, 1965; K. Emery et al., *Bikini and Nearby Atolls, Marshall Islands*, USGS Prof. Pap. no. 260-A, 1954; F. R. Fosberg (ed.), *Atoll Res. Bull.*, nos. 1–117, Pacific Science Board, National Academy of Sciences, 1951–1966; and nos. 118–119, Smithsonian Institution, 1967; O. A. Jones and R. Endean (eds.), *Biology and Geology of Coral Reefs*, vol. 1: *Geology*, 1973; H. Ladd and S. Schlanger, *Drilling Operations on Eniwetok Atoll*, Geol. Soc. Amer. Prof. Pap. no. 260-Y, 1960; H. Ladd et al., Drilling on Midway Atoll, Hawaii, *Science*, 156(3778): 1088–1094, 1967; H. Menard and H. Ladd, Oceanic islands, seamounts, guyots and atolls, in M. N. Hill (ed.), *The Sea*, vol. 3, 1963; S. D. Romasko, *Living Coral and Other Inhabitants of the Reef*, 2d ed., 1976; H. Wiens, *Atoll Environment and Ecology*, 1962.

Aurora

A light occurring in the Earth's upper atmosphere, seen most often along the outer realms of the Arctic and Antarctic regions, where it is called aurora borealis (Fig. 1) and aurora australis, respectively. The aurora is usually confined to northern and southern zones lying roughly 20 to 25° from the geomagnetic axis points. Strong auroras occasionally extend to middle latitudes and are accompanied by small but important changes in the Earth's magnetism. Usually, the lower edge is about 100 km high and has a vertical extent between 20 and 100 km, although some auroras occur as high as 1000 km.

Auroras are usually yellowish-green, from forbidden emission of atomic oxygen at 5577 angstroms (A). Exceptionally high auroras are red, from O lines at 6300 and 6364 A. Occasionally, high auroras seen near twilight are sunlit and appear violet from fluorescence in N_2^+ bands. Very bright displays sometimes reach as low as 65 km and show red lower borders from N_2 and O_2^+ emissions. Atmospheric ionization associated with the aurora often produces anomalous reflection or absorption of radio waves traversing the ionosphere.

A network of all-sky cameras, deployed over

Fig. 1. Photograph of aurora borealis taken at College, Alaska. (*V.P. Hessler, University of Alaska*)

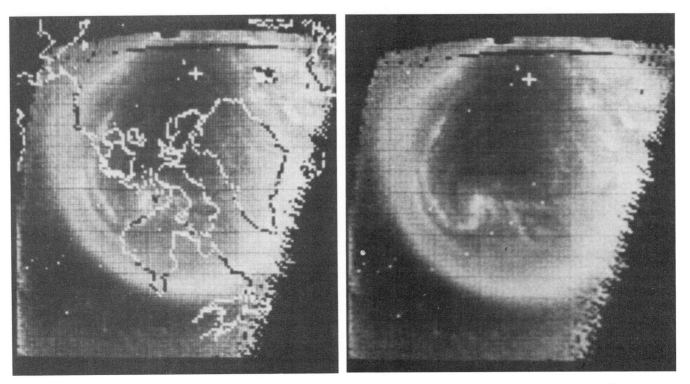

Fig. 2. The auroral distribution around the northern polar cap as seen on Dec. 22, 1971, by *ISIS*-2. The left picture is the same as the right, except that major con-tinental boundaries are superimposed. Midnight is at the bottom, and noon at the top. (*Courtesy of W. Saw-chuk, University of Calgary*)

high and middle latitudes during the International Geophysical Year (1957–1958), established that auroras often extend thousands of kilometers along an oval-shaped curve centered approximately on the geomagnetic pole. Since 1971, satellite pictures have allowed the global pattern of auroras to be reconstructed much more accurately than in the past. The common arc structure of bright auroras shows clearly in space photographs (Fig. 2). Auroral displays often become very active, with rapid lateral motions, sudden brightenings, and flickering vertical rays. This activity typically peaks about midnight and appears on the synoptic, global picture as distorted knots extending pole-ward from an otherwise fairly smooth line of arcs.

A large portion of the auroral light is diffuse and usually faint. It can be detected by long spectro-scopic exposures of the night sky, through the characteristic auroral spectrum, but it is not often noticed from the ground because of the absence of features. The synoptic satellite pictures show the diffuse aurora in a ring outside the bright auroral arcs, completely surrounding the polar cap.

Auroras are the direct result of high-speed bombardment of the rarefied upper atmosphere by ionized hydrogen gas (protons and electrons). That much is known; indeed, the hydrogen emission in the auroral spectrum has Doppler-shifted wave-lengths indicating entry velocities of several thou-sand kilometers per second. But just how and why bombardment occurs has been an elusive problem. Because auroral occurrences are loosely associ-ated with solar activity, early thinking attributed auroral particles to matter ejected a few days earlier from the Sun. Although that may still be true, it is now known that hot plasma (gas com-posed of electrically charged particles) permeates the geomagnetic field from above the atmosphere out to distances of several Earth radii. Hence it seems likely that auroras are caused by distur-bances in this outer "magnetosphere" by some interaction with the solar wind. Presumably the physical mechanism for discharging the magneti-cally trapped plasma into the atmosphere, pro-ducing the vivid luminescence, is one or more of many possible "plasma instabilities." However, the precise processes and the situations that trigger them are not yet understood. *See* ATMO-SPHERE; GEOMAGNETISM; IONOSPHERE; MAG-NETOSPHERE.

[JOSEPH W. CHAMBERLAIN]

Bibliography: J. W. Chamberlain, *Physics of the Aurora and Airglow*, 1961; A. V. Jones, *Aurora*, 1974; A. Olmholt, *Optical Aurora*, 1971.

Backing wind

A wind which changes direction in a counterclock-wise sense, for example, a south wind changing to an east wind or an east wind changing to northerly. At a particular place, the wind gradually backs with time when a cyclone passes eastward on a path south of the observer's location. The wind characteristically backs with increasing height, on the west side of an extratropical cyclone. The meaning, in terms of changes of cardinal direc-tions, is reversed in the Southern Hemisphere. This term is opposite in sense to veering wind.

[CHESTER W. NEWTON]

Baltic Sea

An intracontinental, Mediterranean-type sea, connected with the North Sea by Skagerrak, Kattegat, and the Danish Sounds, which include Little Belt (Lillebaelt), Great Belt (Storebaelt), and the Sound (Öresund). The total area, including the Gulf of Bothnia and the Gulf of Finland, is 154,000 mi² (400,000 km²). The mean depth is only 55 m and the greatest depth, off Landsort, is 459 m (Figs. 1 and 2).

Geologic history. The Baltic Basin is an old depression in Fennoscandian primary rocks. Its waters were part of the ocean several times before the last glacial age. Among the postglacial stages are the Baltic Ice Lake, Yoldia Sea, Ancylus Lake, and Littorina Sea (Fig. 3). The Baltic Ice Lake was a meltwater lake dammed up by the edge of the retreating ice sheet and by the Belt and Sound region, which was higher then than at present. Yoldia Sea was formed when the ice sheet retreated north of central Sweden. Lake water drained across the present great lake region, after which salt water entered the basin from the ocean. Later this sea was connected with the White Sea. Ancylus Lake came into existence when the basin waters were again separated from the ocean by land upheaval and basin tilting, but with an outlet westward across the lake region of Sweden. Littorina Sea, the early stage of the present Baltic, was initiated about 6000 or 5000 B.C., as crustal sinking produced the Danish Sounds. The waters of this sea were much more saline than those of the present Baltic.

Circulation and salinity. The area drained by rivers emptying into the Baltic is about four times the size of the sea area. Annual runoff and precipitation at sea, less evaporation loss, is equal to 1/40 of the Baltic Sea's volume. This fresh-water outflow, in the upper layers, leaves the Baltic through the Sounds. Beneath it the more saline and denser

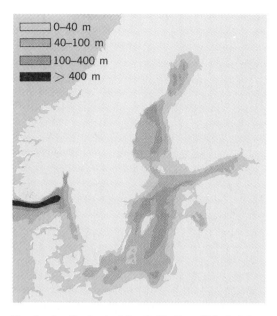

Fig. 2.   Depth chart of the Baltic Sea. (*Adapted from Geographical Society of Finland, A General Handbook on the Geography of Finland, Fennia 72, 1952*)

water from Kattegat penetrates the basin along the deeper parts. Difference in surface and bottom salinity is thus very marked in most parts of the Baltic (see table). A distinct gradient of salinity is found at depths of about 60 m. In addition, a summer thermocline develops every year. In winter convectional processes reach a depth of about 60 m. The annual range in temperature in surface waters is 0–18°C and in deeper waters 2–5°C. *See* OCEAN CURRENTS; THERMOCLINE.

Water level. Tidal ranges are small in the Baltic: off Copenhagen about 6 in., off Stockholm less than 1 in., and off Leningrad 6 in. Drift currents are generally about 0.2 knot, but speeds of 1.5 knots may occur. Storms cause abrupt changes in water level, as much as 10 ft in the northern part of the Gulf of Bothnia and even more at Leningrad.

Ice. In winter severe storms make navigation hazardous among the numerous shallows and rocks. Navigation also is impeded by sea ice from early December to early May. Ice disappears in the northern Baltic in early June. In very severe winters the whole Baltic is covered with ice. On the average, the area of maximum coverage is about 83,000 mi² (214,000 km²). In the northern Baltic, fast ice may be 2 1/2 ft thick.

Ecology. The Baltic Sea is the largest brackish-water area of the world. Its low productivity may be attributed in part to the narrowness of the Danish Sounds and to the low nutrient content of northern rivers. Commerical catches in the Baltic

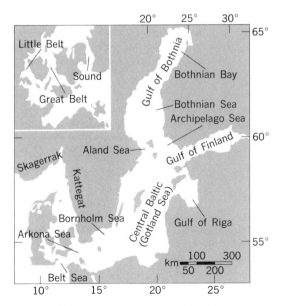

Fig. 1.   Areas of the Baltic Sea. Inset shows connections with the North Sea.

Salinity ranges, parts per thousand, in Baltic Sea

Area	Surface	Bottom
Kattegat	15–20	30–36
Inside Danish Sounds	8–10	15–20
Middle Baltic	6–7	9–12
Innermost estuaries	1–3	3–4

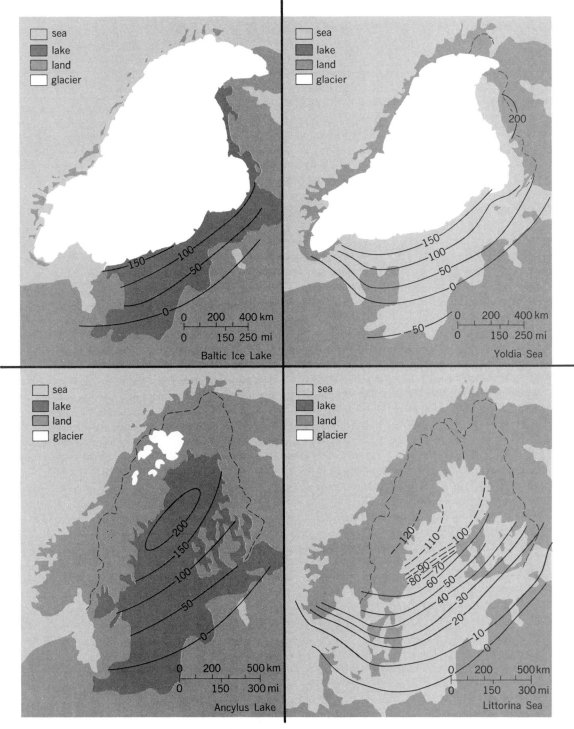

Fig. 3. Successive water bodies in the Baltic Sea basin during the last deglaciation. Isobases (contour interval in meters) show present positions of the related strand lines with respect to existing sea level. (*After E. Fromm, Atlas over Sverige, 1953, in R. F. Flint, Glacial and Quaternary Geology, 1971; reprinted by permission of John Wiley & Sons, Inc.*)

consist of herring, salmon, sprat, and fresh-water fish (whitefish, pike, perch, and trout) in the northern gulfs; herring, cod, salmon, and flounder in the middle Baltic; and herring, cod, eel, salmon, flounder, and plaice in the southwest Baltic.

[ILMO HELA]

Bibliography: J. W. Hedgpeth (ed.), *Treatise on Marine Ecology and Paleoecology*, Geol. Soc. Amer. Memo. no. 67, vol. 1, 1957; L. A. Zenkevich, *Biology of the Seas of the U.S.S.R.*, 1963.

Baroclinic field

A distribution of atmospheric pressure and mass such that the specific volume, or density, of air is a function not solely of pressure. When the field is baroclinic, solenoids are present, there is a gradient of air temperature on a surface of constant pressure, and there is a vertical shear of the geostrophic wind. Significant development of cyclonic and anticyclonic wind circulations typically occurs only in strongly baroclinic fields. Fronts represent baroclinic fields which are locally very intense. *See* Front; Geostrophic wind; Meteorological solenoid; Storm; Wind. [frederick sanders]

Barometer

An absolute pressure gage specifically designed to measure atmospheric pressure.

The basic meteorological barometer is a well-type liquid-column gage which has been filled with mercury and the vertical leg sealed. As atmospheric pressure, acting on the open surface of mercury in the well, allows the mercury level to drop in the vertical leg, a vacuum is created above the mercury. The mercury height is a direct measure of the pressure difference between the full vacuum and the atmosphere, hence of the absolute pressure of the atmosphere.

Apparent mercury column height must be corrected for temperature deviation from standard conditions (0°C or 32°F), for which purpose a mercury-in-glass thermometer is mounted on the barometer. For consistent comparison of atmospheric pressure readings over the Earth's surface, barometer readings must be corrected for elevation of the barometer site above mean sea level, where standard atmospheric pressure is taken to be 760 mm, or 29.92 in., of mercury.

The scale is calibrated in millimeters or inches of mercury and is equipped with a vernier scale for precise reading. A micrometer adjusts the scale to compensate for changes in height of mercury in the reservoir. Mercury barometers with a vernier scale may be accurate to within 0.005 in.

Aneroid indicating and recording barometers using metallic diaphragm elements are usually less accurate, though often more sensitive, devices.

[b. hainsworth; h. payne]

Barotropic field

A distribution of atmospheric pressure and mass such that the specific volume, or density, of air is a function solely of pressure. When the field is barotropic, there are no solenoids, air temperature is constant on a surface of constant pressure, and there is no vertical shear of the geostrophic wind. Significant cyclonic and anticyclonic circulations typically do not develop in barotropic fields. Considerable success has been achieved, paradoxically, in prediction of the flow pattern at middle-tropospheric elevations by methods which are strictly applicable only to barotropic fields, despite the fact that the field in this region is definitely not barotropic. *See* Baroclinic field; Geostrophic wind; Meteorological solenoid; Weather forecasting and prediction; Wind.

[frederick sanders]

Bass Strait

Strait which lies between the Australian mainland and the island of Tasmania. The Bass Strait is an important seaway connecting eastern and western Australian ports and is the center of the principal barracuda and school shark fisheries of Australia. It has an approximate area of 45,000mi², with an average depth of 60 m. There is a deep channel (75–80 m) through the central and northwest sections (Fig. 1). During the winter and early spring a general movement of Southern Ocean waters into the Bass Strait occurs through the western approaches. By November to December, however, contrary sets, especially between King Island and the Victorian mainland, allow the ingress from the east of subtropical waters from southern New

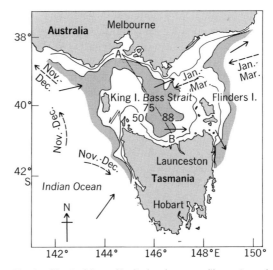

Fig. 1. Chart of Bass Strait showing prevailing sets and direction and months of contrary sets. (*Division of Fisheries and Oceanography, Cronulla, Australia*)

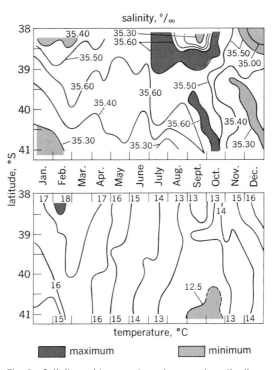

Fig. 2. Salinity and temperature changes along the line A-B in Fig. 1 during 1957. (*Division of Fisheries and Oceanography, Cronulla, Australia*)

South Wales and eastern Tasmania. To some extent these changes in circulation determine the seasonal salinity-temperature changes (Fig. 2). During April to June, Bass Strait waters have the salinity (35.40–35.50 ‰) and temperature (14–16°C) typical of Southern Ocean waters west of Tasmania at that time. From July to October, the highest salinity waters (=35.60 ‰) flow into the Bass Strait from South Australia. Low-salinity waters (35.00–35.30 ‰) occur during the spring and summer along the Victorian coast, but in general their effect on the Bass Strait is slight. The discharge of large quantities of water, retained until early summer as snow and ice in the mountain regions of the major Tasmanian rivers, forms a large body of Tasmanian coastal water, which has a marked effect on salinity values in the Bass Strait during December and January. *See* INDIAN OCEAN; PACIFIC OCEAN.   [DAVID J. ROCHFORD]

Bering Sea

A water body north of the Pacific Ocean, 2,268,000 km² in area, bounded by Siberia, Alaska, and the Aleutian Islands. A basin, with depths greater than 3000 m, occupies the southwest portion of the sea. To the northeast is a wide, extremely flat continental shelf with depths less than 100 m. The main water connections with the Pacific are to the west of the Aleutians, the 2000 m-deep pass between Attu and Komandorskie Islands and the 4400 m-deep pass between the Komandorskies and Kamchatka. Aleutian passes with the largest cross-sectional area are Amchitka (45.7 × 10⁶ m², 1155 m sill depth); Buldir (28.0 × 10⁶ m², 640 m); and Amutka (19.3 × 10⁶ m², 430 m). The Bering Sea connection with the Arctic Ocean is Bering Strait, 85 km wide and 45 m deep.

In summer warm southwesterly winds and cloudy skies prevail. The sea is generally ice-free from July to October. From May through August

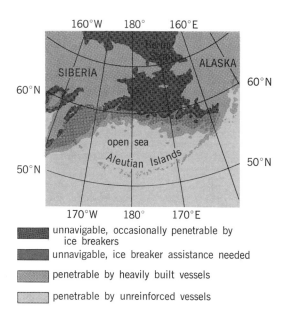

Fig. 1.   Southern limits of landfast and sea ice, February. (*Adapted from U.S. Navy, Ice Atlas of the Northern Hemisphere, H.O. Publ. no. 550, 1946*)

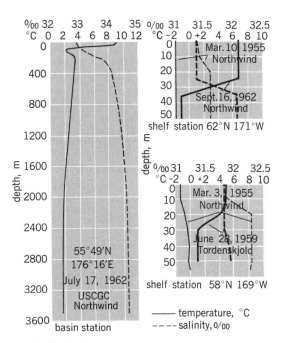

Fig. 2.   Vertical temperature and salinity distributions.

snowmelt adds large quantities of fresh water along the continents, accounting for the low salinity of the coastal waters.

In winter the area is subjected to strong northerly and variable winds, low temperatures, and variable cloudiness. Coastal ice begins to form in late October. By February coastal ice is found in the Aleutians and sea ice extends south to 58°N (Fig. 1). *See* SEA ICE.

Temperature and salinity. There is large seasonal variation in the water structure on the shelf and top 100 m of the basin. In winter the temperatures of this well-mixed layer are cold (<0–3°C); under the ice they approach the freezing point (−1.7°C)(Fig. 2). Salinities range between 31–33 parts per 1000 (‰). In summer temperatures of the surface rise to 6–10°C and occasionally higher near the coasts, while the salinities change little over the open sea but may be markedly reduced (to 25‰ or less) near the coasts. The shelf waters typically become stratified with an upper layer, reflecting the summer warming, overlying a layer with characteristics from the previous winter.

In the basin the waters deeper than 100 m show little seasonal variation. A temperature maximum (3.5–4.0°C) is located between 200–400 m, below which the temperature decreases to bottom values of 1.5–1.6°C. In the level 100–200 m there is a halocline in which the salinities increase from the low surface values to over 34‰. The deep water exhibits a gradual increase in salinity to bottom values of 34.6–34.7‰. *See* HALOCLINE.

At the top of the halocline, about 100 m, is a temperature minimum in summer. This may be partly water cooled locally during the previous winter and unaffected by summer warming. Water of this density is also produced in quantity on the shelf and in embayments, notably the Gulf of Anadyr, in winter and is advected over the basin underneath the surface layer.

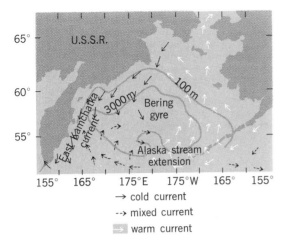

Fig. 3. Surface water circulation, summer.

Tides and currents. Tides in the Bering Sea are semidiurnal with a strong diurnal inequality typical of North Pacific tides. The mean range varies from about 1.5 m in the Aleutians to 30 cm in Bering Strait. Tide ranges are much larger (>4 m) in Bristol Bay. Tidal currents over most of the shelf and basin are not strong (<50 cm sec^{-1}), but in the narrow Aleutian passes and toward the head of Bristol Bay speeds of 200–300 cm sec^{-1} are common.

Nontidal currents (Fig. 3) tend to flow in a counterclockwise pattern during the period of April through October. Surface water entering the sea through the Aleutians as far west as 175°E turns eastward, and the speed in the current along the north side of the islands reaches 50 cm/sec. On the shallow Alaskan shelf the waters flow northward at speeds of 10–20 cm/sec, but reach speeds up to 50 cm/sec near the coast. A strong current flows northward through Bering Strait, where speeds up to 200 cm/sec have been observed near the eastern side of the Strait. The total transport northward averages >1.0 × 10^6 m^3/sec. Along Kamchatka the waters flow southwestward out of the Bering Sea. This outflow receives additional water from currents which set northward into the Bering Sea between 167 and 175°E, the circulation over the basin being a cyclonic gyre. The circulation in winter is essentially unknown.

[LAWRENCE K. COACHMAN]

Bibliography: L. K. Coachman et al., *Bering Strait: The Regional Physical Geography*, 1976; A. J. Dodimead, F. Favorite, and T. Hirano, *Review of Oceanography of the Subarctic Pacific Region*, Int. N. Pac. Fish. Comm. Bull. no. 13, pt. 2, 1963; M. A. Sayles et al., *Oceanographic Atlas of the Bering Sea Basin*, 1978; M. Uda, Oceanography of the subarctic Pacific Ocean, *J. Fish. Res. Board Can.*, vol. 20, no. 1, 1963; U.S. Navy, *Ice Atlas of the Northern Hemisphere*, H.O. Publ. no. 550, 1946.

Bioclimatology

A study of the effects of the natural environment on living organisms. These effects may be direct, such as the influence of ambient temperature on body heat, or indirect, such as the influences on composition of food. Only direct effects are discussed in this article. Natural and artificial elements cannot be sharply distinguished. For example, smoke from a lightning-induced forest fire is natural, but smoke from a chimney is artificial.

Bioclimatology encompasses biometeorology, climatophysiology, climatopathology, air pollution, and other fields. The interplay between disciplines, such as meteorology and physiology, is emphasized. Concerning the time scale, the study of events enduring for hours to days is called biometeorology; for years to centuries, bioclimatology; for millennia and more, paleobioclimatology. For intrinsic reasons bioclimatology should serve as an overall term. Bioclimatology falls naturally into the two areas of plants and of animals and man.

In plant bioclimatology solution of any problem involves study of the natural climatic values, such as air temperature, precipitation, and wind speed and its variations; the transfer mechanism, such as the eddy diffusivity; and the effect of these agents on the plant. The most important mechanisms are (1) photochemical effects, such as photosynthesis and (blue) phototaxis; photosynthesis needs blue and red light, water from the roots (and possibly from the air), and CO$_2$ from the air; all these components vary with weather and climate; (2) evapotranspiration through integument and stomata of plants, a process depending highly on availability of water, transfer of liquids from root to leaf, temperature of the leaf, water vapor of the air, and ventilation; (3) picking up compounds of N, K, P, Ca, and others from the ground; and (4) avoidance of destructive conditions, such as freezing, drying out of leaves (if water supply is smaller than evaporative demand), and overheating. *See* AGRICULTURAL METEOROLOGY; MICROMETEOROLOGY.

Although considerable work has been done on bioclimatology of domestic animals, the following discussion focuses on humans.

Photochemical bioclimatology. Essentially this is the investigation of the effects of light from the Sun and sky. Sunburn of the skin and cornea of the eye is initiated by denaturization of nucleic acid and skin proteins, causing a local histaminelike action. In nature the effect is restricted and is induced by ultraviolet radiation with wavelengths around 0.3–0.31 μ. Sunburn is delayed or prevented by three screeening agents in the skin: pigment, horny layer, and urocanic acid. The pigment of permanently brown or black races, as well as the radiation-induced pigment in variably colored races (the so-called white race), acts as a protective filter. The horny layer of the skin reportedly grows in thickness by ultraviolet exposure. It is still debated how much protection is offered this way. The third screen is urocanic acid, a substance derived by enzymatic action from the animo acid L-histidine in the sweat. A 1-mm sweat layer absorbs 50% of the natural ultraviolet at 300 mm and nearly 100% of the mercury lamp ultraviolet at 250 mm. Artificial ultraviolet sources, such as the cold mercury lamp, act very differently from these sources. Solar erythema can be seen within minutes of exposure. A simple rule to avoid solar overexposure is this: As soon as the dividing line between exposed and shielded (clothed) areas can be discerned, stop sun bathing.

There are two kinds of solar pigmentation, late and direct. The late seems more prevalent in fair races; it occurs in conjunction with sunburns days later. The direct type is found more in southern Europeans and Japanese. It occurs, without sunburn, during a 30-min exposure, and is caused by long-ultraviolet ($0.32-0.4$ μ) and possibly visible light.

Frequent exposure over years leads to skin elastosis (sailor's skin) and finally skin cancer. Skin carcinomas occur more frequently on facial skin exposed to sun. Proof that exposure to sun is carcinogenic comes from statistical evidence and from the results of animal experiments, in which rodents were exposed to very strong artificial ultraviolet radiation.

Although bacteria are easily killed with artificial ultraviolet radiation of short wavelength, natural sunlight probably has very little bactericidal action because it lacks these short wavelengths.

Vitamin D is produced from natural sterols in plant and animal foodstuffs and in human skin by short-wavelength solar ultraviolet radiation (0.3 μ).

It has been claimed that solar ultraviolet radiation favorably affects blood circulation and general health, but sunbather faddism has not been conductive to serious research. It is difficult to separate the effects of ultraviolet radiation, thermohygric exposure, and mental stimuli on a sunbather.

Most sunlight and sky light received by the eye is that reflected by clouds and surfaces. The intensity of the incoming light, its angle of incidence, and the amount and kind (specular or diffuse) of albedo control the amount of light received by the eye. The specular reflections of water, ice, metals, snow, clouds, and white sand are bright enough to irritate the eye. Most sunglasses dampen the whole visible spectrum uniformly and eliminate ultraviolet and infrared rays.

All the important ultraviolet effects mentioned above occur at about 0.3 μ, near the end of the solar spectrum. This ultraviolet radiation is controlled by absorption in stratospheric ozone and scattering by air molecules, clouds, and smog. Dependence of ultraviolet radiation on solar and geometric altitude is pronounced. Usually, the scattered sky ultraviolet radiation exceeds that of the Sun. Snow reflects both ultraviolet rays and visible light to cause snow blindness and sunburn below the chin. No other natural substance reflects more than a few percent of natural ultraviolet; however, metals reflect highly.

Natural penetrating ionizing radiation is composed of β- and γ-rays from radioactive minerals, including their emanations and secondary rays from cosmic radiation, mainly mesons and neutrons. Ionization by both effects at altitudes below 18 km is less than 40 milliroentgens per day, probably an insignificant figure.

Air bioclimatology. Gaseous air constitutents, such as oxygen and water vapor, influence body chemistry and heat balance. The oxygen partial pressure (pO_2) of inhaled air is vital for blood oxygenation and depends mainly on altitude. At or above an altitude of about 3000 m, the pO_2 is low enough to cause air sickness, or mountain sickness. Residence at an altitude of $1-2$ km is supposed to benefit circulation, but may be harmful for some heart patients.

Low water vapor pressure may cause drying of the skin and of the mucous membranes of upper respiratory organs. Water vapor pressure, even indoors, is especially low when outdoor air temperatures are low. Many skin and respiratory complaints in winter are caused more by dryness than by cold. Water vapor influence on heat balance is discussed below.

Carbon dioxide, CO_2, water, H_2O, and organic vapors emitted by man contribute to the unpleasantness of crowded and ill-ventilated rooms, which may also be overheated.

Man's industries, volcanoes, fires, dust storms, and other more or less violent events spew large amounts of matter into the air known as fallout, smog, air pollution, and so forth. No bioclimatic problem in this field can be reported as solved unless source, transfer mode, change in transfer, concentration near the sink, the mode of intake by the sink, and the biological and chemical action of the sink are clear.

Ozone, O_3, may serve as an example. It is produced by solar ultraviolet radiation in the higher stratosphere, by lightning, and by sunlight falling on smog (Los Angeles). Stratospheric ozone is brought down by turbulence, and might reach toxic levels for crews of airplanes at about 15-km altitude. Ozone is blamed for some smog-induced injuries, for example, ocular pain in Los Angeles. A very important sink action is the inhalation into the respiratory organs, where especially the lung's surface is changed by oxidation. This causes coughing at first and tiredness later. Amounts of O_3 less than 0.5 part per million are toxic. *See* ATMOSPHERIC OZONE.

Carbon monoxide, unburned hydrocarbons, NO, NO_2, SO_2, and other by-products are prevalent where there is incomplete combustion. Many by-products are also highly hygroscopic, and serve as nuclei of condensation to cause the low-humidity type of smog, such as that occurring in Los Angeles. This smog condition is further complicated by the occurrence of light-induced reactions between hydrocarbons and NO_2 and other pollutants. The high-moisture type of smog seems typical for London. *See* SMOG.

Very few gases, in fact very little of any gas, pass the skin, with the exception of more easily permeated areas such as chapped lips, scrotum, and labia.

Aerosol bioclimatology. Solid or liquid suspensions in the air affect breathing organs or skin. Widely discussed elements of the atmospheric aerosols are the ions, which are particles containing one or more electrical charges. For a long time beneficial or dangerous effects of an abundance of particles of one charge type have been claimed. These electrical space charges reportedly have influenced results of physiological tests, growing of living cells, and severity of hay fever attacks. As a rule, negative space charge of the order of $1000-10,000$ ions/cm^3 is reportedly beneficial. These results are expected to explain some observations of statistical bioclimatology. *See* ATMOSPHERIC CHEMISTRY; ATMOSPHERIC POLLUTION.

Thermal bioclimatology. This subdiscipline concerns the heat balance of man as controlled by his environment. The basic relation is shown by Eq. (1), where M is metabolic heat production; C is

$$M - C d\theta_b/dt + H_h = H_r + H_a + H_e + R \quad (1)$$

body heat capacity; H is heat exchanged through skin; R is respiratory heat exchange; θ is temperature (°C); b refers to total body; h refers to solar radiation; r refers to infrared radiation; s refers to skin; a refers to air (convection); e refers to evaporation; and t is time. The H terms are defined in Eq. (2), where S is visible and near-infrared radiant

$$H_h = S \cdot \epsilon_h \cdot A_h \qquad H_r = h_r \cdot A_r (\theta_s - \theta_r)$$
$$H_a = h_a A_a (\theta_s - \theta_a) \qquad H_e = h_e \cdot A_e (e_s - e_a) \quad (2)$$

heat flow from the Sun, sky, and environment; and ϵ_h is absorptivity of skin (0.6 for white and 0.9 for black skin). The A factors, that is, the respective surface areas, depend on body posture and on the process involved, such as radiation or convection; these areas are always smaller than the geometric surface. The terms e_s and e_a are the vapor pressures of skin and air, respectively.

The h factors are heat conductances at absolute temperature T, as shown by Eq. (3), where ϵ_r is

$$h_r = \epsilon_r \sigma (T_s^4 - T_r^4)/(\theta_s - \theta_r) \quad (3)$$

0.98 (infrared absorptivity of skin), and σ is Stefan's constant, 4.9 kcal-m²/(hr)(deg⁴). From experiments with men, and using kilocalorie units (kcal), $h_a = 6.3\sqrt{v}$ kcal/(m²)(hr)(deg) for wind velocity v m/sec; further, $h_a = 3.3$ in calm. Equation (4)

$$h_e = 1.63 h_a \text{ kcal/(m²)(hr)(mb)(deg)} \quad (4)$$

follows from h_a. These data are valid for supine adults at sea level, and refer to the heat- or vapor-exchanging area.

Equation (5) applies for the clothed body, if h_e is

$$\theta_s - \theta_a = M/A_a h_c + (M + H_h)/A_a (h_a + h_r) \quad (5)$$

conductance of clothes, and if $\theta_r = \theta_a$, $A_r = A_a$, $H_c = 0$, and $d\theta_b/dt = 0$. The absorptivity of the clothing is inserted for ϵ_h. If $h_a \gg h_e$ (strong wind, thick clothes), solar heating H_h becomes unimportant.

H_e can be measured directly with a good balance. The relative skin humidity $r_s = e_s/e_s^*$, where e_s^* is saturation vapor pressure at skin temperature, can be measured with a hair hygrometer or can be derived from H_e, h_e, e_a, and e_s. During sweating r_s is 100%; otherwise, it is usually below 60%. If skin relative humidity is below 10%, the skin may crack.

Skin water transfer is accomplished by sweating and diffusion. The latter corresponds usually to 1 kcal/(m²)(hr); it may reverse, so that water or vapor is transferred into the skin. Sweat is a powerful emergency measure capable of producing $H_e = 600$ kcal/(m²)(hr) or more. Bioclimatic sweat control works two ways, via body temperatures and via water coverage. The prime control for the amount secreted seems to be the temperature of a section of the hypothalamus, skin temperature acting as a moderator. These temperatures are of course bioclimatologically controlled. Skin totally covered with sweat or bath water lowers its sweat water loss strongly, about 4:1. The effect is absent in saline or with sweat highly concentrated by evaporation.

The respiratory heat and vapor loss is small except during hyperventilation at reduced pressure. Ordinarily, this loss varies little because the exhaled temperature drops as θ_a and e_a fall.

Of the climatic elements, S, h_a (wind), and precipitation are easiest to control. High values of θ_a, θ_r, and e_a are much harder to influence than low values. To ensure a constant θ_b, the body alters the peripheral blood flow and thus alters θ_s. In low temperatures M is raised by shivering or work; in a hot environment H_e is raised by sweat. These emergency regulations are effective only for certain periods. Short-time limits are set by variations of θ_b and finally θ_s (skin burn or freezing). All body controls vary with age, sex, health, exercise, and adaptation. Reportedly, frequent limited exposures to adverse thermal conditions, particularly cold, invigorate many body functions, especially if exposure is combined with exercise.

The most important climatic element is local air temperature; in the cold, wind increases the rate of body cooling. For a well-insulated man or house, Eq. (5), air temperaute alone is the deciding factor. The house heating bill is proportional to the number of degree days, that is, time multiplied by $(\alpha - \theta_a)$, where α is the preferred room temperature, about 22°C. No usable formula for the cooling effect of open air on average individuals can be derived because clothing styles change.

At temperatures greater than 25°C sweating starts. Its evaporation is restricted by high values of e_a. Hence, the combination of θ_a and e_a describes livability in the heat. Tests on sensation experienced by young men exposed to different pairs of θ_a and e_a led to the effective temperature (ET), which approximately equals (θ_a + dew-point temperature)/2. The area of high ET or extreme summer discomfort in the United States is the Gulf

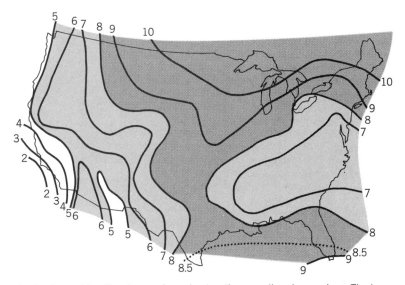

Fig. 1. Sum of heating degree days plus two times cooling degree days. The larger the sum, the larger is the integrated year-round outdoors discomfort, and also total fuel and power bill for heating and air conditioning. Units 1000°F times days. (*From K. J. K. Buettner, Human aspects of bioclimatological classification, in S. W. Tromp, ed., Biometeorology, Pergamon Press, 1962*)

Coast, especially the lower Rio Grande Valley (average ET, 26–27°C). In the Indus Valley ET is 28.5°C; and the world maximum is in Zeila, Somali Coast, where ET is 29.7°C. Figure 1 shows a United States map of the sum of heating and cooling degrees days. Since cooling also involves drying, the cooling degree days are multiplied by two. Large numbers of the sum mean climatic hardship for the outdoors man and high bills for heating fuel and cooling wattage. *See* SENSIBLE TEMPERATURE.

The microclimate can enhance discomfort. The values of θ_a and θ_r are much below normal on calm clear nights, especially when the terrain is concave. The Sun can raise the surface temperatures of such features as sand, walls, and vehicles to as much as 40° above the air temperature.

Extreme climates and microclimates. Certain climatic conditions can be tolerated only for a limited time, since they either overrun the physiological defense mechanism or injure and destroy body parts directly. The first mechanism usually leads to a breakdown of temperature regulation and of the body circulation, and the latter usually involves skin injuries.

The safe survival time depends primarily on air and wall temperature, if there is no artificial ventilation and no radiation other than the walls. Other conditions may suitably be converted into equivalent temperature data. Figure 2 contains such data, listing safe times between seconds (in a fire) to life-span. Some case studies are listed below.

1. The hot desert: Temperatures are in the fifties, vapor pressure is 10–15 mb, and there is sunshine and wind. Over a period of a few hours no equilibrium of body data is reached, and especially heart rate and core temperatures rise. The skin is dry and salt covered. The skin water loss is much below the level needed to compensate by evaporation for the large heat input by solar and infrared radiation and convection.

2. The jungle: Normal jungle is the habitat of a large segment of people. However, extreme jungle conditions are around 35°C and 40 mb vapor pressure. The skin is totally wet, but evaporation becomes insufficient due to external moisture. The effective temperature is a good measure for this danger signal. Worst in this respect are locally wet areas surrounded by a hot desert.

3. Fire: A forest or house fire has flame and glowing fuel temperatures of 600–900°C. Injury comes from skin burns via contact, convection, or radiation. The latter can be warded off by aluminum-covered suits. Near a large fire radiant heat of 40,000 kcal m^{-2} hr^{-1} may cause skin pain in 1 sec and burns in a few seconds. This heat transfer is superior to flame contact. Inhaling of hot air and lack of O_2 are frequently minor problems compared to poisoning by CO.

4. Frostbite and freezing fast: Strong convective cooling by wind and low temperatures may overrun the local defenses, causing defective circulation in the skin and breakdown of small vessels and blood particles. Developing of ice crystals may puncture cells. Skin touching metals below −30°C freezes on contact, causing mechanical loss of skin layers upon removal.

5. Weather accident and disaster: Yearly average of fatalities by weather catastrophes for the United States during 1901–1960 are as follows: hurricanes, 100; tornadoes, 150; lightning, 175; floods, 80; and snowstorms, blizzards, and ice storms, 200. There are probably deaths of similar numbers from heat waves and smog periods. Weather-caused farming disasters have produced many mass-killing famines. These figures fortunately have declined, as have death figures from hurricanes and tornadoes and, probably, lightnings. All others mentioned here have risen.

Statistical bioclimatology. This aspect of bioclimatology investigates correlation between weather and climate phenomena and their effects on man. The most important work concerns clinical data correlated with daily weather. Two weather types seem to be involved: Fronts, central low, and ascending air coincide with frequent attacks of angina pectoris and embolism; incidence of foehn and cyclogenesis to the west are found to correlate with increasing circulatory disorders, mental troubles, increase of accidents, and kidney colic. It has been claimed that there is a correlation between increased solar activity and all deaths in big cities, especially deaths from mental diseases. *See* CLIMATOLOGY; WEATHER.

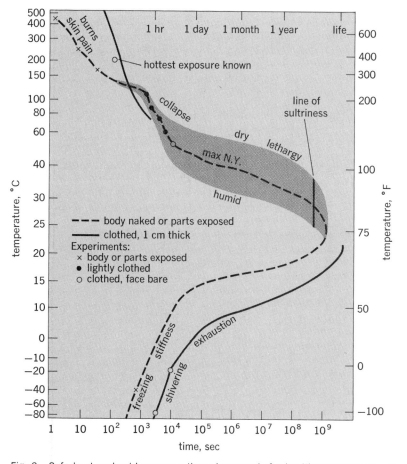

Fig. 2. Safe heat and cold exposure times in seconds for healthy, normal men at rest, with body wholly or partly exposed. Room is free from artificial ventilation and radiation; air and walls have identical temperature. Humidity influence is shown by shading. Max N.Y.= highest data of temperature and humidity in New York City. *(From K. J. K. Buettner, Space medicine of the next decade as viewed by an environmental physicist, U.S. Armed Forces Med. J., 10:416, 1959)*

Climatotherapy. This health treatment is more venerated than exact. Going to a resort is intimately connected with nonclimatic changes of housing, mode of living, food, exercise or rest, and phychological stimuli. The vacation effect itself may be quite helpful.

A health climate location should avoid extremes of temperature, especially high effective temperature felt as sultriness. Air should be free of smoke, smog, allergens, ozone, car exhausts, and industrial and volcanic effluents.

Solar ultraviolet and natural radioactive substances and rays are generally harmful. Beneficial outdoor living usually involves exposure of parts of the skin to solar ultraviolet, and the ensuing pigment is then taken as proof of health, a rather doubtful conclusion. A certain moderate thermal stimulation by wind, waves, and temperature changes may be helpful to most people. Adaptation to the new time zone may severely interfere with adaptation of intercontinental East-West travelers.

If a physician prescribes a very balmy climate for a weak elderly patient, the patient should be aware of the difference between the dry summer heat of for example, El Paso, Texas, and the moist heat of the Lower Rio Grande.

Paleobioclimatology. This deals with possible environmental influences on the development of species, especially of man. The local microclimate has to be considered. A cave, for example, has a very constant temperature and humidity and no sunlight. Fire, as an adjunct, was introduced in China and Hungary more than 500,000 years ago. In a tropical jungle one is not exposed to solar rays. The reduction of fur in man seems correlated with the perfection of eccrine sweating as part of a well-developed cranial temperature control. It might also permit man to utilize his skin temperature sensors to function as infrared "eyes" in the dark of a cave.

In most animals skin and fur color seem to promote camouflage or its opposite. Should this be true for man, one would find dark skin in the tropical jungle; while in a temperate maritime clime white in winter and brown in summer seems adequate. Thermally, a black skin is less suited for sunny hot climate than is a lighter one. Fair-skinned but unprotected man at a latitude of 40–50° may seasonally adapt himself to the varying solar ultraviolet. Sunburn is unlikely if daily exposure is routine. During the ultraviolet-rich summer season a sufficient deposit of vitamin D can be produced in the body to last for the darker seasons. There might have been sufficient vitamin D in the food of Stone Age man. So neither sunburn nor vitamin D seems to have had an effect on evolution.

Ideal bioclimate. There is no ideal climate for all men. One's physiological and psychological preference seems to be set in childhood. History starts, long before effective room heating, in such subtropical climates as Egypt, Mesopotamia, and southern China, or in tropical lowlands, for example, Yucatan, or in tropical highlands, as Peru. It moves in the Old World to higher latitudes, where Romans developed heating. Later, civilized power centers around the 45° latitude, where the house microclimate was adequate the year around, even before air cooling.

Statements that culture best develops in a particular climate are of little value.

In the United States people entirely free to settle where they like the climate seem to prefer the Mediterranean climate of Southern California and the desert of Arizona.

[KONRAD J. K. BUETTNER/H. E. LANDSBERG]

Bibliography: F. Becker et al., *A Survey of Human Biometeorology*, World Meteorol. Organ. Tech. Note no. 65, 1964; K. Buettner, *Physikalische Bioklimatologie*, 1938; K. Buettner et al., Biometeorology today and tomorrow, *Bull. Amer. Meteorol. Soc.*, 48:378–393, 1967; G. E. Folk, *Introduction to Environmental Physiology*, 2d ed., 1974; H. E. Landsberg, *Weather and Health*, 1969; S. Licht (ed.), *Medical Climatology*, Physical Medicine Library, vol. 8, 1964; J. H. Prince, *Weather and the Animal World*, 1975; S. W. Tromp and W. H. Weihe (eds.), Biometeorology Two, in *Proceedings of the 3d International Bioclimatological Congress*, pts. 1 and 2, 1967.

Black Sea

The Black, Azov, Caspian, and Aral seas form a series of bodies of water extending far into the Eurasian landmass. As the distance from the Mediterranean increases, these seas (or lakes) freshen and the composition of their salts departs more widely from that of the oceans. The Black Sea is tenuously connected with the world ocean through the Bosporus, the Sea of Marmara, the Dardanelles, and the Mediterranean Sea.

Runoff and precipitation on the Black Sea exceed losses by evaporation, causing a net outward flow through the surface layers of the Bosporus and a compensating inflow of Mediterranean water along the bottom. The upper layers of the Black Sea are so diluted that the density increases rapidly with depth (Fig. 1), preventing effective vertical mixing. Because of this and the shallowness of the Bosporus (40 to 90 m), the deep waters of the Black Sea are stagnant. The general features of the Black Sea are given in the table.

Circulation. The surface currents form two counterclockwise gyres (Fig. 2) with complicated countercurrents between them, and a current sets outward through the Bosporus. This gyral circulation is reflected in the upward doming of surfaces of equal density, equal chemical concentrations (Fig. 1), and biological zonation in the approximate centers of the gyres.

The Bosporus undercurrent introduces water denser than that in the Black Sea, but this inflow is so small that it would take 2500 years for it completely to replace the water below a depth of 30 m.

The tides of the Black Sea are semidiurnal and of small amplitude: about 9 cm along the central-eastern and central-western coasts and only 2 to 3 cm on the Crimean coast. Their effects on water level are frequently masked by winds and responses to barometric pressure variations.

Physical and chemical characteristics. The Black Sea is distinctly less saline (maximum salinity 22.7°/oo) than the open ocean (about 35°/oo), but its salt has nearly the same composition as that of ocean salt, indicating that the Black Sea is a part

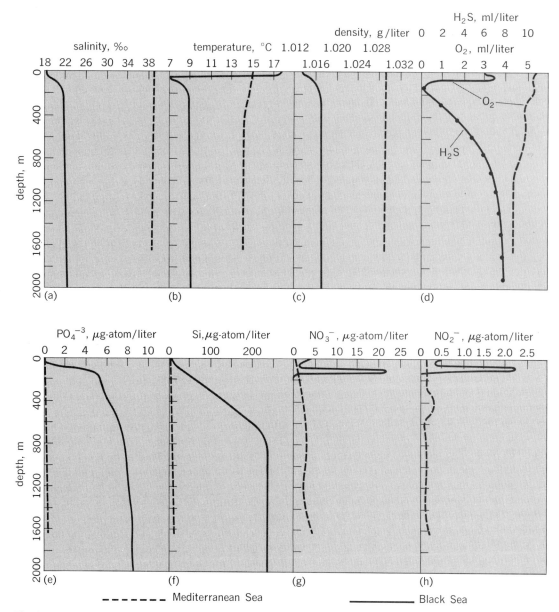

Fig. 1. Vertical distribution of chemical properties and of density in the Black Sea with comparative values from the Mediterranean Sea. The Black Sea values were observed at 41°58'N, 30°23'E on Sept. 29, 1964; the Mediterranean values at 35°54'N, 25°15.5'E on Mar. 30, 1948. (a) Salinity, in grams/kilogram. (b) Celsius temperature. (c) Density of the water at the temperature given in b if brought to the sea surface. (d) Dissolved oxygen and hydrogen sulfide concentrations, both in milliliters/liter. (e) Phosphate-phosphorus concentrations, in microgram atoms/liter (millimoles/liter). (f) Soluble silicate-silicon concentrations, in microgram atoms/liter. (g) Nitrate-nitrogen concentrations, in microgram atoms/liter. (h) Nitrite-nitrogen concentrations, in microgram atoms/liter.

of the ocean system, not a saline lake. The salinity of the surface layers and at 100 m are shown in Fig. 3. The salinity increases rapidly with depth to around 200 m, but at depths greater than 300 to 400 m the salinity increases only slightly with depth (Fig. 1).

River inflow lowers the salinities in the northwestern part of the sea, which allows the formation of ice as much as 70 mi offshore in exceptionally severe winters.

Surface temperatures vary from the freezing point in winter to over 22°C in summer, but at depths greater than 150–200 m, the temperature is remarkably constant at around 8.75°C.

The stagnation of the Black Sea permits the entrapment and decomposition of unusually large amounts of organic matter in the deeper layers, with the following consequences.

1. Dissolved oxygen is absent and hydrogen sulfide, H_2S, is present at all depths greater than about 200 m. The level at which the oxygen disappears is also domed upward in the centers of the gyres, where it may be as shallow as 125 m. The H_2S reaches concentrations of 6 to 8 ml/liter in the depths (Fig.1).

2. Nitrites and nitrates are absent from the oxy-

General features of the Black Sea

Latitude: 40°55′N to 46°32′N	
Longitude: 27°27′E to 41°42′E	
Maximum length:	1149 km
Maximum breadth:	610 km
Area: Black Sea	423,000 km²
Sea of Azov	38,000 km²
Total	461,000 km²
Volume: Black Sea	537,000 km³
Sea of Azov	300 km³
Total	537,300 km³
Depths: Black Sea	
Average	1197 m
Maximum	2245 m
Sea of Azov	
Maximum	13.5 m
Approaches	
Kerch Strait	5 m
Bosporus (max.)	90 m

Water balance estimates:

Main river inflow	
Danube	203 km³/year
Dnieper and Bug	55 km³/year
Dniester	8 km³/year
Don and Kuban	26 km³/year
Others	28 km³/year
Total	320 km³/year
Outflow through Bosporus	388 km³/year
Inflow through Bosporus	188 km³/year
Excess outflow	200 km³/year
Precipitation on Black Sea	234 km³/year
Runoff and river inflow	320 km³/year
Evaporation	354 km³/year
	(84 cm/year)

*The water balance can be expected to vary widely from year to year.

gen-free zone, where denitrification has produced free dissolved nitrogen, although the quantities of free nitrogen from this source are small compared with the amounts from the solution of air.

3. Anaerobic fermentation of organic matter produces up to 0.2 ml/liter of methane in water.

4. The decomposition of organic matter yields phosphate ions, carbon dioxide, ammonia, and soluble silicates, all of which build up to high concentrations in the deep waters. The production of sulfides appears to account for the high alkalinity of the deep water, and the introduced CO_2 reduces the pH from surface values of 8.0–8.2 to about 7.7 in the depths.

5. The oxidation-reduction potential of the water gradually decreases with depth from over +400 millivolts (mv) near the surface until the appearance of sulfides, when it drops abruptly to around −110 mv and then continues to decrease with increasing H_2S concentrations to −172 mv.

6. Only anaerobic bacteria live in the H_2S zone.

7. Metals, such as copper, are presumably removed from solution by sulfide precipitation, but soluble iron and manganese are more concentrated in the sulfide zone than in the oxygenated zone.

Geologic history and features. The Black Sea is a tectonic basin that was, in the early Tertiary, part of a marine system that connected the Atlantic and Indo-Pacific oceans. Later, the eastern end of the system was closed, and in the late Miocene it was cut off from the ocean to the west. Tectonic changes later divided this sea into isolated basins;

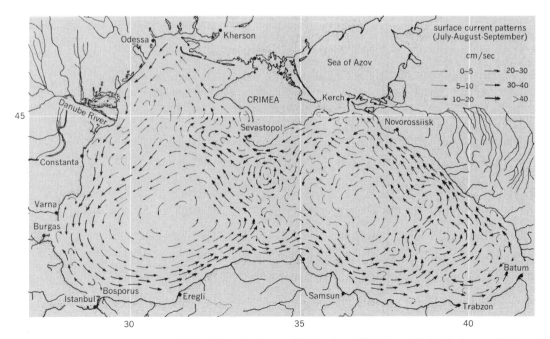

Fig. 2. Surface-current patterns of the Black Sea in midsummer. (*From G. Neumann, 1942, as used in J. W.* Hedgpeth, ed., *Treatise on Marine Ecology and Paleoecology, Geol. Soc. Amer. Mem. no. 67, 1957*)

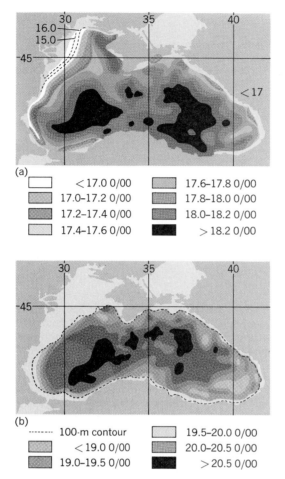

(a)

☐ <17.0 0/00	▨ 17.6–17.8 0/00
▨ 17.0–17.2 0/00	▨ 17.8–18.0 0/00
▨ 17.2–17.4 0/00	▨ 18.0–18.2 0/00
▨ 17.4–17.6 0/00	■ >18.2 0/00

(b)

------ 100-m contour	▨ 19.5–20.0 0/00
▨ <19.0 0/00	▨ 20.0–20.5 0/00
▨ 19.0–19.5 0/00	■ >20.5 0/00

Fig. 3. Salinity distribution in the Black Sea. (a) Salinity of the upper surface layer from July through September. (b) Salinity at 100 m. (*From G. Neumann, 1943, as used in J. W. Hedgpeth, ed., Treatise on Marine Ecology and Paleoecology, Geol. Soc. Amer. Mem. no. 67, 1957*)

the present Caspian, Black, and Mediterranean seas were sometimes connected and sometimes separated. These changes were accompanied by large salinity changes which are reflected in the fossil faunal assemblages of the region.

The present Black Sea is a relatively regular oval, somewhat indented in the south and deeply indented in the north by the Crimean peninsula

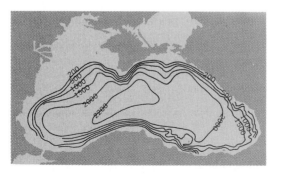

Fig. 4. Bottom topography of the Black Sea. The depths are indicated in meters. (*After L. Zenkevitch, Biology of the Seas of the U.S.S.R., Wiley, 1963*)

(Fig. 4). To the east of the Crimea, the Black Sea and the shallow Sea of Azov are connected by the narrow Kerch Strait. The connection with the Mediterranean was last opened some 6000 years ago, increasing the salinity to its present levels.

Except in the southeast and northwest, the sea deepens abruptly offshore (Fig. 4). The northwestern part is shallow and this coast is characterized by numerous drowned valleys, "limans," whose degree of connection with the sea varies widely. The Sea of Azov is essentially a large liman.

The sediments of the Black Sea tend to grade from sandy coastal material extending to 30 m depths, to highly calcareous muds, some with layers of gray clay, toward the centers of the gyres. All the deep sediments contain relatively large quantities of various forms of ferrous sulfide. The rivers introduce large amounts of terrigenous materials.

Biology. The Black Sea has fewer species of plants and animals than the Mediterranean, but more than the Azov, Caspian, or Aral seas. The 1500 animal species are mainly a depleted Mediterranean fauna with a few relict Caspian types and, near the Danube delta, a few fresh-water forms.

The phytoplankton is characterized by a large diatom bloom that reaches a maximum in May to June. Later, dinoflagellates reach a lesser peak. There are only about 70 species of zooplankton organisms, presumably because of the low salinity. Because of the low oxygen content, only four plankton species regularly occur deeper than 50 m.

There are only 221 algal species in the Black Sea (413 in the Mediterranean), but there are vast fields of the red alga *Phyllophora rubens* var. *nervosa* in the northwestern part and extensive stands of eel grass (*Zostera marina*) along the northwestern coast. There are extensive populations of bivalves in this part of the sea, but the bottom fauna is made up of greatly reduced numbers of species of Mediterranean forms. There are around 170 fish species, of which over 100 are Mediterranean immigrants; 20% of the fishes are of commercial use: among the most important of these are mackerel, bonito, anchovies, herring, mullets, carp, and sturgeon. Most of the fishery is nearshore, and of the 15% which is carried out on the open sea, porpoises make up 90% of the catch. *See* MARINE MICROBIOLOGY; MARINE SEDIMENTS; MEDITERRANEAN SEA; OCEANS AND SEAS; SEA WATER.

[FRANCIS A. RICHARDS]

Bibliography: R. W. Fairbridge (ed.), *The Encyclopedia of Oceanography*, Encyclopedia of Earth Sciences Series, vol. 1, 1975; J. W. Hedgpeth (ed.), *Treatise on Marine Ecology and Paleocology*, Geol. Soc. Amer. Mem. no. 67, vol. 1, 1957; L. Zenkevitch, *Biology of the Seas of the U.S.S.R.*, 1963.

Brocken specter

An illusory magnification to gigantic proportions of an observer's shadow when it is projected on cloud surfaces, particularly when the surfaces extend upward about mountain peaks. The name Brocken specter is derived from the Brocken, the highest peak of the Harz Mountains in Saxony, where the

Fig. 1. A specter image, surrounded by a semblance of its glory, seen on the surface of the cloud between the shadows of two radio towers. (*P. A. Allee, ESSA*)

Fig. 2. View of proportionately huge (seven times that of Fig. 1) specter image resulting from advance of the cloud surface to the observer's position. Note the well-developed rings of glory (in rainbow hues on original color photograph). (*P. A. Allee, ESSA*)

phenomenon is sometimes seen. Rings of colored light known as the Brocken bow, or glory, are to be seen surrounding the Brocken specter.

The apparent magnification of the observer's shadow occurs when he overestimates the distance between himself and his shadow. In Fig. 1 the specter, surrounded by its glory, can be seen on the surface of the cloud between the shadows of two radio towers. The cloud bank is a considerable distance beyond the trees, which are about 50 yd away. As the cloud surface approaches the observer, the specter increases rapidly in size. The radio towers are behind the observer, and spacing of their shadows increases at a slower rate.

In Fig. 2 the cloud surface is at the site of the observer and the specter is seven times larger. The apparent size of the specter is 30 ft in height when compared to the known height and spacing of the radio towers. The camera could not glimpse, as does the eye by means of peripheral vision, the closer and apparently even larger specter shadow. This latter shadow creates an occasional illusion of a nearby giant because the eye focuses at a distance on large objects, such as trees, which appear quite small when compared to the specter shadow glimpsed by peripheral vision. *See* METEOROLOGICAL OPTICS.

[PAUL A. ALLEE]

Capillary wave

Capillary waves, or ripples, occur at the interface between two fluids, in which the principal restoring force is controlled by surface tension. Ripples generated by wind at the interface between air and water on oceans and lakes are of importance to the friction of air flowing over water, and to the reflection and scattering of electromagnetic and sound waves.

The formulas relating the phase velocity c and frequency f to the wavelength λ of low-amplitude sinusoidal waves, in the absence of wind forces, are as shown below:

$$c = c_m\left(\frac{\lambda}{2\lambda_m} + \frac{\lambda_m}{2\lambda}\right)^{1/2} \qquad f = \frac{c}{\lambda}$$

where

$$c_m = \left(2\frac{\rho_2 - \rho_1}{\rho_2 + \rho_1}\right)^{1/2}\left(\frac{gT}{\rho_2 - \rho_1}\right)^{1/4}$$

$$\lambda_m = 2\pi\left[\frac{T}{(\rho_2 - \rho_1)g}\right]^{1/2}$$

T is the surface tension, ρ_1 and ρ_2 are the densities of upper and lower fluids, respectively, and g is the acceleration of gravity. For air over water at 15°C, c_m is 23 cm/sec and λ_m is 1.7 cm. The phase velocity is a minimum at $c = c_m$ when $\lambda = \lambda_m$ (see illustration). Shorter waves are ripples, longer waves are gravity waves. *See* OCEAN WAVES.

In nature, ripples are observed to grow rapidly when the wind blows and to die away rapidly when the wind stops. When the water surface is uncontaminated, ripples die away to e^{-1} of their original amplitude in a time $t_o = \lambda^2/(8\pi^2\nu)$, where ν is the kinematic viscosity of water. For $\lambda = 1.7$ cm and $\nu = .01$ cm²sec, $t_o = 3.8$ sec. When the water surface is contaminated, as by an oil film or other surface-active agent, ripples are damped still more rapidly because the contaminated surface acts as an inextensible film against which the water motions due to the ripples must rub. For a perfectly inextensible film, t_o is equal to $(\pi^3\nu f)^{-1/2}$. For the example treated above, t_o becomes 0.86 sec. For moderate winds, the increased damping of ripples almost completely inhibits their growth; the surface appears smooth and is called a slick. It has been observed that even gravity waves grow at an inappreciable rate under such conditions; the interpretation here is that the fine scale of roughness presented to the wind by a rippled surface is necessary for the formation of gravity waves. On the other hand, ripples have been observed to be formed, in the absence of wind, by momentary nonlinear interactions of steep gravity waves; consequently the formation of both gravity waves and ripples is an interconnected process.

The surface profile of ripples of large amplitude which move without change of form has been calculated by G. D. Crapper. The profile changes from sinusoidal to one with sharper troughs than crests as amplitude increases. [CHARLES S. COX]

Bibliography: G. D. Crapper, An exact solution for progressive capillary waves of arbitrary amplitude, *J. Fluid Mech.*, 2:532–540, 1957; M. E. McCormick, *Ocean Engineering Wave Mechanics*, 1973; A. H. Schooley, Profiles of wind-water waves in the capillary-gravity transition region, *J. Mar. Res.*, 16(2):100–108, 1958; R. L. Wiegel, *Oceanographical Engineering*, 1964.

Caribbean Sea

A deep ocean basin roughly 1500 n mi (2800 km) long and up to 800 n mi (1500 km) wide, bordered by Central and South America and the West Indies island chain (see illustration). Together with the Gulf of Mexico, it is sometimes called the American Mediterranean Sea. Two major basins, the large Caribbean Basin and the smaller and deeper Cayman Basin, are separated by the Jamaica Rise. The mild and equable climate of the Caribbean Sea may be marred in the summer and early autumn by occasional hurricanes.

The basic circulation is from east to west with North Equatorial Current water entering through passages between the eastern islands and exiting through the Yucatan Channel to the Gulf of Mexico, and eventually through the Florida Straits to become an important portion of the Gulf Stream system. The volume of this flow is about 30,000,000 m³/sec.

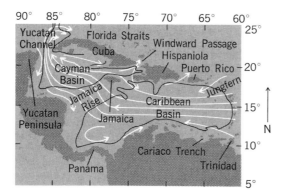

Caribbean Sea. Arrows show general surface circulation.

Bottom water renewal has been the subject of much conjecture over the years. Measurements reported by M. C. Stalcup, W. G. Metcalf, and R. Johnson and by W. Sturges in 1975 showed a small net inflow of deep water of 50,000 to 60,000 m³/sec over the Jungfern Sill (1815 m), which may be sufficient to maintain the present lower oxygen and higher silicate concentrations found in the deep Caribbean Basin as compared with the values just outside in the Atlantic Ocean. Slightly higher oxygen and lower silicate levels in the Cayman Basin suggest that the renewal flow over the Windward Passage Sill (1560 m) may be slightly more vigorous.

Cold water lies at relatively shallow depths under very warm surface water in this area. Proposals are being considered for generating commercially economical quantities of electricity by taking advantage of this sharp thermal gradient. *See* AT-LANTIC OCEAN; GULF OF MEXICO; OCEAN CURRENTS; SEA WATER.   [WILLIAM G. METCALF]

Bibliography: M. C. Stalcup, W. G. Metcalf, and R. Johnson, Deep Caribbean inflow through the Anegada-Jungfern Passage, *J. Mar. Res.*, vol. 33, suppl. 1975; E. M. Nairn and F. G. Stehli (eds.), *The Gulf of Mexico and the Caribbean*, Ocean Basins and Margins Series, vol. 3, 1975; K. Pringle, *Waters of the West*, Caribbean Series, 1976; W. Sturges, Mixing of renewal waters flowing into the Caribbean Sea, *J. Mar. Res.*, vol. 33, suppl., 1975.

Chinook

A mild, dry, extremely turbulent westerly wind on the eastern slopes of the Rocky Mountains and closely adjoining plains. The term is an Indian word which means "snow-eater," appropriately applied because of the great effectiveness with which this wind reduces a snow cover by melting or by sublimation. The chinook is a particular instance of a type of wind known as a foehn wind. Foehn winds, initially studied in the Alps, refer to relatively warm, rather dry currents descending the lee slope of any substantial mountain barrier. The dryness is produced by the condensation and precipitation of water from the air during its previous ascent of the windward slope of the mountain range. The warmth is attributable to the previous release of latent heat of condensation in the air mass and to the turbulent mixing of the surface air with the air of greater heat content that occurs aloft.

In winter the chinook wind sometimes impinges upon much colder stagnant polar air along a sharp front located in the foothills of the Rocky Mountains or on the adjacent plain. Small horizontal oscillations of this front have been known to produce several abrupt temperature rises and falls of as much as 50°F at a given location over a period of a few hours.

In the Alpine regions adverse psychological and physiological effects have been noted in humans during prolonged periods of foehn wind. These phenomena have been referred to as "foehn sickness." *See* FRONT; ISENTROPIC SURFACES; PRECIPITATION; WIND; WIND STRESS.

[FREDERICK SANDERS]

Clear-air turbulence

According to a 1966 definition by the National Committee for Clear Air Turbulence, CAT (as the phenomenon is known) "comprises all turbulence in the free atmosphere of interest in aerospace operations that is not in, or adjacent to, visible convective activity [cumulus clouds]. This includes turbulence found in cirrus clouds not in, or adjacent to, visible convective activity."

CAT thus defined considers only those irregular atmospheric motions which cause bumpy flight conditions in aerospace vehicles (airplanes, rockets, VTOLs, and so on) outside and away from cumulus clouds. Rough flying conditions in the turbulent mixing layer above the ground, in which terrain conditions and solar heating of the Earth's surface cause irregular up- and downdrafts, are expected by the experienced pilot and are therefore usually excluded from CAT statistics. More dangerous is CAT at high levels (in the upper troposphere and in the stratosphere), where it strikes without visible warning and may cause serious

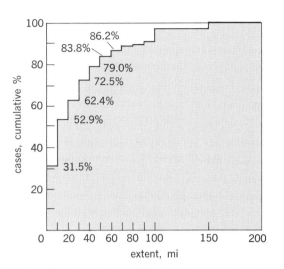

Fig. 1. Horizontal extent of CAT patches encountered in Project Jet Stream. (*After N. W. Cunningham*)

structural damage to the airframe and injury to passengers and crew. *See* AERONAUTICAL METEOROLOGY.

Effects. Sudden up- and downdrafts, if they are strong enough, will cause vertical accelerations of an aircraft which are felt as bumps. A missile, on the other hand, which rises vertically through the atmosphere, may be shaken about when traversing atmospheric layers in which winds blow at different speeds or from different directions or both. Usually the guidance system of the rocket will take care of such variable winds. Nevertheless, in counteracting the changing wind effects, the missile will be subject to stresses which may be undesirable or even dangerous.

Aircraft bumpiness depends not only on the magnitude of atmospheric up- or downdrafts or both, but also on the suddenness with which they are traversed by the aircraft. Therefore large eddies in the atmosphere will have less effect on an airplane than small eddies with the same wind velocities, because in the former the vertical drafts are experienced much more gradually, and the plane will have more time to ride them out. A slow plane generally will experience less CAT than a fast one for the same reason. A heavy plane will ride more smoothly than a light one, because drafts of equal magnitude will cause less vertical acceleration of the former. A plane with flexible wings will experience less turbulence than one with rigid wings, because by its flexibility the airframe is able to absorb much of the beating which it receives from atmospheric eddies. However, it does not take this beating unpunished, because the continued flexing and bending processes will lead to material fatigue which will limit the useful life of an aircraft. Thus CAT becomes an economic problem for military and civilian aircraft operators.

From the foregoing it is obvious that the occurrence of CAT not only depends on atmospheric conditions but also on the type of aircraft and its flight conditions (airspeed, attitude, and so on). In general, eddies of horizontal dimensions of approx-

imately 20 to 200 m will affect subsonic jet aircraft as CAT. Smaller eddies will cause only minor skin vibration, while the aircraft will ride out larger eddies smoothly.

Occurrence. In the subsequent discussion the individual aircraft response effects will be discounted. Instead, emphasis will be on atmospheric conditions as they would be encountered by one specific type of aircraft, a subsonic jet.

Patches of CAT are usually, but not always, relatively small, producing only short periods of bumpiness. Figure 1 shows that 52.9% of CAT cases encountered had horizontal dimensions of 20 mi or less; 83.8% of all cases were 70 mi or less in diameter. For a jet aircraft flying at 500 mph such turbulent patches will last only for a few minutes of flying time.

CAT occurs most frequently near jet streams in the regions of strong vertical wind shear above and below the core of highest wind speeds. This includes the stratospheric regions (above the tropopause) near jet streams. It has also been found that CAT is more readily encountered over mountains than over flat terrain or over oceans, especially if low-level winds blow directly against the mountain range and produce standing lee waves. The latter are often manifested in lenticular cloud patterns which are displayed downstream from the mountain range. Figure 2 shows schematically the formation of lee waves and of a turbulent layer above the tropopause. *See* JET STREAM.

Nature of CAT. Since conventional aircraft are more sensitive to vertical gusts than to horizontal ones, an understanding of the former will suffice to describe the phenomenon of CAT. (As has been

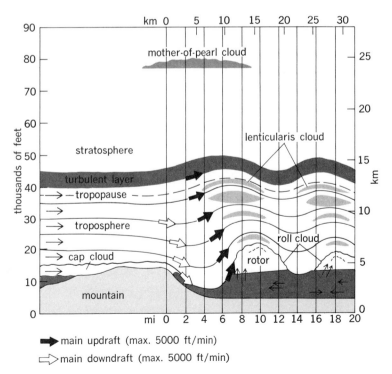

Fig. 2. Atmospheric flow and turbulence conditions associated with typical mountain wave. (*After C. S. Jenkins*)

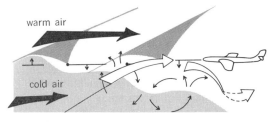

Fig. 3.   Schematic diagram of airflow in the vicinity of an idealized wave pattern caused by vertical wind shear.

mentioned earlier, this approach will not suffice for missiles, which react to wind changes with altitude. These wind changes, however, are not necessarily due to turbulent flow conditions in the atmosphere, but may be due to the vertical structure of atmospheric flow, which usually changes only slowly in the course of a few hours; this flow is more easily measured than CAT in aircraft and will be disregarded here.) Vertical motions of a horizontal scale to be felt as CAT may be caused by convective motions in shallow, thermally unstable layers. If these layers are deep, vertical motions are usually accompanied by cloud formation. Ensuing turbulence would not be classified as CAT. Turbulence in the atmospheric layers near the Earth's surface is usually caused by such convective up- and downdrafts.

Near the jet stream and in the stratosphere CAT occurs mainly in or near thermally stable layers (warm air stratified on top of cold air). Vertical wind shear, if it is strong enough, will cause laminar flow to break up into eddies of all sizes. Those measuring about 20 to 200 m in diameter will be felt as CAT if their motion is violent enough, that is, if they have enough kinetic energy to accelerate the aircraft flying through them.

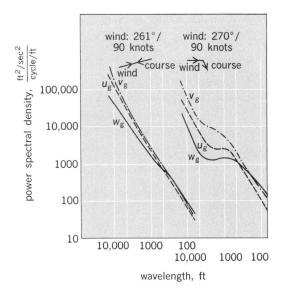

Fig. 4.   Distribution of kinetic energy in turbulent flow as a function of wavelength plotted on logarithmic coordinate scales. Data obtained over Australia by the Royal Aircraft Establishment, Farnborough, England. 1 ft = 0.3048 m;   1[ft²/sec²]/[cycle/ft] = 0.3048 [m²/sec²]/[cycle/m]; 90 knots = 46 m/sec.

Figure 3 shows schematically how waves in a transition layer between cold and warm air flowing at different speeds (indicated by the length of the heavy arrows) break down into turbulent eddies (small black arrows). The kinetic energy of those eddies normally varies directly with their diameter. If these characteristic diameters (that is, the "wavelengths") of the eddies are plotted against the kinetic energy per unit of mass contained per unit of wave number—dimensions $[ft^2/sec^2]/[cycle/ft]$—one may see very clearly this increase of energy with increasing wavelength. In Fig. 4 such a plot is presented, using logarithmic scales along the coordinates: u_g, v_g, and w_g are the eddy velocity components respectively parallel to, normal to, and vertical to the flight direction of the measuring air plane.

Even though the kinetic energies of large eddies are considerable, subsonic jet aircraft do not experience them as bumpiness, because they ride over the waves or eddies smoothly. Supersonic aircraft, however, will react to larger-size eddies than do slower aircraft, and may experience CAT where subsonic commercial jets do not.

From Fig. 4 one can also see that the energy in the w component of waves longer than approximately 1000 ft appears to be smaller when the aircraft flies normal to the wind than when it flies with the wind. Figure 3 schematically accounts for this observation. An aircraft riding against the wind (large, heavy arrows) will experience not only the small-scale CAT eddies (thin arrows) but also the larger-scale up- and downdrafts of the longer waves against which it flies. If the aircraft turns parallel to crests of these waves (according to Fig. 3, this would be a flight perpendicular to the wind arrows), it still would experience CAT from small-scale eddies but no effect from longer waves.

Although the physical nature of CAT is better understood as a result of many investigations, forecasting methods and advance warning devices still need to be improved. *See* METEOROLOGY.

[ELMAR R. REITER]

Bibliography: E. R. Reiter, *Jet Streams: How Do They Affect Our Weather*, 1967; E. R. Reiter, *Jet-Stream Meteorology*, 1963; Society of Automotive Engineers, *Proceedings of the National Meeting on Clear Air Turbulence*, 1966.

Climate modification

Human activity can have a marked effect on the atmospheric environment. This is particularly true of the shallow air layer near the ground and in restricted local areas. Some of these microclimatic changes are brought about deliberately for specific agricultural and engineering purposes. Others are inadvertent and may even be detrimental. Attempts to change climate on a larger scale have been discussed. The possibility of climatic changes on a global scale caused by human pollution of the atmosphere is under investigation.

Climate results from interaction between the general atmospheric circulation with the given geographic surface configuration. Delicate feedback mechanisms exist between the atmospheric variables and continents, oceans, their currents, mountains, and snow and ice surfaces. No complete and realistic mathematical model of climate exists as yet; hence most approaches to climatic modification have been empirical and qualitative. *See* CLIMATOLOGY.

The most successful and beneficial results of modification have been achieved in agriculture and horticulture. Nearly all of these deal with changes in the heat capacity of the soil surface, its emissivity and albedo, moisture balance, and aerodynamic roughness. In their simplest form these measures consist of mulching for conservation of moisture; shielding by glass or plastic to prevent nocturnal heat losses; albedo (reflectivity) changes either to absorb more heat from the Sun or to reflect more light on plants to increase photosynthesis; and shading to protect plants and livestock from radiative heat stress. Protective measures against adverse weather include artificial fog production as a frost precaution, fans to destroy surface temperature inversions, sprinkling to gain the heat of fusion, and the direct heating of orchards. Only direct heating procedure will raise nocturnal temperatures by 2–3°C. *See* MICROMETEOROLOGY.

Shelter belts. Belts, usually consisting of trees and shrubs, are used to reduce wind speeds. Thus the rate of evapotranspiration and plant damage from strong winds is reduced, and also moisture is increased by trapping of snow. Figure 1 shows the wind reduction downwind from a shelter belt with distances measured in units of height of the shelter belt. Maximum protection is offered at about 3–6 belt heights to the lee. Temperature increases in crops during summer may be up to 3°C; evaporation is 20–25% reduced; crop yields are increased 10–35% compared to unsheltered areas. The same principle is used in snow fences to protect highways from blowing snow. Both shelter belts and snow fences show optimal efficiency if they are about 50% permeable to air. Solid walls offer poor protection.

Artificial irrigation. For more than 5000 years artificial irrigation has been practiced to enhance moisture for crops. It also changes the energy balance by increasing net radiation gain and reducing sensible heat loss. Temperatures are lowered by 1–2°C in plant cover and the vapor pressure increased by about 1 millibar. Irrigation effects do not extend to more than 3–4 m height. Even large artificial lakes or reservoirs show an "oasis" effect only to about 1 km downwind. Artificial suppression of evaporation on ponds and reservoirs by use of long-chain alcohols (Hexadecanol) has been quite effective, resulting in a 10–30% reduction of water loss. This procedure has also been successfully applied to rice paddies. *See* HYDROMETEOROLOGY.

Urbanization. Inadvertent climatic changes have taken place locally as a consequence of urbanization. Impermeable surfaces have changed the moisture balance drastically. With a 50% impervious cover, runoff has increased 200% compared to rural conditions, and peak stream discharges 300%. Flood volumes and propagation speeds of flood waves have increased. As result of the rapid drainage, humidities in cities have decreased. Changes in the surface have also altered

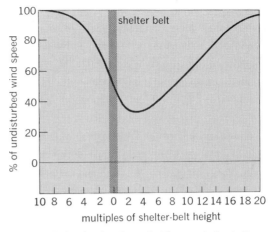

Fig. 1. Wind reduction downwind from a shelter belt.

the heat balance. Combined with various combustion and metabolic processes, they have raised urban temperatures. Although the fossil fuel consumption of 10^{20} cal/year amounts to only $1/2 \times 10^{-6}$ of the solar radiation intercepted by the Earth, artificial heat is 10–15% of solar heating in United States cities and up to one-third of the insolation in some European cities. The result is higher temperatures of 1–2°C for the daily maxima and 1–9°C for the daily minima compared to rural environs. At wind speeds of more than 7 m/sec these effects tend to vanish. Usually wind speeds are reduced by 20–30% because of the increased aerodynamic roughness of cities. The result is an urban heat island (Fig. 2), which also tends to increase the length of the freeze-free season and decrease the duration of snow cover in cities at moderate and higher latitudes.

Because many combustion processes add moisture and because convective activity is higher over cities, there is a tendency for greater cloudiness and higher rainfall. This effect can be enhanced by suitable nuclei that, added from various sources of pollution, stimulate coalescence processes in cloud droplets. The precipitation increase is prob-

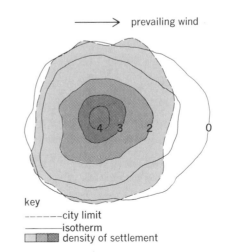

Fig. 2. Urban heat island. Isothermal lines indicate temperature differences from rural conditions.

ably about 5–10% of the annual total. In some spectacular cases a plume effect has been noted downwind from major industrial centers. For example, there is good evidence that the effluents from the industrial complex at the south end of Lake Michigan have caused in the last few decades a 30% increase in rainfall at La Porte, Ind., 50 km downwind. Thunderstorm activity and hail frequency there have also increased.

The city effect on climate is epitomized by air pollution. In larger United States cities the dust load alone averages 60–120 $\mu g/m^3$. This leads to monthly dust falls of 10–40 tons/km² in metropolitan areas. The smoke plume of New York City has been noted by pilots as far as 800 km downwind. Sunshine is reduced by about 20% and visibility is frequently reduced to less than 10 km. High concentrations of condensation nuclei have been suspected to interfere with normal precipitation processes. Plant damage in metropolitan areas has been heavy. Health hazards to humans have become probable even under regular conditions, and are certain when prolonged stagnation of air prevails because of reduced wind and temperature inversions. See ATMOSPHERIC POLLUTION; BIOCLIMATOLOGY.

Atmospheric heat balance. Global climatic effects have been ascribed to increased carbon dioxide from burning of fossil fuel and suspended dust. The two effects counteract each other in their influence on world temperatures. Carbon dioxide, since the last century, has increased about 10%, and a 0.2°C world temperature rise has been ascribed to it because of its absorption of outgoing infrared terrestrial radiation ("greenhouse effect"). A doubling of atmospheric CO_2 content has been predicted for the next half century. Under certain assumptions for world cloudiness and atmospheric moisture, this could lead to a 1°C temperature increase. Yet it is estimated that backscattering of solar radiation to space by man-caused dust suspended in the air would reduce temperature by 1–2°C. Probably more serious than these effects are insolation reductions by condensation trails produced by high-flying aircraft. Under certain atmospheric conditions these persist and reflect 40% of incoming radiation, while not seriously interfering with outgoing terrestrial radiation. In terminal areas this can materially affect energy balance. Measurements have shown 20% reduction in insolation at the surface when persistent condensation trails prevail. See GREENHOUSE EFFECT; TERRESTRIAL ATMOSPHERIC HEAT BALANCE.

Weather and climate modification. There has been much talk of causing climatic changes through various weather modification schemes. Some evidence indicates that systematic and persistent cloud seeding in subcooled, upslope cloud covers could increase rainfall or snowpack by about 10%. There is some hope, backed by limited field experimentation, that lightning strokes to the ground and hail damage can be reduced. Although this modification may be of value locally in mitigating forest fire danger and crop damage, it can hardly be rated as climatic change. Substantial success has been achieved in reducing subcooled water fogs at airports by seeding with dry ice. This has decreased fog hours at some airports by 60–80% during the cold season, but is a microclimatic change. See WEATHER MODIFICATION.

Proposals for inducing climatic changes cover a wide range. Most of them are either technologically or economically not feasible. Those that are within present capabilities and can claim some scientific validity involve changes of the heat balance of the surface. On a small scale, this modification involves the covering of desert areas near shorelines with asphalt. This, it is hoped, will induce strong convection and cause cloud formation with further release of latent heat, creating sufficiently thick cloud layers to produce rain. On a large scale, it has been seriously proposed to change the albedo of the arctic sea ice through the spreading of absorbent black dust. At present only 18 of the available 73 kcal/cm²/year in the Arctic are absorbed. The dust cover would at least triple the amount of heat that might become available for melting of ice. Once the ice melted, it is unlikely that in the current climatic regime permanent pack ice would reform on the Arctic Ocean. What consequences this would have on the various climates of the Northern Hemisphere are not entirely forseeable. It will undoubtedly cause a milder temperature regime in northern countries, reduce the intensity of the general circulation and, possibly, thereby reduce precipitation in lower middle latitudes. Ecologists have raised warning voices against attempting such large-scale climatic modifications. They have characterized them as unwise endeavors which, while perhaps producing limited local benefits, may cause widespread undesirable changes.   [H. E. LANDSBERG]

Bibliography: Committee on Atmospheric Sciences, *Weather and Climate Modification*, 1973; R. M. Hagan et al., *Successful Irrigation: Planning, Development, Management*, 1969; W. N. Hess, *Weather and Climate Modification*, 1974; W. H. Matthews et al., *Man's Impact on the Climate*, 1971; U.S. Department of Agriculture, *Climate and Man*, 1975; World Meteorological Organization, *Proceedings of the Scientific Conference on Weather Modification*, 1975.

Climatic change

The long-term fluctuations in precipitation, temperature, wind, and all other aspects of the Earth's climate. The Earth's climate, like the Earth itself, has a history extending over several billion years. Climatic fluctuations have occurred at time scales ranging from the longest observable (10^8–10^9 years) to interdecadal variability (10^1 years) and interannual variability (10^0 years). Processes in the atmosphere, oceans, cryosphere (snow cover, sea ice, continental ice sheets), biosphere, and lithosphere, and certain extraterrestrial factors (such as the Sun), are part of the climate system.

The present climate can be described as an ice age climate, since large land surfaces are covered with ice sheets (Antarctica, Greenland). The origins of the present ice age may be traced, at least in part, to movement of the continental plates. With the gradual movement of Antarctica toward its present isolated polar position, ice sheets began to develop about 30,000,000 years ago. Within the past several million years, the Antarctic ice sheet reached approximately its present size, and ice sheets appeared on the lands bordering the northern North Atlantic Ocean. During the past million years of the current ice age, about 10 glacial-interglacial fluctuations have been documented. The most recent glacial period came to an end between about 15,000 and 6000 years ago with the rapid melting of the North American and European ice sheets and the associated rise in sea level. The present climate is described as interglacial. The scope of this article is limited to a discussion of climatic fluctuations within the present interglacial period and, in particular, the climatic fluctuations of the past 100 years—the period of instrumental records.

Evidence. Instrumental records of climatic variables such as temperature and precipitation exist for the past 100 years in many locations and for as long as 200 years in a few locations. These records provide evidence of year-to-year and decade-to-decade variability but are completely inadequate for the study of century-to-century and longer-term variability. Even for the study of short-term climatic fluctuations, instrumental records are of limited usefulness, since most observations were made from the continents (only 29% of the Earth's surface area) and limited to the Earth's surface. Aerological observations which permit the study of atmospheric mass, momentum and energy budgets, and the statistical structure of the large-scale circulation are available for only about 20 years. Again, there is a bias toward observations over the continents. It is only with the advent of satellites that global monitoring of the components of the Earth's radiation budget (planetary albedo, from which the net incoming solar radiation can be estimated; and the outgoing terrestrial radiation) has begun. *See* METEOROLOGICAL SATELLITES; TERRESTRIAL ATMOSPHERIC HEAT BALANCE.

There remain important gaps in the ability to describe the present state of the climate. For example, precipitation estimates, especially over the oceans, are very poor. Oceanic circulation, heat transport, and heat storage are only crudely estimated. Also, the solar irradiance is not being monitored to sufficient accuracy to permit estimation of any variability and evaluation of the possible effect of fluctuations in solar output upon the Earth's climate. Thus, although climatic fluctuations appear in instrumental records, defining the scope of these fluctuations and diagnosing potential causes are at best difficult and at worst impossible.

In spite of the inadequacy of the instrumental records for assessing global climate, there is considerable evidence of regional climatic variations. For example, there is evidence of climatic warming in the polar regions of the Northern Hemisphere during the first 4 to 5 decades of the 20th century. During the 1960s, on the other hand, there is evidence of cooling in the polar and mid-latitude regions of the Northern Hemisphere; and, especially in the early 1970s, there were drier conditions along the northern margin of the monsoon lands of Africa and Asia.

Under the auspices of the World Meteorological Organization and the International Council of Scientific Unions, the Global Atmospheric Research Program (GARP) is developing plans for detailed observation and study of the global climate system—especially the atmosphere, the oceans, the sea ice, and the changeable features of the land surface.

Causes. Many extraterrestrial and terrestrial processes have been hypothesized to be possible causes of climatic fluctuations. A number of these processes are listed and described below.

Solar irradiance. It is possible that variations in total solar irradiance could occur over a wide range of time scales ($10^0 - 10^9$ years). If these variations did take place, they would almost certainly have an influence on climate. Radiance variability in limited portions of the solar spectrum has been observed, but not linked clearly to climate variability. *See* SOLAR ENERGY.

Orbital parameters. Variations of the Earth's orbital parameters (eccentricity of orbit about the Sun, precession, and inclination of the rotational axis to the orbital plane) lead to small but possibly significant variations in incoming solar radiation with regard to seasonal partitioning and latitudinal distribution. These variations occur at time scales of $10^4 - 10^5$ years.

Lithosphere motions. Sea-floor spreading and continental drift, continental uplift, and mountain building operate over long time scales ($10^5 - 10^9$ years) and are almost certainly important in long-term climate variation. *See* MARINE GEOLOGY.

Volcanic activity. Volcanic activity produces gaseous and particulate emissions which lead to the formation of persistent stratospheric aerosol layers. It may be a factor in climatic variations at all time scales.

Internal variability of climate system. Components of the climate system (atmosphere, ocean, cryosphere, biosphere, land surface) are interrelated through a variety of feedback processes operating over time scales from, say, 10^0 to 10^9 years. These processes could, in principle, produce fluctuations of sufficient magnitude and variability to explain any observed climate change. For example, atmosphere-ocean interactions may operate over time scales ranging from 10^0 to 10^3 years, and atmosphere-ocean-cryosphere interactions may operate over time scales ranging from 10^0 to 10^5 years. Several hypotheses have been proposed to explain glacial-interglacial fluctuations as complex internal feedbacks among atmosphere, ocean, and cryosphere. (Periodic buildup and surges of the Antarctic ice sheet and periodic fluctuations in sea ice extent and deep ocean circulation provide examples.) Atmosphere-ocean interaction is being studied intensively as a possible cause of short-term climatic variations. It has been observed that anomalous ocean temperature patterns (both equatorial and mid-latitude) are often associated with anomalous atmospheric circulation patterns. Although atmospheric circulation plays a dominant role in establishing a particular ocean temperature pattern (by means of changes in wind-driven currents, upwelling, radiation exchange, evaporation, and so on), the anomalous ocean temperature distribution may then persist for months, seasons, or longer intervals of time because of the large heat capacity of the oceans. These anomalous oceanic heat sources and sinks may, in turn, produce anomalous atmospheric motions. *See* OCEAN-ATMOSPHERE RELATIONS.

Human activities. Forest clearing and other large-scale changes in land use, changes in aerosol loading, and the changing CO_2 concentration of the atmosphere are often cited as examples of possible mechanisms through which human activities may influence the large-scale climate. Because of the large observational uncertainties in defining the state of the climate, it has not been possible to establish the relative importance of human activities (as compared to natural processes) in recent climatic fluctuations. There is, however, considerable concern that future human activities may lead to large climatic variations (for example, continued increase in atmospheric CO_2 concentration due to burning of fossil fuels) within the next several decades. *See* ATMOSPHERIC POLLUTION; CLIMATE MODIFICATION.

It is likely that at least several of the above-mentioned processes have played a role in past climatic fluctuations (that is, it is unlikely that all climatic fluctuations are due to one factor). In addition, certain processes may act simultaneously, or in various sequences. Also, the climatic response to some causal process may depend on the particular initial climatic state, which, in turn, depends upon previous climatic states because of the long time constants of oceans and cryosphere. True equilibrium climates may not exist, and the climate system may be in a continual state of transience.

Modeling. Because of the complexity of the real climate system, simplified numerical models of climate are being used to study particular processes and interactions. Some models treat only the global-average conditions, whereas others, particularly the atmospheric models, simulate detailed patterns of climate. These models are still in early stages of development but will undoubtedly be of great importance in attempts to understand climatic processes and to assess the possible effects of human activities on climate. *See* ATMOSPHERIC GENERAL CIRCULATION; CLIMATOLOGY.

[JOHN E. KUTZBACH]

Ocean-atmosphere interaction. The atmosphere and the oceans have always jointly participated in climatic change, past and contemporary. Some of the contemporary changes can be investigated in the modern records of climatic anomalies in the atmosphere and the oceans.

The most important source of climatic change surpassing 1-year duration seems to be located along the equatorial zone of the Pacific Ocean. The prevailing winds there are easterly and maintain a westward wind drift of the surface water which diverges, under influence of the Earth's rotation, to the right of the wind direction north of the Equator and to the left south of the Equator. The resulting equatorial upwelling of cold water, and subsequent lateral mixing, ordinarily maintains a belt of cold surface water several hundred kilometers wide straddling the Equator from the coast of South America about to the dateline, about one earth quadrant farther to the west. *See* UPWELLING.

Analogous processes are found in the equatorial belt of the Atlantic, but the upwelling water there covers a much smaller area and is also less cold than in the Pacific. The Indian Ocean has no steady easterlies and thus no equatorial upwelling.

The equatorial upwelling process varies in intensity with the equatorial easterlies. Since that wind system is mostly fed by way of the southerlies along the west coasts of South Africa and South America, it is likely that anomalies in the Southern Hemisphere atmospheric circulation frequently are transmitted to the tropical belt. Once an impulse, for example, a strengthening of the Pacific equatorial easterlies, has occurred, the new anomaly has a built-in tendency of self-amplification, because it makes the upwelling strengthen too and thus increases the temperature deficit of the Pacific compared to the persistently warm Indonesian and Indian Ocean tropical waters. This in turn feeds back into further strengthening of the easterlies which started the anomaly in the first place.

The observational proof of this feedback system can be seen in the statistically well-substantiated "southern oscillation," which exhibits opposite contemporaneous anomalies of atmospheric pressure over the tropical parts of the Pacific Ocean on the one side and the Indonesian and Indian Ocean tropical waters on the other (nodal line on the average at 165°E). The periodicity is rather irregular, so the term oscillation should not be taken too literally. The average length of the cycles is 2–3 years.

The cycles of tropical precipitation of more than a year's length by and large agree with those of pressure wherever special local conditions do not interfere. Satellite photos confirm that in the phase of the southern oscillation with positive pressure anomaly over the Pacific, along with strong equatorial easterlies and strong upwelling, most of the Pacific equatorial belt is arid; while in the opposite phase the western and central part of that belt experiences heavy rainfall. In extreme "El Niño" years this rainfall can also extend to the normally arid coast of northern Peru.

When there is more than normal rainfall at the Equator, the general circulation of the atmosphere is supplied with more-than-normal total heat convertible into kinetic energy. The remote effect of this phenomenon, particularly in the winter hemisphere, is the occurrence of stronger-than-normal tradewinds and midlatitude westerlies. At the opposite extreme, the tradewinds are weak and the westerlies meandering. This produces cold winters in the longitude sectors with winds out of high latitudes and mild winters interspersed at longitudes where wind components from low latitudes prevail. Again, it is in the Pacific longitude sector that these teleconnections are most clearly seen, because the interannual variability of sea temperature up to a range of 3°C over a large equatorial area is found only in the Pacific.

[JACOB BJERKNES]

Bibliography: H. P. Berlage, *The Southern Oscillation and World Weather*, Kon. Ned. Meterol. Inst. Meded. Verh. no. 88, 1966; J. Bjerknes, A possible response of the Hadley circulation to variations of the heat supply from the equatorial

Pacific, *Tellus*, 18:820–829, 1966; R. A. Bryson and T. J. Murray, *Climates of Hunger: Mankind and the World's Changing Weather*, 1977; M. I. Budyko, *Climate Changes*, 1977; R. W. Fairbridge (ed.), *Solar Variations, Climatic Change, and Related Geophysical Problems*, Ann. N.Y. Acad. Sci. no. 95, 1961; H. H. Lamb, *Climate: Present, Past and Future*, vol. 1: *Fundamentals and Climate Now*, 1972; *Long-Term Climatic Fluctuations: Proceedings of the WMO-IAMAP Symposium*, 1975; S. H. Schneider and R. E. Dickinson, Climate modelling, *Rev. Geophys. Space Phys.*, 12:447–493, 1974; H. Shapley (ed.), *Climatic Change*, 1953; Study of man's impact on climate, in *Inadvertent Climate Modification*, 1971; C. Tickell, *Climatic Change and World Affairs*, 1977; U.S. National Academy of Sciences, *Understanding Climatic Change: A Program for Action*, 1975; G. Walker, *World Weather VI*, Mem. Roy. Meteorol. Soc., vol. 4, no. 39, 1937.

Climatic prediction

The estimation of the chances that specified climatic conditions will obtain within a certain interval in the future. There are two basic contradictory postulates. One assumes that there are no significant secular trends or cyclic variations (beyond the annual cyclic period). This postulate underlies the estimation of climatic values required for the solution of many practical problems, as in estimating expected rainfall intensities. The problem is essentially one of determining the distribution of the variable (Gaussian, Poisson) and then of estimating the probability p after allowing for persistence. Where a secular trend or cyclic variation is postulated, p must be modified through introduction of linear or curvilinear terms unless the form of the basic distribution is thought to change, in which case entirely different expressions must be developed. *See* CLIMATOLOGY; WEATHER FORECASTING AND PREDICTION.

[DAVID I. BLUMENSTOCK]

Climatology

That branch of meteorology concerned with climate, that is, with the mean physical state of the atmosphere together with its statistical variations in both space and time, as reflected in the totality of weather behavior over a period of many years. Climatology encompasses not only the description of climate but also the physical origins and the wide-ranging practical consequences of climate and of climatic change. Thus it impinges on a wide range of other sciences, including solar system astronomy, oceanography, geography, geology and geophysics, biology and medicine, agriculture, engineering, economics, social and political science, and mathematical statistics. *See* CLIMATE MODIFICATION; CLIMATIC CHANGE.

Like meteorology, climatology is conveniently resolved into subdisciplines in a way that recognizes a hierarchy of geographical scales of climatic phenomena and their governing physics. Macroclimatology refers to the largest (planetary) scale of regimes and phenomena; regional climatology to the scale of continents and subcontinental areas; mesoclimatology to the scale of individual physio-graphic features such as a mountain, lake, or urban area; and microclimatology to the smallest scale, for example, a house lot or the habitat of an insect. Climatology is also resolved in another way that distinguishes between the theoretical, descriptive, and applied aspects of the science: physical and dynamic climatology, concerned with the governing physical laws; descriptive and synoptic climatology, concerned with comparisons of climatic norms and anomalies, respectively, as concurrently observed in different places; and applied climatology, concerned with the practical utilization of climatological data in engineering design, operations strategy, and activity planning. Other important subdisciplines are climatography, concerned with the comprehensive documentation of climate by means of data summaries, maps, and atlases; and statistical climatology, concerned mostly with the estimation of climatological expectancies, risks of extreme events, and probability distributions of climatic variables (including joint distributions of combinations of variables) as needed to solve the problems encountered in applied climatology.

Origins and pattern of global climate. The climate of the Earth as a whole, including its geographical and seasonal variations, is fundamentally prescribed by the disposition of solar radiant energy that is intercepted by the atmosphere. This disposition depends in part on astronomical factors and in part on terrestrial factors, such as atmospheric composition, the distribution of land and sea, and the physiographic nature of the land. At the mean Earth-Sun distance, solar energy arrives in the vicinity of the Earth at the rate of approximately 1.95 g-cal/(cm²)(min), as measured outside the atmosphere perpendicular to the incident rays. This value, known as the solar constant, has been measured to an accuracy of about ±1% and is believed to vary by not more than a fraction of 1% from year to year. When reckoned per unit horizontal area, however, incoming solar radiation (or insolation) is intercepted at different rates over different parts of the Earth. The rotation of the Earth causes this rate to change locally with the time of day. In addition, the shape of the Earth, together with the seasonal changes in orbital distance and axial tilt of the Earth relative to the Sun, causes solar radiation (per unit horizontal area and per unit time) to vary with latitude and with the time of year. *See* INSOLATION.

Terrestrial heat balance. The disposition of solar energy after it penetrates the atmosphere is a complex process that locally depends on the solar elevation angle and on various properties of both the atmosphere and its underlying surface. Averaged over the whole Earth and over all seasons of the year, the disposition is as shown in Fig. 1. The figure represents the planetary average heat balance. This term is appropriate because in the observed absence of large changes of atmospheric temperature from year to year the income and outgo of heat energy through the Earth's surface, within the atmosphere, and through the top of the atmosphere must balance closely. In the annual mean for the whole planet, only about half of the solar energy incident on the top of the atmosphere

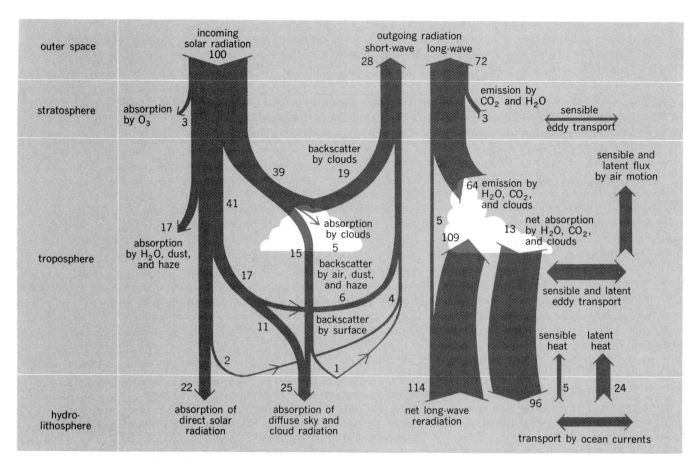

Fig. 1. Mean annual heat balance of whole Earth, showing magnitudes of principal items in percent of total solar radiation arriving at top of atmosphere. (*Based mostly on estimates for Northern Hemisphere by M. I. Budyko and K. Y. Kondratiev, from R. M. Rotty and J. M. Mitchell, Jr., in Symposium Series Volume on Air Pollution Control and Clean Energy, American Institute of Chemical Engineers, 1975).*

penetrates as far as the Earth's surface. Of this amount, more than 90% is absorbed in the surface, which is used in about equal shares to heat the surface material and to evaporate water, primarily from the oceans. Approximately 28% of the incident solar energy is returned unused to space, mostly by backscatter from clouds; this corresponds to the planetary albedo. The remainder of the incident solar energy (25%) is absorbed on the way down through the atmosphere by water vapor, ozone, dust, haze, and clouds. This provides a direct source of heat to the atmosphere.

More than twice as much heat is added indirectly to the atmosphere by way of the surface. Nearly half of this indirect heating is supplied by a net upward transfer of thermal (infrared) radiation energy that is constantly being exchanged between the surface and atmospheric water vapor, carbon dioxide, and clouds. The remainder consists mainly of the latent heat content of water evaporated from the oceans, which is made available to the atmosphere at the point of condensation as clouds and precipitation. Finally, balance is achieved by a flow of thermal (infrared) radiation energy back to space (72% of the incident solar energy), nearly all of it by emission from water vapor, carbon dioxide,

and the tops of clouds in the atmosphere. Less than one-tenth of this thermal radiation loss to space originates at the Earth's surface because the atmosphere is almost completely opaque to infrared radiation except in relatively narrow-wavelength bands (notably between 8 and 12 μ). *See* TERRESTRIAL RADIATION.

The fact that the atmosphere intercepts most of the thermal radiation from the Earth's surface, which in a transparent atmosphere would flow unimpeded into space, is very important to climate because it leads to the maintenance of much higher surface temperatures than those otherwise possible. This warming (the greenhouse effect) is largely attributable to the presence of water vapor and carbon dioxide in air. The magnitude of the warming effect importantly depends on the variable concentration of water vapor, and thus it changes with latitude, time of year, and atmospheric conditions generally. *See* GREENHOUSE EFFECT.

Role of atmospheric circulation. In regions of intense solar radiation, primarily in the tropics, there is a net surplus of radiation energy. Similarly in regions of relatively little solar radiation, primarily in the polar regions, there is a net deficit of radia-

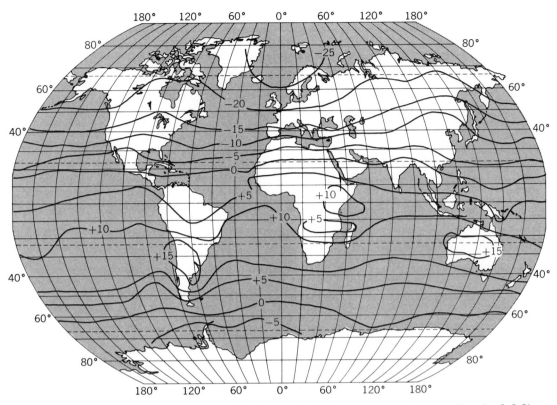

Fig. 2.   Net radiation over Earth, January. Values in kg-cal/(cm²) (month). Regions of net surplus and deficit of radiation energy are shown. (*Modified after G. C. Simpson; base map copyright Denoyer-Geppert Co., Chicago*)

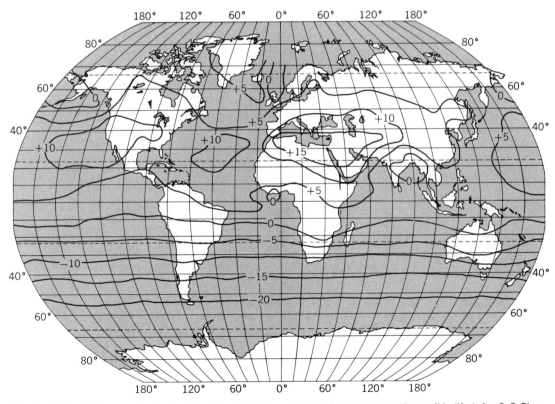

Fig. 3.   Net radiation over the Earth, July. Values in kg-cal/(cm²) (month). Regions of net surplus and net deficit of radiation energy are shown. (*Modified after G. C. Simpson; base map copyright Denoyer-Geppert Co., Chicago*)

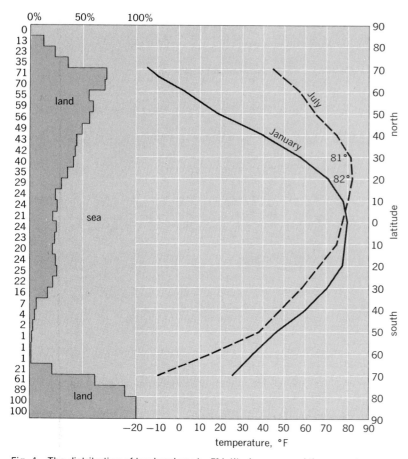

Fig. 4. The distribution of land and sea by 5° latitude zones and the mean temperature by latitude circles between 70°N and 70°S for January and July. (*Land-sea percentages after E. Kossinna; temperatures after J. Hann*)

tion energy (Figs. 2 and 3). These radiative imbalances lead to differential heating and cooling of the
atmosphere, and thereby generate a large-scale
circulation of air that transports heat and moisture
from areas of surplus to areas of deficiency, generally into higher latitudes. In this way global air
currents (along with ocean currents) importantly
modify the zonation of world climate otherwise dictated by the astronomical factors. The characteristic unsteadiness of these air currents causes the
familiar changeability of weather whose statistics
are another important aspect of climate. The pattern of atmospheric circulation, as reflected at
the Earth's surface by differences of atmospheric
pressure and their associated winds, is discussed
in fuller detail below. *See* ATMOSPHERIC GENERAL
CIRCULATION.

Land- and sea-surface influences. The climatic
significance of the distribution of land and sea lies
in the contrast between these two major kinds of
surfaces regarding heating and cooling. Except in
regions of snow cover or inland water bodies, the
lands are heated to higher temperatures than are
the seas, both diurnally during the daytime and
seasonally during the summer. Obversely, during
the nighttime and during the winter the lands are
cooled to lower temperatures than are the seas. As

a direct result, both diurnal and annual temperature ranges are greater over the land than over the
sea. *See* MARINE INFLUENCE ON WEATHER AND
CLIMATE; OCEAN-ATMOSPHERE RELATIONS.

The distribution of land and sea by 5° latitude
zones is shown in Fig. 4. The importance of this
factor is evident from the companion curves, showing mean temperatures by latitude circles between
70°N and 70°S for January and for July. Note, for
example, that the January temperature is 19°F at
50°N, where land occupies more than half the latitude circle; and the temperature is 38° in the corresponding month of July at 50°S, where there is
no land.

Land-sea distribution and the distribution of
radiation are two of the three major factors that
determine the broad features of the distribution of
temperature. The third factor is the circulation of·
the atmosphere and of the oceans. In the mean,
both circulations transfer energy from lower to
higher latitudes. Of the transport, 65% is by the
atmospheric circulation, which carries energy both
in the form of sensible heat and latent heat (as
water vapor that will give up heat upon condensing). The high-latitude regions of great net radiation loss (Figs. 2 and 3) are in large part fed by energy derived from latent heat. Of the transport,
another 35% is accomplished through the circulation of ocean waters, with warm water moving
poleward along the western sides of the oceans
and cool water moving equatorward along the eastern sides. This circulation is reflected in the mean
temperature maps for January and July (Figs. 5
and 6), as in the northward bulge of the isotherms
in the Iceland-Spitsbergen-Norway area. Other
notable features of these maps are the location of
the cold poles in the interior of the great landmasses and the displacement of the warmest zone
northward from the Equator, a displacement associated with the great landmass of Africa in latitudes 10–30°N. *See* OCEAN CURRENTS; WIND.

Relations with air pressure and wind. The global
distribution of temperature is related to the distribution of surface air pressure and of wind. These
relationships are neither simple nor direct,
but there tends to be an inverse relationship
between mean temperature and mean surface
pressure over the continents, with the hot desert
areas sustaining thermal lows and the cold polar
areas sustaining thermal highs. The mean low- and
high-pressure areas over the oceans are quite
different. In the first instance they represent the
summation of moving, dynamic lows (extratropical
cyclones), which are particularly common synoptic
features in the Aleutian and Icelandic areas, as
well as in the waters north of Antarctica. In the
second instance they represent the summation of
subtropical high cells, which are persistent synoptic circulation features over the subtropical
oceans, even though they vary continuously in
size, intensity, and location. *See* AIR PRESSURE:
METEOROLOGY.

The mean pressure patterns are approximately
matched by the patterns of dominant windflow.
The most constant winds are those of the trades,
which lie on the equatorward side of the subtropi-

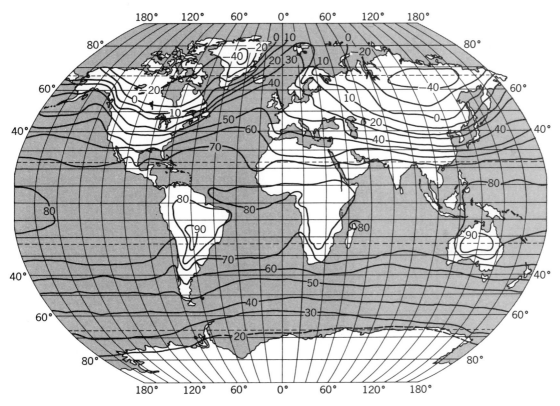

Fig. 5. Mean sea-level temperatures, °F, January. Note location of cold poles in interior of great landmasses. (*Modified from Sir Napier Shaw et al.; base map copyright Denoyer-Geppert Co., Chicago*)

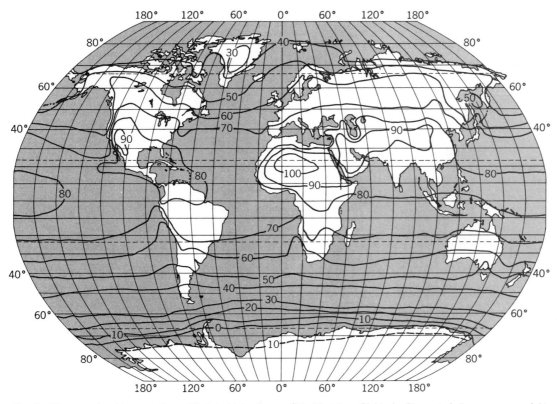

Fig. 6. Mean sea-level temperatures, °F, July. Note displacement of warmest zone northward from Equator. (*Modified from Sir Napier Shaw et al.; base map copyright Denoyer-Geppert Co., Chicago*)

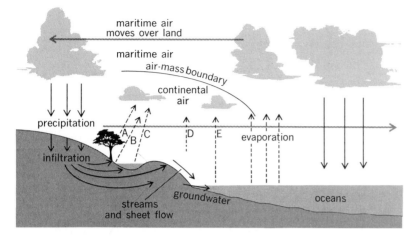

Fig. 7.   Diagram of hydrologic cycle. Water returns chiefly to dry continental air through transpiration (A), evaporation from soil (B), lakes and ponds (C), and streams (D). Continental air moves over ocean to become more moist (E) with conversion to maritime air with precipitation over the oceans. (*Adapted from B. Holzman*)

cal oceanic highs, and those of the monsoon circulation in the Asiatic-Australian area. In this great monsoon circulation, winds blow outward from Asia and inward onto Australia during the Northern Hemisphere winter, with a reversal of the circulation during the summer. Other major wind flow regions are those of the westerlies, to poleward of the oceanic highs, and of the polar easterlies, off Antarctica and Greenland. Neither the westerlies nor the polar easterlies are as constant as the trades or the major monsoon winds because both occur in areas that are frequently the scene of moving cyclones, which may bring winds from any direction.

Marine atmospheric influences. Air that moves long distances across the oceans acquires great quantities of moisture through evaporation from the ocean surface. Because evaporation is greatest where cold dry air moves across much warmer ocean water, the most rapid flux of moisture from sea to air occurs on the western sides of the northern oceans in middle latitudes during the winter season, and over the waters off Antarctica during winter. For example, at 35°N in the western Atlantic an average of about 600 g-cal/(cm²) (day) of thermal energy is expended for evaporation of sea water. In contrast, at latitudes 0–20°N in the eastern Atlantic, the value does not exceed 200 g-cal during any season.

Whatever the variations in detail, the oceans everywhere make a large net contribution of water to the air (except when there is ice). Subsequently, this moisture is carried onto the land by maritime air masses and there it is partly precipitated as rain, snow, hail, dew, or frost. Thereafter, the water returns to the oceans in streams, as sheet wash along the margins of the lands, as underground water, or in moving air (chiefly dry continental air) that acquires water from the land through evaporation and transpiration. This worldwide water cycle is diagrammed in Fig. 7. *See* HYDROLOGY.

Continental precipitation patterns. Figure 11 shows the mean annual precipitation over the continents. Appreciable precipitation occurs only when moist air is forced to rise and this takes place largely through convection, through orographic lifting (the forced lifting of air upslope, as up the flanks of a mountain), or through convergence and the forced ascent of air within an eddy, particularly within an extratropical or tropical cyclone. Hence the precipitation patterns can be viewed in terms of the relative frequency with which moist maritime air is present and the frequency with which the air is forced to rise to appreciable heights. The vertical structure of the air with reference to temperature and moisture is also important, because the structure may be stable and so resist vertical movement, or it may be conditionally unstable, so that when forced lifting has produced condensation, the release of latent heat causes the air to rise to still greater heights. *See* CONTINENTALITY.

Annual precipitation is very high on the west coasts of the continents in high middle latitudes, in the major monsoon areas, and in equatorial areas. On the west coasts of the continents, the high totals are associated with frequent and prolonged cyclonic precipitation and with the orographic lifting of maritime air. In these areas most of the precipitation is in the winter half-year, when cyclonic storms are most common. In contrast, the monsoon areas have their maximum rains during summer, with an influx of very moist, conditionally unstable air. Here, as in the wet equatorial areas, the rainfall mechanisms are convection, orographic lifting, and convergence in minor eddy systems. However, in the equatorial areas the rainfall is fairly well distributed throughout the year.

The extremely dry areas are characterized by infrequent invasions of moist unstable air and relatively little cyclonic activity. These dry areas are the west coast deserts, such as the Sahara; the desert basins that are shielded from fresh maritime air by high mountains, such as the Tarim Basin; and the polar lands.

In absolute terms, the widest swings in annual precipitation from year to year occur in the wettest areas of the monsoon and the equatorial regions. In these regions it is not uncommon during a 20-year period for the annual rainfall to range from a minimum of 60–80 in. to a maximum of 200–300 in. In percentage terms, the greatest variability is in the driest areas—the deserts and dry polar regions. Here the minimum is less than 2 in. in many localities with maximum annual amounts reaching 15 in. or more in so-called wet years. *See* TROPICAL METEOROLOGY.

The water circumstance of a locality is not only a function of the annual precipitation and its variability but even more of the distribution of precipitation throughout the year. Figure 8 shows the seasonal precipitation regimes at eight selected localities. Also shown are the extreme monthly precipitation amounts during the 20-year period which is covered by the graphs. The general form of these graphs is in each instance broadly representative of the precipitation regime throughout the regions which are specified in the diagrams. The variability as shown by the extreme values is typical.

Climatic regions. A wide variety of schemes is available for the definition of climatic regions. These schemes fall into two classes, those intended to bring out one or another aspect of relationships and processes within the atmosphere, and those intended to bring out relationships between areal variations in climatic conditions and corresponding variations in phenomena related to climate.

The first class of schemes is the major radiation regions of the Earth. The distinction is among regions in which the monthly radiation balance is always positive (in the mean), always negative, never strongly positive, never strongly negative, or highly variable and ranging from strongly positive to strongly negative.

The second class of schemes is represented (1) by the major climatic regions of the Earth according to W. Koeppen's classification, and (2) by Fig. 9. The Koeppen classification was intended to bring out the general coincidence between the distribution of natural vegetative formations and climatic conditions. It also serves, however, as a useful classification for purposes of general description, in terms of quantitative definitions of precipitation and temperature conditions. The major aspects of Koeppen's classification scheme are summarized in the table, whose arrangement is taken from H. E. Landsberg. Koeppen and his followers revised his classification twice and carried it beyond the broad scheme given here. For example, cold and hot deserts were distinguished, isothermal regions (mean annual temperature range less than 9°F) were identified, and subregions where fog was common were designated. Clearly a climatic classification scheme such as this could be extended in any desired way to show areal variations in greater and greater detail.

Quite different from Koeppen's scheme is a

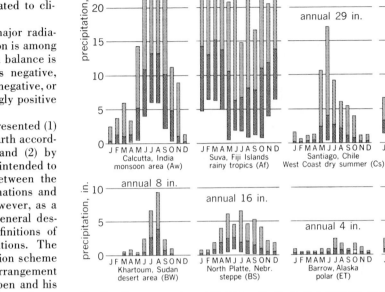

Fig. 8. Mean annual precipitation, with mean and extreme monthly precipitation bar graphs, at selected stations. Top of each bar shows extreme high value; bottom of each bar shows extreme low value; mean indicated by horizontal line within each bar. Letter symbols according to Koeppen type of climate. (*Based on 20-year data for 1921–1940, World Weather Records, Smithsonian Miscellaneous Collections*)

class of schemes illustrated by Fig. 9. This is a classification developed to solve a particular problem in applied climatology, that of designing life rafts for use at sea. The regions were defined pri-

Principal quantitative definitions of major climatic regions*

Symbol	Name of region	Mean temperature, °F	Precipitation	
			Main season	Amount, formula for inches†
Af	Tropical rain forest	>64.4‡	All	Driest month >2.4 Annual >40
Aw	Tropical savanna	>64.4‡	Summer	Driest month <2.4 Annual >40
BW	Desert		Winter	Annual <0.22t − 7
			Summer	Annual <0.22t − 1.5
			Even	Annual <0.22t − 4.25
BS	Steppe		Winter	Annual <0.44t − 14
			Summer	Annual <0.44t − 3
			Even	Annual <0.44t − 8.5
Cf	Humid mesothermal	<64.4 >26.6‡	Even	Annual >0.44t − 8.5
Cw	Humid mesothermal, winter dry	<64.4 >26.6‡	Summer	Annual >0.44t − 3
Cs	Humid mesothermal, summer dry	<64.4 >26.6‡	Winter	Annual >0.44t − 14
Df	Humid microthermal	<26.6‡ >50§	Even	Annual >0.44t − 8.5
Dw	Humid microthermal, winter dry	<26.6‡ >50§	Summer	Annual >0.44t − 3
ET	Tundra	<50 >32§		
EF	Perpetual frost	<32§		

*After W. Koeppen, as outlined in G. T. Trewartha, *An Introduction to Climate*, McGraw-Hill, 3d ed., 1954. Arrangement modified from one by H. E. Landsberg.

†All temperature values are in °F (including *t*, which is mean value). Precipitation values are in inches.

‡Temperature of coldest month.   §Temperature of warmest month.

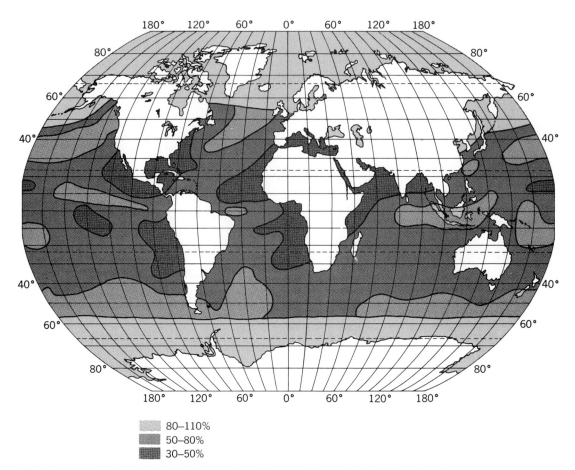

80–110%

50–80%

30–50%

Fig. 9. Climatic regions for fresh-water supply on the oceans, December-February, according to scheme for problem solving. How well climatic conditions meet re-quirements of rainfall and sunshine, as related to use of solar stills, is indicated in percentages. (*After W. C. Jacobs; base map copyright Denoyer-Geppert Co., Chicago*)

marily through consideration of the rainfall proba-bilities and the effective sunshine, the latter being related to the efficiency of portable solar stills. In applied climatology, special classifications of these kinds abound, each designed to serve a practical purpose.

It is feasible and instructive to distinguish minor climatic regions, not on a global scale, but within much smaller areas. Where such regions are defined upon the lands, the topographic factor becomes very important. This is especially true where the area is of the order of a few square miles or less.   [J. MURRAY MITCHELL, JR.]

Bibliography: I. P. Danilina (ed.), *Meteorology and Climatology*, vol. 2, 1977; H. Dickson, *Climate and Weather*, 1976; E. S. Gates, *Meteorology and Climatology*, 1972; J. F. Griffiths, *Applied Climatology*, 1976; H. H. Lamb, *Climate: Present, Past, and Future*, vol. 1, 1972; H. E. Landsberg, *World Survey of Climatology: General Climatology*, vol. 1, 1977; World Meteorological Organization, *Physical and Dynamic Climatology*, 1975; W. D. Sellers, *Physical Climatology*, 1965.

Cloud

Suspensions of minute droplets or ice crystals produced by the condensation of water vapor (the ordinary atmospheric cloud). Other clouds, less commonly seen, are composed of smokes or dusts.

See ATMOSPHERIC POLLUTION; DUST STORM.

This article presents an introductory outline of cloud formation upon which to base an under-standing of cloud classifications. For a more tech-nical consideration of the physical character of atmospheric clouds, including the condensation and precipitation of water vapor *see* CLOUD PHYSICS.

Rudiments of cloud formation. A grasp of a few physical and meteorological relationships aids in an understanding of clouds and skies. First, if water vapor is cooled sufficiently, it becomes satu-rated, that is, in equilibrium with a plane surface of liquid water (or ice) at the same temperature. Further cooling in the presence of such a surface causes condensation upon it; in the absence of any surfaces no condensation occurs until a substan-tial further cooling provokes condensation upon random large aggregates of water molecules. In the atmosphere, even in the apparent absence of any surfaces, they are in fact always provided by invisible motes upon which the condensation pro-ceeds at barely appreciable cooling beyond the state of saturation. Consequently, when atmos-pheric water vapor is chilled sufficiently, such motes, or condensation nuclei, swell into minute water droplets and form a visible cloud. The total concentration of liquid in the cloud is con-trolled by its temperature and the degree of

Fig. 1. Cirrus, with trails of slowly falling ice crystals at a high level. (*F. Ellerman, U.S. Weather Bureau*)

Fig. 2. Small cumulus. (*U.S. Weather Bureau*)

Fig. 3. An overcast of stratus, with some fragments below the hilltops. (*U.S. Weather Bureau*)

Fig. 4. View of cloud formations from Mount Wilson, Calif. High above is a veil of cirrostratus, and below is the top of a low-level layer cloud. (*F. Ellerman, U.S. Weather Bureau*)

Fig. 5. Cirrocumulus, high clouds with a delicate pattern. (*A. A. Lothman, U.S. Weather Bureau*)

chilling beyond the state in which saturation occurred, and in most clouds approximates to 1 g in 1
m³ of air. The concentration of droplets is controlled by the concentrations and properties of the
motes and the speed of the chilling at the beginning of the condensation. In the atmosphere these
are such that there are usually about 100,000,000
droplets/m³. Because the cloud water is at first
fairly evenly shared among them, these droplets
are necessarily of microscopic size, and an important part of the study of clouds is concerned with
the ways in which they often become aggregated
into drops large enough to fall as rain.

The chilling which produces clouds is almost
always associated with the upward movements of
air which carry heat from the Earth's surface and
restore to the atmosphere that heat lost by radiation into space. These movements are most pronounced in storms, which are accompanied by
thick, dense clouds, but also take place on a
smaller scale in fair weather, producing scattered
clouds or dappled skies. *See* STORM.

Rising air cools by several degrees Celsius for
each kilometer of ascent, so that even over equatorial regions temperatures below 0°C are encountered a few kilometers above the ground, and

Fig. 6. Altocumulus, which occurs at intermediate levels. (*G. A. Lott, U.S. Weather Bureau*)

Fig. 7. Altostratus, a middle-level layer cloud. Thick layers of such cloud, with bases extending down to low levels, produce prolonged rain or snow, and are then called nimbostratus. (*C. F. Brooks, U.S. Weather Bureau*)

numbers may be vanishingly small. Consequently, at these temperatures, clouds of unfrozen droplets are not infrequently encountered (supercooled clouds). In general, however, ice crystals occur in very much smaller concentrations than the droplets of liquid clouds, and may by condensation alone become large enough to fall from their parent cloud. Even small high clouds may produce or become trails of snow crystals, whereas droplet clouds are characteristically compact in appearance with well-defined edges, and produce rain only when dense and well-developed vertically (2 km or more thick).

Classification of clouds. The contrast in cloud forms mentioned above was recognized in the first widely accepted classification, as well as in several succeeding classifications. The first was that of L. Howard (London, 1803), recognizing three fundamental types: the stratiform (layer), cumuliform (heap), and cirriform (fibrous). The first two are indeed fundamental, representing clouds formed respectively in stable and in convectively unstable atmospheres, whereas the clouds of the third type are the ice clouds which are in general higher and more tenuous and less clearly reveal the kind of air motion which led to their formation. Succeeding classifications continued to be based upon the visual appearance or form of the clouds, differentiating relatively minor features, but later in the 19th century increasing importance was attached to cloud height, because direct measurements of winds above the ground were then very difficult, and it was hoped to obtain wind data on a great scale by combining observations of apparent cloud motion with reasonably accurate estimates of cloud height, based solely on their form.

WMO cloud classification. The World Meteorological Organization (WMO) uses a classification which, with minor modifications, dates from 1894 and represents a choice made at that time from a

clouds of frozen particles prevail at higher levels. Of the abundant motes which facilitate droplet condensation, very few cause direct condensation into ice crystals or stimulate the freezing of droplets, and especially at temperatures near 0°C their

Cloud classification based on air motion and associated physical characteristics

Kind of motion	Typical vertical speeds, cm/sec	Kind of cloud	Name	Characteristic dimensions, km		Characteristic precipitation
				Horizontal	Vertical	
Widespread slow ascent, associated with cyclones (stable atmosphere)	10	Thick layers	Cirrus, later becoming: cirrostratus altostratus altocumulus	10^3	1–2	Snow trails
			nimbostratus	10^3	10	Prolonged moderate rain or snow
Convection, due to passage over warm surface (unstable atmosphere)	10^2	Small heap cloud	Cumulus	1	1	None
	10^3	Shower- and thunder-cloud	Cumulo-nimbus	10	10	Intense showers of rain or hail
Irregular stirring causing chilling during passage over cold surface (stable atmosphere)	10	Shallow low layer clouds, fogs	Stratus Stratocumulus	10^2 $<10^3$	<1	None, or slight drizzle or snow

Fig. 8. Cumulonimbus clouds photographed over the upland adjoining the upper Colorado River valley. Note the rain showers which appear under some of the clouds. (*Lt. B. H. Wyatt, U.S.N., U.S. Weather Bureau*)

number of competing classifications. It divides clouds into low-level (base below about 2 km), middle-level (about 2 to 7 km), and high-level (between roughly 7 and 14 km) forms within the middle latitudes. The names of the three basic forms of clouds are used in combination to define 10 main characteristic forms, or "genera."

1. Cirrus are high white clouds with a silken or fibrous appearance (Fig. 1).

2. Cumulus are detached dense clouds which rise in domes or towers from a level low base (Fig. 2).

3. Stratus are extensive layers or flat patches of low clouds without detail (Fig. 3).

4. Cirrostratus is cirrus so abundant as to fuse into a layer (Fig. 4).

5. Cirrocumulus is formed of high clouds broken into a delicate wavy or dappled pattern (Fig. 5).

6. Stratocumulus is a low-level layer cloud having a dappled, lumpy, or wavy structure. See the foreground of Fig. 4.

7. Altocumulus is similar to stratocumulus but lies at intermediate levels (Fig. 6).

8. Altostratus is a thick, extensive, layer cloud at intermediate levels (Fig. 7).

9. Nimbostratus is a dark, widespread cloud with a low base from which prolonged rain or snow falls.

10. Cumulonimbus is a large cumulus which produces a rain or snow shower (Fig. 8).

Classification by air motion. Modern detailed studies of clouds were stimulated by the discovery of cheap methods of seeding supercooled clouds with artificial motes, promoting ice-crystal formation, aimed at stimulating or increasing snowfall.

These studies show that the external form of clouds gives only indirect and incomplete clues to the physical properties which determine their evolution. Throughout this evolution the most important properties appear to be the air motion and the size-distribution spectrum of all the cloud particles, including the condensation nuclei. These properties vary significantly with time and position within the cloud, so that cloud studies demand the intensive examination of individual clouds with expensive facilities such as aircraft and radar. The interplay of physical processes in atmospheric clouds is very complicated, and until it is better understood, no satisfactory physical classification will be possible. From a general meteorological point of view a classification can be based upon the kind of air motion associated with the cloud, as shown in the table.

[FRANK H. LUDLAM]

Bibliography: H. R. Byers, *General Meteorology*, 4th ed., 1974; C. E. Koeppe and G. C. DeLong, *Weather and Climate*, 1958; R. S. Scorer and H. Wexler, *A Colour Guide to Clouds*, 1964; World Meteorological Association, Manual on the Observation of Clouds and Other Meteors, rev. ed., WMO no. 47, 1970.

Cloud physics

The study of the physical and dynamical processes governing the structure and development of clouds and the release from them of snow, rain, and hail (collectively known as precipitation).

The factors of prime importance are the motion of the air, its water-vapor content, and the numbers and properties of the particles in the air which act as centers of condensation and freezing. Because of the complexity of atmospheric motions and the enormous variability in vapor and particle content of the air, it seems impossible to construct a detailed, general theory of the manner in which clouds and precipitation develop. However, calculations based on the present conception of laws governing the growth and aggregation of cloud particles and on simple models of air motion provide reasonable explanations for the observed formation of precipitation in different kinds of clouds.

Cloud formation. Clouds are formed by the lifting of damp air which cools by expansion under continuously falling pressure. The relative humidity increases until the air approaches saturation. Then condensation occurs (Fig. 1) on some of the wide variety of aerosol particles present; these exist in concentrations ranging from less than 100 particles per cubic centimeter (cm^3) in clean, maritime air to perhaps $10^6/cm^3$ in the highly polluted air of an industrial city. A portion of these particles are hygroscopic and promote condensation at relative humidities below 100%; but for continued condensation leading to the formation of cloud droplets, the air must be slightly supersaturated. Among the highly efficient condensation nuclei are the salt particles produced by the evaporation of sea spray, but it now appears that particles produced by man-made fires and by natural combustion (for example, forest fires) also make a major contribution. Condensation onto the nuclei continues as rapidly as the water vapor is made available by cooling of the air and gives rise to droplets of the order of 0.01 millimeter (mm) in diameter. These droplets, usually present in concentrations of a few hundreds per cubic centimeter, constitute a nonprecipitating water cloud.

Mechanisms of precipitation release. Growing clouds are sustained by upward air currents, which may vary in strength from a few centimeters per second (cm/sec) to several meters (m) per second. Considerable growth of the cloud droplets (with falling speeds of only about 1 cm/sec) is

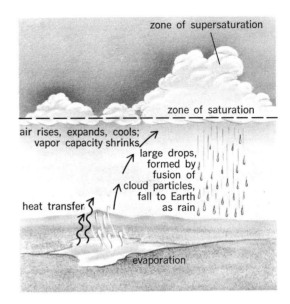

Fig. 2. Diagram of the steps in the formation of rain. (*Based on photodisplay in Willetts Memorial Weather Exhibit, Hayden Planetarium, New York City*)

therefore necessary if they are to fall through the cloud, survive evaporation in the unsaturated air beneath, and reach the ground as drizzle or rain. Drizzle drops have radii exceeding 0.1 mm, while the largest raindrops are about 6 mm across and fall at nearly 10 m/sec. The production of a relatively few large particles from a large population of much smaller ones may be achieved in one of two ways.

Coalescence process. Cloud droplets are seldom of uniform size for several reasons. Droplets arise on nuclei of various sizes and grow under slightly different conditions of temperature and supersaturation in different parts of the cloud. Some small drops may remain inside the cloud for longer than others before being carried into the drier air outside.

A droplet appreciably larger than average will fall faster than the smaller ones, and so will collide and fuse (coalesce) with some of those which it overtakes (Fig. 2). Calculations show that, in a deep cloud containing strong upward air currents and high concentrations of liquid water, such a droplet will have a sufficiently long journey among its smaller neighbors to grow to raindrop size. This coalescence mechanism is responsible for the showers that fall in tropical and subtropical regions from clouds whose tops do not reach the 0°C level and therefore cannot contain ice crystals which are responsible for most precipitation. Radar evidence also suggests that showers in temperate latitudes may sometimes be initiated by the coalescence of waterdrops, although the clouds may later reach to heights at which ice crystals may form in their upper parts.

Initiation of the coalescence mechanism requires the presence of some droplets exceeding 20 microns (μ) in diameter ($1 \mu = 0.0001$ cm). Over the oceans and in adjacent land areas they may well be supplied as droplets of sea spray, but in the interiors of continents, where so-called giant salt particles of marine origin are probably scarce, it

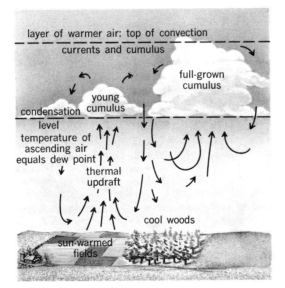

Fig. 1. Conditions leading to birth of a cumulus cloud. (*Based on photodisplay in Willetts Memorial Weather Exhibit, Hayden Planetarium, New York City*)

may be harder for the coalescence mechanism to begin.

Ice crystal process. The second method of releasing precipitation can operate only if the cloud top reaches elevations where temperatures are below 0°C and the droplets in the upper cloud regions become supercooled. At temperatures below −40°C the droplets freeze automatically or spontaneously; at higher temperatures they can freeze only if they are infected with special, minute particles called ice nuclei. As the temperature falls below 0°C, more and more ice nuclei become active, and ice crystals appear in increasing numbers among the supercooled droplets. But such a mixture of supercooled droplets and ice crystals is unstable. The cloudy air, being usually only slightly supersaturated with water vapor as far as the droplets are concerned, is strongly oversaturated for the ice crystals, which therefore grow more rapidly than the droplets. After several minutes the growing crystals will acquire definite falling speeds, and several of them may become joined together to form a snowflake. In falling into the warmer regions of the cloud, however, the snowflake may melt and reach the ground as a raindrop.

Precipitation from layer-cloud systems. The deep, extensive, multilayer-cloud systems, from which precipitation of a usually widespread, persistent character falls, are generally formed in cyclonic depressions (lows) and near fronts. Such cloud systems are associated with feeble upcurrents of only a few centimeters per second, which last for at least several hours. Although the structure of these great raincloud systems, which are being explored by aircraft and radar, is not yet well understood, it appears that they rarely produce rain, as distinct from drizzle, unless their tops are colder than about −12°C. This suggests that ice crystals may be responsible. Such a view is supported by the fact that the radar signals from these clouds usually take a characteristic form which has been clearly identified with the melting of snowflakes.

Production of showers. Precipitation from shower clouds and thunderstorms, whether in the form of raindrops, pellets of soft hail, or true hailstones, is generally of greater intensity and shorter duration than that from layer clouds and is usually composed of larger particles. The clouds themselves are characterized by their large vertical depth, strong vertical air currents, and high concentrations of liquid water, all these factors favoring the rapid growth of precipitation elements by accretion.

In a cloud composed wholly of liquid water, raindrops may grow by coalescence with small droplets. For example, a droplet being carried up from the cloud base would grow as it ascends by sweeping up smaller droplets. When it becomes too heavy to be supported by the vertical upcurrents, the droplet will then fall, continuing to grow by the same process on its downward journey. Finally, if the cloud is sufficiently deep, the droplet will emerge from its base as a raindrop.

In a dense, vigorous cloud several kilometers deep, the drop may attain its limiting stable diameter (about 5 mm) before reaching the cloud base

and thus will break up into several large fragments. Each of these may continue to grow and attain breakup size. The number of raindrops may increase so rapidly in this manner that after a few minutes the accumulated mass of water can no longer be supported by the upcurrents and falls out as a heavy shower. The conditions which favor this rapid multiplication of raindrops occur more readily in tropical regions.

In temperate regions, where the 0°C level is much lower in elevation, conditions are more favorable for the ice-crystal mechanism. However, many showers may be initiated by coalescence of waterdrops.

The ice crystals grow initially by sublimation of vapor in much the same way as in layer clouds, but when their diameters exceed about 0.1 mm, growth by collision with supercooled droplets will usually predominate. At low temperatures the impacting droplets tend to freeze individually and quickly to produce pellets of soft hail. The air spaces between the frozen droplets give the ice a relatively low density; the frozen droplets contain large numbers of tiny air bubbles, which give the pellets an opaque, white appearance. However, when the growing pellet traverses a region of relatively high air temperature or high concentration of liquid water or both, the transfer of latent heat of fusion from the hailstone to the air cannot occur sufficiently rapidly to allow all of the deposited water to freeze immediately. There then forms a wet coating of slushy ice, which may later freeze to form a layer of compact, relatively transparent ice. Alternate layers of opaque and clear ice are characteristic of large hailstones, but their formation and detailed structure are determined by many factors such as the number concentration, size and impact velocity of the supercooled cloud droplets, the temperature of the air and hailstone surface, and the size, shape, and aerodynamic behavior of the hailstone. Giant hailstones, up to 10 cm in diameter, which cause enormous damage to crops, buildings, and livestock, most frequently fall not from the large tropical thunderstorms, but from storms in the continental interiors of temperate latitudes. An example is the Nebraska-Wyoming area of the United States, where the organization of larger-scale wind patterns is particularly favorable for the growth of severe storms.

The development of precipitation in convective clouds is accompanied by electrical effects culminating in lightning. The mechanism by which the electric charge dissipated in lightning flashes is generated and separated within the thunderstorm has been debated for more than 200 years, but there is still no universally accepted theory. However, the majority opinion holds that lightning is closely associated with the appearance of the ice phase, and the most promising theory suggests that the charge is produced by the rebound of ice crystals or a small fraction of the cloud droplets that collide with the falling hail pellets. *See* LIGHTNING.

Basic aspects of cloud physics. The various stages of the precipitation mechanisms raise a number of interesting and fundamental problems in classical physics. Worthy of mention are the supercooling and freezing of water; the nature,

CLOUD PHYSICS

(a) thin hexagonal plate

(b)   needles

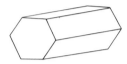

(c) hexagonal prismatic
column

(d) dendritic star-shaped
crystal

Fig. 3. Ice crystal types
formed in various
temperature ranges (°C);
(a) 0 to −3°, −8 to −12°,
−16 to −25°; (b) −3 to −5°;
(c) −5 to −8°, below −25°;
(d) −12 to −16°.

origin, and mode of action of the ice nuclei; and the mechanism of ice-crystal growth which produces the various snow crystal forms.

It has now been established how the maximum degree to which a sample of water may be supercooled depends on its purity, volume, and rate of cooling. The freezing temperatures of waterdrops containing foreign particles vary linearly as the logarithm of the droplet volumes for a constant rate of cooling. This relationship, which has been established for drops varying between 10 μ and 1 cm in diameter, characterizes the heterogeneous nucleation of waterdrops and is probably a consequence of the fact that the ice-nucleating ability of atmospheric aerosol increases logarithmically with decreasing temperature.

When extreme precautions are taken to purify the water and to exclude all solid particles, small droplets, about 1 μ in diameter, may be supercooled to −40°C and drops of 1 mm diameter to −35°C. Under these conditions freezing occurs spontaneously without the aid of foreign nuclei.

The nature and origin of the ice nuclei, which are necessary to induce freezing of cloud droplets at temperatures above −40°C, are still not clear. Measurements made with large cloud chambers on aircraft indicate that the most efficient nuclei, active at temperatures above −10°C, are present in concentrations of only about 10 in a cubic meter of air, but as the temperature is lowered, the numbers of ice crystals increase logarithmically to reach concentrations of about 1 per liter at −20°C and 100 per liter at −30°C. Since these measured concentrations of nuclei are less than one-hundredth of the numbers that apparently are consumed in the production of snow, it seems that there must exist processes by which the original number of ice crystals are rapidly multiplied. Laboratory experiments suggest the fragmentation of the delicate snow crystals and the ejection of ice splinters from freezing droplets as probable mechanisms.

The most likely source of atmospheric ice nuclei is provided by the soil and mineral-dust particles carried aloft by the wind. Laboratory tests have shown that, although most common minerals are relatively inactive, a number of silicate minerals of the clay family produce ice crystals in a supercooled cloud at temperatures above −18°C. A major constituent of some clays, kaolinite, which is active below −9°C, is probably the main source of highly efficient nuclei.

The fact that there may often be a deficiency of efficient ice nuclei in the atmosphere has led to a search for artificial nuclei which might be introduced into supercooled clouds in large numbers. Silver iodide is a most effective substance, being active at −4°C, while lead iodide and cupric sulfide have threshold temperatures of −6°C for freezing nuclei.

In general, the most effective ice-nucleating substances, both natural and artificial, are hexagonal crystals in which spacings between adjacent rows of atoms differ from those of ice by less than 16%. The detailed surface structure of the nucleus, which is determined only in part by the crystal geometry, is of even greater importance. This is strongly indicated by the discovery that several

complex organic substances, notably steroid compounds, which have apparently little structural resemblance to ice, may act as nucleators for ice at temperatures as high as −1°C.

The collection of snow crystals from clouds at different temperatures has revealed their great variety of shape and form. By growing the ice crystals on a fine fiber in a cloud chamber, it has been possible to reproduce all the naturally occurring forms and to show how these are correlated with the temperature and supersaturation of the environment. With the air temperature along the length of a fiber ranging from 0 to −25°C, the following clear-cut changes of crystal habit are observed (Fig. 3):

Hexagonal plates — needles — hollow prisms — plates — stellar dendrites — plates — prisms

This multiple change of habit over such a small temperature range is remarkable and is thought to be associated with the fact that water molecules apparently migrate between neighboring faces on an ice crystal in a manner which is very sensitive to the temperature. Certainly the temperature rather than the supersaturation of the environment is primarily responsible for determining the basic shape of the crystal, though the supersaturation governs the growth rates of the crystals, the ratio of their linear dimensions, and the development of dendritic forms.

Artificial stimulation of rain. The presence of either ice crystals or some comparatively large waterdroplets (to initiate the coalescence mechanism) appears essential to the natural release of precipitation. Rainmaking experiments are conducted on the assumption that some clouds precipitate inefficiently, or not at all, because they are deficient in natural nuclei; and that this deficiency can be remedied by "seeding" the clouds artificially with dry ice or silver iodide to produce ice crystals, or by introducing waterdroplets or large hygroscopic nuclei. In the dry-ice method, pellets of about 1-cm diameter are dropped from an aircraft into the top of a supercooled cloud. Each pellet chills a thin sheath of air near its surface to well below −40°C and produces perhaps 10^{12} minute ice crystals, which subsequently spread through the cloud, grow, and aggregate into snowflakes. Only a few pounds of dry ice are required to seed a large cumulus cloud. Some hundreds of experiments, carried out mainly in Australia, Canada, South Africa, and the United States, have shown that cumulus clouds in a suitable state of development may be induced to rain by seeding them with dry ice on occasions when neighboring clouds, untreated, do not precipitate. However, the amounts of rain produced have usually been rather small.

For large-scale trials designed to modify the rainfall from widespread cloud systems over large areas, the cost of aircraft is usually prohibitive. The technique in this case is to release a silver iodide smoke from the ground and rely on the air currents to carry it up into the supercooled regions of the cloud. In this method, with no control over the subsequent transport of the smoke, it is not possible to make a reliable estimate of the concentrations of ice nuclei reaching cloud level, nor is it known for how long silver iodide retains its nucleating ability in the atmosphere. It is usually these unknown factors which, together with the impossibility of estimating accurately what would have been the natural rainfall in the absence of seeding activities, make the design and evaluation of a large-scale operation so difficult.

In published data, little convincing evidence can be found that large increases in rainfall have been produced consistently over large areas. Indeed, in temperate latitudes most rain falls from deep layer-cloud systems whose tops usually reach to levels at which there are abundant natural ice nuclei and in which the natural precipitation processes have plenty of time to operate. It is therefore not obvious that seeding of these clouds would produce a significant increase in rainfall, although it is possible that by forestalling natural processes some redistribution might be effected.

Perhaps more promising as additional sources of rain or snow are the persistent supercooled clouds produced by the ascent of damp air over large mountain barriers. The continuous generation of an appropriate concentration of ice crystals near the windward edge might well produce a persistent light snowfall to the leeward, since water vapor is continually being made available for crystal growth by lifting of the air. The condensed water, once converted into snow crystals, has a much greater opportunity of reaching the mountain surface without evaporating, and might accumulate in appreciable amount if seeding were maintained for many hours.

The results of trials carried out in favorable locations in the United States and Australia suggest that in some cases seeding has been followed by seasonal precipitation increases of about 10%, but rarely have the effects been reproduced from one season to the next, and overall the evidence for consistent and statistically significant increases of rainfall is not impressive. Experiments of improved design will probably have to be continued for several more years before the effects of large-scale cloud seeding can be realistically assessed.

Attempts to stimulate the coalescence process by spraying waterdroplets or dispersing salt crystals into the bases of incipient shower clouds have been made in places as far apart as Australia, the Caribbean, eastern Africa, and Pakistan. The results of these experiments, though encouraging, are not yet sufficient to allow definite conclusions.

On a more optimistic note, it is relatively easy to clear quite large areas of the supercooled cloud and fog by seeding and converting it into ice crystals, which grow at the expense of the waterdrops and fall out to leave large holes.

The science and technology of cloud modification is, however, still in its infancy. It may well be that further knowledge of cloud behavior and of natural precipitation mechanisms will suggest new possibilities and improved techniques which will lead to developments far beyond those which now seem likely. See WEATHER MODIFICATION.

[BASIL J. MASON]

Bibliography: N. H. Fletcher, The Physics of Rainclouds, 1962; B. J. Mason, Clouds, Rain and Rainmaking, 2d ed., 1975; B. J. Mason, The Physics of Clouds, 2d ed., 1971.

Continental shelf and slope

The continental shelf is the zone around the continent, extending from the low-water line to the depth at which there is a marked increase in slope to greater depth. The continental slope is the declivity from the edge of the shelf extending down to great ocean depths. The shelf and slope comprise the continental terrace, which is the submerged fringe of the continent, connecting the shoreline with the $2\frac{1}{2}$-mi-deep (4-km) abyssal ocean floor (see illustration).

Continental shelf. This comparatively featureless plain, with an average width of 45 mi (72 km), slopes gently seaward at about 10 ft/mi (1.9m/km). At a depth of about 70 fathoms (128 m) there generally is an abrupt increase in declivity called the shelf break, or the shelf edge. This break marks the limit of the shelf, the top of the continental slope, and the brink of the deep sea. However, some shelves are as deep as 200–300 fathoms (180–550 m), especially in past or presently glaciated regions. For some purposes, especially legal, the 100-fathom line or 200-m line is conventionally taken as the limit of the shelf. Characteristically, the shelves are thinly veneered with clastic sands, silts, and silty muds, which are patchily distributed. Geologically, the shelf is an extension of, and in unity with, the adjacent coastal plain. The position of the shoreline is geologically ephemeral, being subject to constant prograding and retrograding, so that its precise position at any particular time is not important. Genetically, the origin of the shelf seems to be primarily related to shallow wave cutting (waves cut effectively as breakers and surf only down to about 5 fathoms (9 m), the depth of vigorous abrasion), shoreline deposition, and oscillations of sea level, which have been especially strong during the Pleistocene and Recent. Although worldwide in distribution and comprising 5% of the area of the Earth, shelves differ considerably in width. Off the east coast of the United States the shelf is about 75 mi wide (120 km), while off the west coast it is about 20 mi wide (32 km). Especially broad shelves fringe northern Australia, Argentina, and the Arctic Ocean. As along the eastern United States, continental shelves commonly acquire a prism of sediments as the continental margin downflexes. Such capping prisms appear to be nascent miogeoclines.

Continental slope. The drowned edges of the low-density "granitic" or sialic continental masses are the continental slopes. The continental plateaus float like icebergs in the Earth's mantle with the slopes marking the transition between the low-density continents and the heavier oceanic segments of the Earth's crust. Averaging $2\frac{1}{2}$ mi high (4 km) and in some places attaining 6 mi (10 km), the continental slopes are the most imposing escarpments on the Earth. The slope is comparatively steep with an average declivity of 4.25° for the upper 1000 fathoms (1.8 km). Most slopes resemble a straight mountain front but are highly irregular in detail; in places they are deeply incised by submarine canyons, some of which cut deeply into the shelf. Usually the slope does not connect directly with the sea floor; instead there is a transitional area, the continental rise, or apron, built by

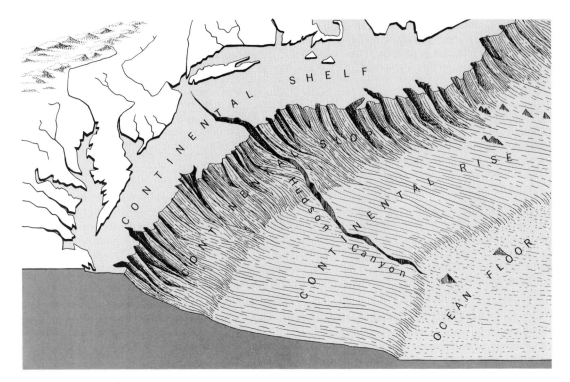

Continental margin off Northeastern United States.

the shedding of sediments from the continental block. *See* MARINE GEOLOGY; SUBMARINE CANYON.

[ROBERT S. DIETZ]

Bibliography: J. L. Culliney, *The Forests of the Sea: Life and Death on the Continental Shelf*, 1977; R. McQuillin, *Exploring the Geology of Shelf Seas*, 1975; F. P. Shephard, *Geological Oceanography*, 1977; J. Waters, *The Continental Shelves*, 1975.

Continentality

The attributes of notably greater temperature ranges with associated weather characteristics that make inland and continental regions distinctly different from marine areas or areas bordering large bodies of water. Unlike the latitudinal uniformity of climate which might result from exposure of a uniform surface by the Earth to the Sun's rays, some of the most outstanding differences are those between continental and oceanic areas, particularly in the middle latitudes. These latitudes, designated temperate zones by the Greeks of the classical period, actually contain some of the least temperate weather and climate of the world over continental areas, as well as truly temperate moderation in maritime provinces.

The reasons are associated with the greatly different thermal behavior of water and land surfaces. The soil surfaces of the continents can be heated or cooled comparatively rapidly because the thermal effects are limited to a shallow surface layer. The heat received at the ground is used to warm the top layers, which consequently become hot in summer. In winter, by contrast, the surface becomes notably cold since there is little conduction of heat from subsurface depths. As a result of

the responsiveness of the surface air temperature to the temperature of the underlying soil surface, the air in contact with the ground undergoes temperature fluctuations of large amplitude, lagging slightly behind the annual and the diurnal periods of insolation (see illustration). *See* MARINE INFLUENCE ON WEATHER AND CLIMATE.

On the other hand, the water surfaces which make up the oceans are heated or cooled much less rapidly, partly because the thermal effects extend to greater depths and partly because much of the solar energy absorbed at the sea surface is utilized for evaporation rather than heating. As a result, the upper layers of the ocean are not heated as strongly as the land surfaces in summer or cooled as strongly by radiation in winter. It follows that large annual and diurnal ranges in temperature with hot summers and cold winters are characteristic of the continental interiors of middle and high latitudes. In contrast, the summers over the oceans are cool, while the winters are comparatively mild.

Because the increase in the annual range in temperature inland over the continents is the most striking effect of the continental surface on climate, climatologists have taken the annual range in temperature (ΔT) as a measure of the continentality. The most frequently used formula is the one shown below, advanced by O. V. Johansson. Here

$$K = \frac{1.6\,\Delta T}{\sin \phi} - 14$$

the index of continentality is K (in percent, a purely oceanic climate presenting a value of 0% and a completely continental climate one of 100%), ΔT is annual range of monthly mean temperature in °C,

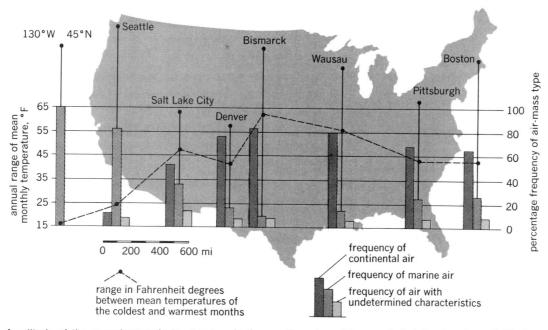

0   200   400   600 mi

range in Fahrenheit degrees
between mean temperatures of
the coldest and warmest months

frequency of
continental air

frequency of marine air

frequency of air with
undetermined characteristics

Amplitude of the annual range in temperature in the United States related to continental location and to frequencies of the marine and the continental air masses.

(Based on data compiled at Headquarters, Air Weather Service, Climatic Center)

and ϕ is latitude. W. Gorczynski and D. Brunt have presented similar formulas, although the one by Brunt, while of greater physical significance than the other two, is somewhat difficult to evaluate because it involves the use of solar radiation data as well as temperature data.

[WOODROW C. JACOBS]

Bibliography: B. Haurwitz and J. M. Austin, *Climatology*, 1944; G. T. Trewartha, *An Introduction to Climate*, 4th ed., 1968.

Coriolis acceleration and force

Two alternative concepts which pertain to the difference between the acceleration of motion as referred to and viewed from a coordinate system which does not partake of the Earth's rotation (absolute acceleration) and the acceleration of motion as referred to and viewed from a coordinate system fixed on the surface of the Earth (relative acceleration). Part of this difference is the centripetal acceleration of a fixed point on the Earth's surface toward the axis of rotation of the Earth. The sum of this acceleration and the true gravitational acceleration (which is directed toward the center of mass of the Earth) is equal to the apparent gravitational acceleration (which is observed to act normal to the ellipsoidal surface of the Earth). The remainder of the difference between the absolute and relative accelerations depends upon the motion relative to the Earth and is called the Coriolis acceleration.

The direction of the Coriolis acceleration is perpendicular to the plane formed by the northward-directed polar axis of the Earth and the direction of motion relative to the Earth. Its sense is such that, if one faces in the direction of this acceleration, the smallest rotation required to bring the axis of the Earth into coincidence with the direction of

relative motion is clockwise. The magnitude of the Coriolis acceleration is equal to twice the product of the speed of motion relative to the Earth times the angular speed of rotation of the Earth about its axis times the sine of the angle between the polar axis and the direction of relative motion. If the vertical velocity component is small, as it is in the large-scale motions of the atmosphere, the horizontal component of the Coriolis acceleration is normal to the horizontal component of relative motion and is directed to the left of the direction of this component on the Northern Hemisphere and to the right on the Southern Hemisphere.

The magnitude of the horizontal component of the Coriolis acceleration is then given by twice the product of the speed of the horizontal component of relative motion (or wind) times the angular speed of rotation of the Earth times the sine of the latitude. Thus it is equal to zero at the Equator and, for a given relative motion, reaches a maximum magnitude at the poles. Though negligible in many physical problems, the Coriolis effect must be considered in meteorological problems relating to the dynamics of large-scale wind currents since its horizontal component is typically as large as any other horizontal force acting.

A simple illustration of the Coriolis acceleration is afforded by consideration of the special case of air motion, or wind, at the poles. Suppose that an air parcel crosses the North Pole and moves directly south so that its motion, to an observer on the Earth, is in a straight line. This same motion, to an observer fixed in space, would be curved and would display a continual acceleration toward the east because of the rotation of the meridian down which the air parcel is moving. This acceleration is the Coriolis acceleration.

The necessity of considering the Coriolis effect

stems from the fact that Newton's second law is valid only when the motions and accelerations are those observed in a coordinate system fixed in space. Thus the sum of the forces must be equated to the sum of the Coriolis acceleration and the acceleration of motion as determined by an observer on the surface of the Earth (relative acceleration). It is often most convenient, however, to think in terms of an equality of the sum of the forces and the relative acceleration alone. In this instance the forces must include a fictitious force equal in magnitude to the Coriolis acceleration and opposed to it in direction. This fictitious force is the Coriolis force. The alteration in concept from the Coriolis acceleration to the Coriolis force is equivalent to the mathematical device of moving the former from the acceleration side of the equation of motion to the force side, with a change of sign. *See* GEOSTROPHIC WIND; GRADIENT WIND.

[FREDERICK SANDERS]

Bibliography: A. H. Gordon, *Elements of Dynamic Meteorology*, 1962; J. R. Holton, *An Introduction to Dynamic Meteorology*, 1972.

Cosmic rays

Electrons and the nuclei of atoms—largely hydrogen—that impinge upon Earth from all directions of space with nearly the speed of light. These nuclei with relativistic speeds are often referred to as primary cosmic rays, to distinguish them from the cascade of secondary particles generated by their impact against air nuclei at the top of the terrestrial atmosphere. The secondary particles shower down through the atmosphere and are found with decreasing intensity all the way to the ground, and below. It was V. Hess's observation of the increasing intensity with height which first established that the particles were of extraterrestrial origin rather than from Earth itself.

Cosmic rays are studied for a variety of reasons, not the least of which is a general curiosity over the process by which nature can produce such energetic nuclei. Apart from this, the primary cosmic rays, being deflected by the interplanetary and geomagnetic fields, are currently used as probes to determine the nature of these fields far out from Earth. Cosmic rays have long served as an inexpensive source of high-energy particles for the study of nuclear interactions and production of the so-called strange particles. The initial identification of the positron, the muon, the π-meson, and certain of the K-mesons and hyperons were made from studies of cosmic rays, which are still the only source of particles with energies of 400 Gev and higher. The observed long flight time of the unstable muons, produced by the primary cosmic rays at the top of the atmosphere, demonstrates directly the time dilatation of special relativity.

Cosmic-ray detection. Cosmic rays are usually detected by instruments which classify each incident particle as to type, energy, and in some cases time and direction of arrival. A convenient unit for measuring cosmic-ray energy is the electron volt (ev), which is the energy gained by a unit charge (such as an electron) accelerating freely across a potential of 1 volt. One electron volt equals about 1.6×10^{-19} joules. For nuclei it is usual to express the energy in terms of electron volt/nucleon, since as a function of this variable the relative abundances of the different elements are nearly constant. Indeed, only since about 1970 have instruments sensitive enough to detect deviations from constancy been in operation. Two nuclei with the same energy/nucleon have the same velocity.

The number of particles measured is generally expressed by dividing the total number seen by the effective size or "geometry factor" of the measuring instrument. Calculation of the geometry factor requires knowledge of both the sensitive area (in square centimeters) and the angular acceptance (in steradians) of the detector, as the arrival directions of the cosmic rays are randomly distributed to within 1% in most cases. A flat detector of any shape but with area of 1 cm² has a geometry factor of π cm²-sr if it is sensitive to cosmic rays entering from one side only. The total flux of cosmic rays in the vicinity of the Earth but outside the atmosphere is about 0.3 nuclei/cm²-sec-sr. Thus a quarter dollar, with a surface area of 4.5 cm², lying flat on the surface of the Moon will be struck by $0.3 \times 4.5 \times 3.14 = 4.2$ cosmic rays/sec.

Energy spectrum. The flux of cosmic rays varies as a function of energy. This type of function is called an energy spectrum, and may refer to all cosmic rays or to only a selected element or group of elements. Since cosmic rays are continuously distributed in energy it is meaningless to attempt to specify the flux at any one exact energy. Normally one speaks of an integral spectrum, in which the function gives the total flux of particles with energy greater that the specified energy (in particle/cm²-sec-sr), or a differential spectrum, in which the function provides the flux of particles in some energy interval (typically 1 Mev/nucleon-wide) centered on the specified energy (in particles/cm²-sec-sr-Mev/nucleon). The basic problem of cosmic-ray research is to measure the spectra of the different components of cosmic radiation and to deduce from them and other observations the nature of the cosmic-ray sources and the details of how and where the particles travel on their way to Earth.

Types of detectors. All cosmic-ray detectors are sensitive to moving electrical charges. Neutral cosmic rays (neutrons, gamma rays, and neutrinos) are studied by observing the charged particles produced in the collision of the neutral primary with some type of target. At low energies the ionization of the matter through which they pass is the principal means of detection. Such detectors include cloud chambers, ion chambers, spark chambers, Geiger counters, proportional counters, scintillation counters, solid-state detectors, and photographic emulsions. One technique has developed from the observation that certain mineral crystals are permanently damaged by individual highly charged cosmic-ray nuclei (usually iron or heavier). The resulting tiny, submicroscopic flaws can be enlarged by chemical etching and the resulting pits measured to determine the type of particle which caused them. Examination of crystals found in meteorites and lunar rock samples yields data on the intensity and composition of cosmic rays over the history of the solar system.

At energies above about 500 Mev/nucleon it is no longer practical to stop cosmic rays completely in the detector. Cerenkov detectors and the deflection in the field of large superconducting magnets provide the best means of studying energies up to a few hundred Gev. Above about 10^{12} ev, direct detection of individual particles is no longer possible since they are so rare. Such particles are studied by observing the large showers of secondaries they produce in Earth's atmosphere. These showers are detected either by counting the particles which survive to strike ground-level detectors or by looking at the flashes of light the showers produce in the atmosphere with special telescopes and photomultiplier tubes. It is not possible to say what kind of particles produce these showers, but the energy can be measured with fair accuracy—the record is about 10^{20} ev (16 J).

Atmospheric cosmic rays. The primary cosmic-ray particles coming into the top of the terrestrial atmosphere make inelastic collisions with nuclei in the atmosphere. The collision cross section is essentially the geometrical cross section of the nucleus, of the order of 10^{-26} cm². The mean free path for primary penetration into the top of the atmosphere is given in Table 1. (Division by the atmospheric density in g/cm³ gives the value of the mean free path in centimeters.)

When a high-energy nucleus collides with the nucleus of an air atom, a number of things usually occur. The abrupt stopping of the incoming nucleus leads to production of π-mesons with positive, negative, or neutral charge; this meson production is closely analogous to the generation of x-rays, or bremsstrahlung, produced when a fast electron is stopped by impact with the atoms in a metal target. The mesons, like the bremsstrahlung, come off from the impact in a narrow cone in the forward direction. Anywhere from 0 to 30 or more π-mesons may be produced, depending upon the energy of the incident nucleus. The ratio of neutral to charged π-mesons is ~0.75.

A few protons and neutrons (in about equal proportions) may be knocked out with energies a significant fraction of that of the incoming nucleus. They are called knock-on protons and neutrons.

All these protons, neutrons, and π-mesons generated by collision of the primary cosmic-ray nuclei with the nuclei of air atoms are the first stage in the development of the secondary cosmic-ray particles observed inside the atmosphere. Since several secondary particles are produced by the collisional stopping of each primary, the total number of energetic particles of cosmic-ray origin will increase with depth, even while the primary density is decreasing. Since electric charge must be conserved and the primaries are positively charged, the positive particles outnumber the negative particles in the secondary radiation by a factor of about 1.2, called the positive excess.

Cascade processes. The uncharged π^0-mesons from the stopping primary nuclei decay into two γ-rays with a life of ~8×10^{-17} sec. The decay is so rapid that π^0-mesons are not directly observed among the secondary particles in the atmosphere.

Table 1. Mean free paths for primary cosmic rays in the atmosphere

Charge of primary nucleus	Mean free path in air, g/cm²
$Z = 1$	60
$Z = 2$	44
$3 \leqq Z \leqq 5$	32
$6 \leqq Z \leqq 9$	27
$10 \leqq Z \leqq 29$	21

The two γ-rays together have the rest energy of the π^0, about 140 Mev, plus the π^0 kinetic energy. They each produce a positron-electron pair, which cannot have individual mean energies of less than 35 Mev. Upon passing sufficiently close to the nucleus of an air atom deeper in the atmosphere, the electrons and positrons convert their energy into bremsstrahlung. The bremsstrahlung in turn create new positron-electron pairs, and so on. This cascade process continues until the energy of the initial π^0 has been dispersed into a shower of positrons, electrons, and photons with insufficient individual energies ($\leqq 1$ Mev) to continue the pair production. The shower, then being unable to reproduce its numbers, is dissipated by ionization of the air atoms. The electrons and photons of such showers are referred to as the soft component of the atmospheric (secondary) cosmic rays, reaching a maximum intensity at an atmospheric depth of 150–200 g/cm² and then declining by a factor of about 10^2 down to sea level.

Mesons. The $\pi^\pm$-mesons produced by the primary collisions have a life of about 2.5×10^{-8} sec before they decay into muons:

$$\pi^\pm \rightarrow \mu^\pm + \text{neutrino}$$

With a life of this order a $\pi^\pm$ possessing enough energy (> 10 Gev) to experience significant relativistic time dilatation may exist long enough to interact with the nuclei of the air atoms. The cross section for $\pi^\pm$ nuclear interactions is approximately the geometrical cross section of the nucleus, and the result of such an interaction is essentially the same as for the primary cosmic-ray nuclei, already discussed. The low-energy $\pi^\pm$-mesons decay into muons before they have time to undergo nuclear interactions.

The muons will not interact with nuclei, and are too massive (207 electron masses) to produce bremsstrahlung. They can lose energy only by the comparatively feeble process of ionizing an occasional air atom as they progress downward through the atmosphere. Because of this ability to penetrate matter they are called the hard component. Their life is 2×10^{-6} sec before they decay into an electron or positron and two neutrinos. Hence, with the time dilatation of their high energy, some 5% of the muons reach the ground. Their interaction with matter is so weak that they penetrate deep into the ground, where they are the only particles of cosmic-ray origin to be found. At a depth equivalent of 300 m of water the muon intensity has decreased from that at ground level only by a factor of 20; at an equivalent depth of 1400 m it has decreased by a factor of 10^3.

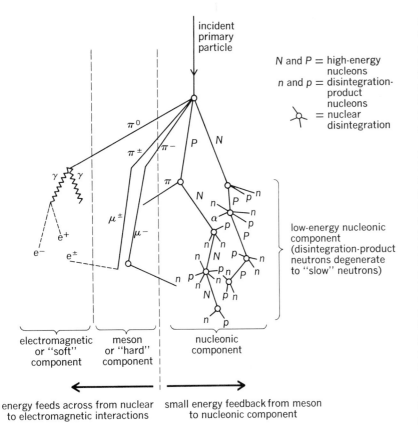

Fig. 1. Cascade of secondary cosmic-ray particles in the terrestrial atmosphere.

Nucleonic component. The high-energy nucleons—the knock-on protons and neutrons—produced by the primary-particle collisions and a few $\pi^\pm$-meson-produced collisions in the atmosphere, proceed on down into the atmosphere with a mean free path of about 60 g/cm² (same path as for the primary protons). They produce nuclear interactions of the same kind as the primary nuclei, though of course with diminished energies. This cascade process constitutes what is called the nucleonic component of the secondary cosmic rays.

Nuclear stars. A nucleus struck by a proton or neutron of the nucleonic component with an energy greater than approximately 300 Mev may have its internal forces momentarily disrupted so that some of its nucleons are free to leave with their original nuclear kinetic energies of about 10 Mev. The nucleons freed in this fashion appear as protons, deuterons, tritons, α-particles, and even somewhat heavier clumps, radiating outward from the struck nucleus. In photographic emulsions the result is a number of short prongs radiating from the point of collision, and for this reason is called a nuclear star. When the nucleon energy falls below about 100 Mev, star production and further knock-ons can no longer be produced. At the same time the protons are rapidly disappearing from the cascade because their ionization losses in the air slow them down before they can make a nuclear interaction. The neutrons are already dominant at 3500 m, about 300 g/cm² above sea level, where they outnumber the protons four to one. Thus the final stages in the lower atmosphere are given over almost entirely to neutrons in a sequence of low-energy interactions which convert them to thermal neutrons (neutrons of kinetic energy of about 0.025 ev) in a path of about 90 g/cm². These thermal neutrons are readily detected in boron trifluoride (BF_3) counters. The nucleonic component increases in intensity down to a depth of about 120 g/cm², and thereafter declines in intensity, with a mean absorption length of about 200 g/cm².

The various cascades of secondary particles in the atmosphere are shown schematically in Fig. 1. Note that about 48% of the initial primary cosmic-ray energy goes into charged pions, 25% into neutral pions, 7% into the nucleonic component, and 20% into stars. The nucleonic component is produced principally by the lower-energy (~5 Gev) primaries. Higher-energy primaries put their energy more into meson production. Hence in the lower atmosphere, a Geiger counter responds mainly to the higher-energy primaries (~15 Gev) because it counts the muons and electrons, whereas a BF_3 counter detecting thermal neutrons responds more to the low-energy primaries. This fact is important for investigations of the effects of interplanetary and geomagnetic fields on the primary cosmic rays, since the lowest-energy primaries are the most drastically affected.

Geomagnetic effects. The magnetic field of Earth is described approximately as that of a magnetic dipole of strength 8.1×10^{25} cgs (centimeter-gram-second) units located near the geometric center of Earth. Near the Equator the field intensity is 0.3 gauss (1 gauss = 10^{-4} tesla), falling off in space as the inverse cube of the distance to the Earth's center. In a magnetic field which does not vary in time, the path of a particle is determined entirely by its rigidity, or momentum per unit charge; the velocity simply determines how fast the particle will move along this path. Momentum is usually expressed in units of ev/c, where c is the velocity of light, because at high energies, energy and momentum are then numerically almost equal. By definition, momentum and rigidity are numerically equal for singly charged particles. The unit so defined is normally called the volt, but should not be confused with the standard and nonequivalent unit of the same name. Table 2 gives examples of these units as applied to different particles with rigidity of 1 Gv. This corresponds to an orbital radius in a 10^{-5}-gauss magnetic field of approximately 10 times the distance from the Earth to the Moon.

The minimum rigidity of a particle able to reach the top of the atmosphere at a particular geomagnetic latitude is called the geomagnetic cutoff rigidity at that latitude, and its calculation is a complex numerical problem. Fortunately, for an observer near the ground obliquely arriving sec-

Table 2. Properties of particles when all have a rigidity of 1 Gv

| Particle | Charge | Nucleons | Kinetic energy | | Momentum, |
			Mev	Mev/nucleon	Mev/c
proton	1	1	430	430	1000
³He	2	3	640	213	2000
⁴He	2	4	500	125	2000
¹⁶O	8	16	2000	125	8000

ondary particles, produced by the oblique primaries, are so heavily attenuated by their longer path to the ground that it is usually sufficient to consider only the geomagnetic cutoff for vertically incident primaries, which is given in Table 3. Around the Equator, where a particle must come in perpendicular to the geomagnetic lines of force to reach Earth, particles with rigidity less than 10 Gv are entirely excluded, though at higher latitudes where entry can be made more nearly along the lines of force, lower energies can reach Earth. Thus, the cosmic-ray intensity is a minimum at the Equator, and increases to its full value at either Pole—this is the cosmic-ray latitude effect. At sea level the variation with latitude is obviously small, ~15%, as seen with Geiger counters, but it is very large when observed with BF_3 counters, amounting to something on the order of a factor of two, as shown in Fig. 2 for latitudes > 40°.

In addition to perturbations in the dipole structure of the geomagnetic field due to turbulence in the iron core of Earth, which must be considered in accurate studies of the geomagnetic cutoff, the distortion of the field by the pressure of ionized gases (for example, the solar wind) in space must often be accounted for, particularly at high latitudes.

This correction can vary rapidly with time because of sudden bursts of solar activity and because of the rotation of Earth. Areas with cutoffs of 400 Mv during the day can have no cutoff at all during the night. The day-night effect is confined to particles with energies so low that neither they nor their secondaries reach the ground, and is thus observed only on high-altitude balloons or satellites.

Since the geomagnetic field is directed from south to north above the surface of Earth, the incoming cosmic-ray nuclei are deflected toward the east. Hence an observer finds some 20% more particles incident from the west. This is known as the east-west effect. *See* GEOMAGNETISM.

Solar modulation. Figure 3 presents portions of the proton and alpha-particle spectra observed near the Earth but outside of the magnetosphere in 1973. Above 20 Gev/nucleon the intensity is steady in time to a few percent and the energy spectra of all elements vary about the −2.7 power of the energy up to a few hundred Gev/nucleon. Above this energy individual elements are no longer resolved, but the combined spectrum seems to have the same energy dependence up to 10^{18} ev. Below 20 Gev/nucleon the cosmic-ray intensity varies markedly with time. S. Forbush was the first to show that the cosmic-ray intensity was low

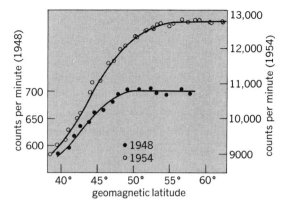

Fig. 2. Latitude variation of the neutron component of cosmic rays in 80°W longitude and at a height corresponding to an atmospheric pressure of 22.5 cm of mercury in 1948, when the Sun was active, and 1954, when the Sun was deep in a sunspot minimum.

during the years of high solar activity and sunspot number, which follow an 11-yr cycle. This effect is clearly seen in the data of Fig. 2 and has been extensively studied with ground-based and spacecraft instruments. While this so-called solar modulation is now understood in general terms, it has not been calculated in detail in large part because of the lack of any direct measurements of conditions in the solar system out of the ecliptic plane, to which all spacecraft are confined because of limitations on the power of rockets.

The primary cause of solar modulation is the solar wind, a highly ionized gas (plasma) which boils off the solar corona and propagates radially from the Sun at a velocity of about 400 km/sec. The wind is mostly hydrogen, with typical density of 5 protons/cm³. This density is too low for collisions with cosmic rays to be important. Rather, the high conductivity of the medium traps part of the solar magnetic field and carries it outward. The rotation of the Sun and the radial motion of the plasma combine to create the observed Archimedean spiral pattern of the average interplanetary magnetic field. Turbulence in the solar wind creates fluctuations in the field which often locally obscure the average direction and intensity. This complex system of magnetic irregularities propagating outward from the Sun deflects and sweeps the low-rigidity cosmic rays out of the solar system.

In addition to the bulk sweeping action, another effect of great importance occurs in the solar wind, adiabatic deceleration. Because the wind is blowing out, only those particles which chance to diffuse upstream fast enough are able to reach Earth. However, because of the expansion of the wind, particles diffusing in it lose energy. Thus, particles observed at Earth with 10-Mev/nucleon energy actually started out with several hundred Mev/nucleon in nearby interstellar space, and those with only 100−200-Mev/nucleon initial energy probably never reach Earth at all. This is particularly unfortunate because at these lower energies the variation with energy of nuclear-reaction probabilities would allow much more detailed investigation of cosmic-ray history. The

Table 3. Geomagnetic cutoff

Geomagnetic latitude	Vertical cutoff, Gv
0°	15
±20°	11.5
±40°	5
±60°	1
±70°	0.2
±90°	0

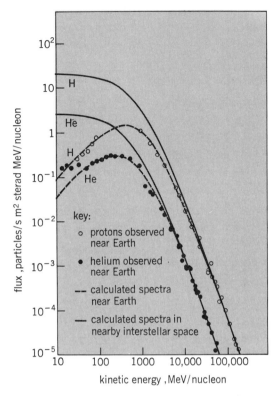

Fig. 3. Spectra of cosmic-ray protons and helium at Earth and in nearby interstellar space, showing the effect of solar modulation. Observations were made in 1973, when the Sun was quiet. (*Courtesy of Dr. M. Garcia-Munoz, University of Chicago*)

region in space where solar modulation is important is probably a sphere of radius 20–50 astronomical units (AU), although this is not at all certain. Spacecraft to Jupiter (5 AU) have not seen a drastic change in the level of cosmic-ray activity. Changes in the modulation with solar activity are caused by the changes in the pattern of magnetic irregularities rather than by changes in the wind velocity, which are quite small.

The amount of modulation can be measured by observation of cosmic-ray electrons. Electrons are light enough to emit synchrotron radiation as they are deflected by the 10^{-6}-gauss galactic magnetic field. This radiation is measured by radio telescopes and allows calculation of the approximate spectrum of electrons in interstellar space, which may be compared with the spectrum observed at Earth to determine the amount of modulation. Knowledge of the amount of modulation permits the calculation of the true interstellar spectra of the nuclear cosmic rays. Curves in Fig. 3 show the result of one such calculation which yielded the surprising result that low-energy helium and protons must have different sources.

There are several phenomenological classes of cosmic-ray variation besides the 11-year variation already mentioned. They are associated with the short-term variations of the solar wind.

There is also the Forbush-type decrease, where the primary intensity around the world may drop in an irregular way as much as 20% in 15 hr or 8% in

3 hr, slowly recovering in the days or weeks that follow. Often, but not always, the Forbush decrease and geomagnetic storms accompany each other. Striking geographical variation is to be seen in the sharper fluctuations during the onset of a Forbush decrease.

Since the original description of the Forbush-type cosmic-ray-intensity decrease, two distinct types of decrease can now be classified as the Forbush type. In one, a sudden outburst at the Sun, at the time of a large flare, yields a cosmic-ray decrease in space which extends to very high energies, 50 Gev or so. The magnetic fields carried in the blast wave from the flare sweep back the cosmic rays, causing a decrease in their intensity in space of up to 40%. In contrast to such events, there is a second type, the recurring decrease, which occurs with each 27-day rotation of the Sun, often for several rotations in succession. The time variation of these decreases often resembles the classical Forbush decrease, but they do not extend to particles much above 10 Gev. The recurring decrease is the result of the higher solar wind velocity from a long-lived, hot, active region in the corona. The recurring decreases are more prominent during the years of declining solar activity, whereas the isolated outburst is more important during the rising and peak years. *See* SUN.

Composition of cosmic rays. Nuclei ranging from protons to uranium have been identified in the cosmic radiation. The relative abundances of the different elements are shown in Fig. 4, together with the best estimate of the "universal abundances" obtained by combining measurements of solar and stellar spectra, lunar and terrestrial rocks, meteorites, and so forth. Most obvious is the similarity between the two distributions. However, a systematic deviation is quickly apparent: the elements lithium-boron and scandium-manganese as well as most of the odd-charged nuclei are vastly overabundant in the cosmic radiation. This effect has a simple explanation: the cosmic rays travel great distances in the galaxy and occasionally collide with atoms of interstellar gas — mostly hydrogen and helium — and fragment. This fragmentation, or spallation as it is called, produces lighter nuclei from heavier ones but does not change the energy/nucleon very much. Thus the energy spectra of the secondaries are similar to those of the primaries. Calculations involving reaction probabilities determined by nuclear physicists show that the overabundances of the secondary elements can be explained by assuming that cosmic rays pass through an average of about $5/cm^2$ of material on their way to Earth. Although an average pathlength can be obtained, it is not possible to fit the data by saying that all particles of a given energy have exactly the same pathlength; furthermore, results indicate that higher-energy particles traverse less matter in reaching the solar system, although their original composition seems energy independent.

When spallation has been corrected for, differences between cosmic-ray abundances and solar-system or universal abundances still remain. The most important question is whether these differences are due to the cosmic rays having come from a special kind of material (such as would be produced in a supernova explosion), or simply to the fact that some atoms might be more easily accelerated than others. It is possible to rank almost all of the overabundances by considering the first ionization potential of the atom and the rigidity of the resulting ion, although this calculation gives no way of predicting the magnitude of the enhancement expected. It is also observed that the relative abundances of particles accelerated in solar flares are far from constant from one flare to the next. The possibility of such preferential acceleration is one of the reasons why much cosmic-ray study is concentrated on determining the isotopic composition of each element, as this is much less likely to be changed by acceleration. It is apparent that the low-energy helium data in Fig. 3 do not fit the calculated values. Since it is known that this low-energy helium is nearly all ^{4}He, whereas the higher-energy helium contains 10% ^{3}He, one can be fairly certain that the deviation is due to a local source of energetic ^{4}He within the solar system rather than a lack of understanding of the process of solar modulation. Similarly, a low-energy enhancement of nitrogen is pure ^{14}N, whereas the higher-energy nitrogen is almost 50% ^{15}N.

Age of radiation. Another important result which can be derived from detailed knowledge of cosmic-ray isotopic composition is the "age" of cosmic radiation. Certain isotopes are radioactive, such as ^{10}Be with a half-life of 2.2×10^6 years. Since Be is produced entirely by spallation, study of the relative abundance of ^{10}Be to the other Be isotopes, particularly as a function of energy to utilize the relativistic increase in this lifetime, will yield a number related to the average time since the last nuclear collision. Measurements show that ^{10}Be is nearly absent at low energies and yield an estimate of the age of the cosmic rays of a few tens of millions of years. This is in disagreement with the conventional age estimate of the comsic rays of a few million years determined by stating that the observed 5-g/cm² pathlength is obtained by moving at the particle velocity through a medium having a density equal to the average density of interstellar gas, about 1 atom/cm³.

Electron abundance. Cosmic-ray electron measurements pose other problems of interpretation, partly because electrons are nearly 2000 times lighter than protons, the next lightest cosmic-ray component. Protons with kinetic energy above 1 Gev are about 100 times as numerous as electrons above the same energy, with the relative number of electrons decreasing slowly at higher energies. But it takes about 2000 Gev to give a proton the same velocity as a 1-Gev electron. Viewed in this way electrons are several thousand times more abundant than protons. (Electrical neutrality of the galaxy is maintained by lower-energy ions which are more dense than cosmic rays although they do not carry much energy.) It is thus quite possible that cosmic electrons have a different source entirely from the nuclei. It is generally accepted that there must be direct acceleration of electrons, because calculations show that more positrons than negatrons should be produced in collisions of cosmic-ray nuclei with interstellar

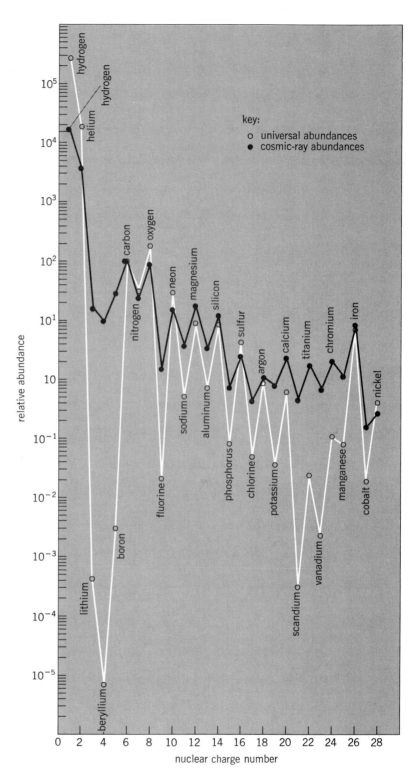

Fig. 4. Cosmic-ray abundances compared to the universal abundances of the elements. Carbon is set arbitrarily to an abundance of 100 in both cases. (*Courtesy of Dr. M. Garcia-Munoz, University of Chicago*)

gas. Measurements show, however, that only 10% of the electrons are positrons. As the number of positrons seen agrees with the calculated secondary production, added confidence is gained in the result that there is indeed an excess of negatrons. Furthermore, these sources of electrons must be in or at least very near to the Milky Way Galaxy, as electrons cannot travel long distances in the 2.7 K blackbody radiation field which permeates the universe. Electrons of energy 15 Gev lose a good portion of their energy in 10^8 yr by colliding with photons via the (inverse) Compton process.

Origin. Although study of cosmic rays has yielded valuable insight into the structure, operation, and history of the universe, their origin has not been determined. The problem is not so much to devise processes which might produce cosmic rays, but instead rather to decide which of many possible processes do in fact produce them.

In general, analysis of the problem of cosmic-ray origin is broken into two major parts: origin in the sense of where the sources are located, whatever they are, and origin in the sense of how the particles are accelerated to such high energies. Of course these questions can never be separated completely.

Location of sources. It is thought that cosmic rays are produced by mechanisms operating within galaxies and are confined almost entirely to the galaxy of their production, trapped by the galactic magnetic field. The intensity in intergalactic space would only be a few percent of the typical galactic intensity, and would be the result of a slow leakage of the galactic particles out of the magnetic trap. It has not been possible to say anything about where the cosmic rays come from by observing their arrival directions at Earth. At lower energies (up to 10^{15} ev) the anisotropies which have been observed can all be traced to the effects of the solar wind and interplanetary magnetic field. The magnetic field of the galaxy seems to be completely effective in scrambling the arrival directions of these particles. At energies around 10^{18} ev, on the order of 100 events had been observed by 1975. There is some indication that these particles may have preferred arrival directions. At these energies, particles have a radius of curvative which is not negligible compared to galactic dimensions, and thus their arrival direction could be related to where they came from.

Only one or two particles with energy above 10^{20} ev have been seen, making the question of their isotropy meaningless. A proton of 10^{20} ev in galactic magnetic field of 10^{-5} gauss would have a radius of curvature of about 10^4 parsec (1 parsec $= 3.08 \times 10^6$ m), which exceeds the distance from the Sun to the galactic center. Containment of such a proton in the galaxy is not possible, but it still might originate within the galaxy and be on its way out. An iron nucleus with the same total energy could be contained because its high charge would cause it to have a much lower rigidity and a smaller radius of curvature. Collision with photons limits the travel of 10^{20}-ev protons between galaxies in the same way as it does that of electrons at lower energy.

Direct detection of cosmic rays propagating in distant regions of the galaxy is possible by observing the electromagnetic radiation produced as they interact with other constituents of the galaxy. Measurement of the average electron spectrum using radiotelescopes has already been mentioned. Proton intensities are mapped by studying the arrival directions of gamma rays (at about 50 Mev)

produced as they collide with the interstellar gas. Unfortunately, the amount of radiation in these processes depends upon both the cosmic-ray flux and the magnetic-field intensity or density of interstellar gas. Hence areas where cosmic rays are known to exist can be pointed out because the radiation is observed. But where no radiation is seen, it is not known whether its absence results from lack of cosmic rays or lack of anything for them to interact with. In particular, very little radiation is seen from outside the galaxy, but there is also very little gas or magnetic field there. There is therefore no direct evidence either for or against galactic containment.

A major difficulty with the concept of cosmic radiation filling the universe is the large amount of energy needed to maintain the observed intensity in the face of an expanding universe—probably more energy than is observed to be emitted in all other forms put together. Continuous acceleration of cosmic rays, rather than a simple decay of intensity from some primordial level, is indicated by the observations of the near constancy of the flux over the last 2×10^9 yr and also by the presence of heavier nuclei such as carbon, oxygen, and iron. It is thought that these nuclei could not be produced in the so-called Big Bang, but must have been fromed in stars and then accelerated. At least this heavy cosmic-ray component has been generated under conditions not too unlike the present and thus is probably continuously produced.

Confinement mechanisms. Three possible models of cosmic-ray confinement are under investigation. All assume that cosmic rays are produced in sources, discrete or extended, scattered randomly through the galactic disc. Most popular is the "leaky box" model which proposes that the particles diffuse about in the magnetic field for a few million years until they chance to get close to the edge of the galaxy and escape. This predicted age is not consistent with the age of a few tens of millions of years inferred from the radioactive isotope ^{10}Be. Another difficulty is that straightforward explanations of how the particles escape predict that this escape would happen either much quicker or much slower than a few million years.

A way around these difficulties is the concept of the galactic halo. In this model the particles are not confined to the galactic disc, but may propagate throughout a larger region of space—possibly corresponding to the halo or roughly spherical but extremely sparse distribution of stars and other materials which typically surrounds a galaxy. This model predicts the long life and is still consistent with the 5 g/cm² pathlength, as the particles spend most of their life in very low-density regions. However, it is hard to explain how such strong magnetic fields could exist so far from the main structure of the galaxy.

A third model assumes that there is almost no escape, that is, cosmic rays disappear by breaking up into protons which then lose energy by repeated collision with other protons. To accept this picture, one must consider the apparent 5-g/cm² mean pathlength to be caused by a fortuitous combination of old distant sources and one or two close young ones. Basically, the objections to this model stem from the tendency of scientists to accept a simple theory over a more complex (in the sense of having many free parameters) or specific theory when both explain the data.

Acceleration mechanisms. Although the energies attained by cosmic-ray particles are extremely high by laboratory standards, their generation can probably be understood in terms of known astronomical objects and laws of physics. Even on Earth, ordinary thunderstorms generate potentials of billions of volts, which would accelerate particles to respectable cosmic-ray energies (a few Gev) if the atmosphere were less dense. Consequently, there are many theories of how the acceleration could take place, and it is quite possible that more than one type of source exists. Two major classes of theories may be identified—extended-acceleration regions and compact-acceleration regions.

Acceleration in extended regions (in fact the galaxy as a whole) was first proposed by E. Fermi, who showed that charged particles could gain energy from repeated deflection by magnetic fields carried by the large clouds of gas which are known to be moving randomly about the galaxy. The original compact-acceleration theory was that particles are accelerated in supernova explosions. One reason for the popularity of this theory is that the energy generated by supernovas is of the same order of magnitude as that required to maintain the cosmic-ray intensity in the leaky box model. Subsequent observations indicated that the acceleration could not take place in the explosion itself. As has been mentioned, the composition of cosmic rays is similar to that of ordinary matter, and is different from the presumed composition of the matter which is involved in a supernova explosion. At least some mixing with the interstellar medium must take place. Another problem with an explosive origin is an effect which occurs when many fast particles try to move through the interstellar gas in the same direction: the particles interact with the gas through a magnetic field which they generate themselves, dragging the gas along and rapidly losing most of their energy. In more plausible theories of supernova acceleration, the particles are accelerated gradually by energy stored up in the remnant by the explosion or provided by the intense magnetic field of the rapidly rotating neutron star or pulsar which is formed in the explosion.

Such acceleration of high-energy particles is clearly observed in the Crab Nebula, the remnant of a supernova observed by Chinese astronomers in AD 1054. This nebula is populated by high-energy electrons which radiate a measureable amount of their energy as they spiral about in the magnetic field of the nebula. So much energy is released that the electrons would lose most of their energy in a century if it were not being continuously replenished. Pulses of gamma rays also show that bursts of high-energy particles are being produced by the neutron star—the gamma rays coming out when the particles strike nuclei in the atmosphere of the neutron star. Particles of cosmic-ray energy are certainly produced in this object, but it is not known whether they escape from the trapping magnetic fields in the nebula and join the freely propagating cosmic-ray population.

The study of energetic particle acceleration in the solar system is valuable in itself, and can give insight into the processes which produce galactic cosmic rays. Large solar flares, about one a year, produce particles with energies in the Gev range, which can be detected through their secondaries even at the surface of the Earth. It is not known if such high-energy particles are produced at the flare site itself or are accelerated by bouncing off the shock fronts which propagate from the flare site outward through the solar wind. Nuclei and electrons up to 100 Mev are regularly generated in smaller flares. X-ray, gamma-ray, optical, and radio mapping of these flares is used to study the details of the acceleration process. By relating the arrival times and energies of these particles at detectors throughout the solar system to the observations of their production, the structure of the solar and interplanetary magnetic fields may be studied in detail.

In addition to the Sun, acceleration of charged particles has been observed in the vicinity of the Earth, Mercury, and Jupiter—those planets which have significant magnetic fields. Again, the details of the acceleration mechanism are not understood, but certainly involve both the rotation of the magnetic fields and their interaction with the solar wind. Jupiter is such an intense source of electrons below 30 Mev that they dominate other sources at the Earth when the two planets lie along the same interplanetary magnetic field line of force. Although the origin of the enhanced flux of ^{4}He has not been identified with certainty, it may be generated by the interaction of the solar wind with interstellar gas in the regions of the outer solar system where the wind is dying out and can no longer flow smoothly.

Direct observation of conditions throughout most of the solar system will be possible in the next few decades, and with it should come a basic understanding of the production and propagation of energetic particles locally. This understanding will perhaps form the basis of an understanding of the problem of galactic cosmic rays, which will remain for a very long time our only direct sample of material from the objects of the universe outside our solar system.

[PAUL EVENSON]

Bibliography: K. M. Apparao, *Composition of Cosmic Radiation*, 1975; S. Hayakawa, *Cosmic Ray Physics*, 1969; F. B. McDonald and C. E. Fichtel (eds.), *High Energy Particles and Quanta in Astrophysics*, 1974; J. L. Osborne and A. W. Wolfendale (eds.), *Origin of Cosmic Rays: Proceedings of the NATO Advanced Study Institute*, 1975; G. D. Rochester and A. W. Wolfendale (eds.), A discussion on the origin of the cosmic radiation, *Phil. Trans. Roy. Soc. Lond.*, ser. A, 277:317–501, 1974; K. Sakurai, *Physics of Solar Cosmic Rays*, 1974; A. W. Wolfendale (ed.), *Cosmic Rays at Ground Level*, 1977.

Crop micrometeorology

Crop micrometeorology deals with the interaction of crops and their immediate physical environment. Especially, it seeks to measure and explain net photosynthesis (photosynthesis minus respiration) and water use (transpiration plus evaporation from the soil) of crops as a function of meteorological, crop, and soil moisture conditions. These studies are complex because the intricate array of leaves, stems, and fruits modifies the local environment and because the processes of energy transfers and conversions are interrelated. As a basic science, crop micrometeorology is related to plant anatomy, plant physiology, meteorology, and hydrology. Expertise in radiation exchange theory, boundary-layer and diffusion processes, and turbulence theory is needed in basic crop micrometeorological studies. A practical goal is to provide improved plant designs and cropping patterns for light interception, for reducing infestations of diseases, pests, and weeds, and for increasing crop water-use efficiency. Shelter belts are modifications that have been used in arid or windy areas to protect crops and seedlings from a harsh environment. *See* AGRICULTURAL METEOROLOGY; MICROMETEOROLOGY.

Unifying concepts. Conservation laws for energy and matter are central to crop micrometeorology. Energy fluxes involved are solar wavelength radiation, consisting of photosynthetically active radiation (0.4–0.7 μm) and near-infrared radiation (0.7–3 μm); far-infrared radiation (3–100 μm); convection in the air; molecular heat conduction in and near the plant parts and in the soil; and the latent heat carried by water vapor. The main material substances transported to and from crop and soil surfaces are water vapor and carbon dioxide. However, fluxes of ammonia, sulfur dioxide, pesticides, and other gases or pollutants to or from crop or soil surfaces have been measured. These entities move by molecular diffusion near the leaves and soil, but by convection (usually turbulent) in the airflow. During the daytime generally, and sometimes at night, airflow among and above crops is strongly turbulent. However, often at night a stable air layer forms because of surface cooling caused by emission of far-infrared radiation back to space, and the air flow becomes nonturbulent. Fog or radiation frosts may result. The aerodynamic drag and thermal (heat-absorbing) effects of plants contribute to the pattern of air movement and influence the efficiency of turbulent transfer.

Both field studies and mathematical simulation models have dealt mostly with tall, close-growing crops, such as maize and wheat, which can be treated statistically as composed of infinite horizontal layers. Downward-moving direct-beam solar and diffuse sky radiation are partly absorbed, partly reflected, and partly transmitted by each layer. Less photosynthetically active radiation than near-infrared radiation is transmitted to ground level and reflected from the crop canopy because photosynthetically active radiation is strongly absorbed by the photosynthetic pigments (chlorophyll, carotenoids, and so on) and near-infrared radiation is only weakly absorbed. The plants act as good emitters and absorbers of far-infrared radiation. Transfers of momentum, heat, water vapor, and carbon dioxide can be considered as composed of two parts; a leaf-to-air transfer and a turbulent vertical transfer. As a bare minimum, two mean or representative temperatures are needed for each layer: an average air temperature and a representative plant surface temperature. Because some leaves are in direct sunlight and some are shaded, a re-

presentative temperature is difficult to obtain. Under clear conditions, traversing solar radiation sensors show a bimodal frequency distribution of irradiances in most crop communities; that is, most points in space and time are exposed to either high irradiances of direct-beam radiation or low irradiances characteristic of shaded conditions. Models of radiation interception have been developed which predict irradiance on both shaded leaves and on exposed leaves, depending on the leaf inclination angle with respect to the rays. The central concept of both experimental studies and simulation models is that radiant energy fluxes, sensible heat fluxes, and latent heat fluxes are coupled physically and can be expressed mathematically. This interdependence applies to a complex crop system as well as to a single leaf.

Photosynthesis. Studies of photosynthesis of crops using micrometeorological techniques do not consider the submicroscopic physics and chemistry of photosynthesis and respiration, but consider processes on a microscopic and macroscopic scale. The most important factors are the transport and diffusion of carbon dioxide in air to the leaves and through small ports called stomata to the internal air spaces. Thence it diffuses in the liquid phase of cells to chloroplasts, where carboxylation enzymes speed the first step in the conversion of carbon dioxide into organic plant materials. Solar radiation provides the photosynthetically active radiant energy to drive this biochemical conversion of carbon dioxide. Progress has been made in understanding the transport processes in the bulk atmosphere, across the leaf boundary layer, through the stomata, through the cells, and eventually to the sites of carboxylation. Transport resistances have been identified for this catenary process: bulk aerodynamic resistance, boundary-layer resistance, stomatal diffusion resistance, mesophyll resistance, and carboxylation resistance. All these resistances are plant factors which control the rate of carbon dioxide uptake by leaves of a crop; however, boundary-layer resistance and especially bulk aerodynamic resistance are determined also by the external wind flow.

Carbon dioxide concentration and photosynthetically active radiation are two other factors which control the rate of crop photosynthesis. Carbon dioxide concentration does not vary widely from about 315 microliters per liter. Experiments have revealed that it is not practical to enrich the air with carbon dioxide on a field scale because it is rapidly dispersed by turbulence. Therefore carbon dioxide concentration can be dismissed as a practical variable. Solar radiation varies widely in quantity and source distribution (direct-beam or diffuse sky or cloud sources). Many species of crop plants have leaves which can utilize solar radiation having flux densities greater than full sunlight (tropical grasses such as maize, sugar cane, and Burmuda grass, which fix carbon dioxide through the enzyme phosphoenolpyruvate carboxylase). Other species have leaves which may give maximum photosynthesis rates by individual leaves at less than full sunlight (such as soybean, sugarbeet, and wheat, which fix carbon dioxide through the enzyme ribulose 1,5-diphosphate carboxylase).

However, in general, most crops show increasing photosynthesis rates with increasing irradiances for two reasons. First, more solar energy would become available to the shaded and partly shaded leaves deep in the crop canopy. Second, many of the well-exposed leaves at the top of a crop canopy are exposed to solar rays at wide angles of inclination so that they do not receive the full solar flux density. These leaves will respond to increasing irradiance also. Furthermore, increased diffuse to direct-beam ratios of irradiance (which could be caused by haze or thin clouds) may increase the irradiance on shaded leaves and hence increase overall crop photosynthesis.

If crop plants lack available soil water, the stomata may close and restrict the rate of carbon dioxide uptake by crops. Stomatal closure will protect plants against excessive dehydration, but will at the same time decrease photosynthesis by restricting the diffusion of carbon dioxide into the leaves.

Transpiration and heat exchange. Transpiration involves the transport of water vapor from inside leaves to the bulk atmosphere. The path of flow of water vapor is from the surfaces of cells inside the leaf through the stomata, through the leaf aerodynamic boundary layer, and from the boundary layer to the bulk atmosphere. Sensible heat is exchanged by convection directly from plant surfaces; therefore there is no stomatal diffusion resistance associated with this exchange. Stomatal diffusion resistance does affect heat exchange from leaves, however, because when stomata are open wide (low resistance) much of the heat exchanged from leaves is in the form of latent heat of evaporation of water involved in transpiration.

Small leaves, such as needles, convect heat much more rapidly than large leaves, such as banana leaves. Engineering boundary-layer theory suggests that boundary-layer resistance should be proportional to the square root of a characteristic dimension of a leaf and inversely proportional to the square root of the airflow rate past a leaf. Experiments support these relationships.

Under high-irradiance conditions, low air humidity, high air temperature, and low stomatal diffusion resistance will favor high transpiration, whereas high air humidity, low air temperature, and high stomatal diffusion resistance will favor sensible heat exchange from leaves. The function of wind is chiefly to enhance the transport rather than determine which form of convected energy will be most prominent. In arid environments, the latent energy of transpiration from crops may exceed the net radiant energy available, because heat from the dry air may actually be conducted to crops which will cause transpiration to increase. In those areas, crop temperature is lower than air temperature.

Flux methods. At least three general methods have been employed to measure flux density of carbon dioxide, water vapor, and heat to and from crop surfaces on a field scale. These methods are restricted to use in the crop boundary layer immediately above the crop surface, and they require a sufficient upwind fetch of a uniform crop surface free of obstructions. Flux densities obtained by these methods will reflect the more detailed interactions of crop and environment, but will not explain them.

The principle of the energy balance methods is to partition the net incoming radiant energy into energy associated with latent heat of transpiration and evaporation, sensible heat, photochemical energy involved in photosynthesis, heat flux into the soil, and heat stored in the crop. Measurement of net input of radiation to drive these processes is obtained from net radiometers, which measure the total incoming minus the total outgoing radiation. The most important components—latent heat, sensible heat, and photochemical energy—are determined by average vertical gradients of water vapor concentration (or vapor pressure), air temperature, and carbon dioxide concentration.

The principle of the bulk aerodynamic methods is to relate the vertical concentration gradients of those transported entities to the vertical gradient of horizontal wind speed. The transports are assumed to be related to the aerodynamic drag (or transport of momentum) of the crop surface. Corrections are required for thermal instability or stability of the air near the crop surface.

The eddy correlation methods are direct methods which correlate the instantaneous vertical components of wind (updrafts or downdrafts) to the instantaneous values of carbon dioxide concentration, water vapor concentration, or air temperature. Under daytime conditions, turbulent eddies, or whorls, transport air from the crop in updrafts, which are slightly depleted in carbon dioxide, and conversely, turbulence transports air to the crop in downdrafts which are representative of the atmospheric content of these entities. More basic and applied research is being done on eddy correlation methods because they measure transports through direct transport processes.

Plant parameters. The stomata are the most important single factor in interactions of plant and environment because they are the gateways for gaseous exchange. Soil-to-air transfers are also very important while crops are in the seedling stage until a large degree of ground cover is attained. Coefficients of absorption, transmission, and reflection by leaves of photosynthetically active, near-infrared, and long-wavelength infrared radiation are not very different among crop species, but the geometric arrangement and stage of growth of plants in a crop may affect radiation exchange greatly. The crop geometry also interacts with radiation-source geometry (diffuse to direct-beam irradiance, solar elevation angle). Crop micrometeorology attempts to show how the plant parameters interact with the environmental factors in crop production and water requirements of crops under field conditions. *See* EVAPOTRANSPIRATION.   [L. H. ALLEN, JR.]

Bibliography: J. Goudriaan, *Crop Micrometeorology*, 1977; E. Lemon, D. W. Stewart, and R. W. Shawcroft, The sun's work in a cornfield, *Science*, 174:371–378, 1971; J. L. Monteith (ed.), *Vegetation and the Atmosphere*, vol. 1, 1975, vol. 2, 1976; R. E. Munn, *Biometeorological Methods*, 1971; N. J. Rosenberg, *Microclimate. The Biological Environment*, 1974; W. D. Sellars, *Physical*

Climatology, 1965; L. P. Smith (ed.), *The Application of Micrometeorology to Agricultural Problems*, 1972; O. G. Sutton, *Micrometeorology*, 1977.

Cyclone

A low-pressure region of the Earth's atmosphere with roundish to elongated oval ground plan, inmoving ground air currents, centrally upward (commonly spiraling) air movement, and generally outward movement at various higher elevations in the lower layer (troposphere) of the terrestrial atmosphere. This type of air circulation is called cyclonic and develops in many variations, but it is distinctly different and approximately the reverse of anticyclonic circulation, which is developed in similarly shaped air portions of high atmospheric pressure.

The unqualified term cyclone is used by meteorologists to designate atmospheric disturbances of considerable size in tropical, middle, or high latitudes. *See* AIR PRESSURE; FRONT; PRECIPITATION; STORM; WIND.

Certain small and usually violent cyclonic disturbances of the tropical and middle latitudes are designated by special names. *See* HURRICANE; TORNADO; WATERSPOUT.

[CHARLES V. CRITTENDEN]

Deep-sea fauna

The deep sea may be regarded as that part of the ocean below the upper limit of the continental slopes (Fig. 1). Its waters fill the deep ocean basins, cover about two-thirds of the Earth's surface, have an average depth of about 4000 m, and provide living space for communities of animals that are quite different from those inhabiting the land-fringing waters which overlie the continental shelves (neritic zone).

The systematic exploration of the deep sea began with the voyage of the HMS *Challenger* (1872–1876). Since that time there have been numerous large-scale, deep-sea expeditions. In 1948 the Swedish Deep Sea Expedition in the Atlantic developed new techniques for trawling and the winch used by later expeditions.

The deep-sea fauna consists of pelagic animals

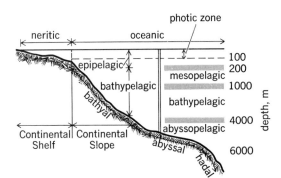

Fig. 1. Classification of marine environments. Right side of diagram illustrates the proposal to divide the bathypelagic zone into mesopelagic, bathypelagic, and abyssopelagic zones. Division of benthic region into bathyal, abyssal, and hadal zones also is shown.

(swimming and floating forms between the surface and deep-sea floor) and below these the benthos, or bottom dwellers, which live on or near the ocean bottom. Pelagic animals can be divided into the usually smaller forms that tend to drift with the currents (zooplankton) and the larger and more active nekton, such as squids, fishes, and cetaceans. Pelagic, deep-sea animals are frequently termed bathypelagic in contrast to the epipelagic organisms of the surface waters (Fig. 1).

Bathypelagic fauna. All animal life in the sea, pelagic and benthic, depends on the growth of microscopic plants (phytoplankton). From the surface down to a maximum depth of about 100 m there is sufficient light for photosynthesis and vigorous phytoplanktonic growth. This layer is known as the photic zone. In the deep sea, plants can exist only as saprophytes. The productivity of the plants, however, is reflected down to the deepest parts of the sea through complex food chains. These consist of zooplankton that graze on phytoplankton, carnivorous species that feed on zooplankton, and large predators that eat the other animals. The typical bathypelagic animals (Fig. 2) begin to appear below about 200 m.

Zooplankton. The planktonic or drifting forms of animal life in the ocean include the Protozoa, larval stages of deep-sea fishes, and even larger organisms with limited powers of movement.

1. Protozoa. Included in this group are various families of Foraminiferida and Radiolaria, such as Challengeriidae and Tuscaroriidae, the skeletons of which form an important part of the deep-sea sediments.

2. Coelenterata. Scyphomedusae such as *Atolla* and *Periphylla* are not uncommon. Other jellyfishes include various Trachymedusae (*Crossota* and *Colobonema*) and Narcomedusae. Siphonophorida, particularly the diphyids, are found down to depths of at least 3000 m but are more common in the upper few hundred meters. One family of Ctenophora (comb jellyfish), the Bathyctenidae, is entirely bathypelagic.

3. Nemertea. This group of worms has bathypelagic species belonging to some 10 families.

4. Crustacea. In numbers of species and individuals, the small Copepoda (0.3–17 mm in length) are the dominant group of crustaceans in the ocean. There are numerous bathypelagic species. Certain of the Ostracoda (*Gigantocypris*) are purely bathypelagic, as are some of the Amphipoda (the gammarid genera *Cyphocaris* and *Hyperiopsis* and most species of the hyperiid families Scinidae and Lanceolidae).

The larger and more active pelagic crustaceans (Euphausiacea, various Mysidacea, and prawns) are usually classed as plankton but might well be called "micronekton," a group intermediate between thrusting nekton and feebler-swimming plankton. Of the euphausiid shrimps, *Benthenphausia* and various species of *Thysanopoda*, *Nematoscelis*, and *Stylocheiron* have centers of abundance in the bathypelagic zone. Deep-water genera of mysids include *Gnathophausia*, *Lophogaster*, and *Eucopia*, while the prawn families Hoplophoridae and Sergestidae have numerous bathypelagic representatives.

5. Chaetognatha. Certain species of arrow-worms, such as *Eukrohnia fowleri* and *Sagitta macrocephals*, and predominantly bathypelagic.

6. Echinodermata: Holothuroidea. The genera *Pelagothauria*, *Enypiastes*, and *Galatheathuria* are bathypelagic, the first two being medusalike.

7. Protochordata: Thaliacea. While they are more abundant in the surface layers, the salps, doliolids, and pyrosomes have been fished down to 3000 m.

Deep-sca nekton. This group consists largely of squids, octopods, and fishes. The sperm whale also enters the deep sea, where it finds some of the squid upon which it feeds.

1. Mollusca: Cephalopoda. Several families of squids (Histioteuthidae, Spirulidae, Lycoteuthidae, Chirotheuthidae, and Enoploteuthidae) form an important part of the deep-sea nekton together with a few octopods, such as *Cirrothauma*, *Amphitretus*, *Vitreledonella*, and *Vampyroteuthis* (Vampyromorpha).

2. Fishes. Apart from a few squaloid sharks, the bathypelagic fish fauna consists of teleosts. The most diverse groups are the stomiatoids (Elupeiformes), with about 300 species; Myctophidae (lantern fishes; Salmoniformes), about 250 species; and the ceratioid anglerfishes (Lophiiformes), about 90 species. The few species forming the orders Anguilliformes (gulper eels) and Cetomimiformes (whale fishes) are entirely bathypelagic, as are certain of the eels (Cyemidae, Nemichthyidae) and Beryciformes (for example, Melamphaidae).

Distribution. The bathypelagic fauna is most diverse in the tropical and temperate parts of the ocean. Numerous species are found in all three temperature zones, but many appear to have a more limited distribution.

Each species also has a definite vertical occurrence. Present findings suggest that there are three main vertical zones, each with a characteristic community. Here the term bathypelagic is used for the fauna between about 1000 and 2000 m, that above (between 200 and 1000 m) being called mesopelagic and that below 2000 m, abyssopelagic (Fig. 1). The typical forms of the mesopelagic fauna (stomiatoids and lantern fishes) live in the twilight zone of the deep sea (between the 20 and 10°C isotherms), while the bathypelagic species (ceratioid angler fishes and *Vampyroteuthis*) are found in the dark, cooler parts below the 10°C isotherm.

Lastly, numerous species of mesopelagic animals, such as euphausiids, prawns, squids, and fishes (particularly lantern fishes), undertake extensive diurnal, vertical migrations, moving upward into the productive surface layers to feed at night. Toward sunrise they begin to descend to their daytime levels. *See* SCATTERING LAYER.

Bioluminescence. Perhaps the most conspicuous feature of pelagic deep-sea life is the widespread occurrence of luminescent species bearing definite light organs (photophores). Many of the squids and fishes have definite patterns of such lights, as do some of the larger crustaceans (hoplophorid and sergestid prawns and euphausiids). Investigations in 1958 with a bathyphotometer showed that

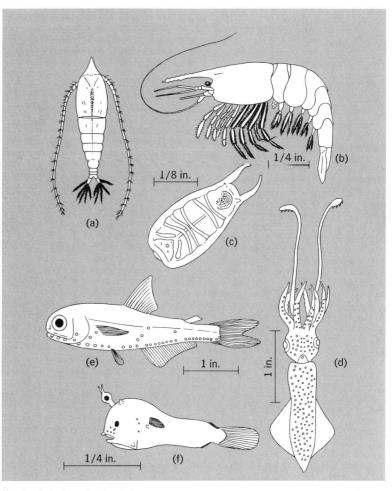

Fig. 2. Pelagic animals of the deep sea. (a) Copepod (*Haloptilus acutifrons*), length 1–8 mm. (b) Prawn (*Acanthephyra multispina*). (c) Salp (*Salpa*, or *Thalia*, *democratica*). (d) Squid (*Abraliopsis morisii*). (e) Lantern fish (*Myctophym punctatum*). (f) Anglerfish (*Lophodolus acanthognathus*). Scale varies as shown.

flashes from luminescent organisms could be detected down to depths of 3750 m.

Benthic fauna. There are two main ecological groups of bottom-living animals (Fig. 3) in the ocean: organisms that attach to the bottom and those that freely move over the bottom.

Attached benthic organisms. This group consists of species that attach themselves to the sediments, rocks, or to other organisms. The more typical forms are included in the following list.

1. Porifera. Hexactinellida (glass sponges), about 375 species.

2. Coelenterata. Certain hydroids, gorgonians, pennatulids (sea fans), antipatharians (black corals), actiniarians (sea anemones), and madrepore corals (*Lophohelia* and *Amphihelia*).

3. Crustacea. Cirripedia (barnacles), such as *Scalpellum* and *Verruca* sp.

4. Echinodermata: Crinoidea (sea lilies). Numerous species of stalked crinoids live in the deep sea together with a number of unstalked forms.

5. Protochordata. Pogonophora (beard-bearers) and certain ascidians (*Culeolous* sp.).

Benthic crawlers and swimmers. This group comprises the freely moving animals, those that

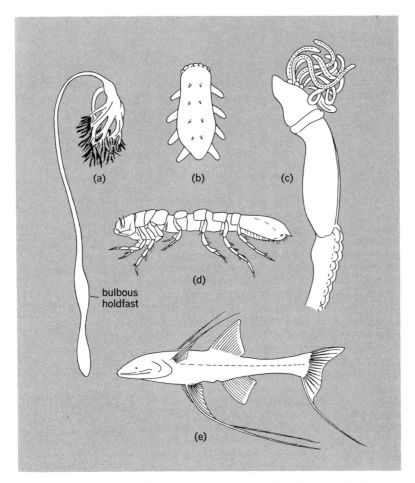

Fig. 3. Benthic animals of the deep sea. (a) Sea-pen (*Umbellula*) showing bulbous holdfast. (b) Sea cucumber (*Elpidia glacialis*), an elasipod holothurian. (c) Head and forepart of the trunk of a pogonophoran (*Birsteinia witjasi*). (d) Isopod crustacean (*Macrostylus hadalis*). (e) Bathypteroid fish (*Benthosaurus*).

swim or crawl over the bottom or burrow into the sediments, the upper layer of which has a rich bacterial flora.

1. Annelida. A few species of Polychaeta (bristle worms).

2. Gephyrea. Certain species of echiuroid and sipunculoid worms.

3. Crustacea. In numbers of species and individuals the most important group of benthic deep-sea crustaceans is the Peracarida, represented by various species of cumaceans (*Bathycuma*, *Macrocylindrus*), isopods (*Ischnomesus* and *Eurycope*), amphipods, and tanaids (*Apseudes*, *Neotanais*). Of the Eucarida, the most prominent groups are the penaeid prawns and the Eryonidae. There are also a number of crabs (*Platymaia*, *Geryon*, *Ethusa*, and *Scyramathia*) and hermit crabs (numerous species of Axiidae).

4. Pycnogonida (sea spiders). Numerous species of the families Colossendeidae and Nymphonidae.

5. Mollusca. Certain of the Octopodidae (octopuses) and the cirromorph octopods live on the deep-sea floor, as do various gastropods, scaphopods, and lamellibranchs. A small, limpetlike mollusk (named *Neoplina galatheae*) was dredged

for the first time on the *Galathea* expedition in 1951.

6. Echinodermata. These form an important part of the benthic fauna, particularly the sea cucumbers (Holothuroidea) of the orders Elasipodida and Molpadida. Among the sea urchins (Echinoidea), the order Cidaroida and the suborder Meridosternata mainly consist of deep-sea species. There are also various brittle stars (Ophiuroidea) and star-fishes (Asteroidea).

7. Fishes. The species of one group of cartilaginous fishes (Holocephali) live over the continental slope. The main groups of benthic, deep-sea teleosts are the Bathypteroidae (Salmoniformes); Halosauridae and Notacanthidae (Notacanthiformes); Macrouridae (rattails) and Morinae (deep-sea cods) (Gadiformes); and Brotulidae, Liparidae, and Zoarcidae (Perciformes).

Distribution. The benthic fauna is most diverse in the temperate and tropical ocean, although the arctic and antarctic areas have their characteristic species. As in the pelagic fauna, certain species occur in all three oceanic zones, while others appear to have a more restricted occurrence.

While a number of species—particularly among the polychaete worms, gastropod mollusks, and the brittle stars (Ophiuroidea)—range from littoral to abyssal regions, most forms tend to live within smaller ranges of depth. Present data suggest that there are typical communities of animals over the continental slopes (Fig. 1) extending down to about 3000 m (bathyal zone); others occur below this in the abyssal zone. Danish and Russian exploration also suggests that the deep-sea trenches (with depths over 7000 m) form another ecological zone (hadal zone) having certain characteristic species—those capable of living under pressures of 700–1000 atm (barophilic species). This work also showed that life could exist at the very bottom of the ocean (down to depths of more than 10,000 m) and that species of certain groups, such as sea anemones, echiuroid and polychaete worms, bivalves, isopod and amphipod crustaceans, sea cucumbers, and Pogonophora, occurred at depths beyond 9000 m.

Lastly, there is a decrease in the numbers of species and individuals with depth. Russian biologists found that at depths of 8000–10,000 m the weight of animals per square meter of sea floor was about one-fifth to one-fifteenth the weight at depths of 1000–4000 m. As the deep-sea benthic fauna is dependent on organic matter originating in the upper, plant-bearing waters and as the amount reaching the bottom must decrease with depth, the above findings are comprehensible. It is also interesting that there are very few carnivorous animals, such as crabs, brittle stars, and starfishes, below a depth of 7000 m. It is the particle-catchers, such as the Pogonophora, and ooze-eaters, such as sea cucumbers and echiuroid worms, that make up most of the hadal fauna. *See* MARINE ECOSYSTEM; SEA-WATER FERTILITY.

[NORMAN B. MARSHALL]

Bibliography: J. Barnard et al., *Abyssal Crustacea*, 1962; A. F. Bruun, Animals of the abyss, *Sci. Amer.*, 197(5):50–57, 1957; A. F. Bruun, Deep sea and abyssal depths, *Geol. Soc. Amer. Mem.*, 67(1):

641–672, 1957; *The Galathea Deep Sea Expedition, 1950–1952*, 1956; C. P. Idyll, *Abyss: The Deep Sea and the Creatures That Live in It*, 2d ed., 1971; A. G. Macdonald, *Physiological Aspects of Deep Sea Biology*, 1975; N. B. Marshall, *Aspects of Deep Sea Biology*, 1977.

Degree-day

A unit used in estimating energy requirements for building heating and, to a lesser extent, for building cooling. It is applied to all fuels, district heating, and electric heating. Origin of the degree-day was based on studies of residential gas heating systems. These studies indicated that there existed a straight-line relation between gas used and the extent to which the daily mean outside temperature fell below 65°F (18.3°C).

The number of degree-days to be recorded on any given day is obtained by averaging the daily maximum and minimum outside temperatures to obtain the daily mean temperature. This procedure was found to be adequate and less time-consuming than the averaging of 24 hourly temperature readings. The daily mean so obtained is subtracted from 65°F and tabulated. Monthly and seasonal totals of degree-days obtained in this way are available from local weather bureaus. They use the 30-year period 1931–1960 as a basis.

The base temperature of 65°F was established by observation of the outside temperatures at which the majority of homeowners started their gas heating systems in the early fall. Interpreted in another way, the internal heat gains in residences, which appear to be about 5000 Btu/hr, are sufficient to provide heat down to a balance temperature of 65°F.

Provided that the efficiency of the heating equipment remains constant, the fuel usage of a given house will be proportional to the number of degree-days. For purposes of comparing one heating season with another, constant efficiency is a reasonable assumption. However, efficiency of the heating equipment decreases markedly in the warmer portions of the heating season. Therefore, the fuel use per degree-day will be higher in spring and fall than in winter. The same considerations apply to use of degree-day data in warmer climates compared to cooler regions.

A frequent use of degree-days for a specific building is to determine before fuel storage tanks run dry when fuel oil deliveries should be made.

Number of Btu which the heating plant must furnish to a building in a given period of time is

$$\text{Btu required} = \text{Heat rate of building} \times 24 \\ \times \text{degree-days}$$

where "Btu required" is the heat supplied by the heating system to maintain the desired inside temperature. "Heat rate of building" is the hourly building heat loss divided by the difference between inside and outside design temperatures.

To determine fuel requirement from this relationship, two decisions are needed. First, the correct seasonal efficiency of the heating system must by obtained or estimated; second, appropriate base temperature for the degree-day data must be calculated. Seasonal efficiency is generally less than the test efficiency of the heating unit although the load efficiency curves are surprisingly level down to quite low loads. Base temperature for degree-day values may be chosen from consideration of internal heat gains of the building. (As already discussed, these steps need not be taken for residences, which tend to be supplied with relatively small yet similar internal gains.) When the estimating procedure is applied to buildings with high levels of internal heat gains, as in a well-lighted office building, then degree-day data on other than a 65°F basis are required. Balance temperatures for both day operation and night operation may also need to be taken into consideration. Then it would be proper to replace the term degree-days in the equation with the new term adjusted degree-days. *See* CLIMATOLOGY; PSYCHROMETRICS.

[C. GEORGE SEGELER]

Delta

A deposit of sediment at the mouth of a river or tidal inlet. The name, from the Greek letter Δ, was first used by Herodotus (5th century B.C.) for the triangular delta of the Nile. It is also used for storm washovers of barrier islands and for sediment accumulations at the mouths of submarine canyons. *See* NEARSHORE PROCESSES.

Structure and growth. The shape and internal structure of a delta depend on the nature and interaction of two forces: the sediment-carrying stream from a river, tidal inlet, or submarine canyon, and the current and wave action of the water body in which the delta is building. This interaction ranges from complete dominance of the sediment-carrying stream (still-water deltas) to complete dominance of currents and waves, resulting in redistribution of the sediment over a wide area (no deltas). This interaction has a large effect on the shape and structure of the delta body.

Most of the sediment carried into the basin is deposited when the inflowing stream decelerates. If there is little density contrast, this deceleration is sudden and most sediment is deposited near the mouth of the river. If the inflowing water is much lighter than the basin water, for example, fresh water flowing into a colder sea, the outflow spreads at the surface over a large distance away from the outlet. If the inflow is very dense, for instance, cold

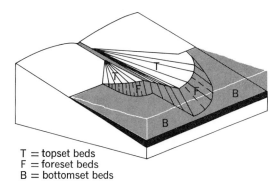

T = topset beds
F = foreset beds
B = bottomset beds

Fig. 1. Schematic diagram showing two stages of growth of a typical Gilbert-type delta. (*Adapted from P. H. Kuenen, Marine Geology, Wiley, 1950*)

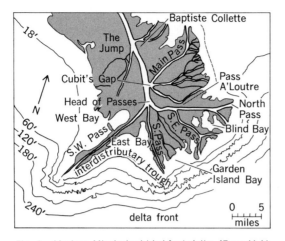

Fig. 2. Modern Mississippi bird-foot delta. *(From H. N. Fisk et al., Sedimentary framework of the modern Mississippi delta, J. Sediment. Petrol., 24:76–99, 1954)*

muddy water in a warm lake, it may form a density flow on or near the bottom, and the principal deposition may occur at great distance from the outlet. A good example is Lake Mead, where 134 ft of sediment have been deposited against Hoover Dam at 75 mi from the inlet of the Colorado River. However, not all long-distance transport of river sediment in the sea can be attributed to deep density flow; almost half of the total load of the Amazon River is deposited more than 1000 mi away near the Orinoco delta and in the southeastern Caribbean because of longshore transport by currents and waves.

Three principal components make up the bodies of most deltas in varying proportions: topset, foreset, and bottomset beds. They were defined originally by G. K. Gilbert for lake and shallow marine deltas; there, sizable differences in the depositional slopes of the three units, which form part of the definition, are readily seen (Fig. 1). The concept has been redefined to apply to most deltas, and as now understood, the topset beds comprise the sediments formed on the subaerial delta: channel deposits, natural levees, floodplains, marshes, and swamp and bay sediments. The foreset beds are those formed in shallow water, mostly as a broad platform fronting the delta shore, and the bottomset beds are the deep-water deposits beyond the deltaic bulge. In marine deltas the fluviatile influence decreases and the marine influence increases from the topset to the bottomset beds. Concurrently, the mean grain size and lithological variability also decrease. The topset beds are variable in grain size and composition over short distances, ranging from coarse channel deposits through layered levee silts and clays to the fine clays of the floodplains and marshes. The foreset beds consist of a uniform blanket of laminated silts and silty clays reflecting the seasonal influence of currents and waves on the shallow platform, while the bottomset beds are homogeneous silty clays, often characterized by abundant plant fragments. The boundary between bottomset beds and neritic shelf clays is completely arbitrary. Similarly, the faunas incorporated in the sediments range from great local variability in the topset beds, with domi-

nance of barren sediments and local pockets of rich brackish faunas, to the almost completely marine fossils uniformly distributed in the bottomset beds.

Through progressive outbuilding, the delta can become overextended with long river courses and very low gradients. Eventually, shorter and steeper paths to the sea will be developed and the existing subdelta will be abandoned in favor of a shorter course. The Mississippi delta shows good examples of such subdelta migration, and the present active delta would be abandoned for a new one off the Atchafalaya River, if artificial control did not keep the flow in check. The abandoned delta body gradually submerges and is eroded by waves and wind. The submergence results from two factors: temporary continuation of subsidence caused by loading of the deltaic sediments on the substrate, and compaction of the fine sediments. These are overcompensated during delta growth, but cause it to sink below sea level after abandonment. Winnowing of the sediments by waves produces a lag deposit of sand at the advancing edge of the sea, which ultimately results in a thin and discontinuous blanket of sand, often bordered by a string of sand bars and small islands near the previous delta shore. This sequence is called the destructive phase of delta formation; a good example is the Breton Sound-Chandeleur Island area of the Mississippi delta, which marks the site of the old St. Bernard subdelta.

In a different way, deltas can be viewed as being composed of three structural elements: (1) a framework of elongate coarse bodies (channels, river-mouth bars, levee deposits), which radiate from the apex to the distributary mouths (sand fingers); (2) a matrix of fine-grained floodplain, marsh, and bay sediments; and (3) a littoral zone, usually of beach and dune sands which result from sorting and longshore transport of river-mouth deposits by waves, currents, tides, and wind. The relative proportions of these components vary widely. The Mississippi delta consists almost en-

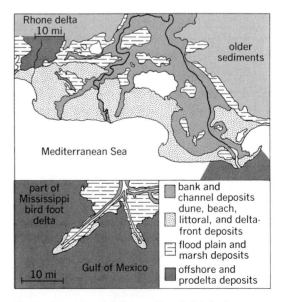

Fig. 3. Examples of deltaic sediment distribution.

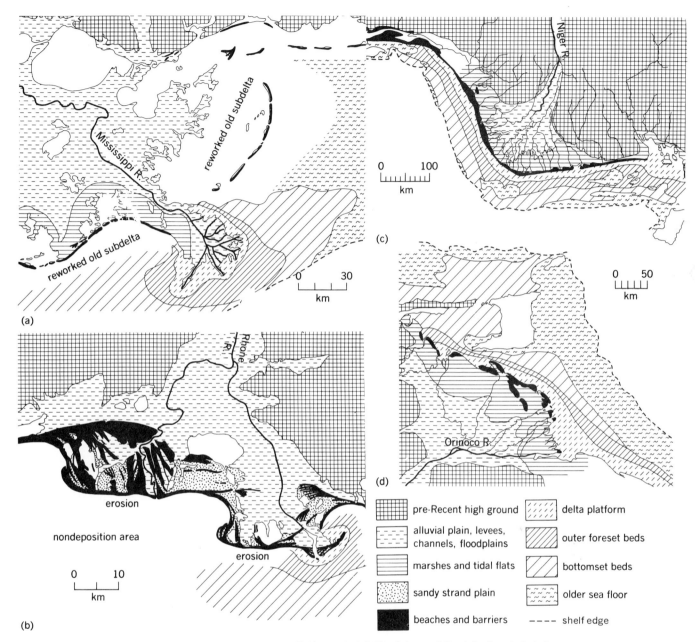

Fig. 4. Sketch maps of the sediments of the (a) Mississippi, (b) Rhone, (c) Niger, and (d) Orinoco deltas. (After T. H. van Andel, The Orinoco delta, J. Sediment. Petrol., 37(2):297–310, 1967)

Legend:
pre-Recent high ground — delta platform — alluvial plain, levees, channels, floodplains — outer foreset beds — marshes and tidal flats — bottomset beds — sandy strand plain — older sea floor — beaches and barriers — shelf edge

tirely of framework and matrix (Fig. 2; Fig. 3, bottom); its rapid seaward growth is the result of deposition of river-mouth bars and extension of levees, and the areas in between are filled later with matrix. This gives the delta its characteristic bird-foot outline. A different makeup is presented by the Rhone delta (Fig. 3, top), where the supply of coarse material at the distributary mouths is slow, and dispersal by wave action and longshore drift fairly efficient, so that nearly all material is evenly redistributed as a series of coastal bars and dunes across a large part of the delta front. This delta advances as a broad lobate front, while the present Mississippi delta grows at several localized and sharply defined points.

Pattern of deposition. Many different types of deltas exist, few of which have been examined in detail. It appears that most of them can be explained by the interplay, in various proportions, of two major controlling forces: the coarseness and the rate of sediment supply by the river, and the rate of reworking and dispersal along the delta front by marine currents and waves. Other factors, such as the shape of the preexisting coastline and of the sea floor, seasonal fluctuations of marine and river forces, the tidal range, and so on, cannot as yet be evaluated, but appear to be secondary. The Mississippi delta (Fig. 4a) is dominantly fine-grained and consists in large part of thick sand fingers with intervening matrix clays. It builds rap-

Statistics pertaining to modern deltas*

River	Dimension of subaerial delta, statute mi		Amount of sediment discharged		Annual extension of subaerial delta	
	Length	Breadth	River water by weight (avg), ppm	Annual volume of sediment, mi³	Measurement period, years	Approximate distance, ft
Mississippi's present bird-foot delta	12	30	550	0.068	1838 – 1947	250
Hwang-Ho	300	470†	50,000‡		1870 – 1937	950
Ganges-Brahmaputra	220	200	870	0.043 (Ganges only)		
Rhone into Mediterranean Sea	30	47	400 – 590	0.005	1737 – 1870	190
Danube	46	46	310	0.008		40
Nile (prior to barrages)	96	145	1600	0.001	1100 – 1870	45
Colorado above Hoover Dam	43	0.05 – 0.6	8300	0.032	1936 – 1948	3.6 mi (gorge)
Euphrates-Tigris	350	90			1793 – 1853	180

*1 statute mi = 1.609 km; 1 mi³ = 4.168 km³; 1 ft = 0.3048 m.   †Includes 100 mi of nondeltaic Shantung Peninsula.
‡Maximum is 400,000 ppm.

idly outward, is almost entirely controlled by river action, and has a typical bird-foot shape.

In the Rhone delta (Fig. 4b) the rate of sediment supply is much lower, but the sediment is coarser, and the rate of reworking and dispersal by waves and currents is higher. Consequently, this delta is more sandy, and its principal growth takes place through the accretion over a broad front of bars and beaches, resulting from winnowing of sand at the river mouth and dispersal by wind and waves. The advance of the delta is slow, but occurs over a broadly lobate front, and the bulk of the delta body consists of the deposits of the sandy strand plain. Matrix and framework deposits are restricted to the topset beds in the upper delta. Littoral marine forces control to a large extent the shape of this delta.

A broad coastal plain is also found in the Orinoco delta (Fig. 4d). The large supply of very fine sediment of this river is augmented by Amazon mud brought along the shore from the east. It builds rapidly in a sheltered sea with little wave action, but longshore drift to the northwest is marked. Consequently, the coastal plain, although also littoral marine, is not wave-built and sandy, but formed by tidal depositon on mud flats. The delta, as a result of the longshore currents, advances over its entire perimeter, and the action of the river plays an insignificant part in its development.

Delta advance over the entire perimeter also takes place in the Niger delta (Fig. 4c), which resembles the Orinoco delta in shape and distribution of sediments. However, enough sediment is supplied by this river to form, under the influence of moderate wave action on an exposed shelf, a marginal zone of sand bars, beaches, and barrier islands. Behind this barrier, tidal sedimentation of a muddy coastal plain takes place, and framework-matrix units are restricted to the upper delta. The delta advances over a broad front as a result of redistribution of sediments by currents. In a sense it is intermediate between the Orinoco and Rhone deltas.

Characteristics of modern deltas. While over 150 major deltas are formed today, not all rivers, or

even all major ones, have deltas. This is the result of a rise in sea level following the last glacial period, which produced deep estuaries in many parts of the world which have not yet been filled (for example, the Amazon estuary). Delta thicknesses vary widely: The Nile is depositing a layer 50 ft thick in a shallow embayment, whereas the Mississippi, building out in deeper water, has constructed a delta body more than 850 ft thick. This thick wedge exerts pressure on the underlying beds, causing downbending of the shelf and producing mudlumps in areas where the underlying sediment is soft (see table).

Engineering problems. Despite difficult engineering problems, many cities, such as Calcutta, Shanghai, Venice, Alexandria (Egypt), and New Orleans, were constructed on deltas. These problems include shifting and extending shipping channels; lack of firm footing for construction except on levees; steady subsidence, which may reach a rate of 5 ft per century; poor drainage; and extensive flood danger. Submergence to a depth of 15 ft during hurricanes or typhoons is not an uncommon occurrence. Moreover, in certain deltas the tendency of the main flow to shift away to entirely different areas, with resulting disappearance of the main channels used for water traffic, is a constant problem that is difficult and costly to counter. See ESTUARINE OCEANOGRAPHY; MARINE SEDIMENTS.

[TJEERD H. VAN ANDEL]

Bibliography: C. C. Bates, Rational theory of delta formation, *Bull. Amer. Ass. Petrol. Geol.,* 37(9):2119 – 2162, 1953; J. M. Coleman and L. D. Wright, *Analysis of Major River Systems and Their Deltas,* 1971; H. N. Fisk et al., Sedimentary framework of the modern Mississippi delta, *J. Sediment. Petrol.,* 24:76 – 99, 1954; R. Russel, *River Plains and Sea Coasts,* 1967; P. C. Scruton, Delta building and the deltaic sequence, in F. P. Shepard (ed.), *Recent Sediments, Northwestern Gulf of Mexico,* Amer. Ass. Petrol. Geol. Spec. Publ. 82 – 102, 1960; T. H. van Andel, The Orinoco delta, *J. Sediment. Petrol.,* 37(2):297 – 310, 1967.

Dew

Drops or films of water formed by condensation of water on outside exposed surfaces, especially of vegetation. Hoarfrost is the corresponding phenomenon at temperatures below freezing. In most cases not enough water is collected to be recorded as precipitation. But at certain locations east of the Mediterranean Sea the annual total has been estimated at around 200 mm (8 in.), enough to permit growth of summer crops that would otherwise require irrigation.

Dew forms on clear nights when there is cooling by radiation, provided exposed surfaces cool below the dew point of the air. Moisture then condenses on the exposed surfaces. If the ground is wet, moisture is evaporated, which adds to the supply and increases the deposit of dew on vegetation. Fog also deposits moisture which, unlike dew, collects inside trees and similar places as readily as on surfaces exposed to the sky. Deposits similar to those produced by fog may form when cold air is replaced suddenly by warmer and moister air. *See* DEW POINT; HUMIDITY; PRECIPITATION; TERRESTRIAL RADIATION.

[J. R. FULKS]

Bibliography: H. J. Critchfield, *General Climatology*, 3d ed., 1974; R. Geiger, *The Climate near the Ground*, 4th ed., 1965.

Dew point

The temperature at which air becomes saturated when cooled without addition of moisture or change of pressure. Any further cooling causes condensation; fog and dew are formed in this way.

Frost point is the corresponding temperature of saturation with respect to ice. At temperatures below freezing, both frost point and dew point may be defined because water is often liquid (especially in clouds) at temperatures well below freezing; at freezing (more exactly, at the triple point, +.01°C) they are the same, but below freezing the frost point is higher. For example, if the dew point is −9°C, the frost point is −8°C. Both dew point and frost point are single-valued functions of vapor pressure.

Determination of dew point (or frost point) can be made directly by cooling a flat polished metal surface until it becomes clouded with a film of water or ice; the dew point is the temperature at which the film appears. In practice, the dew point is usually computed from simultaneous readings of wet- and dry-bulb thermometers. *See* DEW; FOG; HUMIDITY.

[J. R. FULKS]

Bibliography: R. J. List (ed.), *Smithsonian Meteorological Tables*, 6th ed. rev., 1951; M. Thaller, Instrument Development Inquiry, 2d ed., 1977.

Diving

Skin diving, scuba diving, saturation diving, and "hard hat" diving are techniques used by scientists to investigate the underwater environment. Skin diving is usually without breathing apparatus and is done with fins and faceplate. The diver's underwater observation is limited to the time he can hold his breath (1−2 min). Diving with scuba (self-contained underwater breathing apparatus) and "hard hat" provides the diver with a breathable gas, thus expanding the submerged time and the depth range of underwater observations. This type of diving is limited by the physiology of man and his reaction to the pressure and nature of the breathing gas. Saturation diving permits almost unlimited time down to depths of 30 m. *See* DIVING PHYSIOLOGY.

Man has used compressed-air diving as a tool in marine research, salvage, and construction for over 2½ centuries. However, because of a lack of knowledge of diving physiology, the early diving history is a story of danger and full of examples of crippling accidents and deaths. One of the earlier records of scientific diving and its problems is recorded on a plaque at the entrance to Old Port Royal, Jamaica, West Indies: "Lucas Barrett, F.G.S; L.L.S−1837−1862−Died while conducting underwater research off Port Royal on December 19, 1862−25 years old−Geological Society of Jamaica."

Fortunately, efficient training programs, a much better understanding of diving physiology, and reliable equipment have removed many of the early

Fig. 1. A diver wearing a mixed-gas scuba, exposure suit, and face mask. Depth 300 ft (90 m) on the upper lip of Scripps Submarine Canyon. (*U.S. Navy photograph, Naval Electronics Laboratory Center*)

dangers of diving. Diving is now an effective and safe means of conducting marine research. Although some scientific diving was accomplished prior to 1940, it was not until scuba (such as J. Y. Cousteau's Aqualung) became commercially available in 1950 as inexpensive equipment for the sportsman that diving became a widespread means of conducting marine research.

All current methods of diving deliver a breathable gas mixture to the diver's lungs and body cavities at the same pressure as that exerted by the overlying water column. Thus there is no difference in pressure between the interior and exterior parts of the diver's body, and he has no feeling of pressure from the overlying water, regardless of depth. The method of supply and the duration of submergence varies with the type of equipment. Equipment can be divided into two basic types: free diving (scuba), where the diver carries his supply of compressed gas on his back; and tethered diving ("hard hat"), with a gas supply furnished to the diver via a hose from the surface or a habitat.

Scuba diving. Scuba is extensively used by trained personnel as a tool for direct observation in marine research and underwater engineering. This equipment is designed to deliver through a demand-type regulator a breathable gas mixture at the same pressure as that exerted on the diver by the overlying water column. The gas which he breathes is carried in high-pressure cylinders (at starting pressures of 2000–3000 psi or 14–21 megapascals) worn on his back. Because the scuba diver is free from surface tending, he has much

more maneuverability and freedom than the conventional "hard hat" diver.

Scuba can be divided into three types: closed-circuit, semiclosed-circuit, and open-circuit. In the first two, which use pure oxygen or various combinations of oxygen, helium, and nitrogen, exhaled gas is retained and passed through a canister containing a carbon dioxide absorbent for purification. It is then recirculated to a bag worn by the diver (Fig. 1). During inhalation additional gas is supplied to the bag by various automatic devices from the high-pressure cylinders. These two types of equipment are much more efficient than the open-circuit system, which does not take advantage of the unused oxygen in the exhaled gas. The exhaled gas is discharged directly into the water after breathing. The closed- and semiclosed-circuit types are, however, more complicated and can be dangerous if they malfunction or are used by inexperienced divers. Most open-circuit systems use compressed air because it is relatively inexpensive and easy to obtain. Although open-circuit scuba is not as efficient as the other types, it is preferred by most scientists and engineers who work underwater because of its safety, the ease in learning its use, and its relatively low cost.

Free diving or scuba diving (Fig. 2) using compressed air as the breathing gas is used to some extent in most current marine scientific and engineering investigations in all environments, including the Arctic. For physiological reasons the underwater observer is limited to about 50 m of water depth when obtaining observations which require him to think clearly and solve problems while submerged. Below this depth when using compressed air as a breathing gas, he is limited, not by his equipment, but by the complex temporary changes which take place in his body chemistry while breathing gas (air) under high pressure. Nitrogen enters the tissue or the body at higher rates and reaches higher concentrations when breathed under pressure, the solubility going up with increased partial pressure. Under these higher concentrations in the body tissues, nitrogen has a narcotic effect which dulls the diver's reactions and reduces his ability to concentrate.

The practical depth limit to diving with air in clear, warm water is about 80 m. In addition to the narcotic effect of breathing at high pressures, the diver must allow his body to desaturate from the excess nitrogen contained in his body tissue (decompress) before returning to the surface. Too rapid a return to the surface will allow the excess nitrogen, which saturates the body cells and blood, to pass beyond the "bubble point." Small bubbles of gas form in the body tissue, causing the bends, or caisson disease, one of the worst maladies of diving. Tables containing permissible rates of ascent and time allowable at different depths are available from the U.S. Navy Experimental Diving Unit, Washington, DC, and from the Manned Undersea Science and Technology Office of the National Oceanic and Atmospheric Administration (NOAA), Rockville, MD. Tables are also available through most commercial diving supply outlets. Breathing a mixture of helium, instead of nitrogen, with oxygen can increase the diver's ability to

Fig. 2. Underwater scientific investigation by scuba diving, using compressed air. (*U.S. Navy photograph, Naval Electronics Laboratory Center*)

think underwater, but because of the expense and difficulties of handling the specialized diving gear and breathing mixtures needed for this type of deep diving, it is not, at the present time, practical for most scuba use.

By using scuba, scientists are no longer restricted to the sea surface but are able to investigate underwater problems at first hand. Free diving or scuba diving is used as a tool in offshore mapping of geologic formations, biological studies in ecology, underwater engineering, and detailed oceanographic measuring of the water column. An excellent manual for the diver has been published by NOAA that covers all aspects of scientific underwater investigation. *See* UNDERWATER PHOTOGRAPHY.

Saturation diving. Another type of diving has been introduced that permits long periods of submergence (1–2 weeks). Called saturation diving, it allows the diver to take advantage of the fact that at a given depth the body will become fully saturated with the breathing gas and then, no matter how long the submergence period, the decompression time needed to return to the surface will not be increased. Using this method, the diver can live on the bottom and make detailed measurements and observations, and work with no ill effects. This type of diving requires longer periods of decompression in specially designed chambers to free the diver's body of the high concentration of breathing gas, but the longer periods of submergence are worth this inconvenience in many working and scientific dives. Decompression times of days or weeks (the time increases with depth) are common on deep dives of over 60 m. However, when compared with diving from the surface, this type of diving has been shown to be much more efficient since it permits approximately three times the amount of work to be done for the amount of time spent on a project. Furthermore, the marine fauna become accustomed to seeing the divers living among them, and behavioral studies are greatly improved. While saturation diving is an exciting method of diving that greatly extends the time spent underwater, the long periods of time needed to decompress and the limited number of underwater laboratories capable of supporting saturated divers have kept it from becoming a popular tool for the average scientist and engineer. Its primary use has been in the servicing of offshore oil-drilling rigs, in maintenance and repair of dams, and in scientific studies in Federally supported research projects, such as Hydro-Lab, Tektite, and Sea Lab. Factors such as cold water at depth, lack of light, the requirement for bulky and expensive equipment, and the presence of strong bottom currents place a practical limit to this type of diving at about 195 m at the present time. The relatively inexpensive operation of the Hydro-Lab off Freeport, Bahamas, costs less than the operation of most oceanographic surface research vessels. Although saturation diving has been proven an efficient and superior way to study many marine problems, it has received only limited support from Federal funding agencies. *See* OCEANOGRAPHIC SUBMERSIBLES.   [ROBERT F. DILL]

Diving physiology

The study of the effects of increased pressure on physiological functions encountered in various forms of diving, such as breath-hold diving, diving with a self-contained underwater breathing apparatus (scuba), hard-hat diving, and saturation excursion diving developed in connection with underwater explorations. With the rapid advances being made in diving equipment and artificial breathing mixtures, man stands on the threshold of conquering the hostile depths of the ocean bottoms.

Scuba, or self-contained underwater breathing

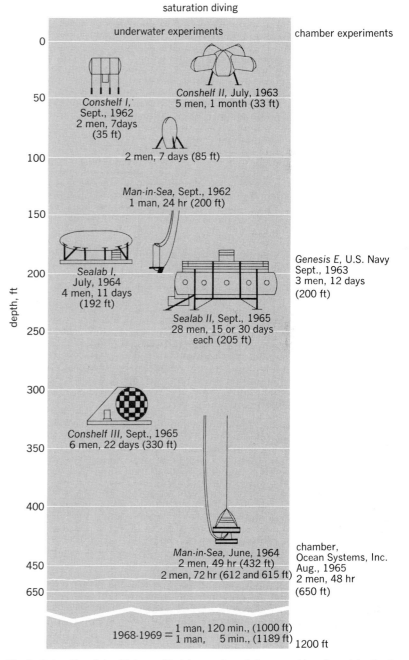

Fig. 1. Saturation diving history, with date, number of divers, and length and depth of underwater and chamber experiments preparatory to actual programs; 1 ft = 0.3 m.

apparatus, developed in 1943 by Capt. J. Y. Cousteau and E. Gagnan of France, was the major innovation in diving technology following the diver's suit and helmet. Now widely in use, scuba permits the diver to swim freely, with a sense of freedom surpassing that in helmet diving, for substantial periods at depths of 100–150 ft, depths unattainable in skin diving. Scuba gave impetus to the furtherance of undersea exploration. The idea of establishing underwater habitations was the next logical step. The pioneering underwater experiments of Cousteau (Conshelf), Capt. G. F. Bond (Sealab I, II), and E. Link (Man-in-Sea), shown in Fig. 1, have demonstrated that man can live underwater, under high pressure, breathing an artificial atmosphere for prolonged periods of time. A record combination of depth and time was achieved in 1964 in one of the Man-in-Sea series when R. Stenuit and J. Lindbergh, were lowered in a cylinder to a depth of 432 ft onto the ocean floor, where they swam to an inflated tent and lived there for two full days.

In 1967 commercial diving operations were undertaken by A. R. Krasberg for Westinghouse and C. H. Hedgepeth for Ocean Systems in the open sea to depths of 612.7 ft and 615 ft. Both diving operations involved several saturation excursion dives and lasted for 72 hr. In 1968 physiological and psychological tests were performed by Ocean Systems during open-sea dives to 650 ft. The average work load of the divers at this depth was $5\frac{1}{2}$ hr per day in water of 56–58°F. They were wearing hot-water suits which kept them warm throughout the dive. No significant impairment of performance and physiological functions was found. However, the divers were not able to carry out the tasks underwater with terrestrial ease.

In 1968 and 1969 a number of simulated deep dives were undertaken in chambers to depths around 1000 ft. No detrimental effects on physiological functions and performance were observed in rest and exercise. Dives were made by the following, with depths and times listed: Dr. A. A. Buehlmann, Zurich, 1000 ft (120 min); U.S. Navy, Capt. G. F. Bond, 825–1025 ft (5 min); IUC, A. Galerne, and U.S. Navy, Dr. K. E. Schaefer, 800–1100 ft and 1000 ft (2 hr 47 min at 1000 ft); and Duke University, Dr. T. Salzano and his coworkers, and U.S. Navy, 1000 ft (3 days). Dr. R. W. Brauer (Duke University and Wrightsville Marine Biology Laboratory) participated in a simulated deep dive at Comex, Marseille. He remained at 1189 ft for 5 min and reported symptoms of intermittent somnolence at this depth.

None of the experiments of the 1960s would have been possible without the progress made in underwater physiology, the central consideration in deep-sea diving. In artificial atmospheres, the gaseous breathing mixtures investigated by the U.S. Navy and others, particularly the substitution of helium for nitrogen, were the key to subsequent deeper and more prolonged dives.

Pressure effects. When a skin diver descends into the sea, the pressure of the surrounding water increases rapidly and compresses the gas-filled space of the lungs. At some depth between 100 and 300 ft, depending on the individual's lung capacity

and shift of blood into the chest, a threshold is reached where the thorax and lungs cannot withstand this pressure and collapse.

With the aid of diving equipment, helmet or scuba, air is supplied to the diver under high pressure, thus preventing the lungs from collapsing. The effects of increased pressure observed on divers in a water environment are different from those found in caisson workers, who are exposed to a dry, pressurized environment in which hydrostatic pressure differences are not present.

Boyle's law. Underwater the pressure increases 1 atmosphere (atm) with each 10 m or 33 ft (fresh water, 33 ft; sea water, 32 ft). According to Boyle's law, the volume of a gas decreases in direct proportion to the increase in pressure, or $P_1V_1 = P_2V_2$. Thus, at 33 ft of water where the pressure is 2 atm, 1 atm produced by the pressure of air above the water and 1 atm by the weight of the 33-ft water column, the volume of gas is compressed to one-half of the original volume at sea level. At 66 ft the pressure is 3 atm, and the volume is therefore compressed to one-third.

Partial pressures of mixed gases. The pressures exerted by individual gases in a gas mixture are independent of the pressures of other gases in the mixture, according to Dalton's law. The total gas pressure in a mixture is, therefore, the sum of the partial pressures of the individual gases in the mixture. The partial pressure of a gas is the product of concentration of the individual gas and the total gas pressure.

A diver is subject to increasing partial pressure of nitrogen, oxygen, and carbon dioxide the deeper he goes. For example, at 1 atm the partial pressure of oxygen measured at 15% concentration in the diver's lungs is 107 mm Hg, whereas at 4 atm overpressure, corresponding to a depth of 132 ft below sea level, the partial pressure of oxygen, now increased in concentration to 20.8%, is 623 mm Hg. (Since the concentration of oxygen here is a dry measurement, the total atmospheric pressure is reduced by 47 mm Hg, the pressure of water vapor in the alveolar air at normal body temperature.)

Nitrogen narcosis. A diver breathing air at a pressure equivalent to that at 300 ft of water is subject to certain changes in personality and performance, ascribed by A. R. Behnke in 1935 to nitrogen narcosis. The subjective symptoms and mental reactions consist of euphoria, overconfidence accompanied by dulling of mental ability, and difficulty in assimilating facts and making quick and accurate decisions. These symptoms are similar to alcohol intoxication. Cousteau has called it "rapture of the deep."

Mild symptoms of euphoria and irresponsibility begin to occur at a depth of 130 ft. The maximum safe depth for scuba divers lingering 15 min has been set at 130 ft.

The mechanism of nitrogen narcosis is not yet fully understood. It was first explained in accord with the Meyer-Overton hypothesis of gas anesthesia, which relates the narcotic action to the lipoid (a fatlike substance) solubility and oil-water partition coefficients of inert gases. A theory advanced by L. Pauling and S. Miller is based on the correlation of the anesthetic activity and the dissociation pressure of gas hydrates formed from the inert gases in the nervous system. The chances of a diver developing nitrogen narcosis are increased as the amount of alveolar carbon dioxide increases because of breathing resistance.

Helium effects. The use of nitrogen-oxygen gas mixtures limited the maximal depth of diving to about 300 ft because of nitrogen narcosis and resistance to breathing. Helium was found superior to nitrogen as an inert gas diluent for oxygen, and it is used for divers below 150 ft. It extended practical diving operations to 500 ft. The world's open-sea record dive to a depth of 1000 ft, established in 1962 by Hannes Keller off the California coast, was accomplished with a helium-oxygen breathing mixture. Helium itself probably does not exert significant narcotic effects up to a depth of 1500 ft. Based on the assumption that helium is about one-ninth as narcotic as nitrogen, Edward Lanphier compared the degrees of impairment of performance produced by nitrogen and helium at different depths (Fig. 2). However, other factors might influence this projection, which suggests that at 100-ft depth helium produces a barely noticeable effect. Peter Bennett has noted tremor, nausea and vomiting, and a deterioration of arithmetical performance at depths of 600 and 800 ft breathing at 95% helium and 5% oxygen mixture, which he ascribed to ventilatory depression and resulting carbon dioxide intoxication under these conditions.

Helium is lighter than nitrogen, thus reducing the resistance to breathing. Moreover, helium is less soluble in the tissues than nitrogen. Therefore, smaller quantities of helium are absorbed in the body during a dive, and less gas is eliminated from the tissues during decompression to result in bubble formation.

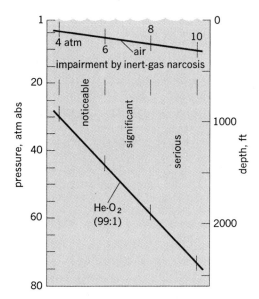

Fig. 2. The degrees of impairment by inert-gas narcosis. The impairment experienced in diving with air is projected into greater depths for a 99% He–1% O_2 mixture on the assumption that helium is one-ninth as narcotic as nitrogen. (From E. H. Lanphier, in Proceedings of the 3d Symposium on Underwater Physiology, 1967)

Nevertheless, in spite of these advantages, helium-oxygen diving has not lessened incidents of bends during decompression. The amount of helium absorbed to the saturation point in the body tissue is only 40% of that of nitrogen. However, because of the faster rate of uptake and elimination of helium by the tissues, bubbles can be more easily formed so that the safe ratio of body gas pressure to external pressure is only 1.7 to 1, compared with 3 to 1 for nitrogen. Diving with helium-oxygen atmospheres requires special equipment and special decompression tables.

Oxygen toxicity. Inhalation of oxygen concentrations above 65% at 1 atm, corresponding to a partial pressure of oxygen above 494 mm Hg, has deleterious effects in most warm-blooded animals; it leads eventually, even after days or weeks, to death due to pulmonary oxygen toxicity. Inhalation of oxygen at a pressure of approximately 2 atm results in generalized convulsions. With increasing depth the toxicity symptoms develop after shorter exposure times. Under higher partial pressures oxygen toxicity manifests itself primarily in the central nervous system.

The symptoms of oxygen poisoning, based on studies of a large number of exposures in a wet chamber, are as follows:

Tingling of fingers and toes
Visual disturbances
Acoustic hallucinations
Sensations of abnormality
Confusion
Muscle twitching
Unpleasant respiratory sensations
Nausea
Vertigo
Lip twitching
Convulsion, epileptiform

There is no typical sequence to these symptoms. It is possible that most of them occur prior to the onset of convulsions. On the other hand, convulsions may develop suddenly without any warning, leading to death for the diver deeply submerged. Interestingly, convulsions stop as soon as the excessive partial pressure of oxygen is removed. No permanent damages resulting from oxygen convulsions in man have been reported. At high pressure exercise greatly increases the susceptibility to oxygen toxicity, and it also produces frequent signs of pulmonary irritation. The most frequent symptom reported in underwater swimmers who contracted oxygen toxicity was dyspnea, or labored breathing.

Carbon dioxide intoxication. With properly designed diving equipment and sufficient ventilation, one would expect the danger of CO_2 accumulation to be minimal. However, because of increased resistance to breathing under higher pressures, CO_2 retention does occur in diving operations. Inhaling CO_2 can cause symptoms of dyspnea, headaches, dizziness, warmth, restlessness, or increased motor activity.

In breath-hold diving CO_2 intoxication does not present a major hazard, since CO_2 is taken up from the tissues, and the levels of alveolar CO_2 and blood CO_2 are rather low during the dive. More-over, they can be easily controlled by the speed of ascent. In scuba diving CO_2 intoxication was found to be significant. Several cases of unexplained loss of consciousness were observed in divers using equipment in which canisters were employed for CO_2 removal. Carbon dioxide intoxication was considered the most likely cause of the "shallow-water blackout" in these cases. Follow-up studies have shown that marked CO_2 intoxication in scuba diving does not necessarily produce severe dyspnea. This means that the depressant effects of CO_2 on the central nervous system can occur without the warning signs of respiratory distress. In deep-sea diving in which suit and helmet are used, a vigorous ventilation of air or gas mixture is needed to prevent the accumulation of CO_2. It is necessary to pump increased amounts of air from the surface to ventilate the helmet of the diver at greater depths. The volumes of air required by a working diver at different depths are listed in Table 1.

Even with sufficient ventilation breathing resistance at greater depths might become so large that CO_2 could accumulate in the alveolar air. Studies by the U.S. Navy Experimental Diving Unit of experienced deep-sea divers breathing oxygen-nitrogen mixtures showed that CO_2 retention was due to increasing breathing resistance. When helium-oxygen mixtures were used, breathing resistance decreased, and CO_2 retention was absent.

Decompression sickness. A diver who has been exposed to high pressure for a long period will have accumulated a large amount of dissolved nitrogen in his tissues. If he were brought back to the surface immediately, say from the pressure of 130 ft (5 atm), nitrogen dissolved at a tension of 4 atm overpressure would suddenly be forced to come into equilibrium with a tension of 1 atm. The result would be bubble formation, which can cause minor or severe damage, depending on the quantity of bubbles and the organs in which they form. It is fortunate that the blood and tissues can hold a certain amount of gas in supersaturation without significant bubble formation. As mentioned earlier, under ordinary circumstances the nitrogen pressure inside the body can rise to three times the pressure outside the body without causing serious gas bubble decompression sickness.

If bubbles are liberated from supersatured tissues, they can exert pressure on nerves and result in local pain to muscles, joints, or bones. Bubbles formed in the spinal cord can produce paralysis. If gas bubbles develop in the brain, they may cause dizziness, blindness, paralysis, unconsciousness, or convulsions, depending on quantity and location. Bubbles formed in the blood can interfere with the pulmonary blood flow and result in asphyxia and the "chokes." The first symptom is often pain below the sternum during deep breathing, followed by shallow and rapid breathing. The onset of symptoms varies greatly, occurring within 1–18 hr following decompression. Incidences of more severe symptoms have been reported in about 5% of divers afflicted with decompression sickness (Table 2). Prevention of decompression sickness is accomplished by minimizing the supersaturation of the tissues. J. S. Haldane in 1907 introduced the idea of stage decompression, which is based on the finding that a man can be safely decompressed to a pressure not less than half the total pressure he was exposed to. This fact was found to be independent of the length of time exposed to high pressure. In other words, if the tension of nitrogen in the body is not greater than twice the ambient pressure, nitrogen bubbles will not form. Decompression involves a series of raises and stops in which the diver is held long enough to desaturate sufficiently to proceed to the next stage and so on, until he has made his way to the surface. The rate at which a diver can be decompressed depends on the depth at which he operated and the time he spent at this depth. If he stayed only a short time at some depth and his tissues did not become saturated, the decompression time is shorter.

Breathing oxygen during part of the decompression period will reduce decompression time. It will hasten the nitrogen elimination from the tissues, since the nitrogen pressure in the lung alveoli is reduced to practically zero. However, administration of pure oxygen is limited to shallower depths because of the danger of oxygen toxicity.

Practical experience has shown that a diver can tolerate for several minutes a strong supersaturation of the tissues without developing bends. This has led to the implementation of surface decompression. The diver is immediately brought to the surface from depth and placed in a deck decompression chamber, where final decompression is achieved following the appropriate decompression table.

Breath-hold diving. Breath-hold diving, an ancient profession, is still practiced by Greek, Polynesian, and Japanese sponge and pearl divers. In

Table 1. Volume of air supply needed by divers at different depths to prevent carbon dioxide accumulation

Depth, ft	Air volume, ft³
Sea level	1.5
33	3
66	4.5
100	6
200	10.5
300	15

Table 2. Frequency of symptoms occurring in decompression sickness*

Symptom	Frequency, %
Local pain ("bends")	89.0
Lower extremity	70.0
Upper extremity	30.0
Skin rash, with itching	11.0
Visual disturbances	6.0
Motor paralysis or weakness	5.5
Vertigo ("staggers")	5.0
Numbness	4.8
Respiratory distress ("chokes")	2.4
Headache	1.9
Unconsciousness	1.5
Aphasia	1.2
Nausea	0.9

*From A. W. Dewey, Jr., *New Engl. J. Med.*, vol. 267, October, 1962.

this diving the organism is exposed to unique stresses. The exchange of CO_2 and O_2 in the lungs was studied by K. E. Schaefer in dives to 90 ft by instructors at the 100-ft escape training tank, U.S. Naval Submarine Base, New London, Conn. During descent the CO_2 tension in the lungs was found to rise quickly above the CO_2 gradient from the lungs to the blood. At 90 ft approximately 50% of the predive CO_2 content of the lungs disappears and is taken up by the blood and tissues. During ascent a normal CO_2 gradient is reestablished. Also, at 90 ft the alveolar oxygen level is rather high, and the CO_2 tension does not rise to dangerous levels (because of the uptake in tissues). Breath-holding time under these conditions is considerably prolonged. This might encourage a diver to stay too long at depth. The real danger, however, develops during ascent, following a prolonged breath-hold at depth. With the reduction of the ambient water pressure, the alveolar oxygen tension falls to a very low level when the diver reaches the surface. He might lose consciousness because of anoxia and drown. Figure 3 shows the changes in CO_2 and O_2 content of the lungs during dives to 90 ft.

Breath-hold divers develop certain physiological adaptations. They show a higher tolerance to increased CO_2 and lower oxygen. Studies on instructors at the escape training tank in New London demonstrated after 1 year of water work an increase in total lung volume commensurate with a decrease in residual-air lung volume, which resulted in a larger compression ratio and an extension of the maximal diving depth from 87 to 112 ft.

Submarine escape. Capt. G. F. Bond of the U.S. Navy was the first to carry out a successful buoyant ascent (equipped with a life jacket only) from a submarine at a depth of 300 ft. The ascent rate was 340 ft/min. No respiratory distress was observed. Based on studies of the alveolar gas exchange during buoyant ascent from 90 ft, a theoretical depth-time curve for buoyant ascent has been established by A. DuBois, according to whom it is possible to ascend from 600 ft while avoiding the hazards of CO_2 narcosis, nitrogen narcosis, and anoxia (Fig. 4).

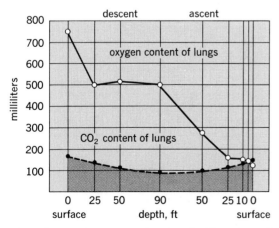

Fig. 3. Changes in oxygen and carbon dioxide content of the lungs during dives to 90 ft. Volumes calculated for standard pressure dry (STPD). (*From K. E. Schaefer and C. R. Carey, Science, 137:1051; copyright 1962 by the American Association for the Advancement of Science*)

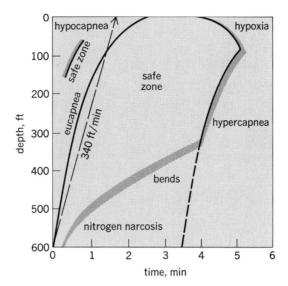

Fig. 4. Submarine escape. Depth-time limits curve for buoyant ascent from 600 ft. (*A. B. DuBois, G. F. Bond, and K. E. Schaefer, J. Appl. Physiol., 18:509, 1963*)

Scuba diving. In the aqualung Cousteau and Gagnan adapted a cylinder-mounted inhalation valve close to the exhalation valve at the level of the diver's lung. This arrangment has made it possible for the diver to inhale and exhale at nearly identical hydrostatic pressures without encountering marked breathing resistance.

In open-circuit demand systems, such as the aqualung, air is exhaled into the water, creating a trail of bubbles. This feature makes open circuits less desirable for military purposes than closed-circuit systems, in which pure oxgyen is breathed and carbon dioxide is absorbed by a canister. Because of the danger of oxygen toxicity, however, the use of closed-circuit systems is limited to a depth of 30 ft at 40 min duration.

The scuba diver must know the time he can remain at a certain depth and return directly to the surface without being endangered by decompression sickness. The no-decompression-limits data, shown in Fig. 5, from the U.S. Navy air tables, are based upon a rate of ascent not exceeding 60 ft/min. The horizontal scale of Fig. 5 shows the bottom time, and the vertical scale the actual depth of the dive. It can be seen that a scuba diver can stay for 15 min at 130 ft and 2 hr at 40 ft without any decompression effects. The recommended depth limit is 130 ft with a bottom time of 10 min. For repetitive dives it is necessary to determine beforehand the exact time and depth of the initial dive, the surface intervals, and the depth and time of the subsequent dives.

Saturation excursion dives. Before 1962 all diving operations started and ended at sea level, and diving tables were based on these conditions accordingly. Since 1962 underwater stations were erected on the ocean floor (Fig. 1). The experiments involved saturation diving, in which the tissues exposed to high pressure for 24 hr became saturated with inert gases, thereby reaching a new equilibrium state. As exposure is continued beyond 1 day, whether it be for days, weeks, or months, no inert gas is further absorbed. The de-

DIVING PHYSIOLOGY

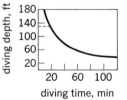

Fig. 5. Limits of no-decompression for compressed air diving. (*From Submarine Medicine Practice, U.S. Navy Bureau of Medicine and Surgery*)

Fig. 6. Sealab II, an underwater station for deep-sea divers; 57 ft long, 12-ft diameter. (*U.S. Navy*)

compression time is therefore the same no matter how much longer the diver remains at depths. For divers based in underwater stations and making excursions to greater depths, a knowledge of the

depth-time limitations for saturation is needed to avoid decompression. Preliminary studies have indicated that saturation permits the diver to go down to greater depths at longer bottom times within the no-decompression limits, as compared to surface-based diving.

Figure 6 shows Sealab II, which was placed at 205 ft beneath the surface of the ocean near Scripps Canyon, off La Jolla, Calif. Three teams of aquanauts with 10 divers in each lived and worked at this depth for periods of 15 days, including two men who stayed for 30 days, breathing an artificial gas mixture of 80% helium, 16% nitrogen, and 4% oxygen. They sustained no ill effects. Persons living in such pressurized environment for a prolonged period do undergo, nevertheless, minor physiological changes, such as reduction of maximum breathing capacity, due to increased breathing resistance. Figure 7 shows the decrease in maximum breathing capacity measured in three subjects in a preparatory experiment for Sealab II (Genesis E).

Three major problems confronted the divers as they climbed out of Sealab II into the cold underwater environment. There were difficulties in orientation, speech communication degraded by helium ("Donald Duck" sound), and excessive heat loss that resulted in chilling, thus aborting operations. Better-heated diving suits will have to be developed to overcome the last problem.

Liquid breathing. Johannes Kylstra advanced the idea of liquid breathing under high pressure and was able to show that mice, rats, and dogs lived for prolonged periods of time submerged in salt solutions equilibrated with oxygen at high pressures. Sixty times more work is necessary to pump equal volumes of water instead of air through the lungs, which seriously restricts carbon dioxide elimination. This method is, however, only of theoretical interest.

Other hazards. Aside from narcosis, toxicity, and decompression sickness, the scuba, helmet, and breath-hold divers are subject to the effects of unequal pressure differences (barotrauma) across their air-containing structures, such as middle ears, sinuses, lungs, and gastrointestinal tract. Swimmers and divers often have difficulty in equalizing pressure in the middle ear during descent because of blockage of the Eustachian tube. They will experience pain and upon further descent will rupture the eardrum. If the sinuses do not equalize during descent, transudation and hemorrhage can develop. For a better illustration the gas-containing system of the body has been divided into rigid and collapsible chambers, as shown in Fig. 8. The rigid chambers must remain open during dives. If negative pressure develops in the rigid chambers with respect to the surrounding hydrostatic pressure, congestion and bleeding will result.

During ascent the air contained in the lungs of the diver expands in relation to the diminishing ambient pressure. If the diver uses a breathing apparatus and breathes normally during ascent, the excess of expanding lung air is discharged through the exhaust valve. But if the diver panics and holds his breath, overexpansion of lung tis-

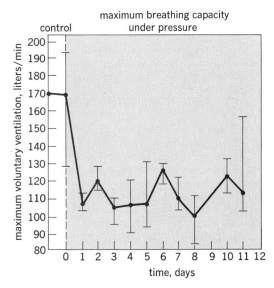

Fig. 7. Effect of prolonged exposure to 7 atm while breathing a helium-oxygen-nitrogen mixture on maximum breathing capacity. Preparatory experiment for Sealab II at the Submarine Medical Research Laboratory in Groton, Conn. (*From G. P. Lord, G. F. Bond, and K. E. Schaefer, J. Appl. Physiol., 21:1833–1838, 1966*)

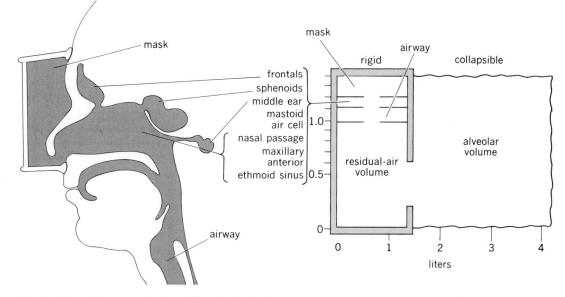

Fig. 8. Schematic representation of the rigid and collapsible parts of the lungs, various rigid chambers of the upper airways, and face mask of a diver. Values given on the right are based on data obtained from Japanese women divers, or amas. (*From H. Rahn, Nat. Acad. Sci.– Nat. Res. Counc. Publ., no. 1341, 1965*)

sues may cause air embolism from gases entering the pulmonary blood vessels from adjoining tissues. Air embolism frequently results in death due to blockage of circulation. It develops when the pulmonary pressure (alveolar-pleural pressure difference) rises above 70 mm Hg.

Through the physiological hazards of the sea are manifold at depth, man is accelerating his attempts to cope with them. He is preparing to reap the great treasures—foodstuffs, fuel, and minerals—of the continental shelf by inhabiting the ocean floor. He has learned that these riches cannot merely be fished out by surface techniques. The acquisition requires his bodily presence with proper life-supporting equipment. Thus, today's scientists have two frontiers to explore: outer space and the ocean depths, or inner space.

[KARL E. SCHAEFER]

Bibliography: P. B. Bennet, Performance impairment in deep diving due to nitrogen, helium, neon and oxygen, in C. Lambertsen (ed.), *Proceedings of the 3d Symposium on Underwater Physiology*, 1967; B. A. Hills, *Decompression Sickness: The Biophysical Basis of Prevention and Treatment*, 1977; J. A. Kylstra, Advantages and limitations of liquid breathing, in C. Lambertsen (ed.), *Proceedings of the 3d Symposium on Underwater Physiology*, 1967; E. H. Lanphier, Interactions of factors limiting performance at high pressures in C. Lambertsen (ed.), *Proceedings of the 3d Symposium on Underwater Physiology*, 1967; K. E. Schaefer, Adaptation to breath-hold diving, *Physiology of Breath-Hold Diving and the Ama of Japan*, Nat. Acad. Sci.–Nat. Res. Counc. Publ. no. 1341, 1965; K. E. Schaefer, Circulatory adaptation to the requirement of life under more than one atmosphere of pressure, *Handbook of Physiology*, vol. 3 1964; K. E. Schaefer and C. R. Carey, Alveolar pathways during 90-ft breath-hold dives, *Science*, 137:1051, 1962.

Drought

A general term implying a deficiency of precipitation of sufficient magnitude to interfere with some phase of the economy. Agricultural drought, occurring when crops are threatened by lack of rain, is the most common. Hydrologic drought, when reservoirs are depleted, is another common form. The Palmer index has become popular among agriculturalists to express the intensity of drought as a function of rainfall and hydrologic variables.

The meteorological causes of drought are usually associated with slow, prevailing, subsiding motions of air masses from continental source regions. These descending air motions, of the order of 200 or 300 m/day, result in compressional warming of the air and therefore reduction in the relative humidity. Since the air usually starts out dry, and the relative humidity declines as the air descends, cloud formation is inhibited—or if clouds are formed, they are soon dissipated. The area over which such subsidence prevails may involve several states, as in the 1962–1966 Northeast drought or in the dust bowl drought of the 1930s over the Central Plains.

The atmospheric circulations which lead to this subsidence are the so-called centers of action, like the Bermuda High, which are linked to the planetary waves of the upper-level westerlies. If these centers are displaced from their normal positions or are abnormally developed, they frequently introduce anomalously moist or dry air masses into regions of the temperate latitudes. More important, these long waves interact with the cyclones along the Polar Front in such a way as to form and steer their course into or away from certain areas. In the areas relatively invulnerable to the cyclones, the air descends, and if this process repeats time after time, a deficiency of rainfall and drought may occur. In other areas where moist air is frequently forced to ascend, heavy rains occur. Therefore, drought in one area, say the northeastern United States, is usually associated with abundant precipitation elsewhere, like over the Central Plains.

After drought has been established in an area, there seems to be a tendency for it to persist and expand into adjacent areas. Although little is known about the physical mechanisms involved in this expansion and persistence, some circumstantial evidence suggests that numerous "feedback" processes are set in motion which aggravate the situation. Among these are large-scale interactions between ocean and atmosphere in which variations in ocean-surface temperature are produced by abnormal wind systems, and these in turn encourage further development of the same type of abnormal circulation. Then again, if an area, such as the Central Plains, is subject to dryness and heat in spring, the parched soil appears to influence subsequent air circulations and rainfall in a drought-extending sense. *See* CLIMATIC CHANGE.

Finally, it should be pointed out that some of the most extensive droughts, like those of the 1930s dust bowl era, require compatibly placed centers of action over both the Atlantic and the Pacific.

In view of the immense scale and complexity of drought-producing systems, it will be difficult for man to devise methods of eliminating or ameliorating them. *See* PRECIPITATION.

[JEROME NAMIAS]

Bibliography: R. A. Bryson and T. J. Murray, *Climates of Hunger: Mankind and the World's Changing Weather*, 1977; J. Namias, *Factors in the Initiation, Perpetuation and Termination of Drought*, Int. Union Geod. Geophys. Ass. Sci. Hydrol. Publ. no. 51, 1960; W. C. Palmer, *Meteorological Drought*, U.S. Weather Bur. Res. Pap. no. 45, 1965; M. A. Weinberg, *Plants Are Water's Factories: A Book about Drought*, 1976.

Dust storm

A strong, turbulent wind carrying large clouds of dust. In a large storm, clouds of fine dust may be raised to heights well over 10,000 ft and carried for hundreds or thousands of miles.

Sandstorms differ by the larger mass, more rapid settling speeds of the particles involved, and the stronger transporting winds required. The sand cloud seldom rises above $1-2$ m and is not carried far from the place where it was raised.

Dust storms cause enormous erosion of the soil, as in the dust bowl disasters of $1933-1937$ in the Great Plains of the United States. Besides causing acute physical discomfort, they present a severe hazard to transportation by reducing the visibility to very low ranges. Conditions required are an ample supply of fine dust or loose soil, surface winds strong enough to stir up the dust, and sufficient atmospheric instability for marked vertical turbulence to occur.

Mechanics of dust raising. Dust (or sand) is initially raised when particles become dislodged by aerodynamic stresses of the strong winds upon exposed grains. The larger particles fall obliquely after attaining considerable horizontal speed, bombarding other particles on the surface which in turn become dislodged and further the process.

R. A. Bagnold classifies as dust the particles having diameters of $10^{-3}-10^{-2}$ mm, with free-fall speeds ranging $10^{-2}-2$ cm/sec; sand particles, $0.1-1$ mm in diameter, have fall speeds $40-600$ cm/sec. Dust can be readily carried upward by ordinary turbulent eddies in an unstable air mass, but these are generally too feeble to sustain large sand particles, which attain only small heights by bouncing. In any dust storm, particles of various sizes are raised, the smallest being carried to greatest heights from which they may take days or weeks to settle while remaining as a dust haze.

Soil factors. Soil condition is the most decisive criterion for development of dust storms. This depends on vegetative cover and upon binding of soil by moisture, both factors being dependent upon prior rain- or snowfall. Dust storms are most frequent in spring, when in semiarid regions the Earth is least covered by vegetation. Loosening of soil and overturning of humus by spring plowing and overgrazing of grasslands are prime contributors to setting up soil conditions favorable for dust storms.

Meteorological factors. Surface wind speeds required vary according to soil characteristics. In some desert regions sandstorms occur with winds of $15-20$ mph. Extensive dust storms in North America usually require winds of $25-30$ mph or more. Such winds are present over large areas in the circulations of many well-developed cyclones, which may raise dense dust clouds several hundred miles across if soil conditions are right.

A further requisite is thermodynamic instability (strong decrease of temperature with height), necessary for development of the vertical eddies required to transport dust aloft from surface layers. Most dust storms occur in daytime, particularly in the afternoon when the air is warmest at the ground, hence most unstable just above. Major dust storms in the United States are almost exclusively confined to maritime polar air masses from the Pacific Ocean, which are characteristically unstable to great heights. *See* AIR MASS.

Dust storms associated with large-scale wind systems are also common in the Sahara and Gobi deserts. More local but often severe dust storms resulting from thundersqualls are common in all the desert regions. *See* SQUALL.

Optical and electrical effects. Small dust particles increase scattering of light, mainly in short (blue) wavelengths. The Sun often appears a deep orange or red when seen through a dust cloud; however, optical effects are variable. Large particles are effective reflectors, and an observer in an aircraft above a dust storm may see a solid sheet with an apparent dust horizon.

Because of friction with air or ground, dust particles acquire appreciable electrostatic charges, and on striking radio antennas may cause severe static. Visible electrical discharges sometimes occur within the dust cloud.

[CHESTER W. NEWTON]

Bibliography: R. A. Bagnold, *The Physics of Blown Sand and Desert Dunes*, 1965; S. A. Trimble et al., *Great Sand Dunes: The Shape of the Wind*, 1975.

Earth sciences

Sciences primarily concerned with the atmosphere, the oceans, and the solid Earth. They deal with the history, chemical composition, physical characteristics, and dynamic behavior of solid Earth, fluid streams and oceans, and gaseous atmosphere. Because of the three-phase nature of the Earth system, Earth scientists generally have to consider the interaction of all three phases —solid, liquid, and gaseous—in most problems that they investigate.

The geosciences (geology, geochemistry, and geophysics) are concerned with the solid part of the Earth system. Geology is largely a study of the nature of Earth materials and processes, and how these have interacted through time to leave a record of past events in existing Earth features and materials. Hence, geologists study minerals, rocks, ore deposits, mineral fuels and fossils, and the long-term effects of terrestrial and oceanic waters and of the atmosphere. They also investigate present processes in order to explain events that took place in the past.

Geochemistry involves the composition of the Earth system and the way that matter has interacted in the system through time. For example, by studying the behavior of radioactive substances it is possible to determine how old the substances are and how much energy has been released through time as a result of decay of radioactive compounds.

Geophysics deals with the physical characteristics and dynamic behavior of the Earth system and thus concerns itself with a great diversity of complex problems involving natural phenomena. For example, earthquakes, vulcanism, and mountain building throw light on the structure and constitution of the Earth's interior and lead to consideration of the Earth as a great heat engine. Study of the magnetic field involves considering the Earth as a self-sustaining dynamo. *See* GEOPHYSICS.

The atmospheric sciences, commonly grouped together as meteorology, are concerned with all chemical, physical, and biological aspects of the Earth's atmosphere. Although the study of weather used to be the chief occupation of meteorologists, man's entry into the space age calls for a vast increase in knowledge of the environment through which vehicles and ultimately living things will go and return. Consequently, many aspects of the Earth's atmosphere are now being studied intensively for the first time. As an example, great planetary currents in the atmosphere, and also in the oceans, are now being investigated not only for the light they may shed on a better understanding of weather but also as a basis for understanding more fully the motion of the entire atmosphere and oceans. *See* METEOROLOGY.

Oceanography encompasses the study of all aspects of the oceans—their history, composition, physical behavior, and life content. Before World War II little was known about the oceans of the world. During that war many important characteristics of the ocean were discovered, and since then, with instruments and facilities developed during the war, oceanographic research has been going on at a quickened pace. *See* OCEANOGRAPHY.

[ROBERT R. SHROCK]

Bibliography: K. B. Krauskopf, *The Third Planet: An Invitation to Geology*, 1974; Scientific American Editors, *Scientific American Resource Library: Readings in the Earth Sciences*, 3 vols., 1973; H. Takeuchi et al., *Debate about the Earth*, 1970.

Earth tides

Cyclic motions, sometimes over a foot in height, caused by the same lunar and solar forces which produce tides in the sea. These forces also react on the Moon and Sun, and thus are significant in astronomy in evaluations of the dynamics of the three bodies. For example, the secular spin-down of the Earth due to lunar tidal torques is best computed from the observed acceleration of the Moon's orbital velocity. In oceanography, earth tides and ocean tides are very closely related.

Efforts to measure earth tides date from the end of the 19th century; accounts are found in George Darwin's *Scientific Papers* and in other writings on natural philosophy about 1890. In 1919 A. A. Michelson and H. G. Gale measured the daily tidal tilt near Geneva, Wisconsin, by recording the changes in water level at the ends of 500-ft-long (152 m) horizontal pipes buried in the ground. Because of its precision and significance their work may be regarded as initiating modern earth tide measurements. By far the most widely used earth tide instruments are the tiltmeter and the gravimeter. Both instruments have the merits of portability, high potential precision, and low cost. Thus they are able to advance economically an important mission—the global mapping of earth tides and ocean tides, both procedures still being in a rudimentary stage.

Tide-producing potential. Tidal theory begins simply, by evaluating the tidal forces produced on a rigid, unyielding spherical Earth by the Moon and Sun. In this discussion, the Earth, Moon, and Sun will be regarded as mass points at their respective centers. The results of this idealized assumption will be modified later to relate earth tide observations to deformable Earth models.

In the expansion of the total gravitational potential of the satellite as a sum of solid spherical harmonics, the tide-producing potential consists only of those terms which vary within the Earth and are the source of Earth deformations. The omitted terms are the constant which is arbitrarily chosen to produce null potential at the Earth's center and the first-degree term expressing the uniform acceleration of every part of the Earth toward the satellite (thus also entailing no distortion). Accordingly, the tide-producing potential U is expressible as the sum of solid spherical harmonics of degree 2 and higher in the notation which follows.

Let the mass of the satellite be denoted by m its distance from the Earth's center by R, the distance of the observing point from the Earth's center by r, and the geocentric zenith angle measured from the line of centers by θ. Then for points $r < R$, Eq. (1) holds, where G is Newton's constant ($6.67 \times 10^{-8} c^3 g^{-1} s^{-2}$).

$$U = U_2 + U_3 + \cdots = GmR^{-1}\left[\left(\frac{r}{R}\right)^2 P_2\,(\cos\theta)\right.$$
$$\left. + \left(\frac{r}{R}\right)^3 P_3\,(\cos\theta) + \cdots\right] \quad (1)$$

On an idealized oblate Earth of equatorial radius r_e and flattening constant f, a point on the surface at geocentric latitude ϕ has radial distance $r = r_e C(\phi)$, where $C(\phi) \equiv 1 - f\sin^2\phi$. The vertical (upward) component of tidal gravity $\partial U/\partial r$ and the horizontal component $r^{-1}\partial U/\partial\theta$ in the azimuth direction away from the satellite, with $P_n^1(\cos\theta) = -\partial P_n(\cos\theta)/\partial\theta$, are given by Eqs. (2) and (3).

$$g_r = \frac{\partial U}{\partial r} = GmC\alpha^3 r_e^{-2}\sum_{n=2}^{\infty} n(\alpha C)^{n-2}P_n(\cos\theta) \quad (2)$$

$$g_\theta = r^{-1}\frac{\partial U}{\partial\theta} =$$
$$-GmC\alpha^3 r_e^{-2}\sum_{n=2}^{\infty}(\alpha C)^{n-2}P_n^1(\cos\theta) \quad (3)$$

Here $\alpha_m \equiv r_e/R_m$ and $\alpha_s \equiv r_e/R_s$, for the Moon and Sun, respectively. R. A. Broucke and coworkers provided a computer program for computing g_r and g_θ in Eqs. (2) and (3) for the Moon within a tested precision of ±0.004 microgal (1 gal = 1 cm $\cdot$ s^{-2}); a similar precision is available for the solar tide.

Harmonic constituents. Equations (2) and (3) indicate the simple nature of the dependence of tidal gravity on the rigid Earth upon the distance R and the zenith angle θ of the satellite. But the time variations of R and θ are complex, because of the Earth's rotation and the complexity of the orbital motions. It is in the analysis of these complexities that earth tide studies show their inheritance from the extensive studies of ocean tides of the last century. For predictions at the world's harbors a year or more in advance, the ocean tides were analyzed in the Darwin-Doodson system as a sum of many harmonic constituents, −386 in A. T. Doodson's 1922 analysis.

However, the objective in earth tide observations is not prediction but assistance in determining present relevant properties of the Earth. For this purpose, only the major tidal constituents in each class provide the critical evidence. Table 1 lists the periods and amplitude factors b of the larger tidal harmonic constituents over the broad

Table 1. Parameters for some equilibrium tides

Species	Symbol	Period	b
Long period	Lunar	18.6 yr	0.066
	Sa	1 yr	0.012
	SSa	½ yr	0.073
	MSm	31.85 day	0.016
	Mm	27.55 day	0.083
	MSf	14.77 day	0.014
	Mf	13.66 day	0.156
	−	13.63 day	0.065
Diurnal	O_1	25.82 hr	0.377
	P_1	24.07 hr	0.176
	K_1	23.93 hr	0.531
Semidiurnal	N_2	12.66 hr	0.174
	M_2	12.42 hr	0.908
	S_2	12.00 hr	0.423
	K_2	11.97 hr	0.115

range of tidal periods. This discussion has been shortened by deemphasizing the purely geometric complexities essential in applied tidal theory and by adopting, when feasible, the simple zenith-centered satellite coordinate system used in Eqs. (1)–(3). Thus this discussion omits mention of the geographic variation with latitude, which differs for the three species in Table 1, and of the tides resolved into their important spectral constituents. Such resolution is especially fertile in the interpretation of earth tide residuals in terms of their ocean tide causes. It should be remembered that the importance of the subject of earth tides is greatly enhanced because of the content of the wide range of known periodicities and known excitation forces.

Tidal torques. The orbital acceleration $\dot{n}_m$ of the Moon is the critical quantity in Eq. (4), giving the

$$N_m = -\tfrac{1}{3}[M_e M_m/(M_e + M_m)]r_m^2\dot{n}_m \quad (4)$$

lunar torque N_m, which slows the Earth's spin. (Here M_e and M_m are masses of the Earth and Moon respectively, and r_m is the distance to the Moon.) In 1939 H. Spencer-Jones, by using observations made subsequent to 1680, inferred that $\dot{n}_m$ has been constant at $22''.4 \pm 1''.0$ century^{-2}. W. H. Munk and G. J. F. McDonald, in adopting this value, found $N_m = -3.9\times10^{23}$ dyne-cm, to which corresponds the energy loss rate $\dot{E}$, shown in Eq. (5), and the relative spin-down, $-\dot{\omega}/\omega = N/c\omega =$

$$\dot{E} = N(\omega - n_m) = 2.7\times10^{19}\text{ erg/s} \quad (5)$$

0.21 eon^{-1} (1 eon = 10^9 yr). (Here ω is Earth's rotational velocity, and c its principal moment of inertia.) The factor $\dot{n}_m$ is not as firm as might be desired. K. Lambeck reported values from several sources which are almost twice the Spencer-Jones value used by Munk. Concerning such differences in $\dot{n}_m$, W. M. Kaula proposed that the matter be ignored for several years until the time base is long enough to get a good figure from laser ranging to the Moon.

Tidal loss in the solid earth. The gravest free mode of the Earth, period 0.9 hr, has the same external form and nearly the same internal geometry as the M_2 tidal bulge. Its observed $Q = 350 \pm 100$ was used by Munk to estimate the loss rate $\dot{E}_B$ in the bodily M_2 tide (Q^{-1} is the loss rate number, and M_2 is the principal semidiurnal tide). The result, $\dot{E}_B = 7\times10^{17}$ erg/s, is only 3% of the required total in Eq. (5). According to Lambeck, the rate at which the Earth dissipates the total lunar-solar tidal energy is $5.7 \pm 0.5\times10^{19}$ erg/s, which is about double Munk's estimate of loss rate due to the Moon alone. The share attributed by Lambeck to the oceans is $5.0 \pm 0.3\times10^{19}$ erg/s. Commenting about the small difference, 7×10^{18} erg/s, he says that "if not all, at least a very major part of the secular change in the Moon's mean longitude is caused by dissipation of tidal energy in oceans, and we do not have to invoke significant energy sinks in the Earth's mantle and core."

Symmetric standard earth. The precision in generating symmetrical oceanless Earth models based upon seismic travel times and the periods of the Earth's numerous observed free vibrations had led to remarkable uniformity in the Love numbers of these models. W. E. Farrell performed a test aimed at producing differences in the Love numbers by changing the upper 1000 km of the Earth. Farrell substituted in the Gutenberg-Bullen A Earth model first a 1000-km upper layer of oceanic crustal structure, then the same thickness of continental shield structure, and compared the three models in respect to the computed Love numbers. Table 2 shows the results for the second-degree Love numbers. Differences exist chiefly in the fourth decimal place. Any of several modern Earth models would serve as a suitable reference for earth tide reductions. Presumably such a standard will be named. For the present, the values $\delta = 1.160$, $\gamma = 0.690$, and $l = 0.0840$ will serve, with an assumed zero phase lag of the tidal bulge. (This lag certainly is small, probably less than $0°.1$, but as noted present estimates of tidal energy dissipation in the solid Earth are imprecise.)

Instrumental dimensionless amplitude factors. The following paragraphs show how the reading of an earth tide meter is altered by the fact that it is anchored to a yielding Earth. The Earth chosen is assumed to be the symmetric standard of the previous paragraph.

Gravimeter. Basically, a gravimeter consists of a mass generally supported by a spring. (The superconducting model uses a magnetic field.) Variations in gravity are measured by the extension of the spring or in the null method by the small corrections required to restore the original configuration. A satellite affects the mean local value of gravity in three ways: (1) by its direct attraction, (2) by the tidal change in elevation of the observing station, and (3) by the redistribution of mass in the deformed Earth. Of these, the value of the first is expressed by Eq. (2), namely, $\Delta g_1 = \partial U / \partial r = 2U(r_0) r_0^{-1}$. An increase in elevation Δr diminishes the Earth's downward gravity field, that is, it increases the upward component (which is taken as positive in the present context). This is expressed as $\Delta g_2 = (\partial g / \partial r) \Delta r = +2g_0 r_0^{-1} \Delta r$, where g_0 is local gravity. For Δr the following expression defining the Love number h is substituted, $\Delta r = hg_0^{-1}U(r_0)$, thus obtaining $\Delta g_2 = 2hr_0^{-1}U(r_0)$. Finally, the change in gravity caused by the redistribution of mass is expressed by introducing another proportionality factor, the Love number k, defined by $V_0 = kU(r_0)$, where V_0 is the potential of the altered mass distribution at points on the Earth's surface. Outside the Earth, Eq. (6) holds.

$$V = kU(r_0) r_0^3 r^{-3} \tag{6}$$

The contribution Δg_3 at the surface $r = r_0$ is $\Delta g_3 = -3kr_0^{-1} U(r_0)$. Total change is per Eq. (7).

$$\Delta g = \Delta g_1 + \Delta g_2 + \Delta g_e = 2U(r_0)r_0^{-1}(1 + h - \tfrac{3}{2}k) \tag{7}$$

The factor $1 + h - \tfrac{3}{2}k$ is called the gravitational factor δ, but obviously the pertinent geophysical

quantity is $\delta - 1 = h - \tfrac{3}{2}k$, which alone characterizes the tidal deformation. On a rigid Earth, $h = k = 0$. On an Earth covered with a fluid layer, the equilibrium displacement Δr is determined by the requirement that the deformed surface remain equipotential, or "level," at its original value W_0. Thus $U(r_0) + (\partial W_0 / \partial r) \Delta r + kU(r_0) = 0$. Using the relations $\Delta r = hg_0^{-1}U$ and $\partial W_0 / \partial r = -g_0$, one finds that Eq. (8) holds for such an Earth. Furthermore,

$$1 - h + k = 0 \tag{8}$$

for a homogeneous incompressible Earth of density ρ, it is easy to compute the external potential due to the superficial mass distribution associated with the displacement $\Delta r = hg_0^{-1}U_2$. This potential is in fact given by Eq. (9), which at $r = r_0$ is $kU(r_0)$

$$V(r,\theta) = \tfrac{4}{5}\pi G\rho r_0^4 r^{-3} hg_0^{-1} U(r_0,\theta)$$
$$= \tfrac{3}{5}g_0 r_0 r^{-1} hg_0^{-1} U(r_0,\theta) \tag{9}$$

by definition. Thus $k = \tfrac{3}{5}h$ for a homogeneous Earth, fluid or solid. For a homogeneous incompressible fluid Earth, in view of Eq. (8), $h = {}^5/_2$ and $k = \tfrac{3}{2}$. For this case, $\delta = 1.25$. Values of $\delta > 1.25$ deduced from observations must be due to errors or to extreme local conditions.

Tiltmeters. The tiltmeter measures changes in the angle of tilt between the tiltmeter's foundation and the local vertical, both subject to tidal variations. Several types of tiltmeters are in use. In the horizontal pendulum the rotation axis is fixed at a small angle with the vertical to produce high sensitivity to tilts. In the level tube, which may be several hundred meters or more long, the difference in elevation of a fluid is measured at the two ends. In this way a sample of the tilt of a large region is obtained, but the horizontal pendulum sometimes also samples a large volume, as in A. Marussi's installation in the Grotto Gigante, Trieste, where these pendulums are 75 m high. Generally two orthogonal tiltmeters are used to deduce the total tilt angle. In terms of the Love numbers h and k, the tilt observations have the following significance. The ground itself is elevated an amount $\Delta r = hg^{-1}U$ by the tidal potential, so that in the direction of maximum tilt the ground's tilt angle ψ is given by Eq. (10).

$$\tan \psi \doteq \psi = hg^{-1}r^{-1}\frac{\partial U}{\partial \theta} \tag{10}$$

This tilt of the solid Earth, for which h is about 0.61, is clearly less than that of the local level (that is, fluid) surface for which h exceeds unity (that is, $h = 1 + k$). Since the tilt of the level surface is $(1 + k)g^{-1}r^{-1}(\partial U / \partial \theta)$, the net tilt angle observed (ψ) is given by Eq. (11). This quantity achieves its maxi-

$$\psi = g^{-1}r^{-1}\frac{\partial U}{\partial \theta}(1 + k - h) \tag{11}$$

mum value when the satellite's zenith angle is 45°, attains an equal minimum at $\theta = 135°$, and is zero when $\theta = 0$, 90°, or 180°. The factor $1 + k - h$ is called the tilt number γ but, in analogy with the previous case, the significant part is only $1 - \gamma = h - k$. In magnitude $h - k$ is about twice the corresponding gravitational quantity $\delta - 1 = h - \tfrac{3}{2}k$.

Extensometer. The extensometer, or linear strainmeter, measures the change in distance between two reference points. In the Benioff design

Table 2. Earth tide parameters for different upper mantles

Model	k	h	l	δ	γ
Gutenberg-Bullen Earth model	0.3040	0.6114	0.0832	1.1554	0.6926
Oceanic mantle	0.3055	0.6149	0.0840	1.1567	0.6906
Shield mantle	0.3062	0.6169	0.0842	1.1576	0.6893

these positions are connected with a quartz tube to bring the points into juxtaposition for precise measurements of relative displacement. By using laser beams, the measurements may be made over distances of kilometers, but the small mechanical meters still have their use. They offer considerably greater utility per unit cost than other equivalent instruments.

The total horizontal component of displacement s toward the satellite serves to define the dimensionless proportionality constant, or Shida's number, l in accordance with Eq. (12), where g is the

$$s = lg^{-1}\frac{\partial U}{\partial \theta} \qquad (12)$$

local value of gravity. However, the absolute displacement s is not measured, but instead the relative displacement between two points, namely a strain component. While the general specification of the strain in an isotropic solid requires six numbers, on the Earth's free surface the two components of shear stress and their associated strain components vanish. Further, a horizontal strainmeter is immune to vertical strains, so that, in all, only three constituents of strain remain effective. To determine the horizontal areal strain (ratio of change in area to original area), since this is independent of azimuth, only two rectangular strain components are needed. These two strain components are sufficient to determine the combination $h - 3l$ of the Love and Shida numbers. With a third horizontal strainmeter, l and h may be independently obtained.

Indicated and true phase lags. The phase indications of a gravimeter or tiltmeter are only a fraction of the true angular lag of the tidal bulge. The tidal bulge is assumed to retain its no-loss equilibrium form, but with axis carried forward from the Moon's zenith by the small angle ϵ required to produce the Moon's known orbital acceleration. In the expressions for the dimensionless amplitude factors, $\delta = 1 + (h - \frac{3}{2}k)$ and $\gamma = 1 + (k - h)$, the tidal losses are produced only by the terms in parentheses. On a rigid, lossless Earth, $\delta = \gamma = 1.0$. Figure 1a is a vector diagram illustrating the relation between a phase angle ϕ observed with a gravimeter and the bulge lag angle ϵ (geometrically a lead angle with respect to the Moon's zenith, but in time, a lag). For small angles, $\epsilon = \delta\phi/(\delta - 1) = 7.25\phi$, if $\delta = 1.16$. The similar relation between true bulge lag ϵ and the phase lag ψ observed in tilt measurements is shown in Fig. 1b. Here $\epsilon = -\gamma\psi/(1 - \gamma) = -2.23\psi$, if $\gamma = 0.69$. Strainmeters, of course, record strain-amplitude components directly, and the true phase lag ϵ of the tidal bulge on the idealized model.

Residuals. The observations of a gravimeter, tiltmeter, or strainmeter are vector quantities, often written in dimensionless form. $\vec{\delta}_0 = \delta_0\exp(i\phi_0)$, $\vec{\gamma}_0 = \gamma_0\exp(i\psi_0)$, $\vec{l}_0 = l_0\exp(i\lambda_0)$, where δ_0, γ_0, and l_0 are the dimensionless amplitudes, and ϕ_0, ψ_0, and λ_0 are the observed phase angles. The residuals, $\vec{\delta} = \vec{\delta}_0 - 1.16$, $\vec{\gamma} = \vec{\gamma}_0 - 0.69$, and $\vec{l} = \vec{l}_0 - 0.084$, referred to the standard symmetrical Earth, contain all the nontrivial information. In practice they are resolved into their harmonic constituents and displayed as functions of these known frequencies. These resolved residuals are the essence of the observational subject.

Earth tides and ocean tides. A major part of research in earth tides relates to ocean tides. Where local earth tides are large and well known, as in the Bay of Fundy, the Irish Sea, and the Alaskan area, the ocean loads afford a tempting opportunity to investigate the local crustal response to such loads and perhaps to revise ideas about its mechanical structure. Observations by gravimeters may be appreciably affected by distant ocean tides. Near the coastline, ocean loads are commonly responsible for 10% of the gravity tide, 25% of the strain tide, and 90% of the tilt tide. Studies of earth tides and ocean tides have become mutually supportive. Solutions for mean values of ocean tides which are free of the local distortions from effects of shallow water can be provided by earth tides. W. E. Farrell produced evidence of the precision of the transcontinental gravity tide profile of J. T. Kuo and coworkers by showing that agreement was produced only when the omitted condition of conservation of mass in the Atlantic ocean tides was introduced. Global coverage with precise earth tide measurements is a major continuing subject of study.

Nearly diurnal resonance. H. Jeffreys and R. O. Vicente, and M. S. Molodensky deduced theoretically the existence of a variation in the amplitudes of the nearly diurnal earth tides due to resonance in the coupling of the Earth's mantle and liquid core. M. G. Rochester noted that either the observations are too inaccurate to determine the location of the resonant frequency or the resonance-band structure is more complicated than expected. With this nearly diurnal resonance should also be observed a strangely neglected nutation, rediscovered by A. Toomre in 1974, of period 460 days, which thus far has equally strangely escaped observation. The continuation of the observations by earth tide meters at well-selected samples of different geographic locations should provide the needed additional evidence. Because the maximum tilt amplitudes in the diurnal constituents occur theoretically at the South Pole, tilt observations in bore holes in the ice have been proposed.

Dilatation. In some seismically active areas it has been observed that slow changes in the dilatation of crustal rocks and their associated absorption of additional interstitial water occur prior to earthquakes. Such variations have been detected by noting changes in the observed ratio of the seismic-wave velocities of the compressional and shear waves. C. Beaumont and J. Berger produced theoretical arguments which show that tidal strains and tidal tilts should be somewhat more strongly affected by the dilatations than are seismic waves. The records of strainmeters, and especially of tiltmeters in the preferred new sites in drill holes, should be monitored in the search for changes in dilatation, hithertofore detected only by seismic waves.

High-latitude stations. At Longyearbyen, Spitzbergen (78°20′N, 15°52′E), observations with three Askania gravimeters and 3 north-south and 3 east-west Verbaandert Melchior quartz pendulums were taken during June 1969 to July 1970. For the gravity factor, δ_{mf} of the fortnightly tide the weighted mean was 1.142 ± 0.014, but the spread in the values obtained by the three gravimeters seems

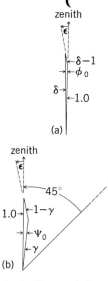

Fig. 1. Observed phase lags in relation to the true lag ϵ of the tidal bulge. (a) ϕ_0 observed by a gravimeter. (b) ψ_0 observed by a tiltmeter.

large, 1.124 to 1.183. For the semiannual tide (SSa) the gravity factor $\delta_{ss} = 1.094 \pm 0.045$ was obtained, with a phase lag of 3 or 4 days.

At the South Pole, latitude $-90°$, elevation 2810.4 m, observations have been carried on since 1967. The observations during 107 days in 1970 gave the value $\delta = 1.153 \pm 0.002$ for the fortnightly and monthly tides (unresolved), and $\lambda = 0°\!.60 \pm 0°\!.40$ for the phase lag for the fortnightly period. B. V. Jackson and L. B. Slichter reported observations of the "forbidden" daily tides, with maximum amplitude in the diurnal band $= 0.606 \pm 0.003$ microgal (K_1), and in the semidiurnal band $= 0.341 \pm 0.0015$ (M_2), as shown in Fig. 2. Tests concerning the presence of a persistent inner-core oscillation showed that no such oscillation producing a signal of amplitude greater than 0.008 microgal could have been present.

The advantages of the South Pole location are the following: (1) The amplitudes of all the long-period gravity tides are theoretically maximal, whereas the short-period gravity tides, which theoretically vanish on a symmetric Earth, have small observed amplitudes in the fractional microgal range. (2) Concerning the observations of tilts, the long-period and semidiurnal amplitudes vanish in theory, but the diurnal constituents, significant in the display of nearly diurnal resonances of the liquid core, all have maximum amplitude. (3) In the observation of free modes, the splitting by rotation

vanishes to first order. The records should be simpler. (4) The corrections for ocean tides are small and will be increasingly better known, aided by more abundant ocean tide measurements in southern seas. The location is good for operating earth tide meters designed for highest precision.

Instrumentation. Improvements in the sensitivity, portability, and economy of earth tide instruments are always in progress. The use of boreholes as stable sites for tilt pendulums has been pioneered. G. Cabanias has reported results on the M_2 tides from three Arthur D. Little borehole tiltmeters installed in holes 100 m apart at Bedford, MA. The three instruments agreed within 2% in amplitude and 1° in phase. Work by J. C. Harrison in the Poorman mine shows the importance there of the mine cavity and of surface topography upon measurements of tidal tilt and strain. A minimodel of the low-drift-rate superconducting gravimeter developed by J. Goodkind is designed to increase to several months the intervals between replenishing the helium supply. It is hoped that this meter will become a widely used standard. At the South Pole the plumb-bob suspensions of the two LaCoste-Romberg gravimeters preserve true levels conveniently. Careful corrections were introduced there in 1971 for the direct and load effects of barometric changes, and for changes in drift rate. This station produces readings of superior quality, significant at the nanogal level in the diurnal and shorter-period range.

[LOUIS B. SLICHTER]

Bibliography: C. Beaumont and J. Berger, Earthquake prediction: Modification of the earth tide tilts and strains by dilatancy, *Geophys. J. Roy. Astron. Soc.*, 39:111–122, 1974; M. Bonatz and T. Chojnicki, International Astro-Geo-Project Spitsbergen 1969/70, *Berechnung langperiodischer Gezeitenwellen für die Gravimeterstation Longyearbyen, Mitteilungen ars dem Institut für Theoretische Geodasie der Universität Bonn*, vol. 7, 1972; R. A. Broucke, W. E. Zürn, and L. B. Slichter, Lunar tidal acceleration on a rigid Earth, in H. C. Heard et al. (eds.), *Flow and Fracture of Rocks*, Geophys. Monogr. 16, American Geophysical Union, 1972; W. E. Farrell, Deformation of the Earth by surface loads, *Rev. Geophys. Space Phys.*, 10:761–797, 1972; J. Goodkind and W. Prothero, Tidal measurements with a superconducting gravimeter, *J. Geophys. Res.*, 77:926–937, 1972; J. C. Harrison, Cavity and topographic effects in tilt and strain measurements, *J. Geophys. Res.*, 81: 319–328, 1976; M. C. Hendershott, Ocean tides, *E. O. S.*, 54:76–86, 1973; B. V. Jackson and L. B. Slichter, The residual daily earth tides at South Pole, *J. Geophys. Res.*, 79:1711–1715, 1974; J. T. Kuo, Earth tides, *Rev. Geophys. Space Phys.*, 13:260–262, 1975; J. T. Kuo et al., Transcontinental gravity profile across the U. S., *Science*, 168: 968–971, 1970; P. L. Lagus and D. L. Anderson, Tidal dissipation in the Earth and planets, *Phys. Earth Planet Interiors*, 1:57–62, 1968; K. Lambeck, Effects of tidal dissipation in the oceans on the Moon's orbit and the Earth's rotation, *J. Geophys. Res.*, 80:2917–2925, 1975; P. Melchior, *Earth Tides*, 1966; P. Melchior, Earth tides and polar motions, *Technophysics*, 13:361–372, 1972; W. H. Munk, Once again—tidal friction, *Geophys.*

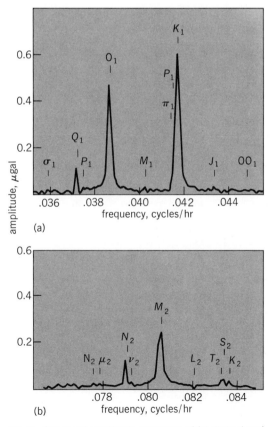

Fig. 2. Observed amplitude spectrum of (a) diurnal and (b) semidiurnal constituents at South Pole. (*From B. V. Jackson and L. B. Slichter, The residual daily earth tides at the South Pole, J. Geophys. Res., 79(11):1711–1715, copyright 1974 by the American Geophysical Union*)

J. Roy. Astron. Soc., 9:352–375, 1968; W. H. Munk and G. J. F. MacDonald, *The Rotation of the Earth*, rev. ed., 1975; P. H. Sydenham, 2000 hr comparison of 10 m quartz-tube and quartz-catenary tidal strainmeters, *Geophys. J. Roy. Astron. Soc.*, 38:377–387, 1974; A. Toomre, On the "nearly diurnal wobble" of the Earth, *Geophys. J. Roy. Astron. Soc.*, 38:335–348, 1974.

Echo sounder

A marine instrument used primarily for determining the depth of water by means of an acoustic echo. A pulse of sound sent from the ship is reflected from the sea bottom back to the ship, the interval of time between transmission and reception being proportional to the depth of the water.

Echo sounders, sometimes called fathometers, are used by vessels for navigational purposes, not only to avoid shoal water, but as an aid in fixing position when a good bathymetric chart of the area is available. Some sensitive instruments are used by commercial fishermen or marine biologists to detect schools of fish or scattering layers of minute marine life. Oceanographic survey ships use echo sounders for charting the ocean bottom. Figure 1 shows an echo-sounder record obtained by ocean-

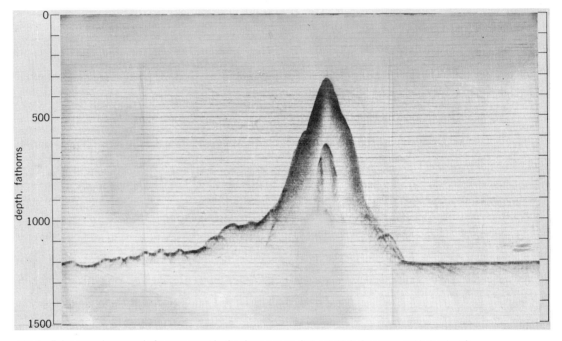

Fig. 1.   Echo-sounder record of a seamount in the deep ocean. (*Woods Hole Oceanographic Institute*)

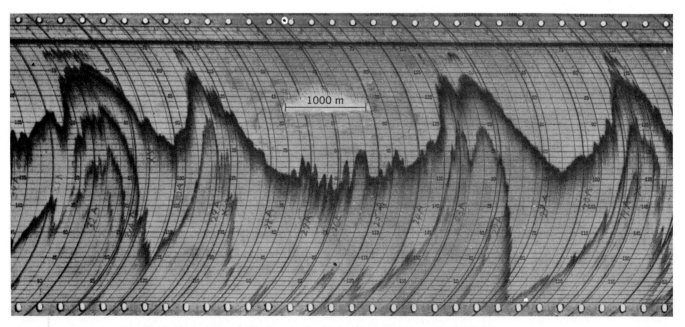

Fig. 2.   Typical record of bottom profile obtained by an echo sounder. (*USCGS*)

ographers of a seamount (undersea mountain). *See* SCATTERING LAYER.

An echo sounder is really a type of active sonar. It consists of a transducer located near the keel of the ship which serves (in most models) as both the transmitter and receiver of the acoustic signal; the necessary oscillator, receiver, and amplifier which generate and receive the electrical impulses to and from the transducer; and a recorder or other indicator which is calibrated in terms of the depth of water. An echo sounder actually measures time differences, so some average velocity of sound must be assumed in order to determine the depth. The frequency generally employed is in the low ultrasonic range (20,000–30,000 Hz). The depth display may be given by a trace-type recorder which supplies a continuous permanent record (Fig. 2) or, as is the case with less expensive commercial instruments, it may be a dial-type display, giving instantaneous depth of water. *See* MARINE GEOLOGY; SONAR; UNDERWATER SOUND.

[ROBERT W. MORSE]

Estuarine oceanography

The study of the physical, chemical, biological, and geological characteristics of estuaries. An estuary is a semienclosed coastal body of water which has a free connection with the sea and within which the sea water is measurably diluted by fresh water derived from land drainage. Many characteristic features of estuaries extend into the coastal areas beyond their mouths, and because the techniques of measurement and analysis are similar, the field of estuarine oceanography is often considered to include the study of some coastal waters which are not strictly, by the above definition, estuaries. Also, semienclosed bays and lagoons exist in which evaporation is equal to or exceeds fresh-water inflow, so that the salt content is either equal to that of the sea or exceeds it. Hypersaline lagoons have been termed negative estuaries, whereas those with precipitation and river inflow equaling evaporation have been called neutral estuaries. Positive estuaries, in which river inflow and precipitation exceed evaporation, form the majority, however.

Topographic classification. Embayments are the result of fairly recent changes in sea level. During the Pleistocene ice age, much of the sea water was locked up in continental ice sheets, and the sea surface stood about 100 m below its present level. In areas not covered with ice, the rivers incised their valleys to this base level; and during the ensuing Flandrian Transgression, when the sea level rose at about 1 m per century, these valleys became inundated. Much of the variation in form of the resulting estuaries depends on the volumes of sediment that the river or the nearby coastal erosion has contributed to fill the valleys.

Where river flow and sediment discharge were high, the valleys have become completely filled and even built out into deltas. Generally, deltas are best developed in areas where the tidal range is small and where the currents cannot easily redistribute the sediment the rivers introduce. They occur mainly in tropical and subtropical areas where river discharge is seasonally very high. The distributaries, or "passes," of the delta are gener-

ally shallow, and often the shallowest part is a sediment bar at the mouths of the distributaries. The Mississippi and the Niger are examples of this type of delta.

Where sediment discharge was less, the estuaries are unfilled, although possibly they are still being filled. These are drowned river valleys or coastal plain estuaries, and they still retain the topographic features of river valleys, having a branching, dendritic, though meandering, outline and a triangular cross section, and widening regularly toward the mouth, which is often restricted by spits. River discharge tends to be reasonably steady throughout the year, and sediment discharge is generally small. These estuaries occur in areas of high tidal range, where the currents have helped to keep the estuaries clear of sediment. They are typical of temperate regions such as the eastern coast of North America and northwestern Europe, examples being the Chesapeake Bay system, the Thames, and the Gironde.

In areas where glaciation was active, the river valleys were overdeepened by glaciers, and fiords were created. A characteristic of these estuaries is the rock bar or sill at the mouth that can be as little as a few tens of meters deep. Inside the mouth, however, they can be at least 600 m deep and can extend hundreds of kilometers inland. Fiords are typical of the Norwegian and the Canadian Pacific coasts.

There is another estuarine type, called the bar-built estuary. These are formed on low coastlines where extensive lagoons have narrow connecting passages or inlets to the sea. Within the shallow lagoons the tidal currents are small, but the deep inlets have higher currents. Again, a sediment bar is generally present across the entrance. In tropical areas the lagoons can be hypersaline during the hot season. They are typical of the southern states of the United States and of parts of Australia.

Estuaries are ephemeral features since great alterations can be wrought by small changes in sea level. If the present ice caps were to melt, the sea level would rise an estimated 30 m, and the effect on the form and distribution of estuaries would be drastic.

Physical structure and circulation. Within estuaries, the river discharge interacts with the sea water, and river water and sea water are mixed by the action of tidal motion, by wind stress on the surface, and by the river discharge forcing its way toward the sea. The difference in salinity between river water and sea water—about 35 parts per thousand—creates a difference in density of about 2%. Even though this difference is small, it is sufficient to cause horizontal pressure gradients within the water which affect the way it flows. Density differences caused by temperature variations are comparatively smaller. Salinity is consequently a good indicator of estuarine mixing and the patterns of water circulation. Obviously, there are likely to be differences in the circulation within estuaries of the same topographic type which are caused by differences in river discharge and tidal range. The action of wind on the water surface is an important mixing mechanism in shallow estuaries, particularly in lagoons; but generally its

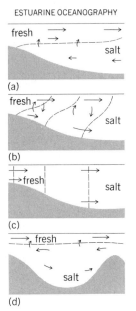

Fig. 1. Diagrammatic representation of mixing in estuaries. (*a*) Salt-wedge type (*b*) Partially mixed type. (*c*) Well-mixed type. (*d*) Fiord.

effect on estuarine circulation is only temporary, although it can produce considerable variability and thus make interpretation of field results difficult. *See* SEA WATER.

Salt-wedge estuaries. Fresh water, being less dense than sea water, tends to flow outward over the surface of sea water, which penetrates as a salt wedge along the bottom into the estuary (Fig. 1). This creates a vertical salinity stratification, with a narrow zone of sharp salinity change, called a halocline, between the two water masses which can reach 30 parts per thousand in 1/2 m. If the sea is tideless, the water in the salt wedge is almost motionless. However, if the surface layer flowing toward the sea has a sufficiently high velocity, it can create interfacial waves on the halocline. These waves break, ejecting small parcels of salt water into the fresher surface layer; this process is called entrainment, and it occurs all along the halocline. No fresh water is mixed downward; thus the salinity within the salt wedge is almost constant along the estuary. However, the salt wedge loses salt water which is mixed into the surface layer and discharged into the sea. Consequently, for this loss to be replaced, there must be a compensatory flow of salt water toward the head of the estuary within the salt wedge, but of a magnitude much less than that of the flow in the surface layer. There is a considerable velocity gradient near the halocline as a result of the friction between the two layers. Consequently, the position of the salt wedge will change according to the magnitude of the flow in the surface layer, that is, according to the river discharge. The Mississippi River is an example of a salt-wedge estuary. When the flow in the Mississippi is low, the salt wedge extends more than 100 mi (160 km) inland, but with high discharge, the salt wedge only extends a mile (1.6 km) or so above the river mouth. Some bar-built estuaries, in areas of restricted tidal range and at times of high river discharge, as well as deltas, are typical salt-wedge types.

Partially mixed estuaries. When tidal movements are appreciable, the whole mass of water in the estuary moves up and down with a tidal periodicity of about 12-1/2 hr. Considerable friction occurs between the bed of the estuary and the tidal currents, and this causes turbulence. The turbulence tends to mix the water column more thoroughly than entrainment does, although little is known of the relationship of the exchanges to the salinity and velocity gradients. However, the turbulent mixing not only mixes the salt water into the fresher surface layer but also mixes the fresher water downward. This causes the salinity to decrease toward the head of the estuary in the lower layer and also to progressively increase toward the sea in the surface layer. As a consequence, the vertical salinity gradient is considerably less than that in salt-wedge estuaries. In the surface, seaward-flowing layer, the river discharge moves toward the sea; but because the salinity of the water has been increased by mixing during its passage down the estuary, the discharge at the mouth can be several times the river discharge. To provide this volume of additional water, the compensating inflow must also be much higher than that in the salt-

wedge estuary. The velocities involved in these movements are only on the order of a few centimeters per second, but the tidal velocities can be on the order of a hundred centimeters per second. Consequently, the only way to evaluate the effect of turbulent mixing on the circulation pattern is to average out the effect of the tidal oscillation, which requires considerable precision and care. The resulting residual or mean flow will be related to the river discharge, although the tidal response of the estuary can give additional contributions to the mean flow. The tidal excursion of a water particle at a point will be related to the tidal prism, the volume between high- and low-tide levels upstream of that point; and the instantaneous cross-sectional velocity at any time will be related to the rate of change of the tidal prism upstream of the section. In details, the velocities across the section can differ considerably. It has been found that in the Northern Hemisphere the seaward-flowing surface water keeps to the right bank of the estuary, looking downstream, and the landward-flowing salt intrusion is concentrated on the left-hand side (Fig. 2). This is caused by the Coriolis force, which deflects the moving water masses toward the right. Of possibly greater importance, however, is the effect of topography, because the curves in the estuary outline tend to concentrate the flow toward the outside of the bends. Thus, in addition to a vertical circulation, there is a horizontal one, and the halocline slopes across the estuary. Because the estuary has a prismatic cross section, the saline water is concentrated in the deep channel and the fresher water is discharged in the shallower areas.

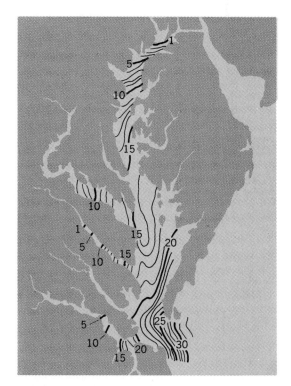

Fig. 2. Typical surface salinity distribution in Chesapeake Bay. (*From D. W. Pritchard, Estuarine Hydrography, in H. E. Landsberg, ed., Advances in Geophysics, vol. 1, Academic Press, 1952*)

Examples of partially mixed estuaries are the rivers of the Chesapeake Bay system.

Well-mixed estuaries. When the tidal range is very large, there is sufficient energy available in the turbulence to break down the vertical salinity stratification completely, so that the water column becomes vertically homogeneous. In this type of estuary there can be lateral variations in salinity and in velocity, with a well-developed horizontal circulation; or if the lateral mixing is also intense, the estuary can become sectionally homogeneous (also called a one-dimensional estuary). Because there is no landward residual flow in the sectionally homogeneous estuary, the upstream movement of salt is produced during the tidal cycle by salty water being trapped in bays and creeks and bleeding back into the main flow during the ebb. This mechanism spreads out the salt water, but it is probably effective only for a small number of tidal excursions inland.

Fiords. Because fiords are so deep and restricted at their mouths, tidal oscillation affects only their near-surface layer to any great extent. The amount of turbulence created by oscillation is small, and the mixing process is achieved by entrainment. Thus fiords can be considered as salt-wedge estuaries with an effectively infinitely deep lower layer. The salinity of the bottom layer will not vary significantly from mouth to head, and the surface fresh layer is only a few tens of meters deep. When the sill is deep enough not to restrict circulation, the inflow of water occurs just below the halocline, with an additional slow outflow near the bottom. When circulation is restricted, the replenishment of the deeper water occurs only occasionally, sometimes on an annual cycle; and between the inflows of coastal water, the bottom layer can become anoxic.

The descriptive classification of estuaries outlined above depends on the relative intensities of the tidal and river flows and the effect that these flows have on stratification. A quantitative comparison between estuaries can be made using the diagram of Fig. 3, which is based on a stratification and a circulation parameter.

Flushing and pollution-dispersal prediction.
Much research into estuarine characteristics is aimed at predicting the distribution of effluents discharged into estuaries. Near the mouth of a partially mixed estuary, the salt water is only slightly diluted, and in order for a volume of fresh water equivalent to the river flow to be discharged, a much greater volume of mixed water must flow seaward. Consequently, estuaries are more effective in diluting and removing pollutants than rivers. It has been observed that increased river flow causes both a downstream movement of the saline intrusion and a more rapid exchange of water with the sea. The latter effect occurs because increased river discharge increases stratification; increased stratification diminishes vertical mixing and enhances the flow toward the sea in the surface layer. Thus, increased river discharge has the effect of increasing the volume of fresh water accumulated in the estuary, but to a lesser extent than the increase of the discharged volume. Obviously, it takes some time for the fresh water from

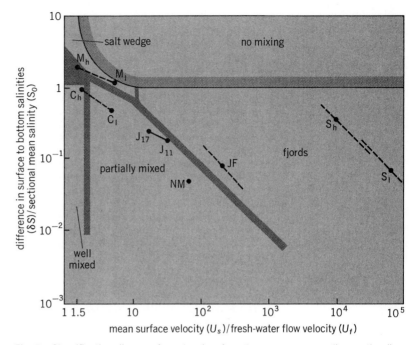

Fig. 3. Classification diagram for estuaries. An estuary appears as a line on the diagram; the upper reaches are less well mixed than the lower sections. Subscript letters refer to high (h) and low (l) river discharge; subscript numbers are distances from the mouth. J = James River; M = Mississippi; C = Columbia River; NM = Narrows of the Mersey; S = Silver Bay; JF = Strait of Juan de Fuca. (*After D. V. Hansen and M. Rattray, Jr., New dimensions in estuary classification, Limnol. Oceanogr., 11:319–326, 1966*)

the river to pass through the estuary. The flushing time can be determined by dividing the total volume of fresh water accumulated in the estuary by the river flow. For most estuaries the flushing time is between 5 and 10 days.

If a conservative, nondecaying pollutant is discharged at a constant rate into an estuary, the effluent concentration in the water moving past will vary with the tidal current velocity, and will spread out by means of turbulent mixing. The concentrations will be increased during the next half cycle as the water passes the discharge point again. After several tidal cycles a steady-state distribution will be achieved, with the highest concentration near the discharge point. Concentrations will decrease downstream, but not as quickly as they decrease upstream. However, the details of the distribution will depend largely on whether the discharge is of dense or light fluid and whether it is discharged into the lower or upper layer. Since its movement will be modified by the estuarine circulation, the effluent will be more concentrated in the lower layer upstream of the discharge point, and it will be more concentrated in the upper layer downstream. To obtain maximum initial dilution, a light effluent would have to be discharged near the estuary bed so that it would mix rapidly as it rose.

For nonconservative pollutants, such as coliform sewage bacteria, prediction becomes more difficult. The population of bacteria dies progressively through the action of sunlight, and concentrations diminish with time as well as by dilution. The faster the mixing, the larger the populations at any distance from the point of introduction, since less decay occurs.

Because of the poor mixing of fresh water into a salt-wedge estuary, an effluent introduced in the surface layer will be flushed from the estuary before it contaminates the lower layer, provided that it is not too dense.

Mathematical modeling. Increasingly, mathematical modeling is being used, with reasonable success in many instances, to predict effluent dispersal with the minimum amount of field data. Although the governing mathematical equations can be stated, they cannot be solved in their full form because there are too many unknowns. Consequently, to reduce the number of unknowns, various assumptions are made, including some form of spatial averaging to reduce a three-dimensional problem to two dimensions or even one dimension. The exchange ratios, about which little is known, are assumed constant, or are considered as a simple variable in space, and are altered so that the model fits the available prototype data.

The first step is usually to model the flow and salinity distribution. Because the density field is important in determining the flow characteristics, density and flow are interlinked problems. Then, for pollutant studies, the pollutant is assumed to act in the same manner as fresh or salt water, or the flow parameters are used with appropriate exchange coefficients to predict the distribution. Simple models consider the mean flow to be entirely the result of river discharge, and tidal flow to be given by the tidal prism. Segmentation is based on simple mixing concepts and crude exchange ratios. Salinity and pollutant concentrations can then be calculated for cross-sectionally averaged and vertically homogeneous conditions by using the absolute minimum of field data. These models are known as tidal prism models. One-dimensional models are very similar, but use a finer grid system and need better data for validation. Two-dimensional models either assume vertical homogeneity and allow lateral variations, or vice versa. There are difficulties in including the effects of tidally drying areas and junctions; the models become increasingly costly and require extensive prototype data, but they are more realistic. The ideal situation of modeling the flow and salinity distribution accurately simply on the basis of knowledge about the topography, the river discharge, and the tidal range at a number of points is still a long way off.

Estuarine environments. Estuarine ecological environments are complex and highly variable when compared with other marine environments. They are richly productive, however. Because of the variability, fewer species can exist as permanent residents in this environment than in some other marine environments, and many of these species are shellfish that can easily tolerate short periods of extreme conditions. Motile species can escape the extremes. A number of commercially important marine forms are indigenous to the estuary, and the environment serves as a spawning or nursery ground for many other species.

River inflow provides a primary source of nutrients such as nitrates and phosphates which are more concentrated than in the sea. These nutrients are utilized by plankton through the photosynthetic action of sunlight. Because of the energetic mixing, production is maintained throughout, in spite of the high levels of suspended sediment which restrict light penetration to a relatively thin surface layer. Plankton concentrations can be extremely high, and when they are, higher levels of the food web—filter-feeding shellfish and young fish—have an ample food source. The rich concentrations provide large quantities of organic detritus in the sediments which can be utilized by bottom-feeding organisms and which can be stirred up into the main body of the water by tidal action. For a more complete treatment of the ecology of estuarine environments from the biological viewpoint *see* MARINE ECOSYSTEM.

There is a close relationship between the circulation pattern in estuaries and the faunal distributions. Several species of plankton peculiar to estuaries appear to confine their distribution to the estuary by using the water-circulation pattern; pelagic larvae of oysters are transported in a similar manner. The fingerling fish (*Micropogon undulatus*), spawned in the coastal waters off the eastern coast of the United States, are carried into the estuarine nursery areas by the landward residual bottom flow.

Estuarine sediments. The patterns of sediment distribution and movement depend on the type of estuary and on the estuarine topography. The type of sediment brought into the estuary by the rivers, by erosion of the banks, and from the sea is also important; and the relative importance of each of these sources may change along the estuary. Fine-grained material will move in suspension and will follow the residual water flow, although there may be deposition and re-erosion during times of locally low velocities. The coarser-grained material will travel along the bed and will be affected most by high velocities and, consequently, in estuarine areas, will normally tend to move in the direction of the maximum current.

Fine-grained material. Fine-grained clay material, about 2 μm in size, brought down the rivers in suspension, can undergo alterations in its properties in the sea. Base (cation) exchange with the sea water can alter the chemical composition of some clay minerals; also, because the particles have surface ionic charges, they are attracted to one another and can flocculate. Flocculation depends on the salinity of the water and on the concentration of particles. It is normally complete in salinities in excess of 4 parts per thousand, and with suspended sediment concentrations above about 300 ppm (mg l^{-1}), and has the effect of increasing the settling velocities of the particles. The flocs have diameters larger than 30 μm, but effective densities of about 1.1 g cm^{-3} because of the water closely held within. If the material is carried back into regions of low salinity, the flocculation is reversed, and the flocs can be disrupted by turbulence. In sufficiently high concentrations, the suspended sediment can suppress turbulence. The sediment then settles as layers which can reach concentrations as high as 300,000 ppm and which are visible as a distinctive layer of "fluid mud" on echo-sounder recordings. At low concentrations, aggregation of particles occurs mainly by biological action.

Turbidity maximum. A characteristic feature of partially mixed estuaries is the presence of a turbidity maximum. This is a zone in which the suspended sediment concentrations are higher than those either in the river or farther down the estuary. This zone, positioned in the upper estuary around the head of the salt intrusion and associated with mud deposition in the so-called mud reaches, is often related to wide tidal mud flats and saltings. The position of the turbidity maximum changes according to changes in river discharge, and is explained in terms of estuarine circulation. Suspended sediment is introduced into the estuary by the residual downstream flow in the river. In the upper estuary, mixing causes an exchange of suspended sediment into the upper layer, where there is a seaward residual flow causing downstream transport. In the middle estuary, the sediment settles into the lower layer in areas of less vigourous mixing to join sediment entering from the sea on the landward residual flow. It then travels in the salt intrusion back to the head of the estuary. This recirculation is a very effective mechanism for sorting the sediment, which is of exceedingly uniform mineralogy and settling velocity. Flocs with low settling velocities tend to be swept out into the coastal regions and onto the continental shelf. The heavier or larger flocs tend to be deposited.

The concentrations change with tidal range and during the tidal cycle, and fluid muds can occur within the area of the turbidity maximum if concentrations become sufficiently high. During the tidal cycle, as the current diminishes, individual flocs can settle and adhere to the bed, or fluid muds can form. The mud consolidates slightly during the slack water period, and as the current increases at the next stage of the tide, erosion may not be intense enough to remove all of the material deposited. A similar cycle of deposition and erosion occurs during the spring-neap tidal cycle. Generally, there is more sediment in suspension in the turbidity maximum than is required to complete a year's sedimentation on the estuary bed.

Mud flats and tidal marshes. The area of the turbidity maximum is generally well protected from waves, and there are often wide areas of mud flats and tidal marshes (Fig. 4). These areas also exchange considerable volumes of fine sediment with sediment in suspension in the estuary. At high water the flats are covered by shallow water, and there is often a long stand of water level which gives the sediment time to settle and reach the bottom, where it adheres or is trapped by plants or by filter-feeding animals. The ebb flow is concentrated in the winding creeks and channels. At low water there is not enough time for the sediment to settle, and it is distributed over the tidal flats during the incoming tide. Thus there is a progressive movement of fine material onto the mud flats by a process that depends largely on the time delay between sediment that is beginning to settle and sediment that is actually reaching the bed. The tidal channels migrate widely, causing continual erosion. Consequently, there is a constant exchange of material between one part of the marsh-

Fig. 4. Aerial photograph of tidal flats showing the areas of pans, marshes, and vegetation between the channels, Scolt Head Island, England. (*Photograph by J. K. St. Joseph, Crown copyright reserved*)

es and another by means of the turbidity maximum. As the muds that are eroded are largely anaerobic, owing to their very low permeability, the turbidity maximum is an area with reduced amounts of dissolved oxygen in the water.

Coarse-grained material. Coarser materials such as quartz sand grains that do not flocculate travel along the bed. Those coming down the river will stop at the tip of the salt intrusion, where the oscillating tidal velocities are of equal magnitude at both flood and ebb. Ideally, coarser material entering from the sea on the landward bottom flow will also stop at the tip of the salt intrusion, which becomes an area of shoaling, with a consequent decrease of grain size inland. However, normally the distribution of the tidal currents is too complex for this pattern to be clear. Especially in the lower part of the estuary, lateral variations in velocity can be large. The flood and ebb currents preferentially take separate channels, forming a circulation pattern that the sediment also tends to follow. The channels shift their positions in an apparently consistent way, as do the banks between them. This sorts the sediment and restricts the penetration of bed-load material into the estuary.

Salt-wedge patterns. In salt-wedge estuaries the river discharge of sediment is much larger, though generally markedly seasonal. Both suspended and bed-load material are important. The bed-load sediment is deposited at the tip of the salt wedge, but because the position of the salt wedge is so dependent on river discharge, the sediments are spread over a wide area. At times of flood, the whole mass of accumulated sediment can be moved outward and deposited seaward of the mouth. Because of the high sedimentation rates, the offshore slopes are very low, and the sediment has a very low bearing strength. Under normal circumstances, the suspended sediment settles through the salt wedge, and there is a zonation of decreasing grain size with distance down the salt wedge, but changes in river flow seldom allow this process to occur.

Fiord sediments. Sedimentation often occurs only at the heads of fiords, where river flow introduces coarse and badly sorted sediment. The sediment builds out into deltalike fans, and slumping on the fan slopes carries the sediment into deep water. Much of the rest of the fiord floor is bare rock or only thinly covered with fine sediment.

Bar-built estuary sediments. Bar-built estuaries are a very varied sedimentary environment. The high tidal currents in the inlets produce coarse lag deposits, and sandy tidal deltas are produced at either end of the inlets, where the currents rapidly diminish. In tropical areas, the muds that accumulate in the lagoons can be very rich in chemically precipitated calcium carbonate.   [K. R. DYER]

Bibliography: R. S. K. Barnes and J. Green (eds.), *The Estuarine Environment*, 1972; Council on Education in the Geological Sciences and F. F. Wright, *Estuarine Oceanography*, 1974; K. R. Dyer, *Estuaries: A Physical Introduction*, 1973; A. T. Ippen (ed.), *Estuary and Coastline Hydrodynamics*, 1966; G. H. Lauff (ed.), *Estuaries*, Amer. Ass. Advan. Sci. Publ. no. 83, 1967; B. W. Nelson (ed.), *Environmental Framework of Coastal Plain Estuaries*, Geol. Soc. Amer. Mem. no. 133, 1973; C. B. Officer, *Physical Oceanography of Estuaries (and Associated Coastal Waters)*, 1976.

Evapotranspiration

The total process of water vapor transfer into the atmosphere from vegetated land surfaces. Evaporation is the change of liquid water into gaseous water vapor; it occurs from bodies of water such as lakes and streams or from other wet surfaces such as wet soil surfaces. Transpiration is the process whereby water is absorbed by plant roots, transported through the plant, and evaporated from plant surfaces, especially from leaves into the air above. The energy required to change liquid water into water vapor comes from sources traceable to the Sun. Energy for evapotranspiration may come from the transfer of heat from warm air to a cooler plant or soil surface, but the major source of energy is generally net radiation. Net radiation is the difference between the incoming and the outgoing shortwave and long-wave radiation streams. Incoming shortwave radiation is composed of direct and diffuse solar radiation, and outgoing shortwave radiation is that amount of solar radiation which is reflected by the surface. The incoming and outgoing long-wave radiation streams are the radiation emitted by the atmosphere and the Earth's surface, respectively, as a function of the atmospheric or the surface temperature.

Rate and amount. The evapotranspiration rate is governed by the total quantity of absorbed energy, by the amount of moisture that is available both at the soil surface and within the plant root zone, and sometimes by the plants themselves. In humid regions the maximum evapotranspiration rate will seldom exceed the net amount of radiative energy available. However, warm dry air emanating from arid regions often becomes a major additional source of energy for evapotranspiration.

Net radiation is also generally greater in arid than in humid regions because of less cloud cover. These two factors combine to cause evapotranspiration rates in arid areas to generally exceed those in humid areas, unless, of course, available water becomes limiting. Evapotranspiration rates in humid areas seldom exceed about 7 mm per day, but rates as high as 12–14 mm per day have been reported from arid, semiarid, and subhumid regions.

In general, the amount of evapotranspiration will be less than or equal to the amount of precipitation that falls in a given region. Exceptions arise when precipitation is supplemented with irrigation, or in instances when plant roots may extract water from underground water tables located within a meter or two of the soil surface. Some plants, particularly those classified as phreatophytes, have rooting systems that can extend several meters, perhaps as deep as 20 to 30 m, into the soil and can draw water from these depths. Phreatophytes are generally located adjacent to streams, canals, or riverbeds and are responsible for nonbeneficial losses of extremely large quantities of water. This is especially serious in arid regions where water is at a premium for irrigation.

It has been estimated that approximately 1.1×10^{10} m³ (3×10^{12} gal) of water are returned daily to the atmosphere by evapotranspiration from the continental United States. Reduction in the evapotranspiration rate by only 1% for a single day would result in saving of enough water to supply the yearly needs of a city of 2,000,000 people. Much current research is aimed at finding techniques to reduce evapotranspiration losses or to improve the efficiency with which water is utilized by plants. Use of windbreaks, proper scheduling of irrigation, minimum tillage practices, and planting at optimal plant populations in proper row widths are among those techniques which have been shown to conserve moisture or increase the water-use efficiency. Water-use efficiency is the amount of dry matter produced divided by the amount of water used.

Measurement. Direct measurement of evapotranspiration is made with lysimeters. Lysimeters are large containers of soil in which plants are grown; loss of water from the container is due to evapotranspiration and is determined by weighing or by accounting for the quantities of water needed to replenish the lysimeter. Some lysimeters are very accurate and can detect evapotranspiration losses as small as 0.01 mm of water over periods as short as 15 min.

There are also many mathematical models developed to estimate evapotranspiration losses. These consider the effects of various weather factors, principally radiation, temperature, humidity, and wind. The models vary in complexity and in the accuracy with which they estimate actual evapotranspiration rates. Some provide estimates that are reliable only on a monthly basis, and others provide acceptable information for periods shorter than 1 hr. Refinements continue to be made in evapotranspiration models, especially those which can make use of information acquired through satellites and other modern systems for collecting meteorological data. *See* ATMOSPHERIC WATER VAPOR.   [BLAINE L. BLAD]

Bibliography: J. L. Monteith, *Vegetation and the Atmosphere*, vol. 1, 1975, N. J. Rosenberg, *Microclimate: The Biological Environment*, 1974; B. Yaron, E. Danfors, and Y. Vaadia, *Arid Zone Irrigation*, 1973.

Fog

Water droplets or, rarely, ice crystals suspended in the air so as to reduce visibility appreciably; a cloud resting on the ground. Fog is most common near seacoasts or large lakes, on mountains, in small valleys of nonarid regions, and over oceans where the water is relatively cold, as in summer off Newfoundland or California. On land, fog is most frequent in the early morning.

Related phenomena. Mist, generally known in the United States as light fog, can be differentiated from true fog by its horizontal visibility. This visibility is more than 1000 m (5/8 mi), according to internationally accepted definition. Also, mist may be wet haze (damp haze), composed of wet hygroscopic particles, such as sea salt, which absorb water in moist but not necessarily saturated air. Dry haze is composed of minute dry particles, and may include smoke or dust, though these are classified separately when possible. Fog, mist, and wet haze are hydrometeors, a general term for all atmospheric water droplets or ice particles. A combination of fog with smoke (and often other pollutants) is called smog. Low stratus clouds may develop when fog is lifted by turbulence or by warming of the ground; in turbulent conditions low stratus clouds may form initially instead of fog.

Formation of fog. Fog results either when air is cooled to its dew point and below or from an increase of moisture through evaporation from water that is warmer than the air. Causes of cooling are the especially rapid nighttime radiation that develops with clear skies; advection (air moving over a colder surface); and expansional (adiabatic) cooling, as when air moves up a mountain slope. For radiation to cause fog, winds must be light or calm; otherwise, mixing prevents fog or produces low clouds. Rapid evaporation into cold air moving over warmer water produces, for example, arctic sea smoke, a thin, wispy fog. Evaporation from warmer raindrops falling through colder air often causes fog.

Before fog can form, there must be sufficient condensation nuclei in the air. This condition is usually met, but ice nuclei are often lacking in sufficient quantity so that many fogs below freezing are water fogs. In the western plateau of the United States, supercooled water fog, known locally as pogonip, sometimes forms over a snow cover with temperatures a few degrees below freezing; it leaves a heavy deposit of rime on all outside objects. Fog with snow cover, however, is common only near freezing, and again, at temperatures below about −40°C (−40°F). *See* CLOUD; CLOUD PHYSICS; DEW; DEW POINT; HUMIDITY; PRECIPITATION; SMOG.   [J. R. FULKS]

Bibliography: F. Cross, Jr., and D. Furehand (eds.), *Air Pollution Meteorology*, 1976; T. F. Malone (ed.), *Compendium of Meteorology*, 1951; S. Petterssen, *Weather Analysis and Forecasting*, vol. 2, 2d ed., *Weather and Weather Systems*, 1956.

Front

A sloping surface of discontinuity in the troposphere, separating air masses of different density or temperature. The passage of a front at a fixed location is marked by sudden changes in temperature and wind and also by rapid variations in other weather elements, such as moisture and sky condition. *See* AIR MASS.

Although the front is ideally regarded as a discontinuity in temperature, in practice the temperature change from warm to cold air masses occurs over a zone of finite width, called a transition or frontal zone. The three-dimensional structure of the frontal zone is illustrated in Fig. 1. In typical cases the zone is about 3000 ft (1 km) in depth and 100 mi (100–200 km) in width, with a slope of approximately 1/100. The cold air lies beneath the warm in the form of a shallow wedge. Temperature contrasts are generally strongest at or near the Earth's surface. In the middle and upper troposphere, frontal structure tends to be diffuse, though sharp, narrow fronts of limited extent are common in the vicinity of strong jet streams. Upper-level frontogenesis is often accompanied by a folding of the tropopause and the incorporation of stratospheric air into the upper portion of the frontal zone. *See* JET STREAM.

The surface separating the frontal zone from the adjacent warm air mass is referred to as the frontal surface, and it is the line of intersection of this surface with a second surface, usually horizontal or vertical, that strictly speaking constitutes the front. According to this more precise definition, the front represents a discontinuity in temperature gradient rather than in temperature itself. The boundary on the cold air side is often ill-defined, especially near the Earth's surface, and for this reason is not represented in routine analysis of weather maps. In typical cases about one-third of the temperature difference between the Equator and the pole is contained within the narrow frontal zone, the remainder being distributed within the warm and cold air masses on either side. *See* WEATHER MAP.

The wind gradient, or shear, like the temperature gradient, is large within the frontal zone and discontinuous at the boundaries. An upper-level jet stream normally is situated above the zone, the

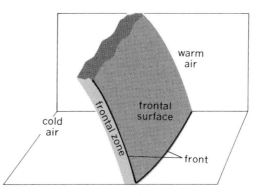

Fig. 1. Schematic diagram of the frontal zone, angle with Earth's surface much exaggerated.

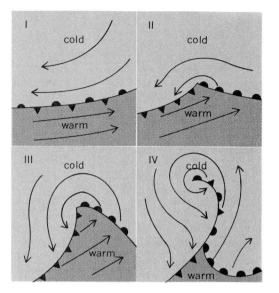

Fig. 2. The life cycle of the wave cyclone, surface projection. Arrows denote airflow. Patterns depicted are for the Northern Hemisphere, while their mirror images apply in Southern Hemisphere.

strong winds of the jet inclining downward along or near the warm boundary.

Frontal waves. Many extratropical cyclones begin as wavelike perturbations of a preexisting frontal surface. Such cyclones are referred to as

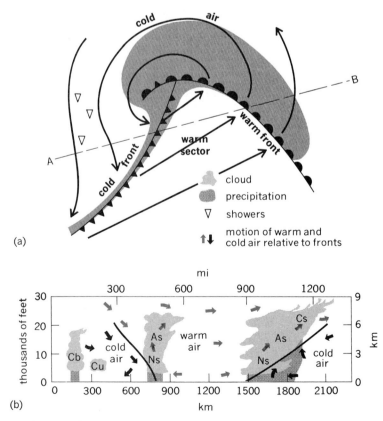

(a)

(b)

Fig. 3. Relation of cloud types and precipitation to fronts. (a) Surface weather map. (b) Vertical cross section (along A-B in diagram a). Cloud types: Cs, cirrostratus; As, altostratus; Ns, nimbostratus; Cu, cumulus; and Cb, cumulonimbus.

wave cyclones. The life cycle of the wave cyclone is illustrated in Fig. 2. In stage I, prior to the development, the front is gently curved and more or less stationary. In stage II the front undergoes a wavelike deformation, the cold air advancing to the left of the wave crest and the warm air to the right. Simultaneously a center of low pressure and of counterclockwise wind circulation appears at the crest. The portion of the front which marks the leading edge of the cold air is called the cold front. The term warm front is applied to the forward boundary of the warm air. During stage III the wave grows in amplitude and the warm sector narrows. In the final stage the cold front overtakes and merges with the warm front, forming an occluded front. The center of low pressure and of cyclonic rotation is found at the tip of the occluded front, well removed from the warm air source. At this stage the cyclone begins to fill and weaken. *See* STORM.

Cases have been documented in which the occluded structure depicted in panel IV (Fig. 2) forms in a different manner than described above. In such cases, sometimes referred to as pseudo-occlusions, the low-pressure center is observed to retreat into, or form within, the cold air and frontogenesis takes place along a line joining the low center and the tip of the warm sector. Cloud observations from meteorological satellites have provided visual evidence of this process. Since the classical occlusion process, in which the cold front overtakes and merges with the warm front, has never been adequately verified, it is possible that most occlusions form in this other way.

A front moves approximately with the speed of the wind component normal to it. The strength of this component varies with season, location, and individual situation but generally lies in the range of 0–50 mph; 25 mph is a typical frontal speed.

Cloud and precipitation types and patterns bear characteristic relationships to fronts, as depicted in Fig. 3. These relationships are determined mainly by the vertical air motions in the vicinity of the frontal surfaces. Since the motions are not unique but vary somewhat from case to case and, in a given case, with the stage of development, the features of the diagram are subject to considerable variation. In general, though, the motions consist of an upgliding of the warm air above the warm frontal surface, a more restricted and pronounced upthrusting of the warm air by the cold front, and an extensive subsidence of the cold air to the rear of the cold front. *See* CLOUD; CLOUD PHYSICS.

Fast-moving cold fronts are characterized by narrow cloud and precipitation systems. When potentially unstable air is present in the warm sector, the main weather activity often breaks out ahead of the cold front in prefrontal squall lines. *See* SQUALL.

Polar front. A front separating air of tropical origin from air of more northerly or polar origin is referred to as a polar front. Frequently only a fraction of the temperature contrast between tropical and polar regions is concentrated within the polar frontal zone, and a second or secondary front appears at higher latitudes. In certain locations such a front is termed an arctic front.

In winter the major or polar frontal zones of the

Northern Hemisphere extend from the northern Philippines across the Pacific Ocean to the coast of Washington, from the southeastern United States across the Atlantic Ocean to southern England, and from the northern Mediterranean eastward into Asia. An arctic frontal zone is located along the mountain barriers of western Canada and Alaska. In summer the average positions of the polar frontal zones are farther north, the Pacific zone extending from Japan to Washington and the Atlantic zone from New Jersey to the British Isles. In addition to a northward-displaced polar front over Asia, an arctic front lies along the northern shore and continues eastward into Alaska.

The polar frontal zone of the Southern Hemisphere lies near 45°S in summer and slightly poleward of that latitude in winter. Two frontal bands, more pronounced in winter than in summer, spiral into the main zone from subtropical latitudes. These bands, originating east of the Andes and northeast of New Zealand, merge with the main frontal zone after making a quarter circuit of the globe.

[RICHARD J. REED]

Frontogenesis. The formation of a front or a frontal zone requires an increase in the temperature gradient and the development of a wind shift. The frontogenesis mechanism operates even when the front is in a quasi-steady state; otherwise the turbulent mixing of heat and momentum would rapidly destroy the front.

The transport of temperature by the horizontal wind field can initiate the frontogenesis process as is shown for two cases in Fig. 4. The two wind fields shown would, in the absence of other effects, transport the isotherms in such a way that they would become concentrated along the A-B lines in both cases. Since the temperature gradient is inversely proportional to the spacing of the isotherms, it is clear that frontogenesis would occur along the A-B lines. Vertical air motions will modify this frontogenesis process, but these modifications will be small near the ground where the vertical motion is small. Thus, near the ground the temperature gradients will continue to increase in the frontal zones as long as the horizontal wind field does not change. As the temperature gradient increases, a circulation will develop in the vertical plane through C-D in each case. The thermal wind relation is valid for the component of the horizontal wind which blows parallel to the frontal zone. The thermal wind, which is the change in the geostrophic wind over a specific vertical distance, is directed along the isotherms, and its magnitude is proportional to the temperature gradient. As the frontogenesis process increases the temperature gradient, the thermal wind must also increase. But if the thermal wind increases, the change in the actual wind over a height interval must also increase (since the thermal wind closely approximates the actual wind change with height). This corresponding change in the actual wind component along the front is accomplished through the action of the Coriolis force. A small wind component perpendicular to the front is required if the Coriolis force is to act in this manner. This leads to the circulation in the vertical plane that is shown

in Fig. 5. *See* CORIOLIS ACCELERATION AND FORCE.

This circulation plays an important role in the frontogenesis process and in determining the frontal structure. The rising motion in the warm air and the sinking motion in the cold air are consistent with observed cloud and precipitation patterns. The circulation helps give the front its characteristic vertical tilt which leaves the relatively cooler air beneath the front. Near the ground the circulation causes a horizontal convergence of mass. This speeds up the frontogenesis process by increasing the rate at which the isotherms move together. The convergence field also carries the momentum lines together in the frontal zone in such a way that a wind shear develops across the front. This wind shear gives rise to the wind shift which is observed with a frontal passage. Eventually the front reaches a quasi-steady state in which the turbulent mixing balances the frontogenesis processes. Other frontogenesis effects are important in some cases, but the mechanism presented above appears to be the predominant one.

[ROGER T. WILLIAMS]

Bibliography: J. G. Charney, Planetary Fluid Dynamics, in P. Morel (ed.), *Dynamic Meteorology*, 1973; H. Dickson, *Climate and Weather*, 1976; J. R. Holton, *An Introduction to Dynamic Meteorology*, 1972; B. A. Hoskins, Atmospheric frontogenesis models: Some solutions, *Quart. J. Roy. Meteorol. Soc.*, 97:139–153, 1971; E. Palmén and C. W. Newton, *Atmospheric Circulation Systems*, 1969.

Frost

A covering of ice in one of several forms produced by the freezing of supercooled water droplets on objects colder than 32°F. The partial or complete killing of vegetation, by freezing or by temperatures somewhat above freezing for certain sensitive plants, also is called frost. Air temperatures below 32°F sometimes are reported as "degrees of frost"; thus 10°F is 22 degrees of frost.

Frost forms in exactly the same manner as dew except that the individual droplets that condense in the air a fraction of an inch from a subfreezing object are themselves supercooled, that is, colder than 32°F. When the droplets touch the cold object, they freeze immediately into individual crystals. When additional droplets freeze as soon as the previous ones are frozen, and hence are still close to the melting point because all the heat of fusion has not been dissipated, amorphous frost or rime results.

At more rapid rates of condensation, the drops form a film of supercooled water before freezing, and glaze or glazed frost ("window ice" on house windows, "clear ice" on aircraft) generally follows. Glaze formation on plants, buildings and other structures, and especially on wires sometimes is called an ice storm, or a silver frost storm, or thaw.

At slower deposition rates, such that each crystal cools well below the melting point before the next joins it, true crystalline or hoar frosts form. These include fernlike assemblages on snow surfaces, called surface hoar; similar feathery plumes in cold buildings, caves, and crevasses, called depth hoar; and the common window frost or ice flowers on house windows.

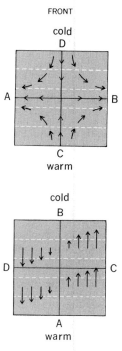

Fig. 4. Two horizontal wind fields which can cause frontogenesis. Broken lines represent isotherms and arrows show wind directions and speeds.

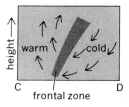

Fig. 5. The circulation in the vertical plane through C-D for both cases of Fig. 4

So-called killing frosts occur on clear autumn nights, when radiative cooling of ground, air, and vegetation causes plant fluids to freeze. At such times, the air temperature measured in a shelter 5–7 ft above ground usually is at least 5°F below freezing, but such standard level temperatures are poor indicators of frost severity. Air temperature varies greatly with height in the first few feet above the ground, and also with topography and vegetation around the shelter.

When wind is absent, the air layer immediately above the ground, rather than the ground itself, loses the most heat by radiation. Then the lowest temperature is 2–6 in. above, and about 1°F colder than, a bare ground surface. Above this near-ground minimum, an inversion of temperature develops for several to many feet thick. Plants radiate their heat faster than the air or ground, and may be colder than this air minimum temperature.

Frost damage can be prevented or reduced by heating the lowest air layers, or by mixing the cold surface air with the warmer air in the inversion above the tops of plants or trees.

Valley bottoms are much colder on clear nights than slopes, and may have frost when slopes are frost-free. In some notable frost pockets or hollows the air temperature may be 40°F colder than at nearby stations on higher ground, because of cold air drainage and the greater radiative cooling of level areas than of slopes. *See* AIR TEMPERATURE; CLOUD PHYSICS; DEW; DEW POINT; PRECIPITATION; TEMPERATURE INVERSION; WEATHER MODIFICATION.

[ARNOLD COURT]

Geomagnetic storm

A violent, transient disturbance of the Earth's magnetic field having durations from 3 hr to several days. The natural magnetic field measured at an observatory results from a mixture of sources internal and external to the Earth's surface. The rapid field variations are due to exterior sources arising in the ionosphere and magnetosphere. G. Graham discovered these variations in 1722, when he observed the small motions of a compass needle under a microscope. The large storms are approximately one-hundredth of the main internal field strength.

H component. The H (magnetic north) component of the field change typically shows the greatest amplitude variation. Many storms have a similar appearance in the H that is divisible into three parts: sudden commencement, or initial, phase; main phase; and recovery phase. The sudden commencements are recorded simultaneously about the Earth when the accuracy of the local clock and the amplitude-frequency response of the observatory magnetometers are suitably sensitive. Typically, there is a sudden onset and increase in the northward field strength which continues for as long as several hours. The H component decreases during the main phase, which often lasts a little longer than the initial phase and is several times its amplitude, or more. The recovery to the quiet-time level takes longer than the other two phases and may extend to several days. The more intense storm shows an increase of both amplitude

and duration. At some stations only the main phase is observed. There, it is called a magnetic bay because of the trace appearance—like a coastline on a map.

Geographic occurrence. The storms are most intense at the night-side, auroral zone latitudes, where they can be about six times larger than at middle latitudes. They show minimum amplitudes in the region of 20° latitude and have a secondary maximum at the Equator. At middle latitudes about 10 storms per year attain over 50 gammas (gamma $= 10^{-9}$ tesla) in magnitude; about 1 per year attains over 150 gammas, and about 1 in 10 years attains over 250 gammas. There is an increase in activity during the equinoxes, and the amplitudes are slightly larger in the winter hemisphere. The number and intensity of storms vary with sunspot number and lag the 11-year solar cycle by about a year or two. During recent sunspot maximum years there were about 5 to 10 storms per year that had amplitudes over about 80 gammas at middle latitudes.

Initial phase. The sudden onset and initial phase of a storm occur when the shock from the material blown out from the Sun as a solar wind encounters the Earth's magnetic field boundary. The boundary facing the Sun is at an average distance from the Earth's center of about 11 Earth radii (R_e). On extremely quiet days this boundary may expand to 17 R_e, and on extremely disturbed days it may be compressed to 7 R_e. Special field-line interactions occur when the solar wind field is directed opposite (southward) to the earth's field at the interface. The sudden compression of the magnetosphere that follows the arrival shock causes a sudden increase in H about the Earth. This compression recovers slowly. However, on the magnetic records the development of the negative H of the main-phase storm feature soon equals and then overrides the compression record. *See* GEOMAGNETISM; SOLAR WIND.

Main phase. The main phase of the storm results from the combined effects of the magnetospheric ring current, the near-Earth tail current, and the ionospheric disturbance current. The first of these is a current in the equatorial plane of the Earth at a distance of about 4–6 R_e, and is formed as material caught from the solar wind is energized and guided into the Earth's environment by the magnetospheric tail. As this westward electric current of charged particles grows, the resulting field at the Earth's surface is depressed. Selected equatorial stations are used to determine an hourly index, Dst, that responds mainly to the magnetospheric compression and ring current. The magnetospheric tail sheet current effects are seen mainly on the night side of the Earth at the antisolar, winter hemisphere locations. The closest approach of these complex tail currents is probably about one to three times more distant than the sun-side compression of the magnetosphere. Tail currents increase in intensity and proximity to the Earth during the height of the storm activity and thereby cause a decrease in H. *See* IONOSPHERE; MAGNETOSPHERE.

Polar substorm. The principal energy in the main phase of a magnetic storm is carried by the intensified westward electrojet currents in the auroral zone ionosphere which flow near the midnight hours about 100 km above the Earth's surface and connect to the magnetosphere by means of field-aligned currents near the dawn and dusk meridians. These currents are the major feature of a polar magnetic substorm disturbance that proceeds through three stages; the growth stage (several tens of minutes), representing the time of injection of energy into the night-side magnetospheric tail; the expansive or explosive stage (several minutes or more), when the disturbance rises to its maximum in amplitude and effective area; and the decay stage (up to an hour or two). Substorm durations are usually less than an hour, but consecutive substorms can blend together to form several hours of geomagnetic disturbance. A peak of substorm activity is usually restricted to a region of less than 5° in latitude to 20 times this in longitude at the Earth's night-side ionosphere. The ratio of high- to low-frequency components of the magnetic variation decreases rapidly with distance from the disturbance center. An hourly index, AE, is a measure of this auroral electrojet current. The conducting ionosphere provides a route for some of the closing current to flow far from the center of substorm activity and leads to an enhancement in the equatorial regions. Related substorm phenomena are the precipitation of low-energy electrons into the ionosphere, auroras, hydromagnetic emissions, and radiowave absorption. *See* AURORA.

Recovery phase. The recovery phase of the geomagnetic storm is a combination of the recovery of the substorm auroral electrojet currents, the magnetospheric compression and tail sheet currents, and the ring current (in order of increasing duration). The recovery may last for several hours to many days, several times the duration of the main phase. *See* SUN; TRANSIENT GEOMAGNETIC VARIATIONS.   [WALLACE H. CAMPBELL]

Bibliography: S. I. Akasofu, *Polar and Magnetospheric Substorms*, rev. ed., 1976; S. Matsushita and W. H. Campbell, *Physics of Geomagnetic Phenomena*, 1967; G. Rostoker, Polar magnetic substorms, *Rev. Geophys. Space Phys.*, 10:157–211, 1972.

Geomagnetism

A term signifying both the magnetism of the Earth and the branch of science that deals with the Earth's magnetism. The term geomagnetism is now given some preference over the older and longer term terrestrial magnetism.

Main geomagnetic field. This is specified at any point O by its vector magnetic intensity F. Its direction is that of the line $P'OP$ from the negative end P' to the positive end P of a magnetized needle $P'P$, perfectly balanced before it is magnetized, and freely pivoted about O, when in equilibrium. The positive pole P is the one that at most places on the Earth takes the more northerly position. Over most of the Northern Hemisphere, P is below O; the needle is said to dip below the horizontal by an angle I, called the magnetic inclination.

Over about half the Earth, however, P will be above O; the inclination I (or magnetic dip) is then reckoned as negative. This is the case over most of

the Southern Hemisphere. The value of I thus ranges from 90 to −90°. A point where $I=\pm90°$ is called a magnetic pole of the Earth.

Magnetic poles and Equator. There are two main magnetic poles; their approximate positions in 1975 were 76.1°N, 100°W and 65.8°S, 139°E. In a few places, near strongly magnetized mineral deposits, there may be local magnetic poles.

The distribution of I over the Earth's surface can be indicated on a globe, or on a map (with any kind of projection), by lines called isoclinic lines or isoclines, along each of which I has the same value. The isocline for which $I=0$ (where the balanced magnetized needle rests horizontal) is called the magnetic or dip equator. Figure 1 shows isoclines over a large part of the Earth for the epoch 1975.

Magnetic declination. A compass needle is magnetized and pivoted at a slightly noncentral point so as to rest and move in the horizontal plane. The deviation of its direction from geographic (gg) north is called the magnetic declination D. This is reckoned positive to the east and negative to the west; alternatively, declinations can be specified by a positive (eastward) value ranging up to 360°.

Over the greater part of the Earth D is numerically less than 90°, but along any small circuit around the magnetic poles it takes all values. At the magnetic poles themselves the compass needle takes no definite direction. At the gg poles the needle takes a definite direction, but as the northward direction changes through a whole revolution along any small circuit around the pole, D likewise takes all values around the circuit.

The distribution of the declination over the Earth's surface can be indicated by lines along each of which D is constant. These are called isogonic lines, or isogones. For the reasons just stated, the two magnetic poles and the two geographical poles are points toward which isogones converge, for all values of D. Figure 2 shows isogones for a large part of the Earth, for the epoch 1975. The isogones for which $D=0$ are called agonic lines. Along these lines, the compass points to true north.

The distribution of D over the Earth can also be indicated by magnetic meridians, which at each point P have the direction of the compass needle at P. These lines extend from the south magnetic pole to the north and have no complication (as do the isogones) at the gg poles. They indicate, more clearly than the isogones, the general distribution of the compass direction over the Earth. But the isogonic map is more convenient in enabling the compass direction at any point to be read or estimated without using an angle measurer. Hence it is the one used by navigators and travelers on sea and land and in the air.

Figure 3, which shows both isogones and magnetic meridians in the region around the north geographic and magnetic poles, illustrates this contrast between them.

Intensity patterns. The strength F of the vector intensity F is the third quantity which, together with I and D, completely specifies F. It is expressed in a unit which in geomagnetic literature

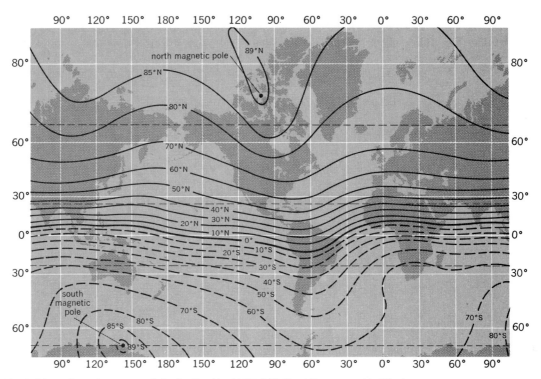

Fig. 1. Lines of equal geomagnetic inclination *I* for 1975. (*U.S. Naval Oceanographic Office*)

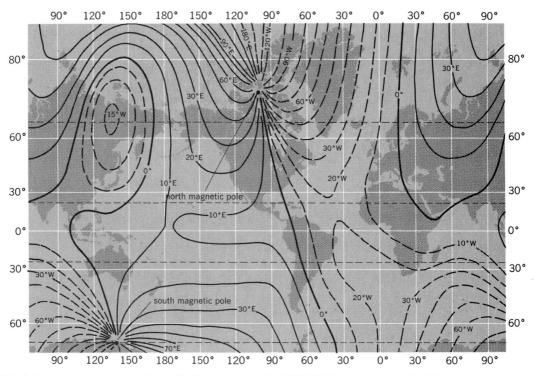

Fig. 2. Lines of equal geomagnetic declination (isogones) for 1975. (*U.S. Naval Oceanographic Office*)

is called the gauss, after the German mathematician, astronomer, and physicist K. F. Gauss (1777–1855), who first showed how F could be measured in units of length, mass, and time. This was the first nonmechanical entity so to be measured; Gauss' innovation was a landmark in the history of physical measurement. Physicists and electrical engineers distinguish the gauss from the oersted, the unit by which they would express the magnetic force. The numerical difference between the two, in geomagnetism, is negligible.

The symbol for the gauss is Γ; much use is made in geomagnetism of a unit called the gamma (symbol γ), which is $10^{-5}\ \Gamma$. The International Sys-

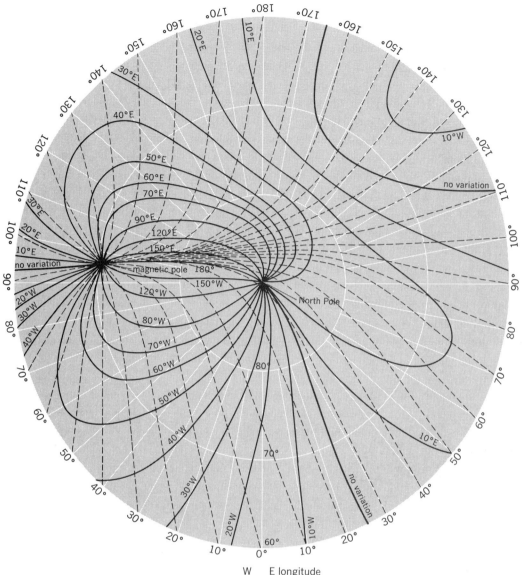

W   E longitude

Fig. 3. Lines of equal declination (full lines) and magnetic meridians (broken lines), for 1922, for the Arctic region north of 60° latitude. The agonic lines (isogones for which magnetic declination equals zero) are marked "no variation." (After H. Spencer Jones, in S. Chapman and J. Bartels, Geomagnetism, 2 vols., Oxford, 1940)

tem of Units (SI) was adopted in 1960, defining magnetic flux density in terms of a tesla and magnetic field strength as amperes per meter. [One oersted = 7.96 amperes per meter; 1 gauss (Γ) = 10^{-4} tesla; and 1 γ = 1 nanotesla (nT).]

The distribution of the magnetic intensity F over the Earth can be indicated, as for I and D, by lines of constant F. They are called isodynamic lines. They are shown for the epoch 1975 in Fig. 4. The value of F in general increases with increasing latitude north or south; its distribution, however, is not simple. In low latitudes it has a minimum of about 0.23 Γ off the east coast of South America. In the Northern Hemisphere there are two maxima, each about 0.6 Γ, one in northern Canada and the other in Asia. The highest intensity, near 0.7 Γ, occurs near the southern dip pole.

The vector intensity $\mathbf{F}$ can also be specified by its vertical component Z and its horizontal vector component $\mathbf{H}$; the latter can be specified by its direction (given by D) and the horizontal intensity H or, alternatively, by its north and east components X, Y. Clearly $Z = F \sin I$, $H = F \cos I$, $X = H \cos D$, and $Y = H \sin D$; Z, like I, is thus reckoned positive when downward, and Y, like D, is positive when eastward.

Geomagnetic elements. All the magnitudes F, H, X, Y, Z, I, and D are called geomagnetic elements. Lines on a map along which any element has a constant value are called isomagnetic lines; the map is called an isomagnetic chart. Figure 5 shows an isomagnetic chart for H for the epoch 1975.

Isomagnetic maps for the whole Earth or any large part of it cannot show the finer details of the field distribution. Maps for smaller areas, such as a European country or a state of the United States, can show such details if the areas have been closely surveyed magnetically. In some parts of

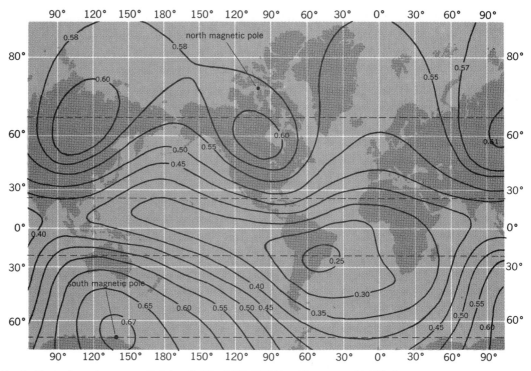

Fig. 4. Lines of equal geomagnetic intensity *F* for 1975. (*U.S. Naval Oceanographic Office*)

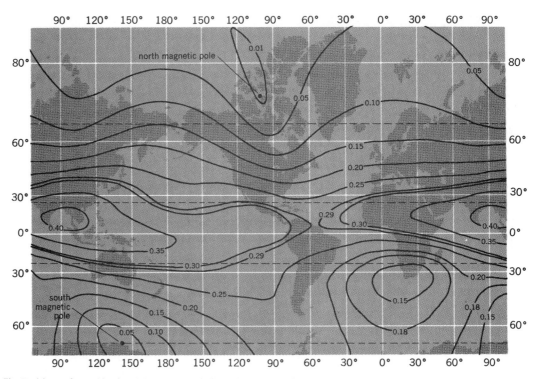

Fig. 5. Lines of equal horizontal geomagnetic force for 1975. (*U.S. Naval Oceanographic Office*)

such maps the lines may depart considerably from the smooth spacing indicated on world maps, for example, in regions of local magnetic anomaly, where the field is disturbed by magnetic minerals not far below the Earth's surface. The most striking local anomalies occur in two narrow strips in the Kursk region of the Soviet Union. They are 60 km apart and run in parallel from NE to SW. The most disturbed part of the major (northerly) strip is only 2 km wide, although the strip is 250 km long; Z is everywhere above normal and ranges up to $1.9\ \Gamma$.

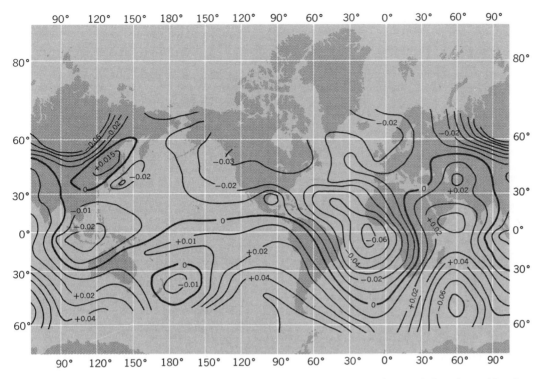

Fig. 6.   Lines of equal rate of change (isopors) of the vertical geomagnetic force *Z*, over the interval from 1885 to 1922. The isopors are drawn at 500-γ intervals; increase of downward magnetic force is considered positive. (*After A. G. McNish, in S. Chapman and J. Bartels, Geomagnetism, 2 vols., Oxford, 1940*)

Spherical harmonic analysis. The Earth's surface magnetic field can be expressed in mathematical terms by a process called spherical harmonic analysis. The data used are magnetic survey measurements of the magnetic elements at a sufficient number of points on the Earth's surface. The analysis provides separate expressions for the parts of the surface field that originate, respectively, within the Earth and above it. It proves that the main field comes almost entirely from within. The part that is of external origin has not been accurately determined, but is less than 100 γ at the Earth's surface and may average only about 20 γ.

The main term in the spherical harmonic expression corresponds to the field of a uniformly magnetized sphere, or of a point dipole at the Earth's center. Hence this part of the geomagnetic field is called the dipole field. The intensity of magnetization for a sphere the size of the Earth, to give this dipole field, would be about 0.08 Γ. The field could also be produced by a surface distribution of electric current around the Earth, of total amount 1.5×10^9 amp, distributed proportionately to the cosine of the latitude. The moment of the equivalent dipole is 8.1×10^{25} Γ-cm³. These three possible systems that could produce the dipole field all differ from the actual source of the field.

The Earth's diameter along the direction of the dipole (which is also the direction of magnetization for the alternative model of a uniformly magnetized sphere) is called the dipole (dp) axis. Its ends on the Earth's surface are called the dp or axis poles. The polarity of the northern pole is negative; that of the southern is positive. On maps the northern dp pole is conveniently marked *B* and the southern, *A*; these letters may stand for boreal and austral. The gg position of *B* is 78.5°N, 69°W, and of *A*, 78.5°S, 111°E. The obliquity of the dp to the gg axis is thus 11.5°. The dp pole *B* is 1280 km from the N gg pole and 1160 km from the north magnetic pole; the corresponding southern distances are 1280 km and 1350 km. The dp axis has shown no change of direction large enough to be reliably estimated since it was first determined nearly 150 years ago. The position of any point *P* may be expressed in dp or dipole coordinates relative to the dp axis: the dp colatitude is the angle *BOP* (*O* being the Earth's center); the dp longitude is measured eastward from the part of the common gg and dp meridian through *B* that lies south of *B*. The dp equator, from which dp latitude is measured north toward *B* or south toward *A*, is in the diametral plane perpendicular to the dp axis.

A closer approximation to the Earth's surface field is the field of a point dipole not at the Earth's center. The eccentric dipole must have the same direction and magnetic moment as the centered dipole, but is displaced from *O* by about 460 km, toward a point in the Pacific Ocean north of New Guinea; it is moving NNW. The eccentric dipole field is only a moderate improvement on the centered dipole field, as an approximation to the actual surface field; there still remain regional deviations from it, amounting in several places to over 0.1 Γ in the horizontal component.

The spherical harmonic expression for the potential of the geomagnetic field enables the field to be calculated at any point above the Earth, except insofar as electric currents in the ionosphere and beyond modify the field. These currents can be explored by rockets and satellites, which confirm that the field does not extend to infinity, varying inversely as the cube of the distance, as formerly supposed. Instead it is confined to a region—the magnetosphere—within a continuing but variable flow of plasma from the Sun. Also, a ring current flows within this region and greatly modifies the field during geomagnetic storms. This current may always be present, though weak when the plasma flow is gentle. The origin of the geomagnetic field is discussed later. *See* GEOMAGNETIC STORM; MAGNETOSPHERE.

Secular magnetic variation. This slow change of the Earth's field necessitates continual redrawing of the isomagnetic maps. In any magnetic element at a particular place, the variation may be an increase or a decrease; its rate is unpredictable and is not constant in either magnitude or sign. The distribution of the rate for any element can be indicated on maps called isoporic, by lines (isopors) along which the rate is constant.

The pattern of such isopors is more complex than that of the isomagnetic lines for the same element. There may be several regions of maximum or minimum change, producing numerous oval systems of isopors (see Fig. 6, giving Z isopors based on an interval of 38 years). Whereas the main field is a planetary property, the secular variation is regional. Moreover isoporic maps change much more than isomagnetic maps from one decade to the next. Regions of maximum change move, disappear, or newly appear. An example of long-continued changes of the field direction is given in Fig. 7, for London. From 1576 to about 1800 the compass there turned from 11°E to 24°W, while the dip first increased from 71° by 3°, until 1700; since 1700 it has decreased again, to about

67°. Since 1800 the compass has swung eastward through about 16°. Figure 7 shows these changes, with some directions and trends for earlier centuries (indicated by numbers, that is, 15 indicates 15th century), estimated from the magnetization of bricks in kilns whose period can be dated. This is an example of archeomagnetic research, which aims to extend knowledge of the past geomagnetic changes by centuries or even by millennia.

Figure 7 suggests a possible cycle of change of direction at London, with a period of 4–5 centuries. But the changes elsewhere are inconsistent with any regular period of secular geomagnetic change.

Another long-continued secular trend is the movement of the agonic line, which in Fig. 2 crosses the Equator at about 74°W. The course of this agonic line has been followed longer than that of the more complicated looped agonic line that traverses Asia. In 1550 it crossed the Equator at about 20°E, and since then it has moved continuously westward through about 90° in about 4 centuries. This is one token of a noticeable tendency—the westward drift—of isomagnetic and isoporic features. Another large-scale feature of the secular magnetic variation is a decrease of the Earth's magnetic moment by about 5% during the last century.

The study of paleomagnetism, the remanent magnetization of rocks and sediments, extends man's knowledge (though with some uncertainties) of the Earth's magnetic history to past geologic ages and indicates that the field has been reversed more than once.

Source of field. The source of the main geomagnetic fild and of its secular variation is the Earth's core, according to accepted theories. Although slow by some standards, the secular variation is very rapid by geologic standards. It cannot reasonably be ascribed to changes in or not far below the Earth's crust. It seems probable that the main field is caused by electric currents in the Earth's liquid core, below about 2900 km (1800 mi). By slow convective movements, electric currents are produced in the core; these maintain the magnetic field, as in a self-exciting dynamo. Rather large-scale eddies in the convective motion produce the regional features of the main field, and their changes produce the secular magnetic variation. These changes have no apparent relation to the broad features of geography and geology.

According to this theory, the geomagnetic field will remain of mainly dipole character nearly down to the surface of the core, with the lines of magnetic force lying nearly in planes through the geomagnetic axis. The magnetic intensity will increase inversely as the cube of the distance from the center, to nearly 5 Γ at the axis poles of the core; but the regional irregularities will become more prominent as the core is approached.

Inside the core the lines of force are probably much twisted around the Earth's axis, and the intensity may exceed 100 Γ. There will be a reaction between the magnetic field and the electrical currents in the core, tending to produce a westward drift of the core (and hence of the secular magnetic

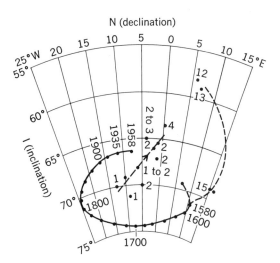

Fig. 7. Geomagnetic direction at London, as observed from 1580 to 1958; the magnetic declination and the dip are shown, mostly at 20-year intervals. In addition, points are shown indicating the direction for earlier centuries, inferred from studies of magnetized bricks in old Roman-British and later kilns. Broken lines are conjectural; the straight broken line shows the estimated changes of direction in Roman times. (*After L. A. Bauer, R. M. Cook, and J. C. Belshe*)

variations) relative to the outer solid part of the Earth.

The estimates of the electrical resistivity σ of the core required by these theories range from 10^{-1} to 10^{-3} or even less, in ohm-cm units. This is far less than that of the upper (dry) layers of the solid Earth, for whose materials σ ranges from 10^4 to 10^{10}. For sea water σ ranges from 15 to 40, depending on the salinity and temperature. At a depth of 600–700 km, σ appears to begin to decrease from about 10^4 to a value of about 0.5 at the core-mantle boundary.

Geomagnetic measurements. To keep the knowledge of the Earth's magnetic field abreast of its secular changes, the world must be magnetically surveyed from time to time. In the past this has generally been done, on land and sea, by measuring D, H, and I. To measure D, it is necessary to determine true north by astronomical observation; for I, the horizontal direction must be determined by a good level. Then a weighted compass needle (declinometer) or a dip circle with balanced needle will indicate D and I; for real accuracy, great care and suitable reversals are required. To measure H, one uses a magnetometer. The original method devised by Gauss was likewise by means of a magnet, whose oscillations and power to deflect another magnet were observed.

Another form of inclinometer to measure I which has almost entirely displaced the dip circle has a rotating coil whose axis of rotation is turned until rotation in the Earth's field produces no induced alternating current in the coil. The axis then lies along the field direction; this enables D and I to be measured. This instrument is called a dip inductor, earth inductor, induction inclinometer, or earth inductor compass.

[SYDNEY CHAPMAN]

Geomagnetic field surveys for geophysical exploration or geologic studies are now usually made with optical pumping or nuclear precession magnetometers, which have an accuracy of the order of a nanotesla. Such instruments give no readings of the field direction, but only of its absolute intensity. These devices have been towed by aircraft, near the ocean surface by ships, and even near the ocean floor. The process of reducing these data to anomaly maps representing only the proportion of the field due to the crustal sources is still more of an art than a science. The reason is that the fraction of observed field that comes from the core and the portion that is due to induced or remanent magnetization of the rocks are yet unsettled.

World magnetic survey. A effort to map the global magnetic field was undertaken by the International Association of Geomagnetism and Aeronomy as a project to be carried out during the International Years of the Quiet Sun (1964–1965). The World Magnetic Survey (WMS) Board was established to coordinate this activity, which eventually involved participation by scientists in 41 member countries.

One significant product of the WMS was the agreement on a magnetic reference field for epoch 1965.0. The WMS group met in Washington, DC,

in 1968, evaluated the various analyses of the magnetic survey data made by different organizations, and adopted an international geomagnetic reference field (IGRF). This standard is one of the more ambitious attempts at scientific coordination since the field changes in unpredictable ways and must be revised every few years.

Mapping the filed. The purpose of the WMS was to obtain data to be used in compiling charts and models for practical applications and for studies of the sources of the field. The objecive then was to map the smooth "main" field by eliminating from survey data the contribution both from the local anomalies in the Earth's crust and the transient changes caused by the ionosphere and magnetosphere. During and prior to the WMS, significant contributions to the magnetic survey data base were made by such organizations as the U.S. Navy and the Dominion Observatory (Canada) performing aeromagnetic surveys, and by the Soviet Union using the nonmagnetic ship *Zarya*. Many more detailed contributions were made by countries which took magnetic surveys within and near their borders.

However, even with this organized international effort, accurate global mapping was not possible in a time short enough to take a crisp "snapshot" of the field before the secular variation blurred the picture. For this reason, one of the first suggested scientific uses of orbiting spacecraft was the mapping of the geomagnetic field.

Satellite surveys. The pioneering attempt to make a satellite survey was by the Soviet experimentalist Sh. Dolginov, who flew flux-gate (vector) magnetometers on *Sputnik 3* in 1958. The vector-component data were never successfully reduced, so that the final analysis was made combining the three observations into a total field of about 100-nT accuracy. The main reason that such vector data are difficult to analyze is that the orientation, or attitude, of the instrument must be known very accurately. For example, a 1° error in attitude yields a 1000-nT error in components measured in a 57,000-nT field. Improvements in spacecraft attitude determination using star cameras now allow the possibility of making accurate vector measurements, and a new survey was planned by the United States in 1979 or 1980.

However, to date, all low-altitude satellites that have attempted absolute magnetic measurements, except the *Sputnik 3*, have carried only total field magnetometers. Key details of these surveys are given in the table. The Soviet *Cosmos 49* and the American *OGO-2* were the two flights made especially for the United States–Soviet Union bilateral agreements and as contributions to the World Magnetic Survey.

Analysis of magnetic surveys. The techniques of spherical harmonic analysis noted earlier have been significantly improved since 1960 due to the use of high-speed computers. The first application of modern equipment to this problem was made by D. C. Jensen and J. C. Cain in 1962 when they made a direct least-squares analysis of all survey data taken since 1940. They used the classical representation of the field expressed in spherical

Low-altitude satellite geomagnetic measurements 1958–1975

| Spacecraft | Orbit | | Interval | Magnetometer | Coverage |
	Inclination, degrees	Altitude range, km			
Sputnik 3	65	440–600	May–June 1958	Flux gates	Soviet Union
Vanguard 3	33	510–3750	Sept.–Dec. 1959	Proton precession	Near ground stations
Cosmos 26	49	270–403	Mar. 1964	Proton precession	Whole orbit
Cosmos 49	50	261–488	Oct.–Nov. 1964	Proton precession	Whole orbit
1964-83C	90	1040–1089	Dec. 1964–June 1965	Rubidium vapor	Near ground stations
OGO-2	87	413–1510	Oct. 1965–Sept. 1967	Rubidium vapor	Whole orbit
OGO-4	86	412–908	July 1967–Jan. 1969	Rubidium vapor	Whole orbit
OGO-6	82	397–1098	June 1969–June 1971	Rubidium vapor	Whole orbit
Cosmos 321	71	270–403	Jan.–Mar. 1970	Cesium vapor	Whole orbit

harmonic coefficients. The magnetic intensity **F** is given by $-\Delta V$, where the scalar potential function V is computed from

$$a \sum_{n=0}^{n*} \left(\frac{a}{r}\right)^{n+1} \sum_{m=0}^{n} (g_n^m \cos m\phi + h_n^m \sin m\phi) P_n^m(\theta)$$

where $a = 6371$ (average radius of Earth in kilometers); $r =$ radial distance from Earth's center; $\theta =$ geographic colatitude; $\phi =$ east longitude; P represents Schmidt's spherical functions; g and h are the spherical harmonic coefficients determined in the analysis; and $n*$ is the truncation level of the expansion. The smallest scale size of the field structure represented by this expression is given by $2\kappa/n*$, or in distance by the circumference of Earth (40,000 km) divided by $n*$.

Since the present satellite surveys are only of total field, the analysis depends on having a very accurate set of observations to produce a good vector model. Fortunately this is possible from spacecraft since they operate above the altitudes where local anomalies add a high noise level (about 200–1000 nT) to the observations. One model was developed which matched a 5-year span of Orbiting Geophysical Observatory (OGO) data everywhere to better than 10 nT. Since the model is constrained only by the strength of the field, there is no direct limit to the error in the component perpendicular to the field. However, due to the properties of a source-free field, it is not possible to distort field direction at one point without changing field intensity elsewhere. It has been learned that the transverse error in the models based only on satellite data can be an order of magnitude greater than the longitudinal error. Thus such a model accurate to 50 nT in intensity could have transverse errors of 500 nT, which would correspond to a 1° error in angle near the Equator. A further consideration regarding the use of the results of satellite data is that errors at a few hundred kilometers' altitude magnify by a factor of 2 or more (depending on $n*$) when the model is extrapolated to the surface.

In spite of these problems, comparisons with surface data taken near the same epoch indicate that satellite models have generally been at least as accurate as those computed using only surface observations. For example, E. Dawson and P. Serson concluded that satellite-derived models agree with the Canadian and Scandinavian aeromagnetic survey data as well as do local surfaces fit to these data themselves.

One of the most interesting and unexpected results from satellite surveys is the discovery of long-wavelength crustal features in the magnetic field. This was unexpected because it was previously thought that geologic anomalies could not produce structures more than about 100 km in size. Part of the reasoning lay in the knowledge that the crust cannot sustain magnetization at temperatures above the Curie point. Such temperatures are reached at least by 10 km under the oceans and 50 km under the continents. However, analysis of the field models from the OGO data appear to show that the field due to the harmonics above $n = 11$ or 12 (wavelengths shorter than 3500 km) are caused by magnetization of rocks in the lithosphere, whereas the longer structure comes from the Earth's core. In addition to these features, other intense anomalies such as the one previously mentioned at Kursk are also detected at satellite altitude. Although work is continuing in this area, it appears that most of this structure is seen over continents or regions of the ocean which may contain sunken continental material.

Secular variation and the dipole. The use of satellite observations has made it possible to determine the secular variation over the whole Earth for a short interval of time. This has been done with the same least-squares analysis which is used to determine the g and h coefficients; for example, $g = g_0 + \dot{g}(t - t_0)$, where t_0 is the epoch of the coefficients, and $\dot{g}$ is the secular change of g in nT/year.

One such analysis of satellite data has determined the global secular variation (SV) to an accuracy of 4 nT/year over a 5-year period. Prior analyses using the year-to-year differences of accurate observations were made at magnetic observatories or special sites called repeat stations. The other studies gave detailed and accurate information concerning the secular variations in the regions of the stations. However, these observing points are located mainly in inhabited land areas, leaving large gaps, such as at the poles and oceans, where data are sparse or nonexistent.

Such studies of global SV revealed that there is a relation between the behavior of the eccentric dipole and the rate of the Earth's rotation. It appears that an irregularity in the length of day causes a disturbance in the flow of material near the core-mantle interface which then propagates in about 7 years to the Earth's surface and changes the rate of westward drift of the eccentric dipole. Such changes may be related to the occurrence of major earthquakes.

[JOSEPH C. CAIN]

Bibliography: J. C. Cain, Structure and secular change of the geomagnetic field, *Rev. Geophys. Space Phys.*, 13:203–204, 1975; S. Chapman, *Solar Plasma, Geomagnetism and Aurora*, 1964; S. Chapman and J. Bartels, *Geomagnetism*, 2 vols., 1940; S. Matsushita and W. H. Campbell (eds.), *Physics of Geomagnetic Phenomena*, 2 vols., 1967; H. Odishaw, *Research in Geophysics*, vol. 1, 1964; *Rev. Geophys. Space Phys.*, 13:35–51, 174–241, 1975; A. J. Zmuda (ed.), *The World Magnetic Survey*, IAGA Bull. no. 28, Paris, 1971.

Geophysics

Those branches of earth science in which the principles and practices of physics are used to study the Earth. Geophysics is considered by some to be a branch of geology, by others to be of equal rank. It is distinguished from the other earth sciences largely by its use of instruments to make direct or indirect measurements of the parts of the Earth being studied, in contrast to the more direct observations which are typical of geology.

SUBDIVISIONS OF GEOPHYSICAL SCIENCE

Geophysics consists of several principal fields plus parallel and subsidiary divisions. These are commonly considered to include plutology, with geodesy, geothermometry, seismology, and tectonophysics as subdivisions; hydrospheric studies, mostly hydrology (groundwater studies) and oceanography; atmospheric studies, meteorology and aeronomy; and several fields of geophysics which overlap one another, including geomagnetism and geoelectricity, geochronology, geocosmogony, and geophysical exploration and prospecting. Planetary sciences, the study of the planets and satellites aside from the Earth, are usually considered a branch of geophysics because the techniques used for such studies are, until man can land on each body, instrumental rather than direct observation.

Plutology. This is a general term covering geophysical methods of studying the solid part of the Earth. The field has the following four major subdivisions.

Geodesy. The science of the shape and size of the Earth is geodesy. This includes consideration of the Earth's mass distribution as determined by measurements of the gravitational-centrifugal field. Because of man's interest in space travel, geodesy is concerned also with the distribution of the gravitational field above the Earth's surface out to the limit of the Earth's detectable effect.

Geothermometry. This is the science of the Earth's heat. It includes the study of temperature variation and of heat generation, conduction, and loss in the Earth, and their effects on the materials of which the Earth is composed. Volcanology, the science of volcanoes, uses some of the principles and techniques of geothermometry.

Seismology. The science of earthquakes and other ground vibrations is the field of seismology. Seismology has made particularly important contributions to man's knowledge of the Earth's interior. Study of the times of passage of seismic waves through the Earth gives information on the distribution of different types of rock.

Tectonophysics. Sometimes called geodynamics, this is the science of the deformation of rocks. It consists of tectonics, the study of the broader structural features of the Earth and their causes, as in mountain building; and rock mechanics, the measurement of the strength and related physical properties of rocks.

Hydrospheric studies. Geophysical study of the hydrosphere has two main branches, hydrology and oceanography.

Hydrology. This groundwater science also includes glaciology, the study of groundwater in the form of snow and ice. *See* GLACIOLOGY; GROUNDWATER; HYDROLOGY.

Oceanography. The scientific study of the oceans includes the study of the shape and structure of the ocean basins; the physical and chemical properties of sea water; ocean currents, waves, and tides; thermodynamics of the oceans; and the relation of these to the organisms which live in the sea. *See* MARINE GEOLOGY; OCEANOGRAPHY; OCEANS AND SEAS.

Atmospheric studies. Because of extending interest from the face of the Earth toward outer space, aeronomy has come to be recognized as a distinct science separate from meteorology.

Meteorology. The science of the Earth's atmosphere yields meteorological information that not only is used in climatology and weather prediction, but also is important in understanding the problems of aircraft flight and air pollution. *See* METEOROLOGY.

Aeronomy. This field is concerned with the phenomena of the upper atmosphere above 100 km, where its electrical behavior is important because the air is strongly ionized. There, where the mean free path of an atom or electron is long, the physical behavior of the material is controlled at least as much by its electrical properties as by its density and other mass properties. The outer boundary of the atmosphere is not a distinct surface; there is instead a gradual transition to the relative emptiness of interplanetary space. In this transition zone there is continuous interaction between the matter of the Earth's atmosphere and the radiations and particles arriving from outside. The Sun, the source of most of these, is a major object of geophysical investigation. The Earth moves within the outer fringe of the Sun's atmosphere. Aeronomy is much concerned with the behavior of the Sun, especially as it influences conditions on the Earth. Understanding of solar phenomena is essential for the safe flight of man beyond the lower reaches of the Earth's atmosphere. *See* AIRGLOW; ATMOSPHERE; AURORA; COSMIC RAYS; IONOSPHERE; MAGNETOSPHERE; SUN; UPPER-ATMOSPHERE DYNAMICS.

Planetary sciences. Astronomical data on the planets and their satellites come from photographs and analysis of the spectra of their reflected light. Radar reflections give additional information. Observing instruments landed on the Moon and nearby planets or flown past their surfaces radio back data of many types. Although these methods may ultimately be replaced by exploration parties and manned observatories, much of the knowledge of the more distant members of the solar system will depend greatly upon such indirect observations for a long time.

Overlapping fields of geophysics. In addition to the regional subdivisions, there are several other fields of geophysics which overlap and concern all the others.

Geomagnetism, geoelectricity. The science of the Earth's magnetic field and magnetic properties is termed geomagnetism. Geoelectricity is the science of electrical currents in the Earth and of its electrical properties. Geomagnetism and geoelectricity are closely related because the magnetic field is due largely to electrical currents in the solid earth and in the atmosphere. *See* GEOMAGNETISM.

Geochronology. This field deals with the dating of events in the Earth's history. The principal technique used is based on radioactive disintegration. The proportions of parent to daughter elements in a mineral or rock are a measure of the age of the material. Other methods depend on the red shift of the spectrum of distant stars, the rate of recession of the Moon, and the rates of erosion and sedimentation.

Geocosmogony. This is the study of the origin of the Earth. The many hypotheses proposed fall into two groups, those which postulate that the Earth is primarily an aggregate of once smaller particles, and those which claim that is is largely a fragment of a larger body. Current speculation favors the former theory. Geocosmogony is intimately linked with the origin of the solar system and our galaxy. Many lines of evidence suggest that the formation of the Earth was a typical minor event in the evolution of the Milky Way or of the universe as a whole, occurring $5-8 \times 10^9$ years ago.

Exploration and prospecting. Geophysical techniques are widely used not only to study the general structure of the Earth but also in prospecting for petroleum, mineral deposits, and groundwater and in mapping the sites of highways, dams, and other structures. Seismic methods are the most widely used, but electrical, electromagnetic, gravity, magnetic, and radioactivity surveying methods are also well developed. Many types of geophysical surveys can be made by lowering measuring apparatus into bore holes.

Technical literature. Some of the principal geophysical publications largely in English are the *Journal of Geophysical Research* of the American Geophysical Union; *Geophysics,* published by the Society of Exploration Geophysicists; the *Bulletin of the Seismological Society of America*; the *Geophysical Journal of the Royal Astronomical Society*; the *Journal of Meteorology* (now the *Journal of Atmospheric Sciences* and the *Journal of Applied Meteorology*); the *Bulletin of the American Meteorological Society*; the *Quarterly Journal of the Royal Meteorological Society*; the *Journal of Atmospheric and Terrestrial Physics*; and the *Bulletin of the Earthquake Research Institute of Tokyo*. Geophysical papers are commonly

found also in the *Bulletin of the Geological Society of America*; *Geochemica and Cosmochemica Acta*, the journal of the Geochemical Society; the *Transactions of the American Institute of Mining, Metallurgical, and Petroleum Engineers*; and the *Proceedings of the National Academy of Sciences of the United States of America*.

[BENJAMIN F. HOWELL, JR.]

PROGRAMS OF GEOPHYSICS

A fundamental feature of geophysics is the necessity to record, collect, exchange, collate, analyze, and synthesize large quantities of data from many sites over extended periods of time. This is a complex and difficult task requiring considerable international cooperation and coordination, large staffs of technicians and scientists, and extensive field programs using sophisticated instrumentation as well as laboratory activities supported by the fastest electronic computers.

The various disciplines of geophysics have one further feature in common, namely, the phenomena they embrace are of fundamental importance to mankind. Whether in the domestic or business environment, rarely a day goes by without the need for an administrative decision based on a meteorological or other geophysical event. As a consequence, geophysical studies are the concern of nations as well as of scientists, and the planning of geophysical programs is becoming a major function of governmental and nongovernmental bodies, particularly in the international sphere. On the international governmental side, specialized agencies of the United Nations have become involved in such programs, especially the United Nations Educational, Scientific, and Cultural Organization (UNESCO) in the fields of hydrology and oceanography and the World Meteorological Organization (WMO) in the field of atmospheric science. On the international nongovernmental side, such activities are coordinated by the International Council of Scientific Unions (ICSU), through its various components, primarily the scientific unions such as the International Union of Geodesy and Geophysics (IUGG), as well as the special scientific and interunion committees which have special responsibilities for individual geophysical programs. A noticeable trend in these programs has been the increasing cooperation between governmental and nongovernmental bodies, often stimulated by specific resolutions from the United Nations General Assembly.

Both the number and extent of large international geophysical programs have increased, and this can be ascribed, at least in part, to the tremendous success of the International Geophysical Year (IGY, 1957–1958). For one thing, IGY captured the interest and imagination of the nonscientific public, in part because it was a truly international program in which virtually all nations collaborated and in part because its more spectacular phases represented readily recognizable advances of major importance. To the scientist, IGY was a success since it led to quasi-global and quasi-simultaneous observational programs; moreover, geophysical studies became the purpose and not merely a sideline of expeditions. The vertical and horizontal probing of the Antarctic continent and of its atmosphere was (and, fortunately, still is) an excellent example of geophysical study being the main purpose of expeditions.

While one may include various justifiable and worthwhile projects in an international geophysical program, one should realize that such a program is not merely an attempt to exploit the enhanced availability of financial and personnel support but is rather a deliberate attempt to schedule, on a global basis, the orderly advance of technology in research and related fields. Because of the requirement for simultaneous or at least coordinated studies, it is simply not effective or economical to introduce new or expanded programs haphazardly in space or in time. This is precisely why each international geophysical program appears to lay major emphasis on relatively new types of observational studies, in order that these programs (whose introduction would have been inevitable, in any case) may be instituted in such a manner (global and coordinated) that significant results should be achievable almost from the start. At the same time, it is necessary to examine existing programs and identify their inadequacies (instrumentation, networks, training, and so on), particularly if they are of fundamental importance for the newer studies, so that a concerted drive to remove or minimize these weaknesses may take place also during the chosen period of observation and study. The international geophysical program has the added advantage of permitting all countries to participate; those which are not prepared to enter the newer (and, generally more expensive) fields will almost certainly be able to discover relevant more conventional fields which are undersubscribed.

Solar-oriented programs. The IGY, from July, 1957, to December, 1958, was an internationally accepted period of concentrated and coordinated geophysical exploration, primarily of the solar and terrestrial atmospheres. It was marked by intense solar activity as well as by intense human activity, and enjoyed an unqualified success partly because of both these factors. It was apparent by 1960 that there would be considerable merit in a second international period of scope, concentration, and coordination, at least comparable to that of the IGY, to take place during the time of the following minimum of solar activity. The period called the International Years of the Quiet Sun (IQSY) was therefore carried out from January, 1964, to December, 1965, to permit a comparison between phenomena and parameters at times of maximum and minimum solar activity.

The IQSY program drawn up by the ICSU special committee for the IQSY on the basis of discussions in nations, in unions, and in the IQSY assemblies comprised activities in the following disciplines: meteorology (large-scale physical and dynamic characteristics of atmosphere above 10 miles), geomagnetism, aurora, airglow, ionosphere (vertical incidence, absorption, and drift studies, plus rocket and satellite data), solar activity, cosmic rays, space research, and aeronomy. Seventy-one nations participated in the IQSY, and the results have appeared in hundreds of research papers in scientific journals and are summarized in the seven-volume *Annals* of the IQSY.

Following the IQSY, cooperation in international and national projects relating to solar-terrestrial disciplines came under the aegis of the Inter-Union Commission on Solar-Terrestrial Physics (IUCSTP), a component body of the ICSU. Initial emphasis was on the period of maximum solar activity (1968–1970), which is referred to as the International Years of the Active Sun (IASY). For that period, specific coordinated programs have been planned in the following fields: monitoring of the solar-terrestrial environment, proton flares, disturbances of the interplanetary magnetic field configuration, characteristics of the magnetosphere, conjugate-point experiments, magnetic storms, low-latitude auroras, upper-atmosphere structure and dynamics, ionospheric chemistry, and ionospheric disturbances.

Solid-earth geophysical programs. One highly effective program, oriented toward solid-earth as well as space problems, has been the world magnetic survey, coordinated on behalf of the ICSU by the World Magnetic Survey Board of IUGG. The aim of this program was to define, with a prior agreement as to precision and detail, the magnetic field of the Earth at the epoch 1965.0. Both surface and airborne data were extrapolated to 1965.0 whenever possible, and many special observation programs were undertaken. A definitive framework was established so that any subsequent observations of geomagnetic phenomena either at the surface or within 1000 km of the surface (by rocket or satellite) could be classified readily as either normal or abnormal.

For pure solid-earth disciplines, the requirement for coordinated programs is considerably less than for other branches of geophysics. Nevertheless, from time to time it becomes apparent that a specific field can be advanced rapidly only if certain studies are carried out in many representative areas. Such was the case for the Upper Mantle Program, with peak activity in the mid-1960s, which was coordinated on behalf of the ICSU by the Upper Mantle Committee of the IUGG, in association with six other ICSU unions. The emphasis was on indirect probing of the upper mantle of the Earth, which is found tantalizingly close to the Earth's surface.

International hydrological decade. Despite its tremendous economic significance, hydrology has long been one of the lesser developed geophysical sciences. To rectify this imbalance and permit hydrology to make a meaningful contribution to global water problems, including especially those of the developing nations, UNESCO took the initiative in launching the International Hydrological Decade (IHD, 1965–1974). A coordinating council, under UNESCO, is the central agency for planning and promoting the Decade, with scientific advice from ICSU's Scientific Committee on Water Research (COWAR). The coordinating council established working groups for various facets of the overall program to deal with network planning and design, global water balance, hydrological maps, representative and experimental basins, hydrology of fractured limestone terrains, nuclear techniques to determine water content in saturated and unsaturated zones, influence of man in the hydrological cycle, floods and their computation, and exchange of information, including publication, standardization problems, and education and training.

Oceanographic programs. The primary coordinating body for oceanographic geophysical programs is the Intergovernmental Oceanographic Commission (IOC), which is an autonomous organization within UNESCO. Scientific advice to IOC is provided by ICSU's Scientific Committee on Oceanic Research (SCOR). A number of international oceanographic programs of somewhat limited extent such as the International Indian Ocean Expedition (IIOE, 1961–1966), studies of the tropical Atlantic and the Kuroshio Current, and so on have been undertaken. A patchwork program of this type, however, is rather unsatisfactory, since priorities for various activities are hard to establish. As a result, SCOR has produced a document, entitled General Scientific Framework for World Ocean Study, which puts regional studies into a global perspective. Almost simultaneously, the United Nations General Assembly adopted Resolution 2172 on Resources of the Sea, which has provided new impetus for oceanic research. As a consequence, international organizations are giving serious consideration to an expanded, accelerated, long-term, and sustained program of exploration of the oceans and their resources.

Atmospheric science programs. Vigorous programs in meteorology were carried out in connection with the IGY and IQSY, but subsequent developments, primarily in space technology, have thrust the atmospheric sciences into the international limelight. Two key United Nations General Assembly resolutions (1721 and 1802) have called upon international governmental (WMO) and nongovernmental (ICSU) organizations to develop significant programs in the atmospheric sciences, both operational (World Weather Watch, WWW) and research (Global Atmospheric Research Program, GARP). A history-making precedent was set in October, 1967, when the ICSU and WMO agreed on the characteristics of GARP and set up a joint organizing committee to plan the program.

The WWW is a global data collecting, processing, and dissemination system whose aim is to provide optimum service to all peoples through collective action within WMO. There are many aspects of WWW that would require research either in the laboratory or in the atmosphere, on a local or global scale. Numerical weather prediction and general circulation research suggest that extended-range prediction should be possible in some detail provided that (1) sufficient initial data on the atmosphere and on the air-Earth boundary are available, (2) all significant atmospheric and boundary processes are adequately included in the physical formulation of the problem, and (3) extremely sophisticated computers and highly accurate and stable numerical procedures are utilized. It is the aim of GARP to test this hypothesis. Preliminary subprograms will tackle individual problem areas and individual geographic areas, with the main phase of GARP embracing a truly global and complete atmospheric observation program, to produce the raw data for detailed testing of atmospheric models and atmospheric predictability.

[WARREN L. GODSON]

Bibliography: S.-I. Akasofu and S. Chapman, *Solar-Terrestrial Physics*, 1972; E. Fluegge (ed.), *Encyclopedia of Physics*, vol. 49, pt. 4: *Geophysics Three*, ed. by K. Rawer, 1972; R. R. Jones, *The Unsettled Earth*, 1975; C. M. Minnis (ed.), *Annals of the IQSY*, vols. 1–7, 1968–1970; H. Odishaw (ed.), *Research in Geophysics*, vol. 1: *Sun, Upper Atmosphere, and Space*, vol. 2: *Solid Earth and Interface Phenomena*, 1964; F. D. Stacey, *Physics of the Earth*, 1977.

Geostrophic wind

A hypothetical wind based upon the assumption that a perfect balance exists between the horizontal components of the Coriolis force and the horizontal pressure gradient force per unit mass, with the implication that viscous forces and accelerations are negligible. Application of the geostrophic wind facilitates an approximation of the wind field from the pressure data over vast regions in which few wind observations are available.

Bases of the approximation. The geostrophic wind blows parallel to the isobars (lines of equal pressure) with lower pressure to the left of the direction of the wind in the Northern Hemisphere and to the right in the Southern Hemisphere. Its speed is given by the equation below, where ρ is

$$V_{\text{geo}} = \frac{1}{2\rho\Omega \sin\phi} \frac{\partial p}{\partial n}$$

the density of air, Ω is the angular speed of rotation of the Earth, p is the atmospheric pressure, ϕ is the latitude, and n is a coordinate normal to the isobars and directed toward higher pressure. The approximation now known as the geostrophic wind was first derived empirically by C. H. D. Buys-Ballot in 1857 and has been known as Buys-Ballot's law.

The geostrophic wind represents a good approximation to the actual wind at elevations greater than about 3000 ft, except in instances of strongly curved flow and in the vicinity of the Equator.

Thermal wind. This is a term denoting the net change in the geostrophic wind over some specific vertical distance. This change arises because the rate of change of pressure in the vertical is different in two air columns of different air density, so that the horizontal component of the pressure gradient force per unit mass varies in the vertical. The thermal wind is directed approximately parallel to the isotherms of air temperature with cold air to the left and warm air to the right in the Northern Hemisphere, and vice versa in the Southern Hemisphere. Thus, for example, the increasing predominance of westerly winds aloft may be viewed as a consequence of the warmth of tropical latitudes and the coldness of polar regions. *See* CORIOLIS ACCELERATION AND FORCE; GRADIENT WIND; WIND; WIND STRESS.   [FREDERICK SANDERS]

Glaciology

The study of existing or modern glaciers in their entirety, involving all related scientific disciplines. As glaciology embraces so many interconnecting facets, it is considered a master science. It is largely concerned with present glacial characteristics and processes, as opposed to studies in glacial geology which relate to the nature and effects of former glaciation.

A glacier is a naturally accumulating mass of ice that moves in the process of discharging from head or center to its margins or a terminal dissipation zone. Glaciers are nourished in areas of snow accumulation that lie above the mean climatological or orographical snow line, which on the glacier surface is referred to as the névé line or firn line. The most active glaciers are generally found in regions receiving the heaviest snowfall, such as the maritime flanks of high coastal mountain ranges. Exemplifying this are the great westerly facing glaciers of Mount Saint Elias and the Saint Elias Mountains in south coastal Alaska which are nourished by the heavy precipitation brought in by warm, cyclonic air masses moving across the warm waters of the Gulf of Alaska. Similarly, in New Zealand the vigorous glaciers of the southern Alps lie in the storm tracks of the prevailing westerly wind which brings much greater accumulation to the western than to the eastern slopes. Other such maritime glacial bodies are the Patagonian ice field in the southern Andes; the small ice sheet and ice caps of Iceland; the mountain ice fields of Norway and the Kebnekaise region of Sweden; and the mountain glaciers of eastern Siberia. There are many other glaciers in the high mountains of the middle and equatorial latitudes (as in Peru and East Africa); however, 96% of the world's glacial ice is represented by the vast continental ice sheets of Antarctica and Greenland. Together, these regions contain at least 5,600,000 mi² (14,300,000 km²) of ice cover, or nearly 10% of the total land area of the globe. *See* SNOW LINE.

The total volume of the world's glaciers, ice fields, and ice sheets can only be estimated, but is at least 24,000,000 km³. This mass of frozen water, if melted and returned to the sea, would raise the sea level 160–200 ft (±60 m). R. F. Flint in 1957

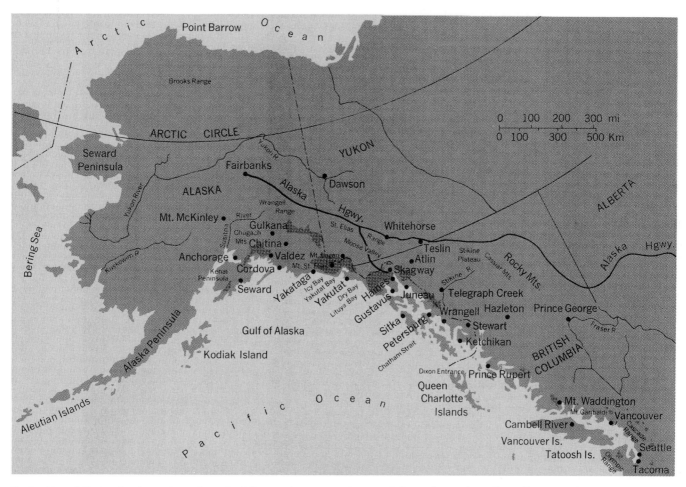

Fig. 1. Map of Alaska showing main centers of existing glaciers. Note that these centers lie along the south and the southeastern coasts, where there is a ready source of moisture and high ranges of mountains, resulting in heavy annual accumulations of snow. In northern Alaska it is too cold and dry to produce the requisite snowfall for glaciation under present climatic conditions. (*Foundation for Glacier and Environmental Research*)

estimated that, during the Ice Age, this volume was probably 300–400% greater. *See* TERRESTRIAL FROZEN WATER.

Glaciation and deglaciation. In many regions glacier activity has formed a continuous series of events throughout the Pleistocene glacial-pluvial epoch, and in postglacial (post–Ice Age) and modern times. In order to simplify the terminology of past and present glaciation, regions which are presently glaciated are also sometimes referred to as glacierized, glacier-covered, or ice-covered. These terms are used instead of the ambiguous and unqualified term glaciated for reference to all areas formerly or currently ice-covered. In this context, for any one region, the glaciation limit is represented by the lowest elevation mountain summit carrying an existing glacier.

A glacier at any one time presents only partial patterns of its long-term regime. The complex study of glacial regimen, the processes and consequences of growth and decay, is being actively pursued in glacial regions throughout the world today. Other glaciological studies embrace research in many allied disciplines, such as geomorphology, meteorology and climatology, physics of ice deformation (glacier mechanics or continuum mechanics), thermodynamics, survey and mapping, glacier geophysics, lichenometry, palynology, and plant ecology.

Materials. Glaciers are composed of three substances: snow, firn, and ice. The main material of glaciers is bubbly glacier ice of a specific gravity approximating 0.88–0.90 g/cm³. This glacier ice is composed of myriads of interlocking crystals, hence it is a polycrystalline material containing air pockets and entrapped water bubbles. Because of the proportion of bubbly glacier ice, a mean bulk specific gravity may be taken as 0.90 g/cm³, as opposed to a specific gravity of 0.917 g/cm³ in dense, solid, and unaerated ice. Below the late-summer glacier snow line (névé line, as defined below), only bubbly glacier ice is exposed. Above the névé line, the other categories of snow and firn (and firn ice) exist to depths of a few to hundreds of feet. It is deeper in polar (colder) firn packs. (Firn is a consolidated granular transition of snow not yet changed to glacier ice. The process of transformation of snow to firn is termed firnification.) The density gradation between firn and ice is usually asymptotic, but with a relatively sharp line of demarcation between a seasonal snowpack and underlying firn. Arbitrary and approximate densities of new snow are 0.1–0.3; old snow, 0.3–0.45; firn, 0.45–0.75; and firn ice, 0.75–0.88. Firn ice is not considered to be a separate stage of metamorphosis but rather the result of a mixture of partially altered firn and bubbly glacier ice. The processes by which new snow is transformed into bubbly and dense glacier ice are complex and varied. In middle-latitude regions the refreezing of percolated meltwater, compaction, and flow deformation play substantial roles in the metamorphosis. In polar regions wind packing, mechanical compaction, and flow recrystallization are the prime factors producing glacier ice.

Terminology. The geographical term névé refers to the area covered by perennial snow or firn, that is, the area lying entirely within the zone of accumulation. The word firn refers only to the substance of the material itself. The terms snowpack and firn pack refer to the volume of snow or firn, respectively, at any one point on the névé of a glacier's surface and connote thickness or depth characteristic rather than area. Snow cover and firn cover refer to the blanket of snow or firn over a névé, a glacier, or a bedrock surface, and have areal rather than depth connotations.

The term névé line (firn line) represents the elevation of the periphery of a névé at any point on a glacier's surface. More specifically, the elevation of the névé line's most stable position over a period of several years is referred to as the semipermanent névé line. The term glacier snow line or transient snow line describes the transient outer limit of retained winter snow cover on a glacier. Its elevation gradually rises until the end of the annual ablation season, by which time the old snow has become firn and the glacier snow line becomes the seasonal névé line. In regions such as coastal Alaska, where glaciers descend to sea level, the summer snow line on intervening rock ridges and peaks is often much higher than the snow line on the valley glacier, and in most instances is more irregular and indefinite. If it has not disappeared completely from the bedrock surfaces by the end of summer, the lowest limits of retained snow may be connected as an irregular limit and be termed the orographical snow line since it is primarily controlled by local conditions and topography. Variations in the position of the orographical snow line are so great from year to year that only from records over a long period can a meaningful trend be discerned. This snow line is important in the development of nivation hollows and protalus ramparts in deglaciated cirque beds. Of greater present-day significance is the average regional level of the orographical snow line rather than its lower limit. This mean position, based on observations over a number of years, is called the regional snow line or climatological snow line. On adjacent ice sheets and glaciers this coincides with the mean névé line. In regions of extensive present glaciation the regional snow line thus equates to the mean névé line. Specifically, the mean névé line is the statistical average of consecutive annual positions of semipermanent névé lines over a period of at least 10 years.

In glaciers of the polar regions a different term is used to delineate areas of net accumulation from those of net wastage, in consequence of the refreezing of meltwater and drainage on surfaces down-glacier from the transient snow line. The position where this refreezing ceases is referred to as the equilibrium line, a theoretical line separating the area of net gain from the area of net loss. On temperate glaciers this coincides with the névé line. It is an important concept in glacier velocity considerations because it represents the position or zone least subject to seasonal variations in flow resulting from excessive accumulation or ablation.

Morphological categories. Glaciers develop numerous forms. Basically, they are of the mountain (alpine) type and of the plateau (polar) type. Mountain glaciers generally are moderate to small

Fig. 2.   Oblique air photograph of 4-mi wide tidewater terminus of the Hubbard Glacier, a prototypical valley glacier which reaches tidewater from a source area in the St. Elias Mountains of Alaska and the Yukon. This glacier has advanced several miles since 1894. (*Photograph by H. B. Washburn, August, 1938*)

in size and include valley glaciers (main ice streams), icefall glaciers, cirque glaciers, basin glaciers, hanging glaciers, cliff glaciers, and glacierets. These terms are self-descriptive and for the most part relate to strong and varied relief of the kind found along the mountainous southern coast of Alaska (Fig. 1). Valley (ice stream) types are the most common (Fig. 2). They are often in the form of glacier systems fed by cirque-headed tributary valleys and serve as outlets from ice fields, such as those found in the St. Elias and Boundary ranges between Alaska and Canada (Figs. 2 and 3). They are also the main type of glacier in the Alps, the southern Andes, New Zealand, the Caucasus, and the Himalayas. The longest valley glacier in the temperate regions is the Hubbard Glacier (Fig. 2), with a length of about 100 mi in the Alaska-Yukon border area. The Vaughan Lewis Glacier and the Upper Herbert (Camp 16) Glacier on Alaska's Juneau Icefield are typical icefall glaciers derived from high névé basins or plateaus (Figs. 3 and 4). Plateau-type glaciers, dominantly of the ice-sheet and ice-cap form, are characterized usually by vast size with relatively flattened surfaces or low relief. These are typical of the Greenland and Antarctic ice sheets, and are sometimes termed inland or continental ice. Often valley glaciers extend outward as distributary tongues. Intermediate between valley glaciers and ice sheets are piedmont glaciers. These occupy broad lowlands bordering a glacial highland. The best known of this category is the Malaspina Glacier near Yakutat, Alaska, with an area of 1400 mi² (Fig. 1).

Geophysical types. Glaciers are classified geophysically into two major groups, polar and temperate, and two transitional groups, subpolar and

Fig. 3.   Part of Vaughan Lewis Glacier, Juneau Icefield, Alaska, showing surface bulges and wave ogives in apron area just below icefall. Similar to view looking downvalley in Fig. 4. Series of medial moraines visible upper right. (*Photograph by M. M. Miller, August, 1968*)

Fig. 4. Oblique air photograph of Upper Herbert (Camp 16) Glacier on the Juneau Icefield, Alaska, showing its plateau névé or accumulation zone at 5000-ft elevation, as well as the ice cascade, seracs, and wave bulges on the apron at the base of the icefall. (*U.S. Forest Service photograph, October, 1962*)

subtemperate (Fig. 5). The temperature of a polar glacier is perennially subfreezing, except for a shallow surface zone which may be warmed for a few weeks of each year by seasonal atmospheric variations. The extreme polar condition depicted in Fig. 5 is found in the heart of the Antarctic continent at the South Pole. In temperate glaciers the temperature below a recurring winter chill layer is always at the pressure melting point. This situation is typical of middle-latitude glaciers, such as those in southern Alaska. Because these terms are thermodynamic in meaning but geographical in connotation, it should be pointed out that glaciers of the geophysically polar type can still exist at relatively low latitudes, and that geophysically temperate glaciers are even found at latitudes to the north of the Arctic Circle.

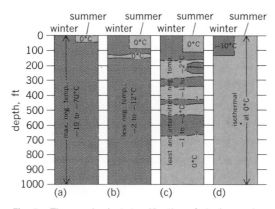

Fig. 5. Thermophysical classification of glaciers or ice sheets. (*a*) Polar. (*b*) Subpolar. (*c*) Subtemperate. (*d*) Temperate. Arbitrary depth-temperature values at surface illustrate seasonal variations. Dark tone, "cold" subfreezing state; light tone, "warm" (0°).

Of the transitional categories, in subpolar glaciers the penetration of seasonal warmth is restricted to a relatively shallow surface layer, but is greater than in polar glaciers. The transitional subtemperate glacier is characterized by a relatively deep zone of annual warming. These subordinate terms are useful because the former may refer to transitional glaciers of dominantly polar character which still have certain temperate characteristics, and the latter to dominantly temperate glaciers having a tendency towards polar characteristics. The significance of such differentiation relates to the close control ice temperatures exert on flow deformation, and so also to glacial fluctuations. This includes a possible relationship to kinematic surges, which are described later. Each geophysical category (polar, subpolar, subtemperate, and temperate ice) can be found in any one glacier system if there is sufficient range in latitude or elevation for the requisite climatological factors. Present-day ice fields and ice sheets which are thermally temperate in some sectors and grade through to thermally polar geophysical conditions in other sectors are referred to as polythermal. Such probably was the geophysical character of the continental ice sheets during the waxing and waning phases of the Pleistocene Epoch. A glacier which is geophysically temperate (0°C) throughout is usually referred to as isothermal.

Thermodynamics. The internal thermodynamic and heat-transfer character of glaciers is more complex than suggested by the foregoing geophysical differentiations. This is because individual glaciers vary much in their structural makeup and related regime histories. Some of the complication is shown by the table of thermal constants for snow, firn, and ice at 0°C.

The problem of thermal changes within glaciers thus requires the approach of the physicist. One supplemental effect, however, must be kept in mind. This relates to the fact that, although normal thermodynamical processes pertain in temperate glaciers, infiltrating meltwater also plays an important role. In polar glaciers the fundamental diffusivity relationship shown in this table dominates, as there is only minimal meltwater effect restricted to the surface zone.

Irregularities in thermal dissipation occasioned by the presence of mobile water and by various physical inhomogeneities in glaciers usually preclude close quantitative agreement with hypothetical temperature curves calculated from assumed bulk diffusivity. Computed values are near to observed conditions only in purely polar glaciers, or in those sections of temperate glaciers in which during the winter months there is no liquid water to prevent changes in internal temperature from being controlled by conduction. The detailed mathematical analysis of the heat transfer is not considered here, beyond mention of the fundamental thermodynamic properties pertaining to understanding and appreciation of field observations.

In snow and firn, internal temperature changes may be considered as occurring in a semi-infinite, homogeneous, isotropic solid (although individual crystals are structurally anisotropic) influenced by fluctuating external temperatures which are a

Thermal constants for snow, firn, and ice at 0°C*

Material	Conductivity, cal °C⁻¹ cm⁻¹ sec⁻¹	Specific heat, cal °C⁻¹ g⁻¹	Density, g cm⁻³	Thermal diffusivity, cm² sec⁻¹	Relative diffusivity to ice, approx ratio
New snow	0.0003	0.5	0.20	.0030	0.27
Old snow	0.0006	0.5	0.30	.0040	0.36
Average firn	0.0019	0.5	0.55	.0070	0.64
Firn ice	0.0038	0.5	0.75	.0100	0.91
Ice	0.0050	0.5	0.92	.0110	1
Water, 0°C	0.0014	1.0	1.00	.0014	0.13
Rubber	0.0005	0.40	0.92	.0014	0.13
Steel, mild	0.1100	0.12	7.85	.12	11
Aluminum	0.4800	0.21	2.70	.86	78
Copper	0.9300	0.09	8.94	1.14	104

*The given conductivities for snow and average firn are based on data from U.S. Army Corps of Engineers, Snow, Ice and Permafrost Research Establishment, *Review of the Properties of Snow and Ice*, rep. no. 4, 1951. The other data for snow and ice are from M. M. Miller, *Glaciothermal Studies on the Taku Glacier, Southeastern Alaska*, Tome 4, Publ. no. 39 de l'Association Internationale d'Hydrologie, Union Internationale de Geophysique et Geologie, Assemblée generale de Rome, 1954.

harmonic function of time. The basic physical factors controlling development of the sinusoidal cold wave are the thermal conductivity k of the medium, its specific heat c, and its density ρ. It can be shown that the transmission of heat actually depends on the diffusivity K, defined by a combination of these quantities in the relationship, $K = k/c\rho$. Here the specific heat is taken as constant at the value for ice. The density and conductivity, however, are variable, depending on the age of the medium and related genetic factors. The resulting diffusivity is therefore a function of these factors.

The relative diffusivity between dense firn and ice is not so large that mass heat transfer and temperature changes in firn are greatly altered by the presence of a few ice strata. On the other hand, a glacier with a substantial covering of snow will display important differences in surface heat transfer compared to one with only solid ice exposed. The general relationships are illustrated in the table. The acute sensitivity of glacier flow to changes in internal temperature, as discussed under structure and movement below, underscores the importance of appreciating the role of these thermal constants in circumstances of pronounced and long-term climatic change.

Regime. Upon the regime, or the annual state of health of a glacier system, depends the glacier's eventual growth or decay, that is, its mass balance. The controlling factors are accumulation (gain) and ablation (loss) over the whole system. On temperate glaciers the critical sector is the névé line, or more properly a névé-line zone which at the end of the annual melt season separates the névé area (accumulator) from the bare-ice area (dissipator). As already noted, it equates to the lower elevation limit of retained winter snowpack as observed at the end of summer. On polar glaciers the equivalent critical sector is, of course, the equilibrium line.

The height of the seasonal névé line, or yearly equilibrium line, shifts greatly according to the regional climatological situation and local conditions. With respect to the Juneau Icefield in southeastern Alaska, this critical line (Fig. 6) varied over

a 20-year period between 2400 and 3800 ft (mean 3100 ft). In high-latitude polar regions the equilibrium line is usually at, or close to, sea level. On alpine glacier systems a higher than average névé line means a tendency toward a negative regime (more loss than gain), whereas a lower than average névé line is associated with an increase in the accumulation area and hence a positive regime. This kind of interrelationship is illustrated by the accumulation trend in Fig. 6 and by the typical mass-balance relationship depicted for a healthy glacier situation in Fig. 7. The specific net budget curve in Fig. 7 shows the difference between accumulation and ablation per unit area at every

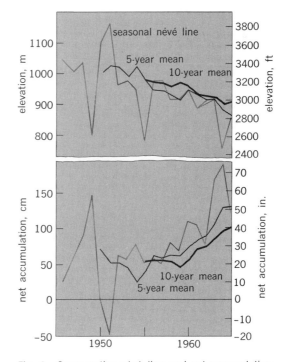

Fig. 6. Comparative névé line and net accumulation trends on the Taku Glacier, Alaska, during 1946–1965. (*Juneau Icefield Research Program*)

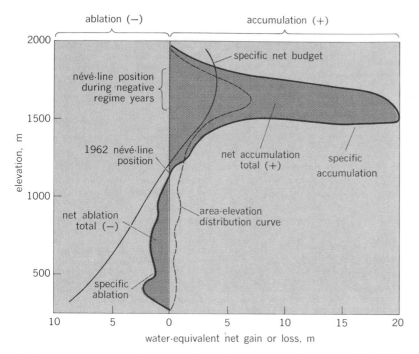

ablation (−)   accumulation (+)

néve-line position
during negative
regime years

1962 néve-line
position

net ablation
total (−)

specific
ablation

specific net budget

net accumulation
total (+)

specific
accumulation

area-elevation
distribution curve

elevation, m

water-equivalent net gain or loss, m

Fig. 7. Mass balance relationship on Nigardsbreen Glacier, Norway, in 1961–1962 budget year. This glacier normally has a negative regime, but it was surplus in this year, calculated to be about 95,000,000 m³ water equivalent. (*After G. Ostrem, Nigardsbreen Hydrologi, 1962, Norsk. Geog. Tidskr., vol. 18, pp. 156–202, 1963*)

specific elevation. To obtain the actual or total budget (mass balance), this figure is multiplied by the total glacier area at that specific elevation. In alpine glaciers the area ratio of accumulator to dissipator is on the order of 4:1, whereas in the Greenland ice sheet it is in excess of 100:1, and in the Antarctic as much as 1000:1. Also, in the accumulation zone there is an increase of retained accumulation in any one year with an increase in elevation until the height of maximum snowfall is attained. It should be noted that this is not always at the highest elevation of land, because maximum snowfall is controlled by the mean freezing level, that is, the greatest snowfall occurs at or near 0°C. Thus, in mountain regions or on large ice sheets with a great elevation range at very high elevations, there is commonly less snowfall, the conditions there being generally too cold for much snow.

The rise and fall of the névé line over a period of years also parallels vertical shifts in the zone of maximum accumulation. This vertical shift may be

measured by means of soundings, borings, test-pit studies, and observations on crevasse walls. In a healthy glacier the average level of maximum snowfall lies at the elevation of greatest glacier area (Fig. 7). A glacier in a poor state of health has usually experienced a rise in its level of maximum snowfall to a height above the area of the main névé. Also to be considered as an additional increment of positive net accumulation in a glacier system is that portion of summer meltwater in geophysically subpolar or polar (cold) névés recaptured by freezing as it percolates to depth. In contrast, meltwater percolation in temperate glaciers drains away almost entirely in subglacial drainage channels, eventually flowing out at the snout of the glacier as an increment of net loss.

Structure and movement. Because of the many structures in ice comparable to those in sedimentary and metamorphic rocks, a glacier is an ideal field laboratory for the structural geologist. A few such structures are: primary stratification (bedding strata), secondary fracture structures, discontinuous marginal or basal tectonic foliation, (Fig. 8), ablation surfaces overthrust surfaces, faults, folded structures, and a varied group of deformed sedimentary and structural bands, including the wave-ogive bands and surface bulges illustrated in Figs. 3 and 4. There are also subsurface diagenetic structures of both stratiform and transverse ice which result from refreezing of downward percolating meltwater in subfreezing firn. Cross-cutting transverse types are manifest at the surface as ice columns and dikes. Tension and shear fractures are also common as crevasses and bergschrunds (the latter exhibiting an overhanging upper lip), as well as moulins (glacier mills, or deep rounded holes caused by water action on embedded stones), cryoconite holes (thermal pits produced by the inmelting of organic or rock fragments), sastrugi (wind-scoured features), and other surface features. All of these are representations of the combination of processes which affect and control the surface regime and the dynamics of the internal structure and movement of glaciers.

Glacier deformation is a composite of internal and external movement. The internal movement is dominated by a continuous plastic creep (flow deformation), and the external in some cases by a fracture type of discontinuous movement at the bed or near the glacier margins. As a general rule, flow deformation, hence the rate of movement, is greater in the upper portion of a glacier than in its basal section (Fig. 9). In glaciers with a healthy or strongly positive regime in the accumulation zone, a dominant proportion of the total mass transfer may be expected to be via actual sliding of the glacier over its bed. Such glaciers are characterized by rectilinear, pluglike, or Block-Schollen velocity profiles. In plan view the greatest amount of movement is expressed in a broad central area. The surface velocity profiles in Fig. 10 were surveyed at stated intervals up-glacier from the terminus and represent the general flow characteristics of this strongly advancing glacier during the 1950s. Sliding glaciers often contain sheared basal ice heavily entrained with rock fragments and serving as effective agents of erosion. In glaciers of equi-

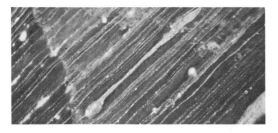

Fig. 8. Basal tectonic foliation (discontinuous laminar shear bands) with entrained fragments of pebble-size rock and fines in sole of Moltke Glacier, a geophysically polar glacier east of Thule. (*Photograph by M. M. Miller*)

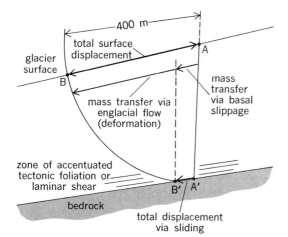

Fig. 9. Sketch of hypothetical case to demonstrate the proportion of total mass transfer across a given profile which can be ascribed to englacial flow deformation plus the proportion resulting from erosive slippage or bed sliding. This case is typical of plug flow or Block-Schollen mass transfer in a healthy advancing glacier.

librium or negative regime in the névé, the dominant mode of movement is expressed as internal streaming flow. In plan view such velocity profiles have a smooth, arched, parabolic form with associated streamlines. All gradations of movement from continuous laminar (streaming) flow to discontinuous Block-Schollen, or plug flow, may occur in any one glacier system. The relative proportions depend upon the névé regime pattern in recent decades, on internal temperature characteristics of the ice, and on configuration of the bedrock channel.

During the 1950s laboratory and field experiments by M. Perutz, J. Glen, M. Miller, J. Nye, S. Steinemann, and others ushered in a new set of concepts in glaciology, which related rheological and mathematical models of glacier mechanics (deformation and flow) to the phenomena of glacier movement and mass transfer observed in nature. The field measurements were largely based on the deformation of glacier tunnels and elongated metal pipes drilled into the glacier perpendicular to its surface. As a result of this pioneering work, the nature of internal deformation of glaciers is treated as a problem in the physics of shear, with reference to plasticity theory. Thus it is expressed by an exponential relationship between gravitational stress and deformation per unit time (creep velocity) with a simplified power law equation, Eq. (1).

$$\frac{d\gamma}{dt} = k\tau^n \qquad (1)$$

Here γ is the shear strain, τ is the shear stress in bars, k is a constant for any given temperature, and n is an empirical constant, depending in large measure on the physical character of the ice. Also, the exponent n probably depends to some degree on the magnitude of stress, a factor judged as probably significant only in very deep or otherwise highly stressed ice.

Under temperate valley glacier conditions, the factor n in nature is found to be close to a value of

3. In the flow law n represents the slope of the double logarithmic plot of the fundamental shear stress–shear strain relationship. The constant k is obtained from the stress-strain formula at the stress of 1 bar. Since the logarithm of 1 is zero, k is, in fact, the ordinate at the point where the abscissa of any log stress–log strain rate line is zero.

Also $d\gamma/dt$ or $\dot{\gamma}$ is the strain rate per year, usually calculated in radians of angular deformation per year, because it is found to be numerically equivalent to the tangent of the changing tilt angle in slowly deforming englacial pipes from which vertical velocity profiles have been measured. Several examples of field measurements are given in Fig. 11.

It should be noted that the power law used with a universal constant n represents only an average for low-gradient temperate glaciers, and therefore can give no more than a good approximation of the true strain rate in a glacier at depth. Nonetheless, by application of this law, at least for glaciers of simple configuration, a reasonable determination can be made of the movement within a glacier. And by combining the englacial movement data with surface velocities obtained by periodic surveys of across-glacier stakes and with geophysically determined depth records, an assessment can be made of the relative proportion of sliding or slippage on the bed (Fig. 9).

The shear stress at any depth with a glacier is calculated from relation (2) in dynes cm⁻², where D is

$$\tau = D\rho g \sin \alpha \qquad (2)$$

the depth in centimeters, ρ the bulk specific gravity of the overlying mass in grams per cubic centimeter, α the surface gradient, and g the acceleration of gravity (980 cm sec⁻²). In relation (2) the very critical control exercised by slight variations in gradient is well revealed.

Thus in the flow law with all factors determinable in nature, and with n and k calculated for any particular velocity profile, it is possible to extrapolate the differential flow all the way to the base of a glacier. In this way determination is made of the proportion of total mass transfer within a glacier. For this calculation the difference is found between the quasi-plastic flow velocity at any given depth U_D and the surface velocity U_O at the glacier surface. The relation is expressed by combining the empirical constants k and n with the stress at the indicated depth D as in Eq. (3), which was formulated by Nye.

$$U_O - U_D = \frac{k}{n+1} D\tau^n \qquad (3)$$

In this manner the approximate value of basal sliding is determined. A hypothetical case is shown in Fig. 9. The actual extent to which bottom sliding takes place depends not only on the surface gradient but on the slope and roughness of the underlying topography, on the width and thickness of the glacier, and on any other blocking factors involved. All these factors vary in their effect on the total stress distribution in different sectors of a glacier, and they are expressed at any particular point in the ice by the creep law, which integrates the gravitational load and englacial temperature.

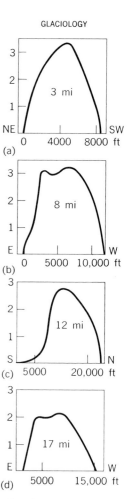

Fig. 10. Horizontal across glacier surface velocity profiles in ft/day from terminus up into névé zone of Taku Glacier, Alaska. (a, c) Partial parabolic streaming. (b, d) Block-Schollen (plug flow).

It must be kept in mind that such analyses over-simplify the situation in nature, because ice is not actually a homogeneous plastic material. Instead it should be considered as a quasi-elastoplastic substance, whose behavior under stress is still not fully understood.

In glacier ice when the stress conditions are great and substantial bottom sliding takes place, there is also much marginal shearing along the edges of the glacier. This is suggested by the form of transverse surface-velocity curves (Fig. 10) and the field measurements of vertical velocity profiles which have been noted (Fig. 11). The accentuation of laminar shearing or tectonic foliation in the sole of a glacier, observed as ribbon structures along the glacier's margins and bed (Fig. 8), indicates that slippage more usually takes place in a zone of highly sheared basal ice carrying entrained rock fragments and debris, which in turn serve as the key erosive tools of the glacier.

Erosion and transportation. Like rivers, glaciers have distinct regions of erosion and transportation. For example, in the high cirque basins of alpine glaciers, rock fragments fall from cliffs or slide into bergschrunds or marginal moats and become entrained in the ice for transport down the valley. These transported fragments thus become the critical tools of erosion along the ice-bedrock interface (Fig. 8). The larger angular rocks scratch the bedrock and form grooves and striae (Fig. 14), while the finer clastic detritus serves as the abrasive to smooth and polish. In such areas the presence of impact structures, striae, grooves, and crescentic features are not only proof of former glaciation but also reveal the former direction of flow.

Other geomorphological characteristics support the role of bottom slippage in erosion, such as deeply entrenched and U-shaped valleys reflecting the parabolic form of englacial creep in an actively advancing glacier, and the vast quantity of rock flour (clay size) carried out from beneath many glaciers by subglacial and proglacial drainage streams. Neither of these conditions could possibly have developed without considerable abrasion from basal sliding of a debris-entrained sole of the glacier.

Plucking is also an important and related agent of erosion. This process involves the penetration of ice or rock wedges into subglacial niches, crevices, and joints in the bedrock. As the glacier moves, it plucks off pieces of jointed rock and incorporates them as supplemental agents of abrasion and further plucking. Down-valley ends of jointed hummocks in the bedrock are produced in this manner and are known as roches moutonnées. Figure 14 shows the glacial grooves and oriented crescentic erosional features on top of a roche moutonnée. A sequence of such bedrock bosses produces a steplike longitudinal profile on a glacial valley floor. The steps often coincide with particularly resistant lithologies or selective low-angled joint surfaces.

Surging glaciers and kinematic waves. Research has suggested that glacier surges, expressed as sudden catastrophic advances or raising and lowering of the ice surface, are relatively common phenomena in some regions. Such abnormal surges are characterized by marked and seemingly anomalous increases in flow velocity and often by a rapid transfer of ice from the névé to the terminus, with much increased pinnacling and

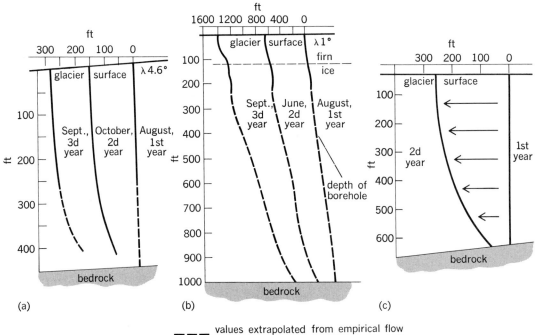

Fig. 11. Vertical velocity profiles, based on measurements made in the field, showing annual changes on the (a) Aletsch Glacier, Switzerland (after M. Perutz); (b) Taku Glacier, Alaska (after M. Miller); and (c) Saskatchewan Glacier, Canada (after M. Meier).

crevassing. The upper glacier surface may sink or be substantially lowered with this volume loss, expressed as a comparable thickening in the lower valley sector. Such surges are believed to be kinematic in nature, in that the wave moves through the glacier at a substantially faster rate than the actual discharge of ice. Often much crevassing and geometric folding of in-ice structures and medial moraines occur, as well as strong buckling, shearing, and surface lowering along the valley walls (Fig. 12). Striking examples are the Dusty, Lowell, Walsh, and Steele glaciers in the Yukon, where in 1965–1969 surge velocities of up to 60 ft per day were reported.

Surging glaciers were first reported by R. F. Tarr and L. Martin in Yakutat Bay, Alaska, in 1910–1913, with later reports of vigorous surging in this same area in 1965–1968. Surges of spectacular nature have been reported on many glaciers, including the Black Rapids Glacier, Alaska; Bruarjokul Glacier, Iceland; Medvezhii Glacier in the Pamirs, Soviet Union; and Muldrow Glacier on Mt. McKinley, Alaska. Others have been reported in Spitsbergen, Ellesmere Land, and even in the Karakoram and Himalayas.

Although this phenomenon is not adequately understood, much research is under way. It is at least known to relate to development of a dynamic instability that usually attenuates within 1 to 5 years, after which the lower glacier area affected by such an advance begins to stagnate. It is not the same phenomenon producing normal discharge wave-bulges and advance, as these may be presumed to relate to annual accumulation varia-

Fig. 12. Massive marginal shearing and lowering of ice along the valley walls of surging Walsh Glacier, in the Yukon Territory of Canada. Note the strongly related development of conjugate shear crevasses. This glacier moves forward tens of feet per day. (*Photograph by H. B. Washburn, August, 1966*)

tions and to gradual climatic change. One possible origin of the catastrophic surge phenomenon is a very unusual and sudden change in load stress via thickening of the glacier in its upper névés or high-level nourishment basins, with a consequent energy release via a kinematic wave.

Of course such thickening may be associated with abnormal increases of snowfall by way of what appears to be sudden climatic change. In certain unique situations, the suddenness of such change can be accentuated by significant horizon-

Fig. 13. Massive earthquake avalanche of rocks and debris covering approximately 10 mi² of the Schwann Glacier in the Eastern Chugach Range, Alaska, was caused by the 1964 Good Friday earthquake. (*Photograph by M. M. Miller*)

tal or vertical shifts in the zone of maximum snow-fall across the critical névé level. But there are several other possibilities. One may relate to warming of the englacial ice itself, a factor which probably caused the Brasvalsbreen Glacier in Spitsbergen to slide forward catastrophically in the late 1940s (a 5-mi advance in less than 5 months). This may not have been a true surge but a response to substantial changes in englacial temperature, since it is known that the mean winter temperatures in the northeastern Arctic rose about 20°F during 1920–1960.

The third possibility is the sudden imbalance produced by earthquake-caused rock and ice slides. A number of huge slides were reported on certain Alaskan glaciers following the 1964 Good Friday earthquake. One of these is pictured in Fig. 13, showing apparent flow effects in a deformed medial moraine below a 10-mi² debris slide on Schwann Glacier in the Copper River region, Alaska. A fourth, and probably minor, possibility is some actual effect of the earthquake shock in the disarticulation of the glacier structure, which could abet the "sudden-slip" character of a subsequent advance. This would probably be true when the epicenter of the earthquake lies in the vicinity of the glaciers involved. A further possibility is a buildup of excessive pressure melt-ing on a glacier's bed, or in some cases an intensification of geothermal heat at the sole of the glacier, resulting in abnormal quantities of lubricating water (slush) which could accentuate the effects of basal slip. In some cases the most dynamic surges could be a result of a combination of two or more of these factors. However, glaciologists must remain cautious, because many glaciers surge and even express seemingly less kinematic effects, such as surface buckling and locally increased crevassing, without any evidence of avalanche material being dumped on them. At least one correlation which is of interest is that some of the most spectacular surging glaciers have been reported in regions of the most active tectonic-earthquake activity.

The greatest bed erosion takes place when the regime of a glacier is healthy and its mass transfer substantial. A vigorously advancing glacier exhibiting Block-Schollen or plug flow is the most effective erosive body. This adds credence to the concept that glacial erosion is an indirect consequence of glacioclimatological oscillations affecting the growth and decay, thickening and thinning, and advance and retreat of glaciers.

As to the direct cause of high ratios of bed slippage and the occasional development of anomalous surges, scrutiny of the shear stress equation reveals that flow within a low gradient glacier or ice sheet (for example, at slopes of 1–3°) is so small that in many cases stresses other than those explained by pure gravitational shear must be in effect. Hydrostatic pressure is ruled out on the basis that it is as negligible in ice as in liquids. The most significant supplemental stress may be attributed to a strong down-glacier longitudinal force superimposed on the gravitational stress in quasi-elastoplastic ice whose yield limit is already exceeded. Such supplemental stressing can be the result of any one factor or a combination of the factors which come into play in advancing and surging glaciers. To such stress tensors, the strain effects of which are sometimes accentuated by an increase in temperature within the ice, the phenomenal glacial advances in Alaska (group IV, Fig. 15) may be ascribed.

The repeated oscillation of glaciers in mountain cirques and over the floors of outlet valleys results in effective scouring, transportation, and removal of material. Such continuous sequences of process produce the wide-strath highland glacial basins and deep U-shaped outlet valleys so common in the Alps, the Cascades, and the Rocky Mountains. Turbid glacier rivers and moraines (lateral, medial, and terminal) are living examples of the immense eroding and transporting power of glaciers. Festooned arrays of terminal moraines also testify to continuous glacier activity, because they express repeated oscillations most usually related to cyclic changes of accumulation and ablation in the distant névés many years before.

Recent and current fluctuations. Following the post-Wisconsinan maximum about 10,000 years ago, the Ice Age entered its latest waning phase. The result was an almost complete retreat and disappearance of glaciers in the mid-latitudes and tropical regions of the Earth. Coincident with this

Fig. 14. Glacial pavement on granitic bedrock surface in recently deglaciated mountain region of Alaska, showing orientated abrasion structures resulting from basal sliding of the glacier over its bed. In this view can be seen crescentic gouges (concave in up-glacier direction), lunate furrows (concave in down-glacier direction), grooves (elongate linear furrows), and striae (fine lines of scratches). The direction of the former ice movement is from foreground toward valley in distance. (*Foundation for Glacier and Environmental Research*)

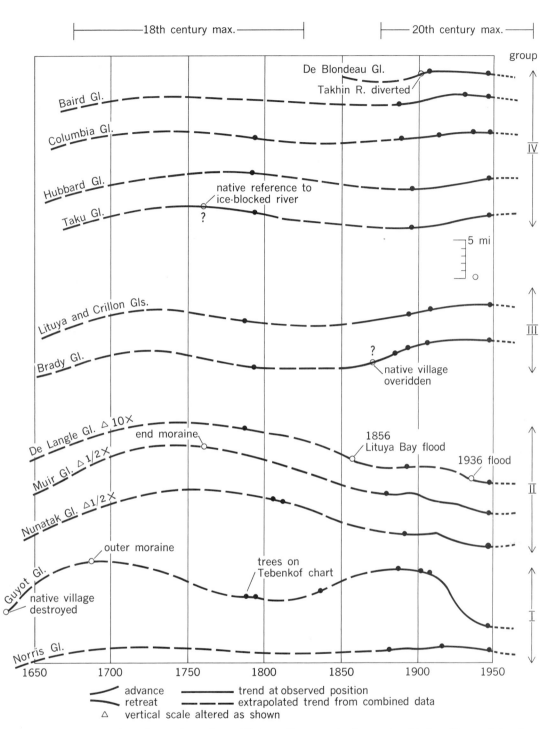

Fig. 15. Two-phased regional fluctuation pattern of the Alaskan Little Ice Age showing how, at any one time, glaciers in a region may reveal patterns of strong different advance simultaneous with significant retreat of others. This out-of-phase pattern reflects the role of differences in size, geographical position, and elevation or orientation of névés, as well as differences in flow lag via differences in size, configuration, and length of the outlet valley glaciers.

disappearance was the Thermal Maximum (Hypsithermal Interval or Climatic Optimum) culminating about 5000 years ago. Between 1000 B.C. and A.D. 1 a worldwide recrudescence of glaciers began to take place. This is termed the Neoglaciation and is associated with a return to harsher climatological conditions, which still prevail today. This temperature fluctuation culminated in a colder condition at the beginning of the Christian era. The evidence suggests that in the 5th and 7th centuries the polar seas were freer of ice than they are today, as far north as the pole. Peripheral waters thus remained relatively clear through the 10th century. The records of Norse settlers in

Greenland indicate that the climate remained relatively mild until the 14th century. Then, about 5 centuries ago, there began another worldwide expansion of glaciers and a thickening of polar ice, called the Little Ice Age. Technically, the Little Ice Age refers to this latest major phase of reglaciation in the latter quarter of the Neoglacial Age.

Little Ice Age fluctuations are double-phased in nature, having produced a worldwide growth of temperate glaciers which reached their culminations in the early to mid-18th century, and again in the late 19th to mid-20th centuries. The Alaska Little Ice Age pattern is illustrated in Fig. 15. This reveals that at any one time the terminal regime of glaciers in adjoining areas may be quite out of phase; that is, one may advance while another simultaneously retreats. The latest advances on a small percentage of high-level trunk glaciers have continued into the 20th century, in spite of a general diminution of ice cover around the periphery of some of the lower Alaskan ice fields. Recent fluctuations in Scandinavia and Patagonia are quite similar to this Little Ice Age pattern and have been shown to have a teleconnectional similarity via upwards of a dozen recessional moraines over the past 200 years. Such evidence supports the global nature of the causal factor, and the acute sensitivity of glaciers as excellent historians of secular climatic change.

In contrast to the behavior of temperate glaciers, the polar glaciers of Antarctica exhibit a fairly stable regime. This suggests that significant volume changes in middle-latitude glaciers, including the sub-Arctic and the sub-Antarctic, and noticeable increases in their internal temperatures have been instrumental factors in accentuating the fluctuation patterns of Neoglacial time.

[MAYNARD M. MILLER]

Bibliography: J. Andrews, *Glacial Systems*, 1975; L. A. Bayrock, *Catastrophic Advance of the Steele Glacier, Yukon, Canada*, Boreal Inst. Univ. Alberta Occasional Publ. no. 3, 1967; J. A. Gerrard, M. F. Perutz, and A. Roch, Measurement of the velocity distribution along a vertical line through a glacier, *Proc. Roy. Soc. London Ser. A*, 213:546–558, 1952; J. W. Glen. The creep of polycrystalline ice, *Proc. Roy. Soc. London Ser. A*, 228:519–538, 1955; A. E. Harrison, Ice surges on the Muldrow Glacier, Alaska, *J. Glaciol.* 5(39): 365–368, 1964; *J. Glaciol.*, 1947 onward; B. Kamb, Glacier geophysics, *Science*, vol. 146, no. 3642, Oct. 16, 1964; M. M. Miller, Alaska's mighty rivers of ice, *Nat. Geogr. Mag.*, 131(2):194–217, February, 1967; M. M. Miller, Phenomena associated with the deformation of a glacier borehole, *Trans. Int. Union Geod. Geophys.*, vol. 4, 1958; M. M. Miller, 1965–1968 Studies of surge activity and moraine patterns on the Dusty Glacier, St. Elias Mountains, Yukon Territory, *Proc. 19th Alaska Sci. Conf.*, American Association for the Advancement of Science, 1968; J. F. Nye, The flow law of ice from measurements in glacier tunnels: Laboratory experiments and the Jungfraufirn borehole experiment, *Proc. Roy. Soc. London Ser. A*, 219:477–489, 1953; P. W. S. Patterson, *The Physics of Glaciers*, 1970.

Gradient wind

A hypothetical wind based upon the assumption that the sum of the horizontal components of the Coriolis force and the atmospheric pressure gradient force per unit mass is equivalent to a wind acceleration which is normal to the direction of the wind itself (centripetal acceleration), with the implication that there are no viscous forces acting. The direction of the gradient wind is the same as that of the geostrophic wind. Its speed is determined by the equation below, where V_{geo} is the

$$V_{grad} = \frac{V_{geo}}{\frac{1}{2} + \sqrt{\frac{1}{4} + \frac{V_{geo}}{2R\Omega \sin \phi}}}$$

speed of the geostrophic wind, Ω is the angular speed of rotation of the Earth, ϕ is the latitude, and R is the radius of curvature of the air trajectory, considered positive when the trajectory is curved in the cyclonic sense (center of curvature on low-pressure side) and negative when the curvature is anticyclonic. The gradient wind speed is less than the geostrophic speed when the air moves in a cyclonically curved path and greater when the air moves in an anticyclonically curved path. The gradient wind is a good approximation of the actual wind and is often superior to the geostrophic wind, particularly when the flow is strongly curved in the cyclonic sense. *See* CORIOLIS ACCELERATION AND FORCE; GEOSTROPHIC WIND; STORM; WIND.

[FREDERICK SANDERS]

Greenhouse effect

The Earth's atmosphere acts as the glass walls and roof of a greenhouse in trapping heat from the Sun. Like the greenhouse, it is largely transparent to solar radiation, but it strongly absorbs the longer-wavelength radiation from the ground. Much of this long-wave radiation is reemitted downward to the ground, with the paradoxical result that the Earth's surface receives more radiation than it would if the atmosphere were not between it and the Sun.

The absorption of long-wave (infrared) radiation is effected by small amounts of water vapor, carbon dioxide, and ozone in the air and by clouds. Clouds actually absorb about one-fifth of the solar radiation striking them, but unless they are extremely thin, they are almost completely opaque to infrared radiation. The appearance even of cirrus clouds after a period of clear sky at night is enough to cause the surface air temperature to increase rapidly by several degrees because of radiation from the cloud. *See* ATMOSPHERIC OZONE; TERRESTRIAL RADIATION.

The greenhouse effect is most marked at night, and usually keeps the diurnal temperature range below 20°F. Over dry regions such as New Mexico and Arizona, however, where the water-vapor content of the air is low, the atmosphere is more transparent to infrared radiation, and cool nights may follow very hot days.

[LEWIS D. KAPLAN]

Groundwater

The water in the zone in which the rocks and soil are saturated, the top of which is the water table. The zone of saturation is the source of water for wells, which provide about one-fifth of the water supplies of the United States. It is also the source of the water that issues as springs and seeps, and maintains the dry-weather flow of perennial streams. The saturated zone is a natural reservoir which absorbs precipitation during wet periods and releases it during dry periods, thus tempering the severity of floods and droughts. The amount of groundwater stored in the rocks in the United States is estimated to be several times as great as that stored in all lakes and reservoirs, including the Great Lakes.

A knowledge of geology is essential to an understanding of the occurrence of water. For this reason, the study of groundwater is sometimes called hydrogeology or geohydrology.

Subterranean water. Water beneath the land surface occurs in the zone of aeration above the water table and in the zone of saturation below the water table. Water in the zone of aeration, also called vadose water, is divided into soil water, intermediate vadose water, and water of the capillary fringe.

Water in the capillary fringe is connected with the zone of saturation and is held above it by capillary forces. The lower part of the fringe may be saturated, but is not a part of the zone of saturation because the water is under less than atmospheric pressure and will not flow into a well. When the well reaches the zone of saturation, water will begin to enter it and will stand at the level of the water table.

Rock formations capable of yielding significant volumes of water are called aquifers. Some wells are artesian; that is, the water rises above the top of the water-bearing bed. Other wells encounter water above the saturated zone and lose their water if extended through the impermeable bed upon which the water rests. Such bodies of water are said to be perched.

The total volume of water available and the rate at which it can be released to a well are determined by the number and size of the openings in rocks and soils and the manner in which they are interconnected. Openings are practically absent in some igneous rocks. They are numerous but microscopic in clay. They are large and interconnected in many sands and gravels. There are huge caverns and tubes in many limestones and lavas. The distribution and types of openings are as diverse as the geology itself, so that general statements about them applicable to one area may be incorrect for another.

Openings in rocks. Primary openings are those which existed when the rock was formed, and secondary, those which resulted from the action of physical or chemical forces after the rock was formed. Primary openings are found in sedimentary rocks such as sand and clay and certain kinds of limestone composed of triturated shells. Openings in lava formed at the stage when the lava is partly

liquid and partly solid are also considered primary. Most rocks containing primary openings are geologically young. Rocks which contain primary openings large enough to carry useful amounts of water are represented, for example, by the seaward-dipping strata of the Atlantic and Gulf coastal plains, including the coquina limestone of Florida, the intermontane valleys of the western United States, the glacial deposits of the North-Central states, and the lava rocks of the Pacific Northwest.

Secondary openings are common in older rocks. Sand and gravel that have been cemented by chemical action, limestone indurated by compression or recrystallization, schist, gneiss, slate, granite, rhyolite, basalt and other igneous rocks, and shale generally contain few primary openings; but they may contain fractures that will carry water. Limestone is subject to solution which, beginning along small cracks, may develop channels ranging from openings a fraction of an inch across to enormous caverns capable of carrying large amounts of water.

Porosity. The property of rocks for containing voids, or interstices, is termed porosity. It is expressed quantitatively as the percentage of the total volume of rock that is occupied by openings. It ranges from as high as 80% in newly deposited silt and clay down to a fraction of 1% in the most compact rocks.

Permeability. This is the characteristic capability of rock or soil to transmit water. The porosity of a rock or soil has no direct relation to the permeability or water-yielding capacity. This capacity is related to the size and degree to which the pores or openings are interconnected. If the pores are small, the rock will transmit water very slowly; if they are large and interconnected, they will transmit water readily. Permeability is expressed as the volume of flow through unit cross-sectional area under a hydraulic gradient of 100% (1 meter of head loss per meter of water travel) in unit time. It is a velocity usually expressed in feet or meters per day. Permeability varies with the viscosity of the water, hence with water temperature.

The Meinzer unit, used by the U.S. Geological Survey, is defined as the rate of flow of water at 60°F (15.6°C) in gallons per day through a cross section of 1 ft² (1 gallon per day through a cross section of 1 ft² = 4.716×10^{-7} m³/s through a cross section of 1 m²), under a hydraulic gradient of 100%. Under field conditions, the adjustment to standard temperature is commonly ignored.

Transmissibility. Transmissibility expresses the rate at which water moves through a saturated body of rock. It is expressed as the rate of flow of water at the prevailing temperature, in gallons per day through a vertical strip of aquifer 1 ft (0.3048 m) wide, extending the full saturated height of the aquifer under a hydraulic gradient of 100%.

Controlling forces. Water moves through permeable rocks under the influence of gravity from places of higher head to places of lower head, that is, from areas of intake or recharge to areas of discharge, such as wells or springs. Water moving through rocks is acted upon also by friction and by molecular forces. The molecular forces are the attraction of rock surfaces for the molecules of water (adhesion) and the attraction of water molecules for one another (cohesion). When wetted, each rock surface is able to retain a thin film of water despite the effect of gravity. In very-fine-grained rocks, such as clay and fine silt, the interstices may be so small that molecular attraction extends from one side of a pore to the opposite side. Molecular force then becomes dominant, and water moves through the rock very slowly under the gradients typical of natural conditions.

The amount of water that drains from a saturated rock under the influence of gravity, expressed as a percentage of the total volume of the rock, is called the specific yield. Specific yield is often called effective porosity because it represents the pore space that will surrender water to wells. The term porosity is poorly defined, and its use should be discontinued. A part of the water stored in the rocks and soil is held by molecular forces and may have only a small share in supplying springs or wells. This latter portion is of special interest to the agriculturalist because it sustains plant life. A soil that is highly permeable permits water to pass through it easily, and little is retained for the nourishment of plant life, whereas a soil that is relatively impermeable retains much of its water until it is extracted by plants or by evaporation.

Sources. The chief means of replenishment of groundwater is downward percolation of surface water, either direct infiltration of rainwater or snowmelt or infiltration from bodies of surface water which themselves are supplied by rain or snowmelt. Evidence on the replenishment of groundwater is furnished by analysis of data on the downward movement of precipitation through the soil and subsoil, the rise and fall of groundwater levels and spring discharge in response to precipitation and seepage losses from streams, and the slope of the hydraulic gradient from known areas of intake to areas of discharge. Some groundwater may originate by chemical and physical processes that take place deep within the Earth. Such water is called juvenile water to indicate that it is reaching the Earth's surface for the first time. Such water is always highly mineralized. Some water is stored in deep-lying sedimentary rocks and is a relic of the ancient seas in which these rocks were deposited. It is called connate water. The total quantity of water from juvenile and connate sources that enters the hydrologic cycle is insignificant when compared with the quantities of water derived from precipitation (meteoric water). It is balanced to some extent by withdrawal of water from the hydrologic cycle by such processes as deposition of minerals that include water in their crystalline structure. See HYDROLOGY.

Infiltration. Replenishment of water in the zone of saturation involves three steps: (1) infiltration of water from the surface into the rock or soil that lies directly beneath the surface, (2) downward movement through the zone of aeration of the part of the water not retained by molecular forces, and (3) entrance of this part of the water into the zone of saturation, where it becomes groundwater and moves, chiefly laterally, toward a point of discharge. Infiltration is produced by the joint action of molecular attraction and gravity. The rate of infiltration is a function of the permeability of the soil. Under conditions of unsaturated flow in the zone of aeration, it varies with moisture content as well as with pore size. It varies also with the geology. For example, in the Badlands, South Dakota, where the soil and rocks are of low permeability, the infiltration capacity of the soil is low and is reached quickly after rainfall or snowmelt begins. Hence, there is not much infiltration, and any excess of precipitation or snowmelt over infiltration runs off over the surface and enters the streams. If the excess is large, serious floods and erosion result. On the other hand, the soils of the Sand Hills, Nebraska, and the glacial outwash deposits of Long Island, New York, are so permeable that they absorb the water of the most violent storms and permit little or no direct runoff.

The permeability of the rock materials beneath the soil zone also is important. Since the soil is commonly formed by weathering of the underlying rock, the permeability of the rocks is generally comparable to that of the soil.

Water-table and artesian conditions. Water that moves downward through the soil and subsoil in excess of capillary requirements continues to move downward until it reaches a zone whose permeability is so low that the rate of further downward movement is less than the rate of replenishment from above. A zone of saturation then forms, its thickness depending on the opportunity for lateral escape of water in relation to replenishment. The top of this zone is the water table (see illustration). Under these circumstances, the water is said to be unconfined, or under water-table conditions. However, since much of the crust of the Earth has a more or less well-defined layered structure in which zones of high and low permeability alternate, situations are common in which groundwater moving laterally in a permeable rock passes between layers of relatively low permeability. Although the permeable layer contains unconfined groundwater in the area where there is no impermeable layer (confining bed) above, the part of the layer, or aquifer, that passes beneath the confining bed contains water that is pressing upward against the confining bed, and if a well is drilled in this area the water in it will rise. It tends to rise to the level of the water in the unconfined area, but fails to reach that level by the amount of pressure head lost by friction as the water moves from the unconfined area to the well. Confined water is also called artesian water, and wells tapping it are called artesian whether or not their head is sufficient for them to flow at the land surface.

Chemical qualities. Water is said to be the universal solvent. When it condenses and falls as rain or snow, it absorbs small amounts of mineral and organic substances from the air. After falling, it continues to dissolve some of the soil and rocks through which it passes. Thus, no groundwater is chemically pure. Its most common mineral constituents are the bicarbonates, chlorides, and sulfates of calcium, magnesium, sodium, and potassium, in ionized or dissociated form. Silica also is an im-

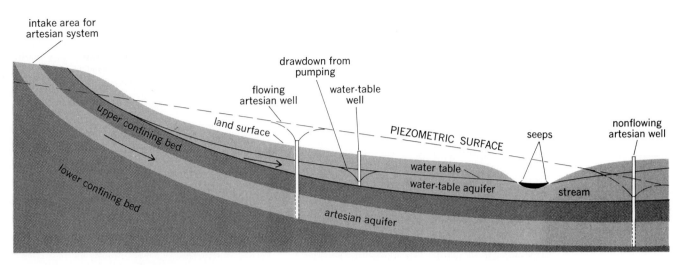

Water-table and artesian conditions.

ortant constituent. Common also are small, but significant, concentrations of iron, and manganese, fluoride, and nitrate. The concentration of the dissolved minerals varies widely with the kind of soil and rocks through which the water has passed. Ordinarily, water that contains more than 1000 mg/liter of dissolved solids is considered unfit for human consumption, and water that contains more than 2000 mg/liter, unfit for stock. However, both human beings and animals may become accustomed to greater concentrations.

Pollution of groundwater. The dissolved minerals commonly found in groundwater are not harmful to humans unless present in excessive amounts. Groundwater has been generally considered safe for human consumption without treatment. However, pollution of groundwater is an increasing problem, and measures are being taken to protect major aquifers from pollution. Pollutants can reach an aquifer as a result of unwise waste disposal, overpumping the groundwater, or chemicals distributed on the land. Sanitary landfills, mine waste dumps, disposal areas for sludges from wastewater treatment or air-pollution control equipment, and industrial waste lagoons can all be sources from which pollutants, including toxic chemicals, can be leached into the underlying groundwater. Fertilizers, herbicides, and pesticides used in agriculture and forestry can be leached into the groundwater. Recharge of the groundwater with treated wastewater may transfer dissolved pollutants to the groundwater, and disposal of hazardous wastes into injection wells may, inadvertently, lead to pollution of an aquifer used as a potable water source. Overpumping of groundwater lowers the water table or piezometric surface. In coastal aquifers, sea water may be permitted to enter the aquifer. Some inland aquifers may be polluted by flow from an adjacent saline aquifer. Lowered groundwater levels may also augment seepage from any of the surface sources of pollution. Pollution, once having entered an aquifer, may be slow to disappear because of the low flow velocities encountered in some aquifers. In the hard-rock aquifers, velocity of flow in the primary openings may range from a few meters per day to a few meters per year, leaving the possibility that a pollutant, once introduced into an aquifer, might remain there for years. If pollution is induced by excessive withdrawals, inflow of pollutants will continue until water levels have risen sufficiently to reverse the gradient of flow.

[RAY K. LINSLEY]

Bibliography: H. Bouwer, *Groundwater Hydrology*, 1978; S. N. Davis and R. J. M. DeWiest, *Hydrogeology*, 1966; R. J. Kazmann, *Modern Hydrology*, 1965; W. C. Walton, *Groundwater Resource Evaluation*, 1970; R. C. Ward, *Principles of Hydrology*, 1964.

Gulf of California

A deep, elongated trough in northwestern Mexico, 1500 km long and 100–200 km wide, separated from the Pacific Ocean by the mountainous peninsula of Baja California. A central constriction and a group of islands separate a shallow (<200 m) northern section from a deep region to the south which contains a series of basins 1000–3600 m deep (Fig. 1).

The southern part of the Gulf has a relatively thin crust, where the Mohorovičić discontinuity (Moho) is 10–11 km below sea level; the Moho is 23–25 km deep beneath Baja California and the mainland. The Gulf is therefore part of the Pacific Ocean and not a sunken block of the continent. It lies where the East Pacific Rise meets the continent. The topography of the Gulf floor suggests that it was formed by lateral movements along northwest-southeast-trending fractures which are probably associated with the San Andreas Fault system in California.

Climate and water relations. The Gulf is an arid region over most of its expanse. Baja California has an annual rainfall of 10–15 cm; no permanent streams exist. On the eastern side rainfall increases from 10 cm in the northwest to 85 cm in the southeast; permanent streams occur only in the southern half. Surface water temperatures in the northern Gulf range from 16°C in winter to 30°C in summer. In the southern Gulf the water is

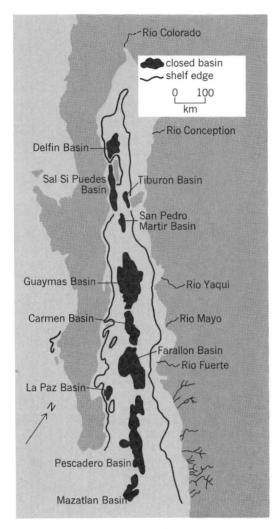

Fig. 1. Principal relief features of Gulf of California.

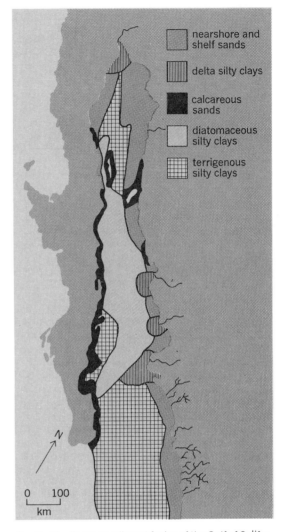

Fig. 2. The Recent sediment facies of the Gulf of California. (*Adapted from T. H. van Andel and G. G. Shor, eds., Marine Geology of the Gulf of California, Amer. Ass. Petrol. Geologists Mem. no. 3, 1964; and S. E. Calvert, Bull. Geol. Soc. Amer., 77:569–596, 1966*)

Equatorial Pacific with a marked oxygen minimum between 300 and 900 m.

Surface water circulation is governed by seasonal wind regimes. In winter water is driven out of the Gulf; this is replenished by a subsurface flow from the Pacific. In summer water is blown into the Gulf, while a subsurface flow leaves the Gulf. Upwelling occurs along the eastern side in winter and along the western side in summer, resulting in high plankton productivity.

Bottom sediments. The northern Gulf has sediments derived predominantly from the Colorado River (Fig. 2). The southern Gulf receives sediments from the eastern side; a wide coastal plain and shelf indicate high rates of supply. Very little sediment is supplied from the peninsula, and the western shallow-water sediments consist of calcareous skeletal debris.

The deep-water sediments are land-derived silty clays in the southern Gulf and diatom-rich silty clays in the central Gulf, reflecting the high plankton production. Rates of sedimentation are high (1–5 mm/year). Considerable quantities of silica are permanently extracted from Pacific Ocean waters as diatom frustules. Some of the diatomaceous sediments are finely laminated, the laminae

consisting of alternating diatom-rich and clay-rich layers. Each pair, averaging 2–5 mm in thickness, is a yearly deposit, or varve. *See* PACIFIC OCEAN.

[STEPHEN E. CALVERT]

Bibliography: T. H. van Andel and G. G. Shor (eds.), *Marine Geology of the Gulf of California*, Amer. Ass. Petrol. Geologists Mem. no. 3, 1964; S. E. Calvert, Accumulation of diatomaceous silica in the sediments in the Gulf of California, *Bull. Geol. Soc. Amer.*, 77:569–596, 1966; S. E. Calvert, Origin of the diatom rich sediments in the Gulf of California, *J. Geol.*, 74:546–565, 1966; *1940 E. W. Scripps Cruise to the Gulf of California*, Geol. Soc. Amer. Mem. no. 43, 1950.

Gulf of Mexico

A semienclosed marginal sea forming the southern marine boundary of the United States, the scene of extensive marine fisheries and important offshore gas, oil, and sulfur production. Covering an area of 1,700,000 km², with wide continental shelves on

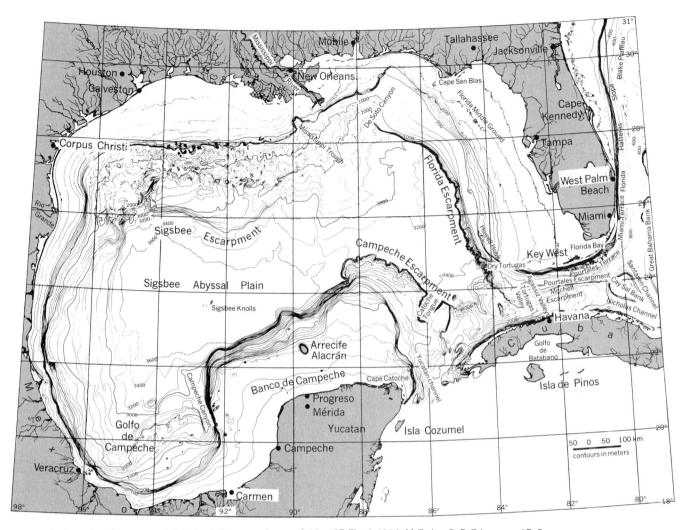

Fig. 1. Bathymetry. (*Courtesy of E. Uchupi. Sources of data: submarine topography—U.S. Coast and Geodetic Survey Hydrographic Surveys; U.S. Naval Oceanographic Office Charts 1126 and 2056; J. S. Creager, Texas A. and M.* *ref. 58–12F, Fig. 6, 1956; M. Ewing, D. B. Ericson, and B. C. Heezen, Habitat of Oil, Amer. Ass. Petrol. Geologists, Fig. 10, 1958; base—U.S. Coast and Geodetic Survey)*

the northern and eastern boundaries, the basin has a large central area (Sigsbee Abyssal Plain) with depths in excess of 3500 m. This central area has a modified oceanic crust with the Mohorovičić discontinuity at as little as 16 km below sea level.

Submarine structures and deposits. The Florida shelf and the Campeche bank are carbonate areas with active reef formation currently taking place on some of the Florida Keys and the outer reefs of Campeche. Both meet the deep ocean in abrupt escarpments that are probably the result of reef building rather than faulting. The remainder of the Gulf has detrital sediments (mainly montmorillonite), deriving at present principally from the Mississippi and the Atchafalaya rivers but in earlier periods from sources progressively nearer to the northwest Gulf. *See* MARINE SEDIMENTS.

The Texas-Louisiana shelf and slope display a great number of salt domes similar to those occurring under the adjacent land areas. Associated petroleum pooling has made this a productive area for oil and gas recovery. Beyond the narrow west-

ern shelf the slope shows a subsurface ridge structure indicating a basement formation that is probably made up of salt or other evaporites. In the deep central region a series of diapiric structures (Sigsbee Knolls) suggests that here, too, the basement rocks may be evaporites (Fig. 1). *See* MARINE GEOLOGY.

Currents. In the southeast the island of Cuba separates the two connections with the Atlantic— one via the Caribbean through the Yucatán Strait with controlling sill depth of 1500 m and the other through the Straits of Florida with a controlling depth of 800 m. The Yucatán current carries some 30,000,000 m³ of water per second northward in an intense flow with speeds typically 2–4 knots close to the western side of the Strait. This flow generally loops to the north and exits as the Florida Current, which is continuous with, and after augmentation becomes, the Gulf Stream. Anticyclonic eddies, semiattached or broken off from this loop, at times extend as far north as the offing of the Mississippi Delta. Surface-layer circulation in the

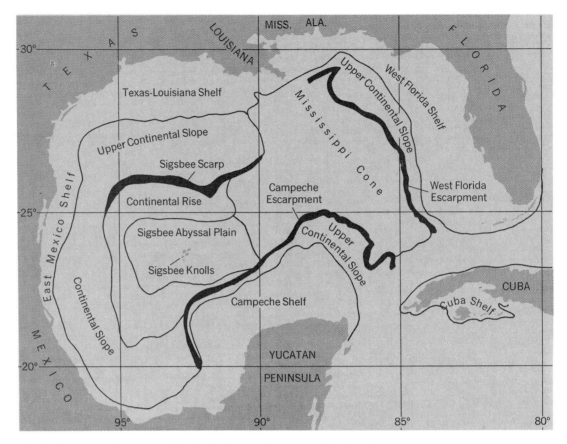

Fig. 2. Physiographic provinces in the Gulf of Mexico. (*Courtesy of W. R. Bryant*)

west-central Gulf is dominated by an elongate anti-cyclonic gyral with the most intense flow in the north, roughly paralleling the continental slope (Fig. 2). *See* CARIBBEAN SEA; GULF STREAM.

Salinity and temperature. The major contributions of fresh water to the Gulf come from the Mississippi, Atchafalaya, and the lesser rivers of Texas. The low-salinity coastal waters hold to the shelf with a westward flow. Although the shallow Texas bays at times may be brackish, restricted circulation and local evaporation make them hypersaline over extended periods.

Much of the basin water lies below the still depth at Yucatán Strait and shows remarkably uniform characteristics: potential temperatures, 4.0–4.15°C; salinity, 34.965–34.975°/oo; and dissolved oxygen, 4.8–5.2 ml/liter. Above this a layer derived from sub-Antarctic Intermediate Water shows a salinity minimum (34.86–34.89°/oo) at a temperature about 6.5°C. This core is typically at a depth of 700–800 m and has lost its identity through mixing by the time of exit in the Florida Current. A salinity maximum (as high as 36.7°/oo) occurs at depth from 100 to 200 m and has its origin in the Subtropical Underwater of the Atlantic Ocean. Except for the loop current region, surface layers are locally modified and show regional, seasonal, and year-to-year variability (Fig. 3). *See* SEA WATER.

Astronomical tides are of low amplitude (1–2 ft) with a dominant diurnal component. The wind set-up tends to be a more important factor in determining coastal sea level. During hurricanes "storm tides" may raise sea level at the coast by as much as 15 ft.

The most valuable of the food fisheries are for shrimp and oysters. Several high-quality fin fishes are also landed in quantity. A fish-meal industry is supported on landings of menhaden and thread herring. There are large inshore and offshore sports fisheries. *See* STORM SURGE.

[HUGH J. MC LELLAN]

Bibliography: Gulf Coast Association of the Geologic Society, *Proceedings of Symposium on*

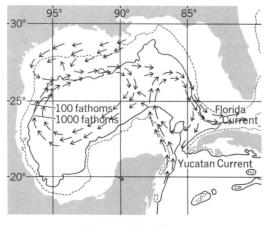

Fig. 3. Basic surface layer circulation.

the *Geological History of the Gulf of Mexico*, vol. 17, 1967; H. J. McLellan and W. D. Nowlin, Some features of the deep water in the Gulf of Mexico, *J. Mar. Res.*, 21(3):233–245, 1963; W. D. Nowlin and H. J. McLellan, A characterization of the Gulf of Mexico waters in winter, *J. Mar. Res.*, 25(1):29–59, 1967; U.S. Fish and Wildlife Service, Gulf of Mexico: Its origin, waters, and marine life, *Fisheries Bull.*, no. 89, 1954.

Gulf Stream

A great ocean current transporting about 70,000,000 tons of warm water per second northward from the latitude of Florida to the Grand Banks off Newfoundland, an amount of water 1000 times the discharge of the Mississippi River. Before the days of scientific oceanography it was supposed that the origin of the water in the Gulf Stream was the Gulf of Mexico—as indeed the name implies; but now the origin has been definitely traced much farther upstream, through the Gulf of Mexico and Caribbean, to the Great North and South Equatorial currents of the Atlantic (Fig. 1). In fact, it seems much more reasonable to think of the Gulf Stream as a portion of a great horizontal circulation in the ocean, where each particle of water executes a closed circuit, sometimes moving slowly in mid-ocean regions and other times rapidly in strong currents like the Gulf Stream. Thus there can scarcely be a meaning to the concept of the beginning and end of the Stream other than simple arbitrary geographical limits.

Description. The Stream is as deep as 610 m, is 20–40 mi wide, and moves at velocities up to 4 mph. It does not stay in a fixed path but meanders like a river. The meanders generally are 100–200 mi long and move downstream at a rate of about 5 mi per day. The large-scale meanders first become noticeable after the Stream leaves the coast at Cape Hatteras, and increase in amplitude until the pattern of flow has become extremely disordered and complex in the longitude of the Grand Banks. Farther to the east the field of motion is not clearly defined. There are small-scale irregularities that have been detected by aerial surveys; and large-scale, long-period irregularities and some fluctuations evidently connected with seasonal changes of the winds have been detected from analysis of tide gage records.

Over the tropical and subtropical Atlantic the upper layers of water are warm (10–27°C), whereas the deep waters are as cold as 2.2°C (Fig. 2). There is rather an abrupt transition separating these two layers, the so-called main thermocline. *See* THERMOCLINE.

Along the United States coast and along a roughly defined line between Cape Hatteras and the Faeroes the thermocline rises to the surface, and north of it all the water is cold, even at the surface. The Gulf Stream flows along this boundary immediately on the warm-water southern side, and the Coriolis force acting upon the Gulf Stream (to the right of its direction of flow) balances the northward pressure gradients that otherwise would force the warm water above the thermocline to flow much farther northward. In this sense the Gulf Stream acts as a "dynamic dam" to prevent

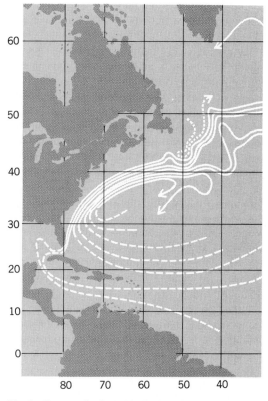

Fig. 1. Sources (indicated by broken lines) and the flow pattern (indicated by solid lines) of the Gulf Stream system. (*Adapted from C. O'D. Iselin, Papers Phys. Oceanogr. Meteorol., Mass. Inst. Technol. and Woods Hole Oceanogr. Inst., 4(4):1–99, 1936*)

warm surface waters from further flow northward, so that an increase in Gulf Stream discharge might actually be construed as leading to cold northern European climate rather than the reverse, as is so often asserted in lay literature.

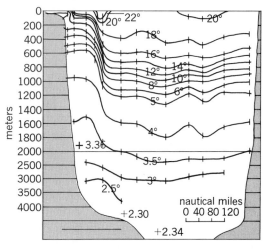

Fig. 2. Temperature (°C) section across the Gulf Stream from Chesapeake Bay to Bermuda, Apr. 17–23, 1932. The vertical exaggeration is ×370 above 2000 m, and ×148 below 2000 m. (*Adapted from C. O'D. Iselin, Papers Phys. Oceanogr. Meteorol., Mass. Inst. Technol. and Woods Hole Oceanogr. Inst., 4(4):1–99, 1936*)

Theory. In other oceans there are also strong currents on the western sides, such as the Kuroshio in the North Pacific and the Agulhas in the South Indian. The marked tendency toward an east-west asymmetry in the horizontal circulations of all the oceans (an intensification of currents in the west) has been a subject of theoretical investigation in the decade 1948–1958. The theoretical explanation of broad features of the general oceanic circulation now seems to have been established. In the particular case of the Gulf Stream it appears that about half of the transport is due to the prevailing wind system over the North Atlantic, whereas the other half is tied in with a worldwide system of currents driven by thermal differences and is fed by surface waters from all over the globe. Associated with the thermal part is a deep cold countercurrent under the Gulf Stream, most of which crosses the Equator into the South Atlantic and thence is distributed into the cold deep waters of the Indian and Pacific oceans. *See* AT-LANTIC OCEAN; CARIBBEAN SEA; GULF OF MEXI-CO; OCEAN CURRENTS.   [HENRY STOMMEL]

Bibliography: T. F. Gaskell, *The Gulf Stream*, 1974; H. Stommel, *The Gulf Stream: A Physical and Dynamical Description*, 2d ed., 1965 (reprint, 1977).

Hail

Ice balls, occurring mainly in thunderstorms, composed usually of alternate layers of rime and hard ice, but varying between all hard ice (ice pellets) and all rime (soft hail or snow pellets). Small hail has a soft center and a single outer coat of hard ice. Hailstones are spherical or irregular, often conical; a typical diameter is 1/2 in. (1 cm), but sizes range to more than 5 in. in rare cases.

Hail in the United States is most frequent in a belt roughly from the southwest corner of North Dakota into the Texas Panhandle, and also from the southwest corner of Montana southward through western Utah. Such frozen precipitation is uncommon on the Pacific Coast and in Florida, and it is generally uncommon throughout the tropics. Hail does extensive damage to crops and sometimes buildings, especially windows. Hailstorm paths vary greatly; a typical size is 20 mi long and 1 mi wide. The conditions for true hail are (1) warm moist air in low levels; (2) thunderstorms with turbulence extending to great heights, sometimes over 50,000 ft; and (3) the freezing height roughly 10,000–12,000 ft above ground. Hail grows by alternate accumulation of rime and of liquid water which freezes to form hard ice. *See* CLOUD PHYSICS; PRECIPITATION; SNOW; THUN-DERSTORM.   [J. R. FULKS]

Bibliography: R. List, New developments in hail research, *Science*, 132(3434):1091–1098, 1960.

Halo

A system of luminous rings or arcs around the Sun or Moon, caused by refraction or reflection of light rays illuminating ice crystals floating in the air. Since the ice crystals have the form of hexagonal pyramids or prisms, the rays of minimum deviation in the prismatic refraction on the sides and the top or bottom of these prisms form a cone centered around the direction toward the Sun. These rays are seen as luminous circles around the Sun with a radius of either 22° (inner halo) or 46° (outer halo). Because the minimum deviation of the emerging rays is smaller in the red than in the blue, the inner and outer halos show a distinct red tint on the boundary toward the Sun. Besides the inner and outer halos several similar arcs, luminous arcs, and bright spots are occasionally seen in the sky as a result of refraction or reflection of light rays illuminating ice crystals in the atmosphere. All these phenomena are usually considered parts of a more complex system, called the halo. The inner halo is, however, most frequently observed. *See* SUNDOG.
          [ZDENEK SEKERA]

Halocline

A layer of sea water within which a marked salinity gradient is present. A permanent halocline of decreasing salinity usually coincides with the top of the permanent thermocline. Although weak and generally less important than the thermocline, the oceanic halocline furnishes significant clues to the origin and movement of water masses. Haloclines are extremely important in estuarine oceanography, however, as their depth, strength, and shape greatly influence circulation, mixing, and flushing of these embayments. Estuarine haloclines have strong positive gradients which are formed as fresh water flows seaward over sea water entering along the bottom. *See* ESTUARINE OCEANOGRAPHY; SEA WATER; THERMOCLINE.
          [JOHN J. SCHULE, JR.]

Halogen atmospheric chemistry

The halogens include the elements fluorine, chlorine, bromine, and iodine and occur in the atmosphere as both aerosols and gases. Natural sources include the dispersion of sea water through the action of breaking bubbles and the emanation of vapors from the sea surface. Pollution sources include automotive exhaust emissions and the release of synthetic organohalogen compounds. The halogens are marked by their chemical reactivity in the atmosphere, both in the troposphere and in the stratosphere. Although they occur only at trace levels, they play in important role in the total balance of chemical reactions in the atmosphere.

Sea salt particles. Sea salt is transferred to the atmosphere by bubbles of trapped air that rise to the sea's surface and break. Several jet drops of a diameter equal to one-tenth of the bubble diameter are ejected from the center of the bursting bubble and transfer material, initially accumulated within the bubble surface film, to heights of several centimeters in the atmosphere. An additional shower of smaller droplets from the breaking film cap is also transferred to the atmosphere. Once airborne, the droplets of less than 10 μm in diameter, such as are formed from small bubbles, form an aerosol and are carried far from their point of origin. They may ultimately be removed by precipitation or by fair weather deposition processes.

The composition of sea salt particles is modified from that of sea water by fractionation in the bubble-breaking process. Materials initially concentrated in the sea surface microlayer and in bubble films are preferentially transferred into the atmosphere. Measurements of chlorine, bromine, and iodine in sea spray show the weight ratio of bromine to chlorine to be approximately equal to the sea water ratio of 0.0034, although varying significantly with particle size. The ratio of iodine to chlorine, on the other hand, may be 1000 times greater in the aerosol than in sea water, especially in the smallest aerosols. The iodine enrichment is believed to be due to a combination of the concentration of surface-active organic compounds which contain iodine, in airborne droplets produced by bubble bursting and of volatile compounds of iodine emanating from the sea surface. The latter combine with aerosols by physical and chemical processes in the atmosphere.

The relative abundance of halogens at higher levels in the atmosphere is determined by their ratio in the source material and by their residence times in the atmosphere. A longer residence time, favored for elements largely in the gas phase which are removed only slowly from the atmosphere, results in a higher steady-state concentration in the atmosphere. Observations show that the ratio of bromine to chlorine is substantially greater for particles in the stratosphere than for those in the troposphere, suggesting that vapor-to-particle transfer of bromine must be important at higher altitudes.

Automotive pollution particles. The combustion of leaded gasoline generates lead chlorobromide particles and is now the major source of atmospheric particulate bromine found in populated areas. Unlike the sea salt particles, which are more abundant in larger particle-size classes than in small because of the particles' dispersive origin, automotive lead halide particles are very small, mostly less than 0.5 μm in diameter, because of their formation by condensation and coagulation of vapors in the exhaust system. Such small particles are not removed readily from the atmosphere by sedimentation, impaction, or water vapor condensation and rainout and therefore may persist in the atmosphere with longer residence times before being deposited at the Earth's surface. Similarly, small lead halide particles penetrate deeply into the human respiratory tract and may be deposited preferentially in the pulmonary instead of the upper tracheobronchial region. It has been observed that the weight ratio of bromine to lead in air is generally severalfold lower than that in leaded gasoline (0.4) and decreases with the residence time of airborne particles. Consequently, bromine from automotive pollution is a reactive element under natural conditions and is gradually released from particles as a vapor. Its subsequent chemical reactions in the atmosphere are not fully understood. See ATMOSPHERIC POLLUTION.

Tropospheric halogen gases. The concentrations of gaseous compounds of chlorine, bromine, and iodine are found to be very much greater than particulate concentrations in locations remote from urban areas. By use of special filter techniques, which can trap inorganic and organic halogen gases, the gas-to-particle ratio has been measured in air over North America, distant from the main source. During 1974 chlorine was found to be 20 to 50 times more abundant in the gas phase than in aerosols; bromine 4 to 40 times, and iodine 6 to 15 times. Most of the gaseous halogens are believed to be organic compounds rather than inorganic vapors. Photochemical reactions may be important in their formation and transformation in the atmosphere. See TROPOSPHERE.

Stratospheric halogen gases. Tropospheric trace gases are transferred into the stratosphere by a complex pattern of transport processes. In the stratosphere they may remain with residence times of the order of years before being transported down into the troposphere and removed at the Earth's surface. In the stratosphere, many trace gases take part in photochemical and free-radical reactions linked to the formation or destruction of ozone and therefore may influence the radiation balance of the stratosphere. Halogenated hydrocarbon compounds, derived from both natural and pollution sources in the lower atmosphere, may be dissociated in the stratosphere by the absorption of ultraviolet radiation and thereby yield reactive halogen atoms with catalytic effects on the photochemical cycle of ozone. In 1975 attention was focused on the possible significance of widespread releases of chlorofluoromethane compounds to the atmosphere. In the stratosphere, chlorine atoms formed by photochemical dissociation of chlorinated hydrocarbons may enter into reactions of the type: $Cl + O_3 \rightarrow ClO + O_2$ and $ClO + O \rightarrow Cl + O_2$.

When combined with other reactions which influence the balance of O_3 in the stratosphere, the effect of added chlorine is expected to be a decrease in the O_3 concentration. Moreover, the chlorine acts as a catalyst that is regenerated after each reaction and may transform a large number of O_3 molecules to O_2. Concern over the possibility that anthropogenic chlorofluoromethanes are the major source of chlorine atoms now in the stratosphere has stimulated research interest in the atmospheric chemistry of the halogens. See ATMOSPHERIC CHEMISTRY; STRATOSPHERE.

[JOHN W. WINCHESTER]

Bibliography: G. G. Desaedeleer et al., Bromine and lead relationships with particle size and time along an urban freeway, *Trans. Amer. Nucl. Soc.*, 21:36–37, 1975; R. A. Duce, J. W. Winchester, and T. W. Van Nahl, Iodine, bromine, and chlorine in the Hawaiian marine atmosphere, *J. Geophys. Res.*, 70:1775, 1965; R. Guderian, *Air Pollution: Phytotoxicity of Acidic Gases and Its Significance in Air Pollution Control*, 1977; K. A. Rahn, R. D. Borys, and R. A. Duce, Tropospheric halogen gases: Inorganic and organic components, *Science*, 190:549–550, 1976; F. S. Rowland and M. J. Molina, Chlorofluoromethanes in the environment, *Rev. Geophys. Space Phys.*, 13:1, 1975; J. W. Winchester et al., Lead and halogens in pollution aerosols and snow from Fairbanks, Alaska, *Atmos. Environ.*, 1:105, 1967; World Meteorological Association, *Climatological Aspects of the Composition and Pollution of the Atmosphere*, 1975.

Humidity

Atmospheric water-vapor content, expressed in any of several measures, especially relative humidity, absolute humidity, humidity mixing ratio, and specific humidity. Quantity of water vapor is also specified indirectly by dew point (or frost point), vapor pressure, and a combination of wet-bulb and dry-bulb (actual) temperatures. See DEW POINT.

Relative humidity is the ratio, in percent, of the moisture actually in the air to the moisture it would hold if it were saturated at the same temperature and pressure. It is a useful index of dryness or dampness for determining evaporation, or absorption of moisture. See PSYCHROMETRICS.

Human comfort is dependent on relative humidity on warm days, which are oppressive if relative humidity is high but may be tolerable if it is low. At other than high temperatures, comfort is not much affected by high relative humidity. See TEMPERATURE-HUMIDITY INDEX.

However, very low relative humidity, which is common indoors during cold weather, can cause drying of skin or throat and adds to the discomfort of respiratory infections. The term indoor relative humidity is sometimes used to specify the relative humidity which outside air will have when heated to a given room temperature, such as 72°F, without addition of moisture. It always has a low value in cold weather and is then a better measure of the drying effect on skin than is outdoor relative humidity. This is even true outdoors because, when air is cold, skin temperature is much higher and may approximate normal room temperature.

Absolute humidity is the weight of water vapor in a unit volume of air expressed, for example, as grams per cubic meter or grains per cubic foot.

Humidity mixing ratio is the weight of water vapor mixed with unit mass of dry air, usually expressed as grams per kilogram. Specific humidity is the weight per unit mass of moist air and has nearly the same values as mixing ratio.

Dew point is the temperature at which air becomes saturated if cooled without addition of moisture or change of pressure; frost point is similar but with respect to saturation over ice. Vapor pressure is the partial pressure of water vapor in the air. Wet-bulb temperature is the lowest temperature obtainable by whirling or ventilating a thermometer whose bulb is covered with wet cloth. From readings of a psychrometer, an instrument composed of wet- and dry-bulb thermometers and a fan or other means of ventilation, values of all other measures of humidity may be determined from tables. See HYGROMETER; MOISTURE-CONTENT MEASUREMENT; PSYCHROMETER.

[J. R. FULKS]

Hurricane

A tropical cyclone whose maximum sustained winds reach or exceed a threshold of 33 meters per second (m/s). In the western North Pacific ocean it is known as a typhoon. Many tropical cyclones do not reach this wind strength.

Maximum surface winds in hurricanes range up to about 200 mph (320 km/hr). However, much greater losses of life and property are attributable to inundation from hurricane tidal surges and riverine or flash flooding than from the direct impact of winds on structures.

Tropical cyclones of hurricane strength occur in lower latitudes of all oceans except the South Atlantic and the eastern South Pacific, where combinations of cooler sea temperatures and prevailing winds whose velocities vary sharply with height prevent the establishment of a central warm core through a deep enough layer to sustain the hurricane wind system.

In the United States, property losses resulting from hurricanes have climbed steadily in the last 2 decades because of the increasing number of seashore structures. However, the loss of life, which has been huge in many storms, has decreased markedly. This is due mainly to the fact that warnings, aided by a more complete surveillance from aircraft and satellite, and extensive programs of public education, have become more accurate and more effective. Improvements in methodology for hurricane prediction have reduced the error in pinpointing hurricane landfall and have greatly reduced the probability of large errors in prediction.

Structure and movement. At the Earth's surface the hurricane appears as a nearly circular vortex 400–800 km in diameter. Its dynamic and thermodynamic properties, however, are distributed about the vortex asymmetrically. The cyclonic circulation (counterclockwise in the Northern Hemisphere) extends through virtually the entire depth of the troposphere (15 km). In lower layers (2–4 km in depth) winds spiral inward and accelerate toward lower pressure, reaching peak velocities in a narrow annulus surrounding the pressure center at a radial distance of 20–30 km. Here there is a near balance between the pressure forces acting radially inward and the centrifugal and Coriolis forces acting outward, so that the air, no longer able to move radially, is forced upward, transporting with it the horizontal momentum it acquired in lower layers. Traveling upward through the warm core, the air parcels soon reach layers at which horizontal pressure forces begin to diminish (hydrostatically). The imbalance that results causes the air to spiral outward and join environmental circulations, carrying with it a canopy of cloud debris known as the outflow cloud shield. This circulation of mass—in, up, and out of the vortex—at the rate of some 2,000,000 metric tons per second constitutes a kind of atmospheric heat pump. It uses as fuel the latent heats of condensation and fusion released as the water vapor, brought in with the environmental air and augmented by fluxes from the ocean surface, rises, cools, and generates tall cumulus clouds in or near the annulus of maximum winds. These cumuli are the conduits for transporting mass upward to the outflow level, and in a mature hurricane they merge into a wall of nimbostratus encircling a small benign weather area known as the eye, where cloudiness is minimal, rain is absent, and winds are relatively light. The circulation in the vortex not only generates the eye-wall clouds but also a family of spiral rain bands, which move cyclonically around the vortex center (Fig. 1).

The hurricane system moves through its environment in response to the interactions between the circulations of the environment and the vortex. These interactions create systematic imbalances between the amount of mass flowing in and that flowing out of selective quadrants or sectors and thereby produce the pressure changes that cause the center to move. The steering of the hurricane thus involves the interaction of vortex with environment from the bottom of the inflow to the top of the outflow layers. Most tropical cyclones reach hurricane strength in the belt of sluggish but steady tradewinds and are propelled westward at 10–15 knots (5–7.7 m/s) to the extremity of the subtropical anticyclone, where they tend to recurve northward into the vigorous west winds of midlatitudes and are carried eastward, often at speeds greater than 30 knots (15.4 m/s).

Energy sources and transformations. The process by which the tall cumuli in the eye wall maintain the warm, light air in the hurricane core, which in turn generates the pressure forces that determine the strength of hurricane winds, is not a simple matter of releasing latent heat within the clouds. Early diagnostic computer models of the hurricane encountered this difficulty when simulations led to run-away development and unrealistic pressure forces. This led to the development of procedures for parameterizing cloud effects. Most procedures assumed that clouds heat the adjacent air by direct turbulent transfer of their own excess warmth.

William Gray hypothesized that the warm core of a hurricane is maintained primarily by adiabatic heating of air descending between individual clouds. If true, this would mean that the energy driving the hurricane is supplied indirectly and would call for a much different kind of parameterization. Gray's hypothesis gained support from recent studies based on observations from research aircraft flights in hurricanes, which showed that the warm core in cases studied was indeed maintained by a combination of cumulus ascent and of air descending between clouds. Nevertheless, energetically the ultimate strength developed by a hurricane still depends on the amount of latent heat liberated in the eye wall, which in turn depends on the heat content of the air rising in the eye wall and the rate of vertical transport. The rate of vertical transport depends on environmental circulations that may constrain the inflow at low levels or the efflux at the top of the vortex.

While more than three-fourths of the water vapor that fuels the hurricane is imported from outside the storm boundaries, the critical contribution to its energetics is from the flux of latent and sensible heat it derives from the warm ocean over which it travels. This is a critical source, because it is precisely this contribution that is responsible for converting a tropical cyclone from gale strength to hurricane strength. This energy flux from sea to air is sometimes constrained by an important feedback mechanism. It has been shown theoretically and experimentally that cooling of sea surface temperatures occurs in the wake of hurricanes because of the mixing and upwelling caused by hurricane wind stresses on the sea. The amount of cooling is a maximum (2–5°C) when the hurricane travels at speeds less than that of gravity waves in the ocean thermocline (4–8 knots; 2–4 m/s), which is enough to substantially reduce the sea-air heat flux and the strength of the wind system. *See* OCEAN-ATMOSPHERE RELATIONS.

The parameterization of latent heat releases in hurricanes remains one of the biggest challenges

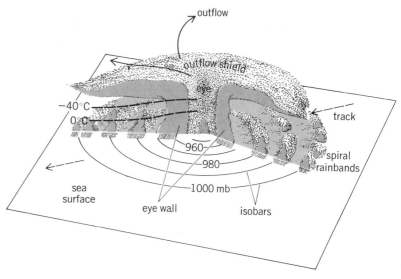

Fig. 1. Model of a hurricane circulation and cloud structure.

for modelers and is still unresolved. However, important progress in simulating other aspects of hurricane structure has been made, most notably by R. Anthes and by S. Rosenthal. Anthes's three-

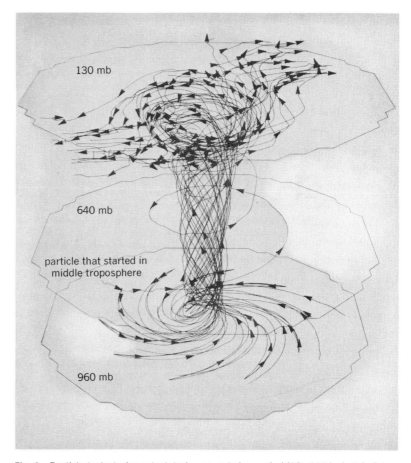

Fig. 2. Particle trajectories calculated over an 8-day period (90–282 hr in 9-hr intervals) in an experiment with a three-dimensional model hurricane. All particles start in the lower atmospheric boundary layer except one, which is started in the middle troposphere. (*From R. A. Anthes, S. L. Rosenthal, and J. W. Trout, Preliminary results from an asymmetric model of the tropical cyclone, Mon. Weather Rev., 99:744–758, 1971*)

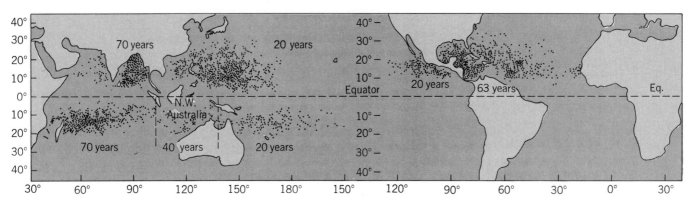

Fig. 3. Points on the globe where tropical cyclones were first detected by weather observers. (*From W. Gray, Global view of the origin of tropical disturbance, Mon. Weather Rev., 96(10):670, 1968*)

dimensional asymmetric hurricane model (Fig. 2) simulated a hurricane circulation in three dimensions, including a derived eye wall and the spiral rain bands. Such successes are necessary prerequisites to the development of an effective dynamic prediction model and lend encouragement that this task is achievable.

Hurricane formation. An average of about 70 tropical cyclones develop gale or hurricane strength somewhere on the globe each year, a figure some 15 to 20% higher than climatological records indicated before satellite weather surveillance became available in the early 1960s. About 9 occur in the North Atlantic ocean, 16 in the eastern North Pacific, and 24 in the western North Pacific. They also occur in the central and western South Pacific, the Bay of Bengal, the Arabian Sea, and the South Indian Ocean (Fig. 3). *See* METEOROLOGICAL SATELLITES.

The disturbances, sometimes referred to as seedlings, that breed tropical cyclones of hurricane strength originate mainly in tropical latitudes where the trade winds invade the equatorial trough (sometimes referred to as the intertropical convergence zone, or ITCZ). Some have their origins over continental areas. Seedling disturbances, comprising an agglomeration of convective clouds 300–500 km in diameter, often move more than 2000 km across tropical oceans as benign rainstorms before developing closed circulations and potentially dangerous winds. *See* TROPICAL METEOROLOGY.

In the North Atlantic, approximately 100 seedling disturbances are tracked across the tropical belt each year, more than half of which emanate from the African continent north of the ITCZ. A few form in the subtropics from cold lows, and still others from the trailing edge of old cold fronts or shear lines in subtropical latitudes. Regardless of the sources, however, it is notable that the average of 100 seedlings per year has a standard deviation of only 8, one of the most dependable recurrences of a meterological event. It is even more notable and difficult to explain why, with 100 opportunities for tropical cyclone development each year, an average of only 9 seedlings become dangerous wind storms, and the standard deviation from year to year is 4.

Two well-known factors constrain the develop-

ment of seedlings. One is the temperature of the sea environment. It was established in the early 1950s that to supply the heat flux from sea to air needed to generate a hurricane wind system initially, the sea surface must be warmer than about 26°C. The second factor is the variation of prevailing winds with height. If winds in the lower troposphere move at appreciably different velocities from those in the upper troposphere, the heat released by convection, or the warming by subsidence between convective clouds, cannot be stored in vertical columns of sufficient height to produce the pressure drop needed to support hurricane winds. Nevertheless, on many occasions, cyclogenesis still fails to occur when these constraints are absent.

Work by Michael Garstang based on observations from the BOMEX and GATE expeditions in 1969 and 1974 point to some additional constraints. These researches show that aggregate convection, even in the presence of a forcing influx of air from the ocean environment, can be self-defeating for several reasons. Vertical mixing by convection brings drier, sometimes cooler, air down to the layers below cloud bases, diluting the heat and moisture content. This process also increases the thermal stability of the subcloud layer. The net result is that the initial convection frequently cannot survive. This deadly cycle can be broken only if the convective mixing occurs in an environment that allows air of substantially higher momentum to be brought to the surface to increase the circulation of mass to the convection. Curiously, the latter condition seems to associate itself with a unique spacing of the convective elements of the disturbances. Further research into the environmental factors that control this spacing may supply the answer to one of the most puzzling questions in meteorology: Why are there so few hurricanes?

Prediction. Most methods for hurricane prediction are addressed exclusively to the movement and landfall of the center. In part, this is because the landfall position is the most important element in issuing warnings and in evacuating residents from areas subject to flooding. The change in strength, often a dramatic one as a hurricane approaches land, is so intimately related to cumulus parameterization and to nonlinear interactions between environment and vortex that these predictions continue to depend mainly on monitoring by reconnaissance aircraft and satellites. *See* STORM DETECTION.

Today, hurricane forecasting draws heavily on machine prediction models that combine the output of dynamical short-period predictions of environmental changes with statistical and analog information on expected hurricane behavior to determine the most probable future position of the center in 12-hr increments, for up to 72 hr. The latest version of such a technique, currently in use at the National Hurricane Center in Miami, is the NHC-73 method developed by Hope and Neumann. Since the 1950s, researchers have been trying to evolve dynamical models that predict hurricane movement. The most successful of these, and the only one in operation, is known as SANBAR (Sanders barotropic model), a single-layer barotropic filtered model which uses as input the mean wind velocities for the layer 1000–100 mb. The first three-dimensional asymmetric prediction model to show useful skill was developed by B. I. Miller.

However, several problems have restricted the operational usefulness of nearly all dynamical models. Foremost is the initial value problem, which requires either that the vortex be replaced by an ersatz circulation or point vortex, or a machine analysis with high resolution using very small mesh lengths. For the latter, adequate initial data are not available. Second, if adequate physics are included, computation time for a 24-hr forecast

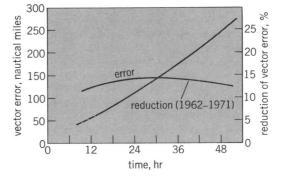

Fig. 4. Growth of the absolute magnitude of the vector error with length of forecast. The broken line is the percentage reduction in this error from 1962 through 1971. *(From R. H. Simpson, Hurricane prediction, Science, Sept. 7, 1973, copyright 1973 by the American Association for the Advancement of Science)*

is enormous, even on the largest computer systems.

The outlook for substantial increases in expected skill for predicting movement (Fig. 4), while uncertain, is not clearly promising. What some improved models may be able to do, whether they are dynamically or statistically founded, is to decrease the chances of large, disastrous errors. *See* WEATHER FORECASTING AND PREDICTION.

Mitigation. Systematic experimentation aimed at reducing the destructiveness of hurricanes by

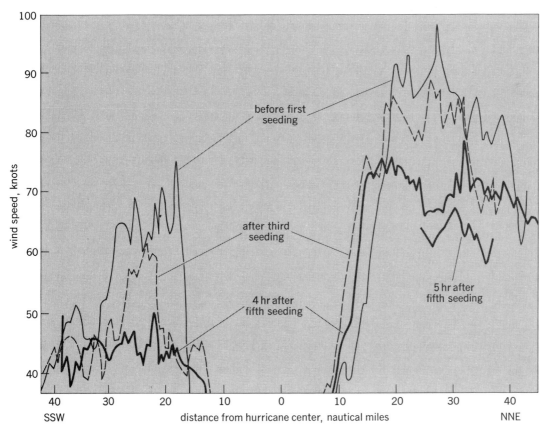

Fig. 5. Changes with time of wind speeds at 12,000 ft in Hurricane Debbie on Aug. 18, 1969.

strategic cloud seeding began in 1961 when Atlantic hurricane Esther was seeded with silver iodide crystals on 2 days. Additional experiments were conducted in Beulah (1963), Debbie (1969), and Ginger (1971). The objectives of these experiments, sponsored by the United States government under the project name of Stormfury, were to reduce the destructive winds of the hurricane by dispersing silver iodide crystals in the eye wall. Stormfury employs a strategy that seeks a selective release of latent heat of fusion in the eye wall, designed to induce the eye wall and its annulus of maximum winds to expand, or to reform at a greater distance from the center, the results of which, from conservation of absolute angular momentum, would be a reduction in maximum wind speeds.

The initial hypothesis, advanced by Robert Simpson and Joanne Simpson, argued on hydrostatic grounds that the strategic release of latent heat of fusion in the eye wall would reduce surface pressures outward from the annulus of maximum winds, and in doing so would reduce the pressure gradient. This, it was argued, would cause a circulation imbalance and would induce the eye wall to migrate outward. After exhaustive experiments with computer simulations of this process by Stanley Rosenthal, Cecil Gentry presented a revised hypothesis that, in seeking the same basic objectives, proposed a seeding procedure designed to stimulate growth of cumulus clouds located radially outward from the annulus of maximum winds. The increases in buoyancy resulting from seeding, it was argued, should cause these clouds to penetrate and transport mass to the outflow level with the result that the eye wall should tend to reform at a greater distance from the center. The expected result, as in the earlier hypothesis, is that maximum winds would be reduced. This revision has remained the basis for Stormfury experiments in recent years, although a number of alternative and imaginative suggestions for mitigating the damage potential of hurricanes remain untested. Most notable of these is the proposal by William Gray to use carbon black seeding as a radiation sink for altering the convection strategically and thus altering the mass circulation in the vortex.

The results of each of the Stormfury seeding experiments has been a response of the kind predicted by the hypothesis, although in all but perhaps hurricane Debbie, which showed a dramatic reduction (31%) in maximum winds after a succession of seeding runs at 90-min intervals (Fig. 5), the results were inconclusive, since the changes were not of sufficient magnitude to rule out the possibility that they were caused by natural variations common in hurricanes.

The biggest handicap in conducting the Stormfury experiments has been the limited number of hurricane cases suitable for experimentation. The National Oceanic and Atmospheric Administration, in 1976, planned to resume Stormfury experiments in 1977. The biggest challenges in these follow-up experiments are (1) to determine whether there is ordinarily enough unfrozen cloud water present to generate the results predicted by the hypothesis; and (2) to establish whether the dynamical results sought would cause any significant

change in storm surge levels, since inundation is the greatest source of losses of both life and property. *See* CLOUD PHYSICS; STORM.

[ROBERT SIMPSON; JOANNE SIMPSON]

Bibliography: G. E. Dunn and B. I. Miller, *Atlantic Hurricanes*, 1964; W. Hess (ed.), *Weather and Climate Modification*, chaps. 14 and 15, 1974; W. J. Kotsch and E. T. Harding, *Heavy Weather Guide*, 1965; E. Palmen and C. W. Newton, *Atmospheric Circulation Systems*, chap. 15, 1969; R. H. Simpson, Hurricane prediction, *Science*, 181:899–907, 1973.

Hydrology

The science that treats of the waters of the Earth, their occurrence, circulation, and distribution; their chemical and physical properties; and their reaction with their environment, including their relation to living things. The domain of hydrology embraces the full life history of water on the Earth.

Hydrologic cycle. The central concept of hydrology is the hydrologic cycle, a term used to describe the circulation of water from the oceans through the atmosphere to the land and back to the oceans over and under the land surface (see Fig. 1). Water vapor moves over the sea and land. It condenses into clouds and is precipitated as rain, snow, or sleet when one air mass rises to pass over another or over a mountain. Some of that which falls is evaporated while still in the air, or is intercepted by vegetation. Of the precipitation that reaches the ground surface, some evaporates quickly, some penetrates the soil, and some runs off over the land surface into streams, lakes, or ponds. Of that which penetrates the soil, some is held for a time and then returned to the atmosphere by evaporation or plant transpiration. The remainder penetrates below the soil zone to become a part of the groundwater. Study of the water in the atmosphere, although closely related to hydrology, lies more properly in the field of meteorology. The science relating to the oceans also touches closely on hydrology but is regarded as a separate science. *See* ATMOSPHERIC WATER VAPOR; METEOROLOGY; OCEANOGRAPHY.

Water is important for domestic, agricultural, and industrial uses. Thus the study of water and the means by which it may be obtained and controlled for use is of utmost importance to the welfare of mankind. Water is also a destructive agent of awesome force. Great floods periodically inundate valleys, causing death and destruction. Less dramatic are the rising water tables which, especially in irrigated areas, cause deterioration of the soil and make worthless large areas that would otherwise produce crops. Erosion of soil by flowing water and ultimate deposition of the sediment in lakes, reservoirs, stream channels, and coastal harbors is also a problem for hydrologists.

Functions. Hydrology is concerned with water after it is precipitated on the continents and before it returns to the oceans. It is concerned with measuring the amount and intensity of precipitation; quantities of water stored as snow and in glaciers, and rates of advance or retreat of glaciers; discharge of streams at various points along their courses; gains and losses of water stored in lakes

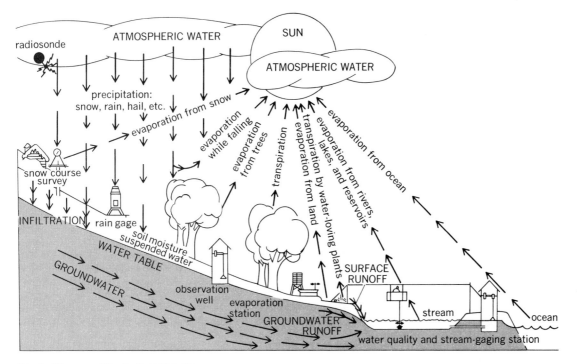

Fig. 1. Diagram of the hydrologic cycle.

and ponds; rates and quantities of infiltration into the soil and movement of soil moisture; changes of water levels in wells as an index of groundwater storage; rate of movement of water in underground reservoirs; flow of springs; dissolved mineral matter carried in water and its effects on water use; quantities of water discharged by evaporation from lakes, streams, and the soil and vegetation; and sediment transported by streams. In addition to devising methods for making these diverse measurements and storing the data in usable form, hydrology is concerned with analyzing and interpreting them to solve practical water problems. Rigorous studies of all the basic data are required to determine principles and laws involved in the occurrence, movement, and work of the water in the hydrologic cycle.

Engineering hydrology studies are basic to the design of projects for irrigation, water supply, flood control, storm drainage, and navigation. The hydrologist must estimate the probability of floods and droughts, volumes of water available for use, and probable maximum floods for spillway design. Often these estimates are required at locations where streamflow has not been measured. Digital computers are playing an increasing role in these studies and electrical analog computers are widely used in groundwater studies. *See* GROUNDWATER.

[ALBERT N. SAYRE/RAY K. LINSLEY]

Snow management. Many aspects of snow management have received attention in recent years due to a continuing need for additional water supplies and a need for better forecasts of existing supplies. Significant advances in snow management have occurred relative to instrumentation, snowpack modeling, and snow augmentation techniques.

Instrumentation. Improvements in the use of radioactive isotopes for monitoring the snowpack, together with a widespread use of telemetry equipment which provides direct contact with computer facilities, have made possible more detailed monitoring of the snowpack at remote locations. The use of radioactive isotopes requires the preseason installation of a gamma-radiating source in a sealed tube plus a scintillation crystal detector in another tube. A mechanism is provided to raise and lower both in their respective tubes by remote control. Each time the installation is interrogated by radio control, a profile of the snowpack water content, density, and depth is obtained. Consequently, vital forecast information can be obtained as frequently as necessary without the need for a visit to the site.

Adaptation of the lysimeter to snowpack measurements has also aided in evaluating water yields from the snow. The type of lysimeter referred to here generally consists of some type of weighing platform which supports a column of soil or snow. Provisions are made for measuring water entering or leaving the column, thus giving a measure of water gained, stored, or lost relative to the starting condition of the column. The lysimeter has been used for years to evaluate water movement through soil; however, application of the technique to snow conditions has been difficult. This is due to constant changes in the snowpack, both in dimension and internal structure, which affects the flow of water from the pack. These changes prevent an accurate computation of unit area above the lysimeter which is necessary in relating water flow from the pack with contributing area.

The snow pillow, which has been used for several years, is in effect a borderless lysimeter. That is, no vertical restraints are applied to the snow column above the surface area measured. The pillow is frequently used to obtain a continuous record of snowpack water content for forecast purposes. However, this is somewhat unreliable for determining the amount and rate of water loss as runoff from the pack.

Development of a universal surface precipitation gage has effectively taken the snow pillow one step further by including a recorder system for measuring runoff from the bottom of the snowpack. Consequently, evaluations of snowpack water content, precipitation additions, and runoff losses are continuously carried out. Problems of bridging and ice lense formation still affect unit area calculations on slopes. A simple snowmelt lysimeter has been developed for use on slopes which utilizes flexible plastic sheets to form a border around the area measured. The border is moved up or down with changes in the snowpack depth. This allows control over the unit area measured by ensuring collection of all meltwater from the area directly above the plot. Maintaining the border flush with the snow surface is necessary to prevent abnormal effects on the snow surface.

One use of the snow lysimeter has been to evaluate effects of the forest canopy on water yields from the snowpack. In the past, effects of the forest canopy on snow accumulation and distribution have been measured by monitoring the water content of the snowpack; however, such measurements have led to false interpretation of how much water is yielded from the pack. Accumulation of snow in the crowns of trees suggests a loss to the snowpack, while large openings appear to have more snow, but melt faster in the spring. Measurements of the water content of the snowpack alone suggests that at peak accumulation (approximately April 1), the most snow exists in small openings, followed by large openings, then deciduous forests, and finally by coniferous forests. However, lysimeter tests indicate that coniferous forests produce more runoff by the time maximum accumulation occurs. The water for early runoff from the snow under a conifer stand is obtained from melting snow intercepted in the crowns of the trees and melting of the snowpack under the trees due to the heat absorbed by the dark foliage of the trees. Consequently, the snowpack under the conifer stands has less water but has already contributed more to runoff. As the season progresses, the melt rates increase in large openings due to a lack of shading and advective heat from wind movement. Consequently, the snowpack disappears in large openings first, followed by deciduous forests, then coniferous forests, and finally in small openings. The small openings appear to exhibit the best conditions for maximum accumulation of snow and delayed melt. The forest around the small opening contributes windblown snow from its crown, and shade to delay melt.

Modeling. Mathematical models of dynamic systems such as a snowpack are useful tools for predicting changes in such systems over a period of time. Because such a large number of variables are present which affect the accumulation and melt of a snowpack, use of a model was virtually impossible before computers became available to handle the computations involved. Today analog-to-digital

computers make such a model very practical. The input data such as precipitation, temperature, and snowpack conditions, in the form of variable electrical resistances (analog data), are applied to the computerized model and digitized to provide discrete (digital) output in terms of runoff, stored water, or other forms of forecast information.

One useful snowpack model accounts for gains or losses of water content by a series of heat balance equations, along with precipitation input information. The snowpack can be considered a dynamic heat reservoir, which makes it possible to account for water gains or losses by balancing heat gains or losses. This is possible since water losses are primarily a result of melting or evaporation which are heat-related processes. The addition of heat and water from the atmosphere must also be accounted for once the snowpack begins to form. Once the heat balance of the pack reaches 0°C, it is considered isothermal; and further additions of heat result in melted snow, forming runoff.

Augmentation. Management of the snowpack for increased water yields has developed along two primary lines. One approach is to manipulate the forest canopy to increase the snow catch locally or alter the melt rate for a more favorable timing of runoff. The second approach is to increase winter precipitation through cloud seeding.

In the techniques of vegetation manipulation, improvements have resulted through the application of modern logging methods, such as helicopter, balloon, or skyline yarding, which provide more flexibility in the cutting patterns that are followed. This is significant on a local basis, but does not have a significant effect on basin-wide water demands.

Cloud-seeding techniques have reached the stage of providing operational systems for basin-wide increases in water yield, which have very significant ramifications on both the land in the seeded area and on downstream areas where the additional water would be used. Cloud seeding for snowpack augmentation basically consists of forming a smoke cloud of silver iodide crystals which becomes the nucleus for ice crystal formation, producing snow when the moisture-laden cloud is orographically lifted and cooled (see Fig. 2). A detailed explanation of this process has been described by M. Neiberger.

Basically, snow augmentation represents an increased water content of the snow, deeper snow, a prolonged period when snow is on the ground, increased runoff during melt, a longer period of high peak flows, and lower stream temperatures during the early summer period. The additional snow is obtained by increasing snowfall during periods that normally would have produced only light snowfall.

The upper Colorado River basin is an excellent example of how a snowpack augmentation program may be applied. The target area is the zone where increased snowfall is expected, which in this case would be the high Rocky Mountains of Colorado. A heavy snowpack normally occurs in this area. Additional water in the Colorado River system would be beneficial, primarily to downstream users from Colorado to California. It is estimated that snow augmentation could produce 20–

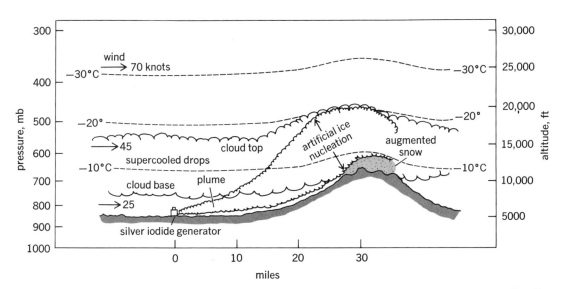

Fig. 2. Idealized model showing meterological conditions that will result in increased snowfall if clouds are seeded with silver iodine particles. (*From L. W. Weisbecker, comp., Technology Assessment of Winter Orographic Snowpack Augmentation in the Upper Colorado River Basin, vol. 2, Stanford Research Institute, Menlo Park, CA, 1972*)

25% more water in the Colorado River system, or approximately 2,300,000 acre-feet of water. The water-rich target area would not benefit from additional water, but could benefit from the snow as a winter sports enhancement. Skiing and associated activities attract many people to this area of Colorado and adjacent states. Basin-wide snow augmentation is not without potential side effects. Changes in ecological balances, hydrologic conditions, and economic opportunities both favorable and unfavorable are possible. A detailed discussion of such potential changes has been reported by L. W. Weisbecker. *See* SNOW; SNOW GAGE.   [ROBERT D. DOTY]

Water balance in North America. Demand for water has customarily been balanced against the supply on a predominantly local basis. However, in recent years water shortages have occurred in many localities, including some which have had abundant supplies in the past, and, as the population grows, more areas can be expected to experience such shortages. At present neither the North American continent as a whole nor any of its component countries has an overall water shortage. It would seem possible, therefore, to balance the supply and the demand on a larger scale so that local water shortages could be prevented or eliminated. An examination of the availability of water and the factors involved in the increasing demand for it is given below, followed by a discussion of the various means of achieving a balance between supply and demand.

Availability of water. The availability of water in any area is determined by the balance of certain elements in the hydrologic cycle. The International Hydrological Decade has undertaken the task of studying the balance and the long-term changes among these elements throughout the world. For the purposes of this discussion, however, the long-term average runoff to the oceans is a satisfactory measure of the surplus water available for man's needs in North America.

Table 1 shows estimated water availability in terms of the long-term average runoff from the North American continent as a whole and from some selected subdivisions. The last column gives the availability on a per capita basis. The figures in Table 1 are approximate; better data should be available when the work of the Decade has been completed. The values given are long-term averages of quantities which can vary widely from day to day and from season to season, but somewhat less from year to year. The variability is greater in some areas than in others because natural and artificial storage tends to regulate discharge.

Interestingly, the average per capita availability of water is greater in Texas and New Mexico, areas

Table 1. Estimated water availability for North America

Area	Average runoff, millions of cubic feet per second*	Population, millions*	Water availability, gallons per capita per day*
NORTH AMERICA	7.3	263	18,000
Canada	3.6	20	115,000
Conterminous United States	1.8	200	5,800
Mexico	0.4	43	6,000
Eastern seaboard from Virginia to New York	0.16	47	2,200
Texas and New Mexico	0.08	12	4,300
Southern California	0.003	10	200
Mexico, central and southern plateau	0.015	16	600
Mexico, between plateau and Gulf Coast	0.1	6.5	10,000

*Rounded values.

which are generally associated with water shortage, than it is on the heavily populated east coast of the United States, which is usually thought of as adequately supplied. Table 1 also shows that a large per capita water supply exists when the average is taken over the entire continent or even over each individual country. This is because of the large runoff in relatively lightly populated areas such as Alaska, northern Canada, northwestern United States, and southern Mexico.

Demand for water. The demand for water is not based on man's minimal needs alone; it is a complicated function of need, availability, and custom. Demand breaks down into two categories — withdrawal uses and nonwithdrawal uses. Water withdrawals for the United States are summarized in Table 2, and these are probably not much larger, on a per capita basis, than would be the corresponding statistics for Canada and Mexico. Only a small part of the water withdrawn is lost, a little less than 10%, principally through evaporation and transpiration in connection with irrigation. The rest is returned to the lakes and streams, frequently degraded in quality. Nevertheless, much of it can be and is used again many times over, and some of the withdrawals given in Table 2 include reused water. These withdrawal data can serve as a measure of demand in the absence of a better indicator.

It is worth noting that the average water withdrawal rate per capita exceeds the availability in some areas of the United States. However, because water is reused, only a few areas have real shortages. For example, along the Eastern seaboard of the United States, where there is heavy industrial use of water, treatment of waste water generally yields enough water for reuse to satisfy all demands. On the other hand, areas with heavy demand for water for irrigation have little water for reuse, both because of losses sustained from evaporation and transpiration and because of the increased salinity of the recovered water.

Nonwithdrawal uses include hydroelectric power generation, navigation, transport of waste products, storage of floodwaters, recreation, and enhancement of aesthetic features of the landscape. In general, these uses do not degrade the quality or affect the quantity of the water in serious measure; however, some evaporation losses and quality changes may be associated with impoundments, and quality changes certainly accompany transport of waste products. There is no general way to measure the demand for nonwithdrawal uses because the demand depends so heavily on such factors as how much waste of what type will be transported, how deep the navigation channels will be, and how much water must flow in a river to make it aesthetically pleasing. These factors determine how much of the available water passes unused through a given river basin. Thus, even though the water availability in Table 1 exceeds the withdrawal use in Table 2, there still may be water shortages.

Although the availability of water to a given area is fairly well fixed when averaged over long time periods, the demand is not. The last column in

Table 2. Estimated water withdrawal in the United States

Use	Billions of gallons per day	Gallons per capita per day*	Annual increase, %
Public water supplies	26	130	} 3.0
Rural domestic supplies	7	35	
Irrigation	155	775	1.8
Steam electric cooling	127	635	2.7
Industrial and other	81	405	2.8
Total	396†	1980	2.4

*Based on 200,000,000 people.
†Includes 70,000,000,000 gal/day of groundwater.

Table 2 presents estimated increases in withdrawals. These increases are a product of the increased standard of living. Thus, whereas availability per capita is decreasing, demand per capita is increasing, and sooner or later some water-rich areas will become water-poor.

When the demand factors are considered and weighed against the availability of water, the water-short areas of North America can be identified as parts of the Canadian prairie provinces, the northern plains and the southwestern part of the United States, and northern Mexico including the central and southern plateau. Other areas, such as the heavily populated part of the Eastern seaboard of the United States, may fall into this category in the foreseeable future.

Meeting the demand. Where demand threatens to outreach available water supplies, several courses of action are possible. Such courses, listed in their most probable order of application on the North American continent, include taking measures to conserve water, controlling industrial growth and population, introducing new sources of supply, and transferring water from one basin to another.

Possible conservation measures include the metering of, and adequate charge for, all water used, detection and sealing of leaks in conduits and reservoirs, control of plant growth to retard transpiration, more efficient application of water in irrigation and in industrial processes, and recycling of waste waters from industrial plants and public water supplies.

It may become desirable to plan the abandonment of certain industries in some water-short areas in favor of others that require less water. For example, it has been suggested that irrigation be abandoned in certain parts of the southwestern United States. At the same time, shortages of water may motivate certain industries which cannot accommodate themselves to conservation measures to move to areas with water surpluses, and population shifts would accompany these industrial shifts.

Possible methods of adding to the water supply include mining groundwater (as opposed to simply removing it from storage during one season or during dry years and replacing it at other times), desalinating sea water or brackish water, and mining glaciers or towing icebergs to water-short coastal areas. Controlling precipitation by seeding clouds could also be mentioned in connection with increasing the water supply; however, this process might also be characterized as a method of interbasin transfer. Indeed, the question of whether water so produced is really new water or simply water removed from another basin may present insurmountable political and legal obstacles to the practical use of precipitation control.

The data in Table 1 point toward the possibility of transferring water from areas of high per capita availability to those of low availability. Several grand schemes for such interbasin transfer of water have been proposed. Some smaller transfer projects, such as New York City's Delaware River supply and the Colorado–Big Thompson project across the Continental Divide, are already in operation. The California Water Project, which is nearing completion, and the Texas Water Plan, now stalled, are examples of larger projects. The grand schemes, however, visualize the transport of water from the northern parts of Canada, where there is a large surplus, to the central and southwestern parts of the United States and into northern areas of Mexico.

It is likely to be many years before any of these large-scale projects reaches fruition, since reasonable cost figures will need to be settled and many international political and social problems overcome. However, lesser interbasin transfer projects will probably continue to develop, and it may be through sequential interconnection of these that one of the grand schemes will eventually be realized.

[EDWARD SILBERMAN]

Bibliography: R. del Arenal C., Water resources of Mexico, *Water Resour. Bull.*, 5(1):19–38, 1969; V. T. Chow, *Handbook of Applied Hydrology*, 1964; L. M. Cox, *Proceedings of the 39th Western Snow Conference*, pp. 84–87, 1974; S. N. Davis and R. J. M. DeWiest, *Hydrogeology*, 1966; H. F. Haupt, *USDA For. Serv. Res. Pap.*, no. INT-114, 1972; R. J. Kazmann, *Modern Hydrology*, 2d ed., 1972; A. H. Laycock (ed.), *Proceedings of the Symposium on Water Balance in North America*, American Water Resources Association, 1969; W. G. McGinnies and B. J. Goldman (eds.), *Arid Lands in Perspective* (including AAAS Papers on Water Importation into Arid Lands), 1969; M. F. Meier, *J. Amer. Water Works Ass.*, 61(1):8–12, 1969; D. H. Miller, *Water at the Surface of the Earth: An Introduction to Ecosystem Hydrodynamics*, 1977; M. Neiberger, *Meteorol. Org. Tech. Note*, no. 105, 1969; J. L. Smith, H. G. Halverson, and R. A. Jones, *U.S. Dep. Commerce Nat. Tech. Inform. Serv. Rep.*, no. TID-25987, 1972; M. Tribus, Physical view of cloud seeding, *Science*, 168:201–211, 1970; R. C. Ward, *Principles of Hydrology*, 1967; L. W. Weisbecker (comp.), *Technology Assessment of Winter Orographic Snowpack Augmentation in the Upper Colorado River Basin*, Stanford Research Institute, Menlo, Park, CA, 1972; A. Wilson and K. T. Iseri, *River Discharge to the Sea from the Shores of the Conterminous United States, Alaska, and Puerto Rico*, USGS Atlas no. HA-282, rev. ed., 1969; G. Young, Dry lands and desalted water, *Science*, 167:339–343, 1970.

Hydrometeorology

The study of the occurrence, movement, and changes in the state of water in the atmosphere. The term is also used in a more restricted sense, especially by hydrologists, to mean the study of the exchange of water between the atmosphere and continental surfaces. This includes the processes of precipitation and direct condensation, and of evaporation and transpiration from natural surfaces. Considerable emphasis is placed on the statistics of precipitation as a function of area and time for given locations or geographic regions.

Water occurs in the atmosphere primarily in vapor or gaseous form. The average amount of vapor present tends to decrease with increasing elevation and latitude and also varies strongly with season and type of surface. Precipitable water, the mass of vapor per unit area contained in a column of air extending from the surface of the Earth to the outer extremity of the atmosphere, varies from almost zero in continental arctic air to about 6 g/cm^2 in very humid, tropical air. Its average value over the Northern Hemisphere varies from around 2.0 g/cm^2 in January and February to around 3.7 g/cm^2 in July. Its average value is around 2.8 g/cm^2, an amount equivalent to a column of liquid water slightly greater than 1 in. in depth. Close to 50% of this water vapor is contained in the atmosphere's first mile, and about 80% is to be found in the lowest 2 mi.

Atmospheric water cycle. Although a trivial proportion of the water of the globe is found in the atmosphere at any one instant, the rate of exchange of water between the atmosphere and the continents and oceans is high. The average water molecule remains in the atmosphere only about 10 days, but because of the extreme mobility of the atmosphere it is usually precipitated many hundreds or even thousands of miles from the place at which it entered the atmosphere.

Evaporation from the ocean surface and evaporation and transpiration from the land are the sources of water vapor for the atmosphere. Water vapor is removed from the atmosphere by condensation and subsequent precipitation in the form of rain, snow, sleet, and so on. The amount of water vapor removed by direct condensation at the Earth's surface (dew) is relatively small.

A major feature of the atmospheric water cycle is the meridional net flux of water vapor. The average precipitation exceeds evaporation in a narrow band extending approximately from 10°S to 15°N lat. To balance this, the atmosphere carries water vapor equatorward in the tropics, primarily in the quasi-steady trade winds which have a component of motion equatorward in the moist layers near the Earth's surface. Precipitation also exceeds evaporation in the temperate and polar regions of the two hemispheres, poleward of about 40° lat. In the middle and higher latitudes, therefore, the atmosphere carries vapor poleward. Here the exchange occurs through the action of cyclones and anticyclones, large-scale eddies of air with axes of spin normal to the Earth's surface.

For the globe as a whole the average amount of

evaporation must balance the precipitation. The subtropics are therefore regions for which evaporation substantially exceeds precipitation. The complete meridional cycle of water vapor is summarized in Table 1. This exchange is related to the characteristics of the general circulation of the atmosphere. It seems likely that a similar cycle would be observed even if the Earth were entirely covered by ocean, although details of the cycle, such as the flux across the Equator, would undoubtedly be different.

Complications in the global pattern arise from the existence of land surfaces. Over the continents the only source of water is from precipitation; therefore, the average evapotranspiration (the sum of evaporation and transpiration) cannot exceed precipitation. The flux of vapor from the oceans to the continents through the atmosphere, and its ultimate return to atmosphere or ocean by evaporation, transpiration, or runoff is known as the hydrologic cycle. Its atmospheric phase is closely related to the air mass cycle. In middle latitudes of the Northern Hemisphere, for example, precipitation occurs primarily from maritime air masses moving northward and eastward across the continents. Statistically, precipitation from these air masses substantially exceeds evapotranspiration into them. Conversely, cold and dry air masses tend to move southward and eastward from the interior of the continents out over the oceans. Evapotranspiration into these continental air masses strongly exceeds precipitation, especially during winter months. These facts, together with the extreme mobility of the atmosphere and its associated water vapor, make it likely that only a small percentage of the water evaporated or transpired from a continental surface is reprecipitated over the same continent. *See* ATMOSPHERIC GENERAL CIRCULATION; EVAPOTRANSPIRATION; HYDROLOGY.

Precipitation. Hydrometeorology is particularly concerned with the measurement and analysis of precipitation data. Since 1950 increasing attention has been paid to the use of radar in estimating precipitation. By relating the intensity of radar echo to rate of precipitation, it has been possible to obtain a vast amount of detailed information concerning the structure and areal distribution of storms. *See* METEOROLOGICAL INSTRUMENTATION; RADAR METEOROLOGY; STORM DETECTION.

Deficiencies in the observational networks over

Table 1. Meridional flux of water vapor in the atmosphere

Latitude	Northward flux, 10^{10} g/sec
90°N	0
70°N	4
40°N	71
10°N	−61
Equator	45
10°S	71
40°S	−75
70°S	1
90°S	0

the oceans and over the more sparsely inhabited land areas of the Earth are now being bridged through the use of meteorological satellite observations. Progress in the 1970s toward development of methods for estimating rainfall amounts from satellite observations of cloud type and distribution is of particular significance to hydrometeorology. *See* METEOROLOGICAL SATELLITES.

Precipitation occurs when the air is cooled to saturation. The ascent of air towards lower pressure is the most effective process for causing rapid cooling and condensation. Precipitation may therefore be classified according to the atmospheric process which leads to the required upward motion. Accordingly, there are three basic types of precipitation: (1) Orographic precipitation occurs when a topographic barrier forces air to ascend. The presence of significant relief often leads to large variations in precipitation over relatively short distances. (2) Extratropical cyclonic precipitation is associated with the traveling regions of low pressure of the middle and high latitudes. These storms, which transport sinking cold dry air southward and rising warm moist air northward, account for a major portion of the precipitation of the middle and high latitudes. (3) Air mass precipitation results from disturbances occurring in an essentially homogeneous air mass. This is a common precipitation type over the continents in mid-latitudes during summer. It is the major mechanism for precipitation in the tropics, where disturbances may range from areas of scattered showers to intense hurricanes. In most cases there is evidence for organized lifting of air associated with areas of cyclonic vorticity, that is, areas over which the circulation is counterclockwise in the Northern Hemisphere or clockwise in the Southern Hemisphere.

The availability of data from geosynchronous meteorological satellites, together with surface and upper-air data acquired as part of the Global Atmospheric Research Program (GARP), is leading to significant advances in the understanding of the character and distribution of tropical precipitation. *See* HURRICANE; STORM; TROPICAL METEOROLOGY.

Precipitation may, of course, be in liquid or solid form. In addition to rain and snow there are other forms which often occur, such as hail, snow pellets, sleet, and drizzle. If upward motion occurs uniformly over a wide area measured in tens or hundreds of miles, the associated precipitation is usually of light or moderate intensity and may continue for a considerable period of time. Vertical velocities accompanying such stable precipitation are usually of the order of several centimeters per second. Under other types of meteorological conditions, particularly when the density of the ascending air is less than that of the environment, upward velocities may locally be very large (of the order of several meters per second) and may be accompanied by compensating downdrafts. Such convective precipitation is best illustrated by the thunderstorm. Intensity of precipitation may be extremely high, but areal extent and local duration are comparatively limited. Storms are sometimes observed

Table 2. Record observed point rainfalls*

Duration	Depth, in.	Station	Date
1 min	1.23	Unionville, Md.	July 4, 1956
8 min	4.96	Füssen, Bavaria	May 25, 1920
15 min	7.80	Plumb Point, Jamaica	May 12, 1916
42 min	12.00	Holt, Mo.	June 22, 1947
2 hr 45 min	22.00	Near D'Hanis, Texas	May 31, 1935
24 hr	73.62	Cilaos, La Reunion (Indian Ocean)	Mar. 15–16, 1952
1 month	366.14	Cherrapunji, India	July, 1861
12 months	1041.78	Cherrapunji, India	August, 1860 to July, 1861

*From R. K. Linsley, M. A. Kohler, and J. L. H. Paulhus, *Hydrology for Engineers*, 2d ed., 1975.

in which local convective regions are embedded in a matrix of stable precipitation. *See* PRECIPITATION.

Analysis of precipitation data. Precipitation is essentially a process which occurs over an area. However, despite the experimental use of radar, most observations are taken at individual stations. Analyses of such "point" precipitation data are most often concerned with the frequency of intense storms. These data are of particular importance in evaluating local flood hazard, and may be used in such diverse fields as the design of local hydraulic structures, such as culverts or storm sewers, or the analysis of soil erosion. Intense local precipitation of short duration (up to 1 hr) is usually associated with thunderstorms. Precipitation may be extremely heavy for a short period, but tends to decrease in intensity as longer intervals are considered. Several record point accumulations of rainfall are shown in Table 2.

A typical hydrometeorological problem might involve estimating the likelihood of occurrence of a storm of given intensity and duration over a specified watershed to determine the required spillway capacity of a dam. Such estimates can only be obtained from a careful meteorological and statistical examination of large numbers of storms selected from climatological records. In the United States the U.S. Army Corps of Engineers, in cooperation with the National Weather Service, has embarked on a continuing program of analysis to make such historical depth-area-duration data available to the practicing engineer.

Evaporation and transpiration. In evaluating the water balance of the atmosphere, the hydrometeorologist must also examine the processes of evaporation and transpiration from various types of natural surfaces, such as open water, snow and ice fields, and land surfaces with and without vegetation. From the point of view of the meteorologist, the problem is one of transfer in the turbulent boundary layer. It is complicated by topographic effects when the natural surface is not homogeneous. In addition the simultaneous heating or cooling of the atmosphere from below has the effect of enhancing or inhibiting the transfer process. Although the problem has been attacked from the theoretical side, empirical relationships are at present of greatest practical utility. *See* METEOROLOGY; MICROMETEOROLOGY.

[EUGENE M. RASMUSSON]

Bibliography: R. D. Fletcher, Hydrometeorology in the United States, in T. F. Malone (ed.), *Compendium of Meteorology*, 1951; R. K. Linsley, M. A. Kohler, and J. L. H. Paulus, *Hydrology for Engineers*, 2d ed., 1975; E. N. Lorenz, *The Nature and Theory of the General Circulation of the Atmosphere*, 1967; W. D. Sellers, *Physical Climatology*, 1965.

Hydrosphere

Approximately 74% of the Earth's surface is covered by water, in either the liquid or solid state. These waters, combined with minor contributions from groundwaters, constitute the hydrosphere:

World oceans	1.3×10^9 km³
Fresh-water lakes	1.3×10^5 km³
Saline lakes and inland seas	1.0×10^5 km³
Rivers	1.3×10^3 km³
Soil moisture and vadose water	6.7×10^4 km³
Groundwater to depth of 4000 m	8.4×10^6 km³
Icecaps and glaciers	2.9×10^7 km³

The oceans account for about 97% of the weight of the hydrosphere, while the amount of ice reflects the Earth's climate, being higher during periods of glaciation. (Water vapor in the atmosphere amounts to 1.3×10^4 km³.) The circulation of the waters of the hydrosphere results in the weathering of the landmasses. The annual evaporation of 3.5×10^5 km³ from the world oceans and of 7.0×10^4 km³ from land areas results in an annual precipitation of 3.2×10^5 km³ on the world oceans and 1.0×10^5 km³ on land areas. The rainwater falling on the continents, partly taken up by the ground and partly by the streams, acts as an erosive agent before returning to the seas. *See* GROUNDWATER; HYDROLOGY; RIVER; SEA WATER; TERRESTRIAL FROZEN WATER.

[EDWARD D. GOLDBERG]

Bibliography: R. L. Nace, Water resources, *Environ. Sci. Technol.*, 1:550–560, 1967.

Hydrospheric geochemistry

The oceans of the world constitute a principal reservoir for substances in the major sedimentary cycle, which involves the processes of transport of material from the Earth's crust to the sea floor. The cycle begins with the precipitation of water, acidified by the uptake of carbon dioxide in the atmosphere, onto the continents. This results in the physical and chemical breakdown of exposed surfaces. A part of the weathered material, in dis-

Table 1. Oceanic and land-drainage areas

Ocean	Area, 1000 km²	Land area drained, 1000 km²	Ratio, area drained/ ocean area
Atlantic	98,000	67,000	0.684
Indian	65,500	17,000	0.260
Antarctic	32,000	14,000	0.440
Pacific	165,000	18,000	0.110
Interior drainage		32,000	

solved or solid states, is borne by the rivers to the oceans. Evaporation at the oceanic surfaces provides atmospheric water, which precipitates in part upon the continents. This latter process completes the cycle. Table 1 gives the quantitative details by contrasting the marine areas and the respective land areas draining into the world's oceans. Clearly, per unit area, the Atlantic receives the weathering products from an integrated drainage area six times as large as that of the Pacific. The interior drainage areas are responsible for such water bodies as the Great Salt Lake, the Caspian Sea, and the Dead Sea.

Oceanic waters. The reactivities of chemical species in the oceans are reflected in the average times spent there before precipitation to the sea floor. Those elements with short residence time in the oceans engage more readily in chemical reactions that result in the formation of solid phases than those elements with long residence times.

Residence times. The calculations of residence times are based upon a simple reservoir model of the oceans, whose chemical composition is assumed to be in steady state; that is, the amount of a given element introduced by the rivers per unit time is exactly compensated by that lost through sedimentation. An elemental residence time may be defined then by $t = A/(dA/dt)$, where A is the amount of the element in the oceans and (dA/dt) is the rate of introduction or the rate of removal of the element from the marine hydrosphere. Table 2 gives values for a representative group of elements.

The element of longest residence, sodium, has a residence time within an order of magnitude of the age of the oceans, several billion years. Similarly, the more abundant alkali and alkaline-earth metals all have residence times in the range of 10^6 to 10^8

Table 2. Residence times of elements in the oceans

Element	Residence time, years	Element	Residence time, years
Na	2.6×10^8	Mg	4.5×10^7
Ca	8.0×10^6	Li	1.2×10^7
U	6.5×10^5	K	1.0×10^7
Cu	6.5×10^4	Rb	6.1×10^6
Si	1.0×10^4	Ba	4.0×10^4
Mn	7.0×10^3	Zn	2.0×10^4
Ti	1.6×10^2	Pb	4.0×10^2
Al	1.0×10^2	Ce	3.2×10^2
Fe	1.4×10^2	Th	1.0×10^2

years, resulting from the relative lack of reactivity of these elements in marine waters.

Elements which pass rapidly through the marine hydrosphere in the major sedimentary cycle, such as titanium, aluminum, and iron, not only enter the oceans in part as rapidly settling solids but also are reactants in the formation of the clay and ferromanganese minerals. For a discussion of the inorganic regulation of the composition *see* SEA WATER.

Although the relatively low values of these residence times are significant, the absolute values are probably unrealistic inasmuch as they are in conflict with an assumption used in their derivation. In treating the oceans as a simple reservoir, the mixing times of the oceans are assumed to be much less than the residence times of the elements. Yet the oceans are believed to mix in times of the order of thousands of years. Nonetheless, it is significant that one may expect to find the concentration of such elements varying from one oceanic water mass to another.

Those elements with residence times of intermediate length, periods of the order of tens and hundreds of thousands of years, are probably of nearly uniform concentration in the oceans of the world, as are the chemical species of longer marine lives. Typical members of the group are such metals as barium, lead, zinc, and nickel, elements in extremely dilute solution but actively involved in the inorganic chemistries of the seas. Such behavior is confirmed by the observation that these elements are in states of undersaturation in oceanic waters.

Photosynthesis. The presence of the large photosynthesizing biomass in the oceans gives rise to dramatic concentration changes. The amount of photosynthesis in the oceans, calculated to be of the order of 1×10^{17} tons of carbon dioxide consumed per year, compares with estimates for land of 2×10^{16} tons/year. The depth of the photosynthetic zone can extend downward from the surface to depths of 100 meters (m) or so, depending upon the transparency of the water, season of the year, and latitude. In waters of active plant growth, carbon dioxide and oxygen (the intake and release gases of photosynthesis) are often observed in states of depletion and supersaturation, respectively.

The photosynthesizing plants require a group of dissolved chemical species, the nutrients, which are necessary for growth and multiplication. Ions of the orthophosphoric acids, nitrate, nitrite, and ammonia, as well as monomeric silicic acid, have very low concentrations in the regions of plant productivity as compared to regions of lesser fertility. Certain other substances concerned with plant growth, such as vitamins and trace-metal ions, may have their marine concentrations governed by biological activity in surface waters, but as yet no definite relationships have been established.

The primary production of plant material furnishes the basis of nutrition for the animal domain of the oceans. The plant material which is removed from the photosynthetic zones but is not consumed by higher organisms, together with the organic debris resulting from the metabolic waste products or death of members of the marine biosphere, is oxidized principally in the oceanic waters below but adjacent to the photosynthetic zone. This results not only in low values of dissolved gaseous oxygen at such depths but also in high concentrations of the nutrient species which are released subsequent to the combustion of the organic matter (see illustration).

The dissolved organic substances, arising from life in the sea, are of the order of 0.3 mg of carbon per liter. Higher amounts of such materials are found in surface or coastal waters. *See* SEA-WATER FERTILITY.

Salt content. For many problems in the physics of the sea and in engineering, the significant parameter is the total salt content, which governs the density as a function of pressure and temperature, rather than elemental, ionic, or molecular compositions.

The total salt content is expressed by either the salinity or the chlorinity, both terms given in units of parts per thousand °/₀₀. The salinity is defined as the weight in grams in vacuo of solids which can be obtained from a water weight of 1 kg in vacuo under the following conditions: (1) the solids have been dried to a constant weight at 480°C; (2) the carbonates have been converted to oxides; (3) the organic matter has been oxidized; and (4) the bromine and iodine have been replaced by chlorine. A weight loss from the solid phases results from such chemistries, and hence the salinity of a given sample of sea water is somewhat less than its actual salt content.

In practice, the salt content is ascertained by the precipitation of the halogens with silver nitrate or by such physical methods as electrical conductivity, sound velocity, and refractive index, with the first of these techniques having widespread use. In the chemical titration with silver nitrate the mass of halogens contained in 1 kg of sea water, assuming the bromine and iodine are replaced by chlorine, is designated as the chlorinity. The values so obtained are dependent upon the atomic weights of both silver and chlorine. Inasmuch as changes in the salt content of water are of interest when taken over long time periods, the chlorinity has been made independent of atomic weight changes by redefining it in terms of the weight of silver precipitated, as in Eq. (1). Chlorinity deter-

$$Cl\ °/_{00} = 0.3285234\ Ag \qquad (1)$$

minations, based either on chemical or physical methods, are related to standard sea water, a water of known salinity which is obtainable from the Hydrographic Laboratory in Copenhagen, Denmark. Chlorinity is related to the salinity, Eq. (2).

$$S\ °/_{00} = 0.030 + 1.8050\ Cl\ °/_{00} \qquad (2)$$

Salinity of open ocean waters varies regionally between 32 and 37 °/₀₀. Areas where evaporation exceeds precipitation, such as enclosed basins, are characterized by higher values. Salinities of 38–39 °/₀₀ are representative of the Mediterranean Sea, while the northern part of the Red Sea has values ranging from 40 to 41 °/₀₀. Coastal bays, sub-

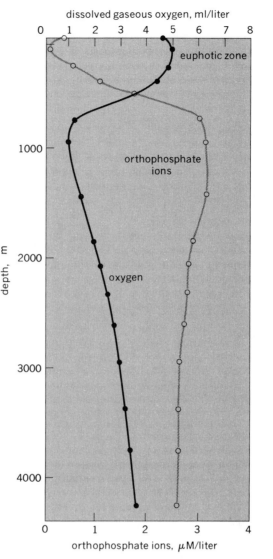

Distribution of dissolved gaseous oxygen (low values) and nutrient species (high concentrations) of orthophosphate ions at 26°22′.4N and 168°57′.5W in the Pacific Ocean. (*Data from Chinook Expedition of 1956 of the Scripps Institution of Oceanography*)

ject to land drainage, and those waters which mix with meltwaters from cold regions, possess salinities which have all degrees of dilution comparable to those of the open ocean. *See* OCEAN-ATMOSPHERE RELATIONS.

Although sea waters exhibit marked regional and depth differences in salt content, the ratios of the major dissolved constituents to one another, listed below, are almost invariable. (Ratios are grams of given element per kilogram of seawater divided by chlorinity in parts per thousand.)

Constituents	Ratio	Constituents	Ratio
Na/Cl	0.5555	S/Cl	0.0466
Mg/Cl	0.0669	Br/Cl	0.0034
Ca/Cl	0.0213	Sr/Cl	0.00040
K/Cl	0.0206	B/Cl	0.00024

Development of the hydrosphere. Many hypotheses on the origin and evolution of the Earth's hydrosphere have been advanced over the past 50 years. Most of them can be placed in one of two categories: (1) the hypothesis of an original ocean, which proposes that the present ocean has had much the same size and composition since the beginning of the geologic record; and (2) the hypothesis of continuous accumulation, which considers that the ocean has been growing continuously, but not necessarily uniformly, since the time of its inception.

Such considerations have certain common assumptions. First, the rock-forming species in sea water, sodium, potassium, silicon, iron, magnesium, and others, have been derived from the weathering of the Earth's surface, whereas the marine quantities of water and the anionic constituents, such as chlorine and sulfur, cannot be adequately supplied by such a mechanism. These latter substances, as gases or dissolved species, have apparently evolved from the Earth's interior by the degradation of the surface rocks. This proposal has been reached through the following arguments. The abundances of the noble gases—neon, nonradiogenic argon, and krypton—are many orders of magnitudes less on the Earth than in the universe relative to other elements. It is thus assumed that they were lost by the Earth during its formative period. Therefore, substances which existed as gases and had comparable molecular weights were similarly not retained.

The hypothesis of the permanency of the oceans through geologic time complements the hypothesis of the permanence of the continents. Since old rocks are found in the basement of the central shield areas, with greater thicknesses of younger rocks at the edges, it has been assumed that the continents have grown laterally. Consequently, the reduction in area occupied by the oceans is compensated for by an increase in average depth. However, the calculated amounts of weathered materials that would form by the action of an initial ocean containing all the various anionic constituents as acids are far greater than those estimated by geologists to have been decomposed over all of geologic time. This hypothesis has been modified by some geochemists to one of an initial ocean of water with the gradual accretion of anionic substances.

The hypothesis of the slow growth of the oceans gains strength with the following observations. If only 1% of the hot-spring water is juvenile, the amounts found today, if extrapolated over geologic time, give a sufficient volume to produce the present oceans. Further, the ratios of the major anionic constituents in sea water are similar to those found in plutonic gases.

Fresh waters and rain. Fresh continental waters show enormous variations in total salt content and in the relative concentrations of their various components, and such parameters in any given water body vary seasonally as well. Chemical analyses of a number of representative waters are given in Table 3.

The factors governing the composition of freshwater bodies are many, and some are poorly understood. Significant amounts of the dissolved phases appear to come from rock weathering, organic-decomposition products, atmospheric dusts, fuel combustion, air-borne particles from the marine environment, and volcanic emanations. Such materials, with the exception of the first, carried from their sources in the atmosphere, can be taken from the air to the river and lakes below by rain. Soil and rock weathering provide the major supply of ions to rivers and lakes. This can be seen from Table 3, where rainwaters have less dissolved solids than river or lake waters by an order of magnitude.

Most fresh waters can be characterized as bicarbonate waters, with HCO_3^- exceeding all other anions, although in some instances chloride or sulfate is the dominant anion.

The sodium, chlorine, and often magnesium are dominantly of marine origin. They leave the oceans as sea spray and are air-borne to the continents. These elements show dramatic decreases in concentration in surface waters going from the coastal to the interior regions. Exceptions can be found in certain waters which derive their salts mainly by the denudation of igneous areas.

Calcium, and sometimes magnesium, can originate from drainage areas, as in the case of the Wisconsin fresh waters which drain over ancient magnesian limestones. Sulfate not only has sources in the marine environment but is also produced by the combustion of sulfur-containing fuels and from the oxidation of sulfur dioxide which results from the atmospheric burning of hydrogen sulfide, a product of the decomposition of organic matter.

The average salt content of fresh waters is of the order of 120 parts per million (ppm). Lower amounts of dissolved solids (50 ppm and less) are found in waters draining igneous rock beds, while open lakes and rivers carrying high salt contents (200 ppm and above) normally result either from the leaching of salt beds or from contamination by man.

Closed basins. The chemical compositions of closed basins, water bodies in which evaporation is the mechanism for the loss of water, are illustrated

Table 3. The chemical compositions of fresh waters and rain (values in parts per million)

Substance	Rivers*	Irish lakes†	English rain‡
Ca^{++}	15	4.0	0.1 – 2.0
Na^+	6.3	8.6	0.2 – 7.5
Mg^{++}	4.1	1.0	0.0 – 0.8
K^+	2.3	0.5	0.05 – 0.7
CO_3^{--}	58.4	8.8	0.0 – 2.7
SO_4^{--}	11.2	5.2	1.1 – 9.6
Cl^-	7.8	14.7	0.2 – 12.6
SiO_2	13.1		
NO_3^-	1		
Fe	0.67		

*Daniel A. Livingston, *Data of Geochemistry*, USGS Prof. Pap. no. 440–G, 1963.

†E. Gorham, The chemical composition of some western Irish fresh waters, *Proc. Roy. Irish Acad.*, vol. 58B, 1957.

‡E. Gorham, On the acidity and salinity of rain, *Geochim. Cosmochim. Acta*, vol. 7, 1955.

Table 4. The composition of waters from chloride, sulfate, and carbonate lakes*

Substance	Dead Sea	Little Manitou Lake	Pelican Lake, Ore.
Na^+	11.14	16.8	29.25
K^+	2.42	1.0	3.58
Mg^{++}	13.62	10.9	2.62
Ca^{++}	4.37	0.48	2.27
CO_3^{--}	Trace	0.47	30.87
SO_4^{--}	0.28	48.4	22.09
Cl^-	66.37	21.8	7.97
SiO_2	Trace	0.009	1.21
Al_2O_3 Fe_2O_3		0.21	0.02
Salinity	226,000	106,851	1983

*Data from G. E. Hutchinson, *A Treatise on Limonology*, vol. 1, Wiley, 1957.

in Table 4, which contains representative examples of the three classical types, the carbonate, sulfate, and chloride waters. These classes appear in sequence during the removal of water from a system with the composition of average river or lake water. The carbonate types exist up until evaporation leads to the precipitation of calcium carbonate and to liquid phases enriched in sulfate and chloride ions. Further removal of water results in the precipitation of calcium carbonate and the subsequent precipitation of gypsum, $CaSO_4 \cdot 2H_2O$. The residual waters hence contain chloride as the dominant anion. *See* SALINE EVAPORITE.

[EDWARD D. GOLDBERG]

Bibliography: M. N. Hill (ed.), *The Sea*, vol. 2, 1963; G. E. Hutchinson, *A Treatise on Limnology*, vol. 1, 1957.

Hygrometer

An instrument for giving a direct indication of the amount of moisture in the air or other gas, the indication usually being in terms of relative humidity as a percentage the moisture present bears to the maximum amount of moisture that could be present at the location temperature without condensation taking place. There are three major types of hygrometers: mechanical, electrical, and cold-spot or dew-point. *See* DEW POINT; HUMIDITY.

In a simple mechanical type of hygrometer the sensing element is usually an organic material which expands and contracts with changes in the moisture in the surrounding air or gas. The material used most is human hair. Other materials may be paper, animal tissues such as goldbeater's skin, and wood. As illustrated in Fig. 1, the bundle of hair is held under a slight tension by a spring, and a magnifying linkage actuates a pointer. The usually designed operating range of this type of hygrometer is from about 40 to 100°F. For very short periods these hygrometers may be used down to 32°F and up to about 140°F. When used within the range of usual room conditions, these instruments show very little drifting of the indications over periods of a year or more, but if exposed for any length of time to extremes of humidity or temperature, a permanent change in the sensitive material may result. The time required to respond fully to a sudden wide change of humidity may be as much as 5 min.

In an electrical hygrometer the change in the electrical resistance of a hygroscopic substance is measured and converted to percent relative humidity. In one procedure a pair of fine gold or platinum wires are wound as a helix on a glass or polystyrene cylinder and the spaces between the wires bridged with a film or a hygroscopic salt, such as lithium chloride (Fig. 2). At a constant temperature the logarithm of the electrical resistance between the wires varies almost linearly with the logarithm of the relative humidity. The resistance may be measured with a Wheatstone bridge or a series milliammeter and converted to relative humidity. In order to avoid polarization of the salt film, it is best to use an alternating current or a direct current through a reversing commutator. Also, since a single element is temperature-sensitive and applicable within a limited temperature range, it is usual practice to have a number of elements to cover the humidity range from 10 to 90%.

In a third group of hygrometers, commonly called dew-point apparatus, the dew-point temperature is determined; this is the temperature at which the moisture in the gas is at the point of sat-

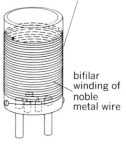

HYGROMETER

plastic form coated with hygroscopic material

bifilar winding of noble metal wire

Fig. 2. Electrical hygrometer. (*From D. M. Considine, ed., Process Instruments and Controls Handbook, McGraw-Hill, 2d ed., 1974*)

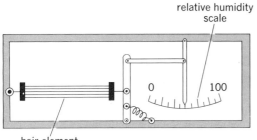

relative humidity scale

0   100

hair element

Fig. 1. Hygrometer which uses hair as the sensing element. (*From D. M. Considine, ed., Process Instruments and Controls Handbook, McGraw-Hill, 2d ed., 1974*)

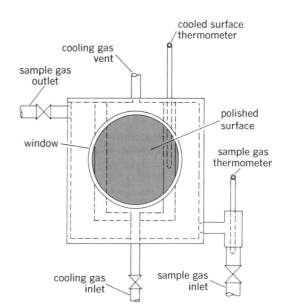

cooled surface thermometer

cooling gas vent

sample gas outlet

polished surface

window

sample gas thermometer

cooling gas inlet

sample gas inlet

Fig. 3. Dew-point type of hygrometer.

uration, or 100% relative humidity. The usual procedure is to chill a polished surface until dew or a film of moisture just starts to appear and to measure the temperature of the surface. In Fig. 3 the polished surface is the outer surface of an expansion chamber through which a gas such as ether, carbon dioxide, or ethane is caused to expand slowly, thereby cooling the chamber. This chamber is enclosed in an outer casing into and out of which the gas under test is circulated slowly. A window in the outer casing permits viewing the polished surface, and a thermometer in the wall of the expansion chamber gives the temperature at which dew first appears, or starts to disappear as cooling is stopped. With this temperature, together with the temperature of the incoming sample gas and a table of dew-point temperatures, relative humidity of the incoming gas may be determined.

With each of the three types of hygrometers described there are, of course, a number of commercial instruments available, including both direct reading and recording.   [HOWARD S. BEAN]

Bibliography: M. R. Diskson, *A Survey of Hygrometry Methods*, 1963; H. Spencer-Gregory and E. Rouke, *Hygrometry*, 1957; C. E. Terrell (ed.), *AGA Gas Measurement Manual*, 1963; A. Wexler and W. G. Brombacher, *Methods of Measuring Humidity and Testing Hygrometers*, U.S. Department of Commerce, Nat. Bur. Stand. Circ. no. 512, 1951.

Ice island

One of the massive bodies of floating ice in the Arctic Ocean, from 20 to 200 ft thick, irregular in shape, and from a few square miles to 300 mi² in area. Ice islands originate in the landfast ice along the high-latitude northern shores of the Canadian archipelago and Greenland. This distinguishes them from the smaller and more rugged icebergs which originate in the glaciers of Greenland. The unbroken appearance of the ice islands contrasts greatly with the surrounding pack ice, which normally almost completely covers the Arctic Ocean with a maximum thickness of 20 ft. It was the unbroken appearance and the elevation of ice islands above the pack ice which first attracted the attention of U.S. Air Force weather reconnaissance planes in 1946 (Fig. 1). Since that time about 100 ice islands have been observed, mostly in the numerous bays and straits of the Canadian archipelago.

Character and formation. Ice islands are tabular features without the pressure ridges found on pack ice. Long shallow drainage channels are most evident in summer when they are filled with meltwater. The presence of rock piles on the surface and of dust and dirt layers within the ice, as well as plant and animal remains, testify that the ice islands were close to land at one time.

The principal source of ice islands is the Ward Hunt Ice Shelf, a floating body of landfast ice about 60 by 10 mi in size located between McClintock and Markham bays on northern Ellesmere Island. Extensively studied, it has been found to be similar in all respects to the ice islands. The Ward Hunt Ice Shelf has decreased considerably in area since it was first visited by G. S. Nares in

1875 and R. E. Peary in 1906. A dramatic formation of ice islands occurred during the winter of 1961–1962 when the entire northern part (some 200 mi²) of the Ward Hunt Ice Shelf broke off to form five large ice islands. The largest of these, WH-5 (about 11 by 5 mi), drifted eastward, eventually turning southward and entering Robeson Channel. It continued southward through Baffin Bay and eventually disintegrated off Labrador and Newfoundland in 1964. The other four islands drifted westward and southward, skirting the northern edge of the Canadian islands.

The ice islands probably break from the shelf during years of generally warm Arctic climate. The growth of this landfast shelf ice to thicknesses of 100 ft or more takes centuries and depends on general climatic conditions. *See* ICEBERG; SEA ICE.

Scientific endeavor. In March, 1952, a landing was made on one of these islands, designated T-3 (T for radar target), also called Fletcher's Ice Island for Col. J. O. Fletcher, who was in charge of the first operations there. It was occupied from March, 1952, to May, 1954; April, 1955, to September, 1955; and from April, 1957, to September, 1961, when it ran aground on the continental shelf 80 mi northwest of Point Barrow, Alaska. It was reoccupied in February, 1962, after floating free again and has been continually occupied since then (as of 1969). In 1952 a landing was made on T-1, the first island discovered and the largest of the known ice islands, but no permanent facilities were established. The Soviet Union located one of its North Pole stations, NP-6, on an ice island from 1956 until 1959, without encountering fracturing of the ice or danger to the camp, thus showing the usefulness of ice islands as research platforms; eight other Soviet stations on pack ice were reestablished a total of 57 times. After 1961 the operation of United States drifting ice stations in the Arctic Ocean shifted from the Air Force to the Navy. In 1961 ARLIS II (Arctic Research Laboratory Ice Station II) was established by the Naval Arctic Research Laboratory of Barrow, Alaska, on an ice island that measured 3.25 by 2.0 mi and 20–80 ft in thickness.

The ice islands are excellent platforms for Arc-

ICE ISLAND

Fig. 1. Ice Island T-3. (*Official photograph, USAF Cambridge Research Center, Mass.*)

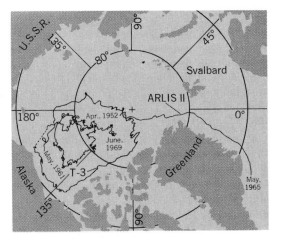

Fig. 2. Drift of T-3 and ARLIS II during occupancy.

tic scientific studies. They drift under the influence of winds and currents in an erratic course through the Arctic Ocean (Fig. 2). There are two patterns of drift: a clockwise drift around the western Arctic, for example, T-3, or a direct drift across the basin and out into the Greenland Sea, for example, ARLIS II. T-3 has completed two revolutions around the western Arctic Basin since 1952, averaging about 9 years for one revolution. ARLIS II drifted from the area north of Alaska across the Arctic Ocean and down the coast of Greenland. It was evacuated by icebreaker in May, 1965, in Denmark Strait between Iceland and Greenland, and its breakup was later observed in the waters off southern Greenland.

Camps on ice islands have been extensively used as scientific research stations. T-3, ARLIS II, and NP-6 have been bases for investigations in the meteorological, oceanographic, geophysical, and upper atmospheric phenomena of the north polar regions. T-3 was the only ice island being used as a research base in 1969. The research program conducted from T-3 in the Arctic Ocean is similar to that on an oceanographic research vessel in other oceans: The ice island position is determined by celestial methods and with the U.S. Navy Satellite Navigation System; echo soundings of ocean depths are recorded continuously; the Earth's magnetic and gravity fields are recorded regularly; surface and upper-air meteorological observations are taken several times per day; the ocean is sampled with nets and water bottles lowered from a winch on a cable through a hole cut in the pack ice near the island; and sediment cores and photographs of the ocean floor are obtained regularly. *See* ARCTIC OCEAN.

[KENNETH L. HUNKINS]

Bibliography: H. Landsberg (ed.), *Advances in Geophysics*, vol. 3, 1956; J. Sater (ed.), *Arctic Drifting Stations: A Report on Activities Supported by the Office of Naval Research*, 1968; B. Staib, *On Skis Toward the North Pole*, 1965; T. Weeks and R. Mather, *Ice Island: Polar Science and the Arctic Research Laboratory*, 1965.

Iceberg

A large mass of glacial ice broken off and drifted from parent glaciers or ice shelves along polar seas. Icebergs should be distinguished from polar pack ice which is sea ice, or frozen sea water, though rafted or hummocked fragments of the latter may resemble small bergs. *See* GLACIOLOGY; SEA ICE.

Characteristics and types. The continental or island icecaps of both Arctic and Antarctic regions produce icebergs where the icecaps extend to the sea in the form of glaciers or ice shelves. The "calving" of a large iceberg is one of nature's greatest spectacles, considering that a Greenland berg may weigh over 1,000,000 tons and that Antarctic bergs are many times larger. An iceberg consists of glacial ice which is compressed snow having a variable specific gravity that averages about 0.89. This results in an above-water mass of from one-eighth to one-seventh of the entire mass. However, spires and peaks of an eroded or weathered berg will result in height to depth ratios

of between 1–6 and 1–3. Tritium age experiments with melted Greenland berg ice indicate these bergs may be of the order of 50,000 years old. Minute air bubbles imprisoned in glacial ice impart to bergs a snow-white color and cause it to effervesce when immersed. *See* SEA WATER.

Icebergs are classified by shape and size. The terms used are arched, blocky, dome, pinnacled, tabular, valley, and weathered for berg discription, and bergy-bit and growler for berg fragments ranging smaller than cottage size above water. The lifespan of an iceberg may be indefinite while the berg remains in cold polar waters, eroding only slightly during summer months. But under the influence of ocean currents, an iceberg that drifts into warmer water will disintegrate rapidly, its life being measured in weeks in sea temperatures between 40–50°F and in days in sea temperatures over 50°F. A notable feature of icebergs is their long and distant drift which may carry them into steamship tracks, where they become hazards to navigation. The normal extent of iceberg drift is shown in Fig. 1.

Arctic icebergs. In the Arctic, icebergs originate chiefly from glaciers along Greenland coasts. It is estimated that a total of about 16,000 bergs are calved annually in the Northern Hemisphere, of which over 90% are of Greenland origin; but only about half of these have a size or source location to enable them to achieve any significant drift. The majority of the latter stem from some 20 glaciers along the west coast of Greenland between the 65th and 80th parallels of latitude. The most productive glacier is the Jacobshavn Glacier at latitude 68°N, calving about 1400 bergs yearly, and the largest is the Humboldt Glacier at latitude 79° with a seaward front extending 65 mi. The remainder of the Arctic berg crop comes from East Greenland and the island icecaps of Ellesmere Island, Iceland, Spitzbergen, and Novaya Zemlya, with almost no sizable bergs produced along the Eurasian or Alaskan Arctic coasts. No icebergs are discharged or drift into the North Pacific Ocean or its adjacent seas, except a few small bergs each year that calve from the piedmont glaciers along the Gulf of Alaska. These achieve no significant drift. *See* ARCTIC OCEAN; ATLANTIC OCEAN.

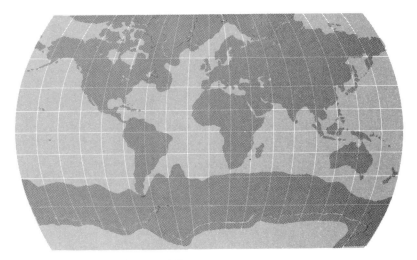

Fig. 1. Normal extent of iceberg drift.

Fig. 2. Arctic iceberg, eroded to form a valley or dry-dock type; grotesque shapes are common to the glacially produced icebergs of the North. Note the brash and small floes of sea ice surrounding the berg.

Ocean currents of the Arctic and adjacent seas determine the drift and ultimate distribution of icebergs, wind having little effect except on small, sail-shaped fragments. The dominant drift along the East Greenland coast is southward around the tip of Greenland and then northward along the west coast. Here the drifting bergs join the main body of West Greenland bergs and drift in a counterclockwise gyral across Davis Strait and Baffin Bay. The bergs are then swept southward along the coasts of Baffin Island, Labrador, and Newfoundland by the Labrador Current. This drift terminates along the Grand Banks of Newfoundland, where the waters of the Labrador Current mix with the warm Gulf Stream and even the largest of bergs melt within 2–3 weeks. Freak iceberg drifts have been reported where bergs or remaining fragments were sighted off Scotland, Nova Scotia, Bermuda, and even the Azores Islands. Such reports, however, are extremely rare. About 400 bergs each year are carried past Newfoundland as survivors of the estimated 3-year journey from West Greenland. The remainder become stranded along Arctic coasts and shoals and are ultimately destroyed by wave erosion and summer melting.

Icebergs in the Northern Hemisphere rarely reach proportions larger than 2000 ft in breadth or 400 ft in height above the water (Fig. 2). However, true glacial ice islands several miles in extent are occasionally found and have even served as

Fig. 3. Antarctic iceberg, tabular type. Such bergs develop from great ice shelves along Antarctica and may reach over 100 mi in length. The U.S. Coast Guard icebreaker *Westwind* is in the foreground.

floating bases for scientific studies. The origin of these rare counterparts of the common Antarctic type is uncertain but is thought to be an ice shelf along northern Ellesmere Island. *See* ICE ISLAND.

Antarctic icebergs. In the Southern Ocean, bergs originate from the giant ice shelves all along the Antarctic continent. These result in huge, tabular bergs (Fig. 3) or ice islands several hundred feet high and often over a hundred miles in length, which frequent the entire waters of the Antarctic seas. The most active iceberg-producing regions are the Ross and Filchener ice shelves in the Ross and Weddell seas. The large size of these bergs and influence of the Antarctic Circumpolar Current give them an indeterminant life-span. When weathered, Antarctic icebergs attain a deep bluish hue of great beauty rarely seen in the Arctic. *See* ANTARCTIC OCEAN. [ROBERTSON P. DINSMORE]

Indian Ocean

The smallest and geologically the most youthful of the three oceans. It differs from the Pacific and Atlantic oceans in two important aspects. First, it is landlocked in the north, does not extend into the cold climatic regions of the Northern Hemisphere, and consequently is asymmetrical with regard to its circulation. Second, the wind systems over its equatorial and northern portions change twice each year, causing an almost complete reversal of its circulation. During the International Indian Ocean Expedition from 1960 to 1965, the ocean was explored systematically by more than 50 ships of 18 nations.

Size and bathymetry. The eastern and western boundaries of the Indian Ocean are 147 and 20°E, respectively. In the southeastern Asian waters the boundary is usually placed across Torres Strait, and then from New Guinea along the Lesser Sunda Islands, across Sunda Strait and Singapore Strait. The surface area within these boundaries is 75,900,000 km², or about 21% of the surface of all oceans. The volume of the Indian Ocean is 293,000,000 km³, and the average depth is about 3850 m.

The ocean floor is divided into a number of basins by a system of ridges (Fig. 1). The largest is the Mid-Ocean Ridge, the greater part of which has a rather deep rift valley along its center. It lies like an inverted Y in the central portions of the ocean and ends in the Gulf of Aden. More recently discovered is the Ninety East Ridge. Most of the ocean basins separated by the ridges reach depths in excess of 5000 m. The Sunda Trench, stretching along Java and Sumatra, is the only deep-sea trench in the Indian Ocean and contains its deepest observed depth of 7455 m. The Andaman Basin, with a maximum depth of 4360 m, is separated from the open ocean by a 1400-m sill. The Red Sea has a maximum depth of 2835 m, but its entrance at the Strait of Bab-el-Mandeb is only 125 m deep. East of the Mid-Ocean Ridge, deep-sea sediments are chiefly red clay; in the western half of the ocean, globigerina ooze prevails and, near the Antarctic continent, diatom ooze. *See* MARINE GEOLOGY; MARINE SEDIMENTS.

Wind systems. Atmospheric circulation over the northern and equatorial Indian Ocean is character-

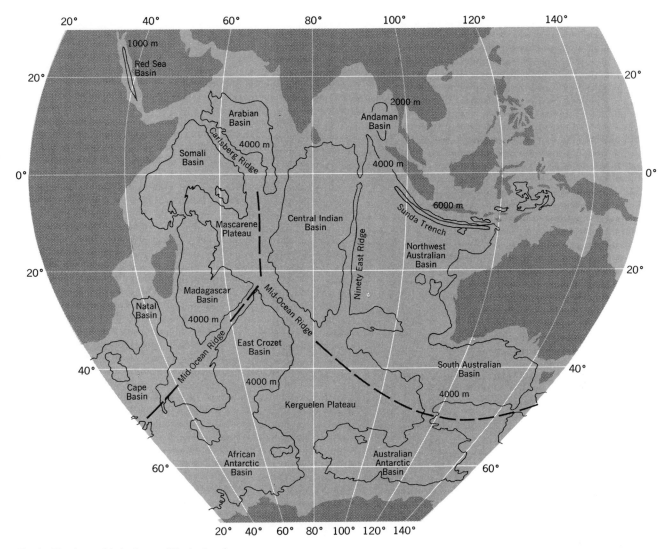

Fig. 1. Physiographic features of the Indian Ocean.

ized by the changing monsoons. In the southern Indian Ocean atmospheric circulation undergoes only a slight meridional shift during the year (Fig. 2). During winter in the Northern Hemisphere, from December through February, the equatorial low-pressure trough is situated at about 10°S and continues into a low-pressure system over northern Australia. A strong high-pressure system lies over Asia. This situation causes the Northeast Monsoon to blow everywhere north of the Equator. The winds cross the Equator from north to south and then usually become northwest winds before reaching the Intertropical Convergence. The subtropical high-pressure ridge of the Southern Hemisphere is situated near 35°S. North of it, the Southeast Trades blow. South of 40°S, winds from the west prevail and are associated with cyclones traveling around Antarctica.

During summer in the Northern Hemisphere, from June through August, a low-pressure system is developed over Asia with the center around Iran. The subtropical high-pressure ridge of the Southern Hemisphere has shifted slightly northward and continues into a high-pressure system over Australia. The Southeast Trades are more strongly developed during this season, cross the Equator from south to north, and become the Southwest Monsoon, bringing rainfall and the wet season to India and Burma. Atmospheric circulation during June through August is much stronger and more consistent than during February. In the Southern Hemisphere the West Wind Belt has shifted about 5° to the north, with westerly winds starting from 30°S and storms becoming stronger and more frequent during winter in that hemisphere.

Circulation. The surface circulation is caused largely by winds and changes in response to the wind systems (Fig. 3). In addition, strong boundary currents are formed, especially along the western coastline, as an effect of the Earth's rotation and of the boundaries created by the landmasses. In the Southern Hemisphere south of 35°S, a general drift from west to east is found as a result of the prevailing west winds. Near 50°S, where west winds are strongest, the Antarctic Circumpolar Current is embedded in the general West Wind Drift. In the subtropical southern Indian Ocean, circulation is anticyclonic. It consists of the South Equatorial Current flowing west between 10 and 20°S, a flow to the south along the coast of Africa and Madagascar, parts of the West Wind Drift, and flow to the north in the eastern portions of the ocean, especially along the coast of Australia. The flow to the south between Madagascar and Africa is called the Mozambique Current. It continues along the coast of South Africa as the Agulhas Current, most of which turns east into the West Wind Drift. The circulation in the Southern Hemisphere south of 10°S changes only slightly with the seasons.

North of 10°S the changing monsoons cause a complete reversal of surface circulation twice a year. In February, during the Northeast Monsoon, flow north of the Equator is mostly to the west and the North Equatorial Current is well developed. Its water turns south along the coast of Somaliland and returns to the east as the Equatorial Countercurrent between about 2 and 10°S. In August, during the Southwest Monsoon, the South Equatorial Current extends to the north of 10°S; most of its water turns north along the coast of Somaliland, forming the strong Somali Current. North of the Equator flow is from west to east and is called the Monsoon Current. Parts of this current turn south along the coast of Sumatra and return to the South Equatorial Current.

Transports of many of these current systems have been determined. The Antarctic Circumpolar Current and the West Wind Drift carry between

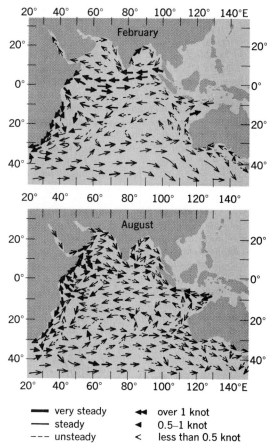

very steady   ◄◄ over 1 knot
steady   ◄   0.5–1 knot
unsteady   <   less than 0.5 knot

Fig. 3. Surface currents of the Indian Ocean.

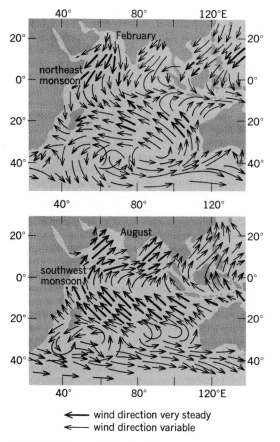

⟵ wind direction very steady
⟵ wind direction variable

Fig. 2. Winds over the Indian Ocean.

100×10^6 and 140×10^6 m³/sec to the east. The South Equatorial Current transport ranges between 20×10^6 and 30×10^6 m³/sec, and that of the Agulhas Current is about 40×10^6 m³/sec. During the full development of the Southwest Monsoon, the Somali Current and the Monsoon Current transport between 30×10^6 and 40×10^6 m³/sec. Although the Antarctic Circumpolar Current reaches to great depths, probably to the bottom, most of the other currents are much shallower. The Agulhas Current reaches to approximately 1200 m, the Somali Current to 800 m, and the remainder of the circulation is limited to the upper 300 m of the ocean. Below these depths, movements are sluggish and irregular, and the spreading of water properties is a slow process accomplished chiefly by mixing. At the Equator, during the time of the Northeast Monsoon, an Equatorial Undercurrent is found flowing as a subsurface current near the 150-m depth from west to east. This current is weaker than the corresponding currents in the Pacific and Atlantic oceans. *See* OCEAN CURRENTS.

Surface temperature. The pattern of sea-surface temperatures changes considerably with the seasons (Fig. 4). During February, when the Intertropical Convergence is near 10°S, the heat equator is also in the Southern Hemisphere and most of the area between the Equator and 20°S has temperatures near 28°C. The water in the northern

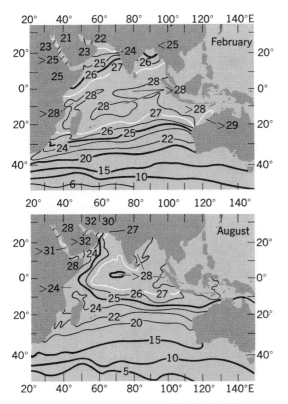

Fig. 4.   Sea-surface temperature (°C) of Indian Ocean.

parts of the Bay of Bengal and of the Arabian Sea is much cooler, and temperatures below 20°C can be found in the northern portions of the Persian Gulf and the Red Sea. In the Southern Hemisphere temperatures decrease gradually from the tropics to the polar regions. Surface circulation affects the distribution of temperature, and warm water spreads south along the coast of Africa and cool water north off the west coast of Australia, causing the isotherms to be inclined from west to east.

During August high temperatures are found in the Northern Hemisphere and in the equatorial region. The Somali Current advects cool water along the coast of Africa to the north. In the Southern Hemisphere isotherms are almost 10° of latitude farther north.

Surface salinity. The distribution of surface salinity is controlled by the difference between evaporation and precipitation and by runoff from the continents (Fig. 5).

High surface salinities, which are greater than above 36 parts per thousand (°/oo), are found in the subtropical belt of the Southern Hemisphere, where evaporation exceeds rainfall. In contrast, the Antarctic waters are of low salinity because of heavy rainfall and melting ice. Another area of low salinities stretches from the Indonesian waters along 10°S to Madagascar. It is caused by the heavy rainfall in the tropics. The Bay of Bengal has very low salinities, often less than 30°/oo, as a result of the runoff from large rivers. In contrast, because of high evaporation the Arabian Sea has salinities as high as 36.5°/oo. High salinities are also found in the Persian Gulf (>38°/oo) and in the Red Sea

(41°/oo), representing the arid character of the landmasses surrounding them. The salinity distribution changes relatively little during the year; however, south of India from the Bay of Bengal to the west, a flow of low-salinity water, caused by the North Equatorial Current, can be noticed during February.

Surface water masses. The different climatic conditions over various parts of the Indian Ocean cause the formation of characteristic surface water masses. The Arabian Sea water is of high salinity, has a moderate seasonal temperature variation, and can be classified as subtropical. The water in the Bay of Bengal is of low salinity and always warm, representing tropical surface water. Another type of tropical surface water stretches from the Indonesian waters to the west and is called the Equatorial Surface Water. Subtropical Surface Water with salinities in excess of 35.5°/oo and a seasonal temperature variation of between 15 and 25°C is found in the subtropical regions of the Southern Hemisphere. Its southern boundary is the Subtropical Convergence coinciding with temperatures of about 15°C. From there, temperature and salinity decrease in the area of the transition water to about 4°C and 34°/oo at the Antarctic Polar Front. South of the Antarctic Polar Front, Antarctic Surface Water of low salinity (<34°/oo) is found; its temperature is near the freezing point (−1.9°C) in winter and is approximately 2°C in summer.

Subsurface water masses. The vertical distributions of temperature, salinity, oxygen content,

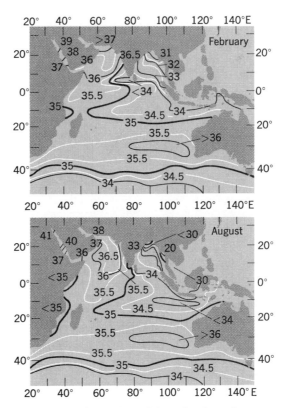

Fig. 5.   Sea-surface salinity of the Indian Ocean in parts per thousand by weight (°/oo).

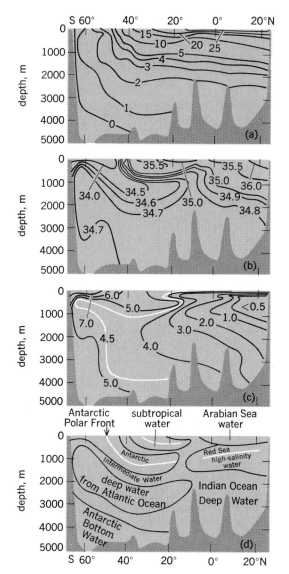

Fig. 6. The Indian Ocean from the Arabian Sea to Antarctica showing (a) temperature (°C), (b) salinity (°/oo), (c) oxygen content (ml/liter), and (d) water masses.

and main water masses in the Indian Ocean are shown in Fig. 6, along a section from the Arabian Sea to Antarctica. Warm water of more than 15°C occupies only a very thin surface layer of less than 300 m depth. This layer contains the tropical and subtropical water masses. Subtropical water of the Southern Hemisphere spreads as a subsurface salinity maximum toward the Equator at depths between 100 and 200 m.

Water of high salinity formed in the Red Sea and in the Persian Gulf leaves these basins and spreads as a subsurface layer of high salinity throughout the Arabian Sea at depths between 600 and 1000 m. To the south it can be traced as far as Madagascar, and to the east as far as Sumatra. Water of low salinity and a temperature of approximately 4°C sinks near the Antarctic Polar Front and spreads north as a salinity minimum between 600 and 1500 m. It is called the Antarctic Intermediate Water.

The deep and bottom water of the Indian Ocean is of external origin. South of Africa, water of rather high salinity, originating south of Greenland in the North Atlantic Ocean, enters the Indian Ocean, filling the layers between 1500- and 3000-m depths. The water below the 3000-m depth is Antarctic Bottom Water originating in the Weddell Sea. Its temperature is lower than 0°C.

Because of their origin at the sea surface, both the Antarctic Intermediate Water and the Antarctic Bottom Water are of rather high oxygen content (>5 ml/liter). Since their residence time is long, the water masses of the northern Indian Ocean below the surface layer have a very low oxygen content. An oxygen minimum is associated with the Red Sea water and extends south between the Antarctic Intermediate Water and the North Atlantic Deep Water. *See* ANTARCTIC OCEAN; RED SEA; SEA WATER.

Upwelling. Several areas of upwelling are found along the shores of the Indian Ocean, but all of them are seasonal in character, in contrast to upwelling areas in the other oceans. They are especially developed during the Southwest Monsoon season from May through September, when upwelling takes place in the Banda Sea, south of Java, along the coast of Somaliland, and off Arabia. The strongest upwelling is found along the Somaliland coast, where surface temperatures during this season may be as low as 14°C. Water of high nutrient content ascends at a rate of approximately 1 m/day from as deep as 150 m and is integrated into the surface flow. These areas have a high biological production. During the Northeast Monsoon, some upwelling takes place in the Andaman Sea and off the west coast of India. *See* UPWELLING.

Tides. Both semidiurnal and diurnal tides occur in the Indian Ocean. The semidiurnal tides rotate around three amphidromic points situated in the Arabian Sea, southeast of Madagascar, and west of Perth. The diurnal tide also has three amphidromic points: south of India, in the Mozambique Channel, and between Africa and Antarctica. It has more the character of a standing wave, oscillating between the central portions of the Indian Ocean, the Arabian Sea, and the waters between Australia and Antarctica. *See* TIDE.

Along the coast of Africa, in the Bay of Bengal, and along the northwest coast of Australia semidiurnal tides prevail, with two high waters each day. Mixed tides are found around the Arabian Sea and along the coasts of Sumatra and Java. Prevailing diurnal tides with only one high water each day occur only along southwest Australia. The ranges of spring tides are high in the Arabian Sea, 2.5 m at Aden and 5.7 m at Bombay. The Bay of Bengal has high tidal ranges, chiefly along the coast of Burma, with 7 m at Rangoon. Moderately high tides between 1 and 2 m are found along the coast of Sumatra and Java, but spring tides along the northwest coast of Australia are between 6 and 10 m. Along the coast of South Africa, tidal ranges are between 1.5 and 2 m, but in the Mozambique Channel they increase to 3–5 m. The islands in the central Indian Ocean usually have spring tidal ranges between 1 and 1.5 m.   [KLAUS WYRTKI]

Bibliography: J. G. Colborn, *The Thermal Structure of the Indian Ocean*, 1975; G. Dietrich, *Gen-*

eral Oceanography, 1963; International Indian Ocean Expedition, *Collected Reprints*, vols. 1–4, UNESCO, 1965–1967; *Monatskarten fur den Indischen Ozean*, Ger. Hydrogr. Inst. Publ. no. 2242, 1960; G. Schott, *Geographie des Indischen und Stillen Ozeans*, 1935; H. U. Sverdrup, M. W. Johnson, and R. H. Fleming, *The Oceans: Their Physics, Chemistry, and General Biology*, 1946; G. B. Udintsev (ed.), *Geological-Geophysical Atlas of the Indian Ocean*, 1976.

Industrial meteorology

The commercial application of weather information to the operational problems of business, industry, transportation, and agriculture in a manner intended to optimize the operation with respect to the weather factor. The weather information may consist of past weather records, contemporary weather data, predictions of anticipated weather conditions, or an understanding of physical processes which occur in the atmosphere. The operational problems are basically decisions in which weather exerts an influence.

The evolution of applied meteorology from primarily a governmental function to today's mixed economic approach has taken place largely since World War II. In that period there has been a gradual development in the application of meteorological information. Some of the applications are: professional meteorologists serving television or radio stations to explain to the public the changes in weather which either have occurred or are about to occur; forecasters working with airline dispatchers to utilize optimum flight paths and to avoid equipment tie-ups during lengthy adverse weather conditions; consultants who help utilities locate future power plant sites with minimal environmental impact; forecasters who advise shipping lines regarding ocean routing paths which avoid storms and decrease by a few hours the travel time for long-distance ocean freight movement; and consultant advisories which help industrial firms in the marketing of weather-sensitive products. In many cases the need for meteorological information is sufficiently important that a complete department within the operational segment of a company is established to serve the needs of that weather-affected firm.

One specialized consulting service involves snowstorm forecasting to aid in keeping city streets, highways, and especially toll roads usable. The city administrator seeks to match the use of worker-power and equipment to the size and duration of the storm. In addition to forecasting the beginning time of the snowstorm, the consultant firm typically makes additional contacts to keep the client advised of the storm's intensity, rate of movement, and total duration. The costs of such weather advisory services are nominal compared with costs of overreacting when only light snow begins, and with street or highway blockage when a little lead time could have been used to enlist supplemental equipment or worker-power.

Contract research work by professional meteorologists meets a need for intermittent special requirements of governmental agencies. Studies of this type might require the determination of atmospheric airflow frequencies related to the dispersion of air pollution throughout a broad area as it might relate to governmental regulatory policies. A distinct advantage exists when a government agency can meet a nonrecurring need with contract consulting effort.

Along with the advances of communication capacity, use of weather satellite photography, and the expanded international marketing of goods and services, consulting meteorologists within the United States have found a demand for their services among commodity-trading firms. Knowledge of recent past weather or current weather throughout the wheat-growing areas of the world is useful information to wheat users, wheat growers, and speculators in commodity futures who help establish a market value for wheat well in advance of the actual harvest.

Climatological records can be used effectively in developing statistical odds for extreme weather conditions. Climatological records also can be used for postanalyses of sales records of weather-sensitive items. For example, the sales of room air conditioners move up sharply in the months of May and June when early-season extreme hot spells occur throughout the northern areas of the United States. A series of several such hot spells will guarantee a peak sales season for all manufacturers. By contrast, a cold May and June in the same marketing areas will guarantee a carry-over of large numbers of such equipment until the next peak marketing season. Postanalyses of actual daily sequences of weather in a given season can help both the manufacturer and the marketing firms in their planning for subsequent years. An extremely cold season is not likely to be followed a year later with similar low sales.

Meteorological consulting firms range from individual consultants who operate similarly to individual consulting engineers, to large firms having nearly 100 professional meteorologists on their staff. In early 1979 the Professional Directory in the *Bulletin of the American Meteorological Society* carried a listing of 84 different firms. A high fraction of these firms serve multiple clients as compared with full-time staff members within individual companies. Approximately 10% of the 8000 professional meteorologists in the United States are engaged in some facet of industrial meteorology.

Some measure of the interest in applications of meteorology is indicated by the number of industrial corporate members within the American Meteorological Society. The nearly 100 corporate members include industrial firms, equipment suppliers, utilities, and insurance companies.

Special applications of meteorological knowledge are required in the field of weather modification. Fog dispersal at some commercial airports and field projects to increase winter snowpack on the western slopes of mountain barriers throughout the United States are examples of such use.

Forensic meteorology, the application of atmospheric information to legal cases, requires careful attention to postanalyses of factual information. Since accidents seldom happen in the immediate vicinity of weather observing stations, professional opinions are useful in determining the sequence of weather-affected events that are subject to litigation. Expert-witness testimony is the full-time emphasis of some meteorological consultants and a part-time effort by many consultants.

The National Weather Service recognizes the importance of the role of professional meteorologists serving industry. The Special Assistant for Industrial Meteorology in the National Oceanic and Atmospheric Administration is charged with coordinating the efforts of the governmental data collection service and the many users of such information, whether on a current basis or from the archives of weather records at the National Climatic Center. National Weather Service personnel are encouraged to recognize the needs of private professional practitioners. Industrial meteorologists represent an indispensable part in the government's effort to bring tailored meteorological service to those segments of commerce and industry affected by weather.

The development of weather sensing equipment fitted to the needs of industrial firms, particularly as related to air quality, has led to expanded use of professional personnel who install and maintain such measuring equipment. Weather equipment sales to private industrial firms now exceed sales to governmental agencies.

In 1968 a nonprofit organization, the National Council of Industrial Meteorologists (NCIM), was formed to further the development and expansion of industrial meteorology. In a recent 3-year period the membership of the NCIM served the following: 210 industrial firms, 52 Federal and state agencies, 296 local government agencies, and 14 university subcontract projects.

Annual conferences on industrial meteorology are sponsored by the Committee on Industrial Meteorology of the American Meteorological Society. That society has established a program for the certification of consulting meteorologists who meet rigorous standards of knowledge, experience, and adherence to ethical practice. *See* AERONAUTICAL METEOROLOGY; AGRICULTURAL METEOROLOGY; WEATHER FORECASTING AND PREDICTION.

[LOREN W. CROW]

AMS Bull., vol 59, no. 8, August 1978; W. J. Maunder, *The Value of Weather*, 1971; Papers presented at Session 4 of the 56th Annual Meeting of the American Meteorological Society, Philadelphia, Jan. 20, 1976, *AMS Bull.*, 57(11):1318–1342, November 1976; J. A. Taylor (ed.), *Weather Forecasting for Agriculture and Industry: A Symposium*, 1972; J. C. Thompson and G. W. Brier, The economic utility of weather forecasts, *Mon. Weather Rev.*, 83(11):249–254, 1955.

Insolation

The amount of solar radiation which reaches a unit horizontal area of the Earth. It is the latitudinal variation of insolation which supplies energy for the general circulation of the atmosphere. Insolation outside the Earth's atmosphere depends on the angle of incidence of the solar beam and on the solar constant. The solar constant is the amount of energy which, in unit time, reaches a unit plane surface perpendicular to the Sun's rays outside the Earth's atmosphere when the Earth is at its mean distance from the Sun. Insolation is measured in langleys (ly) or calories per centimeter squared

(ly = cal/cm^2). Outside the Earth's atmosphere a marked gradient exists, from ~900 ly/day near the Equator to zero in the dark polar areas; but in summer the insolation is quite uniform—over 1000 ly/day near the poles compared with about 800 ly/day near the Equator.

The extraterrestrial solar energy is modified by the air, the clouds, and the land and water surface of the Earth. Energy of wavelengths shorter than about 2900 A does not reach the surface but is absorbed high in the atmosphere mainly by nitrogen, oxygen, and ozone.

Air molecules also scatter energy in accordance with Rayleigh's law, and absorption by ozone, water vapor, and other gases further reduces the solar energy transmitted.

Particles always present in the lower atmosphere scatter and absorb energy, too. However, clouds affect the extraterrestrial insolation more than any other atmospheric factor. The reflectivity (or albedo) of clouds varies from less than 10 to over 90% of the insolation on them. The albedo will be higher in visible light than in total solar energy, for clouds absorb mainly in the near infrared. The cloud albedo depends on the drop sizes, liquid water content, water vapor content, and thickness of the cloud; it also depends on the Sun's zenith distance Z. The smaller the drops and the greater the liquid water content, other things being the same, the greater the cloud albedo.

Clouds also absorb solar energy. Consider a cloud with a given liquid water content; if the drops are large and therefore relatively few in number, solar energy can deeply penetrate the cloud. If the cloud is also warm (high water vapor content) and occurs in deep vertical layers, it will absorb more than other cloud types.

The solar energy measured underneath clouds also depends on the reflectivity of the surface. An overcast which transmits 0.4 of the energy incident on it when the cloud is over a forest will transmit 0.7 of the incident energy when it is located over a highly reflecting snow surface. This is caused by the multiple reflections between the snow and the cloud, but the forest still absorbs more energy than the snow.

All these atmospheric effects will modify the distribution of the solar energy which reaches the surface. For example, H. G. Houghton computed the annual values given in the table, which shows that the latitudinal gradients of insolation are appreciably smaller at the Earth's surface than at the outside of the atmosphere. The high value of surface insolation near latitude 30°N is due in part to the small cloudiness in the large high-pressure areas at those latitudes. Additional longitudinal variations in insolation depend on cloudiness. Thus near latitude 40°N in summer, the average insolation exceeds 700 ly/day in desert areas and falls to 250 ly/day in the cloudy areas off Japan. *See* TERRESTRIAL RADIATION.

The energy absorbed at the surface depends also on the surface albedo. The reflectivity of the solid Earth varies with ground cover. Forests absorb nearly all the energy incident on them. Fresh snow, on the other hand, reflects up to 90% of the incident energy. In middle latitudes, snow albedos may decrease to less than 40% with time, especial-

Latitudinal variation of insolation

Latitude	0°N	30°N	60°N	90°N
		(annual mean, ly/day)		
Extraterrestrial	850	740	470	350
At surface				
With clear sky	570	520	320	220
With normal cloud				
cover	410	440	200	150
Total absorption (atmos-				
phere and surface)	570	530	260	120

ly after warm spells have modified the surface. Grass, fields, and other surfaces will have intermediate albedos.

The reflectivity of the water surface of the Earth with cloudless skies depends strongly on the Sun's zenith angle Z. The albedo for the direct solar beam is only 0.02 for $Z = 0°$ and 0.13 for $Z = 70°$. It increases to 0.35 for $Z = 80°$ and to 1.00 for $Z = 90°$. For these larger values of Z, the wind may cause the albedo to decrease. An albedo value of 0.17 for $Z = 80°$ with a wind of 15 knots has been suggested. *See* ALBEDO.

When all the factors are considered, the absorption by the atmosphere and Earth's surface can be estimated, as shown in the last line of the table. It is interesting to note that the gradient of absorbed radiation, which is the driving force of the circulation, is similar to the gradient of the extraterrestrial insolation.

[SIGMUND FRITZ]

Bibliography: F. A. Berry, Jr., E. Bollay, and N. R. Beers (eds.), *Handbook of Meteorology*, 1945; K. L. Coulson, *Solar and Terrestrial Radiation*, 1975; S. Fritz, Solar energy on clear and cloudy days, *Sci. Mon.*, 84:55–65, 1957; G. Partridge and C. Platt, *Radiative Processes in Atmospheric Science*, 1976; N. Robinson (ed.), *Solar Radiation*, 1966.

Internal wave

Internal waves are wave motions of stably stratified fluids in which the maximum vertical motion takes place below the surface of the fluid. The restoring force is mainly due to gravity; when light fluid from upper layers is depressed into the heavy lower layers, buoyancy forces tend to return the layers to their equilibrium positions. Internal waves have been found in the atmosphere as lee waves (waves in the wind stream downwind from a mountain) and as waves propagated along an inversion layer (a layer of very stable air). They are also associated with wind shears at the lower boundary of the jet stream. In the oceans internal oscillations have been observed wherever suitable measurements have been made (see Fig. 1), but it is not completely certain that all of these oscillations are manifestations of internal waves rather than turbulent eddies. The observed oscillations can be analyzed into a spectrum with periods ranging from a few minutes to many days. At a number of locations in the oceans, internal tides, or internal waves having the same periodicity as oceanic tides, are prominent.

The vertical distribution of motions and phase velocity of internal waves depends on the vertical gradient of density in the fluid and the frequency of the generating forces. There is a simple density distribution which is illustrative: The fluid consists of two homogeneous layers, a lighter one on top of a heavier one, such as kerosine over water. The internal waves in this system are sometimes called boundary waves because the maximum vertical motion occurs at the discontinuity of density at the boundary between the two fluids. Let the thickness of the layers be h_1 and h_2, let g be the acceleration of gravity, and let $\delta\rho/\rho$ be the fractional change of density across the boundary. (In the ocean $\delta\rho/\rho$ is of the order of 0.1%, and squares of this small quantity can be neglected.) Then the phase velocity of internal waves of wavelength long compared to $h_1 + h_2$ is given by expression (1). This is to be compared with expression (2), the phase velocity

$$[(\delta\rho/\rho)(gh_1h_2)/(h_1+h_2)]^{1/2} \qquad (1)$$

$$[g(h_1+h_2)]^{1/2} \qquad (2)$$

for surface waves of great length. Because of the factor $(\delta\rho/\rho)^{1/2}$, the internal waves move at a slow speed, of the order of a few knots in the deep oceans. The effect of the rotation of the Earth is to increase the phase velocity of waves having periods long enough to approach one pendulum day.

When there is a continuous distribution of density in the fluid, as in the ocean or atmosphere,

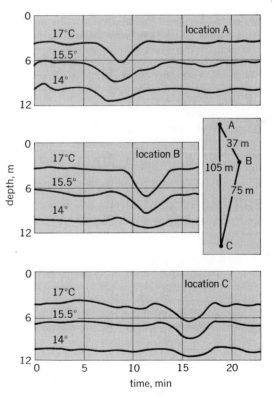

Fig. 1. Diagrams showing passage of an internal wave in the ocean off Mission Beach, CA. Curves indicate depth of certain isotherms recorded at locations A, B, and C as functions of time. Prominent trough recorded at successively later times at the three locations represents a solitary internal wave trough traveling at 23 cm/sec. Average depth to sea floor was 20 m. Insert shows relative location of recorders in horizontal plan.

Aug. 24, 1970

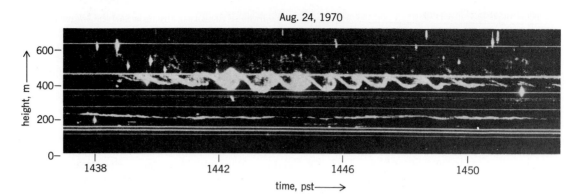

Fig. 2. Internal waves detected in the atmosphere. A narrow radar beam is pointed directly upward, and the height from which echoes are produced is recorded. The prominent waves evident just above 400 m elevation are characteristic of internal waves along a temperature inversion and show signs of breaking induced by wind shear. Radar methods can also detect internal waves at elevations above 10 km.

internal waves are possible only for frequencies lower than the value of expression (3), called the

$$(2\pi)^{-1}\left[g\frac{d}{dz}(ln\rho)-\frac{g^2}{c^2}\right]^{1/2} \qquad (3)$$

Väisälä-Brunt frequency, where $d(ln\rho)/dz$ is the maximum downward rate of increase of the logarithm of density and c is the velocity of sound in the fluid. In the ocean this maximum frequency occurs in the thermocline, where it commonly amounts to about 1/5 cycle per minute. At any frequency lower than this limit, there is an infinity of possible modes of internal waves. In the first mode, the vertical motion has a single maximum somewhere in the body of the fluid; in the second mode, there are two such maxima (180° out of phase), with a node between; and so on. The actual motion usually consists of a superposition of modes. See THERMOCLINE.

Internal waves in the atmosphere have been detected by a variety of instruments: microbarographs and wind recorders at ground level, and long-term recordings of the scattering of radar or sonar beams by sharp density gradients in the high atmosphere (see Fig. 2). In the ocean, internal waves have been found by recording fluctuating currents in mid-depths by moored current meters and by studies of the fluctuations of the depths of isotherms as recorded by instruments repeatedly lowered from shipboard or by autonomous instruments floating deep in the water.

Internal waves are thought to be generated in the sea by variations of the wind pressure and stress at the sea surface, by the interaction of surface waves with each other, and by the interaction of tidal motions with the rough sea floor. The importance of internal waves is that they can transmit energy and momentum throughout the ocean, not only laterally but also vertically. They can, therefore, transmit energy from the surface to all depths. In this way the otherwise sluggishly moving water at great depths can be agitated.

[CHARLES S. COX]

Bibliography: J. L. Cairns, Internal wave measurements from a midwater float, J. Geophys. Res., 80(3):299–306, 1975; N. P. Fofonoff, Spectral characteristics of internal waves in the ocean, Deep-Sea Res., 16(suppl.):58–71, 1969; C. Garrett and W. Munk, Space-time scales of internal waves: A progress report, J. Geophys. Res., 80(3):291–293, 1975; E. E. Gossard and W. H. Hooke, Waves in the Atmosphere, in Developments in Atmospheric Science, vol. 2., 1975; W. J. Gould, Spectral characteristics of some deep current records from the eastern North Atlantic, Phil. Trans. Roy. Soc. London Ser. A., 270:437–450, 1971.

Ionosphere

That part of the upper atmosphere which is sufficiently ionized by solar ultraviolet radiation so that the concentration of free electrons affects the propagation of radio waves. Existence of the ionosphere was suggested simultaneously in 1902 by O. Heaviside in England and A. E. Kennelly in the United States to explain the transatlantic radio communication that was demonstrated the previous year by G. Marconi, and for many years it was commonly referred to as the Kennelly-Heaviside layer. The existence of the ionosphere as an electrically conducting region had been postulated earlier by Balfour Stewart to explain the daily variations in the geomagnetic field. The first direct observation was accomplished in 1924, when E. W. Appleton and M. A. F. Barnett in England and G. Breit and M. A. Tuve in the United States independently observed the direct reflection of radio waves by the ionosphere.

The ionosphere has been extensively explored. The earliest technique involved the ionosonde, which utilizes a pulsed transmitter to send radio signals vertically upward while slowly sweeping the radio frequency. For normal incidence upon the ionosphere, a pulse is reflected at that level in the ionosphere where the plasma frequency equals the radio frequency. Pulses reflected by the ionosphere are received at the transmitter and recorded; the elapsed time between pulse transmission and reception can be interpreted as an apparent distance to the point of reflection. The resulting presentation is a curve of apparent height versus plasma frequency. This technique is still used extensively. Many rockets have flown through the ionosphere to record data, and satellites have orbited in and above the ionosphere for similar purposes. Ionosondes in satellites have been utilized to explore the topside of the ionosphere (that portion above the region of maximum electron concentration) and have provided extensive data.

Data from ground-based ionosondes showed that the ionosphere was structured in the vertical direction. It was first thought that discrete layers were involved, referred to as the D, E, F_1, and F_2 layers; however, rocket measurements have shown that the "layers" merge with one another to such an extent that they are now normally referred to as regions rather than layers. Since a vertically incident radio wave is reflected from the ionosphere at that level where the natural frequency of the plasma equals the radio frequency, it is possible to identify the electron concentration at the point of reflection in terms of the radio frequency according to the relation in Eq. (1), where N_e is the elec-

$$N_e = 1.24 \times 10^{10} f^2 \qquad (1)$$

tron concentration per cubic meter and f is the radio frequency in megahertz. The maximum electron concentrations in the various regions are designated N_mD, N_mE, N_mF_1, and N_mF_2. The critical frequency is the frequency that will just penetrate a given region; for the F_2 peak this is designated f_oF_2. From a careful analysis of the indices of refraction of the ionosphere at various levels, the apparent heights h'_mD, h'_mE, h'_mF_1, and h'_mF_2 of the maximum electron concentration in each region can be converted to true heights h_mD, h_mE, h_mF_1, and h_mF_2.

The ionosphere shows important geographic and temporal variations; the latter include regular diurnal, seasonal, and sunspot-cycle components and irregular day-to-day components associated mainly with variations in solar activity and atmospheric motion. Typical electron concentration profiles through the atmosphere are shown in Fig. 1. Curves are shown for daytime and nighttime conditions near the maximum and the minimum of the sunspot cycle. These curves are typical of temperate latitude conditions.

D region. The D region is the lowest ionospheric region, extending approximately from 60 to 85 km.

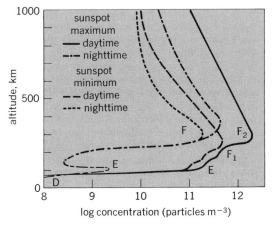

Fig. 1. Typical mid-latitude electron concentration profiles for daytime and nighttime conditions near maximum and minimum of sunspot cycle.

The upper portion is normally caused mainly by the ionization of nitric oxide by hydrogen Lyman-alpha radiation in sunlight, and the lower portion is mainly due to ionization by cosmic radiation. The daytime electron concentrations are about $10^8 - 10^9$ per cubic meter. The region virtually disappears at night even though the cosmic-radiation source of ionization continues, because attachment of electrons to molecules (that is, formation of negative ions) quickly removes free electrons; this effect is suppressed during the daytime by photodetachment. This is the only ionospheric region in which negative ions are thought to be of significance.

The collision frequency for electrons with heavier particles in the D and lower E regions is relatively high, and this condition causes absorption of energy from radio signals traveling through the regions. This severely limits radio propagation and is responsible for the very limited daytime range for stations in the normal broadcast band, where daytime propagation by ionospheric reflection is rendered ineffective by the strong D-region absorption.

The D region is susceptible to disturbance during certain solar events. Sudden ionospheric disturbances (SIDs) occur over the daylight hemisphere with some solar flares; these disturbances last about 1/2 hr and are apparently caused by bursts of solar x-rays with wavelengths near 5 A. These penetrate into the lower D region and cause enhanced electron concentrations there. Polar-cap absorption events (PCAs) are caused by solar cosmic rays that are occasionally emitted by solar flares; these events last for a few days and occur only at high latitudes because of the guiding properties of the geomagnetic field on charged particles. A third type of D-region disturbance occurs regularly near active auroras. All these disturbances are characterized by an abnormally intense absorption of radio waves. *See* COSMIC RAYS; SUN.

E region. Soft x-rays and the more penetrating portions of the extreme ultraviolet radiation from the Sun are absorbed in the altitude region from 85 to 140 km, where they cause daytime electron concentrations of the order of 10^{11} per cubic meter in what is known as the E region. This is the region from which ionospheric reflections were first identified. The principal ions have been observed to be O_2^+, O^+, and NO^+, the last presumably being formed in chemical reactions involving the primary ions. The soft x-rays that are principally responsible for the formation of the E region must also produce N_2^+ ions; however, these are not observed because they are removed very rapidly by the reactions in Eq. (2a) and (2b). Of these two reactions,

$$N_2^+ + O \rightarrow NO^+ + N \qquad (2a)$$
$$N_2^+ + O_2 \rightarrow O_2^+ + N_2 \qquad (2b)$$

the former is the more important, but both dominate over direct recombination of N_2^+ ions with electrons.

F region. The most strongly absorbed of the solar extreme ultraviolet radiations ($200 A < \lambda < 900 A$) produce the F region, which is the region above 140 km. These radiations are most strongly absorbed near 160 km, where they produce a daytime peak in ionization referred to as the F_1 peak or ledge; the F_1 region extends approximately from 140 to 200 km. Above the region of maximum photoionization, the loss rate for electrons and ions decreases with altitude more rapidly than does the source of ionization, so the equilibrium electron concentrations increase with altitude. This decrease in loss rate occurs because O^+ ions recombine directly with electrons only very slowly in a two-body radiative-recombination process, and another loss process predominates instead, an ion-atom interchange reaction, shown in Eq. (3), followed by a dissociative recombination, shown in Eq. (4). The loss rate is controlled by the ion-atom

$$O^+ + N_2 \rightarrow NO^+ + N \tag{3}$$
$$NO^+ + e \rightarrow N + O \tag{4}$$

interchange reaction, and it is the decreasing N_2 concentration with altitude that is responsible for lowering the loss rate so rapidly as to cause equilibrium concentrations of O^+ to increase with altitude above the region of maximum production.

The increase in O^+ and electron concentration with altitude finally stops because the atmospheric density becomes so low that there is a rapid downward diffusive flow of ions and electrons by ambipolar diffusion in the gravitational field. The region of maximum ionization concentration normally occurs near 300 km, and it is known as the F_2 peak. Most of the photoionization that occurs above the peak involves atomic oxygen and it is lost by downward diffusion through the peak into the denser atmosphere below, where the ionization can be lost by the ion-atom interchange reaction followed by dissociative recombination, as described above.

The peak electron concentration in the F_2 region is in the vicinity of 10^{12} ions per cubic meter. Above the peak, the distribution of ionization is in diffusion equilibrium in the gravitational field. When a single ion species is dominant, its distribu-

tion with altitude is a negative exponential, but with a lesser rate of fall with altitude than for neutral particles of the same mass. The gradient of the logarithm of the concentration is given by Eq. (5),

$$\frac{d}{dz} \ln n = -\frac{mg}{k(T_i + T_e)} \tag{5}$$

where m is the ion mass, g the acceleration of gravity, k Boltzmann's constant, T_i the temperature of the ions, and T_e the temperature of the electrons. The corresponding expression for neutral particles is Eq. (6), where T is the neutral particle tempera-

$$\frac{d}{dz} \ln n = -\frac{mg}{kT} \tag{6}$$

ture and the neutral particles are assumed to have the same mass as the ions. If $T = T_i = T_e$, the rate of decrease in the logarithm of the ion concentration with altitude is only half as rapid as for neutrals. However, due to the energy with which photoelectrons are released, it is frequently the case that $T_e > T_i > T$, in which case the rate of fall of ion concentration is still less rapid.

There is an anomaly in F-region behavior, known as the winter anomaly, in that the electron concentrations in winter are higher than in summer, especially if compared under comparable conditions of solar illumination. This is due to a change in atmospheric composition, in particular an increase in atomic oxygen concentration relative to molecular nitrogen. The effect of this change is to reduce the rate of recombination, which is controlled by reaction (3), thus allowing ion and electron concentrations to increase. The change in atmospheric composition results from complications associated with large-scale atmospheric circulation at ionospheric altitudes.

Heliosphere and protonosphere. Helium and atomic hydrogen are important constituents of the upper atmosphere, and their ions are also important at levels above 500 km or so. These gases become important, and finally predominant, constituents of the upper atmosphere because of their low masses. In diffusion equilibrium, each neutral gas is distributed in the gravitational field just as if the other gases were not present, and the lighter ones therefore finally come to predominate over the heavier ones above some sufficiently high altitude. Diffusion equilibrium for ions is more complicated than for neutrals. If the ions and electrons were free to act as independent gases, the electrons would fall off hardly at all with altitude, because of their very small mass. However, even a very small tendency in this direction would set up an electric field sufficient to stop further charge separation, thus requiring that the ions and electrons adopt almost identical distributions. The electric field that is established is sufficient to support half the ion mass and hold the electrons down in the same degree as if they had masses equal to half the ion mass. Therefore, ions and electrons are both distributed as if they had half the ion mass (assuming that $T_e = T_i$; if $T_e > T_i$, the effective mass is even less). Now, if there is in addition, a minor constituent lighter ion, it will experience the same electric field that is established by the major constituent

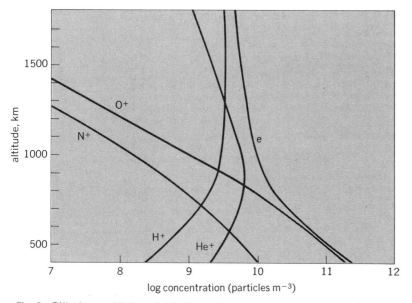

Fig. 2. Diffusion equilibrium distributions of N^+, O^+, He^+, and H^+ ions. Curve e is the total ion concentration, which equals the electron concentration.

heavier ion and that is sufficient to support half the mass of the heavier ion. This field may be more than sufficient to support the entire mass of the lighter ion.

For example, when the predominant ion is O^+, the field is sufficient to support a mass of 8 atomic mass units with a single charge on it. A helium ion will therefore experience a net upward force equivalent to 4 m_p, where m_p is the mass of a hydrogen atom; a hydrogen ion will experience a net upward force of 7 m_p; and these ions will therefore actually increase in concentration with altitude in a diffusion equilibrium situation. This increase with altitude will continue until the lighter ions finally come to predominate, at which point the electric field decreases and the lighter ions adopt a distribution in which their concentration falls with altitude. This is illustrated in Fig. 2. See UPPER-ATMOSPHERE DYNAMICS.

The terms heliosphere and protonosphere are sometimes used to designate the regions in which helium and hydrogen ions respectively are predominant. Sometimes, there may be no region in which helium ions dominate; the transition being from oxygen ions to hydrogen ions.

Magnetosphere and plasmasphere. The magnetosphere is that region in which the movement of ionized plasma is dominated by the geomagnetic field. It extends roughly from the F_1 region upward, as the movement of ions below about 150 km is dominated by collisions with neutral particles. Electrons are not dominated by collisions with neutral particles except below about 90 km.

The outer limit of the magnetosphere is set by interaction of the geomagnetic field with the solar wind, which compresses the magnetic field on the daytime side and limits its outer extension to about 10 earth radii. At that point, the compressed geomagnetic field is sufficiently strong to turn aside the solar wind, and this causes a shock front to develop in the solar wind, as indicated in Fig. 3. See GEOMAGNETISM.

On the nighttime side of the Earth, the magnetosphere extends far out into space as a long tail. A notable feature is the separation of the magnetic tail into two segments that are separated by a current sheet; such a current system must exist between two regions of oppositely directed magnetic field. The bundle of field lines in each section of the tail reaches the Earth's surface in a somewhat smaller area than that encompassed by the auroral zone. See AURORA; MAGNETOSPHERE.

There is a sudden decrease in hydrogen ion concentration that occurs at a distance of about 4 Earth radii from the Earth's center in the equatorial plane. Ion concentrations fall by a factor of about two orders of magnitude, from values near 10^8 to about 10^6 per cubic meter. Since ions can move more or less freely along magnetic-field lines but not across them, this boundary is projected downward along magnetic-field lines and reaches the region of the F_2 peak near 60° geomagnetic latitude. The region of relatively high ion concentration within this boundary is known as the plasmasphere, and the boundary itself is called the plasmapause. The plasmapause is not entirely regular, as it moves in and out diurnally and with geomagnetic activity.

Spread F and sporadic E. Since the ionosphere is produced by solar radiation, one might expect it to be very uniform in its properties, but this is not so. There are many irregularities in the ionosphere with scale sizes varying at least from a few meters to a few kilometers. The cause of these is not known. In the F_2 region and above, the irregularities must have a structure aligned along the magnetic field. The phenomenon of spread F is apparently due to these kinds of irregularities, and the phenomenon is so named because it gives rise to a widened echo trace on the ionosonde record.

Sporadic E is a condition in which there is a thin region of greatly enhanced electron concentration—a region a kilometer or two thick in which the concentration increases by a factor of two or more. This is apparently caused by a strong wind shear that causes ionization to converge on the shear region by transport from above and below. If the ionization transported into the sporadic E layer includes several constituents, the species with higher recombination coefficients will be preferentially removed by recombination, and the species with the smallest recombination coefficients will accumulate. Sporadic E layers have been observed at times to consist of metallic ions, and it is thought that these are of meteoric origin. Such ions are probably always present as minor constituents in the E region, and they have lesser recombination coefficients than the normal atmospheric ions.

Electrodynamic drift. An important physical process in the atmosphere is electrodynamic drift and dynamo effects. Winds and the resultant movement of ionization in the geomagnetic field give rise to induced currents, which in turn perturb the geomagnetic field. These dynamo effects are recognizable in geomagnetic records made at the Earth's surface.

The magnetosphere is an active magnetoelectrodynamic system driven in part by electric fields generated by atmospheric motions in the dynamo region and in part by the solar wind. Although the processes of energy and momentum transfer from the solar wind are not satisfactorily understood in detail, there is no doubt about the facts that a general pattern of magnetospheric motion or convection exists, that it is due mainly to the solar wind, and that it can be described in terms of electric fields. Auroras and some ionospheric currents constitute a load on this system. Magnetospheric convection is probably responsible for the reduction in ion concentration outside the plasmasphere. The plasmasphere corotates with the Earth and is not involved in the pattern of magnetospheric convection.

Two important ionospheric effects have been recognized as resulting from the transport of ionization under influence of electric fields produced by winds and dynamo action at lower altitudes. One effect is the production of the equatorial anomaly in the F_2 region. The ionization maximum in the afternoon does not occur at the Equator or at the subsolar latitude; instead it is divided into two maxima lying about 10° north and south of the geomagnetic equator. The cause of this has been identified as an eastward-directed field that causes an upward drift of ionization at near equatorial latitudes. This is followed by diffusion downward along the magnetic-field lines under action of gravity. The net effect is a transport of ionization from near the geomagnetic equator to the region of the maxima about 10° north and south of it.

The other important effect is the maintenance of the nighttime F region. Recombination should cause a much more rapid decay than is actually observed; the lesser decay is due to upward electrodynamic drift into a region in which the loss processes are less rapid, due to the lesser concentrations of molecular species there. *See* ATMOSPHERE.

[FRANCIS S. JOHNSON]

Bibliography: P. M. Banks and G. Kocharts, *Aeronomy*, 1973; A. D. Danilov, *Chemistry of the Ionosphere*, 1970; K. Davies, *Ionospheric Radio Propagation*, 1967; E. R. Dyer (ed.), *International Symposium on Solar-Terrestrial Physics, Leningrad, 1970*, 1972; A. P. Mitra, *Ionospheric Effects of Solar Flares*, 1974; J. A. Ratcliffe, *An Introduction to the Ionosphere and Magnetosphere*, 1972; H. Rishbeth and O. K. Garriott, *Introduction to Ionospheric Physics*, 1969; R. C. Whitten and I. G. Poppoff, *Fundamentals of Aeronomy*, 1971.

Isentropic surfaces

Surfaces along which the entropy and potential temperature of air are constant. Potential temperature, in meteorological usage, is defined by the relationship

$$\theta = T\left(\frac{1000}{P}\right)^{\frac{c_p - c_v}{c_p}}$$

in which T is the air temperature, P is atmospheric pressure expressed in millibars, c_p is the specific heat of air at constant pressure and c_v is the specific heat at constant volume. Since the potential temperature of an air parcel does not change if the processes acting on it are adiabatic (no exchange of heat between the parcel and its environment), a surface of constant potential temperature is also a surface of constant entropy. The slope of isentropic surfaces in the atmosphere is of the order of 1/100 to 1/1000. An advantage of representing meteorological conditions on isentropic surfaces is that there is usually little air motion through such surfaces, since thermodynamic processes in the atmosphere are approximately adiabatic. *See* ATMOSPHERIC GENERAL CIRCULATION.

[FREDERICK SANDERS]

Isobar

A line passing through points at which a constant value of the air pressure exists, within a specified surface of reference. Central regions of closed isobars on the globe reveal systems of relative high and low pressure as shown on synoptic weather charts based upon simultaneous barometric observations at many stations. For such charts the surface of reference is usually the geoid (mean sea level). In this case the data represent pressures reduced to sea level, which yield unreal isobars over land. Systems of relatively high and low pressure are called anticyclones and cyclones, respectively.

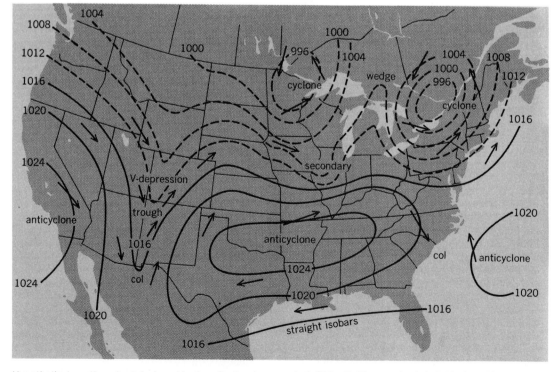

Hypothetical weather chart designed to show the fundamental configurations of isobars. Arrows fly with surface wind. (*After R. Abercromby, in L. P. Harrison, Meteorology, National Aeronautics Council, Inc., 1940*)

Horizontal pressure gradients determined from real isobars correlate well with the wind velocity about 300–800 m above surface. The illustration presents a fictitious weather chart designed to portray the configurations of isobars generally observed. *See* AIR PRESSURE.   [LOUIS P. HARRISON]

Isopycnic

The line of intersection of an atmospheric isopycnic surface with some other surface, for instance, a surface of constant elevation or of constant atmospheric pressure. An isopycnic surface is a surface in which the density of the air is constant. Such surfaces are also designated isosteric surfaces because these surfaces, in which the specific volume is constant, coincide with the reciprocal conditions of isopycnic surfaces since density is the reciprocal of specific volume. On a surface of constant pressure, isopycnics coincide with isotherms, because on such a surface, density is a function solely of temperature. On a constant-pressure surface, isopycnics lie close together when the field is strongly baroclinic and are absent when the field is barotropic. *See* BAROCLINIC FIELD; BAROTROPIC FIELD; METEOROLOGICAL SOLENOID.

[FREDERICK SANDERS]

Isotach

A line along which the speed of the wind is constant. Isotachs are customarily represented on surfaces of constant elevation or atmospheric pressure, or in vertical cross sections. The closeness of spacing of the isotachs is indicative of the intensity of the wind shear on such surfaces. In the region of a jet stream the isotachs are approximately parallel to the streamlines of wind direction and are closely spaced on either side of the core of maximum speed. The term isovel is used synonymously with the term isotach. *See* JET STREAM; WIND.

[FREDERICK SANDERS]

Isothermal chart

A map showing the distribution of air temperature (or sometimes sea surface or soil temperature) over a portion of the Earth's surface, or at some level in the atmosphere. On it, isotherms are lines connecting places of equal temperature. The temperatures thus displayed may all refer to the same instant, may be averages for a day, month, season, or year, or may be the hottest or coldest temperatures reported during some interval.

Maps of mean monthly or mean annual temperature for continents, hemispheres, or the world sometimes show values reduced to sea level to eliminate the effect of elevation in decreasing average temperature by about 3.3°F/1000 ft (see illustration). Such adjusted or sea-level maps represent the effects of latitude, continents, and oceans in modifying temperature; but they conceal the effect of mountains and highlands on temperature distributions. The first isothermal chart, prepared by Alexander von Humboldt in 1817 for low and middle latitudes of the Northern Hemisphere, was the first use of isopleth methods to show the geographic distribution of a quantity other than elevation.

These maps are now varied in type and use. Isothermal charts are drawn daily in major weather

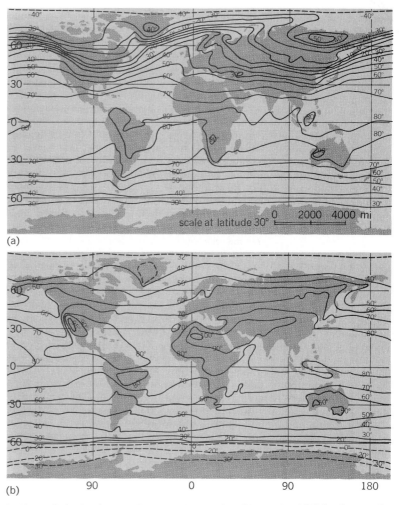

(a)

(b)

Isothermal charts of mean air temperature reduced to sea level (°F) for the months of (a) January and (b) July. (*From G. T. Trewartha, Introduction to Climate, 4th ed., McGraw-Hill, 1968*)

forecasting centers; 5-day, 2-week, and monthly charts are used regularly in long-range forecasting; mean monthly and mean annual charts are compiled and published by most national weather services, and are presented in standard books, on, for example, climate, geography, and agriculture. *See* AIR TEMPERATURE; TEMPERATURE INVERSION.

[ARNOLD COURT]

Jet stream

A relatively narrow, fast-moving wind current flanked by more slowly moving currents. Jet streams are observed principally in the zone of prevailing westerlies above the lower troposphere and in most cases reach maximum intensity, with regard both to speed and to concentration, near the tropopause. At a given time, the position and intensity of the jet stream may significantly influence aircraft operations because of the great speed of the wind at the jet core and the rapid spatial variation of wind speed in its vicinity. Lying in the zone of maximum temperature contrast between cold air masses to the north and warm air masses to the south, the position of the jet stream on a given day usually coincides in part with the regions of greatest storminess in the lower troposphere, though portions of the jet stream occur over regions which are entirely devoid of cloud. *See* CLEAR-AIR TURBULENCE.

Characteristics. The specific characteristics of the jet stream depend upon whether the reference is to a single instantaneous flow pattern or to an averaged circulation pattern, such as one averaged with respect to time, or averaged with respect both to time and to longitude.

If the winter circulation pattern on the Northern Hemisphere is averaged with respect to both time and longitude, a westerly jet stream is found at an elevation of about 13 km near latitude (lat) 25°. The speed of the averaged wind at the jet core is about 148 km/hr (80 knots). In summer this jet is displaced poleward to a position near lat 42°. It is found at an elevation of about 12 km with a maximum speed of about 56 km/hr (30 knots). In both seasons a speed equal to one-half the peak value is found approximately 15° of latitude south, 20° of latitude north, and 5–10 km above and below the location of the jet core itself.

If the winter circulation is averaged only with respect to time, it is found that both the intensity and the latitude of the westerly jet stream vary from one sector of the Northern Hemisphere to another. The most intense portion, with a maximum speed of about 185 km/hr (100 knots), lies over the extreme western portion of the North Pacific Ocean at about lat 22°. Lesser maxima of about 157 km/hr (85 knots) are found at lat 35° over the east coast of North America, and at lat 21° over the eastern Sahara and over the Arabian Sea. In summer, maxima are found at lat 46° over the Great Lakes region, at lat 40° over the western Mediterranean Sea, and at lat 35° over the central North Pacific Ocean. Peak speeds in these regions range between 74 and 83 km/hr (40–45 knots). The degree of concentration of these jet streams, as measured by the distance from the core to the position at which the speed is one-half the core speed, is only slightly greater than the degree of concentration of the jet stream averaged with respect to time and longitude. At both seasons and at all longitudes the elevation of these jet streams varies between 11 and 14 km.

Variations. On individual days there is a considerable latitudinal variability of the jet stream, particularly in the western North American and western European sectors. It is principally for this reason that the time-averaged jet stream is not well defined in these regions. There is also a great day-to-day variability in the intensity of the jet stream throughout the hemisphere. On a given winter day, speeds in the jet core may exceed 370 km/hr (200 knots) for a distance of several hundred miles along the direction of the wind. Lateral wind shears in the direction normal to the jet stream frequently attain values as high as 100 knots/300 nautical miles (185 km/hr/556 km) to the right of the direction of the jet stream current and as high as 100 knots/100 nautical miles (185 km/hr/185 km) to the left. Vertical shears below and above the jet core are often as large as 20 knots/1000 ft

(37 km/305 m). Daily jet streams are predominantly westerly, but northerly, southerly, and even easterly jet streams may occur in middle or high latitudes when ridges and troughs in the normal westerly current are particularly pronounced or when unusually intense cyclones and anticyclones occur at upper levels.

Insufficiency of data on the Southern Hemisphere precludes a detailed description of the jet stream, but it appears that the major characteristics resemble quite closely those of the jet stream on the Northern Hemisphere. The day-to-day variability of the jet stream, however, appears to be less on the Southern Hemisphere.

It appears that an intense jet stream occurs at high latitudes on both hemispheres in the winter stratosphere at elevations above 20 km. The data available, however, are insufficient to permit the precise location or detailed description of this phenomenon. *See* AIR MASS; ATMOSPHERE; GEOSTROPHIC WIND; STORM.   [FREDERICK SANDERS]

Lightning

The large spark produced by an abrupt discontinuous discharge of electricity through the air generally under turbulent conditions of the atmosphere.

Cumulonimbus clouds, which produce lightning, contain strongly upward-moving air and frequently extend to altitudes higher than 40,000 ft. Within these clouds, various factors act to cause the creation and separation of electric charge. Conflicting theories have been proposed to explain the electrification. For the most part, they involve cloud and precipitation particles and lead to essentially the same charge distribution. The upper parts of thunderstorms are positively charged, whereas the central and lower parts are predominantly negative. In some clouds, after rain has formed, a small positive charge center is found in the lower part of the cloud.

When cloud electrification has proceeded to the point where the potential gradient between adjacent charge centers, or between the cloud and ground, reaches the breakdown value, the large spark called lightning occurs. A breakdown poten-

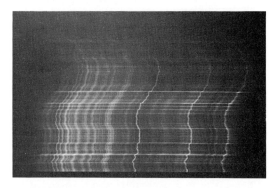

Fig. 1. Slitless spectrum of a ground flash, obtained by placing a diffraction grating in front of a camera lens with focal length 61 cm. Range of wavelengths is 3900 A (left) to 6165 A (right). (*From L. E. Salanave, Lightning and Its Spectrum, University of Arizona Press, copyright 1980; used with permission*)

tial of about 10,000 volts/cm has been suggested as an appropriate value for air containing water droplets.

Leaders and return strokes. It has been established that a lightning stroke to ground is initiated by a step leader. A surge of electrons moves downward about 50 m in about 1 μsec and then stops. After a pause of perhaps 50 μsec another step takes place. This sequence is repeated until the leader reaches the ground. When this occurs, a surge of charge moves rapidly up the path taken by the step leader. Whereas the average speed of the step leader is of the order of 10^6 m/sec, the main stroke moves up the preionized path at a speed of the order of 10^7 m/sec. The luminosity of this so-called return stroke is quite intense.

After an interval of the order of 0.01 sec, another stroke usually occurs in the same channel. These secondary return strokes are preceded by downward-moving dart leaders which do not show step characteristics. As many as 30–40 strokes have been observed to occur in the same channel.

Cloud-to-cloud strokes also involve a leader and main return stroke.

The current in the return strokes increases in an average of about 6 μsec to a maximum value frequently exceeding 30,000 amp, but sometimes as high as 200,000 amp. It decreases to about one-half the maximum value in an average of about 24 μsec. Average currents during the strokes are of the order of 10,000 amp, and about 500 amp may flow through the ionized channel in the interval between strokes. A reasonable estimate of the diameter of the channel appears to be about 10 cm. The high currents lead to thunder and intense heating. For a brief instant the temperature is probably near 30,000°C. The high currents and temperatures are mainly responsible for the damage caused by lightning. *See* CLOUD PHYSICS; THUNDERSTORM.

[LOUIS J. BATTAN]

Optical spectrum. Lightning is a thermal source of radiation emitting a very broad spectrum of electromagnetic frequencies. If boundaries are set by practical considerations of atmospheric transmission and the spectral sensitivity of photographic film, optical observations extend from approximately 3000 A (ozone absorption) to nearly 10,000 A (limit of ordinary infrared photographic film). This is the "spectrum" that will be discussed here.

Ground flashes in the clear air below a cloud provide the best opportunity to study the optical properties of lightning. In such cases the source is a narrow line at a considerable distance from the observer, and the technique of slitless spectroscopy can be used to advantage. In this method of observation, the dispersive element (a prism or diffraction grating or both) is placed in front of a camera focused at infinity.

Figure 1 is the reproduction of a slitless spectrum of lightning from 3900 to 6200 A obtained on panchromatic film. With a few minor exceptions, the spectrum in this region is composed of emissions from singly ionized nitrogen. When the spectrum is recorded on film with extended red sensitivity, one finds a very strong line due to hydrogen,

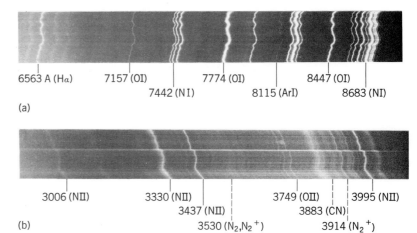

6563 A (Hα)   7157 (OI)   7774 (OI)   8447 (OI)   8683 (NI)
7442 (N I)   8115 (ArI)
(a)

3006 (NII)   3330 (NII)   3749 (OII)   3995 (NII)
3437 (NII)   3883 (CN)
(b)   3530 (N₂,N₂⁺)   3914 (N₂⁺)

Fig. 2. Lightning spectra. (*a*) Vertical section of a red and infrared slitless spectrum. Very strongly exposed spectra show absorption features due to molecular oxygen and water vapor; the A band of oxygen near 7600 angstrom units is especially prominent. (*b*) The ultraviolet spectrum. Atomic lines are due to ionized nitrogen and oxygen, emitted when the temperature (and the current) is near its peak.

termed the Hα line, with a wavelength of 6563 A. The atomic hydrogen is derived from water vapor completely dissociated by the electrical discharge. Still further extensions with film sensitive to infrared reveal a spectrum dominated by strong lines of neutral nitrogen and oxygen and a few faint lines due to neutral argon (Fig. 2*a*).

The photographic ultraviolet spectrum of lightning extends from the visible limit (approximately 4000 A) to slightly below 3000 A, where absorption bands due to ozone become significant and eventually cut down the transmitted radiation to very low levels. (Ozone is generated around the electri-

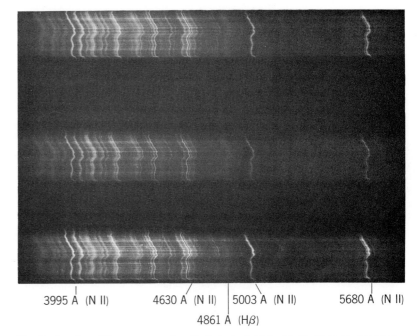

3995 A (N II)   4630 A (N II)   5003 A (N II)   5680 A (N II)
4861 A (Hβ)

Fig. 3. Spectra of three consecutive strokes of a lightning flash. Time sequence upward; intervals 45 and 55 msec, respectively. (*From L. E. Salanave, Lightning and Its Spectrum, University of Arizona Press, copyright 1980; used with permission*)

cal discharge channel and in the intervening air path during thunderstorms.) Emission lines due to singly ionized nitrogen and oxygen predominate, as in the visible spectrum, but in addition there may occur emission bands due to molecules of cyanogen (CN) and nitrogen (N_2 and N_2^+). These molecular features are notably variable in their relative intensities from flash to flash; their excitation is presumably dependent upon the amount of electrical current and its duration (Fig. 2b).

Bright horizontal streaks are produced by the comparatively weak continuous spectrum of heated air—enhanced where short sections of the discharge channel lie along the direction of dispersion or in a line toward the spectrograph. Some of the faintest streaks may be caused by regions (beads) of intrinsic brightening along the channel, but this is not established with certainty.

The spectrum of each stroke in a flash can be obtained by placing a mask with a narrow, horizontal slot close to the photographic film which is moving vertically (usually by means of a rotating drum). Since strokes typically occur 50 msec apart or more, a film transport of only 20 cm/sec suffices with a masking slot 1 cm high.

Figure 3 shows the time-resolved spectrum of a three-stroke flash. The similarity among the component spectra shows that the discharge attained substantially the same temperature at each stroke, though the overall luminosity of individual strokes clearly differ.

In some cases a stroke continues for 50 msec or more at a very low discharge rate, called a continuing current. This produces a continuing luminosity whose spectrum is dominated by Hα and N_2^+ emissions, with lesser contributions from neutral oxygen and nitrogen.

An application of a high-speed drum camera and extremely sensitive film produces time-resolved spectra of individual strokes, on which changes occurring in microseconds can be measured.

A comparison of the relative intensities of certain emissions of ionized nitrogen (N II) shows that temperatures of the order of 25,000–30,000°C are attained in a few microseconds, dropping to half that in about 30 μsec. Measurements of the broadened profile of Hα indicate that electron densities are somewhere between 10^{17} and 10^{18} cm^{-3} (almost complete ionization) at peak temperature, at which time the pressure inside the lightning channel is of the order of 10 atm. The physical basis for these rather spectacular values can be intuitively grasped in terms of the very high, rapidly rising currents described in the previous section.

[LEON E. SALANAVE]

Bibliography: E. P. Krider, Lightning spectroscopy, *Nucl. Instrum. Methods*, vol. 110, July 1973; R. E. Orville and L. E. Salanave, Lightning spectroscopy: Photographic techniques, *Appl. Opt.*, vol. 9, August 1970; M. A. Uman, *Understanding Lightning*, chap. 11, 1971.

Magnetosphere

The region of the Earth in which the geomagnetic field plays a dominant part in controlling the physical processes that take place. The magnetosphere

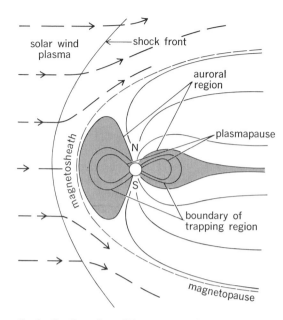

Fig. 1. Configuration of the magnetosphere in the plane containing the Sun-Earth line and the geomagnetic axis. The regions of open magnetic field lines (tail) are unshaded, and the region of closed magnetic field lines (doughnut) is shaded. Trapped energetic particles exist throughout the doughnut, but are relatively unstable outside the plasma-pause and tend to be precipitated into the atmosphere, where they produce auroras.

is usually considered to begin at an altitude of about 100 km, corresponding to the ionospheric E region, and to extend outward to a distant boundary, termed the magnetopause, that marks the transition to the interplanetary medium. The shape of the magnetosphere is similar to that of a comet in that it has a long tail extending away from the Sun and is rather blunt on the sunward side (Fig. 1). *See* IONOSPHERE.

The geocentric distance to the nose of the magnetosphere is usually about 10 R_e (R_e = earth radius $\simeq$ 6400 km), although it is known to vary between 6 and 14 R_e, depending on conditions in the interplanetary medium. Along the dawn and dusk meridians and over the poles of the Earth, the geocentric distance to the magnetopause is more typically 15 R_e. The magnetosphere gradually expands laterally with distance along the tail, and at the distance of the Moon (60 R_e) it has a radius of approximately 20 R_e. The magnetosphere gradually expands with distance along the Sun. Nevertheless its influence has been detected at a distance of 800 R_e in observations made from the space probe *Pioneer 6*. On the other hand, observations of solar electron events made within the tail of the magnetosphere suggest that its length does not exceed 2×10^7 km ($\simeq$ 3000 R_e). There are theoretical reasons for believing that the length of the tail varies between a few hundred and a few thousand earth radii and this appears to be in reasonable accord with the observations.

Shape. The shape of the magnetosphere is determined by stresses exerted at the magnetopause by the solar wind, which constitutes the interplanetary medium. On the sunward side the normal stress is most important, and adequate calculations of the shape can be made by equating the pressure exerted by the solar wind outside the magnetopause to the magnetic pressure inside. Since the solar wind is highly supersonic, a shock wave must stand on the upstream side of the magnetosphere, and in the relatively thin transition region between the shock and the magnetopause the pressure should be approximately equal to the local ram pressure of the solar wind. Thus Eq. (1)

$$p \approx nm V^2 \cos^2 \theta \qquad (1)$$

holds, where p is the pressure in the transition region, n the number density, V the speed of the solar wind in front of the shock, m the mean mass of solar-wind ions, and θ the angle between the solar-wind direction and the normal to the shock and magnetopause. The magnetic field strength (B_1) just inside the magnetopause on the forward side of the magnetosphere is approximately double that of the Earth's dipole, the enhancement being due to currents flowing in the magnetopause itself. An example of the calculated shape, obtained by putting $p = B_1{}^2/8\pi$, is shown in Fig. 2. The theoretical and observed shapes are in good agreement, as shown in Fig. 3; here the lower curve is the calculated mean position of the magnetopause, based on observed mean values of n and V; the black dots represent the observed position found by instruments on the satellite *Explorer 18 (IMP 1)*.

Existence of the shock wave which stands in the solar wind on the sunward side of the magnetosphere (Fig. 1) was predicted because the solar wind is supersonic and a drastic change in its properties would be required to direct the flow around the obstacle represented by the magnetosphere. Using standard aerodynamic procedures, the expected position of the shock wave can be calculated, provided the shape of the magnetosphere is known. Observed and calculated shock-wave positions are shown in Fig. 3, and it can be seen that there is good agreement. The most remarkable feature of this particular shock wave is that particle-particle collisions are not involved, since the mean free path in the solar wind at the Earth's orbit is of the order of 10^{13} cm and the thickness of the shock wave is in fact not more than a few

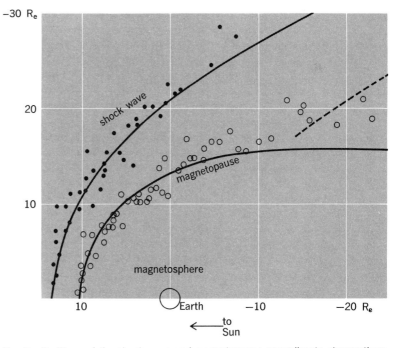

Fig. 3. Positions of the shock wave and magnetopause according to observations from the satellite *Explorer 18 (IMP 1)*. Solid lines represent the computed positions of the shock wave and magnetopause for average solar-wind conditions. Broken line represents the region of space where the magnetopause was observed on several occasions by the satellite *Explorer 10*. (After N. F. Ness)

hundred kilometers, which is roughly the gyro-radius of a proton in the interplanetary magnetic field near the Earth. This is the first clear example of a collision-free shock wave to be found either in nature or in the laboratory.

Tail. The geometry of the tail of the magnetosphere is not determined by simple pressure balance. If it were, it would be weak and composed of the small part of the dipole magnetic field of the Earth able to bulge out into the shadow region cast in the solar wind by the stronger fields in the forward part of the magnetosphere. In fact, the total amount of magnetic flux contained in the tail is quite large and involves all the field lines that enter or leave the Earth within about 1500 km of the geomagnetic poles. The magnetic field lines in the tail do not curve gently around and connect the two polar regions of the Earth, but instead lie almost parallel to the Earth-Sun line. In the northern half of the tail, the direction of the magnetic field is toward the Earth, and in the southern half, away from the Earth. The two halves of the tail are separated by a thin neutral sheet of very low magnetic field strength which contains plasma with enough pressure to prevent linkage of the oppositely directed magnetic fields on each side. Some of the magnetic field lines in the tail do connect across the neutral sheet, but most are linked with the interplanetary magnetic field. Thus, if one were to trace by compass a magnetic field line starting from the geomagnetic south pole, it would normally lead not to the geomagnetic north pole but out into the interplanetary medium and eventually to the Sun or possibly out into the Galaxy beyond the solar system. It is essentially the fol-

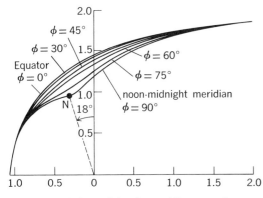

Fig. 2. Calculations of the shape of the magnetopause in planes containing the Sun-Earth line and making an angle ϕ with the geomagnetic equatorial plane. The scale is approximately in units of 10 R_e.

lowing effect that confirms the above description of the tail most decisively: Low-energy (40-kev) electrons emitted by solar flares have been observed in the tail moving toward the Earth from the antisolar direction, and since these particles are effectively constrained to move parallel to magnetic field lines, this can only mean that the tail is linked magnetically with the interplanetary medium.

The tail must be produced by transverse or "viscous" stresses exerted at the magnetopause by the solar wind. The mechanism involved is probably linkage of geomagnetic field lines with the interplanetary magnetic field on the upstream side of the magnetosphere and their subsequent transport downstream (away from the Sun) by the solar wind. By this process, solar-wind energy is converted to magnetic energy in the tail, and the total magnetic flux in the tail tends to grow at the expense of the flux in the forward part of the magnetosphere, which is continually eroded away. However, the tail does not grow indefinitely because the magnetic field lines in the tail can be reconnected in the neutral sheet and, on contracting, once more become looped geomagnetic field lines which close near the Earth. Thus a general circulation of the magnetic field must take place through the magnetosphere with the tail as a natural byproduct of the process that causes the circulation, namely, erosion of the front of the magnetopause by the solar wind. The circulation implies the existence of a quasi-steady electric field, and this is believed to be the main cause of several more readily apparent magnetospheric phenomena, notably the auroras, geomagnetic storms, and the trapped radiation belts. *See* AURORA; GEOMAGNETIC STORM; VAN ALLEN RADIATION.

"Doughnut" region. One can consider the magnetosphere as a doughnut-shaped region of closed field lines that emanate from regions near the Earth with geomagnetic latitude (Λ) less than $70-75°$, together with bundles of open field lines that emanate from the polar regions ($\Lambda \gtrsim 70-75°$). However, there is a considerable interchange of magnetic flux between the open and closed regions, and the division, on occasion, can fluctuate from perhaps $\Lambda \simeq 65°$ to $\Lambda \to 90°$ in some sectors at least. The doughnut is considerably stretched on the night side of the Earth and somewhat flattened on the day side, but the field is on the whole dipolelike, especially in the region $\Lambda \lesssim 60°$. Energetic charged particles can be trapped within the doughnut for relatively long periods, and it is in this region that the so-called radiation belts of the Earth are found. In contrast, the open magnetic field lines in the tail cannot act as a trap, and any particles found in here must be flowing either toward or away from the Earth. Energetic particles are often found in the vicinity of the neutral sheet in the tail, but this probably indicates that the cusplike outer boundary of the doughnut on the night side has temporarily become enlarged.

Magnetospheric plasma. A low-energy plasma with particle energies less than a few electron volts permeates the entire doughnut region. It has been detected directly by spacecraft observations and by ground-based observations of whistlers. The latter are very-low-frequency (1–10 kHz) radio signals that originate in lightning flashes and propagate in ducts following the geomagnetic field lines; they move from one hemisphere to the other many times before eventually decaying. It appears that the magnetospheric plasma is simply an extension of the ionosphere and consists mostly of light ions (H^+, He^+) produced by photoionization of neutral atmospheric atoms by sunlight. However, the plasma exhibits a remarkable feature in that the density drops quite sharply (by a factor 10 or more) on a shell defined by geomagnetic field lines that intersect the Earth at $\Lambda \simeq 60°$ (Fig. 1). Within this shell, called the plasmapause, the plasma seems to be in diffusive equilibrium with the ionosphere, but outside the shell, the density is very low (less than 10 particles per cubic centimeter near the equatorial plane of the magnetosphere), and there is certainly no such equilibrium. The probable explanation of this effect is that the plasmapause marks the inner boundary of those geomagnetic field lines, now part of the doughnut, which have recently been open. While the field lines form part of the tail, the plasma can stream freely out into space and be lost. It takes a day or so for fresh ions and electrons produced in the ionosphere to replenish the field lines that have been drained of plasma. Thus one can regard the existence of the plasmapause as very good evidence for the interchange of geomagnetic field lines between the tail and the doughnut and, accordingly, as a confirmation of the existence of a quasi-steady electric field in the magnetosphere. The changes of shape and size undergone by the plasmapause are consistent with this explanation, as are many other observations.

Van Allen radiation belts. The energetic-particle belts that reside in the doughnut were discovered in 1958 by James Van Allen and colleagues from the University of Iowa. In fact, these early experiments were capable of detecting only the most penetrating components of the trapped particles which compare with the visible portion of an iceberg as far as the whole population of particles is concerned. The first belt of particles found contains protons with energies in the range 1–100 Mev; this very stable belt is centered at an altitude of about 4000 km above the Equator, and is known as the inner Van Allen belt. The second belt is composed of electrons with energies in the range 1–10 Mev and is centered at a geocentric distance of about 4 R_e. It is known as the outer Van Allen belt and varies in time and to some extent in position.

Both the inner and outer Van Allen belts are only minor components of the entire population of trapped particles, which fills the doughnut completely and involves particles of all energies down to the few electron volts typical of the thermal plasma described above. The various components have quite different characteristics as far as temporal and spatial variations are concerned, and it is not easy to make general statements about the behavior of the trapped particles as a whole. However, since these particles provide the major part of the pressure exerted by the magnetospheric plasma, one can be certain that, whenever the magnetosphere is observed to be inflated by internal pressure, the total energy contained in trapped particles is large. Such an inflation of the magnetosphere is easily recognized because it corresponds to a reduction in the geomagnetic field strength observed on the ground at low and middle latitudes; it takes place during geomagnetic storms that occur following flares and other activity on the visible disk of the Sun. *See* SUN.

Geomagnetic storm. It has been shown that the reduction in the geomagnetic field strength (ΔB) occurring during the main phase of a storm can be related directly to the total energy (E, in ergs) of trapped particles. Thus Eq. (2) holds, where $1\gamma =$

$$\Delta B = 2 \times 10^{-21} E\, \gamma \qquad (2)$$

10^{-5} gauss. For a moderate geomagnetic storm, $\Delta B \simeq 200\gamma$; hence $E \simeq 10^{23}$ ergs. It is known that the bulk of the particles that cause the magnetosphere to become inflated are protons with energies of the order of 10 kev. The inflation begins first on the night side of the Earth, possibly as a result of the reconnection of formerly open magnetic field lines in the tail of the magnetosphere. The resulting asymmetric bulge of the doughnut builds up over a period of a few hours (corresponding to an energy input rate of about 10^{19} ergs sec^{-1}), and then spreads throughout the whole doughnut region outside the plasma-pause in a series of bursts, each of which is associated with a violent eruption of auroral activity, called a polar substorm.

During substorms, energy is dissipated in the upper atmosphere at a rate of about 3×10^{18} ergs sec^{-1}, partly as a result of heating by intense electric currents which flow in the ionosphere at such times and partly as a result of direct bombardment of the atmosphere by electrons with typical energies of about 10 kev, which produce most of the auroral luminosity. The details of the substorm mechanism are not understood. However, the overall effect appears to be a collapse of the bulge produced by injection of 10-kev electrons and protons on the night side of the closed region of the magnetosphere. The electrons are precipitated into the atmosphere to produce auroras, while the

protons move deeper into the magnetosphere, compressing the plasmapause and causing a more general inflation of the field. Eventually, during the recovery phase of the geomagnetic storm, the trapped protons also leak out of the magnetosphere, partly by being absorbed in the atmosphere and partly by escaping into the interplanetary medium. This recovery phase may take several days and is associated with the appearance of enhanced fluxes of the more energetic particles which make up the outer Van Allen belt (although geomagnetic field deflation implies that the total pressure of trapped particles must decrease).

Small geomagnetic storms may begin gradually, but major storms begin with an abrupt increase of the geomagnetic field strength observed at ground level and usually occur a day or so after a large solar flare. The increase, known as the "sudden commencement" of the storm, is produced when a blast wave caused by the flare impinges on the magnetosphere. The blast waves precedes an enhancement of the solar wind velocity, density, and ram pressure, which can keep the magnetosphere compressed for several hours (the initial phase of the storm) before fresh plasma injected into the magnetosphere via the neutral sheet produces the main phase. The complete sequence of events composing a geomagnetic storm is summarized in Fig. 4.

Apart from the gross variations of the structure of the magnetosphere that take place during geomagnetic storms and to some extent at all times, smaller fluctuations also occur. Oscillations of the magnetospheric plasma and fields are possible at all frequencies ranging from ultralow frequency (≈ 0.01 Hz and up), where hydromagnetic (Alfvén) waves are involved, up to very low frequency (≈ 10 kHz), where the predominant waves are whistlers. Even higher frequency emissions have been observed, but these are of lesser importance as far as other phenomena are concerned. Most emissions are believed to be generated by the trapped particles in the radiation belts, the lower frequencies (~ 1 Hz) being associated with protons and the higher frequencies ($\sim 1 - 10$ kHz) with electrons. From every point of view, the magneto-

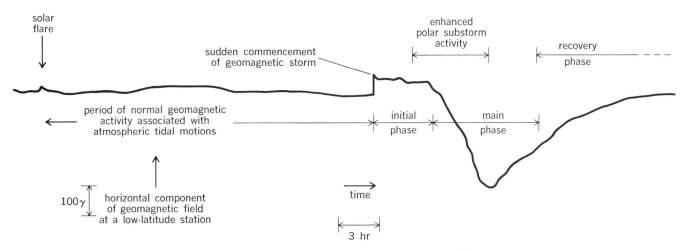

Fig. 4. Schematic summary of events following the occurrence of a large solar flare on the visible disk of the Sun.

sphere constitutes a remarkable natural plasma laboratory, which, together with its inherent importance, makes it a highly worthwhile subject for investigation.   [W. I. AXFORD]

Bibliography: S. I. Akasofu, Electrodynamics of the magnetosphere: Geomagnetic storms, *Space Sci. Rev.*, 6:21, 1966; V. C. A. Ferraro and E. N. Parker, The worldwide magnetic storm phenomenon, *Handbuch der Physik*, 1968; S. Matsushita and W. H. Campbell, *Physics of Geomagnetic Phenomena*, vols. 1 and 2, 1968; V. Vacquier, *Geomagnetism in Marine Geology*, 1972.

Marine biological sampling

The collection and observation of living organisms in the sea. This article deals primarily with the methods of collecting marine organisms as part of ecological investigations and points out some of the difficulties that arise in obtaining data for quantitative investigations. The biological survey of the ocean depends to a large extent on vessels equipped with nets and other sampling devices. Only in intertidal regions at low tide is it possible to observe and collect marine organisms without special apparatus. *See* OCEANOGRAPHY.

Objectives. The foremost aim of sampling is to reveal the different kinds of organisms that live in the ocean, a task that is far from complete; many new species of marine animals and plants are still being found. Because marine organisms range in size from bacteria to the 100-ft blue whale (*Balaenoptera musculus*), many kinds of gear are needed. Another objective of sampling is to discover the varying quantities of each species in time and space; this aim further complicates the design of gear.

Because of the diverse nature of these objectives and the variety of sampling methods, marine biological sampling can best be considered in categories based on size of the organisms, beginning with the smallest. For a description of the organisms themselves and of their habitats *see* DEEP-SEA FAUNA; MARINE ECOSYSTEM; MARINE MICROBIOLOGY.

Bacteria. Since assessment of the numbers of bacteria in the sea and bottom deposits depends on culturing a measured subsample and then counting the numbers of colonies that grow in the culture medium, whole samples must be taken under sterile conditions. Samples from mid-water levels can be taken in a sterilized pressure-withstanding bottle, which is lowered on a wire to the desired depth. A sealed glass inlet tube is then opened by sliding a brass messenger down the wire to operate a breaking device. The main difficulty of mid-water sampling is that a great many bacteria live on the surfaces of living organisms and decaying matter; these bacteria, therefore, are not included in the estimate. The quantities of particulate organic material can be assayed by drawing a sea-water sample through a membrane filter. Benthic bacteria are obtained by taking a sterile subsample from the center of a core of sediment.

Phytoplankton. Planktonic plants range in size from about 1 to 2000 μ. They can be collected by any of various types of hydrographic water bottle or with a fine-meshed silk net.

Water bottle samples. Phytoplankton is collected in water bottles at standard depths (0.5, 10, 25, 50, 75, 100, 150, 200, and 300 m). The sample is then concentrated, counted, and identified. One method is to strain the sample through fine-meshed filters. Millipore filters, through which a known volume of sea water (1/2 to 2 liters) is drawn by suction, are particularly useful if some estimate is to be made of the smaller species. The filters are made transparent by treatment with cedar oil, after which the plants become visible for microscopic examination.

Phytoplankton collected in a water bottle may also be precipitated by placing a known volume in an ultracentrifuge, but even at high speeds (15,000 rpm) complete sedimentation is not obtained. Plants in a subsample may also be left to settle in a special counting chamber in a dark room. After a day or so they are identified and counted by means of an inverted (Utermohl's) microscope. As an alternative, the sample may be preserved in formalin and allowed to settle for several days. A preservative other than formalin is necessary if the naked flagellates are not to be destroyed.

Instead of identifying and counting the plants in a sample, an estimate of total plant matter can be made by obtaining a measure of the chlorophyll content. After filtering or centrifuging a measured volume of sea water, the precipitate is treated with a standard amount of acetone, which dissolves the chlorophyll. Indication of the pigment density is then obtained by optical or spectroscopic means.

Net samples. Fine-meshed conical nets are widely used to collect phytoplankton. The conical filtering part of the net is generally fastened to a canvas foresection, which is lashed to a metal ring bearing three towing bridles. A suitable diameter for the ring is about 50 cm. The filtering part, which leads to a terminal bucket, is made of fine silk, having, for instance, 200 meshes to the linear inch. The net may be hauled vertically through the phytoplankton-bearing part of the sea, which in deep waters is between the surface and a depth of about 100 m.

There are two main problems with net samples. First, even when the finest silk is used, the smaller forms of plants easily pass through the meshes. Second, when phytoplankton crops are heavy, the meshes may become clogged, thereby reducing their filtering capacity. Further, the entire water column cut by the metal ring may not pass through the net, even when filtering at its maximum efficiency. To overcome these problems, it is essential to fit a calibrated flow meter at the entrance of the net. The meter has a propeller counting device, the revolutions of which give an indication of the volume of water that has passed through the net.

The foregoing methods simply give a measure of the standing crop of plants, the numbers (or volume or weight) present at the moment and place of sampling. If plant productivity (usually expressed as the amount of carbon assimilated below 1 m² of sea surface in 1 day) is to be measured, other means are needed. *See* SEA-WATER FERTILITY.

Zooplankton. Fine-meshed phytoplankton nets are also suitable for catching the smallest members of the zooplankton, such as protozoans and small larval stages of worms, mollusks, and crustaceans (the first nauplius stages of copepod crustaceans may be no longer than 0.1 mm). Adequate sampling of these small forms is still a major problem.

Zooplankton nets may be towed horizontally, obliquely, or vertically. The towing speed is usually about 1 knot. Nets with a mouth diameter of about 70 cm are suitable for collecting small- and medium-sized zooplankton, copepods, ostracods, pteropods, appendicularians, and others. Mesh size may range from 40 to 100 meshes to the linear inch. Larger zooplankton species, such as euphausiid shrimps, prawns, arrow worms, jellyfish, young fishes, and so on, are sampled better in a larger net with a mouth diameter of about 100 cm. This type of net is made of coarser material, such as stramin (with 11 or 12 meshes to the linear inch).

If some measure of the vertical distribution of species is being sought, a mechanism for closing the net is essential. The net may be hauled vertically from one depth to the next (for example, 100–50 m or 1500–1000 m), after which the net is closed and hauled to the surface.

The net is closed usually by sliding a brass messenger down the towing wire to trigger a spring mechanism. In the Nansen-type device, the spring disengages the towing bridles, collapsing the net, which is then closed by a line running from the mechanism to a throttling band (around the forepart of the net). For oblique or horizontal tows, a closing net should be fitted with a depth gage. Several layers may be sampled at one towing by attaching a series of nets to the tow line (Fig. 1).

Such types of plankton nets must be towed slowly if they are to fish efficiently. For taking samples from ships under way, special gear is needed. High-speed samplers, such as the one developed at the Scripps Institution of Oceanography, consist of a streamlined metal tube containing a plankton filter and a receptacle for the catch. In the Scripps model there is a depth-flow recorder at the rear of the instrument. The filter section is made of Monel metal mesh, having 23 mesh openings to the centimeter. Adequate and well-preserved samples can be taken at speeds of 8–10 knots.

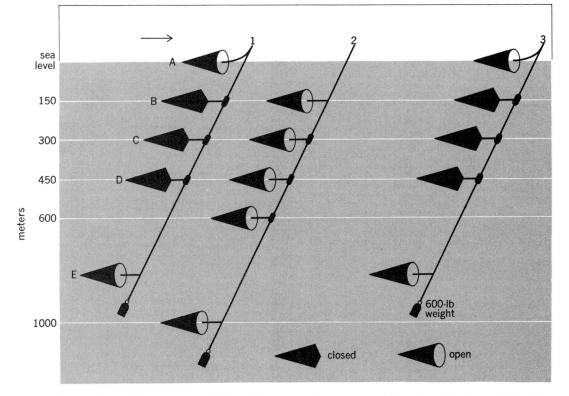

Fig. 1.   Diagram illustrating an opening-closing net tow series as taken on the Trans-Pacific Expedition. Nets A and E were standard quantitative nets. Nets B, C, and D were opening-closing nets. At point 1, just before net A entered the water, a messenger was sent down to open nets B, C, and D; then the entire series was lowered 150 m, at point 2. Between points 2 and 3 the series was raised 75 m and towed at that depth for approximately 20 min. At point 3 the series was raised another 75 m, net A was removed from the water, and a second messenger was sent down to close nets B, C, and D. The entire series was then removed from the water. Nets A, B, C, and D each fished a stratum 150 m thick while net E fished from the surface to depths of 1000 m or more. The ship maintained a speed of 1–1.5 knots during these tows. The entire procedure took approximately 2 hr. (*Scripps Institution of Oceanography*)

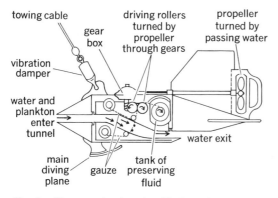

Fig. 2. Diagram showing simplified section of Hardy plankton recorder. Plankton are caught on moving gauze and covered by second gauze band as it leaves water tunnel. Gauze with plankton is wound on spool in tank of preserving fluid. (*From H. Barnes, Oceanography and Marine Biology, Allen and Unwin, 1959*)

The Hardy plankton recorder (designed by Sir Alister Hardy) is towed from merchant ships on routes across the North Sea and North Atlantic. This instrument, fitted with a forward diving plane, fishes at a constant depth of 10 m and takes a continuous sample of the plankton. An inlet at the front of the instrument leads to a tunnel at the end of which is a slowly moving silk gauze (driven by a mechanism that is turned by a propeller at the tail of the instrument). On leaving the tunnel the gauze and trapped plankton are covered by another moving gauze, the two passing back to a spool in a

preserving tank (Fig. 2). While the zooplankton is somewhat crushed and comes only from a fixed depth, the great advantage is the continuous record of plankton distribution over long distances. Monthly records can be taken on the shorter routes, thereby yielding data on seasonal changes in the distribution of both plants and animals of the plankton.

Nekton. For catching the more active fishes and cephalopods and the larger crustaceans, large-sized nets are essential. The Isaacs-Kidd mid-water trawl (Fig. 3), developed at the Scripps Institution of Oceanography and now used by research vessels of several nations, is a particularly suitable instrument. It can be towed at speeds of 4–5 knots, thus catching forms that might evade a slower-moving net. This net has a leading diving-vane or depressor at the foot of the opening and a spreader bar at the top. By these means the bridles are kept from obstructing the inlet of the trawl, which is 10–15 ft wide. The trawl is stable and easily handled and tends to catch larger nektonic animals than other nets of comparable size. When a closing device is developed for use with the trawl, its value will be much enhanced. A net that will catch the larger squids and fishes has yet to be designed.

Benthos. Organisms that live on the bottom are sampled by three main kinds of gear: grab, dredge, and trawl. For a discussion of bottom-sampling devices used to collect sediments from the ocean floor *see* MARINE SEDIMENTS.

Grab samplers. The Petersen grab is designed to sample 0.1 m² of the bottom surface and is suitable for biological work in both shallow and deep waters. The iron grab weighs about 45 kg and is dropped with the jaws in the open position. As it hits the bottom, the shock releases a catch and closes the grab. The sample is washed through a set of three sieves consisting of coarse, medium, and fine meshes. It is then weighed to give an indication of the biomass. The van Veen grab has the same weight and sampling capacity as the Petersen model, but each jaw is connected to a long lever, enabling the grab to bite deeply into the bottom deposit. This gear is most suitable for work in shallow waters with hard level bottoms.

Dredge samplers. Dredges have a bag of coarse-meshed netting attached to a rectangular wrought-iron frame which forms the mouth of the net. The towing line is attached to two bridles, movably connected to the short sides of the rectangular opening. Each of the two long sides is fashioned into an obliquely set blade which digs into the sea bed to scoop up organisms living just below the surface of the sediments as well as those that are exposed. Dredges are suitable for sampling attached and slowly moving invertebrates but not for the more active forms. Most dredges are 2–3 ft wide.

Bottom trawl. The otter trawl is the most suitable if fishes as well as invertebrates are to be caught. It is a large baglike net, tapering to a cod end. The mouth is kept open by a pair of otter boards, to which the two towing warps are attached. The head rope is buoyed by a series of floats, while the foot rope is weighted so that it

Fig. 3. Isaacs-Kidd mid-water trawl. (*Scripps Institution of Oceanography*)

drags along the bottom. The Agassiz or Sigsbee trawl is less suitable for taking fishes but yields good catches of invertebrates. This net is attached to an iron frame with a hoop at each end so that the trawl will fish whichever side is up when it falls on the sea bed.

If some measure is to be made of the density of animals on the sea floor, a counting device can be fitted to a dredge or an Agassiz trawl. The device is rotated as the gear is dragged, to indicate the distance of tow. There is no suitable attachment on an otter trawl. The catch is hauled to the surface, washed, and then weighed.

Methods of observation. Direct observation from bathyscaphs or by specially designed cameras and television equipment often provides valuable supplementary and corroborative data to those obtained by nets. If bathyscaphs can be made more mobile and equipped with sampling gear, the range and accuracy of marine biological sampling should be greatly improved. *See* SCATTERING LAYER; UNDERWATER PHOTOGRAPHY; UNDERWATER TELEVISION. [NORMAN B. MARSHALL]

Bibliography: H. Barnes, *Oceanography and Marine Biology*, vol. 14, 1976; E. A. Drew et al., *Underwater Research*, 1975; A. C. Hardy, *The Open Sea: Its Natural History*, 2d ed., 1971; J. W. Hedgpeth (ed.), *Treatise on Marine Ecology and Paleoecology*, Geol. Soc. Amer. Mem. no. 67, vol. 1, 1957; B. Kullenberg, The technique of trawling, in *Galathea Deep Sea Expedition 1950–52*, 1956; N. B. Marshall, *Aspects of Deep Sea Biology*, 2d ed., 1977; H. U. Sverdrup, M. W. Johnson, and R. H. Fleming, *The Oceans*, 1942; U.S. Navy, Obtaining biological specimens, in *Instruction Manual for Oceanographic Observations*, H. O. Publ. no. 607, 2d ed., 1955.

Marine ecosystem

Marine ecology comprises the ecology of the world's oceans with their shores and estuaries. Individual organisms, groups of organisms, and the ocean environments in which they exist consti-

Fig. 1. A community of ribbed mussels attached to a decayed log in the intertidal zone.

Fig. 2. A marsh grass–mussel community, representing a small marine ecosystem of the intertidal zone.

tute ecological systems, termed ecosystems. Examples vary in size from a particular seaweed ecosystem, to a tidepool ecosystem, to an estuarine bay ecosystem (Figs. 1 and 2). The size limits, or boundaries, of a particular ecosystem that distinguish it from others may depend upon physical barriers to community dispersal, such as a submarine mountain; environmental factors that restrict the area of the biotic community, such as the salinity factor; the productivity cycles within the community; or the reproductive and dispersal potentials of the community. This conceptual unit of ecological science, whether large or small, possesses individuality, a degree of stability and permanence, characteristic functional cycles, and readily recognizable components, either living or nonliving or both.

The nonliving, or abiotic, materials in a marine ecosystem cycle comprise not only a variety of water-soluble inorganic nutrient salts, such as the phosphates, nitrates, and sulfates of calcium, potassium, and sodium and the dissolved gases oxygen and carbon dioxide, but also organic compounds, such as the various amino acids, vitamins, and growth substances. Most of the solid material dissolved in the sea originated from the weathering of the crust of the Earth.

MARINE ENVIRONMENT

The environment of the marine ecosystems is a subject of study by oceanographers who investigate the physical, chemical, and biological properties of ocean waters, ocean currents, and ocean basins. Although it is by far the world's largest inhabitable environment, much of its area and volume is relatively unproductive of life. Yet, it contains representatives of a few major taxonomic groups not found in terrestrial or fresh-water environments. *See* OCEANOGRAPHY.

Chemical factors. The chemical properties of ocean waters constitute an important aspect of the marine environment. Such chemical factors as the dissolved solids and gases in the ocean waters have been the subject of study. *See* SEA WATER.

Dissolved solids. The total quantity of dissolved solids in the waters of the world's oceans or in the marine ecosystem as a whole approximates 5×10^{16} metric tons, enough to form a layer 153 in. thick over the land area of the Earth. Typical ocean water contains about 34.9 grams (g) per liter of dissolved materials, in which there are 19.0 g of chlorine, 10.50 g of sodium, 1.35 g of magnesium, 0.885 g of sulfur, 0.400 g of calcium, and 0.380 g of potassium. Some of the most important nutrient elements are present in very small amounts, for example, the phosphates needed for the growth of algae. Phosphorus exists in ocean waters as phosphate ions. It is an essential component of living things, and in ocean water the amount present may limit the production of plants. The inorganic form occurs in amounts varying from zero concentration to over 0.10 mg per liter. Phytoplankters, principally microscopic algae, may absorb inorganic phosphorus and reduce the amount remaining in the water to a minimum. Some is permanently lost to recycling by being bound in nonsoluble forms and deposited on the sea bottom. By their death and decay, phosphorus is returned to the aquatic environment. Certain of the 40 or more elements in sea water which, like phosphorus, exist in extremely small amounts are concentrated to important degrees. For example, macroscopic algae concentrate potassium and iodine in relatively large amounts. Of special importance is silicon, utilized by diatoms and other silica-secreting organisms.

Dissolved gases. The essential gas, carbon dioxide, is readily dissolved in sea water, but in equilibrium with the air it would contain only about 0.5 milliliters (ml) of free CO_2 per liter at 0°C. Actually, sea water contains about 47 ml of CO_2 per liter because appreciable amounts are present in the form of carbonate and bicarbonate ions. This gives great importance to the marine ecosystem as the world's reservoir of CO_2 although only about 1% is in the free form. Since, in relation to the atmosphere, its CO_2 content is 50 times greater, oceans regulate the concentration in the atmosphere. The CO_2 content of sea water originates not only from the atmosphere but also from biochemical processes in the sea—respiration and decay—and in the soil of the ocean bottom and coastal shores. The free carbon dioxide in water exists in simple solution and in the form of carbonic acid, H_2CO_3. Combined CO_2 is in the bicarbonate ions (HCO_3^-) and the carbonate ions (CO_3^{--}). These and other ions of weak acids constitute the buffer system of sea water that stabilizes its hydrogen-ion concentration, or pH, at about 8–8.4 at the surface and at 7.4–7.9 at the deeper levels. The high buffer capacity of sea water is a dominant environmental characteristic. Chemical and biochemical interactions and actions of biological agents which would otherwise produce pH changes that could seriously modify living conditions are largely negated by this buffer system. The ocean environment is thus chemically stable. Marine plants, the producers, may utilize primarily either free CO_2, or some free CO_2 and some combined CO_2. In any case, lack of CO_2 is not a limiting factor to growth of ecosystems in the sea. High concentrations, however, are a limiting factor to many marine animals such as fish.

The dissolved oxygen content in the sea-water environment varies from 0 to over 8.5 ml per liter. The oxygen in water comes from the atmosphere by diffusion and from aquatic plants by their photosynthetic action. Low temperatures and low salinities of water favor high solubilities of oxygen gas. There are low-oxygen strata in ocean waters, but in general the supply is adequate for animals. The considerable oxygen content of the deeper water layers of the ocean reached these layers when they were in the photosynthetic surface strata.

Physical factors. The physical aspects of marine ecological systems, temperature and salinity, are especially significant because of their degree of constancy away from land. The mean annual temperatures in different latitudes on the Earth remain unchanged. Seasonal variations in the ocean are small compared to those on land. Also, the entire temperature range in ocean waters is within the tolerance limits of numerous plants and animals. In polar and tropical seas, temperatures do not vary more than about 5°C during the year. Temperate seas commonly vary about 10–15°C. The extent of seasonal variations decreases in the deeper strata. In temperate and tropical regions a permanent thermal gradient, the thermocline, in which temperature decreases rapidly with depth, lies between the surface mixed layer and the deep unmixed strata. Low temperatures, around 3°C, in the deep and bottom waters of the ocean exist because the waters of greatest density, formed in high latitudes, sink to the bottom or to levels of similar density, then spread out and move toward warmer latitudes, to form a pattern of oceanic circulation. The salinity of the world's oceans varies only a few parts per thousand and averages around 35°/oo. Major latitudinal differences in density re-

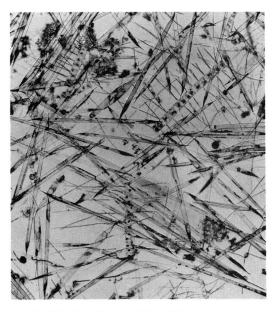

Fig. 3. Plankton diatoms. Their light spiny shells are suitable for floating on the surface of the water. *(Photograph by P. Conger, Smithsonian Institution)*

sult from temperature differences. As expected, there is a vast system of oceanic circulation involving all depths and modified by large water masses contributed by adjoining seas. Thus, this offshore environment of the unit marine ecosystem is characterized by relative constancy of salinity over exceedingly large areas. The massive and relatively constant properties of this environment are reflected in the distribution, abundance, and morphological traits of its biota.

Biota. The organisms that comprise the biota of the marine ecosystem are characterized by a lack of diversity among the plants, in contrast to the animals. This generalization applies to both microscopic and macroscopic organisms. With few exceptions, all types of animals inhabiting land and fresh-water environments occur in the marine ecosystem. Six major animal groups, including ctenophores, starfishes, and certain worms, are restricted to the marine environment. Other major taxonomic units are predominantly marine. The kinds of plants and animals in the marine environment fall into three major groups: organisms of the plankton, nekton, and benthos.

Plankton organisms. These organisms are small, mostly microscopic, and have little or no power of locomotion, being distributed by water movements. There are two main types, the phytoplankton and zooplankton. The former includes all of the floating plants, such as the small algae, fungi, and sargassum weeds. Of these, the most important in the economy of the sea are algae—diatoms and dinoflagellates. They are the major producers in marine plankton (Fig. 3). Diatoms are microscopic, unicellular plants, some of which form chains. They possess characteristic shells composed of translucent silica, and have a great variety of form and sculpture. The shell structure consists of two nearly equal valves, one of which fits over the other and hence may be compared to a box with a telescoping lid. The valves are joined by connecting bands. The protoplasm within the shell is exposed by a slit or by small pores to permit metabolic interchanges with the environment. During reproductive division of the protoplasm of the diatom, one of the two protoplasmic daughter cells retains the larger epivalve, or lid valve, the other the hypovalve, or box valve. The daughter cells then lay down the needed complementary valves. This simple binary form of division is the most common one. It permits rapid production of vast populations in favorable environments of an ecosystem. During successive binary divisions, the size attainable by individuals is progressively reduced until a minimum limit is reached when, usually, the protoplasmic content of the shells escapes from the parted valves. It is enclosed in a flexible pectin membrane and is called an auxospore. These specialized spores grow in size and finally form the characteristic two valves. Diatoms possess one or more chromatophores, ranging in color from yellow to brown.

Diatoms occur as fossil, siliceous shell deposits, called diatomaceous earth, and as living producers in practically all habitats of the broad marine ecosystem. They are found floating in water, attached to the bottom, on larger plants, on animals, and, as

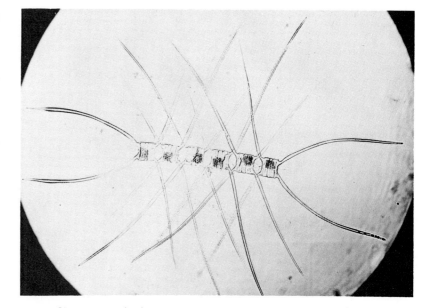

Fig. 4. *Chaetoceros atlanticus* Cleve, a branched-type oceanic diatom. (*Photograph by P. Conger, Smithsonian Institution*)

spores, enclosed in Arctic ice. Free-floating diatoms possess structural adaptations that permit adjustments in depth. The bladder-type diatom is relatively larger, and in some forms such as *Planktoniella*, the shape is disklike, so that a zigzag course is followed in sinking. The hair type, such as *Rhizosolenia*, is a long and slender diatom, which sinks slowly when the long axis is horizontal to the pull of gravity, but more rapidly when oriented vertically. The ribbon-type cells, such as *Fragillaria*, are broad and flat, and are attached to form chains. The most abundant diatoms in the offshore oceanic waters are of the branched type, such as *Chaetoceros*. They possess numerous spiny projections that resist sinking (Fig. 4).

Dinoflagellates possess whiplike flagella that provide a slight degree of locomotion. Like diatoms, they possess structural modifications that indicate adaptation to environmental conditions. Some, such as *Dinophysis*, possess winglike structures that favor suspension; some have cellulose plates; others are naked cells. Many kinds are luminescent.

Phytoplankton organisms are much more abundant in nutrient-rich coastal waters than in offshore oceanic waters. They are the primary producers upon which large and small marine animals feed. The primary determinant of phytoplankton productivity is the Sun's radiant energy. Yet, less than 0.3% of the incident solar energy is converted by the diatoms and dinoflagellates to provide the basic food of the world's oceans.

Zooplankton organisms are the floating or weakly swimming animals, which include the eggs and larval stages as well as adult forms. The principal kinds include numerous Protozoa, such as Foraminifera and Radiolaria; great numbers of small Crustacea, such as ostracods and copepods, with their various larval stages; various jellyfishes; numerous worms; a few mollusks; and also the eggs and early developmental stages of most of the

Fig. 5. A benthic community of brittle stars and isopods at a depth of 1200 m in the San Diego Trough, off California, 32°54′ N and 117°36′ W. (*Photograph by G. A. Shumway, U.S. Naval Electronics Laboratory*)

nonplanktonic organisms in the sea. Plankton organisms are grouped on a basis of size. The smallest organisms range from about 5 to 60 μ, the next largest range up to about 1 mm in size, the next up to 1 cm, and the largest plankton group over 1 cm.

Nekton organisms. These are the actively swimming animals of the marine ecosystem. They comprise adult stages of such familiar forms as crabs, squids, fish, and whales. Some of these undergo long horizontal migrations over hundreds of miles; some migrate periodically to great depths; and a few live mostly in deep waters, for which habitat they possess marked adaptations.

Benthos organisms. These inhabit the bottom, and range from high-tide level on shore to the deep-sea bottom. There are relatively few kinds and numbers of animals on the deep-sea floor. They are mainly mud dwellers, possessing characteristic structures permitting life in a quiet, dark, muddy environment where food is scarce. Certain isopods, sponges, hydroids, brittle stars, sea urchins, and shrimp are typical animals of the deep-sea biota (Fig. 5). Inshore bottom communities at depths less than 50 m are rich in plants and animals. Although mosses and ferns are entirely absent in the sea, there are approximately 30 species of marine flowering plants. Of these, eelgrass (*Zostera*) is a characteristic shore plant. It is a perennial flowering plant, not actually a grass, that is worldwide in distribution, and is most abundant on soft bottoms in protected, coastal areas. Large brown algae, such as *Fucus* and *Ascophyllum*, are widely distributed conspicuous plants of exposed rock surfaces in the intertidal zone. *Ulva*

and *Enteromorpha* are typical green algae of mud flats in quiet waters. Red algae, such as dulse (*Rhodymenia*), are commonly found below the intertidal zone. These are important food plants for the bottom-living animals of the coasts.

Conspicuous members of the benthic animal communities of the sea coast are barnacles, snails, mussels, clams, oysters, sea anemones, sea urchins, sea cucumbers, and starfish. Shore corals are largely restricted to warmer seas. The bottom animals of the marine ecosystem have structural modifications facilitating adhesion, burrowing, feeding, and protection.

MAJOR DIVISIONS

The marine ecosystem can be divided into two large areas, the pelagic and benthic divisions. Each of these consists of various zones.

Pelagic division. This division embodies all the waters of the oceans and their adjacent salt-water bodies. It is divisible into the neritic zone, extending offshore to the edge of the continental shelf to a depth of over 200 m, and the oceanic zone, embracing the remaining offshore waters. The neritic zone is rich in plant nutrients, especially phosphates and nitrates, which originate from coastal tributary waters and from bottom deposits carried upward to surface water layers by upwelling, diffusion, turbulence, or convection. The water is more variable in density and in chemical content than oceanic waters. It is also far more productive of plankton, fish, and shellfish. Many of the inshore, coastal forms are adapted to withstanding brackish waters of coastal tributaries. The oceanic zone has a well-populated upper, lighted, 200-m

Fig. 6. A littoral community with abundant sand dollars. This photograph was taken at a depth of 4 m in Mission Bay Channel, San Diego, Calif. (*Photograph by R. F. Dill, U.S. Naval Electronics Laboratory*)

stratum, and deeper, relatively dark, and sparsely populated layers, characterized by great pressures, animals modified for life in darkness and under great pressures, and very few bottom animals. Since there is less plankton and suspended organic material in the water, light pentrates further than in the neritic zone. Also the water is usually very transparent. Its salt content is uniformly high and not subject to major variations in time and space. In the upper photosynthetic zone, or zone of productivity, plant nutrients are much less concentrated, and the cycle for replenishment is much longer than in coastal waters that receive nutrients and organic detritus from the land.

Benthic division. This division of the marine environment embraces the entire ocean floor, both coastal bottom and deep-sea bottom, properly termed the littoral and deep-sea systems, respectively. The littoral system consists of the eulittoral zone and the sublittoral zone.

Eulittoral zone. This extends from high-tide level to about 60 m, the approximate depth below which attached plants do not grow abundantly. This is probably the richest zone of the marine ecosystem in respect to numbers and kinds of organisms, as well as in variety of ecological types and habitat modifications. The upper intertidal portion of the zone extends between high- and low-water marks, a vertical distance that varies on different continental shores from over 12 m to a few centimeters. Changes in such environmental factors as light, temperature, salinity, and time of exposure vary tremendously within short vertical distances across the zone. These variations are reflected in the shapes, movements, tolerances,

and life histories of the characteristic animals and plants. Numerous small, partly independent ecosystems thrive in this area because the substratum includes a variety of rock exposures, gravel, sand, and mud types admixed in all degrees (Fig. 6). The lower, permanently submerged portion of the eulittoral zone is characterized by abundant sessile plants, such as conspicuous rock weeds (*Fucus*), bladder wrack (*Ascophyllum*), green sea lettuce (*Ulva*), and kelp (*Laminaria* and *Nereocystis*), many of which are common to both portions of the zone. Certain of those algae, such as the giant kelp, form productive algal forests that extend down below the typical eulittoral zone into the sublittoral zone. Here, they utilize the dimly lighted, lower, and least productive portion of the littoral system.

In tropical littoral waters, the coral reef communities abound with characteristic types and forms of both plants and animals. Visible algal growth is sparse, animals often greatly predominate over plants, and depth range is to about 60 m. Sparsely populated animal benthos extends to greater depths into the sublittoral zone, terminating at depths varying in different latitudes between about 200 to 400 m, depending upon light and temperature factors that modify the distribution of benthic animals and plants.

Deep-sea system. This portion of the benthic division is subdivided into an upper part, the archibenthic zone or, more meaningfully, the continental deep-sea zone, extending from the edge of the continental shelf (200–400 m) to depths of about 800–1100 m, and the abyssal-benthic zone that embraces the remainder of the benthic deep-sea system (Fig. 5). The zones have little or no light, relatively constant conditions of salinity and temperature, and steadily decreasing numbers and kinds of organisms. In the abyssal regions of great depth, perpetual darkness, and extremely low temperature (5 to −1°C), the extreme in monotony of environment prevails. However, bacteria and at least some animals exist at about all known depths. The remains of plants and particulate organic detritus continuously rain down to settle over the bottom as a light coating, utilized by the bottom dwellers for food. The pelagic food supply from above decreases with the increase in offshore distance for two reasons: The offshore oceanic waters contain less particulate matter than inshore waters, and the longer period required to descend the deeper strata results in greater disintegration while sinking. *See* DEEP-SEA FAUNA.

Estuaries. These are coastal adjuncts of the marine ecosystem. They embrace bodies of water which, by virtue of their position, are directly subject to the combined action of river and tidal currents. Compared with offshore ocean waters, they lack constancy, possessing characteristic horizontal and vertical gradients in physical, chemical, and biological properties. These gradients are subject to pronounced changes in space and time. Estuarine waters are characterized by rapidity of response to changing external conditions. Their temperature, salinity, and turbidity conditions are distinctly not uniform. They are usually rich in plant nutrients of land origin, which they transport to coastal waters, thus fertilizing these waters. Estuarine waters contain an environmentally selected biota containing representative fresh-water and salt-water forms. *See* ESTUARINE OCEANOGRAPHY.

Marshes. These transitional land-water areas, covered part of the time at least by estuarine or coastal waters, comprise parts of the peripheral area of the marine ecosystem. Mangroves are characteristic woody marshes in tropical tidal waters of flat, muddy shores. More generally, coastal marshes are dominated by grasses and sedges. Marshes have characteristic biota for a certain latitude. In temperate regions, they are inhabited by characteristic birds such as bitterns and rails, and by crustaceans such as sand hoppers and fiddler crabs. Typical marsh animals and plants have tolerances to fresh water and salt water that differ from those of related forms living in fresh-water or salt-water habitats.

Sediments. Inanimate, particulate materials of organic and inorganic origin that have settled on the bottom in aquatic environments comprise the sediments. Vast quantities of particulates, eroded from land surfaces by natural waters, are carried to the sea by rivers and estuarine waters. They settle to the bottom and provide food for benthic animals. Sediments may be carried to the surface, or photosynthetic stratum, either by upwelling of coastal waters or by ocean currents, and thus enter a biochemical cycle of the marine ecosystem. Marine sediments are made up of microscopic fragments of weathered rock, partly decomposed plant and animal remains, skeletal remains of various organisms, inorganic precipitates from sea water, terrestrial particulates, and particulate material

of volcanic origin. *See* MARINE SEDIMENTS.

[CURTIS L. NEWCOMBE]

Bibliography: J. Fraser, *Nature Adrift: The Story of Marine Plankton*, 1962; E. D. Goldberg, The oceans as a chemical system, in *The Sea: Ideas and Observations on Progress in the Study of the Seas*, vol. 2, pp. 3–25, 1963; C. P. Idyll, *Abyss: The Deep Sea and the Creatures That Live in It*, 1964; H. B. Moore, *Marine Ecology*, 1958; C. L. Newcombe, Mussels, *Turtox News*, vol. 25, no. 1, 1947; E. P. Odum, *Fundamentals of Ecology*, 1971; H. U. Sverdrup, M. W. Johnson, and R. H. Fleming, *The Oceans: Their Physics, Chemistry, and General Biology*, 1942.

Marine fisheries

The harvest of animals and plants from the ocean to provide food and recreation for people, food for animals, and a variety of organic materials for industry. In 1976 the world marine harvest was about 66,000,000 metric tons—the highest ever recorded—plus an undetermined catch by recreational and subsistence fishermen. It is now generally agreed that, with present fishing gear and methods, the world catch is approaching a maximum, which may be less than 100,000,000 metric tons. If methods can be devised to harvest smaller organisms not currently used because they are too costly to catch and process, it has been estimated that the yield could perhaps be increased severalfold. The Soviet Union is said to have succeeded in developing an acceptable human food product from Antarctic krill, an abundant small shrimplike animal which is the principal food of the blue whale. This could lead to a substantial increase in the harvest of the sea.

Harvest of the sea. The world marine commercial fish catch grew more than 6% per year from the end of World War II to 1967. In 1969, however, the catch dropped slightly, and in the period since 1970 the average rate of increase was only about 1% per year. Another way of putting it is that in 1960–1967 the world catch increased by nearly 60% (Fig. 1), but in the following years grew scarcely at all. Much of this decline in rate of growth was caused by the virtual collapse of the Peruvian anchovy fishery, which at its peak accounted for about 20% of the total world catch from the sea (see Fig. 1).

Whales are not included in these catches be-

Table 1. World whale catch in 1973 in numbers of whales*

Whaling area	Fin	Sei and Bryde	Minke	Sperm	Others
Southern Hemisphere					
Antarctic Ocean†	1,288	4,392	7,713	4,927	
Pacific Ocean	11	19		3,227	311
Atlantic Ocean	1	497	650	2,363	
Indian Ocean	41	10	175	3,099	1
Northern Hemisphere					
Pacific Ocean and Bering Sea	460	2,585		8,568	215
Atlantic and Arctic Oceans	342	139	2,445	613	10
Totals	2,143	7,642	10,983	22,797	537

*From *International Whaling Statistics LXXIII*, Bureau of International Whaling Statistics, Oslo, 1974.
†1973–74 season in the Antarctic.

cause they are recorded by number rather than by weight (Table 1). The 1973 world whale catch yielded about 128,000 tons of whale meat; 38,000 tons of meal; 6000 tons of solubles; 18,000 tons of other products; and 5,600,000 units of vitamin A, in addition to 860,000 barrels of oil.

Most of the world fish catch still comes from the Northern Hemisphere, but the catch in the Southern Hemisphere was growing until the 1970s. In the late 1950s nearly 90% of the world catch came from north of the Equator. By 1967 this had dropped to about 70%, largely through the phenomenal development of the Peruvian anchovy fishery. The decline of this fishery in the 1970s caused the total Southern Hemisphere catch to drop also. By 1975 the Northern Hemisphere produced about 77% of the world total. The Northern Hemisphere includes only about 43% of the total area of ocean, but it contains most of the world's estuaries and continental shelves. It is in these rich and relatively shallow waters that most marine fish and shellfish resources are concentrated.

Sixteen countries landed 1,000,000 metric tons or more of fishery products (live weight) in 1973. These countries accounted for about 75% of the world catch (Table 2). In the late 1950s the United States ranked second in weight of fishery landings; in 1976 it ranked sixth. This has been cited by many as evidence that the United States is declining as a world fishing power, but like many popular views about fisheries this is a gross oversimplification. Total domestic fishery production has remained about level for many years, while catches by several other nations have been growing. Americans have obtained increasing amounts of fish and shellfish by imports. With less than 7% of the world population, the United States consumes more than

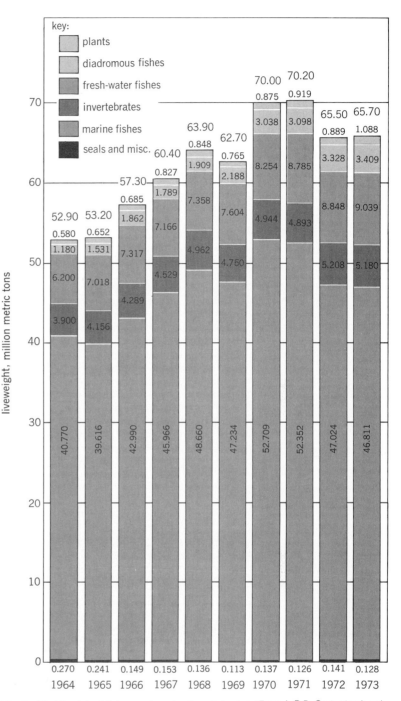

Fig. 1. World catch of fishes and marine invertebrates. (*From L.P.D. Gertenbach, ed., Yearbook of Fishery Statistics 1973, Food and Agriculture Organization, 1974*)

12% of world fishery production, and thus is the most attractive fishery market in the world.

The most remarkable fishery development was the climb by Peru from a position of insignificance among fishing nations in the late 1950s to first place by weight landed in the late 1960s. The subsequent decline of this major fishery, based almost entirely on a single species of anchovy, illustrates the inherent instability of most fishery resources. This densely schooling herringlike fish owed its tremendous abundance to the high biological fertility of the Humboldt Current. Fluctuations in

Table 2. Leading commercial fishing nations in 1976*

Nation	Weight,† 10^3 metric tons	Percent of world catch
Japan	10,620	14.5
Soviet Union	10,134	13.8
People's Republic of China	6,880	9.4
Peru	4,343	5.9
Norway	3,435	4.7
United States	3,004	4.1
Republic of Korea	2,407	3.3
India	2,400	3.3
Denmark	1,912	2.6
Thailand	1,640	2.2
Spain	1,483	2.0
Indonesia	1,448‡	2.0
Philippines	1,430	1.9
Chile	1,264	1.7
Canada	1,136	1.5
South Vietnam	1,014	1.4
Totals	54,550	74.3

*From *Fisheries of the United States, 1977*, U.S. Department of Commerce, National Oceanic and Atmospheric Administration, C.F.S. no. 7500, 1978.

†Includes weight of clam, oyster, scallop, and other mollusk shells.

‡Data estimated by the Food and Agriculture Organization.

anchovy catches are related mainly to fluctuations in the Humboldt Current. *See* OCEAN CURRENTS; SEA-WATER FERTILITY.

Most marine fisheries are located close to coasts. Only a small part of the world catch is taken more than 100 mi (160 km) from shore.

No one knows exactly how many different kinds of marine life are used by people. In official statistics many species are lumped together. Several thousand species are included in the world catch, but a surprisingly small number make up the bulk of the catch. For example, most of the domestic catch in the United States, over 83% by weight, is made up of only 12 kinds of fish and shellfish (Table 3). Omitting the considerable part of the world marine fishery catch which is unsorted and unidentified (9,900,000 metric tons), 20 kinds of fish, shellfish, and plants make up about 99% by weight of the world catch. The 8 most important in order of weight landed in 1976 were herring, cod, jack, redfish, mollusks, tuna, crustaceans, and flounder. They made up about 66% of the total commercial harvest.

Changing patterns. Before World War II few fishermen ventured very far away from their home shores in search of marine fish. There were some notable exceptions, however. Some European fishermen had been lured years before to the rich banks of Newfoundland, Nova Scotia, and New England; whaling had begun in the productive waters surrounding Antarctica; the Japanese had started their march across the ocean in search of salmon, tuna, and other species; and the United States tuna fleet was already developing its fishery off Central and South America. But these developments were halted during the war. After the war the need for animal protein stimulated several countries to develop distant water fisheries. In the forefront were Japan and the Soviet Union. These two countries have developed modern, self-sufficient fleets of factory ships, catcher boats, and supply vessels which can and do operate anywhere in the world ocean. Japan more than tripled its catch from 1938 to 1973, from about 3,500,000 to 10,700,000 metric tons. In the same period the Soviet Union more than quintupled its catch, from about 1,600,000 to 8,600,000 metric tons. Almost 99% of the Japanese catch and about 93% of the Soviet catch come from the ocean. Both countries are net exporters of fishery products.

In contrast, the United States fish catch has remained almost static for many years. It rose slightly after the war, reaching an all-time high of about 3,000,000 metric tons (live weight) in 1962. Since 1962 the total catch has dropped about 12%. About 4½% of the United States commercial catch comes from fresh water. The United States is a net importer of fishery products. In 1974 the United States imported about 2,226,000 metric tons (live weight) and exported less than 4% of that amount.

United States fishing industry. The reasons why the United States supplies most of its demand for fishery products by importing, whereas other major fishing nations produce more than they consume are complicated, and only a brief outline can be given in a short article. First, it must be understood that, although there are a few large fishing companies in the United States, most commercial fishing is conducted by a large number of small, independent operators. Most of them are in competition with each other, either to make the catch or to purchase the raw material from the fishermen. These segments of the industry can be classified in various ways, but the important distinction is between the fisherman and the processing-distributing segment of industry.

Problems of fishermen. Almost all United States fishermen are independent operators. Some are prosperous, but many are struggling to make a living. In many fisheries there are more fishermen and units of gear than are necessary to make the catch. They are hemmed in by laws and regulations, many enacted in the name of conservation, but most merely increasing the cost of catching fish. The living resources fluctuate widely in abundance from natural causes, and their migration patterns change from time to time. Most fishermen in the United States lack the flexibility to shift from one fishery to another in response to these changes. They pay more for boats and gear than do their foreign competitors. Foreign fishermen are liberally subsidized in various ways by their governments, and substantial quantities of this subsidized catch are offered in the United States at prices lower than American fishermen are willing to accept. Foreign fishermen, with their highly organized and efficient fleets, are crowding the independent United States fisherman off traditional fishing grounds, or making incidental catches of resources which support traditional American fisheries. Many of the oldest fishery resources in the United States are fully utilized. Others have been overfished, and attempts to rehabilitate them are being made. These obstacles are almost overwhelming to many fishermen in the United States.

Table 3. Major kinds of fish and shellfish in the 1977 United States commercial marine fishery catch*

Kinds of fish and shellfish	Landed value,† 10^6 dollars	Weight, 10^3 lb (metric tons)‡	
Shrimp (less than 10 spp.)	355.2	476,654	(216,207)
Salmon (5 spp.)	221.8	335,642	(152,245)
Tuna (5 spp.)	135.8	345,229	(156,593)
Crab (less than 10 spp.)	202.5	398,539	(180,774)
Menhaden (2 spp.)	68.1	1,796,115	(814,704)
Flounder	59.5	169,603	(76,931)
Oysters	52.5	46,026	(20,877)
Clams (less than 10 spp.)	74.3	96,160	(43,617)
Cod	18.9	86,481	(39,227)
Scallops (3 spp.)	46.0	27,826	(12,622)
Lobster (2 spp.)	67.3	37,191	(16,870)
Herring (2 spp.)	11.6	155,847	(70,691)
Totals	1313.5	3,971,313	(1,801,357)
Grand totals of commercial marine catch	1515.1	5,198,100	(2,357,818)

*From *Fisheries of the United States, 1977*, U.S. Department of Commerce, National Oceanic and Atmospheric Administration, C.F.S. no. 7500, 1978.

†Amount paid to fishing crews, not retail value.

‡Weight of mollusk shells not included.

Most Americans believe that elimination of foreign fishing off their shores would solve all the major problems of the domestic fisheries. It is probable that Congress will draft legislation declaring domestic jurisdiction over a coastal zone extending 200 mi (320 km) seaward from the shores. This will remove one source of difficulty for domestic fisheries, but it has not been clearly recognized that it will create other problems. Extended jurisdiction, whether it is accomplished by unilateral action or by international agreement, will focus attention on the remaining domestic fishery management problems. Preoccupation with foreign fishing has caused some exceedingly difficult domestic problems to go almost unnoticed. Many of these problems are long-standing, and have resisted solution. The pertinent states may find that they can no longer afford virtually to ignore fishery management responsibilities.

The United States tuna and shrimp industries generally, and some other segments of the American fishing industry, have been able to avoid most of these problems. One reason is that tuna and shrimp are highly popular seafoods in the United States, and this demand has helped these industries to meet their competition. But as the harvest reaches equilibrium these once prosperous segments of the American fishing industry are having problems. Japan, the Soviet Union, and other nations that fish in distant waters have been able to expand because their flexibility and self-sufficiency allow them to move toward new resources as the catch on older fishing grounds declines. With few exceptions the American fisherman does not have this flexibility.

Problems of processor and distributor. These segments of the United States fishing industry do not usually have the same difficulties as the fishermen. Those who rely upon a single species, as the now defunct California sardine industry did, are at the mercy of a fluctuating supply of raw material and, when the total catch begins to drop, they are likely to encourage fishermen to increase their fishing effort to maintain the total catch at a level that will protect capital investment. Such a policy leads almost inevitably to overfishing and, possibly, destruction of the resource. A reasonable solution is to have alternative resources to turn to as the abundance of a species drops. Unfortunately, however, no two kinds of fish behave exactly alike, and it requires new techniques, and often other types of fishing gear, to catch another kind of fish economically. Thus, while the principal resource is abundant, the industry has no interest in seeking alternate resources, and when the principal resource declines, the capital to develop other fishing methods is hard to find.

Other fish processors in the United States stabilize their supply of raw material by importing partially processed or processed fish in quantity. There has been a growing tendency to merge with large food-processing companies. By diversifying operations and source of supply the processor or distributor of fish can avoid many economic problems.

Sport fisheries. Sport fishing in the ocean is often ignored in discussions of marine fisheries. This is probably because there are not good records of sport fish catches and because the individual sportsman's catch is small. Information on sport catches is gathered by interviews or questionnaires that represent rather small samples of the total number of fishermen. Although these provide estimates rather than precise counts, there is general agreement that sport fishermen in the United States catch large numbers of marine fish and some shellfish. The total catch by weight is at least 11%, and may be as great as 25 to 30%, of the total domestic commercial catch, and it is growing steadily. The catch of some species is as large as, or even larger than, the commercial catch of the same species. Thus, the marine sport fisheries are an important force in determining the condition and the yield of many of the United States coastal fishery resources. Sport fisheries must be considered in any fishery management plan. Marine sport fishing is an important activity in other countries but is probably largest in the United States.

A considerable industry has developed around the marine sport fisheries. The investment in manufacturing and retailing establishments for tackle, boats, motors, bait, fuel, and all the other necessities of the fisherman is large. Operators of marinas, fishing piers, and other establishments in the coastal area may derive all or a considerable part of their income from sport fishing and associated activities.

Fishery management. The aim of modern biological fishery management is maintaining the resource at the level of maximum sustainable yield, which means reaching a balance between the capacity of the resource to renew itself and the harvest that man may safely take. A fishery resource can reach an equilibrium at almost any level of fishing intensity. After fishing begins, the catch increases in proportion to the fishing effort. There is a limit, however, to the total amount that may be caught. If the intensity of fishing increases beyond that point, the total catch will begin to drop because the capacity of the resource to renew itself has been reduced. The catch per fisherman will drop more rapidly than the total catch, for more fishermen will be sharing a smaller catch. Actually, the catch per fisherman will begin to drop before the maximum sustainable yield is reached, for as the fishery grows fishermen begin to compete for the available fish. Many economists believe that the amount of fishing effort should be limited at the point of maximum economic yield, which is reached before the catch reaches a maximum, and which is the point at which the value of the catch over the cost of taking it is at a maximum. From a conservationist's point of view such a restriction would have advantages, for limiting the catch at a level below the maximum biological yield would provide a safety factor against overfishing. A relationship between fishing intensity, total catch, and the numbers or weight of fish in the resource is illustrated in Fig. 2.

Very few marine fisheries are being regulated to maintain the maximum sustainable yield. The classic examples were the Pacific halibut fishery and the fur seal industry on the Pribilof Islands. Both resources had been restored from a condition of overfishing and were producing approximately the maximum sustainable yield. The joint Canadian–

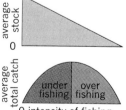

Fig. 2. Relations among fishing intensity, stock size, and average sustainable total catch from a fish population (not to scale).

United States halibut management program has been affected adversely by incidental catches of foreign trawlers fishing for other species. The North American Pacific halibut catch is now less than half of what it was before intensive foreign fisheries developed in the Gulf of Alaska. In the northwest Atlantic a unique development in the mid-1970s was international agreement on a total catch quota for all species combined, which is less than the sum of the quotas for individual major species and stocks. This approach forces the fishing fleets to make major strategy decisions in advance of the fishing season. The plan was designed to relieve pressure on important species, such as haddock and yellowtail flounder, which have been seriously overfished. Fishery management is made difficult by local traditions, natural fluctuations in abundance, difficulty of surveillance and enforcement of laws, and domestic or international disagreements as to the condition of the resource.

International management. Where fishermen of two or more nations are harvesting stocks of fish jointly, various international arrangements have been developed to deal with mutual problems. The 1958 Geneva Convention on Fishing and Conservation of the Living Resources of the High Seas provides general guidelines for international fishing activities, but developments in the 1970s made that convention obsolete. It was hoped that the round of Law of the Sea Conferences, which convened in Caracas late in 1974, reassembled in Geneva in the spring of 1975, and continued in New York in the fall of 1975, would resolve some major international fishery problems. As already mentioned, however, the United States Congress, impatient at the slow pace of international negotiations, might take matters into its own hands. Unilateral action by the United States and other nations is likely to add complications rather than improve the situation.

The Food and Agriculture Organization of the United Nations, through its Department of Fisheries, has established a number of regional fishery councils and commissions. Several international fishery conventions have been negotiated to deal with single joint-fishery agreements or special fishery problems. Some involve only two nations, such as the International Pacific Salmon Fisheries Convention between Canada and the United States. Others have many members, such as the International Convention for the Regulation of Whaling, which has 15. Often, when special problems develop, as when one nation is fishing off the coast of another, bilateral agreements of relatively short duration are negotiated. Such agreements have been necessary in the continental shelf region between Cape Cod and Cape Hatteras as foreign fishing activities there increased. This region is outside the regulatory area defined by the International Convention for the Northwest Atlantic Fisheries. Although these and other international arrangements for fishery management are far from perfect, they have been considerably more effective in resolving conflicts and promoting rational management than individual nations have been with marine fishery management in their own waters.   [J. L. MC HUGH]

Bibliography: *Fisheries of the United States, 1977*, U.S. Department of Commerce, National Oceanic and Atmospheric Administration, C.F.S. no. 7500, 1978; Food and Agriculture Organization, *Expanding the Utilization of Marine Fishery Resources for Human Consumption*, 1976; J. A. Gulland, *The Management of Marine Fisheries*, 1974; G. Pontecorvo (ed.), *Fisheries Conflicts in the North Atlantic: Problems of Management and Jurisdiction*, 1974; *Report of the Committee on Fisheries, 9th Session, Rome, Oct. 15–22, 1974*, FAO Fisheries Rep. no. 154, 1976; B. Rothschild (ed.), *World Fisheries Policy: Multidisciplinary Views*, 1972.

Marine geology

The study of the portion of the Earth beneath the oceans. More than 70% of the Earth's surface is covered by marine waters. Of the oceanic area (361×10^6 km²) approximately 300×10^6 km² is contributed by the deep-sea floor; the remaining 60×10^6 km² represents the submerged margins of the continents. The distribution of elevations on the Earth is shown in Fig. 1.

Soundings. Soundings are measurements of ocean depth made from ships. Early soundings were made with a lead attached to a hemp line; about 1875 the hemp line was replaced by piano wire. Since about the middle of the 1920s, virtually all deep-sea soundings have been made by echo sounding. The echo-sounding machine sends out a sound pulse (10–20 Hz) and then times the interval from the sound pulse to the returning echo. The early sounders required manual operation, but since about 1935 automatic recording sounders, which plot a graph of depth versus time or distance, have been used almost exclusively. Since 1953 precision, high-resolution echo sounders have been used in increasing numbers. *See* ECHO SOUNDER.

Sounding corrections. Wire and hemp line soundings require a correction for wire angle, for stretch of the wire, and for calibration of the metering counters used. Echo soundings require a correction for sound velocity, since the average vertical velocity is not constant, and for slope of

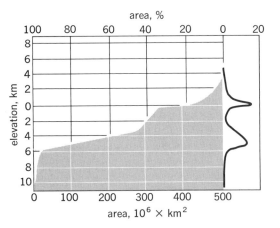

Fig. 1. Hypsographic curve showing area of Earth's solid surface above any given level of elevation or depth. Curve at the right shows frequency distribution of elevations and depths for 2-km intervals.

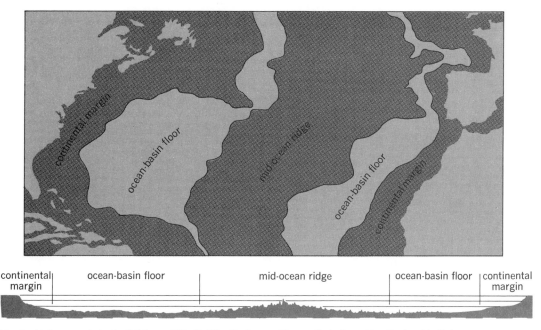

Fig. 2. Major morphologic divisions of North Atlantic Ocean. The profile is from New England to Sahara.

the bottom, as the point from which the first echo returns is not always directly beneath the ship. In addition, corrections for inaccuracies of timing and mechanical imperfections must be made for most sounders. The position of the sounding lines is generally determined by standard astronomical fixes and dead reckoning. Errors of a few miles are the rule in deep-sea sounding surveys. *See* UNDERWATER SOUND.

PHYSIOGRAPHIC PROVINCES

The Earth relief lies at two dominant levels (Fig. 1); one, within a few hundred meters of sea level, represents the normal surface of the continental blocks; the other, between 4000 and 5000 m below sea level, and comprising more than 50% of the Earth's surface, represents the ocean-basin floor. The topographic provinces beneath the sea can be included under three major morphologic divisions: continental margin, ocean-basin floor, and mid-oceanic ridge. These are indicated on a typical transoceanic profile taken from the North Atlantic in Fig. 2. Each of these major divisions can be further divided into categories of provinces, and those in turn into individual physiographic provinces (Fig. 3).

Continental margins. The continental margin includes those provinces associated with the transition from continent to ocean floor. The continental margin in the Atlantic and Indian oceans is generally composed of continental shelf, continental slope, and continental rise. A typical profile off the northeastern United States is shown in Fig. 4.

Gradients on the continental shelf average 1:1000, while on the continental slope gradients range from 1:40 to 1:6, and occasionally local slopes approach the vertical. The continental rise lies at the base of the continental slope. Continental rise gradients average 1:300, but individual slope segments may be as low as 1:700 or as steep

as 1:50. The continental slopes are cut by many submarine canyons. Some of the larger canyons such as the Hudson extend across the continental rise (Fig. 4). Submarine alluvial fans extend out from the seaward ends of the larger canyons. *See* CONTINENTAL SHELF AND SLOPE; SUBMARINE CANYON.

The continental margin can be divided into three categories of provinces. Category I includes the continental shelf, marginal plateaus, and shallow epicontinental seas, all slightly submerged portions of the continental block. Category II includes the continental slope, marginal escarpments, and the landward slopes of marginal trenches, all expressions of the outer edge of the continental block. Category III includes the continental rise, the ridge-basin complex, and the ridge-trench complex. The continental slope of the northeastern United States can be traced directly into the marginal escarpment (Blake Escarpment) off the southeastern United States (Fig. 5) and the landward slope of the Antilles marginal trench (Puerto Rico). The continental rise off New England can be traced into the Antilles Outer Ridge. Seismic refraction studies show that a trench filled with sediments and sedimentary rocks lies at the base of the continental slope off New England. Thus the main difference in morphology between the trenchless continental margins and continental margins with a marginal trench is that in the former the trench has been filled with sediments.

In the continental margins of the Atlantic, Indian, Arctic, and Antarctic oceans and the Mediterranean Sea, the continental rise generally represents the category III provinces. The Pacific, however, is bounded by an almost continuous line of marginal trenches. The high seismicity, vulcanism, and youthful relief of the Pacific borders suggest a very recent origin. In contrast, the nonseismic, nonvolcanic character, as well as the

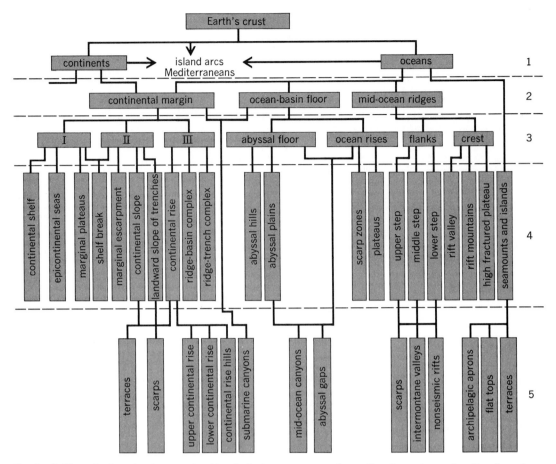

Fig. 3. Outline of submarine topography. Line 1, first-order features of the crust; line 2, major topographic features of the ocean; line 3, categories of provinces and superprovinces; line 4, provinces; line 5, subprovinces and other important features.

lower relief, of the Indo-Atlantic margins suggests a greater age. Thus on the old, stable, continental margins the deposition of sediment derived from the land has filled the marginal trench and produced the continental rise. The local relative relief on the continental margin rarely exceeds 20 fathoms, with the major exception of submarine canyons and occasional seamounts.

Submerged benches. Submerged marine beach terraces have been identified throughout the world. Since the beaches seem to correlate well between areas of vastly different tectonic development, it has been concluded that those which are listed in the accompanying table represent submerged late Pleistocene beaches.

Structural benches. Structural benches, the topographic expression of outcropping beds, have been identified on the continental slope. Near

Depth in fathoms of prominent continental-shelf terraces*

Placentia Bay, Newfoundland	Norfolk, Va.	Charleston, S.C.	Bimini, B.W.I.	St. Vincent, Cape Verde Is.	Dakar, Senegal	San Pedro, Calif.
10		12	10	8	10	10
				15	15	15
20	18	20	20	24	20	20
	30	30	28	28	28	28
35	35	35		32		
40				38	38	38
42		45	42	42	45	45
	50					
55	58			54	55	55
				60		
68		68	65			
80	80	80	85	80	78	80

*Each column based on a single nonprecision echogram.

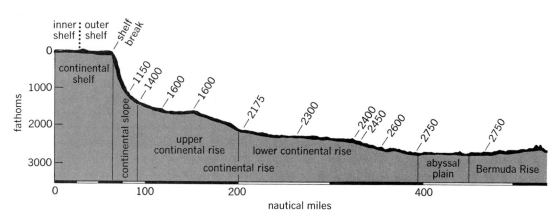

Fig. 4. Continental margin provinces: type profile off northeastern United States.

Cape Hatteras, N.C., the structural benches have been dated by extrapolating data obtained in several test borings near the coastline (Fig. 6). Benches on Georges Bank have been dated from bottom samples obtained by dredging. Through the action of slumps, bottom currents, and turbidity currents, sediments are continually removed from the continental slope. Thus, there is no cover of recent sediments to obscure the outcrops of the ancient formations. *See* TURBIDITY CURRENT.

Ocean-basin floor. Excluding the marginal trenches and mid-oceanic ridges, the deepest portions of the ocean are included in this division. Approximately one-third of the Atlantic and three-fourths of the Pacific fall under this heading. The ocean-basin floor can be divided into three categories of provinces—the abyssal floor, oceanic rises, and seamounts and seamount groups—which are discussed below.

Abyssal floor. The abyssal floor includes the broad, deep areas of the central portion of the ocean. In the Atlantic, Indian, and northeast Pacific oceans, abyssal plains occupy a large part of the abyssal floor. An abyssal plain is a smooth portion of the deep-sea floor where the gradient of the bottom does not exceed 1:1000. Abyssal plains adjoin all continental rises and can be distinguished from the continental rise by a distinct change in bottom gradient. At their seaward edge, most of the abyssal plains gradually give way to abyssal hills. Individual abyssal hills are 50–200

fathoms high and 2–6 mi wide. In the Atlantic, the abyssal hill provinces only locally exceed 50 mi in width. Abyssal plains in the same area range from 100 to 200 mi in width. Core samples of sediment obtained from the Atlantic abyssal plain invariably contain beds of sand, silt, and gray clay intercalated in the red or gray pelagic clay which is generally characteristic of the deep-oceanic environment. These deep-sea sands were transported by turbidity currents from the continental margin. Some of the currents probably descended along a broad front, while others certainly followed the submarine canyons and spread out fanwise from their submarine alluvial cones. *See* MARINE SEDIMENTS.

The abyssal hills are thought to represent tectonic or volcanic relief of a type identical with that buried beneath the abyssal plains. Abyssal plains are also found in the marginal trenches, marginal basins, and in epicontinental marginal seas. Features of exactly the same morphology and origin are found in some lakes. Of similar origin are archipelagic aprons, which spread out from the base of oceanic islands. *See* OCEANIC ISLANDS.

Oceanic rises. Oceanic rises are areas slightly elevated above the abyssal floor which do not belong to the continental margin or the mid-oceanic ridges. In the North Atlantic, the Bermuda Rise is the best-known example (Fig. 7). In contrast to the mid-oceanic ridges, oceanic rises are nonseismic; their relief is more subdued, and they are asym-

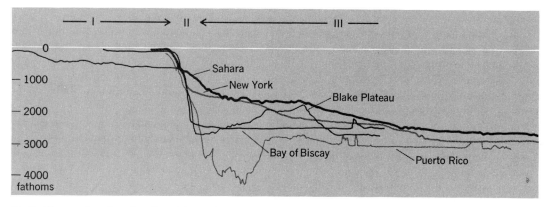

Fig. 5. Three categories of continental-margin provinces.

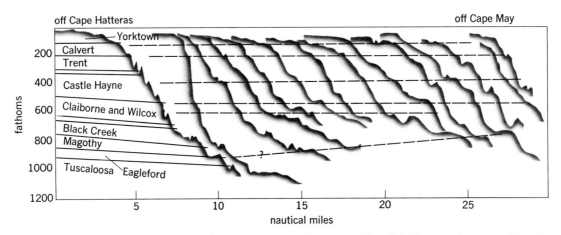

Fig. 6. Correlation of structural benches (outcropping beds) on the continental slope from Cape Hatteras, N.C., to Cape May, N.J. The soundings were taken by the U.S. Coast and Geodetic Survey.

metrical in cross section. The western and central Bermuda rise is characterized by gentle, rolling relief. The average depth gradually decreases toward the east. In the eastern third, the rise is cut by a series of scarps, 500–1000 fathoms in height, from which the sea floor drops to the level of the abyssal plain on the east. The series of eastward-facing scarps suggest block faulting. Situated approximately in the center of the Bermuda Rise is the volcanic pedestal of Bermuda. A small archipelagic apron surrounds the pedestal. The turbidity-current origin of the smooth apron is supported by cores containing shallow-water carbonate clastic sediments in depths of 2300 fathoms. In the Pacific, extending for more than 3000 mi west of Cape Mendocino, Calif., is an asymmetrical rise with a southward-facing scarp, which has been named the Mendocino Fracture Zone. The Bermuda Rise is less than one-fourth as long as the Mendocino Rise, but otherwise the relief of both features is quite similar. Although the circum-Pacific seismic belt crosses its trend, the Mendocino Escarpment is nonseismic. Other nonseismic fracture zones, which probably can be classified as oceanic rises, have been reported from the eastern Pacific. The Rio Grande Rise of the South Atlantic and the Mascarene Ridge of the Indian Ocean are similar in form.

Seamounts and seamount groups. A seamount is any submerged peak more than 500 fathoms high. This discussion, however, is limited to the larger, more or less conical peaks more than 1000 fathoms in height. Seamounts are distributed through all the physiographic provinces of the oceans. Seamounts sometimes occur randomly scattered but more often lie in linear rows. It seems safe to conclude that virtually all conical seamounts are extinct or active volcanoes. The Kelvin seamount group, a line of seamounts 800 mi long, stretches out from the vicinity of the Gulf of Maine toward the Mid-Atlantic Ridge. The Atlantis–Great Meteor seamount group extends for 400 mi along a north-south line, south of the Azores. In the southwestern Pacific, many lines of islands and seamounts crisscross the ocean. In the mid-Pacific, southwest of Hawaii, is a large area of seamounts whose flat summits range from 50 to 850 fathoms beneath sea level. These tablemounts have been termed guyots. From the flat summits, shallow-water fossils of Cretaceous age have been dredged. Such sunken islands are not limited to the Pacific. Several of the Kelvin seamounts are flat-topped at 650 fathoms, and the seamounts of the Atlantis–Great Meteor group have flat summits at 150–250 fathoms. *See* SEAMOUNT AND GUYOT.

Mid-Oceanic Ridge. The middle third of the Atlantic, Indian, and South Pacific oceans is occupied by a broad, fractured swell known as the Mid-

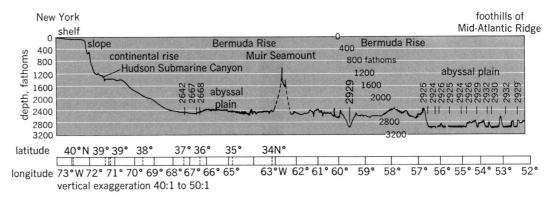

Fig. 7. Precision-depth-recorder profile between Mid-Atlantic Ridge and New York.

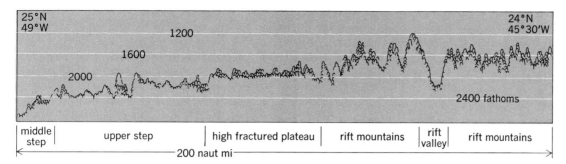

Fig. 8. Tracing of a precision-depth-recorder record showing crest and western flank of Mid-Atlantic Ridge.

Oceanic Ridge. In the Atlantic, it is known as the Mid-Atlantic Ridge; in the southern Indian Ocean, the Mid-Indian Ridge; in the Arabian Sea, the Carlsberg Ridge and Murray Ridge; and in the South Pacific, the Easter Island Ridge.

The Mid-Atlantic Ridge can be divided into distinctive physiographic provinces which can be identified on most transatlantic profiles (Figs. 2 and 8).

Crest provinces. The rift valley, rift mountains, and high-fractured plateau which constitute this category form a strip 50–200 mi wide. The rift valley is bounded by the inward-facing scarps of the rift mountains. The floor of the rift valley lies 500–1500 fathoms below the adjacent peaks of the rift mountains, which drop abruptly to the high fractured plateau, lying at depths of 1600–1800 fathoms on either side of the rift mountains. The topography of the crest provinces is the most rugged submarine relief. An earthquake belt accurately follows the rift valley through a distance of over 40,000 mi. Heat-flow measurements in the crest provinces give values several times greater than have been obtained in normal ocean or continental areas. A large, positive magnetic anomaly and a moderate (−20 mgal) negative gravity anomaly are associated with the rift valley. Seismic refraction measurements indicate a crust intermediate in composition between the oceanic crust and the mantle. The crest provinces of the Mid-Oceanic Ridge can be traced directly into the rift valleys, rift mountains, and high plateaus of Africa. These features of African geology are clearly the result of extensional forces in the Earth's crust. The Mid-Oceanic Ridge is probably similar in all essential characteristics, including origin, to the African rift valley complex.

Flank provinces. The flank provinces of the Mid-Oceanic Ridge can be divided into several steps or ramps, each bounded by scarps somewhat larger than those which characterize the entire area.

Parts of the flank provinces, particularly the Upper Step south of the Azores, are characterized by smooth-floored intermontane valleys. Photographs, cores, and dredging indicate that the crest of the Mid-Atlantic Ridge north of the Azores is being denuded of its sediments. As these sediments are eroded from the crest provinces and deposited on either side, they are gradually filling the intermontane basins and smoothing the relief of the flanks.   [BRUCE C. HEEZEN]

UNDERLYING STRUCTURE

Because approximately 70% of its surface is covered by the oceans, the typical structure of the Earth is found in the oceanic and not in the land areas. Statistical examination shows that most of the Earth's solid surface is either at the elevation of the ocean floors or at the elevation of the continents. The anomalous areas, those of extreme or of intermediate elevation, are long, narrow features—the mountain ranges, island arcs, deep-sea trenches, and continental margins. With the exception of a few intermediate areas, such as the Red Sea, the crustal structure of the ocean basins is distinctly different from the crustal structure of the continents, and it appears that this has been the case for most of the Earth's history.

To have a rough model of a section through the Earth, one draws a circle about 5 in. in diameter and a concentric one of about half that diameter. Inside the smaller circle is the core, probably of a nickel-iron composition. The part between the circles is the mantle, a crystalline, basic rock with density about 3.3 g/cm³. The line forming the outer circle, if made with an average pencil point, will include all of the crust of the Earth. The crust has a density of about 2.7 g/cm³. The oceanic crust is about 6 km thick compared to about 36 km for the continents. The boundary between the crust and the mantle is called the M-discontinuity (Mohorovičić discontinuity) by seismologists.

Instruments and techniques. Unlike the continental areas, the ocean floors cannot be studied directly; hence, most of the information about the structure of the Earth beneath the oceans comes from geophysical measurements, from samples dredged or cored from the ocean floor, and, to a smaller extent, from observations made from deep submersibles. Principally employed geophysical techniques are earthquake seismology, explosion seismology, and measurements of the variations in the Earth's gravitational and magnetic fields. Seismology is the study of the propagation of sound waves or elastic waves in the Earth. The source of these waves can be natural, which is the case in earthquake seismology, or man-made. Commonly used man-made sound sources include explosions, high-energy electrical sparkers, and pneumatic devices. The speed of sound waves traveling in the various layers, considered together with gravity and magnetic measurements, makes it

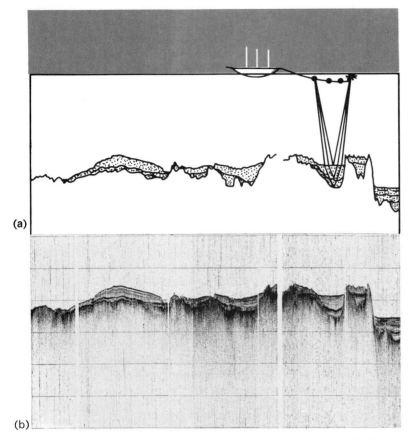

(a)

(b)

Fig. 9. Continuous seismic profiling: (a) technique and (b) typical record. Record shows approximately 200 km of traverse. The average thickness of these ocean-bottom sediments is about 500 m. Vertical exaggeration × 25.

possible to estimate certain physical properties such as density and elastic constants. These in turn suggest the type of rock constituting each layer. The travel time of sound waves reflected or refracted by a particular layer provides a measure of the depth and thickness of the layer. In marine investigations, a relatively new device for measuring the thickness and structure of the sediments is the continuous seismic reflection profiler. A sound source, towed behind the ship, emits periodic pulses of acoustic energy. These pulses travel through the water and are reflected back to the surface from the interfaces between the layers of sediment and rock. The reflected pulses are received by a towed array of hydrophones, amplified, and printed by a scanning recorder. This tech-

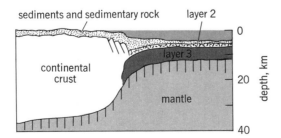

Fig. 10. Structure section across Atlantic-type (rifted) margin.

nique produces a cross-sectional view of the sediments on the ocean floor and usually shows the topography of the basement beneath the sediments as well (Fig. 9). See ECHO SOUNDER; MARINE SEDIMENTS.

The sea gravimeter is essentially a highly damped, extremely sensitive spring balance. Changes in the displacement of the springs reflect minute variations in gravitational force. Total magnetic field intensity is measured with a magnetometer towed well astern to minimize the magnetic effects of the ship. The gravitational and magnetic anomalies (variations) when combined with seismological data are indications of compositional changes, or structural features such as folds or faults, and have provided much information about the structure of the Earth.

Paleomagnetism is the study of the remanent magnetization in rocks. The direction of remanent magnetization in sedimentary rocks is parallel to the Earth's magnetic field at the time that the rock was deposited. The remanent magnetic vector in igneous rocks indicates the direction of the Earth's field at the time the rock cooled through the Curie temperature. Paleomagnetic studies of rocks of various ages and wide geographic distribution appear to indicate that the magnetic poles have shifted with respect to the geographic poles through geologic time or that the continents have shifted with respect to each other. Data have been accumulated that prove there have been many reversals of the Earth's magnetic field. These reversals have been used to explain linear magnetic anomaly patterns in some parts of the oceans and to support the hypothesis of continental drift by spreading of the sea floor away from the axes of the mid-ocean ridges.

Ocean basins and continental margins. Figure 10 is a structure section across an Atlantic type (rifted) margin based on an interpretation of geologic and geophysical data within the framework of plate tectonics and sea-floor spreading. On the continent and continental shelf, the rocks beneath the sedimentary layers are mainly of the acidic type, such as granite, gneiss, or schist. Beneath the rocks is an intermediate layer believed to be gabbroic or basaltic. The mantle is probably composed of ultramafic rocks, such as peridotite, enstatite, or eclogite. This is the most prominent layer in the Earth, extending from near the surface approximately halfway to the center. There are some variations in the upper parts of the mantle between continental and oceanic areas, but the most apparent difference is in the thickness and composition of the crust. The continental crust is six or seven times thicker than the oceanic crust and contains almost all of the acidic rocks, such as granites, whereas the oceanic crust is almost entirely composed of basic rock.

The average depth of the ocean basins is about 4.8 km. The topography of the ocean floor is rough in the majority of the explored parts, although there are broad areas, particularly in the Atlantic, where the bottom is almost completely flat. These abyssal plains are thought to have been formed by turbidity current deposition where sediments, set in motion during underwater landslides and thrown into suspension in the water, flow to the deepest parts of the basins.

Away from the abyssal plains and continental rises (see section above on physiographic provinces) the sediment layers are mainly composed of clay-sized particles (less than 2μ) and of the skeletons of plankton that lived and died in the waters above. Pelagic sediments of this type generally form an approximately uniform blanket over preexisting topography. In some areas the sea-floor topography is shaped by nonuniform deposition and erosion by bottom currents, particularly by the flow of the cold, dense water generated in the polar regions. The thickness of the unconsolidated sediments varies from tens of meters in areas which receive no turbidities and where the planktonic population is sparse to thousands of meters in other areas. The average is less than 1 km.

Layer 2, immediately below the layer-1 sediments, has a seismic velocity between 4 and 6 km/s. This range of velocities encompasses those appropriate to compacted or metamorphosed sediments or volcanic rocks. Shallow penetrations into layer 2 by deep-sea drilling indicate that its upper part consists of a mixture of pillow basalts, dikes, volcanic debris, and sediments. The lower part is probably massive basalt or metabasalt or both.

The principal layer of the oceanic crust, layer 3, shown by check marks in the structure sections (Figs. 10–12), has been found by most of the numerous seismic refraction measurements made in the Atlantic, Pacific, and Indian oceans. The average velocity of sound in this layer is near 6.7 km/s, and the thickness is about 5 km. Variations from these average are often found in the neighborhood of anomalous areas such as seamounts, trenches, or continental margins. Laboratory sound-velocity measurements of various rock types dredged from sea-floor escarpments suggest that layer 3 is composed of gabbro and metagabbro. Whether or not a similar layer exists under the continents is in dispute. There is much evidence that the velocity in the lower part of the continental crust is intermediate between that in the upper part and that in the mantle, and this could indicate the presence of material similar to that in the oceanic crust.

Continental shelves. These are the submerged borders of the continents. The water depth is on the order of a few hundred meters, and the width of the shelf varies from a few kilometers in some places to a few hundred kilometers in others. The thickness of the crust is intermediate between that of the continents and oceans, and its composition is continental. In some places, such as parts of the east coast of North America and South America, the continental shelf and continental slope are broad areas where erosion of the continental masses has resulted in the deposition of many thousands of meters of sediment during the past several million years. In other areas which do not receive much sediment, notably the west coast of the Americas, there is only a narrow continental shelf.

Submarine ridges. There are two types of oceanic ridges, seismic and aseismic. The mid-ocean ridge system is of the former type and is the largest single feature on the Earth. It is more than 80,000 km long and completely encircles the globe. In many places the ridge is offset by large fracture zones, along which seismic activity is generally higher than elsewhere on the ridge system. Over its entire length the ridge is characterized by a narrow zone of shallow-focus earthquake epicenters. Usually associated with this zone are a rift valley and a large positive magnetic anomaly. In addition, parallel linear bands of alternating positive and negative magnetic anomalies are found on the flanks in many areas. The sediment cover is typically thin and in some places in the crest region is entirely absent in a zone 100–500 km wide. On the flanks an appreciably thicker cover is found; in some places it covers the basement rock more or less uniformly, while in others it is collected in pockets separated by exposed basement peaks or ridges.

The tops and flanks of seismic ridges have been dredged and have yieled basaltic volcanic rocks with some inclusions of gabbroic or ultramafic materials. Figure 11 shows a structure section across the Mid-Atlantic Ridge based on seismic refraction and gravity measurements.

In addition to the mid-ocean ridge system, there are other oceanic ridges which are seismically inactive. Typical examples of these are the Walvis Ridge off southwestern Africa and the Hawaiian Ridge, Emperor Seamount Chain, and Line Island Chain in the Pacific. These ridges are thought to have been formed by extrusion of large volumes of volcanic material as the oceanic crust drifted over "hot spots" in the lower part of the mantle. Another class of aseismic ridge is exemplified by the Lomonosov Ridge in the Arctic Ocean and the Jan Mayen Ridge in the Norwegian Sea. These ridges appear to be thin slivers of continental rocks which were separated from the larger continental masses by rifting associated with sea-floor spreading.

Deep-sea trenches. Deep-sea trenches are important structural features associated with some continental margins, island arcs, earthquake belts, and areas of volcanic activity. Most of them are confined to the margins of the Pacific Ocean, although there are trenches in the Atlantic and Indian oceans. The greatest depths in the oceans are found in these trenches, the deepest in the Pacific being about 10.7 km in the Marianas Trench, and the deepest in the Atlantic about 8.4 km in the Puerto Rico Trench. The trench is formed by the depression of the high-velocity crustal layer and the mantle by several kilometers. In the bottom

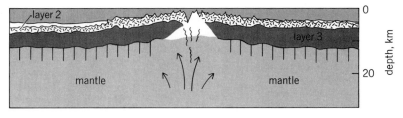

Fig. 11. Structure section across Mid-Atlantic Ridge, indicating crustal accretion and divergence.

are layers of sediments which are generally thickest landward of the trench axes. The great depth of the underlying dense layers causes a pronounced deficiency of gravity, a characteristic feature of all deep-sea trenches. Earthquake foci tend to lie on inclined planes dipping landward from the trenches.

Several hypotheses have been advanced to account for the existence of island arcs and the associated trenches and to relate them to a major processes going on in the Earth. The theory receiving much attention currently is that there are convection cells in the mantle which cause upwelling of new crustal material along the mid-ocean ridges and convergence at the trenches. The convergent flow causes the oceanic crust to be thrust underneath the continents or island arcs, creating the negative gravity anomalies, the earthquake foci along the shear zones, and the volcanic and orogenic activity landward of the trench itself. Figure 12 illustrates the concept of trench structure at a convergent margin.

Continental drift and sea-floor spreading. The parallel Atlantic borders of the Americas and Africa first led scientists to speculate that these continents were originally a single unit which was split apart by drifting. Alfred Wegener first popularized the idea in 1910. Throughout this century, the theory has been the subject of much debate, and although the early evidence was confined to the continents themselves, an impressive body of geological, climatological, and paleontological observations was brought to bear on the theory, including the existence of truncated mountain ranges and similar fossils and mineral assemblages found on opposite sides of the Atlantic. Paleomagnetic data also have been interpreted as indicating that magnetic poles have moved and that the continents have drifted relative to each other. Opposition to the theory has been due largely, but not entirely, to the difficulty in finding a suitable mechanism for drift that was not in conflict with concurrent theories on the composition of the Earth's interior.

Convection cells. A concept of continental drift by spreading of the sea floor has attracted much attention. An important concept in the sea-floor spreading hypothesis is that heat generated in the Earth causes convective overturn of mantle material in a number of convection cells. The mid-ocean ridges are thought to lie along the juncture of the upwelling limbs of separate cells. At the surface, the upwelling material diverges and flows away from the ridge axes, carrying the continental

blocks along with it until they come to rest over or near a downflowing convergence. It is suggested that reversals of the magnetic field produce areas of reversely magnetized crustal rock, symmetrical with respect to the ridge axis. The times of reversals are known for the last 5,000,000 years, and on the basis of this hypothesis values of 1–4 cm/year have been deduced for spreading rates. The descending flow is thought to create the deep-sea trenches and earthquake fault planes mentioned previously, and the resulting compression to be responsible for the mountain building and volcanism associated with the trenches. This mechanism has also been offered as an explanation for the fact that no sediments or rocks older than Jurassic have yet been found in the ocean basins, based on the assumption that all older ones have been swept into the mantle at the trenches. Such a description is greatly simplified, and many complexities are encountered in attempting to fit the entire surface of the Earth into a plausible pattern and in visualizing convection cells with the great horizontal dimensions implied by this theory and the small vertical dimensions implied by mantle stratification. But there are many arguments in favor of it.

Mid-ocean ridge system. Modern investigations in submarine geology and geophysics have supplied many new facts with which to test the theory. Among these was the discovery of the mid-ocean ridge system and the fact that, particularly in the Atlantic Ocean, the ridge lies almost exactly equidistant between the continents on either side. In addition to the median position of the ridge in the Atlantic, several other features about it have been discovered which seem to support the concept of continental drift. The great majority of earthquake epicenters in the oceans are concentrated in the axial zone of the ridge. An axial rift valley characterizes much of its length and appears to be a tensional feature, such as would be produced if the crust were being pulled apart. In many places long fracture zones are found, perpendicular to the ridge axis, which may represent flowlines of the continents' movements. There are distinct concentrations of earthquake activity near the intersections of the fracture zones and the ridge axis. Directions of first motion for these earthquakes appear to be best explained by transform faults associated with crustal spreading away from the ridge axis. Measurements of the thermal gradient in the bottom sediments have shown that the crest of the ridge is, in many places, characterized by abnormally high heat flow, which would be expected if convection cells were bringing new, hot material to the surface at this point.

Rock magnetism. Widely acclaimed evidence cited in support of sea-floor spreading has come from the observed bands of alternating positive and negative magnetic anomalies parallel to and symmetric about the ridge axis. Figure 13 shows an example of this magnetic pattern on the Mid-Atlantic Ridge southwest of Iceland. It has been proposed that the positive anomaly bands (including the central one) correspond to rocks that cooled through the Curie temperature with the Earth's magnetic field in its present polarity.

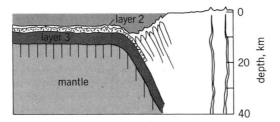

Fig. 12. Structure section across Pacific-type (convergent) margin.

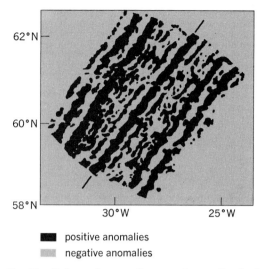

positive anomalies

negative anomalies

Fig. 13. Pattern of magnetic anomalies on the Reykjanes Ridge southwest of Iceland. (*After Heirtzler, Le Pichon, and Baron, 1966*)

The negative bands correspond to rocks magnetized by a reversed field. According to the sea-floor spreading hypothesis, the banded pattern was formed by upwelling and outward spreading of new material at the ridge crest, each newly generated strip of crust acquiring permanent magnetization in the direction of the ambient field. The spatial arrangement of the positive and negative bands near the ridge crest, normalized for spreading rate, is in agreement with that predicted by the history of field reversals.

Sediment distribution. Since the mid-1960s many hundreds of thousands of kilometers of seismic reflection measurements have shown the general pattern of sediment thickness in the oceans. On and near the crests of the mid-ocean ridges the sediment cover is very thin, becoming progressively thicker down the ridge flanks and into the basins. The total amount of sediment on the ridges varies from area to area because sediment productivity varies, but the increase of thickness with increasing distance from the ridge crest is a general characteristic of the pattern. The deep-sea sediment drilling program has now provided the additional evidence that the age of the sediment lying on the igneous basement (that deposited earliest) increases with increasing distance from the ridge crest. Thus, both the patterns of sediment thickness and sediment age are consistent with a young crust and progressively older flanks, concepts inherent in the sea-floor spreading hypothesis.

[MAURICE EWING; JOHN EWING]

Bibliography: C. A. Burk and C. L. Drake (eds.), *The Geology of Continental Margins*, 1974; R. E. Craig, *Marine Physics*, 1972; C. King, *Introduction to Marine Geology and Geomorphology*, 1975; A. E. Maxwell (ed.), *The Sea*, vol. 4, pts. I and II: *New Concepts of Sea Floor Evolution*, 1970; W. C. Pitman and J. R. Heirtzler, Magnetic anomalies over the Pacific-Antarctic Ridge, *Science*, 154:1164–1171, 1966; L. R. Sykes, Mechanism of earthquakes and nature of faulting on the mid-oceanic ridges, *J. Geophys. Res.*, 72:2131–2153, 1967; Symposium on Continental Drift, *Philosophical Transactions of the Royal Society of London, Ser. A*, 258(1088): 41–75, 1965; F. J. Vine, Spreading of the ocean floor: New evidence, *Science*, 154:1405–1415, 1966; T. J. Wilson, A new class of faults and their bearing on continental drift, *Nature*, 267:343–347, 1965.

Marine influence on weather and climate

The moderating, energizing, and redistributing effects that bodies of water have on air temperature, weather, and climate by releasing energy and heat to the atmosphere in a number of characteristic ways.

To a greater extent than is commonly realized, the climates of the oceans determine the climates of the Earth. In meteorological literature, however, discussions are mostly limited to a few words on the moderating effects of oceans on the climate of continents. It is often overlooked that oceans constitute the prime source of energy for all water-vapor-induced atmospheric processes. The latent heat liberated by the condensation of water vapor evaporated from the oceans constitutes the prime source of energy for tropical cyclone formation; and such water vapor from the oceans is otherwise a primary stage in the energy cycle of the general atmospheric circulation. *See* STORM.

When the climatologist speaks of the marine influence on climate, he is usually referring to the moderating effect of the oceans rather than to their role in the energy cycle of the atmospheric circulation. The usual explanation which is given for the moderating effect of the oceans is that sea water has a higher specific heat than does the soil surface of the continents. This factor, however, is only one among a number of other factors of equal or greater importance that serve to impose a lag in the heating of the surface layers of the atmosphere.

These factors include: (1) The largest fraction of the solar energy absorbed at the sea surface is utilized in evaporating sea water; a very much smaller fraction (average about 10%) is utilized in directly heating the atmosphere in contact with the sea surface. (2) A large proportion of the surplus solar energy absorbed at the sea surface during summer at mid-latitudes is stored by the oceans, to be released to the atmosphere during the colder seasons when there is a deficiency in solar energy. By the same mechanism much of the solar energy absorbed at low latitudes is transported by ocean currents and released to the atmosphere at higher latitudes where receipt of solar energy is also comparatively deficient. (3) The energy absorbed at the sea surface is mixed throughout a relatively deep layer, whereas the energy absorbed at a soil surface is confined to heating a shallow surface layer. Neither superheated nor subcooled surface zones remain unaltered for any length of time over the open ocean. (4) Shorter wavelengths of energy which reach the sea surface are transmitted to considerable depths because of the relative transparency of sea water to such radiation; but this is of secondary importance in the energy budget of the oceans. (5) Less solar energy is available for absorption over the sea than over land because of the higher albedo of the ocean areas. The high albedo results primarily from the greater average cloudiness rather than from a high reflectivity of the sea surface. *See* ALBEDO.

The oceans affect weather and climate, then, because they are a tremendous reservoir for the storage of solar energy. This energy is absorbed during periods of the year (or in regions) of surplus solar energy. It is then stored or transported by ocean currents, to be released to the atmosphere at times (or in regions) of deficiency in solar energy. In this sense the oceans become equalizers or moderators of climates. *See* CONTINENTALITY.

It has been shown that nonperiodic regional variations in the ocean circulation (and thus sea-surface temperature and energy transfer) will produce changes in regional weather and climate. Much of the present research on the effects of the oceans on weather and climate has to do with the effects of short-term variations in the oceanic circulation upon weather and climate.

[WOODROW C. JACOBS]

Bibliography: J. Bjerknes, Climatic change as an ocean-atmosphere problem, *Proceedings of UNESCO*, 1961; W. C. Jacobs, Large-scale aspects of air-sea interactions, *International Dictionary of Geophysics*, 1964; W. C. Jacobs, On the energy exchange between sea and atmosphere, *J. Mar. Res.*, 1942; J. S. Malkus, Large-scale interactions, in M. N. Hill (ed.), *The Sea*, vol. 1, 1962; J. Namias, The sea as a primary generator of short-term climatic anomalies, *Proceedings of the World Meteorological Organization/International Association of Meteorology and Atmospheric Physics Symposium*, 1975.

Marine microbiology

The study of the microscopic organisms living in the sea. Marine microorganisms were probably among the first (cellular) living entities and have played a tremendous part in geology and in the biology of the oceans. They are still of paramount importance as transformers of organic and inorganic substances and as food for other, often larger, organisms.

Marine microbes may be considered as (1) autotrophs and (2) heterotrophs or phagotrophs. Autotrophs live on inorganic materials and use carbon dioxide as the sole carbon source, obtaining their energy from sunlight (phototrophs) or from chemical reactions (chemoautotrophs). The heterotrophs and phagotrophs (bacteria, heterotrophic algae and fungi, and protozoa) require a source of carbon more complex than carbon dioxide; this they obtain, along with energy, by breaking down or ingesting other microbes, plants, and animals.

Phototrophic microorganisms. These constitute the most important plants in the sea, since all but the macroscopic seaweeds are microscopic and the latter occupy only about 2% of the surface area in the sea, though they probably contribute more than 2% of the primary productivity. The microscopic components are known as phytoplankton and occur in the upper waters of the oceans, the photic zone, where light can penetrate. This zone varies in depth depending on the incident angle of sunlight and on the turbidity of the water. Most phytoplanktonic organisms are motile (for example, the flagellate protozoa), asymmetrically shaped, or provided with processes (for example, the diatoms; see illustration) so that they tend to rotate in a way that brings them back toward the surface whenever they begin to sink. Many have oil as their storage product, which aids their buoyancy. *See* MARINE ECOSYSTEM.

The phytoplankton is made up of a number of taxonomic groups, including diatoms, blue-green algae, and flagellates, all of course containing chlorophyll. The flagellates comprise chlamydomonads, dinoflagellates, coccolithophores, chrysomonads, cryptomonads, and silicoflagellates. The organisms range in size from 1 mm long (a few diatoms) to 2 μ. In numbers, diatoms, blue-green algae, and the smaller flagellates are usually the most abundant. Diatoms are most numerous in boreal regions, blue-green algae are abundant in the tropical waters of the Pacific and Indian oceans, coccolithophores and microflagellates in the warm and temperate waters, armored dinoflagellates in inshore (neritic) waters, and naked dinoflagellates in warm oceans. "Red tides" appear when some organisms locally grow to such

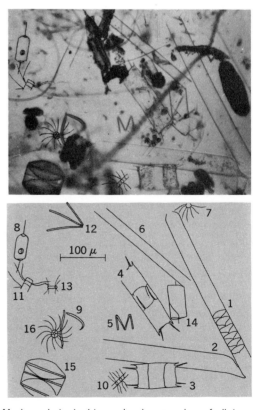

Marine phytoplankton, showing species of diatoms. (1,6) *Rhizosolenia* sp.; (2) *Rhizosolenia alata* var. *indica*; (3,4) *Biddulphia chinensis*; (5) *Thalassiothrix nitzschioides*; (7, 16) *Bacteriastrum varians*; (8) *Ditylum brightwellii*; (9) *Chaetoceros coarctatum* (fragment); (10, 11) *Chaetoceros* sp.; (12) *Thalassiothrix frauenfeldii*; (13) *Chaetoceros laeve*; (14) *Ditylum sol*; (15) *Hyalodiscus stelliger*.

a density that the water actually becomes colored; the reddish hue is due to accessory pigments. The Red Sea owes its name to the heavy blooms of a blue-green alga, *Oscillatoria* (*Trichodesmium*) *erythraea*, which also occurs in huge masses in the Indian Ocean, the Gulf of Mexico, the tropical Atlantic, and the tropical Pacific. Dinoflagellates form red tides in coastal waters and often contain a neurotoxin which may poison fish and other marine animals, causing heavy mortality. Such dinoflagellate blooms may produce large deposits of dead organic matter, as at Walvis Bay, South Africa; these have been considered to represent a possible source of petroleum, resulting from the bacterial reduction of the organic matter to hydrocarbons.

Chemoautotrophs. These are important only on the sea floor. They utilize energy-producing (exothermic) reactions such as the aerobic oxidation of sulfides (*Thiobacillus*), ammonia (*Nitrobacter* and *Nitrosomonas*), or hydrogen (*Hydrogenomonas*), and the anaerobic oxidation of hydrogen coupled with the reduction of sulfates to sulfides (*Desulfovibrio*). Thus they are important in the sulfur and nitrogen cycles of the sea bed, but are less common in the sea itself, except in shallow waters. In estuaries, bacteria of the sulfur cycle influence and sometimes control the biology of the overlying water.

Heterotrophs. While some diatoms and many dinoflagellates can live either as autotrophs or heterotrophs, and many flagellates are obligate heterotrophs, lacking chlorophyll, the most important heterotrophs in the sea are probably bacteria. These aid in the decomposition of organic matter in the sea and in the returning of some of this organic matter to the inorganic state, releasing carbon dioxide, nitrogen, phosphate, sulfur, and other essential elements in a form available for use by photosynthetic plants. Their importance in this area has been questioned, and it is suggested that protozoa and metazoa also play a large part in nutrient regeneration. Even refractory substances, such as cellulose and chitin, are decomposed by marine bacteria. Heterotrophic bacteria are not numerous in the open sea, except in the vicinity of plankton swarms, and their scarcity may be due to grazing by animals.

Phagotrophs. Phagotrophic protozoa ingest other phytoplankton organisms, bacteria, or detritus, and behave as do the other animals of the zooplankton. They include some flagellates, among them dinoflagellates; the tintinnids and other ciliates; and the foraminiferidans and radiolarians (rhizopods).

Interrelationships of microorganisms. Life in the sea is a vast symbiosis or metabiosis, and no organisms can exist independently of others. Moreover, the sea is never without microorganisms even in a particular region. There are always enough nearby to start another bloom if conditions are favorable. Many phytoplankton organisms require growth factors, and it is believed that these are frequently supplied by marine bacteria; some phytoplankton organisms produce toxic substances which inhibit the growth of other organisms. Bacteriophages have been detected in the

Black Sea, the North Sea, and elsewhere. Phosphate is accumulated and stored by diatoms and is released by bacterial or autolytic action after their death; some types of phytoplankton (for example, the diatom *Fragilaria*) require nitrates, others (for example, the blue-green algae) prefer or require ammonia, and a few can fix atmospheric nitrogen.

Role of microorganisms. Marine microorganisms are the food of a large number of marine animals, including many crustacea of the zooplankton, mollusca, and plant-eating fish. The phytoplankton that can be counted or measured represents merely the residue that is left after the animals have grazed. Thus, phytoplankton forms the primary food in the sea. Zooplankton animals appear to be selective in their feeding and to reject certain diatoms, such as *Chaetoceros* and *Rhizosolenia*. *Chlorella* may be ingested by cyprid larvae but passes whole into the feces, and the larvae die on this diet. Selectivity in grazing and facultative heterotrophy of some phytoplankton organisms create difficulties in assessing phytoplankton production.

Many plankton microbes have calcareous or siliceous skeletons. Foraminiferida and coccolithophores have, as a group, calcareous skeletons, and diatoms, radiolaria, and silicoflagellates have siliceous skeletons; these may accumulate on the sea bottom to form vast beds which can persist through geological ages, for example, foraminiferal limestones, diatomaceous oozes, and radiolarian jaspers.

Bacteria, and possibly diatoms, live on the sea floor, even at the greatest depths (10,000 m), with hydrostatic pressures up to 1000 atm. This bottom (benthic) flora forms a biosphere very distinct from the phytoplankton biosphere. *See* SEA WATER; SEA-WATER FERTILITY.

[EDWARD J. FERGUSON WOOD]

Bibliography: C. D. Litchfield, *Marine Microbiology*, 1976; J. M. Sieburth, *Microbial Seascapes: A Pictorial Essay on Marine Microorganisms and Their Environments*, 1975; E. F. Wood, *The Living Ocean*, 1975; E. F. Wood, *Microbiology of Oceans and Estuaries*, 1967.

Marine mining

The process of recovering mineral wealth from sea water and from deposits on and under the sea floor. Unknown except to technical specialists before 1960, undersea mining is receiving increasing attention. Frequent references to marine mineral resources and marine minerals legislation by national and international policy makers, increasing activity in marine minerals exploration, and the launching of major new seagoing mining dredges for South Africa and Southeast Asia all indicate the beginnings of a viable and expanding industry.

There are sound reasons for this sudden emphasis on a previously little-known source of minerals. While the world's demand for mineral commodities is increasing at an alarming rate, most of the developed countries have been thoroughly explored for surface outcroppings of mineral deposits. The mining industry has been required to advance its capabilities for the exploration and exploitation of low-grade and unconventional sources of ore. Corresponding advances in oceanology have highlighted the importance of the ocean as a source of minerals and indicated that the technology required for their exploitation is in some cases already available.

There is a definite realization that the venture into the oceans will require large investments, and the trend toward the consortium approach is very noticeable, not only in exploration and mining activities but in research. Undersea mining has become an important diversification for many major oil and aerospace companies, and a few mining companies appear to be taking an aggressive approach. In the late 1960s over 80 separate exploration activities were reported in coastal areas worldwide. In the 1970s interest turned more to the deep-seabed deposits, subject to much negotiation at the United Nations 3d Law of the Sea Conference. Anticipating agreement, several countries, including the United States and West Germany, prepared draft interim legislation which would permit nationals of their countries and reciprocating states to mine the deep seabeds for nodules containing manganese, copper, nickel, and cobalt. Five major consortia were active in testing deep-seabed mining systems in the late 1970s including companies from the United States, Canada, United Kingdom, West Germany, France, and Japan.

While mineral resources to the value of trillions of dollars do exist in and under the oceans, their exploitation is not simple. Many environmental problems must be overcome and many technical advances must be made before the majority of these deposits can be mined in competition with existing land resources.

The marine environment may logically be divided into four significant areas: the waters, the deep sea floor, the continental shelf and slope, and the seacoast. Of these, the waters are the most significant, both for their mineral content and for their unique properties as a mineral overburden. Not only do they cover the ocean floor with a fluid medium quite different from the earth or atmosphere and requiring entirely different concepts of ground survey and exploration, but the constant and often violent movement of the surface waters combined with unusual water depths present formidable deterrents to the use of conventional seagoing techniques in marine mining operations.

The mineral resources of the marine environment are of three basic types: the dissolved minerals of the ocean waters; the unconsolidated mineral deposits of marine beaches, continental shelf, and deep-sea floor; and the consolidated deposits contained within the bedrock underlying the seas. These are described in Table 1, which shows also the subclasses of surficial and in-place deposits, characteristics which have a very great influence on the economics of exploration and mining. *See* MARINE GEOLOGY; SEA WATER.

As with land deposits, the initial stages preceding the production of a marketable commodity include discovery, characterization of the deposit to assess its value and exploitability, and mining, including beneficiation of the material to a salable product.

Exploration. Initial requirements of an exploration program on the continental shelves are a thorough study of the known geology of the shelves and adjacent coastal areas and the extrapolation of known metallogenic provinces into the offshore areas. The projection of these provinces, which are characterized by relatively abundant mineralization, generally of one predominant type, has been practiced with some success in the localization of certain mineral commodities, overlain by thick sediments. As a first step, the application of this technique to the continental shelf, overlain by water, is of considerable guidance in localizing more intensive operations. Areas thus delineated are considered to be potentially mineral-bearing and subject to prospecting by geophysical and other methods.

A study of the oceanographic environment may

Table 1. Marine mineral resources

| Dissolved | Unconsolidated | | Consolidated | |
	Surficial	In place	Surficial	In place
Metals and salts of: Magnesium Sodium Calcium Bromine Potassium Sulfur Strontium Boron Uranium And 30 other elements Fresh water	Shallow beach or offshore placers Heavy mineral sands Iron sands Silica sands Lime sands Sand and gravel Authigenic deposits Manganese nodules (Co, Ni, Cu, Mn) Phosphorite nodules Phosphorite sands Glauconite sands Deep ocean floor deposits Red clays Calcareous ooze Siliceous ooze Metalliferous ooze	Buried beach and river placers Diamonds Gold Platinum Tin Heavy minerals Magnetite Ilmenite Rutile Zircon Leucoxene Monazite Chromite Scheelite Wolframite	Exposed stratified deposits Coal Iron ore Limestone Authigenic coatings Manganese oxide Associated Co, Ni, Cu Phosphorite	Disseminated massive, vein, or tabular deposits Coal Iron Tin Gold Sulfur Metallic sulfides Metallic salts

indicate areas favorable to the deposition of authigenic deposits in deep and shallow water. Some deposits may be discovered by chance in the process of other marine activities.

Field exploration prior to or following discovery will involve three major categories of work: ship operation, survey, and sampling.

Ship operation. Conventional seagoing vessels are used for exploration activities with equipment mounted on board to suit the particular type of operation. The use of submersibles will no doubt eventually augment existing techniques but they are not yet advanced sufficiently for normal usage.

One of the most important factors in the location of undersea minerals is accurate navigation. Ore bodies must be relocated after being found and must be accurately delineated and defined. The accuracy of survey required depends upon the phase of operation. Initially, errors of 1000 ft or more may be tolerated.

However, once an ore body is believed to exist in a given area, maximum errors of less than 100 ft are desirable. These maximum tolerated errors may be further reduced to a few feet in detailed ore body delineation and extraction.

There are a variety of types of electronic navigation systems available for use with accuracies from 3000 ft down to approximately 3 ft. Loran, Lorac, and Decca are permanently installed in various locations throughout the world. Small portable systems are available for local use that provide high accuracy within 30–50-mi ranges. For deep-ocean survey, navigational satellites have completely revolutionized the capabilities for positioning with high accuracy in any part of the world's oceans.

During sampling and mining operations, the vessel must be held steady over a selected spot on the ocean floor. Two procedures that have been fairly well developed for this purpose are multiple anchoring and dynamic positioning.

A three-point anchoring system is of value for a coring vessel working close to the surf. A series of cores may be obtained along the line of operations by winching in the forward anchors and releasing the stern anchor. Good positive control over the vessel can be obtained with this system, and if conditions warrant, a four-point anchoring system may be used. Increased holding power can be obtained by multiple anchoring at each point.

Dynamic positioning is useful in deeper water, where anchoring may not be practical. The ship is kept in position by use of auxiliary outboard propeller drive units or transverse thrusters. These can be placed both fore and aft to provide excellent maneuverability. Sonar transponders are held submerged at a depth of minimum disturbance, or the system may be tied to shore stations. The auxiliary power units are then controlled manually or by computer to keep the ranges at a constant value. *See* OCEANOGRAPHIC SUBMERSIBLES.

Survey. The primary aids to exploration for mineral deposits at sea are depth recorders, subbottom profilers, magnetometers, bottom sampling devices, and subbottom sampling systems. Their use is dependent upon the characteristics of the ore being sought.

For the initial topographic survey of the sea floor, and as an aid to navigation, in inshore waters, the depth recorder is indispensable. It is usually carried as standard ship equipment, but precision recorders having a high accuracy are most useful in survey work.

In the search for marine placer deposits of heavy minerals, the subbottom profiler is probably the most useful of all the exploration aids. It is one of several systems utilizing the reflective characteristics of acoustic or shock waves.

Continuous seismic profilers are a development of standard geophysical seismic systems for reflection surveys, used in the oil industry. The normal energy source is explosive, and penetration may be as much as several miles.

Subbottom profilers use a variety of energy sources including electric sparks, compressed air, gas explosions, acoustic transducers, and electromechanical (boomer) transducers. The return signals as recorded show a recognizable section of the subbottom. Shallow layers of sediment, configurations in the bedrock, faults, and other features are clearly displayed and require little interpretation. The maximum theoretical penetration is dependent on the time interval between pulses and the wave velocity in the subbottom. A pulse interval of 1/2 sec and an average velocity of 8000 ft/sec will allow a penetration of 2000 ft, the reflected wave being recorded before the next transmitted pulse.

Penetration and resolution are widely variable features on most models of wave velocity profiling systems. In general, high frequencies give high resolution with low penetration, while low frequencies give low resolution with high penetration. The general range of frequencies is at the low end of the scale and varies from 150 to 300 Hz, and the general range of pulse energy is 100–25,000 joules for nonexplosive energy sources. The choice of system will depend very much on the requirements of the survey, but for the location of shallow placer deposits on the continental shelf the smaller low-powered models have been used with considerable success.

With the advent of the flux gate, proton precession, and the rubidium vapor magnetometer, all measuring the Earth's total magnetic field to a high degree of accuracy, this technique has become much more useful in the field of mineral exploration.

Anomalies indicative of mineralization such as magnetic bodies, concentrations of magnetic sands, and certain structural features can be detected. Although all three types are adaptable to undersea survey work, the precession magnetometer is more sensitive and more easily handled than the flux gate, and the rubidium vapor type has an extreme degree of sensitivity which enhances its usefulness when used as a gradiometer on the sea surface or submerged.

Once an ore body is indicated by geological, geophysical, or other means, the next step is to sample it in area and in depth.

Sampling. Mineral deposit sampling involves two stages. First, exploratory or qualitative sampling locate mineral values and allow preliminary judgment to be made. For marine deposits this will involve such simple devices as snappers, drop corers, drag dredges, and divers. Accuracy of positioning is not critical at this stage, but of course is dependent on the type of deposit being sampled. Second, the deposit must be characterized in sufficient detail to determine the production technology requirements and to estimate the profitability of its exploitation. This quantitative sampling requires much more sophisticated equipment than does the qualitative type, and for marine work few systems in existence can be considered reliable and accurate. However, in particular cases, systems can be put together using available hardware which will satisfy the need to the accuracy required. Specifically, qualitative sampling of any mineral deposit offshore can be carried out with existing equipment. Quantitative sampling of most alluvial deposits of heavy minerals (specific gravity, less than 8) can be carried out at shallow depths (less than 350 ft overall) using existing equipment but cannot be carried out with reliability for the higher-specific-gravity minerals such as gold (specific gravity 19). Quantitative sampling of any consolidated mineral deposit offshore can be carried out within limits.

Any system that will give quantitative samples can be used for qualitative sampling, but in many cases heavy expenses could be avoided by using the simpler equipment.

To obviate the effects of the sea surface environment, the trend is toward the development of fully submerged systems, but it should be noted that the deficiencies in sampling of the heavy placer minerals are not due to the marine environment. Even on land the accuracy of placer deposit evaluation is not high and the controlling factors not well understood. There is a prime need for intensive research in this area.

Evaluation of surficial nodule deposits on the deep seabed is carried out by using a combination of photography, television, or acoustic imagery and sampling by dredge, corer, or clam, either towed or attached to free-fall self-surfacing devices. Analysis of nodules is done on board the vessel.

Exploitation. Despite the intense interest in undersea mining, new activities have been limited mostly to conceptual studies and exploration. The volume of production has shown little change, and publicity has tended to overemphasize some of the smaller, if more newsworthy, operations. All production comes from nearshore sources, namely,

sea water, beach and nearshore placers, and nearshore consolidated deposits.

Minerals dissolved in sea water. Commercial separation techniques for the recovery of minerals dissolved in sea water are limited to chemical precipitation and filtration for magnesium and bromine salts and solar evaporation for common salts and fresh-water production on a limited scale. Other processes developed in the laboratory on pilot plant scale include electrolysis, electrodialysis, adsorption, ion exchange, chelation, oxidation, chlorination, and solvent extraction. The intensive interest in the extraction of fresh water from the sea has permitted much additional research on the recovery of minerals, but successful commercial operations will require continued development of the combination of processes involved for each specific mineral.

As shown in Table 2, three minerals or mineral suites are extracted commercially from sea water: sodium, magnesium, and bromine. Of these, salt evaporites are the most important. Japan's total production of salt products comes from the sea. Magnesium extracted from sea water accounts for 75% of domestic production of this commodity in the United States, and fresh water compares with bromine in total production value.

Unconsolidated deposits. Unconsolidated deposits include all the placer minerals, surficial and in place, as well as the authigenic deposits found at moderate to great depths.

The mining of unconsolidated deposits became widely publicized with the awareness of the potential of manganese nodules as a source of manganese, copper, nickel, and cobalt, and because of the exciting developments in the exploitation of offshore diamonds in South-West Africa in the late 1960s. Despite the fact that there are presently no operations for nodules and the offshore diamond mining operations have been suspended, unconsolidated deposits have for some years presented a major source of exploitable minerals offshore.

So far the methods of recovery which have been used or proposed have been conventional, namely, by dredging using draglines, clamshells, bucket dredges, hydraulic dredges, or airlifts. All these methods (Fig. 1) have been used in mining to maximum depths of 200 ft, and hydraulic dredges for digging to 300 ft are being built. Extension to depths much greater than this does not appear to

Table 2. Production from dissolved mineral deposits offshore

Mineral	Location	Number of operations	Annual production	Value (millions of dollars)*
Sodium, NaCl	Worldwide	90+	10,000,000 tons	46
Magnesium, metal	United States, United	2	221,000 tons	156
Mg, MgO, Mg(OH)₂	Kingdom, Germany, Soviet Union	25+	800,000 tons	532
Bromine, Br	Worldwide	7	102,000 tons	612
Fresh water	Middle East, Atlantic region, United States	150+	†	†
Heavy water	Canada	1	†	†
Total		275+		1346

*Reported 1973.   †Not reported.

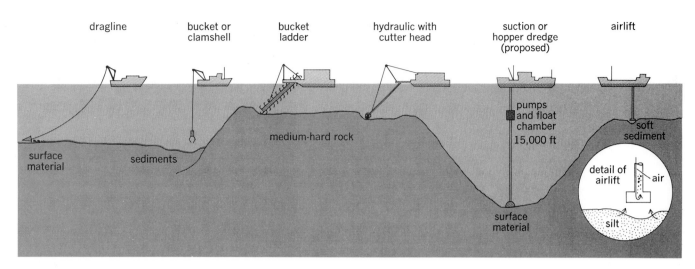

Fig. 1. Methods of dredging used in the exploitation of unconsolidated mineral deposits offshore.

present any insurmountable technical difficulties.

More than 70 dredging operations were active in the 1970s, exploiting such diverse products as diamonds, gold, heavy mineral sands, iron sands, tin sands, lime sand, and sand and gravel (Table 3). The most important of all of these commodities is the least exotic; 60% of world production from marine unconsolidated deposits, or about $100,000,000 annually, is involved in dredging and mining operations for sand and gravel. Other major contributors to world production are the operations for heavy mineral sands (ilmenite, rutile, and zircon), mostly in Australia, and the tin operations in Thailand and Indonesia, which account for more than 10% of the world's tin.

Economics of these operations are dictated by many conditions. Table 4 shows a comparative range of throughput and cost for typical dredging operations both onshore and offshore. The spectacular range of costs offshore is indicative of the effects of different environmental conditions. In general, it may be said that offshore operations are more costly than similar operations onshore.

The operations of Marine Diamond Corporation are of considerable historical interest. The first pilot dredging commenced in 1961 with a converted tug, the *Emerson K*, using an 8-in. airlift. The

operation expanded until 1963, with the fleet consisting of 3 mining vessels, all using an air- or jet-assisted suction lift, 11 support craft, and 2 aircraft. Production totaled over $1,700,000 of stones during that year from an estimated 322,000 yd³ of gravel. At that time, the estimation of mining cost was $2.33/yd³, showing a profit of nearly $3/yd. Subsequent unexpected problems, including severe storms, operating difficulties, and loss of one of the mining units, led to a reduction of profits and transfer of company control. In the year ending June 30, 1965, the company reported an operating loss of $2.02/yd³ in the treatment of 220,000 yd³ of gravel, valued at $28/yd.

Production operations of Marine Diamond Corporation fluctuated considerably. *Diamantkus*, a vessel designed to produce 7000 yd/hr, was withdrawn as uneconomic after only 30 months of service. Only two mining units, *Barge III* and *Colpontoon*, were in operation in 1968, both converted pipe-laying barges using combination airlifts and suction dredging equipment. A third and larger unit, the *Pomona*, a multiple-head suction dredge, was commissioned in March 1967, but was damaged by storms on the first trial run. The characteristics of the bedrock, with its many potholes and extremely irregular surface, added to the difficulty

Table 3. Production from unconsolidated mineral deposits offshore

Mineral	Location	Number of operations	Annual production	Value (millions of dollars)
Diamonds	South-West Africa	1	221,500 yd³	8.9
Gold	Alaska	1	—	—
Heavy mineral sands	North America, Europe, Southeast Asia, Australia	15	1,307,000 tons	13.1
Iron sands	Japan	3	36,000 tons	3.6
Tin sands	Southeast Asia, United Kingdom	4	10,000 tons concentrates	24.2
Lime shells	United States, Iceland	9	20,000,000 yd³	30.0
Sand and gravel	Britain, United States	38	100,000,000 yd³	100.0
Total		71		179.8

Table 4. Range of throughput and costs for typical dredging operations (1967)*

	Onshore	Offshore
Yd³/month	100,000–500,000	2,500–350,000
Dollars/yd³	0.08–0.25	0.15–74.5

*Values within these ranges are dependent on system characteristics.

in recovering the maximum amount of diamonds. Mining operations ceased in May 1971, but large areas remained to be explored.

Liberal offshore mining laws introduced in 1962 in the state of Alaska resulted in an upsurge of exploration activity for gold, particularly in the Nome area. In 1968 a mining operation was attempted, using a 20-in. hydrojet dredge, in submerged gravels about 60 mi east of Nome. No production was reported.

Over 70% of the world's heavy mineral sand production is from beach sand operations in Australia, Ceylon, and India. Only two oceangoing dredges are used. The majority being pontoon-mounted hydraulic dredges, or draglines, with separate washing plants.

The Yawata Iron and Steel Co. in Japan used a 10.5-yd³ barge-mounted grab dredge and a hydraulic cutter dredge to mine iron sand from the floor of Ariake Bay in water depths of 50 ft. These operations were suspended finally in 1966, the reason being given that the reserves had not been accurately surveyed and the cost of mining had not been competitive with Yawata's alternate sources of supply.

An interesting comparison between a clamshell dredge from Aokam Tin and a bucket-ladder dredge from Tongka Harbor, working offshore on the same deposit in Thailand was made. The clamshell was set up as an experimental unit using an oil tanker hull. It was designed for a digging depth of 215 ft, and mobility and seaworthiness were prime factors in its favor. However, in practice, it was never called upon to dig below 140 ft, its mobility was superfluous, and the ship hull proved very unsatisfactory in terms of usable space. Although it was able to operate in sea states which prevented the operation of the neighboring bucket dredge, its mining recovery factor was low and its operating costs much higher than anticipated. It

was withdrawn from service after only 9 years, in favor of the bucket dredge.

Another major operation is run by Indonesian State Mines off the islands of Bangka and Belitung. The operations are as far as 3 mi from shore in waters which are normally calm. They do have storms, however, which necessitate delays in the operation and the taking of precautions unnecessary onshore. The operations employ 12 dredges, one of which, the *Belitung I*, constructed in 1979, is the world's largest mining bucket-line dredge. With a maximum digging depth of 150 ft it is designed to dig and treat 650,000 yd³ per month of 600 hr and produce over 1500 tons of tin metal per year, depending on the richness of the ground. The dredge is working up to 15 mi offshore.

Lime shells are mined as a raw material for portland cement. Two United States operations for oyster shells in San Francisco Bay and Louisiana employed barge-mounted hydraulic cutter dredges of 16- and 18-in. diameter in 30–50 ft of water. The Iceland Government Cement Works in Akranes uses a 150-ft ship to dredge sea shells from 130-ft of water, with a 24-in. hydraulic drag dredge.

In the United Kingdom, hopper dredges are used for mining undersea reserves of sand and gravel. Drag suction dredges up to 38 in. in diameter are most commonly used with the seagoing hopper hulls. Similar deposits have been mined in the United States, and the same type of dredge is employed for the removal of sand for harbor construction or for beach replenishment. Some sand operations use beach-mounted drag lines for removal of material from the surf zone or beyond.

For deep-seabed mining, at depths below 15,000 ft, systems have been tested using airlift and hydraulic suction lift as well as a continuous dragline (CLB system), all towed from surface vessels. One other system incorporated a mobile gathering device on the bottom. All the tests carried out in the 1970s were ⅕ scale for 1,000,000–3,000,000 tons/year production models.

Consolidated deposits. The third and last area of offshore mineral resources has an equally long history. As Table 5 shows, the production from in-place mineral deposits under the sea is quite substantial, particularly in coal deposits. Undersea coal accounts for almost 30% of the total coal production in Japan and just less than 10% in Great Britain.

Extra costs have been due mainly to exploration,

Table 5. Production from consolidated mineral deposits offshore

Mineral	Location	Number of operations	Annual production	Value (millions of dollars)
Iron ore	Finland, Newfoundland*	2	1,700,000 tons	17.0
Coal	Nova Scotia, Taiwan, United Kingdom, Japan, Turkey	57	33,500,000 tons	335.0
Sulfur	United States	1	600,000 tons†	15.0
Total		60		367

*Closed June 1966.   †Estimated.

with mining and development being usually conventional. In the development of the Grand Isle sulfur mine off Louisiana, some $8,000,000 of the $30,000,000 expended was estimated to be due to its offshore location. There is no doubt that costs will be greater, generally, but on the other hand, in the initial years of offshore mining as a major industry, the prospects of finding accessible, high-grade deposits will be greater than they are at present on land.

Some of the mining methods are illustrated in Fig. 2. For most of the bedded deposits which extend from shore workings a shaft is sunk on land with access under the sea by tunnel. Massive and vein deposits are also worked in this manner. Normal mining methods are used, but precautions must be taken with regard to overhead cover. Near land and in shallow water a shaft is sunk at sea on an artificial island. The islands are constructed by dredging from the seabed or by transporting fill over causeways. Sinking through the island is accompanied by normal precautions for loose, waterlogged ground, and development and mining are thereafter conventional. The same method is also used in oil drilling. Offshore drilling and in-place mining are used only in the mining of sulfur, but this method has considerable possibilities for the mining of other minerals for which leaching is applicable. Petroleum drilling techniques are used throughout, employing stationary platforms constructed on piles driven into the sea floor or floating drill rigs.

In summary, Table 6 shows offshore mining production, valued at $1.8 billion annually. Though this is only a fraction of world mineral production, estimated at $700 billion, the results of the extensive exploration activity taking place off the shores of all five continents may alter this considerably in the future.

The future. Despite the technical problems which still have to be overcome, the future of the undersea mining industry is without doubt as potent as it is fascinating.

Deposits of hot metalliferous brines and oozes enriched with gold, silver, lead, and copper have been located over a 38.5-mi^2 area in the middle of the Red Sea at depths of 6000–7000 ft. Similar deposits are indicated over vast areas of the East Pacific Rise.

Major problems of dissolved mineral extraction must be solved before their exploitation, and significant advances must be made in the handling of

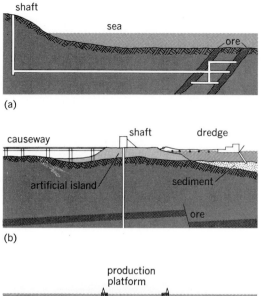

(a)

(b)

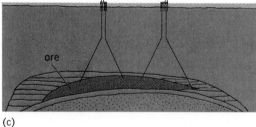

(c)

Fig. 2. Methods of mining for exploitation of consolidated mineral deposits offshore. (a) Shaft sunk on land, access by tunnel. (b) Shaft sunk at sea on artificial island. (c) Offshore drilling and in-place mining.

these sometimes corrosive media at such depths and distances from shore.

The mining of unconsolidated deposits will call for the development of bottom-sited equipment to perform the massive earth-moving operations that are carried out by conventional dredges today. The remarkable deposits of Co-Ni-Cu-Mn nodules covering the deep ocean floors will require new concepts in materials handling, and while some initial attempts are being made to mine them from the sea surface, it is almost certain that future operations will include some form of crewed equipment operating on the sea floor.

Table 6. Summary of production from mineral deposits offshore

Type	Minerals	Number of operations	Annual value (millions of dollars)
Dissolved minerals	Sodium, magnesium, calcium, bromine	275+	1346
Unconsolidated minerals	Diamonds, gold, heavy mineral sands, iron sands, tin sands, lime shells, sand and gravel	71	179.8
Consolidated minerals	Iron ore, coal, sulfur	60	367.0

land 10 mi distant

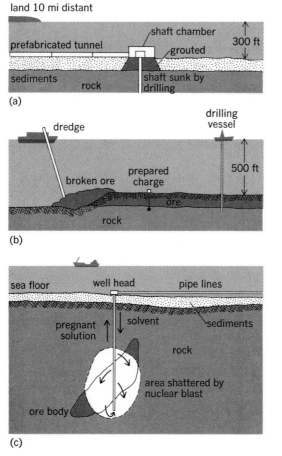

(a)

(b)

(c)

Fig. 3. Possible future methods of mining consolidated mineral deposits under the sea. (*a*) Shaft sinking by rotary drilling from tunnels laid on the sea floor. (*b*) Breaking by nuclear blasting and dredging. (*c*) Shattering by nuclear blast and solution mining.

Consolidated deposits may call for a variety of new mining methods which will be dependent on the type, grade, and chemistry of the deposit, its distance from land, and the depth of water. Some of these methods are illustrated in Fig. 3. The possibility of direct sea floor access at remote sites through shafts drilled in the sea floor has already been given consideration under the U.S. Navy's Rocksite program and will be directly applicable to some undersea mining operations. In relatively shallow water, shafts could be sunk by rotary drilling with caissons. In deeper water the drilling equipment could be placed on the sea floor and the shaft collared on completion. The laying of large-diameter undersea pipelines has been accomplished over distances of 25 mi and has been planned for greater distances. Subestuarine road tunnels have been built using prefabricated sections. The sinking of shafts in the sea floor from the extremities of such tunnels should be technically feasible under certain conditions.

Submarine ore bodies of massive dimensions and shallow cover could be broken by means of nuclear charges placed in drill holes. The resulting broken rock could then be removed by dredging. Shattering by nuclear blast and solution mining is a method applicable in any depth of water. This method calls for the contained detonation of a nuclear explosive in the ore body, followed by chemical leaching of the valuable mineral. Similar techniques are under study for land deposits.

There are many other government activities which may have a direct bearing on the advancement of undersea mining technology but possibly none as much as the International Decade of Ocean Exploration. The discovery of new deposits brings with it new incentives to overcome the multitude of problems encountered in marine mining.

[MICHAEL J. CRUICKSHANK]

Bibliography: C. F. Austin, In the rock: A logical approach for undersea mining of resources, *Eng. Mining J.*, August 1967; G. Baker, *Detrital Heavy Minerals in Natural Accumulates*, Australian Institute of Mining and Metallurgy, 1962; H. R. Cooper, *Practical Dredging*, 1958; M. J. Cruickshank, Mining and mineral recovery, *U.S.T. Handbook Directory*, pp. A/15–A/28, 1973; M. J. Cruickshank, *Technological and Environmental Considerations in the Exploration and Exploitation of Marine Mineral Deposits*, Ph.D. thesis, University of Wisconsin, 1978; M. J. Cruickshank et al., Marine mining, in *Mining Engineering Handbook*, AIME Society of Mining Engineers, sec. 20, pp. 1–200, 1973; M. J. Cruickshank, C. M. Romanowitz, and M. P. Overall, Offshore mining: Present and future, *Eng. Mining J.*, January 1968; R. H. Joynt, R. Greenshields, and R. Hodgen, Advances in sea and beach diamond mining techniques, *S.A. Mining Eng. J.*, p. 25 ff., 1977; J. L. Mero, *The Mineral Resources of the Sea*, 1965; C. G. Welling and M. J. Cruickshank. Review of available hardware needed for undersea mining, *Transactions of the Second Annual Marine Technology Society Conference*, 1966.

Marine resources

The oceans, which cover two-thirds of the Earth's surface, contain vast resources of food, energy, and minerals, and are also invaluable in many indirect ways. Capabilities to harvest these resources, and to diminish them, have become so significant that the world's nations have been forced into difficult negotiations toward an internationally accepted legal regime for the oceans, and abiding concerns have been raised about the oceans' ecological status and capacity to serve humankind.

Food. World fisheries expanded steadily from World War II to 1970, particularly the fisheries of Japan and Eastern-bloc countries. Annual harvest seems to have become stationary near 60,000,000 tons (54 teragrams); an additional 10,000,000 or perhaps 20,000,000 tons (9 or 18 teragrams) comes from fish culture in lakes, ponds, and rivers, principally in China and Southeast Asia. The ocean harvest is a very selective one, with 1% of the known oceanic species supplying nine-tenths of the total. It divides into about equal parts for food and industrial products, the latter including meal for livestock feed, fertilizers, and oils. *See* MARINE FISHERIES.

Seafood resources show a distinct geographic pattern, favoring the eastern edge of oceans because of geostrophic upwelling, the high latitudes through seasonal mixing and recycling of nutrients, and the proximity of large rivers. Estimates of the potential yield vary widely, from a technically optimistic high of 2,000,000,000 tons (1.8 petagrams) yearly to a pragmatically pessimistic low near the current level. The former envisages raising the productivity of wide areas by cultivation; the latter, a sharpening of institutional and environmental difficulties.

Unwanted species. Both human preferences and natural factors account for the pause in fisheries expansion. Conservative tastes and dietary preferences prevent the use of many unfamiliar species and of those whose names evoke psychological rejection. Horse mackerel, redfish, spider crabs, and dogfish are examples of undesired food that became highly accepted as Pacific mackerel, ocean perch, Alaska king crab, and grayfish. The marketing of squids, seals, jellyfishes, crabs, and other under- or unexploited species could probably benefit from similar rechristening strategies. Particularly abundant is Antarctic krill, the food of baleen whales; its harvesting and marketing are being explored by the Soviets and West Germans. Large populations of sable and lantern fishes in deep water are also attracting fishing effort. Such refocusing measures would counter the present domination of ecosystems by left-behind unwanted species.

Species diversity. The California sardine–northern anchovy imbalance seems to be a case in point, except that changes in oceanic conditions—the "climate" in terms of physical and chemical factors—have also been implicated in the ascendancy of the anchovy over the sardine. The study of this fluctuating climate and its effect on species diversity and abundance is a main thrust of oceanographic research.

Competition for resources. It also seems that the insistence on short-term profitability is incompatible with long-term yield and profitability. In whaling, the fleets have shifted from the large to the small whales, endangering the survival of most of the cetacean species as well as of the industry. Had the industry been content with a balanced and slightly reduced harvest after the lull of World War II, both the whales and fleets would still thrive. Herring, cod, tuna, and salmon fisheries also suffer from overcapacity and international competition. In both national and international arenas, means must be found to regulate the competition for clearly limited and sometimes fragile stocks and to effect a competent management of the living resources and their use.

Mariculture. To help increase the small share that seafood has in world nutrition—less than 20% in animal protein, about 1% in carbohydrates—great effort is currently invested in mariculture in the United States and Japan. There exists, of course, much skill in raising of fish and shellfish in fresh and brackish water, and techniques have been developing for thousands of years. Present work under the U.S. Sea Grant College Program emphasizes their cultivation in sea water, a sensitive process, and the hope is eventually to relieve pressure on such overexploited species as the salmon. The Japanese have been successful in the artificial propagation of the coastal sea bream and yellowtail and the eel and salmon, and also of food

and pearl oysters, abalone, and scallops. Mussels are extensively cultured in some bays in Italy and France. Although natural stocks of marine organisms are not overly affected by moderate pollution and do not suffer lasting damage from most oil spills, such events would wipe out the sensitive culture work. Achievement of the goal of large-scale market penetration by the culturing industry requires consideration of its environmental and juridical needs with reference to existing uses.

Energy. The potential for energy from the sea's motion and processes has long been apparent, and designs for wave- and tidal-power devices can be traced back for hundreds of years. Wind was enlisted, of course, by the first sailor. The period from mid-1800 to mid-1900 was particularly fertile for ocean-energy technology and included discovery of the power potential from ocean-thermal gradients. World War II and the period of cheap oil and gas following it pushed many of these ideas aside. With the 1973 Oil Producing and Exporting Countries (OPEC) embargo and recognition of the limited reserves of fossil fuels, efforts into the use of alternate, renewable energy resources, including the oceanic ones, have redoubled.

Categorizing resources. The categorizing of ocean energy is somewhat arbitrary. The following are not truly marine: Offshore oil and gas are in principle not different from onshore oil and gas but require greater rigor in exploration and production; about 15% of the world's oil is produced offshore, and extraction capabilities are advancing. Coal deposits, known as extensions of land deposits, are mined under the sea floor in Japan and England. Geothermal resources are known to exist offshore; they are presently not being used, and their prospects are only dimly perceived. Biomass energy in the form of methane from giant kelp is under active investigation. Another extensive energy resource is potentially available in the fissionable and fusible elements contained in sea water; they may provide power for 10^5 and 10^9 years, respectively.

In a strict sense, ocean energy is expressed in the processes of the ocean, such as in the currents, tides, waves, thermal gradients, and the only recently recognized salinity gradients. Solar energy drives them, except the tides, which are fueled by the fossil kinetic energy of the Earth-Moon system. Estimates of the intensities of the processes are given in Fig. 1. Figure 2 shows the size of the resources.

Currents. It is evident that currents constitute the smallest and weakest resource. Low-pressure turbines of 170-m diameter have been proposed for the center of the Gulf Stream, but it is doubtful that ocean currents can be profitably harnessed. A few straits possess fast tidal currents—the Seymour Narrows in British Columbia and the Apolima Straits in Western Samoa, for instance—and offer better prospects. *See* OCEAN CURRENTS.

Tides. Feasible tidal power (see Fig. 2) is limited to a few sites with high tides. Tides now provide the only source of commercial ocean power in the Rance River tidal plant in Brittany, France. Since 1968 it has produced moderate amounts of power in its bank of twenty-four 10-MW turbines, and is still being fine-tuned for greater efficiency. The

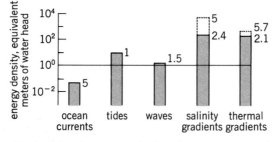

Fig. 1. Intensity or concentration of energy expressed as equivalent head of water. "Ocean currents" shows the driving head of major currents. For tides, the average head of favorable sites is given. For waves, the head represents a spatial and temporal average. The salinity-gradients head is for fresh water versus sea water, the dotted extension for fresh water versus brine (concentrated solution). The thermal-gradients head is for 12°C, and that for 20°C is dotted; both include the Carnot efficiency. *(From G. L. Wick and W. R. Schmitt, Prospects for renewable energy from the sea, Mar. Technol. Soc. J., 11(546):16–21, 1977)*

plant is successful technically and, increasingly, economically, and the experience gained there is being considered in renewed studies for tidal-power development in Australia, England, Canada, India, the United States, and the Soviet Union. The Soviets have an experimental 400-kW plant near Murmansk, and the Chinese have a number of very small plants. *See* TIDE.

Waves. Considerable efforts are being directed toward power from waves in England, Japan, Sweden, and the United States. Advanced lab and model testing is taking place in the British Isles, which are subjected to some of the most powerful waves of the Atlantic; the British hope to supply as much as 30% of their electricity needs from this source. As shown in Fig. 2, wave power is a unique resource in that it could expand under use, because in the open sea the waves could be

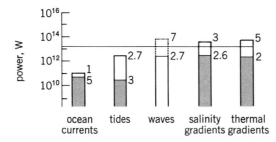

Fig. 2. On the ocean-currents bar, the shading represents the power contained in concentrated currents such as the Gulf Stream. Estimated feasible tidal power is shaded. The dotted extension on "waves" indicates that wind waves are regenerated as they are cropped. "Salinity gradients" includes all gradients in the ocean; the large ones at river mouths are shown by shading. Not shown is the undoubtedly large power if salt deposits are worked against fresh or sea water. On "thermal gradients," the shading indicates the unavoidable Carnot-cycle efficiency. The line at 1.5×10^{13} W is a projected global electricity consumption for the year 2000. *(From G. L. Wick and W. R. Schmitt, Prospects for renewable energy from the sea, Mar. Technol. Soc. J., 11(546):16–21, 1977)*

rebuilt by the winds that cause them. *See* OCEAN WAVES.

Salinity gradients. Salinity-gradient energy is present between aqueous solutions of different salinities. Vast amounts of this power are being dissipated in the estuaries of large rivers. Some conversion processes involve semipermeable and ion-selective membranes; they will require considerable development, however. One conversion process does not require membranes and is akin to an open-cycle thermal-gradient process; advances there will thus help. At the moment, salinity-gradient energy appears to be the most expensive of the marine options. Its high intensity and potential, especially when brines and salt deposits are included (see Figs. 1 and 2), which yield energy by the same principle, should, however, spur a strong level of research.

Thermal gradients. Thermal-gradient power, which would use the reservoirs of warm surface water and cold deep water, was first worked on by the French. Georges Claude spent much time and his own money on it in the 1920s, with generally bad luck. Now the United States government has made it the cornerstone of its ocean-energy program because, unlike tidal and wave power, it has the capacity for supplying base-load power. It also has a high intensity and potential. Floating tethered units of 100 megawatts electric are foreseen for locations in the Gulf of Mexico and the Hawaiian Islands. The design as much as possible employs conventional and proved technology, but the continuous operation of a complex system in a corrosive, befouling, and hazardous environment poses a formidable challenge.

Problems. There are some problems common to most marine alternatives. The principal one is transferring the power to shore. Indirect power harvest, such as by tanked or piped hydrogen or by energy-intensive products, from distant units or arrays is often suggested. But the detailed evaluation in the United States' thermal-gradient program points to electricity, the highest-value product, to justify the investment and to build up capacity and experience. Other problems pertain to corrosion in saline water, befouling by marine organisms, and deployment in the often violent sea.

Environmental impact. Environmental effects of ocean-energy conversion are generally negligible since the individual sources are large compared with initial demand. Only under extensive conversion would some impacts possibly be felt, such as downstream climatic changes from major ocean-current use, bay and estuarine ecosystem upsets from tidal-basin use, possible diminishing of dissolved oxygen from extensive use of waves, and short-circuiting of the ocean's internal heat transfer from thermal-gradient use. Conflicts would also arise with present uses of the sea through hazards to navigation and fisheries operations. Juridical issues will further complicate wide ocean-energy development. Some inadvertent effects may be beneficial, however. Upwelling of nutrients by thermal-gradient plants and the sheltering of organisms by ocean structures could enhance the productivity of fish and invertebrate stocks and their harvest, for instance.

Minerals. Annual production of marine minerals is worth one-tenth that of seafood or offshore oil and one-hundredth that of such nonextractive uses as maritime or military traffic. These minerals fall into two broad categories: geologic ones, or those deposited on or in the seabed; and chemical ones, or those dissolved in sea water. *See* MARINE MINING.

Geologic resources. Among the deposited resources, the manganese nodules have currently captured the spotlight in technology development and juridical deliberations, although commercial production had not commenced as of mid-1979. Far more important is the dredging of sand and gravel for construction materials, a billion-dollar-a-year industry. Produced also are aragonite in the Bahamas, monazite off Australia, and diatomaceous earth, while production of diamonds off Namibia was halted in 1971. Southeast Asia produces tin from shallow water, and phosphorite nodule deposits constitute potential fertilizer reserves off Florida and Morocco. Moreover, some hard-rock minerals are won by shaft mining, such as coal off Japan and the British Isles, and by drilling, such as barytes off Alaska.

Knowledge of the extent of mineral deposits is very scant and spotty. About 0.1% of the United States' continental shelf is surveyed in detail. The shelves of Europe, Southeast Asia, Japan, and South Africa are also explored. Often the surveys belong to private industry, and little is publicly disclosed. If and when shortages and economic prospects stimulate interest, resource surveys and extraction technology progress rapidly, as in the case of oil and gas and the manganese nodules.

The prospects are further enhanced by the work of the United States–operated international Deep Sea Drilling Program, which has drilled and cored about 500 sites thus far (mid-1979). The resultant insights have revolutionized knowledge of the geologic history and physical dynamics of the sea floor and have advanced understanding of the formation of mineral deposits and ores.

Chemical resources. Among the many elements and compounds dissolved in sea water and taken up by marine organisms, a few have become economically important. Where natural evaporation rates are high, salt and magnesium are won by distillation. Bromine yields to an oxydation-fractionation process. Currently the Japanese are planning to extract uranium, but the metals that some marine organisms strongly concentrate—strontium, copper, zinc, nickel, and vanadium—remain unused. Some marine plants are harvested for valuable compounds—a seaweed for agar, kelp for algin, Irish moss for carrageenin. Sponges, pearls, and ambergris are further products elaborated by organisms. Sea water itself is made to yield water for human use, mainly in rich arid countries and on some ships, but large-scale desalination for cities and crops seems to have become the victim of high-energy costs. Rather, the immense tabular icebergs of the Antarctic are now under study as water supplies for some arid countries such as Saudi Arabia. One modest-sized Antarctic berg could supply all of California with domestic and industrial water and absorb the waste heat of its power plants for a year.

Artifacts. In some sense, past and present artifacts that rest on the bottom of the sea may also be counted as mineral resources. These are often of very high quality in the form of processed ores, trapped oil, finished products and metals, and archeological treasures—coins, jewelry, art objects. Capabilities have grown in recent years in archeological, commercial, and military salvage, allowing increasing access to the wrecks of the past.

Passive uses. Often overlooked in resource assessments are nonextractive uses, since they are difficult to quantify. In this category belong such invaluable and indispensable items as inspiration and recreation, commercial and military traffic, and the disposal of waste materials and low-grade heat.

Besides inspiring artists and laymen, the sea has a strong recreational appeal. The crowding on coasts derives in part from the opportunities for sailing, fishing, and water sports, in part from those for trade and defense. Two-thirds of present oil production moves in tankers. Countries use the oceans for cooling water and as a receptacle for liquid and solid wastes: power plants are shore-sited wherever feasible; metropolitan areas discharge wastewater to sea and dump solid materials on a large scale. Such practices have, of course, raised concerns about environmental impact and are now under scrutiny and control; of particular concern is the planned disposal of radioactive wastes in the sub-sea floor.

Conclusions and prospects. The sea is important for intangible and practical ends. It sustains life with its materials and processes. Its organisms satisfy nearly one-fifth of the human demand for animal protein; its content of minerals helps overcome critical shortfalls in terrestrial resources; and it is beginning to yield some of its immense energy flux.

At present the overriding benefit, however, comes from its just being there. It sustains the continents by the spreading of its floor, purifies water and air by the hydrologic cycle, modifies climate zonation by its circulation, inspires humans' vision and carries their trade, and absorbs many of their wastes. It is thus crucial that the sea suffer relatively little from abuse and mismanagement often associated with development and growth, so that it can continue to provide the varied and balanced uses humankind has come to expect from it.   [WALTER R. SCHMITT]

Bibliography: Committee on Merchant Marine and Fisheries, U.S. House of Representatives, *Aquaculture*, 1977; Committee on Science and Technology, U.S. House of Representatives, *Energy from the Ocean*, 1978; J. D. Isaacs, The nature of oceanic life, *Sci. Amer.*, vol. 221, no. 3, 1969; J. D. Isaacs, The sea and man, *Explorers J.*, vol. 46, no. 4, 1968; National Research Council– National Academy of Sciences, *Priorities for Research in Marine Mining Technology*, 1977; National Research Council–National Academy of Sciences, *Supporting Papers, World Food and Nutrition Study*, vol. 1, 1977; G. L. Wick and W. R. Schmitt, Prospects for renewable energy from the sea, *Mar. Technol. Soc. J.*, vol. 11, no. 546, 1977.

Marine sediments

The accumulation of minerals and organic remains on the sea floor. Marine sediments vary widely in composition and physical characteristics as a function of water depth, distance from land, variations in sediment source, and the physical, chemical, and biological characteristics of their environments. The study of marine sediments is an important phase of oceanographic research and, together with the study of sediments and sedimentation processes on land, constitutes the subdivision of geology known as sedimentology. *See* MARINE GEOLOGY; OCEANOGRAPHY.

ENVIRONMENTS OF DEPOSITION

Traditionally, marine sediments are subdivided on the basis of their depth of deposition into littoral (0–20 m), neritic (20–200 m), and bathyal (200–2000 m) deposits. This division overemphasizes depth. More meaningful, although less rigorous, is a distinction between sediments mainly composed of materials derived from land, and sediments composed of biological and mineral material originating in the sea. Moreover, there are significant and general differences between deposits formed along the margins of the continents and large islands, which are influenced strongly by the nearness of land and occur mostly in fairly shallow water, and the pelagic sediments of the deep ocean far from land.

Sediments of continental margins. These include the deposits of the coastal zone, the sediments of the continental shelf, conventionally limited by a maximum depth of 100–200 m, and those of the continental slope. Because of large differences in sedimentation processes, a useful distinction can be made between the coastal deposits on one hand (littoral), and the open shelf and slope sediments on the other (neritic and bathyal). Furthermore, significant differences in sediment characteristics and sedimentation patterns exist between areas receiving substantial detrital material from land, and areas where most of the sediment is organic or chemical in origin.

Coastal sediments. These include the deposits of deltas, lagoons, and bays, barrier islands and beaches, and the surf zone. The zone of coastal sediments is limited on the seaward side by the depth to which normal wave action can stir and transport sand, which depends on the exposure of the coast to waves and does not usually exceed 20–30 m; the width of this zone is normally a few miles. The sediments in the coastal zone are usually land-derived. The material supplied by streams is sorted in the surf zone; the sand fraction is transported along the shore in the surf zone, often over long distances, while the silt and clay fractions are carried offshore into deeper water by currents. Consequently, the beaches and barrier islands are constructed by wave action mainly from material from fairly far away, although local erosion may make a contribution, while the lagoons and bays behind them receive their sediment from local rivers. The types and patterns of distribution of the sediments are controlled by three factors and their interaction: (1) the rate of continental runoff and sediment supply; (2) the intensity and direction of marine transporting agents, such as waves, tidal currents, and wind; and (3) the rate and direction of sea level changes. The balance between these three determines the types of sediment to be found. *See* DELTA.

On the Texas Gulf Coast, rainfall and continental runoff decrease gradually in a southwesterly direction. The wind regime favors considerable wave action and a southwesterly drift of the nearshore sand from abundant sources in the east. Since sea level has been stable for several thousand years, the conditions have been favorable for the construction of a thick and nearly closed sand barrier which separates a large number of bays from the open Gulf. This barrier is constructed by marine forces from sediments from distant sources and varies little in characteristics along its length. The bays, on the other hand, receive local water and sediment. In the east, the supply of both is fairly abundant, and since the streams are small, the sediment is dominantly fine; the bays have muddy bottoms and brackish waters. Conditions are fairly stable, and a rich, but quantitatively not large fauna is present, including oyster banks. At the southwestern end, continental runoff and sediment supply are negligible.

The only sediment received by the bays comes from washovers from the barriers and is therefore mainly sandy, and the virtually enclosed bays with no runoff are marine to hypersaline. Locally, this yields chemical precipitates such as gypsum and calcium carbonate, and is also conducive to the development of a restricted but very abundant fauna, which produces significant deposits of calcareous material. The sediments of bays and lagoons are often more stratified than those of the open sea as a result of fluctuating conditions. The textural and compositional characteristics depend on local conditions of topography, shore development, and wave and current patterns. They range from coarse gravel and cobbles on rocky beaches fronting the open sea to very fine clayey silt in the interior of quiet lagoons. *See* ESTUARINE OCEANOGRAPHY.

The effects of sea level changes are imperfectly known, but it is easily comprehended that the development of coastal sediments is to a large extent a function of the duration of this environment in a particular place. Thus if sea level rises or falls rapidly, there is no time for extensive development of beach, barrier, and lagoon deposits, and discontinuous blankets of nearshore sands, with muds behind them, are formed. As the rate of change decreases, open barriers, consisting of widely spaced low sand islands, tend to form, which imperfectly isolate open shallow lagoons in which essentially marine conditions prevail. A prolonged stability is required to produce thick, closed barriers and completely isolate the lagoon environment.

Entirely different nearshore deposits are found on shoals where supply of sediment and fresh water from the land is absent, either because land areas are small (Bahamas), the drainage is directed elsewhere (southern Cuba), or there is no rainfall (Arabian Peninsula). Calcareous muds and coarse calcareous sands then make up the lagoon and beach deposits. If, in addition, the shoal borders directly on the deep ocean without transitional shelf, as is the case in Fig. 1, cool water is driven onto the shoal, where it warms up and precipitates calcium carbonate. In the turbulent water of the shoal this precipitation either takes place in the form of oolites or it cements organic debris together in small aggregates (grapestone). At the edge of the shoal, the presence of cool, nutrient-rich ocean water is favorable for the growth of coral and algal reefs which are bordered by a zone of skeletal sand derived from broken calcareous organisms. The inner sheltered portions receive the finest calcareous sediments. The types of sediment and their distribution patterns are mainly controlled by the shape of the shoal, in particular the position of its edge, by the prevailing wind, and by the location of sheltered or somewhat deeper quiet areas.

Shelf and slope sediments. The continental shelf is a gently seaward sloping plain of greatly varying width, ranging from less than a mile along steep rocky coasts to several hundred miles, for example, in the western Gulf of Mexico. A distinct break in slope at 100–500-m depth marks the transition to the continental slope which descends somewhat more steeply (5–10°) to the deep-sea floor. In areas of active tectonism, for example, off the coast of southern California, the shelf is narrow and separated from the continental slope by a wide zone of deep basins alternating with shallow banks and islands. Submarine canyons cut the edge of the continental shelf in many places and sometimes reach back into the nearshore zone. *See* CONTINENTAL SHELF AND SLOPE; SUBMARINE CANYON.

During the Pleistocene, the continental shelf was subjected to repeated transgressions and regressions. During each interglacial, sea level was high and the shoreline was located near its present position; during each glacial period, much water was withdrawn from the ocean and the shoreline occurred near the edge of the shelf. The last low sea level stand occurred approximately 19,000 years ago, and the present shoreline was established as recently as 3000–5000 years ago. On most shelves, equilibrium has not yet been fully established and the sediments reflect to a large extent the recent rise of sea level. Only on narrow shelves with active sedimentation are present environmental conditions alone responsible for the sediment distribution.

Sediments of the continental shelf and slope belong to one or more of the following types: (1) biogenic (derived from organisms and consisting mostly of calcareous material); (2) authigenic (precipitated from sea water or formed by chemical replacement of other particles, for example, glauconite, salt, and phosphorite); (3) residual (locally weathered from underlying rocks); (4) relict (remnants of earlier environments of deposition, for example, deposits formed during the transgression leading to the present high sea level stand); and (5) detrital (products of the weathering and erosion of the land, supplied by streams and coastal erosion, such as gravels, sand, silt, and clay).

On shelves with abundant land-derived sediment, the coastal zone is composed of deltas, lagoons, bays, and beaches and barriers. Outside the beaches and barriers, a narrow strip of wave-transported sand, usually less than 2 or 3 mi wide, fringes the coast. On the open shelf, the sediment deposited under present conditions is a silty clay, which, near deltas, grades imperceptibly into its bottomset beds. Usually, the silty clay, which results from winnowing near the coast, is carried no more than 20–30 mi offshore by marine currents, so that the zone of active deposition is restricted. If the shelf is narrow, all of it will fall into this zone, but if it is wide, the outer part will be covered by relict sediments resulting from the recent transgression. These relict sediments were deposited near the migrating shoreline and consist of beach sands and thin lagoonal deposits. They have been extensively churned by burrowing animals and wave action, resulting in a mottled structure, and authigenic glauconite has formed in them.

On many shelves, small calcareous reefs (shelf-edge reefs) occur at the outer edge. These reefs apparently depend on the presence of deep water

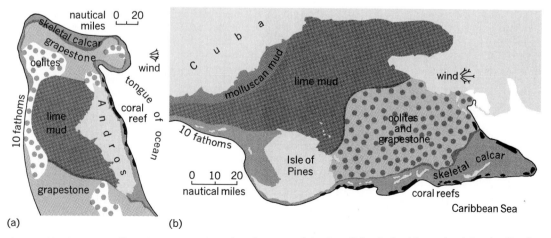

Fig. 1. Shoal water sediments as examples of sedimentation without supply of land-derived sediment. (a) The Bahamas. (b) The Gulf of Batabano, which is in southwestern Cuba. Both of these shoals border directly on the deep ocean.

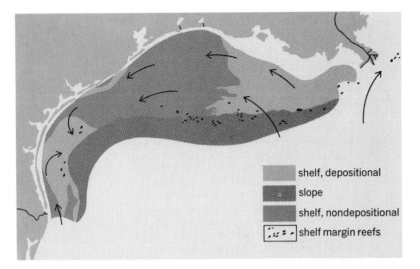

Fig. 2. Sediment distribution of the northwestern Gulf of Mexico as an example of the sediments of a shelf with abundant land-derived sediment. Shelf depositional sediments and slope deposits are silty clays; nondepositional area is covered with relict sediments. Arrows show generalized circulation.

which provide a sediment source far out on the shelf, as in the Mississippi delta; or by exposure to unusually vigorous wave action which prevents fine sediments from being deposited, even though a supply is present.

The fine-grained deposits that are being formed tend to be deposited more rapidly near the source than farther away, and as a result contain more biogenous material with increasing distance from the source, so that they become more calcareous. An example of shelf sediments with abundant supply of land-derived material is shown in Fig. 2. On shelves with little or no land-derived material, the only available sources of sediment are biogenous and authigenic. These sources provide far less material than rivers do, and as a result sedimentation rates are much lower. Even on shelves with abundant supply of land-derived material, the areas of nondeposition are extensive, often 40–50% of the total area. On the calcareous shelves, relict sediment may cover up to 75% of the entire area. Near the shelf edge these relict sediments are shallow-water deposits formed during the last low stand of the sea, and consist of small algal reefs and oolites as described for the Bahamas.

Landward of this zone, the deposits, formed when sea level rose rapidly, are thin blankets of calcareous debris, consisting of shell material, bryozoa, coral debris, and so forth (skeletal calcarenites). The only active sedimentation zones occur very near the shore, where calcareous and sometimes land-derived sand, silt, and clay are being deposited at present, and at the outermost shelf margin and continental slope, where a blanket of fine-grained calcareous mud very rich in planktonic Foraminiferida is found (foraminiferid calcilutite and calcarenite). Figure 3 shows two examples of calcareous shelves; similar are the Persian Gulf and many of the shelves of Australia.

The rise of sea level has, in many regions, severely restricted the supply of sediment to the continental shelf. Along the eastern coast of the United States, the valleys of many rivers have been flooded, and during the present stable sea level, barriers have been built across them which restrict the escape of sediment from the estuaries. As a result, sedimentation on this shelf is slow, and relict transgressive sands occur nearly everywhere at the surface.

Much of the fine-grained sediment transported into the sea by rivers is not permanently deposited on the shelf but kept in suspension by waves. This material is slowly carried across the shelf by currents and by gravity flow down its gentle slope, and is finally deposited either on the continental slope or in the deep sea. If submarine canyons occur in the area, they may intercept these clouds, or suspended material, channel them, and transport them far into the deep ocean as turbidity currents. If the canyons intersect the nearshore zone where sand is transported, they can carry this material also out into deep water over great distances.

for their growth, although it is not certain that they are growing vigorously at the present time. In the Gulf of Mexico, where they are particularly abundant, they mark the tops of salt domes in the subsurface. Beyond the reefs begins the zone of slope deposition, where in deeper and quiet water silty clays with abundant calcareous remains of open water organisms are being slowly deposited. Thus, there are in principle four parallel zones on each shelf: an inner sandy zone; an intermediate zone of clay deposition; an outer shelf zone of no deposition, where relict sediments occur, terminating in edge reefs; and a slope zone of calcareous clays. This parallel zonation is often strongly modified by special current patterns, which carry fine sediments farther out across the shelf, as in the western Gulf of Mexico; by rapidly advancing deltas

Complex sediment patterns form in areas of considerable relief, for example, the borderland off southern California, where very coarse relict and residual sediments on shallow banks alternate rapidly with silty clays and calcareous deposits in the

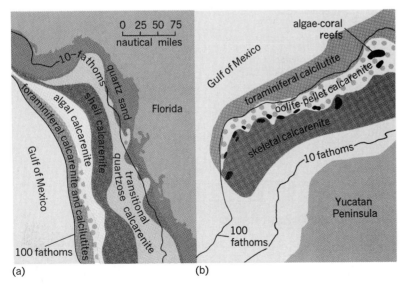

Fig. 3. Sediments on shelves with little land-derived sediment supply. (a) Off western coast of Florida. (b) Off northwestern Yucatan Peninsula.

deep troughs. Such cases, however, are rare along the continental margins. Unimportant, but striking and geologically interesting, are the calcareous sediments associated with coral reefs and atolls. Usually, they occur on islands in mid-ocean, where clear water with abundant nutrients is available, and land-derived sediments are absent; but fringing and barrier reefs with associated calcareous sediments also occur along coasts with low sediment supply. *See* REEF.

[TJEERD H. VAN ANDEL]

Deep-sea sediments. Sediments covering the floor of the deep sea were first systematically described and classified during the late 19th century by J. Murray and A. F. Renard (1884, 1891) after their observations during the Challenger Expedition (1872–1876). Their classification included two principal sediment types, terrigenous (sediments deposited near to and derived from continental areas) and pelagic (sediments, principally fine grained, accumulated slowly by settling of suspended material in those parts of the ocean farthest from land). Both categories include biogenic and nonbiogenic material, as well as sediment derived from continents, that is, terrigenous, making the classification, at best, difficult to apply. Some of the terms, such as pelagic red clay, however, have remained in general usage but with meanings somewhat modified from the original definitions. Some of the more widely quoted classifications have been proposed subsequently by H. U. Sverdrup and coworkers (1942), R. R. Revelle (1944), P. H. Kuenen (1950), F. P. Shepard (1963), and E. D. Goldberg (1964). A problem with most classifications has been in distinguishing descriptive categories, for example, red clay, from genetic categories, that is, those that include an interpretation of sediment origins, for example, volcanic mud. In addition, classifications are difficult to apply because so many deep-sea sediments are widely ranging mixtures of two or more end-member sediment types. The following sections will briefly describe the most important end members, their manner of origin, and some of the factors that control their distribution.

Biogenic sediments. Biogenic sediments, those formed from the skeletal remains of various kinds of marine organisms, may be distinguished according to the composition of the skeletal material principally either calcium carbonate or opaline silica. The most abundant contributors of calcium carbonate to the deep-sea sediments are the planktonic foraminiferids, coccolithoforids and pteropods. Organisms which extract silica from the sea water and whose hard parts eventually are added to the sediment are radiolaria, diatoms, and to a lesser degree, silicoflagellates and sponges. The degree to which deep-sea sediments in any area are composed of one or more of these biogenic types depends on the organic productivity of the various organisms in the surface water, the degree to which the skeletal remains are redissolved by sea water while settling to the bottom, and the rate of sedimentation of other types of sediment material. Where sediments are composed largely of a single type of biogenic material, it is often referred to as an ooze, after its consistency in place on the ocean

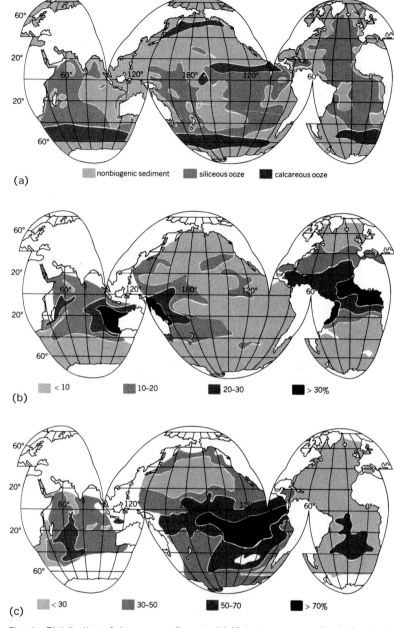

Fig. 4. Distribution of deep-sea sediments. (*a*) Major types generalized after H. U. Sverdrup and coworkers, 1942, and K. K. Turekian, 1964. (*b*) Kaolinite concentration. (*c*) Montmorillonite concentrations. Parts *b* and *c* are generalized after J. J. Griffin and coworkers, 1968, and P. E. Biscaye, 1965. The concentrations are in the $<2\text{-}\mu$ size fraction on a carbonate-opaline-silica-free basis.

floor. Thus, depending on the organism, it may be called a foraminiferal ooze (or globigerina ooze, after one of the most abundant foraminiferid genera, *Globigerina*), coccolith ooze, pteropod ooze, or for the siliceous types, diatomaceous or radiolarian ooze. The distribution map of major sediment types (Fig. 4*a*) does not distinguish the several types of organism, but indicates that the siliceous oozes are most abundant at high latitudes and in the highly productive areas of upwelling such as the Equatorial Pacific. Calcareous oozes are also abundant beneath zones of upwelling and along topographic highs such as the Mid-Atlantic Ridge. Their relative abundance in shallower water is due to an increased tendency for calcium carbonate (calcite) to redissolve at ocean depths greater than about 4000 m because of lower water temperature, greater pressure, and increased dissolved carbon dioxide in the water

Nonbiogenic sediments. The nonbiogenic sediment constituents are principally silicate materials and, locally, certain oxides. These may be broadly divided into materials which originate on the continents and are transported to the deep sea (detrital constituents) and those which originate in place in the deep sea, either precipitating from solution (authigenic minerals) or forming from the alteration of volcanic or other materials. The coarser constituents of detrital sediments include quartz, feldspars, amphiboles, and a wide spectrum of other common rock-forming minerals. The finer-grained components also include some quartz and feldspars, but belong principally to a group of sheet-silicate minerals known as the clay minerals, the most common of which are illite, montmorillonite, kaolinite, and chlorite. The distributions of several of these clay minerals have yielded information about their origins on the continents and, in several cases, clues to their modes of transport to the oceans. For example, the distribution map of kaolinite concentrations in the less than 2-μ fraction (Fig. 4b) shows that this mineral occurs in the deep sea principally adjacent to continental areas, mostly at low latitudes, where it is a common constituent of tropical soils. Its distribution indicates that it occurs only in the deep sea as a detrital phase and, in the western Equatorial Atlantic, appears to be carried to the ocean only by rivers. On the eastern side of the Equatorial Atlantic, however, its more widespread distribution corresponds with a region of frequent dust storms and suggests that its transport there is both by rivers and by the Northeast Trade Winds.

One of the common clay minerals, montmorillonite, has both a detrital and volcanogenic origin in different parts of the deep sea. In the Atlantic Ocean its distribution is controlled by its relative abundance in the soils of adjacent continental areas, whereas in the South Pacific Ocean its great abundance results from its formation, along with authigenic minerals called zeolites, as an alteration product of widespread volcanic glass (Fig. 4c). The reason these materials—both the volcanic glass shards and their alteration products—are so abundant in the South Pacific is that they are relatively undiluted by detrital constituents, the South Pacific being so large and rimmed by numerous deep trenches which act as traps for seaward-moving detrital sediments.

An authigenic constituent of deep-sea sediments that has received much recent attention, because of the potential economic value of trace elements that occur with it (principally copper), is manganese oxide. This material occurs in many areas as manganese-iron nodules. These are spheroidal pellets of MnO_4 and Fe_2O_3, ranging in size from microscopic to tens of centimeters in diameter and covering thousands of square miles of the deep-sea floor, principally in areas where both biogenic and nonbiogenic sedimentation rates are very slow.

Both in the case of biogenic and nonbiogenic sedimentation, the old concept of a slow, gentle "snowfall" of sediment particles to the sea floor where they remain forever buried, has been shown to be untrue in many parts of the ocean. Agencies are at work within the deep sea capable of distributing and redistributing sediment from its point of origin or entry in the ocean. Turbidity currents, high-density clouds of entrained sediment, are responsible for rapidly moving downward vast amounts of sediment from the continental slopes and other topographic highs. In addition, the deep oceanic current systems have been shown to be capable of eroding and transporting sediment along the ocean floor. The evidence for this comes from deep-sea photographs which show current ripple marks and scour marks in many parts of the ocean as well as measurements of relatively turbid zones of water traveling near the bottom (the nepheloid layer) in many parts of the ocean; this research suggests a rather continuous process of sediment erosion and redeposition. This process is discussed in a later section on transport and rate of sedimentation.

Ancient sediments. Because intensive studies of deep-sea sediments are a relatively recent phenomenon in the study of the Earth, and because the surface layer of sediment covers nearly two-thirds of the Earth's surface, investigators have only begun to understand the factors which presently control the distribution of sediments. An entire additional dimension, however, is introduced by extending these studies down through the sediment column, that is, back through geologic time. Samples of sediment have been taken in deep-sea cores back through the Pleistocene, Tertiary, and as old as Jurassic, or more than 130,000,000 years old. The few studies made to date (1969) on these older deep-sea sediments show that sedimentary conditions were very different from those that now obtain. Several studies in the Pacific Ocean and scattered results from the Atlantic Ocean suggest that sediments derived from volcanic sources were possibly much more important in late Mesozoic and during most of Tertiary time, and that the importance of detrital sediment sources is a relatively recent phenomenon, beginning as late as Pliocene or Pleistocene time.

[PIERRE E. BISCAYE]

Bibliography: Tj. H. van Andel, Morphology and sediments of the Sahul Shelf, northwestern Australia, *Trans. N.Y. Acad. Sci.*, 28:81–89, 1965; Tj. H. van Andel and J. R. Curray, Regional aspects of modern sedimentation in northern Gulf of Mexico and similar basins, in F. P. Shepard (ed.), *Recent Sediments, Northwestern Gulf of Mexico*, Amer. Ass. Petrol. Geol. Spec. Publ., 1960; P. E. Biscaye, Mineralogy and sedimentation of Recent deep-sea clay in the Atlantic Ocean and adjacent seas and oceans: *Geol. Soc. Amer. Bull.*, 76:803–832, 1965; K. O. Emery, *The Sea of Southern California*, 1960; D. B. Ericson and G. Wollin, *The Deep and the Past*, 1964; E. D. Goldberg, The oceans as a geological system, *Trans. N.Y. Acad. Sci.*, 27:7–19, 1964; J. J. Griffin, H. Windom, and E. D. Goldberg, The distribution of clay minerals in the world ocean, *Deep-Sea Res.*, 15:433–459, 1968; F. P. Shepard, *Submarine Geology*, 2d ed., 1963; H. U. Sverdrup, M. W. Johnson, and R. H. Fleming, *The Oceans*, 1942; K. K. Turekian, The geochemistry of the Atlantic Ocean Basin, *N.Y. Acad. Sci. Trans.*, ser. 2, 26:312–330, 1964.

PHYSICAL PROPERTIES

Physical properties of marine sediments, such as density and the elastic constants, depend on many factors. These include grain shapes, sizes, and compositions; the amount of interstitial fluid and its properties; the nature of grain-to-grain contacts; the degree of compaction and consolidation; and the age. Measurable physical quantities appear to be affected to a much greater extent by the fractional volume of fluid in the sediment (porosity) than by sediment type. This is to be expected since density and elastic constants do not differ much for the principal constituents of sediments (silica, calcium carbonate, clay minerals). Water depth affects physical properties only to a slight extent.

Some measurements of physical properties are made on samples recovered from the ocean bottom by coring devices or in sufficiently shallow water by divers. Such observations are limited to sediments lying within a few tens of feet of the water-sediment interface. Properties of deeper-lying sediments are known from seismic refraction measurements of the velocities of elastic waves, by inference from dispersion of surface waves, or from gravity data. A few typical measurements are given in the table. These are merely illustrative and are not in any sense average or most likely values.

Definitions and some useful interrelationships among measurable quantities are given by Eqs. (1)–(5), where ρ_1 and ρ_2 are fluid and average parti-

$$\rho = \rho_1\phi + \rho_2(1-\phi) \quad \text{(bulk density)} \quad (1)$$

$$\alpha = \sqrt{(k + 4/3\mu)/\rho} \quad \text{(compressional wave velocity)} \quad (2)$$

$$\beta = \sqrt{\mu/\rho} \quad \text{(shear wave velocity)} \quad (3)$$

$$(\alpha/\beta)^2 = 2(1-\sigma)/(1-2\sigma) \quad (4)$$

$$k = 1/C \quad \begin{array}{l}\text{(incompressibility}\\ = 1/\text{compressibility)}\end{array} \quad (5)$$

cle densities, respectively, ϕ is the porosity ($\phi = 1$ at 100% fluid, $\phi = 0$ at 0% fluid), μ is rigidity, and σ is Poisson's ratio.

Several general conclusions may be drawn from observations. Density and porosity are nearly linearly related and for most ocean sediments observations lie between the two lines, as in Eq. (6).

$$\rho_{\text{cgs}} = 1.03 + (1.67 \pm 0.05)(1-\phi) \quad (6)$$

Compressional wave velocities for $\phi > 0.6$ agree well with the predictions, or equation, of A. B. Wood. This is obtained by inserting into Eq. (2) $\mu = 0$ and $C = 1/k = C_1\phi + C_2(1-\phi)$, where C_1 and C_2 are compressibilities of fluid and particles, respectively. At smaller porosities compressional velocity rises more steeply, the $\phi = 0$ limit being near 6 km/sec. Poisson's ratio may be expected to vary from a value near 0.25 at zero porosity to 0.5 at $\mu = 0$. Good agreement of α with the Wood equation for $\phi > 0.6$ indicates that σ reaches its upper limit near $\phi = 0.6$.

Seismic refraction measurements indicate that compressional velocity increases with depth in the sedimentary column, the gradient being from 0.5 to 2.0/sec. Thus ρ, μ, k, β should also increase with depth and ϕ and σ decrease. [JOHN E. NAFE]

Bibliography: E. Bullard, The flow of heat through the floor of the Atlantic Ocean, *Proc. Roy. Soc. London, Ser. A*, 222:408–429, 1954; E. L. Hamilton et al., Acoustic and other physical properties of shallow-water sediments off San Diego, *J. Acoust. Soc. Amer.*, 28:1–15, 1956; A. S. Laughton, Sound propagation in compacted ocean sediments, *Geophysics*, 22:233–260, 1957; J. E. Nafe and C. L. Drake, Variation with depth in shallow and deep water marine sediments of porosity, den-

Typical measurements of selected properties of marine sediments

Property measured	Fine sand, 17-station average[a]	Clayey fine silt[a]	Gray clay or silt[b]		Cream calcisiltite[c]		Gray clay[c]	Artificially compacted globigerina ooze pressure[d], kg/cm²		
								512	768	1024
Medium grain diameter, mm	0.19	0.02			[0.01][e]	[0.01]				
ρ, g/cm³	1.93	1.60	1.72	1.58	1.60	1.71	1.46	2.14	2.22	2.26
ϕ, %	46.2	65.6	[56]	[65]	65	57	74	[32]	[28]	[26]
α, km/sec	[1.68]	[1.46]			1.59	1.68	1.49	2.68	2.89	3.06
β, km/sec								1.20	1.42	1.57
σ (Poisson's ratio)	0.44[f]	0.50[f]						0.37	0.34	0.32
k, 10^{-11} dyne/cm²	[0.472]	[0.342]						1.13	1.25	1.38
μ, 10^{-11} dyne/cm²	[0.06][g]	[0.00][g]						0.31	0.45	0.56
Thermal conductivity, 10^{-4} cal/(cm)(°C)(sec)		26.8	23.1							
Approximate water depth, fathoms	15	15	1000	1000	2300	1600	2550			

[a]After E. L. Hamilton et al., 1956.   [b]After E. Bullard, 1954.   [c]After G. H. Sutton et al., 1957.   [d]After A. S. Laughton, 1957.   [e]Brackets indicate conversion of units from those used in the original publication.   [f]Lower limit.   [g]Upper limit.

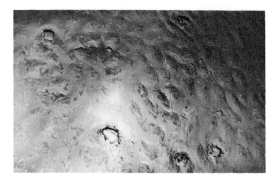

Fig. 5. On southern margin of the Burdwood Bank, in northern Drake Passage. Rhomboidal ripple marks in foraminiferid ooze attest to current velocities in excess of 1 knot. Scour around ice-rafted boulders occurs on down-current side. Lat. 55-59°S; long. 61-43°W; 4060 m.

sity and the velocities of compressional and shear waves, *Geophysics*, 22:523–552, 1957; C. B. Officer, Jr., A deep-sea seismic reflection profile, *Geophysics*, 20:270–282, 1955; G. H. Sutton, H. Berckhemer, and J. E. Nafe, Physical analysis of deep sea sediments, *Geophysics*, 22:779–821, 1957; A. B. Wood, *A Textbook of Sound*, 3d ed., 1955.

TRANSPORT AND RATE OF SEDIMENTATION

Rivers, glaciers, wind, and ocean waves and currents carry particles from continents and continental margins to the various environments of deposition in the ocean. The rates of sedimentation can be determined by study of the micropaleontology, radioactivity, or paleomagnetism of the sediments.

Transport of sediments. Sedimentary particles derived from the continents by weathering are brought to the sea by rivers and streams. Most of these particles never reach the deep sea but are (1) trapped in estuaries; (2) deposited in or near deltas; (3) concentrated on beaches or other littoral deposits; or (4) carried to and deposited on the inner continental shelves by waves and longshore currents. A small fraction of the clay-sized particles, especially near the mouths of major rivers, is carried off of the continental shelf by currents. *See* DELTA; ESTUARINE OCEANOGRAPHY; NEARSHORE PROCESSES.

Low stands of the sea during glacial epochs exposed only the inner margins of deeper continental shelves to erosion, while their outer portions became zones of littoral deposition. Rivers crossed shallower shelves through valleys cut in the shelf

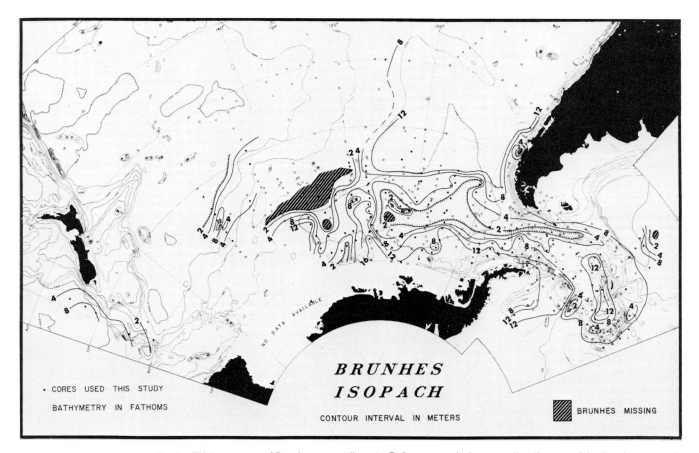

Fig. 6. Thickness map of Brunhes-age sediments. Reference horizons are sediment-water interface and Brunhes-Matuyama polarity reversal boundary 700,000 years ago. The area where the Brunhes-age sediments are missing, as well as the area of the Brunhes-age sediments outlined by 2-m thickness or less, across the Southern Pacific is a zone of currents, documented by bottom photography, as in Fig. 5.

plain and debauched directly down the continental slopes. As glaciation waned and sea level rose, the littoral deposits of the outer shelf were sometimes reworked and their finer clasts carried inward by wave action to contribute, in part, to present coastal sediments.

On most continental shelves the outer sections are receiving no contemporary continental sedimentation; therefore, they retain relict littoral sedimentary deposits which document former low stands of the sea. Some continental shelves in high latitudes are floored with glacial deposits: terminal moraines, recessional moraines, or glacial marine sediments deposited during Pleistocene glacial epochs. These deposits, too, are relict, except for the Antarctic and Greenland continental shelves where present glaciers terminate on the shelf and icebergs raft glacially derived sediments seaward before melting. Tropical shelves which lie between the 18°C sea-water isotherms often have carbonate sediments, in part derived by wave and current action from barrier and fringing reefs at the shelf margin, and in part biochemically precipitated by other invertebrates associated with carbonate reef-bank ecologies. Under optimum conditions reefs grow upward at rates up to 50 m/1000 years. *See* GLACIOLOGY; REEF.

Sediments accumulate on the continental slopes until slumping occurs to carry the deposits down to the abyssal floor. These slumps often occur by rotational faulting, but sometimes the slump transforms by mixing with water to become a mud flow (landslide) or density current. The transport of very fine-grained material by density currents, also known as turbidity currents, has been hypothesized as being a major transportive mechanism from continental slopes to abyssal plains, especially in the Northern Hemisphere. During the low sea level stands of glacial epochs when rivers discharged directly down the continental slopes, density currents were much more prevalent and may have been instrumental in eroding submarine canyons, forming the great submarine alluvial cones which occur at their bottoms and building abyssal valleys which show natural levies hundreds of miles from the continental margins. *See* SUBMARINE CANYON; TURBIDITY CURRENT.

There is increasing evidence that geostrophic ocean currents are responsible for considerable sediment transport, even of sand-sized sedimentary particles, in the deep sea (Fig. 5). Such currents have been demonstrated to cause erosion, not only from submarine ridges and sea mounts, but also across abyssal plains, and to deposit sediments on the protected side of obstructions and in submarine valleys. *See* OCEAN CURRENTS.

Large portions of the ocean floor that are far from land and lie under zones of high surface biological productivity are covered by biological deposits which consist of the calcareous or siliceous tests of plankton. These deposits constitute the organic oozes which cover more than 60% of the world's ocean floors (Fig. 4).

Much smaller contributions to abyssal sediments are made by wind transport of silt, clay, and volcanic ash from the continents, cosmic dust and

micrometeorites, and precipitation of inorganic constituents directly out of sea water. Of still less importance is biological transport of detritus by mammals, fish, birds, and kelp.

Rates of sedimentation. Early measures of rates of sedimentation on nearshore deposits were estimated from historical filling of bays and estuaries, and progradation and erosion of coastlines. Direct observation in the deep sea was limited to the apparent depth of burial of such objects as deep-sea cables, which had been on the sea floor for several years. Since such objects could have partially sunk into the sediment, rates of burial derived from sedimentation alone were conjectural.

Other early attempts revolved about the recognition in deep-sea cores of the end of the last glacial epoch as documented by temperature-sensitive pelagic Foraminiferida. However, such climatic changes are usually not synchronous worldwide and in tropical regions may not even have occurred. Furthermore, the time of the end of the last glacial epoch was not established until recently, and in the deepest parts of the sea, solution of the calcareous tests erased the record.

Modern methods of calculating rates of sedimentation necessitate the determination of time boundaries in deep-sea cores, using radioactivity or paleomagnetism. Radioactive determinations utilize the rate of change of surface disequilibrium to equilibrium-with-depth in cores in isotope pairs of the U^{238}, U^{235}, or Th^{232} families. These methods depend upon numerous assumptions, including

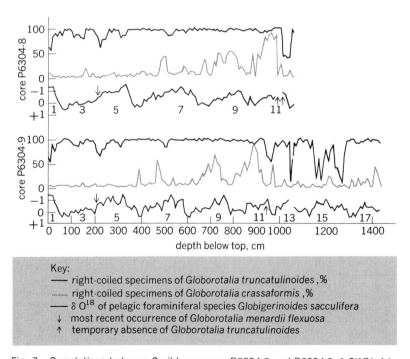

Key:
— right-coiled specimens of *Globorotalia truncatulinoides* ,%
--- right-coiled specimens of *Globorotalia crassaformis* ,%
— δ O^{18} of pelagic foraminiferal species *Globigerinoides sacculifera*
↓ most recent occurrence of *Globorotalia menardii flexuosa*
↑ temporary absence of *Globorotalia truncatulinoides*

Fig. 7. Correlations between Caribbean cores P6304-8 and P6304-9. δ O^{18}(‰) is with respect to the Chicago standard PDB-1 (δ O^{18} = R/R standard −1 where R = $(O^{18})/(O^{16})$). The numbers identify the different core stages. The downward arrows indicate the most recent occurrence of *Globorotalia menardii flexuosa*. The upward arrows identify the core layers in which *Globorotalia truncatulinoides* becomes rare. (*After C. Emiliani, J. Geol., 74:109–126, 1966*)

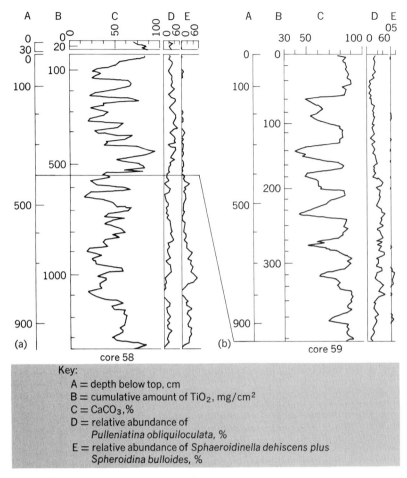

Key:

A = depth below top, cm
B = cumulative amount of TiO₂, mg/cm²
C = CaCO₃,%
D = relative abundance of
    Pulleniatina obliquiloculata, %
E = relative abundance of *Sphaeroidinella dehiscens plus
    Spheroidina bulloides*, %

Fig. 8. Correlation between two cores from the eastern equatorial Pacific ocean bottom. (*a*) Core 58. (*b*) Core 59. (*After C. Emiliani, Science, 154:851–857, copyright © 1966 by the American Association for the Advancement of Science*)

products within the sediment column. Theoretically, these methods cover the time span back to more than 300,000 years. Other methods have used C¹⁴, Be¹⁰, Al²⁶, or Si³¹, all of which are produced in the outer atmosphere by cosmic ray bombardment. Carbon-14 can be used only in carbonate sediments and has a limited half-life. The remainder occur in very low quantities, requiring sophisticated ion-exchange techniques for concentration or low-level counting for long periods.

Sedimentation rates based on paleomagnetic chronology depend upon the determination of the paleomagnetic polarity of deep-sea sediments and upon the correlation of these polarities with the paleomagnetic polarity time scale developed from terrestrial lavas. It has been demonstrated that fine particles settle through a water column to align themselves with the Earth's field, rather than doing so hydrodynamically, and acting as tiny magnets, they represent the Earth's field at the time they are deposited. If only the vertical orientation of a core is known (that is, which end is up), the direction of the vertical vector of the detrital remanent magnetism is sufficient to determine whether or not the sediment is normally magnetized (deposited during a time when the Earth's magnetic field is oriented N–S, as it is today), or if there is reversed magnetism (sediment deposited when the Earth's field was in the opposite sense). Inasmuch as the times when the Earth's magnetic field has reversed are known with some precision back to almost 3,500,000 years, reversals in remanent magnetism in sedimentary cores provide time lines for correlation, against which rates of sedimentation can be measured. Reversals in the Earth's magnetic field are worldwide and geologically synchronous and therefore provide a chronological datum everywhere in fine-grained deep-sea deposits. Where sufficient cores are available, it has been possible to draw maps of the thickness of sediment deposited during any selected paleomagnetic time interval. These maps are also maps of rates of sedimentation (Fig. 6).

Rates of sedimentation around deltas and in

the chemical separation in sea water of uranium from thorium by the precipitation of thorium hydroxide into the sediment, as well as the retention and nonmigration of radiodisintegration daughter

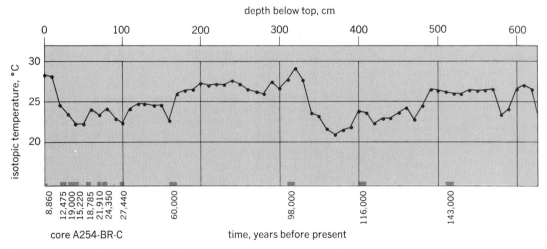

Fig. 9. Core A254-BR-C (central Caribbean Sea). Paleotemperatures measured on pelagic foraminiferid species *Globigerinoides sacculifera*. Absolute ages by C¹⁴ and Pa²³¹/Th²³⁰ measurements. (*After C. Emiliani, Bull. Geol. Soc. Amer., 75:129–144, 1964*)

estuaries are as high as 50,000 cm/1000 years; on continental slopes and rises, 100 cm/1000 years; and on the abyssal floors, 0.01–2 cm/1000 years. These rates have varied considerably in the past, depending upon climate, tectonism, and organic productivity in the ocean. Rates have been highest when the continents and mountain ranges were high, during glacial stands of low sea level when continental drainage was dumped directly onto abyssal floors, and when the wind and ocean circulation patterns were such as to ensure maximum biological productivity. Rates have been lowest when widespread epeiric seas have inundated the continents, when the continental margins consisted of basins or geosynclines which trapped sediments, and prior to the advent of calcareous Foraminiferida in the late Mesozoic.

[H. G. GOODELL]

Bibliography: G. Arrhenius, Pelagic sediments, in M. N. Hill (ed.), *The Sea*, vol. 3, 1963; R. A. Bagnold, Beach and nearshore processes, pt. 1: Mechanics of marine sedimentation, in M. N. Hill (ed.), *The Sea*, vol. 3, 1963; A. Cox, Geomagnetic reversals, *Science*, 163:237–245, 1969; K. O. Emery, Organic transportation of marine sediments, in M. N. Hill (ed.), *The Sea*, vol. 3, 1963; H. G. Goodell and N. D. Watkins, The paleomagnetic stratigraphy of the Southern Ocean: 20° west to 160° east longitude, *Deep-Sea Res.*, 15:89–112, 1968; B. C. Heezen, Turbidity currents, in M. N. Hill (ed.), *The Sea*, vol. 3, 1963; D. L. Inman and R. A. Bagnold, Beach and nearshore processes, pt. 2: Littoral processes, in M. N. Hill (ed.), *The Sea*, vol. 3, 1963; F. F. Koczy, Age determination in sediments by natural radioactivity, in M. N. Hill (ed.), *The Sea*, vol. 3, 1963.

CLIMATIC RECORD IN SEDIMENTS

An undisturbed section of pelagic sediments contains important information on the geophysical and environmental conditions prevailing in the Earth's interior, on the ocean floor, in the water column above, at the ocean surface, in the atmosphere, on adjacent continents, and even in outer space, during the time of sediment deposition. One of the principal and better-studied parameters is temperature. Temperature variations occurring both at the ocean surface and on the ocean floor are recorded by variations in the relative abundance of the tests of different species of pelagic and benthonic organisms, and by variations of the O^{18}/O^{16} ratio in the calcium carbonate of the tests of these organisms. Pelagic and benthonic Foraminiferida in deep-sea cores of globigerina-ooze facies have been primarily used for these studies.

Methods of study. In order to obtain a meaningful temperature record, the sediment cores must be shown to contain continuous and undisturbed stratigraphic records. This can be done by intercorrelating different cores by means of different parameters, both temperature-dependent and temperature-independent, together with absolute dating by radiometric methods and unique micropaleontological events. If concordance is obtained, the cores must be undisturbed. Only a few cores representing a substantial portion of Pleistocene time have been found to meet these stringent cri-

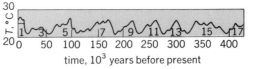

Fig. 10. Generalized temperature curve for surface water of central Caribbean. Numbers identify temperature cycles. (*After C. Emiliani, J. Geol., 74:109–126, 1966*)

teria. Among these are two cores from the Caribbean and a suite of cores from the eastern equatorial Pacific. These cores have been studied by G. Arrhenius, C. Emiliani, and L. Lidz, using geochemical, isotopic, and micropaleontological techniques.

Figures 7 and 8 show different curves, representing mutually independent parameters, for the two cores from the central Caribbean and two from the eastern Pacific. By using the different curves simultaneously, the two cores within each set can be unequivocally correlated with each other, leading to the conclusion that both represent continuous and undisturbed sediment sections.

Accurate geochemical, isotopic, and micropaleontological analysis of these cores and other undisturbed deep-sea cores of globigerina-ooze facies, together with carbon-14 and protactinium-231/thorium-230 age measurements (Fig. 9), revealed quasi-sinusoidal temperature oscillations having an apparent amplitude of 9–10°C (uncorrected for glacial-interglacial changes in the isotopic composition of sea water) in the Caribbean and the equatorial Atlantic, and an average wavelength of about 50,000 years (see isotopic curves in Figs. 7 and 9). The oxygen isotopic curves of the various Atlantic and Caribbean cores and the C^{14} and Pa^{231}/Th^{230} age measurements (together with reasonable extrapolations from the latter) have been used to construct a generalized temperature curve (Fig. 10). In constructing this curve, the O^{18}/O^{16} results were corrected for the glacial-interglacial changes in the isotopic composition of sea water. Because of the uncertainty involved in this correction, the isotopic temperatures cannot be considered as absolute, but only as reasonable

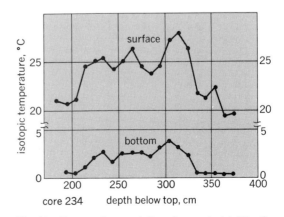

Fig. 11. Temperature variations in equatorial Atlantic during last interglacial (about 100,000 to 70,000 years ago). (*After C. Emiliani, J. Geol., 66:264–275, 1958*)

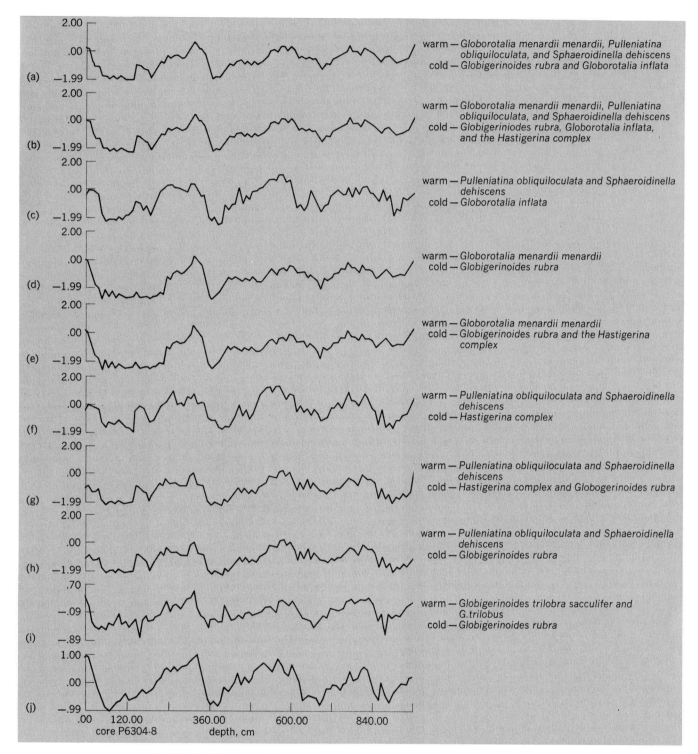

Fig. 12. Caribbean core P6304-8. (a–i) Ratios of the relative abundance of planktonic foraminiferid species or groups of species diagnostic of warm and cold periods. (j) O¹⁸/O¹⁶ ratio in pelagic foraminiferid species *Globigerinoides triloba sacculifera.* Vertical scales (in inches) are proportional to original micropaleontological and isotopic ratios. (*After L. Lidz, Science, 154:1448–1452, copyright © 1966 by the American Association for the Advancement of Science*)

approximations. The generalized temperature curve, which extends from the present to about 425,000 years ago, shows eight temperature cycles which, with the single exception of the cycle extending 65,000–15,000 years ago, have approximately the same amplitude.

Some of the Pacific cores mentioned (for example, core 59) exhibit very clear and well-defined oscillations in the percentages of carbonate, while others (for example, core 58) do not, largely because of postdepositional solution (Fig. 8). While the shallower pelagic foraminiferid species are insufficiently abundant in these cores for isotopic analysis, analysis of the deeper subspecies *Globorotalia menardii tumida* confirmed the inverse relationship between carbonate content and temperature predicted by Arrhenius. Thus, each carbonate cycle in the Pacific cores is likely to be the equivalent of a temperature cycle in the Atlantic and Caribbean cores. As a result, core 59, containing about nine carbonate cycles, should include sediments about 440,000 years old.

The rate of accumulation of TiO_2 in core 58, when averaged across several tens of thousands of years, remains almost constant (Fig. 8, column B), indicating an almost constant rate of bulk sedimentation. As a result, the age of about 440,000 years for the layer at 400 cm below the top may be used for estimating an age of about 1,100,000 years for the bottom of the core (991 cm below the top).

The carbonate cycles continue through core 58, and one of the highest carbonate peaks, presumably representing a strong glaciation, occurs near the bottom of the core, at an estimated age of about 1,100,000 years.

Figure 11 shows the isotopic temperature variations during the last interglacial period for both the surface and bottom water of the equatorial Atlantic. Assuming that the temperature of the bottom water remained constant across the glacial and interglacial ages of the Pleistocene (a very conservative assumption), the minimum amplitude of the glacial-interglacial temperature variation of the surface water would be given by the difference between the two curves of Fig. 11, or about 6°C. This value has been used to establish the amplitude of the generalized temperature curve shown in Fig. 10.

Different species of pelagic Foraminiferida occupy different habitats, with respect to both latitude and depth. Of 31 species listed by A. W. H. Bé, 22 occur mainly in the tropical-subtropical areas, 5 in the temperate seas, and 1 in the polar zones. The majority of the species live close to the surface, but at least five are known to live at considerable depth during a portion of their life cycles. Also, different species have different seasonal growth habits in regions where appreciable seasonal temperature variations occur. It is evident that the ecology and paleoecology of the species used for temperature reconstructions must be known and accurately taken into consideration.

W. Schott, R. Todd, and F. Phleger have done extensive micropaleontological work on deep-sea cores, directed to obtaining estimates of past temperature variations. They have used the classical method of recording the relative abundances of different species having different temperature habitats. D. Ericson has relied mainly on the relative abundance of the *Globorotalia menardii* complex, a method which was shown to be highly unreliable. The micropaleontological techniques were refined by Lidz, who used ratios of warm to cold species or groups of species to amplify and clarify the signal. The results for one of the Caribbean cores are shown in Fig. 12, together with the O^{18}/O^{16} curve. The agreement between isotopic and micropaleontological paleotemperatures is excellent.

Temperature oscillations. The Caribbean, Atlantic, and Pacific deep-sea cores indicate the

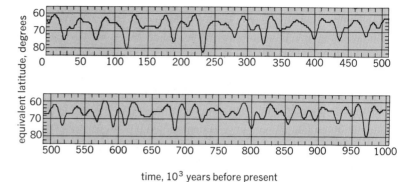

Fig. 13. Variation of summer insolation at 65°N during the past million years (in degrees of equivalent latitude), based on data calculated by D. Brouwer and A. J. J. van Woerkom. (*From C. Emiliani, J. Geol., 63:538-578, 1955*)

occurrence of about 20 major temperature oscillations during the past 1,000,000 years, with each colder and warmer interval lasting about 25,000 years. This conclusion, which is in sharp contrast with the generally accepted, classical view of four major glaciations separated by four interglacials, each lasting 60,000–200,000 years, is supported by recent evidence from various sources. In particular, the fossil pollen record in the lake deposits at Leffe, Lombardy, showed the occurrence of eight major temperature cycles during the "Donau-Mindel" interval; two of the classical interglacials of Europe have been shown, by varve analysis, to have lasted only about 30,000 years each; and age measurements by the K^{40}/Ar^{40} method have dated the Günz glaciation at about 400,000 years ago, and have revealed the repeated occurrence of glaciation back to at least 2,700,000 years ago.

Variations in the astronomically induced semiannual insolation in the high latitudes, which have a quasi-periodicity of about 50,000 years (Fig. 13), have been repeatedly proposed by J. Croll, M. Milankovitch, and others as a cause of glaciation. In the modern development proposed by C. Emiliani and J. Geiss, glaciation was induced by Tertiary cooling caused by albedo increase associated with the alpine orogenesis; the successive glaciations

Fig. 14. Bottom dredge, showing bottom sample. (*U.S. Navy Hydrographic Office*)

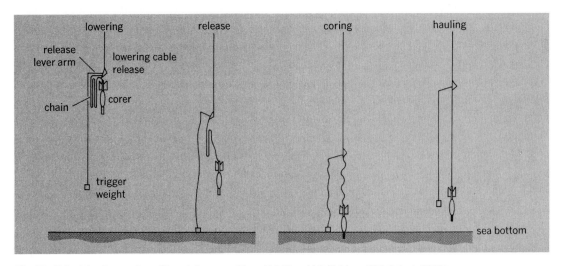

Fig. 15.   Principle of operation of free-fall corers. (*From U.S. Navy H.O. Publ. no. 607, 2d ed., 1955*)

were initiated by the quasi-periodic reductions of summer insolation at the high northern latitudes; the further development of each glaciation was largely a self-sustaining process; and complete deglaciation (Antarctica and Greenland excepted) at the end of each glacial cycle resulted from surface freezing of the northern North Atlantic and time-delay effects introduced by plastic flow of glaciers, heat absorption by ice melting, and crustal warping. On the basis of the frequency of the past temperature variations, it is possible to predict the beginning of a new glaciation within a few thousand years, reaching its maximum about A.D. 20,000. *See* CLIMATIC CHANGE.

[CESARE EMILIANI]

Bibliography: G. Arrhenius, Sediment cores from the east Pacific, in *Reports of the Swedish Deep-Sea Expedition, 1947–1948*, vol. 5, 1952; A. W. H. Bé, *Foraminifera: Families Globigerinidae and Globorotaliidae*, Conseil Perm. Intern. Explor. Mer, Fiches Ident. Zooplankton, no. 108, 1967; C. Emiliani, Isotopic paleotemperatures, *Science*, 154:851–857, 1966; C. Emiliani, Paleotemperature analysis of core 280 and Pleistocene correlations, *J. Geol.*, 66:264–275, 1958; C. Emiliani, Paleotemperature analysis of the Caribbean cores A254-BR-C and CP-28, *Bull. Geol. Soc. Amer.*, 75:129–144, 1964; C. Emiliani, Paleotemperature analysis of the Caribbean cores P6304-8 and P6304-9, and a generalized temperature curve for the past 425,000 years, *J. Geol.*, 74:109–126, 1966; C. Emiliani, Pleistocene temperatures, *J. Geol.*, 63:538–578, 1955; D. B. Ericson et al., Atlantic deep-sea sediment cores, *Bull. Geol. Soc. Amer.*, 72:193–286, 1961; F. L. Parker, Eastern Mediterranean Foraminifera, *Reports of the Swedish Deep-Sea Expedition, 1947–1948*, vol. 8, pp. 217–283, 1958; F. B. Phleger, F. L. Parker, and J. F. Peirson, North Atlantic Foraminifera, *Reports of the Swedish Deep-Sea Expedition, 1947–1948*, vol. 7, fasc. 1, p. 122, 1953; W. Schott, Die Foraminiferen in dem aequatorialen Teil des Atlantischen Ozeans, *Wiss. Ergeb. Deut. Atl. Exped. "Meteor,"* 1925–1927, vol. 3, pp. 43–134, 1935; R. Todd, Foraminifera from western Mediterranean deep-sea cores, *Reports of the Swedish Deep-Sea Expedition, 1947–1948*, vol. 8, pp. 167–215, 1958.

SAMPLING AND CORING DEVICES

The bottom-sampling devices used to collect sediments from the ocean floor are of three main types: snappers, dredges, and coring tubes. Snapper-type samplers, with closing jaws actuated by a tension spring and trigger mechanism, are used to obtain small samples from the sediment surface. Snapper, or grab, samplers are generally used in shallow waters where it is desirable to gather a large number of samples as rapidly as possible. The dredge is essentially a bag made of steel links; its mouth is held open by a rectangular frame provided with a bail to which the ship's trawl wire is shackled. When it is dragged along the bottom, the dredge collects relatively large

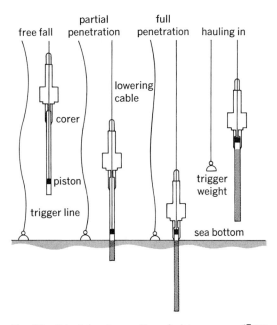

Fig. 16.   Principle of operation of piston corers. (*From U.S. Navy H.O. Publ. no. 607, 2d ed., 1955*)

objects such as manganese oxide and phosphate nodules, sharks' teeth, and ice-rafted rocks lying on or near the sediment-water interface (Fig. 14).

Coring devices are used to obtain samples of bottom material in place. This device consists essentially of a steel tube, a glass or plastic liner that may be removed without disturbing the sample, attached weights, and a core-catcher ring and cutting edge which fit the bottom, or penetration edge, of the tube. The amount of sample collected depends on the length of the corer, its weight, and the penetrability of the bottom. Core samples up to 1.2 m in length may be obtained when the smaller types of corer are used. Modifications in coring devices include the piston corer and free-fall release mechanism. Both are utilized in the heavier coring equipment which is used to obtain longer cores.

Piston corer. The piston corer, a steel tube measuring about 60 mm in inside diameter, is capable of recovering undistorted vertical sections of sediment as much as 20 m in length. The corer is driven into the sediment by a free fall of about 5 m and by 500–1000 kg of lead attached to the upper end. Free fall is effected by a trigger weight attached by a wire line to the end of a trigger arm, which rises and releases the coring tube when the trigger weight comes to rest on the sediment surface (Fig. 15). Connection between the ship's cable and the apparatus is maintained by a length of wire rope which passes from the end of the ship's cable down through the coring tube to a piston which is initially at the bottom of the tube. The length is so adjusted that the wire rope becomes taut just when the bottom of the coring tube reaches the sediment surface (Fig. 16). As the tube sinks into the sediment, the piston, immobilized by the cable to the ship, makes hydrostatic pressure outside the tube effective in overcoming frictional resistance between the entering sediment and the inner wall of the tube.

The apparatus is raised by the wire rope attached to the piston which is retained by shoulders near the top of the tube. The bottom of the tube is armed with a removable steel cutting edge. Just above this is the core catcher, a circular set of leaves of spring bronze which close as the tube is withdrawn from the sediment.

The original piston corer of B. Kullenberg (Sweden) is provided with a metal liner in which the core is removed. A simplified corer designed by M. Ewing is used without a liner, and the core must therefore be pushed out of the tube with a plunger and rod (Fig. 17). A pilot core about 30 cm long is usually taken with each long core by means of a small tube with plastic liner attached to the trigger weight.

Bottom contact. In bottom-sampling operations it is necessary to know when bottom contact is made because any excessive wire that is payed out usually kinks. Close attention must be given to the sonic depth, the wire angle, and the length of wire payed out.

In shallow waters contact with the bottom may be detected by a slack in the wire and a jerk in the meter wheel. In deeper waters a spring scale or dynamometer may be used to register variations in

Fig. 17. Lowering the Ewing simplified piston corer. (*U.S. Navy Hydrographic Office*)

wire tension. A bottom-signaling device, or ball-breaker, is sometimes used for this purpose. Upon contact with the bottom a small glass ball is crushed (Fig. 18). The resulting implosion signal may be seen on the pen trace of the echo sounder

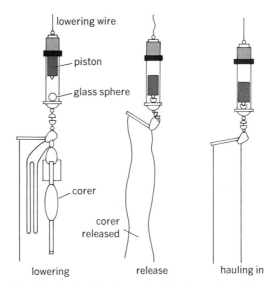

Fig. 18. The ball-breaker principle of operation. (*From U.S. Navy H.O. Publ. no. 607, 2d ed., 1955*)

or heard as an audible signal on suitable monitoring equipment.

[DAVID B. ERICSON]

Bibliography: H. Barnes, *Oceanography and Marine Biology*, vol. 14, 1976; J. B. Hersey, Acoustically monitored bottom coring, *Deep-Sea Res.*, 6:170–172, February, 1960; *Instruction Manual for Oceanographic Observations*, U.S. Navy H.O. Publ. no. 607, 2d ed., 1955; P. D. Komar, *Beach Processes and Sedimentation*, 1976; B. Kullenberg, The piston core sampler, *Svenska Hydrograf. Biol. Komm. Skrifter*, [3]1(2):1–46, 1945; D. A. McManus, A large-diameter coring device, *Deep-Sea Res.*, 12:227–232, 1965; D. G. Moore, The freecorer: Sediment sampling without wire and winch, *J. Sediment. Petrol.*, 31:627–630, 1961; A. F. Richards and H. W. Parker, Surface coring for sheer strength measurements, in *Proceedings of the Conference on Civil Engineering in the Oceans*, 1968; T. F. Yen, *Chemistry of Marine Sediments*, 1977.

Mediterranean Sea

The Mediterranean Sea lies between Europe, Asia Minor, and Africa. It is completely landlocked except for the Strait of Gibraltar, the Bosporus, and the Suez Canal. The Strait of Gibraltar is 13 mi wide at the boundary between the Atlantic and the Mediterranean (Great Europa Point, Gibraltar, to Punta Almina, Ceuta), and the Bosporus is as narrow as 1/2 mi, widening to 2 mi at the Black Sea boundary (Rumeli Burnu, European Turkey, to Anadolu Burnu, Asiatic Turkey). The Suez Canal, a man-made channel connecting the Mediterranean with the Red Sea, is 87½ mi long, about 300 ft wide, and 37 ft deep. In latitude the Mediterranean extends from 30°15′N in the Gulf of Sidra to 45°47′N in the Gulf of Trieste; in longitude, from 5°21′W at Great Europa Point to 36°12′E in the Gulf of Alexandretta.

The Mediterranean is conveniently divided into an eastern basin and a western basin, which are joined by the Strait of Sicily and the Strait of Messina. The Strait of Sicily, between Cap Bon, Tunisia, and Sicily, is 77 mi wide, but the Strait of Messina, between Punta Sottile, Sicily, and the Italian mainland, is only 1.7 mi across.

The major subdivisions of the western basin are the Alboran Sea, which is the portion west of Cabo de Gata, Spain, and is named for the Isla de Alboran; the Balearic or Iberian Sea, between Spain and the Islas Baleares; the Ligurian Sea, the gulf north of Corsica; and the Tyrrhenian Sea, lying between Corsica, Sardinia, Sicily, and the Italian mainland. The eastern basin includes the Adriatic Sea, between Italy down to the heel of the boot (Capo Santa Maria de Leuca) and the Balkan Peninsula north of Corfu; the Ionian Sea, which lies south of the Adriatic as far as the southern tips of Sicily and the Greek mainland; the Aegean Sea, the waters north of Kithera, Andikithera, Crete, Karpathos, and Rhodes, as far as the Dardanelles; and the Sea of Marmara, which extends from the Dardanelles to the Bosporus.

The Mediterranean Sea has numerous islands, particularly in the Ionian Sea (which has the alternate designation of The Archipelago) and the eastern Tyrrhenian and Adriatic. The largest islands are Sicily, 25,710 km²; Sardinia, 24,040 km²; Cyprus, 9250 km²; Corsica, 8720 km²; and Crete, 8380 km².

The total water area of the Mediterranean is 2,501,000 km², and its average depth is 1536 m. The greatest depth in the western basin is 3719 m, in the Tyrrhenian Sea. The eastern basin is deeper, with a greatest depth of 5530 m in the Ionian Sea about 35 mi off the Greek mainland. The Adriatic Sea mainly overlies the Continental Shelf, but there is a basin in the southern part with a greatest depth of 1315 m. The sill depth in both the Strait of Gibraltar and the Strait of Sicily is about 300 m.

The Atlantic tide disappears in the Strait of Gibraltar. The tides of the Mediterranean are predominantly semidiurnal. The eastern and western basins have standing wave systems, with nodal lines extending roughly from Barcelona to Bougie, and from Kerme Korfezi to Kithera to Tobruch, respectively. The tide range elsewhere is about 1 ft. At the ends of the straits of Sicily and Messina, high water occurs in exactly opposite phase, giving rise to strong hydraulic currents, particularly in the narrow Strait of Messina, the legendary seat of Scylla and Charybdis. The Adriatic has a progressive tide, with range as great as 3 ft in the northern end, radiating about an amphidromic point near 43°N, 15°E.

Since evaporation over the whole Mediterranean greatly exceeds the supply of water from precipitation and river runoff, the surface salinity is higher than that of the Atlantic, increasing from a value of less than 36.50‰ in the Strait of Gibraltar to over 39.00‰ in the extreme east. Surface temperatures reach 27 or 28°C in the southern and eastern gulfs in the summer, elsewhere ranging from 21 to 25°C. In winter the northern Adriatic may be as cold as 8°C, but off Egypt and Syria the temperature may not drop below 16°C.

Its subsurface waters are all formed within the Mediterranean, owing to the shallow sill separating it from the Atlantic. To a depth of about 600 m, these waters are characterized by a salinity maximum of 38.60–39.90‰ and temperature of 14°C or greater; they are formed at the surface, where evaporation is most significant. The deeper waters are formed in winter along the northern shores of the sea, where winter cooling has its greatest effect; they are about 1°C colder than the inter-

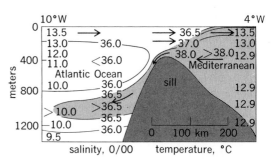

Vertical distribution of salinity and temperature in west-east section through Strait of Gibraltar along 36°N, according to G. Schott, 1944. Arrows show principal spreading of the waters. (*From G. Dietrich and K. Kalle, Allgemeine Meereskunde, eine Einführung in die Ozeanographie, Gebrüder Borntraeger, Berlin, 1957*)

mediate waters and 0.20°/₀₀ or 0.30°/₀₀ lower in
salinity. In turn, the deeper water of the eastern basin has a temperature about 0.5°C higher
and salinity 0.25°/₀₀ higher than that of the western
basin.

The excess of evaporation within the Mediterranean requires that a surface current of Atlantic
water enter the Straits of Gibraltar. The layer of
intermediate water, since it has a greater density
than Atlantic water at the same depth, gives rise to
an outgoing countercurrent which pours down over
the Continental Slope and whose waters can be
detected for thousands of miles in the Atlantic.
The volumes of water in the currents are about 15
times as great as the actual evaporation; thus,
some 1.75×10^6 m³/sec enter the Strait of Gibraltar, or enough to renew the entire 3.842×10^6 km³
in about 70 years (see illustration).

The International Commission for the Scientific
Exploration of the Mediterranean Sea, with mem-
bership including most of the countries engaged in
Mediterranean oceanographic studies, has its
headquarters at the Musée Océanographique,
Monaco. *See* ATLANTIC OCEAN; RED SEA.

[JOHN LYMAN]

Mesosphere

An atmospheric layer stretching from the stratopause, at an altitude of about 50 km, to the mesopause, at about 85 km. Like the troposphere, from
which it is separated by the stratosphere, it is a
layer in which the temperature generally decreases with height. Although much remains to be
learned about the mesosphere, it is clear that many
of the atmospheric variations encountered therein
are linked to complicated processes in the underlying layers. In contrast, the region above the
mesosphere (the thermosphere) undergoes large
variations of temperature and density which are
directly related to variations in solar activity.

Special features of the mesosphere include a region of extremely strong winds centered near 65
km, tidal oscillations and gravity waves of appreciable amplitude, meteors, and noctilucent clouds.
The clouds are observed in middle and high latitudes near an altitude of 80 km, in summer, when
the Sun is at a small angle below the horizon. From
satellite photometer measurements in 1969, it was
deduced that a light-scattering cloud prevails near
the mesopause, over the summer pole. It has been
suggested that noctilucent clouds represent extensions of this phenomenon to lower latitudes. Meteors, especially numerous near 95 km, commonly
burn up before reaching the stratopause. Also in
the domain of the mesosphere is the greater part of
the D region, the lowest ionized layer of the atmosphere. *See* ATMOSPHERE; IONOSPHERE; METEO-
ROLOGY; NOCTILUCENT CLOUDS.

Observations. Both direct and indirect methods
of measurement are used to determine the properties of the mesosphere. Rocket techniques are
used semiroutinely to obtain height profiles of
temperature, pressure, density, and wind; and infrequently the concentration of minor constituents
such as ozone has been measured. Ground-based
methods of determining wind, temperature, and
density include radar tracking of meteor trails,
partial reflection of medium-frequency radio

waves, and lidar. The capability for remote soundings of temperature and minor constituents with
the aid of satellite-borne infrared radiometers was
extended in 1975 to mesospheric altitudes, and it
was also demonstrated that such soundings could
be made from microwave measurements. *See*
METEOROLOGICAL ROCKETS; METEOROLOGICAL
SATELLITES.

Temperature structure. The absorption of solar
ultraviolet radiation by ozone accounts for a temperature maximum at the stratopause, averaging
about 270 K. In high latitudes during winter, the
stratopause temperature may be much lower, and
indeed the distinction between stratosphere and
mesosphere nearly disappears. The ozone concentration decreases rapidly with height, from
a maximum in the stratosphere, and the lack of
absorbing gases in the mesosphere should largely
account for the general altitudinal decrease of
temperature to an average annual value of about
190 K at the mesopause. Many photochemical
processes affecting the upper mesosphere and
lower thermosphere are under study. Just above
the mesopause, the dissociation of molecular
oxygen by solar ultraviolet rays at wavelengths
shorter than those reaching the stratopause accounts for a rapid temperature increase with
height.

In the middle to high latitudes, the mesospheric
thermal structures in summer and winter are remarkably different. In summer, the temperature
usually falls off smoothly with height at the rate of
3–5°C per kilometer (mesopause temperatures as
low as 130 K have been measured), whereas in
winter alternating layers of decreasing and increasing temperature are commonly encountered,
with a net decrease to temperatures usually in the
range 180–250 K near 85 km. Moreover, there is
evidence that on a coarse vertical scale the wintertime thermal structure changes coherently
throughout the stratosphere and mesosphere:
when there is strong warming at some level, warming also occurs at a level some 40–50 km higher
(or lower), with cooling at a level in between. Such
temperature perturbations are thought to be linked
to the upward propagation, from the lower atmosphere, of horizontal atmospheric waves of quasi-
planetary scale. (Large-scale waves may be generated *within* the upper atmosphere, however, under
certain conditions.) Intense wintertime perturbations, with temperature changes of about 30–
80°C within 1 week, are termed sudden warmings
and have been best studied in the stratosphere.

In the mean, the horizontal temperature gradient is directed poleward in winter below 65 km
(temperature lower in high latitudes) and equatorward above 65 km. The opposite pattern prevails
in summer. Various theories have been suggested
to explain the relatively high temperature of the
upper mesosphere in winter, which exceeds the
value based on radiative heating and cooling alone.
The final explanation must doubtless require consideration of "dynamically" induced temperature
changes in horizontal- and vertical-flow patterns
associated with the perturbations mentioned.
See ATMOSPHERIC OZONE; STRATOSPHERE.

Wind systems. From about 20 km through the
stratosphere and into the mesosphere, there is in

winter a vast system of cyclonic winds from the west (westerlies), roughly concentric about the pole, often referred to as the polar night vortex. In summer, this system is replaced by an anticyclonic system of winds from the east (easterlies). These systems arise from an approximate balance between the Coriolis force and the prevailing horizontal pressure gradient, which itself is largely determined in the mesosphere by the horizontal temperature gradients below 65 km. Moreover, the reversal of the temperature gradient near 65 km accounts for a strong wind maximum in the lower mesosphere, in accordance with the thermal wind equation. The mean speed at 65 km is greatest in middle latitudes, about 80 m/s in winter (westerly) and 60 m/s in summer (easterly); extreme winter velocities may reach 100–200 m/s.

This simplified picture of the mesospheric circulation omits such features as the distortion of the winter polar vortex by subpolar, large-scale anticyclonic waves; and the lowering of the altitude of maximum wind in sudden warming events. At some level above this altitude, the wind may even become easterly. Indeed, if sufficiently intense polar warming occurs at stratospheric altitudes, the polar vortex may undergo a general circulation reversal from westerly to easterly. The temporary establishment of midwinter easterlies in the mesosphere has been simulated in theoretical studies and observed at heights up to about 100 km. *See* ATMOSPHERIC GENERAL CIRCULATION.

Special variations. Superimposed on the gross variations described above may be changes with various time scales, from minutes to hours in the case of gravity waves, to 12 and 24 hr in tidal variations, to semiannual changes. The last, which are not well understood, are pronounced near 50 km in the tropics but can also be discerned at higher latitudes and altitudes, with a phase reversal in wind above 80 km. Observational evidence and theoretical predictions indicate a generally increasing amplitude with height in the case of the shorter-term variations, so that near the mesopause they constitute a large part of the overall atmospheric variability.   [RODERICK S. QUIROZ]

Bibliography: R. A. Craig, *The Edge of Space,* 1968; R. A. Craig, *The Upper Atmosphere Meteorology and Physics,* 1965; R. J. Murgatroyd, The physics and dynamics of the stratosphere and mesosphere, *Rep. Prog. Physics,* 33:817–880, 1970; R. S. Quiroz (ed.), Meteorological investigations of the upper atmosphere, *Meteorol. Monogr.,* 9(1), 1968; R. S. Quiroz, A. J. Miller, and R. M. Nagatani, A comparison of observed and simulated properties of sudden stratospheric warmings, *J. Atmos. Sci.,* 32:1723–1736, 1975; W. L. Webb, *Structure of the Stratosphere and Mesosphere,* 1966.

Meteorological instrumentation

Apparatus and equipment used to obtain quantitative information about the weather. This information includes the state of the atmosphere in such aspects as motion, energy exchange, gaseous composition, solid- and liquid-particle content, electrical activity, and others, as well as their composite effects in various combinations. Air temperature and pressure, which describe the state of the atmosphere, are obtained with simple instruments such as thermometers and barometers. However, visibility influenced by the amount and size distribution of liquids or solids in the atmosphere and the amount and direction of illumination is an example of a meteorological parameter which is not so easy to measure directly, but is expressed as the distance to which objects can be recognized. Any instrument able to measure useful properties of the atmosphere is potentially a meteorological instrument. Ordinarily, however, the expression is used to indicate instruments in use for routine weather observations.

Classes of observations. Four types of observation are made: climatological, synoptic, operational, and research.

Climatological observations. These measure the broad features of the weather at a given location over a long period of time.

Synoptic observations. These are taken over an area, ideally at a number of altitudes over the entire globe, and at a number of times during the day. The resulting data are the basis of weather forecasting, whether from maps or from high-speed computers. *See* WEATHER FORECASTING AND PREDICTION.

Operational observations. Special observations are used to control some activity; for example, measurement of wind and temperature and visibility and cloud ceiling over an airport are necessary for efficient and safe operation of aircraft.

Research observations. Observations used in research may be any of the above or special ones taken with newly developed instruments.

Problems of instrument exposure. In addition to those generally desirable characteristics of instruments, such as accuracy, reliability, and ease of maintenance, another requirement common to all meteorological instruments is care in exposure. A rain gage whose catchment area is 10 in. in diameter often represents the single rainfall measurement for an area more than 10 mi in diameter. Atmospheric properties are greatly affected by such factors as surface cover, proximity to trees or buildings, and elevation, to mention only a few. Because errors due to exposure cannot be completely eliminated, uniformity of exposure from station to station is essential. Often it is more difficult to obtain a representative sample than to secure high accuracy in the instrument itself.

The requirements for proper exposure are particularly stringent for climatological stations. The records of a climate station might make it appear that the climate has slowly warmed, when actually the environment of the station has changed because of the growth of a city around it. Likewise, it might appear that wind velocities are gradually becoming less—all caused by growth of trees in the vicinity. So important is the question of proper instrument exposure that the U.S. Weather Bureau has established a number of "bench-mark" stations carefully selected for likely constant conditions. Changes in a station record can then confidently be taken to be changes in the general climate and not changes in the environment. *See* MICROMETEOROLOGY.

Accessibility and conversion of data. Still another feature necessary in meteorological instruments is rapid access to the pertinent data, particularly for synoptic observations. If many hours are required to convert a particular set of measurements into meaningful meteorological data, much of its value is lost. It arrives too late to be much use in the next forecast. This has been a major problem in meteorological use of satellites. The sheer volume of worldwide detailed observation is formidable. *See* METEOROLOGICAL SATELLITES.

Surface observational practices. Considerable information on the common use of instruments for surface observations of various elements of weather can be found in separate articles. For examples on temperature *see* AIR TEMPERATURE; on atmospheric pressure *see* AIR PRESSURE; BAROMETER; on humidity *see* HUMIDITY; HYGROMETER; PSYCHROMETER; on wind *see* WIND; WIND MEASUREMENT; on rain and other forms of precipitation *see* PRECIPITATION; PRECIPITATION GAGES; SNOW; SNOW GAGE.

Upper-air instruments. Modern meteorology began in the late 1930s with the advent of regular soundings of the atmosphere from many stations on a broad geographical scale. Isolated explorations of the atmosphere by using kites were made as early as 1749, and with manned and free balloons during the 19th century. In the 1920s routine ascents with airplanes equipped with aerometeorographs, which recorded air temperature, pressure, and humidity, gave a valuable first look at the three-dimensional picture of the weather.

Air soundings and radiosonde. Only with the development in the mid-1930s of the radiosonde and the subsequent improvements into the radiosonde-radiowind system was much progress possible in gaining a three-dimensional understanding of the atmosphere. The radiosonde is a light-weight miniature weather station carried aloft by balloons with a radio link to the ground. A radiosonde-radiowind system consists of a balloon, the radiosonde, ground equipment to receive and record the signals from the radiosonde, and a radio theodolite, which measures the elevation and azimuth angles of the balloon, used to obtain measurements of the upper winds.

Meteorological balloons. Used for lifting radiosondes to high altitudes, these balloons are made of neoprene rubber of high quality. To reach altitudes of 100,000 ft, the rubber must stretch enough to enclose a volume 100 times greater than it had at the Earth's surface. To reach even higher altitudes, the volume must double for roughly every 15,000 ft of additional altitude. This can be done by using large 10-kg balloons only partly inflated at the ground. They become fully extended at 30,000 ft, and then stretch to their maximum volume as they ascend to bursting altitudes. These larger balloons weigh much more than the payload they carry. One reaches a point of very rapidly diminishing returns at altitudes of 140,000–150,000 ft. Indeed, no further increase in altitude is possible even with zero payload because of the weight of the balloon. Routine radiosonde ascents are made with smaller 1-kg balloons, which reach to about 90,000 ft about 85% of the time. Very large balloons constructed from thin plastic films, such as polyethylene or Mylar, make it possible to carry loads of tons to 100,000 ft or higher. These balloons have a fixed volume, in contrast to the variable volume of the rubber balloons. To prevent bursting, they are either vented or are built strong enough to resist the slight superpressure inside. They can float at a fixed altitude for days. The vented balloon must be automatically ballasted to compensate for heating and cooling during the day and night. Balloons made of Mylar are strong enough to withstand the increased internal pressure caused by solar heating, and therefore require no complex ballast system. These are ideal platforms for fixed-altitude radiosondes, called transosondes, because they make it possible to obtain upper-air observations over vast ocean areas. The maximum altitude of plastic balloons is about the same as that for large rubber balloons. To carry instruments to higher altitudes, rockets must be used. For additional information on this technique *see* METEOROLOGICAL ROCKETS.

Data conversion and transmission. Radiosonde flight equipment, in general use by meteorological services of different countries, can be divided into two basic systems: (1) those in which temperature and humidity sensors produce dimensional changes which are mechanically coded for transmission to the ground, and (2) those in which sensors produce electrical changes directly.

1. Mechanical-change type. The thermometer in the first group is usually a bimetal element or a stretched wire. The humidity element is a hair or goldbeater's-skin hygrometer. The pressure element is an aneroid capsule. Several schemes for coding these mechanical motions into meaningful radio signals are used. In the Väisälä radiosonde, the sensors are coupled to variable capacitors which shift the carrier frequency of the radio transmitter. A wind-driven switch connects to the transmitter each sensor-variable capacitor and two fixed capacitors in sequence. The two fixed capacitors, one large and the other small, are needed for in-flight calibrations. Another system in common use converts the mechanical motions into variable time intervals. A clockwork or constant-speed electric motor turns a helix or two-turn flat spiral, which forms one contact. Contact levers, mechanically coupled to the meteorological sensors, make contact with the helix or spiral each revolution. The time a contact is made during the revolution is a function of the position of the sensor contact. The beginning of each revolution is given by still another fixed-reference contact. The relative position in time of each meteorological contact compared with the reference contact is a measure of the position of each meteorological element.

Photoengraving and photoetching techniques make possible substitution of a disk or drum containing many groups of Morse code signals for the spiral or helix. The radio signals from the radiosonde are then readable as regular Morse code signals. In still another type of radiosonde, the mechanical motion is used to move μ-metal slugs to vary the inductance of small coils. These are made part of an audio-frequency oscillator. Changes in the sensors cause changes in the audio tone transmitted to the ground. A precision oscillator or audio-frequency meter at the ground station is necessary to interpret the signals.

2. Direct electrical-change type. The radiosonde system in use by the U.S. Weather Bureau

and military service employs a mechanical bellows-operated commutating switch and variable resistors for the temperature and humidity elements. The thermistor temperature element is a rod of ceramic material which possesses a large negative temperature coefficient of resistance, roughly 4%/°C. The resistance of the element varies from 20,000 ohms at 60°C to 2,000,000 ohms at −90°C. The diameter of the rods is made as small as practicable so that the elements will respond rapidly to changes in temperature. They are coated with a reflective paint or are shaded in a duct to reduce heating from solar radiation.

The humidity element consists of a polystyrene strip provided with two electrodes on the edges. The space between the electrodes is coated with a weak solution of lithium chloride held in a binder of polyvinyl acetate and alcohol. The resistance of the strip varies with changes in vapor pressure and temperature. During flight the pressure-actuated commutating switch operates a relay. In the normally closed position, the relay connects the temperature element into an electronic blocking-oscillator measuring circuit. In the alternate position, the relay connects the humidity element into the circuit. The value of pressure is determined by the number of times the humidity, temperature, and a fixed-value resistor have been connected into the circuit, whereas humidity and temperature are determined from the pulse repetition frequency of the blocking oscillator.

A notable feature of the radiosonde in use in the United States is the simple linear frequency versus electrical conductance characteristic of the blocking-oscillator circuit. The frequencies of the meteorological sensor signals can be compared with the frequency generated by a known calibration resistance included in the sensor switching sequence. Because of this feature, the manufacture and calibration are greatly simplified. Careful matching of temperature-thermistor elements and humidity elements to a specific blocking oscillator is unnecessary. Within reasonable limits any temperature element will work with any blocking oscillator, and a simple preflight calibration establishes the ratio for a particular combination. The pulses from the blocking oscillator modulate a 1680-MHz cavity oscillator transmitter to provide the radio link to the ground station.

The ground equipment includes not only a radio receiver and recording audio-frequency meter for reception of temperature, pressure, and humidity signals, but also an automatic tracking antenna. The latter, a parabolic dish about 10 ft in diameter, is capable of measuring the elevation and azimuth angle to the airborne radiosonde to a few hundredths of 1°. This radiotheodolite angle data and the height of the balloon available from the radiosonde data allow the computation of upper winds to maximum altitude, even though the balloon is hidden by cloud. To improve the accuracy, particularly at high altitude and long range, a ranging unit is sometimes used.

For the ranging unit a tiny radio receiver is added to the balloon-borne equipment to make a transponder. This permits measurement of the round-trip time for a radio signal, from the ground station to the balloon and back again, to give the slant range. The slant range multiplied by the cosine of the elevation angle gives the horizontal distance to the point under the balloon, whereas when the height of the balloon is used to obtain the distance, the product with the cotangent of the elevation angle is required. Because at low angles of elevation for a given error in angle measurement the cosine varies much less than the cotangent of the angle, the accuracy of the wind measurement at high altitude and long range is considerably enhanced by use of the ranging device.

Ground evaluation and dissemination. The data from a typical radiosonde receiver flight is evaluated in about 1 hr, coded for rapid transmission by teletype, and transmitted to weather stations throughout the world.

Radar wind observation. Radar is also used to obtain data on upper winds. The radar beam is reflected by a passive target which is part of, or attached to, the balloon. The balloon itself may be coated with a thin metallic film to make it a conductor. To obtain a strong return signal, a corner reflector constructed of aluminum foil with a light framework is used. A corner reflector tends to reflect the incident electromagnetic energy back toward the radar antenna from whence it came, much as a billiard ball moving from one corner of a billiard table toward the opposite corner tends to return to the starting corner. Because the radar has excellent angular and range capability, the wind determination by radar is usually quite good. When an active transponder is used instead of a passive reflector, wind measurements at great range and altitude are possible.

Radar storm tracking. Radar is of even greater utility for storm detection. Strictly speaking, radar detects the precipitation associated with the storms. It presents a graphic picture of the precipitation area in the three dimensions together with all the changes that occur. The radar echo information is usually presented on an oscilloscope as a horizontal map with the location of the radar set in the center. These data appear on the plan-position indicator (PPI scope, or PPI). A vertical picture of the precipitation area at a fixed azimuth angle is presented on the range-height indicator (RHI scope). The antenna on a tracking radar is automatically locked onto the balloon target, whereas on a search radar used for storm detection the antenna rotates continuously to illuminate a horizontal sheet for the PPI presentation and moves up and down in angle of elevation for the RHI picture.

In addition to providing the location of a precipitation area, the strength of the echo signal is related to intensity of the precipitation. In thunderstorms the turbulent cells are usually associated with the heaviest precipitation. Radar-equipped aircraft are able to detect and avoid these dangerous parts of a storm, hence the name weather-avoidance radar. There is no simple relationship between the strength of the radar echo and the amount of precipitation in the cloud (rainfall if it all falls to the ground). The strength of the echo signal is related, among other parameters, to the number of drops in a given volume and the sixth power of the drop diameters. The amount of liquid water is, however, proportional to the number of drops and the third power of the drop diameters. Hence, to obtain the amount of rainfall, the drop-size distribution must be known independently. Attenuation of the radar signal by other rain between the target in question and the radar set complicates the determination further. The power scattered by a given drop size is greater for short wavelengths, typically 3-cm wavelength, than for larger wavelengths, typically 10 cm. Thus it is possible to detect light rainfall with short-wavelength radar, but further penetration into a storm requires 10-cm radar. The optimum wavelength for weather radar is near 5–7 cm. Despite these seeming shortcomings, weather radar has been of enormous benefit, especially through forecasting. Violent thunderstorms, tornadoes, and hurricanes have been located and tracked on a routine basis by the weather services of many countries. *See* STORM DETECTION. [VERNER E. SUOMI]

Meteorological optics

A branch of atmospheric physics or physical meteorology in which optical phenomena occurring in the atmosphere are described and explained. Meteorological optics can be divided into two main parts. In the first part, the atmosphere is considered as a continuous medium in which the speed of propagation of electromagnetic waves (light) varies with the density. In the second part, attention is given to the presence of particulate matter in the atmosphere, such as air molecules, dust and haze particles, cloud and raindrops, and ice crystals, and to the phenomena resulting from the interaction of light with these particles. In a third, minor part, applications to measurement of atmospheric properties are considered.

Variations of refractive index. The first part of meteorological optics discusses optical phenomena in the atmosphere caused by the variation of the speed of light propagation, or of the refractive index, as a result of changes in the air density, such as (1) astronomical and terrestrial refraction, for instance the bending of light rays as they penetrate the denser layers of the lower atmosphere, causing the apparent increase of the angular elevation of celestial or terrestrial objects above the horizon; (2) mirages, in which the bending of the light rays is due to abnormal vertical distribution of air density; (3) scintillation, a condition of rapid changes of images of stars or other distant bright objects caused by rapid variations in the air density; (4) green flash, or the green color of the last visible segment of the setting Sun, which results from the dispersion and attenuation of the horizontal rays. *See* MIRAGE.

Interaction with particulate matter. The interaction of light with particulate matter can be twofold. The radiant energy, received by a molecule of air or by a particle floating in the air, can be changed totally or in part into another type of energy such as heat or photochemical energy. This process of energy change is called absorption, and the amount of energy transformed represents a loss of radiant energy. In the visible range of the spectrum of electromagnetic waves, however, the atmospheric gases, except ozone, do not absorb light. Much more important is the second type of interaction of light with the particulate matter. In this process, called scattering, the energy received is completely reradiated in all directions around the scattering center, so that no part of the radiant energy is transformed into another type of energy and thus lost. The different types of scattering as well as the important consequences of light scattering are discussed below.

Light scattering by air molecules. This type of scattering is governed by a law, first formulated by J. W. Strutt, Lord Rayleigh (1871). It applies to particles of any shape, the dimensions of which are negligible with respect to the wavelength of the incident light. The intensity of light scattered by such a particle is inversely proportional to the fourth power of the wavelength, being more intense in the violet than in the red part of the spectrum. The scattered light is polarized; that is, when observed through an analyzer such as a nicol prism or Polaroid plate, it shows variations in the intensity while the analyzer is rotated around its axis (two maxima and two minima when rotated by 360°). The blue color of the sky and the polarization of skylight in a pure atmosphere containing only air molecules can be explained as consequences of light scattering according to the Rayleigh law. Another important consequence of light scattering is the attenuation or extinction of light due to scattering. The radiant energy which is scattered in all directions by the scattering center can be noticed as a loss of the radiation in the direction of the incident light. The amount of attenuated radiant energy is also inversely proportional to the fourth power of the wavelength. Therefore the light rays of a long path length will contain less violet light than red. The loss of the violet and blue light is particularly noticeable in the change of color of the setting Sun, from white to red.

Mie scattering. The scattering of light by particles of size comparable to or larger than the wavelength of the incident light is characterized by greater intensity of the scattered light in the forward direction than in the backward direction. For a dielectric sphere the law of scattering was developed by G. Mie and is sometimes called Mie scattering. The light scattered by dust and haze particles in the air varies with the wavelength much less than in the case of molecular scattering, and may even be independent of the wavelength. Thus the intensity of the scattered light is much larger in the long wavelength compared with that for molecular scattering. The presence of dust and haze particles in the atmosphere can be noted by the change of the deep blue color of the sky in a pure atmosphere into a whitish blue color of the sky in a turbid atmosphere, with a bright white region around the Sun, called the aureole. The attenuation of light due to the scattering by dust and haze particles is very nearly independent of the wavelength, so that the setting Sun, seen through a dense layer of haze or mist, does not change its white color.

Large-particle optics. The interaction of light with large particles, such as water droplets in fog, cloud, or rain, is governed by the laws of diffraction and geometrical optics. Interference of rays passing very close to the edge of the drops gives rise to a series of rings of the maximum and minimum intensity called corona and visible as a system of rings of different colors around the illuminating bodies, Sun or Moon. A similar interference effect occurs in the backward direction, around the antisolar point, and becomes noticeable as a system of colored rings around the shadow cast by the Sun on a layer of cloud or fog, the anticorona or glory. The rays reflected once or twice inside the sphere cause the colored phenomena of the primary and secondary rainbow. *See* RAINBOW.

Ice crystals appear in the atmosphere in the form of hexagonal prisms. Part of the light incident on the ice crystal is refracted as by a prism with the sides forming an angle of 60 or 90°. As a consequence of such refraction, two luminous rings appear around the Sun or Moon, as part of a complex phenomenon called the halo. On both sides of the Sun two bright spots, called parhelia, occasionally appear, provided that the hexagonal prisms have a prevailing vertical orientation of their axes. *See* HALO; SUNDOG.

Visibility and related features. Closely connected with the scattering of light in the atmosphere is the problem of visibility. The light scattered into the observer's eye by particles situated along the line of sight to a distant object diminishes the contrast of the object and its background, thus decreasing its visibility.

By a combination of the effects of terrestrial refraction and of light scattering the twilight phenomena can be explained, particularly the different colors of the sky, the twilight arcs, seen above the Sun horizon as well as in the opposite direction, above the rising Earth shadow.

Applications to related problems. Several phenomena, some of which are enumerated above, can be used for indirect measurement of certain physical quantities. For example, air density may be judged by measurement of the amount of lifting of distant objects due to the anomalous refraction of horizontal rays. Similarly, the measurements of the skylight intensity and its polarization, as well as of the attenuation of the Sun radiation, can be used for the determination of atmospheric turbidity. The measurement of the angular distances from the Sun of different rings in the corona can be used for estimating drop sizes in a cloud. For the same purpose similar measurements can be applied to the anticorona or to the rainbow. The measurements of the attenuation of Sun radiation, combined with the measurements of the intensity of skylight in the ultraviolet, were successfully used for the determination of the total amount and vertical distribution of ozone in the upper atmosphere.

Several technical problems can be solved by direct application of the results of meteorological optics, such as the amount of illumination of objects on the ground or high in the atmosphere and the conditions of their visibility under different atmospheric conditions.   [ZDENEK SEKERA]

Bibliography: D. Deirmendjian, *Electromagnetic Scattering on Spherical Polydispersions*, 1969; S. Fluegge (ed.), *Handbuch der Physik*, vol. 48, 1957; J. M. Greenberg and H. C. Van de Hulst (eds.), *Interstellar Dust and Related Topics: Proceedings of the Symposium Organized by the International Astronomical Union, SUNY at Albany, May–June 1972*, 1973; E. J. McCartney, *Optics of the Atmosphere: Scattering by Molecules and Particles*, 1976; R. A. Tricker, *Introduction to Meteorological Optics*, 1971.

Meteorological rockets

Synoptic exploration of the stratospheric circulation (25–80 km altitude) through use of small rocket systems has matured since 1960 into a highly productive source of information on atmospheric structure and dynamics. More than 25,000 small meteorological rockets have been fired in a coordinated investigation of the wind field and the temperature and ozone structures in the 25–55 km altitude region of the middle atmosphere.

These new data have produced dramatic changes in the scientific view of this frontier region of the atmosphere, with a resulting alteration of the structural concepts—which had previously developed without adequate measurements—into a new space-age atmospheric model which is primarily characterized by intense dynamics.

The Meteorological Rocket Network (MRN) has roughly doubled the observed atmospheric volume, and its conception and operation has epitomized the international cooperation which has characterized most meteorological efforts. The MRN began at North American missile ranges in 1959 with equipment developed by the Atmospheric Sciences Laboratory at White Sands Missile Range. Under sponsorship of the United States National Academy of Sciences through the Committee on Space Research of the International Council of Scientific Unions, the MRN was expanded into a global network in 1968.

Meteorological Rocket Network. Meteorological rocket launch sites from which data have been contributed are illustrated on the world map in Fig. 1, with the numbers near each station identifying the name and coordinate data presented in the table. The productivity of these 56 sites and six ships has been highly variable, with some stations conducting only a few firings for studies of special phenomena and others firing on almost a daily basis over much of the period. Nine of the stations have fired more than 1000 rocketsondes each, with the maximum number fired at one station being above 5000 for White Sands Missile Range. The current MRN rate of accumulation of soundings is approximately 2000 rocketsondes per year.

Rocketsonde development has followed a complex path. The most used system as of 1975 is illustrated in Fig. 2. This PWN-8B rocketsonde system uses a motor of 1.8 sec burning time to propel its dart to roughly 70 km altitude, where a parachute and radiosonde payload are expelled. The meteorological measurements are made during the sensor system descent.

The Starute parachute sensor system is illustrated in Fig. 3. The 30-cm-long transmitter-sensor payload is enlarged in Fig. 4. The overall weight of these rocketsondes is 15.4 kg, and the length is 2.87 m. The rocketsondes have proved to be highly efficient in the MRN application and, more importantly, have produced the most accurate and sensitive information on the wind and temperature structure of the stratospheric region thus far ob-

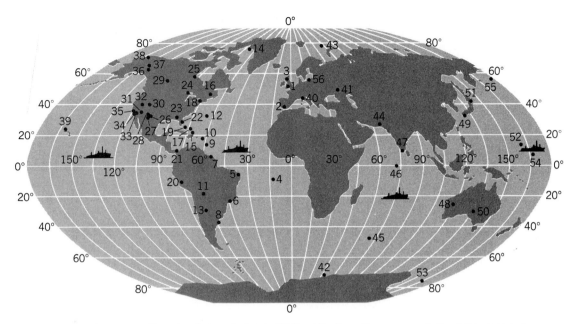

Fig. 1. Sites which have contributed rocketsonde data to MRN publications. Four representative ships are shown.

Meteorological Rocket Network launch sites shown in Fig. 1

Name	Coordinates		Name	Coordinates	
1. Aberporth	52°N	4°6'W	29. Primrose Lake	54°45'N	110°3'W
2. El Arenosillo	37°6'N	6°44'W	30. Green River	38°56'N	110°4'W
3. West Geirinish	57°21'N	7°22'W	31. Yuma	32°52'N	114°19'W
4. Ascension Island	7°59'S	14°25'W	32. Tonopah	38°0'N	116°30'W
5. Natal	5°55'S	35°10'W	33. Pt. Mugu	34°7'N	119°7'W
6. Marambaia	23°S	44°W	34. San Nicolas Island	33°14'N	119°25'W
7. Kourou	5°12'N	52°43'W	35. Vandenberg	34°40'N	120°36'W
8. Mar Chiquita	37°45'S	57°25'W	36. Fort Greely	64°0'N	145°44'W
9. Harp Seawell	13°6'N	59°37'W	37. Poker Flat	65°8'N	147°29'W
10. Antigua	17°9'N	61°47'W	38. Point Barrow	71°21'N	156°59'W
11. Tartagal	22°46'S	63°49'W	39. Barking Sands	22°2'N	159°47'W
12. Bermuda	32°21'N	64°39'W	40. Sardinia	40°N	10°E
13. Chamical	30°22'N	66°17'W	41. Volgograd	48°41'N	44°21'E
14. Thule	76°33'N	68°49'W	42. Moledezhnaja	67°40'S	45°51'E
15. Grand Turk	21°26'N	71°9'W	43. Heiss Island	80°37'N	58°3'E
16. Highwater	45°1'N	72°27'W	44. Sonmiani	25°11'N	66°44'E
17. San Salvador	24°7'N	74°27'W	45. Kerguellen Island	48°50'S	70°0'E
18. Wallops Island	37°50'N	75°29'W	46. Gan	0°41'S	73°9'E
19. Eleuthera Island	25°16'N	76°19'W	47. Thumba	8°32'N	76°52'E
20. Chilca	12°30'S	76°48'W	48. Carnarvon	25°S	114°E
21. Fort Sherman	9°20'N	79°59'W	49. Uchinoura	31°15'N	131°5'E
22. Cape Canaveral	28°27'N	80°32'W	50. Woomera	30°57'S	136°31'E
23. Eglin	30°23'N	86°42'W	51. Ryori	39°2'N	141°50'E
24. Keweenaw	46°26'N	87°42'W	52. Eniwetok	11°26'N	162°23'E
25. Fort Churchill	58°44'N	93°49'W	53. McMurdo	77°51'S	166°39'E
26. Holloman	32°51'N	106°6'W	54. Kwajelein	8°44'N	167°44'E
27. White Sands	32°23'N	106°29'W	55. Shemya	52°43'N	174°6'E
28. ZURF	33°46'N	106°36'W	56. Ustoka	55°N	18°E

tained. A similar system, with the addition of an ozone sampler, is available for special studies.

The total observational system includes a ground station composed of a radar for tracking the metallized Starute and a receiving station for the temperature and ozone telemetry data. Again, the equipment of these stations over the MRN is highly variable, the best being the precision tracking and telemetry systems found at the missile ranges. *See* METEOROLOGICAL INSTRUMENTATION.

Results. MRN synoptic exploration of the stratospheric circulation has revealed an upper atmosphere which is markedly different from the quiescent, static, drab, and uninteresting place characterized in assumed models before the scientific inventory included real data of accuracy and sensitivity.

A prime example of the synoptic-scale dynamical systems which originate in the stratospheric region is illustrated by the "explosive warmings"

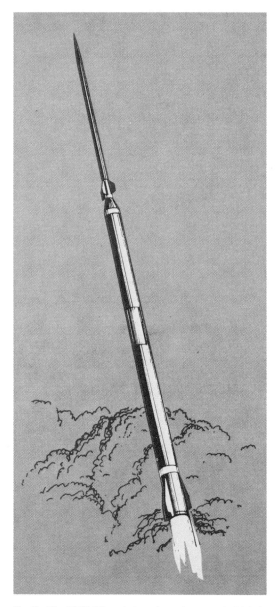

Fig. 2.   The PWN-8B rocketsonde system used widely in the MRN.

Fig. 3.   The Starute parachute system used to measure the wind profile and carry a transmitter-sensor payload.

titude. These data were in strong contradiction to observations obtained with less sensitive systems and to most theoretical considerations. R. L. Lindzen's tidal theory cannot be reconciled with the MRN temperature data, very probably because of his assumption of negligible viscosity and neglect of additional important heating and cooling mechanisms.

Synoptic-scale circulation systems in the upper atmosphere are demonstrated by the MRN data to be very obviously keyed to the geographic and orographic structures of the Earth's surface. In winter, oceanic regions characteristically have poleward extensions of ridges of high pressure, and continental regions have shifty troughs of low pressure extending equatorward over them. This intimate relationship between the surface and 50 km

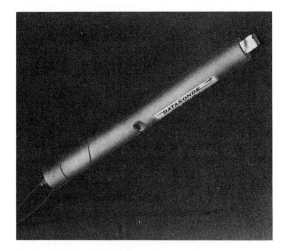

Fig. 4.   The transmitter–bead thermistor payload for the Starute.

first detected in high-altitude balloon flights over Berlin by R. Scherhag. These major disturbances of the stratosphere have been studied extensively through use of the MRN data, the strongest thus far observed being a 93°C temperature increase at 43 km over Fort Churchill in January 1973.

Possibly of equal significance at the other end of the turbulence spectrum is the fact that the sensitive sensors of the MRN have revealed a tremendous amount of small-scale variability in the upper atmosphere. These data make clear that assumptions of an inviscid fluid in the atmosphere are highly suspect; the values of the viscous terms in the equations of motion are probably orders of magnitude larger than has been assumed in prior analysis.

An early discovery of MRN exploration was a clear picture of the diurnal temperature and wind variations of the stratopause region near 50 km al-is most likely the direct result of turbulent energy transport in the vertical direction, and a total understanding of the entire atmospheric system cannot be realized until these factors are incorporated. *See* ATMOSPHERIC GENERAL CIRCULATION; STRATOSPHERE; UPPER-ATMOSPHERE DYNAMICS.

[WILLIS L. WEBB]

Bibliography: H. N. Ballard, *A Guide to Stratospheric Temperature and Wind Measurements*, COSPAR Technique Man. Ser., 1967; N. J. Beyers and B. T. Miers, Diurnal temperature change in the atmosphere between 30 and 60 km over White Sands Missile Range, *J. Atm. Sci.*, 22(3):262–266, 1965; E. Hesstvedt, On the effect of vertical eddy transport on atmospheric composition in the mesosphere and lower thermosphere, *Geofysiske Publikasjoner*, 27(4):1–35, 1968; R. L. Lindzen, Thermally driven diurnal tide in the atmosphere, *Quart. J. Roy. Meteorol. Soc.*, 93:18–42, 1967; B. T. Miers, Wind oscillation between 30 and 60 km over White Sands Missile Range, New Mexico, *J. Atm. Sci.*, 22(4):382–387, 1965; R. Scherhag, *Die Explosionsartigen Stratosphärewarmunger des Spätwinters 1951–52*, Berlin Deutschland Wetterdients, U.S. Zone, 38:51–63, 1952; W. L. Webb, Dynamic structure of the stratosphere and mesosphere, in K. Rawer (ed.), *Handbuch der Physik*, 1976; W. L. Webb, *Stratospheric Circulation*, 1969; W. L. Webb, *Structure of the Stratosphere and Mesosphere*, 1966; W. L. Webb, *Thermospheric Circulation*, 1972; World Data Center A, Ashville, NC, *Meteorological Rocket Network Firings*, vols. 1 ff.

Meteorological satellites

Meteorological satellites are Earth-orbiting spacecraft carrying a variety of instruments for measuring visible and invisible radiation from the Earth and its atmosphere. The instruments of a single satellite can view the entire Earth twice each 24 hr—a striking contrast to the small fraction of the global atmosphere that is sampled by the combination of all other methods by all nations.

Only the United States and the Soviet Union operate their own satellite systems, but many nations are using satellite data. One type of United States satellite transmits pictures in a form that can be recorded at receiving stations with relatively simple equipment. Hundreds of such stations have been installed all over the world. Pictures of cloud systems help detect and track storms, depict snow cover, and reveal the extent of pack ice. Infrared (heat) radiation delineates clouds at night and provides heat budget information. *See* WEATHER FORECASTING AND PREDICTION.

Satellite data are also important to research. Clouds are sensitive indicators of atmospheric circulation, providing visual evidence of weather systems. Some of these systems can be observed in no other way. Infrared measurements reveal the distribution of water vapor and other constituents of the atmosphere, all sensitive indices of weather. *See* ATMOSPHERIC GENERAL CIRCULATION; METEOROLOGY.

First 15 years. The space age of meteorology began when *Tiros 1* (television and infrared ob-

servation satellite) was launched on Apr. 1, 1960. Within a few days experimental cloud analyses were transmitted from the receiving stations in New Jersey and Hawaii.

Before the end of 1975 the United States had launched 42 satellites which produced a prodigious flow of weather data. Most satellites provide information for routine analysis; others are intended for research and development. The early instruments surveyed only a fraction of the Earth because they did not pass near the poles and did not always look earthward. A significant innovation was made early in 1965. *Tiros 9* was placed into a near-polar orbit; its instruments viewed the Earth several times each minute; and the orbit precessed to bring the vehicle across each latitude at the same local time each day (a Sun-synchronous orbit). This opened the door to a full-coverage system. After one more Tiros, the operational series, known as ESSA (environmental survey satellite), began in February, 1966. Since then the Earth has been photographed completely every day. A new ESSA was launched when its predecessor wore out. *ESSA 9* entered service in February, 1969, and in December, 1970, a new series commenced with the launch of *NOAA 1*. The second of this series, *NOAA 2* in October, 1972, carried the first operational sounder to measure the vertical profile of temperature in the atmosphere. *NOAA 4* was operating at the end of the fifteenth year of the meteorological space age.

Nine research and development satellites were flown successfully during this period—six of the Nimbus series and three applications technology satellites (ATS). Nimbus, larger and more sophisticated than the operational spacecraft, serves as the space platform for testing new instrument systems. Frequently, new sensors were tested on Nimbus before entering regular service on the ESSA and NOAA series.

The ATS series, although not primarily for meteorological purposes, marked the next major advance. These geostationary satellites remain fixed, approximately 36,000 km above the Equator. *ATS 1*, launched in December, 1966, carried a new device, the spin-scan camera system, which sweeps the Earth's scene with 2000 scan lines and produces a high-quality picture despite its great distance from Earth. Wind-driven clouds that are tracked on ATS picture sequences reveal airflow at various levels—wherever clouds happen to be. *ATS 3*, launched about a year later, carried a three-channel scan camera that transmitted color pictures. *ATS 6* in 1974 was used primarily for the relay of television signals, but it also bore a camera for meteorological application. Rather than spinning about its axis as the earlier ATS, it was stabilized about all three axes and carried a different, very-high-resolution camera. These satellites also serve as communication and relay stations for collecting and distributing all types of meteorological data.

The experimental geosynchronous spacecraft have been succeeded by *SMS 1* and *SMS 2* (stationary meteorological satellites), launched May, 1974, and February, 1975. Besides taking pictures in visible light, SMS spin-scan cameras are sensi-

Fig. 1.   Single picture taken from *ESSA 7* at 06:06 GMT on Oct. 17, 1968. Coastlines and latitude-longitude grid superimposed by computer after reception from satel-lite. The white spiral in the upper right quadrant is the cloud pattern of typhoon Gloria. (*NOAA*)

tive to infrared radiation from the Earth and clouds, taking pictures at half-hour intervals both day and night. Their operational use is discussed later.

The Department of Defense developed its own meteorological satellite system (DMSP) tailored to its special worldwide needs. The first DMSP was launched in August, 1973; the second in March, 1974; and the third in August, 1974. These satel-lites carry temperature sounders, sensors of space environment, and imagers in both infrared and visible light. Unique to these spacecraft are very-low-light cameras. So sensitive are they that cloud pictures can be taken by moonlight.

Soviet scientists followed closely the develop-ments in the United States, and their studies with published data of Tiros paralleled research in the United States. Then, just after the mid-1960s, the Soviet Union initiated its own weather satellite program. Scientific interchange led to the opening in 1964 of a communication link between the Unit-ed States National Environmental Satellite Service (NESS) and the Soviet center in Moscow. Satellite

measurements and weather maps flow both ways across the Atlantic. By 1975 the Soviet Union had launched its twenty-first COSMOS satellite, which is designated the METEOR series.

Operational system. Pictures from NOAA spacecraft are assembled with a sophisticated communication-computer system. Pictures are tape-recorded while the vehicle transits remote regions. Once within (FM) radio range of a receiv-ing station (at Fairbanks, Alaska, or at Wallops Islands, Va.), the tape is read out and the signals sent by land lines to the NESS near Washington, D.C. There they are fed into a high-speed comput-er for assembling and mapping. Pictures displayed for photographing on a cathode-ray tube are over-laid with latitude-longitude lines and coastlines. These are fitted to the picture by a computer that calculates their shape and placement from the sat-ellite message, and from the known height and position of the camera at picture time. *See* WEATH-ER MAP.

Figure 1 is a single picture so produced. A total of 12 such frames cover a swath along a single south-to-north pass of the satellite from pole to pole—13 passes cover the Earth. Each 24 hr, after the whole Earth has been photographed, the computer transforms the oblique perspective pictures to the shapes and scales of two standard map projections and assembles them into montages (Figs. 2 and 3).

The United States also operates a nonstorage camera, the APT (automatic picture transmission) system. This system snaps and transmits pictures every few minutes. Transmitted slowly, these pictures can be recorded on quite simple equipment. University students have constructed their own APT station. Every such station can receive several pictures from two or three adjacent passes to obtain pictures of a huge surrounding area. The station at Berlin, for example, records pictures of Newfoundland in the west and North Africa in the south. The Australians monitor typhoon formation north of the Fiji Islands and track storms in the Indian Ocean. Initially this camera operated only in sunlight, but in January, 1970, an infrared image scanner was tested on a NASA satellite, and since December of that year, with the start of the NOAA series, the operational APT has transmitted only infrared pictures. APT receivers anywhere on Earth can assemble pictures of their part of the world each 12 hr.

Infrared data. While visible pictures are irreplaceable for some purposes, the 1970s witnessed an accelerated use of infrared images and data. Infrared radiation, heat energy, is highly dependent on the temperature of the radiating material, a characteristic that makes it possible to deduce a myriad of information. By measuring at wavelengths where water vapor and other atmospheric gases radiate only weakly, the instrument "sees" through the atmosphere and measures radiation from the Earth, oceans, and clouds. Various intensities of radiation are transposed into contrasting densities on film so that images are produced similar to those formed by reflected light. But visible and infrared images exhibit an important difference, as illustrated in Fig. 4. Thick clouds, regardless of their altitude, reflect most of the sunlight that falls on their upper surface so that they are very bright when photographed from space. On the other hand, low clouds, lying in the warmer atmosphere, and high clouds, in the cold upper atmosphere, radiate at quite different intensities and therefore exhibit different brightnesses on film. Figure 4a, made from visible light, was taken simultaneously with Fig. 4b, an infrared image. Areas of light gray and white on Fig. 4b represent cold clouds—they have a similar appearance on both images. By contrast, the clouds immediately west of the United States are nearly white on the visible image in Fig. 4a but are dark on the infrared image. These are low-lying clouds with warm tops. Quantitative measurements of infrared radiation portrays much more than clouds and their temperatures, however.

Ocean-surface temperatures, critical to both meteorology and oceanography, can be deduced from infrared measurements in cloudless regions. Sophisticated computer programs analyze infra-

red data from all ocean areas each day, deducing which measurements were made in cloud-free areas, using only these for computing sea-surface temperatures. Other sensors respond to water vapor in the atmosphere, while those measuring a broad spectrum of wavelengths observe the total heat lost to space. This debit side of the Earth's heat budget is a critical variable for long-range forecasting.

Nimbus 3, flown in the spring of 1969, carried some new experiments that have marked the beginning of another advance in observing the atmosphere from space. SIRS (satellite infrared spectrometer) and IRIS (infrared interferometer) both measure radiation from carbon dioxide in the atmosphere. From these data are deduced temperatures of several layers of the atmosphere. Clouds are nearly opaque to radiation in the carbon dioxide spectrum, however, so that temperatures are unobtainable below clouds by this method. To overcome this difficulty, *Nimbus 5* carried additional detectors sensitive to long wavelengths (microwaves) which penetrate clouds. Oxygen radiates strongly in this wave band, and so temperature of oxygen (and that of the other atmospheric gases it is mixed with) can be deduced. Combining these measurements with observations of carbon dioxide radiation provides soundings above and below clouds.

Fig. 2. *ESSA 7* montage of about 28 pictures taken Oct. 17, 1968, transposed to Mercator projection. Typhoon Gloria (Fig. 1) appears above picture center. (*NOAA*)

These means of observing temperature profiles cannot completely replace routine balloon measurements (radiosondes) of temperature, pressure, and humidity in the free atmosphere. Satellite soundings provide rather smoothed profiles, while balloon soundings measure sharper variations of temperature and humidity. Forecasters need both the detail provided by radiosondes and the tremendous areal coverage available only with satellites.

Another new experiment lofted by *Nimbus 3* was IRLS, a system for locating and communicating with unmanned sensors anyplace on Earth. This system can collect and relay weather reports from automatic weather stations located on buoys and in remote areas. More importantly, IRLS can determine the position of free-floating balloons. Balloons that float at a constant elevation have already been tested extensively. Some have circled the Earth several times before descending. Larger constant-level balloons, called carrier balloons, have been flown bearing many dropsondes which are parachuted on command. They contain instruments much like the radiosondes to measure pressure, temperature, and humidity as they fall through the atmosphere, relaying their measurements via the satellite. Winds can be deduced from the path of the high-level balloon, and lower winds are measured from the dropsonde.

Nimbus 4 added a back-scattered ultraviolet

(BUV) experiment. Ozone in the upper atmosphere scatters back ultraviolet light from the Sun, providing a means to monitor the distribution of ozone around the world. *See* ATMOSPHERIC OZONE.

Nimbus 5 measured yet another part of the spectrum—the longer wavelengths of microwaves that penetrate both clouds and atmosphere. The electron-scanning microwave radiometer (ESMR) has been used to map sea ice in cloud-covered polar regions.

Using pictures for weather analysis. Before the era of satellites storms sometimes existed undetected until a ship happened into their path; now every intense cyclone on Earth can be watched as it moves across remote regions. Such features and many more of vital interest to forecasters can be seen in Fig. 3. In addition to the typhoon approaching the Philippine Islands, two mature cyclones can be seen in the North Pacific, one just east of Kamchatka, the other in the Gulf of Alaska. The cold front from the latter swept toward the southwest and produced rain a few hundred miles north of Hawaii at the time the picture was taken. Similar storms appear in the North Atlantic, and scores of other weather disturbances are shown. Pictures such as these enable the analyst to prepare accurate weather maps, a critical step in weather forecasting.

Figure 4 is one pair of a sequence of pictures taken by *SMS 2* from its fixed position at 115°W longitude on the Equator. Similar sequences are obtained with *SMS 1* at 75°W. An enormous area is viewed by this combination. The view from *SMS 1* sweeps from the bulge of equatorial Africa, across the North and South Atlantic, across North and South America, into the eastern Pacific. *SMS 2* overlaps part of this expanse, picturing the western United States at more favorable angles for West Coast forecasters, and monitors tropical disturbances as far west as the International Date Line.

The unique feature of these spacecraft is their capability to photograph clouds frequently so that their motion can be measured. Clouds act as tracers which are carried along by the winds. Clouds that are both cold and bright are high clouds which follow the circulation of the upper atmosphere, while low clouds, which appear as darker shades of gray on infrared images, reveal the flow of the lower atmosphere. Analysts use digital values of the measured infrared radiance to determine cloud temperatures and, thereby, their altitude. Hundreds of wind measurements are made daily by means of this technique, and messages are transmitted on international communication circuits which provide wind speed, direction, location, and height. The directions of some clouds are shown in Fig. 4*b*.

Wind and change of wind with height are closely related to the pressure and temperature of the atmosphere. Therefore the wind patterns deduced from such pictures enable meteorologists to derive pressure and temperature over vast areas that were previously unobserved.

It is also important to detect short-period changes of cloud patterns. Hundreds of weak dis-

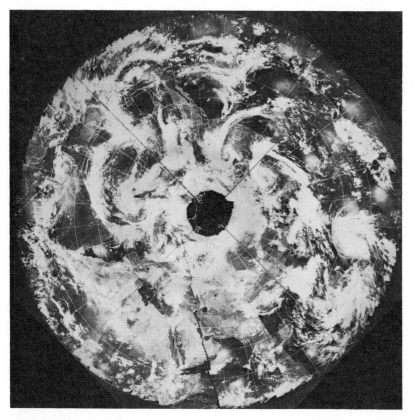

Fig. 3. *ESSA 7* montage of about 80 pictures taken Oct. 16–17, 1968, transposed to polar stereographic projection of Northern Hemisphere. Asia and western Pacific at bottom; outline of North America at top. (*NOAA*)

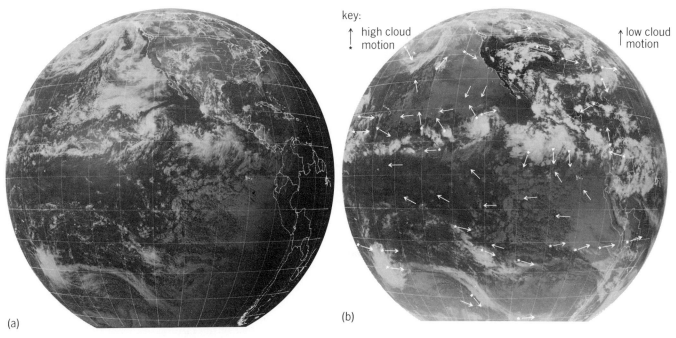

(a)

(b)

key:

↑ high cloud
⋮ motion

↑ low cloud
⋮ motion

Fig. 4. Simultaneous pictures from about 36,000 km above the eastern Pacific with *SMS 2* (*a*) visible and (*b*) infrared spin-scan radiometer, on Aug. 11, 1975. Winter sunset has already darkened South America in *a*, but in *b* clouds are imaged by their infrared radiation. (*NOAA*)

turbances, for example, drift across tropical oceans each summer, but only a few become hurricanes. Transformation of a squally center into a hurricane is signaled by formation of a pattern similar to that in Fig. 1. The continuous surveillance that is possible from geostationary vehicles permits early storm detection and more timely warnings. *See* STORM DETECTION.

The dramatic success of geosynchronous satellites, the advances made in measuring the vertical profile of atmospheric temperature, and the sophisticated computer programs that analyze huge quantities of data every day have inaugurated a new meteorological era. Added to the earlier types of observations, scientists can now measure the workings of the atmosphere to an extent unimagined only a few years ago. Meteorologists have always been hampered by their inability to determine the present state of this heat engine—the atmospheric shell that covers the globe. Adequate laboratory models cannot be constructed. Even the most advanced numerical models incorporate crude assumptions to fill gaps left by scientists' inability to measure the atmosphere. Satellites provide the means to determine the state of the atmosphere over the entire globe.

[LESTER F. HUBERT]

Bibliography: R. K. Anderson, E. W. Ferguson, and V. J. Oliver, *The Use of Satellite Pictures in Weather Analysis and Forecasting*, World Meteorol. Organ. Tech. Note Ser. no. 124, 1973; A. W. Johnson, Weather satellites, pt. 2, *Sci. Amer.*, 220(1):52–68, 1969; World Weather Watch, *Information on the Application of Meteorological Satellite Data on Routine Operations*, 1978.

Meteorological solenoid

In meteorological usage, solenoids are hypothetical tubes formed in space by the intersection of a set of surfaces of constant pressure (isobaric surfaces) and a set of surfaces of constant specific volume of air (isosteric surfaces). The isobaric and isosteric surfaces are such that the values of pressure and specific volume, respectively, change by one unit from one surface to the next. The state of the atmosphere is said to be barotropic when there are no solenoids, that is, when isobaric and isosteric surfaces coincide. The number of solenoids cut by any plane surface element of unit area is a measure of the torque exerted by the pressure gradient force, tending to accelerate the circulation of air around the boundary of the area. *See* BAROCLINIC FIELD; BAROTROPIC FIELD; ISOPYCNIC.

[FREDERICK SANDERS]

Bibliography: H. R. Byers, *General Meteorology*, 1974; S. L. Hess, *Introduction to Theoretical Meteorology*, 1959.

Meteorology

The science concerned with the atmosphere and its phenomena. Meteorology is primarily observational; its data are generally "given." The meteorologist observes the atmosphere—its temperature, density, winds, clouds, precipitation, and so on—and aims to account for its observed structure and evolution (weather, in part) in terms of external influence and the basic laws of physics.

Empirical relations between observed variables, as those between the patterns of wind and weather, are developed to pose more effectively

Table 1. Components of dry air

Gas	Symbol	% vol
Nitrogen	N_2	78.09
Oxygen	O_2	20.95
Argon	Ar	0.93
Carbon dioxide	CO_2	0.03
Neon	Ne	1.8×10^{-3}
Helium	He	5.2×10^{-4}
Krypton	Kr	1×10^{-4}
Hydrogen	H_2	5×10^{-5}
Xenon	Xe	8×10^{-6}
Nitrous oxide	N_2O	3.5×10^{-5}
Radon	Rn	6×10^{-16}
Methane	CH_4	1.5×10^{-4}

the problems to be investigated and explained and to provide essential material for the application of the science. Weather forecasting serves as an example of such application because theory still remains insufficiently developed to provide more certain applications. Little controlled experiment has been made on the atmosphere, but more is probable. *See* WEATHER FORECASTING AND PREDICTION; WEATHER MODIFICATION.

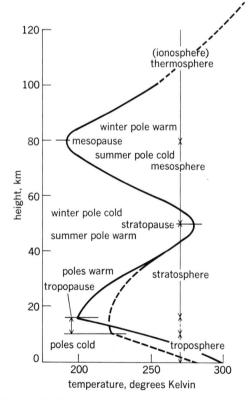

Fig. 1. Vertical structure and nomenclature of atmospheric layers in relation to temperature. Values of temperatures shown are approximate; the form of the temperature variation with height determines the nomenclature. The two curves shown for troposphere and stratosphere refer to lower (solid line) and higher (dashed line) latitude conditions. The sense of the meridional temperature gradient at various levels is shown by entries "poles cold" and "poles warm." A change of meridional gradient appears to coincide with a change of vertical gradient (pause) only at the tropopause.

This background article has a threefold organization. The first portion presents a summary of the general physics of the air; this has been the principal approach and basis for meteorological science and its applications, such as to weather phenomena (the condition of the atmosphere at any time and place) and climate (a composite generalization of weather conditions throughout the year). A second portion, synoptic meteorology, discusses the character of the atmosphere on the basis of simultaneous observations over large areas. Concurrently, and at an accelerating pace, dynamic principles (thermodynamic and hydrodynamic) are being applied to meteorological investigations. This study of naturally produced motions in the atmosphere is forming much of the scientific basis for modern weather forecasting and physical climatology. Hence, the third portion of this article deals with dynamical meteorology. *See* CLIMATOLOGY.

GENERAL PHYSICS OF THE AIR

The components of dry air, excluding ozone, and their relative volumes (mol fractions of gases) up to a height of at least 50 km are given in Table 1. Some of the rarer gases, such as CO_2, are continually entering and leaving the atmosphere through the Earth's surface, and the fractions quoted are thus mean values. The effective molecular weight of the mixture is 28.966 and its equation of state, to 1 in 10^4, is $p = R\rho T$, where p is pressure, ρ density, T absolute temperature, and R, the specific gas constant, is 2.8704×10^6 erg/(g)(°K). *See* AIR; ATMOSPHERE.

Above 50 km, O_2 becomes progressively dissociated to O, which is probably dominant above about 150 km. N_2 probably dissociates at appreciably higher levels. There is no good evidence for diffusive gravitational separation of the lighter from the heavier gases below about 100 km—nor indeed is this likely. Above 100 km, however, such separation is probable and in the levels of escape (exosphere), at several hundreds of kilometers, helium and hydrogen may be dominant with proportions varying during the solar 11-year cycle.

Water vapor and ozone are highly variable additional components of air. The fraction of the former commonly decreases rapidly with height, from about 10^{-2} at sea level to 10^{-6} or 10^{-7} at about 16 km, with dissociation at much higher levels. Ozone, the product of photochemical action in sunlight at high levels, has a maximum fraction ($\sim 10^{-8}$) at about 25 km. The importance of these two constituents is far greater than their fractions might suggest because (1) both are radiatively active, water vapor and ozone in the infrared of terrestrial emission, ozone in the near ultraviolet of solar emission; and (2) water vapor condenses to the liquid or solid, giving cloud, which reflects upward a major part of solar radiation incident upon it, and in condensing releases a large latent heat to the air. Carbon dioxide, CO_2, is the only other constituent with strong infrared activity.

Thermal structure. It is convenient to divide the atmosphere into a number of layers on the basis of its thermal structure. The first layer (Fig. 1) is the troposphere, in which temperature on average decreases with height 6°C/km (the "lapse rate") everywhere except near the winter pole. The troposphere is about 16 km deep in the tropics and about 10 km deep, with substantial variations, in higher latitudes. At any one level in the layer, the mean temperature decreases from Equator to pole, by about 40°C in winter and 25°C in summer.

The second layer is the stratosphere, in which the temperature varies at first very little with height and then increases to near the surface temperature at about 50 km. In this layer temperature increases poleward, except quite near the winter pole. The boundary between troposphere and stratosphere is termed the tropopause, which may drop abruptly in altitude at about 30° lat and again in higher latitudes.

From 50 km to 80 km the temperature again decreases with height to an absolute minimum of about −80°C, lower in summer than in winter, and in summer associated with noctilucent clouds. This layer is called the mesosphere, and its lower and upper boundaries are called the stratopause and mesopause, respectively. *See* MESOSPHERE; STRATOSPHERE; TROPOSPHERE.

Above 80 km the temperature again increases with height, at a rather uncertain rate. The layer is strongly ionized by solar ultraviolet and other radiations and is variously termed the ionosphere or thermosphere. *See* IONOSPHERE.

Radiation and thermal structure. During the passage of solar radiation downward through the atmosphere, the following, in broad outline, takes place. The shorter ultraviolet waves are absorbed by the thermosphere, more above than below, so that temperature may be expected to fall along the path of the beam. Entering the mesosphere, O_3 (ozone) is encountered, and new absorption of energy takes place in the near ultraviolet. The increase of ozone concentration along the path outweighs the depletion of the beam energy by O_3 at upper levels so that the temperature increases along this portion of the path. Beneath the stratopause, however, depletion in the wavelengths concerned has become large and the O_3 concentration itself ultimately falls so that temperature decreases along the path. There is little absorption in the troposphere—the slight near-infrared absorption by water vapor is practically uniform along the path because of the increasing concentration of vapor—but there is substantial backscatter by air molecules, haze, and clouds. The Earth's surface, except snow, absorbs the incident solar radiation strongly, and this leads to the heating of the atmosphere from below by various kinds of convection and to a lapse of temperature in the troposphere. *See* INSOLATION; TERRESTRIAL RADIATION.

Because the atmosphere and underlying surface maintain their temperature over the years, the terrestrial emission of radiation to space by water vapor, carbon dioxide, and ozone in the far infrared, must nearly balance the absorption of solar energy. The balance is achieved globally but not locally, the winds providing the necessary transport of heat from the regions of net absorption (lower latitudes) to those of net emission (higher latitudes). While the broad form of the temperature profile of Fig. 1 is determined by the pattern of solar absorption, actual temperatures are also due to terrestrial emission and wind transport. *See* TERRESTRIAL ATMOSPHERIC HEAT BALANCE.

Pressure and wind. The pressure at any point in the atmosphere is the weight of a column of air of unit cross section above that point. Therefore pressure decreases with height. At any level it decreases more rapidly where the air is cold and dense than where it is warm and light. The pressure at sea level is generally a little over 1 kg/cm², varying by a few percent in space and time because of the inflow of air over some regions and its outflow over others. The horizontal pressure pattern necessarily changes with height when, characteristically, the temperature varies horizontally.

In an unheated, nonrotating atmosphere, air would flow horizontally from high pressure to low to remove the pressure difference. Other forces arise in the actual atmosphere to provide a flow which is more nearly along, than perpendicular to, the isobars (lines on a map connecting points of equal barometric pressure). Therefore pressure patterns persist for days, with gradual modification, and are translated with speeds often comparable with those of the winds blowing in them. The effect of the Earth's rotation on the wind is nearly always dominant, except very near the Equator. If the horizontal force arising from the Earth's rotation (called the Coriolis force) precisely balances the pressure force, the wind blows with a speed V along the isobars (for an observer with his back to the wind, low pressure is to the left in the Northern Hemisphere). The speed is given by the equation below, where $\partial p / \partial n$ is the horizontal

$$V = \frac{(\partial p / \partial n)}{2\omega \rho \sin \phi}$$

pressure gradient, ω the angular velocity of the Earth, ρ the air density, and ϕ the latitude. This wind V is called the geostrophic wind and is a good approximation to the actual except near the surface, where friction always intervenes. Since, as seen above, the pressure gradient generally varies with height, so do both V and the actual wind. The change of V with height is proportional to the horizontal temperature gradient, low temperature being to the left of the vector change of V in the Northern Hemisphere. *See* AIR PRESSURE; WIND.

General atmospheric circulation. The mean wind in nearly all parts of the atmosphere is predominantly along the latitude circles. The pattern of this zonal motion in latitude and height is shown in Fig. 2 and is the basis of the general circulation of the atmosphere. Surface easterlies (trade winds) are found in the tropics, surface westerlies in middle latitudes, and surface easterlies again near the poles. But everywhere in the troposphere, except near the Equator, winds increase in strength from the west with height; in the tropics and polar regions easterlies decrease before giving place aloft to increasing westerlies. Above the tropopause the westerlies decrease with height except over the winter polar cap where, at around 40 km, there occurs a winter westerly jet stream. This pattern of winds is consistent with the meridional gradient of temperature in the troposphere and its reversal (except over the winter polar cap) in the stratosphere, since the winds are quasi-geostrophic. The absolute maximum of zonal wind shown in Fig. 2 is found in the upper troposphere at about 30° latitude and is the core of the subtropical jet

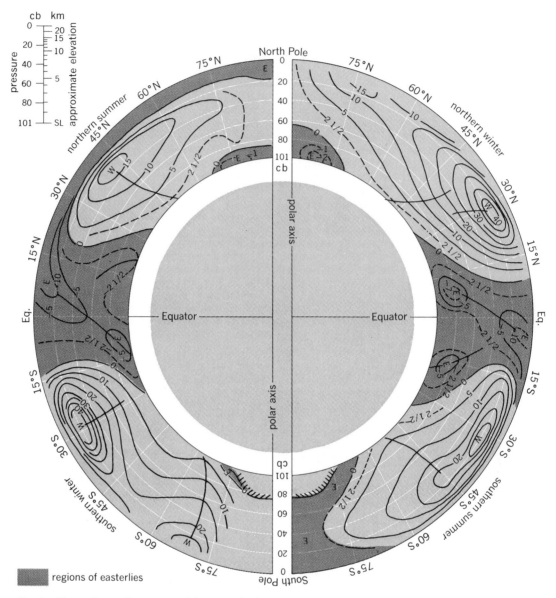

Fig. 2. The pattern of mean zonal (east-west) wind speed averaged over all longitudes as a function of latitude, height, and season (*after Y. Mintz*). Height, greatly enlarged relative to Earth radius, is shown on a linear pressure ($\propto$ mass) scale with geometrical equivalent given at upper left. The mean zonal wind (westerly positive, easterly negative) has the same value along any one line and is shown in meters per second (1 m/sec $\cong$ 2 knots) on the line. Note subtropical westerly jet in high troposphere at about 30° lat (or more in northern summer). This jet should not be confused with the polar front jet of higher latitudes; the latter is migratory and so does not appear in the mean.

stream. *See* GEOSTROPHIC WIND; JET STREAM.

The general circulation is maintained against frictional dissipation by a "boiler-condenser" arrangement provided by the heat absorbed, directly or indirectly, from the Sun and that emitted to space by the Earth. The wind thereby generated transports heat from source to sink to maintain thermal balance, disturbances on the mean motion being important in the process. *See* ATMOSPHERIC GENERAL CIRCULATION.

Water vapor, cloud, and precipitation. The motion of the air is rarely quite horizontal; it is commonly moving upward or downward at a few centimeters per second over large areas, while locally in thunderclouds the vertical velocity may

be several meters per second. Air which rises is cooled by expansion into lower pressure and that which descends is correspondingly compressed and warmed. Since air is always more or less moist, a sufficient rise will cause the temperature to drop to the saturation point and cloud will then form on nuclei present in the air. Large-scale ascent leads to extensive sheets of cloud called stratus: cirrostratus at high, ice-forming levels, altostratus at medium levels, nimbostratus at lower levels. Local, strong ascent produces heap clouds (cumulus or cumulonimbus). If the cloud base is sufficiently warm or the cloud depth sufficiently great, or both, the condensed water or ice in the cloud falls out as rain, snow, or other precipitation.

Certain lenticular clouds are due to local ascent and descent brought about by hills, and these clouds do not move with the wind. Cloud formation is practically confined to the troposphere because this alone contains sufficient water vapor and sufficiently sustained vertical motion. *See* CLOUD; CLOUD PHYSICS; PRECIPITATION.

Atmospheric disturbances. The mean motion, described under the general circulation of the atmosphere, is disturbed by many patterns of motion of varying scale. A chart of the flow well above the surface over the Northern Hemisphere will generally show, on any day, large meanderings of the air poleward and equatorward superposed on a general westerly flow. These are the long waves of the westerlies, with a few thousand kilometers between the turning points. They are generally associated with a jet stream in the upper troposphere. Another class of large-scale atmospheric disturbances, seasonal in nature and most apparent in the lower troposphere, is designated the monsoons; these are most developed in the South Asian cyclone of summer and the Siberian anticyclone of winter. *See* MONSOON; UPPER SYNOPTIC AIR WAVES.

Proceeding downward in scale are the traveling depressions (cyclones) and anticyclones, ridges and troughs of extratropical latitudes, 1000–2000 km in horizontal extent, and then the somewhat shorter waves in the easterlies (trades). Next in size is the tropical cyclone (∼100 km)—the hurricane or typhoon, according to location—which is mainly confined to the tropical oceans and eastern seaboards of continents. Still smaller (∼1 km) is the tornado, mainly confined to land and associated with cumulonimbus cloud, and the waterspout associated with similar cloud over the sea. The smallest "revolving storm" is the dust devil (∼10 m), which occurs immediately above a hot, dry surface in a very light general wind. *See* CYCLONE; FRONT; STORM; TORNADO.

In addition to the above well-defined patterns of flow there are randomly distributed fluctuations of flow over several orders of magnitude of scale from about 1 cm upward, particularly evident in the bottom kilometer of the atmosphere and in and around cumulus. They are related to friction, convection, and wind shear and are important agents of vertical transfer of heat, matter (water vapor, dust, ozone), and momentum.

Air masses and fronts. The horizontal variations of air temperature referred to in previous paragraphs are commonly concentrated into narrow zones, or even discontinuities, with low gradients in intervening areas. These narrow zones or discontinuities are called fronts, and the air in a region of small temperature gradient is called an air mass. If a front moves so that a warm air mass replaces a cold air mass, it is called a warm front, and conversely a cold front. Fronts may also be stationary. They slope upward at a slope ratio of about 1 in 100, with the cold air as a wedge beneath the warm air.

Most extratropical depressions first appear as dents or waves on a frontal surface, warm air rising slowly over an extended area in the forward part of the wave. This results in stratiform cloud. A stronger updraft immediately ahead of the rear part of the wave yields cumulonimbus cloud.

The tropopause is higher and its temperature lower above a warm air mass than above a neighboring cold air mass, and a break, or offset, appears at the frontal boundary with a jet stream in the warm air mass near the break in the tropopause. *See* AIR MASS; TROPOPAUSE.

Optical phenomena. Scattering, refraction, and reflection of light by the air or by particulate matter (dust, cloud particles, or rain) in the air give rise to a variety of optical phenomena. *See* METEOROLOGICAL OPTICS.

Electrical properties. The atmosphere and the underlying Earth are like a leaky electrical capacitor. Positive charge is separated vertically from negative charge in thunderclouds and some other areas of disturbed weather; the net result is that the Earth's surface is left in fine-weather areas with an average negative charge σ of 2.7×10^{-4} esu/cm² to which corresponds a vertical field F_0 ($= 4\pi\sigma$) at the surface of 100 volts/m. The other "plate" of the capacitor is the highly conducting upper atmosphere, and the leak arises from the small conductivity of the air between the plates, produced by ionization by radioactive matter in the soil and air, and by cosmic radiation; the conductivity increases with height. The resistance R of a 1-cm² column of atmosphere is about 10^{21} ohms and an air-Earth current $i = V_\infty/R$ (V_∞ being the potential of the upper conducting layer) of about 2×10^{-16} amp/cm² flows as a discharge current. V_∞ is thus about 2×10^5 volts above Earth. The increase of conductivity with height implies a proportionate decrease of the field with height, and this in turn implies a small positive ionic space charge in the air.

The air-Earth current would discharge the capacitor in about 1/2 hr if V_∞ (or σ) were not maintained by the charge separation in thunderclouds. This separation is more than adequate, the excess providing lightning flashes within the cloud—a shorting of the generator. *See* ATMOSPHERIC ELECTRICITY.

[P. A. SHEPPARD]

SYNOPTIC METEOROLOGY

This branch of meteorology comprises the knowledge of atmospheric phenomena connected with the weather, applied mainly in weather forecasting, and based on data acquired by the synoptic method. This method involves the study of weather processes through representations of atmospheric states determined by synchronous observations at a network of stations, most of which are at least 10 km apart. By international agreement, the data taken from the Earth's surface and aloft at certain international hours of observation are inserted on weather maps, upper-air maps, vertical cross sections, time sections, and sounding diagrams with international symbols according to fixed rules. These crude representations are then analyzed and critically evaluated in accordance with the knowledge of existing structure models in the (lower) atmosphere, in order to ascertain the best approximation to a three-dimensional image of the true atmospheric state at the hour of observation. From one such representation, or a series of them, future atmospheric states

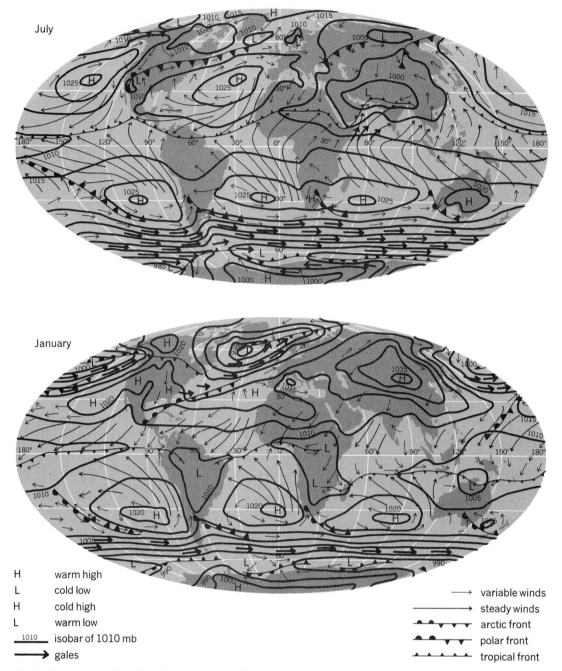

H	warm high
L	cold low
H	cold high
L	warm low
—1010—	isobar of 1010 mb
⟶	gales

⟶	variable winds
⟶	steady winds
▬●▬▼▬	arctic front
▬●▬▼▬	polar front
▬▲▬▲▬	tropical front

Fig. 3. Air masses and fronts as links in the general atmospheric circulation.

are then derived with the aid of empirical knowledge of their behavior and by application of the theoretical results of dynamic meteorology.

General atmospheric circulation. Figure 3 shows some of the main constituents of the average state in the bottom layer of the atmosphere as to flow and pressure patterns, main air masses and fronts. The illustration also indicates how these large-scale mechanisms form part of a general atmospheric circulation with trade winds, monsoons, high-reaching middle-latitude westerlies, and shallow polar easterlies (see the vertical cross section of Fig. 2).

The planetary high-pressure belt at 30°N and S

is split into subtropical high-pressure cells mainly as a result of the joint dynamic and thermodynamic effect of the great continents and mountain ridges of the Earth, especially the Cordilleran highlands of South and North America. These cells again determine the formation and average position of the different polar fronts and air masses at the Earth's surface.

Other general structures in the atmospheric circulation of importance to weather are the quasi-horizontal tropopause layers at different heights within different air masses and the tropical fronts, maintained within the Zones of Intertropical Convergence (ITC). The former generally mark the top of any considerable convection or upglide motion and of clouds in the atmosphere, but represent no store of potential energy. The latter form at the meeting of two opposite trade wind systems in the doldrums and have functions partly similar to those of the polar fronts, although they are much weaker and less distinct.

Air masses and front↔jet systems. Air masses may be classified in two distinctly different ways, thermodynamically and geographically.

Thermodynamic classification. This classification is based on their recent path and life history, distinguishing mainly the two opposite cases: the air being warmer or colder than the Earth's surface, resulting in warm mass and cold mass. The warm mass, usually flowing poleward, is much warmer than the seasonal normal of the region, at least aloft. With the cooling from below, it gradually acquires a stable stratification, which damps turbulence and vertical mixing. Thus, the wind is relatively steady, the visibility low, and advection fog or stratus clouds often form within it, at least at sea, sometimes even yielding drizzle. This air mass is most typically found on the poleward side of the warm subtropical highs (Figs. 3 and 8), where the air is subsiding. These highs may get displaced poleward (Fig. 8) and will then bring periods of steady and rather dry weather—very warm in summer on land—to middle and higher latitudes. *See* WIND.

The cold mass, mostly flowing equatorward, is by definition colder than normal, at least aloft. Due to heating from below it rapidly acquires an unstable stratification which favors turbulence and vertical mixing or convection. Thus, the wind is gusty, visibility is good, and usually where the air is moist enough convective clouds form: cumulus → cumulonimbus with showers → thunderstorms and even hail. This air mass is most typically found within (upper) cold lows (Fig. 10), where, if there is sufficient moisture, the instability, general convergence, and lifting tendency of the air may favor the formation even of nonfrontal rain areas, which are discussed later. Therefore, such a low usually brings a period of wet and stormy weather to the equatorward part of middle latitudes, and in winter to the subtropics. However, at night over land, even this air may be stabilized so efficiently that radiation fog occurs within it.

Geographic classification. The second classification is based on the geographical origin or position of the air and the values of characteristic properties (such as temperature and specific humidity).

Tropical air occupies all the space between the polar-front↔jet systems of both hemispheres; aloft it may reach far into the polar region. At the midtroposphere this air is 10–20°C warmer and much more humid than the polar air at the same level and latitude. At low levels, the tropical air emerges from the quasi-stationary subtropical highs, reaching middle latitudes from the southwest, particularly within the warm sectors of migrating cyclones (Fig. 7), as a mild or warm, moist and hazy air current (the stable warm mass). Aloft, it appears in high latitudes within the warm highs (Fig. 8). Below, the tropical air also flows equatorward and westward within the northern

and southern trade wind systems. When approaching the tropical front (Fig. 3) it appears, at sea, as an unstable cold mass with intense convection and shower activity on both sides of the front: the equatorial rains. Over land, for example, in North Africa, only the southwestern monsoon (that is, the southeastern trade wind that has invaded the other hemisphere), undercutting the very hot and dry northeast trade wind, is moist enough to produce any convective clouds and rain.

Polar air is found on the polar side of the polar fronts, as shown in Fig. 3. Below, it emerges in winter from continental subpolar highs of 40–60°N, and in summer from the polar basin. The polar masses, and still more so the arctic and antarctic air masses, are mainly characterized by very low temperatures and low specific humidities aloft.

Arctic air is produced over the ice- and snow-covered parts of the arctic region during the colder

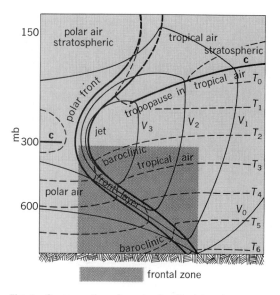

Fig. 4. Cross section of a polar-front↔jet system.

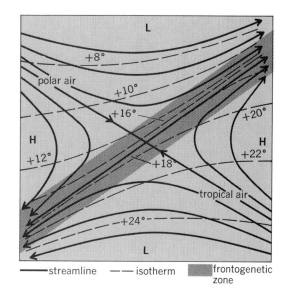

Fig. 5. Frontogenesis in a field of deformation.

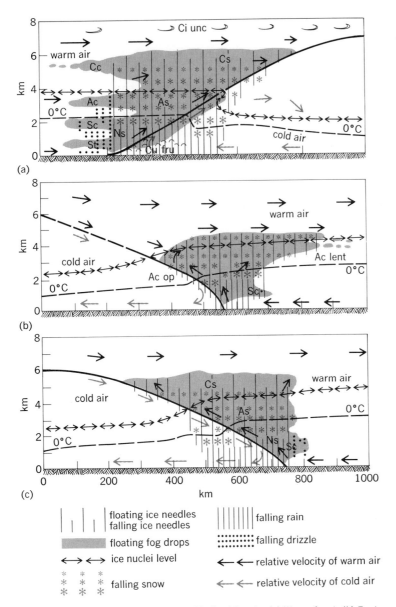

Fig. 6. Schematic cross sections of three kinds of fronts. (a) Warm front. (b) Fast-running cold front. (c) Slow-moving cold front. Ac, altocumulus; Ac lent, altocumulus lenticularis; Ac op, altocumulus opacus; As, altostratus; Ci unc, cirrus uncinus; Cc, cirrocumulus; Cs, cirrostratus; Cu fru, cumulus fractus; Ns, nimbostratus; St, stratus; and Sc, stratocumulus.

Table 2. Atmospheric waves

Name	Type*	Wavelength, km
Ultrasound	C	$<2 \times 10^{-5}$
Ordinary sound (tones)	C	2×10^{-5} to 10^{-2}
Explosion waves	C	10^{-2} to 10^{-1}
Helmholtz waves (Sc und, Ac und)	G	10^{-1} to 1
Short lee-waves (Ac lent, Cc lent)	G	1 to 2×10
Long lee-waves (nacreous clouds, precip.)	G(I)	2×10 to 10^{2}
Short jet-waves (frontal)	GI(V)	5×10^{2} to 5×10^{3}
Long jet-waves	V	5×10^{3} to 10^{4}
Tidal waves	(G)	$\leqq 2 \times 10^{4}$

*C, compression, longitudinal; G, gravitational; I, inertial; and V, vorticity-gradient; the last three are transversal.

or rising pressure behind. Frictionally produced convergence, or static and inertial instability, or both, preferably within the warmer air, favors the ascent of air along the front. As a result, a vast cloud mass, a cloud system, may form, with an area of continuous precipitation near the front (Fig. 6).

The main fronts reach into the stratosphere (Fig. 4) and have a horizontal extension of several thousand kilometers. They originate as links in the atmospheric circulation, within frontogenetic zones, between two stationary anticyclones (highs) and two cyclones (lows) when the axis of stretching, or confluence, of this deformation field (Fig. 5) has some extension west to east. This pattern is in its turn determined by the large-scale orography of the Earth. Because of the north-south temperature contrast, air masses of different temperature, specific humidity, and density are then brought together in a front zone along this axis. On the warm side of the polar fronts, which together more or less encircle the hemisphere, the thermal wind, as discussed in dynamical meteorology below, corresponding to the mass distribution of the fronts, implies a narrow band of intense flow near the tropopause level, a jet stream. The thermal wind is most pronounced at middle and higher latitudes, where the front is marked, steep, and usually reaches into the stratosphere (Fig. 5). There, a main front and its jet together represent a zone of maximum potential and kinetic energy. Below, in lower latitudes, the frontal tilt is often small, and there is an outflow of polar air into the tropics.

A front may appear as warm or cold. At a warm front, the warmer air gains ground and slides evenly upward above the retreating cold-air wedge (tilt or slope 1:200 to 1:100), producing a wedge-shaped upglide cloud system (Fig. 6a). At the approach of a marked warm front, therefore, a typical cloud sequence invades the sky: cirrostratus → altostratus → nimbostratus, the last yielding continuous and prolonged precipitation ahead of the front line.

At a cold front the rather steep cold-air wedge (with a slope of 1:100 to 1:50) pushes forward under the warm air and forces it upward according to one of the two flow patterns shown in Fig. 6b and c. At the approach of a cold front of the more

Key (Fig. 6):

floating ice needles / falling ice needles — falling rain
floating fog drops — falling drizzle
ice nuclei level — relative velocity of warm air
falling snow — relative velocity of cold air

seasons, when the Earth's surface is a marked cold source. In the North American sector its seat, for dynamic as well as thermal reasons, lies on the average near Baffin Island, being separated from the polar air south of it by the American arctic front (Fig. 3).

Fronts and frontogenesis. Because of the Earth's rotation a surface of density discontinuity, a front, seeks a tilted position of dynamic equilibrium. A front is defined as a dynamically important, tilting layer of transition between two air masses of markedly different origin, temperature, density, and motion. Colder air lies as a wedge below the warmer; therefore, the front will coincide with a trough or bend in the (moving) isobaric system, at whose passage the wind will veer, and as a rule with falling pressure ahead and stationary

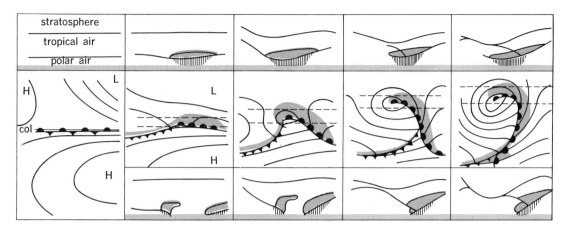

Fig. 7. Life history of cyclones from wave to vortex. The vertical sections (top and bottom) show clouds and precipitation along the two dashed lines in the map sequence. (*From C. L. Godske et al., Dynamic Meteorology and Weather Forecasting, American Meteorological Society, Waverly Press, 1957*)

common type (Fig. 6*b*) there is, therefore, another typical cloud sequence: altocumulus, partly lenticular, rapidly thickening into nimbostratus. The precipitation is usually more intense but of shorter duration than with the warm front. *See* CLOUD; CLOUD PHYSICS.

Waves and vortices (the weather). The actual atmospheric states affecting the weather may be regarded as composed of the general circulation and its disturbances. Together they form a multitude of weather mechanisms, some of which have already been described, in which the water-vapor cycle (evaporation → transport and lifting → condensation → precipitation) evidently has a fundamental role. The atmospheric disturbances consist of a spectrum of waves of different wavelengths λ (Table 2), corresponding circulations (vortices), or both. Only the larger of these are studied synoptically. By their size the circulations may be classified as planetary (or geographical), secondary, or tertiary:

1. The long waves forming in the front↔jet zone, that is, the region of maximum energy, may represent a steady state (see section on dynamical meteorology) when their wavelength corresponds to about four circumpolar waves at low, three at medium, and possibly two at high latitudes. The planetary waves have $\lambda \sim 10,000$ km. Thereby they partly determine the shape of the general circulation (Fig. 3). The shorter long waves $(3000 < \lambda < 8000$ km) propagate slowly eastward and, because of inertial instability, mostly develop as shown in Figs. 8 and 10 through the stages wave → tongue → cutoff vortex. Since the air aloft flows rapidly through these quasi-stationary long-wave patterns, isotherms will approach coincidence with the streamlines and isobars. Thus, equatorward tongues and cutoffs will be cold and will tend to coincide with lows (at least aloft); the poleward ones will tend to be warm and coincide with highs. The former will contain a polar-air hourglass (at least in higher latitudes) or dome, possibly with an arctic-air dome inside it (Fig. 9). Correspondingly, the poleward tongues and cutoff vortices will contain tropical air and a much higher tropopause.

2. The secondary waves are the short waves $(1000 < \lambda < 3000$ km). The short waves in a front↔jet system, when unstable, will form secondary circulations, developing according to the scheme of Fig. 7, that is, initial front wave → young cyclone → initial occlusion → backbent occlusion. Important secondary circulations also form outside the front ↔ jet systems: the easterly wave, the tropical hurricane, and the convective system, occurring primarily in lower latitudes, the last two without an obvious preceding wave stage.

3. The tertiary circulations, barely observable by the ordinary synoptic network, require a mesoscale network $(d < 10$ km), time sections, or both, for a detailed study and forecast. Weather mechanisms of this size are land and sea breezes, mountain and valley winds, katabatic and other local winds (foehn, chinook, bora), local showers, tornadoes, orographic cloud and precipitation systems, and lee waves. Moreover, numerous different orographic factors and the daily period of radiation affect most meteorological elements and weather mechanisms. Their detailed study is therefore basic for both climatology and weather forecasting.

Life cycle of extratropical cyclones. An extratropical cyclogenesis starts as a wavelike front bulge. Ahead of the wave the front becomes a warm front, behind it a cold front. These two sections then flank a warm tongue (Fig. 7*b*) moving along the front. At first there is only a shallow low around the tip of this warm sector, but even the passage of such an initial frontal wave may cause sudden, severe, and unexpected weather changes. If the front wave has an appropriate size $(\lambda \sim 1500$ km, amplitude ≥ 200 km), it will usually be unstable, narrow gradually, and at last overshoot as does a breaking sea wave. The cold front overtakes the warm front at the ground, and the warm-sector air is lifted and spreads aloft. As a result, the common center of gravity of the system sinks, and potential energy is transformed into the kinetic energy of an increasing cyclonic circulation. As long as this front occluding continues, the circulation increases, and the ensuing low deepens, whereas when the warm sector has disappeared from the interior of the main cyclone, the latter will weaken, and the low fill. These features serve as good, physically comprehensible, prognostic rules. The occluded front is called an occlusion. The occluding process gives the clue to the life history of cyclones and anticyclones (Fig. 7). At its last phase the occluded front often lags behind and bends (Fig. 7d and e), a "false warm sector" forming between the direct and the backbent occlusion. Several cyclones (and lows), together constituting a cyclone series, may form on one and the same front, moving with the upper, steering current, the first one being in the most advanced stage of development (Fig. 8).

Important front↔jet systems in the Northern Hemisphere are the North Pacific polar front and the North Atlantic polar front (Fig. 3). The former extends in winter on the average from near the Philippines north of Hawaii to the northwest coast of the United States; the cyclone series forming in it then brings wet and unsteady weather to the northwestern part of North America. The latter front, extending in winter on an average from near the Bermudas to England, plays the analogous role for almost all Europe, and during certain periods for eastern United States. In summer both these

systems are much weaker and lie on an average farther north. The North American arctic front and the European arctic front (Fig. 3) are fundamental for the weather in their vicinity. The former accounts for many severe weather vagaries, such as the blizzards and glaze storms of northeastern United States, and the most severe cold waves and their killing frosts farther south; it is thus of special importance to North American weather forecasting. The production of real arctic air as defined here seems to cease in summer, or at least in July. On the other hand, the corresponding air mass in the Southern Hemisphere, antarctic air, and the antarctic fronts, exist throughout the year, the antarctic ice plateau being an enormous cold source even in summer. Evidently most migrating extratropical precipitation and storm areas originate at front↔jet systems, separating air masses of radically different motion and weather type. Therefore, these systems are of utmost importance to weather forecasting, being the real atmospheric zones of danger and main sources of the salient aperiodic weather changes of synoptic size outside the tropics.

Large-scale tropical weather systems. In the tropics (and in the subtropics in summer) nonfrontal convective and other weather phenomena may grow to hundreds of kilometers in width; consequently, they can be studied and forecast individually by synoptic methods. Both on land and at sea, short waves (λ ~ 500 km) form in the frontless trade winds outside the doldrums. These easterly waves propagate slowly westward, showing a characteristic intensification of the shower activity in their eastern part.

Over tropical seas in late summer and early fall (Northern Hemisphere, August–October; Southern Hemisphere, February–April), conditions exist that favor the formation of tropical hurricanes. Tropical hurricanes are cyclonic vortices (smaller but much more intense than the extratropical ones) of 100–400-km width; they have wind velocities often exceeding 50 m/sec (100 knots), 100–500 mm total rainfall, and very low central pressure. They move on the whole poleward, often recurring around a subtropic high. Necessary conditions for their formation seem to be (1) air (and sea-surface) temperature above, say, 27°C, implying enough lability energy to drive such large-scale convections, (2) a preexisting cyclonic motion and frictional inflow (either at the tropical front or in an easterly wave), needed to order, and possibly trigger, the convections, (3) a divergence mechanism aloft to dispose of the air that converges and rises in their interior, possibly also triggering their formation, (4) sufficient Coriolis force (that is, the hurricanes cannot form too near the Equator), and (5) no disturbing land surface within the area of formation. Whether these conditions are sufficient is not certain. The widespread destructive power of hurricanes (shore flooding, wind pressure, downpour) makes their study a major task of tropical synoptic meteorology and weather forecasting. Tracking those already formed can be done with radar, reconnaissance flights, and satellites, but discovering their imminent formation is still an unsolved problem. *See* HURRICANE.

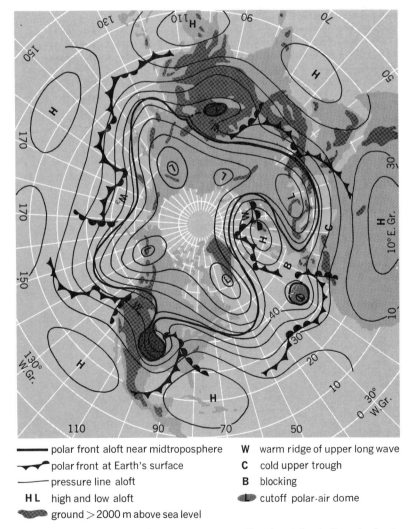

———— polar front aloft near midtroposphere   W   warm ridge of upper long wave

⌄⌄⌄ polar front at Earth's surface   C   cold upper trough

———— pressure line aloft   B   blocking

H L   high and low aloft   ◖ cutoff polar-air dome

▰▰▰ ground > 2000 m above sea level

Fig. 8. Schematic circumpolar upper pressure pattern in relation to the polar fronts at the Earth's surface. (*After E. Palmén*)

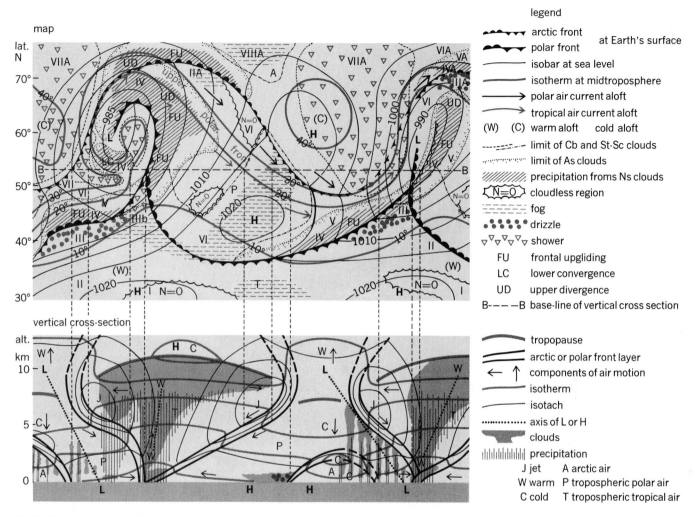

legend

arctic front ⎫
polar front ⎬ at Earth's surface
isobar at sea level
isotherm at midtroposphere
⟶ polar air current aloft
⟶ tropical air current aloft
(W)   (C) warm aloft   cold aloft
------- limit of Cb and St-Sc clouds
.... limit of As clouds
//////// precipitation froms Ns clouds
⟨N=0⟩ cloudless region
≈≈≈ fog
•••• drizzle
▽▽▽▽ shower
FU frontal upgliding
LC lower convergence
UD upper divergence
B———B base-line of vertical cross section

tropopause
arctic or polar front layer
← ↑ components of air motion
isotherm
isotach
.... axis of L or H
clouds
precipitation
J jet   A arctic air
W warm   P tropospheric polar air
C cold   T tropospheric tropical air

Fig. 9.   Model of middle-latitude front↔jet disturbances and their weather regions (fall and spring).

Over land, conditions 1 and 2 never lead to hurricane formation but may instead—within the tropics and also in warmer seasons farther poleward—favor the formation of a migrating convective system, with a forerunning pseudo cold front (in the United States also called a squall line) at its outer edge; these systems have roughly the same extent and precipitative power as a tropical hurricane. Outside the tropics they mainly form in the warmer seasons in cyclonic warm sectors, especially in the midwestern United States, where they provide the main water supply. Because of the flood hazards, soil erosion, and other aspects, and the sudden violent squalls, or even tornadoes, sometimes attending the pseudo cold front, these systems, therefore, form another major problem of weather forecasting. Within the tropics, migrating convective systems, often in the easterly waves, will cause more variation of weather from day to day than is generally recognized, whereas the small-scale showers usually have a daily period; together they constitute the equatorial rains.

Apart from the polar-front jet (meandering between 30 and 70° lat), there is a subtropical jet, and a corresponding front, at about 30° lat near the tropopause level (~15 km altitude). As a rule, it makes only small meridional excursions. Therefore, it shows up (instead of the polar-front jet) as the main wind maximum aloft (at 30° lat) in the average meridional cross section of Fig. 2. But for the same reason, and because its front is confined to the tropopause, its influence on daily weather, at least in the tropics, seems slow and diffuse.

Circumpolar aerology. Since 1940, the technical facilities of meteorology have undergone an explosive development, partly due to the great exigencies during World War II. The vast gaps formerly in the meteorological network over the oceans are partly bridged by stationary weather ships, or ocean weather stations, with complete air-sounding equipment. This huge technical improvement has been followed by equally outstanding scientific achievements in the understanding of the dynamics and thermodynamics of the free atmosphere. The weather mechanisms and weather processes described above, which before 1940 had been mainly observed and studied within the lower half of the troposphere, can now be treated in their entirety and fitted into their general relationships with the rest of the atmosphere. On the whole, an intimate interaction takes place between the disturbances seen on the ordinary

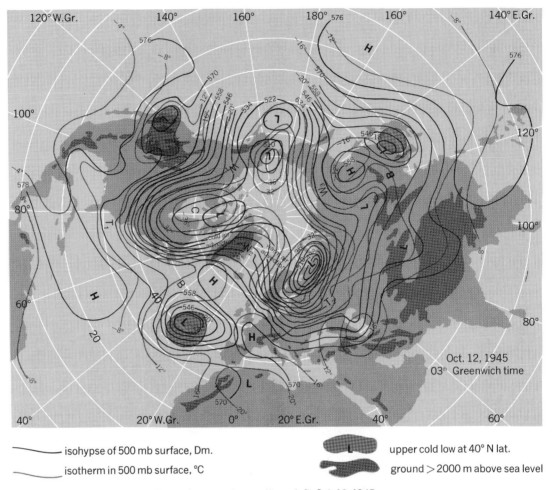

Fig. 10. Circumpolar pressure, flow and temperature pattern aloft, Oct. 12, 1945.

synoptic map (surface map) and the upper large-scale patterns shown by Fig. 8 and described above. Figure 9 shows this interdependence three dimensionally and in detail for disturbances of front↔jet systems (including the arctic front). Particularly, it shows the eight weather regions that are conditioned by such systems, these naturally being of fundamental importance to synoptic meteorology and weather forecasting.

In winter, the polar-front jet lies on an average at 25–45° lat, in summer at 40–60° lat, at 7–11 km alt; it is only a few hundred kilometers broad, with wind speed surpassing 100 m/sec (200 knots). It displays two fundamentally different patterns: the zonal or high index type, where the jet has waves of only small amplitudes; and the meridional or low index type, where its meridional excursions are huge and often are cut off. Figure 10 gives an example of the isothermal, isobaric, and flow patterns at the midtroposphere, represented by the isotherms and isohypses ("contour lines") of the 500-mb surface. It shows three long waves w_1 w_2 w_3, one deep trough T, and four cutoff lows $L_1 L_2 L_3 L_4$. Where the isotherms diverge markedly from the parallelism with the streamlines and "contours," there will be a considerable horizontal advection of colder and denser air or of warmer air (Fig. 10), and ensuing advective changes of pres-

sure and wind at most levels. These observations gave another powerful tool for quantitatively forecasting the thermal changes of flow patterns. Calculating the vertical motion by the divergence and vorticity equations has proved even better for this and similar purposes (as discussed below in dynamical meteorology).

Since the planetary waves represent a rather stable steady state, they offer no real forecast problem. The shorter long waves will move east, conserving most of their absolute vorticity, and may therefore be forecast numerically, using a very simplified general model of the atmosphere, the barotropic model, where production and annihilation of kinetic energy is excluded. The propagation and development of the short front↔jet waves, that is, the weather mechanisms, will to a certain extent be steered by this long-wave pattern. However, in these mechanisms frontal potential energy is transformed into cyclonic kinetic energy, and the short jet waves will often react upon the large-scale pattern. Such processes, which may lead to intense frontal cyclogenesis and the formation of large-scale cold tongues and cutoffs, confront numerical forecasting with considerable difficulties. However, the fact that the upper steering flow patterns are so simple and move much more slowly than the lower ones has proved a great help to short-range forecasting and has made an extended forecasting (2–5 days) possible on a synoptic-physical basis. Attempts are being made, with some success, to use baroclinic models of the atmosphere as well, and to incorporate the effect of heat sources and large-scale orography into the numerical short-range and extended forecasting. A main obstacle to real long-range forecasting (7–30 days) is the fact that the causes for a definitive change from the zonal to the meridional upper flow pattern are not yet known.

Aided by radar, continuous detailed mapping and tracking of rain areas is gradually being introduced. Analogously, photographic and television cloud surveys from satellites are used in studying the large-scale development of weather regions. *See* METEOROLOGICAL SATELLITES; RADAR METEOROLOGY. [TOR BERGERON]

DYNAMICAL METEOROLOGY

This branch of meteorology is the science of naturally produced motions in the atmosphere and of the related distributions in space of pressure, density, temperature, and humidity. Based on thermodynamic and hydrodynamic theories, it forms the main scientific basis of weather forecasting and climatology. *See* CLIMATOLOGY; WEATHER FORECASTING AND PREDICTION.

Thermodynamic properties of air. Atmospheric air contains water vapor and sometimes suspended water droplets or ice particles or both. In dynamical meteorology it suffices to use a simplified cloud physics; thus it may be assumed that the water vapor pressure cannot exceed the saturation pressure over a plane water surface and that the vapor pressure equals this saturation pressure whenever the air contains water (cloud) droplets. The distinction between water and ice is ignored most of the time.

Nonsaturated air. Nonsaturated air contains no condensation products and behaves nearly as a single ideal gas. Its pressure p, density ρ, and absolute temperature T satisfy very closely gas equation (1). Its enthalpy per unit mass h is

$$\frac{p}{\rho} = RT \qquad (1)$$

given by Eq. (2), and its entropy per unit mass may be written $s = c_p \ln \theta$. The potential temperature θ is given by Eq. (3), where $p_0 = 1000$ mb. In these

$$h = c_p T + \text{constant} \qquad (2)$$

$$\theta = T\left(\frac{p_0}{p}\right)^{R/c_p} \qquad (3)$$

expressions the gas constant per unit mass R and the specific heat at constant pressure c_p both depend slightly on the specific humidity or water-vapor mass concentration m. It is often sufficiently accurate to ignore this dependence and to use the values which hold for dry air; these are $R = 287$ m^2 sec^{-2}C$^{\circ -1}$ and $c_p = 1004$ m^2 sec^{-2}C$^{\circ -1}$.

Saturated air. A sample of nonsaturated air becomes saturated if its pressure is reduced or heat is removed or both while its water-vapor concentration m is kept constant. If this process continues, cloud droplets form by condensation at a rate just sufficient to maintain the vapor pressure at the saturation value. If the process is reversed, evaporation takes place. During such phase transitions the release or binding of latent heat causes temperature changes which are important for many motion processes. This property of saturated air is expressed in the formula for its enthalpy per unit mass, Eq. (4). Here L is the latent heat, and

$$h = c_p T + L m_s(p,T) + \text{constant} \qquad (4)$$

m_s is the mass concentration of saturated water vapor, which is a known function of p and T. On the other hand, the liquid (or solid) phase contributes very little to the total mass or volume of the air, so that the gas equation (1) holds with good approximation also for saturated air. Note, however, that the approximate equations (1) and (4) are not fully consistent with the second law of thermodynamics.

Thermodynamic state. The thermodynamic state of an air sample, whether saturated or not, is defined by three state variables, for example, pressure p, density ρ, and mass concentration m of the water component (in all phases). It is often convenient to use T instead of ρ. Comparison between m and the saturation vapor concentration $m_s(p,T)$ shows whether the sample is saturated, and if so, the mass of the condensed phase.

There are many motion phenomena in which the released latent heat is unimportant. When dealing with such phenomena, the air may be considered dry and its thermodynamical state as defined by the two variables p and ρ (or T) only.

Instantaneous state of atmosphere. The instantaneous state of the atmosphere may be characterized by the distributions in space of the thermodynamic variables p, ρ, and possibly m and three components of the air velocity v relative to the solid Earth. These basic quantities are regarded as functions of three space coordinates and time, and as such they characterize a sequence of states or a motion process in the atmosphere.

Pressure and density are free to vary independently in space, and the density is usually variable in isobaric surfaces of constant pressure; the density field is then said to be baroclinic. Only in special cases is the density field barotropic, that is, constant in each isobaric surface.

A basic difficulty in dynamical meteorology results from the complexity of the field of motion in the real atmosphere. Superimposed upon the large-scale motion systems revealed by weather maps is a fine structure of motions of all scales down to small eddies of millimeter size. Such motion fields cannot be dealt with in all details; it is necessary to deal instead with smoothed motion fields, where details smaller than a certain scale have been left out. As a consequence of the nonlinearity of the basic equations, these details still exert a certain influence upon the larger-scale motions. This influence can be taken into account only in a statistical sense, in the form of so-called eddy terms which express the transport of momentum, enthalpy, and water substance by small-scale eddies.

Equations. The dependent variables p, ρ, m, and $\mathbf{v}$ satisfy Eqs. (5)–(8), which express, respectively,

$$\frac{\partial \rho}{\partial t} + \operatorname{div}(\rho \mathbf{v}) = 0 \tag{5}$$

$$\rho \frac{D\mathbf{v}}{Dt} = -\mathbf{k} \times f\rho \mathbf{v} - \rho g \mathbf{k} - \operatorname{grad} p - \frac{\partial \mathbf{F}_M}{\partial z} \tag{6}$$

$$\rho \frac{Dh}{Dt} - \frac{Dp}{Dt} = -\frac{\partial}{\partial z}(F_{\text{rad}} + F_h) + \Delta \tag{7}$$

$$\rho \frac{Dm}{Dt} = -\frac{\partial}{\partial z}(F_{\text{precip}} + F_m) \tag{8}$$

the conservation of total mass (continuity equation), Newton's second law (equation of motion), the thermodynamic energy equation (or first law), and the conservation of mass of the water component.

Here Eqs. (9) and (10) apply, and the notation

$$f = 2\Omega \sin \Phi \tag{9}$$

$$\frac{D}{Dt} = \frac{\partial}{\partial t} + \mathbf{v} \cdot \operatorname{grad} \tag{10}$$

is as follows: t denotes time; D/Dt individual time derivative or rate of change as experienced by a moving air particle; $\mathbf{k}$ a vertical unit vector; f Coriolis parameter; Φ latitude; $\Omega = 0.7292\ 10^{-4}$ sec^{-1} angular velocity of the Earth's rotation; g acceleration of gravity; $\mathbf{F}_M$ vertical eddy-flux density of momentum; F_{rad} vertical radiative-heat-flux density; F_h vertical eddy-flux density of enthalpy; F_{precip} vertical flux density of water substance by precipitation; F_m vertical eddy-flux density of water substance; and Δ viscous dissipation. To these equations one must add boundary conditions, expressing the kinematic constraint at the Earth's surface, and the various fluxes into the atmosphere from below and above.

In Eqs. (6)–(8) horizontal flux densities (that is, fluxes through vertical surfaces) have been neglected. Moreover, in Eq. (6) the expression for the Coriolis force has been simplified by including only the effect of the vertical component of the Earth's rotation ($\Omega \sin \Phi$). These simplifications are common in dynamical meteorology.

In order that Eqs. (5)–(8) with boundary and initial conditions constitute a closed system of equations, it is necessary that $\mathbf{F}_M$, F_{rad}, F_h Δ, F_{precip}, and F_m be expressible in terms of the dependent variables. The radiative flux F_{rad} depends upon the distribution of temperature, water vapor, cloud, and the properties of the underlying surface; it can be calculated quite accurately when these quantities are known, although the calculation is time-consuming. A major difficulty is the determination of eddy-flux densities $\mathbf{F}_M$, F_h, and F_m. It is usually assumed that a major part of these fluxes is contributed by turbulence in the atmospheric boundary layer, and this part of the fluxes can be estimated from turbulence theory. Thus the last term of Eq. (6) represents the force of turbulent friction. The calculation of F_{precip} presents another difficulty, since it depends strongly on small-scale eddy motions and microprocesses within the clouds.

Note that Eqs. (5)–(8) are nonlinear, since the operator Eq. (10) produces product terms when applied to the dependent variables. Nonlinearity may also enter in the relations between eddy fluxes and dependent variables, and in the relation between condensation heat and dependent variables.

Equations (5)–(8) can be treated either analytically or numerically, and one can distinguish between analytical and numerical dynamic meteorology.

Analytical dynamical meteorology. It is not possible to find analytical solutions of Eqs. (5)–(8) which represent motion processes of the composite kind found in the real atmosphere. Instead, one deals with various kinds of idealized motion systems, for which simplifications of the equations can be justified on the basis of scale analysis and simplified geometry.

Motion without heat sources and friction. A common simplification consists in ignoring all heat sources, including condensation heat, so that Eq. (7) may be integrated to give Eq. (11) for isen-

$$\theta = \text{constant in time for each air particle} \tag{11}$$

tropic motion. From Eq. (3) there is then for each particle a relation between p and ρ or T; such a fluid is said to be piezotropic. Another simplification consists in ignoring the force of eddy friction ($-\partial \mathbf{F}_M/\partial z$) in Eq. (6). The bulk of the work in analytical dynamical meteorology has been done under the assumption that heat sources and friction are absent.

Equilibrium. In the absence of friction and heat sources and disregarding the slight centripetal acceleration which exists because an air current is bound to follow the Earth's curvature, Eqs. (5)–(8) permit a particularly simple solution which represents a steady, straight, horizontal current with parallel streamlines. With vanishing particle accelerations, Eq. (6) expresses balance of forces. In the vertical direction there is balance between gravity and vertical pressure force. This hydrostatic equilibrium is expressed by Eq. (12),

$$-g\rho - \frac{\partial p}{\partial z} = 0 \tag{12}$$

where z represents height. In the horizontal direction the Coriolis force must balance the horizontal pressure force $\operatorname{grad}_h p$. Thus this geostrophic equilibrium can be expressed by Eq. (13).

$$-\mathbf{k} \times f\rho \mathbf{v} - \operatorname{grad}_h p = 0 \tag{13}$$

Ignoring the Earth's curvature, one may place a cartesian coordinate system with the z axis pointing upward and the x axis along the horizontal velocity u. Then Eq. (13) gives Eq. (14). Thus the

$$u = -\frac{1}{f\rho}\frac{\partial \rho}{\partial y} \tag{14}$$

geostrophic wind velocity is directed along the isobars (lines of constant pressure) in level surfaces, with pressure increasing to the right (or left) in the Northern (or Southern) Hemisphere (Fig. 11). At the Equator $f = 0$, and the formula breaks down. The dependent variables u, p, and ρ

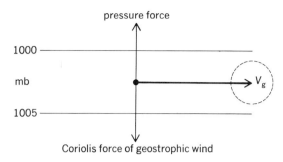

Fig. 11. The geostrophic balance. (*From S. Petterssen, Introduction to Meteorology, 3d ed., McGraw-Hill, 1969*)

may depend upon y and z, but not on x if the current is steady. From Eqs. (12) and (14) one may derive Eq. (15), showing that the rate of change of u

$$\frac{\partial u}{\partial z} = \frac{g}{f}\frac{1}{\rho}\left(\frac{\partial \rho}{\partial y}\right)_p = -\frac{g}{f}\frac{1}{T}\left(\frac{\partial T}{\partial y}\right)_p \qquad (15)$$

with height is proportional to the baroclinicity (the subscript p indicates derivative in a surface of constant pressure). Thus in a barotropic current the velocity is constant with height; in particular, a state of persistent rest is necessarily barotropic.

Atmospheric boundary layer. Next to the Earth's surface is a turbulent boundary layer approximately 1 km thick, inside which the force of turbulent friction $(-\partial \mathbf{F}_M/\partial Z)$ is strong enough to disturb significantly the geostrophic equilibrium. The vertical eddy-flux density of momentum $\mathbf{F}_M$ is given by Eq. (16), where the coefficient of eddy

$$\mathbf{F}_M = -K_M \partial \mathbf{v}_h/\partial z \qquad (16)$$

viscosity K_M depends upon the intensity of the turbulence. In the lowest 20–30 m, K_M increases rapidly with height, and the wind velocity is typically a logarithmic function of height; this wind profile is modified if the eddy flux of heat F_h is numerically large.

Overlaying this shallow layer is a much deeper layer, in which K_M changes less with height. Assuming balance between the Coriolis force, the horizontal pressure force, and turbulent friction, one obtains for constant K_M the Ekman spiral, which represents the variation of wind velocity with height (Fig. 12). As z increases, the wind approaches the geostrophic value.

Stability. A steady state is termed stable if any small disturbance remains small indefinitely and unstable if a significant deviation from the steady state can develop from an initial disturbance, however small. In this definition it is usually understood that the motion is isentropic, as in Eq. (11).

For a state which deviates only slightly from a known steady state, the basic nonlinear equations (5)–(8) turn into a set of linear perturbation equations in the small perturbation quantities which define the deviations of the variables $(\mathbf{v}, p, \rho)$ from their steady-state values. These linear equations can be treated analytically; they have solutions representing various kinds of wave motions, whose amplitudes may either remain small or grow indefinitely. Existence or nonexistence of growing wave solutions is assumed to correspond to in-

stability or stability of the steady state, respectively. Several types of stability may be identified.

Static stability. When in an atmosphere in equilibrium a nonsaturated air particle is displaced vertically with unchanged potential temperature $(D\theta = 0)$, its temperature changes at a rate $DT/Dz = -0.01\,°\text{C } m^{-1}$ (the dry-adiabatic lapse rate). Therefore, if the distribution of temperature with height in the atmosphere is such that at all levels $\partial\theta/\partial z > 0$ or $\partial T/\partial z > -0.01\,°\text{C } m^{-1}$, then a particle displaced upward (or downward) becomes colder (or warmer) than its environment, and the sum of the buoyancy and the weight acting on the particle represents a net restoring force. The state is then said to be statically stable. A rigorous proof of the stability can be given by showing that the sum of potential and internal energy of a bounded part of the atmosphere is a minimum in the state of rest if $\partial\theta/\partial z > 0$.

If $\partial\theta/\partial z < 0$ (or $\partial T/\partial z < -0.01\,°\text{C } m^{-1}$) at some levels, then the state is statically unstable. Saturated air is less stable than nonsaturated air with the same temperature distribution, because the temperature of a saturated particle which is displaced vertically changes at a rate which is slower than the dry-adiabatic lapse rate, the rate depending upon p and t.

Static instability is frequently caused by heating the air from the underlying surface and produces convection currents, whose horizontal scale is of the same order as the depth of the unstable layer. Above a certain condensation level, condensation takes place in the rising currents which then become visible as cumuliform clouds.

Inertial stability. In a steady current satisfying Eqs. (12) and (14), inertial stability results from excessive Coriolis forces acting on a chain of particles parallel to the current which have been displaced along the isentropic surface normal to the current. The criterion of stability, valid in both hemispheres, is $f[f - (\partial u/\partial y)_\theta] > 0$, where the subscript θ indicates that the derivative is taken in an isentropic surface $(\theta = \text{constant})$ and f is reckoned positive in the Northern, negative in the Southern Hemisphere.

Shearing instability. If the vertical shear has a sufficiently pronounced maximum at some level, shearing instability results. The layer of maximum shear breaks up into a system of equally spaced vortices with horizontal axes normal to the flow.

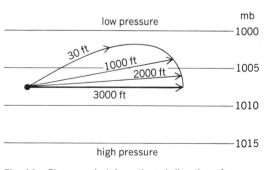
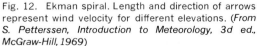

Fig. 12. Ekman spiral. Length and direction of arrows represent wind velocity for different elevations. (*From S. Petterssen, Introduction to Meteorology, 3d ed., McGraw-Hill, 1969*)

Condensation may occur in the upper part of the vortices and form what is known as billow clouds.

Wave motions. In a statically and inertially stable atmosphere without shearing instability there is a spectrum of possible wave motions, ranging from high-frequency acoustic waves through gravity waves at intermediate frequencies to inertia-gravity waves at frequencies comparable with the Coriolis parameter *f*. Typical gravity waves have orbital periods which are short compared with a pendulum day (12 hr/sin Φ) and can therefore be studied without taking the Earth's rotation into account. Their frequency cannot exceed the Väisälä-Brunt frequency $[(g/\theta)\partial\theta/\partial z]^{1/2}$, which in the troposphere mostly corresponds to a minimum orbital period of about 8 min.

Organized stationary gravity wave motions occur in the atmosphere when a stably stratified air current blows over hilly terrain. Their properties can be explained quite well from linear theory. The terrain corrugations act to draw kinetic energy from the current and transform it into wave energy, and the process may be looked upon as a radiation of wave energy upward and horizontally downstream. Depending upon the wavelength and the distributions of temperature and wind with height, the waves may either be transmitted to very high levels, absorbed, or trapped at low levels. Condensation may occur in the wave crests making them visible as stationary clouds, even on satellite photographs; their spacing is mostly 10–50 km. Such waves cause a drag (wave resistance) on the mountains and a corresponding braking force on the airstream.

Organized long gravity-inertia waves occur as forced tidal waves, the predominant forcing effect being the diurnal heating and cooling. To some extent gravity-inertia waves are also generated, as disorganized noise, by nonlinear interaction with other motion types.

Kinematics of motions in atmosphere. Apart from sound waves, which may be treated as a separate phenomenon, all motions in the atmosphere have the character of circulations along closed streamlines. The distribution of horizontal velocity $\mathbf{v}_h$ in a horizontal surface may be characterized by the fields of two scalar quantities, namely, the relative vertical vorticity $\zeta = \mathbf{k} \cdot \text{curl } \mathbf{v}$ and the horizontal divergence $\delta = \text{div}_h \mathbf{v}_h$. The former represents twice the angular velocity of an air particle around a vertical axis relative to the solid Earth (reckoned positive when counterclockwise). Since the solid Earth rotates around the local vertical with the angular velocity $f/2$, it follows that the absolute vorticity of the air particle relative to a nonrotating frame is $f + \zeta$. The horizontal divergence δ represents the relative rate of expansion of an infinitesimal horizontal area moving with the air.

The field of $\mathbf{v}_h$ in a horizontal surface may be broken up into one field $\mathbf{v}_\zeta$ which carries all of the vorticity but no divergence, and another field $\mathbf{v}_\delta$ which carries all of the divergence but no vorticity. This is expressed by Eqs. (17).

$$\mathbf{v}_h = \mathbf{v}_\zeta + \mathbf{v}_\delta \qquad \zeta = \mathbf{k} \cdot \text{curl } \mathbf{v}_\zeta \qquad \delta = \text{div}_h \mathbf{v}_\delta \quad (17)$$

The continuity equation may be written approximately as Eq. (18), where ρ^* represents a standard

$$\rho^*\delta + \frac{\partial(\rho^* w)}{\partial z} = 0 \qquad (18)$$

density distribution, depending upon z only. It follows that the horizontal field $\mathbf{v}_\delta$ together with the vertical velocity w forms a system of circulations in the vertical, whereas $\mathbf{v}_\zeta$ represents a horizontal circulatory motion. These two parts of the motion field behave very differently; for an individual air particle, δ may change quickly as a result of divergent pressure forces, whereas ζ changes more slowly. For this reason the motions in the atmosphere may be classified into two categories: predominantly vertical circulations, in which $\mathbf{v}_\delta$ is the dominating component, and $\mathbf{v}_\zeta$ is of secondary importance; and predominantly horizontal circulations, in which $\mathbf{v}_\zeta$ is the dominating component, and $\mathbf{v}_\delta$ is secondary. All types of gravity-inertia waves and small-scale convection currents belong to the first category; in the atmosphere these motions generally have small amplitudes and carry little energy, and they are mostly too weak or of too small scale to show up on weather maps. The large-scale motion systems revealed by weather maps, which contain the bulk of the energy and are the carriers of weather systems, are all predominantly horizontal circulations and thus belong to the second category above.

Quasi-static approximation. For motion systems whose vertical scale is small compared to the horizontal scale, the vertical motion is weak and its momentum can be neglected. Disregarding also vertical eddy friction, the hydrostatic equation, Eq. (12), holds everywhere and replaces the vertical component of Eq. (6). In the atmosphere, where the vertical scale of motion systems is limited by the high static stability of the stratosphere, this quasi-static approximation applies to motion systems of horizontal scale larger than about 200 km.

Large-scale quasi-horizontal circulations. Equations (5)–(8) apply equally well to motion systems of all types. In the study of the important large-scale quasi-horizontal circulations, it is convenient to derive equations which apply specifically to such motions, whereas predominantly vertical circulations have been eliminated at the outset. This is achieved by a filtering approximation: The horizontal momentum due to the velocity component $\mathbf{v}_\delta$ in Eqs. (17) is neglected; that is, the horizontal acceleration is approximated by $D\mathbf{v}_\zeta/Dt$. It follows that the horizontal divergence of the forces on the right-hand side of Eq. (6) must balance. This gives a relation between $\mathbf{v}_\zeta$ and the pressure field, which for small Rossby numbers U/fL (U is characteristic velocity and L is horizontal scale) reduces to the geostrophic relationship, Eq. (13). In extratropical latitudes the Rossby number of the large-scale quasi-horizontal circulations is of the order 10^{-1}, and these motions are therefore also quasi-geostrophic, within 10–20% error.

The evolution of the velocity field is approximately determined by the vorticity equation, which is obtained from Eq. (6) by taking the vertical vorticity. In a somewhat simplified form, Eq. (19),

$$D(f + \zeta)/Dt = -(f + \zeta)\delta \qquad (19)$$

it expresses the familiar mechanical principle that

the absolute rotation of an air particle around a vertical axis speeds up when the particle contracts horizontally.

In a current moving up a mountain slope, air columns must shrink vertically and hence, by Eq. (18), diverge horizontally ($\delta > 0$); from Eq. (19) it follows that the absolute vorticity ($f + \zeta$) must decrease numerically. The opposite process takes place when the motion is downslope. This explains the observed predominance of anticyclonic vorticity ($\zeta < 0$ in the Northern Hemisphere) over large mountain ranges which are crossed by air currents.

Over level country there is no such net vertical stretching or shrinking through the whole atmospheric column; therefore, in the mean for the column there is a tendency toward conservation of absolute vorticity. This tendency shows up at the middle level of the atmosphere about 5 km above sea level, where in a crude approximation Eq. (20)

$$f + \zeta = \text{constant in time for each particle} \quad (20)$$

holds (the barotropic model). Since f increases northward, Eq. (20) requires ζ to decrease for particles moving northward and increase for particles moving southward. As a consequence, large-scale motion systems have a tendency to move westward relative to the air.

The barotropic model has been used to study the properties of a broad zonal current, such as the westerlies in middle latitudes. It has been found that barotropic instability (a kind of shearing instability) occurs if the absolute vorticity of the undisturbed current has a pronounced maximum at some latitude; in that case large-scale vortices may grow spontaneously, feeding upon the kinetic energy of the zonal current. If the zonal current does not possess a vorticity maximum, the current is barotropically stable. In the case of a uniform current, perturbations propagate westward relative to the air as permanent Rossby waves. Such waves may be studied approximately in a cartesian coordinate system, where f increases with y (northward) at a rate $\beta = df/dy$. A wave of wavelength L, superimposed upon a uniform westerly current u, will be propagated at a speed c, satisfying the Rossby formula, Eq. (21).

$$c = u - \frac{\beta}{4\pi^2} L^2 \quad (21)$$

In spherical coordinates one obtains instead the more accurate Rossby-Haurwitz formula, Eq. (22),

$$\gamma_n = \omega - \frac{2(\omega + \Omega)}{n(n+1)} \quad (22)$$

which gives the angular velocity of propagation around the Earth's axis (γ_n) for a horizontal motion system where the stream function is a spherical harmonic of the order n (with any longitudinal wave number), assuming that the atmosphere has a solid rotation ω relative to the solid Earth. Some of the large-scale motion systems of the atmosphere are of this kind, and there is a satisfactory agreement between their observed propagation and Eq. (22).

The barotropic model describes only a very special mode of motion; for instance, it cannot describe conversion of potential and internal energy

into kinetic energy or vice versa. A more general use of the filtered equations has revealed other types of large-scale motion systems. In the case of a zonal current which is sufficiently baroclinic, the linearized perturbation equations have solutions representing growing disturbances, which are similar to extratropical cyclones with respect to size, growth rate, and structure. Such a zonal current is said to possess baroclinic instability, and it is characteristic that the growing disturbances feed on the potential and internal energy of the zonal current.

Extratropical cyclones are observed to form in the zone of middle-latitude westerlies, characterized by pronounced baroclinicity and also by a strong maximum of cyclonic wind shear. Baroclinic and barotropic instability may both operate to cause the sudden growth of cyclones that is often observed.

Numerical dynamic meteorology. High-speed electronic computers have made it possible to find approximate numerical solutions to equation systems which are untractable by analytical methods. This has opened up a new branch of dynamical meteorology. It has become possible to drop many of the simplifying assumptions which were necessary for the analytical treatment; in particular, it is no longer necessary to linearize the equations.

Weather prediction. Starting from an initial state of the atmosphere which is known from observations, later states may be calculated from Eqs. (5)–(8) by numerical integration in time. To achieve this, the fields are represented by their values in a grid of points in space, and the differential equations are approximated by finite difference equations; thus one obtains a numerical model of the atmosphere. The total amount of computation for a given integration period depends on the approximation used in Eqs. (5)–(8), on the spacing of grid points in horizontal and vertical direction, and on the area over which the computation is extended. The quasi-static approximation is always used; the filtering approximation is used in some cases, but the equations may also be applied in nonfiltered, "primitive" form.

For use in practical weather forecasting, it is necessary that the computation proceed considerably faster into the future than real weather. This strongly restricts the size of the numerical model and makes considerable simplifications necessary, depending upon the available computer capacity.

The simplest possible model is the barotropic model, which is obtained when only one grid point is used along each vertical, at the 500-mb level. The set of governing equations then reduces to Eq. (20) or a slight modification thereof, and this equation may be expressed in terms of a stream function as the single dependent variable; all other physical processes are ignored. Although simplified to the extreme, the barotropic model can give predictions of practical value up to about 3 days ahead.

By increasing the number of grid points along the vertical, one obtains baroclinic models of increasing fidelity, requiring increasing amounts of computer time. A simplification which is frequently made in such models designed for weather prediction consists in neglecting friction and heat sources (also latent heat) and omitting humidity as a variable. The error due to this simplification grows relatively slowly with the prediction period, so that predictions of value may be obtained up to about 4 days ahead. Although humidity has been omitted, such models can still be used to predict precipitation, because this is known to occur wherever there is ascending motion.

Influences in the atmosphere propagate at the speed of sound and reach one from the remotest parts of the globe in less than 20 hr. Strictly speaking, the integration should therefore be extended to the entire atmosphere, even in a prediction for only 1 day ahead. This is prohibitive in practice, partly because the amount of computation would be too large, and partly because data which define the initial state of the entire atmosphere are not yet available. Fortunately, the bulk of the influences do not propagate at sound speed, and integrations of value can be carried out for a limited part of the world.

Integrations in time carried out with baroclinic numerical models of the atmosphere have confirmed and extended many of the results of analytical dynamic meteorology, such as the growth of cyclonelike disturbances in a baroclinically unstable zonal current and the formation of the associated frontal systems.

General circulation of atmosphere. The motions in the atmosphere are caused by solar heat. The combined process of absorption and emission of radiation causes a distribution of heat and cold sources which continually disturbs the equilibrium and maintains the motion. During this process the volume and pressure of individual air particles change with time, so that expansion takes place on the average at higher pressure than contraction. As a result, heat is converted into mechanical energy, just as in a man-made heat engine. The bulk of the mechanical energy thus released is used to overcome friction within the atmosphere itself; only a very small fraction is spent to do work on the ocean surfaces, thus maintaining ocean currents and waves.

The composite motion of the entire atmosphere is termed the general circulation. It has become possible to simulate the general circulation by numerical models. Some requirements which a general-circulation model must satisfy are: (1) It must cover the entire globe (or possibly one hemisphere, assuming symmetry at the Equator); (2) have enough vertical resolution so that all energy conversions can be represented; (3) contain radiative heat sources; and (4) contain turbulent eddy fluxes of momentum, enthalpy, and water vapor.

A numerical model for study of the general circulation must necessarily be exceedingly complicated. However, it is not intended for weather prediction, and it is therefore not necessary that the numerical integration proceed faster than real time; and it is not necessary to start the integration from a real state determined by global observations. By performing long-term numerical integrations with general-circulation models, starting from a constructed initial state, it has been possible to reproduce the main statistical characteristics of the real atmosphere, including the distribution of climate. *See* ATMOSPHERIC GENERAL CIRCULATION.

Special motion systems. Another important branch of numerical dynamic meteorology is the numerical study of various mesoscale and small-scale motion systems, such as hurricanes, fronts, squall lines, cumulus clouds, and land and sea breezes.   [ARNT ELIASSEN]

Bibliography: H. R. Byers, *Elements of Cloud Physics*, 1965; H. R. Byers, *General Meteorology*, 4th ed., 1974; I. P. Danilina (ed.), *Meteorology and Climatology*, vol. 2, 1977; A. Eliassen and K. Pederson, *Meteorology*, 2 vols., 1977; R. Goody and J. C. Walker, *Atmospheres*, 1972; J. Gribbin, *Forecast, Famines, and Freezes*, 1977; G. J. Haltiner and F. L. Martin, *Dynamical and Physical Meteorology*, 1957; S. L. Hess, *Introduction to Theoretical Meteorology*, 1959; G. I. Marchuk, *Numerical Methods in Weather Prediction*, 1973.

Micrometeorology

A branch of atmospheric dynamics and thermodynamics which deals primarily with the interaction between atmosphere and ground; the interchange of masses, momentum, and energy at the earth-air interface; in short, with the lower boundary conditions of atmospheric processes. Thus, there are strong interconnections with those sciences that deal with the media underlying the atmosphere. Micrometeorology, along with other branches of meteorology, is also concerned with investigating and predicting the transport and dispersion of pollution from such sources as smokestacks and automotive exhausts in the lower atmosphere. Micrometeorology typically deals with the transport (or flux) of air properties in the vertical direction (conduction and convection) and the vertical variation of these fluxes. *See* ATMOSPHERIC POLLUTION; OCEAN-ATMOSPHERE RELATIONS; SEA WATER.

Scale of focus. Micrometeorology differs from general meteorology (in particular, meso- and macrometeorology) in scale and with respect to the features of the atmosphere studied. All of the various branches of meteorology deal with the temporal-spatial variations of the weather elements (such as air density, temperature, components of momentum, and mixing ratios). Meso- and macrometeorological studies are typically based on standard synoptic observations (from a nationally and internationally organized network with about 20 to 200 km distance between stations), and of primary interest are large-scale air currents which serve to transport "conservative" atmospheric properties (such as heat, admixtures, and absolute vorticity) over relatively large horizontal trajectories. *See* METEOROLOGY.

Micrometeorology is concerned with the great variety of atmospheric processes which are incompletely covered by the synoptic network. Examples of such small-scale processes are mountain and valley circulations, sea breezes, and airflow past mountains and islands. The prefix micro is justified by the fact that the detailed study of the temporal-spatial variations of air properties in the lower atmosphere requires close spacing of instruments of relatively small size and low inertia (lag time).

Instrumental needs. The development and use of nonstandard mast-supported thermometers, anemometers, and other sensing instruments for the measurement of mean vertical profiles and gradients of temperature, wind speed, and other air properties is a basic requirement of micrometeorological research. Further miniaturization of sensing instruments is necessary for the study of short-time fluctuations of meteorological variables. A significant range of such fluctuations occurs with frequencies between several cycles per second to several cycles per minute.

Another basic requirement is the development and use of instruments for the direct measurement of boundary fluxes (such as evaporation, surface stress, and energy transfer, including net radiation). Micrometeorological studies are frequently concerned with the diurnal cycle of heating and cooling, the hydrologic cycle, and the energy dissipation in large-scale air currents due to surface stress (friction).

Micrometeorology is sometimes referred to as fair-weather meteorology. This appellation can be justified by the fact that of the total solar energy intercepted in space by the Earth about 70% reaches the ground under clear skies, with only about 10% scattered back to space (the remainder being absorbed in the air), while under an average cloud deck the corresponding percentages are about 30% reaching the ground, with nearly 60% being scattered back to space. Solar energy arriving at the bottom of the atmosphere is partly reflected ("albedo" radiation, normally about 15% of incoming radiation but varying between extremes of about 5–90%, for very dark-colored ground to very bright, fresh snow cover) and partly absorbed. A primary task of micrometeorology is to analyze and predict the distribution and transformation of the absorbed solar irradiation by establishing a complete energy budget of the air-submedium interface. The respective energy balance equation takes into consideration the vertical fluxes of terrestrial (or long-wave) radiation, of sensible heat conducted into the submedium as well air, and of latent energy (most important, phase transformations of H_2O such as evaporation, sublimation, and melting, but also photosynthesis wherever significant). The partitioning of energy and the subsequent changes in surface temperature, in direct response to variation in intensity of the solar forcing function, depend strongly on the strength of the existing overall air motion in the region. Because of prevailing nonuniformity in the physical structure of natural ground and its various (vegetative or nonvegetative) covers, and the relatively strong energy flux density of solar radiation, the atmosphere within a few meters of the ground is a surprisingly complex and variable structure. Strong fluctuations and gradients of temperature, density, and moisture affect the propagation of sound and light, creating irregular scattering and refraction phenomena which account for optical mirages, shimmer, and "boiling," as well as "ducting" of radio and radar beams. *See* METEOROLOGICAL INSTRUMENTATION.

Significant applications. Micrometeorological research is important in that it supplies detailed information about the physical processes in the region of the atmosphere where life is most abundant. It closes the gaps of information from synoptic networks. It produces results useful for applications in various fields such as climatology, oceanography, soil physics, agriculture, biology, chemical warfare, and air pollution. Among the branches of atmospheric physics, micrometeorology is the one whose subjects are most amenable to fairly complete experimental description, and to testing of the theoretical models.

The characteristic scale is small enough that it is both feasible and economical to attempt control of natural processes in the lower atmosphere, for example, by artificial changes of albedo or other surface characteristics such as aerodynamic roughness; by thermal admittance of the submedium (through mulching of soil or other means); by windbreaks; by the use of sunshades and smoke screens; by heat supply and artificial stirring of air; and by irrigation. *See* AGRICULTURAL METEOROLOGY; CLIMATOLOGY; CLIMATE MODIFICATION; CROP MICROMETEOROLOGY; INDUSTRIAL METEOROLOGY; WEATHER MODIFICATION.

Turbulence research and applications. The mechanism of the vertical flux is essentially one of eddy mixing or turbulent exchange of air properties. It is convenient to distinguish two methods of approach: One deals with the mean vertical gradients caused by turbulence, while the other is based on statistical treatment of turbulent fluctuations recorded by low-inertia (fast-response) instruments. The connecting link between the two approaches is presented by the Reynolds expression of the mean flux of an air property as the average covariance of the fluctuations of vertical wind speed and the property considered.

A powerful tool for research is the spectrum analysis of fluctuations by which the distribution of the variance of a meteorological element over the various frequencies is measured. According to J. Van der Hoven the power spectrum (in the range 0.001–1000 cycles per hour) of horizontal wind speed recorded at the upper levels of the meteorological tower at Brookhaven, Long Island, N.Y., shows two major peaks of energy, one at about 1 cycle per 4 days, another at about 1 cycle per minute. A broad minimum of energy at frequencies of about 1–10 cycles per hour seems to exist under varying terrain and synoptic conditions. This spectral gap appears to separate objectively the macrometeorological from the micrometeorological scale in the atmosphere.

Owing to the relatively high Reynolds number of atmospheric flow, all air motions are turbulent, which means that individual particles do not describe straight and parallel trajectories (as in truly laminar flow) even though the originating force (pressure gradient force) may be constant and uniform and the ground smooth and flat. Turbulence is sometimes visible in the shapes assumed by smoke, and felt in the variable pressure and cooling power of the wind. The intensity of atmospheric turbulence (as measured by the ratio standard deviation of a wind component divided by mean wind speed) is normally between 0.2 and 0.4, which is significantly larger than that of wind tunnel turbulence.

Low-level turbulence derives its energy basically from large-scale air motions and depends on the surface roughness. This mechanical turbulence is intensified by surface heating and damped by nocturnal cooling. Individual gusts can be ascribed to a disturbance, or eddy, of a certain extent. There is a wide range of eddy sizes (eddy spectrum). Near the ground, eddy sizes increase in proportion to distance from the ground, then decrease with further increase in height.

Atmospheric turbulence affects the flight of airplanes and ballistic missiles (rough air). It is especially troublesome when intensified by surface heating (thermal turbulence) and additional buoyancy forces generated by latent heat release in updrafts that ascend above the condensation level (cumulus convection). There are many features of turbulence research that are of common interest to micrometeorology and aeronautics. *See* CLEAR-AIR TURBULENCE.   [HEINZ H. LETTAU]

Bibliography: R. Geiger, *The Climate near the Ground*, 4th ed., 1965; H. E. Landsberg (ed.), *Advances in Geophysics*, suppl. 1: *Descriptive Micrometeorology*, by R. E. Munn, 1966; H. Lettau and B. Davidson (eds.), *Exploring the Atmosphere's First Mile*, 2 vols., 1957; C. H. Priestley, *Turbulent Transfer in the Lower Atmosphere*, 1974; P. Schwerdtfeger, *Physical Principles of Micrometeorological Instruments*, 1976; O. G. Sutton, *The Challenge of Atmosphere*, 1961; O. G. Sutton, *Micrometeorology*, 1977; J. Van der Hoven, Power spectrum of horizontal wind speed in the frequency range from 0.0007 to 900 cycles per hour, *J. Meteorol.*, 14:160, 1957.

Millibar

A unit of pressure commonly used in meteorology, a force of 100 pascals (100 newtons/m² or 1000 dynes/cm² of surface). One millibar (mb) is 1/1000 of a bar. The normal atmospheric pressure, a 760-mm column of pure mercury at 0°C in the gravity field at mean sea level and 45° latitude, is equal to 1013.25 mb. One millibar equals 0.2953 in. or 0.750062 mm of a mercury column. The approximation 1 mb = 3/4 mmHg gives an error of only 8×10^{-5}. The International System (SI) unit of pressure, the pascal, is equal to 1×10^{-2} mb. Its dimensions are $M L^{-1} T^{-2}$. *See* ATMOSPHERE.   [VERNER E. SUOMI]

Mirage

A name for a variety of unusual images of distant objects seen as a result of the bending of light rays in the atmosphere during abnormal vertical distribution of air density. If the air closer to the ground is much warmer than the air above, the rays are bent in such a way that they enter the observer's eyes along a line lower than the direct line of sight. The object is then seen below the horizon, the inferior mirage. If the air closer to the ground is much colder than the air above, the rays are bent in the opposite direction, arriving at the observer's eyes above the line of sight; the object then seems to be elevated or floating in the air, the superior mirage. More complicated or irregular stratification of air density causes apparent multiple reflection of the object and vertical or horizontal distortion of the image. Mirages can be seen most frequently along an overheated highway surface;

the inferior mirage of the sky gives the impression of water reflection over a wet pavement, which disappears upon a closer viewing. *See* METEOROLOGICAL OPTICS.   [ZDENEK SEKERA]

Moisture-content measurement

Measurement of the ratio or percentage of water present in a gas, a liquid, or a solid (granular or powdered) material. Nearly all materials contain free water, the relative amount being dependent upon the physical and chemical properties of the material. The primary purpose of determining and maintaining moisture contents within specified limits can usually be traced to economic factors, trade practices, or legal requirements.

Moisture content has a number of synonymous terms, many of which are specific to certain industries, types of product, or material. The water content in solid, granular, or liquid materials is usually referred to as moisture content on either the wet or dry basis; the wet basis is common to most industries. Specifically, moisture content on the wet basis refers to the quantity of water per unit weight or volume of the wet material. A weight basis is preferred. The textile industry uses the dry basis for moisture content of textile fibers. Often referred to as regain moisture content, the dry basis or regain refers to the quantity of water in a material expressed as a percentage of the weight of the bone-dry (thoroughly dried) material. The relationship between the wet and dry moisture-content basis is shown in Fig. 1.

The moisture content in air is referred to as humidity, either absolute or relative. Absolute humidity is the number of pounds of water vapor associated with 1 lb of dry air, also called just humidity. Relative humidity is the ratio, usually expressed as a percentage, of the partial pressure of water vapor in the actual atmosphere to the vapor pressure of water at the prevailing temperature. Relative humidity (RH) is customarily reported by the U.S. Weather Bureau because it essentially describes the degree of saturation of the air. However, air which is saturated (100% RH) at 50°F is quite dry (19% RH) when heated to 100°F. A changing basis of this type is not convenient for many purposes such as computations used in air-conditioning, combustion, or chemical processing; therefore absolute units, such as dew point or grains of water per pound of dry air, are more acceptable. Dew point is the temperature at which a

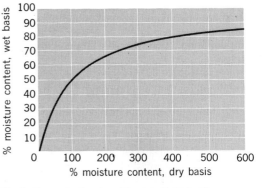

Fig. 1.   Dry versus wet moisture-content basis.

given mixture of air and water vapor is saturated with water vapor.

GASES

The measurement of water content in gases and mixtures of air and gases is important in industry. A number of commercially manufactured instruments are available for these measurements; their principles of operation include condensation, used in dew- or fog-point indicators; dimensional change, used by hygrometers; thermodynamic equilibrium, used by wet-bulb psychrometers; and absorption methods, which serve as the basic principle for gravimetric and electric conductivity or dielectric types.

The importance of humidity in relation to personal comfort is well known. As a result the air-conditioning industry has grown to considerable proportions by producing equipment to maintain comfortable conditions of temperature and humidity. Considerable industrial air conditioning is also done for process reasons. The textile industry makes wide use of humidity control in rooms for weaving, carding, spinning, and other processes, because the amount of moisture held or absorbed by the textile fibers affects these operations. Paper manufacturers face problems similar to those of the textile industry, as do certain chemical, plastic, and allied processing industries in which control of humidity is important for product quality.

Control of humidity is important in the preservation of materials, especially those which are hygroscopic, and in the storage of food products. In many instances, humidities must be maintained at high levels, as in the storage of apples and vegetables, whereas humidity is maintained at low levels in the storage of dried milks, eggs, and similar products.

Human comfort is affected by high humidities, because the air is so close to its saturation content that it cannot absorb moisture from the surface of the skin and thus cool the individual by evaporation. The higher the temperature of the air, the greater the amount of moisture it can hold. A rule of thumb for human discomfort is that any combination of temperature and relative humidity totaling 130 or higher is uncomfortable. *See* HUMIDITY.

Psychrometers. A psychrometer is a device for measuring moisture content of air or gases by means of two thermometers. One thermometer bulb is covered with a wick and maintained wet (the wet bulb); the other bulb is exposed directly to the air or gas (the dry bulb). The evaporation of water from the moistened wick of the wet bulb produces a lowering of its temperature, and by observation of the difference in temperature between the two bulbs the absolute or relative humidity can be determined. For accurate results, the gas or air must have a velocity of 15–20 ft/sec past the wet bulb. Psychrometric charts or tables are used with the readings obtained from the two thermometers to determine the moisture content of the air or gas. Instruments utilizing this principle are often called wet- and dry-bulb thermometers. *See* PSYCHROMETER; PSYCHROMETRICS.

Hygrometers. Hygrometers measure humidity by the change in dimensions of a hygroscopic material, such as human hair, organic membranes,

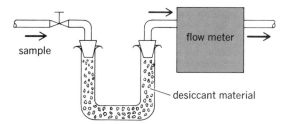

Fig. 2. Gravimetric absorption method.

wood, and plastics. Their most dependable range of operation is from 15 or 20% to 85 or 90% relative humidity at temperatures of 0 to approximately 160°F. Stability of these instruments is better when they are not subjected to extremes of temperature or humidity. There is considerable time lag in the system response to changing humidity conditions when operated at low temperatures. Accuracies of ±3% relative humidity can be expected at normal room temperatures.

Salt conductivity type. Another type of hygrometer utilizes changes in the electrical conductivity of a hygroscopic salt as its operating principle. The conductivity of a solution is dependent upon its concentration (amount of water if the salt content is constant) and its temperature. *See* HYGROMETER.

Gravimetric type. The change in weight of an absorbing material can be used to measure its moisture content, a method known as gravimetric hygrometry. The measurement of moisture content in gases is obtained by passing a known volume of the gas through a suitable desiccant, such as phosphorus pentoxide, silica gel, or similar material, and observing its change in weight. This method is considered a primary standard and is often used in the exact calibration of instruments. It is necessary to make certain that all of the gas or atmosphere has been in contact with the desiccant for a sufficient time to ensure complete absorption of the water vapor. Exact initial and final weighings are also necessary. Figure 2 shows the basic principle of this instrument.

Hygroscopic materials may also be employed directly to determine changes in water vapor in air. The Aldrich regain indicator makes use of a loose ball of cotton attached on one arm of a sensitive balance; the other arm of the balance serves as a pointer, as in Fig. 3. It is important that the sensing material change weight only with change in relative humidity of the surrounding atmosphere.

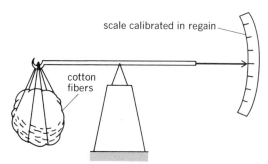

Fig. 3. Aldrich regain indicator.

Dew-point indicators and recorders. When water vapor is cooled, a temperature is reached at which the phase changes to a liquid or solid. This temperature is known as the dew point. The classical method of determining the dew point consists of slowly cooling a polished metal surface until condensation takes place; the temperature at which the first droplet appears is taken as the dew point. The manually operated dew cup is one of the simplest forms of dew-point apparatus. This consists of a polished metal cup containing ether or another volatile liquid into which is placed an accurate mercury-in-glass thermometer. Air is bubbled through the volatile liquid, lowering its temperature and the temperature of the metal cup until dew (condensation) forms on the cup (Fig. 4). *See* DEW POINT.

These instruments are capable of high accuracy, provided correct techniques are employed. This technique is widely used for measuring water vapor in flue gases, gasoline vapors, furnace gases, compressed gases, and others. More refined instruments utilize refrigeration and closed systems to contain the sample while the condensation is viewed through inspection windows. Others automatically record the condensation point by means of photocells, which alternately control heat or refrigeration to the target area. An adaption of the dew-point technique makes use of the Wilson cloud chamber principle. The gas sample is compressed and then vented to atmosphere, producing cooling by adiabatic expansion. By repeated trials a pressure ratio can be obtained at which a cloud or fog forms. Dew point is computed from the ratio of the initial to final pressure.

Miscellaneous methods. Several alternate methods exist for determining the amount of water vapor present in gases or air.

The difference in thermal conductivity of dry air and air with water vapor may be determined by measuring the difference in the electrical resistance (temperature) of a hot wire sealed in a small cell. This method is affected by changes in gas composition. Usually a bridge circuit is used, with a hot-wire cell containing dry air as the reference and a cell with the sample to be tested as the unknown.

Spectroscopic methods and index of refraction have been used experimentally, as has the measurement of pressure or volume after the absorption of the water vapor from a sample.

LIQUIDS AND SOLIDS

Instrumentation for the measurement and control of moisture in liquids and solids has grown rapidly during the past 20 years. The development of these instruments has been due in a large part to the great need that exists in many processes where the control of a precise moisture is critical. The desirability of a specific moisture content in a product during its preliminary manufacturing process is often required. In general, however, the rigid control of moisture content occurs most frequently in the final product to assure its quality and the fulfillment of legal or trade practices for the individual product.

Instrument types. Instruments suitable for the measurement of moisture content are available from a large number of manufacturers. These instruments may be classified as periodic and continuous. In general, only those instruments offering continuous measurement are practical for the automatic control of moisture content in a product. The periodic instrument types are generally automated versions of conventional laboratory moisture-analysis procedures. The speed of response for the periodic (intermittent) instruments is typically 2 min or longer, often 15–20 min, making this type of instrument impractical for automatic control. The initial cost of the periodic or intermittent sampling instrument is usually less than for continuous types.

Moisture measuring instruments may also be classified by operating principle. Those instruments employing electrical conductivity (either dc or ac), absorption of electromagnetic energy (radio-frequency regions), electrical capacitance (dielectric constant change), and infrared energy radiations are more readily adapted to continuous measurements inasmuch as the response of these instruments to moisture changes is very fast. Those instruments employing automatic oven drying, chemical titrations (Karl Fischer technique), equilibrium hygrometric methods, distillation methods, and so forth are usually of the intermittent, or periodic, type; the time for analysis is usually 2 min or longer. Most of the more popular methods presently utilized are outlined in following pages.

Electrical conductivity. These methods are based on the relationship between dc resistance and moisture content for such materials as wood, textiles, paper, grain, and similar products. Specific resistance plotted against moisture content results in an approximate straight line up to the moisture saturation point. Beyond the saturation point, where all of the cells and intermediate spaces are saturated with free water, conductivity methods are not reliable. This point varies from approximately 12 to 25% moisture content, depending on the type of product.

The sample under test is applied to suitable electrodes in the form of needle points for penetrating into woods, plaster, and similar products or

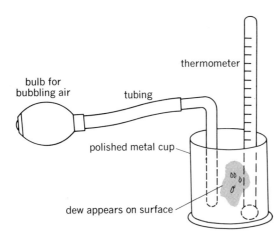

Fig. 4. Manually operated dew-point unit.

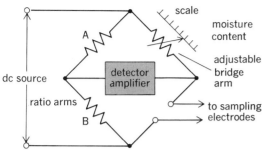

Fig. 5.   Basic bridge circuit for conductivity method of moisture-content measurement.

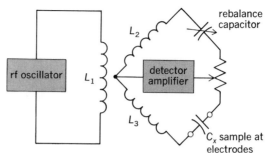

Fig. 6.   Basic bridge circuit for capacitance method of moisture-content measurement.

flat plates for sheet materials. Granular or fibrous materials may make use of electrodes in the form of a cup or clamp arrangement to confine the material to a fixed volume. The electrodes and sample under test make up one arm of a Wheatstone bridge, as shown in Fig. 5. The high sensitivity required of the detector dictates the use of electronic amplifiers. The range of resistance values corresponding to normal moisture contents varies from less than 1 to 10,000 megohms or higher, depending upon material, electrode design, and moisture content. Increasing moisture content results in decreasing resistance values.

Electrical capacitance. These methods are based on the principle of the change in dielectric constant between dry and moist conditions of a material. The dielectric constant of most vegetable organic materials is 2–5 when dry. Water has a dielectric constant of 80; therefore the addition of small amounts of moisture to these materials causes a considerable increase in the dielectric constant. The material being measured forms part of a capacitance bridge circuit, which has radio-frequency power applied from an electronic oscillator (Fig. 6). Electronic detectors measure bridge unbalance or frequency change, depending on the method employed.

Electrode design varies with the type of material under test. Parallel-plate types are used for sheet materials, whereas cylindrical electrodes are usually adapted for liquids or powders. Modifications of the parallel-plate capacitor into a rectangular enclosure are used with granular, fibrous, or powdered materials. Range of moisture contents measurable is from 2 or 3% to 15 or 20%, varying with the product.

Radio-frequency absorption. Instruments utilizing the attenuation of electromagnetic energy when passed through the material have been developed and tested during the last decade. The radio-frequency types are operated at frequencies below 10 MHz and, for best results, are used with products composed of polar materials. The radio-frequency energy is passed through the polar material and the water molecules absorb some of the energy as molecular motion.

Figure 7 shows the basic block diagram of a typical radio-frequency moisture instrument. These instruments must be calibrated to the particular product under test. By means of suitable electrodes for the sampling technique the instrument

can be applied to solid or sheet materials; granular materials are placed into cell-type electrodes. The test results will be affected by any polar material in the sample; however, water of hydration in a product will not be detected. These instruments are mostly available as periodic types but can be adapted for continuous measurements on many products. Moisture content is dependent on the material and ranges from 0.1 to 60%; for certain materials the range will be quite narrow.

Microwave absorption. Instruments employing very-high-frequency electromagnetic energy have been developed and are in limited use for both grab samples and continuous measurement applications. The frequency of the electromagnetic energy is in the region of 1 GHz and higher. The 2.45-GHz "S" band, 8.9–10.68 GHz "X" band, and 20.3–22.3 GHz "K" band frequencies have all been used for microwave moisture-analysis instruments.

The principle of operation is based upon the fact that the water molecule greatly attenuates the transmitted signal with respect to other molecules in the material in the S and X band frequencies. In the case of the K band microwave frequencies, the water molecule produces molecular resonance. There are no other molecules that respond to this particular resonant frequency, making this frequency most specific to the moisture (free water) in paper products. The wavelength of the K band frequencies are approximately 1.35–1.5 cm long; at these very short wavelengths, the electromagnetic energy may be guided (transmitted) by means of metallic pipes, usually of rectangular shape, called waveguides. Thus the output of a microwave oscillator can be coupled to a horn-shaped opening on the end of the waveguide, propagated through the sample material, and collected by means of an input or receiver horn. The output

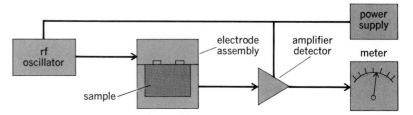

Fig. 7.   Radio-frequency-absorption moisture indicator.

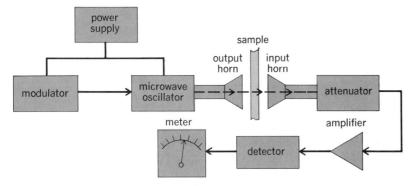

Fig. 8. Microwave-absorption moisture indicator.

and input horn are physically separated so that a noncontacting (nondestructive) continuous measurement of the material is possible. The basic block diagram of a microwave moisture instrument is shown in Fig. 8. The detection of energy absorption can be made by attenuation (loss) or by phase shift methods, or both measurements may be made simultaneously. Moisture content ranges of less than 1% to 70% or higher can be made. At low moisture contents the S and X bands require a minimum mass of material for practical operation. Accuracies of these instruments are within ±0.5% of indication up to approximately 15% moisture content.

Infrared absorption. These instruments operate on the principle of absorption of infrared radiation when passed through the sample material. The water molecule becomes resonant at certain infrared frequencies and thus the amount of energy absorbed by the water absorption band is a measure of the moisture content.

One instrument, used for measuring the moisture content of sheet paper, makes use of infrared energy at two different wavelengths by measuring the differential absorption (Fig. 9). Infrared radiation is passed through a rotating disk chopper containing two filters, one filter passing a wavelength of 1.94 μ (which covers the water absorption band)

and the second filter passing a band of 1.80 μ (which is not significantly affected by moisture content). The paper sheet is exposed to the dual-frequency narrow-band radiation and the reflected radiation is collected in an integrating sphere, where it activates a lead sulfide detector. The signals are amplified and the ratio of the two narrow-band signals is read out directly as moisture content. This type of measuring system is quite immune to the effect of paper variables such as basis weight, coatings, density, and composition. The useful range is from 1 to 12% and can be used as either contacting or noncontacting.

An infrared instrument useful for gas and liquid measurements operates on the principle of selective energy absorption of the sample when compared to a reference material having little or no absorption. The instrument contains two infrared sources (Fig. 10). One source is directed through a cell containing the sample material, and the second source is directed through the reference cell, which in gas analysis is filled with an inert gas, usually air. A detector unit is present which consists of two cavities separated by a metallic diaphragm that acts as a variable capacitor in respect to a fixed plate within the cavity. The operation consists of chopping the infrared energy to the sample and reference cells at 10 Hz. The gas or fluid in the sample cell absorbs the wavelengths which are characteristic of the material and proportional to its concentration. The reference cell does not absorb any appreciable infrared energy. The detector, which has two equal volumes separated by a diaphragm, is filled with gas or vapor of the component of interest. The infrared energy passes through the cells and into the detector, which causes the temperature of gas in the detector to rise; however, if some of the energy has already been absorbed by the sample cell, the sample side of the detector receives less energy than the reference half of the detector. The temperature rise of the gas in the detector causes the pressure to increase and, if the temperature is unequal, the pressure increase is unequal, causing the diaphragm to move in a direction that will tend to equalize the two pressures. This in turn causes the capacitance existing between the diaphragm and its sensing plate to change; this change is used to modulate an oscillator.

The chopper alternately blocks and then passes the energy from each infrared source. Each time the energy is blocked, the pressures equalize in the detector compartments and again the energy is permitted to enter the detector cavities. The modulating frequency is 10 Hz and the magnitude of the capacitance change is directly related to the amount of energy absorbed in the sample cell. The signal from the oscillator is amplified and demodulated for meter indication or applied to a recorder. The range of moisture in gases is from 1000 ppm to 100% water vapor, and in fluids from 6 to 100% water. Response speed is 0.5 sec for 90% of full-scale output.

Equilibrium. The equilibrium moisture content of the air at the surface of a material is representative of the moisture within the material. This is particularly true of hygroscopic granular or fibrous

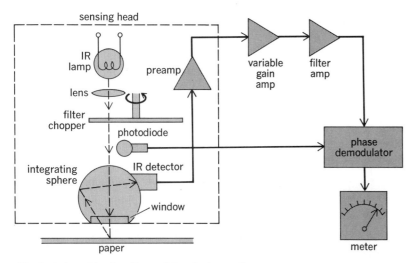

Fig. 9. Infrared-backscatter moisture instrument.

materials. Humidity-measuring instruments are employed by inserting the measuring element into the material. Moisture content is determined from data of relative humidity plotted against moisture content for the material under test.

Absorption. The water content of liquid organic compounds can be determined by the use of spectrophotometers. The sample is placed in a suitable cell and monochromatic infrared radiation is passed through it. Water vapor absorbs the infrared radiation, but the material absorbs little radiant energy. Thus the radiation varies with the water content. The radiation is focused on a sensitive thermocouple and the resulting voltage amplified. The indication of a sensitive galvanometer can be calibrated in terms of moisture content. This method has been used for accurately detecting 0–10 ppm of water in Freon-12.

Chemical methods. A widely used chemical titration method is the Karl Fischer technique. The method offers extreme sensitivity with good accuracy and covers a wide range of materials. Small samples can be analyzed readily. End points can be determined either visually or electrometrically. The Karl Fischer reagent is added in small increments to a glass flask containing the sample until the color changes from yellow to brown or a change in potential is observed at the end point. Dark-colored solutions require electrical end-point measurement. This method does not differentiate between free and combined water.

Various other methods involve chemical reactions with the free water in liquids or solids by (1) the evolution of free acidic or basic compounds, (2) the evolution of an inert gas, or (3) the formation of an insoluble precipitate. In general, these methods fail to distinguish active hydrogen from water in many materials and are reliable only under laboratory control.

Distillation. A representative sample of the material can be placed in a suitable flask with an excess of liquid which usually has a boiling point higher than water but is immiscible with water. Heat is applied to the flask, and the water and some of the liquid are distilled off. The combined vapors are condensed and collected. The water is separated and measured volumetrically to calculate the moisture content. A variety of laboratory glass apparatus is available for distillation of liquids which are either lighter or heavier than water. The measuring tubes are etched with graduations in cubic centimeters corresponding to grams of water. The moisture content is calculated from the weight of the original sample and the weight of water. This method is slow, and the accuracy depends upon the care exercised and the apparatus used. The method was widely used in the analysis of grains but has largely given way to electrical or oven-drying methods.

Oven drying. The oldest and most common analytical method of determining moisture consists of heating the sample to ensure complete drying. Moisture is calculated on the basis of loss in weight between original and dried sample. The method is applicable to many solids and some liquids, and does not require unusual operator skill. Semiautomatic drying and weighing ovens are

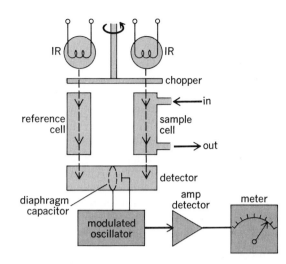

Fig. 10.   Infrared gas or liquid analyzer.

available; the moisture content is indicated directly by the weighing scale built into the oven. Problems involve making certain that the material has not lost some of its weight by loss of volatile products other than water, that the material is completely dry at the time of final weighing, and that accurate weighings are made. The oven method is widely used and often serves as the primary standard for calibration of electrical and other indirect methods.   [LEE E. CUCKLER]

Bibliography: D. M. Considine (ed.), *Process Instruments and Control Handbook*, 1974; D. M. Considine and S. D. Ross (eds.), *Handbook of Applied Instrumentation*, 1964; A. Pande, *Handbook of Moisture Determination and Control: Principles, Techniques, Applications*, 1974.

Monsoon

Large-scale wind system which predominates or strongly influences the climate of large regions, and in which the direction of the wind flow reverses from winter to summer. The outstanding example of a monsoon circulation is the vast wind system which dominates the weather over most of the Asian continent, especially the southern and eastern parts, the north Indian Ocean, and the extreme western portion of the northern Pacific Ocean. The monsoon influence is felt more or less distinctly on all the continental landmasses outside equatorial latitudes, except Antarctica. The monsoonal circulation in the lower troposphere is cyclonic around the margins of a continental area in summer, and anticyclonic in winter.

General development. These seasonally reversing circulations are attributable to differences in heating or cooling of the atmosphere over continental landmasses in contrast to that over the surrounding oceanic areas.

The monsoon influence is absent or feeble in equatorial regions because of the small differences in atmospheric heating from season to season, as well as between the cloudy land and oceanic regions. In Antarctica the monsoon influence is not felt because the presence of a permanent ice and snow cover precludes the possibility of relative heating of the atmosphere over the continent even in summer. In many other regions the monsoon circulation is effectively superposed upon the zonal wind systems and serves mainly to distort them.

Great Asiatic monsoons. The Asiatic system dominates the continent south of approximately latitude 45°. In winter the anticyclonic circulation is centered on the average in Siberia near 50°N and 100°E. The mean pressure in this area in the month of January exceeds 30.40 in. (reduced to sea level), the highest such value to be found on the Earth. Persistent northerly winds flow from the interior to produce bitterly cold weather far south along the eastern coast of Asia. When the Asiatic anticyclone is displaced westward from its average position, easterly or southeasterly winds bring outbreaks of cold air from the continental interior into eastern Europe and Scandinavia, and occasionally as far as the British Isles. Prevailing southwesterly flow on the northern and northwestern flanks of the anticyclone produces a relatively mild winter climate in these coastal portions of Asia. Flow of cold air from the north into India and adjacent regions is blocked by the massive mountain barrier along their northern boundaries. At elevations comparable to that of the Tibetan plateau the prevailing wind is part of the circumpolar upper-level westerly circulation. A branch of this circulation descends the mountain slopes and moves across the southern regions in winter as a relatively mild but quite persistent northeasterly wind.

Winter precipitation is sparse in most regions dominated by the Asiatic monsoon circulation, because of the small water vapor content of the continental air mass and because of the predominant sinking motion of the atmosphere. Precipitation is plentiful, however, in the East Indies and on the islands of the north Indian Ocean, because the northeasterly wind current experiences a long trajectory over a warm oceanic surface before arriving at these locations with a considerably enhanced water vapor content. This wind current is so firmly established that the intertropical convergence zone is driven well south of the Equator. Along this zone rainfall is abundant.

In summer the pattern of the Asiatic monsoon circulation is cyclonic. It is centered about a region of low pressure extending from Arabia to India between latitudes 25 and 30°N. In this region in July the mean pressure reduced to sea level is near 29.50 in. The intertropical convergence zone is displaced far to the north, as a warm and extremely moist southwesterly or southerly wind current prevails over the north Indian Ocean and the extreme western portion of the North Pacific Ocean. Heavy precipitation falls as this current encounters the rugged terrain of southern and eastern Asia. The world's heaviest annual rainfall occurs on the southern slopes of the Himalayas, where, locally, amounts in excess of 400 in., on the average, fall entirely during the summer season. The southwest monsoon begins abruptly on the Indian Peninsula in May or June. In a particular year the time of onset of the southwest monsoon and the amount of rain that falls may have a crucial effect upon the agriculture of these regions.

Another portion of the Asiatic summer monsoon system is a persistent northerly wind throughout much of the Near and Middle East, the eastern Mediterranean Sea, and extreme southeastern Europe. This current traverses predominantly continental regions and is therefore associated with extremely hot, dry weather.

Effect on North America. On the other continents the monsoon influence is less clear-cut. In North America, for example, monsoon influences include the persistent northwesterly winds in summer along the coast of California; the penetration of warm, moist air from the Gulf of Mexico far to the north in the central and eastern United States during summer with associated copious precipitation amounts; the southward penetration of cold dry air masses in winter in these same regions; and a shift of the prevailing wind from southwest in summer to northwest in winter along much of the Atlantic seaboard. *See* AIR MASS; PRECIPITATION; WIND.

[FREDERICK SANDERS]

Naval meteorology

Those aspects of meteorology that are relevant to naval operations. However, in more general terms, naval meteorology refers to marine or maritime meteorology, which is the study of weather phenomena over the oceans and other large bodies of water, including the interactions between the atmosphere and the oceans, particularly the exchange of energy.

Weather over the sea. Some weather phenomena which occur mainly over the sea include fog, tradewinds, tropical storms, and waterspouts.

Fog. Although fog is prevalent over the land and sea, certain types of fog are more extensive over the sea. Steam fog results from evaporation of relatively warm water into overlying cooler air followed by almost immediate condensation into minute liquid droplets, much like steam from a pan of hot water. Because the air is warmed by the sea surface and tends to rise in convective currents, steam fog is usually not as dense as other types of fog. In high latitudes, steam fog is referred to as arctic sea smoke or, if the air temperature is below freezing, as arctic frost smoke. Advection fog may occur when relatively warm, moist air moves over colder water, cooling the air, saturating it, and eventually forming fog. Extensive areas of advection fog may develop when warm air streams poleward over colder water, especially over cold ocean currents, reducing visibility drastically and making navigation hazardous. This danger has been considerably reduced in modern times through the use of radar to detect other ships, icebergs, and the shore. *See* FOG.

Tradewinds. Tradewinds, or simply trades, refer to an extensive tropical wind system found primarily over oceanic areas between the subtropical high-pressure belts centered near 30°N and 30°S latitude. On the average, the trades blow from the northeast or east-northeast in the Northern Hemisphere and from the southeast or east-southeast in the Southern Hemisphere, converging more or less toward the Equator into a rather narrow belt of sometimes intense cumulus convection and

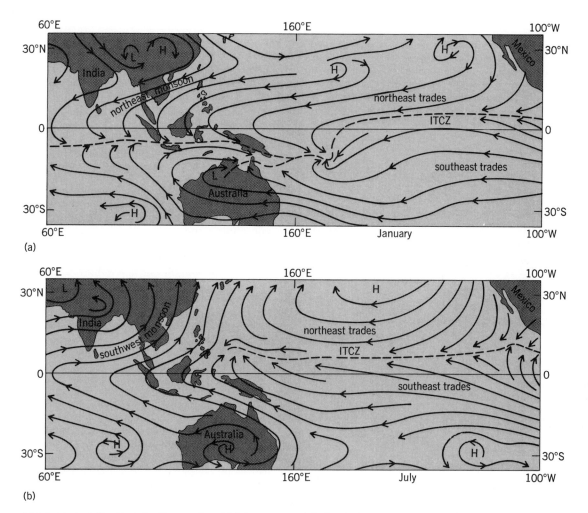

(a)

(b)

Low-level wind directions for the months of (a) January and (b) July over the Pacific and Indian oceans. The intertropical convergence zone (ITCZ) is shown with a broken line.

showers known as the intertropical convergence zone (ITCZ). The trades are best developed in the lower 5000 ft (1525 m) of the atmosphere and are gradually replaced by more complex wind systems aloft. In the extreme western Pacific and Indian oceans, the trades largely give way to a strong monsoon circulation induced by the continent of Asia and to some extent by Australia. The southwesterly summer monsoon pours warm, moist air into India and southeast Asia in general, whereas in winter the wind comes from the northeast out of Asia. A somewhat similar but less intense monsoon circulation is found over Mexico and Central America. Also, northern Australia experiences a general pattern of onshore winds in summer and offshore winds in winter. The ITCZ is normally slightly north of the Equator in the eastern Pacific year-round, but becomes more diffuse and tends to shift markedly into the summer hemisphere in the western Pacific and Indian oceans. For the most part, solar heating and terrain influences distort the airflow over tropical land areas, making the wind systems complex, less steady, and more subject to local conditions (see illustration).

Tropical storms. Warm-center storms (cyclones or low-barometric-pressure centers) develop in low latitudes and derive their energy from warm tropical waters. The winds are nearly circular, spiraling inward to form a vortex rotating in counterclockwise direction in the Northern Hemisphere and clockwise in the Southern Hemisphere. In intense storms, known as hurricanes in the Atlantic, Gulf of Mexico, and the eastern North Pacific, and as typhoons in the western North Pacific, the winds may reach 150 knots (75 m/s) or more, with a central calm area or eye roughly 10–40 mi (15–70 km) in diameter. Tropical storms, which have many other local names, usually develop in latitude bands 5–20° north and south of the Equator over the warmest ocean waters. These storms generally move from east to west and often curve poleward into middle latitudes; however, many tracks are erratic, making forecasting difficult. Movement inland leads to dissipation; however not without damage and sometimes loss of life due to the strong winds, heavy rains, and flooding storm tides. *See* HURRICANE.

As an aid in providing early warnings of severe tropical storms and also for routine observations, aircraft reconnaissance was developed after World War II by the U.S. Navy and Air Force. Incipient storms were detected, and full-blown hurricanes and typhoons were penetrated to provide accurate positions of storms from which their speed and direction of movement could be calculated. This information enabled forecasters to make better predictions of the future movement of a storm and its intensity. The National Oceanic and Atmospheric Administration (NOAA) has partially taken over reconnaissance duties from the military in the Atlantic area, which are now conducted on a reduced scale because of the information provided by meteorological satellites. Tropical and middle-latitude cyclones as well are identifiable with the satellite radiometers by the tell-tale vortical cloud patterns. *See* METEOROLOGICAL SATELLITES; STORM DETECTION; WEATHER FORECASTING AND PREDICTION.

Waterspouts. In addition to the tropical cyclones, much smaller vortices known as waterspouts occur over the sea, mainly in tropical and subtropical waters. These whirlwinds are associated with cumulus clouds and have a tornadolike appearance, but are less intense and often similar in size and intensity to dust devils over land. Dense spray is generated at the sea surface by a waterspout.

Observations. To support weather forecasting over the globe, observations of pressure, temperature, wind, and humidity are needed both at the Earth's surface and aloft, even to heights of 20 mi (32 km) or more. Over populated land areas there is a fairly dense observational network; however, over the oceans and uninhabited lands observations are scarce and expensive to make, except over the major sea lanes and air routes. Currently, over the Northern Hemisphere oceans there are about 700 surface ship observations per 6-hr period, only about a half dozen or so upper-air observations taken with rawinsondes (radar-tracked balloons carrying instruments) every 12 hr, and, including land areas, about 800–1000 aircraft observations of wind and temperature per 12-hr period (data counts from the Navy Fleet Numerical Weather Central). The number of observations over the sea in the Southern Hemisphere is far less and does not add greatly to the aforementioned totals.

A great boon to the weather forecaster over data-sparse areas is the meteorological satellites. In the early days of satellites only TV camera pictures showing clouds (from a distance of several hundred miles or more in space) were available, which were nevertheless invaluable to meteorologists for observing weather systems on a grand scale for the first time in history. Modern satellites have radiometers with better resolution, microwave sounders measuring moisture, and infrared spectrometers measuring terrestrial radiative fluxes from the atmosphere, from which a vertical profile of temperature may be constructed. Successive satellite pictures also enable analysts to make wind estimates by observing the movement of identifiable cloud elements.

Despite these advances, data remain scarce over the oceans and are inadequate not only for daily forecasting but also for meteorological research over ocean areas. To help support such research, various data-gathering projects have

been funded on occasion. An important project, the First Global GARP (Global Atmospheric Research Project) Experiment (FGGE), sponsored by the World Meteorological Organization with the collaboration of many nations, is planned for 1978–1979 with the objective of obtaining a fairly complete coverage of surface- and upper-air meteorological observations over the entire globe for 12 consecutive months. Novel observing platforms such as free-floating balloons, ocean buoys, and additional satellites will be used as well as conventional observations from sounding balloons, aircraft, ships, land stations, and so on. These global data will permit study of meteorological phenomena and weather forecasting on a global scale hitherto impossible. Previous international experiments, such as the GARP Atlantic Tropical Experiment (GATE), have been very successful.

Air-sea interaction. There is a two-way interaction between the atmosphere and the sea (or land) which results in the exchange of energy in the form of heat, water vapor, and momentum. A consequence of the interactions is the formation of boundary layers from which energy is eventually transmitted to the deepest parts of the ocean, the outer limits of the atmosphere, and space. *See* OCEAN-ATMOSPHERE RELATIONS.

Sensible heat. When the water is warmer than the air, heat is added to the air (referred to as sensible heat), and then turbulence and convective currents carry the energy to higher levels in the atmosphere. Simultaneously the water is cooled and sinks to lower levels in the ocean. Air warmer than the water cools, and fog may develop. The rate at which sensible heat is transferred is directly proportional to the sea-air temperature difference and the wind speed.

Latent heat. Another form of heat energy exchange is a consequence of evaporation from, or condensation on, the sea surface. When the vapor pressure at the sea surface exceeds the partial pressure exerted by the water vapor molecules in the air, evaporation occurs. The conversion of the sea water to vapor requires considerable energy, referred to as latent heat, which is furnished by the sea and carried into the atmosphere with the evaporated water. Thus the atmosphere receives latent heat energy which is later released to the air when the water vapor condenses back to liquid in the form of clouds and rain. The opposite process, that is, condensation of water vapor from the air on the sea surface, takes place when the vapor pressure in the air exceeds that at the water surface. When relatively cold, dry air moves over warmer water, both sensible and latent heat are rapidly transmitted to the air, and play a very important role in weather phenomena both locally and on a planetary scale. For example, sensible and latent heat are the primary sources of the energy driving tropical storms.

Momentum flux. The third form of energy exchange is through momentum flux. Winds blowing over the sea surface generate waves and ocean currents, resulting in a gain of kinetic energy by the oceans and a loss by the air. This is a primary mechanism for the dissipation of atmospheric kinetic energy which is constantly being generated by the unequal solar heating between the poles and the Equator. Also the drag of relatively persistent wind systems such as the trades over the water generates not only surface waves but also ocean currents. In fact, the winds are both directly and indirectly responsible for most of the major ocean currents.

Ocean waves. The surface waves generated by the wind vary in height, length, and period, depending on the wind speed, the length of water over which the wind is blowing in a more or less uniform direction (fetch), and the duration of time the wind has blown. Generally, the sea is "confused," with many wavelengths present, but at times a particular length and period predominates. The wave action mixes a layer of water near the surface, tending to produce a uniform temperature. The depth of this mixed layer has important implications with respect to sound propagation from sonars, which play an important role in detection of surface ships, submarines, and even schools of fish.

Ocean waves can be very destructive to ships at sea and also when the surf strikes the shore. As a consequence, oceanographers and meteorologists have developed wave forecasting techniques which are utilized primarily by government, navy, and private meteorologists to prepare wind wave, swell (waves that have left the generation area), and surf forecasts for a variety of purposes, including merchant marine operations, navy ship maneuvers, refueling at sea, amphibious landing exercises, oil drilling operations, and search and rescue. *See* OCEAN WAVES.

Ship routing. The headway made by a ship at sea depends to a great extent on the state of the sea, as well as other factors. Head waves retard a ship's progress the most, but following and beam waves also slow a ship's speed, and the larger the waves the slower the headway. Hence it is clearly desirable to avoid heavy seas not only to escape damage and possible injuries but also to save time in crossing. Empirical data gathered from ships' logs have enabled researchers to construct curves relating ship speed to the sea state. By using these curves and Huygens' principle for light propagation through a refracting medium, it is possible to construct a least-time track for a ship traversing the ocean. However, this technique requires a detailed prediction of winds for the entire period of the voyage, perhaps up to 3 weeks, which is beyond current skill in forecasting. As a result, preliminary tracks are recommended based on 5–7-day predictions and climatology and then modified, if need be, as the voyage progresses. Other factors, such as the sensitivity of the cargo and passenger comfort, are also taken into account in addition to the time factor. An interesting point here is that the shortest distance between two ports, that is, the great-circle route, usually is not the least-time track or the safest route to follow. Ship routing based on weather and sea conditions has been found to improve the efficiency and reduce the cost of shipping; consequently, it has become widely used for both commercial and naval ships of the United States and other countries.

Ice accretion. Ice accretion may occur on ships (as on aircraft) and not only decreases the efficiency of operation but may be extremely hazardous, especially to small vessels. The accumulation of ice reduces freeboard and, more importantly, the stability of the ship by making it top-heavy, especially when ice collects on the superstructure. It can also interfere with radar and radio antennas and render them inoperable. Ice formation may occur in various ways, for example, from freezing rain, arctic frost smoke, and freezing spray. When the air temperature is well below freezing, arctic frost smoke composed of supercooled water droplets can freeze on contact, forming rime ice. Accumulations of a foot or more of rime ice on the sides of a ship and a half foot on the superstructure have occurred in a 12-hr period. Rime ice is porous and more easily dislodged than clear ice. The most dangerous accumulations of ice develop with freezing spray which can occur when the air temperature is several degrees or more below freezing. The colder the air and the sea and the stronger the wind, the more rapid is the accumulation of ice. As much as 50 tons (45 metric tons) and more of ice have accumulated in 24 hr, even on small ships of 500–1000 tons (1 register ton = 100 ft^3 or 2.83 m^3), creating a serious hazard to the ship and crew.

[GEORGE J. HALTINER]

Nearshore processes

The processes that shape the shore features of coastlines and begin the mixing, sorting, and transportation of sediments and runoff from land; particularly those interactions among waves, winds, tides, currents, and land that relate to the waters, sediments, and organisms of the continental shelf and nearshore areas.

The energy for nearshore processes comes from the sea and is produced by the force of winds blowing over the ocean, by the gravitational attraction of Moon and Sun acting on the mass of the ocean, and by various impulsive disturbances at the atmospheric and terrestrial boundaries of the ocean. These forces produce waves and currents that transport energy toward the coast. The configuration of the land mass and adjacent shelves modifies and focuses the flow of energy and determines the intensity of wave and current action in coastal waters. Rivers and winds transport erosion products from the land to the coast, where they are sorted and dispersed by waves and currents.

The dispersive mechanisms operative in the nearshore waters of oceans, bays, and lakes are all quite similar, differing only in intensity and scale, variables that are determined primarily by the nature of the wave action and the dimensions of the surf zone. The most important mechanisms are the orbital motion of the waves, which is the basic mechanism by which wave energy is expended on the shallow sea bottom, and the currents of the nearshore circulation system that produce a continuous interchange of water between the surf zone and offshore. The dispersion of water and sediments near the coast and the formation and erosion of sandy beaches are some of the more common manifestations of nearshore processes.

Erosional and depositional nearshore processes

play an important role in determining the configuration of coastlines. Erosion is usually dominant off headlands and along coastal sections backed by alluvium and other unconsolidated material, whereas deposition is most common along indentations between headlands. The overall effect of such processes is usually a straightening and smoothing of the coastline. However, this is not always the case; differential wave erosion may cause a rapid erosion of the material between headlands and thus produce irregularities in the coastline.

Whether deposition or erosion will be predominant in any particular place depends upon a number of interrelated factors: the amount of available beach sand and the location of its source; the configuration of the coastline and of the adjoining ocean floor; and the effects of wave, current, wind, and tidal action. The establishment and persistence of natural sand beaches are often the result of a delicate balance among a number of these factors, and any changes, natural or man-made, tend to upset this equilibrium.

Waves. Waves and the currents that they generate are the most important factor in the transportation and deposition of nearshore sediments. Waves are effective in moving material along the bottom and in placing it in suspension for weaker currents to transport. In the absence of beaches, the direct force of the breaking waves erodes cliffs and sea walls.

Wave action along most coasts is seasonal in nature in response to the changing wind systems over the waters where the waves are generated. The height and period of the waves depend on the speed and duration of the winds generating them and the fetch, or length, over which the wind blows. Consequently, the nature and intensity of wave attack against coastlines varies considerably with the size of the water body, as well as with latitude and exposure. Waves generated by winter storms in the Southern Hemisphere of the Pacific Ocean may travel more than 5000 mi before breaking on the shores of California, where they are common summer waves for the Northern Hemisphere. Using sensitive instruments, W. H. Munk and others have recorded waves off San Clemente Island, Calif., that have traveled more than 10,000 nautical miles from their generation by storms in the South Indian Ocean.

The profiles of ocean waves in deep water are long and low, approaching a sinusoidal form. As the waves enter shallow water the wave velocity and length decrease, the wave steepens, and the wave height increases, until the wave train consists of peaked crests separated by flat troughs. Near the breaker zone the process of steepening is accelerated, so that the breaking waves may attain a height several times greater than the deep-water wave. This transformation is particularly pronounced for long-period waves from a distant storm. The profiles of local storm waves and the waves generated over small water bodies such as lakes show considerable steepness even in deep water, so that the shallow-water steepening is not as pronounced as in the case of ocean swell.

The shallow-water transformation of waves commences at the depth where the waves "feel bottom." This depth is equal to one-half the deep-water wave length, where the wave length is the horizontal distance from wave crest to crest. The deep-water wave length is given by the relationship $L = gT^2/2\pi$, where g is the acceleration of gravity and T is the wave period in seconds. Upon

Fig. 1. Longshore currents, generated when waves approach the beach at an angle. In this photograph at Oceanside, Calif., the longshore current is flowing toward the observer. (*Department of Engineering, University of California, Berkeley*)

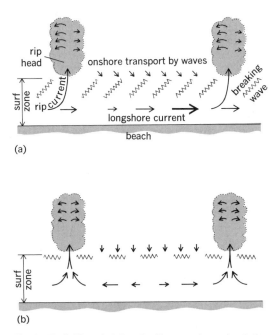

Fig. 2. Definition sketches for the nearshore circulation cells. (*a*) Asymmetrical cell for breakers oblique to the shore. (*b*) Symmetrical cell for breakers parallel to the shore.

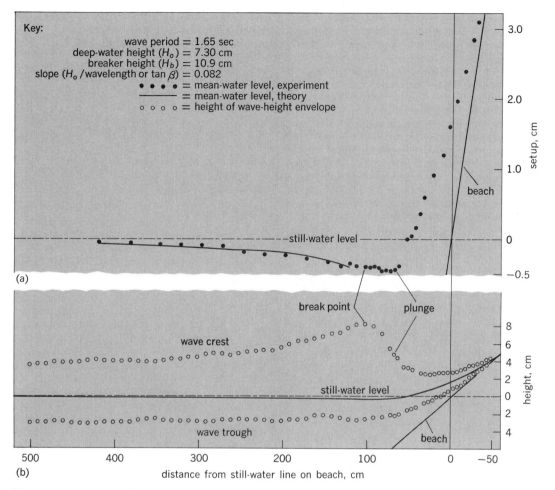

Key:
wave period = 1.65 sec
deep-water height (H_o) = 7.30 cm
breaker height (H_b) = 10.9 cm
slope (H_o/wavelength or tan β) = 0.082
• • • • = mean-water level, experiment
——— = mean-water level, theory
o o o o = height of wave-height envelope

setup, cm

3.0

2.0

1.0

beach

still-water level

0

—0.5

(a)

break point   plunge

8
6
wave crest

4

still-water level

2
0

beach

2

wave trough

4

500   400   300   200   100   0   —50

(b)   distance from still-water line on beach, cm

height, cm

Fig. 3. Measurements of (*a*) wave setup and (*b*) envelope of wave height for waves breaking on a plain beach in the laboratory. The measurements are relative to still-water level and show the decrease in mean-water level near the break point of the waves (setdown) and increase in level over the beach face (setup). The vertical exaggeration is ×80 for *a* and ×10 for *b*. (*Measurements from A. J. Bowen, D. L. Inman, and V. P. Simmons, J. Geophys. Res., 73(8), 1968; theoretical curve from M. S. Longuet-Higgins and R. W. Stewart, Deep-Sea Research, vol. 11, © 1964 by Pergamon Press; reprinted with permission*)

entering shallow water, waves are also subjected to refraction, a process in which the wave crests tend to parallel the depth contours. For straight coasts with parallel contours, this decreases the angle between the approaching wave and the coast and causes a spreading of the energy along the crests. The wave height is decreased by this process, but the effect is uniform along the coast (Fig. 1). A submarine canyon or depression causes waves to be refracted, or bent, in such a manner that waves over the canyon will diverge and decrease in height and the line of wave crests will be convex toward the shore. Waves will converge on either side of the canyon over a ridge, causing the wave height to increase and the line of wave crests to be concave toward the shore. The amount of wave refraction and the consequent change in wave height and direction at any point along the coast is a function of wave period, direction of approach, and the configuration of the bottom topography.

Waves from distant storms may have periods as great as 20 sec or more when they reach ocean coasts. Since refraction commences when waves reach a depth equal to one-half the wave length, these long waves will be refracted by topographic features on the ocean floor in depths as great as 300 m. Thus, the formation of beaches and the effect of waves on a coastline may be influenced by the topography of the bottom many miles from the coast. When submarine ridges cause a concentration of wave energy at certain points along a coast, severe erosion and damage to coastal structures often result. This occurs periodically along the California coast when the wave period, deep-water direction of approach, and height are such as to focus energy on coastal structures. *See* OCEAN WAVES; SEA STATE.

Currents in the surf zone. When waves break so that there is an angle between the crest of the breaking wave and the beach, the momentum of the breaking wave has a component along the beach in the direction of wave propagation. This results in the generation of longshore currents that flow parallel to the beach inside of the breaker zone (Fig. 2*a*). After flowing parallel to the beach as longshore currents, the water is returned seaward along relatively narrow zones by rip currents.

The net onshore transport of water by wave action in the breaker zone, the lateral transport inside of the breaker zone by longshore currents, the seaward return of the flow through the surf zone by rip currents, and the longshore movement in the expanding head of the rip current all constitute the nearshore circulation system. The pattern that results from this circulation commonly takes the form of an eddy or cell with a vertical axis. The dimensions of the cell are related to the width of the surf zone; the spacing between rip currents is usually two to eight times the width of the surf zone.

When waves break with their crests parallel to a straight beach, the flow pattern of the nearshore circulation cell becomes symmetrical (Fig. 2b). Longshore currents occur within each cell, but there is no longshore exchange of water from cell to cell.

The nearshore circulation system produces a continuous interchange between the waters of the surf and offshore zones, acting as a distributing mechanism for nutrients and as a dispersing mechanism for land runoff. Offshore water is transported into the surf zone by breaking waves, and particulate matter is filtered out on the sands of the beach face. Runoff from land and pollutants introduced into the surf zone is carried along the shore and mixed with the offshore waters by the seaward-flowing rip currents.

There is no evidence of undertow in the surf zone, other than the instantaneous motion occurring as the backwash flows under a wave breaking on the face of a steep beach. The principal danger to swimmers is from rip currents that may carry them seaward unexpectedly. Longshore currents may attain velocities in excess of 5 knots, while rip current velocities in excess of 3 knots have been measured.

Periodicity or fluctuation of current velocity and direction is a characteristic of flow in the nearshore system. This variability is primarily due to the grouping of high waves followed by low waves, a phenomenon called surf beat that gives rise to a pulsation of water level in the surf zone.

Formation of circulation cells. Nearshore circulation cells result from differences in mean water level in the surf zone associated with changes in breaker height along the beach. Waves transmit momentum in the direction of their travel, and their passage through water produces second-order pressure fields that significantly change the mean water level near the shore. Near the surf zone the presence of the pressure field produces a decrease in mean water level termed wave setdown. The setdown is proportional to the square of the wave height and for waves near the surf zone has a maximum value that is about one-sixteenth that of the breaker height. Shoreward of the break point the onshore flux of momentum against the beach produces a rise in mean water level over the beach face referred to as setup. A profile of the envelope of wave height and the experimental and theoretical values of wave setdown and setup are shown in Fig. 3.

If the wave height varies along a beach, the setup will also vary, causing a longshore gradient in mean water level within the surf zone. Longshore currents flow from regions of high water to regions of low water and thus flow away from zones of high waves. The longshore currents flow seaward as rip currents where the breakers are lower. Pronounced changes in breaker height along beaches usually result from wave refraction over irregular offshore topography. However, on a smaller scale, uniformly spaced zones of high and low breakers occur along straight beaches with parallel offshore contours. It has been shown that these alternate zones of high and low waves are due to the interaction of the incident waves traveling toward the beach from deep water with one of the many possible modes of oscillation of the nearshore zone known as edge waves. Edge waves are trapped modes of oscillation that travel along the shore (Fig. 4). Circulation in the nearshore cell is enhanced by edge waves having the period of the incident waves, or that of their surf beat, because these interactions produce alternate zones of high and low breakers whose positions are stationary along the beach. It appears that the edge waves can be either standing or progressive. In either

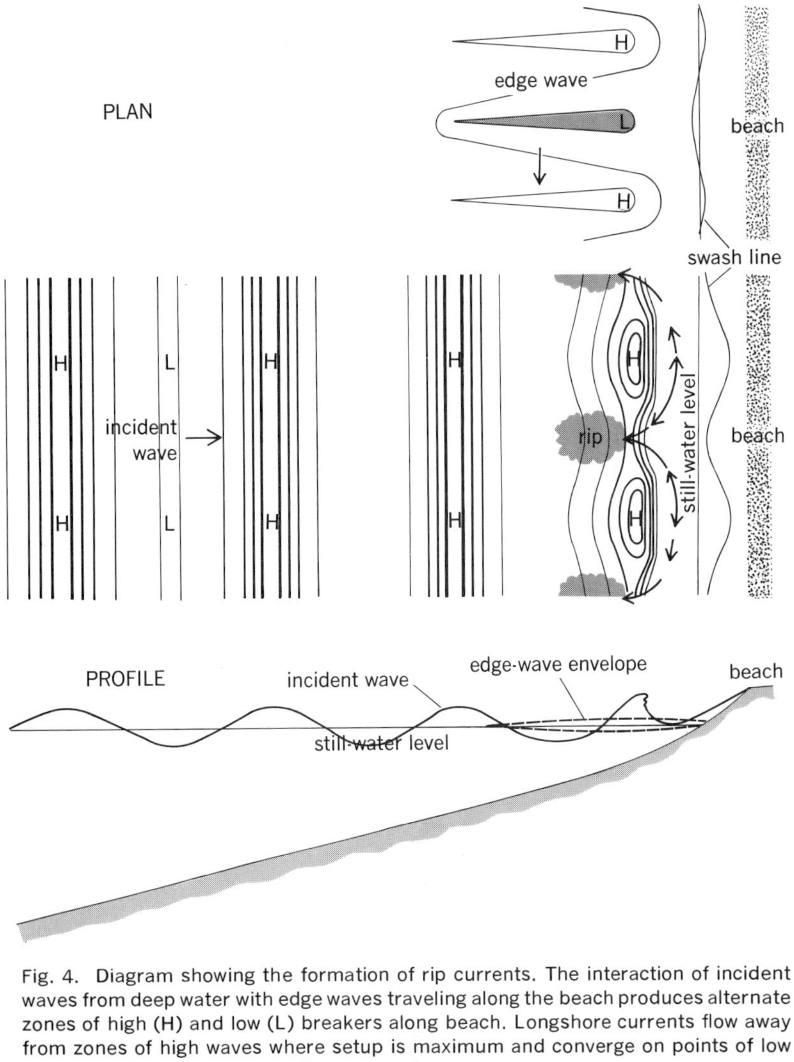

Fig. 4. Diagram showing the formation of rip currents. The interaction of incident waves from deep water with edge waves traveling along the beach produces alternate zones of high (H) and low (L) breakers along beach. Longshore currents flow away from zones of high waves where setup is maximum and converge on points of low waves, causing rip currents to flow seaward.

case, the spacing between zones of high waves (and hence between rip currents) is equal to the wave length of the edge wave.

The position and spacing of rip currents can be predicted under certain controlled conditions in the laboratory if the height and period of the incident wave and the slope and length of the beach are known. However, because of variability in incident waves and in the beach configuration, prediction of rip current spacing on natural beaches is uncertain. Where beaches are irregular, or where offshore topography produces irregularities in the wave refraction pattern, the location of cells is also dependent upon the regional gradation of breaker height along the beach.

Points, breakwaters, and piers all influence the circulation pattern and alter the direction of the currents flowing along the shore. In general, these obstructions determine the position of one side of the circulation cell. In places where a relatively straight beach is terminated on the down-current side by points or other obstructions, a pronounced rip current extends seaward. During periods of large waves having a diagonal approach to the shore, these rip currents can be traced seaward for 1 mi or more.

Beach types. Beaches consist of transient clastic material (unconsolidated fragments) that reposes near the interface between the land and the sea and is subject to wave action. The material is in dynamic repose rather than a stable deposit, and thus the width and thickness of beaches is subject to rapid fluctuations, depending upon the amount and rigor of erosion and transportation of beach material. Beaches are essentially long rivers of sand that are moved by waves and currents and are derived from the material eroded from the coast and brought to the sea by streams.

The geometry of beaches is also dependent on coastal history, and there is a close relationship between beach characteristics and type of coast. Long straight beaches are typical of low sandy coasts; shorter crescent-shaped beaches and small pocket beaches are more common along mountainous coastlines. The coast may be cliffed as shown in Fig. 5, or it may contain a ridge of windblown sand dunes and be backed by marshes and water. Along many low sandy coasts, such as the East and Gulf coasts of the United States, the beach is separated from the mainland by water or by a natural coastal canal. Such beaches are called barrier beaches. A beach that extends from land and terminates in open water is referred to as a spit, while a beach that connects an island or rock to the mainland or another island is a tombolo.

While differing in detail, beaches the world over have certain characteristic features which allow application of a general terminology to their profile (Fig. 5). The beach or shore extends landward from the lowest water line to the effective limit of attack by storm waves. The region seaward is termed the offshore; that landward, the coast. The beach includes a backshore and foreshore. The backshore is the highest portion and is only acted upon by waves during storms. The foreshore extends from the crest of the berm to the low-water mark and is the active portion of the beach traversed by the

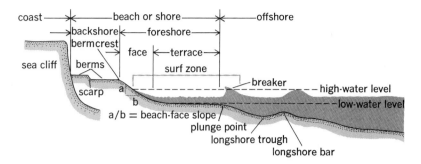

Fig. 5. Beach profile, showing characteristic features.

uprush and backwash of the waves. The foreshore consists of a steep seaward dipping face related to the size of the beach material and the rigor of the uprush and of a more gentle seaward terrace, sometimes referred to as the low-tide terrace or step, over which the waves break and surge. In some localities the foreshore face and terrace merge into one continuous curve; in others, there is a pronounced discontinuity at the toe of the beach face. The former condition is characteristic of fine sand beaches and of coasts where the wave height is equal to or greater than the tidal range; the latter is typical of coastlines where the tidal range is large compared with the wave height, as along the Patagonian coast of South America and portions of the Gulf of California.

The offshore zone frequently contains one or more bars and troughs that parallel the beach; these are referred to as longshore bars and longshore troughs. Longshore bars commonly form on the bottom at the plunge point of the wave, and their position is thus influenced by the breaker height and the nature of the tidal fluctuation.

Beach cycles. Waves are effective in causing sand to be transported laterally along the beach by longshore currents and in causing movements of sand from the beach foreshore to deeper water and back again to the foreshore. These two types of transport, although interrelated, are more conveniently discussed in separate sections. The offshore and onshore transport of sand is closely related to the beach profile and to the cycles in beach width and will be discussed here and under the mechanics of beach formation.

Along most coasts there is a seasonal migration of sand between the beaches and deeper water, in response to the changes in the character and direction of approach of the waves. In general, the beaches build seaward during the small waves of summer and are cut back by high winter storm waves. There are also shorter cycles of cut and fill associated with spring and neap tides and with nonseasonal waves and storms. Bottom surveys indicate that most offshore-onshore interchange of sand occurs in depths less than 10 m but that some effects may extend to depths of 30 m or more.

Figure 5 shows the profile of a typical summer beach which has been built seaward by low waves. During stormy seasons the beach foreshore is eroded, frequently forming a beach scarp. Subsequent low waves build the beach foreshore seaward again. The beach face is a depositional feature and its highest point, the berm crest, represents the maximum height of the run-up of water on the face of the beach. The height of wave run-up usually varies between one and three times the height of the breaking wave. Since the height of the berm depends on wave height, the higher berm, if it is present, is sometimes referred to as the winter or storm berm, and the lower berm as the summer berm.

The entire beach may be cut back to the country rock during severe storms. Under such conditions the waves erode the coast and form the well-known sea cliffs and wave-cut terraces, which are frequently preserved in the geologic record and serve as markers to the past relations between the levels of the sea and the land.

Beach cusps. A series of regularly spaced scallops, called beach cusps, sometimes forms on the beach foreshore (Fig. 6). Beach cusps consist of short transverse valleys formed in the beach face and sepaated by ridges with cuspate points. The spacing between cuspate points ranges from several to several hundred meters, while the depths of the valleys range from a few centimeters on fine sand beaches to several meters on pebble and cobble beaches. The formation of cusps appears to be associated with a wave-wave interaction similar to that operative in the formation of the nearshore

Fig. 6. The effect of headlands on the accretion of beach sand, shown in photograph of Point Mugu, Calif. The point forms a natural obstruction which interrupts the longshore transport of sand, causing accretion and a wide beach to form (foreground). The regularly spaced scallops are beach cusps. (*Department of Engineering, University of California, Berkeley*)

circulation cell. In the case of cusps the run-up on the beach face and the longshore flow modify the beach face where the longshore flow converges, causing a depression or valley. When low waves break directly on narrow, steep beaches, the cusps assume the dimensions of the nearshore circulation cell from which they are derived (Fig. 6). However, where the surf zone is wide, cusps that form on the beach face are small compared to the spacing between rip currents. In this case it seems likely that the formation of cusps is in response to an interaction of the bore from the breaking wave with an edge wave disturbance traveling within the surf zone. Cusps occur with greatest frequency during neap tides when the water level remains nearly constant.

Mechanics of beach formation. Wherever there are waves and an adequate supply of sand or coarser material, beaches form. Even man-made fills and structures are effectively eroded and reformed by the waves. The initial and most characteristic event in the formation of a new beach from a heterogenous sediment is the sorting out of the material, with coarse material remaining on the beach and fine material being carried away. Concurrent with the sorting action, the material is rearranged, some being piled high above the water level by the run-up of the waves to form the beach berm, some carried back down the face to form the foreshore terrace. In a relatively short time, the beach assumes a profile which is in equilibrium with the forces generating it.

The back-and-forth motion of waves in shallow water produces stresses on the bottom that place sand in motion. The interaction of the wave stresses with the bottom also induces a net boundary current flowing in the direction of wave travel. The most rapidly moving layer of water is near the bed, and for waves traveling over a horizontal bed the interaction of wave stresses and the boundary current produces a net transport of sand in the direction of wave travel. Thus waves traveling toward the shore tend to contain sand against the shore.

The action of waves on an inclined bed of sand eventually produces a profile that is in equilibrium with the energy dissipation associated with the oscillatory motion of the waves over the sand bottom. If an artificial slope exceeds the natural equilibrium slope, an offshore transport of sand will result from the gravity component of the sand load, until the slope reaches equilibrium. Conversely, if an artificial slope is less than the natural equilibrium slope, a shoreward transport of sand will result and the beach slope will steepen. Thus, the equilibrium slope is attained when the up-slope and down-slope transports are equal.

Since the equilibrium slope steepens with increasing energy dissipation and with increasing speed of the boundary current, it is usually gentle in deeper water where wave dissipation is slight and currents are low, and steeper near the surf zone where both dissipation of energy and the currents are greatest. This dependence of slope on wave stress and boundary currents causes the slight steepening of slope found just outside of the surf zone in many beach profiles (Fig. 5).

The slope of the beach face is related to the dissipation of energy by the swash and backwash over the beach face. Percolation of the swash into a permeable beach reduces the amount of flow in the backwash and is thus conducive to deposition of the sand transported by the swash. If in addition the beach is dry, this action is accentuated. Coarse sands are more permeable and consequently more conducive to deposition and the formation of steep beach faces. Large waves elevate the water table in the beach. When the beach is saturated, the backwash has a higher velocity, a condition that is conducive to erosion. From the foregoing it follows that the slope of the beach face, sometimes referred to as the foreshore slope, increases both with increasing sediment size and with decreasing wave height. Some typical average values for the face slopes of ocean beaches are given in the table.

Oscillatory ripples, which form on the sandy bottom, are an important factor in the mechanism of transportation and sorting of beach sands. Because turbulence and lift force are most intense at the crest of the ripple, only the coarsest material is deposited there. The fine material is placed in suspension and removed from the area, while the coarse material moves with the ripple toward the beach. As a consequence of the sorting action by ripples and by the uprush and backwash on the beach face, beach sands are the best sorted sedimentary deposits.

The beach face is frequently characterized by laminations (closely spaced layers) that show slight differences in the size, shape, or density of the sand grains. The laminations parallel the beach face and represent "shear sorting" within the granular load as it is transported by the swash and backwash of the waves over the beach face (Fig. 7). Detailed examination shows that each lamina of fine-grained minerals delineates the plane of shear between moving and residual sand. The mechanism concentrating the heavy fine grains at the shear plane is partly the effect of gravity acting during shearing, causing the small grains to work their way down through the interstices between larger grains, and partly the dependence of the normal dispersive pressure upon grain size. When grains are sheared, the normal dispersive pressure between grains, which varies as the square of the grain diameter, causes large grains to drift toward the zone of least shear strain, that is, the free surface, and the smaller grains toward the zone of greatest shear strain, that is, the shear plane.

Longshore movement of sand. The movement of sand along the shore occurs in the form of bed load (material rolled and dragged along the bottom) and suspended load (material stirred up and carried with the current). Suspended load transportation occurs primarily in the surf zone where the turbulence and vertical mixing of water are most effective in placing sand in suspension and where the longshore currents which transport the sediment-laden waters have the highest velocity. Concentrations of suspended sand in the surf zone, as high as 30 g/liter, have been measured along the coast of the Black Sea, suggesting that suspension is one of the most important processes in littoral drift. On the other hand, it is probable that suspension becomes less important for very

Fig. 7. Laminations in the beach face at La Jolla, Calif. Black laminae are heavy minerals and the light ones are quartz. Scale is 15 cm long. (*From D. L. Inman, Beach Erosion Board, U.S. Eng. Corps Tech. Mem. no. 39, 1953*)

coarse sand and cobble beaches.

The volume of littoral transport along oceanic coasts is usually estimated from the observed rates of erosion or accretion, most often in the vicinity of coastal engineering structures such as groins or jetties. In general, beaches build seaward upcurrent from obstructions and are eroded on the current lee, where the supply of sand is diminished. Such observations indicate that the transport rate varies from almost nothing to several million cubic meters per year, with average values of 150,000–600,000 m³/year. Along the shores of smaller bodies of water, such as the Great Lakes of the United States and the Mediterranean Sea, the littoral transport rate can be expected to range about 7000–150,000 m³/year. In general these are conservative estimates, since the volume of material moved commonly exceeds that indicated either by deposition or erosion.

The large quantity of sand moved along the shore and the pattern of accretion and erosion that occurs when the flow is interrupted pose serious problems for the coastal engineer. The problem is particularly acute when jetties are constructed to stabilize and maintain deep navigation channels through sandy beaches. A common remedial procedure is to dredge sand periodically from the accretion on the up-current side of the obstruction and deposit it on the eroding beaches in the cur-

Relationship between size of beach sediment and slope of beach face

Beach sediment		Slope of beach face, degrees
Type	Size, mm	
Very fine sand	1/16–1/8	1
Fine sand	1/8–1/4	3
Medium sand	1/4–1/2	5
Coarse sand	1/2–1	7
Very coarse sand	1–2	9
Granules	2–4	11
Pebbles	4–64	17
Cobbles	64–256	24

rent lee. Another method is the installation of permanent sand bypassing plants which continually remove the accreting sand and transport it by hydraulic tube to the beaches in the lee of the obstruction.

Mechanics of longshore transport of sand. The longshore transport of sediment is caused by the action of waves and currents. Waves and surf dissipate power on a granular bed by the stress of the motion of the water over the bed. The immersed weight of sediment placed in motion is proportional to the power dissipated by the waves. The presence of any net current, such as the longshore current, will produce a net transport of sediment in the direction of the current. In the absence of a current, the wave power is simply expended in a back-and-forth motion of sediment, and there is no net transport of sediment. Thus the transport rate should be proportional to the product of the power dissipated by the waves and the speed of the current. The power dissipated by waves can be approximated by assuming that all of the wave energy is expended in the surf zone. Unfortunately, reliable relations for predicting the longshore current have yet to be developed.

The longshore transport rate of sand is directly proportional to the longshore component of wave power estimated at the breaker point. Thus, the longshore transport rate of sand can be estimated from a knowledge of the total budget of wave energy incident upon the beach. The longshore transport rate is given by the relation $I = KP$, where I is the immersed weight transport rate in dynes per second, K is a constant of proportionality having a dimensionless value of about 0.77, and P is the longshore component of wave power per unit of beach length computed at the breaker point of waves in ergs per second per centimeter of beach length.

Source of beach sediments. The principal sources of beach and nearshore sediments are the rivers, which bring large quantities of sand directly to the ocean; the sea cliffs of unconsolidated material, which are eroded by waves; and material of biogenous origin (shell and coral fragments and skeletons of small marine animals). Occasionally sediment may be supplied by erosion of unconsolidated deposits in shallow water. Beach sediments on the coasts of the Netherlands are derived in part from the shallow waters of the North Sea. Wind-blown sand may be a source of beach sediment, although winds are usually more effective in removing sand from beaches than in supplying it. In tropical latitudes many beaches are composed entirely of grains of calcium carbonate of biogenous origin. Generally the material consists of fragments of shells, corals, and calcareous algae growing on or near fringing reefs. The material is carried to the beach by wave action over the reef. Some beaches are composed mainly of the tests of foraminifera that live on sandy bottom offshore from the reefs.

Streams and rivers are by far the most important source of sand for beaches in temperate latitudes. Cliff erosion probably does not account for more than about 5% of the material on most beaches. Wave erosion of rocky coasts is usually slow, even where the rocks are relatively soft shales. On the other hand, retreats greater than 1 m a year are not uncommon in the unconsolidated sea cliffs. C. Lyell in 1873 showed that the Ganges and Brahmaputra rivers carry a volume of sediment into the Bay of Bengal each year that is 780 times greater than the material eroded by wave action from the 36-mi stretch of cliffs in the vicinity of Holderness, England. The Holderness cliffs, which are on the exposed North Sea coast, are noted for their rapid rate of erosion. According to J. A. Steers, they are 40 ft high and recede at the rate of about 7 ft/year.

Surprisingly, the contribution of sand by streams in arid countries is quite high (Fig. 8). This is because arid weathering produces sand-size material and results in a minimum cover of vegetation, so that occasional flash floods may transport large volumes of sand. The maximum sediment yield occurs from drainage basins where the mean annual precipitation is about 30 cm/year, as has been shown by W. B. Langbien and S. A. Schumm.

Following initial deposition at the mouths of streams entering the ocean, much of the sand-size fraction of terrestrial sediments is carried along the coast by longshore currents. The sand carried by these littoral currents may be deposited in continental embayments, or it may be diverted to deeper water by submarine canyons which traverse the continental shelf and effectively tap the supply of sand. Recent observations by H. W. Menard suggest that most of the deep sediments on the abyssal plains along a 250-mi section of the California coastline are derived from two submarine canyons, Delgada Submarine Canyon in northern California and Monterey Submarine Canyon in central California. *See* DELTA; MARINE SEDIMENTS.

Biological effects. The rigor of wave action and the continually shifting substrate make the sand beach a unique biological environment. Because few large plants can survive, the beach is occupied largely by animals and microscopic plants. Much of the food supply for the animals consists of particulate matter that is brought to the beach by the nearshore circulation system and trapped in the

Fig. 8.   Sand delta at Rio de la Concepción on the arid coast of the Gulf of California. Such deltas are important sources of sand for beaches. (*Courtesy of D. L. Inman*)

sand. The beach acts as a giant sand filter that strains out particulate matter from the water that percolates through the beach face.

Since the beach-forming processes and the trapping of material by currents and sand are much the same everywhere, the animals found on sand beaches throughout the world are similar in aspects and habits, although, according to E. Dahl, different species are present in different localities. In addition, since the slopes and other physical properties of beaches are closely related to elevation, the sea animals also exhibit a marked horizontal zonation. Organisms on the active portion of the beach face tend to be of two general types, insofar as the procurement of food is concerned: those that burrow into the sand, using it for refuge while they filter particulate matter from the water through siphons or other appendages that protrude above the sandy bottom, and those that remove organic material from the surface of the sand grains by ingesting them or by "licking." There are usually few species, but those which are present may be very abundant; for example, *Thoracophelia nucronata*, a beach worm which ingests sand grains, was estimated to have a population of $100,000/m^3$ in the fine sand beach at La Jolla, Calif., and the bean clam *Donax gouldii* has peak populations of more than $25,000/m^2$ on the same beach.

A black layer is frequently found at depths of 5–75 cm below the surface of the beach foreshore. Chemically this is a reducing layer which has pH values greater than 8.0. J. R. Bruce has shown that the discoloration is caused by presence of ferrous sulfide which oxidizes to a reddish yellow on exposure to air. The formation of the layer is apparently related to the activity of bacteria on the organic material in the beach. This reducing layer is conducive to the deposition of calcium carbonate and may play an important role in cementing the beach sand and forming beach rock and nodules.

In tropical seas the entire shore may be composed of the cemented and interlocking skeletons of reef-building corals and calcareous algae. When this occurs, the nearshore current system is controlled by the configuration of the reefs that the organisms form. Where there are fringing reefs, breaking waves carry water over the edge of the reef, generating currents that flow along the shore inside of the reef and then flow back to sea through deep channels between reefs. Under such conditions beaches are usually restricted to a berm and foreshore face bordering the shoreward edge of the reef. *See* REEF.

[DOUGLAS L. INMAN]

Bibliography: R. A. Bagnold, *An Approach to the Sediment Transport Problem from General Physics*, USGS Prof. Pap. no. 422–1, 1966; A. J. Bowen, D. L. Inman, and V. P. Simmons, Wave set-down and set-up, *J. Geophys. Res.*, 73(8), 1968; J. S. Fisher and R. Dolan (eds.), *Beach Processes and Coastal Hydrodynamics*, 1977; D. L. Inman and R. A. Bagnold, Littoral processes, in M. N. Hill (ed.), *The Sea: Ideas and Observations*, vol. 3, 1963; D. L. Inman and A. J. Bowen, Flume experiments on sand transport by waves and currents, in J. W. Johnson (ed.), *Proceedings of the Eighth Conference on Coastal Engineering*, 1963; D. L. Inman, G. C. Ewing, and J. B. Corliss, Coastal sand dunes of Guerrero Negro, Baja California, Mexico, *Geol. Soc. Amer. Bull.*, 77(8), 1966; D. L. Inman and J. D. Frautschy, Littoral processes and the development of shorelines, *Coastal Engineering*, Santa Barbara Conference of the American Society of Civil Engineers, 1965; M. S. Longuet-Higgins and R. W. Stewart, Radiation stress in water waves: A physical discussion with application, *Deep-Sea Research*, vol. 11, 1964; F. P. Shepard, *Submarine Geology*, 2d ed., 1963; U.S. Army, *Shore Protection, Planning, and Design*, Coastal Eng. Res. Center Tech. Rep. no. 4, 1966; R. L. Wiegel, *Oceanographical Engineering*, 1964.

Noctilucent clouds

High-altitude clouds which are seen at the mesopause (about 260,000 ft) and are still illuminated by the Sun when it is already well below the observer's horizon during astronomical twilight. Noctilucent clouds occur only during the summer months at latitudes from about 45 to 70°. Generally, they resemble high cirrus clouds with pronounced band or wave structures and are usually so tenuous that stars can be seen through them; they are always pure white to blue-white. Rocket measurements and data from the worldwide observer network have lead to major revisions in hypotheses regarding the nature and formation of noctilucent clouds.

Data network. As a result of the establishment of the worldwide network for their observation, it is known that noctilucent clouds appear with about equal frequency over landmasses around the world in both the Northern and Southern hemispheres. Data over the oceans are still needed to establish whether the clouds and their wave structure may be associated with terrain features such as mountain ranges.

For the use of theoreticians and for analysis, the data from the existing network of stations are being processed into punched-card form at three world data centers—one at the Tartu Observatory in Estonia, Soviet Union; another at the University in Edinburgh, Scotland; and the third at the Meteorology Branch of the Canadian Department of Transport. An internationally accepted *Observer's Handbook* issued in 1968 contains information needed for observing and reporting the data. Exchange of information has also been facilitated by the organization of a subcommission on noctilucent clouds as part of the International Commission of the Meteorology of the Upper Atmosphere of the International Union of Geodesy and Geophysics.

Rockets. Since 1962 rocket measurements of noctilucent clouds and associated meteorological phenomena have been performed with increasing frequency. The results of the first rocket sampling experiments which showed a high concentration of nonvolatile residue particles on the collection plates that were exposed on passage through these altitudes appear to have been anomalous. Five subsequent firings have shown no appreciable increase over background levels in nonvolatile particulate concentration measured when the clouds were in evidence. It has not yet been established whether these findings are a result of the extreme variability of particulate concentration within cloud displays or the ability of noctilucent clouds to form in the absence of solid condensation nuclei. Complete analysis of data from 1968 and 1969 rocket-sampling measurements may help to clarify this issue. Temperature soundings by rockets in the presence and absence of noctilucent clouds have shown that there exists extremely low temperatures (130°K) at the mesopause during the arctic summer. This temperature structure is extremely uniform and stable in the presence or absence of noctilucent clouds and is quite different from the highly variable, irregular, much warmer temperatures (about 200°K) measured at the same heights during the arctic winter months. *See* Meteorological rockets.

Formation. It is now agreed that the cold temperatures are a necessary, but not a sufficient, condition for noctilucent-cloud formation. It is also generally accepted that the clouds are composed of ice crystals or water droplets which condense from water vapor brought up to the mesopause by the turbulence present during the arctic summer. Since cosmic dust nuclei are not always present in sufficient quantity, theories have been formulated which use ions as the condensation nuclei. One theory in particular hypothesizes the growth of the cloud particles by chemical attachment to iron hydride ions which result from meteoric ablation and subsequent atmospheric reactions that occur at 260,000–300,000-ft altitudes. This hypothesis would explain why noctilucent clouds are sometimes said to appear with higher probability on certain summer dates following meteoric showers,

even when no cosmic dust residue is found in the cloud particles. While it is generally concluded that the wave structure of noctilucent clouds is a manifestation of gravity waves in the atmosphere, the source of these gravity waves is still unknown. A recent theory associates the formation of the clouds and the wave structure with gravity waves propagated up from the ageostrophic component of the tropospheric jet stream. Such a mechanism would not only demonstrate the relationship between upper and lower atmosphere processes but, if valid, would lead to a criterion for predicting the occurrence of noctilucent clouds. *See* Cloud physics; Jet stream.   [robert k. soberman]

Bibliography: B. Fogle and B. Haurwitz, *Noctilucent Clouds*, in D. Reidel (ed.), *Space Science Reviews VI*, 1966; J. Kaunieks and P. Greenberg (eds.), *Physics of Mesospheric (Noctilucent) Clouds*, 1974; I. A. Khvostikov and G. Witt (eds.), *Noctilucent Clouds*, 1967; R. K. Soberman, *Noctilucent Clouds*, in S. L. Valley (ed.), *Handbook of Geophysics and Space Environments*, 1965; R. K. Soberman, S. A. Chrest, and R. F. Carnevale, *Rocket Sampling of Noctilucent Cloud Particles During 1964 and 1965*, in A. P. Mitra, L. G. Jacchia, and W. S. Newman (eds.), *Space Research VIII*, 1968.

North Sea

The North Sea overlies the European continental shelf between latitudes 51°N and 61°N. Its waters circulate freely with those of the northeast Atlantic Ocean between Scotland and Norway and through the Straits of Dover. The North Sea is a prolific fishery region. Numerous fishing grounds have dis-

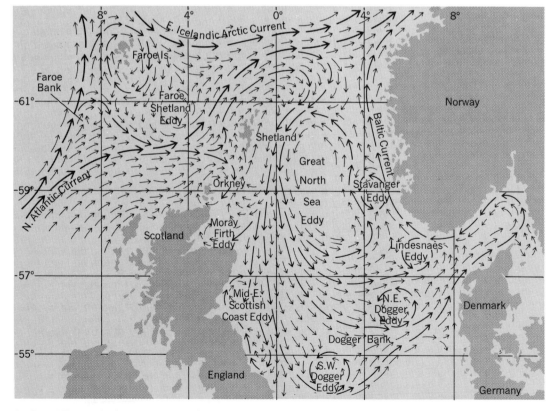

Surface drift currents in northern North Sea.

tinctive names, such as Dogger Bank and Fladen Ground. A wide range of clupeoids, gadoids, flatfishes, and crustaceans are commercially fished.

The southern half of the North Sea floor is a plateau, mostly less than 40 m deep. The northern half is a basin which deepens northward to the edge of the continental shelf at a depth of about 200 m. There is a narrow submarine valley along the Norwegian coast with depths ranging from about 240 to 350 m.

Atlantic oceanic water and continental (mainly Baltic) waters constantly flow into the North Sea. The Atlantic Ocean water is warmer and saltier than that entering the North Sea from the Baltic. In autumn and winter these waters mix to produce a characteristic water mass with intermediate conservative properties. During other seasons a halothermocline exists at 30–40 m depth. Occasionally, waters with Mediterranean or Arctic water mass characteristics invade the region. Each of these water-mass types has a characteristic fauna, chiefly plankton forms, by which it can be recognized. *See* ATLANTIC OCEAN; HALOCLINE; THERMOCLINE.

The inflow of oceanic water through the Straits of Dover is relatively small; a greater volume enters from the north and becomes part of the prevailing circulation. The surface current system in the northern part of the North Sea is shown in the illustration. Subsurface and bottom currents set in similar directions, except along the Norwegian coast. There a deep oceanic current moves south beneath a north-setting surface current. Dynamically and in their physicochemical characteristics these several water masses vary seasonally, fluctuate annually, and undergo other changes over longer periods of time. Catastrophic variations, or those which cause unusual mortality among fish, are known to occur.

Mean minimum surface temperatures in February range from about 7°C in the northwest part of the North Sea to less than 2°C in the southeast part. Mean maximum surface temperatures in August range from 11.5° to over 17°C from northwest to southeast, respectively. Extreme conditions of salinity do not coincide exactly with extreme conditions of temperature. Salinity values for February (expressed in parts per thousand) range from nearly 35.25°/oo to less than 32.00°/oo, and for August from more than 35.25°/oo to less than 31.00°/oo. Other less conservative properties, such as phosphate, nitrate, and oxygen contents, vary seasonally and according to biological activity. *See* SEA WATER.

Marked tidal forces in the North Sea in conjunction with atmospheric disturbances produce surges which cause damage on neighboring coasts, particularly in the United Kingdom and the Low Countries. *See* STORM SURGE.

[JOHN B. TAIT]

Bibliography: K. F. Bowden, Storm surges in the North Sea, *Weather*, 8:82–84, 1953; K. Chapman, *North Sea Oil and Gas*, 1976; J. H. Fraser, The plankton of the waters approaching the British Isles in 1953, *Marine Research Service, Scottish Home Department*, no. 1, pp. 1–12, 1955; J. R. Lumby and G. T. Atkinson, On the unusual mortality amongst fish during March and April 1929, in the North Sea, *J. cons., Cons. perma. intern. explor. mer*, 4(3):309–332, 1929; I. McClean, *North Sea Information Sources*, 1976; M. M. Sibthorp, *The North Sea: Challenge and Opportunity*, 1975; J. B. Tait, Hydrography of the Faroe-Shetland Channel, 1927–1952, *Marine Research Service, Scottish Home Department*, no. 2, 1957; J. B. Tait, The surface water drift in the northern and middle areas of the North Sea and in the Faroe-Shetland Channel, Fishery Board of Scotland, *Sci. Invest.*, no. 1, 1937.

Northwest Passage

The northern sea route between the Atlantic and Pacific oceans through the Canadian Archipelago. The entire route is frozen over in winter. As the ice cap retreats northward in summer, two routes are opened (see illustration). The eastern approach through Davis and Hudson straits is navigable in June, at which time the partially melted ice floes drift into the Atlantic Ocean on the Labrador Current. This approach remains open until November.

In the Canadian Archipelago the ice is landlocked. These passages remain closed until the ice melts. The tortuous southern route, close along the Canadian mainland coast, is usually passable from mid-August to mid-October. In this period it has been traversed by a number of ships. The more direct northern route, through Lancaster Sound and the Barrow and McClure straits, has never been found entirely ice-free. This passage has only been made by Wind-class icebreakers.

The southern Beaufort Sea is usually passable from mid-June to mid-October. The passage around Point Barrow is only possible while the arctic ice floes are held offshore by the summer southeast winds. Usually this condition prevails from early August to late September. The distribution of sea ice in the Chuckchi Sea and Bering Strait is controlled by the same winds. However, the ice retreats from these parts in late June and does not close them until October. The Bering Sea approach is navigable from early June to December. *See* BERING SEA.

Perhaps the most critical part of the passage is around Cape Barrow. Ships entering the Beaufort Sea must ensure their escape around the Cape, which may be closed by wind-drifted ice anytime after the first week of September. The approach to McClure Strait, along the west side of Banks Island, closes at the same time. With these routes closed, the ships must be prepared to retreat eastward to the Atlantic Ocean before the passages through the Archipelago become frozen. Icebreakers can pass through Prince of Wales Strait to the eastward of Banks Island and traverse the shorter northern passage. Other ships must take the longer southern passage where the freeze-up occurs a few weeks later. The best route for surface vessels is from west to east by the southern passage, which is used for most service to the DEW line radar stations.

In 1958 the Northwest Passage was used as a submarine route when the nuclear powered submarine USS *Nautilus* made the first submarine passage from the Pacific to the Atlantic Ocean. The first attempt in June was frustrated in the

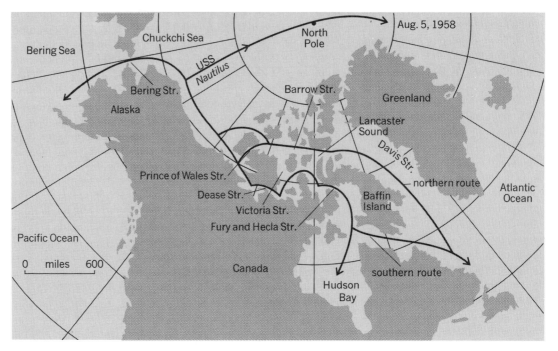

Map of Northwest Passage, showing northern and southern routes. A west-to-east shipping route has been proposed from the open waters in the western Arctic through Coronation Gulf and Dease, Victoria, Fury, and Hecla straits into the Atlantic Ocean or Hudson Bay.

shallow (160-ft) Chuckchi Sea by an ice ridge which extended 80 ft below the surface. The remaining depth did not provide safe passage for the 50-ft height of the ship. Returning through Bering Strait on July 29, the ship encountered the ice ridge farther north. This time the ice ridge was avoided by passing close around Point Barrow and entering the Arctic Ocean through the Barrow Canyon. This trench provides a minimum of 400-ft depth through the continental shelf.

During September, 1969, the super oil tanker SS *Manhattan,* converted to the world's largest icebreaker, made a test cruise from Boston, Mass., through the Northwest Passage to Prudhoe Bay, Alaska. Escorted by the Canadian icebreaker CGS *Sir John A. MacDonald* and with two aircraft to scout ahead, the ship undertook the direct passage from Baffin Bay through Lancaster Sound to McClure Strait. An attempt to exit into the Arctic Ocean at the western end of the strait was frustrated by heavily rafted (ridged) pack ice. After being extricated from the ice, the *Manhattan* turned south along the eastern side of Banks Island through the usually ice-free Prince of Wales Strait to the Beaufort Sea and reached Prudhoe Bay without incident. After taking on a token cargo of a gilded barrel of oil, the ship retraced the route.

Since the early 1950s many commercial vessels, including towed barges, have made all or part of the Northwest Passage with icebreaker escort. The passage of the *Manhattan* confirmed the feasibility of the route for an annual round-trip passage by adequate vessels during autumn months. With present technology the route is not feasible throughout the year. *See* ARCTIC OCEAN; SEA ICE.

[JOHN P. TULLY]

Ocean-atmosphere relations

This field of investigation is concerned with the boundary zone between sea and air and the dynamic relationships between oceanographic and meteorologic studies. When it is considered that the largest fraction of the heat energy the atmosphere receives for maintaining its circulation is derived from the condensation of water vapor originating primarily from oceanic evaporation, it becomes evident that an understanding of processes occurring at the air-sea boundary is fundamental to an understanding of atmospheric behavior. The oceanic energy supply to the atmosphere is highly regionalized because of the character of ocean currents, which in turn implies that the atmospheric circulation itself (and resulting weather) is greatly influenced by the oceanic circulation. Conversely, the oceanic circulation represents a state of equilibrium in which the effects of the frictional stresses of the wind on the sea surface are balanced by changes in the distribution of density of oceanic waters. These compensating density changes are in turn related to time and space variations in radiation, heat conduction, evaporation, and precipitation. It is therefore equally manifest that an understanding of the oceanic circulation (and the resulting distribution of properties within the ocean) requires a thorough knowledge of atmospheric processes at the air-sea boundary.

The conclusion is that neither ocean nor atmosphere should be treated independently, but that they should be considered together as a single dynamical-thermodynamical system. However, the interaction of ocean and atmosphere is so complicated that it is not yet possible to completely separate cause and effect. Therefore separate discussions are presented for each of the major classes of atmospheric influences upon the ocean, as well as the oceanic influences upon the atmosphere.

Effects of wind on ocean surface. The frictional stresses of the wind on the surface of the sea produce ocean waves (and storm surges) and ocean currents. The former are transitory phenomena and are not discussed in the present article because they are of little direct meteorological interest, even though waves at sea, coastal breakers, and storm tides are of considerable maritime, as well as oceanographic, importance. *See* SEA STATE; STORM SURGE.

Wind-induced ocean currents, on the other hand, are of large-scale significance from both the oceanographic and meteorological points of view. The wind exerts a twofold effect upon the surface layers of the ocean. In the first instance the stress of the wind leads to the formation of a shallow surface wind drift. The resulting transport of surface water by the wind drift leads in the second instance to pressure variations with depth and a changed distribution of mass (density) throughout the ocean. In the final analysis it is the resulting fields of density which account for the major current systems of the oceans. The total transport due to the wind drift is directed to the right of the wind (in the Northern Hemisphere), but the final density (slope) current which results from the sloping sea surface tends to flow in the direction of the prevailing wind, except where coastal configuration prevents the realization of such flow. Nevertheless, it should again be emphasized that the wind effect is not the sole meteorological factor that serves to determine the distribution of mass or the slope of isobaric surfaces in the oceans; heating and cooling, freezing and thawing, and evaporation and precipitation all exert their influences. *See* WIND STRESS OVER SEA.

Major ocean current systems. For the reasons just outlined, the major ocean currents of the world conform closely with the prevailing anticyclonic wind circulatons of the oceans. With the exception of the northern Indian Ocean, warm currents flow poleward to about 40° lat in the western portion of all oceans, with easterly flow in the higher latitudes, equatorward drift (relatively cold) in the eastern portions of the oceans, and westerly flowing currents near the Equator. The low temperature of the waters in the eastern portions of the oceans is due partly to the high-latitude origin of the currents and partly to coastal upwelling of cold subsurface waters. Of all the currents the poleward-flowing warm currents of the western portions of the oceans are the best developed and the most important. Examples are the Gulf Stream of the North Atlantic and the Kuroshio of the North Pacific, each of which transports a tremendous volume of warm tropical water into higher latitudes. *See* OCEAN CURRENTS; UPWELLING.

Hydrologic and energy relations. For the Earth as a whole and for the entire year, the amounts of energy received and lost through radiation are in balance for all practical purposes. However, this is not true for any given portion of the Earth's surface, particularly during any given fraction of the annual solar cycle. The ratio of insolation to outgoing radiation decreases from the Equator toward the poles. Between the Equator and the 35th parallel, the Earth receives more energy through radiation than it loses; the reverse is true poleward from about the 35th parallel.

Because observations indicate that the lower latitudes are not becoming progressively warmer and the higher latitudes colder, it must be assumed that considerable heat is transported from lower to higher latitudes by both atmosphere and ocean. According to H. U. Sverdrup (see bibliography, G. P. Kuiper, 1954), the meridional transport of energy in the Northern Hemisphere reaches a maximum a little north of latitude 35°N. At latitude 30°N, Sverdrup computes the total energy transport across the latitude circle to be 6.5×10^{16} cal/min of which 1.9×10^{16} cal/min (or 29%) is accomplished by ocean currents, principally by the Gulf Stream and Kuroshio. The remaining energy transport is accomplished by the atmosphere.

The largest fraction of radiant energy absorbed by the oceans is utilized in evaporating sea water. A much smaller fraction, about 10%, is utilized in more direct heating of the atmosphere in contact with the sea surface. The latent energy of vaporization subsequently becomes available to the atmosphere as either sensible heat or gravitational potential energy when condensation takes place, often in a region far removed from the area where the evaporation occurred. The precipitation resulting from the condensation of atmospheric water vapor is then returned to the ocean, either directly as rainfall or snowfall, or indirectly as runoff and discharge from land areas. The hydrologic and energy cycle is thereby completed. *See* HYDROLOGY; HYDROSPHERIC GEOCHEMISTRY; MARINE INFLUENCE ON WEATHER AND CLIMATE; TERRESTRIAL ATMOSPHERIC HEAT BALANCE.

The maximum evaporation and heat exchange between sea and atmosphere take place where cold air flows over warm-water surfaces. The ideal locations for maximum moisture or energy transfer are therefore those areas where cold continental air flows out over warm poleward-moving ocean currents. Such ideal conditions exist during winter off the eastern coasts of the continents and over warm currents such as the Gulf Stream and Kuroshio. Radiant energy that was absorbed and stored by the oceans at lower latitudes is given off to the atmosphere by this process at places and during seasons of marked deficiency in radiative energy. Any change in the oceanic transport by ocean currents must be reflected in corresponding changes in the rates of evaporation and must finally have significant effects upon the atmospheric circulation. *See* CLIMATIC CHANGE.

Relations with sea-water salinity. In the absence of horizontal flow, the surface salinity of any portion of the ocean is mainly determined by three processes: decrease of salinity by precipitation, increase of salinity by evaporation, and change of salinity by vertical mixing. Salinities thus tend to be high in regions where evaporation exceeds precipitation and low where precipitation exceeds evaporation. However, horizontal transport of surface waters by wind-induced ocean currents serves to displace the areas of maximum or minimum salinity in the direction of surface flow away from the areas of maximum differences (positive or negative) between evaporation and precipitation. The conclusion, of course, is that the distributions of surface salinities (as well as other properties) in the ocean are determined almost completely by atmospheric circumstances. *See* SEA WATER.

Other factors. In addition to the transfer of momentum, heat, and the latent energy involved in the evaporation-condensation-precipitation process, the oceans and the atmosphere interact in a multitude of other ways. These include radiation effects, flow of electrical charges, sea salts (a source of atmospheric condensation nuclei), and the flow of gases across the air-sea interface. Of the gases, the exchange of carbon dioxide between sea and atmosphere is perhaps the most important. Since the ocean absorption of carbon dioxide increases with salinity and decreases with temperature, it is suggested that in low latitudes the air is enriched with carbon dioxide given up by the ocean and that the general atmospheric circulation carries the carbon dioxide into higher latitudes, where it again dissolves in the sea water which in time brings it back toward the Equator. The combustion of fossil fuels has added considerable excess of carbon dioxide to the atmosphere during the past 100 years. In view of the great importance of carbon dioxide in determining the radiational balance of Earth and space, it is extremely important to determine whether this output of carbon dioxide has caused a significant increase in the total content of carbon dioxide in the atmosphere or whether most of it has been absorbed by the oceans. This is a controversial matter that is still under investigation by meteorologists and oceanographers.

[WOODROW C. JACOBS]

Bibliography: B. Bolin and E. Eriksson, Changes in the carbon dioxide content of the atmosphere and sea due to fossil fuel combustion, in B. Bolin (ed.), *The Atmosphere and Sea in Motion*, pp. 130–142, 1959; T. E. Graedel, The kinetic photochemistry of the marine atmosphere, *J. Geophys. Res.*, January 1979; W. C. Jacobs, The energy exchange between sea and atmosphere and some of its consequences, *Bull. Scripps Inst. Oceanogr. Univ. Calif.*, 6:27–122, 1951; G. P. Kuiper (ed.), *The Solar System*, vol. 2, 1954; P. S. Liss and P. G. Slater, Flux of gases across the air-sea interface, *Nature*, 247:181–184, 1974; J. S. Malkus, *Large-scale Interactions*, in M. N. Hill (ed.), *The Sea*, vol. 1, 1962; G. Newman and W. J. Pierson, *Principles of Physical Oceanography*, 1966; H. U. Roll, *Physics of the Marine Atmosphere*, 1965; H. U. Sverdrup, *Oceanography for Meteorologists*, 1942.

Ocean currents

The general circulation of the ocean. The term is usually understood to include large-scale, nearly steady features, such as the Gulf Stream, as well as current systems which change seasonally but are persistent from one year to the next, such as the Davidson Current, off the northwestern United States coast and the equatorial currents in the Indian Ocean. Since 1965, it has been learned that a great number of energetic motions have periods of a month or two and horizontal scales of a few hundred kilometers—a very-low-frequency turbulence, collectively called eddies. These are transient features, but locally have all the characteristics of currents. They are ubiquitous features in the ocean, although their energy varies widely, being highest near Western Boundary Currents. Energetic motions are also concentrated near the local inertial period (24 hr, at 30° latitude) and at the periods associated with tides (primarily diurnal and semidiurnal). *See* TIDE.

The greatest single driving force for currents, as for waves, is the wind. Furthermore, the ocean absorbs heat at low latitudes and loses it at high latitudes. The resultant effect on the density distribution is coupled into the large-scale wind-driven circulation. Some subsurface flows are caused by the sinking of surface waters made dense by cooling or high evaporation. *See* OCEAN WAVES.

This discussion treats the dynamics of ocean currents, characteristics of the surface and deep circulation, and methods of measuring currents. For a discussion of water movements within estuaries, the distribution of physical properties in sea water, the characteristics of water masses, and mixing and stirring processes *see* ESTUARINE OCEANOGRAPHY; SEA WATER.

BASIC EQUATIONS

The wind at the sea surface acts through frictional drag to exert a stress on the water's surface. Except for this effect, the direct effects of friction play a very small role in ocean circulation. The other important forces are gravity and pressure-gradient forces. The conservation of momentum, in the x, y, and z directions (a cartesian coordinate system), is expressed in Eqs. (1)–(3), where x is to

$$\frac{du}{dt} - fv = -\frac{1}{\rho}\frac{\partial p}{\partial x} + F_x \qquad (1)$$

$$\frac{dv}{dt} + fu = -\frac{1}{\rho}\frac{\partial p}{\partial y} + F_y \qquad (2)$$

$$\frac{dw}{dt} + g = -\frac{1}{\rho}\frac{\partial p}{\partial z} + F_z \qquad (3)$$

the east, y is to the north, and z is positive upward, with velocity components u, v, w in these directions; p is pressure; ρ is density; and F represents (schematically) the effects of frictional stresses in each component equation. The Coriolis parameter, f, is $2\Omega \sin \phi$, where ϕ is latitude and Ω is the angular rotation rate of the Earth. The magnitude of f is about 10^{-4} s at middle latitudes. The terms fu and fv are apparent accelerations and result from the rotation of the Earth. *See* CORIOLIS ACCELERATION AND FORCE.

For low-frequency motions (periods of about a week or longer) the middle two terms in Eqs. (1) and (2) balance each other to a good approximation (better than 5%) away from lateral boundaries and from the surface and bottom boundary layers so that, for the x equation [Eq. (1)] only, Eq. (4) is ob-

$$-fv = -\frac{1}{\rho}\frac{\partial p}{\partial x} \qquad (4)$$

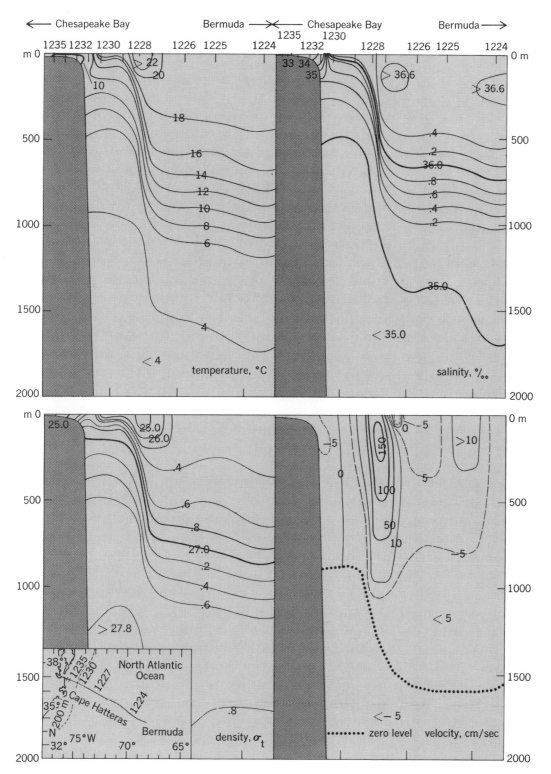

Fig. 1. Cross section of Gulf Stream showing vertical distribution of temperature, salinity, density, and velocity. Current flow in southwest-northeast direction. Inset shows locations of cross section and stations. (*After* G. Dietrich, *General Oceanography*, *Wiley*, 1963, *based on section by RV Atlantis, Woods Hole Oceanographic Institution, Stations 1224–1235*)

tained. This is the (x component of the) geostrophic equation. Most open-ocean currents are nearly in geostrophic balance, except for the upper 50–100 m which is influenced directly by wind stress.

Except in (high-frequency) surface waves, the middle two terms in Eq. (3) balance each other, giving the hydrostatic equation, which is a better approximation than the geostrophic balance.

The right-hand side of Eq. (4) is the horizontal pressure-gradient force per unit mass, in the x direction. Its magnitude can be determined, for a typical case, from the following considerations. It is known that sea level near New York or Chesapeake Bay is about 1 m lower than near Bermuda, on the other side of the Gulf Stream. This change in level takes place in a distance of roughly 100 km. The magnitude of this term is then 10^{-2} cm/s. With $f = 10^{-4}$ s, an average value for the northward-flowing stream speed, v, is 100 cm/s. The peak velocities in the Gulf Stream are $2-2\frac{1}{2}$ times this amount ($4-5$ knots, or $2-2.5$ m/s). A cross section of properties is shown in Fig. 1. Looking downstream, it may be seen that the thermocline rises sharply in the region of the stream. On the offshore side, the water is warmer than on the inshore side at the same depth. Thus, since the warmer water is lighter, the sea surface must [through the hydrostatic Eq. (3)] stand higher on the offshore side if the horizontal pressure gradients vanish in the deep water. (The Gulf Stream extends only to about 3000 m depth.)

This last idea leads to a calculation that is fundamental to much of physical oceanography—knowing the density distribution, which can be measured, to calculate the pressure field, which is almost never measured. The equation for vertical shear [combining Eqs. (1) and (3)] is Eq. (5). Note

$$\frac{\rho f}{g}\frac{\partial v}{\partial z} = \frac{\partial \rho}{\partial x} \qquad (5)$$

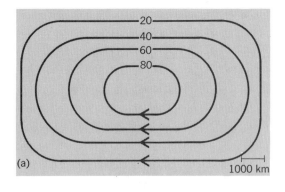

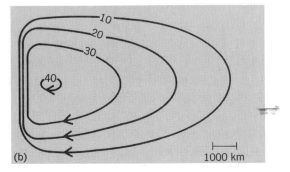

Fig. 3. Streamlines showing currents for (a) an ocean on a nonrotating globe and (b) an ocean on a uniformly rotating globe in which the Coriolis forces increase with the geographic latitude. (*After H. Stommel, The Gulf Stream, University of California Press, 1965*)

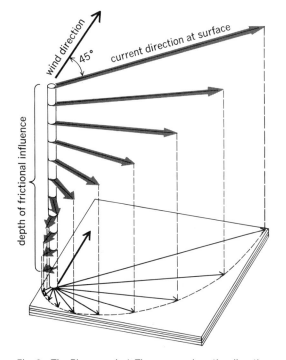

Fig. 2. The Ekman spiral. The arrows show the direction and magnitude of the purely wind-driven current, as it changes with depth. The angle between the wind and the surface current is a function of the vertical mixing by turbulent motions; if this mixing is uniform with depth, the angle is 45° as shown.

that only the vertical variation in speed (shear) can be computed; hence the need, on a density section such as in Fig. 1, to measure the velocity field directly at some depth (such as at 2000 m). The velocity at 2000 m, for example, in the Gulf Stream is about 10 cm/s. Lack of exact knowledge of the deep velocities, therefore, introduces an uncertainty of about 10% in the calculation of surface speeds in a strong current, but introduces large uncertainty in total mass transport calculations.

Many theories of ocean circulation deal with vorticity, by taking the curl of Eqs. (1) and (2). The conservation of vorticity, especially in the vertically integrated form, is in several ways simpler to deal with than the conservation of momentum.

Wind stress. Air does not merely glide along the surface of the water, but exerts a frictional effect, or wind stress, which causes the surface water to be carried along with it. The movement of this thin layer on the surface of the water is conveyed by an internal turbulent friction to the deeper levels. The eventual result of such interaction, in a limitless homogeneous sea under the influence of a steady wind, would be a pure drift current, the theory of which was developed by V. W. Ekman. The resulting current distribution is illustrated by the so-called Ekman spiral (Fig. 2). There is a current at the sea surface at an angle to the right of the direction of the wind in the Northern Hemisphere. With increasing depth, the current turns farther toward the right and gradually subsides. When the direction of this current reaches an angle of 180° to the

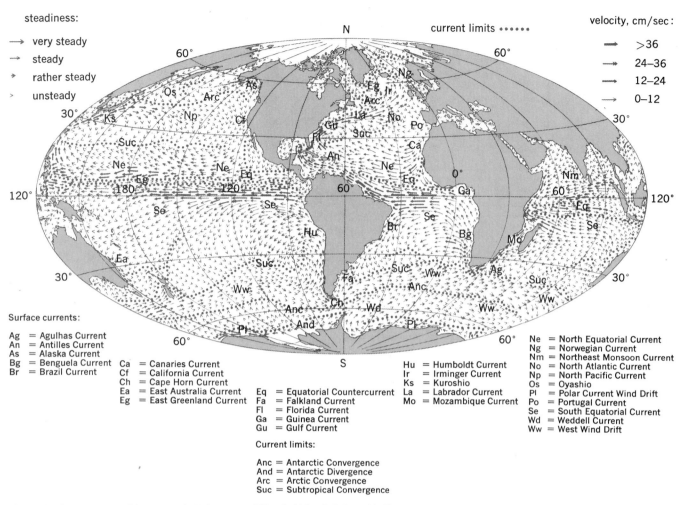

steadiness:

→ very steady

→ steady

↗ rather steady

> unsteady

current limits ••••••

velocity, cm/sec:

→ >36

→ 24–36

→ 12–24

→ 0–12

Surface currents:

Ag = Agulhas Current
An = Antilles Current
As = Alaska Current
Bg = Benguela Current
Br = Brazil Current

Ca = Canaries Current
Cf = California Current
Ch = Cape Horn Current
Ea = East Australia Current
Eg = East Greenland Current

Eq = Equatorial Countercurrent
Fa = Falkland Current
Fl = Florida Current
Ga = Guinea Current
Gu = Gulf Current

Hu = Humboldt Current
Ir = Irminger Current
Ks = Kuroshio
La = Labrador Current
Mo = Mozambique Current

Ne = North Equatorial Current
Ng = Norwegian Current
Nm = Northeast Monsoon Current
No = North Atlantic Current
Np = North Pacific Current
Os = Oyashio
Pl = Polar Current Wind Drift
Po = Portugal Current
Se = South Equatorial Current
Wd = Weddell Current
Ww = West Wind Drift

Current limits:

Anc = Antarctic Convergence
And = Antarctic Divergence
Arc = Arctic Convergence
Suc = Subtropical Convergence

Fig. 4. Surface currents of the oceans in February and March. (*After G. Schott, 1943*)

flow on the sea surface, the speed of the current is only 1/23 that of the surface water. This depth is called the depth of frictional influence. For example, at a latitude of 50° this depth amounts to about 60 m when the wind speed reaches 7 m/s (13.8 knots).

Observations show that the angle between wind and surface velocity is usually between 30° and 45°; the surface velocity is about 2% of the wind speed. Both these factors depend on how strongly the upper layers are stratified and how effectively the vertical turbulent motions transmit the stresses downward.

The wind-driven currents are in addition to any other currents that are present—from the large-scale density distribution, eddies, tides, and so on. One important feature of the wind-driven currents that is independent of stratification or turbulent intensity is that the total volume transport is at right angles to the wind. This can be seen from the vertical integral of Eqs. (1) and (2), if the time-dependent and pressure-gradient terms are neglected. It is understandable, therefore, that upwelling of cold deep water occurs at a coast when the wind blows parallel to the coast, with the coast on the left-hand side of the wind. *See* WIND STRESS OVER SEA.

An outstanding feature of ocean currents is that

intense flows (such as the Gulf Stream) occur on the western sides of the oceans. These flows were shown by H. Stommel to result from the variation of Coriolis parameter with latitude (Fig. 3). *See* GULF STREAM.

Surface currents. Except in western boundary currents, and in the Antarctic Circumpolar Current, the system of strong surface currents is restricted mainly to the upper 100–200 m of the sea. The mid-latitude anticyclonic gyres, however, are coherent in the mean well below 1000 m. The average speeds of the open-ocean surface currents remain mostly below 20 cm/s (0.4 knot). Exceptions to this are found in the western boundary currents, such as the Gulf Stream, and in the Equatorial Currents of the three oceans, all of which have velocities of 1–2 m/s (2–4 knots). Knowledge of the surface currents is based in part on direct measurements of the current, and more generally on dead reckoning from ships.

The primary causes of surface currents are wind stress and internal pressure forces resulting from the density distribution. Frictional forces and the Coriolis acceleration influence the surface currents. The effect of the wind is greatest when the direction and strength of the wind are steady; this is the case in lower and middle latitudes. In these latitudes an anticyclonic (clockwise, in the Northern Hemisphere) current system corresponds to the anticyclonic wind system (Fig. 4). Surface currents which flow in a westerly direction in the lower latitudes are part of this system (the North and South Equatorial Currents). The continuation of these currents is found along the eastern sides of these continents in narrow and strong surface currents directed toward the poles—the Western Boundary Currents; the Gulf Stream, Brazil Current, and Somali Current (only in the summer of the Northern Hemisphere); and the Agulhas Current, Kuroshio, and East Australia Current. In middle latitudes these currents turn and flow in an easterly direction (North Atlantic Current, North Pacific Current, and West Wind Drift of the Southern Hemisphere). On the eastern sides of the oceans this pattern contains surface currents directed toward the Equator—the Eastern Boundary Currents; the Canary, Benguela, West Australia, California, and Humboldt currents.

Embedded in the system of the North and South Equatorial Currents, which flow to the west, are the Equatorial Countercurrents, flowing to the east. These are found about 5° north of the Equator in all three oceans. An additional easterly flowing countercurrent is sometimes found in the eastern Pacific Ocean, south of the South Equatorial Current.

A large subsurface current is found centered on the Equator. This Equatorial Undercurrent is about 400 km wide but only 200 m thick (centered at a depth of 150–200 m), flowing to the east with peak speeds of 2–3 knots (1–1.5 m/s). This current is absent in the Indian Ocean in Northern Hemisphere summer; in the Pacific it is sometimes called the Cromwell Current.

In higher latitudes the currents tend to follow the coasts and shelf edges; in the Northern Hemisphere the continents lie to the right-hand side of the current when one looks in the downstream direction. Examples are the Norwegian, East Greenland, West Greenland, Labrador, and Alaska currents. Islands in this fashion are, so to speak, surrounded by currents moving in a clockwise direction (for example, Iceland by the Irminger, North Iceland, and East Iceland currents). It is for this reason that the western sides of continents in these latitudes are bordered by comparatively warm waters coming from lower latitudes, whereas off the eastern coasts of the same latitudes there are cold waters from higher latitudes. For example, in the Norwegian Current the surface temperature in summer is 10°C higher than in the East Greenland Current; both are at the same latitude. Thus surface currents are of great climatic importance.

All surface currents contain vertical components which vary from region to region. These vertical current components are influenced by converging or diverging winds, acceleration of the currents, and other factors. The vertical components are certainly very small (for example, a large value would be a speed of 0.003 cm/s in the upwelling area of the California Current), but in the long run they are of great importance when considering the balance of heat and all sea-contained substances. *See* MARINE INFLUENCE ON WEATHER AND CLIMATE.

One important idea about the large mid-ocean gyres is that their transport is determined not simply by the wind stress but by the latitudinal variation of the east-west winds (the curl of the wind stress). The curl of the vertically integrated Eqs. (1) and (2), for steady flow, is the Sverdrup relation, Eq. (6), where M_y is the total vertically integrated

$$\frac{\partial f}{\partial y} M_y = \mathrm{curl}\,\tau \qquad (6)$$

mass transport (per unit width) in the y direction. If the winds are all easterly or westerly, the only component of interest is given in Eq. (7).

$$\mathrm{curl}\,\tau = \frac{\partial \tau_x}{\partial y} \qquad (7)$$

The maximum transport in the Florida Current occurs in June or July, almost 6 months out of phase, or delayed from the winter maximum in the curl of the wind stress over the North Atlantic. The seasonal variation is from about 27 to 33×10^6 m³/s in volume transport in the Straits of Florida; this transport increases to approximately 100×10^6 m³/s as the Stream passes the longitude of Cape Cod.

Current variability. Most of the major ocean currents have been known fairly well for several decades. However, advances have been made in trying to understand the variations in ocean currents. Figure 5 shows the instantaneous path of the Gulf Stream, as determined by observations of its sharp surface-temperature gradient (see Fig. 1). The instantaneous path is much more sinuous than the long-term average path; the wavelike path variations will change entirely in a period of roughly 2 weeks to a month. Some wavelike meanders grow to have very large amplitudes, which become un-

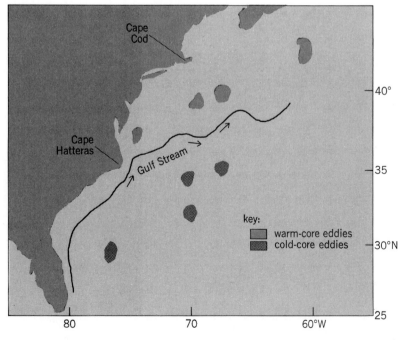

Fig. 5.   The position of the Gulf Stream at the beginning of October 1975.

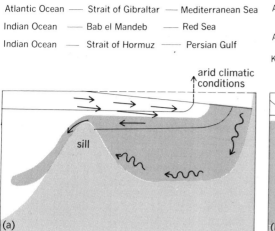

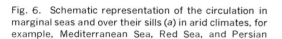

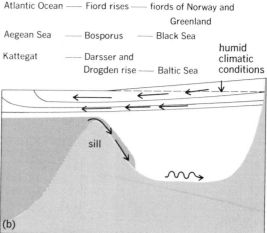

Fig. 6. Schematic representation of the circulation in marginal seas and over their sills (a) in arid climates, for example, Mediterranean Sea, Red Sea, and Persian Gulf; (b) in humid climates, for example, Black Sea, Baltic Sea, and fiords of Norway and Greenland.

stable and "pinch off" to form "Gulf Stream rings," or eddies, as also shown in Fig. 5. The water in the center of a ring is cold or warm, depending merely upon the phase of the meander that pinched off. These eddies usually drift to the west or southwest.

South of Cape Hatteras, the Gulf Stream may be found going 45 to 90° away from its "normal" path. Downstream from Cape Hatteras, the meanders may grow to hundreds of miles away from the mean path, which becomes convoluted; a ship traveling perpendicular to the Stream, for example, could cross it three times.

Another major source of variability in the central ocean arises from planetary waves, in which the motion is very nearly geostrophic; these motions resemble the ringlike eddies in many ways. Planetary motions seem to have periods of about a month or two, and horizontal scales of several hundred kilometers. It is not very well understood how these eddylike motions interact with the large-scale mean circulation.

The idea has been advanced by several researchers that a substantial portion of the transport of the Western Boundary Currents does not continue to the northeast but returns southerly in a weak countercurrent just offshore.

DEEP CIRCULATION

The deep circulation results in part from the wind stress and in part from the internal pressure forces which are maintained by the budgets of heat, salt, and water. Both groups of forces are dependent upon atmospheric influences. Apart from Coriolis and frictional forces, the topography of the sea bottom exercises a decisive influence on the course of deep circulation.

Marginal seas. The deep circulation in marginal seas depends largely on the climate of the region, whether arid or humid.

Arid climates. Under the influence of an arid climate, evaporation is greater than precipitation. The marginal sea is therefore filled with relatively salty water of a high density. Its surface lies at a

lower level than that of the neighboring ocean. Examples of this type are the Mediterranean Sea, Red Sea, and Persian Gulf. Figure 6a shows a schematic cross section of such a marginal sea which has a sill at its entrance. At the connection between the two seas there is water of a slightly lower density from the ocean flowing in at the surface. The water from the marginal sea flows over the sill into the ocean, where it sinks to a level at which it finds water corresponding to its density. Substantial vertical mixing takes place during this initial flow. At the deeper level it then spreads horizontally. The waters from the Mediterranean Sea and the Red Sea, because of their high salinity, can be followed far out into the Atlantic and Indian oceans, respectively. *See* INDIAN OCEAN; MEDITERRANEAN SEA.

Humid climates. The deep circulation of marginal seas in humid climates shows a different pattern, however, shown schematically in Fig. 6b. The level of the sea is higher than in the neighboring ocean. Therefore, the surface water with its lower density and accordingly its lower salinity flows outward, and the relatively salty ocean water of higher density flows over the sill into the marginal sea. Examples of this circulation are the Baltic Sea with the shallow Darsser and Drogden rises, the Norwegian and Greenland fiords, and the Black Sea with its entrance through the Bosporus.

The Black Sea is an example of a special case. The sill depth always remains in the water of low density—that is, the outflowing upper layer. The renewal of deep water, and with it the deep circulation, comes to a complete halt. The result is that the oxygen is entirely used up, and poisonous hydrogen sulfide is generated. Below depths of 200 m, only anaerobic organisms live in the Black Sea. *See* BLACK SEA.

If the sill depth interferes only occasionally with the lighter water of the upper levels, as in the entrances to the Baltic Sea, the renewal of deep water is interrupted at intervals. In the Baltic Sea these interruptions sometimes last for several years. *See* BALTIC SEA.

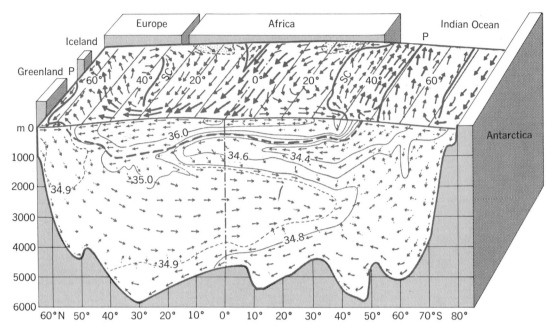

Fig. 7. Schematic representation of the surface and deep circulation in the Atlantic Ocean. All arrows show current directions; on the surface thin arrows indicate speeds of 5–40 cm/sec (0.1–0.8 knot), and thick arrows indicate speeds of 40–150 cm/sec (0.8–2.9 knots). SC indicates convergence of surface currents in subtropical waters. P indicates oceanic polar front where cold-water masses from polar and subpolar geographical latitudes meet relatively warm waters of temperate zone. In vertical section heavy broken line shows division between warm- and cold-water spheres, and other lines indicate equal salinities. (*After G. Wüst, 1949*)

Oceans. The deep circulation in the oceans is more difficult to perceive than the circulation in the marginal seas. In addition to the internal pressure forces, determined by the distribution of density and the piling up of water by the wind, there are also the influences of Coriolis forces and large-scale turbulence. There are areas in tropical latitudes in which the surface water, as a result of strong evaporation, has a relatively high density. In thermohaline convection, the water sinks while flowing horizontally until it reaches a density corresponding to its own, and then spreads out horizontally. In this way the colder and deeper levels of the oceans take on a layered structure consisting of the so-called bottom water, deep water, and intermediate water. In the Atlantic Ocean the deep-water circulation is strong on the western side of the ocean, where measurable speeds — roughly 10 cm/s — are found. Figure 7 is a schematic representation of the surface and deep circulation in the Atlantic Ocean as viewed from the western side of the ocean. From the lines indicating equal salinity, the origin of the water masses in the various strata can be inferred. The Bottom Water comes from very cold water masses which have sunk along the edge of Antarctica. In the layer immediately above, or Deep Water, the water masses have their origin in the far North Atlantic. In the next layer above, or Intermediate Water, the water south of about 30°N comes from the Southern Hemispheric polar front.

There are five major areas where the surface water becomes denser and sinks.

1. The Norwegian Sea, where the water with the highest density of the world oceans is formed. This supplies the North Polar Sea with cold deep water, as well as the North Atlantic Ocean with cold bottom water to a latitude of approximately 50°N.

2. The Antarctic continental slope in the Weddell Sea, where water temperatures beneath the winter pack ice are near freezing (for sea water, −1.9°C). Favored by the topography of the ocean floor, this water flows northward in the western Atlantic to the foot of the Grand Banks, as well as through the Romanche Deep on the Equator (through the Mid-Atlantic Ridge) into the eastern side of the Atlantic. It also spreads northward into the Indian and Pacific oceans, flowing toward the Equator on the western side of each ocean.

3. In the Labrador and Irminger seas, where the Deep Water (between depths of 1000 and 4000 m) of the Atlantic Ocean originates. Apart from low temperatures, this water is distinguished by its richness in oxygen. This water has been traced around South America and into the Pacific Ocean.

4. The polar front at latitude of 50°S. Here cool water with low salinity is formed, supplying the Antarctic Intermediate Water of the Atlantic, Indian, and Pacific oceans.

5. The polar front in the North Pacific Ocean, where North Pacific Intermediate Water is formed. The distribution of this water in the North Pacific Ocean has been studied extensively.

The deep-sea circulation of the Atlantic Ocean appears to be the most active in comparison with the Indian and Pacific oceans, because the important sources of thermohaline convection are found in the Atlantic. In addition, the continental barrier of South America seems to force water from the surface currents of the South Atlantic into the

North Atlantic Ocean. To compensate for the loss of surface water in the South Atlantic, there is a more active deep-water circulation, in which North Atlantic Deep Water flows southward into the South Atlantic Ocean. Another way of looking at the spreading and mixing of the Antarctic Bottom Water is shown in Fig. 8. *See* ATLANTIC OCEAN; INDIAN OCEAN; PACIFIC OCEAN.

MEASUREMENT OF CURRENTS

Current-measuring devices used for direct measurements are of several functional types. The first utilizes the drift of a free body such as a drogue, a drift bottle, a ship, or a mid-depth neutrally buoyant pinger. Speed and direction are determined by observing the distance and direction the body drifts in a given time interval. The second method is based on drag effects on a fixed body. Information can be obtained when the current rotates a propeller, twists a vane, tilts an instrument case, or creates a pressure difference in a pitot tube. In both methods it is necessary to know the actual motion of the instrument or to know that it is stationary. Hence the navigational problem of knowing accurately the position of the current meter or ship is an important part of an ocean-current measurement. A basic problem arises, because the speed of deep-ocean currents is usually only about 10 cm/s, or 1/5 knot (but they may be as great as 1 knot or more). Currents having such speeds in the central ocean are transient flows, however, having

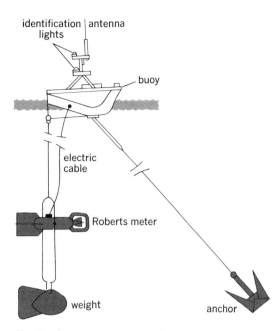

Fig. 9. Buoy-operated Roberts radio current meter. Buoy contains battery-operated radio transmitter and antenna. When meter is lowered directly from ship, there is no need for transmitter and receiver as electric cable is connected directly to relay box. (*Adapted from U.S. Navy Hydrographic Office Publ. no. 607, 2d ed., 1955*)

periods of perhaps a month. The mean speeds are not well known but are probably 1 cm/s or less. Only in a few great currents, such as the Gulf Stream, are there speeds as great as 2–4 knots — or in special regions of strong tidal currents.

If the ship is close enough to land to be in an electromagnetic navigational network, such as loran, positions can be known to within 1 mi (1.6 km) or even 100 yd (91 m). Special systems, in limited areas, allow navigation to a few meters. Many vessels now have navigation by artificial satellites of the U.S. Navy's Transit System, giving a fix about every 2 hr. Buoys are routinely anchored in very deep water, but anchoring a ship in the deep ocean is costly and time-consuming. Since oceanographers have learned the need for long data records, the usual method is to anchor a current-meter mooring, using subsurface floats. The ship returns in 3 to 6 months and recovers the mooring.

Drift of a free body. Most ocean surface currents have been discovered because of their effect on the course and speed of ships.

The neutrally buoyant float, developed by Sir John Swallow, is a mid-depth drifting device consisting of a pressure case which is designed to be less compressible than sea water and floats at a predetermined depth. It emits acoustic pings which can be heard for several miles from a ship equipped with appropriate sound gear. The sensitivity of this method can approach 0.01 knot (0.005 m/s) if the buoy is followed for several days. It provided the first way to measure mid-depth currents. Large floats of this type are now tracked from shore-based stations.

It should be noted that observations obtained by drifting objects are Lagrangian observations,

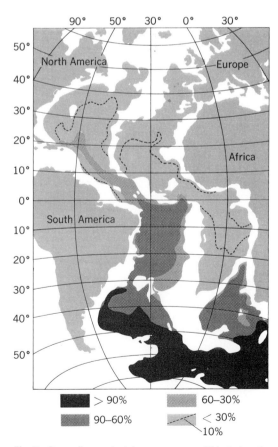

Fig. 8. Spreading and mixing processes within Antarctic Bottom Water. (*After G. Wüst*)

which require very careful interpretation if they are to be interpreted within an Eulerian framework.

Effects on a fixed body. In this method the effect of the current on a fixed instrument is measured. Most subsurface currents have been measured with self-contained meters, which record the revolutions of a propeller in a given time, and the direction has been established by use of a compass. There are propeller-type meters which are self-contained and meters which telemeter their data to the ship or a surface buoy. The Roberts meter is an early version of this type (Fig. 9). In weak currents a Savonius rotor is less affected by vertical motion than is a conventional propeller. Some instruments employing a propeller enclose it in a horizontal housing. The Savonius rotor, however, has poor response characteristics for periods near 1 s, which is a typical surface-wave period.

Most of the quantitative knowledge of deep currents has been gained from moored current meters using Savonius rotor-type current meters, of a type developed by W. S. Richardson. Wave action on a surface buoy (as in Fig. 9) makes a deep mooring strum like a banjo string; modern moorings employ subsurface floats (often glass spheres) and can be deployed for a maximum, at this writing, of 6 to 9 months. These each have typically three to six current meters, with accompanying temperature recorders. The discovery of the large, energetic, eddylike motions has shown the need for very long observations (several years) in order to allow significant conclusions to be drawn. A deep mooring represents a capital investment of about $50,000; a single day of ship time costs about $5,000; a cruise may last 2 to 3 weeks, and launch up to a dozen or more moorings. *See* OCEANOGRAPHY.

[WILTON STURGES]

Bibliography: G. Dietrich, *General Oceanography*, 1963; G. L. Pickard, *Descriptive Physical Oceanography*, 1976; J. L. Reid, Jr., *Intermediate Waters of the Pacific Ocean*, Johns Hopkins Oceanographic Studies, vol. 2, 1965; Melvin E. Stern, *Ocean Circulation Physics*, 1975; H. Stommel, *The Gulf Stream: A Physical and Dynamical Description*, 1977; H. Stommel and K. Yoshida (eds.), *Kuroshio: Physical Aspects of the Japan Current*, 1972.

Ocean waves

The irregular moving bumps and hollows on the ocean surface. Winds blowing over the ocean, in addition to producing currents, create surface water waves called waves or a "sea" (Fig. 1). The characteristics of these waves (or the state of the sea) depend on the speed of the wind, the length of time that it has blown, the distance over which it has blown, and the depth of the water. If the wind dies down, the waves that remain are called a dead sea. Waves can travel hundreds and thousands of miles from where they were generated into areas where the wind is light. These waves are called swell. The rise and fall of the water as a function of time at a fixed point can be recorded. Such a record, as taken by the ocean weather ship *Weather Explorer* with a shipborne wave recorder, is shown in Fig. 2. The highest wave in this record is 46 ft from crest to trough. *See* SEA STATE.

This article treats the generation of ocean waves, the mathematical theory of forecasting ocean waves, and the instruments used to measure ocean waves. For a discussion of wave characteristics *see* CAPILLARY WAVE; INTERNAL WAVE; NEARSHORE PROCESSES; SEICHE; STORM SURGE; TSUNAMI.

Generation. The physical processes by which waves are generated have been studied theoretically by O. M. Phillips and J. W. Miles and partially verified by numerous investigators in terms of observations in wind tunnels and in nature. The discrepancies between theory and observation are substantially less than they were a few years ago. Wave spectral components first grow linearly in resonance with turbulent pressure components advected by the winds of the atmosphere. The components then grow exponentially by extracting energy from the wind profile. As saturation at a particular frequency is reached, the formation of whitecaps limits further growth.

Although not accepted by all scientists, there is considerable evidence that, if the wind blows long enough at constant velocity over a large enough area, the spectrum of the waves will depend only on wind speed. Formula (1) has been suggest-

$$S(\mu) = \frac{\alpha g^2}{\mu^5} e^{-\beta(\mu_0/\mu)^4} \qquad (1)$$

ed ($\alpha = 8.1 \times 10^{-3}$, $\beta = 0.74$, $\mu_0 = g/y$, wind measured at 19.5 m above the surface).

The rate at which waves are generated depends very much on whether or not there is a background spectrum present when the wind speed increases. With no background, a fairly long time is required to generate a fully developed sea. With a background, the time required is some variable fraction, often one-half or less, of the time required from a truly zero initial condition. Variable background conditions imply that the spectrum at all points over the ocean must be kept track of accurately so as to determine changes accurately. *See* WIND STRESS OVER SEA.

Equations of motion. The mathematical equations governing the motions of the irregular wavy surface have never been completely solved as they are nonlinear. If potential flow can be assumed, and if the effect of surface tension is neglected, the equations governing the motions of the water are described by the potential equation, Eq. (2); Bernoulli's equation, Eq. (3); the kinematic boundary condition, Eq. (4); and the condition that the

$$\phi_{xx} + \phi_{yy} + \phi_{zz} = 0 \qquad (2)$$

$$\frac{p}{\rho} + gz - \phi_i + \frac{1}{2}(\phi_x^2 + \phi_y^2 + \phi_z^2) = 0 \qquad (3)$$

$$\eta_t = -\phi_z + \phi_x \eta_x + \phi_y \eta_y \qquad (4)$$

potential $\phi_z \to 0$ as $z \to -\infty$ if the water is deep (subscripts denote partial differentiation). The equations governing the motions of the air above should be considered, but are not given here. The position of the boundary between the water and the air is described by $z = \eta(x,y,t)$, the free surface, where z is positive upward. Equations (2) and (3) apply at $z = \eta(x,y,t)$ and below the surface. Equation (4) applies at $z = \eta(x,y,t)$.

With the wind blowing, the pressure p at $z =$

Fig. 1. Aerial photograph of sea waves. (*Courtesy of Capt. D. B. MacDiarmid, U.S. Coast Guard*)

$\eta(x,y,t)$ will be a function of x, y, and t. If p at $z = \eta$ is set equal to zero and if the squared and product terms of the above equations are omitted, the equations become linear. Numerous solutions exist. The important problem is to find solutions that approximately represent waves as found in nature.

Wave statistics and stochastic models. With complicated computations and sufficiently detailed recording techniques, the wavy surface over any finite area could be represented as closely as desired by a mathematical formula within a linear theory. Wave records of any length, such as the one in Fig. 2, could be represented as closely as desired as a function of time, but such an effort would be wasted. The space pattern could be observed only over an area that is small compared to the dimensions of the oceans. The time record could be observed only for a short period. The waves would never again be exactly like the ones that were studied in such detail.

The waves must therefore be studied by means of statistical models, and given wave records or stereophotogrammetric observations must be analyzed by means of statistical techniques. This permits economy of computation. The analysis of a given record that will never again be repeated then fits into an overall pattern that describes the statistical properties of the waves. Fortunately, such models exist, from the field of electronics in particular, and they have been extended and adapted to the study of ocean waves.

Such a model is given by the ensemble of all possible sea states for a particular power spectrum, $S(\mu, \theta)$, as in Eq. (5).

$$\eta(x,y,t) = \int_0^\infty \int_{-\pi}^\pi \cos\left[\frac{\mu^2}{g} (x \cos\theta + y \sin\theta) \right.$$
$$\left. - \mu t + \epsilon(\mu,\theta) \right] \sqrt{2S(\mu,\theta)\, d\mu\, d\theta} \quad (5)$$

The symbol $\epsilon(\mu, \theta)$ stands for a random phase uniformly distributed over the interval between zero and 2π. The variance spectrum $S(\mu, \theta)$ has the dimensions of cm² sec/radian. If Eq. (5) is represented by an approximating double summation, the sea surface can be represented by a large sum of many simple harmonic progressive waves, each with a different frequency $\mu = 2\pi/T$, a wavelength

determined by $2\pi/\lambda = \mu^2/g (\lambda = gT^2/2\pi)$, a phase speed equal to $c = g/\mu$ ($c = gT/2\pi$), and a direction toward which the wave is traveling determined by θ. The amplitude of the wave is determined by the square root of the volume under $2S(\mu, \theta)$ for the appropriate range of μ and θ. Other representations are also possible.

An ensemble of sea states can be defined by considering $^{(1)}\eta(x, y, t)$, $^{(2)}\eta(x,y,t)$, $\cdots$, $^{(n)}\eta(x,y,t)$, $^{(n+1)}\eta(x,y,t)$, $\cdots$, where the only difference from record to record would be the random phases chosen for each term in each partial sum.

For the ensemble space, the covariance function can be found as in Eq. (6), where E denotes the expected value.

$$E[\eta(x,y,t)\,\eta(x+x^*, y+y^*, t+\tau)]$$
$$= \int_0^\infty \int_{-\pi}^\pi S(\mu,\theta) \cos\left[\frac{\mu^2}{g} (x \cos\theta + y \sin\theta) \right.$$
$$\left. - \mu\tau \right] d\theta\, d\mu \quad (6)$$

It was shown by W. J. Pierson in 1955 that the n random variables $\eta(x_1,y_1,t_1)$, $\eta(x_2,y_2,t_2)$, $\cdots$, $\eta(x_n,y_n,t_n)$ have a multivariate normal distribution $E[\eta(x_j,y_j,t_j)]^2$ given by Eq. (6) with x^*, y^*, and τ equal to zero. The covariances, $E[\eta(x_j,y_j,t_j)\eta(x_k,y_k,t_k)]$, are given by Eq. (6) with $x^* = x_j - x_k$, $y^* = y_j - y_k$, and $\tau = t_j - t_k$.

By virtue of the ergodic theorem, the time and

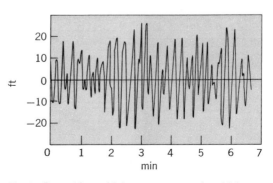

Fig. 2. Record from shipborne wave recorder which was taken on Nov. 16, 1953, at 60.2°N, 14.0°W. Wind was Beaufort force 11. (*National Institute of Oceanography, Wormley, Surrey, England*)

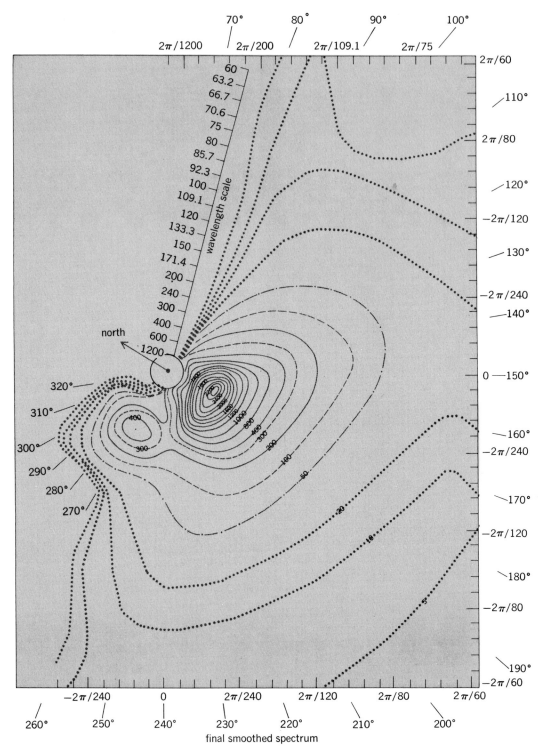

Fig. 3.   Directional spectrum of a wind-generated sea. (*From N.Y. Univ. Meteorol. Pap.*)

space variation of a particular sample function has the same statistical properties as the variations in the ensemble. A given sample function can be analyzed so as to estimate the spectrum and also the required parameters for the various probability density functions that describe the statistical properties of the waves.

With τ equal to zero, the Fourier inversion of

Eq. (6) yields information on the directional spectrum if the transformations $\alpha = \mu^2\cos\theta/g$ and $\beta = \mu^2\sin\theta/g$ are made, where α and β are wave numbers in a cartesian coordinate system. There is a 180° indeterminancy in direction that can be resolved by an analysis of the meteorological conditions that generated the waves. When a finite area is used to obtain such an estimate, the analysis of

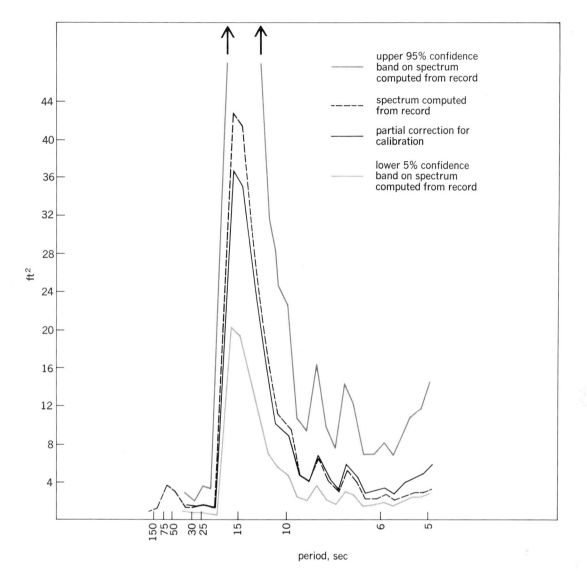

ft²

upper 95% confidence
band on spectrum
computed from record

spectrum computed
from record

partial correction for
calibration

lower 5% confidence
band on spectrum
computed from record

period, sec

Fig. 4. Frequency spectrum of the wave record shown in Fig. 2.

the data cannot be based simply on the above equations. The techniques developed by J. W. Tukey in 1949 for single-variable cases have been extended to the two-variable case.

As shown in Fig. 3, such a directional spectrum has been estimated from stereophotogrammetric measurements of waves generated by a wind of 18.7 knots near the surface. Figure 3 is a smoothed version of such an analysis based on estimates that are distributed according to χ^2 (chi square probability density function) with 19 degrees of freedom. The contours when divided by 10^4 and interpreted in terms of square feet estimate the results of integrating $S(\mu, \theta)$, approximately over a square with a length of side given by the distance between two of the scale marks on the side of the figure. A wide range of directions and of wavelengths from 600 ft down to 60 ft is evident. The secondary peak appears to be due to swell. There are even shorter waves which could not be detected by the process of analysis. If the contour values are halved, Fig. 4 can be interpreted as the resolution of the total variance of the wavy surface into contributions

from different wave numbers. Numerous other methods for estimating directional wave spectra have been developed, as described in the references.

The three-variable process has many interesting properties that were studied in detail and reported upon by M. S. Longuet-Higgins in 1957. Such properties as the probability density function (pdf) of the speeds of contours of constant elevation, the appearance and disappearance of maxima and points of inflection, and the number of relative maxima and minima on the surface have been determined.

If the waves are observed at a fixed point as a function of time, as in Fig. 2, x^* and y^* become zero in Eq. (6) and directional effects are lost, giving Eq. (7).

$$S(\mu) = \int_{-\pi}^{\pi} S(\mu, \theta) \, d\theta \qquad (7)$$

Such a time record becomes a sample from a stationary Gaussian process, and the results of the analysis of such processes can be immediately applied to the study of wave records. The estimate of the spectrum of the record shown in Fig. 2 is given in Fig. 4, for example, after at least partial correction for instrumental response. Each estimate has about 11 degrees of freedom. Also the variation of $\eta(x,y,t)$ along any line in x,y,t space is similarly represented by a one-variable process.

The needed parameters for many theoretical pdf's that describe quantities that can be evaluated from the record of the waves can be determined from the wave spectrum. The average time interval between zeros and the average time interval between maxima in the record can be found. The crest-to-trough wave heights are roughly distributed according to a Rayleigh distribution. For pressure records made below the surface, the pdf of the zero intervals can be found with a fair degree of accuracy.

Such quantities as sea surface slopes and curvatures can also be evaluated. These quantities depend on the various moments of the spectrum, on the values of $E[\eta(x_1,y_1,t_1)\ \eta(x_2,y_2,t_2)]$, and on spectra derived by transformations and differentiations.

Nonlinear considerations. From a probabilistic point of view, the problem area where the greatest need for further work exists is that of accounting for the nonlinear properties of the waves. Contributions have been made by L. J. Tick, M. S. Longuet-Higgins, M. S. Chang, O. M. Phillips, and K. Hasselman. For long crested waves, Tick has obtained the second-order corrections to the wave profile and to the spectrum. Longuet-Higgins has obtained a Gram-Charlier series correction to the normal marginal probability density of points chosen at random from the time history of an ocean wave record. Chang has studied the motion of a particle at the surface of long crested random waves and found second-order effects about seven times more important than the linear effects in the forward drift. Some strange things may occur at third order for gravity waves and at second order for capillary waves. Opinions differ, and these can be investigated through the bibliography.

Wave forecasting. Sea state forecasts are based on a combination of empirical knowledge and certain theoretical relationships of waves and wind. The sea surface wave pattern usually consists of a locally generated sea and superimposed swell which was generated in a distant storm. Forecasts are generally based on oceanic weather maps. After combination of these phenomena, the actual state of the sea surface pattern at a given locality can be estimated. The most rational way of describing the state of a sea is provided by the wave energy spectrum $S(\mu, \theta)$.

The first part of the problem of wave forecasting is to find the energy spectrum as it changes from hour to hour or from place to place as the wind blows over the ocean surface. The growth of the spectrum of the sea in an area of wave generation has been described by various authors. It is agreed that the higher the wind speed, the higher the waves; that the greater the distance over which the wind blows (the fetch), the higher the waves; and that the longer the wind blows (the duration), the higher the waves.

The second part of the problem of forecasting

waves is the problem of describing how the wave height decreases when the wind dies down, and how the waves travel out of the area of generation as swell. Numerical filtering operations (somewhat analogous to electronic filters) that depend on the dimensions of the generating area and on the time of the forecast can be applied to the spectrum of the sea to forecast the spectrum of the swell and the rate at which the waves will die down in the generating area. These filters are based on the fact that the waves are highly dispersive. The spectrum $S(\mu,\theta)$ covers a wide range of frequencies and directions, and therefore the waves spread out over a wide area as they travel out of the generating area.

Attempts are being made to develop ways to describe the function $S(\mu,\theta)$ as it varies from point to point over the ocean. These wave forecasting methods are an extension of the earlier wave forecasting concepts, first the significant wave height and period procedures and then the spectral methods, and many features in common can be found. Also, numerous discrepancies between these different methods have been resolved. The extension beyond these earlier methods is possible only because of the new theories mentioned earlier and the use of high-speed electronic computers. The procedure involves a computation of how the wave spectrum changes every 3 hr on a grid of points over the oceans and can be accomplished by computing how the spectrum will grow with time for components traveling with the wind and dissipate with time for components traveling against the wind, and how the spectrum will change as the spectral components propagate at group velocity.

Data sources and further reading. The bibliography at the end of the article gives references that contain extensive additional references, so the reader can further study the present status of the subject. The references also contain lists of papers that contain data on wave spectra both as a function of frequency alone and as a function of frequency and direction. These results have had considerable application in naval architecture.

[W. J. PIERSON, JR.; GERHARD NEUMANN]

Wave-measuring devices. Instruments designed to measure ocean waves can be grouped into two classifications: those that sense the elevation of the surface water, and those that sense the subsurface pressure fluctuations generated by the waves. Surface elevation sensing gages include (1) surface float devices mechanically connected to a recording mechanism, (2) devices that record the buoyant force on a vertical cylinder, (3) recorders actuated by vertical accelerometers mounted in surface buoys, (4) inverted echo sounders (fathometers) fixed below the surface to echo off the water surface, (5) accurate absolute altimeter recorders operated in aircraft flying at a fixed elevation, (6) stereophotographs, and (7) electrical elements whose resistance or capacitance is a function of the elevation of the water surface.

Surface gages. Of the gages that sense the surface elevation, the step resistance gage shown in Fig. 5 (type 7 above) is the most widely used. Electrical contact points are mounted along a vertical support and connect to a resistance circuit. The values of the resistors connected to the contact points are selected so that the current increases in proportion to the number of contacts shorted along the submerged length of the gage. The alternating current (used to prevent polarization of the contacts) which flows through the gage is converted by means of a rectifier to a proportional direct current to drive a pen recorder.

Wave-measuring instruments that sense subsurface pressure fluctuations rely on hydrodynamic theory to compute the surface wave heights. This is shown in Eqs. (8) and (9), where K is ratio of

$$K = \frac{\cosh \frac{2\pi b}{L}}{\cosh \frac{2\pi d}{L}} \quad (8)$$

$$L = \frac{g}{2\pi} T^2 \tanh \frac{2\pi d}{L} \quad (9)$$

pressure variation expressed as equivalent water heights to surface wave height, d is the depth of the water, b is the height of the instrument above bottom, L is the wave length, T is the wave period, and g is the acceleration of gravity. For example, if $d = 40$ ft, $b = 10$ ft, and the pressure record indicates $T = 10$ sec with an amplitude of 5 ft, $g/2\pi = 5.12$ ft/sec², L would be 329 ft, K would be 0.187, and the surface wave would have an amplitude of $5.0/0.817 = 6.1$ ft.

Subsurface gages. Subsurface pressure signals can be converted into an electrical signal by any of the numerous transducers. A typical pressure gage which utilizes a differential potentiometer-type transducer is shown in Fig. 6. One pressure port of the transducer is connected to the sea pressure by silicone oil and a rubber bellows. The second port is open to air in the compliant chamber.

As a result of the restricted flow of fluid into this chamber through the capillary tube, the pressure is equal to the average pressure. The differential pressure across the transducer is therefore only that produced by the wave-generated pressure fluctuations. The mean pressure (hydrostatic) and

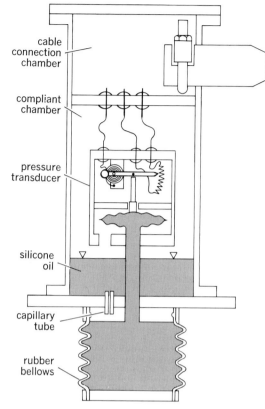

Fig. 6. Subsurface pressure gage.

the slow pressure fluctuations due to tides are canceled and do not appear in the record.

[FRANK E. SNODGRASS]

Bibliography: R. B. Blackman and J. W. Tukey, The measurement of power spectra from the point of view of communications engineering, *Bell Syst. Tech. J.*, vols. 1 and 2, 1958; L. J. Cote et al., The directional spectrum of a wind generated sea as determined from data obtained by the Stereo Wave Observation Project, *N.Y. Univ. Meteorol. Pap.*, 2(6), 1960; B. Kinsman, *Wind Waves: Their Generation and Propagation on the Ocean Surface*, 1965; M. S. Longuet-Higgins, The statistical analysis of a random moving surface, *Phil. Trans. Roy. Soc. London*, ser. A, 249:321–387, 1957; G. Neumann and W. J. Pierson, *Principles of Physical Oceanography*, 1966; O. M. Phillips, *The Dynamics of the Upper Ocean*, 2d ed., 1977; O. M. Phillips, On the generation of waves by turbulent wind, *J. Fluid Mech.*, 3:417, 1957; W. J. Pierson, Jr., Gravity waves, *Trans. A.G.U.*, 48:584–588, 1967; M. J. Tucker, The accuracy of wave measurements made with vertical accelerometers, *Deep-Sea Res.*, 5:185–192, 1959; M. J. Tucker, A shipborne wave recorder, *Trans. Soc. Naval Architects Marine Eng. London*, 98:236–250, 1956.

Oceanic islands

Those islands which rise from the deep-sea floor rather than from shallow continental shelves. Most islands in gulfs and seas that fringe the great ocean basins are geologically similar to the nearby continents. On the other hand, almost all islands that

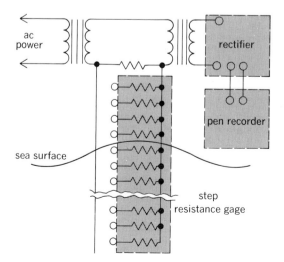

Fig. 5. Step resistance wave-measuring gage.

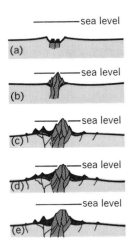

Fig. 1. Development of submarine volcano or ridge. Center areas are initial volcanic extrusion; black areas are later deposits of volcanic material and erosional debris. Sequence (a) to (e) is explained in text.

rise from the ocean basins are volcanoes with or without coral reef and, geologically, bear little relation to the continents. Volcanic islands are only the tops of much larger undersea volcanoes, most of which are associated with great submarine structures, such as submarine ridges and fractures in the Earth's crust (Fig. 1).

Submarine volcanoes. On the deep-sea floor, volcanoes begin as lava flows from fissures in the Earth's crust under 2 or 3 mi of water. Gradually they build upward through the water, but about nine-tenths of the submarine volcanoes become inactive and stop their growth before they reach the sea surface. The others burst from the deep water into a new realm, where wave and subaerial erosion combat their upward growth. At first the volcanoes tend to produce ash and cinders, which are easily eroded. Falcon Island, an active volcano in the Tonga group, has several times been reduced to a submarine bank within a few years after an eruption built up an island. Gradually as the pile becomes broader, volcanoes rise above the waves and more resistant fluid lava flows build a solid island. Where several nearby volcanoes merge together, as in Hawaii, a great island may form. *See* VOLCANO; VOLCANOLOGY.

Volcanoes are active for no more than a few million years, however, and inactive volcanoes are inevitably worn down to shallow submarine banks by erosion which never stops. In addition to these worn-down volcanoes, drowned former islands called guyots or tablemounts have been discovered in all the ocean basins, mostly at depths of 1000–7000 ft. Reef coral as old as the Cretaceous and volcanic erosional debris have been dredged on

some guyots. Also, drilling on atolls shows that the coral is a capping several thousand feet thick on former volcanic islands. *See* ATOLL; SEAMOUNT AND GUYOT.

Associated submarine structures. Most submarine volcanoes are associated with great submarine structures: long, straight, narrow features, such as the Hawaiian Ridge and Murray Fracture Zone; broad oceanic rises, such as the Mid-Atlantic Ridge; and island arcs and trenches, such as the Aleutian Arc. *See* MARINE GEOLOGY.

Long straight structures. Lines of volcanoes occur in the Atlantic and Indian oceans but are relatively rare. A linear group of guyots extends southeast from Cape Cod, and other groups may be undiscovered in the less well-surveyed parts of these oceans.

It is the Pacific, however, that is the type area for linear archipelagoes, and there they are extremely common. Existing linear groups are largely confined to the southwestern and central Pacific (Fig. 2), but, in the past, islands were present in the northern part of the basin (Fig. 3). Some very large archipelagoes, like the Tuamotu Islands, and former archipelagoes, such as the Mid-Pacific Mountains, consist of individual volcanoes only a few thousand feet high rising as peaks above great steep-sided ridges. Other archipelagoes, including the Hawaiian, Samoan, and Marquesas islands, have large volcanoes rising from lower ridges. The occurrence of volcanoes in a straight line suggests an underlying linear fracture in the Earth's crust. The association of volcanoes with a fracture in the crust (Fig. 1a) is clearly demonstrated along the Clarion Fracture Zone, a submarine feature in the

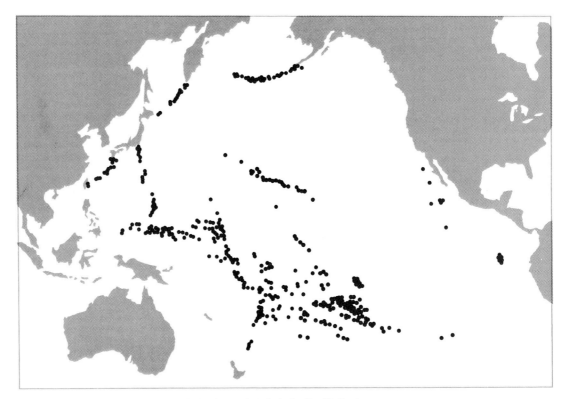

Fig. 2. Distribution of volcanic islands, banks, and atolls in the Pacific Basin.

Fig. 3. Distribution of guyots, or former islands, in the Pacific Basin.

east-central Pacific. A long straight trough forms the western part of the fracture zone. Toward the east it is interrupted by seamounts until the trough disappears and is replaced by a line of shallow banks, and even farther to the east it is replaced by the volcanic Revillagigedo Islands.

The Earth's crust is able at first to support the load of a new volcano or volcanic ridge, but as the structure becomes larger, a point is reached when crustal strength is insufficient and elastic downbowing begins (Fig. 1b). The topographic expression is that of a central ridge surrounded by a depression or moat outside of which an arc may occur. A single volcano may be encircled by a small individual moat. As downbending continues, tension fractures permit volcanism on the arches (Fig. 1c). Volcanism and erosion debris commence filling the marginal depressions and faulting may occur. These depressions may become filled with volcanic material and sediment to form smooth archipelagic aprons sloping away from the island, seamount, or ridge, as around the Marquesas Islands (Fig. 1d). If eroded to a flat bank and relatively sunk below the surface, the seamount becomes a guyot (Fig. 1e); if coral was present at the surface of the seamount and kept pace with the sinking, an atoll forms.

Broad oceanic rises. Submarine ridges or rises are the locus of solitary volcanic islands and seamounts, such as Ascension, Reunion, and Easter islands. Typically volcanoes are located near, but not on, the crests of the broad submarine rises. In addition, volcanic islands occur in clusters. Examples are the Azores and Galapagos islands. The latter group is surrounded by a thin archipelagic apron, but moats have not been found around this type of island cluster.

Island arcs. Groups of islands which follow more or less curved lines are called island arcs. They are associated with deep trenches, large gravity anomalies, and earthquakes; hence they are in a region of great crustal instability. Typically there is an inner arc of active volcanoes and an outer arc of nonvolcanic islands which may contain sediments of the types now found on the deep-sea floor, thus indicating uplift of several miles. Uplift is also shown by raised sea cliffs and deposits of coral. Drowned former islands do not occur along island arcs.

[EDWIN L. HAMILTON; HENRY W. MENARD, JR.]

Bibliography: R. S. Dietz, Marine geology of Northwestern Pacific, *Bull. Geol. Soc. Amer.*, 65: 1199, 1954; E. L. Hamilton, *Sunken Islands of the Mid-Pacific Mountains*, Geol. Soc. Amer. Mem. no. 64, 1956; H. W. Menard, Archipelagic aprons, *Bull. Amer. Assoc. Petrol. Geol.*, 40:2195, 1956; H. W. Menard, *Marine Geology of the Pacific*, 1964; F. P. Shepard, *Submarine Geology*, 3d ed., 1973.

Oceanographic platforms

Supporting structures (stations) for instruments used in sensing and recording the various parameters of the oceanic environment. The platforms come in assorted shapes and sizes (manned and unmanned), ranging from moored buoys to fixed offshore towers and from surface survey ships to deep-diving submersibles. The former group collects a history of data over long periods of time in a

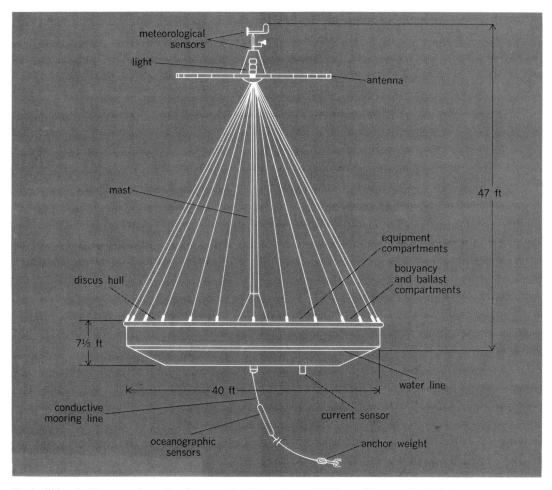

Fig. 1. "Monster" buoy configuration. Buoy weighs 43 tons dry and 100 tons with sea-water ballast.

specific location, whereas the latter group collects data over a large area, usually for a specific instant of time. *See* OCEANOGRAPHY.

Buoys. Buoyant platforms, called buoys, are often used in oceanographic research. They are transported to a selected site where they are moored to the sea floor. Instruments are suspended beneath the buoys at desired data-gathering locations in the water column. The data may be recorded, stored, and recovered periodically or

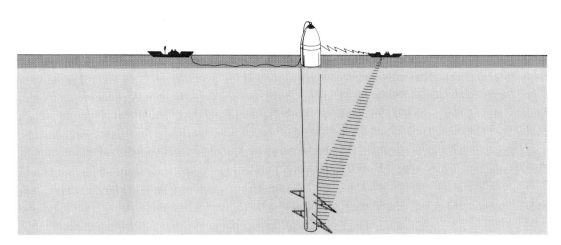

Fig. 2. Schematic of U.S. Navy SPAR (seagoing platform for acoustic research). In the vertical "listening" position SPAR receives both underwater acoustic signals and abovewater radio signals from the transmitter, also called target vessel, depicted on the horizon at the right. The tending tow ship, shown on the left, is tethered to SPAR by a half-mile of power supply and tow cables. Data are processed aboard the tender.

telemetered to a receiving station on a scheduled basis or on command by means of a radio link or both. The "monster" buoy (Fig. 1) was developed by the U.S. Navy and Convair Corp. It is a 40-ft-diameter, disk-shaped buoy with a 100-channel capacity for data, which is sampled once every hour. It contains both short-term and long-term memory devices for telemetering the data storage. The long-term memory stores all collected data during 1 year of unattended operation.

Other notable, large-size buoy platforms developed in the United States are the Pacific-based FLIP (floating instrument platform) and the Atlantic-based SPAR (seagoing platform for acoustic research). Both are stable spar-type buoys with marine instruments mounted on the submerged portions. Nominally 355 ft in length the platforms are towed horizontally into position and then flipped to the upright (vertical) position by selective ballasting. When upright they extend about 55 ft above the water surface. FLIP, launched in 1962, is manned with crew quarters above the water surface. SPAR, launched in the mid-1960s, is unmanned (Fig. 2).

Fixed platforms. Platforms similar to those employed in offshore oil exploration are also used for oceanographic studies. The Argus Island Research Station (Office of Naval Research), constructed in 1960, is a truss-type structured platform supported on four legs. It is located off the coast of Bermuda in 200 ft of water on Plantagenet Bank. Space is provided for electronic equipment, a general shop, maintenance facilities, storage, and living accommodations for about 30 persons. The top of the tower is 65 ft above mean low water. Another Navy tower, known as the NEL tower, was erected offshore near San Diego in 60-ft water depth for oceanographic studies. The special feature of this tower is its railway track system for raising and lowering instruments, which operates between the main deck and the sea floor on three sides of the tower. Many other offshore platforms, such as the U.S. Coast Guard light towers and the earlier Texas towers, conduct oceanographic studies as secondary missions.

Oceanographic research ships are the workhorses at sea. Studies dealing with areal variability in volume or in and about the sea floor rely on these highly mobile platforms. See OCEANOGRAPHIC SUBMERSIBLES; OCEANOGRAPHIC VESSELS.

[J. J. HROMADIK]

Bibliography: J. F. Brahtz (ed.), *Ocean Engineering*, 1968; The monster buoy, *Geol. Mar. Technol.*, vol. 2, no. 4, 1966; B. J. Muga and J. Wilson, *Dynamic Analysis of Ocean Structures*, 1970; R. L. Trillo (ed.), *Jane's Ocean Technology 1976-77*, 1976.

Oceanographic submersibles

Small research vessels for underwater transport of people and equipment and for use as underwater platforms for observation, sampling, measurement, and performing various work tasks. They are also referred to as undersea vehicles or submersibles.

Manned undersea exploration first employed small research submersibles in the 1950s. Early types were employed in general reconnaissance, observation, and photographic missions. Power, control, navigation, and life support functions generally limited submerged endurance to only a few hours. Available structural and buoyancy materials, together with economic restrictions, limited operating depths to a maximum of about 1000 ft (305 m). Despite their limitations, these early vehicles were most successful in creating an interest in the practical usage of small submersible vehicles.

The 1960s saw major United States corporations building advanced-capability research submersibles in the expectation of large United States government support for oceanographic research. Somewhat later, the emerging undersea oil industry provided strong stimuli for developing practical and economical undersea work boats. New materials and designs produced versatile work boats, equally effective in varied assignments and still able to perform specialized tasks with add-on equipment.

There are about 100 manned submersibles available for use around the world. Of these, about 30 submersibles are available in the United States, the leading country in their development and construction (see table). The highest concentration of submersibles is in support of the offshore oil industry, mainly in the North Sea. The leading mission applications are inspection of pipelines and cables, followed by cable burial, salvage, coral harvesting, geology, fisheries, biology, and environmental research.

Design considerations. Undersea vehicles are being utilized more now that experience has confirmed their utility, and systems are being designed in accordance with user requirements. A major trend pertains to designing a completely integrated system, which, in addition to the submersible, includes support ship, handling gear for launch and retrieval, and logistic and maintenance support. The objective is to obtain an effective, high utilization rate under varying weather conditions. The major vehicle operating problem is handling during launch and retrieval in heavy seas. Equipment for conducting efficient deep-water surveys will require the use of accurate navigation and guidance systems. Manipulators with greater dexterity will be needed for human-occupied and unoccupied systems to perform intricate manipulative operations more quickly. Many new vehicles are being developed with large panoramic Plexiglass windows to provide a wider viewing field, very effective in survey and inspection missions.

The ratio of the weight of structure to the seawater weight for an equal volume determines to a great extent the size and payload capabilities of a submerged vehicle. This ratio is dependent on several items, one of the most important of which is the compressive strength-to-density ratio of the pressure hull material. Steel with a yield strength of 100,000 psi (689,500,000 N/m²) is generally employed in vehicles. This provides a strength-to-density ratio about double that normally attainable in the 1950s. Research programs have again doubled this ratio.

Hull buoyancy. Most small submersibles need buoyancy in excess of that produced by their pressure hulls to attain neutral buoyancy while sub-

United States submersibles*

Vehicle	Operator	Depth, ft	Crew	Length/ beam, ft	Weight, lb	Payload, lb
Sea Ranger	Verne Engineering Corp.	600	4	17/8	19,000	2,200
Nemo	Southwest Research Institute	600	2	6/6	2,000	850
PC-3B	International Underwater Contractors, Inc.	600	2	22/4	6,350	1,000
Sea Explorer	Sea Line Inc.	600	2	15/5	3,600	300
PRV-2†	Pierce Subs Inc.	600	3	19/8	15,500	1,000
Margenaut	Margen International	600	8	44/9	108,000	6,000
Nekton Alpha	General Oceanographics	1,000	2	15/4	4,500	300
Nekton Beta	General Oceanographics	1,000	2	15/4	4,700	460
Nekton Gamma	General Oceanographics	1,000	2	15/4	4,700	460
Johnson Sea Link I†	Harbor Branch Foundation	1,000	4	23/8	21,000	1,200
Snooper	Undersea Graphics	1,000	2	15/4	4,500	200
Guppy	Sun Shipbuilding & Drydock Co.	1,000	2	11/8	5,000	400
Opsub	Ocean Systems	1,000	2	18/8	10,400	400
Sea Ray	Submarine R & D Corp.	1,000	2	20/5	9,000	350
Mermaid II	International Underwater Contractors, Inc.	1,000	2	17/6	14,000	1,000
Nemo I	Seaborne Ventures	1,000	3	12/8	20,000	1,200
Diaphus	Texas A&M University	1,200	2	13/5	10,000	225
PC-14C-2	Army Missile Command	1,200	2	13/5	10,000	225
Star II	Deepwater Explorations Ltd.	1,200	2	17/5	10,000	500
PC-17†‡	Perry Oceanographics, Inc.	1,500	4	34/8	38,000	500
Deep View	Southwest Research Institute	1,500	2	16/6	12,000	500
Johnson Sea Link II†	Harbor Branch Foundation	2,000	4	23/8	21,000	1,200
Beaver MK IV†	International Underwater Contractors, Inc.	2,700	5	25/8	34,000	2,000
DSRV-1	U.S. Navy	5,000	4	50/8	75,000	4,300
DSRV-2	U.S. Navy	5,000	4	50/8	75,000	4,300
Sea Cliff	U.S. Navy	6,500	3	26/12	42,000	700
Turtle	U.S. Navy	6,500	3	26/12	42,000	700
Deep Quest	Lockheed Missiles & Space Co.	8,000	4	40/16	115,000	7,000
Alvin	Woods Hole Oceanographic Institute	12,000	3	23/8	32,000	1,500
Trieste II	U.S. Navy	20,000	3	78/19	180,000	2,000

*1 ft = 0.03 m; 1 lb = 0.45 kg.   †With diver lockout.   ‡Under construction.

merged. Low-density solids that may be considered for this purpose include polyethylene, polypropylene, expanded plastics, inorganic foams, and syntactic foam. The foamed plastics and inorganic foams have low strength and high water permeability. The low-density plastics have limited buoyancy as well as other problems. Syntactic foams consisting of extremely small and hollow glass, ceramic, or rigid plastic spheres embedded in a plastic matrix have been effectively used at great depths. Syntactic foam is available with a weight of $36-44$ lb/ft³ (sea water is 64 lb/ft³; 1 lb/ft³ = 0.45 kg/0.028 m³), which will withstand pressures at 10,000-ft (3048 m) submergence with less than 1% water absorption. Its cost per pound of net buoyancy is high, but low compared to that of a pressure hull of titanium or a rigid pressure vessel of higher-strength steel.

Power. As a source of power, the lead-acid storage battery, in use in submarines since the beginning of this century, continues to be most widely employed. It places severe limitations on small submersibles because of its relatively high weight-to-energy ratio. However, its characteristics are well known. It is relatively inexpensive, rugged, and reliable, can be quickly recharged, and has a high cell voltage. These batteries are normally carried external to the hull, either pressurized in oil or dry in cylindrical pods. In order to provide quick access for servicing and replacement, there is a trend to use trays of dry batteries in cylindrical pods. Externally carried batteries can also be used as droppable ballasts in the event of an emergency requiring additional buoyancy to surface. Silver-zinc or silver-cadmium batteries provide three to five times as much energy per pound as lead-acid batteries and are employed where mission requirements justify their added costs. For endurances beyond 50 hr, fuel cells, radio isotopes, and nuclear reactors can be employed at much greater cost. Most submersibles have maximum speeds in the range of $3-5$ knots (1.5–2.6 m/s) and have modest power needs for propulsion. Doubling the speed would require at least eight times more power.

Sensors. These instruments are required on small submersibles to determine position, communicate with other units, and make observations of the environment both for a record and while performing work. Typical sensors on a small oceanographic submersible include radio telephone, underwater telephone, external lighting, magnetic recorder, movie camera (carried internal in the pressure hull for viewing through sight ports), magnetic recorder, directional gyro, depth sounder, and manometer depth indicator.

Life support systems. All submersibles require life support systems, and those on the small oceanographic type take into consideration the following

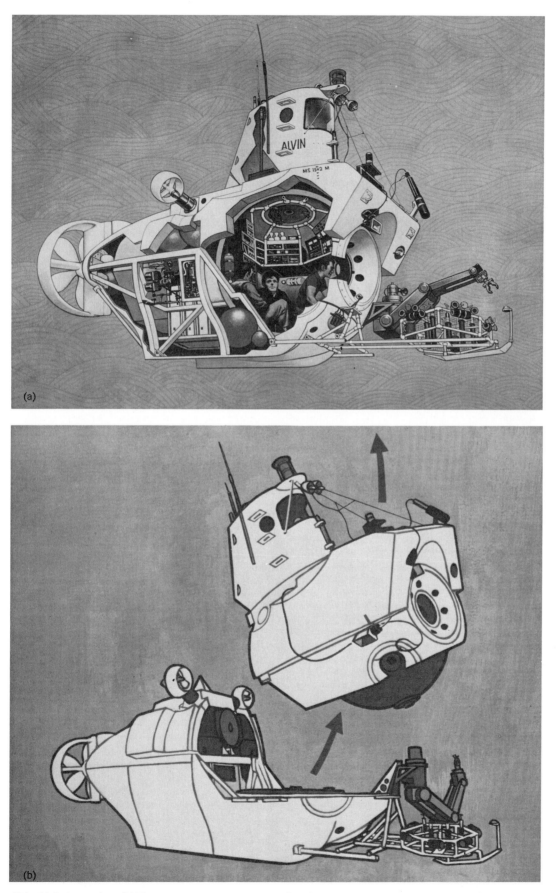

Alvin. (*a*) Cutaway view. (*b*) The passenger capsule and conning tower become disengaged and float to the surface in case of entanglement. (*Courtesy of National Geographic Society*)

services for atmospheric control and monitoring: breathing mixture supply system; carbon dioxide removal system; hydrogen, carbon monoxide, and toxin removal systems; air purification and filtering system; atmospheric monitoring system; and emergency breathing supply system. Most systems on the small submersibles are relatively simple and emphasize the removal of carbon dioxide and the supply of replacement oxygen. Air inside the submersible is drawn continuously through an absorbent system composed of activated charcoal to remove odors and of lithium hydroxide and boralyme to remove CO_2 so that its concentration never exceeds about 0.1%. The internal hull pressure, displayed on an aneroid barometer and sensitive altimeter, decreases slightly as CO_2 is removed. Oxygen is bled in at a point where the air exhausts from the CO_2 removal equipment to maintain the pressure at 1 atm (101,325 N/m²). Both O_2 and CO_2 content are constantly monitored by meters.

Alvin. Since 1965, the Alvin has been probably the best-known submersible (see illustration) and is still operating, with many new capabilities incorporated. By early 1976, it had made over 600 dives, primarily engaged in deep-ocean research in geology and biology. In 1974, it was used in the French-American Mid-Atlantic Ocean Undersea Study (FAMOUS) project together with two well-known submersibles from France, *Archimede* and *Cyana*. Their submersibles were effective in enabling microscale examinations and selective sampling of the deep-ocean features of the Mid-Atlantic Ridge in support of theories on continental spreading.

Alvin is equipped with an extensive instrumentation suite, including navigation and obstacle avoidance sonar, underwater telephone, radios, underwater lights, 35-mm cameras and strobes, underwater television system, and a mechanical manipulator capable of lifting up to 50 lb (22.7 kg) of bottom samples. Sample baskets and racks are carried for transporting collected samples back to the surface. There are provisions both inside and outside the submarine for the installation of specific mission devices or instruments. To power such devices, 30- and 60-V direct current and a limited amount of 115-V 60-cycle alternating current are available. Normal dive duration is 6–8 hr based on rate of battery expenditure. The life support system is capable of supporting three persons for 72 hr. Maximum operating depth is 12,000 ft (3600 m). Normal operations carry a crew of one or two scientific observers.

Alvin is normally supported by its tender, *Lulu*, and is carried on board when not diving. *Lulu* is a 105-ft (32 m) catamaran type and carries a complete shop and repair facility. It also provides berthing and messing facilities and laboratory and work space for six scientists. The usual *Alvin/Lulu* cruise is 10 to 14 days.

A precision navigation system is also available which allows accurate positioning of the submersible at any time during a dive series. This system and other specialized equipment such as hard-rock samplers, magnetometer, precision temperature sensors, and data-logging equipment are available for use with *Alvin*.

Submersibles have proven to be a significant tool in many commercial applications and for scientific research, and their abundance and utilization is steadily increasing. *See* OCEANOGRAPHIC VESSELS.

[D. C. BEAUMARIAGE; J. R. VADUS]
Bibliography: R. D. Ballard, Project FAMOUS, Dive into the Great Rift, *Nat. Geogr.*, May 1975; R. F. Busby, *Manned Submersibles*, Office of the Oceanographer of the U.S. Navy, Spec. Pub. 102, 1976; J. R. Heirtzler, Project FAMOUS, Where the Earth turns inside out, *Nat. Geogr.*, May 1975; M. C. Link, *Windows in the Sea*, 1973; E. H. Shenton, *Diving for Science: The Story of the Deep Submersible*, 1972; R. L. Trillo (ed.), *Jane's Ocean Technology 1974–75*, 1976; National Ocean Policy Committee Print, *Soviet Ocean Activities: A Preliminary Survey*, pt. 4: *Soviet Undersea Research Activities*, Apr. 30, 1975; U.S. Government Printing Office, *Safety and Operational Guidelines for Undersea Vehicles*, bk. 1 and 2, 1974; J. R. Vadus, *International Status and Utilization of Undersea Vehicles*, National Oceanic and Atmospheric Administration, presented at Inter-Ocean 1976, Dusseldorf, June 1976.

Oceanographic vessels

The largest and most basic tools used in the scientific study and exploration of the oceans. The term oceanographic vessels includes both conventional and special-purpose research ships and specialized research vehicles, such as deep submersibles and moored buoys. Uncrewed platforms and equipment, such as buoys, are commonly considered instruments rather than vehicles. *See* OCEANOGRAPHIC SUBMERSIBLES.

The primary purpose of a research ship is to carry scientists and technicians and their equipment to locations at sea where their investigations are to take place. This requires a vehicle which is seaworthy, which has an arrangement that expedites work both on board and over the side, whose position and ability to maneuver can be precisely controlled at all speeds, and which can be navigated accurately.

Conventional oceanographic ships are used for multidiscipline research; their operation requires flexibility in the equipment they can carry and in the space available to accommodate special requirements of various projects. Larger ships can engage in multidiscipline research on a single cruise, but smaller craft with limited space can only handle various disciplines on a successive cruise basis.

Special-purpose oceanographic vessels, such as research submarines, manned buoys, fisheries research ships, special ships to tend and carry deep-submergence vehicles, research drilling ships, and routine survey ships, are often limited by the special requirements of their primary tasks from being economically suitable for multidiscipline operation in oceanographic investigations.

However, scientific investigations at sea are now becoming more specific in nature. This is leading to a noticeable tendency toward more special-purpose ships. An example is the deep-drilling ship *Glomar Challenger* (Fig. 1), in which oil exploration technology has been extended to deep-ocean geo-

Fig. 1. *Glomar Challenger*, a deep-sea drilling project research ship, which is operated by Global Marine; it was placed in service in 1968. The length is 400 ft (122 m).

logical investigations. Another example is the 105-ft (32-m) catamaran research vessel (Fig. 2) designed to carry and tend the deep submersible *Alvin*.

Types. Increased interest in oceanography since 1950 has resulted in a corresponding increase in the world oceanographic fleet. In some countries this has led to new ship construction, but there have been many conversions, particularly of smaller military and merchant vessels built during World War II. Such conversions have met a twofold need: partial satisfaction of the urgent demand for research vessels while new ones are in preparation, and provision for less expensive experience for laboratories and nations which have not previously operated oceanographic ships.

Tradition and geographical conditions are important factors in the type of research ship desired and required by various nations and even by various laboratories. For example, nations with large fishing grounds in home waters, such as Japan, are most apt to develop a fleet of numerous relatively small research vessels, while geographically land-bound nations, such as the Soviet Union, may rely upon larger, more self-sufficient vessels for extended cruises to seas remote from home waters. Nations with fleets developed by private, nongovernmental laboratories, such as the United States, historically have tended to rely on small vessels resulting from, and operated on, reduced budgets. Thus national oceanographic research fleets can be identified as developments from military, fisheries, or hydrographic survey activities of their governments.

Fig. 2. Deep submersible tender *Lulu*, 450 tons (1275 m³), operated by Woods Hole Oceanographic Institution; placed in service in 1964. Length is 105 ft (32 m), beam 46 ft (14 m), draft 10 ft (3 m), speed 6 knots (3 m/s).

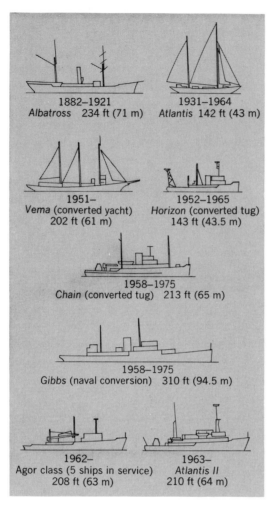

Fig. 3.   The major steps in the development of research ships in the United States.

The history of the development of oceanographic vessels is inseparable from the history of oceanography itself. Figure 3 illustrates the major steps in the development of conventional oceanographic vessels in the United States. The *Albatross*, built and operated by the Federal government, was the only major research ship in service for almost 40 years and was heavily used by academic laboratories and investigators. From 1930 until about 1950, when the major research laboratories were becoming established, the *Atlantis* of Woods Hole carried on as the nation's major seagoing oceanographic vessel. During the 1960s with expansion in the science, major conversions by academic laboratories with the *Horizon*, *Chain*, and *Gibbs* indicated the many possibilities which would result from more capable ships. Since 1960, new, specially designed vessels have been placed in service. These include the *Agor* class ships, the *Atlantis II*, and several smaller ships such as the *Oceanus* class and the *Gyre* class (Figs. 3–5).

The cost of operating a research ship is a large part of the total outlay for oceanographic research. Current expenses are in the order of 40% of total funds. Because the economic indexes of a research ship are dependent upon scientific productivity, which is related to the vessel itself and to the scientific personnel working with it, the efficiency of a research vessel is difficult to measure by parameters appropriate for other vessels.

New construction tended to emphasize replacement of obsolete vessels. In the United Kingdom the 2800-ton (7930-m³) *Discovery* (new) replaced the 1736-ton (4916-m³) steamer *Discovery II* of 1929. In the United States the new 2300-ton (6510-m³) *Atlantis II* (Fig. 6) replaced the 350-ton (990-m³) ketch *Atlantis*, built in 1930. The *Atlantis II* accommodates 26 scientists and a permanent crew

Fig. 4.   *Oceanus*, 960 tons (2720 m³), operated by Woods Hole Oceanographic Institution; placed in service in 1975. Length is 175.2 ft (53.4 m), beam 32.8 ft (10 m), draft 17.4 ft (5.3 m), speed 16 knots (8 m/s). Scientific staff numbers 12, crew 12.

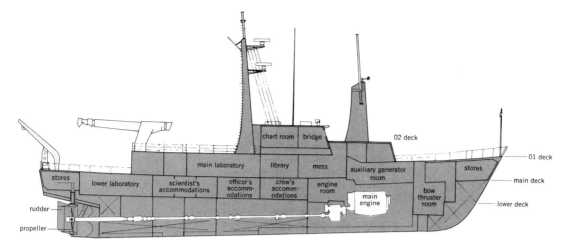

Fig. 5. Inboard profile of the *Oceanus* showing arrangement of spaces and propulsion system.

of 26, and has 3200 ft² (300 m²) of enclosed scientific laboratory area. The original *Atlantis* accommodated 9 scientists and a permanent crew of 20, and had 430 ft² (40 m²) of laboratory area. An interesting comparison between the *Atlantis II* of 1963 and the *Challenger* of 1872 shows that, while their displacements and power are approximately equal, in the first year of operation the *Atlantis II* traveled 41,800 mi (67,300 km), two-thirds of the total for the 3-year voyage of the *Challenger*.

Research requirements. A successful research ship must be sufficiently flexible in design so as not to hinder new developments in research techniques. For example, in France and the United States particularly, the development of deep-submergence vehicles requires that research vessels be capable of lifting such vehicles from the water for servicing, even in high-sea states. The ability to lift vehicles up to 100 tons (91 megagrams) in heavy seas should be possible with future multipurpose but generally conventional research ships. Figure 7 shows a highly maneuverable conventional multipurpose research ship working with a special-purpose vessel, the submersible tender *Lulu*.

Fig. 6. *Atlantis II*, 2300 tons (6510 m³), operated by Woods Hole Oceanographic Institution; placed in service in 1963. Length is 209.3 ft (63.8 m), beam 44.0 ft (13.4 m), draft 16.1 ft (4.9 m). Scientific staff numbers 26, crew 26.

Fig. 7.   Research vessel *Knorr,* 2100 tons (5950 m³), working with deep submersible tender *Lulu.* Length is 236 ft (72 m).

Future trends. The utilization of submarines as primary research ships will be given attention. Crewed ocean stations similar to large buoys have been in service in Monaco and California. Hydrofoil vessels may be efficient in high-speed survey and reconnaissance operations using continuous measurement equipment. *See* OCEANOGRAPHIC PLATFORMS.

Conventional research ships and some special vessels will be called upon to handle a variety of heavy equipment with precise control, such as buoys of greater durability, coring and drilling devices, benthic living experiments, and bottom-supported instruments, and to handle towed instruments of greater weight and heavy strain at depths of 20,000 ft (6100 m) or more. They also are needed to house and operate large automated data-gathering and recording installations with computers for at-sea analysis.   [JONATHAN LEIBY]

Bibliography: J. Dermody, J. Leiby, and M. Silverman, An evaluation of recent research vessel construction in the United States, *Trans. Soc. Nav. Architects Mar. Eng.,* 72:405–444, 1964; IGY World Data Center for Oceanography and the National Oceanographic Data Center, *Oceanographic Vessels of the World,* 1961; T. G. Lang, W. J. Sturgeon, and J. D. Hightower, *The Use of Semisubmerged Ships for Oceanic Research,* Mar. Tech. Soc., San Diego, 75 CHO 995-1 OEC, September 1975; J. Leiby, *History, Design Criteria and Construction of a High Powered Intermediate-Sized Oceanographic Ship,* Mar. Tech. Soc., San Diego, 75 CHO 995-1 OEC, September 1975; S. B. Nelson, *Oceanographic Ships Fore and Aft,* U.S. Government Printing Office, 1971; L. Rosenblatt, The design of modern oceanographic research ships, *Trans. Soc. Nav. Architects Mar. Eng.,* 68:193–263, 1960; J. O. Traung and N. Fujinami, *Research Vessel Design,* Food and Agriculture Organization of the United Nations, 1961; A. C. Vine and J.

Leiby, *Status of Oceanographic Ships in the United States with a View into Future Developments,* Soc. Nav. Architects Mar. Eng. Philadelphia Sec., March 1967; R. P. Voelker and J. K. Kim, *Oceanographic Research Ship Capabilities in Ice,* Mar. Tech. Soc., San Diego, 75 CHO 995-1 OEC, September 1975.

Oceanography

The scientific study and exploration of the oceans and seas in all their aspects, including the sediments and rocks beneath the seas; the interaction of sea and atmosphere; the body of sea water in motion and subject to internal and external forces; the living content of the seas and sea floors and the behavior of these organisms; the chemical composition of the water; the physics of the sea and sea floor; the origin of ocean basins and ancient seas; and the formation and interaction of beaches, shores, and estuaries. Hence oceanography, sometimes called the science of the seas, consists of the marine aspects of several disciplines and branches of science: geology, meteorology, biology, chemistry, physics, geophysics, geochemistry, fluid mechanics, and in its more theoretical aspects, applied mathematics. Oceanography is also an environmental science which describes and attempts to explain all processes in the ocean, and the interrelation of the ocean with the solid and gaseous phases of the Earth and with the universe.

OCEAN RESEARCH

Because of the fluid nature of its contents, which permits vertical and horizontal motion and mixing, and because all the waters of the world oceans are in various degrees of communication, it is necessary to study the oceans as a unit. Further unification results from the technological necessity of studying the ocean from ships. Many phases of oceanic research can be carried out in a laboratory, but to study and understand the ocean as a whole, scientists must go out to sea with vessels adapted or built especially for that purpose. Furthermore, data must be obtained from the deepest part of the ocean and, if possible, scientists must go down to the greatest depths to observe and experiment. Another unifying influence is the fact that many oceanic problems are so complex that their geological, biological, and physical aspects must be studied by a team of scientists. Because of the unity of processes operating in the ocean, and because some writers have separated marine biology from oceanography (implying the term oceanography to embrace primarily physical oceanography, bottom relief, and sediments), the term "oceanology" is sometimes used as embracing all the science divisions of the marine hydrosphere. As used in this article, the term oceanography applies to the whole of sea science. *See* HYDROSPHERE.

Development. The early ocean voyages by Frobisher, Davis, Hudson, Baffin, Bering, Cook, Ross, Parry, Franklin, Amundsen, and Nordenskiold were undertaken primarily for geographical exploration and in search of new navigable routes. Information gathered about the ocean, its currents, sea ice, and other physical and biological phenomena was more or less incidental. Later, the polar expeditions of the Scoresbys, Parry, Markham, Greeley, Nansen, Peary, Scott, and Shackleton were also voyages of geographical discovery, although scientific observations about the sea and its inhabitants were made by some of them. William Scoresby took soundings and observed that discolored water containing living organisms (now known to be diatoms) was related to whale movements. Ross made dredge hauls of bottom-living animals. Nansen contributed to the improvement of plankton nets and suggested the existence of internal waves.

More closely related to the beginning of oceanography as comprehensive study of the seas are the 19th-century activities of naturalists Ehrenberg, Humboldt, Hooker, and Örstedt, all of whom contributed to the eventual recognition of plankton life in the sea and its role in the formation of bottom deposits. Charles Darwin's observations on coral reefs and Müller's invention of the plankton net belong to this phase of developing interest in marine science, in which men began to investigate ocean phenomena as biologists, chemists, and physicists rather than as oceanographers. In this group should also be included such physicists and mathematicians as Kepler, Vossius, Fournier, Varenius, and Laplace, who provided the background for the development of modern theories and investigations of ocean currents and air circulation.

Toward the middle of the 19th century a few scientists began to study the oceans as a whole, rather than as an incidental part of an established discipline. Forbes, as a result of his work at sea, first developed a scheme for vertical and horizontal distribution of life in the sea. On the physical side Matthew Fontaine Maury, developing and extending Franklin's earlier work, made comprehensive computations of wind and current data and set up the machinery for international cooperation. His book *Physical Geography of the Sea* has been regarded as the first text in oceanography.

Forbes and Maury were followed by a distinguished group of men whose interest in oceanography led them to make the first truly oceanographic expeditions. Most famous of these was the three-year around-the-world voyage of HMS *Challenger*, which followed earlier explorations of the *Lightning* and *Porcupine.* Instrumental in organizing these was Wyville Thompson, later joined by John Murray. Later in succession were the Norwegian Johan Hjort and the *Michael Sars* North Atlantic exploration; Louis Agassiz; and Albert Honoré Charles, Prince of Monaco, in a series of privately owned yachts named *Hirondelle I* and *II* and *Princess Alice I* and *II.* Other important contributions were made by Michael and G. O. Sars, Björn Helland-Hansen, Carl Chum, Victor Hansen, Otto Petterson, Gustav Ekman, and the vessels *Valdivia,* the Danish *Dana,* the British *Discovery,* the German *National* and *Meteor,* and the Dutch *Ingold, Snellius,* and *Siboga,* the French *Travailleur* and *Talisman,* the Austrian *Pola,* and the North American *Blake, Bache,* and *Albatross.* Among the North American pioneers were Alexander Agassiz, L. F. de Pourtales, and J. D. Dana. Pioneers in modern oceanographic work are M. Kunelsen, Sven Ekman, A. S. Sverdrup, A. Defant, Georg Wüst, Gerhard Schott, and Henry Bigelow.

Modern oceanography relies less upon single explorations than upon the continuous operation of single vessels belonging to permanent institutions, such as *Atlantis II* of the Woods Hole Oceanographic Institution, *Argo* of Scripps Institution of Oceanography, *Vema* of the Lamont Geological Observatory, the French *Calypso,* and the large Soviet vessels *Vitiaz* and *Mikhail Lomonosov.* Single explorations continue to be made, as exemplified by the Swedish *Albatross* and Danish *Galatea.*

The reduction of data and study of collections from earlier expeditions were carried out generally in research institutions, museums, and universities not solely or primarily engaged in oceanography. The first marine laboratories were interested principally in fishery problems or were designed as biological stations to accommodate visiting investigators. Many of the former have extended their activities to cover chemical and physical oceanography during their growth and development. The latter, often active as extensions of university biological departments, are exemplified by the Naples Zoological Station and the Marine Biological Laboratory of Woods Hole. Visitors to such stations contribute greatly to the development of biology, generally in such fields as embryology and physiology.

The number of institutions devoted to organized oceanographic investigations with permanent scientific staffs has gradually grown. At first the requirements of fishery research provided the stimulus in countries adjacent to the North Sea, but in later years laboratories in other countries wholly or mainly devoted to oceanography have grown considerably in number. A few may be mentioned here. In England, among other important

Fig. 1. Deep-sea drilling vessel *Glomar Challenger,* equipped with satellite navigation equipment and capable of holding position to within 100 ft for several days, using bottom-mounted sonic beacons, tunnel thrusters, and computers.

sity. In Germany oceanographic laboratories are located at Kiel and at Hamburg. In Denmark the Danish Biological Station is at Copenhagen. Other European laboratories include those at Bergen, Norway; Göteborg and Stockholm, Sweden; Helsinki, Finland; and Trieste. Laboratories are located at Tokyo, Japan; Namaimio and Halifax, Canada; and Hawaii. This list is not inclusive and necessarily leaves out a considerable number of important institutions.

Surveys. Oceanographic surveys require careful planning because of high cost. Provision must be made for the proper type of vessel, equipment, and laboratory facilities, adapted to the nature and duration of the survey.

Research ships. Ships of all types and sizes have been gathering information about the oceans since earliest times. Vessels of less than 300 tons displacement seldom range farther than several hundred miles from land, whereas ships larger than 300 tons displacement may work in the open ocean for several months at a time. Research ships of all sizes must be seaworthy and must provide good platforms from which to work (Fig. 1). More specifically, a ship must have comfortable quarters, adequate laboratory and deck space for preliminary analyses, plus storage space for equipment, explosives, samples, and scientific data. Machinery, usually in the form of winches and booms, is necessary for handling the complex and often heavy scientific equipment needed to probe the ocean depths (Fig. 2).

A number of the larger oceanographic vessels are equipped with general-purpose digital computers, including tapes, disk files, and process-interrupt equipment. The result is that data can be reduced on board for experimental work, and all the operations taking place on the vessel can be centralized, including the satellite navigation equipment.

Standard oceanographic equipment includes collecting bottles (Nansen bottles) for obtaining water samples and thermometers (both reversing thermometers and bathythermographs) for measuring temperatures at all depths. In addition, there are various devices for obtaining samples of ocean bottom sediments and biological specimens. These include heavy coring tubes which punch cylindrical sediment sections out of the bottom, dredges which scrape rock samples from submerged mountains and platforms, plankton nets for collecting very small planktonic organisms, and trawls for collecting large free-swimming organisms at all oceanic depths. Echo sounders provide accurate profiles of the ocean floor.

institutions, are the National Institute of Oceanography, the Marine Biological Laboratory at Plymouth, and the Fisheries Laboratory at Lowestoft. In the United States are the Woods Hole Oceanographic Institution in Massachusetts, the Scripps Institution of Oceanography in California, the Lamont-Doherty Geological Observatory in New York, the University of Miami Marine Laboratory in Florida, the Texas A. & M. College Department of Oceanography, and the Oceanography Laboratories of the University of Washington at Seattle, the University of Hawaii, and Oregon State University.

Specialized equipment for oceanic exploration includes seismographs for measuring the Earth's crustal thickness, magnetometers for measuring terrestrial magnetism, gravimeters for measuring variations in the force of gravity, hydrophotometers for measuring the distribution of light in the sea, heat probes (earth thermometers) for measuring the flow of heat from the Earth's interior, deep-sea cameras to photograph the sea bottom, bioluminescence counters for measuring the amount of luminescent light emitted by organisms, salinometers for measuring directly the salinity of sea wa-

Fig. 2. Trawl winch used on research vessel *Vema.* (*Lamont-Doherty Geological Observatory*)

ter, and current meters to clock the speed of ocean currents.

Positioning of a ship is very important for accurate plotting of data and detailed charting of the oceans. Celestial navigation is in wide use now as in the past. Navigational aides such as electronic positioning equipment (loran, shoran, and radar) are increasing the accuracy of positioning to within several tens of yards of the ship's true position. Navigational and radio communications equipment normally is situated near the captain's bridge, but often is duplicated in the scientific laboratories in order that complete communication between the ship's operators, scientists, and other participating ships can be carried on at all times.

Probably the single most important advance in deep-sea oceanography has been the introduction of satellite navigation. Combined with a computer, satellite navigation permits fixes to within several hundred yards while the ship is moving and to within several hundred feet when the ship is located with respect to bottom-mounted beacons, for example, in deep-sea drilling.

Ship's laboratories. Laboratories must be adaptable for a large number of operations. In general they are of two categories, namely, wet and special laboratories. Wet laboratories are provided with an open-drain deck so that surplus ample water can be drained out on deck. Such a laboratory is located near the winches used for running out and retrieving a long string of water-sample bottles (hydrocasts). Adjoining the wet laboratory are special laboratories equipped with benches for measuring chemical properties of the recovered water and for examination of biological and geological samples. Electronics laboratories are either part of, or adjacent to, the special laboratory, depending upon the size of the ship. Here, numerous recording devices, amplifiers, and computers are set up for a variety of purposes, such as measurements of underwater sound, measurement of the Earth's magnetic and gravity fields, and seismic measurements of the Earth's crustal thickness (Fig. 3).

Marine technology. The development of nuclear power plants permits extended voyages without the necessity of refueling. Nuclear power plants, used in conjunction with inertial guidance in the submarines *Nautilus* and *Skate*, made possible the first extended journey under the Arctic ice pack. Uncharted regions of the oceans are within the reach of exploration.

Direct visual observations of the ocean depths are fast becoming a reality, both by "manned" submersibles (bathyscaph) and by television cameras. Deep-sea cameras have been developed to the point whereby motion pictures of even the deepest parts of the sea bottom can be taken (Fig. 4). However, such observations yield no information about the subsurface material. Major crustal features are determined by seismic measurement. The shallower features of the subbottom structure and deposits were not really observed until the advent of the subbottom acoustic probe. This device is a very-high-energy echo sounder capable of penetrating below the sediment-water interface and yielding a continuous profile of the subbottom

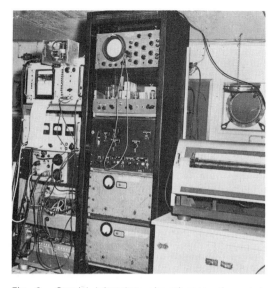

Fig. 3.   Special laboratory aboard research vessel. (*Lamont-Doherty Geological Observatory*)

strata.   *See* OCEANOGRAPHIC   SUBMERSIBLES; UNDERWATER   PHOTOGRAPHY;   UNDERWATER TELEVISION.

Information as to the physical and chemical makeup of the underlying material, however, is dependent upon penetration and actual recovery. Commonly used coring devices rarely penetrate more than 10–20 m (on occasion to about 33 m) below the surface. A new "incremental" coring device has been developed for taking successive 2-m sediment cores to depths of possibly 100 m or more.

Fig. 4.   Remote-controlled underwater television camera mounted in self-propelled vehicle, which can make visual surveys to depths of 1000 ft. Self-buoyant unit moves about or hovers at desired depth in currents or tides of several knots. (*Vare Industries, Roselle, N.J.*)

The most impressive achievement in this area, however, has been in deep-sea drilling. The *Glomar Challenger*, operated by the Global Marine Corp., was designed primarily for the purpose of taking cores throughout the full column of sediment in mid-ocean. It is capable of handling 24,000 ft of drill string in mid-ocean and drilling through more than 2500 ft of sediment, taking 30-ft cores in the process. By mid-1969 the vessel had successfully operated at more than 40 sites in both the Atlantic and Pacific. This operation is sponsored by the National Science Foundation, with scientific guidance supplied by JOIDES (Joint Oceanographic Institutions Deep Earth Sampling), which include the University of Washington; Institute for Marine Sciences, University of Miami; Woods Hole Oceanographic Institution; Lamont-Doherty Geological Observatory of Columbia University; and Scripps Institution of Oceanography. The last-mentioned is the operating institution. Besides incorporating the necessary innovations in drilling and coring, the *Glomar Challenger* is an example of the many important advances in oceanography, such as satellite navigation and positioning.

Oceanographic stations. Work on station consists of sampling and measuring as many marine properties as possible within the limitations of an expedition. Water, sea-bottom, and biological samples are successively collected on the long cables extended to the ocean floor. In some surveys one cable lowering may include samplers for all these items, but this is not the usual procedure. Stations are systematically located at predetermined points along the ship's path. At hydrographic stations, observations of water temperature, salinity, oxygen, and phosphate content are determined upon sample recovery. Seismic stations generally are carried out by two ships, one running a fixed course and dropping explosives while the second remains stationary and records the returning subbottom reflected or refracted sound waves. Biological stations may consist of vertical net hauls or horizontal net tows depressed to sweep the ocean at a fixed depth. Geological stations are usually coring or bottom-dredging operations. Wherever possible the recovered data are given a preliminary reduction aboard ship so that interesting discoveries are not bypassed before sufficient information is obtained. Detailed analyses aboard ship are seldom possible because of the limitations of time, space, and laboratory equipment. Instead, the carefully processed, labeled, and stored material is preserved for intensive study ashore.

Home laboratory. This phase of the work may entail many months of careful examination and detailed analyses. Batteries of sophisticated scientific instruments are often necessary: data computers for reduction of physical oceanography information; spectrographic apparatus consisting of emission units; infrared, ultraviolet, x-ray, and mass spectrometers for chemistry; aquaria, pressure chambers, and chemostats for the biologist; electron microscopes and high-powered optical microscopes for examination of inorganic and organic constituents; radioisotope counters; and numerous standard physical, chemical, geological, and biological instruments. The great variety of measurable major and minor properties is reduced to statistical parameters which may then be integrated, correlated, and charted to increase the knowledge of sea properties and show their relationships. The essentially descriptive properties lead to an understanding of the principles which control the origin, form, and distribution of the observed phenomena. The present knowledge of the oceans is still fragmentary, but the increasing store of information already is being applied to a rapidly expanding number of man's everyday problems.

Research problems. Since the middle of the 19th century, man has learned more and more about that 71% of the Earth which is covered with sea water. The rate of increase of knowledge is being accelerated by improved tools and methods and increased interest of scientists and engineers. Some of the present and future research problems that are attracting scientists are mentioned below.

One of the oldest and still unsolved problems is the motion of ocean waters, involving surface currents, deep-sea currents, vertical and horizontal turbulent motion, and general circulation. New methods such as distribution of radioactive substances, deep-sea current meters and neutrally buoyant floats, high-precision determination of salt and gas content, and hot-wire anemometers for turbulence studies and current measurements have increased present knowledge considerably. The surface movement of water is wind-produced, and a general theory of the motion has been worked out. The deep-sea currents, which are known to be caused in part by variations in the thermohaline circulation, are still an open problem. Superimposed on these movements is turbulent motion, which ranges over the whole spectrum from large ocean eddies transporting millions of cubic meters of water per second to the tiniest vibrations of water particles. Very little is known about turbulence. *See* OCEAN CURRENTS.

The mixing of water masses and the formation of new water masses cannot yet be completely and adequately described, as many of the thermodynamic parameters are not precisely known. Laboratory experiments and measurements of thermal expansion, saline contraction, and specific heat at constant pressure must be carried out. A further problem is the composition of sea water and the extent to which the ratio among the components is constant. In connection with these problems it has been urged that a library of water samples be established. Improved techniques of measuring sound velocity, electrical conductivity, refractive index, and density must be developed to enable scientists to follow many processes in the ocean. It is therefore necessary to study the small variations of these parameters in the sea. *See* SEA WATER.

The tides in the oceans are rather well known at the surface but are almost completely unknown in the deep sea; also, the influence of land boundaries on deep-sea tides is not yet understood. Further research also must be devoted to the interesting phenomonon of internal waves. *See* TIDE.

The study of ocean waves is one of the most advanced topics in oceanography, but the energy exchange between atmosphere and sea surface by friction must be studied further. Another problem is that of the heat exchange between ocean and atmosphere, an important link in the heat mechanism which determines the weather and the oceanic circulation. *See* OCEAN-ATMOSPHERE RELATIONS; OCEAN WAVES.

The climate of the past, in particular that of the last 1,000,000 years, is best studied in the ocean. Isotopic methods in paleoclimatologic research allow the determination of temperature variations in the ocean with a high degree of accuracy. The rapid growth of geochemistry and the increased sampling of deep-sea sediments through improved techniques have solved some of the problems of deep-sea sedimentation. At the same time a number of new ones have been created, such as: Why is the sediment carpet only about 300 m thick? What is the mechanism of sediment transport? What is the history of sea water? Of the ocean basin? What is the cause of the ice ages? *See* MARINE GEOLOGY; MARINE SEDIMENTS.

The results of the Deep Sea Drilling Project have confirmed the expectations of its most optimistic supporters. It has strikingly confirmed the sea-floor-spreading hypothesis of the development of the ocean basins and the newer concepts of plate tectonics. A core to the Moho (about 3 mi deep) may help answer many of the questions about the structure of the Earth's crust. The problem of the mechanism of formation of the ridges and island chains may be near solution.

The age determination of sediments by radioactivity methods, which was thought impossible 30 years ago, is now used on deep-sea sediments older than 10,000,000 years. Very little is known about the formation of minerals on the sea floor, the diffusion and adsorption of elements in and on sediments, and the reaction at slow rates in sediments. Certainly microbiological processes on the sea floor are an important factor, as they seem to produce chemical energy in sediments. *See* HYDROSPHERIC GEOCHEMISTRY.

In marine biology the systematics and ecology remain the major aspects. It is still the science of the "naturalist." The interest of marine biology is many-sided and not grouped around a few central problems. Ocean life offers to the general biologist the best opportunities to study such complex problems as the structure of communities and the flux of energy through these communities. The zonation of animals on the shore and in the open ocean is not yet fully understood; the cause of patchiness in the distribution must be found. On the other hand, the distribution of species by currents and eddies must be studied, and large-scale experiments on behavior must be carried out. Observation at sea has been neglected to a large extent, and therefore the equilibrium between sea observation and laboratory experiment must be restored. The great advances in genetics, biochemistry, physiology, and microbiology also will advance the study of life in the sea. *See* DEEP-SEA FAUNA; MARINE ECOSYSTEM; MARINE MICROBIOLOGY.

Applications of ocean research. Directly and indirectly the ocean is of great importance to man. It is valuable as a reservoir of natural resources, an outlet for waste disposal, and a means of transport and communication. The ocean is also important as a harmful agent causing biological, chemical, and mechanical destruction of life and property. In addition to the peaceful exploration of the oceans, there are many military applications of surface and submarine phenomena. In all of these aspects oceanography provides basic information for engineers who seek to increase its benefits and to avoid its harmful effects. *See* MARINE RESOURCES.

Food resources. The food resources of the ocean are potentially greater than those of the land since its larger area receives a proportionately larger amount of solar radiation, the source of living energy. Nevertheless, this potential is only in part realized. Oceanographic studies provide information which can help to increase fishing yields through improved exploratory fishing, economical harvesting methods, fisheries forecasts, processing techniques at sea, and aquaculture.

Fishes are dependent in their distribution upon food organisms and plankton, vertical and horizontal currents which bring nutrients to the plankton, bottom conditions, and physical and chemical characteristics of the water. A knowledge of the relation of food fishes to these environmental conditions and of the distribution of these conditions in the oceans is vital to successful extension of fishing areas. Satisfactory measurements of the basic organic productivity of the sea may become essential in the selection of regions for extended fishery exploration. *See* SEA-WATER FERTILITY.

The catching of fishes may be facilitated and new and more efficient methods devised through a knowledge of the reaction of fishes to stimuli and of their habits in general. This knowledge may result in better design of nets and in the use of electrical, sonic, photic, and chemical traps or baits. The harvest also may be increased by using improved methods of locating schools of fish by sonic or other means.

The biology of fishes, their food preferences, their predators, their relation to oceanographic conditions, and the fluctuations in these conditions seasonally and from year to year are important factors in forecasting fluctuations in the fisheries. This information will aid in preventing the economic waste of alternating glut and scarcity, and is essential to good management of fisheries and to sound regulation by conservation agencies.

Other anticipated advances which require further oceanographic study include (1) the improvement of fishing by transplanting the young of existing stocks or by introducing new stocks; (2) farming or cultivation of sea fishes (although this does not seem feasible at present, scientific research has improved the cultivation of oysters and mussels in France and Japan); and (3) the use of planktonic vegetation as a source of food or animal nutrition. *See* MARINE FISHERIES.

Mineral resources. Although most of the valuable chemical elements in sea water are in very great dilution, the great volume of water of the oceans (about 300,000,000 mi³) provides a limitless and readily accessible reservoir, if such dilute concentrations can be economically extracted. Magnesium is produced largely from sea water, and bromide also has been extracted commercially. High concentrations of manganese are found in manganese nodules, which are very common on certain areas of the sea floor.

Other elements occur in too great dilution to be extracted by present methods. Possibly a better understanding of the ability of certain marine plants and animals to accumulate and concentrate elements from sea water in their tissues may lead to new methods of recovering these elements. *See* MARINE MINING.

Energy and water source. Sea water contains deuterium and would be a limitless source of this element in the event of successful nuclear fusion developments. Further oceanographic knowledge and advances in engineering may lead to the increased utilization of tidal energy sources, or of the heat energy available from temperature differences in the ocean. The development of new methods for removing salt from sea water offers promise that the sea may become a practical source for potable water.

Disposal outlet. Because of its large volume, the sea is frequently used for the disposal of chemical wastes, sewage, and garbage. A knowledge of local currents and tides, as well as of the bottom fauna, is essential to avoid pollution of beaches or commercial fishing grounds. Radioactive waste disposal in offshore deeps poses problems of the rate of movement of deep waters and the transfer of radioactive materials through migration and food chains of marine organisms. A problem of rapidly growing concern is oil pollution, which is caused in part by the rapidly expanding exploitation of the continental-shelf-oil resources. The Santa Barbara incident in 1969 was one example. There must be considerable development in marine engineering and better ecological understanding if these oil resources are to be fully utilized. A second major cause of oil pollution is the breakup of giant oil tankers, such as the *Torrey Canyon*. Prevention of such accidents requires improved vessels and more stringent navigation controls.

Traffic and communication. The sea still remains an important highway; thus the knowledge and forecasting of waves, currents, tides, and weather in relation to navigation are of great practical importance. New developments include the continuous rerouting of ships at sea in order that they may follow the most economic paths in the face of changing weather conditions. A knowledge of submarine topography, geologic processes, and temperature conditions is important for the satisfactory location, operation, and repair of submarine cables. The use of Sofar in air-sea rescue operations is based upon submarine acoustics.

Defense requirements. Defense aspects of marine research involve not only the navigation of surface vessels but also undersea craft with special navigational problems related to submarine topography, echo sounding, and the distribution of temperature, density, and other properties. Research in submarine acoustics has improved communication between, and detection of, undersea craft. In spite of these advances natural conditions, such as warm water pockets and subsurface magnetic irregularities, can conceal submarines from conventional means of detection. Investiga-

tion of these conditions is essential to any defense against missile-carrying submarines.

Property and life. Damage to docks and ships by marine borers and fouling organisms is controlled by methods that utilize a knowledge of the biology, behavior, and physiology of the destructive organisms, and of the oceanographic conditions which control their distribution. Loss of life caused by the attacks of sharks and other fishes may be reduced through an understanding of their behavior and the development of repellants and other protective devices. The chemical characteristics of sea water pose special problems of corrosion of metals. Beach erosion, wave damage to harbor and offshore structures, the effects of tsunamis and internal waves, and storms cause loss of property and life. Much of this damage may be minimized by the application of oceanographic knowledge to forecasting methods and warning systems. *See* NEARSHORE PROCESSES; STORM SURGE; TSUNAMI.

Indirect benefits of oceanography arise from the application of marine meteorology to weather prediction, not only over the sea areas but also over the land. The study of marine geology and marine ecology aid in the understanding of the character of oil-bearing sedimentary rocks found on land. *See* MARINE INFLUENCE ON WEATHER AND CLIMATE.   [WILLIAM A. NIERENBERG]

Bibliography: H. Barnes (ed.), *Oceanography and Marine Biology*, vol. 14, 1976; M. B. Deacon (ed.), *Oceanography: Concepts and History*, 1977; G. Dietrich, *General Oceanography*, 1963; W. A. Herdman, *Founders of Oceanography and Their Work*, 1923; G. L. Pickard, *Descriptive Physical Oceanography*, 1976; R. G. Pirie (ed.), *Oceanography: Contemporary Readings in the Ocean Sciences*, 1977; Scientific American Editors, *Ocean*, 1969; F. W. Smith and F. A. Kalber (eds.), *Handbook Series in Marine Science*, 2 vols., 1974.

THEORETICAL OCEANOGRAPHY

The basis for theoretical oceanographic studies is the known set of conservation equations for momentum, mass, and energy, supplemented by an equation of state for sea water and a conservation equation for dissolved salts. A general solution to the mathematical system is not possible. The aim of theoreticians is to develop simple mathematical models from the general set to explain observed oceanic features. A model may describe a process, such as the convective overturning of surface waters, or a phenomenon, such as the existence of the Gulf Stream on the western side of the North Atlantic. Whatever the purpose of the model, simplicity is important: the more directly that one can relate a feature to processes that are already-understood, the better.

Oceanic flow. Oceanic flows with a horizontal scale of 100 km or more are strongly affected by the rotation of the Earth. Just as a tilted spinning top wobbles laterally instead of falling directly when acted upon by gravity, the rotation of the Earth causes a fluid flow to be deflected from the direction of the applied force. The deflecting (Coriolis) force is proportional to the angular rate of rotation of the Earth and to the sine of the latitude of the position of the fluid. Therefore, it is larger at high latitudes than at low.

The effect of rotation is easily incorporated into the conservation equation for momentum. When it is made to play a dominant role, the analysis is simplified because flows that are not affected by the rotation are effectively filtered out of the equations. *See* CORIOLIS ACCELERATION AND FORCE.

Wind effects. Simple theoretical analysis shows that the wind directly affects the surface waters of the ocean only in the top hundred meters or so and that rotation causes the net transport of water in this wind-driven layer to flow to the right of the direction of the wind stress. This indirect circulation in the surface layers induces a deeper flow in the direction of the wind, thereby setting up the wind-driven circulation. The Coriolis force exerts a dominating influence on the structure of the flow, and its variation with latitude gives rise to the Gulf Stream and the other observed western boundary currents.

Thermohaline circulation. Solar heating is most intense at low latitudes, and cooling of surface waters occurs in polar regions. The vertical circulation caused by the generated density differences has a global scale because the dense water that fills the abyssal ocean is formed in only a few polar locations. The rotation of the Earth affects the flow pattern of the deep circulation as well, giving rise to strong currents near coasts and underwater ridges that bound the basins on the west and to a relatively weak flow in the remaining areas. Theories of this thermohaline circulation verify the observation that the waters of the ocean are stably stratified. Light surface waters lie above the thermocline, a layer with a relatively sharp density change, and the abyss is filled with nearly homogeneous, dense water. *See* THERMOCLINE.

Waves. Given the stable stratification of the oceans, theory predicts the existence of a variety of large-scale waves. The periods of gravity waves are limited by rotation to be less than or equal to a day, or more precisely, the period of a Foucault pendulum. Longer-period, planetary waves owe their existence to the variation of the Coriolis force with latitude and to the variable depth of the ocean. These waves have periods ranging from several days to several years. Variable depth also gives rise to waves that are trapped in a layer near the bottom, the depth of the layer depending on stratification, rotation, and the scale of the waves. The periods here range from a few days to a few months. All of these waves can be identified in observational records, and some of the observed, nonwavelike features can be described by combinations of such wave motions.

Instability. Simpler wave studies are valid when the wave speeds are substantially larger than the fluid velocity, a condition that is often violated, especially for long-period waves. In such cases the current velocity must be included in the analysis. The energy associated with the currents then becomes available as a possible source of instability. The waves may grow in amplitude by drawing energy from the kinetic energy of the currents or from the potential energy of the stratification. Both types of instability occur in the ocean, though the more important source of energy seems to be the latter, in which case the basic flow is said to be baroclinically unstable. As the waves grow, they interact with each other and alter the current field from which they draw their energy. Eventually a state of statistically steady equilibrium may be achieved in which the basic flow is altered to the point where it provides enough energy to feed waves whose amplitudes are in equilibrium with the altered mean field.

Turbulence. In more extreme situations the instability may become so intense that it leads to turbulence. In that case traditional analysis breaks down, and a statistical-dynamical approach to the problem is required. The effect of turbulence is often parameterized in terms of properties of the mean field, somewhat in the manner in which the net effect of individual particle motions of a gas is parameterized in kinetic theory as a viscosity coefficient. However, the two situations are very different, and in the present setting that approach gives reasonable results only in some cases.

The latter 1970s saw the development of numerical oceanographic models that can resolve the unstable modes and follow their development as the turbulent regime is approached. The constraints of rotation and stratification help to restrict the structure of the motions essentially to two dimensions so that the problem can be handled with present computing facilities. Tentative results offer considerable hope that the turbulent effects can be parameterized. The role of bottom topography may be important in such a description.

The resolution of the large-scale turbulence problem is especially important in the oceanographic context because the fluctuating turbulent motions are normally much stronger than the velocity of the mean circulation. Hence, good understanding of the latter will require a knowledge of the effects of the former. A similar situation exists in the atmosphere, though the relative intensity of the fluctuations is somewhat smaller there.

Time-averaged circulation. A totally different theoretical approach can be used to determine the time-averaged circulation. Ocean water contains a number of dissolved chemicals, such as oxygen and silicates, which can be looked upon as tracers that indicate the flow from known source regions. Ideally one would like to be able to calculate the flow from the tracer distributions, but it is easy to show that this inverse problem generally does not have a unique solution. So far the most effective attacks on this problem have used a known velocity field, one that satisfies at least the major constraints, in the convective-diffusive equation for the tracer. The calculated distribution is then compared with the observed, and parameters are adjusted until an optimal fit is achieved. A more direct approach is to include the tracer in a numerical model of the circulation. [GEORGE VERONIS]

Bibliography: N. P. Fofonoff, Dynamics of Ocean Currents, in *The Sea*, vol. 1, pp. 323–395, 1962; National Academy of Sciences, *Numerical Models of Ocean Circulation*, 1975; P. B. Rhines. The Dynamics of Unsteady Ocean Currents, in *The Sea*, vol. 6 pp. 189–318, 1977; M. Stern, *Ocean Circulation Physics*, 1975; H. Stommel, *The Gulf Stream*, 1965; G. Veronis, Model of World Ocean Circulation, *J. Mar. Res.*, pt. I, 31:228–288, 1973, pt. III, 36:1–44, 1978.

Oceans

The vast salt-water bodies surrounding the continents and filling the great basins of the Earth's crust. Some 70% of the Earth's surface is covered by the oceans, with a total volume of approximately 1.35×10^9 km³. The world ocean can be more aptly described as a hemispheric ocean: that is, the Southern Hemisphere has a greater aerial distribution of sea water than the Northern Hemisphere. In fact, more than two-thirds of all land lies north of the Equator.

The oceanic regions of the Earth comprise not only the bodies of water which fill the large deep-ocean basins, but also the shallower seas that cover the slightly submerged edges and interiors of continents. The distribution of elevations of the Earth's surface relative to sea level can be plotted on a graph called a hypsographic curve. This curve (Fig. 1) indicates that two dominant elevations characterize the Earth's surface: one approximately 100 m above sea level and the other 5000 m below sea level. These two average elevations are separated by a sharp transition zone. The distribution indicates that the oceans form a geographic province entirely distinct from that of the continents. The great oceanic depth is a reflection of the vastly different compositional makeup of the Earth's crust beneath the sea compared with that which underlies the continents.

An understanding of the oceans includes the history of formation, composition, and origin of the ocean basins; the distribution and origin of sediment and mineral deposits; the composition, distribution, and circulation of the water masses; and the study and distribution of the various types of life within the oceans.

Features of ocean floor. Three major morphologic provinces characterize the ocean floor: the continental margins, the deep-ocean basins, and the mid-oceanic ridges (Fig. 2).

Around the Atlantic Ocean, and along other similar ocean margins, the continental margins consist of shelf areas which are submerged edges of the continents and the continental slopes and rises which are transition zones between the continents and deep-ocean basins. Continental shelves comprise about one-sixth of the Earth's surface. Their width varies from several kilometers (southern California) to several hundred kilometers (Gulf Coast, Grand Banks), averaging about 65 km. The shelves dip seaward at a very gentle gradient of about 0.07° and attain depths of about 130 m at their outer margin, called the shelf break. The shelf edge may be a depositional or erosional feature. It is subject to modification by the sculpting and scouring force of strong currents and erosion during low sea-level stands. Active faulting occurs along unstable margins, such as that off the coast of southern California and southern Alaska. Seaward of the shelf break is the more steeply dipping (4°) continental slope, an approximately 20-km-wide zone characterized by either seaward-dipping sediments or slumping of sediments downslope. The continental rise, at the base of the continental slope, has a seaward dip of less than 0.5° and varies in width from 100 to 1000 km. Rises consist of thick wedges of sediment which accumulate by gravitational gliding of semifluidized material down the steeper continental slope or by the deposition of land-derived clastics in the form of great aprons of sediment on the deep-ocean floor, termed submarine fan deposits. *See* CONTINENTAL SHELF AND SLOPE.

Some continental margins are so unlike the conventional Atlantic-type margins (with continental shelves, rises, and slopes) that they constitute an additional oceanic morphological realm, the western Pacific type and Andean types of continental margin. These may include offshore volcanic island arcs like those fringing the western Pacific (Japanese Islands, Philippine Islands, and so on) and eastern Caribbean. These island arcs are paralleled offshore by deep narrow trenches which represent the greatest depths in the oceans. The deepest trench is the Challenger Deep, part of the Marianas Trench in the Pacific Ocean, where a depth of 11,022 m below sea level has been recorded. In general, trenches are hundreds of kilometers wide and 3–4 km deeper than the surrounding ocean floor. They are continuous for thousands of kilometers and are generally V-shaped, with narrow flat floors due to sediment infilling. Small ocean basins which commonly occur behind the trench-bounded chains of volcanic islands (Sea of Japan, Philippine Sea) are called marginal seas. Other small ocean basins such as the Mediterranean Sea, Gulf of Mexico, and Caribbean, Bering, and Black seas, which are partially separated from the main ocean bodies, are often

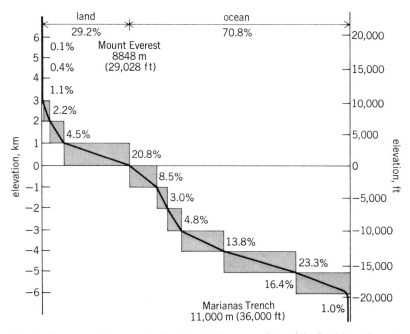

Fig. 1. Hypsographic curve showing the relative proportions of the Earth's surface at elevations above and below sea level. (*From A. N. Strahler, Physical Geography, 2d ed., copyright © 1960 by John Wiley & Sons, Inc.; reprinted by permission*)

Opposite page:
Fig. 2. Physiography of the floor of the North Atlantic Ocean, showing major features. (*From M. G. Gross, Oceanography: A View of the Earth, 2d ed., copyright © 1977 by Prentice-Hall, Inc.; reprinted by permission*)

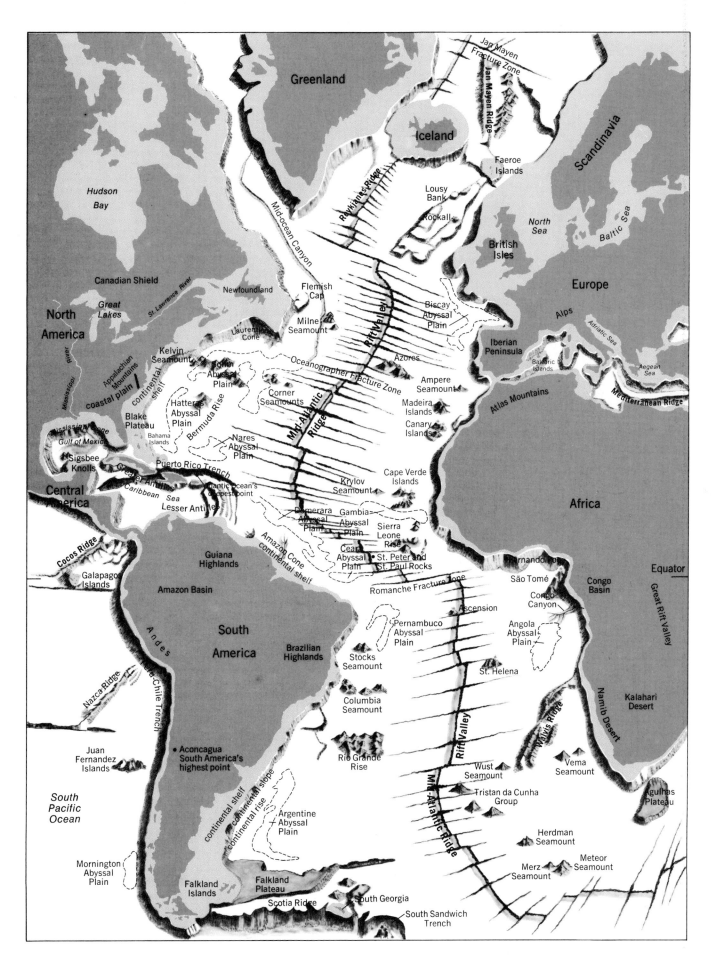

also referred to as marginal seas. *See* MARINE GEOLOGY.

The deeper-realm ocean floor (excluding the continental margins) covers 56% of the Earth's surface. Much of the deeper-realm ocean floor consists of the abyssal plains (30% of the Earth's surface area), which are broad, flat, featureless plains with an average depth below sea level of 4–6 km. Surface slopes do not exceed 0.05 degrees. The subdued relief is produced by the absence of tectonic activity and by the blanketing effect of sedimentary cover.

The topography of the deeper-realm ocean floor is dominated by an extensive mid-oceanic ridge system which is continuous for more than 55,000 km through the Atlantic, Pacific, Arctic, and Indian oceans. The ridges of this system are characterized by a central, 25–50-km-wide rift valley whose average depth is 2.7 km. The rift is bordered by steep mountains standing 1–3 km above the sea floor, peaks of which emerge above sea level as islands (Iceland and Bouvet and St. Paul islands). Numerous fracture zones which trend perpendicular to the ridge and offset the ridge crest (transform faults), and the many volcanoes and relict volcanoes associated with the ridge system produce irregular and rugged morphologic surface in some

portions of the usually monotonous abyssal plain areas. Some volcanic islands (Hawaii, Line Islands) and seamounts (New England and Emperor seamounts) form linear chains of emergent or submergent volcanoes. Submerged seamounts with flat tops are called guyots. These beveled peaks were eroded flat when the seamounts were near sea level. Coral atolls sometimes fringe submerged seamounts or guyots in the equatorial regions. Rises and swells are broad regions of higher elevation on the ocean floors, such as the Bermuda Rise in the Atlantic and the Shatsky Rise in the Pacific. These higher regions presumably reflect an extensive period of past volcanic outpouring over a wide region of the ocean floor. *See* OCEANIC ISLANDS; SEAMOUNT AND GUYOT.

Evolution of ocean floor. The difference in elevation of the continents and ocean basins attests to the differences in composition between continental and oceanic crust. Continental crust is thick (35–60 km) and composed primarily of granitic-type rocks with density = 2.7 (Fig. 3). Oceanic crust is denser (density = 3.0) and only 5–10 km thick, consisting of a carpet of sediment (layer 1) overlying thick piles of extrusive volcanic rocks (layer 2). These volcanic rocks of the ocean crust are basaltic in composition and are characterized by tubular and bulbous "pillow lavas" and ropy-textured plains flooded by "pahoehoe" lavas. The extrusive flows cap a third component of the volcanic crust (layer 3), intrusive-type volcanic rocks composed of densely spaced vertical dikes fed from subjacent plutonic complexes and the dense, underlying, coarsely crystalline rocks of the mantle.

The oceanic crust is also much younger than the continental crust. The bulk of the continental crust is at least 10^9 years old, much is $2–3 \times 10^9$ years old, and continental rocks as old as 3.9×10^9 years have been described. However, the oldest in-place oceanic crust is about 200×10^6 years old. No oceanic crust older than this exists within the ocean basins, although older slices of ultramafic (mantle) rocks, sheeted dike complexes (densely spaced vertical dikes), and basaltic rocks capped by thin, deep marine sediments are found on land. These have been tectonically thrust onto the continents and incorporated into highly deformed mountain belts. Known as "ophiolites," these sequences are interpreted to represent fragments of older oceanic terranes that no longer exist.

The contrasts between the lithology and age of continental and oceanic crust is presently explained largely by the theories of sea-floor spreading and plate tectonics. These theories also provide an interpretative basis for differences in the geological characteristics of continental margin types, the abyssal plains, and the mid-ocean ridge system.

The sea-floor spreading theory postulates that new oceanic crust is continually being produced by the upwelling of magmatic material along the mid-ocean ridge system. As new magma rises and solidifies, older ocean crust is pushed aside and the ocean basins grow. This spreading of the sea floor also causes continental drift because the granitic continental blocks become further and further apart as the ocean floor grows. The high

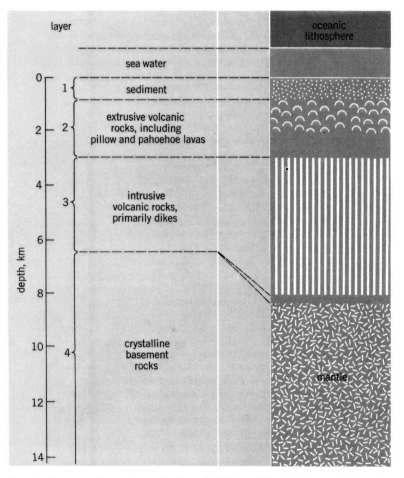

Fig. 3.   Structure of oceanic crust. (*From P. A. Rona, Plate tectonics and mineral resources, Sci. Amer., 229(1):86–95; copyright © 1973 by Scientific American, Inc.; all rights reserved*)

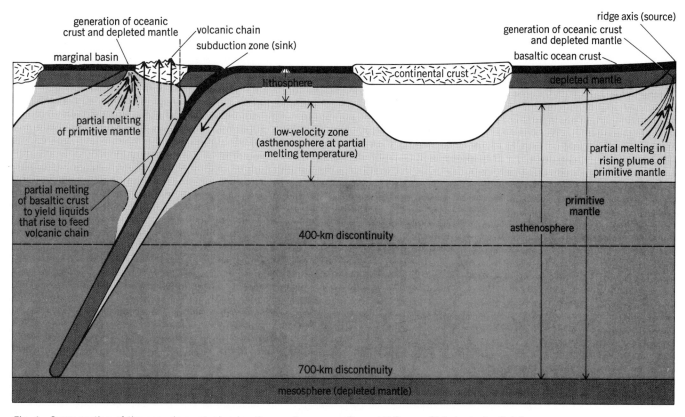

Fig. 4. Cross section of the oceanic crust, showing the mode of formation at mid-oceanic ridges and the subsequent destruction along deep-sea trenches or subduction zones. (*From J. F. Dewey, Plate tectonics, Sci. Amer., 226(5):62; copyright © 1972 by Scientific American, Inc.; all rights reserved*)

topography of the mid-ocean ridge system is directly related to the presence of anomalously hot rocks which expand in volume. As newly formed oceanic crust cools with progressively more time (and further from the ridge crests), oceanic crust volumetrically contracts, producing the topographically lower, subdued abyssal plain areas. Formation of new crust at ridges occurs at the rate of 1–10 cm per year on each side of a ridge.

A variety of data support this model for sea-floor spreading. Magnetometer surveys indicate that the ocean floor exhibits a pattern of linear stripes parallel to the mid-ocean ridges. These stripes are actually belts of magnetic anomalies, that is, alternating strips of crust which exert greater or less magnetism than the average. These magnetic anomaly patterns characterize many major ridge systems. In general, the strips or belts parallel the ridge axes and are symmetric in width across the ridge system. For years they remained a mystery, until in 1963 F. J. Vine and D. H. Matthews proposed that they represented evidence for the continuous generation of new ocean crust at the ridges by the extrusion of liquid basaltic lava. They reasoned that as the lava cooled, it became magnetized in the Earth's magnetic field. Because the Earth's magnetic field apparently alternates through geologic time between normal (as it is now) and reverse polarization, there is a reinforcing or subtractive effect exerted by the basaltic oceanic crust on the present-day magnetic field. Magma extruded during times of normal polarity

operates in coincidence with the present-day magnetic field to produce a stronger than normal, or positive, anomaly. Magma extruded during times of reversed polarity produces a small component of magnetic strength opposite to that at present, and results in a negative (less than normal strength) magnetic anomaly.

A paleomagnetic time scale had been developed by dating continental basalts that displayed magnetic properties that were interpreted as the results of polar reversals through time. These were then correlated with the stripelike magnetic anomalies on parts of the ocean floor. By estimating spreading rates from these data, it was possible to extrapolate the paleomagnetic time scale further into the past and predict the ages of the ocean floors. The oldest oceanic crust is believed to lie in the northwest Pacific and is about 200×10^6 years old. Spreading rates of about 2–20 cm per year are adequate to produce all the ocean basins within this time span.

The exact nature of the forces behind sea-floor spreading and continental drift is not clear. It is now known that long, linear mid-oceanic ridges form along belts of higher-than-average heat flow. The accretion of the oceanic crust is evidently linked to partial melting of the deep primitive mantle of the aesthenospheric layer of the Earth along linear zones which coincide with the mid-ocean ridge system. This melting liberates basaltic composition magma, which rises to the surface and either chills rapidly to form glass in contact with

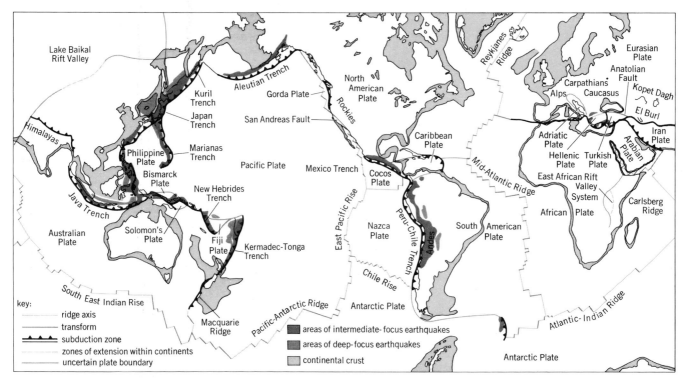

Fig. 5.   Configuration of plates which segment the Earth's crust, showing locations of ocean ridges, trenches or subduction zones, and transform faults. Triangles along subduction zones are on upper, overriding plate.

sea water or moves slowly in dike swarms or through crystal settling in deep cumulate chambers. Local "hot spots" also occur where plumes of hot magma rise from the mantle to produce linear chains of volcanoes as they are overridden by a moving rigid plate of ocean crust (Hawaiian-Emperor Seamount chain).

Because the Earth is believed not to be expanding, the formation of new ocean crust must be accompanied by simultaneous destruction of ocean crust elsewhere (Fig. 4). This allows the origin of trenches, island arcs, and inland seas to be explained in the context of sea-floor spreading. The deep-sea trenches are believed to be areas where older, cooler oceanic crust is consumed as it slips beneath adjacent oceanic or continental crustal blocks. Regions of descending lithospheric slabs are the locus of destructive earthquakes, caused as the downgoing rock internally deforms. These earthquake regions, called Benioff zones, define the configuration of the downgoing lithospheric plate. Partial melting of descending rocks, as well as melting of rocks overlying the descending plate, causes magma rich in sodium and potassium to rise to the surface, forming the chains of volcanic islands which parallel the overlying deep-sea trenches. The frictional drag exerted on subjacent, continentally capped lithosphere probably produces marginal ocean basins. Compressive forces along these zones of convergence also cause deformation of the Earth's crust and the formation of mountain belts, such as the Andes.

Deep-sea trenches and mid-oceanic ridges are the main (convergent and divergent, respectively)

boundaries of the rigid plates of the Earth's crust which move with respect to each other (Fig. 5). A third type of boundary is a transform fault, where the plates on either side are neither created nor destroyed but slide past one another, such as along the San Andreas Fault. Ridge-ridge transform faults of the mid-oceanic ridge system join different portions of a spreading ridge axis. The concept that the Earth's crust consists of plates that are created in one area, move, and are destroyed in another is termed plate tectonic theory.

Sediments, mineral deposits, and energy resources. The ocean floor is a repository (Fig. 6) for a continuous rain of material manufactured in the sea itself (shells, skeletons, teeth, scales, and animal and plant tissue), transported from the surrounding land (gravel, sand, silt, and mud derived from rivers and coastal beaches; windblown dust from deserts; ash and pumice from volcanic eruptions; boulders carried by melting icebergs; and spores and pollen of terrestrial plants), impacted from space (meteorites and tektites), products chemically precipitated from sea water or subsea thermal springs (halite, gypsum, metalliferous shales, polymetallic nodules, and crusts), or derived from alteration or abrasion of the ocean bedrock and previous generations of sediment (volcanic sands, zeolites, some iron/aluminum silicate clay minerals and phosphates). These various kinds of bottom sediments can be grouped into a number of genetically distinct types, of which three are most important: the terrigenous sediments, biogenous sediments, and red clay. Deep-sea sediment without significant amounts of terri-

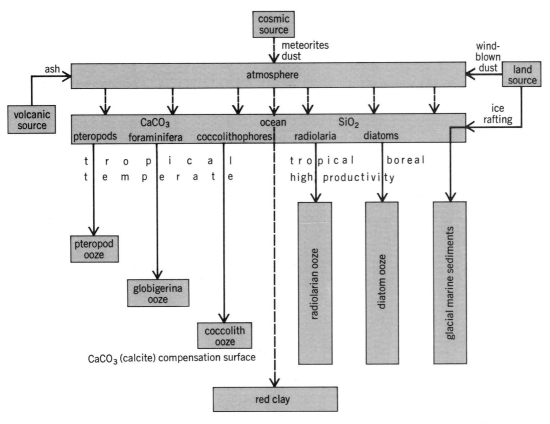

Fig. 6. Schematic representation of pelagic sedimentation in the ocean. (*From W. W. Hay, ed., Introduction, Studies in Paleo-Oceanography, Soc. Econ. Paleontol. Mineral. Spec. Publ. no. 20, pp. 1–5, 1974*)

genous (land-derived) detritus are referred to as pelagic deposits.

Terrigenous sediment. Land-derived material is sometimes carried to the ocean basins in episodic, swiftly flowing sediment suspensions known as turbidity currents which are generated by underwater avalanches, often initiated in the heads of submarine canyons carved into the continental slope. Deposits of turbidity currents, or turbidites, consist of size-graded sand, silt, and clay layers, with the coarsest components at the base of the layer, and are commonplace on continental rises and the adjacent abyssal plains. Fine-grained materials are often carried long distances in currents before settling to the seabed. Clouds of diffuse sediment in suspension in currents hugging the ocean floor are known as nepheloid layers. Violent explosions from land volcanoes in island arcs blast ash and glass shards to heights reaching 30 km in the upper atmosphere. Winds can carry this sand-sized volcanic tephra seaward for distances exceeding 500 km and the finer dust up to several thousand kilometers. Flowing continental glaciers (such as those found today in Greenland and Antarctica) and mountain glaciers (such as those in Alaska and Scandinavia) rip up bedrock and soil at their basal shear planes and incorporate this debris into the ice flow. Upon reaching the land's edge, the glaciers collapse into icebergs which, prior to melting, raft this material, known as erratics, for hundreds of kilometers out to sea. During the last ice age, glacial boulders were rafted in the Northern Hemisphere as far south as the coast of southern Portugal and almost to the vicinity of Bermuda.

Biogenous sediment. Microscopic skeletons of marine plankton make up layers of biogenic ooze which can be either calcareous (composed of pteropods, foraminiferans, coccolithophores, and discoasters) or siliceous (consisting of radiolarians, diatoms, and silicoflagelletes). Siliceous oozes are dominant in regions of high latitude and are abundant in equatorial and coastal regions of strong upwelling. In areas where the ocean is very deep, the calcareous material dissolves in corrosive bottom waters. The level of dissolution, known as the carbonate compensation surface (CCS), is shallowest near the margins of the ocean basins and deepest in the equatorial belt of high ocean-surface fertility. Although the mean depth of the CCS is about 5 km today, it has varied significantly in the past, being several hundred meters deeper in the Pacific during the ice ages and several kilometers shallower prior to about 40×10^6 years ago.

Red clay. Bright reddish-brown to chocolate brown, fine-grained (particle diameters less than 4 μm) pelagic material containing less than 10% calcium carbonate and amorphous silica is referred to as red clay. Red clay is polygenic in origin, consisting of windblown terrigenous and volcanogenic material, cosmic material such as tektites, and fine organic dust derived from the abrasion of biogenous material such as shark teeth.

The areal distribution of the various kinds of bottom sediment on the ocean floor is largely a function of oceanic depth, water chemistry and

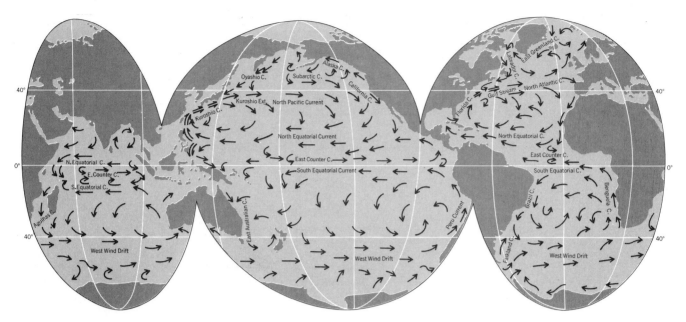

Fig. 7. Map showing major surface currents of the world's oceans. Winter conditions are shown for the Indian Ocean. (*From D. A. Ross, Introduction to Oceanography, copyright © 1970 by Prentice-Hall, Inc.; reprinted by permission*)

temperature, organic productivity, and the distance from terrigenous source areas. Terrigenous sediments form the dominant sediment cover in belts bordering all the major continental masses and along the flanks of some portions of the mid-ocean ridge system. Away from terrigenous sources, biogenous sediments predominate when organic activity is sufficiently high, as is typical in tropical latitudes. Where the water depth in such areas exceeds the critical carbonate compensation depth, the dominant sediment is siliceous ooze. Where the water depth is shallower, calcareous oozes predominate. Abyssal plains which are both remote from terrigenous sources and in areas of low organic productivity are covered with a mantle of red clay.

Minerals. In areas characterized by either very slow deposition or erosion from bottom currents, slabs or nodules of iron and manganese oxides are accreted in concentric shells to eventually construct a nearly continuous indurated pavement (hydrogenous sediments). Some nodules reach 10 cm or more in diameter and contain economically interesting abundances of copper, nickel, and cobalt. Important nodule fields extend over vast tracts of the equatorial Pacific between the Clipperton and Clarion fracture zones, but also occur in the Cape Basin of the South Atlantic and in the Mozambique Basin of the Indian Ocean.

High heat flow in the crestal area of the mid-oceanic ridge induces hydrothermal circulations in permeable strata of layer 2 of the oceanic crust. Leaching of the basaltic rocks in the absence of oxygen scavenges iron and manganese ions into solutions that are ejected in hot thermal springs within the rift valley floor. Under such circumstances, metalliferous shales are precipitated onto the top of the volcanic pile and metallic oxides build hollow cylinders centered over the thermal vents. Unusual animal communities have adapted themselves to the springs by feeding upon sulfate-reducing bacteria in the emissions. Where thick layers of salt and associated evaporites are present such as in the Red Sea, pools of anoxic brine as hot as 60°C become trapped in rift valley depressions.

Persistent winds, such as the trades, blow surface water offshore, allowing cold subsurface water to upwell and replace it. Belts of upwelling (for example, the Benguela Current off southwestern Africa, Canary Current off northwestern Africa, and Peru-Chile Current off western South America) are recognized by their anomalously high nutrient content, great abundances of life, and generally reduced level of dissolved oxygen in underlying intermediate waters. Phosphates are an important economic resource of upwelling areas, and organic rich muds (sapropels) are present only where the sediment carpet of the continental slope intersects water masses which have been depleted in oxygen by the consumptive processes related to high productivity.

Carbonate platforms and barrier reefs (the Bahamas and northeastern Australia) occur at the continent's edge in warm surface waters and in isolation from coastal deltas. The seaward escarpments of these limestone terraces are among the steepest continuous slopes on the Earth's surface.

Gas and oil. Upon deep burial, the organic compounds in marine strata thermally maturate and generate natural gas, liquid petroleum, or both. Under favorable conditions, gas and oil migrate along permeable beds to become stored in geologic reservoirs which can include ancient strandline (beach) sands or reef tracts whose matrices have been enlarged by solution-related processes. Continental shelves and slopes are most likely to provide fossil energy resources.

General ocean circulation. The atmosphere and hydrosphere interact to circulate the ocean's various water masses. Principal energy inputs are

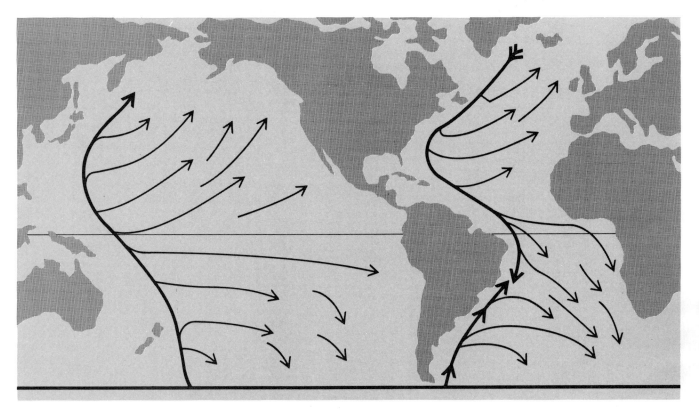

Fig. 8 Approximate direction of deep-water circulation. Details of deep circulation are mostly unknown. (*From R. W. Stewart, The atmosphere and the ocean, Sci. Amer.,* *221(3):76–86; copyright © 1969 by Scientific American, Inc.; all rights reserved*)

wind shear and heat. For the most part, wind shear drives the surface-ocean gyres, and heat, the deep-ocean boundary currents. Complications are introduced by uneven solar insolation as the result of both latitudinal differences and the Earth's orbital tilt and by deflections due to the Earth's rotation (Coriolis forces) which cause moving masses to turn to the right (clockwise) in the Northern Hemisphere and to the left (counterclockwise) in the Southern Hemisphere.

Equatorial regions are dominated by the westerly flow of the North and South Equatorial currents driven primarily by the tradewinds and bisected by a narrow easterly flowing countercurrent (Fig. 7). The equatorial currents form gyres elongated east-west and enclosing the subtropical regions centered on 30°N and 30°S latitude. Prominent easterly directed currents include the West Wind Drift in the Southern Hemisphere and the North Pacific Current and North Atlantic Current (Gulf Stream) in the Northern Hemisphere. Seasonally variable winds in the Indian Ocean dominated by the heating and cooling of surrounding land areas are responsible for the monsoon currents. Small counterclockwise surface gyres are present in the north subpolar and polar regions.

The deep thermohaline circulation (Fig. 8) is a global convective process in which dense cold waters in high latitudes sink and spread toward the Equator. In the Northern Hemisphere most of the deep water is created in the Norwegian Sea and, flowing southward as North Atlantic Deep Water, it hugs the western boundary of the ocean basin,

reaching velocities of 10–20 cm/s. These bottom-hugging, contour-following currents are capable of sculpting the ocean bottom and retransporting and depositing fine-grained sediment in thin laminated deposits termed contourites. Cold, dense water produced around Antarctica forms both the Antarctic Bottom Water and the Antarctic Intermediate Water. The Antarctic Bottom Water flows through the Argentine and Brazil basins in the Atlantic and through the Samoa Passage in the Pacific, and crosses the Equator to spread out onto the Bermuda Rise and reach as far north as the Aleutian Abyssal Plain. Bottom-water circulation is strongly influenced by sea-floor topography.

Chemistry and origin of sea water. Sea water is a weak solution of various substances dissolved in water (see table). The term "salinity" is used to define the total weight of dissolved substances.

Major elemental abundances in sea water

Element	Abundance, mg/liter	Principal species
Oxygen (O)	857,000	H_2O
Hydrogen (H)	108,000	H_2O; O_2; SO_4^{--}
Chlorine (Cl)	19,000	Cl^-
Sodium (Na)	10,500	Na^+
Magnesium (Mg)	1,350	Mg^{++}; $MgSO_4$
Sulfur (S)	885	SO_4^{--}
Calcium (Ca)	400	Ca^{++}; $CaSO_4$
Potassium (K)	380	K^+
Bromine (Br)	65	Br^-
Carbon (C)	28	HCO_3^-; H_2CO_3; CO_3^{--} and organic compounds

Salinity is customarily expressed in grams per kilogram, parts per thousand (ppt), or parts per million (ppm). The average salinity of sea water is 35,000 ppm. The bulk of this salinity, which gives sea water its "saltiness," is dissolved sodium and chlorine.

Speculation on the origin of sea water and the source of sea water salinity is based largely on comparisons between the chemistry of sea water, river runoff, volcanic gases, and geysers. The water itself is thought to represent the accumulation of condensed steam and juvenile gases "exhaled" from the Earth during the early stages of Earth history. This degassing is also thought to have produced the bulk of the constituents in the Earth's atmosphere. Several constituents dissolved in sea water must have had a similar origin, because they are not contributed to the sea by river runoff (examples are chlorine, boron, and bromine).

Many of the other dissolved constituents in sea water (carbonate, sulfate, magnesium, calcium, potassium, and sodium) evidently represent the cumulative buildup of components provided to the sea by the dilute solutions of river runoff. These substances are put into solution by the chemical decomposition which occurs during the weathering of preexisting rocks and minerals. The discrepancy in the proportion of some of these constituents in sea water versus runoff is largely due to likelihood of their extraction once in sea water. Some constituents have a very brief "residence time" in sea water. For example, calcium, carbonate, and silica are readily extracted from sea water by organisms and eventually incorporated as shell fragments in siliceous and calcareous oozes. Still other constituents such as iron and potassium are quickly extracted from solution and absorbed on clay minerals or become involved in diagenetic chemical reactions. Conversely, other constituents, particularly sodium and chlorine, are not organically extracted, nor do the proper conditions for their chemical precipitation occur frequently. Consequently, they tend to accumulate as the most important dissolved constituents in sea water.

The distribution of life in the oceans is controlled by the biolimiting nutrients (oxygen, silica, nitrogen, and phosphorus) dissolved in sea water. Dissolved oxygen is abundant in the surface ocean, depleted at intermediate levels, and enriched in bottom waters advected from the cold polar surface ocean. The Black Sea is the only major ocean basin whose center is totally deficient in oxygen (that is, euxinic), although oxygen levels are extremely low in silled basins on the continental margin of California and in certain coastal fiords. Silica, nitrogen, and phosphorus are removed from the surface ocean by biological factories and are enriched in deep-water masses by the downward settling and decay of dead marine organisms. Enhanced fertility is therefore brought about by upwelling phenomena. Sunlight is attenuated in sea water, and for all purposes is negligible at depths in excess of 250 m.

Life in the ocean. The surface ocean is the most prolific in terms of biomass alone. The sunlit realm, known as the photic region, supports a microscopic plant kingdom, predominantly algae, consumed by floating protozoa (plankton), swimming fish (nekton), and large marine mammals (whales and porpoises). Life on the ocean floor is much less diverse. Of the many major groups of marine organisms, only three dozen have abyssal representatives. Animals that live within the seabed are benthic infauna, and those that crawl on, swim closely above, or are attached to the substratum are benthic epifauna. With increasing depth, the number of animals deriving their nourishment from the sediment carpet, or filtering it from sea water, increases as the number of scavengers and carnivores decreases. Life on the ocean floor is sparse beneath temperate ocean surface gyres, but abundant beneath polar and coastal waters. Whereas 5000 g of living organisms can be found per square meter of inshore sea floor and 200 g might be recovered from a square meter of the continental shelf, in the mid-ocean abyss the seascape is stark and barren. Species diversity in the deep sea is nevertheless remarkably great, possibly because biological disturbances such as predation keep the population of potential competitors at low enough densities to preclude competitive exclusion.

Rates of biological processes in the deep sea are shown to be extremely slow, as supported by studies of growth rates, metabolic activity, and recolonization. *See* OCEANS AND SEAS.

[ELIZABETH L. MILLER; WILLIAM B. F. RYAN]

Bibliography: C. A. Burk and C. L. Drake (eds.), *The Geology of Continental Margins*, 1974; D. B. Ericson and G. Wollin, *The Ever-Changing Sea*, 1970; M. G. Gross, *Oceanography: A View of the Earth*, 2d ed., 1977; B. C. Heezen and C. D. Hollister, *The Face of the Deep*, 1971; B. Mason, *Principles of Geochemistry*, 1958; F. P. Shepard, *Submarine Geology*, 3d ed., 1973.

Oceans and seas

The interconnecting body of salt water that covers 70.8% of the surface of the Earth is called the world ocean, or simply the ocean. Its major subdivisions, corresponding to the continents, are oceans. Subdivisions of oceans in turn are called seas; these range all the way from vague regions with no fixed limits (such as Sargasso Sea) to almost completely landlocked bodies (Black Sea). The terms bight, strait, gulf, and bay are often used interchangeably with sea (Great Australian Bight, Denmark Strait, Gulf of Mexico, Bay of Bengal). Salt lakes lacking outlets to the ocean are also usually called seas (Salton Sea, Dead Sea, Caspian Sea).

The ocean has a total area of 362×10^6 km² and an average depth of 3729 m, with a total volume of 1.35×10^{18} m³. It has a mean temperature of 3.90°C and a mean specific gravity of 1.045, giving a total mass of 141×10^{16} metric tons. Since the mean salinity is 34.75‰ (3.475% by weight), there are 136×10^{16} tons of water and 4.93×10^{16} tons of salt in the ocean.

The world ocean is commonly divided into four oceans (see table). The Arctic Ocean is separated from the Pacific at Bering Strait and from the Atlantic by Fury and Hecla straits, Davis Strait, and a line from Greenland to Svalbard, Bear Island,

Characteristics of the oceans and adjacent seas

Oceans and adjacent seas	Area, 10^6 km^2	Volume, 10^6 km^3	Mean depth, m
Pacific	166.241	696.189	4188
Asiatic Mediterranean	9.082	11.366	1252
Bering Sea*	2.261	3.373	1492
Sea of Okhotsk	1.392	1.354	973
Yellow and East China seas	1.202	0.327	272
Sea of Japan	1.013	1.690	1667
Gulf of California*	0.153	0.111	724
Pacific and adjacent seas, total	181.344	714.410	3940
Atlantic	86.557	323.369	3736
American Mediterranean	4.357	9.427	2164
Mediterranean	2.510	3.771	1502
Black Sea*	0.508	0.605	1191
Baltic Sea*	0.382	0.038	101
Atlantic and adjacent seas, total	94.314	337.210	3575
Indian	73.427	284.340	3872
Red Sea*	0.453	0.244	538
Persian Gulf	0.238	0.024	100
Indian and adjacent seas, total	74.118	284.608	3840
Arctic	9.485	12.615	1330
Arctic Mediterranean	2.772	1.087	392
Arctic and adjacent seas, total	12.257	13.702	1117
Totals and mean depths	362.033	1349.929	3729

*See article by this title.

and the north cape of Norway. It includes the Arctic Mediterranean, consisting of Hudson Bay, Baffin Bay, and the Northwestern Passages or Canadian Straits. The meridian of Cape Agulhas (20°E) divides the Atlantic and Indian oceans. The Indian and Pacific oceans are separated by the Andaman Islands; Indonesia; a line from Timor to Cape Talbot, Australia; the western end of Bass Strait; and the meridian of Southeast Cape, Tasmania (146°52′E). All of Magellan Strait is part of the Pacific; to the south the boundary between Atlantic and Pacific is a line from Cape Horn to the South Shetland Islands and thence to the Antarctic Peninsula.

The Gulf of Mexico and Caribbean Sea together make up the American Mediterranean, and the seas between the Andaman Islands, East Indies, New Guinea, the Phillipines, and Formosa collectively make up the Asiatic Mediterranean. Using these limits, H. W. Menard and R. Smith (1966) computed the characteristics of the oceans and their principal seas listed in the table. *See* ANTARCTIC OCEAN; ARCTIC OCEAN; ATLANTIC OCEAN; CARIBBEAN SEA; GULF OF CALIFORNIA; GULF OF MEXICO; INDIAN OCEAN; MEDITERRANEAN SEA; NORTH SEA; PACIFIC OCEAN.

[JOHN LYMAN]

Pacific Ocean

The Pacific Ocean has an area of 165,000,000 km^2 and a mean depth of 4282 m. It covers 32% of the Earth's surface and 46% of the surface of all oceans and seas, and its area is greater than that of all land areas combined. Its mean depth is the greatest of the three oceans and its volume is 53% of the total of all oceans. Its greatest depths in the

Marianas and Japan trenches are the world's deepest, more than 10 km (Fig. 1).

Surface currents. The two major wind systems driving the waters of the ocean are the westerlies which lie about 40–50° lat in both hemispheres (the "roaring forties") and the trade winds from the east which dominate in the region between 20°N and 20°S. These give momentum directly to the west wind drift (flow to the east) in high latitudes and to the equatorial currents which flow to the west. At the continents there is flow of water from one system to the other and huge circulatory systems result (Fig. 2).

The swiftest flow (greater than 2 knots) is found in the Kuroshio Current near Japan. It forms the northwestern part of a huge clockwise gyre whose north edge lies in the west wind drift centered at about 40°N, whose eastern part is the south-flowing California Current, and whose southern part is the North Equatorial Current.

Part of the west wind drift turns northward into the Gulf of Alaska, thence westward again and into the Bering Sea, returning southward off Kamchatka and northern Japan, where it is called the Oyashio Current.

H. U. Sverdrup (1942) reported estimates of the flow in the North Pacific in the upper 1500 m. The Kuroshio carries about 65,000,000 m^3/sec at its greatest strength off Japan. The west wind drift in mid-ocean carries about 35,000,000, the California Current east of 135°W carries about 15,000,000, and the North Equatorial Countercurrent about 45,000,000.

A gyre corresponding to the Kuroshio–California–North Equatorial current gyre is found in the Southern Hemisphere. Its rotation is counter-

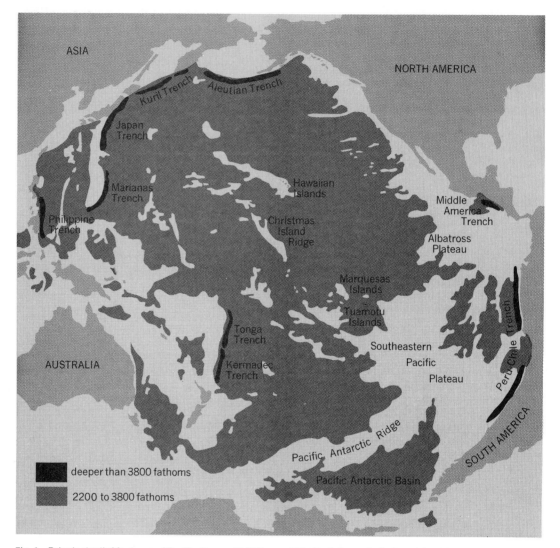

Fig. 1.   Principal relief features of Pacific Ocean. (*F. P. Shepard, The Earth Beneath the Sea, Johns Hopkins, 1959*)

clockwise, with the highest speeds (about 2 knots) in the Southeast Australia Current at about 30°S. The current turns eastward and flows around New Zealand to South America, where it turns northward. Along this coast it is called the Humboldt, or the Peru, Current. It turns westward at the Equator and is known as the South Equatorial Current in its westward flow. It is to be remarked that the northwestern edge of this gyre is severely confused by the chain of islands extending southeastward from New Guinea to New Zealand. This island chain partly isolates the Coral and Tasman seas from the rest of the South Pacific, so that the western equatorial edge of the gyre is not so regular in shape nor so clearly defined as its northern counterpart. *See* SOUTHEAST ASIAN WATERS.

In the region of the west wind drift in the South Pacific the ocean is open both to the Indian and the Atlantic, although the eastward passage to the Atlantic through Drake Passage is narrower and shallower than the region between Australia and Antarctica. Part of the water flows, however, all around Antarctica with the wind behind it, and it receives more momentum than its northern coun-

terpart. The total transport is several times greater.

G. E. R. Deacon (1937) estimated the transport to the east in the South Pacific part of the west wind drift as more than 100,000,000 m³/sec in the upper 3000 m. Sverdrup estimates only about 35,-000,000 m³/sec for the upper 1500 m in the North Pacific west wind drift. The gyre corresponding to the Oyashio–Gulf of Alaska gyre of the North Pacific is thus vaster in transport and area, since much of it passes around Antarctica. *See* ANTARCTIC OCEAN; INDIAN OCEAN.

Between the two subtropical anticyclones (Kuroshio–California–North Equatorial Current and the Southeast Australia–Peru–South Equatorial Current) lies an east-flowing current between about 5 and 10°N, called the Equatorial Countercurrent. Sverdrup estimates flow of 25,000,000 m³/sec for this current. There is also an eastward flow along 10°S from 155°E to at least 140°W.

Within the upper 1000 m the flow of water is generally parallel to the surface flow, but slower. Certain important exceptions occur. Beneath the surface waters of the California and Peru currents,

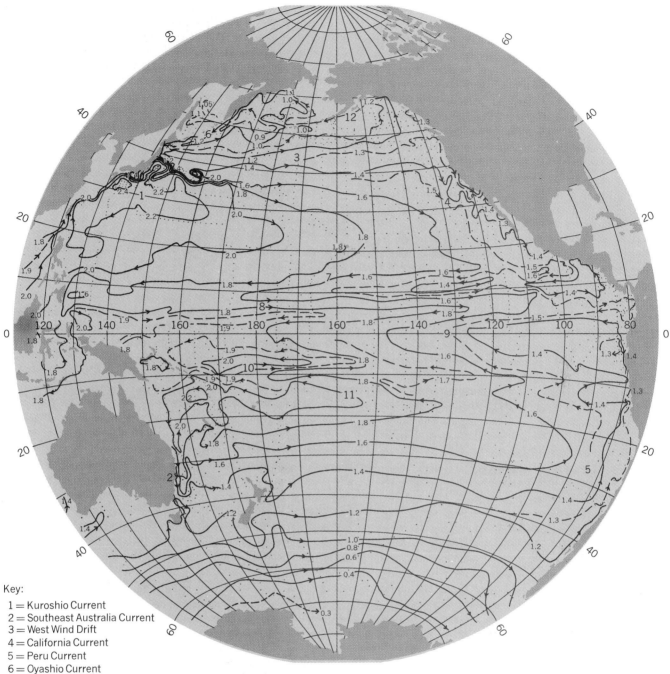

Key:

 1 = Kuroshio Current
 2 = Southeast Australia Current
 3 = West Wind Drift
 4 = California Current
 5 = Peru Current
 6 = Oyashio Current
 7 = North Equatorial Current
 8 = North Equatorial Countercurrent
 9 = Equatorial Current
10 = South Equatorial Countercurrent
11 = South Equatorial Current
12 = Alaska Current

Fig. 2. Principal currents of the Pacific Ocean. Contours give geopotential anomaly in meters at the sea surface with respect to the 1000-decibar surface, from whose gradient the relative geostrophic flow can be calculated. The large whole numbers 1–12 refer to the currents which are listed in the key. (From J. L. Reid, Jr., On the circulation, phosphate-phosphorous content and zooplankton volumes in the upper part of the Pacific Ocean, Limnol. Oceanogr., 7:287–306, 1962)

at depths of 200 m and below, countercurrents are found in which some part of the tropical waters is carried poleward. In the California Current this flow reaches to the surface in December, January, and February, and it is known as the Davidson Current north of 35°N. It is not known whether a surface poleward flow occurs in southern winter in the Peru Current.

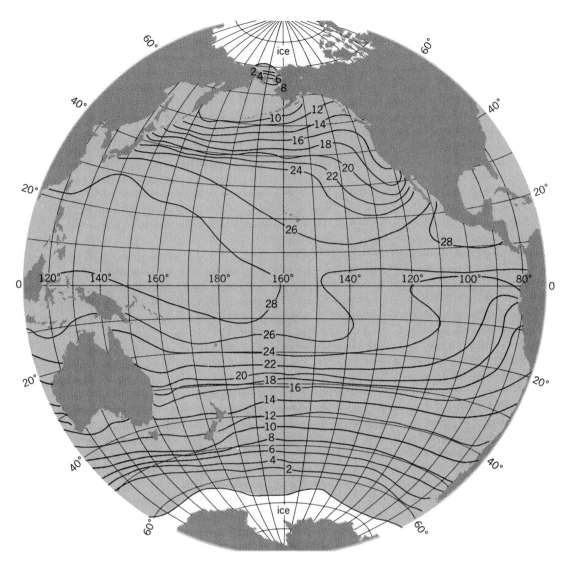

Fig. 3.   Temperature at sea surface in August in °C. (*Adapted from U.S. Navy Hydrographic Office, H.O. Publ. no. 225, 1948*)

Direct measurements at the Equator and 140°W have revealed that a subsurface flow to the east is found at least from 140°W to 90°W, with highest velocities at depths of about 100 m. Speeds as high as 2–3.5 knots to the east were found at this level, while the upper waters (South Equatorial Current) were flowing west at 0.5–1.5 knots. This current has been called the Equatorial Undercurrent, or Cromwell Current after its discoverer. *See* OCEAN CURRENTS.

Temperature at the sea surface. Equatorward of 30° lat heat received from the Sun exceeds that lost by reflection and back radiation, and surface waters flowing into these latitudes from higher latitudes (California and Peru currents) increase in temperature as they flow equatorward and turn west with the Equatorial Current System. They carry heat poleward in the Kuroshio and Southeast Australia Currents and transfer part of it to the high-latitude cyclones (Oyashio–Gulf of Alaska Gyral and Antarctic Circumpolar Current) along the west wind drift. The temperature of the equatorward currents along the eastern boundaries of the subtropical anticyclones is thus much lower

than that of the currents of their western boundaries at the same latitudes. Heat is accumulated, and the highest temperatures (more than 28°C) are found at the western end of the equatorial region (Fig. 3). Along the Equator itself somewhat lower temperatures are found. The cold Peru Current contributes to its eastern end, and there is apparent upwelling of deeper, colder water at the Equator, especially at its eastern end (with temperatures running as low as 19°C in February at 90°W) as a consequence of the divergence in the wind field.

Upwelling also occurs at the edge of the eastern boundary currents of the subtropical anticyclones. When the winds blow strongly equatorward (in summer) the surface waters are driven offshore, and the deeper colder waters rise to the surface and further reduce the low temperatures of these equatorward-flowing currents. The effect of these seasonal variations in the winds is thus to reduce the seasonal range of temperature, since the upwelling occurs in spring and summer. The seasonal range of nearshore surface temperature of the California Current at 35°N is less than 4C° (9–13°C), though the latitudinal mean is about 10°C.

The equatorward winds off Japan occur in winter and the poleward in summer, so that seasonal range is increased at that latitude to more than 16C° (10–26°C).

The temperatures at the surface of the South Pacific Ocean have not been nearly so well documented, since most of the information comes from measurements made by merchant vessels, and the commercial shipping lines cover but a small part of its great extent. It may be reasoned that most of the temperature characteristics of the North Pacific will have analogies in the Southern Hemisphere. The damped seasonal variation of the California Current seems to occur in the Peru Current as well. But the temperatures of the Southeast Australia Current do not seem to vary so widely through the year as those of the Kuroshio, probably because of the restrictions upon the flow which are imposed by the islands.

The limiting temperature in high latitudes is that of freezing. Ice is formed at the surface at temperatures slightly less than −1°C depending upon the salinity; further loss of heat is retarded by its insulating effect. The ice field covers the northern and eastern parts of the Bering Sea in winter, and most of the Sea of Okhotsk, including that part adjacent to Hokkaido (the north island of Japan). Summer temperatures, however, reach as high as 6°C in the northern Bering Sea and as high as 10° in the northern part of the Sea of Okhotsk.

Pack ice reaches to about 62°S from Antarctica in October and to about 70°S in March, with icebergs reaching as far as 50°S. *See* BERING SEA; ICEBERG; SEA ICE.

Salinity at the sea surface. The highest values of salinity observed in the Pacific Ocean are slightly greater than 35.5 and 36.5 parts per thousand (°/oo), found respectively in the surface water of the centers of the north and south subtropical anticyclones (Fig. 4). These anticyclones cover the latitudes in which evaporation exceeds precipitation, and the overlying anticyclonic winds oppose outward flow at the sea surface. Three regions where precipitation most strongly exceeds evaporation are found poleward of 40° latitude and in the eastern tropical Pacific. The result is that the high-latitude cyclones are regions of low salinity (as low as 32.5°/oo in the north, 33.8 in the south) which, through mixing with the anticyclones in the region of the west wind drift, contribute water of low salinity to the eastern boundary currents (off California and South America). The greater part of the effect of the eastern tropical precipitation is found at the surface of the North Equatorial Current and Countercurrent. Near Central America values are less than 33°/oo at 10°N, but they rise nearly to 34.5 near the Philippine Islands.

Dissolved oxygen at the sea surface. Above the thermocline the water is in continual overturn and is thus in constant contact with the atmosphere. Oxygen from the atmosphere dissolves in the water until equilibrium is established, and over most of the Pacific the upper layer is very close to saturation in oxygen content with typical values from about 98 to 103% of the saturation value. *See* THERMOCLINE.

The saturated value of dissolved oxygen rises as both the temperature and salinity fall, but the range of surface temperature in the ocean accounts for a wider variation in saturated value than does that of surface salinity, and it is principally variation in space and time of surface temperature which accounts for the oxygen values at the surface. Values greater than 7 ml/liter are found in cold waters of high latitudes and less than 5 in warm regions near the Equator.

Nutrients at the sea surface. Nutrients such as inorganic phosphate-phosphorus, silicate-silicon, and nitrate generally increase from the sea surface downward, since photosynthesis and growth in the upper mixed layer tend to use such quantities as are there, and diffusion upward from the higher concentrations is limited by the great stability usually found immediately below the surface layer. At the surface phosphate-phosphorus varies from less than 0.25 microgram-atoms (µg-atoms) per liter in the centers of the anticyclones to more than 1.5 in the high-latitude cyclones (Fig. 5). High values are also found in the California and Peru currents. In a manner similar to the low temperatures found there, their high concentrations are partly the result of transport of mixed water from the cyclones and partly the result of upwelling at the coasts under equatorward winds. Values greater than 1 µg-atom/liter are found in both areas during the summer period of upwelling. At the Equator in the eastern Pacific upwelling raises values to more than 1 throughout the year.

Silicate-silicon has not been so extensively measured. It also is low in value at the surface and increases with depth. Surface values range from as high as 40 µg-atoms/liter in the high-latitude cyclones and 12 in the upwelling regions of the California Current, to 4 or less in the center of the anti-

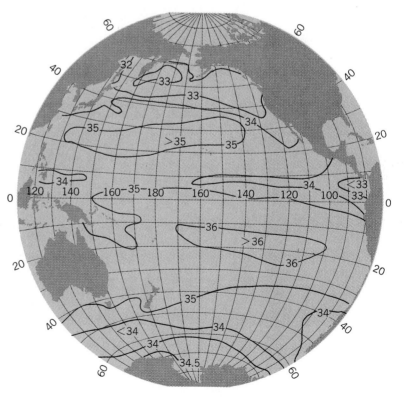

Fig. 4. Salinity at sea surface in northern summer, in parts per thousand by weight.

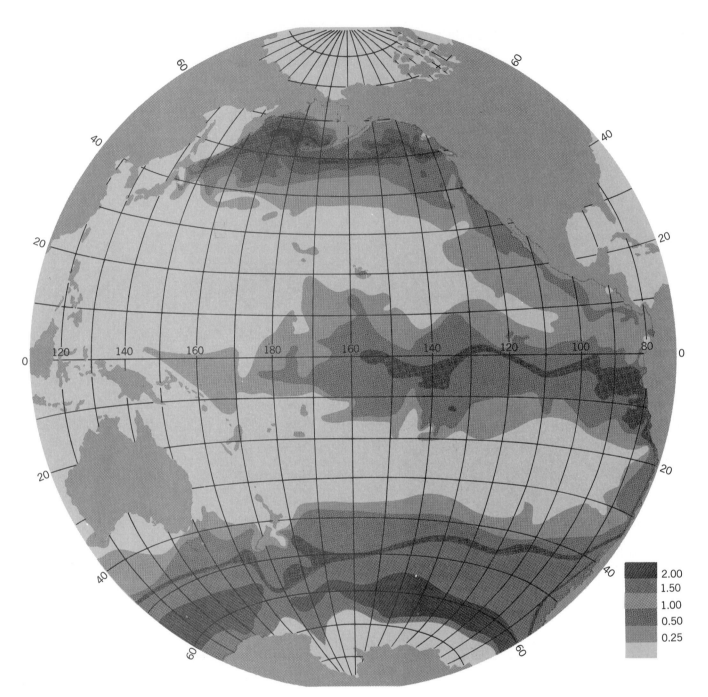

Fig. 5.   Inorganic phosphate at sea surface, in microgram-atoms per liter.

cyclones and to values too small to be detected near the Equator in the eastern region.

Distribution of properties with depth. Surface waters in high latitudes are colder and heavier than those in low latitudes. As a result, some of the high-latitude waters sink below the surface and spread equatorward, mixing mostly with water of their own density as they move, and eventually become the dominant water type in terms of salinity and temperature of that density over vast regions (Fig. 6).

The deep and bottom water of all oceans is believed to be formed in the Atlantic high latitudes off Greenland and in the Weddell Sea and arrives in the South Pacific with temperature less than 2°C, salinity about 34.65–34.7 °/∞, and oxygen about 4.5 ml/liter. The waters filling the Pacific below 3000 m retain these values of temperature and salinity everywhere, but in the northern part the oxygen is reduced to values between 2.5 and 3.5. This would be consistent with depletion by decay and respiration during a slow movement north.

Over most of the North Pacific Ocean the temperature does not decrease all the way to the bottom, but beneath a minimum value at about 3800–4000 m it rises again. The increase in temperature is probably in close balance with that in pressure, since no evidence has been found of instability. Two possible explanations for the temperature maximum at the bottom are the flow of heat upward from the ocean floor and an adiabatic rise in temperature as the water flows downward into the deeper basins.

The most conspicuous water masses formed in the Pacific are the Intermediate Waters of the North and of the South Pacific, which on the vertical sections include the two huge tongues of low salinity extending equatorward beneath the surface from about 55°S and from about 45°N. The southern tongue is higher in salinity and density and lies at a greater depth, since the surface waters of the high-latitude cyclone are more saline in the south than in the north.

It has been observed that the higher salinities are found at the surface in the anticyclones. The highest values are in the equatorward halves of the anticyclones. High values penetrate the base of the mixed layer, and tongues of high salinity extend in the thermocline toward the Equator from both north and south.

Beneath the mixed layer the waters are not in contact with the atmosphere, and the oxygen that is consumed cannot be replaced directly from the atmosphere. The oxygen quickly falls below the saturated value. Even where high values of oxygen accompany the sinking of water masses, as in the South Pacific Intermediate Water, oxygen is not in saturation below 300 m.

Between saturated waters of the surface and cold bottom waters entering from the south lies a minimum value of oxygen. Beneath the tongue of high oxygen associated with the South Pacific Intermediate Water the minimum is only a little less than 4 ml/liter. Beneath the North Pacific Intermediate Water, however, no water as high as 3.5 is found, and in the minimum itself values less than 0.5 occur over large areas. Values of oxygen beneath the surface are the result of consumption by organisms and replenishment by mixing and renewal of the water. Since at any position the values are nearly constant in time, consumption must everywhere equal replenishment by both flow and diffusion.

Phosphate-phosphorus increases rapidly beneath the sea surface to a maximum which usually lies beneath the oxygen minimum. In regions of upwelling and divergence, higher values of phosphate and other nutrients from depth may from time to time be brought to the surface and thus made available to plants. Such regions (California and Peru currents, South Equatorial Current) are highly productive. The values gradually diminish beneath the maximum to about 2.5 μg-atoms/liter at the bottom in the north and slightly less than 2 in the south. The maximum value is greater than 3.5 in the north and between 2 and 2.5 in the south.

The Pacific Ocean is higher in phosphate concentration than the Atlantic or Indian oceans, exceeding the Atlantic by about 1 μg-atom/liter on the average, though the values at the surface do not differ so much. This excess, like the lower value of oxygen, is undoubtedly related to deep circulation.

Nitrate-nitrogen is present in the ratio of approximately 8:1 by weight of phosphate-phosphorus in

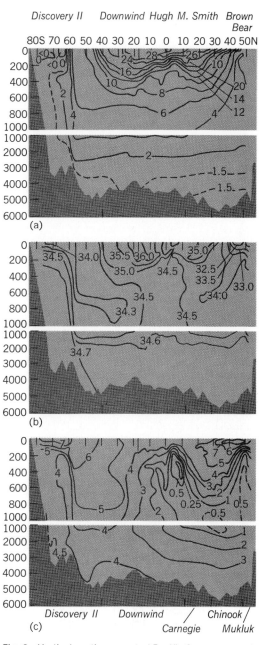

Fig. 6. Vertical sections, central Pacific Ocean, approximately along meridian 160°W, from Alaska (right) to Antarctica (left). (a) Temperature in °C. (b) Salinity in parts per thousand. (c) Dissolved oxygen in milliliters per liter. Depths are in meters. Depth scale is expanded in upper 1000 m. (*From Carnegie expedition; NORPAC and EQUAPAC expeditions; Discovery expeditions; and Chinook, Mukluk, and Downwind expeditions*)

those areas where it has been measured, but the details of its distribution are not so well known.

Silicate-silicon has also not been adequately sampled. Its vertical distribution parallels that of phosphate-phosphorus except that beneath the level of the phosphate maximum it remains nearly constant, at about 170 μg-atoms/liter in the center of the northern anticyclone, as high as 220 in the Bering Sea, about 130 in the center of the southern anticyclone. Silicate-silicon, like phosphate-phosphorus, is more highly concentrated in the Pacific than in the Indian and Atlantic oceans. *See* MA-RINE GEOLOGY; MARINE SEDIMENTS; OCEANOGRA-PHY; SEA WATER; SEA-WATER FERTILITY.

[JOSEPH L. REID]

Bibliography: R. A. Barkley, *Oceanographic Atlas of the Pacific Ocean*, 1969; C. A. King, *Physical Geography*, 1978; G. E. R. Deacon, The hydrology of the southern ocean, *Discovery Reports*, vol. 15, pp. 1–124, 1937; J. A. Knauss, Measurements of the Cromwell Current, *Deep-Sea Res.*, 6:265–286, 1960; H. W. Menard, *Marine Geology of the Pacific*, 1964; R. B. Montgomery and E. D. Stroup, *Equatorial waters and currents at 150°W in July–August 1952*, Johns Hopkins Oceanographic Studies, vol. 1, 1962; NORPAC Committee, *Oceanic Observations of the Pacific: 1955*, NORPAC Atlas, 1960; G. L. Pickard, *Descriptive Physical Oceanography*, 2d ed., 1976; J. L. Reid, Jr., *Intermediate waters of the Pacific Ocean*, Johns Hopkins Oceanographic Studies, vol. 2, 1965; J. L. Reid, Jr., On the circulation, phosphate-phosphorus content and zooplankton volumes in the upper part of the Pacific Ocean, *Limnol. Oceanogr.*, 7:287–306, 1962; G. H. Sutton et al. (eds.), *The Geophysics of the Pacific Ocean Basin and Its Margin*, 1976; H. U. Sverdrup, M. W. Johnson, and R. H. Fleming, *The Oceans*, 1942; W. S. Wooster and G. H. Volkmann, Indications of deep Pacific circulation from the distribution of properties at five kilometers, *J. Geophys. Res.*, 65:1239–1249, 1960.

Polar meteorology

A regional branch of atmospheric science, taking special account of circumstances and phenomena prevailing in the northern and southern polar regions. The atmosphere in these regions has a distinctive character brought about principally by low solar elevation, as well as the given geographical configuration of oceans and land masses.

Radiation balance. Weather and climate in the polar regions are dominated by the annual cycle of insolation. The maximum elevation angle of the Sun at noon, H, is determined by its declination δ and the geographic latitude φ according to $H = \varphi + (90 - \delta)$. Hence, the highest angle under which the Sun's rays can strike the poles (summer solstice) is 23.5°. At winter solstice the Sun is an equal angle below the horizon. While daily amounts of incoming radiation energy in the polar regions during summer can be as large as in the tropics, the prevailing snow and ice surfaces reflect much of it back into space. In addition, infrared blackbody radiation is emitted to space by the Earth's surface and the atmosphere (mainly its clouds and water vapor) throughout the year. The total radiation balance of the Earth-atmosphere system in the Arctic is strongly negative throughout the year except during a brief period in summer. The distribution of total net radiation at the top of the atmosphere (visible and infrared radiation) as observed from satellites, as a function of latitude and time of the year, is shown in Fig. 1. The total heat balance of the Earth-atmosphere system differs little in the two polar regions. However, the cold and centrally located Antarctic continent, surrounded by a warm ocean, produces a much sharper meridional gradient of heating rates between 50° and 60° S. Within the global atmospheric heat engine the polar regions function as heat sinks. *See* TERRESTRIAL ATMOSPHERIC HEAT BALANCE; TERRESTRIAL RADIATION.

One of the consequences of radiational cooling is the formation of inversions, layers in which the temperature increases with height. They can form at the ground, or aloft, when a dense layer of stratus clouds acts as an elevated radiating surface. Inversions are important, as they suppress vertical mixing and are frequently associated with fog. *See* TEMPERATURE INVERSION.

To maintain the near-equilibrium state to which the atmosphere must adjust, the radiative heat loss in the polar region is compensated by the influx of sensible heat and latent heat carried by warmer air masses. These advective fluxes result from the migration of low- and high-pressure systems across the polar regions. On an annual average, barometric surface pressure in the Arctic shows a weak anticyclone, centered over the Beaufort Sea. In the south polar region, surface pressure is not meaningful since more than half of the Antarctic continent lies at an elevation of more than 2000 m (25% lies above 3000 m). The mean 500-mb (1 mb = 100 Pa) surface over Antarctica shows a well-developed cyclonic circumpolar vortex. *See* AIR MASS; ATMOSPHERIC GENERAL CIRCULATION.

Precipitation. Saturation vapor pressure at typical polar temperatures, 0° to −50°C, ranges from 6 to 0.04 mb. Therefore, precipitation in the polar regions is sparse. Some of the high-latitude precipitation, especially during the cold months, is not necessarily associated with clouds but falls as an extremely thin, continuous mist of small ice crystals ("diamond dust"). In Antarctica, precipitation increases from 5 cm/year on the central plateau to about 40 cm/year near the coast. In the Arctic, precipitation is more uniform and lies between 15 and 20 cm/year, except in the Norwegian and Barents seas (50 cm/year). Contrary to the precipitation regime of Antarctica, the Greenland ice cap shows an increase from 20 cm/year near its west coast to 50 cm/year in the interior at 2000 m elevation. By climatological definition, this qualifies the polar regions as deserts. *See* PRECIPITATION.

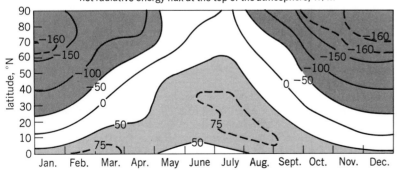

net radiative energy flux at the top of the atmosphere, W/m²

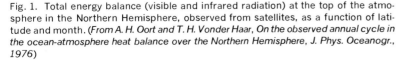

Fig. 1. Total energy balance (visible and infrared radiation) at the top of the atmosphere in the Northern Hemisphere, observed from satellites, as a function of latitude and month. (*From A. H. Oort and T. H. Vonder Haar, On the observed annual cycle in the ocean-atmosphere heat balance over the Northern Hemisphere, J. Phys. Oceanogr., 1976*)

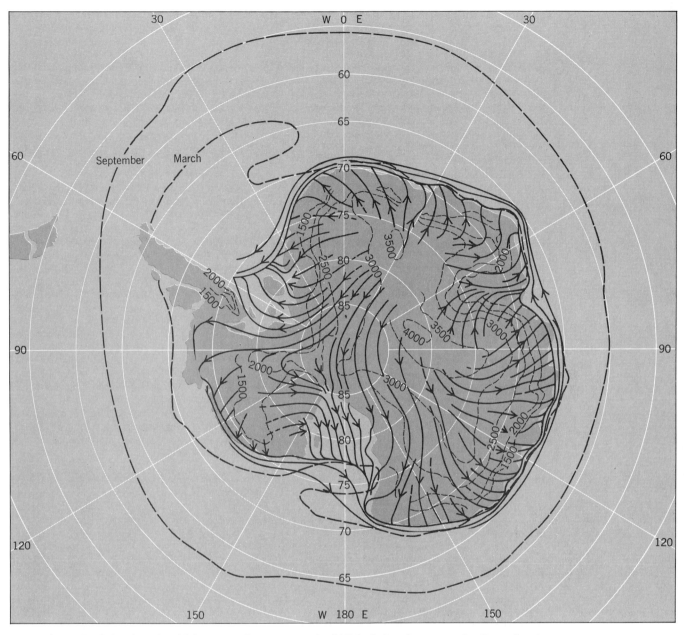

Fig. 2. Antarctic wind and sea ice. (*a*) Average pattern of katabatic surface wind, inferred from predominant wind frequencies at fixed stations and from traverse records. (*b*) Extent of sea ice surrounding the continent, 2,600,000 km² in March and 18,800,000 km² in September.

Even though precipitation is small, massive ice caps several thousand meters thick have been developing on Greenland and Antarctica since the onset of the Quaternary Age. If these ice caps were to melt, the world ocean level would rise by 170 m. In Antarctica there is virtually no snow melt, and in Greenland snow accumulation outweighs ablation. Both ice caps produce an excess of ice that flows into the ocean and turns into drifting icebergs.

Freshly fallen polar snow is dry and light (density 0.1 g/cm³). It is easily picked up by wind of about 10 m/s and transformed into "blowing snow." At stronger winds the number of snow particles in the lower atmosphere boundary layer can become so large that light is scattered uniformly in all directions and all shadows disappear. This "white-out" condition is most dangerous for aircraft operations. During cold, calm weather, a strong temperature gradient develops in the polar snow pack (cold at the surface, warm at depth), causing an upward diffusion of water vapor in the saturated interstitial air. The result is evaporation of ice at some depth and sublimation at the surface. After a few days, this sublimation cements the snow crystals together to form a light but strong crust 10–20 cm thick. The Eskimos have taken advantage of this phenomenon to develop their technique of constructing igloos.

Jet streams. Above the prevailing surface inversions the troposphere in both polar regions shows variations associated with jet streams and fronts similar to, but less vigorous than, the patterns in midlatitudes.

A significant feature of both polar stratospheres above 20 km is a strong horizontal temperature gradient across the twilight zone, causing the so-called polar-night jet. In the Arctic, this westerly jet develops after sunset, begins to meander, and frequently breaks down between late December and February. The associated eddies serve to redistribute ozone, an important minor constituent of the atmosphere, generated by the Sun at lower latitudes between 30 and 60 km altitude. During summer the lower Arctic stratosphere shows a steady easterly flow. Over the geographically more symmetrical Antarctic the stratospheric jet remains strong (cyclonic) throughout the winter and weakens only with the return of the Sun and its direct heating effect (via ozone absorption) on the stratosphere.

Paleoclimatology. Polar ice caps are, along with deep-sea sediments, the most important repositories for paleoclimatic information. Especially the ratio of the stable isotopes of O^{18} and O^{16}, which is a measure of the temperature at which the deposited snow formed in the atmosphere, has been successfully used in reconstructing the temperature history of the Pleistocene. Multiple evidence indicates that, at the end of the last glacial stage, about 12,000 years ago, the temperature in the Arctic region rose by about 10°C within less than 1000 years.

Polar contrasts. Weather and climate in the Arctic are generally milder and less extreme than in the Antarctic. The extreme temperature range in the central Antarctic plateau of 65°C is smaller than the one found in certain continental areas of northern middle latitudes, but central Antarctica has the lowest surface temperatures recorded on Earth: −88°C. Associated with this are inversions of up to 25°C lasting for several months.

While summer surface temperatures stay well below freezing in Antarctica, even near the coast, the Arctic summer shows a completely different pattern. In late May or early June the low-lying land masses surrounding the Arctic Ocean become free of snow. The snow on the pack ice of the Arctic Basin begins to melt in early June, and the entire area maintains a temperature near zero throughout the summer. At the same time the characteristic persistent "Arctic stratus" forms. At 85°N, total cloudiness from June to August is 80–90%.

Unlike the Arctic, whose surface winds are generally weak, variable, and dominated by moving pressure systems, the Antarctic has a pronounced regime of drainage (katabatic) winds. A thin layer of air, cooled to extremely low temperatures by radiative heat loss, drains downslope from the central plateau toward the coast (Fig. 2). Cape Denison, in East Antarctica, recorded 234 days per year on which the average 24-hr wind speed was greater than 18 m/s. These winds, which also occur less strongly developed on the Greenland ice cap, are capable of transporting significant amounts of snow.

One of the most important contrasts between north and south polar regions exists in their sea ice regime. Even though the total mass of sea ice present on the globe is small compared with the mass of its ice caps, the great horizontal extent of sea ice makes it a crucial and variable factor in the heat balance of the Earth. With the formation of floating ice on the sea surface, its reflectivity (albedo) undergoes a drastic change from 0.1 to 0.6, and all forms of energy exchange between ocean and atmosphere are suppressed. In the Arctic the total sea ice area varies from 11,400,000 km² in March to 7,000,000 km² in August. In the Southern Ocean surrounding Antarctica the minimum extent occurs in March and the maximum in September (Fig. 2). Therefore, most Arctic sea ice is several years old. Under the present climatic regime it reaches an average thickness of about 3 m. At that value, seasonal melting and accretion are in balance. Except in the Weddell Sea, all Antarctic sea ice is less than a year old. It has been theorized that a perturbation on the average extent of sea ice might have a positive feedback (self-enhancement) in the complex interaction of atmosphere and ocean in the polar regions, and might result in a profound and lasting effect on the terrestrial climate. *See* ARCTIC CIRCLE; ANTARCTIC CIRCLE; ANTARCTICA.   [NORBERT UNTERSTEINER]

Bibliography: P. Putnins, The climate of Greenland, in S. Orvig (ed.), *World Survey of Climatology*, vol. 14, pp. 3–128, 1970. W. Schwerdtfeger, The climate of the Antarctic, in S. Orvig (ed.), *World Survey of Climatology*, vol. 14, pp. 253–355, 1970; E. Vowinckel and S. Orvig, The climate of the North Polar Basin, in S. Orvig (ed.), *World Survey of Climatology*, vol. 14, pp. 129–252, 1970; G. Weller and S. A. Bowling (eds.), *Climate at the Arctic*, 24th Alaska Science Conference, University of Alaska, 1975; World Meteorological Organization, *Symposium on Polar Meteorology*, 1967.

Precipitation

The fallout of water drops or frozen particles from the atmosphere. Liquid types are rain or drizzle, and frozen types are snow, hail, small hail, ice pellets (also called ice grains; in the United States, sleet), snow pellets (graupel, soft hail), snow grains, ice needles, and ice crystals. In England sleet is defined as a mixture of rain and snow, or melting snow. Deposits of dew, frost, or rime, and moisture collected from fog are occasionally also classed as precipitation.

All precipitation types are called hydrometeors, of which additional forms are clouds, fog, wet haze, mist, blowing snow, and spray. Whenever rain or drizzle freezes on contact with the ground to form a solid coating of ice, it is called freezing rain, freezing drizzle, or glazed frost; it is also called an ice storm or a glaze storm, and sometimes is popularly known as silver thaw or erroneously as a sleet storm. *See* CLOUD.

Most precipitation particles carry an electrostatic charge, either positive or negative, but the origin of the charge and its relation to other problems of atmospheric electricity are not completely understood. *See* ATMOSPHERIC ELECTRICITY; THUNDERSTORM.

Rain, snow, or ice pellets may fall steadily or in

showers. Steady precipitation may be intermittent though lacking sudden bursts of intensity. Hail, small hail, and snow pellets occur only in showers; drizzle, snow grains, and ice crystals occur as steady precipitation. Showers originate from instability clouds of the cumulus family, whereas steady precipitation comes from stratiform clouds.

The amount of precipitation, often referred to as precipitation or simply as rainfall, is measured in a collection gage. It is the actual depth of liquid water which has fallen on the ground, after frozen forms have been melted, and is recorded in millimeters (mm) or inches and hundredths. A separate measurement is made of the depth of unmelted snow, hail, or other frozen forms. *See* PRECIPITATION GAGES.

Liquid types. Rain and drizzle are somewhat arbitrarily differentiated. Raindrops are generally over 0.5 mm in diameter and differ from drizzle mainly in having larger sizes. Drizzle may also be distinguished from rain by the meteorological conditions of its formation; drizzle falls usually from fog or thick stratus clouds, whereas rain comes from clouds of cumulus type or clouds extending well above the freezing level. Raindrops are rarely larger than 5 mm in diameter because larger sizes tend to break up.

The small raindrops are approximately spherical, but large falling drops are flattened (Fig. 1), especially on the bottom side. The following speeds of falling drops, in meters per second, were measured in still air by R. Gunn (drop diameters are in millimeters):

Speed	0.27	2.06	4.03	6.49	8.06	8.83
Diameter	0.1	0.5	1.0	2.0	3.0	4.0

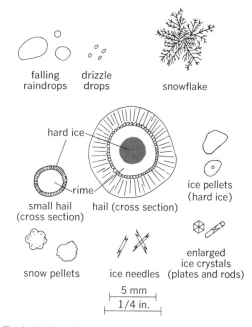

Fig. 1. Various forms of precipitation.

Frozen precipitation. Snowflakes are branched, six-point star crystals, irregular forms such as matted ice needles, or combinations of both; often they are coated by rime. The speed of falling flakes is variable; it is greater when the flakes are rimed and is generally 1–2 m/sec. *See* SNOW.

Hail, mostly seen in thunderstorms, forms with the aid of warm, moist air when clouds build to great heights. Hailstones usually have concentric

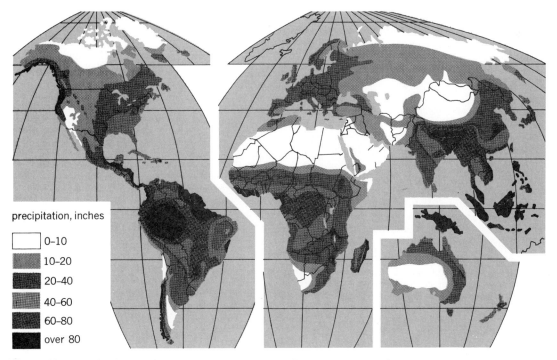

Fig. 2. Map showing the distribution of average annual precipitation over the land areas of the world. (*From N. A. Bengtson and W. Van Royen, Fundamentals of Economic*

Geography, 5th ed., Prentice-Hall, © *1964; reprinted by permission of Prentice-Hall, Inc., Englewood Cliffs, NJ*)

precipitation, inches

- 0–10
- 10–20
- 20–40
- 40–60
- 60–80
- over 80

layers of rime and hard ice, typically 1 cm in diameter but ranging up to more than 10 cm. *See* HAIL.

Ice pellets are of hard ice, either clear or opaque, and may be spherical or irregular. They are formed by the freezing of drops of rain or drizzle.

Snow pellets are of soft, opaque ice, irregularly shaped, sometimes with scalloped edges and often studded with a few oblong or branched crystals. Typical diameters are 2–5 mm. They fall usually in showers, sometimes even when ground temperatures are a few degrees above freezing. U. Nakaya found their specific gravity to be about 0.12. Snow grains are similar but smaller and flatter, with diameters around 1 mm or less; they fall generally in small amounts from stratus clouds or fog.

Ice needles are narrow, pointed crystals, roughly 1–3 mm long and about 1/4 mm in diameter, which fall singly or in clusters, mostly at temperatures near or a little below freezing.

Ice crystals are small rods or plates. They fall in cold, stable air that is often without clouds, and glisten in light as they settle to the ground.

Geographical distribution. Precipitation is part of the hydrologic cycle, the continuous interchange of water between sea, land, and atmosphere, but its distribution over the Earth is uneven (Fig. 2).

Some world-record rainfall amounts are, for 1 min, 1.23 in., Unionville, Md., 1956; 1 hr, 12.00 in., Holt, Mo., 1947; 12 hr, 52.76 in., Belouve, La Réunion, 1964; 24 hr, 73.62 in., Cilaos, La Réunion, 1952; 2 days, 98.42 in., Cilaos, La Réunion, 1952; 31 days, 366.14 in., Cherrapunji, India, 1861; and 1 year, 1041.78 in., 2 years, 1605.05 in., both Cherrapunji, India, 1860–1861.

Precipitation and weather. Precipitation occurs most often in cyclones or tropical disturbances. In weather forecasting, several synoptic (weather map) types of precipitation are recognized: warm front, warm moist air rising over a wedge of cold air; cold front, cold air undercutting and lifting warmer air; convective, caused by local updrafts of moist air; convergent, general lifting of air caused by convergence at low levels; and orographic, moist air forced upward on mountain slopes. Any type may act alone; however, convection or convergence may occur with other types. Lake snow falls when air, at first extremely cold, blows off a bay or off lakes such as the Great Lakes. Minor causes of precipitation are turbulence, contact cooling (air moving over a colder surface), and nighttime radiation from cloud tops, but the amounts from these sources are small.

Condensation in the atmosphere. Basically, this requires that air be cooled to and below its dew point. Then minute water droplets form by condensation, or ice crystals form by sublimation. These droplets, or ice crystals, are cloud particles from which, by a growth mechanism, the larger precipitation particles may form. The only cooling process sufficient to produce appreciable precipitation is adiabatic expansion, which occurs in air rising toward lower pressures. The rate of condensation in rising saturated air is greater the warmer the air and is directly proportional to the speed of ascent. For discussions of condensation nuclei,

formation of precipitation in clouds, and artificial stimulation of precipitation *see* CLOUD PHYSICS; WEATHER MODIFICATION.

For discussions of related topics *see* ATMOSPHERIC GENERAL CIRCULATION; DEW; DEW POINT; FOG; HUMIDITY; HYDROLOGY; HYDROMETEOROLOGY; RAIN SHADOW.   [J. R. FULKS]

Bibliography: T. F. Malone (ed.), *Compendium of Meteorology*, 1951; W. E. Middleton, *History of the Theories of Rain and Other Forms of Precipitation*, 1968; A. Miller, *Meteorology*, 3d ed., 1976; J. L. H. Paulhus, Indian Ocean and Taiwan rainfalls set new records, *Mon. Weather Rev.*, 95:331–335, 1965; S. Petterssen, *Weather Analysis and Forecasting*, vol. 2, 2d ed., 1956.

Precipitation gages

Instruments used to measure the amount of rainfall or snowfall expressed as inches or millimeters depth of water which falls on a level surface. When precipitation gages are equipped with a recorder, the time of occurrence and the rate or intensity are available as well. Other forms of precipitation, such as dew, frost, and moisture absorbed by the soil, are also of considerable interest, but they are not measured by the precipitation gages routinely used by the meteorological services.

General design features. The gage is basically an open container and funnel constructed to minimize any splashing out and mounted 1–3 ft above the surface to prevent splashing into the container, yet not so high that its catch is affected by the wind which normally increases in velocity with height (Fig. 1). From the volume of water calibrated and the area of the opening, the depth of water is easily obtained. Various schemes are used to obtain the water volume. The U.S. Weather Bureau calibrates the catch in a cylinder whose area is one-tenth that of the gage area. The depth is measured by a dip stick calibrated in inches. Gages used in other countries have bottles, graduated cylinders, or other arrangements for obtaining the volume of the catch. The funnel in each of these gages reduces water loss by evaporation.

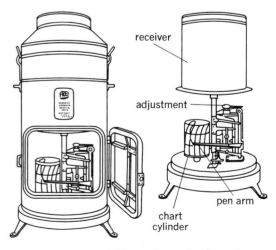

Fig. 1. Fergusson weighing and recording type of gage. (*From F. A. Berry, Jr., E. Bollay, and N. R. Beers, eds., Handbook of Meteorology, McGraw-Hill, 1945*)

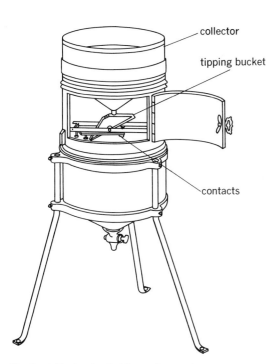

Fig. 2. A tipping-bucket type of gage. (*From F. A. Berry, Jr., E. Bollay, and N. R. Beers, eds., Handbook of Meteorology, McGraw-Hill, 1945*)

Recording instruments. A continuous type of recording rain gage uses a spring balance to record on a clock-driven chart the weight of the precipitation calibrated. High sensitivity is obtained by using a slotted linkage so that large amounts of rainfall cause the recorder pen to travel back and forth several times across the chart. Another type of recording gage uses a shallow V-shaped two-compartment tipping bucket, pivoted at the vertex so that it has two stable positions (Fig. 2). The vertical partition lies just below the opening in the funnel and directs the flow into the other position where the second compartment fills while the first empties into a container. Each tip of the bucket corresponds to 0.01 in. of rainfall, and closes an electrical contact to actuate a remote register. Still other types of gages use a float to actuate the recording pen. Some of these have an automatic siphon to empty the float chamber when a specific amount of rainfall (0.02 in.) is collected.

Gaging snowfall. When a suitable wind screen is used, snowfall can be measured by melting the snow collected in a rain gage (the funnel must be removed). Fallen snow is also measured by taking sample cores. A device for measuring the mass of snowfall uses a cobalt-60 γ-ray source at ground level columnated toward a Geiger counter mounted several feet above the surface. As snow covers the surface, the reduction in counting rate as a result of γ-ray absorption by the snow is used to determine the mass of snow in the column. An attractive feature of the device is that the counts are easily telemetered by radio, making it very useful for mountainous regions. *See* SNOW GAGE; SNOW SURVEYING.

[VERNER E. SUOMI]

Bibliography: F. A. Berry, Jr., E. Bollay, and N. R. Beers (eds.), *Handbook of Meteorology*, 1945; W. E. K. Middleton, *Invention of the Meteorological Instruments*, 1969; W. E. K. Middleton and A. E. Spilhaus, *Meteorological Instruments*, 3d ed., 1953.

Psychrometer

An instrument consisting of two thermometers which is used in the measurement of the moisture content of air or other gases. The bulb or sensing area of one of the thermometers either is covered by a thin piece of clean muslin cloth wetted uniformly with distilled water or is otherwise coated with a film of distilled water. The temperatures of both the bulb and the air contacting the bulb are lowered by the evaporation which takes place when unsaturated air moves past the wetted bulb. An equilibrium temperature, termed the wet-bulb temperature (T_W), will be reached; it closely approaches the lowest temperature to which air can be cooled by the evaporation of water into that air. The water-vapor content of the air surrounding the wet bulb can be determined from this wet-bulb temperature and from the air temperature measured by the thermometer with the dry bulb (T_D) by using an expression of the form $e = e_{SW} - aP (T_D - T_W)$. Here e is the water-vapor pressure of the air, e_{SW} is the saturation water-vapor pressure at the wet-bulb temperature, P is atmospheric pressure, and a is the psychrometric constant, which depends upon properties of air and water, as well as on speed of ventilation of air passing the wet bulb.

Other moisture parameters, such as relative humidity and dew-point temperature, can be conveniently evaluated from the wet- and dry-bulb measurements by means of psychrometric tables. These tables are accurate for a ventilation speed of 4.6 m/s (15 ft/s) when applied to measurements made using standard U.S. Weather Bureau thermometers. Other sensors may require different ventilation rates.

The common sling psychrometer consists of two mercury-in-glass thermometers which are mounted side by side on a metal or plastic frame. The frame is attached to a handle by means of a chain or a suitable bearing in such a manner that the frame can be freely rotated about the handle. A clean muslin patch is tightly mounted over the bulb of one of the thermometers and is thoroughly wetted with distilled water prior to an observation. To make a measurement, the frame is rotated about the handle at a rate of approximately 4 revolutions per second, and this rotation is continued

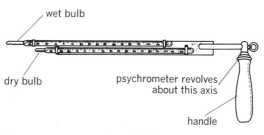

wet bulb

dry bulb   psychrometer revolves about this axis

handle

Sling psychrometer. (*From D. M. Considine, ed., Process Instruments and Controls Handbook, McGraw-Hill, 1957*)

until the wet-bulb temperature decreases to a constant value.

Another form of the psychrometer consists of two temperature-sensing elements, one of which is wetted, mounted in an air duct. A suction fan is used to aspirate air past the sensing elements at a rate sufficient to assure that the wet-bulb temperature will be reached for the particular thermal sensor employed in the instrument. This form of the instrument is used frequently at permanent weather stations.

The error associated with the measurement of atmospheric moisture by means of a psychrometer originates from a number of sources. These can be classified as errors due to (1) inherent deficiencies in temperature sensing, (2) improper ventilation, (3) improper muslin preparation or bulb coverage, (4) impure water, (5) extraneous heat flow to the bulb from radiative or conductive sources, and (6) an uncertainty in the phase of the water when the wet-bulb temperature is below the freezing point.

In general, the accuracy of the instrument is greatest at temperatures above the freezing point of water and decreases substantially at low temperatures. For example, normal calibration errors in the standard sling psychrometer can lead to a measurement error of 25% in air of nominal 50% relative humidity when the dry-bulb temperature is −15°C. The corresponding measurement error at a dry-bulb temperature of 15°C is only 5%.

[RICHARD M. SCHOTLAND]

Bibliography: C. F. Marvin, *Psychrometric Tables*, U.S. Department of Commerce, W.B. 235, 1941; W. F. K. Middleton and A. F. Spilhaus, *Meteorological Instruments*, 1953; A. Wexler (ed.), *Humidity and Moisture*, 1965.

Psychrometrics

A study of the physical and thermodynamic properties of the atmosphere. The properties of primary concern in air conditioning are (1) dry-bulb temperature, (2) wet-bulb temperature, (3) dew-point temperature, (4) absolute humidity, (5) percent humidity, (6) sensible heat, (7) latent heat, (8) total heat, (9) density, and (10) pressure.

The atmosphere in a clean pure state is a mechanical mixture of dry air and water vapor. Each is independent of the other and follows the laws of physics in accordance with its respective physical properties. Normal atmosphere contains impurities such as carbon dioxide, ozone, dust, and dirt in varying quantities.

Graphic representation. Even though dry air and water vapor are independent entities, there are certain relationships which permit the inclusion of these factors on a common chart, called a psychrometric chart (see illustration).

The psychrometric chart is a valuable tool for understanding the relationship of the many variables encountered in the atmosphere and for solving problems in air conditioning.

Dry-bulb temperature. The dry-bulb temperature is the ambient temperature of the air and water vapor as measured by a thermometer or other temperature-measuring device in which the thermal element is dry and shielded from radiation. Temperature scales are usually in degrees

Fahrenheit or degrees Celsius. The vertical lines on the chart are dry-bulb temperature lines.

Wet-bulb temperature. If the bulb of a dry-bulb thermometer is covered with a silk or cotton wick saturated with distilled water and the air is drawn over it at a velocity not less than 1000 ft/min, the resultant temperature will be the wet-bulb temperature.

If the atmosphere is saturated with water vapor, the water on the wick cannot evaporate and the wet bulb will read the same as the dry bulb. If the atmosphere is not saturated, water will evaporate from the wick at a rate dependent upon the percentage of saturation of the atmosphere. The cooling produced by the evaporation will result in a lowering of the temperature of the bulb and the consequent reading is the wet-bulb temperature.

The lines on the psychrometric chart extending from upper left to lower right are wet-bulb temperature lines. Where the dry-bulb and wet-bulb temperatures are the same, the atmosphere is saturated and a line drawn through these points is called the saturation or 100% relative humidity curve.

Dew-point temperature. The dew-point temperature is the temperature at which the water vapor in the atmosphere begins to condense. This is also the temperature of saturation at which the dry-bulb, wet-bulb, and dew-point temperatures are all the same. *See* DEW POINT.

Table 1. Properties of water

Temperature, °F	Pressure, psia	Enthalpy of evaporation, Btu/lb
40	0.122	1070.64
45	0.147	1067.81
50	0.178	1064.99
55	0.214	1062.16
60	0.256	1059.34
65	0.306	1056.52
70	0.363	1053.71
75	0.430	1050.89
80	0.507	1048.07
85	0.596	1045.23
90	0.698	1042.40
95	0.816	1039.56
100	0.950	1036.72

Table 2. Temperature-volume-enthalpy relation of saturated air

Temperature, °F	Volume, ft³/lb of dry air	Enthalpy, Btu/lb of dry air
40	12.70	15.23
45	12.85	17.65
50	13.00	20.30
55	13.16	23.22
60	13.33	27.15
65	13.50	30.06
70	13.72	34.09
75	13.88	38.61
80	14.09	43.69
85	14.31	49.43
90	14.55	55.93
95	14.80	63.32
100	15.08	71.73

The dew-point temperature is a measure of the actual water-vapor content in the atmosphere. The water-vapor content is constant for any dew-point temperature regardless of the dry-bulb or wet-bulb temperature. The dew-point temperatures are shown as the horizontal lines on the psychrometric chart.

Absolute humidity. The actual quantity of water vapor in the atmosphere is designated as the absolute humidity. It is measured as pounds of water vapor per pound of dry air, grains of water vapor per pound of dry air, or grains of water vapor per cubic foot of dry air. One pound of water vapor is equal to 7000 grains. Because the absolute water-vapor content is directly related to dew-point temperature, the vapor content in grains per pound of dry air is shown on the vertical scale on the right-hand side of the psychrometric chart for different dew-point temperatures. *See* HUMIDITY.

Percentage or relative humidity. Percentage or relative humidity is the ratio of the actual water vapor in the atmosphere to the quantity of water vapor the atmosphere could hold if it were saturated at the same temperature. For example, if the atmosphere at 50°F is saturated with water vapor, there would be 53 grains of water vapor per pound of dry air. If the dry-bulb temperature of this atmosphere were then raised to 69°F, it could hold 106 grains of water vapor (corresponding to a saturation temperature of 69°F). Because no moisture was added in changing the temperature from 53°F to 69°F, the actual vapor content is still 53 grains. The percentage humidity would then be 53/106 or 50% at 69°F.

Sensible heat. Sensible heat, or enthalpy of dry air, is heat which manifests itself as a change in temperature. It is expressed in British thermal units (Btu). One Btu will raise the temperature of 1 lb of water 1°F. One Btu will also raise the temperature of 1 lb of dry air approximately 4.25°F at atmospheric temperatures and sea-level pressure.

Latent heat. Latent heat, or enthalpy of vaporization, is the heat required to change a liquid into a vapor without change in temperature. For example, it would require 1054 Btu to change 1 lb of water at 70°F from a liquid to a dry, saturated vapor at 70°F.

Latent heat is sometimes referred to as the latent heat of vaporization and varies inversely as the pressure. The higher the pressure (or saturation temperature) the lower the Btu required to evaporate water from a liquid to a vapor. Table 1 shows the thermodynamic properties of water for various temperatures.

Total heat. The total heat, or enthalpy, of the atmosphere is the sum of the sensible heat, latent heat, and superheat of the vapor above the saturation or dew-point temperature. At saturation, the total heat measured in Btu/lb dry air is measured at the wet-bulb temperature and includes both the sensible heat of the dry air and the latent heat of the water vapor at the temperature measured. For example, at 60°F on the saturation curve the total heat is 26.46 Btu/lb of dry air with water vapor to saturate the air at that temperature.

Total heat is relatively constant for a constant wet-bulb temperature, deviating only about 1.5–

Table 3. Altitude and pressure (standard atmosphere)

Altitude, ft	Pressure, in. Hg
−1,000	31.02
−500	30.47
0 (sea level)	29.92
500	29.38
1,000	28.86
5,000	24.89
10,000	20.58

2% low at relative humidities below 30%. Table 2 shows enthalpy values for saturated air.

Density. The density of the atmosphere varies with both altitude and percentage humidity. The higher the altitude the lower the density, and the higher the moisture content the lower the density.

At sea level (29.92 in. Hg absolute pressure) and 59°F the density is 0.0765 lb/ft³. At 5000 ft elevation and 59°F, the density would be 0.0637 lb/ft³.

At sea level and 65°F saturated, the density is 0.074 lb/ft³. At 65° dry bulb and 30% saturation, the density is 0.0752 lb/ft³. The reciprocal of density (cubic feet per pound of dry air) is usually used rather than the density. Table 2 shows temperature-volume relations for saturated air.

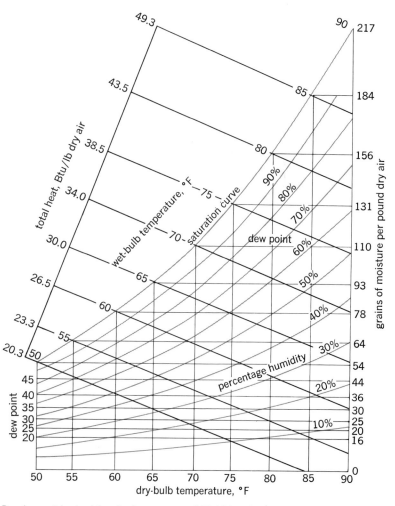

Psychrometric chart for air at a pressure of 29.92 in. of mercury.

Pressure. Atmospheric pressure, usually referred to as barometric pressure, is measured either in inches of mercury (29.92 in. Hg at sea level) or in pounds per square inch absolute (14.7 psia at sea level).

Pressure varies inversely as elevation, as temperature, and as percentage saturation. Pressure decreases with elevation as shown in Table 3. *See* MOISTURE-CONTENT MEASUREMENT; PSYCHROMETER.

[JOHN EVERETTS, JR.]

Bibliography: W. H. Carrier, Rational psychrometric formulae, *Trans. Amer. Soc. Mech. Eng.*, 33:1005–1053, 1911; W. H. Carrier and C. O. Mackey, *A Review of Existing Psychrometric Data in Relation to Practical Engineering Problems*, Amer. Soc. Mech. Eng. Advance Pap., 1936; J. A. Goff, Thermodynamic properties of moist air, *Heat. Piping Air Cond.*, 6(3):117–132, 1934; W. K. Lewis, The evaporation of a liquid into a gas: A correction, *Mech. Eng.*, 55(9):567–568, 1935; A. Pande, *Handbook of Moisture Determination and Control*, 4 vols., 1974–1975; A. Wexler, *Humidity and Moisture*, 1964–1965; D. D. Wile, Psychrometric charts, *Amer. Soc. Heat. Air Cond. Eng. J.*, vol. 1, no. 8, 1959.

Radar meteorology

The study of the scattering of radar waves by all types of atmospheric phenomena and the use of radar for making weather observations and forecasts. In general, radar is useful in meteorology because it is capable of detecting water and ice particles, but it can also be used to observe lightning and regions of the atmosphere which have large gradients of temperature and water vapor. Important problems and applications of radar treated in this article include radar reflectivity from water and ice particles; use of radar for rainfall measurement; study of cloud and precipitation formation; observations of tornadoes and hurricanes; and radar echoes from targets other than water or ice particles.

Echoes from water and ice particles. The radar equation applies to water and ice particles to the same extent as it does to airplanes or ships provided that the term σ, the radar cross section, is properly specified. The radar cross section is defined as the area that would intercept that amount of radiation which, when reradiated isotropically, produces an echo equal to that observed from the water or ice particle. When a radar wave is intercepted by a drop, small fractions of the energy are absorbed and scattered back to the radar while the major part of the energy propagates forward and is intercepted by other drops. The cross section of a spherical water or ice particle whose diameter is small relative to the wavelength is given by the Rayleigh scattering law, Eq. (1), where λ is wave-

$$\sigma = \frac{\pi^5}{\lambda^4} |K|^2 D^6 \qquad (1)$$

length, D is particle diameter, and $|K|^2$ is a term which depends on the dielectric properties of the particle. It is about 0.93 for water and 0.20 for ice and accounts for the fact that a waterdrop reflects about five times more power than does an ice particle of the same size. Although the quantity of pow-

er backscattered by a single particle is very small, the radar echo is caused by all the drops within a region delineated by the beam width and a distance equal to one-half the pulse length of the radar. Since precipitation particles occur in concentrations of perhaps $1000/m^3$, a large number of particles cause backscatter simultaneously.

Because precipitation particles are constantly moving relative to one another, the radar echo intensity fluctuates rapidly from one instant to the next. However, if the particles are randomly dispersed, the time-averaged power is given by the sum of the power from each of the particles. When the entire radar beam is intercepted by a region of rain or snow, the average received power $\overline{W}_r$ is given by Eq. (2), where P_t is the power transmitted,

$$\overline{W}_r = \frac{P_t A_e h}{8\pi R^2} \Sigma \sigma \qquad (2)$$

A_e is the so-called effective antenna area, h is the pulse length, R is the range, and $\Sigma \sigma$ is the sum of the radar cross sections of all the particles in a unit volume.

By combining Eqs. (1) and (2) it is found that the echo power increases rapidly as the particle size increases and as the wavelength decreases. To detect small particles, for example, cloud droplets $50-100~\mu$ in diameter, short wavelengths should be used. Radar sets for measuring cloud bases and tops have been designed to operate at a wavelength of about 1 cm. Unfortunately the short waves suffer strong attenuation by water vapor and waterdrops. As a result, short-wave radar is not suitable for observing heavy rain or for detecting clouds at long range. To detect large water or ice particles associated with moderate to heavy precipitation, wavelengths greater than 5 cm should be employed. Radar sets operating at 10-cm wavelengths are used for observing intense storms at distances exceeding 200 mi.

When dealing with water or ice particles whose diameters are about the same or greater than the radar wavelengths, the Rayleigh law no longer applies, and it is not possible to write a simple expression for the radar cross section. This is the case when large hailstones are involved. Calculations have shown that, in general, the radar cross sections of large ice spheres are greater than those of water spheres of the same diameter and very much greater than the cross sections of raindrops. When small ice particles develop a thin layer of water (of the order of 0.1 mm), they behave on radar almost as if they were composed entirely of water.

Doppler radar. Starting in the early 1960s, meteorologists began to employ pulsed-Doppler radar sets for the study of atmospheric phenomena. Such radar sets, in addition to measuring the quantity of reflected power, also measure the velocity of the scatterers along the radar beam. If the beam is pointed vertically, a Doppler radar measures the vertical velocity of the scattering particles —rain, snow, hail, and so on. In general, a great many scattering particles are viewed at any instant of time. As a result, a Doppler radar observes a spectrum of returned power as a function of the vertical velocity of the particles.

By means of data from a vertically pointing pulsed-Doppler radar, it is possible some of the time to infer the vertical motions of the air. This has led to information on the updrafts in thunderstorms. In cases where the air motion is small, Doppler radar data have been used to calculate the sizes and concentrations of raindrops. Such calculations make it possible to examine raindrop spectra in time and space and investigate the mechanisms influencing rainfall. In some instances Doppler radar may also yield information on hailstone sizes in the free atmosphere.

By scanning azimuthally with the antenna at various elevation angles, a Doppler radar may be used to measure wind velocity, providing there are widespread scatterers such as rain, snow, or even insects. The observations also can reveal information about the turbulent structure of the atmosphere.

If two or three scanning Doppler radars are employed to observe the same storm system, it is possible to obtain the two- or three-dimensional wind patterns. Multiradar observations have also yielded detailed data on three-dimensional air motions within thunderstorms and showers.

Rainfall measurement. Since radar detects the power backscattered by rain and snow particles, it is reasonable to expect that radar should be a good means for measuring rainfall. In order to accomplish this purpose, it is necessary to relate rainfall intensity I to $\Sigma\sigma$, the radar cross section.

Raindrop diameters seldom exceed 5 or 6 mm. Therefore, if a radar operates at a wavelength of about 10 cm, the raindrops will act as Rayleigh scatterers and $\Sigma\sigma$ will be proportional to ΣD^6. Another advantage of employing this relatively long wavelength is that signals may pass through normal rainfall without suffering serious attenuation. This allows one to neglect intervening rain when estimating the rainfall rate in the interior of a storm.

A great many observations of raindrops at the ground have shown that their concentrations decrease exponentially as D increases. Furthermore, the measurements reveal that in general Eq. (3)

$$\Sigma D^6 = aI^b \qquad (3)$$

holds, where a and b have different values for rain and snow, for various types of rain, and for various times within some rainstorms.

By substituting in Eq. (2), Eq. (4) is obtained,

$$\overline{W}_r = \frac{CI^b}{R^2} \qquad (4)$$

where C is a constant depending on the characteristics of the radar set and the phase of the precipitation. The exponent b is often taken to be about 1.6 for rain and about 2.0 for snow.

Interestingly, when snow is dry, its nonspherical nature is of little importance. A snowflake may be treated as if it were a sphere composed of low-density ice.

The equation relating $\overline{W}_r$ and I has been tested experimentally. Investigations conducted in various parts of the world reveal that if extreme care is employed in taking all factors into account, radar may measure I with a probable error of about 20%. Usually errors are considerably larger. The major sources of errors are the uncertainties in the ΣD^6 versus I relationship, and the inaccuracies in measuring the received power and in calibrating the radar set.

The difficulties in obtaining more accurate measurements of rainfall rate have held back the widespread use of radar for the practical measurement of rainfall.

Since the mid-1960s increased attention has been devoted to the development of an operational scheme to measure the total rainfall over a watershed. The techniques offering the most promise involve the following steps. The area of interest is divided into smaller units which may range in size from a few to as many as 18 mi². At intervals of perhaps 30 min, the radar echo intensity from each unit area is read and put in digital form. The data are then transmitted to a computer which could be in a River Forecast Center at some distance from the radar set. Once the computer has the information on echo intensity as a function of position, it may calculate the pattern of precipitation intensity. With measurements at regular time intervals, the total quantities of rain or snow over any area may be calculated.

This procedure has been under development at various research centers, in particular, at the National Severe Storm Laboratory in Norman, Okla. Early tests have yielded encouraging results.

Hopefully, in the near future radar will begin to be used more widely to supply quantitative information on rain and snow falling over large areas.

Cloud and precipitation formation. Radar has played an important role in the study of mechanisms of precipitation formation. Figure 1 is a photograph of a range-height indicator (RHI) showing the precipitation echo from a cumuliform cloud of the type often seen in the summer. Observations such as this show the altitudes at which there are

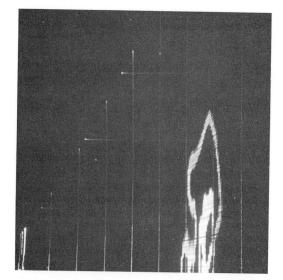

Fig. 1. Isolated thunderstorm echo on the RHI scope of a 10-cm vertically scanning radar set. The white vertical lines are at 10-mi intervals. The white horizontal lines are at 20,000 ft intervals. The echo extends to an altitude of nearly 50,000 ft. The alternate bright and dark areas show differences in radar echo intensity. (*National Severe Storms Laboratory, Norman, Okla.*)

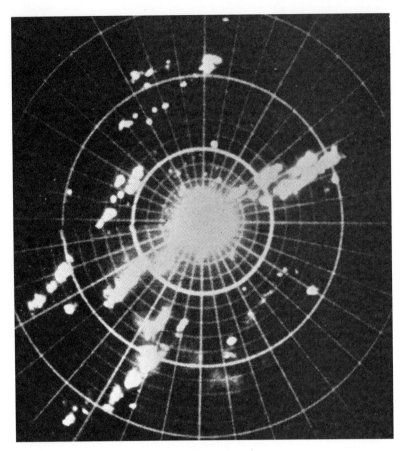

Fig. 2.   Series of squall lines on a PPI scope of a 10-cm radar set. The heavy circular lines are at 50-mi intervals; the light circular lines are at 10-mi intervals. The lines of thunderstorms are oriented roughly northeast-southwest.

large water or ice particles. From a series of observations, the region in which the large particles first form and the rate at which they spread may be determined. By means of such analyses it has been shown that in convective clouds precipitation often forms in the absence of ice crystals, a fact which was once in doubt. *See* CLOUD PHYSICS.

Radar observations have also shown that thunderstorms grow to great altitudes at rates which

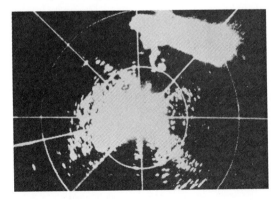

Fig. 3.   Thunderstorm echo northeast of radar station (operating at 3 cm) with a 6-shaped appendage. A tornado formed near the bottom part of the loop at the southerly end of the appendage. (*Photograph by Illinois State Water Survey, Champaign-Urbana, Ill.*)

may exceed 2000 ft/min. Sometimes they may penetrate into the stratosphere to altitudes over 60,000 ft. In order for thunderstorms to grow to these extreme altitudes, there must be very strong updrafts. In such clouds there is likely to be severe turbulence and hail. Also, the possibility of tornado formation is strong. The degree of turbulence can be inferred, to a certain extent, from analyses of the fluctuations of the echo intensity. *See* HAIL; THUNDERSTORM.

The most widespread practical use of weather radar has been to detect thunderstorms and to watch their movements. On the basis of this type of information it is possible to predict their passage over critical areas. For this purpose a plan position indicator (PPI) is the most suitable. Figure 2 shows a number of lines of thunderstorms. By keeping them under observation, warnings may be issued to communities in their paths. A line of thunderstorms may extend for many hundreds of miles, but it will be composed of individual storm cells whose diameters are usually less than 10 mi. Although a squall line may last for many hours, the individual storms usually last for less than an hour or two. The component storms are in a continuous state of development and dissipation. For this reason the forecasting problem is more difficult than it seems, especially if forecasts are to be issued more than an hour or two in advance. *See* SQUALL.

The character of precipitation in large winter storms differs in some important respects from the rain which forms in summer convective clouds. It has been concluded that in such storm systems the precipitation particles first develop in the form of ice crystals at high altitudes. The crystals agglomerate to form snow particles which drift toward the ground and are carried horizontally by the wind. When they fall through the level of 0°C, they begin to melt, and their radar reflectivities rapidly increase to form the bright band that appears at about 6000 ft. After the particles have melted, they fall faster than they did as snowflakes. When the temperature of the ground is below freezing, the snow particles do not melt but reach the ground as snow. *See* HYDROMETEOROLOGY; PRECIPITATION.

Tornadoes and hurricanes. The most violent storm produced by nature, the tornado, is characterized by its small size and short duration. These properties make it difficult to study. Since 1953 radar observations have been made of many thunderstorms which have spawned tornadoes. In some cases the funnel was associated with a 6-shaped appendage such as that shown in Fig. 3. Unfortunately this feature occurs with only a small fraction of the observed tornadoes. When it does occur, it is a sure sign that a tornado either is present or is about to form. As of 1976 there still was not a reliable technique for determining, in every case, whether or not a tornado is present. On the other hand, once a tornado has been spotted, a radar set can accurately track the associated thunderstorm, and warnings can be issued to communities ahead of the storm. Research on the development of Doppler radar techniques for tornado detection has made impressive advances.

Hurricanes do tremendous amounts of damage

every year. As they move toward or over coastlines, they may cause great floods and damaging winds. Radar observations have shown that most often the precipitation patterns consist of spiral bands, as in Fig. 4. In some hurricanes the spirals are replaced by a ring of echoes. Often lines of thunderstorms roughly perpendicular to the direction of travel precede the storm center by several hundred miles. Radar-equipped airplanes and powerful ground-based radar stations are now being employed to locate and track hurricanes so that accurate forecasts may be issued. *See* HURRICANE; STORM DETECTION; TORNADO.

Echoes from other targets. It has been found that the free-electron concentrations following a lightning stroke are sufficiently large to cause a detectable radar echo. Lightning echoes having total lengths over 50 mi have been observed. The reflectivity of the lightning channel increases with increasing wavelength. Because of the rapid recombination of the electrons with positive ions, the durations of lightning echoes average about 0.5 sec. *See* LIGHTNING; SFERICS.

In many instances radar echoes have been received from regions where there were no visible targets. These echoes are usually called angel echoes. Studies have shed considerable light on the origin of these echoes. It is now clear that they may be caused by insects, birds, or by refractive index variations in the atmosphere. Simultaneous observations of angel echoes have been made with sensitive radar sets operating at wavelengths of 3.2, 10.7, and 71.5 cm. Comparisons of the reflected power have led to conclusions that the source of echoes sometimes is swarms of insects. In some cases one bug in a cubic mile can be detected.

Before 1950 scientists observed that sometimes radar echoes were received from regions of highly varying refractive index. They most often are associated with temperature inversions, that is, with layers where the temperature increases with height. When the water vapor content decreases through a temperature inversion, the microwave refractive index may decrease markedly.

If turbulent motions of the air cause the forma-

tion of eddies having different refractive index, it is possible to explain how clear air may reflect sufficient power to produce radar echoes. Layers of clear-air echoes detected by powerful radar sets have been associated with regions of clear-air turbulence.

Wind measurements. Radar is used for wind measurements. This is accomplished by tracking targets carried along by the wind. The targets may be radar reflectors suspended from balloons rising to altitudes approaching 30 km or chaff dispersed from an airplane or rocket. *See* METEOROLOGICAL INSTRUMENTATION.   [LOUIS J. BATTAN]

Bibliography: American Meteorological Society, *Proceedings of Radar Meteorology Conferences*; D. Atlas, *Advances in Radar Meteorology*, in H. E. Landsberg (ed.), *Advances in Geophysics*, vol. 10, 1964; NATO Advanced Study Institute, Gaslor, *Atmospheric Effects on Radar Target Identification and Imaging*, 1976; A. Williams, *The Use of Radar Imagery in Climatological Research*, 1973; World Meteorological Organization, *Hydrological Requirements for Weather Radar Data*, 1969.

Radiation

The emission and propagation of energy; also, the emitted energy itself. The etymology of the word implies that the energy propagates rectilinearly, and in a limited sense, this holds for the many different types of radiation encountered.

The major types of radiation may be described as electromagnetic, acoustic, and particle, and within these major divisions there are many subdivisions.

For example, electromagnetic radiation, which in the most familiar energy ranges behaves in a manner usually characteristic of waves rather than of particles, is classified roughly in order of decreasing wavelength as radio, microwave, visible, ultraviolet, x-rays, and γ-rays. In the last three subdivisions, and frequently in the visible, the behavior of the radiation is more particlelike than wavelike.

Since the energy of a photon (light quantum) is inversely proportional to the wavelength, this classification is also on the basis of increasing photon energy.

Acoustic or sound radiation may be classified by frequency as infrasonic, sonic, or ultrasonic in order of increasing frequency, with sonic being between about 16 and 20,000 Hz. Infrasonic sound can result, for example, from explosions or other sources so loud that exceptional waves are set up because the large amplitudes of the source vibrations exceed the elastic limit of the transmitting medium. Ultrasonic sound can be produced by means of crystals which vibrate rapidly in response to alternating electric voltages applied to them. There is a nearly infinite variety of sources in the sonic range.

The traditional examples of particle radiation are the α- and β-rays of radioactivity. Cosmic rays also consist largely of particles—protons, neutrons, and heavier nuclei, along with β-rays, mesons, and the so-called strange particles. *See* COSMIC RAYS.   [MC ALLISTER H. HULL, JR.]

Fig. 4. Hurricane Donna observed at 0730 EST, Sept. 10, 1960, by a 10-cm radar set located at Miami, Fla. (*Photograph by L. F. Conover, National Hurricane Project, U.S. Weather Bureau, Miami, Fla.*)

Radioactive fallout

The radioactive material which results from a nuclear explosion in the atmosphere. In particular the term applies to the debris which is deposited on the ground, but common usage has extended its coverage to include airborne material as well.

Nuclear explosion. A nuclear explosion results when the fission of uranium-235 or plutonium-239 proceeds in a rapid and relatively uncontrolled way, as opposed to the controlled fission in a reactor. The energy release of an atomic (fission) bomb is usually expressed in terms of thousands of tons of TNT equivalent, and such explosions may have yields into the range of hundreds of kilotons. Still larger explosions can be produced by using the fission device as a trigger for a fusion reaction to produce a thermonuclear explosion. The yield of such thermonuclear devices can range up to hundreds of thousands of kilotons (hundreds of megatons).

The radioactivity from the explosion is produced by fission products and by activation products. The basic reaction in atomic fission is the splitting of an atom of the fissionable material (uranium or plutonium) into two lighter elements (fission products). These lighter elements are all unstable and emit beta and gamma radiation until they reach a stable state. During the fission reaction a number of neutrons are released, and they can interact with the surrounding materials of the device or of the environment to produce radioactive activation products. The fusion process does not produce fission products but does release neutrons which can add to the activation.

The fissioning of a single atom of uranium or plutonium yields almost 200 million electron volts (MeV) instantaneously, and just over 50 g of fissionable material are required to produce a yield of 1 kiloton. The tremendous energy release produces the explosive shock and temperatures ranging upward from 1,000,000°C. The device itself and the material immediately surrounding it are vaporized into a fireball which then rises in the atmosphere. The altitude at which the fireball cools sufficiently to stabilize depends very much on the yield of the explosion. In general, an atomic explosion fireball stabilizes as a cloud high in the troposphere, while a thermonuclear fireball tends to break through into the stratosphere and stabilize at altitudes above 10 mi.

As the fireball cools, the vaporized materials condense to fine particles. If the explosion has taken place high above the Earth, the only material present is that of the device itself, and the fine particulates are carried by the winds and distributed over a wide area before they descend to Earth. Bursts at or near the surface carry large amounts of inert material up with the fireball. A large fraction of the radioactive debris condenses out onto the large inert particles of soil and other material, and many small radioactive particulates attach themselves to the larger inert particles. The larger particles settle rapidly by gravity, and large percentages of the radioactivity may be deposited in a few hours near the site of the explosion. In contrast, the radioactivity from a thermonuclear explosion at high altitude takes months or years to reach the surface. It is also possible to test nuclear explosives deep in the earth. Such underground tests do not produce fallout as long as the explosion is completely contained, and they are not covered by the test ban treaty of 1962.

The first nuclear explosion took place in New Mexico in 1945 and was rapidly followed by the two bombings at Hiroshima and Nagasaki, Japan. Through 1978 six nations have tested a large number of nuclear and thermonuclear weapons. The United States has tested at its sites in Nevada and the Pacific; the United Kingdom has tested in Australia and Christmas Island. The Soviet Union has several areas, including Novaya Zemlya in the Arctic. Atmospheric testing was heaviest in 1954, 1958, and 1961 and was stopped following negotiation of the test ban treaty. France and China did not sign the treaty, however, and have carried out a number of tests since, France in the Sahara and the South Pacific and China at an inland test site at Lop Nor. India has also carried out one underground test.

Radioactive products. The total fission yield for all tests has been about 200 megatons, the total explosive yield being over 500 megatons. The production of long-lived fission products has been about 20 megacuries of strontium-90 and 30 megacuries of cesium-137. Figure 1 shows the time distribution for strontium-90 fallout in New York City, which may be considered typical of the Northern Hemisphere mid-latitudes. Figure 2 shows the latitude distribution as of 1966. Testing through 1978 does not modify this picture because atmospheric explosions since the test ban have added less than 10% to the total.

Local fallout. Local fallout is the deposition of large radioactive particles near the site of the explosion. It may present a hazard near the test site or in the case of nuclear warfare. The smaller particulates, which are spread more widely, constitute the worldwide fallout. The radioactivity comes from the isotopes (radionuclides) of perhaps 60 different elements formed in fission, plus a few others formed by activation and any unfissioned uranium or plutonium. Each of the radioactive iso-

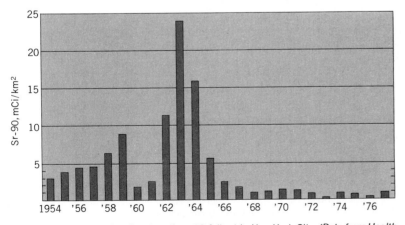

Fig. 1. Time distribution for strontium-90 fallout in New York City. (*Data from Health and Safety Laboratory, U.S. Atomic Energy Commission*)

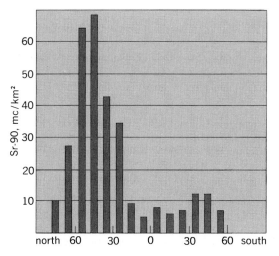

Fig. 2. Latitude distribution of cumulative strontium-90 deposit as of 1966. (*Data from Health and Safety Laboratory, U.S. Atomic Energy Commission*)

topes produced in the fission process goes through three or four successive radioactive decays before becoming a stable nuclide. The radioactive half-lives of these various fission products range from fractions of a second to about 100 years. The overall beta and gamma radioactivity decays according to the minus 1.2 power of the activity at any time. Thus decay is very rapid at first and then slows down as the short-lived isotopes disappear.

The hazard from local fallout is largely due to the external gamma radiation which a person would receive if this material were deposited near him; fallout shelters are designed to shield survivors of an explosion from this radiation. Actually, the effects of an explosion are so great that a shelter which would resist the shock would shield the occupants from fallout radiation.

In so-called clean nuclear weapons the ratio of explosive force to the amount of fission products produced is high. On this basis, small fission weapons are considered "dirty," although the absolute amount of fission products is much less than with a large clean bomb. It is also possible to increase the radioactivity produced, and thus the fallout, by surrounding the device with a material which is readily activated. Cobalt has been frequently mentioned in this connection.

Tropospheric and stratospheric fallout. As previously mentioned, the large particulates from a nuclear explosion are deposited fairly close to the site. The smaller particles are carried by the winds in the troposphere or stratosphere, generally in the direction of the prevailing westerlies. The debris in the troposphere circles the Earth in about 2 weeks, while material injected into the stratosphere may travel much faster. The horizontal distribution in the north-south direction is much slower, and it may take several months for radioactivity to cover the Northern Hemisphere, for example, following a test in the tropics. Transfer between the Northern and Southern hemispheres is slight in the troposphere but can take place in the stratosphere. Radioactive debris in the troposphere is brought down to the surface of the Earth mainly in

precipitation, with perhaps 10–15% of the total amount brought down by dry deposition. Radioactive material remains about 30 days in the troposphere. The mechanism of transfer from the stratosphere to the troposphere is not completely understood, but it is obvious from measurements that the favored region for deposition is the mid-latitudes (30–50°) of the Northern and Southern hemispheres. Since most of the testing took place in the Northern Hemisphere, the deposition there was about three times as great as in the Southern Hemisphere.

Hazards. Once the fallout has actually been deposited, its fate depends largely on the chemical nature of the radionuclides involved. Since the level of gamma radiation from worldwide fallout is negligible, attention has been paid to those nuclides which might possibly present some hazard when they enter the biosphere. Plants, for example, may be contaminated directly by foliar deposition or indirectly by deposition on the soil that is followed by root uptake. Animals may be contaminated by eating the plants, and man by eating plant or animal foods. The inhalation of radioactive material is not considered to be a significant hazard as compared to ingestion.

Based on the various considerations mentioned, the three radionuclides of greatest interest are strontium-90, cesium-137, and iodine-131. Strontium-90 is a beta emitter which follows calcium metabolically and tends to deposit in the bone. The radiation emitted could result in bone cancer or leukemia if the levels were to become sufficiently high. Cesium-137 is distributed throughout soft tissues and emits both beta and gamma radiation. It is considered to be a possible genetic hazard but not as dangerous to the individual as strontium-90. Iodine-131 is relatively short-lived and probably is a hazard only as it appears in tropospheric fallout. It is readily absorbed by cows on pasture and is transferred rapidly to their milk. The iodine in milk in turn is concentrated in the human thyroid and presents a possible high radiation dose, particularly to children, who consume larger amounts of milk and who have smaller thyroids than adults.

Sampling. Radioactive fallout is monitored by many countries. The usual systems involve networks that collect samples of airborne dust, deposition, and milk and other foods. The levels of iodine-131 or cesium-137 can be checked in living individuals by external gamma counting. Strontium-90 on the other hand can be measured only in autopsy specimens of human bone. These national data are also combined and evaluated by the Scientific Committee on the Effects of Atomic Radiation of the United Nations. Their reports are issued at intervals and offer broad reviews and summaries of the available information on levels of fallout and on possible hazards.

The studies of radioactive fallout have also included sampling in the stratosphere by balloons and high-flying aircraft, as well as taking aircraft samples in the troposphere. Very elaborate monitoring systems have been set up, and these have provided considerable scientific information in addition to their original monitoring purposes. The major benefits have been the understanding of stratospheric transfer processes as significant to meteorology and information on the uptake and metabolism of various elements by plants, animals, and man. These studies are usually reported in the specialty journals for meteorology, agriculture, and biology.

[JOHN H. HARLEY]

Bibliography: *Reports of the United Nations Committee on the Effects of Atomic Radiation*, 1958, 1962, 1964, 1966, 1969, 1972, 1977, U.S. Public Health Service, *Radiological Health Data and Reports*, monthly (through 1974).

Rain shadow

An area of diminished precipitation on the lee side of mountains. There are marked rain shadows, for example, east of the coastal ranges of Washington, Oregon, and California, and over a larger region, much of it arid, east of the Cascade Range and Sierra Nevadas. Precipitation on the northern Oregon coast is around 100 in. per year; east of the coastal ranges in the Willamette Valley it is about 40 in.; and east of the Cascades it drops to around 10 in. All mountains decrease precipitation on their lee; but rain shadows, as is shown by annual totals, are sometimes not marked if moist air comes frequently from different directions, as in the Appalachian region.

The causes of rain shadow are (1) precipitation of much of the moisture when air is forced upward on the windward side of the mountains, (2) deflection or damming of moist air flow, and (3) downward flow on the lee slopes, which warms the air and lowers its relative humidity. *See* CHINOOK; PRECIPITATION.

[J. R. FULKS]

Rainbow

Colored arcs seen in the skies when the Sun or Moon is illuminating large numbers of falling raindrops. Such arcs are centered around the antisolar point (180° from the Sun). Among the parallel rays striking a water droplet there is one ray which after one internal reflection leaves the drop at the smallest angle of deviation from the direction of the incident rays. All other rays emerge from the drop, after an internal reflection, at larger angles of deviation. The ray of minimum deviation, also called Descartes ray, has the greatest intensity and thus is more visible than the others. Since the angle of deviation for such a ray of minimum deviation is smaller for the red light and larger for the blue, the rays of minimum deviation from a large number of water drops of the same size are seen as arcs of different colors, red on the upper part, and blue on the lower part of the bow. The angular distance of these arcs from the antisolar point is greater than 42°, and the system of arcs is called the primary rainbow. The rays of minimum deviations, after two internal reflections, form similar colored arcs at the angular distance of 51° from the antisolar point to make the secondary rainbow. In this rainbow the color sequence is reversed, with the red arcs visible on the lower part, the blue on the upper part of the bow. *See* METEOROLOGICAL OPTICS.

[ZDENEK SEKERA]

Red Sea

The Red Sea lies between Arabia and northeastern Africa. It is about 2000 km long and 300 km wide, and has a maximum depth of about 2300 m. Details of its bathymetry remain little known because echo soundings are difficult to obtain along the precipitous slopes and in the narrow valleys of the rugged sea floor. New knowledge has resulted from two sources: geophysical investigation of the underlying rocks, and geochemical studies of hot, dense brines discovered in a few deep basins in the central sector.

Geology and geophysics. Geophysical and geological findings have indicated that the Red Sea basin is a complex system of rifts formed during several periods of crustal movement and accompanying vulcanism. These movements have resulted in the separation of the ancient crystalline rock masses of Africa and Arabia. The Red Sea basin is a segment of the African rift system, which extends from central Africa to the Dead Sea. Negative gravity anomalies and minor magnetic anomalies suggest that the northern Red Sea and the Gulfs of Suez and Aqaba are relatively simple downfaulted basins. In contrast, the central and southern sectors of the Red Sea exhibit positive gravity anomalies on the sides, negative gravity anomalies over the center of the deep rift, and extreme magnetic irregularity, especially over the deep median valley. This has been interpreted to represent a complex series of rifts, wherein basaltic magma has risen to engulf remnants of lighter crustal rock (Fig. 1).

Geological studies have indicated that four crustal blocks, including the Sinai Peninsula and the "Horn of Africa," have taken part in the rotation and separation that formed the Red Sea basin. These studies differ somewhat from the geophysical investigations in concluding that compressional forces may have been active in the northern sector; they agree, however, that tensional forces and associated volcanic emanations probably produced the complex structures in the south. Rift movements began as early as the middle Mesozoic (about 150,000,000 years ago) and have continued intermittently since then. The sedimentary record in the Red Sea basin dates back at least to the Miocene (10,000,000–25,000,000 years ago). Faulting of Pliocene volcanics in the area of the straits of Bab-el-Mandeb in the south suggests that the Red Sea basin was opened to the Indian Ocean perhaps no more than 10,000,000 years ago.

Isotopic studies of sediment cores indicate that the Red Sea was apparently closed off entirely during at least the last sea-level lowering of the Pleistocene, and that conditions of an evaporite basin prevailed during this interval. Previous sedimentologic studies had revealed semiconsolidated layers in the upper few meters of sediment. These layers may be associated with the evaporite conditions that marked the glacial intervals. Glacial sea-level lowering was at least 100 m and probably more.

Oceanography. The water that enters the Red Sea basin is surface water from the Gulf of Aden, which flows through the straits over a sill that is only 125 m deep. No permanent rivers discharge

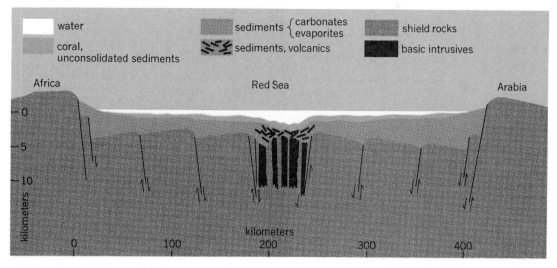

Fig. 1. Schematic diagram of a cross section of the central and southern segments of the Red Sea basin. (*From C. Drake and R. Girdler, A geophysical study of the Red Sea, Geophys. J., 8:473, 1964*)

into the Red Sea, and the Suez Canal affords no appreciable interchange with the Mediterranean. Several oceanographic ships passing through the Red Sea in recent years have collected hydrographic data. A study of the overall circulation during the early summer suggests that evaporation exerts a greater influence on the circulation of the Red Sea than the seasonal wind pattern emphasized by earlier workers. Evaporation rates reach 210 cm per year in the northern sector and, in consequence, an inflow from the Gulf of Aden compensates for these evaporative losses. This inflow is usually at the surface, except in late summer when northwest winds temporarily superimpose a shallow surface outflow through the straits.

Because of a peculiar thermohaline situation set up by the cooling and evaporation of northward-flowing surface water, a shallow density-driven outflow is produced that carries saline water back to the straits, where it spills over the sill and into the Gulf of Aden. This Red Sea outflow can be traced at intermediate levels well into the Indian Ocean. In the early summer this outflow is apparently derived entirely from the upper 100–200 m of the Red Sea, indicating that there are seasonal intervals when the great bulk of water below sill depth does not take part in exchange with the outside.

The well-mixed water that fills the Red Sea basin from 200 to over 2000 m exhibits remarkably uniform properties. Its salinity is nearly constant at 40.6 parts per thousand and its temperature remains near 22°C. It is formed and replenished at the surface in the northern sector during the winter, when cooling and evaporation reach a maximum. Under these conditions, the deepest parts of the basin remain unstratified and well ventilated by mixing with surface-derived water. *See* INDIAN OCEAN; OCEAN CURRENTS; SEA WATER.

Anomalously hot saline water. There is one extraordinary exception to this general situation. The British research vessel *Discovery*, acting on earlier reports of anomalously warm and saline water in a deep isolated central basin, made a preliminary bathymetric reconnaissance of the area and lowered thermometers and water-sampling devices to the bottom of the deepest basin revealed by their survey. What they retrieved sparked a "gold rush" of oceanographic investigation. The hottest, saltiest water yet found in the deep ocean filled the bottom 200 m of this small basin some 2200 m below the surface. The salt content is over 25% by weight in contrast to the normal Red Sea salinity of 4%. Temperatures are as high as 44°C (111°F) in contrast to the already high "normal" bottom temperatures of 22°C. The relative proportions of the constituent salts differ markedly from the constant ratios that characterize normal sea water, as well as brines concentrated from normal sea water.

One theory proposes that the hot brine has a connate origin; that is, it has seeped out of a sedimentary rock reservoir, where it has been entrapped since its incorporation with the sediments at the time they were deposited, perhaps millions of years ago. Within that time the ancient waters may have migrated through sediments other than those that first enclosed them, reacting chemically with minerals, other waters, and perhaps volcanic fluids and vapors. With the burial of the enclosing rocks, the temperature of the brines rose in accordance with the geothermal gradient. The chemical composition of the Red Sea brines compares favorably with oil-well brines of known connate origin. Faulting or vulcanism could have provided a pathway for the upward migration of the brines.

Another theory proposes that these pathways could also provide the downward path. Nevertheless, isotopic studies indicate that the source of the brines is the Red Sea water itself, and that they have entered these deep basins by the process of submarine discharge. Thus, earlier theories of either contemporary surface formation in an evaporating lagoon or an ancient residuum from a time of lowered sea level are in question.

The discovery of anomalous water in these small central Red Sea basins dates back to the Swedish

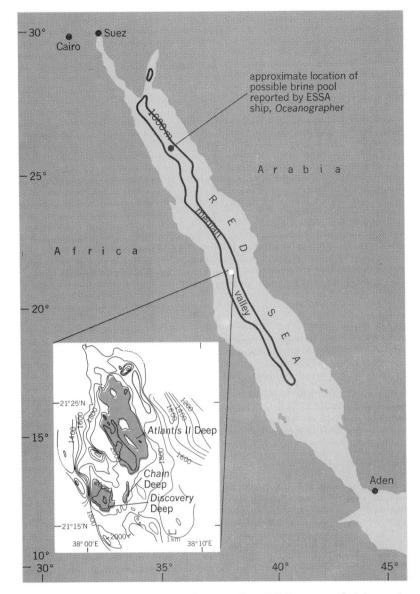

Fig. 2. Brine-filled deeps in central Red Sea. (*From E. T. Degens and D. A. Ross, eds., Hot Brines and Recent Heavy Metal Deposits in the Red Sea, Venlay Inc., 1969*)

which shows that there are three deeps containing hot saline water within an area of 15 by 15 km (Fig. 2). The *Discovery* Deep was found to be separate from the area explored by both the *Atlantis II* and *Atlantis* to the northeast, and a new deep, called *Chain* Deep, was discovered that contained water with a temperature and salinity at least as high as 34°C (94°F) and 7.4%, respectively. Early in 1967 the *Oceanographer*, a research vessel of the U.S. Environmental Sciences Services Administration, observed a reflection on their fathometer which may be the surface of the dense brine lying within a basin which is located 340 mi north of the three known deeps (Fig. 2).

Sediment. Samples obtained from the brine area show high concentrations of iron, manganese, zinc, lead, copper, and silver in both the sediments and the overlying brines. The sediments beneath the brine are in some places a homogeneous black slurry, and in others a series of brightly colored brown, red, orange, and yellow layers. A large portion of the sediment is composed of limonite-like amorphous iron hydroxide, with lesser amounts of montmorillonite, sphalerite, and amorphous silica. Traces of anhydrite and dolomite are also reported. Chemical and mineralogic analyses reveal amorphous iron hydroxides, iron-rich silicates, sulfides, oxides, and carbonates. The sediments are believed to be largely precipitates from the overlying brine. The distribution of these different phases indicates two major processes of precipitation: (1) cooling and precipitation near the brine source area (probably in the *Atlantis II* Deep) of the silicates and sulfides, and (2) oxidation during mixing with the overlying sea water and precipitation of the hydroxide and oxide suites over a large area. The carbonate may result either from a simple loss of CO_2 from the brine or by pelagic carbonate debris reacting with the hot brine.

If a sediment thickness of 10 m is assumed, the concentration of copper, zinc, silver, and gold in the *Atlantis II* Deep has a value of $2,300,000,000, not considering the cost of retrieval. Regardless of the economic value of these deposits, they are of considerable importance to geology because of their similarity to ancient ore deposits. These small Red Sea basins afford a unique opportunity to study in action the same processes that deposited the vast and rich Precambrian iron ores more than 1,000,000,000 years ago.

[A. CONRAD NEUMANN; DAVID A. ROSS]

Bibliography: P. Brewer, J. Riley, and F. Culkin, The chemical composition of the hot salty water from the bottom of the Red Sea, *Deep-Sea Res.*, 12: 497, 1965; E. T. Degens and D. A. Ross (eds.), *Hot Brines and Recent Heavy Metal Deposits in the Red Sea*, 1969; C. Drake and R. Girdler, A geophysical study of the Red Sea, *Geophys. J.*, 8:473, 1964; J. Gevirtz and G. Friedman, Deep sea carbonate sediments of the Red Sea and their implications on marine lithification, *J. Sediment Petrol.*, 36:143, 1966; J. Hunt et al., Red Sea: Detailed survey of hot brine areas, *Science*, 156: 514, 1967; A. Miller et al., Hot brines and recent iron deposits in deeps of the Red Sea, *Geochim. Cosmochim. Acta*, 30:341, 1966; D. Swartz and D. Arden, Jr., Geologic history of the Red Sea area, *Bull. Amer. Ass. Petrol. Geol.*, 44:1621, 1960.

Deep-sea Expedition, which in 1948 found water whose temperature, salinity, and oxygen content were slightly different from normal overlying sea water. Following this initial observation, waters of similar properties were resampled in the same area by the *Atlantis* in 1958, the *Atlantis II* in 1963, and the *Discovery* in 1963. It was the *Discovery*, however, on the second visit on Sept. 11, 1964, which first found and sampled the extremely hot brines that depart so markedly from anything else previously reported. In 1965 the *Atlantis II* returned to the same general area and measured water temperatures as high as 56°C (133°F) and salinities similar to those observed by the *Discovery*. The German oceanographic vessel *Meteor* also made temperature and salinity measurements at about the same time, using newly devised electronic instrumentation. In 1966 the *Chain* from Woods Hole Oceanographic Institution spent 3 weeks in the hot brine region of the Red Sea and compiled a detailed bathymetric map of the area

Reef

A mass or ridge of rock or rock-forming organisms in a water body, a rock trend on land or in a mine, or a rocky trend in soil. Usually the term reef means a rocky menace to navigation, within 6 fathoms of the water surface. Various kinds of calcium carbonate–secreting animals and plants create biogenic, or organic, reefs throughout the warmer seas (Figs. 1 and 2). Naturally cemented sand ridges make reefs along the coast of Brazil and elsewhere. Rocky shores of seas, lakes, and navigable rivers commonly exhibit reefs of rock types similar to those of the adjacent land, for example, the *Felsenriffe* of the Lorelei legend.

Biogenic reefs. Reefs designated as biogenic, or organic, consist of the hard parts of organisms, or of a biogenically constructed frame enclosing detrital particles; the hard parts of free-living organisms; and precipitated calcium carbonate. Most biogenic reefs are made of corals and associated organisms, but some entire reefs and important parts of others consist mainly of lime-secreting algae (Fig. 3), hydrozoans, annelids, oysters, or sponges. In nautical language a rocklike organic mass must be a menace to navigation before it can be classed as a reef. However, the term may also be accurately applied to any sizable biogenic eminence or buildup that grows, or once grew, upward from the floor of a water body, ordinarily the sea.

Coral reef. The most widespread and, volumetrically, the most important kind of biogenic reef is the coral reef, consisting of corals and associated calcium carbonate–secreting organisms. Coral reefs first attracted wide scientific attention through the accounts of Charles Darwin, who divided them into three principal types: fringing reef, barrier reef, and atoll. Darwin considered that these developed in the order named as a result of persistent and profound subsidence. Modern reef theory is more complicated but still retains subsidence as an important feature. In modern seas, coral reefs are important as hazards to navigation, as natural breakwaters surrounding boat passages and harbors, and as sites of complex life associations and high biological productivity. Fossil biogenic reefs are common reservoir rocks for oil.

Corals are exclusively marine, and the typical reef-building types are restricted to shallow warm water because of their symbiotic association with microscopic algae, known as zooxanthellae (Fig. 4). Although fossil representatives are known from Middle Ordovician time, Paleozoic coral reefs were made by organisms quite different from those responsible for the Mesozoic and Cenozoic reefs.

Kinds and origins. A fringing reef growing against the shore may, with subsidence and continued upward growth, become separated from the beach by a lagoon to become a barrier reef (Fig. 5). Continued subsidence and upgrowth can produce an atoll after all preexisting land has disappeared beneath a central lagoon, which is surrounded and defined by the peripheral atoll reef. Filling of an atoll lagoon, independent upward growth, or emergence and planation may produce a table reef,

Fig. 1.   A biogenic reef, the Ine Anchorage Reef, Arno Atoll, Marshall Islands.

Fig. 2. Coral reef at Ine Anchorage, Arno Atoll, Marshall Islands.

Fig. 3. Algal buttresses and surge channels of eastern peripheral reef at Onotoa Atoll, in the Gilbert Islands.

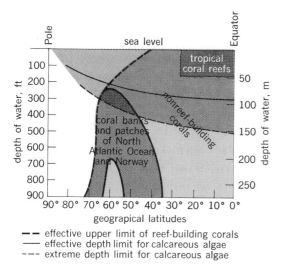

Fig. 4. The generalized depth ranges of the existing corals and calcareous algae. (*After C. Teichert, Cold- and deep-water coral banks, Amer. Ass. Petrol. Geol. Bull., 42(5):1064–1082, 1958*)

which is less common than the other major types. These are all open-sea reef types, characteristically rising from oceanic depths. Small reefs associated with these major categories or forming complex reef communities in quiet shallow waters that lack the larger reef types fall into the general categories of reef patches, pinnacles, and knolls.

Because the typical coral-reef association (Fig. 6) rarely grows with vigor below a depth of 20 m and largely dies out below 100 m, a reef frame thicker than this implies a subsidence of the sea floor or a rise of sea level. It was demonstrated by drilling that Eniwetok Atoll has subsided about 1200 m since first reef growth occurred beneath it. This finding confirms the importance of subsidence in the evolution of central Pacific atolls. Change of sea level with growth and melting of the Pleistocene ice sheets is equally important in the explanation of existing reef features. As an added elaboration, biogenic reefs may also grow upward from, or form a veneer over, existing platforms, even inheriting their general configuration from preceding topographic forms. *See* OCEANIC ISLANDS; SEA-LEVEL FLUCTUATIONS.

Geographic distribution. Existing coral reefs occur widely at tropical latitudes. They are most common in the Pacific and Indian oceans but are also found in the Atlantic Ocean and the Red Sea. They are uncommon on the western sides of continents because of the upwelling of cold water from the depths along these coasts. Coral reefs extend farther north than south and locally range well beyond the tropics because of displacement of the controlling 22°C isotherm. Some calcium carbonate buildups made by corals and brachiopods are also found in cold, relatively deep water in Scandinavian fiords and the northern Mediterranean. The corals that build these reeflike masses are of types not dependent upon the microscopic algal symbionts (zooxanthellae) of typical reef corals.

Geologic range. Coral reefs are known to range from Middle Ordovician to modern times. They became abundant during the Middle Silurian,

when their distribution included regions now within the Arctic Circle—a fact that has been explained in a variety of ways. Devonian reefs ranged across central Europe and much of central and western North America and Asia. Biogenic reefs, but not coral reefs, are prominent Permian features. A great change in the coral faunas and their associates took place near the end of Paleozoic time. No Early Triassic coral reefs or corals are known, but coral-reef associations of modern types are known from the Middle Triassic onward and locally make up much of the rock sequence.

Economic and human significance. Tens of thousands of people live on Pacific and Indian Ocean islands that are built of reef formations or have reef foundations. In the reef waters and on the islands live animals and plants that provide food, shelter, and tools to maintain life, and copra for the markets of the world. The air bases and harbors at many reef islands provide the only stopovers in broad expanses of water. Rich phosphate deposits occur on some elevated reef islands, such as Ocean, Nauru, and Angaur. The porous nature and organically rich environment of reef rock make it a potential source of petroleum.

The fringing reef and the barrier reef are discussed below. For a discussion of the third main form in Darwinian reef evolution, the atoll, *see* ATOLL.

Fringing reef. Fringing reef refers to a coral or other biogenic reef that fringes the edge of the land. A fringing reef is ordinarily divided into a steeply descending seaward front and a flat, broad or narrow pavementlike surface that is awash at low tide. Although surfaces and fronts of such reefs may show vigorous growth of algae, corals, or other lime-secreting organisms, such growth is commonly only a veneer over an erosional sea-level bench. Charles Darwin's idealized sequence of reef development begins with the fringing reef, but studies of fossil and recent reef development show that it need not precede, or be followed by, more complex reef types.

Barrier reef. Barrier reef refers to a reef, ordinarily of corals or other organisms, that parallels the shore at the seaward side of a natural lagoon. The surface may be regularly awash at low tide or may break water only at times of strongest swell.

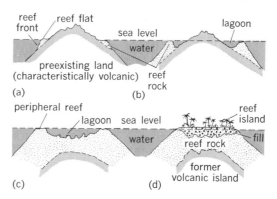

Fig. 5. Diagram showing reef sequence according to Darwin (table reef added). (a) Fringing reef. (b) Barrier reef. (c) Atoll. (d) Table reef.

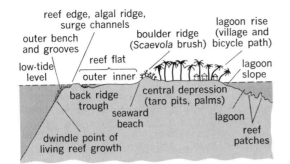

Fig. 6. Diagram of zonation of Pacific windward reef.

Ordinarily, the lagoonward slope is gentle, the seaward slope abrupt. Subsidiary reef patches are common in barrier-reef lagoons and locally beyond the ends or front of the continuous reef. Barrier reefs may develop by continuous upward growth or veneer ridges at the seaward edges of submerged erosional benches, but sinking bottom or rising water is probably essential at some stage of their development.

[PRESTON CLOUD]

Bibliography: L. Barnett, The coral reef, *Life*, The World We Live In, pt. 8, pp. 74–94, Feb. 8, 1954; L. Barnett, The mystery of coral isles, *Life*, Darwin's World of Nature, pt. 7, pp. 54–68, July 20, 1959; P. E. Cloud, Jr., Facies relationships of organic reefs, *Bull. Amer. Ass. Petrol. Geol.*, 36: 2125–2149, 1952; P. E. Cloud, Jr., Nature and origin of atolls, *Proc. 8th Pac. Sci. Congr. Pac. Sci. Ass.*, 3-A:1009–1024, 1958; R. A. Daly, Coral reefs: A review, *Amer. J. Sci.*, 246:193–270, 1948; R. W. Fairbridge et al., *Selected Bibliography on the Geology of Organic Reefs*, Int. Comm. Reef Terminol., Pac. Sci. Board, Nat. Acad. Sci.–Nat. Res. Counc. Circ. no. 3, 1958; H. S. Ladd et al., Drilling on Eniwetok Atoll, Marshall Islands, *Bull. Amer. Ass. Petrol. Geol.*, 37:2257–2280, 1953; F. S. MacNeil, The shapes of atolls: An inheritance from subaerial erosion forms, *Amer. J. Sci.*, 252: 402–427, 1954; W. E. Pugh, *Bibliography of Organic Reefs, Bioherms, and Biostromes*, Seismograph Service Corp., 1950; C. Teichert, Cold and deep-water coral banks, *Bull. Amer. Ass. Petrol. Geol.*, 42:1064–1082, 1958; J. I. Tracey, Jr., *Natural History of Ifaluk Atoll: Physical Environment*, Bernice P. Bishop Mus. Bull. no. 222, 1961.

River

A water stream of natural origin which flows across the surface of a continent or island. A river is part of a river system which drains a topographically related section of land surface known as a river basin (Fig. 1). The system begins in the precipitation which falls on a rock-, soil-, or vegetation-covered surface and immediately becomes surface runoff, or eventually appears as snow and ice meltwater or underground drainage. Such a system may be divided into headwater streams, tributary streams, and the main stem. The headwaters are in springs, marshes, lakes, or small upper streams, generally in the highest relative elevation in a basin. A river ends in a mouth, where it may discharge into a major lake, a dry basin of interior drainage (playa), an inland sea, or the ocean.

Terminology. Like many words which have long been in general use, the term river is somewhat elastic in meaning. In English usage the main stem of a stream system is nearly always designated as a river, but so are all important tributaries and even many secondary tributaries. A tributary may also be known as a fork, branch, or creek, and may have the same volume of flow as other streams called rivers. Smaller headwater streams are usually creeks or brooks.

Rivers flow in channels or watercourses and develop many distinctive valley features by erosion and deposition.

Rivers may be described by the pattern of the system of which they are part and by their length, velocity, volume of discharge, and the nature of water flowing within them. Most rivers are part of a dendritic drainage pattern (Fig. 2), but some, responding to the underlying geologic structure, are in radial, annular, rectangular, or trellised (lattice-like) pattern. In some limestone regions a karst (enclosed depression) drainage may be found, with associated underground rivers. A few rivers, such as the Nile in its lower reaches, are exotic and flow for considerable distances without receiving drainage of any consequence from tributaries. Such river reaches always occur in arid regions.

Regime and flow patterns. The regime is directly dependent on the climate of the region or regions involved. It also is influenced by the size of the drainage basin funneling upon the stream; the direction of flow; the conditions of vegetative cover; and the nature of the surface geology (Fig. 3), topography, and soil conditions in the basin. Few if any streams have completely stable conditions of flow; the rule is variation from day to day,

RIVER

boundaries of secondary drainage basins

boundary of main drainage basin

Fig. 1. Maplike diagram of a drainage basin. Note that such basins are composed of a system of secondary basins.

stream
divide

Fig. 2. Cartographic diagram illustrating stream, divide, and basin patterns in a dendritic drainage system.

season to season, and year to year. Study of these variations and their causes is an important part of the science of hydrology.

In arid regions, intermittent streams are common. The flow of an intermittent stream may fluctuate markedly from nothing to flood stage within a matter of minutes if a storm of sufficient extent and intensity covers part or all of its drainage area. Normally dry channels of these streams are called arroyos or wadis.

Under more humid climatic conditions, the channels of streams of sufficient volume to be called rivers are only occasionally dry. Fluctuations of flow nonetheless are found everywhere. For example, natural flow near the mouth of the Tennessee has varied between 4500 and 500,000 ft³/sec. In middle latitudes the season of low flow is generally summer, when evaporation and transpiration within the basin are greatest. High water may come during autumn, winter, or spring, depending on temperature conditions and the time of heaviest precipitation. Storage of large volumes of water, such as snow over frozen ground, characteristically causes early spring floods in the Great Plains region of the United States when a rapid thaw takes place.

Within high latitude areas of the Northern Hemisphere, high water inevitably occurs in spring on the northward flowing rivers because melting progresses from headwater to mouth, and the flow of water released upstream is barred by ice dams remaining downstream. The rivers of Siberia are notable examples of this condition, with broad flooding over lowland plains.

In low latitude areas, on the other hand, high water is directly related to seasonal maxima of rainfall, but high altitude conditions may complicate the regime in most zones. Where there is a pronounced dry season, as on the Indian peninsula, high water occurs soon after the onset of the rainy season, when the moisture requirements of hitherto dormant vegetation are still low. In all parts of the world, altitudinal conditions may influence the regime of a stream in another manner. Where headwaters are in extensive high mountain areas with heavy winter accumulations of snow and ice, high water occurs at the season of heaviest melt, early summer or midsummer. The Columbia, Ganges, Indus, and Rhine rivers show this influence in their regimes.

Surface materials and the nature of vegetative cover also influence the regime. The more continuous the forest or grass cover, in general the more stable the volume of discharge. Soil conditions which favor easy infiltration also promote more equitable flow, as on the sands of the Atlantic and Gulf Coastal Plain of the United States.

Water qualities. Every river is an agent of erosion, as well as an agent of drainage. Many mineral materials other than water consequently are constantly in motion where a river flows. These materials are transported by water in solution, in suspension, and as bed load.

The high capacity of water as a solvent imparts many different qualities to river water as a solution. The great majority of rivers are fresh water, but a few are saline (relatively high salt content). All rivers, however, contain perceptible amounts

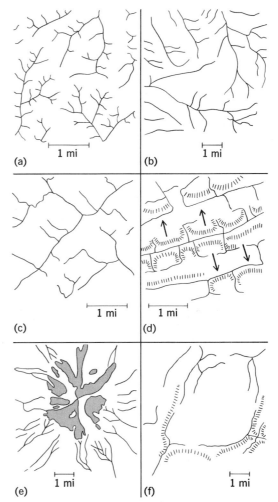

Fig. 3. Stream patterns. (a) Dendritic drainage in horizontal rocks, West Virginia. (b) Dendritic drainage in crystalline rocks, Rocky Mountains. (c) Rectangular drainage in jointed crystalline rocks, Adirondacks. (d) Trellis drainage in folded rocks, Pennsylvania. (e) Radial drainage on a volcano, Mount Hood, Ore. (f) Annular drainage on dome, Turkey Mountain, N.Mex. (After A. K. Lobeck, Geomorphology, McGraw-Hill, 1939)

of mineral material in water solution. In most cases this is calcium, the most common cause of "hard" water, but any of the elements soluble in water may be found, such as magnesium, potassium, sodium, silicon, nitrogen, and the elements which combine with them to form salts. The content of salts in solution is highest in the rivers of regions under desert or semiarid climates, but calcareous materials derived from limestone may yield hard water in humid regions.

Like most other bodies of water on the Earth's surface, a river also is a medium for the support of life, from bacteria and simple forms of plant life to fish, and amphibian, mammal, and bird wildlife. This is related not only to the capacity of water to carry nutrient minerals in solution but also dissolved gases, particularly oxygen.

Management. The characteristics of rivers have made them important to human society. No other natural feature, except the soil, has been more closely tied to the past progress of civilization for

Discharge, basin area, and length of some of the world's major rivers*

River	Avg. discharge, ft³/sec	Basin area, mi²	Length, mi
Amazon	4,000,000	2,772,000	3900
La Plata-Paraná	2,800,000	1,198,000	2450
Congo	1,400,000	1,425,000	2900
Yangtze	770,000	750,000	3100
Brahmaputra	700,000	361,000	1680
Ganges	660,000	450,000	1640
Mississippi-Missouri	620,000	1,243,000	3892
Yenisei	615,000	1,000,000	3550
Orinoco	600,000	570,000	1600
Lena	547,000	1,169,000	2860
St. Lawrence	500,000	565,000	2150
Ob	441,000	1,000,000	2800
Mekong	390,000	350,000	2600
Volga	350,000	592,000	2325
Amur	338,000	787,000	2900
Mackenzie	280,000	682,000	2525
Columbia	256,000	258,200	1214
Zambesi	250,000	513,000	2200
Danube	218,000	347,000	1725
Niger	215,000	584,000	2600
Indus	196,000	372,000	1700
Yukon	180,000	330,000	2100
Huang	116,000	400,000	2700
Nile	100,000	1,293,000	4053
Sâo Francisco	100,000	252,000	1811
Euphrates	30,000	430,000	1700

*1ft³/sec $= 2.832 \times 10^{-2}$ m³/sec; 1 mi² $= 2.590$ km²; 1 mi $= 1.609$ km.

the majority of human beings. Means of counteracting the vagaries of flow have been an important part of civil engineering for centuries. This has been true in part because of the attractiveness of floodplains to agricultural occupance, and the consequent need to avoid natural flooding. It has also followed from man's need for water storage in order to live through drought seasons. In modern times the problem of river management or river control has become much more difficult because of the rapid increase of population, its concentration in dense settlements, the extent to which manufacturing and other economic functions have encroached on floodplains, the vastly increased disposal of wastes in rivers, and the larger number of purposes that rivers must serve simultaneously. The general objects of river management are the conservation of natural flow for release at the times needed by man, the confinement of flood flow to the channel and planned areas of floodwater storage, and the maintenance of water quality at a level which will yield optimum benefit through multiple use. The techniques of river management are well understood; their practice is still very incomplete, in part because the economics of river development is not well known. Domestic river development is an important responsibility of the U. S. Army Corps of Engineers. It also is the central responsibility of the Tennessee Valley Authority and is an important objective of the Bureau of Reclamation of the Department of the Interior.

Of the major rivers in the world (see table) none is yet controlled or managed in the manner which modern engineering, administrative, and biological techniques would permit. The closest approach to such management is made on some medium-sized streams, the Tennessee, the Rhine, and the Rhône, for example. Some other rivers, such as the San Joaquin in California, have been fully developed for a single purpose, irrigation. Commencing in the 1930s, the greatest river-regulation works of all history were undertaken. The United States, the Soviet Union, and (since 1946) France have been foremost in supporting work of this kind. Among the notable achievements have been the series of great dams on the Columbia, Missouri, and Colorado and the regulation of the Tennessee in the United States; the Volga-Don Canal, the lower Volga dams, and the very large dams on the Angara and Yenisei in the Soviet Union; and the Rhône regulation in France.

The greatest and potentially most productive works remain for the future. These include plans for important work on the three largest streams of all the Amazon, the La Plata-Paraná, and the Congo. These basins contain storage and power-generation sites of several times the capacity of the largest hitherto developed. Of the eight rivers having basins of 1,000,000 mi² or more in extent, only the Mississippi and Nile have more than minor control works. Still other great streams offering major possibilities for physical development are the Yenisei, Yangtze, Huang, Amur, Mekong, Chao Phraya, Tigris-Euphrates, Niger, Zambesi, Orinoco, Sâo Francisco, Danube, Mackenzie, and Yukon. The extent and timing of such development will depend upon economic need, availability of investment funds, and political cooperation. The need is patent for development of the Yangtze, Huang, Nile, Niger, Tigris-Euphrates, Danube, Sâo Francisco, and lesser streams in densely settled, underdeveloped areas. It is therefore probable that the latter half of the 20th century will be a period of extending control of these streams, as political conditions permit.

[EDWARD A. ACKERMAN; DONALD J. PATTON]

Bibliography: C. H. Crickmay, *The Work of the River*, 1975; M. Morisawa, *Streams: Their Dynamics and Morphology*, 1968; J. N. Rayner, *Conservation, Equilibrium and Feedback Applied to Atmospheric and Fluvial Processes*, 1972; R. J. Russel, *River Plains and Sea Coasts*, 1967; S. S. Schumm, *The Fluvial System*, 1977; S. A. Schumm (ed.), *River Morphology*, 1972; B. A. Whitton, *River Ecology*, 1975.

River tides

Tides that occur in rivers emptying directly into tidal seas. These tides show three characteristic modifications of ocean tides. (1) The speed at which the tide travels upstream depends on the depth of the channel, $v \approx \sqrt{gh}$, where v is the speed, g the acceleration of gravity, and h the channel depth. (2) The further upstream, the longer the duration of the falling tide and the shorter the duration of the rising tide. (3) The range of the tide decreases with distance upstream. See TIDE.

In a river the difference between the depths of water at high and low tides may be relatively large, leading to a marked difference between the speeds at which high and low tides move. In the Hudson River the low tide (at lower high water) takes 10 min longer than the high tide (at higher high water) to reach Tarrytown, 24 nautical mi from the mouth, whereas the low tide takes 60 min longer than the high tide to reach Albany, 125 nautical mi from the mouth.

The difference in depth between various points on the river also partially explains the second modification, or duration of fall and rise. In addition, the river flow, which may fluctuate widely, helps a falling tide but hinders a rising tide, increasing the difference in duration. At the mouth of the Hudson the average fall lasts 6 hr 22 min, whereas the average rise lasts 6 hr 3 min. At Tarrytown the values are 6 hr 33 min and 5 hr 52 min, and at Albany 7 hr 21 min and 5 hr 4 min, respectively.

The third modification or decrease in tidal range upstream may be accounted for by loss of energy of the water through friction with the sides and bottom of the channel (see illustration). At the mouth of the Hudson the average tidal range is 4.4 ft, whereas at Troy, 131 nautical mi upstream, the range is 3.0 ft. Although friction always saps energy from the tide, if the channel becomes constricted within a short distance, the water may be forced into a smaller space, thus producing a

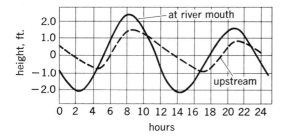

River tide curves at river mouth and upstream.

larger tidal range. For example, Bristol Channel in Great Britain is 40 mi wide at the mouth, where the average tidal range is 20 ft. Within 80 mi the channel narrows to 5 mi, at the mouth of the Avon River, where the tidal range is 33 ft. See TIDAL BORE.

Tides penetrate upstream until they encounter a dam, rapids, or falls. In the Amazon a 10-ft tide at the mouth is detectable 450 nautical miles upstream.   [BLAIR KINSMAN]

Bibliography: A. T. Ippen, *Estuary and Coastline Hydrodynamics*, 1966; C. B. Officer, *Physical Oceanography of Estuaries and Associated Coastal Waters*, 1976; A. C. Redfield, The analysis of tidal phenomena in narrow embayments, *Pap. Phys. Oceanogr. Meteorol.*, 11:1–36, 1950; J. J. Stoker, *Water Waves*, 1957.

Saline evaporite

A sedimentary deposit of soluble salts resulting from the evaporation of a standing body of water. Quantitatively the most important evaporites are anhydrite, $CaSO_4$; gypsum, $CaSO_4 \cdot 2H_2O$; and halite (rock salt), $NaCl$. Other evaporites, of much more limited distribution and volume but of economic significance, include potassium chlorides, sodium carbonates, borates, and nitrates.

Evaporites, being soluble, are rarely exposed at the surface except in arid regions. By far the greater part of the data on evaporites are derived from deep borings, chiefly those drilled in the search for oil and gas, and from underground workings developed to exploit economically valuable deposits. Gypsum, the mineral used in plaster and cement manufacture, is common near the surface but is invariably replaced by anhydrite at greater depths, the latter mineral being vastly more abundant. Other evaporite minerals are not related to depth of burial, except for the effects of near-surface solution.

Anhydrite and halite, the dominant evaporite minerals, occur in bedded deposits ranging from thin laminae to massive beds several tens of feet in thickness. These may be present in vertical succession, separated by partings of shale or carbonates, to make up aggregate thicknesses of several hundreds, or even thousands, of feet. Such large accumulations are commonly interbedded with other strata bearing marine fossils which clearly demonstrate a relationship between major evaporite deposits and marine waters. Typically, the evaporites are found in repeated cycles which approximate this vertical order from base upward: (1) marine fossil-bearing limestone; (2) dolomitized marine limestone; (3) fine-grained, finely laminated, unfossiliferous dolomite; (4) anhydrite; and (5) halite. Many cycles lack the halite member, while in a few localities a sixth member bearing potash minerals is present.

Marine basin evaporites. The geographic distribution of major evaporite deposits and the thickness and character of the other rocks with which they are associated strongly indicate a relationship between evaporites and marine sedimentary basins which tended to subside during deposition of the sediments. It is estimated that more than 95% of the volume of known evaporite deposits occur in such sedimentary basins. Two distinct patterns of

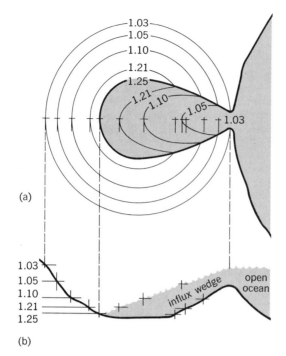

(a)

(b)

Fig. 1. Geometry of model evaporite basin, showing lines of equal brine density at stage of saline equilibrium. (a) Plan view. (b) Cross section. (*From L. I. Briggs, Evaporite facies, J. Sediment. Petrol., 28(1):46–56, 1958*)

basinal evaporite occurrence are noted: basin-center evaporites occupying the interior of sedimentary basins, and basin-margin evaporites localized in a ring at the margin of the basins.

The basin-center occurrences commonly exhibit a concentric pattern in plan view, duplicating in a

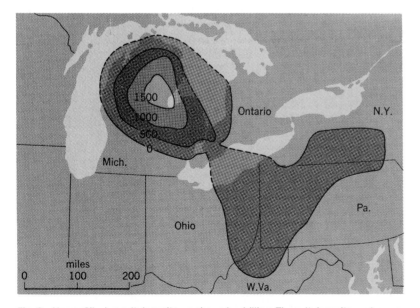

Fig. 2. Upper Silurian salt deposits, as shown by drilling. The salt deposits are known to occur throughout most of the southern peninsula of Michigan, with thickness ranging upward to 1800 ft. Silurian salt beds aggregating 325 ft or less are distributed as indicated by the map. (*Data from USGS and from Fettke, Martens, Pepper, and Alling, Baltimore and Ohio Railway Co.*)

lateral fashion the vertical succession of the evaporite cycle, with marine carbonates (limestone or dolomite) at the outside and halite at the center. The peripheral carbonate belt in many instances includes elongate trends developed parallel to the margin of the basin and forming an encircling barrier. Such a barrier may be either of two types: apparently wave-resistant trends of coral-algal reefs, or bar- or banklike trends of shell fragments, ooliths, and carbonate pellets which appear to have been built by wave action in shoaling water. Basins with extreme subsiding tendencies during evaporite deposition appear to have the greatest possibilities of containing potash salts, found with halite along the axis of maximum subsidence.

Basin-margin evaporites, usually anhydrite without large volumes of halite, occupy zones of varying width at the edge of a sedimentary basin. At greater distances from the basin, these evaporites are typically intercalated with, and pass into, red shales and siltstones. Toward the basin they tend to end relatively abruptly against trends of limestone or dolomite which may form an inner encirclement of the basin in the form of reefs, bars, or banks, as noted above. The interiors of such basins, within the encircling reefs, bars or banks, are filled with marine limestones and shales, commonly dark in color and bearing a fossil fauna lacking the remains of bottom-dwelling organisms.

Theory of origin. Both major types of evaporite occurrence give evidence of forming in partially isolated bodies of water separated from the open sea by barriers established by organic growths (reefs) or by wave action (banks and bars) which restrict and confine the circulation of the waters. In the case of basin-center evaporites, it is thought that evaporation of water in the basin produces an inflow of water from the surrounding or adjacent open sea. The inflowing water, of normal salinity and therefore of low density, flows readily as surface currents through passes in the barrier. Dense, highly saline water produced through evaporation sinks to the bottom and is unable to pass back through the barrier to mix with normal sea water (Fig. 1). Thus sea water can continue to enter; but heavy brines are confined in the basin, where they collect until the saturation points, first of the sulphates, then of the chlorides, are reached and precipitation of the salts takes place.

Basin-margin evaporites appear to represent a reversal of this pattern of openly circulating and restricted waters. In the basin-margin case the waters of the basin interior are in open communication with the sea and are of normal salinity. At the edge of the basin, confined between barriers of reefs, banks, or bars, and a surrounding land area, marginal lagoons form evaporating pans. Here water passing through the barrier from the basin is confined, and evaporation leads to the precipitation of salts, as noted before. It is apparent that basin-margin evaporites require an arid climate in the surrounding land area, since large streams of fresh water flowing into the lagoons would cancel the effects of evaporation.

Shelf evaporites. Although basinal evaporites include by far the greatest volume of these salts, there are extensive deposits which show no relationship to subsiding sedimentary basins. The lateral continuity in thickness and character of relatively thin evaporite strata and the other bedded sediments with which they occur indicates deposition in, or at the margin of, shallow shelf seas covering broad areas of slow, uniform subsidence. Shelf evaporites, because of their wide distribution and the shallow depths at which they may be encountered, include many of the commercially exploited deposits of gypsum.

Two types of shelf evaporite occurrence are recognized; both are known as extensions of basin-margin deposits and also occur at great distances from basinal influences. One type is characterized by repeated thin cycles involving interbedded marine carbonates and evaporites. The second type is intercalated with red silts and shales which may contain remains of land animals and plants. The carbonate association is believed to represent deposition in shallow, ephemeral, evaporating pans at a distance from land areas or adjoining land areas of very low relief. The redbed type is interpreted as an alternation of land-derived detritus and the deposits of a shallow bordering shelf sea.

Nonmarine evaporites. Deposits suggesting precipitation from nonmarine waters are extremely rare in ancient rocks. However, present-day lakes, many of them temporary or seasonal, in areas of arid or semiarid climate are sites of evaporite deposition.

Streams in areas such as these carry the soluble products of rock weathering to closed depressions occupied by ephemeral bodies of water, where the salts are concentrated by evaporation. Nonmarine evaporites include locally important halite deposits and the major occurrences of borates, nitrates, and sodium carbonate.

Geologic distribution. In North America Cambrian evaporites appear to be confined to the arctic areas of Canada. Ordovician strata are notably poor in thick and extensive evaporite deposits. Strata of all subsequent geologic ages include major accumulations in one or more areas of North America. Certain sedimentary basins of long-continued subsiding tendencies contain thick evaporites, representing several successive periods of geologic time (Fig. 2). Examples are the Michigan Basin (Silurian, Devonian, and Mississippian) and the Gulf Coast Basin (Permian ?, Jurassic, Cretaceous, and Tertiary).   [L. L. SLOSS]

Bibliography: H. Borchert and R. C. Muir, *Salt Deposits: The Origin, Metamorphism and Deformation of Evaporites*, 1964; W. C. Krumbein, Occurrence and lithologic associations of evaporites in the United States, *J. Sediment. Petrol.*, 21(2): 63–81, 1951; P. C. Scruton, Deposition of evaporites, *Bull. Amer. Ass. Petrol. Geol.*, 37(11):2498–2512, 1953; L. L. Sloss, The significance of evaporites, *J. Sediment. Petrol.*, 23(3):143–161, 1953.

Saltmarsh

Generally, a maritime habitat characterized by special plant communities, which occur primarily in the temperate regions of the world; however, typical saltmarsh communities can be formed in association with mangrove swamps in the tropics and subtropics. In inland areas where saline springs emerge or where there are salt lakes, typical saltmarsh communities can be found, dominated sometimes by the same species that occur on maritime saltmarsh. The extensive areas of inland salt desert, while exhibiting some features of similarity with saltmarsh, are nevertheless best regarded as a separate entity. Excess sodium chloride is the predominant environmental feature of salt marsh (maritime or inland). In the case of salt deserts sodium chloride is only one of the alkali salts that may occur in excess.

Occurrence of different types. Maritime saltmarsh can be found on stable, emerging, or sinking coastlines. On emerging or sinking coasts the actual extent of saltmarsh depends upon an adequate degree of wave protection and also upon the rate of change of coast level in relation to the rate of silt deposition. Mud or sand flats are raised by silt deposition to a level at which the characteristic phanerogamic plants can colonize. Saltmarshes therefore are common features of estuaries and protected bays, provided the seabed is shallow and does not shelve steeply. They also occur behind spits, barrier beaches, and offshore sand, shell, and shingle islands.

On emerging coastlines the true saltmarsh zone tends to be narrow, though older saltmarsh areas, recognizable by remnant saltmarsh species, have subsequently been invaded and dominated by the local terrestrial species. On sinking coastlines the extent of saltmarsh depends essentially on the rate of accretion from silt deposition in relation to rate of sinking. The greater the former in relation to the latter, the more extensive the saltmarshes are likely to be. Because of the dependence of saltmarshes upon accretion, they are likely to be best developed in association with eroding soft rock coastlines and estuaries of rivers that bring down abundant silt from soft rock upland. Work on accretion rates has shown that previously calculated time scales for marsh development have been too great.

Physiography. Saltmarsh can form on mud, muddy sand, or sandy mud, but not on pure sand because mobile sand does not provide a sufficiently stable substrate. Typical physiographic features associated with saltmarsh are the creeks, which serve as drainage channels, and pools, which are known as pans. The type of creek system depends upon the initial substrate, whether muddy or sandy, local fresh-water drainage channels, and the type of primary colonist, for example, annual species of samphire (*Salicornia*) or clumps of ricegrass (*Spartina*) (Fig. 1). Various types of pan have been recognized, including primary, secondary, and creek (cutoff ends of creeks).

Plant zonation (succession). The colonizing plants are subject to considerable tidal inundation, but with the advent of plants the rate of accretion is hastened and the land level rises. The physical factors of the environment change, there are fewer submergences, and new species invade. A characteristic feature of saltmarsh vegetation, therefore,

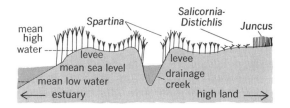

Fig. 1. Diagrammatic cross section of a marine marsh on the southern Atlantic coast of the United States; vertical exaggeration is about 10 to 1.

is the zonation of plant communities associated with changes in the environment. The zonation or succession for maritime saltmarsh is essentially dynamic since the saltmarsh is continuously, albeit slowly, building up toward the sea.

When maritime saltmarsh develops between two lateral ridges with only a narrow creek entry, the full succession may be passed through in a few hundred years, even on a subsiding coastline. The zonation of inland saltmarshes, a static zonation, is related to decreasing salinity in proportion to the distance from the source of salt.

The early stages on any saltmarsh are sufficiently rapid to be observed in 25–50 years. The final stages of the succession depend upon whether the saltmarsh borders sand dune, meadow, or fresh-water inflow. In the last case the succession continues into brackish communities and finally into fresh-water swamp.

Plant species and life forms. The plants that grow on salt marshes must tolerate the excess sodium chloride and are termed halophytes. These plants possess features associated with the halophytic environment, for example, development of succulence, waxy cuticle, and salt-excreting glands; overall there is a tendency for the vegetation to exhibit a drab grayness. The predominant life forms on saltmarshes are the hemicryptophytes and therophytes. Members of the Chenopodiaceae are common (*Salicornia*, *Suaeda*, and *Allenrolfia*) and also the Plumbaginaceae (*Limonium* or *Statice*). Among the grasses *Spartina* and *Puccinellia* are important genera in different parts of the world (Fig. 2). Other genera that are widely represented in this habitat include *Plantago*, *Triglochin*, *Cotula*, *Scirpus*, and *Juncus*.

Fig. 2. View of *Spartina* marsh on the coast of Georgia.

Inasmuch as there are characteristic phanerogams, so also there are characteristic algae associated with saltmarsh vegetation. Particular communities of green and blue-green algae are common, and in the Atlantic – North Sea area there may be extensive communities of free-living brown fucoids. Characteristic red algal communities are dominated by species of *Bostrychia* and *Catenella*. It has been shown that nitrogen fixation by blue-green algae may be significant. Work on the marine soil fungi and bacteria has been undertaken as well as on soil microfauna, but information on these groups is still far from adequate.

As the number of species capable of growing under the specialized conditions is limited, the type of succession tends to be similar for major areas. The following groups have been proposed for maritime saltmarshes: Arctic, European (with four subdivisions), Mediterranean, Atlantic North American (with three subdivisions), South American, Pacific North American, Japanese, and Australasian (with two subdivisions), each with a characteristic succession. Six groups have also been proposed for inland saltmarshes. These are Inland European (with two subdivisions), Inland Asia (with three subdivisions), Inland African (with three subdivisions), North American–Mexican (with five subdivisions), South American, and Australasian. The saltmarsh communities can be well distinguished by using the Montpellier system of classification; and on this basis 12 orders have been recognized.

Environment. The principal feature of the environment is the excess sodium chloride, with the resulting effect of the sodium ion upon the soil colloids and the chloride ion upon plant metabolism. Other specific ions may, in places, be important, and the ratio of chloride to sulfate can be significant. The phosphate ion appears to vary with season. So far as ion uptake by plants is concerned, two mechanisms appear to operate depending on whether there is a low or high salt concentration.

The frequency of tidal inundations is very important, with its effect upon salinity, as is also the water table and soil aeration. At lower marsh levels the existence of an aerated layer seems essential for the growth of the plants, and its absence may inhibit plant colonization. A lowering of the soil salinity at some season also appears necessary for successful seed germination, though temperature can also be involved. In respect to seed germination, the halophytes can be classed into three groups: the glycophyte group, where low salinity is essential; euryhalophyte group, with a wide salinity tolerance; and stenohalophytes, with a narrow salinity tolerance.

Productivity. From the scanty data available, it appears that wild saltmarsh is highly productive biologically and compares favorably with fresh-water reed swamps. Less than 10% of the net production appears to be consumed by grazing herbivores, most of it going into the detritus path of energy flow. For this reason, wild saltmarsh dominated by grasses forms excellent grazing for stock, except in spring tide periods. An evaluation of salt marshes has provided the estimates shown in the table.

Dollar value for various saltmarsh usages

	Annual return/acre	Capital value/acre
Fish and fish food	$ 100	$ 2,000
Oyster culture (maximum)	900	18,000
Sewage effluent treatment	2500	50,000
Life support	4150	83,000

Reclamation. In many parts of the world, salt-marshes have been enclosed by seawalls and converted to valuable agricultural land. About 25% of the saltmarsh on the eastern seaboard of Canada and the United States has been thus treated. In view of the importance of saltmarsh as the primary production agent for the coastal fisheries, concern has been rising over reclamations, and careful thought must be given before reclamation projects are approved. Thermal cooling water from power stations can inflict damage on the vegetation, as may industrial effluents. The use of saltmarsh for disposal of garbage should no longer be tolerated. Finally, before a reclamation is approved, very careful consideration needs to be given to the soil properties, the environmental conditions, and the economic return.   [VALENTINE J. CHAPMAN]

Bibliography: V. J. Chapman, *Coastal Vegetation*, 2d ed., 1976; V. J. Chapman, *Salt Marshes and Salt Deserts of the World*, 2d suppl. reprint ed., 1974.

Sargasso Sea

A region of the North Atlantic Ocean. The boundaries of the Sargasso Sea are clearly defined in the west and north by the Gulf Stream. In the east it extends to 40°W and in the south to 20°N, though these boundaries are less definite (Fig. 1).

The Sargasso Sea gets its name from the indigenous, yellow-brown, floating seaweed *Sargassum*, which is found throughout the sea. The floating masses of *Sargassum* contrast vividly with the deep blue of the water. It has been estimated by A. E. Parr that there are 4,000,000–11,000,000 tons of this weed floating in the Sargasso Sea and its environs (Fig. 2).

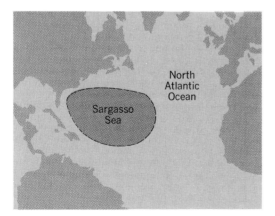

Fig. 1. Location of Sargasso Sea.

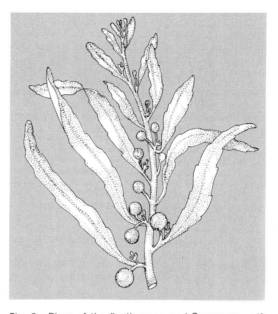

Fig. 2. Piece of the floating seaweed *Sargassum*, gulf weed, with stemlike stipe, leaflike blades, and berrylike bladders, or floats. (*From H. J. Fuller and O. Tippo, College Botany, rev. ed., Holt, 1954*)

Dynamically, the Sargasso Sea is a high cell (clockwise gyre), and the currents revolve anticyclonically about its center. Much of its deep circulation is involved with the Gulf Stream, to which it contributes a volume of about 45,000,000 m³/sec between the Straits of Florida and Cape Hatteras. It is not yet clear how this water is recirculated. *See* OCEAN CURRENTS.

At the surface the contribution is small, and sometimes (especially in the summer months) it is reversed. In consequence the upper layers of the Sargasso Sea have a closed circulation. Water is cooled in the northern end of the sea to 18°C and mixes to a depth of about 350 m at the end of each winter. The excess quantities of this 18°C water flow off to the south, and a distinct wedge of water with a temperature of 18°C can be found throughout the Sargasso Sea at a depth of 300 m. *See* ATLANTIC OCEAN; GULF STREAM.

[L. VALENTINE WORTHINGTON]

Bibliography: M. N. Hill (ed.), The composition of sea water, *The Sea*, vol. 2, 1962; A. E. Parr, Quantitative observations on the pelagic *Sargassum* vegetation of the western North Atlantic, *Bull. Bingham Oceanogr. Collect.*, 6(7):1–94, 1939; J. Teal and M. Teal, *Sargasso Sea*, 1975; L. V. Worthington, The 18° water in the Sargasso Sea, *Deep-Sea Res.*, 5(4):297–305, 1959.

Scattering layer

A layer of organisms in the sea which causes sound to scatter and returns echoes. Recordings by sonic devices of echoes from sound scatterers indicate that the scattering organisms are arranged in approximately horizontal layers in the water, usually well above the bottom. The layers are found in both shallow and deep water.

Shallow water. In the shallow water of the continental shelves (less than 200 m deep), scattering

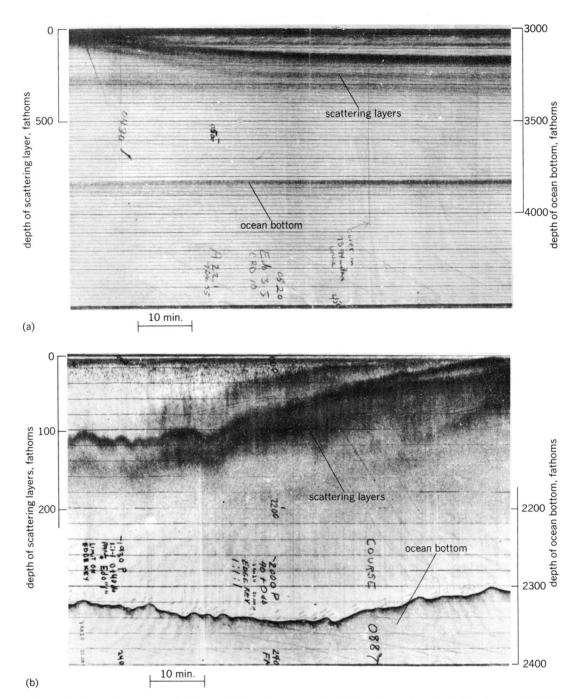

Fig. 1. Scattering layers recorded by a 12-kHz echo sounder. (a) Sunrise descent of deep scattering layers in eastern Pacific off northern Chile. A layer, which appears to have remained at depth throughout the night, is shown near 300 fathoms. (b) Sunset ascent of deep scattering layers in western North Atlantic near 40°30′N, 50°W. (From R. H. Backus and J. B. Hersey, Sound scattering by marine organisms, in M. N. Hill, ed., The Sea, vol. 1, Interscience, 1962)

layers and echoes from individuals or compact groups are very irregularly distributed and are probably made up of a variety of sea animals and possibly plants. While some fishes, notably herring, and some zooplankton have been identified by fishing, many others have not been identified.

Deep scattering layers. In deep water (greater than 200 m) one or more well-defined layers generally are present. Though commonly variable in detail, they are found to be very widely distributed.

They are readily detected by echo-sounding equipment operating in the frequency range 3–60 kHz, the sound spectrum of each layer generally having maximum scattering at a somewhat different frequency than others found at the same place. Commonly, but not universally, the deep-water layers migrate vertically in apparent response to changes in natural illumination. The most pronounced migration follows a diurnal cycle, the layers rising at night, sometimes to the surface, and descending to greater depths during the day (Fig. 1). The common range of daytime depths is 200–800 m. The migration is modified by moonlight and has been observed to be modified during the day by heavy local cloud cover, for example, a squall. Occurence of the layers in deep water was first demonstrated by C. Eyring, R. Christiansen, and R. Riatt. *See* ECHO SOUNDER; UNDERWATER SOUND.

Deep scattering organisms. All animals and plants, as well as nonliving detritus, contrast acoustically with sea water and, hence, any may be responsible for observed scattering in a particular instance. Many animals and plants have as part of their natural structure a gas-filled flotation organ which scatters sound many times more strongly than would be inferred from the sound energy they intercept. These are the strongest scatterers of their size. M. W. Johnson pointed out in 1946 that the layers which migrate diurnally must be animals that are capable of swimming to change their depth, rather than plant life or some physical boundary such as an abrupt temperature change in the water. In 1953 V. C. Anderson demonstrated that some of the deep-water scatterers have a much smaller acoustical impedance than sea water. This fact fits the suggestion by N. B. Marshall in 1951 that the scatterers may be small fishes with gas-filled swim bladders, many of which are known

to be geographically distributed much as the layers are. In 1954 J. B. Hersey and R. H. Backus found that the principal layers in several localities migrate in frequency of peak response while migrating in depth, thus indicating that the majority of scatterers fit Marshall's suggestion (Fig. 2). In deep water, corroboration of Marshall's suggestion has come from several independent combined acoustical and visual observations made during dives of the deep submersibles *Trieste, Alvin, Soucoupe, Deep Star,* and others. Nearly all these dives were made within a few hundred miles of the east or west coasts of the continental United States. The fishes observed most commonly to form scattering layers are the lantern fishes, or myctophids, planktonic fishes that are a few inches long and possess a small swim bladder. The siphonophores (jellyfish) also have gas-filled floats and have been correlated with scattering layers by observation from submersibles. It is not clear whether fishes with swim bladders and siphonophores make up the principal constituents of the deep-water scattering layers in deep-ocean areas, but it is nearly certain that they are not exclusively responsible for all the widely observed scattering. *See* DEEP-SEA FAUNA.

[JOHN B. HERSEY]

Bibliography: R. H. Backus and J. B. Hersey, Sound scattering by marine organisms, in M. N. Hill (ed.), *The Sea*, vol. 1, 1962; N. B. Marshall, Bathypelagic fishes as sound scatterers in the ocean, *J. Mar. Res.*, 10:1–17, 1951.

Sea

The term sea has several meanings: (1) the ocean; (2) a major subdivision of an ocean, (3) a lake lacking an outlet to the ocean, therefore usually salty; and (4) ocean waves still under the influence of the wind that produced them, or a single such wave. *See* OCEAN WAVES; OCEANS AND SEAS; SEA STATE.   [JOHN LYMAN]

Sea ice

Ice formed by the freezing of sea water is referred to as sea ice. Ice in the sea includes sea ice, river ice, and land ice. Land ice is principally icebergs which are prominent in some areas, such as the Ross Sea and Baffin Bay. River ice is carried into the sea during spring breakup and is important only near river mouths. The greatest part, probably

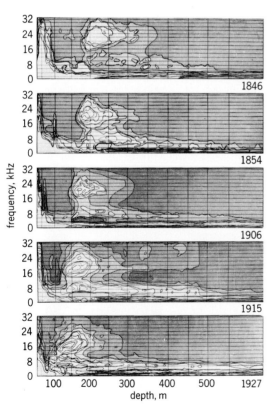

Fig. 2. Sequence of sunset observations showing scattering as a function of depth and frequency. Contours of equal sound level are 2 dB apart, with lightest areas denoting highest levels. Time of day of each observation is indicated by the number at the lower right-hand corner of each record. (*From R. H. Backus and J. B. Hersey, Sound scattering in marine organisms, in M. N. Hill, ed., The Sea, vol. 1, Interscience, 1962*)

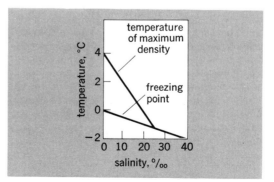

Fig. 1. Change of freezing point and temperature of maximum density with varying salinity of sea water.

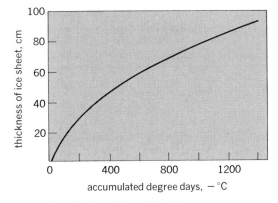

Fig. 2. Growth of undisturbed ice sheet.

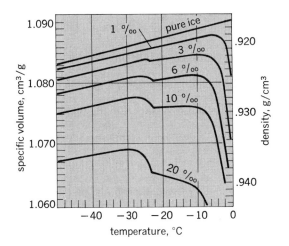

Fig. 3. Specific volume of sea ice for varying salinity and temperature, computed on basis of chemical model. (*By D. L. Anderson, based on data in Arctic Sea Ice, NAS-NRC Publ. no. 598, 1958*)

99% of ice in the sea, is sea ice. *See* ICEBERG; TERRESTRIAL FROZEN WATER.

Properties. The freezing point temperature and the temperature of maximum density of sea water vary with salinity (Fig. 1). When freezing occurs, small flat plates of pure ice freeze out of solution to form a network which entraps brine in layers of cells. As the temperature decreases more water freezes out of the brine cells, further concentrating the remaining brine so that the freezing point of the brine equals the temperature of the surrounding pure ice structure. The brine is a complex solution of many ions. With lowering of temperature below −8°C, sodium sulfate decahydrate ($Na_2SO_4 \cdot 10H_2O$) and calcium sulfate dihydrate ($CaSO_4 \cdot 2H_2O$) are precipitated. Beginning at −24°C, sodium chloride dihydrate ($NaCl \cdot 2H_2O$) is precipitated, followed by precipitation of potassium chloride (KCl) and magnesium chloride dodecahydrate ($MgCl \cdot 12H_2O$) at −34°C, and the remaining ions with further lowering of temperature.

The brine cells migrate and change size with changes in temperature and pressure. The general downward migration of brine cells through the ice sheet leads to freshening of the top layers to near zero salinity by late summer. During winter the top surface temperature closely follows the air temperature, whereas the temperature of the underside remains at freezing point, corresponding to the salinity of water in contact. Heat flux up through the ice permits freezing at the underside. In summer freezing can also take place under sea ice in regions where complete melting does not occur. Surface melt water (temperature 0°C) runs down through cracks in the ice to spread out underneath and contact the still cold ice masses and underlying colder sea water. Soft slush ice forms with large cells of entrapped sea water which then solidifies the following winter.

The salinity of recently formed sea ice depends on rate of freezing; thus sea ice formed at −10°C (14°F) has a salinity from 4 to 6 parts per thousand (°/oo), whereas that formed at −40°C may have a salinity from 10 to 15°/oo. Sea ice is a poor conductor of heat and the rate of ice formation drops appreciably after 4–6 in. are formed. An undisturbed sheet grows in relation to accumulated degree-days of frost. Figure 2 shows an empirical relation between ice thickness and the sum of the mean diurnal negative air temperature (degrees Celsius). The thermal conductivity varies greatly with air bubble content, perhaps between 1.5 and 5.0 × 10^{-3} cal/(cm)(sec)(°C).

The specific gravity of sea ice varies between 0.85 and 0.95, depending on the amount of entrapped air bubbles. The specific heat varies greatly because changing temperature involves freezing or melting of ice. Near 0°C, amounts that freeze or melt at slight change of temperature are large and "specific heat" is anomalous. At low temperatures the value approaches that of pure ice; thus, specific heat for 4°/oo saline ice is 4.6 cal/(g) (°C) at −2°C and 0.6 at −14°C; for 8°/oo saline ice, 8.8 at −2°C and 0.6 at −14°C.

Sea ice of high salinity may expand when cooled because further freezing out occurs with attendant increase of specific volume, for example, ice of salinity 8°/oo at −2°C expands at a rate of about 93 × 10^{-4} cm³/g per degree Celsius decrease in temperature, at −14°C expands 0.1 × 10^{-4}, but at

SEA ICE

Fig. 4. Pancake ice interspersed with blocks of young ice. (*U.S. Naval Oceanographic Office*)

Fig. 5. Honeycombed structure of an overturned rotten block. (*U.S. Naval Oceanographic Office*)

Fig. 6. Hummocky floes that have weathered. (*U.S. Naval Oceanographic Office*)

−20°C contracts 0.4×10^{-4} per degree Celsius decrease. Change of specific volume with temperature and salinity is illustrated in Fig. 3.

Sea ice is viscoelastic. Its brine content, which is very sensitive to temperature and to air bubble content, causes the elasticity to vary widely. Young's modulus measured by dynamic methods varies from 5.5×10^{10} dynes/cm² during autumn freezing to 7.3×10^{10} in winter to 3×10^{10} at spring breakup. Static tests give much smaller values, as low as 0.2×10^{10}. The flexural strength varies between 0.5 and 17.3 kg/cm² over salinity range of $7-16°/\infty$ and temperatures −2 to −19°C. Acoustic properties are highly variable, depending principally on the size and distribution of entrapped air bubbles.

Electrical properties vary greatly with frequency because of ionic migration within the brine cells. For example, for sea ice salinity of $10°/\infty$ at −22°C, the dielectric coefficient is very large, about 10^6 at 20 Hz, and decreases with increasing frequency to about 10^3 at 10 kHz and to 10, or less, at 50 MHz. The effective electrical conductivity decreases with lowering of temperature, for example, from less than 10^{-3} mho/cm at −5°C to 10^{-6} mho/cm at −50°C (frequency 1 to 10 kHz).

Types and characteristics. The sea ice in any locality is commonly a mixture of recently formed ice, old ice which has survived one or more summers, and possibly old ridges of ice that formed against a coast and contain beach material. The various descriptive forms are shown in Figs. 4–7. Except in sheltered bays, sea ice is continually in motion because of wind and current. The weaker parts of the sea ice canopy break when overstressed in tension, compression, or shear, pulling

Fig. 7. Unweathered pressure ridges, formed by rafting of floes. (*U.S. Naval Oceanographic Office*)

apart to form a lead (open water), or piling block on block to form a pressure ridge. Depending on the composition of the ice canopy, the ridges may form in ice of any thickness, from thin sheets (10 cm thick) to heavy blocks (3 m or more in thickness). The ridges may pile 13 m high above and extend 50 m, or more, below the sea surface. Massive ridges become grounded in coastal zones, further producing disruptive forces within the ice canopy. See ICE ISLAND.

[WALDO LYON]

Bibliography: T. Karlsson (ed.), *Sea Ice*, 1972; W. D. Kingery (ed.), *Ice and Snow: Properties, Processes, and Applications*, 1963; National Academy of Sciences, Division of Earth Sciences, *Beneficial Modifications of the Marine Environment*, 1972; *Proceedings of the Conference on Arctic Sea Ice*, NAS–NRC Publ. no. 598, 1958; J. C. Reed and J. E. Sater (eds.), *The Coast and Shelf of the Beaufort Sea*, 1974; World Meteorological Organization, *Sea-Ice Nomenclature*, 1971.

Sea-level datum planes

Sea level is the elevation of the sea surface measured as the vertical distance between the surface and some fixed point on land—a rocky outcrop on a beach, a mountain peak, or a reference point installed by man.

Mean sea level. Mean sea level is the average elevation and is frequently used as a reference level in describing the elevation of points on land, or of depths in the sea. The surface of the sea is by no means a stationary spheroidal surface, so that the accurate determination of its average elevation requires a long series of observations. Usually a recording tide gage is installed, and the records are read each hour for several weeks, months, or years. The recommended length of record, to eliminate as nearly as possible all tidal constituents, is 19 years. However, shorter series are frequently used; the value so obtained can often be adjusted to a more representative value of mean sea level by comparison with the records at a nearby gage (where sea level is accurately known) for the same period of time. Such a mean value, computed for observations for a stated period in time, has been adopted for United States surveying and is called

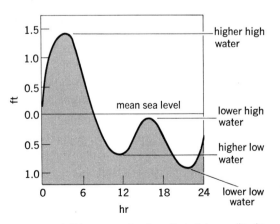

Diagram of tide curve showing diurnal inequality in heights of high waters and low waters.

sea-level datum. See SEA-LEVEL FLUCTUATIONS.

If hourly values cannot be obtained, selected hourly values throughout each day may be used, with some loss of accuracy. On the other hand, if precise daily means are to be determined, the hourly values must be multiplied by weighting factors determined from the known periods of the tidal constituents.

Other datum planes. Other datum planes are used for special purposes. Half-tide level is the average of all the highest and lowest readings during the period (called high and low waters). This datum was in early use before automatically recording gages were generally distributed, and at many localities only the high and low waters were recorded by an observer. Mean low water and mean high water are the averages of all of the low waters and high waters, respectively; half-tide level lies midway between these two means.

When there is a large diurnal inequality in the tide (see illustration), mean lower low water is frequently used as the datum on navigational charts. This is the average of only the lower of the two daily low waters, which makes it a safer reference to operators of vessels in tidal waters. Similarly, mean higher high water is the average of the higher of the two high waters each day and is a more realistic estimate of the daily excursion of water up a beach where the highs are quite different in elevation. Other datum planes with features useful for the particular area are in use. H. A. Marmer (1951) gave a detailed treatment of the definitions, determination, and usage of datum planes. See TIDE.

[JUNE G. PATTULLO]

Bibliography: G. W. Groves, Numerical filters for discrimination against tidal periodicities, *Trans., Amer. Geophys. Union*, 36(6):1073–1084, 1955; H. A. Marmer, *Tidal Datum Planes*, 5th ed., USCGS Spec. Publ. no. 135, 1951; G. L. Picard, *Descriptive Physical Oceanography*, 2d ed., 1976; J. R. Rossiter, Note on methods of determining monthly and annual values of mean water level, *Int. Hydrogr. Rev.*, May, 1958.

Sea-level fluctuations

The sea-level fluctuations first observed by man were surely waves; next he must have noted the rise and fall of the tides. But more concentrated attention (and carefully made records) reveal that the height of sea level is constantly changing and that its average value (mean sea level) will depend critically on the length of the series of observations and the period in time during which they are made. Furthermore, the sea surface is not even level. See SEA-LEVEL DATUM PLANES.

Fluctuations in space. Concrete evidence that the sea surface is not perpendicular to a plumb line has been found by precision leveling along the Atlantic Coast of the United States (Fig. 1).

Effect of Coriolis acceleration. Sea level stands about 30 cm higher along the coast of Maine than it does off Florida, and the steepest slope occurs off Cape Hatteras, N.C. This occurs because the water in the sea is always in motion. The motion of any fluid over the surface of the Earth is influenced by the rotation of the Earth itself; fluids are deflected to the right (looking downstream) in the Northern Hemisphere, to the left in the Southern Hemisphere. This effect results from the Coriolis acceleration. *See* CORIOLIS ACCELERATION AND FORCE.

Offshore from the eastern coast of the United States the Gulf Stream flows northward, but inshore, on the continental shelf, flow is to the south. The water is then deflected to the west and piles up along the coast. More water piles up along the New England and Middle Atlantic states where southward flow is greatest; less, south of Cape Hatteras where the northward-flowing Gulf Stream lies closer to shore.

This effect is not confined to continental coasts. Each of the major oceans contains at least one current gyre where the water flows more or less continually around in a circle. In the Northern Hemisphere, if the flow is clockwise, water will collect in the center of the gyre until the deflecting force of the Earth's rotation is just balanced by the gravitational force urging the water to run downhill out of the center of the gyre. On the other hand, if the flow is counterclockwise there will be a lowering of sea level in the center of the gyre. *See* OCEAN CURRENTS.

In the North Atlantic, for example, sea level must stand higher near Bermuda, which lies near the center of a clockwise gyre, than it does southeast of Iceland, where the water is revolving in a counterclockwise flow. It is not known exactly how large this difference in elevation is because the precision leveling network has not yet been extended out over the open ocean. From oceanographic measurements the difference is estimated as between 1 and 2 m.

If the Coriolis effect were the only important factor, then the sea surface could be visualized as a nearly smooth surface approximately perpendicular everywhere to a plumb line, but with rises of a meter or less near Bermuda, off Japan, and also in middle latitudes of the Southern Hemisphere. There would be dips in the elevation of the sea surface off Iceland, in the Gulf of Alaska, and near the Antarctic Continent, which is girdled by a great clockwise flow.

Effect of atmospheric pressure. Sea level also deviates from the mean, however, under the influence of differential atmospheric pressure. If pressure is high over one area of the ocean and low somewhere else, the water will tend to flow toward the low-pressure area. Thus the sea surface behaves as an inverted barometer; it stands high where air pressure is low, and low where air pressure is high. *See* AIR PRESSURE.

A difference in air pressure of 1 millibar (mb) corresponds very nearly to a difference in sea level of 1 cm. Because of the average distribution of air pressure, this effect tends to reduce the current-induced differences in elevation just considered. That is, off Iceland the air pressure is low. Near Bermuda and the Azores, air pressure is high. The average difference is about 20 mb so that if this effect alone were acting on the sea its surface level would be 20 cm higher off Iceland than near Bermuda and the Azores. The actual difference in level must show the combined effects of the forces related to the relative motions of the Earth and the ocean waters (Coriolis effect) and the force resulting from differences in atmospheric pressure from place to place.

Barriers and density differences. A third effect can be observed where parts of the sea surface are separated from each other by a land barrier. Precise leveling across the continent of North America shows that sea level is about 50 cm higher on the Pacific than on the Atlantic side of the continent. Currents and winds probably cause part of this difference in level, but another important factor is the difference in the density of the sea water.

Water, including sea water, expands and contracts with changing temperature. In the sea the effect is complicated by the presence of the salt, since a mass of water also takes up more or less room as the relative salt content is changed. Furthermore, the two effects are not completely independent, but as treated here they may be considered so. Thus, when sea water is cold and salty it will take up less space per unit mass, and hence stand at a lower level, than when it is warm and fresh. *See* SEA WATER.

Let it be assumed that somewhere very deep in the oceans all isobaric surfaces (surfaces of constant pressure) are level. This is equivalent to saying that any two water columns that are of the same area and contain the same mass of water will have their bases on a common level surface. This is nearly true wherever the columns may be or whatever the temperature and salinity distributions may be. However, if the water in one column is less dense than in the other, its surface will stand higher.

Fluctuations in time. The height of sea level at a given point changes from second to second, hour to hour, even century to century. The slowest regular changes in sea level that are large enough to be readily observed by eye have periods of approximately 12 and 24 hr. *See* TIDE.

Short-period variations. Less regular than tides are fluctuations where the period is several days to a week or two. These are caused by the moving high- and low-pressure systems in the atmosphere; they are another example of the sea in its role as an inverted barometer. As a low-pressure area moves in over the coast, sea level rises; as the low pressure moves inland and a high pressure overlies the coast, sea level falls again. The area of sea surface that rises and falls is controlled by the size of the meteorological disturbance and has a diameter, usually, of a few hundred kilometers. The height of

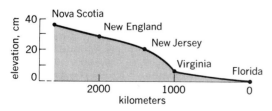

Fig. 1. Elevation of mean sea level from north to south along the eastern coast of North America. Heights refer to the elevation at Florida. (*Based on data from H. U. Sverdrup et al., The Oceans, Prentice-Hall, 1942*)

the variation in level is dependent on the pressure difference in the atmosphere and may be 10 or 20 cm—sometimes greater.

Occasionally very large and destructive rises in sea level that last for one or several days occur in connection with severe atmospheric storms. It has been estimated that more than three-fourths of the loss of life that has resulted in hurricanes has been caused by the inundations from storm waves, rather than by the direct effects of the high winds. These storm waves are partly induced by the low central pressures in the eye of the storm and are partly the result of water being driven onshore (especially over shoaling bottom areas) by the winds. *See* STORM SURGE.

Annual variations. A more strictly periodic variation, although not nearly so striking, is the fluctuation which has an annual period. Sea level is highest in autumn, lowest in spring, over all middle latitude areas of the sea. This means, of course, that when sea level is high along the coast of the United States, Europe, and Japan (northern autumn, southern spring), it is low along South America, South Africa, and Australia.

The average annual change in level is 20 cm, although it varies from almost zero at some equatorial islands to more than 1 m along the north shore of the Bay of Bengal. The oscillation seems to be smaller at mid-ocean near offshore islands (in the open ocean) than at continental coasts, but it has the same phase in either case. That is, in September sea level is high all across the Atlantic Ocean from the United States to Europe, between latitudes 20 to 40°N. At the same time it is high in the same general band of latitude across the North Pacific, but low in the middle latitudes of the Southern Hemisphere.

North of 40°N a similar oscillation of about equal amplitude occurs, but not at the same time. That is, the maximum heights are observed in winter and the lowest levels in summer. Data in the Southern Hemisphere are inadequate for determining whether a similar change of time of maximum occurs at high southern latitudes. The variations in the two zones of the Northern Hemisphere are illustrated in Fig. 2.

There are as yet insufficient data on ocean currents to compute the annual variation in the effect of Coriolis acceleration except at one or two locations. Data from the United States eastern coast and Bermuda indicate the changing speed of the Gulf Stream is probably closely related to differences in the slope between these two points, but this effect cannot explain why the level should rise at the coast and at Bermuda at the same time.

In most locations annual variations in atmospheric pressure are not large enough to account for the annual change in sea level as an inverted barometer effect. (One must be careful to exclude from this computation annual changes in the total air mass over all the oceans of the world. Sea water is almost incompressible, and only relative changes from one part of the sea to another can cause appreciable changes in sea-surface elevation. The known migration of air mass to continents during winter and back over the sea during summer must be excluded.) Even near Japan and Iceland, where pressure changes throughout the year are large, only about one-third of the sea-level change can be attributed to the effects of atmospheric pressure. In many parts of the world, indeed, the air pressure changes should cause the sea surface to move up when it is moving down, and vice versa.

In middle latitudes the density of sea water changes measurably throughout the year. At the end of summer the upper 400 m of the sea water has become several degrees warmer than it is at the end of winter. This warmer water stands some 10–20 cm higher than the same mass of cool water does in winter. This effect accounts for about two-thirds of the measured rise in sea level in middle latitudes.

In the more northerly parts of the northern oceans (where the maximum height occurs 3 months later than in middle latitudes), the seasonal changes in ocean temperature are neither large enough nor at the right time to explain the observed variations.

Seasonal changes in salinity usually have a much smaller effect than the changes in temperature. One area where this statement does not hold is in the Bay of Bengal, where the largest seasonal changes in sea level are observed. Here the summer monsoon is accompanied by very heavy rainfall and onshore winds. These effects combine to form a lens of low-salinity water near the coast in summer which is replaced by cool, high-salinity ocean water in winter. Most of the observed changes in height are due to the low salinity of this fresh-water lens, although its high temperature and the onshore winds that prevail at that season are probably also factors.

A fourth possibility must be considered. In discussing salinity above, it was tacitly assumed that the total mass of water plus salt in the oceans remains constant throughout the year. Certainly if a particular area is considered, this may not be true. After all, changes in salinity are largely brought about by seasonal changes in evaporation or precipitation, such as the removal or addition of fresh

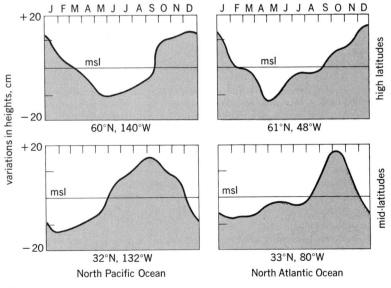

Fig. 2.   Seasonal variations in sea level.

water near the surface; this may not be completely compensated by internal flow.

Furthermore, there is some slight evidence that even the total mass of fresh water in the oceans does not remain constant throughout the year. Both oceanographic evidence and hydrological estimates suggest that there is less water in the oceans at the end of Northern Hemisphere winter in March.

The data indicate that this water is held on the large continents in the form of snow and ground water and gradually returned to the sea during summer months. A similar effect must occur in the Southern Hemisphere, but the land masses in cool regions of the Southern Hemisphere are not nearly so large and therefore not so effective in storing moisture during the Southern Hemisphere winter season.

The amount of water involved is very great, but spread over all the oceans of the world it would change the average height of sea level by only 1 or 2 cm. Because this change in level is so small, measurement from the sea-level data cannot be considered reliable; however, the oceanographic and hydrological estimates agree in both magnitude and phase.

The problem of the causes of the annual variation in sea level cannot be considered solved. As a case in point: If sea level is high in the Northern Hemisphere while it is low in the Southern, what is the force acting to maintain this slope?

Long-period variations. Still slower changes in sea level have been observed, although these become more difficult to discuss quantitatively because continuous observations are not yet available over all oceans.

Long records from recording gages exist only in Europe, the United States, Japan, and Indonesia. These show that sea level may rise more or less continuously for several years, rising perhaps as much as 1 m, and then suddenly or slowly subside again.

Another form of evidence, ancient beach lines, suggests that in earlier centuries sea level was several, perhaps tens of meters, higher than it is today. The causes for these changes cannot be explained until there is more complete information on their geographical extent, when they took place, and how long the changes persisted. For example, slow changes in sea-level elevation may have been caused, and may again be caused, by the melting or refreezing of the gigantic Antarctic and Greenland icecaps, which contain enough water to raise the sea level by several tens of meters. *See* GLACIOLOGY; HYDROLOGY.

[JUNE G. PATTULLO]

Sea state

The description of the ocean surface or state of the sea surface with regard to wave action. Wind waves in the sea are of two types: Those still growing under the force of the wind are called sea: those no longer under the influence of the wind that produced them are called swell. Differences between the two types are important in forecasting ocean wave conditions. Properties of sea and swell and their influence upon sea state are described in this article.

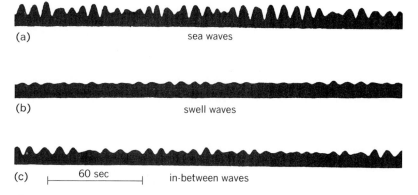

(a)   sea waves

(b)   swell waves

(c) |— 60 sec —| in-between waves

Fig. 1.   Records of surface waves. (*a*) Sea, (*b*) swell, and (*c*) in-between waves. (*Adapted from W. J. Pierson, Jr., et al., Observing and Forecasting Ocean Waves, H.O. Publ. no. 603, U.S. Navy Hydrographic Office, 1955*)

Sea. Those waves which are still growing under the force of the wind have irregular, chaotic, and unpredictable forms (Fig. 1*a*). The unconnected wave crests are only two to three times as long as

Table 1. Sea height code*

Code	Height, ft	Description of sea surface
0	0	Calm, with mirror-smooth surface
1	0–1	Smooth, with small wavelets or ripples with appearance of scales but without crests
2	1–3	Slight, with short pronounced waves or small rollers; crests have glassy appearance
3	3–5	Moderate, with waves or large rollers; scattered whitecaps on wave crests
4	5–8	Rough, with waves with frequent whitecaps; chance of some spray
5	8–12	Very rough, with waves tending to heap up; continuous whitecapping; foam from whitecaps occasionally blown along by wind
6	12–20	High, with waves showing visible increase in height, with extensive whitecaps from which foam is blown in dense streaks
7	20–40	Very high, with waves heaping up with long frothy crests that are breaking continuously; amount of foam being blown from the crests causes sea surface to take on white appearance and may affect visibility
8	40+	Mountainous, with waves so high that ships close by are lost from view in the wave troughs for a time; wind carries off crests of all waves, and sea is entirely covered with dense streaks of foam; air so filled with foam and spray as to affect visibility seriously
9		Confused, with waves crossing each other from many and unpredictable directions, developing complicated interference pattern that is difficult to describe; applicable to conditions 5–8

*Modified from *Instruction Manual for Oceanographic Observations*, H.O. Publ. no. 607, 2d ed., U.S. Navy Hydrographic Office, 1955.

Table 2. Swell-condition code*

Code	Description	Height, ft	Length, ft
0	No swell	0	0
	Low swell	1–6	
1	Short or average		0–600
2	Long		600+
	Moderate swell	6–12	
3	Short		0–300
4	Average		300–600
5	Long		600+
	High swell	12+	
6	Short		0–300
7	Average		300–600
8	Long		600+
9	Confused		

*Instruction Manual for Oceanographic Observations, H.O. Publ. no. 607, 2d ed., U.S. Navy Hydrographic Office, 1955.

the distance between crests and commonly appear to be traveling in different directions, varying as much as 20° from the dominant direction. As the waves grow, they form regular series of connected troughs and crests with wave lengths commonly ranging from 12 to 35 times the wave heights. Wave heights only rarely exceed 55 ft. The appearance of the sea surface is termed state of the sea (Table 1).

The height of a sea is dependent on the strength of the wind, the duration of time the wind has blown, and the fetch (distance of sea surface over which the wind has blown). *See* OCEAN WAVES.

Swell. As sea waves move out of the generating area into a region of weaker winds, a calm, or opposing winds, their height decreases as they advance, their crests become rounded, and their surface is smoothed (Fig. 1*b*). These waves are more regular and more predictable than sea waves and, in a series, tend to show the same form or the same trend in characteristics. Wave lengths generally range from 35 to 200 times wave heights.

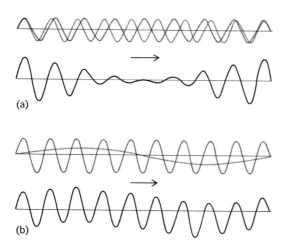

Fig. 2. Wave patterns resulting from interference. (*a*) Interference of waves of equal height and nearly equal length, forming wave groups. (*b*) Interference between short wind waves and long swell. (*From Techniques for Forecasting Wind Waves and Swell, H.O. Publ. no. 604, U.S. Navy Hydrographic Office, 1951*)

The presence of swell indicates that recently there may have been a strong wind, or even a severe storm, hundreds or thousands of miles away. Along the coast of southern California long-period waves are believed to have traveled distances greater than 5000 mi from generating areas in the South Pacific. Swell can usually be felt by the roll of a ship, and, under certain conditions, extremely long and high swells in a glassy sea may cause a ship to take solid water over its bow regularly.

A descriptive classification of swell waves is given in Table 2. When swell is obscured by sea waves, or when the components are so poorly defined that it is impossible to separate them, it is reported as confused.

In-between state. Often both sea waves and swell waves, or two or more systems of swell, are present in the same area (Fig. 1*c*). When waves of one system are superimposed upon those of another, crests may coincide with crests and accentuate wave height, or troughs may coincide with crests and cancel each other to produce flat zones (Fig. 2). This phenomenon is known as wave interference, and the wave forms produced are extremely irregular. When wave systems cross each other at a considerable angle, the apparently unrelated peaks and hollows are known as a cross sea.

Breaking waves. The action of strong winds (greater than 12 knots) sometimes causes waves in deeper water to steepen too rapidly. As the height-length ratio becomes too large, the water at the crest moves faster than the crest itself and topples forward to form whitecaps.

Breakers. As waves travel over a gradually shoaling bottom, the motion of the water is restricted and the wave train is telescoped together. The wave length decreases, and the height first decreases slightly until the water depth is about one-sixth the deep-water wave length and then rapidly increases until the crest curves over and plunges to the water surface below. Swell coming into a beach usually increases in height before breaking, but wind waves are often so steep that there is little if any increase in height before breaking. For this reason, swell that is obscured by wind waves in deeper water often defines the period of the breakers.

Surf. The zone of breakers, or surf, includes the region of white water between the outermost breaker and the waterline on the beach. If the sea is rough, it may be impossible to differentiate between the surf inshore and the whitecaps in deep water just beyond.

[NEIL A. BENFER]

Bibliography: W. Bascom, *Waves and Beaches: The Dynamics of the Ocean Surface*, 1964; H. J. McLellan, *Elements of Physical Oceanography*, 1966; G. Neumann and W. J. Pierson, Jr., *Principles of Physical Oceanography*, 1966.

Sea water

Water is most often found in nature as sea water (≅98%). The rest is found as ice, water vapor, and fresh water. Sea water is an aqueous solution of salts of a rather constant composition of elements. Its presence determines the climate and makes life possible on the Earth. The boundaries of sea water are the boundaries of the oceans, the mediterranean seas, and their embayments. The physical, chemical, biological, and geological events in the hydroplane within these boundaries are the studies which are grouped together and called oceanography. The basic properties of sea water, the distribution of these properties, the interchange of properties between sea and atmosphere or land, the transmission of energy within the sea, and the geochemical laws governing the composition of sea water and sediments are the fundamentals of oceanography.

The discussion of sea water which follows is divided into six sections: (1) physical properties of sea water; (2) interchange of properties between sea and atmosphere; (3) transmission of energy within the sea; (4) composition of sea water; (5) distribution of properties; and (6) sampling and measuring techniques. For further treatment of related aspects of physical character, composition, and constituents see HYDROSPHERIC GEOCHEMISTRY; MARINE RESOURCES; SEA-WATER FERTILITY; UNDERWATER SOUND.

PHYSICAL PROPERTIES OF SEA WATER

Sea water is basically a concentrated electrolyte solution containing many dissolved salts. The ratio of water molecules to salt molecules is about 100 to 1. Since nearly all the salt exists as electrically conducting ions, this means that the ratio of water molecules to ions is about 50 to 1; consequently, the ions are on the average no farther than about 10^{-7}cm from each other, a distance equivalent to the diameter of about five water molecules. Because pure water has a relatively open structure with tetrahedral coordination, the water molecules by virtue of their electric dipole moment (arising from the separation of the positive and negative charges) can be readily oriented or polarized by an electric field. The polarizability manifests itself in the high dielectric constant of pure water. Since electrostatic attraction between ions is inversely proportional to the dielectric constant, a high dielectric constant facilitates ionization of electrolytes because of the reduced forces between ions of opposite sign.

Salinity effects. In the neighborhood of ions, extremely high electric fields exist (around 100,000 volts/cm) and water molecules near them become aligned; water molecules that remain in the vicinity of the ions for a long time constitute a hydration shell, and the ions are said to be solvated. The alignment of water molecules produces local dielectric saturation (that is, no further alignment is possible) around the ions, thereby lowering the dielectric constant of the solution below that of pure water.

Details of ion-ion and ion-solvent interactions and their effects on the physical property of solutions are treated in the theory of electrolyte solutions.

As a consequence of the salts in sea water, its physical properties differ from those of pure water, the difference being closely proportional to the concentration of the salts or the salinity. Salinity measurements (which can be conveniently made using electrical conductivity apparatus), along with pressure and temperature data, are used to differentiate water masses. In studying the movement of water masses in the oceans and their small- and large-scale circulation patterns, including geostrophic flow, properties such as density, compressibility, thermal expansion coefficients, and specific heats need to be known as functions of temperature, pressure, and salinity.

The large value of the osmotic pressure of sea water is of great significance to biology and desalination by reverse osmosis; for example, at a salinity of 35 °/$_{\infty}$ (parts per thousand) the osmotic pressure relative to pure water is around 25 atm. Related to osmotic pressure and very important to the formation of ice is the reversal of the freezing point and temperature of maximum density of sea water compared to pure water: The freezing point

Table 1. Some physical properties of sea water (salinity, 35°/$_{\infty}$) at sea level

Property	Temperature, °C			
	0	10	20	30
Specific volume, cm³/g	0.9726	0.9738	0.9757	0.9784
Isothermal compressibility × 10⁶, bars⁻¹	46.7	44.3	42.7	41.7
Thermal coefficient of volume expansion × 10⁵, °C⁻¹	5.4	16.6	25.8	33.4
Sound speed, m/sec	1449.4	1490.1	1521.7	1545.7
Electric conductivity × 10³, (ohm-cm)⁻¹	29.04	38.10	47.92	58.35
Molecular viscosity, centipoise	1.89	1.39	1.09	0.87
Specific heat, cal/g°C	0.953	0.954	0.955	0.956
Optical index of refraction (n − 1.333,338) × 10⁶, λ = 0.5876 μ	6966	6657	6463	6337
Osmotic pressure, bars	23.4	24.3	25.1	26.0
Molecular thermal conductivity coefficient × 10³, cal/cm sec °C	1.27	1.31	1.35	1.38

Table 2. Sound attenuation coefficient α in sea water*

	Sound frequency, Hz			
	100	1000	10,000	100,000
Attenuation coefficient α, km^{-1}	0.00023	0.0069	0.15	6.3

*Depth $\sim$ 1200 m. temperature $\sim$ 4°C, and salinity = 35°/₀₀.

SOURCE: After W. H. Thorp, Deep ocean attenuation in the sub- and low-kilocycle-per-second region, *J. Acoust. Soc. Amer.*, 38:648–654, 1965.

temperature is lowered to −1.9°C, and the temperature of maximum density is decreased from just below 4°C for pure water to about −3.5°C for 35°/₀₀ salinity sea water. Some other properties which show significant changes between sea water (salinity 35°/₀₀) and pure water at atmospheric pressure are shown in Tables 1 and 2. Although the world oceans show a wide range in temperature and salinity, 75% by volume occurs within a range of 0 to 6°C and 34 to 35°/₀₀ salinity.

Pressure effects. Since the greatest ocean depths exceed 10,000 m and more than 54% of the oceans' area is at pressure above 400 bars, it is necessary to consider the effect of pressure, as well as temperature, on the physical properties of sea water. The pressure corresponding to the maximum ocean depth is about 1100 bars. Table 3 shows the percent change in some properties at 500 and 1000 bars, corresponding to depths of about 5000 and 10,000 m at two temperatures, 0 and 20°C. The unusual pressure dependence of viscosity is a consequence of the open structure of water which is altered by pressure, temperature, and solutes.

Sound absorption. Because electromagnetic radiation can propagate in the ocean for only limited distances, sound waves are the principal means of communication in this medium. The equation for attenuation of the intensity of a plane wave (without geometrical spreading losses) is given by Eq. (1) where I_0 is the initial intensity, and I is the

$$I = I_0 e^{-2\alpha x} \qquad (1)$$

intensity at a distance of x km. Sound absorption in sea water, shown in Table 2, is considerably greater than in fresh water, about 30 times greater between frequencies of 10 and 100 kHz. This arises from a pressure-dependent chemical reaction involving magnesium sulfate with a relaxation frequency around 100 kHz. Another relaxation frequency around 1 kHz has been found in the ocean; the origin of this phenomenon has not been determined.

[F. H. FISHER]

Bibliography: A. Bradshaw and K. E. Schleicher, The effect of pressure on the electrical conductance of sea water, *Deep-Sea Res.*, 12:151–162, 1965; L. A. Bromley et al., Heat capacities of sea water solutions at salinities of 1 to 12% and temperatures of 2° to 80°, *J. Chem. Eng. Data*, 12:202–206, 1967; A. Defant, *Physical Oceanography*, vol. 1, 1961; F. H. Fisher, Ion pairing of magnesium sulfate in sea water: Determined by ultrasonic absorption, *Science*, 157:823, 1967; H. W. Menard and S. M. Smith, Hypsometry of ocean basin provinces, *J. Geophys. Res.*, 71:4305–4325, 1966; R. B. Montgomery, Water characteristics of Atlantic Ocean and of world ocean, *Deep-Sea Res.*, 5:134–148, 1958; G. Neumann and W. J. Pierson, Jr., *Principles of Physical Oceanography*, 1966; W. S. Reeburgh, Measurements of electrical conductivity of sea water, *J. Mar. Res.*, 23:187–199, 1965; M. Schulkin and H. W. Marsh, Sound absorption in sea water, *J. Acoust. Soc. Amer.*, 34:864–865, 1962; E. M. Stanley and R. C. Batten, *Viscosity of Sea Water at High Pressures and Moderate Temperatures*, Nav. Ship Res. Dev. Center Rep. no. 2827, 1968; W. D. Wilson, Speed of sound in sea water as a function of temperature, pressure and salinity, *J. Acoust. Soc. Amer.*, 32:641–644, 1960; W. D. Wilson and D. S. Bradley, Specific volume of sea water as a function of temperature, pressure and salinity, *Deep-Sea Res.*, 15:355–363, 1968.

INTERCHANGE BETWEEN SEA AND ATMOSPHERE

The sea and the atmosphere are fluids in contact with one another, but in different energy states — the liquid and the gaseous. The free surface boundary between them inhibits, but by no means totally prevents, exchange of mass and energy between the two. Almost all interchanges across this boundary occur most effectively when turbulent conditions prevail: a roughened sea surface, large differences in properties between the water and the air, or an unstable air column that facilitates the transport of air volumes from sea surface to high in the atmosphere.

Heat and water vapor. Both heat and water (vapor) tend to migrate across the boundary in the direction from sea to air. Heat is exchanged by three processes: radiation, conduction, and evaporation. The largest net exchange is through evaporation, the process of transferring water from sea to air by vaporization of the water.

Evaporation depends on the difference between the partial pressure of water vapor in the air and the vapor pressure of sea water. Vapor pressure increases with temperature, and partial pressure increases with both temperature and humidity; therefore, the difference will be greatest when the sea (always saturated) is warm and the air is cool and dry. In winter, off east coasts of continents, this condition is most ideally met, and very large quantities of water are absorbed by the air. On the average, 100 g water per square centimeter of ocean surface are evaporated per year.

Since it takes nearly 600 cal to evaporate 1 g

Table 3. Percent change of sea water properties at elevated pressures

Property	500 bars pressure		1000 bars pressure	
	0°C	20°C	0°C	20°C
Electrical conductivity	6.76	3.88	11.19	6.49
Specific volume	−2.16	−1.97	−4.00	−2.28
Compressibility	−13.0	−12.0	−24.0	−22.0
Thermal expansion coefficient	224.0	19.0	386.0	34.3
Sound speed	5.85	5.66	12.02	11.08
Sound absorption coefficient		−47.0		−65.0
Molecular viscosity	−3.8	−0.3	−4.7	0.5

water, the heat lost to each square centimeter of the sea surface averages 150 cal/day. This heat is stored in the atmospheric volume but is not actually transferred to the air parcels until condensation takes place (releasing the latent heat of vaporization) perhaps 1000 mi away and 1 week later.

Radiation of heat from the water surface to the atmosphere and back again are both large—of the order of 800 cal/(cm²)(day) according to E. R. Anderson (1953). However, the net flux is out from the sea; it averages about 100 cal/day.

Conduction usually plays a much smaller role than either of the above; it may transfer heat in either direction, but usually it contributes a small net transfer from sea to air. *See* OCEAN-ATMOSPHERE RELATIONS; TERRESTRIAL ATMOSPHERIC HEAT BALANCE.

Momentum. Momentum can be exchanged between these two fluids by a process related to evaporation, that is, migration of molecules of air or water across the boundary, carrying their momentum with them. However, in natural conditions the more effective mechanism is the collision of "parcels" of the fluids, as distinct from motions of individual molecules. Also, momentum is usually transferred from air to sea, rather than vice versa. Winds whip up waves; these irregular shapes are more easily attacked by wind action than is the flat sea surface, and both waves and currents are initiated and maintained by the push and stress of the wind on the water surface. *See* WIND STRESS OVER SEA.

[JUNE G. PATTULLO]

Isotopic relationships. The isotopic water molecules H_2O^{16}, HDO^{16}, and H_2O^{18} have different vapor pressures and molecular diffusion coefficients and therefore exchange at different rates between the atmosphere and sea. Variations in the relative proportions of the hydrogen isotope deuterium, D, and O^{18} can now be measured mass spectrometrically with high precision, and the isotopic fractionation effects can be studied both at sea and in the laboratory. The isotopic vapor pressures have been measured very accurately, and the ratios of the binary molecular diffusion coefficients for HDO-air and H_2O^{18}-air to that for H_2O^{16}-air have been calculated theoretically and confirmed experimentally. Since the relative transport properties of isotopic molecules are much better known than those of different chemical species, the isotopic variations observed in surface ocean water and atmospheric vapor and in experimental studies on evaporation in small wind tunnels provide a powerful method for the study of the air-sea interface and the molecular and turbulent transport processes controlling the moisture supply to the air.

Precipitation over sea and land varies in isotopic composition because of the effects of fractional condensation of liquid water and variation of the isotopic vapor pressure ratios with temperature. Local equilibrium however, is maintained, and kinetic effects do not occur; the deuterium and oxygen-18 variations in precipitation are linearly correlated, and the concentrations of both heavy isotopes decrease with increasing latitude because of their preferential concentration in the liquid precipitate, which strips them out in lower latitudes. These variations provide a continually varying liquid input into the sea which must be balanced against the direct molecular exchange.

Water vapor over the oceans is never in isotopic equilibrium with surface sea water. The deuterium and oxygen-18 concentrations are always lower than the two-phase equilibrium separation factors given by the vapor-pressure ratios. The deviations from equilibrium are correlated with latitude and go through a maximum in each hemisphere in the trade-winds regions of maximum evaporation to precipitation ratio. The vapor composition relative to surface sea water cannot be understood simply on the basis of multiple equilibrium stage processes in fractional condensation during precipitation, and the isotopic variations reflect the kinetic isotope effects in the molecular exchange of water at the interface.

Two types of kinetic processes affect the isotopic composition of the vapor. There is a fractionation at the interface between liquid and vapor, since the condensation coefficients for the isotopic species are not necessarily the same. (The condensation coefficient may be thought of as the fraction of molecules of a given type striking the liquid surface which actually condense into the liquid structure; conversely, it is also a measure of the probability of a molecule to surmount the energy barrier for evaporation and actually to escape from the liquid.) For such fractionation to occur, the vapor concentration at the liquid-vapor interface must be significantly lower than the equilibrium concentration.

The second kinetic process is molecular diffusion from the interface into the turbulent mixing zone in the atmosphere above the boundary layer. Two types of models have been postulated for the boundary layer. At low wind speeds it is generally postulated that a true laminar layer exists with a fixed thickness and vapor gradient for a given wind speed, the transport through this layer being by steady-state molecular diffusion. At high wind speeds, above a certain critical velocity, the water surface changes from a smooth surface to a hydrodynamically rough surface. H. U. Sverdrup proposed that small turbulent eddies extend down to the actual liquid surface when it is rough. In such a model the vapor flux into these eddies can be postulated to take place by unsteady-state diffusion. With isotopic measurements it is possible to distinguish between these two models, because in the first case the isotopic fractionation is governed by the ratio of molecular diffusion coefficients, whereas in the unsteady-state process the fractionation is governed by the square root of this ratio. The single-stage enrichments in the two processes thus differ by a factor of 2.

Experimentally, the following points are observed: The oxygen-18 fractionation in the exchange process can be predicted from the rough-surface model for conditions under which both smooth and rough surfaces should be present. These results thus indicate that interfacial fractionation (from differences in condensation coefficients) is not important for oxygen and that a "microeddy" transport process operates in all cases; no actual fixed laminar layer exists. Deuterium data from the same experiments show larger separation factors than can be obtained from diffusional effects, so that it appears that there is significant interfacial fractionation for this isotope. This is plausible because of the asymmetric character of the HDO molecule and the large effect of deuterium substitution on vibrational frequencies, both of which should produce marked fractionation effects. The deuterium data therefore indicate that in the experiments which have been made the vapor pressure at the interface is lower than the saturation vapor pressure and that there is a significant thermodynamic disequilibrium. The oceanic data on surface waters and atmospheric vapor and precipitation indicate that kinetic effects of the same type as observed in laboratory experiments are important over the oceans.

These investigations thus provide a way to determine directly the contributions of molecular and turbulent transport processes to the evaporation rate over lakes and oceans and to relate numerical measures of the transport coefficients to physical parameters, such as wind speed. In addition, if the interfacial fractionation factor for deuterium can be determined, a measure of the actual vapor pressure at the interface can be obtained. However, a great deal of work remains to be done on this subject, which is still in its infancy.

Tritium variations. The radioactive hydrogen isotope tritium occurred with an abundance of about 0.4×10^{-12} molecule of HTO per 10^6 molecules of H_2O in oceanic surface waters in the prenuclear era, representing steady-state production of tritium by cosmic rays in the atmosphere. In 1968 the concentration of this molecule was about 10 times higher in surface waters because of nuclear weapon testing. The concentration in deep waters, below the thermocline, has always been too low to be measurable because of the short tritium half-life (12.5 years) compared to oceanic mixing times of the order of hundreds to thousands of years. Studies of tritium variations in present-day ocean waters have been made by Arnold Bainbridge, W. F. Libby, and others. Calculations of mixing rates of surface and deep waters and of atmospheric residence times for tritium can be made from these data. The models used in the various studies of this subject have neglected the molecular exchange of tritium between atmosphere and sea, which is important relative to the simple input by precipitation. Calculations, based on the stable isotope effects in molecular exchange, show that the molecular-exchange input of tritium into the ocean is about 2.5 times greater than the input by precipitation in both the prenuclear and postnuclear epochs. It is therefore necessary to reevaluate the tritium calculations, taking this finding into account.   [HARMON CRAIG]

Projection of droplets. Water, salts, organic materials, and a net electric charge are transferred to the air through the ejection of droplets by bubbles bursting at the sea surface. The exchange of these properties between the sea and the atmosphere is of importance in meteorology and geochemistry. Upon evaporation of the water, the droplet residues are carried great distances by winds. These particles become nuclei for cloud-

drop and raindrop formations and probably represent a large part of the cyclic salts of geochemistry.

Air bubbles are forced into surface waters of the sea by wave action, impinging raindrops, melting snowflakes, and other means. The larger bubbles rise to the surface, burst, and eject droplets. Many of the smaller bubbles dissolve before reaching the surface. The photomicrograph (Fig. 1) shows stages in the collapse of a bubble, and the jet and droplet formations which result. The graph (Fig. 2) shows the approximate relationships between the sizes of the bubbles, the sizes of the ejected droplets, and the weight of sea salt in these droplets.

The amounts of water which become airborne as droplets near the sea surface are not known. The best estimates which can now be made (from the limited information about the number and size of bubbles in the sea) range from about 2 to 10 g/ (m^2) (day) during fresh winds.

The average amounts of sea salt which become airborne at considerable altitudes are shown in Table 4. The total range of observed amounts in individual samples is from about 4×10^{-13} g/ml in a wind of Beaufort force 1 to 10^{-9} g/ml in a wind of force 12. *See* WIND.

Parts of marine organisms are seen in droplets ejected from plankton-rich water. The droplets also become coated with organic monolayers when they arise through contaminated surfaces. During moderate winds in oceanic trade wind areas, organic materials can equal 20 to 30 percent of the airborne sea salt.   [ALFRED H. WOODCOCK]

Electrification of the atmosphere. The traditional view states that the net positive space

Fig. 1. Collapse of air bubble and formation of jet and droplets. (a) High-speed motion pictures of stages in the process. (b) Oblique view of jet and droplets from bubble 1.0 mm in diameter.

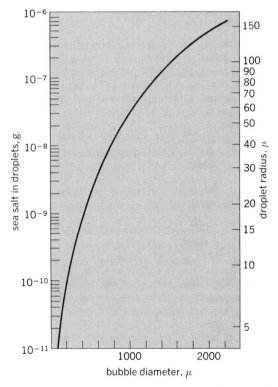

Fig. 2. Approximate relationships between the sizes of the bursting bubbles and the sizes and salt contents of the ejected droplets.

Table 4. Airborne sea salts in relation to wind force

Beaufort wind force	Concentrations,* μg/m³	Flux,* mg/(m²)(day)	Total,† mg/m²
2–3	2.7	0.42	6.0
4–5	9.9	3.9	11.6
6–7	21.3	24.0	21.8

*At about 500 m.

†Integrated through lowest 2000 m.

charge usually found in regions of fair weather is maintained by a charge separation process within thunderstorms. However, research has indicated that a charge separation mechanism at the surface of the ocean may contribute significantly to the atmospheric space charge over the oceans. It appears that this mechanism enables the oceans of the world to supply positive charge to the atmosphere at a rate of at least 10% of that supplied to the atmosphere over the oceans by thunderstorms. See ATMOSPHERIC ELECTRICITY.

The carriers of the charge separated at the ocean surface are the drops that emerge at the collapse of the jet that forms when a bubble breaks at the ocean surface (Fig. 1). Laboratory measurements have shown that the charge on these drops is a function of their size and of the age of the bubble from which they came. For drops in the size range commonly found in the atmosphere over the ocean, the charge per drop is positive and of the order of 10^{-6} electrostatic units (esu). From a consideration of the numbers and sizes of drops and the rate that they leave the sea surface, it has been computed that the net oceanic charge production is roughly proportional to the square of the wind speed, and is 5×10^{-8} esu/(cm²) (sec) for winds of 10 knots. Measurements of space charge made on the windward shore of the island of Hawaii are in good agreement with these calculations. As the winds over the oceans attain a maximum mean speed in each hemisphere at latitudes of 40–60°, a similarly located maximum should be found in the latitudinal distribution of the oceanic charge separation.

The normal oceanic fair-weather potential gradient does not have any significant influence on the magnitude of the charge of the drops from the bubble jet, but intense negative thunderstorm potential gradients of the order of 100 volts/cm at the sea surface can, by the process of induction charging, produce a positive charge on the drops that exceeds by many times the positive charge found on the drops in fair weather. Consequently, the normal positive space charge may be increased considerably in regions near oceanic thunderstorms.

[DUNCAN C. BLANCHARD]

Microlayer. The microlayer is the thin zone beneath the surface of the ocean or any free water surface within which physical processes are modified by proximity to the air-water boundary. It is characterized by suppression of vertical turbulence, a consequent decrease in diffusivity and increase in material, and an increase in kinetic and thermal gradients.

Because the microlayer at a free water surface is at least superficially similar to the boundary layer observed in the tangential flow of any viscous fluid near a rigid surface, it is tempting to identify one with the other. However, in view of the thermohydrodynamic complexity of the free ocean surface, such identification is unwarranted. It is not possible, in the present state of knowledge, to deal analytically with this problem. Consequently, all that can be described of the nature of the microlayer is gleaned from a few scattered observations. For the most part, these have been measurements of the effect of the microlayer on the flux of heat between water and air. The earliest determination appears to have been made by Alfred Merz in 1920. Subsequently A. H. Woodcock and H. Stommel measured thermal gradients at the surface of steaming ponds and estuaries, using a specially designed mercury thermometer of thin-stem diameter. G. C. Ewing and E. D. McAlister, using an infrared radiometer, observed the ocean surface to be as much as 0.6°C cooler than the underlying water. A somewhat related experiment, by W. S. Wise and others (1960), showed a measurable concentration gradient of solute at the surface of evaporating sugar solution. This suggests that a salinity gradient probably exists at the ocean surface, although it has yet to be described in the literature.

A different and striking manifestation of the microlayer can be observed when a gentle breeze blows over calm water. Specks of dust at the surface flow along noticeably faster than those 1 cm or so beneath the surface. The strongly developed vertical shear in the wind-driven flow which is thus revealed is possible only because the small eddy stresses in the microlayer permit the motion to remain nearly laminar at a relatively high Reynolds number. The reduction of the shear at higher wind speeds shows that the microlayer is thinner under these conditions.

The development and stability of the microlayer is enhanced by a contaminating film of surface-active agents, which is characteristically present on natural water surfaces. Such films quickly accumulate on any body of water exposed to the air. Even in the laboratory elaborate precautions are necessary to maintain a truly clean water surface.

The origin and composition of such films are complex. However, it can be asserted on thermodynamic grounds that substances will accumulate in a liquid surface if they reduce the surface tension and, hence, the free energy of the surface. Patches of such film can be observed on the sea under all normal conditions of wind and wave, although they are more strikingly visible at wind speeds under 3 m/sec, when they take the form of long, broad slicks. The most obvious effect of the film is to smooth the smallest ripples, giving the water a shiny appearance. The smoothing results from an altered boundary condition at the interface, substituting a sort of rubber-sheet elasticity for the relatively unrestricted freedom of a clean liquid surface. Such a stabilized surface more nearly approximates a rigid boundary, and therefore the associated microlayer becomes more nearly similar to the familiar boundary layer characterizing flow near a solid surface. Gradients, whether of substance, temperature, or momentum, are thus appreciably enhanced by surface films. In this purely mechanical manner, contaminating films can reduce the convective flux of heat across the air-water boundary independently of any throttling action they may have on the evaporation rate.

The flux of sensible heat across the air-water boundary of the ocean is usually in the upward direction, and hence the microlayer is cooler above than beneath. The sense of the flux results from the circumstance that most of the heating of the ocean is by solar radiation, which penetrates several meters into the sea before being absorbed, whereas heat balance is largely maintained by upward flux of sensible heat from a layer less than a few molecular diameters deep. This is a form of the well-known greenhouse effect. Thus, on the average, the microlayer is heated from below and cooled from above. The net upward flux is of the order of 150 g-cal/(cm²) (day), varying with the latitude, the season, and the time of day. The flux is greatest in the tropics, in the autumn, and in the forenoon; least at the poles, in early summer, and in early afternoon.

The importance of the microlayer resides in the fact that most surface measurements are in reality volume measurements of a thin but finitely thick layer. Consequently, the value recorded depends on the method of measurement employed. For many purposes the differences are trivial and are ignored. However, where precision is required, the only way to arrive at a true value of any parameter at the exact surface is to calculate it from theoretical considerations or to estimate it by extrapolating some measured gradient to the boundary. As an example, one may assume intuitively that the surface temperature of water must approach the psychrometric, or so-called wet-bulb, temperature of the overlying air as a limit. However, the psychrometric temperature itself varies as the boundary is approached and therefore cannot be directly measured at the exact surface.

Hence, from a physical point of view, the concept of a surface is something of an abstraction which has precise meaning only when referred to a specific parameter. An estimate of the surface value of any physical quantity depends on arbitrary assumptions as to the pertinent gradient in the microlayer. The best that can be done at present is to ensure that the assumptions adopted make physical sense.   [GIFFORD C. EWING]

Exchange of gases. A sample of sea water taken from any location and depth in the oceans is found generally to contain in solution all the gaseous constituents of the atmosphere. The concentrations of the dissolved gases depend upon gas exchange processes at the air-sea interface and upon chemical, biological, and physical processes within the body of the oceans. Perhaps the most important fact about the dissolved gases is that their concentrations in near-surface waters are found to differ from saturated values by only a few percent. This makes possible the use of deviations from saturation as clues to the processes indicated above.

In solution, nitrogen, oxygen, and the noble gases do not react chemically with the water. Carbon dioxide, however, tends toward dissociation equilibrium with H_2CO_3, HCO_3^-, and CO_3^{--}. If air and sea water are equilibrated under normal conditions in the laboratory, the concentration of each dissolved gas is proportional to its partial pressure in the air above and independent of the presence of the other gases. The solubility coefficient of each gas (that is, the concentration of that dissolved gas when its partial pressure above the solution is 1 atm) increases rapidly as the temperature is lowered and decreases with rising salinity. For the gases which do not react chemically with water, both the solubility coefficients and their variations with temperature generally increase with molecular weight.

Exchange of gases across the air-sea boundary may occur by molecular diffusion through a thin surface layer or by the motion of aggregates of molecules in the form of bubbles or gases dissolved in water droplets. The nearly saturated state of surface waters implies that molecular diffusion is the primary controlling process. The direction of net diffusion always is toward producing equilibrium. For example, upwelling warm water usually is undersaturated relative to the cold surface, and as the water cools net gas flow is into the water. Net transfer by molecular aggregates may be in either direction.

Surface water may sink and travel great distances beneath the surface. At all depths dissolved oxygen is consumed and carbon dioxide is produced by respiration and organic decomposition. The reverse of these effects is produced by photosynthesis when sufficient light is present. Consequently, these gases vary in concentration more than the others. A widespread characteristic of oxygen is the occurrence of a zone of minimum concentration at intermediate depths. This is accompanied by a maximum in the O^{18}/O^{16} ratio, which implies that O^{16} is consumed relatively more rapidly.

Nitrogen gas is produced in anoxic waters such as the Cariaco Trench, but no clear evidence has been obtained for significant biological influence on dissolved nitrogen in normal ocean water.

The concentration of carbon dioxide in the atmosphere probably plays a significant role in the heat budget of the Earth. Consequently, the partition between air and sea of the CO_2 produced by the combustion of fossil fuels has special importance. Although measurements have shown that a large proportion of the gas is absorbed by the oceans, precise calculations are difficult because of the complex solution chemistry of CO_2, mixing within the oceans, and influence of the biosphere.

Because helium is light and escapes from the Earth's gravitational field, its concentration in the atmosphere is very small (5 ppm). The amount in the sea is correspondingly small. Changes in dissolved He^4 concentrations from radioactive decay within the oceans are not measurable, but efflux through the sea floor from decay within the Earth should be detectable with sensitive techniques. There is controversial evidence for excess dissolved helium in deep waters. The observed world distribution of helium is the result of a steady state between production in the lithosphere and hydrosphere and escape from the atmosphere.

Similarly, radioactive decay of K^{40} produces Ar^{40}

which does not escape but accumulates in the atmosphere. The relatively large abundance of argon in normal sea water precludes direct observation of concentration changes from radioactive decay.

The multiplicity of physical processes, such as atmospheric pressure changes, bubble trapping, upwelling, and internal mixing, complicates the problem of unraveling the history of a sample of water, but the differing solubility characteristics of the gases promise to be useful. For example, because the solubility coefficient of argon is approximately twice that of nitrogen, deviations from surface equilibrium which result from bubble entrainment will be accompanied by enhanced N_2/Ar ratios. Conversely, the physical processes influencing oxygen and argon are almost identical, and the O_2/Ar ratio depends primarily upon the biological history of the sample. For the same reason, the He/Ne ratio offers the most sensitive approach to the helium efflux problem. *See* ATMO-SPHERIC CHEMISTRY.   [BRUCE B. BENSON]

Bibliography: A. E. Bainbridge, Tritium in the North Pacific surface water, *J. Geophys. Res.*, 68: 3785, 1963; B. B. Benson, *Some Thoughts on Gases Dissolved in the Oceans*, Univ. Rhode Island Occas. Publ. no. 3, 1965; D. C. Blanchard, The electrification of the atmosphere by particles from bubbles in the sea, *Progr. Oceanogr.*, 1:71–202, 1963; H. Craig and L. I. Gordon, Deuterium and oxygen-18 variations in the ocean and marine atmosphere, in E. Tongiorgi (ed.), *Stable Isotopes in Oceanographic Studies and Paleotemperatures*, 1965; H. Craig, L. I. Gordon, and Y. Horibe, Isotopic exchange effects in the evaporation of water, *J. Geophys. Res.*, 68:5079, 1963; H. Craig, R. F. Weiss, and W. B. Clarke, Dissolved gases in the Equatorial and South Pacific Ocean, *J. Geophys. Res.*, 72:6165–6181, 1967; W. Dansgaard, Stable isotopes in precipitation, *Tellus*, 16:436, 1964; J. I. Drever (ed.), *Sea Water: Cycles of the Major Elements*, 1977; G. Ewing and E. D. McAlister, On the thermal boundary layer of the ocean, *Science*, 131(3410):1374–1376, 1960; I. Friedman et al., The variation of the deuterium content of natural waters in the hydrologic cycle, *Rev. Geophys.*, 2:177, 1964; C. E. Junge, *Air Chemistry and Radioactivity*, 1963; R. Revelle and H. E. Suess, Gases, in M.N. Hill (ed.), *The Sea*, 1962; A. H. Woodcock, Salt nuclei in marine air as a function of altitude and wind force, *J. Meteorol.*, 10:362–371, 1953.

TRANSMISSION OF ENERGY

Electromagnetic and acoustic energy from various natural sources permeates the sea, supplying it with heat, supporting its ecology, and providing for sensory perception by its inhabitants; artificial sources afford man the means for underwater communication and detection.

Light. The primary source of energy which heats the ocean and supports its ecology is light from the Sun. On a clear day as much as 1 kw of radiant power from the Sun and sky may impinge on each square meter of sea surface. Of this power, 4–8% is reflected and the remainder is absorbed within the water as heat or as chemical potential energy

due to photosynthesis. The peak of the irradiation is close to the wavelength of greatest transparency for clear sea water, 480 mμ, but nearly half of the radiant power is infrared radiation which water absorbs so strongly that virtually none penetrates more than 1 m beneath the surface. As much as one-fifth of the incident power may be ultraviolet (below 400 mμ), and this radiation may penetrate a few tens of meters if little or no "yellow substance" (humic acids and other materials associated with organic decomposition) is present. Only a narrow spectral band of blue-green light, representing less than 10% of the total irradiation, penetrates deeply into the sea. This radiation has been detected by multiplier-phototube photometers at depths of more than 600 m. Visibility, important to predators in the feeding grounds of the sea, is possible chiefly because of this blue-green light.

Irradiance. Irradiance on a flat surface oriented in any manner decreases exponentially with depth, as illustrated by Fig. 3, which depicts experimental values of irradiance on an upward-facing surface. Irradiance on any other surface could be represented by a curve parallel (within 5%) to the one shown; irradiance on downward-facing surfaces is approximately one-fiftieth of the irradiance on upward-facing surfaces at all depths.

Absorption. Light, to be useful for heating or for photosynthesis, must be absorbed. The quantity of radiant power absorbed per unit of volume depends upon the amount of power present and the magnitude of the absorption coefficient; to a useful (5%) approximation power absorbed per unit of volume at any depth can be calculated by multiplying the irradiance at that depth, as in Fig. 3, by the slope of the curve expressed in natural log-units per unit of depth, that is, the attenuation coefficient K. Thus in Fig. 3, at a depth of 64 m where

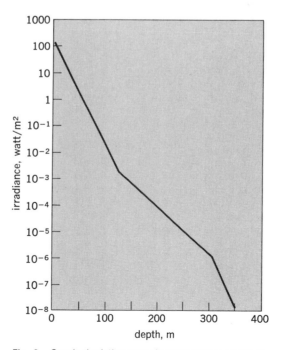

Fig. 3. Graph depicting experimental values of irradiance on an upward-facing surface.

the irradiance is 0.5 watt/m² and is decreasing with depth at the rate of 0.08 natural log-units/m, approximately $0.5 \times 0.08 = 0.04$ watt of radiant power is absorbed by every 1 m³ of sea water.

Visibility. Visibility under water is accomplished by image-forming light (rays) which must pass from the object to the observer without being scattered. The transmission of water for image-forming light is less than for diffused light, since scattering in any direction constitutes a loss of image-forming light, whereas only scattering in rearward directions is a loss for diffused light. Image-forming light is exponentially attenuated with distance, but the attenuation coefficient α averages 2.7 times greater than the attenuation coefficient for irradiance K, defined above. Apparent contrast of an underwater object having deep water as a background is exponentially attenuated with distance, the effective attenuation coefficient being $\alpha + K \cos \theta$, where θ is the inclination angle of the path of sight, and $\cos \theta = 1$ when the observer looks straight down. See discussion of water color and transparency in section on sampling and measuring techniques.   [SEIBERT Q. DUNTLEY]

Compensation intensity and depth. As daylight penetrating into the sea diminishes, the photosynthesis of plants is reduced but respiration remains approximately the same. The light value at which the rates of photosynthesis and respiration are equal is the compensation intensity. The depth at which the compensation intensity is found is the compensation depth. Both of the foregoing have also been termed compensation point, but since ambiguity may occur, it is best to avoid this term. The compensation intensity varies according to the species, the physiological condition of the plants, and other factors, particularly temperatures. Lowered temperature depresses respiration more than photosynthesis. The compensation depth depends upon the intensity of the incident radiation, the transparency of the water, and the period considered, since illumination varies with time. Compensation intensities of 10–200 foot-candles (ftc) have been measured for phytoplankton and of 17–45 ftc for filamentous algae. Compensation depths for 24-hr periods for phytoplankton range from less than 1 m in turbid water to more than 30 m in coastal areas and to 80 m or more in the clearest tropical waters, and for attached plants, to 50 m along the coast and to 160 m in especially clear water, as in the Mediterranean. Generally the compensation depth is found where daylight is reduced to about 1% of its value at the surface for phytoplankton or about 0.3% for bottom plants. The compensation depth is of particular significance since it marks the lower limit of the photic zone within which green plants can carry on primary production necessary as an energy source for the whole marine ecosystem.

[GEORGE L. CLARKE]

Electromagnetic fields. In sea water, as in any conductor, the electromagnetic behavior is determined by the magnetic permeability μ and the electrical conductivity σ. From these one may find the skin depth δ and the characteristic impedance η given by Eqs. (2) and (3).

$$\delta = (\pi f \mu \sigma)^{-1/2} \tag{2}$$

$$\eta = (2\pi f \mu / \sigma)^{1/2} \tag{3}$$

Both δ and η relate to electromagnetic waves of frequency f, for which the wavelength is $2\pi\delta$, the absorption over a path length x reduces the amplitude in the ratio $e^{-x/\delta}$, and the ratio of electric to magnetic field amplitude in a plane wave is η, the former leading in phase by 45°.

For sea water, magnetic permeability μ is nearly the same as for free space, and electrical conductivity σ is given to about 1% by Eq. (4), where t is

$$\sigma = [4.00 + a(t - 12)][1 + .0269(S - 35)] \tag{4}$$

temperature in °C, S is salinity in parts per thousand by weight, and a is .10 for $t > 12$ or $a = .092$ for $t < 12$ (Fig. 4). Using $\sigma = 4.0$ mho/m as a typical value, one obtains Eqs. (5) and (6). These formulas

$$\delta = 250 f^{-1/2} \text{ meter} \tag{5}$$

$$\eta = .0014 f^{1/2} \text{ ohm} \tag{6}$$

are expected to hold for all frequencies below about 900 MHz.

Absorption limits the penetration of a field, either inward from a boundary or outward from an electric or magnetic source, to a small multiple of the skin depth δ. A submerged horizontal dipole source near the surface will, however, have a more extensive field in the air and a shallow layer of water. At .01/Hz, δ is 2.5 km; this is a rough upper frequency limit for field fluctuations (such as those of the geomagnetic field) which can penetrate the entire thickness of the ocean layer. At 10 kHz, δ is 2.5 m. Fields of this and higher frequencies, existing, for example, in the natural atmospheric noise, can penetrate only a thin surface layer. For radio signals, the sea acts as an excellent ground plane, involving lower losses than transmission over land.

[PHILIP RUDNICK]

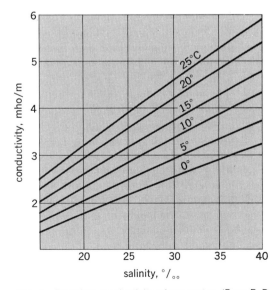

Fig. 4. Electrical conductivity of sea water. (*From B. D. Thomas, T. G. Thompson, and C. L. Utterback, The electrical conductivity of sea water, J. Cons. Perma. Int. Explor. Mer. 9:28–35, 1934*)

Sound. Sound in sea water travels four or five times its speed in air, or about 1500 m/sec. As sound propagates in the sea, its intensity diminishes inversely as the square of the distance from the source, in the absence of appreciable absorption, refraction, reflection, and scattering. Although losses by absorption are small compared with those that occur when sound of the same frequency travels through air, the increase in absorption with higher frequencies limits the effective range of ultrasonic waves, that is, waves having frequencies above those audible to the human ear. This is an important factor in submarine detection, where ultrasonic frequencies are used because of their desirable directional properties. *See* SONAR.

The velocity of sound in sea water varies with temperature, salinity, and pressure. Hence a beam of ultrasonic waves, when transmitted in a horizontal direction, may be refracted and then reflected one or more times from the surface, ocean bottom, or some layer within the vertical water structure. In this manner several different rays may be received at different intervals from a single source. The direct transmission is limited to specific distances, depending on depth of the bottom, and its theoretical velocity may be computed if the temperature and salinity are known. At greater distances the apparent horizontal velocity is less than

the theoretical velocity because of such factors as distance between source and receiver, depth, bottom profile and character, and physical properties of the water.

The vertical velocity is a function of depth (pressure) and the distribution of temperature and salinity. Except in polar latitudes it generally decreases from the surface to some moderate depth (from 500 to 1500 m) because of decreasing temperature. Below this depth the velocity gradually increases again as the effect of increasing pressure becomes dominant. Because most sonic depth-finding instruments are calibrated for a constant velocity, when a very accurate depth reading is needed, it is necessary to correct readings to true depths. *See* ECHO SOUNDER; MARINE GEOLOGY.

Investigations of sound in the ocean offer many promising areas for further study, particularly studies of underwater noises produced by marine life, the relation of these noises to other sounds in the sea, the seasonal rhythm and geographic variation of the noise makers, and the ecological significance of sound-producing organisms. A variety of problems in underwater acoustics may result from the presence of sound-producing organisms. *See* SCATTERING LAYER; UNDERWATER SOUND.   [NEIL A. BENFER]

Bibliography: V. M. Albers, *Underwater Acoustics*, 1965; J. W. Caruthers, *Fundamentals of Marine Acoustics*, 1977; C. S. Clay and H. Medwin, *Acoustical Oceanography: Principles and Applications*, 1977; R. W. Holmes, Solar radiation, submarine daylight, and photosynthesis, in *Treatise on Marine Ecology and Paleoecology*, Geol. Soc. Amer. Mem. no. 67, vol. 1, 1957; R. H. Lien, Radiation from a horizontal dipole in a semi-infinite dissipative medium, *J. Appl. Phys.* 24(1): 1–4, 1953; J. E. Tyler and R. C. Smith, *Measurements of Spectral Irradiation Underwater*, 1970; J. R. Wait, The radiation fields of a horizontal dipole in a semi-infinite dissipative medium, *J. Appl. Phys.*, 24(7):958–959, 1953.

COMPOSITION OF SEA WATER

The concentrations of the various components of sea water are regulated by numerous chemical, physical, and biochemical reactions.

Inorganic regulation of composition. The present-day compositions of sea waters (Table 5) are controlled both by the makeup of the ultimate source materials and by the large number of reactions, of chemical and physical natures, occurring in the oceans. This section considers the nonbiological regulatory mechanisms, most conveniently defined as those reactions occurring in a sterile ocean. For a discussion of the weathered and weathering substances that give rise to the waters of the world *see* HYDROSPHERIC GEOCHEMISTRY.

Interactions between the ions results in the formation of ion pairs, charged and uncharged species, which influence both the chemical and physical properties of sea water. For example, the combination of magnesium and sulfate to form the uncharged ion pair accounts for the marked absorption of sound in sea water. A model accounting for such interactions has been developed for the principal dissolved ions in sea water (Table 6).

Table 5. Chemical abundances in the marine hydrosphere

Element	Concentration, mg/liter	Element	Concentration, mg/liter
H	108,000	Ag	0.0003
He	0.000007	Cd	0.00011
Li	0.17	In	0.000004
Be	0.0000006	Sn	0.0008
B	4.6	Sb	0.0003
C	28	Te	–
N	15	I	0.06
O	857,000	Xe	0.00005
F	1.2	Cs	0.0003
Ne	0.0001	Ba	0.03
Na	10,500	La	1.2×10^{-5}
Mg	1350	Ce	5.2×10^{-6}
Al	0.01	Pr	2.6×10^{-6}
Si	3.0	Nd	9.2×10^{-6}
P	0.07	Pm	–
S	885	Sm	1.7×10^{-6}
Cl	19,000	Eu	4.6×10^{-7}
A	0.45	Gd	2.4×10^{-6}
K	380	Tb	–
Ca	400	Dy	2.9×10^{-6}
Sc	<0.00004	Ho	8.8×10^{-7}
Ti	0.001	Er	2.4×10^{-6}
V	0.002	Tm	5.2×10^{-7}
Cr	0.00005	Yb	2.0×10^{-6}
Mn	0.002	Lu	4.8×10^{-7}
Fe	0.01	Hf	<0.000008
Co	0.0004	Ta	<0.000003
Ni	0.007	W	0.0001
Cu	0.003	Re	0.0000084
Zn	0.01	Os	–
Ga	0.00003	Ir	–
Ge	0.00006	Pt	–
As	0.003	Au	0.00001
Se	0.00009	Hg	0.0002
Br	65	Tl	<0.00001
Kr	0.0002	Pb	0.00003
Rb	0.12	Bi	0.00002
Sr	8.0	Po	–
Y	0.00001	At	–
Zr	0.00002	Rn	0.6×10^{-15}
Nb	0.00001	Fr	–
Mo	0.01	Ra	1.0×10^{-10}
Tc	–	Ac	–
Ru	0.0000007	Th	0.000001
Rh	–	Pa	2.0×10^{-9}
Pd	–	U	0.003

Table 6. Distribution of major cations as ion pairs with sulfate, carbonate, and bicarbonate ions in sea water of chlorinity 19°/₀₀ and pH 8.1*

Ion	Free ion, %	Sulfate ion pair, %	Bicarbonate ion pair, %	Carbonate ion pair, %
Ca^{++}	91	8	1	0.2
Mg^{++}	87	11	1	0.3
Na^+	99	1.2	0.01	—
K^+	99	1	—	—

Ion	Free ion, %	Ca ion pair, %	Mg ion pair, %	Na ion pair, %	K ion pair, %
SO_4^{--}	54	3	21.5	21	0.5
HCO_3^-	69	4	19	8	—
CO_3^{--}	9	7	67	17	—

*From R. M. Garrels and M. E. Thompson, A chemical model for sea water at 25°C and one atmosphere total pressure, *Amer. J. Sci.*, 260:57–66, 1962.

pH and oxidation potential. The pH of surface sea waters varies between 7.8 and 8.3, with lower values occurring at depths. The pH normally goes through a minimum with increasing depth in the ocean, and this depth dependence shows a marked resemblance to the profiles of oxygen.

The oxidation potentials of sea-water systems are determined by oxygen concentration in aerobic waters and by hydrogen sulfide concentration in anoxic waters. For 25°C, 1 atm pressure, and a pH of 8, the oxidation potential is about 0.75 volt for waters containing dissolved oxygen in a state of saturation. In anoxic waters the oxidation potentials can vary from about −0.2 to −0.3 volt.

Solubility. Only calcium, among the major cations of sea water, is present in a state of saturation, and such a situation generally occurs only in surface waters. Here its concentration is governed by the solubility of calcium carbonate. Barium concentrations in deep waters can be limited by the precipitation of barium sulfate. The noble gases and dissolved gaseous nitrogen have their marine concentrations determined by the temperature at which their water mass was in contact with the atmosphere and are in states of saturation or very nearly so.

Authigenic mineral formation. The formation and alteration of minerals on the sea floor apparently are responsible for controlling the concentrations of the major cations Na, K, Mg, and Ca. Such clay minerals as illite, chlorite, and montmorillon-

ite are presumably synthesized from these dissolved species and the river-transported weathered solids (aluminosilicates, such as kaolinite) by reactions of the type shown in Eq. (7). The SiO_2

$$Aluminosilicates + SiO_2 + HCO_3^- + Cations$$

$$= Cation\ aluminosilicate + CO_2 + H_2O \qquad (7)$$

is introduced in part as diatom frustules. Such a process implies control of the carbon dioxide pressure of the atmosphere.

The formation of ferromanganese minerals, sea-floor precipitates of iron and manganese oxides, may govern the concentrations of a suite of trace metals, including zinc, manganese, copper, nickel, and cobalt. These elements are in highly undersaturated states in sea water (Table 7) but are highly enriched in these marine ores, the so-called manganese nodules, which range in size from millimeters to about 1 m in the form of coatings and as components of the unconsolidated sediments.

Cation and anion exchange. Cation-exchange reactions between positively charged species in sea water and such minerals as the marine clays and zeolites appear to regulate, at least in part, the amounts of sodium, potassium, and magnesium, as well as other members of the alkali and alkaline-earth metals which are not major participants in mineral formations. High charge and large radius influence favorably the uptake on cation-exchange minerals. It appears, for example, that while 65% of the sodium weathered from the continental rocks resides in the oceans, only 2.5, 0.15, and 0.025% of the total amounts of potassium, rubidium, and cesium ions (ions increasingly larger than sodium) have remained there. Further, magnesium and potassium are depleted in the ocean relative to sodium on the basis of data obtained from igneous rock.

The curious fact that magnesium remains in solution to a much higher degree than potassium is not yet resolved but may be explained by the ability of such ubiquitous clay minerals as the illites to fix potassium into nonexchangeable or difficultly exchangeable sites.

Similarly, anion-exchange processes may regulate the composition of some of the negatively charged ions in the oceans. For example, the chlorine-bromine ratio in sea water of 300 is displaced to values around 50 in sediments. Such a result may well arise from the replacement of chlorine by bromine in clays; however, the meager amounts of work in this field preclude any unqualified statements.

Physical processes. Superimposed upon these chemical processes are changes in the chemical makeup of sea water by the melting of ice, evaporation, mixing with runoff waters from the continents, and upwelling of deeper waters. The net effects of the first three processes are changes in the absolute concentrations of all of the elements but with no major changes in the relative amounts of the dissolved species.

Changes with time. Changes in the composition of sea water through geologic time reflect not only differences in the extent and types of weathering processes on the Earth's surface but also the relative intensities of the biological and inorganic reac-

Table 7. Observed concentrations of some trace metals in sea water*

Ion	Observed sea-water concentration, moles/liter	Limiting compound	Calculated limiting concentration, moles/liter
Mn^{++}	4×10^{-8}	$MnCO_3$	10^{-3}
Ni^{++}	4×10^{-8}	$Ni(OH)_2$	10^{-3}
Co^{++}	7×10^{-9}	$CoCO_3$	3×10^{-7}
Zn^{++}	2×10^{-7}	$ZnCO_3$	2×10^{-4}
Cu^{++}	5×10^{-8}	$Cu(OH)_2$	10^{-6}

*Calculated on the basis of their most insoluble compound.

tions. The most influential parameter controlling the inorganic processes appears to be the sea-water temperature.

Changes in the abyssal temperatures of the oceans from their present values of near 0 to 2.2, 7.0, and 10.4°C in the upper, middle, and lower Tertiary, respectively, have been postulated from studies on the oxygen isotopic composition of the tests of benthic foraminifera. Such temperature increases would of necessity be accompanied by similar ones in the surface and intermediate waters. One obvious effect from the recent cooling of the oceans is an increase in either or both of the calcium and carbonate ions, since the solubility product of calcium carbonate has a negative temperature coefficient. Similarly, the saturated amounts of gases that can dissolve in sea water in equilibrium with the atmosphere increase with decreasing water temperatures.

[EDWARD D. GOLDBERG]

Biological regulation of composition. In the open sea all the organic matter is produced by the photosynthesis and growth of unicellular planktonic forms. During this growth all the elements essential for living matter are obtained from the sea water. Some elements are present in great excess, such as the carbon of CO_2, the potassium, and the sulfur (as sulfate). Other elements—for example, phosphorus, nitrogen, and silicon—are present in small enough quantities so that plant growth removes virtually all of the supply from the water. During photosynthesis, as these elements are being removed from the water, oxygen is released.

The biochemical cycle. The organic matter formed by photosynthesis and growth of the unicellular plants may be largely eaten by the zooplankton, and these in turn form the food for larger organisms. At each step of the food chain a large proportion of the eaten material is digested and excreted, and this, along with dead organisms, is decomposed by bacterial action. The decomposition process removes oxygen from the water and returns to the water those elements previously absorbed by the phytoplankton.

The distribution of oxygen and essential nutrient elements in the sea is modified by the spatial separation of these biological processes. Photosynthesis is limited to the surface layers of the ocean, generally no more than 100 m or so of depth, but the decomposition of organic material may take place at any depth. Reflecting this separation of processes, the concentration of nutrient elements in the surface is low, rises to maximum values at intermediate depths (300–800 m), and decreases slightly to fairly constant values which extend nearly to the bottom. Frequently, a slight increase in the concentration of essential elements is observed near the bottom. The oxygen distribution is the opposite of the one just described, with high values at the surface, a minimum value at mid-depth, and intermediate values in the deep water. The oxygen-minimum–nutrient-maximum level in the ocean is the result of two processes working simultaneously. In part it is formed by the decomposition of organic matter sinking from the surface, and in part it results from the fact that this water was originally at the surface in high latitudes, where it contained organic matter and subsequently cooled, sank, and spread out over the oceans at the appropriate density levels.

Because of the nearly constant composition of marine organisms, the elements required in the formation of organic material vary in a correlated way. Analyses of marine organisms indicate that in their protoplasm the elements carbon, nitrogen, and phosphorus are present in the ratios of 100:15:1 by atoms. In the production of organic matter these elements are removed from the water in these ratios, and during the decomposition of organic matter they are returned to the water in the same ratios. However, since the decomposition of organic material is not an instantaneous process which releases all elements simultaneously, it is not unusual to find different ratios of concentration of these elements in the sea. Particularly in coastal waters and in confined seas the ratio of concentration of nitrogen to phosphorus, for example, may differ widely from the 15:1 ratio of composition within the organisms.

The biochemical circulation. Unlike the major elements in sea water, the concentrations of these nutrients are widely different in different oceans of the world. Pacific Ocean water contains nearly twice the concentration of nitrogen and phosphorus found at the same depth in the North Atlantic, and intermediate concentrations are found in the South Atlantic and Antarctic oceans. The lowest concentrations for any extensive body of water are found in the Mediterranean, where they are only about one-third of those in the North Atlantic.

These variations can be attributed to the ways in which the water circulates in these oceans and to the effect of the biological processes on the distribution of elements. The Mediterranean, for example, receives surface water, already low in nutrients, from the North Atlantic and loses water from a greater depth through the Straits of Gibraltar. While the water is in the Mediterranean, the surface layers are further impoverished by growth of phytoplankton, and the organic material formed sinks to the bottom water and is lost in the deeper outflow. A similar process explains the low nutrient concentrations in the North Atlantic, which receives surface water from the South Atlantic and loses an equivalent volume of water from greater depths. *See* OCEAN CURRENTS; SEA-WATER FERTILITY.

[BOSTWICK H. KETCHUM]

Buffer mechanism. The constituents of sea water include a number of cations, all of which are weak acids, and a smaller number of anions, some of which are strong bases. Thus sea water is always somewhat on the alkaline side of neutrality, ranging in pH roughly between the limits of 7.5 and 8.3.

In chemical oceanography the term alkalinity is used to denote not the concentration of hydroxyl ions, as might be expected, but the concentration of strong bases. Alkalinity can be defined as the number of equivalents of strong acid required to convert stoichiometrically all the strong bases to weak acids.

The addition or subtraction of weak acids therefore does not affect the alkalinity of sea water, although through the operation of the buffer mechanism it changes the pH.

The principal weak acid in sea water is carbonic, resulting from the hydrolysis of dissolved carbon dioxide. Boric acid is also present in significant amounts. Salts of these two weak acids are the strong bases which make up the alkalinity. Total combined boron, whether as acid or borate, is in virtually constant proportion to chlorinity, the ratio of boron to chlorinity being about 0.00024, which is equivalent to a specific ratio of boric acid to chlorinity of about 0.022 (concentrations in millimoles per liter).

The total alkalinity is more variable in the ocean. Near the surface it can be increased by addition of dissolved carbonates in river discharge or decreased by the precipitation of lime in the formation of coral and shells. In deeper layers it can be increased by the solution of calcareous debris sinking from the surface. F. Koczy (1956) gives a thorough discussion of the variation of specific alkalinity in the oceans. The average ratio of alkalinity to chlorinity is about 0.120 for surface sea water, increasing somewhat with depth (concentrations in milliequivalents per liter).

Even more variable than the alkalinity is the total dissolved CO_2. At the surface, CO_2 moves between the sea and the atmosphere, and in the euphotic zone CO_2 is removed by photosynthesis to be incorporated into organic matter. Below the euphotic zone, CO_2 is regenerated by biological oxidation of organic matter. Total CO_2 typically may vary from 2.0 or 2.1 mM/liter at the surface to 2.8 or more at the depth of the oxygen minimum.

Some of this CO_2 is in physical solution, and some is undissociated carbonic acid; but most of it is in ionic form, mainly bicarbonate ion with some carbonate. The proportions of the various forms are governed by the dissociation constants of carbonic and boric acids, which vary with temperature and pressure, and by the activity coefficients of the various ions concerned, which vary with temperature, pressure, and salinity. In practice, the activity coefficients and the dissociation constants are combined into apparent dissociation constants, which are tabulated by H. W. Harvey (1957) as functions of temperature and salinity at 1 atm pressure.   [JOHN LYMAN]

Bibliography: R. M. Garrels and M. E. Thompson, A chemical model for sea water at 25°C and one atmosphere total pressure, *Amer. J. Sci.*, 260: 47–66, 1962; H. W. Harvey, *Chemistry and Fertility of Sea Waters*, 1957; M. N. Hill (ed.), *The Sea*, vol. 2, 1963; F. Koczy, The specific alkalinity, *Deep-Sea Res.*, 3:279–288, 1956; J. Lyman and R. B. Abel, Chemical aspects of physical oceanography, *J. Chem. Educ.*, 35:113–115, 1958; F. T. MacKenzie and R. M. Garrels, Chemical mass balance between rivers and oceans, *Amer. J. Sci.*, 264:507–525, 1966; D. F. Martin, *Marine Chemistry*, vol. 1: *Analytical Methods*, 1972; A. C. Redfield, The biological control of chemical factors in the environment, *Amer. Sci.*, 46(3):205–221, 1958.

DISTRIBUTION OF PROPERTIES

The distribution of physical and chemical properties in the ocean is principally the result of the following: (1) radiation (of heat); (2) exchange with the land (of heat, water, and solids such as salts) and with the atmosphere (of water, salt, heat, and dissolved gases); (3) organic processes (photosynthesis, respiration, and decay); and (4) mixing and stirring processes. These processes are largely responsible for the formation of particular water types and ocean water masses.

Horizontal distributions. The general distribution of properties in the oceans shows a marked latitudinal effect which corresponds with radiation income and differences between evaporation and precipitation.

Temperature. Heat is received from the Sun at the sea surface, where parts of it are reflected and radiated back. Equatorward of 30° latitude the incoming radiation exceeds back radiation and reflection, and poleward it is less. The result is high sea-surface temperature (more than 28°C) in equatorial regions and low sea-surface temperatures (less than 1°) in polar regions.

Salinity. Various dissolved solids have entered the sea from the land and have been so mixed that their relative amounts are everywhere nearly constant, yet the total concentration (salinity) varies considerably. In middle latitudes the evaporation of water exceeds precipitation, and the surface salinity is high; in low and high latitudes precipitation exceeds evaporation, and dilution reduces the surface salinity.

Open ocean surface salinities range from lows of about 32.5°/∞ in the North Pacific, 34.0 in the Antarctic, 35.0 in the equatorial Atlantic, 34.0 in the equatorial Indian, and 33.5 in the equatorial Pacific, to highs in the great evaporation centers of the middle latitudes of 35.5 and 36.5 in the North and South Pacific, 37.0 in the North and South Atlantic, and 36.0 in the Indian Ocean.

Dissolved oxygen. Dissolved oxygen is both consumed (respiration and decay) and produced (photosynthesis) in the ocean, as well as being exchanged with the atmosphere at the sea surface. Above the thermocline the waters are nearly always near saturation in oxygen content (>7 ml/liter in the cold waters of high latitudes and <5 ml/liter in warm equatorial waters).

Density. The density of sea water depends upon its temperature, salinity, and depth (pressure) but can vary horizontally only in the presence of currents, and hence its distribution depends closely upon the current structure. In low and middle latitudes the effect of the high temperature exceeds that of the high salinity, and the surface waters are lighter than those in high latitudes, with values ranging from less than 1.022 g/ml to more than 1.027. The density at great depth (10,000 m) may exceed 1.065. The greatest vertical gradient is associated with the thermocline and the halocline and is therefore very near to the surface. The heavier surface waters from high latitudes move and mix underneath the lighter water at depths which depend upon their density. The difference in surface density is usually not great, since the high-latitude salinity is low, and the waters usually sink only a few hundred meters, forming intermediate water. But in cases where water of high salinity has been carried into high latitudes by the currents and cooled (as in the North Atlantic) or where water of relatively high salinity freezes and gives up part of its water (as on the continental shelf of Antarctica), deep and bottom waters with temperature less than 1°C are formed. *See* HALOCLINE; THERMOCLINE.

Vertical distributions. The subsurface distribution of properties is controlled largely by external factors, particularly those which influence surface density, and the type of deep-sea circulation.

Temperature. The thermohaline circulation results in a vertical distribution of temperature such that in low and middle latitudes the deeper waters are colder than the surface waters, and at very high latitudes where surface temperatures are low, the deeper waters are as warm as those at the surface, or warmer. Seasonal cooling in high latitudes may cause the temperature to be at a minimum at some intermediate depth, and the circulation may cause a temperature minimum or maximum at intermediate depths. Over a large part of the Pacific Ocean, where no bottom water is formed, the temperature increases downward from about 4000 m. Since the gradient is not greater than the adiabatic, the water is not unstable.

Density and salinity. Since much of the flow of the ocean is geostrophically balanced, surfaces of constant density slope in various ways with respect to the sea surface and other surfaces of constant pressure, and density varies in the east-west as well as the north-south direction. Mixing and movement of intermediate and deep water along these surfaces cause more complex distributions of other variables. The intermediate waters of the North and South Pacific and of the South Atlantic appear to have salinity minima at intermediate depths in middle and low latitudes, since they originate in the high-latitude regions of low salinity and pass between the high-salinity surface waters of middle latitudes and the bottom waters. In the North Pacific this minimum varies from 33.4 to 34.1°/oo and in the South Pacific and South Atlantic, from 34.2 to about 34.6°/oo. In the North Atlan-

tic Ocean there is no intermediate water of low salinity formed, but very saline water (36.5°/oo) flows in from the Mediterranean at depth and results in a more complicated distribution of salinity than is found in the other oceans. For the temperatures and salinities of the bottom waters, which are more homogeneous than the others, see the later discussion on ocean water masses.

Dissolved oxygen. Where the cold surface waters sink in high latitudes, quantities of oxygen are carried downward. Below the compensation depth consumption gradually reduces the concentration. The bottom waters of the Atlantic Ocean contain from 5 to 6 ml/liter of dissolved oxygen; the Indian and South Pacific Ocean, about 4; and the North Pacific, less than 4. Between these bottom waters and those at the surface, which are saturated, smaller values of oxygen are found ranging from less than 0.10 in the eastern tropical Pacific and less than 1.0 in the eastern tropical Atlantic to greater than 4 ml/liter in the South Atlantic.

Nutrients. Other properties such as the nutrients, phosphate and nitrate, have low concentrations in the surface layer, where they are consumed by the plants, and high values at depths, where they are concentrated by the sinking and decay of organisms.

The maximum values of phosphate are found at intermediate depths, usually beneath the layer of minimum oxygen, and vary from less than 1 μg-atom/liter in the North Atlantic to more than 2 in the Antarctic, Indian, and South Pacific, and more than 3.5 in the North Pacific. Surface values vary from less than 0.1 in the North Atlantic and 0.5 in the North Pacific to more than 1.5 in the Antarctic. Bottom values vary from about 1.0 in the North Atlantic and 2.0 in the Antarctic, South Pacific, and Indian oceans to 2.5 in the North Pacific.

Nitrate-nitrogen is present in a ratio of about 8:1 by weight to phosphate-phosphorus.

Silicate has no intermediate maximum but increases monotonically toward the bottom, where

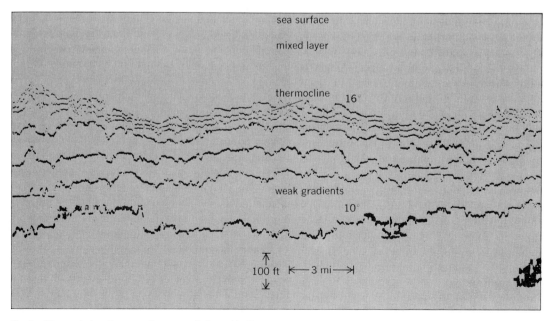

Fig. 5. Thermal structure most commonly found in the sea.

Table 8. Slope of an isotherm*

Slope, minutes from horizontal	Observations, %
0–10	34
10–20	20
20–30	11
30–40	7
40–50	6
50–60	5
>60	17

*Based on 60,000 depth observations.

the values vary from more than 150 μg-atoms/liter in the North Pacific to less than 40 in the North Atlantic. Surface values are generally less than 10.

[JOSEPH L. REID]

Detailed thermal structure. A detailed knowledge of the sea's thermal structure and its relation to oceanic dynamic processes has been accumulated. The details of two-dimensional structure measurement of the sea have been provided by the temperature structure profiler. Although thermal changes occur during the towing of the profiler, the spatial plot is realistic since heat transfer at the surface is negligible, and advective changes proceed more slowly than the 6-knot movement of the ship.

In Fig. 5 the vertical scale represents depth from the surface to 800 ft. The horizontal scale may be interpreted as either time or distance, because the towing ship has a constant speed of 6 knots. Since the vertical scale is about 100 times the horizontal, the isotherm lines appear to be steeper than they are. Each isotherm is a whole-degree Celsius from 16 to 10°. This example reflects the common, mixed layer, without isotherms, found from the surface to a depth, in this case, of 260 ft. Below these are the closely spaced isotherms that compose the sharp thermocline.

Isotherm oscillations. The profiler made possible a determination of detailed horizontal changes in isotherms. An important large-scale feature re-

vealed is the vertical, wavelike undulation in the main thermocline, with wave heights of 50–100 ft and crests about every 12 mi. Smaller oscillations are conspicuous on all isotherms. These are the ever-present internal waves, which occur on the average thermocline at the rate of 1–4 per mile. Their height varies inversely with the strength of the thermocline and usually amounts to 3–20 ft. *See* INTERNAL WAVE.

The two-dimensional operation discloses the complexity of the thermal structure in the sea. From the data acquired, it is possible to derive quantitative information from the slope of isothermal surfaces. By scaling the depth of isotherms every 304 ft, it is possible to obtain the angle that the isotherm makes with the horizontal (Table 8).

The isotherms in the thermocline are unusually flat, with a median slope of 17 min. By use of a power spectrum, it was found that the wavelengths vary widely. There are frequently several peaks in the power spectrum corresponding to wavelengths between 0.25 and 1.0 mi. Repeated peaks occur at 0.3 and 0.7 mi. The greatest power is in the long-period waves. Other analyses from data collected in a single place reveal (Fig. 6) that vertical changes in the isotherms have definite cycles corresponding to the short-period internal waves (5–8 min), seiches (14 and 20 min), tidal phenomena (6, 8, 12, and 24 hr), meteorological (3–5 days), lunar cycles (14 and 28 days), and seasonal cycles (1 year).

It is believed that the internal waves in the open sea move in different and changing directions. Analyses of tows in different directions have provided information on the direction of propagation of dominant waves. Up to 200 mi off the coast of southern California the dominant direction is shoreward.

An attempt has been made to measure a three-dimensional structure. This is done by towing the profiler in circles and box patterns. Preliminary data indicate that the ocean-temperature structure possesses numerous small thermal domes or humps moving in different directions.

Circulation and boundary influences. The detailed vertical thermal sections reveal the dynamic processes going on in the sea. Domes or ridges in the thermal structure, caused by eddies or oppositely flowing adjacent currents, have been observed to be 150 ft in height within a distance of 6 mi. A similar change in the depth and character of a thermocline has been found across major current systems. Areas of upwelling are apparent where the thermocline tilts upward toward the surface and generally becomes weaker.

The structure is also influenced by the boundaries of water masses and land features. A strong horizontal thermocline may serve as a vertical boundary. If one water layer moves in a different direction from another on the opposite side of the thermocline, a large-scale turbulence, with temperature inversion, can occur (Fig. 7).

Vertical water mass boundaries, or fronts, are detected easily by taking detailed two-dimensional profiles. The isotherms extend to the surface to form a horizontal thermal gradient with turbulence along the thermal boundary.

Islands, points of land, and shoals also influence

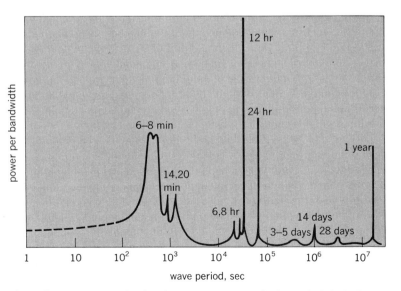

Fig. 6. Power spectrum showing the most common oscillation periods in the internal thermal structure of the ocean.

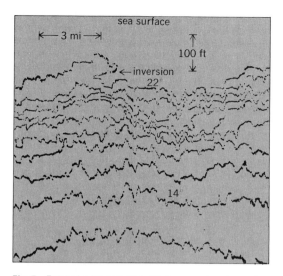

Fig. 7. Extensive turbulence with temperature inversion.

thermal structure and internal waves. The progressive nature of internal waves is slowed down over shoals, where they have a shorter wavelength. This happens when the waves approach shore, where they not only become more closely spaced but also refract and move shoreward with long crests. Another effect on the thermal structure is created by the tide, which causes a large mass of water to move in and out across the continental shelf off California. The colder, heavier water mass coming shoreward along the bottom is affected by the bottom friction to form a wall-like front (Fig. 8). This is essentially an internal tidal bore, which occurs at flooding spring tides. The front of this bore contains turbulence and sometimes weak thermal inversions. It is also evident that the strength of the thermocline changes with tide. When internal waves approach the surface boundary, their crests become flat and their troughs sharp. If they converge with the sea floor, the troughs grow flat, and the crests become steep and sharp.

[E. C. LA FOND]

Stirring and mixing processes. Stirring and mixing are processes of prime importance in determining the distribution of properties and in the formation of water masses in the ocean. Stirring refers here to motions which increase the average magnitude of the gradient of a property throughout a specified region. Mixing is employed in a narrow sense to denote the molecular diffusion or conduction processes by which gradients are decreased. Stirring and subsequent mixing can decrease the gradients in a region much more quickly than molecular processes acting alone. Combined stirring and mixing are often called turbulent diffusion.

Stirring motions involve shearing or more precisely, deformations of the water. Stirring scales range from those of the permanent currents down nearly to molecular scale. Smaller-scale motions, often highly complex, are the most effective in stirring.

With very important exceptions noted below, the preferred direction of stirring is along surfaces of constant potential density. (Potential density is the density of water when brought adiabatically to atmospheric pressure.) The preference is due to the fact that potential density almost always increases with depth. Under this condition, vertical displacement of a water parcel from a potential density surface is resisted by a stabilizing force proportional to the vertical gradient of potential density.

At the sea surface, powerful mechanisms exist which induce stirring across potential density surfaces. These are the wind stresses and the thermal convections resulting from cooling and evaporation. Near the sea surface, they greatly outweigh the stabilizing forces to produce and maintain a shallow, homogeneous top layer over much of the ocean. Stirring induced at the surface penetrates across the potential density surfaces just below the homogeneous layer but is damped out as it extends deeper into the thermocline or halocline layer. Thus, at some depth in these layers, the more usual stirring along potential density surfaces becomes dominant. In great depths and in high latitudes, the dominance is less strong since the vertical gradient of potential density is small. Stirring across potential density surfaces may also be brought about by tidal currents, but these are strong only in the shallow, coastal parts of the ocean.

Because of the variety and complexity of stirring motions, no general method of treating them quantitatively has proved really satisfactory. It is often expedient to assume that stirring and mixing follow rules analogous to those for molecular diffusion. Sometimes analysis of the motions in detail is possible, although difficulties in both theory and observation are great. Statistical treatment of the detailed motion seems to provide the most realistic approach. *See* OCEAN CURRENTS.

[JOHN D. COCHRANE]

Ocean water masses. Ocean water masses are extensive bodies of subsurface ocean water characterized by a relatively constant relationship between temperature and salinity or some other conservative dissolved constituent. The concept was developed to permit identification and tracing of such water bodies. The assumption is made that the characteristic properties of the water mass were acquired in a region of origin, usually at the surface, and were subsequently modified by lateral and vertical mixing. The observed characteristics in place thus depend both on the original proper-

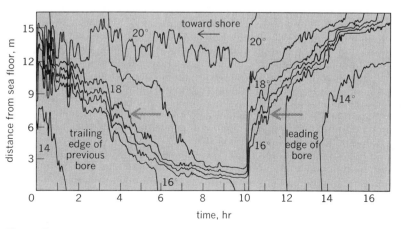

Fig. 8. Thermal structure changes caused by an internal tidal bore.

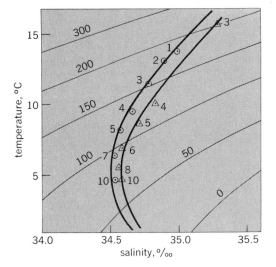

Fig. 9. Temperature-salinity values for Carnegie station 40 (circles) and Dana station 3756 (triangles); see Fig. 11 for locations. Depths of observations in hectometers (100 m). Dashed lines represent definition of Pacific equatorial water. Solid lines represent specific volume as thermosteric anomaly in centiliters (.01 liter) per ton.

ties and on the degree of modification en route to the region where observed.

A water mass is usually defined by means of a characteristic diagram, on which temperature or some other thermodynamic variable is plotted against an expression for the amount of one component of the mixture. A point on such a diagram defines a water type (representing conditions in the region of origin), and the line between points observes, at least approximately, the property of mixtures; that is, on the line connecting the two points the proportion lying between the point representing one water type and the point for any mixture equals the proportion of the second water type in the mixture. The resulting curve for a vertical

water column has been called a characteristic curve, because for a given water mass its shape is invariant, regardless of depth. The existence of such a curve implies continuous renewal of water types, since otherwise mixing would lead to homogeneous water, represented by a point on the diagram.

Temperature-salinity relationships. In oceanography the characteristic diagram of temperature against salinity is usually used in studies of water masses, and the resulting temperature-salinity curve (the *T-S* curve) is used to define a water mass (Fig. 9).

In drawing such a curve, data from the upper 100 m are usually omitted because of seasonal variation and local modification in the surface layer, so that strictly speaking, a water mass as defined extends only to within 100 m of the sea surface. Although ideally a water mass is defined by a single *T-S* curve, because of random errors in field measurements and perhaps fine structure in the water mass itself, in practice an envelope of values provides a more useful definition.

On the *T-S* diagram any property which is a function only of temperature and salinity can be represented by the appropriate family of isopleths (such as values at constant pressure of density expressed as σ_t, or thermosteric anomaly, sound speed, and saturation concentration of dissolved gases). Therefore, the *T-S* diagram with isopleths can be used to determine values of such temperature-salinity dependent functions. Since the ocean is inherently stable (that is, density increases monotonically with depth), examination of the slope of a *T-S* curve (on which depth is indicated) relative to the isopleths of density permits an estimate of the vertical distribution of stability. The diagram is often useful for the detection of faulty observations and as a guide to interpolation on neighboring stations. When a uniform series of data is available, it can also be used for the quantitative representation of the frequency distribution of water characteristics.

Water-mass types. The most important and best-established water masses (characterized by *T-S* curves in Fig. 10; distribution shown in Fig. 11) occur in the upper 1000 m of the ocean. These are of three general types: (1) polar water, present south of 40°S in all oceans, and north of 40°N in the Pacific; (2) central water, occurring at mid-latitudes over most of the world ocean; and (3) equatorial water, present in the equatorial zones of the Pacific and Indian oceans.

Polar waters, including the Subarctic, Subantarctic, and Antarctic Circumpolar water masses, are formed at the surface in high latitudes and thus are cold and have relatively low salinity. Subantarctic water is bounded in the south by the well-defined Antarctic Convergence, south of which circumpolar water is found; Subarctic water has no clearcut northern boundary.

The central water masses appear to sink in the regions of the subtropical convergences (35–40°S and N), where during certain seasons of the year horizontal *T-S* relations at the surface are similar to the vertical distributions characteristic of the various water masses. The great differences in their properties are attributed to differences in the

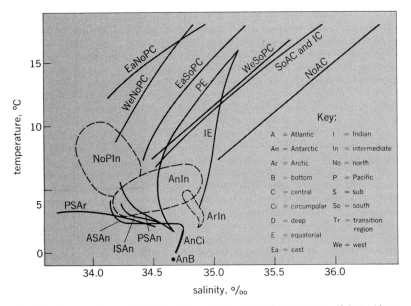

Fig. 10. Temperature-salinity curves for water masses of world ocean. (*Adapted from H. U. Sverdrup, R. H. Fleming, and M. W. Johnson, The Oceans, Prentice-Hall, 1942*)

amounts of evaporation and precipitation, heating and cooling, atmospheric and oceanic circulation, and the distributions of land and sea in the source regions.

The widespread and well-defined equatorial waters (Fig. 9) separate the central water masses of the Indian and Pacific oceans. These equatorial water masses are apparently formed by subsurface mixing at low latitudes, although the place and manner of their formation is not well known.

Intermediate waters underlie the central water masses in all oceans. Antarctic Intermediate Water sinks as a water type along the Antarctic Convergence; the water mass then formed by subsequent mixing is characterized by a salinity minimum. Arctic Intermediate Water, of little importance in the Atlantic, is widespread in the Pacific and is apparently formed northeast of Japan. Other important intermediate water masses are formed in the Atlantic and Indian oceans by addition of Mediterranean and Red Sea water, respectively. Deep and bottom waters of the world ocean are formed in high latitudes of the North Atlantic, South Atlantic (Weddell Sea), and Indian oceans. *See* ANTARCTIC OCEAN; ARCTIC OCEAN; ATLANTIC OCEAN; INDIAN OCEAN; PACIFIC OCEAN.

[WARREN S. WOOSTER]

Isotopic variations. Studies of the isotopic composition of ocean water have characterized the major deep-water masses and investigated their origin and mixing, have investigated isotopic exchange with dissolved ions, and have been concerned with the geological history of the sea. Other studies have been concerned with the exchange of moisture between the atmosphere and the sea, and the nature of the boundary layer at the air-sea interface.

Variations in the ratios of the stable isotopes are uniformly reported in terms of the ratio in an arbitrarily defined standard mean ocean water (SMOW), whose isotopic composition is numerically defined relative to a water standard distributed by the International Atomic Energy Agency in Vienna, Austria. Absolute concentrations of the isotopes are much more difficult to measure than variations in ratio, but the absolute concentrations of SMOW are believed to correspond to a D/H ratio of 1/6328, or 158 parts per million (ppm) of D, and an O^{18}/O^{16} ratio of 1993×10^{-6}, or 1989 ppm of O^{18}. The isotopic variations are reported as δ values relative to SMOW, defined from the relation $R/R_{SMOW} = 1 + \delta$, where R is the ratio D/H or O^{18}/O^{16}. The delta values are always given in parts per thousand ($^\circ/oo$), the same units in which salinity is recorded by oceanographers.

The isotopic variations in surface ocean waters reflect the net effects resulting from the overall balance between precipitation, evaporation, molecular exchange, and mixing with other surface and deep waters. In equatorial latitudes and in high latitudes, precipitation exceeds evaporation, while the reverse is true in the latitudes of the trade winds. The isotopic relationships are best seen by plotting the δ values versus salinity, as shown in Fig. 12.

The δ-S relationships are seen to have different slopes in the North and South Pacific surface waters because of the varying intensities of the

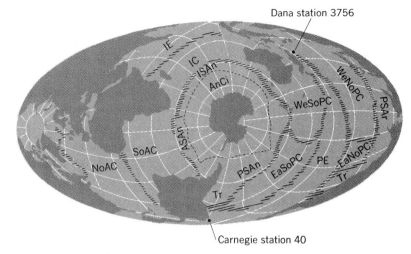

Fig. 11. Distribution of representative water masses of upper 1000 m (symbols as in Fig. 10). Dashed line around Antarctica represents Antarctic Convergence. (*Adapted from H. U. Sverdrup, R. H. Fleming, and M. W. Johnson, The Oceans, Prentice-Hall, 1942*)

different processes operating. At higher latitudes the relationships are approximately linear in regions where precipitation exceeds evaporation, but in low latitudes the relationships become very complicated in passing through the trade-wind regions to the equatorial zone of high precipitation. These variations can be related quantitatively to the net evaporation, precipitation, and mixing rates in local regions if mixing models are used in which the number of equations does not exceed the variables, and preliminary models along these lines have been made.

Deuterium-O^{18} relationships. In equatorial and temperate latitudes, surface ocean waters show simple linear relationships between D and O^{18} of the form $\delta D = M\delta O^{18}$, with zero intercept and a slope M which decreases with increasing ratio of

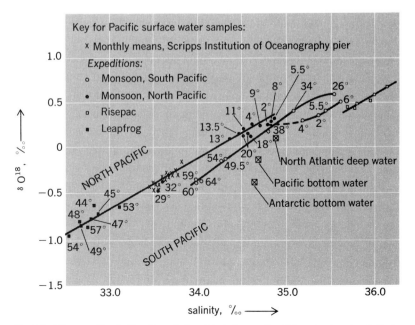

Fig. 12. Diagram illustrating the relationship between the O^{18} content and salinity in the surface and deep waters of the Pacific, Antarctic, and North Atlantic oceans.

evaporation to precipitation in a region. Characteristic values of the slope M are 7.5 in the North Pacific, 6.5 in the North Atlantic, and 6.0 in the Red Sea.

Freezing effects. The freezing of sea water concentrates salt in the liquid so that water of high salinity tends to form and sink. This process is especially important in the formation of Antarctic bottom water on, for example, the Weddell Sea Shelf in the Atlantic portion of Antarctica. The isotopic separation factors in the freezing process are so small (1.0203 for D and 1.0027 for O^{18}) that very little change in isotopic composition is produced. Ocean waters generated by such a process will thus be expected to lie along an almost horizontal line, parallel to the abscissa in a plot, such as Fig. 12, of isotopic composition against salinity. This has been observed in measurements on both Arctic and Antarctic waters.

For example, in Fig. 12 the Antarctic Bottom Water plotted in the diagram is related to the points shown for samples of South Pacific surface waters in latitudes 59–64°S. Monsoon, Risepac, and Leapfrog refer to Scripps Institution of Oceanography (SIO) expeditions. Risepac samples were along an east-west track in the South Pacific between the Society Islands and South America. The Antarctic Bottom Water samples represented were actually taken from the Weddell Sea area. The surface waters in comparable latitudes in the Atlantic plot at the same points shown for the South Pacific surface waters at 59–64°S, because all these waters are restricted to the same isotopic composition and salinity by the rapid circumpolar circulation around Antarctica.

In the Weddell Sea region, however, the points for surface and intermediate waters connecting the high-latitude surface waters and the Antarctic Bottom Water point are parallel to the abscissa in Fig. 12, indicating that the bottom water is directly generated by the freezing process. This is completely in accord with the classical picture worked out by oceanographers. There is no evidence, either from classical or isotopic data, that this process operates to a significant extent anywhere else in the oceans, although it may also operate on the Ross Sea Shelf, where no detailed winter studies have yet been made.

Deep-water relationships. The oxygen-18 values for the core waters in the major deep-ocean-water masses are shown in Fig. 12. Indian Ocean Deep Water is indistinguishable from Pacific Bottom Water, and the Pacific Deep Water has this composition over the entire Pacific Basin, except in very high southern latitudes where mixing with Antarctic Bottom Water is observed. The precision of the isotopic data shown is about ±0.02°/oo. The isotopic data are sufficiently precise that mixing between North Atlantic Deep Water and underlying, northerly flowing Antarctic Bottom Water can be seen directly on such a diagram.

The North Atlantic Deep Water point plots directly on the line relating the salinity and δ values of surface waters in the North Atlantic (not shown). Hence, the convective origin of this water from sinking of cooled surface water (discovered by F. Nansen) is directly verified by the isotopic data. However, Pacific and Indian Ocean deep and bottom waters nowhere plot on the lines relating the surface waters in the δ-S diagram. This is seen directly for the Pacific waters in Fig. 12. Thus, the deep water in these oceans cannot be significantly replenished by convective mixing with surface waters in the Pacific and Indian Ocean basins; they must originate by the mixing of other deep waters. This is in agreement with the classical oceanographic picture based on density and stability considerations.

The data in Fig. 12 further indicate that the Pacific and Indian deep waters do not lie on a line connecting the North Atlantic Deep Water and the Antarctic Bottom Water, the only major sources which can provide deep water for the Pacific and Indian oceans. Pacific and Indian deep waters are almost 50–50 mixtures of North Atlantic Deep Water and Antarctic Bottom Water, but there must be a so-called "third component" in small amounts, which almost certainly represents intermediate water. It is not yet possible to decide whether the third component is intermediate water added in the Atlantic, where the major mixing of the two main sources takes place, or whether it represents Pacific and Indian Ocean intermediate water added locally in these latter oceanic basins; the former interpretation seems more probable and is supported by some measurements, but detailed studies are necessary. In any case, the major characteristics of the deep-ocean mixing processes seem to be understood.

Changes in isotopic composition. During the Pleistocene Epoch, the isotopic composition of the sea has oscillated about its present composition because of the periodic formation and disappearance of continental ice sheets greatly depleted in D and O^{18}. These variations are estimated to be of the order of 1°/oo for O^{18} and 7°/oo for D. On the geological time scale of hundreds of millions of years the changes may have been even greater. It is known that H and D atoms can escape from the Earth's gravitational field, and it is believed that H atoms will diffuse upward and escape at a greater rate than D. These atoms are formed by photolysis of water vapor in the atmosphere, and it thus seems likely that the D/H ratio of the ocean may have been increasing with time because of preferential loss of H atoms. Hydrogen found in igneous rocks generally averages about 80°/oo lower in D than ocean water and is, therefore, at least consistent with the idea that this may be juvenile hydrogen which is of the same composition as the early oceans.

The geochemical cycle of O^{18} is much more complicated because of the variety of oxidation-reduction reactions in oxygen geochemistry. A schematic outline of the isotopic oxygen cycle is shown in Fig. 13. According to present knowledge of oxygen isotope fractionation effects, ocean water of 0°/oo would be in isotopic exchange equilibrium with igneous rocks at about 250°C; at oceanic temperatures it is far removed from equilibrium. The ocean water observed today is therefore not a direct sample of juvenile water which has escaped from the interior of the Earth at high temperatures—such water would be about 9°/oo on the SMOW scale. It is possible, however, that there is a continual input of juvenile water into the ocean,

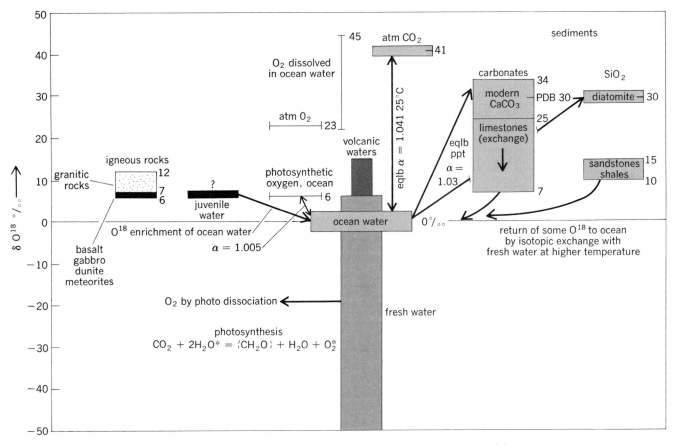

Fig. 13.   The oxygen isotope geochemical cycle. The isotopic variations are the per mil deviations from standard mean ocean water (SMOW) with approximate isotopic abundances $O^{16}=1$, $O^{17}=1/2500$, and $O^{18}=1/500$.

Single-stage isotopic fractionation factors (α) between oxygen in water and in the other substances are noted by connecting arrows.

so that the ocean is changing to, or has reached, a steady state in which the net oxygen removed from the ocean has the composition of the incoming juvenile water.

Calcium carbonate and silica remove oxygen of about $30°/_{oo}$ from the ocean, but some of this O^{18} is cycled back into the sea by isotopic exchange with fresh water on the contentnts. The average oxygen removed in this way is probably about $20°/_{oo}$. It is difficult to see how other processes can account for significantly more back-cycling of O^{18}, and it seems likely that the ocean may have undergone a progressive depletion of O^{18} through time, especially since Cretaceous times when large amounts of limestone began to be deposited in the deep sea by organisms. Studies are currently being made of oxygen isotope ratios in ancient carbonates, phosphates, and sulfates to obtain definitive evidence on this problem. When more is known about the actual variations and secular changes in isotopic compositions of the ocean, this information can in turn be applied to questions of the origin and growth of the ocean itself. *See* MARINE SEDIMENTS.   [HARMON CRAIG]

Bibliography: J. L. Cairns, Asymmetry of internal tidal waves in shallow coastal water, *J. Geophys. Res.*, 72:3563–3565, 1967; J. D. Cochrane, The frequency distribution of water characteristics in the Pacific Ocean, *Deep-Sea Res.*, 5:111–127, 1958; H. Craig, Abyssal carbon and radiocarbon in the Pacific, *J. Geophys. Res.*, 74:5491–5506, 1969; H. Craig, Isotopic composition and origin of the Red Sea and Salton Sea geothermal brines, *Science*, 154:1544, 1966; H. Craig and L. I. Gordon, Deuterium and oxygen 18 variations in the ocean and marine atmosphere, in E. Tongiorgi (ed.), *Stable Isotopes in Oceanographic Studies and Paleotemperatures*, 1965; H. Craig and B. Hom, Deuterium–oxygen 18–chlorinity relationships in the formation of sea ice (abstract), *Trans. Amer. Geophys. Union*, 49:216, 1968; C. Eckart, An analysis of the stirring and mixing processes in incompressible fluids, *J. Mar. Res.*, 7:265, 1948; S. Epstein and T. Mayeda, Variation of O^{18} content of waters from natural sources, *Geochim. Cosmochim. Acta*, 4:213, 1953; I. Friedman et al., Variation of deuterium content of natural waters in the hydrologic cycle, *Rev. Geophys.*, 2:177, 1964; Y. Horibe and N. Ogura, Deuterium content as a parameter of water mass in the ocean, *J. Geophys. Res.*, 73:1239, 1968; E. C. LaFond, in *Encyclopedia of Oceanography*, 1966; E. C. LaFond and K. G. LaFond, Internal thermal structure in the ocean, *J. Hydronautics*, 1:48–53, 1967; E. C. LaFond and K. G. LaFond, in *The New Thrust Seaward*; Transactions of the 3d Annual Marine Technology Society Conference, 1967; R. B. Montgomery, Water characteristics of Atlantic Ocean and of world ocean, *Deep-Sea Res.*, 5:134–148, 1958; D. Rochford, Total phosphorus as a means of identifying

East Australian water masses, *Deep-Sea Res.*, 5:89–110, 1958.

SAMPLING AND MEASURING TECHNIQUES

Observations of conditions in the sea generally must be made in situ and the information transmitted back to the observer, or the result must be recorded in situ and retained for reading when withdrawn from the sea. Consequently, the marine scientist is faced with peculiar problems of technique and the use of special equipment to obtain much of his information about sea water. Some of the more commonly used methods and devices for obtaining observations relative to the physical properties of sea water are described here. For information pertaining to the collection of living organisms in the sea, the sampling of the ocean bottom, and the measurement of subsurface currents, *see* MARINE SEDIMENTS; OCEAN CURRENTS.

Temperature-measuring devices. Temperature measurements in the surface layer of the ocean, to depths of 900 ft (275 m), are usually made with the bathythermograph, or BT, a nonelectric device which gives a continuous record of water temperature and depth as it is lowered and raised. The instrument can be operated at frequent intervals while underway and therefore provides a rapid means of obtaining a detailed picture of temperature distribution within the surface layer.

Reversing thermometers. The most reliable and widely used temperature-measuring device is the deep-sea reversing thermometer. This mercury-in-glass thermometer records the temperature at the time it is inverted. It is reliable to about 0.01°C with proper corrections. These thermometers are usually in a pressure-proof glass tube. In this manner the thermometer is protected from the effect of pressure, and the true water temperature is given. These same thermometers are available with a mercury bulb which can be exposed to sea pressure. The compression of the bulb on these unprotected thermometers causes them to read about 1°C high for each 100 m depth. When protected and unprotected reversing thermometers are used in pairs, they give both temperature and depth (thermometric depth). Reversing thermometers usually are attached to a reversing water bottle which collects a water sample when it is inverted.

Electrical thermometers. Since 1950 a rapidly increasing number of electric temperature recorders have been developed around thermistor beads encased in glass. A typical thermistor will change resistance 4% or about 100 ohms/°C and will have a thermal time constant of a few tenths of 1 sec. Electric temperature recorders are usually made to plot temperature against time. They are often used to study microstructure and may have a sensitivity of 0.01–0.001°C. When measuring elements are to be lowered from a ship, the recorder is usually made to plot temperature against depth.

In 1958 a system was developed utilizing electrical thermometers attached to a special cable which could be towed behind a vessel at 500 ft and at speeds of 10 knots. The thermometers are attached to the cable at 25-ft intervals. A facsimile type of recorder draws continuous isotherms on a depth-distance plot. The depth of each degree or tenth-degree isotherm is plotted every 2 sec.

Radiation thermometers have been built and flown from low-flying aircraft. These measure changes in temperature of the water surface to about 0.1°C. Radiotelemetering buoys permit water and air temperatures to be observed on unattended buoys and transmitted to land. *See* OCEANOGRAPHIC PLATFORMS.

Water samplers. For many chemical and gas analyses it is essential to obtain samples of 100–1000 ml of sea water. These usually are obtained by a series of Nansen bottles attached on $\frac{3}{16}$-in. hydrographic cable. The bottles are designed to flush continuously when lowered in an upright position. A weight messenger is then sent down the cable. As it strikes the tripping device of the Nansen bottle, the bottle is inverted and the lids are closed (Fig. 14). At the same time another messenger is released to trip the next lower bottle and so on. Usually two reversing thermometers are attached to each bottle. The thermometer and water bottle give accurate results, but the method is very time-consuming. To obtain a synoptic picture of temperature and salinity, or density, a number of closely spaced observations must be taken in a relatively short period of time.

In an effort to eliminate contamination from a metallic case, samplers such as the Van Dorn have been made from plastic tubing. Large rubber stoppers on each end are pulled shut by rubber bands. Simple open-tube-type samplers of $\frac{1}{10}$–10-liter capacity can easily be made.

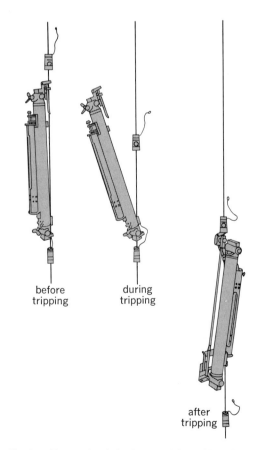

before tripping   during tripping

after tripping

Fig. 14.   Nansen bottle in three positions. (*From Instruction Manual for Oceanographic Observations, H.O. Publ. no. 607, 2d ed., U.S. Navy Hydrographic Office, 1955*)

Measurements such as carbon-14 require samples as large as 200–400 liters. Flushing of such large samplers requires either a large open-ended hose construction or a barrel with two flapper-type ports and water scoops to aid ventilation. After the sample is brought to the surface, some large samplers are retrieved on deck while full. Others are emptied while still in the water by means of a hose.

Continuous electrical temperature recorders that obtain temperatures with a single probe will almost certainly be used in the future. This will increase the need for a single collecting device that will take many water samples when used at the end of the cable.   [ALLYN C. VINE]

Serial observations. These are measurements of temperature, salinity, and other properties at a series of depths at some location in the ocean (an oceanographic station), by which the distribution in space and time of these properties (and others computed from them such as density and geopotential) may be described.

Bottles in series. A number of water samplers with thermometers attached are usually lowered on the same cast. As many as 26 samplers have been lowered at once. The number depends on the number of levels to be sampled, the strength of the wire, and the extent of possible damage to the equipment which may result from the roll of the ship or from dragging the bottom.

After the first (deepest) bottle is attached, the wire is paid out and the next bottle and its messenger are attached. When all bottles have been lowered, it is necessary to wait for the thermometers to approach equilibrium (about 10 min) before releasing a messenger to trip the cast. If the wire is nearly vertical, the messenger falls about 200 m/min. At high wire angles (60° and more have occurred under high wind conditions or strong current shear) the messenger will fall more slowly and may stick. The angle can sometimes be reduced by maneuvering the ship.

After allowing time for the final messenger on the cable to trip the deepest bottle, the wire is pulled in and the bottles removed. Water samples are drawn into laboratory bottles and the thermometers are read.

Thermometric depths. When protected and unprotected thermometers are reversed at the same depth, the unprotected will give a higher reading because of the pressure on its bulb. The difference in the two readings depends upon the pressure at reversal, and since this is proportional to depth the "thermometric depth" can be computed. With information from both protected and unprotected thermometers at several of the levels, the shape of the wire can be estimated, and the depths of the other samplers computed. Unprotected thermometers are ordinarily used at depths greater than 200 m, since the depth of the upper bottles can be computed from the wire angle and length. Depth computations are estimated to be accurate to ±5 m in the upper 1000 m and to about 0.5% of wire length below that.

Standard depths. In 1936 the International Association of Physical Oceanography proposed certain standard levels at which observations should be made or values interpolated in reporting. They are (in meters) 0, 10, 20, 30, 50, 75, 100, 150, 200, (250), 300, 400, 500, 600, (700), 800, 1000, 1200, 1500, 2000, 2500, and 3000 and by 1000-m intervals at greater depths (depths in parentheses being optional). These values are recommended as a convenient standard of comparison, not as being sufficient for measuring the ocean everywhere. Where the precise level of maxima or minima in the various properties is to be determined, the standard depths may not be adequate and more depths must be sampled.   [JOSEPH L. REID]

Analysis of water samples. The development of analytical methods to measure the kind and quantities of dissolved and suspended substances in sea water has paralleled advances in analytical chemistry. In addition to the usual considerations of accuracy, precision, speed, and cost that control the choice of analytical methods in most applications, methods for sea-water analyses are further restricted by the necessities of performing some analyses on shipboard immediately after the samples are obtained and of storing other samples for analyses that can be performed only in a shore-based laboratory. For example, analyses for biologically active substances, especially those present in trace quantities, are performed on shipboard, whereas analyses for which the highest precision and accuracy are demanded, frequently those which require precision weighing, are performed in shore-based laboratories.

In-place techniques. During the 1960s considerable effort was expended in the development of in-place analytical techniques. All of the new techniques depend upon the generation of an electrical signal in which either a voltage or a frequency is made to be proportional to the concentration of the constituent or the intensity of the property being measured. All such methods produce a continuous record in which the property being measured is displayed either as a function of time or of depth of water.

Three kinds of in-place instruments have been used successfully. The first two utilize a transducer lowered on a cable beneath the ship. One version (the older) requires an electrical cable from transducer to ship with recording of the signal on shipboard; the second has the recorder and transducer on the cable end. With the recorder on shipboard there can be continuous monitoring of the signal, and the raising and lowering can be changed to rerecord or emphasize a feature. However, recording on shipboard requires the use of at least one electrical conductor insulated from the sea, a system which requires much maintenance. Placing the recording equipment on the end of the cable with the transducer eliminates the need for long lengths of insulated electrical conductors but introduces the possibility of damaging the entire instrument by flooding with sea water should a seal on the pressure case develop a leak under pressure. A third, the newest type of instrument, is a free-fall device in which all components are in a single container which is dropped into the water with no attachment to the ship. The device, which by itself has positive buoyancy, enters the water with jettisonable weights attached and is carried either to the bottom, where the weights are released, or to some predetermined depth, where a pressure-activated release drops the weights.

Once back on the surface the device can be located with the aid of radar reflectors, flashing lights, and radio beacons. The primary advantage of free-fall devices, other than eliminating winches with long lengths of cable, is the possibility of a rapid survey of a region. Many devices can be set out in a predetermined pattern and later retrieved without the time-consuming "on station" intervals required by the older hydrographic casts with cable and sampling bottles.

In-place devices having transducers for electrical conductivity, temperature, and pressure (depth) are commercially available. Many of the so-called single-ion electrodes being developed and marketed have potential application with in-place devices. Included are electrodes for measuring dissolved oxygen, halides, fluoride, calcium, magnesium, and sulfide.

Examination of Table 5 shows that the dissolved substances in sea water fall into two main groups: the major constituents, which include sodium, potassium, calcium, magnesium, chloride, sulfate, bromide, and the sum of carbonate, bicarbonate, and carbonic acid; and the minor constituents, which include all of the other elements in the table. The major constituents show the unique property of being present in very nearly contant ratio in open sea water. For this reason, direct analyses for these substances are rarely performed today, except when estimates of small variations in concentrations that may be produced by biochemical and geochemical processes are sought. When analyses are made for these constituents (except for the CO_2 system constituents) samples are returned to a shore-based laboratory. Storage containers for such samples must be carefully chosen, because changes in concentration of some constituents may result from the interaction of sea water with the glass in many common, soft glass containers.

The most complete study on record of the concentrations of major sea-water constituents still is that conducted by W. Dittmar on 77 samples taken during the round-the-world cruise of the HMS *Challenger* in 1873–1876. Dittmar's analyses were made by what are now considered classical gravimetric and weight titration procedures: Chloride was analyzed by the Volhard method; calcium, by precipitation of the oxalate followed by ignition to, and weighing of, the oxide; magnesium, by precipi-

tation of magnesium ammonium phosphate followed by ignition to, and weighing of, pyrophosphate; sulfate, by precipitation and weighing of barium sulfate; potassium, by precipitation of potassium chloroplatinate followed by weighing of metallic platinum after reduction with hydrogen; and sodium, indirectly, as the difference between the measured sum of all cations as sulfates and the sum of the magnesium, calcium, and potassium calculated as sulfates. Despite empirical corrections and analyses by difference in Dittmar's study, more recent analyses have produced only small changes in the mean values of the concentrations of the major constituents.

Activation analysis. Several modern analytical techniques are ideally suited for analyses of many of the minor constituents in sea water. Activation analysis is the production of radioactive isotopes of elements present in a sample by controlled exposure of the sample to the neutron flux in a nuclear reactor followed by measurement of the decay properties, such as half-life and rate of decay of the artificial activities produced. This method has been used to provide estimates of the concentrations of some 18 trace elements and minor constituents in both sea and fresh waters.

Two sample preparation procedures have been used in activation analysis studies. One uses scavenging or carrier techniques to concentrate minor elements by precipitating a naturally occurring major constituent. For example, trace quantities of iron can be carried by precipitation of magnesium hydroxide. Carrier techniques provide a means of concentrating minor elements from large volumes of sea water and in some cases allow selective removal of one or more elements.

The second method of sample preparation involves the freeze-drying of a sample and irradiation of all of the solids produced. This method has the advantage of low probability of sample contamination; however, the massive quantities of sodium and halide activities produced require "cooling" for many days to allow the short-lived activities of the major elements to die out before postirradiation chemistry can be conducted. Another distinct advantage of neutron-activation analysis of whole (freeze-dried) samples is that several elements can be determined in one sample and with one irradiation.

Gas chromatography. Gas-chromatographic methods have been adapted to the analyses of dissolved atmospheric gases and to a rapidly growing list of dissolved organic substances. As with the adaption of many other analytical systems to sea-water analysis, sample preparation for gas-chromatographic analyses requires special attention. Not only does water interfere in many gas-chromatographic procedures but, in addition, the concentration of many gaseous and organic constituents is below the useful range of gas-chromatographic instruments. Sample preparation then always requires separation of water from the test substances and usually a concentrating step.

The development of gas-chromatographic methods for the analysis of dissolved oxygen, nitrogen, and argon gives a quick and convenient means of separating indicators of biological and physical processes from one another. This is pos-

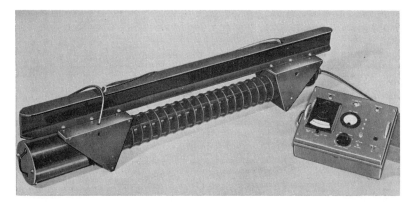

Fig. 15. Hydrophotometer for attenuation coefficient α. (*Visibility Laboratory, Scripps Institution of Oceanography*)

sible because all three gases are atmospheric components, and the sea surface, under most circumstances, can be considered to be in equilibrium with the atmosphere. The concentration of argon in sea water, because it is not involved in any naturally occurring biological or chemical processes, provides a record of the physical conditions which existed at the sea surface when a sample of water now at some subsurface depth was last at the surface. Present-day argon concentration in subsurface samples then allows one to estimate what the oxygen and nitrogen concentrations would have been in the same samples had there been no changes of biological origin. The concentration of oxygen in subsurface waters, on the other hand, responds rapidly to increases by photosynthesis (in the upper 75–100 m) and to decreases by plant and animal respiration and bacteriological decomposition at all depths. Nitrogen concentration may be either increased or decreased during nitrification or denitrification processes, which are much slower than oxygen-controlling processes.

A single 5-mm sample is sufficient for gas-chromatographic analyses for oxygen, nitrogen, and argon. The three test gases are removed from sea water in an attachment to the chromatograph, which acts as a scrubber through which the carrier gas, usually helium, bubbles through the sample and scrubs the other gases from solution. A special sampling bottle, which can be interspersed among the usual Nansen-type samples on a cable and can be attached directly to the gas chromatograph, reduces sampling errors and eliminates the possibility of contamination by contact with the atmosphere.

Studies, using gas-chromatographic analyses, have shown some unique features concerning the distribution of carbon monoxide between the atmosphere and the sea. *See* GAS CHROMATOGRAPHY.

Mass spectrometry. The high sensitivity of the mass spectrometer has been utilized in a method for the analyses of helium, neon, krypton, xenon, and argon in the sea. A single 5.5-ml sample is sufficient for these analyses. The method uses a unique sampling procedure, in which a piece of thin-walled metal tubing is crimped off at both ends at the sampling location. The spacing between the two crimping tools establishes the volume size.

The sample is released in a vacuum chamber within the mass spectrometer system by simply puncturing the thin-walled tubing.

Anodic stripping. Anodic-stripping techniques, which consist of modifications of the older polarographic method of analysis, provide measures of the extent to which some of the biochemically reactive elements—zinc, copper, cobalt, lead, and iron, for example—are complexed with both inorganic and organic ligands. Anodic-stripping methods have useful sensitivities down to $10^{-12}\,M$ for some elements and have shown complex formation to be a common occurrence and, therefore, a significant factor in biochemical and geochemical processes.   [DAYTON E. CARRITT]

Water color and transparency. The physical relationships governing the penetration and ab-

Fig. 16.   Irradiance collector for obtaining experimental values of the diffuse attenuation coefficient *K*. (*Visibility Laboratory, Scripps Institution of Oceanography*)

sorption of light, the color of the water, and the transparency of the sea are of prime importance to physical and biological oceanography. Instruments for measuring the color (transmittance) and transparency of the water are discussed below.

Water color. The color of water in the visual sense is a phenomenon which has both objective and subjective aspects. From an objective point of view the color of water is primarily the result of selective scattering and absorption of visible light by the water itself or by the dissolved or suspended material in the water. The color which is brought about by these basic mechanisms can, however, be drastically altered by the color of the bottom when visible, by surface films, by the color of the sky or the reflected images of other objects, by the spectral quality of the source of light, and by the optical state of the water surface, as well as by subjective phenomena such as the chromatic adaptation of the observer.

From a physical point of view the color of water is the color of the hydrosol only and can be observed on an overcast day in deep water with the aid of a face mask.

Transmittance. The transmittance of water is closely related to its color. In measuring the transmittance of a path length R of water, consideration must be given to the directional distribution of the light, as well as to its spectral composition. For monochromatic light the general equation for transmittance is $T = e^{-CR}$. The attenuation coefficient C varies with the directional distribution of the light, being maximum for collimated light and minimum for completely diffuse light. Modern theory makes use of the attenuation coefficients for both collimated and diffuse light. Their independent measurement is therefore essential. The coefficient α, equal to C for collimated light, is commonly measured with an instrument having a collimated source of light (Fig. 15). The coefficient K, equal to C for diffuse light, is com-

Table 9. Spectral values of diffuse attenuation coefficient, K per meter*

Wavelength, $m\mu$	Near-surface values, 28–56 m	Deep-ocean values, 56–159 m
400	0.121	
420	0.139	0.0902
440	0.133	0.0820
460	0.118	0.0749
480	0.115	0.0611
500	0.106	0.0527
520	0.0995	0.0545
540	0.107	0.0806
560	0.105	0.0852
580	0.110	0.0908
600	0.107	0.0933

*Computed from measurements of relative downwelling irradiance obtained by Dr. J. C. Hubbard, Woods Hole Oceanographic Institution, with a submerged monochromater having a bandwidth of approximately 10 mμ. Location of measurements lat 39°38′N 68°42′W.

monly measured under natural illumination by means of a photo detector and a diffusing plate (Fig. 16) having cosine collecting properties. Downwelling irradiance readings H_1 and H_2 are taken at two depths, d_1 and d_2, and the experimental coefficient K is computed from Eq. (8).

$$H_1/H_2 = e^{K(d_2 - d_1)} \qquad (8)$$

Data on the geographic and chronological variations of α and K are being collected. Some monochromatic data for α are available. Monochromatic data (1958) for K are given in Table 9.

Transparency. Measurements of water transparency or clarity are often made with a Secchi disk, an opaque white disk which is held horizontally and lowered into the water until it disappears. The greatest depth at which it can be visually detected is called the Secchi-disk reading and is related in a complex way to the optical properties of the water. Secchi-disk readings depend on the size and reflectance of the disk, the state of the sea surface, the state of the sky, the light adaptation level of the observer, and the technique of observing, as well as many other minor factors.

When the above factors are controlled, it can be shown that the apparent contrast of the disk is given by Eq. (9), where C_0 is the inherent contrast of

$$C_R = C_0 e^{-(\alpha + K)d} \qquad (9)$$

the submerged disk. C_0 depends on the submerged reflectance r of the disk and the reflectance of the surrounding water r_b as follows: $C_0 = (r - r_b)/r_b$. Definitions of α and K appear in previous sections; d is the depth of the disk below the surface.

If one uses the utmost precaution and careful technique and does not overlook the effect of human eye capabilities in the final computation, it is possible to obtain the sum $\alpha = K$ from a Secchi-disk reading. It is obvious that Secchi-disk measurements taken from the deck of a ship can be used only to describe the surface strata.

[JOHN E. TYLER]

Bibliography: H. Barnes, *Oceanography and Marine Biology: Annual Review*, vol. 14, 1976; D. E. Carritt, Analytical chemistry in oceanography, *J. Chem. Educ.*, 35:119–122, 1958; J. B. Hersey, Electronics in oceanography, *Advan. Electron. Electron Phys.*, 9:239–295, 1957; E. O. Hulburt, Optics of distilled and natural water, *J. Opt. Soc. Amer.*, 35(11):698–705, 1945; J. M. Kamenovich, *Fundamentals of Ocean Dynamics*, 1977; J. J. Myers, C. H. Holm, and R. F. McAllister (eds.), *Handbook of Ocean and Underwater Engineering*, 1969; H. Sverdrup et al., *Oceans: Their Physics, Chemistry and General Biology*, 1942.

Sea-water fertility

The fertility of a given area of the sea may be defined in terms of the production of living organic matter by the organisms which it contains. The primary productive process in the sea, as on land, is the photosynthetic reduction of carbon dioxide by plants, and the rate of this process for unit area or volume, expressed as mass of carbon fixed in unit time, is a measure of fertility. It may be measured directly in several ways, or it may be estimated from chemical changes in the water. The quantity of living organisms present at any time in a given place (standing stock or standing crop), although it may be related to fertility, is not a measure of it, being the result of a balance between production and removal. Coastal and oceanic waters differ considerably in the types of organisms which they support; the chief difference is in the bottom fauna, which is usually abundant in shallow water and sparse in the deep oceans. Along some shores there may be dense growths of kelp which may have exceptionally high rates of carbon fixation. *See* DEEP-SEA FAUNA; MARINE BIOLOGICAL SAMPLING.

Fertility of the oceans. This depends on the production of phytoplankton in the upper layer, called the euphotic zone, which receives ample sunlight for the photosynthetic processes of plants. The plants are mostly diatoms and flagellates which reproduce by division, on average probably less than once per day. If not carried downward by water movement, they tend to remain for some time in the euphotic zone since they often have structures which offer resistance to motion through the water. They sink slowly but are usually eaten by the small zooplankton before reaching the bottom. It has been estimated that about 80% of these plants are so utilized; the remainder, comprising those uneaten, or eaten, partly digested, and excreted by zooplankton, descend to nourish the benthos.

For growth the phytoplankton need radiant energy, carbon dioxide and water, nitrogen, phosphorus, and a range of trace elements of which sea water usually contains enough, with possible occasional exceptions of iron and manganese. Diatoms also need silicon, which is present in the water as monomeric silicic acid, generally in sufficient quantity. There is also a requirement, varying in nature and amount for different species, for accessory organic nutrient factors, but this is not yet fully understood. There is always enough carbon dioxide and water, so that plant growth is normally limited by the supply of radiant energy, nitrogen, and phosphorus.

About one-half of the energy in the solar spectrum can be used for photosynthesis, since only the visible spectrum of 3800–7200 A is used. Carbon fixation (gross production) is proportional to light intensity up to about 20 g-cal/cm² per day (saturation intensity); above this amount photosynthetic efficiency decreases. It is inhibited at intensities about one-third that of full sunlight, so that carbon fixation at the surface may be less than at intermediate depths. Net production is positive when the radiation intensity is sufficient for carbon gained by photosynthesis to exceed that lost by respiration. This compensation intensity has been found to be somewhat less than 100 g-cal/cm² per day, and the maximum depth at which it is attained (compensation depth) varies from a few meters to more than 100 m, depending on the transparency of the water. Compensation depth corresponds very roughly to the depth at which the surface intensity in the middle of the day has been attenuated to about 1%. At low intensities it has been found that to convert 1 mole of carbon dioxide to carbohydrate, the algae require about 10 quanta of radiant energy; this corresponds, when various corrections are made, to a maximum energy-conversion efficiency of about 17%. Making further allowance for decreased efficiency at higher intensities and correcting for loss by respiration, it is possible to calculate the maximum possible yields of organic material at various levels of surface irradiation. These yields range from zero at the compensation intensity of 100 g-cal/cm² per day to some 25 g/m² (day weight) at 600 g-cal/cm² per day, which may be attained in spells of fine summer weather. Over longer periods intensities in most parts of the world average between 200 and 400 g-cal/cm² per day, and the calculated production comes out at some 10–20 g/m² organic matter daily. Such values are in fact approached by the highest values recorded for single days in shallow water. Open-ocean values in spring are usually around 2–5 g/m² per day, and annual averages are 0.4–1.0 g/m² per day.

When irradiation is adequate for any length of time, the limitation on growth is set by the availability of nitrogen and phosphorus. The amount of these elements depends on the locality. Deep water in the oceans contains relatively high concentrations of nitrate and phosphate ions, maintained by decomposition of descending dead organisms. However, these nutrients are hindered from reaching the surface by the stable vertical density gradient in the ocean. Surface concentrations of nitrogen and phosphorus normally range up to about 200 μg N/liter and 30 μg P/liter in temperate latitudes in winter. In spring, however, active plant growth can reduce these concentrations to undetectable levels, which may remain fairly constant throughout summer. New growth is then dependent upon nutrients regenerated from dead plants and from excretory matter from the animals. A similar state of affairs generally exists throughout the year in the tropics, and relatively low apparent rates of carbon fixation are normal. Nevertheless, the total yearly production is comparable with that of other regions since radiation intensities are adequate throughout the whole year and also because high transparency of the water allows light to penetrate deeply.

In regions such as the equatorial divergences and the west coasts of Africa or South America, where physical processes cause nutrient-rich deep water to be brought to the surface, high instantaneous productivities may occur, resulting in large standing stocks of all organisms in the food chain.

The chemical composition of marine phytoplankton, unlike that of land plants, shows high protein (40–50%) and high fat (20–27%) contents and is resembled by that of the animals which succeed it in the food chain. This food chain has more links than most terrestrial ones and is in fact a fairly complex web. Some estimates of production of each species in the food chain have been made in areas, such as the North Sea, where there are important fishing grounds, and it has been possible to make reasonably reliable estimates of fish populations. To account for these estimated rates of production of the various links in the chains leading to pelagic and to demersal fish, it is necessary to postulate transfer efficiencies well over 10%, a suggestion which is supported by experimental evidence. Indeed, feeding experiments with herbivorous zooplankton have shown efficiencies greater than 50%. Compared with those on land, marine food chains appear to be efficient. It is perhaps remarkable that when the annual fishery yield of the North Sea is compared with the estimated primary production, it is found that nearly 0.5% of primary production is available for human use. The productivity of all the oceans has been put at $1.6–15.5 \times 10^{10}$ tons carbon/year, compared with 1.9×10^{10} tons carbon/year for the land areas. The table gives estimates of the annual production of living material in two nearshore environments. The estimates are based on many assumptions.

[FRANCIS A. J. ARMSTRONG]

Productivity and its measurement. The production of organic matter in the sea, as on land, is accomplished by the photosynthetic activity of autotrophic plants. In coastal waters, where sunlight penetrates to the bottom, both rooted plants and benthic algae contribute to this process. In the open sea, organic production is limited to unicellular algae, the phytoplankton, which live suspended in the upper layers of all ocean waters.

Productivity of benthic plant communities may be determined by periodic harvest and measurement of their growth over discrete time intervals. Such direct methods are impossible in the study of the short-lived plankton because of unmeasurable losses from natural death, predation, sedimentation, and advection.

A more satisfactory approach to both benthic and planktonic plant production is through measurement of chemical changes of the water accompanying photosynthesis and growth. These may be followed in the natural environment for periods ranging from 1 day to several weeks, or in the labo-

Annual production of living organisms, g/m²

Organism	English Channel	North Sea
Phytoplankton	730–910	1000
Zooplankton	275	160
Pelagic fish	2.9	6
Demersal fish	1.9	1.7
Benthos	55	50

ratory by exposing representative samples of the plant population to natural conditions for periods not exceeding 1–2 days.

Photosynthetic activity is indicated by the changes in the water of nitrogen and phosphorus salts, oxygen, and carbon dioxide, and by the degree of acidity (pH). Calculations based on natural-environment changes of these indicators must allow for gas exchanges across the water surface and the effects of vertical mixing between surface and deep waters. In such calculations the horizontal advection is generally neglected, and complete chemical recycling between sampling periods cannot be accounted for. For the last reason, the method tends to give conservative estimates of productivity.

Experimental laboratory studies include measurement of oxygen production, carbon dioxide assimilation, and pH change. Both natural-environment and laboratory studies of changes of these properties represent the net effect of photosynthesis and respiration (by both plants and animals), and hence measure net production. Respiration may be measured separately in the laboratory in dark-bottle experiments. This measurement, when added to the net change observed in transparent bottles, gives a measure of real photosynthesis or gross production. Oxygen-bottle experiments lack the necessary sensitivity for use in the open sea, where plankton are sparse; such experiments have been largely replaced by the extremely sensitive method of measuring CO_2 uptake using C^{14} as a tracer. $C^{14}O_2$ uptake appears to be equivalent to net production (photosynthesis minus respiration) by the plant community.

A third method for estimating productivity is based on the premise that photosynthesis is a function of two independent variables, the chlorophyll content of the plants and the light intensity which they receive. Production may be calculated from simultaneous measurements of these factors in the ocean and from their experimentally derived relationship to photosynthesis.

The few existing measurements of dense benthic plant communities indicate that they may produce as much as 20 g organic matter/(m²) (day), an amount equivalent to the best agricultural yields on land. Plankton production seldom if ever attains this level, though values half as great are not uncommon. The productivity of shallow, inshore waters is generally higher than that of the open sea, but the seasonal range of most marine areas includes two orders of magnitude. The mean annual rate of production in the oceans as a whole is a matter of some controversy, but probably lies between 100 and 300 g organic matter/m² sea surface, which represents an efficiency of utilization of 0.1–0.2% of incident, visible solar energy.

[JOHN H. RYTHER]

Geographic variations in productivity. Strictly speaking, variations in productivity imply variations in the rate of entry of carbon into the organic cycle, or gross photosynthesis. The extent to which this takes place in the sea is determined by the amount of photosynthesizing plant tissue present, the temperature, and the available light energy. The net productivity is the rate of plant growth. This is of greater value as a measurement because

it eliminates from the determination the respiratory and excretory losses of the plants and specifies the production of food for the planktonic animals. Variations in net productivity depend on the physiological and oceanographical factors affecting algal growth in the sea, to which the availability of nutrient salts is of prime importance.

The limited penetration of daylight into the sea restricts plant growth to the upper euphotic layer where the nutrients can be assimilated. The plants sink to deeper layers where the nutrients are released by decomposition of plant material by microorganisms and returned to the sea. Acting against this downward transport of nutrients is eddy diffusion (produced by turbulence), which brings nutrients up to the surface from the richer deeper waters. This process is facilitated where the water is homogeneous and it is suppressed by stratification. Currents perform the major transport of nutrients, and where surface divergences occur, upwelling currents bring deeper water to the surface (see illustration). In the illustration the northern Indian Ocean and the China Sea appear under southwestern monsoon conditions; these currents are reversed, with a shift in regions of divergence, in the northeastern monsoon. *See* SOUTHEAST ASIAN WATERS; UPWELLING.

In temperate and higher latitudes there is a pronounced seasonal cycle, with suppressed production in winter because of excessive turbulence and lack of light, a rapid burst of growth in spring, and limitation of this growth by lack of nutrients when the waters stabilize in summer. Frequently in autumn there is a subsidiary flowering before growth is again limited by lack of light.

In tropical and subtropical regions a more permanent stratification limits nutrient supply to the euphotic zone, and there appears to be a low net production rate. Exceptions are found in regions of divergence, principally on the western coasts of the continents and to a lesser extent in the open ocean in the equatorial region, where upwelling of rich deeper waters permits a high productivity, often throughout the year.

The question of relative production in high and low latitudes is still undecided, for the high productivity of the polar seas is of short seasonal duration; the tropics may in fact equal it on an annual basis, although running at a lower instantaneous rate. As yet, measurements of production rates are inconclusive on this point.

Some estimate of the production of higher animals can be obtained from commercial fishing and whaling statistics. The correlation with net production rate appears to be fairly good; however, it is modified by feeding and breeding requirements of the animals in question and by the fact that many higher animals of no commercial interest may be produced in some areas.   [RONALD I. CURRIE]

Size of populations and fluctuations. In temperate waters fish tend to spawn at the same place at a fixed season. The larvae drift in a current from spawning ground to nursery ground, and the adults migrate from the feeding ground to the spawning ground. Adolescents leave the nursery to join the adult stock on the feeding ground. The migration circuit is based on the track of the larval drift, and the population is isolated by its unique time of

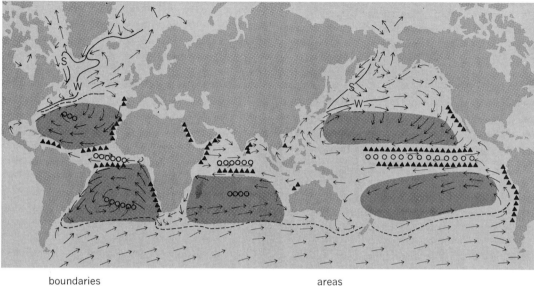

boundaries
—S—— polar waters, summer
—W—— polar waters, winter
-------- temperate waters

areas
▲▲▲▲▲▲ divergence and upwelling
ooooo convergence and sinking
▓ subtropical and tropical areas of low productivity

A world map (Mercator projection) which shows the ocean surface currents, the boundaries between areas of high and low productivity, and the principal regions of divergence and sinking.

spawning. Because larvae suffer intense mortality, numbers in the population are perhaps regulated naturally during the period of larval drift.

Methods of measuring the size of marine populations are of three basic kinds. The first is by a census based on samples which together constitute a known fraction either of the whole population, as in the case of sessile species such as shellfish, or of a particular age range, as in the case of fish which have pelagic eggs whose total abundance can be measured by fine-meshed nets hauled vertically through the water column. The second is by marking or tagging, in which a known number of marked individuals are mixed into the population and the ratio of unmarked to marked individuals is subsequently measured from samples. The third is applicable to commercially exploited populations where the total annual catch is known; the mortality rate caused by exploitation is measured, based on the age composition of the populations, thus establishing what fraction of the population the catch is. The last two methods are most generally used.

The largest measured populations are of pelagic fish, particularly of the herring family and related species, Clupeidae. One of these is the Atlantic herring (*Clupea harengus*) which contains on the order of 1,000,000,000 mature individuals and ranges over hundreds of miles of the northeast Atlantic.

Populations of bottom-living fish tend to fluctuate slowly over long time periods of 50–75 years. However, those of pelagic fish, like the Atlantic herring, may fluctuate dramatically. If fish spawn at a fixed season, they are vulnerable to climatic change. During long periods there are shifts in wind strength and direction, with consequent delays or advancements in the timing of the

production cycle. If the cycle becomes progressively delayed, the fish larvae, hatched at a fixed season, become progressively short of food, and the populations decline with time. Climatic change affects the population during the period of larval drift when isolation is maintained and when numbers are normally regulated.   [D. H. CUSHING]

Biological species and water masses. Biogeographical regions in the ocean are related to the distribution of water masses. Their physical individuality and ecological individuality are derived from partly closed patterns of circulation and from amounts of incident solar radiation characteristic of latitudinal belts. Each region may be described in terms of its temperature-salinity property and of the biological species which are adapted to all or part of the relatively homogeneous physical-chemical environment.

Cosmopolitan species. The discrete distributions of many species are circumscribed by the regions of oceanic convergence bounding principal water masses. Other distributions are limited to current systems. Cosmopolitan species are distributed across several of the temperature-salinity water masses or oceans; their wider specific tolerances reflect adaptations to broadly defined water types. No pelagic distribution is fully understood in terms of the ecology of the species.

A habitat is integrated and maintained by a current system: oceanic gyral, eddy, or current, with associated countercurrents. This precludes species extinction that could occur if a stock were swept downstream into an alien environment. The positions of distribution boundaries may vary locally with seasonal or short-term changes in temperature, available food, transparency of the water, or direction and intensity of currents.

Phytoplankton species are distributed according

to temperature tolerances in thermal water masses, but micronutrients (for example, vitamin B_{12}) are essential for growth in certain species. The cells of phytoplankton reproduce asexually and sometimes persist in unfavorable regions as resistant resting spores. New populations may develop in prompt response to local change in temperature or in nutrient content of the water. Such species are less useful in tracing source of water than are longer-lived, sexually reproducing zooplankton species.

Indicator organisms. The indicator organism concept recognizes a distinction between typical and atypical distributions of a species. The origin of atypical water is indicated by the presumed affinity of the transported organisms with their established centers of distribution.

Zooplankton groups best understood with respect to their oceanic geography are crustaceans such as copepods and euphausiids, chaetognaths (arrow worms), polychaetous annelids, pteropod mollusks, pelagic tunicates, foraminiferids, and radiolarians. Of these the euphausiids are the strongest diurnal vertical migrants (200–700 m). The vertical dimension of euphausiid habitat agrees with the thickness of temperature-salinity water masses, and many species distributions correspond with the positions of the masses. In the Pacific different species, some of which are endemic to their specific waters, occupy the subarctic mass (such as *Thysanoessa longipes*), the transition zone, a mixed mass lying between subarctic and central water in midocean and between subarctic and equatorial water in the California Current (for example, *Nematoscelis difficilis*), the barren North Pacific central (such as *Euphausia hemigibba*) and South Pacific central masses (for example, *E. gibba*), the Pacific equatorial mass (such as *E. diomediae*), a southern transition zone analogous to that of the Northern Hemisphere (represented by *Nematoscelis megalops*), and a circumglobal subantarctic belt south of the subantarctic convergence (such as *E. lucens*).

Epipelagic fishes and other strongly swimming vertebrates are believed to be distributed according to temperature tolerances of the species and availability of food. However, distributions of certain bathypelagic fishes (such as *Chauliodus*) have been related to water mass. See SEA WATER.

[EDWARD BRINTON]

Bibliography: R. J. H. Beverton and S. J. Holt, On the dynamics of exploited fish populations, *Fish Invest.* (London), vol. 19, 1957; R. J. Browning, *Fisheries of the North Pacific*, 1974; D. H. Cushing, Biological and hydrographic changes in British seas during the last thirty years, *Biol. Rev.*, 41:221–258, 1966; A. W. Ebeling, *Melamphaidae, I: Systematics and Zoogeography of the Species in the Bathypelagic Fish Genus Melampheas*, Dana Rep. no. 58, 1962; F. E. Firth, *Encyclopedia of Marine Resources*, 1969; J. A. Gullond, *Population Dynamics of the World Fisheries*, 1972; H. W. Harvey, *The Chemistry and Fertility of Sea Waters*, 2d ed., 1957; M. N. Hill (ed.), *The Sea*, vol. 2: *The Composition of Sea Water, Biological Species, Water-Masses and Currents*, 1963; J. E. G. Raymont, *Plankton and Productivity*

in the Oceans, 1963; J. P. Riley and G. Skirrow, *Chemical Oceanography*, vol. 1, 1965; W. D. Russel-Hunter, *Aquatic Productivity: An Introduction to Some Basic Aspects of Biological Oceanography and Limnology*, 1970.

Seamount and guyot

A seamount is an isolated submarine mountain rising 3000 ft or more above the ocean floor. In the Pacific Basin there are at least 10,000 such mountains, which occur as volcanic peaks on ridges, or rises, or as individual peaks.

Flat-topped seamounts are called guyots, or tablemounts (see illustration). They are present on all ocean floors but are most common in the Pacific. Bottom samples dredged from several guyots include reef corals and rounded volcanic cobbles. Both the coral and volcanic erosion debris indicate that the flattops were once at sea level

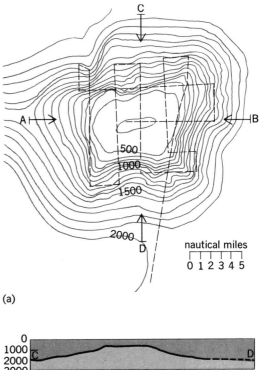

(a)

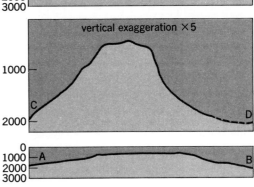

(b)

Pratt Seamount, an isolated flat-topped seamount (guyot) in the Gulf of Alaska, 142°30'W 56°20'N. (a) Plan. Contour interval = 100 fathoms. (b) Profiles.

though they are now 1000–7000 ft below the ocean surface. Thus guyots are ancient islands which were truncated to sea level by erosion. *See* OCEANIC ISLANDS.

Although the relative subsidence of guyots, atolls, and seamounts is established, the causes of subsidence are still subject to speculation. Among the possible causes are local, regional, or general sinking of the sea floor, fluctuations of sea level, and an unusual increase in the volume of the oceans since formation of the features. Although combinations of causes are likely, the most important is probably the elevation and subsidence of broad regions of the sea floor during the evolution of oceanic rises. The guyots of the central Pacific and the Gulf of Alaska may have formed in this manner. Some guyots and seamounts lie in or near regions of tectonic instability and apparently subsided as the result of large-scale faulting. *See* ATOLL; MARINE GEOLOGY; REEF; SEA-LEVEL FLUCTUATIONS.

[EDWIN L. HAMILTON; HENRY W. MENARD]
Bibliography: H. W. Menard, *Marine Geology of the Pacific*, 1964; F. P. Shepard, *Geological Oceanography*, 1977; F. P. Shepard, *Submarine Geology*, 3d ed., 1973.

Seiche

A standing wave (stationary oscillation) that occurs in enclosed or partially enclosed water bodies, such as lakes, bays, gulfs, and harbors, in which the water has a natural period of oscillation depending on the horizontal dimensions and depth of the containing basin. Seiches are commonly generated by wind, atmospheric pressure gradients, tides, or oscillations of adjacent water bodies. Relatively weak external forces can start a prolonged set of damped seiche oscillations that may reach large proportions if the periodicity of the cause approximates the natural seiche period.

As wave length is long relative to the depth of the water body, seiche behavior conforms to long-wave theory. Thus, the velocity is given by $\sqrt{gh}$, where g is the acceleration of gravity and h is the water depth. The natural period of oscillation for uniformly deep, completely enclosed, rectangular basins is given by $2l/n\sqrt{gh}$, where l is the length of the basin and n is the number of nodal lines present (1 for the fundamental case or mononodal seiche and 2 for the binodal seiche). Antinodes exist at the extremities of such a basin, with the node (or nodes) in the central region.

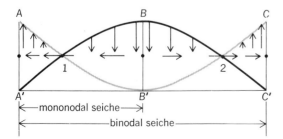

Fig. 1. Seiches in rectangular basins. Figure is exaggerated in that oscillation of water surface is shown to extend to bottom of basin.

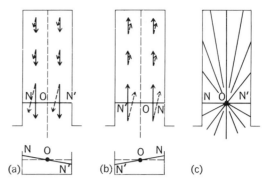

Fig. 2. Schematic representation of transverse oscillations in a bay in the Northern Hemisphere leading to the development of an amphidromic point. (*a*) Ebb current. (*b*) Flood current. (*c*) Cotidal lines.

In Fig. 1 showing the profile of standing waves in the closed basins $ABB'A'$ and $ACC'A'$, vectors indicate the vertical and horizontal components of motion as the water surface oscillates from the position of the solid line to that of the broken line. (Directions are reversed in the return motion.) Points 1 and 2, below which there is maximum horizontal and no vertical motion, are called nodes. Points A, B, and C, below which there is maximum vertical and no horizontal motion, are called antinodes. The seiche in the smaller basin $ABB'A'$, having only one node, is the fundamental mode, or mononodal seiche. The seiche in the larger basin $ACC'A'$ has two nodes, and is a binodal seiche.

The natural period for an open-end basin is $4l/n\sqrt{gh}$, with nodal and antinodal lines at the open and closed margins, respectively. Basin irregularities and Coriolis effects may introduce errors as much as 10% of formula calculations; however, refined procedures exist which make it possible to correct seiche determinations and compensate for these errors.

Coriolis effect. The Coriolis (deflective) effect of the Earth's rotation imposes a secondary effect on the water level changes produced by a seiche. In plan views Fig. 2*a* and *b*, arrows show surface currents at 1/4 and 3/4 cycles respectively, after high water at the channel head. Coriolis deflection, indicated by dotted arrows, causes an increase of water level in the right half of the current in both *a* and *b*, with a corresponding decrease to the left. The dotted line is one of mean level with respect to the deflection. Water profiles along nodal lines are shown beneath Fig. 2*a* and *b*. Point O, called the amphidromic point, is thus the only position of no water change. Line ON connects points having simultaneous high water. Other such "cotidal lines" are shown in Fig. 2*c* for different phases of one seiche oscillation. The superposition of the transverse (deflective) motion on the primary longitudinal motion causes this line to rotate counterclockwise through one complete oscillation, in the Northern Hemisphere. *See* TIDE.

Internal seiches. Internal seiches may develop when vertical variations of temperature or salinity produce a significant variation in density between two water layers. The zone or surface between these layers, will, if the density contrast is sufficient, behave as a free surface, similar to that between water and air. Seiches can be supported by this internal surface where the vertical water motion will be a maximum.   [WILLIAM L. DONN]

Sensible temperature

The temperature at which air with some standard humidity, motion, and radiation would provide the same sensation of human comfort as existing atmospheric conditions. Of the many sensible temperature formulas thus far proposed, none is completely satisfactory or generally accepted. Most are intended for warm, moist conditions; a few, like "wind chill," are for cold weather. Some are purely empirical, modifying the actual temperature according to the humidity; others are theoretical, and express estimated heat loss rather than an equivalent temperature. *See* AIR TEMPERATURE; HUMIDITY.

Heat is produced constantly by the human body at a rate depending on muscular activity. For body heat balance to be maintained, this heat must be dissipated by conduction to cooler air, by evaporation of perspiration into unsaturated air, and by radiative exchange with surroundings. Air motion (wind) affects the rate of conductive and evaporative cooling of skin, but not of lungs; radiative losses occur only from bare skin or clothing, and depend on its temperature and that of surroundings, as well as sunshine intensity.

As air temperatures approach body temperature, conductive heat loss decreases and evaporative loss increases in importance. Hence, at warmer temperature, humidity is the second most important atmospheric property controlling heat loss and hence comfort, and the various sensible temperature formulas incorporate some humidity measure. Most used is C. P. Yaglou's effective temperature, represented by lines of equal comfort on a chart of dry-bulb versus wet-bulb temperature; it relates existing comfort to that in motionless, saturated air.

Effective temperature is approximated by E. Thom's temperature-humidity index (THI), called discomfort index until certain southern United States cities objected, given in °F by Eq. (1), where

$$THI = 15 + 0.4t \qquad (1)$$

t is the sum of the dry-bulb and wet-bulb temperatures. The THI is routinely tabulated at major Weather Bureau stations, and accumulated into "cooling degree days." Similar is humiture, first applied by O. F. Heavener in 1937 to the average of temperature in °F and relative humidity, and redefined by V. E. Lally and B. F. Watson in 1960, as Eq. (2), where t is Fahrenheit temperature and e is

$$Humiture = t + e - 10 \qquad (2)$$

vapor pressure in millibars. Many other "comfort temperatures" and "sultriness indices" have been proposed.

Under cold conditions, atmospheric moisture is negligible and wind becomes important in heat removal. P. A. Siple's wind chill, given by Eq. (3),

$$Wind\ chill = (10.45 + 10\sqrt{v} - v)(33 - t) \qquad (3)$$

does not estimate sensible temperature as such,

but heat loss in kcal/m² hr, for wind speed v in m/sec and air temperature t in °C. It is used extensively by the U.S. Army, government agencies, construction contractors, and others in cold areas.
  [ARNOLD COURT]
  Bibliography: S. Licht (ed.), *Medical Climatology*, 1964; F. G. Sulman, *Health, Weather, Climate*, 1976; S. W. Tromp (ed.), *Medical Biometeorology*, 1963.

Sferics

The electromagnetic radiations originating in atmospheric electrical discharges. Thunderstorms are the principal sources of these discharges. These radiations are also known as atmospherics, but the contracted form sferics is more common. The term sferics is occasionally used for radiations coming to the Earth from outside the Earth's atmosphere, but this usage is not common. *See* ATMOSPHERIC ELECTRICITY.

Sferics are the major cause of static heard in amplitude-modulated radio receivers. This same noise can, however, be put to good use in the detection and tracking of severe storms such as squall lines, tornadoes, snowstorms, and hailstorms. Sferics are important in two major scientific areas: (1) the determination of the characteristics of the source of the radiations; and (2) the study of the propagation of these radiations within and beyond the Earth's atmosphere. An example of the latter use is the determination of the electron concentrations beyond the Earth's atmosphere by studies of the propagation of sferics along the lines of the Earth's magnetic field to the conjugate point on the opposite hemisphere. The sferics which propagate in this manner are in the audiofrequency range and are called whistlers. *See* STORM DETECTION.

Information on the nature of a storm may be obtained from the characteristics (waveform, frequency, rate of occurrence) of the sferics which originate therein. For example, it has been found that when there are numerous sferics characteristic of intracloud discharges, tornado activity is likely to occur. *See* TORNADO.

The frequencies radiated by lightning discharges vary from a few hertz to a few thousand megahertz with the peak of the frequency distribution at about 10 kilohertz. At this peak frequency, electrical storms from as far away as 3000 mi can sometimes be detected. *See* LIGHTNING.

Sferics are commonly detected by means of direction-finding antennas consisting of two loops, one oriented north-south and the other east-west. The signals picked up by these loops are fed into separate amplifiers and then into the deflection plates of a cathode-ray oscilloscope. A straight-line trace is formed on the face of the oscilloscope, the angle of the line indicating the direction of arrival of the sferics. By using three or more such direction-finding stations, the location of the source of the radiations may be pinpointed.
  [CHRISTOS G. STERGIS]
  Bibliography: J. A. Chalmers, *Atmospheric Electricity*, 1967; C. P. Mook (ed.), Symposium on sferics and thunderstorm electricity, *J. Geophys. Res.*, 65(7):1865–1966, 1960; *Sferics* (supplement),

in *Meteorological Abstracts and Bibliography*, vol. 10, no. 4, 1959; M. Uman, *Lightning*, 1969; U.S. Air Force, *Handbook of Geophysics*, 1960; R. C. Wanta, *Sferics*, in T. F. Malone (ed.), *Compendium of Meteorology*, 1951.

Sirocco

A southerly or southeasterly wind current from the Sahara or from the deserts of Saudia Arabia which occurs in advance of cyclones moving eastward through the Mediterranean Sea. The sirocco is most pronounced in the spring, when the deserts are hot and the Mediterranean cyclones are vigorous. It is observed along the southern and eastern coasts of the Mediterranean Sea from Morocco to Syria as a hot, dry wind capable of carrying sand and dust great distances from the desert source. The sirocco is cooled and moistened in crossing the Mediterranean and produces an oppressive, muggy atmosphere when it extends to the southern coast of Europe. Rain that falls in this air is often discolored by the dust or sand which is precipitated along with the water drops. Under sunny conditions, the sirocco can produce temperatures in excess of 100°F in southern Europe. Various other names are used to denote the sirocco in specific localities, such as khamsin in Egypt. *See* AIR MASS; WIND.

[FREDERICK SANDERS]

Smog

Air pollution consisting of smoke and fog. This portmanteau word was coined and first used publicly by H. A. Des Voeux in England in 1905. The main types are London smog and Los Angeles smog. The London kind consists primarily of sulfur compounds, aerosols, and CO, and it leads to bronchial irritation and coughing and, under persistent conditions of atmospheric stagnation, to a marked rise in mortality. The worst air pollution disaster on record occurred as a result of a severe smog during Dec. 5–8, 1952, which led to an estimated total of 4000 deaths in the Greater London area.

Smog of the Los Angeles type is induced by photochemical action of sunlight and consists mainly of organic compounds, nitrogen oxides, O_3, and CO. Aside from temporary eye irritation, the exact nature of the response of individuals to photochemical smog is difficult to specify. For example, people living in less smoggy areas of Los Angeles County survive heart attacks more readily than those who are exposed to more smog, and there is a small but significant relationship between motor vehicle accidents and oxidant levels. Medical groups are concerned about probable long-term effects of photochemical smog, even if such effects have not yet been definitely proved. *See* ATMOSPHERIC CHEMISTRY; ATMOSPHERIC POLLUTION; BIOCLIMATOLOGY.   [E. WENDELL HEWSON]

Snow

The most common form of frozen precipitation, usually flakes of starlike crystals, matted ice needles, or combinations, often rime-coated; also transitions between these forms, or combinations of them with ice columns, plate crystals, snow pellets, snow grains, or ice pellets. Photographs by W. A. Bentley, mostly of carefully selected six-pointed star crystals and hexagonal ice plates, show an infinite variety of beautiful symmetrical patterns. *See* PRECIPITATION.

Geographical distribution. Snow falls at sea level generally poleward from latitudes 35°N or S, or even closer to the Equator in the interior of continents and at high elevations. On west coasts it commonly falls only poleward of latitude 45°. The elevation of perpetual snow on mountains and highlands decreases poleward from the subtropics down to sea level in polar regions. Glaciers develop where the annual accumulation of snowfall exceeds melting, runoff, and evaporation. Along seacoasts, as in Greenland, large ice masses break off glaciers to form icebergs.

The greatest annual snowfalls in the United States attain several hundred inches in some locations. Examples are in the Sierra Nevadas, Cascade Range, and mountains of Colorado; other places with heavy amounts are northern New York near Lake Ontario and eastward, in Michigan along the southern shore of Lake Superior, and in northern New England.

Specific gravity. When newly fallen, snow has a gravity ratio commonly 1:10 but varying greatly. Wet snow is heavier, but dry, fluffy snow may have a specific gravity of 1:30 or less. The density of snow cover is increased by packing, melting, and refreezing, and also because snow changes to ice granules through transfer of water by sublimation from small to larger ice particles. Deep in glaciers entrapped air is compressed and density approaches that of pure ice.

Thermal effects of snow cover. Snow, especially when new, is a good insulator because air is entrapped by the flakes. Thus it often protects low vegetation, such as grains, from freezing. Air temperatures tend to be low over snow because it reflects sunshine, is a good radiator at night, and interferes with conduction of heat from the ground.

Formation of snowflakes. U. Nakaya, who studied the forms, growth, and properties of snowflakes, found in the laboratory that star crystals could be grown only in a narrow temperature range of about −14 to −17°C. Above −7°C only ice needles formed. Plates and columns formed predominantly at about −10 to −22°C with relative humidity 100–110%, whereas other types grew mostly when humidity was 110–130%. B. J. Mason later reported (1959) the following experimentally determined relationships between temperature (°C) and the type of ice crystals grown on a fiber in a cold chamber: 0 to −3, thin hexagonal plates; −3 to −5, needles; −5 to −8, hollow prisms; −8 to −12, hexagonal plates; −12 to −16, dendritic crystals; −16 to −25, plates; −25 to −50, hollow prisms. *See* CLOUD PHYSICS; GLACIOLOGY; HAIL; HYDROLOGY.

[JOE R. FULKS]

Bibliography: W. A. Bentley and W. J. Humphreys, *Snow Crystals*, 1931, reprint, 1962; P. Dalrymple et al., *A Year of Snow Accumulation at Plateau Station: Meteorological Studies at Plateau Station*, 1977; W. D. Kingery, *Ice and Snow: Properties, Processes, and Applications*, 1963.

Snow gage

An instrument for measuring the amount of water equivalent in snow; more commonly known as a snow sampler. In the earliest attempts to determine the amount of snow on the ground a simple pole with an attached scale was employed to read the depth. Since the density of snow varies considerably both as new-fallen snow and as snow packed on the ground, there was a need for a more accurate way of measuring the water equivalent of the snow than using the depth of snow as an indicator. J. E. Church of the University of Nevada first recognized this need in 1910. In Nevada and elsewhere in the western United States, it was important to know the amount of water equivalent in each winter's snow pack, for this is the source of irrigation water.

Tubular samplers. The first snow samplers were hollow steel tubes $1\frac{1}{2}$ in. in diameter. These tubes varied in length and were coupled by threads into one length sufficient to penetrate whatever snow depth was encountered. Later, G. D. Clyde of Utah State College designed a lightweight seamless aluminum tube consisting of easily coupled 30-in. lengths. The bottom section was tipped with a steel circular saw-edged cutter for penetrating hard snow or ice crusts. The inside diameter of these tubes was reduced to 1.485 in. This dimension was chosen because the weight of a cylinder of water of this size is exactly 1 oz avdp/in. of the length. This design change enables any scale that weighs in ounces to be used, and the amount of snow core to be read directly in inches of water.

Over the years several other snow samplers have been developed from material such as fiber glass and plastic for use in shallow snow, deep dense snow, and so forth. All these samplers are used in the same manner as the standard sampler. To obtain a measurement the sampler is pushed vertically through the snow to the ground surface (Fig. 1). The sampler, together with its snow core, is withdrawn and weighed. The water equivalent of the snow layer is obtained by subtracting the weight of the sampler from the total. In addition to any error introduced by the scale, there is usually a 6–8% error in the weight of samples taken in

Fig. 1. Snow surveyors measure deep snow in Utah with snow gage. Scales hang from ski pole in front of door of oversnow vehicle. (*USDA, Soil Conservation Service*)

Fig. 2. Twelve-foot-diameter pressure pillow installed on a snow course in Oregon. The small house located at the right contains riser pipe and instruments for reading increase in pressure as snow covers the pillow. (*USDA, Soil Conservation Service*)

this manner. Research has shown that the bluntness of the cutter is responsible for this error.

Automatic devices. Automatic devices that permit the remote observation of the water equivalent of the snow have been developed. These devices also permit telemetering the data to a central location, eliminating the need for travel to the snowfields to obtain the data. The two principal methods are the radioactive gage developed by R. W. Gerdel in 1954 and the pressure pillow developed by R. T. Beaumont in 1964. The radioactive snow gage employs the attenuation of energy from a source of radioactive material placed at the ground surface. The radioactive gage is calibrated prior to installation, and the attenuated energy is measured by a Geiger counter. Several of these gages have been installed in Europe, Japan, and North America. The pressure pillow snow gage is a mechanical-hydraulic system that weighs the snow. It is circular in shape, usually 12 ft in diameter, and is made of rubber material (Fig. 2). It is filled with a solution of antifreeze and water to a thickness of 3–4 in. As the snow falls and covers the pillow, pressure inside the pillow is increased because of the weight of the snow. Either a pressure transducer or a riser pipe is connected to the pressure pillow to record this increase in pressure. The increase or decrease is directly related to the mass of the snow resting on the pillow. Radios are used to transmit these data to central locations. Some hydrologists have used thin metal plates instead of rubber for construction of the pressure pillows. Several networks of pressure pillows and telemetry systems have been installed in the western United States and send hourly and daily information on the water equivalent of the snowpack for water resources management. *See* SNOW SURVEYING.

[ROBERT T. BEAUMONT]

Bibliography: R. T. Beaumont, Field accuracy of volumetric snow samplers at Mt. Hood, Oregon, in *Physics of Snow and Ice*, vol. 1, 1967; R. T. Beaumont, Mt. Hood pressure pillow snow gage, *J. Appl. Meteorol.*, 4(5), 1965; R. W. Gerdel, Radioactive snow gage, *Weatherwise*, 5(6), 1952; J. L. Smith, D. W. Willen, and M. S. Owens, Measurement of snowpack profiles with radioactive isotopes, *Weatherwise*, 18(6), 1965.

Snow line

A term generally used to refer to the elevation of the lower edge of a snow field. In mountainous areas, it is not truly a line but rather an irregular, commonly patchy border zone, the position of which in any one sector has been determined by the amount of snowfall and ablation. These factors may vary considerably from one part to another. In regions where valley glaciers descend to relatively low elevations, the summer snow line on intervening rock ridges and peaks is often much higher than the snow line on the glaciers, and in most instances it is more irregular and indefinite. If by the end of summer it has not disappeared completely from the bedrock surfaces, the lowest limit of retained snow is termed the orographical snow line, because it is primarily controlled by local conditions and topography. On glacier surfaces it is sometimes referred to as the glacier snow line or névé line (the outer limit of retained winter snow cover on a glacier).

Year-to-year variation in the position of the orographical snow line is great. The mean position over many decades, however, is important as a factor in the development of nivation hollows and protalus ramparts in deglaciated cirque beds. The average regional level of the orographical snow line in any one year is the regional snow line. Since the regional snow line is controlled entirely by climate and not influenced by local conditions, glaciologists sometimes call it the climatological snow line. The average position of the climatological snow line over a decade or more may be termed the mean climatological snow line. On broad ice sheets and glaciers, the climatological snow line is the same as the névé line. The average position of this, over a decade or more of observation, may be termed the mean névé line. *See* GLACIOLOGY; SNOWFIELD AND NÉVÉ.

[MAYNARD M. MILLER]

Snow surveying

An inventory of the winter's accumulation of snow. Snow surveying was first developed by James E. Church, University of Nevada. In Nevada, as in many areas of the western United States, the majority of the runoff is from the snowpack accumulated in the winter. The annual variability of the snowpack led Church to develop the snow gage and the concept of snow surveying to predict the expected runoff from the melting of the snow.

A snow survey is made at the same location each year, and this site is referred to as the snow course. A snow course consists of a series of sampling points, usually 10–20, to give a dependable sampling average. These points are situated at a measured spacing, 50–100 ft apart, along a permanently marked and mapped route (see illustration). The significant information obtained by means of a snow survey is the average water equivalent of all of the samples taken at the snow course. The snow courses may be sampled several times a year, year after year.

Comparison of these data with similar information concerning stream flow shows a relationship between the average water equivalent and the

Recording snow-gage measurements at a marked location along a snow-course line in the Nevada highlands. (*From Univ. Nev. Agr. Exp. Sta. Bull. no. 184, 1949*)

stream flow for several months, usually April through September inclusive. Using this relationship and measuring the snow prior to the melt each spring, a forecast of the expected runoff can be made for the coming irrigation season, which in most cases is also April through September.

Since snow surveys were initially made to forecast water supplies for irrigation, original applications were limited to that purpose. However, with the construction of large multiple-purpose reservoirs in the western United States, forecasts are also used for water-management purposes: flood forecasting, power requirements, and municipal water supplies.

In the western United States and Canada, there are some 1500 snow courses at which surveys are made each year. Newly developed automatic devices for measuring and transmitting such data from remote areas by telemetry systems are being installed at established snow courses so that the data can be obtained without traveling the snow courses. *See* SNOW GAGE.

[ROBERT T. BEAUMONT]

Bibliography: R. A. Work, Measuring snow to forecast water supplies, in *Water*, Yearbook USDA 1955.

Snowfield and névé

The term snowfield is usually applied to mountain and glacial regions to refer to an area of snow-covered terrain with definable geographic margins. Where the connotation is very general and without regard to geographical limits, the term snow cover is more appropriate; but glaciology requires more precise terms with respect to snowfield areas. These terms differentiate according to the physical character and age of the snow cover. Technically, a snow field can embrace only new or old snow (material from the current accumulation year). Anything older is categorized as firn or ice. The word firn is a derivative of the German adjective fern, meaning "of last year," and hence refers to hardened snow not yet metamorphosed to ice which has been retained from the preceding year or years. Thus, by definition, a snowfield composed of firn can be called a firn field. Another term familiar to glaciologists is névé, from the French word for a mass of hardened snow in a mountain or glacier environment. In English, rather than a specific word for the material itself, a descriptive phrase is used such as: consolidated granular snow not yet changed to glacier ice. Because of the need for simple terms, however, it is becoming most acceptable to use the French term névé when specifically referring to a geographical area of snowfields on mountain slopes or glaciers (that is, an area covered with perennial "snow" and embracing the entire zone of annually retained accumulation). For reference to the compacted remnant of the snowpack itself, that is, the material which is retained at the end of an annual melting period on a snowfield or névé, it is appropriate to use the derivative German term firn. See GLACIOLOGY.

[MAYNARD M. MILLER]

Bibliography: L. De Vries and W. E. Clason, *Dictionary of Pure and Applied Physics* (German and English), 1963; W. D. Kingery, *Ice and Snow: Properties, Processes and Applications*, 1963; M. M. Miller, The terms "névé" and "firn," *J. Glaciol.*, 2(12), 1952.

Sofar

The name associated with explosive signals which follow transmission paths through deep ocean layers. Sounds originating moderately deep in the ocean are refracted so that they propagate to large distances with small losses. For example, a small explosive charge set off at a 3000- or 4000-ft depth is detectable many hundreds of miles away. If such an explosive signal is received at two or more points, the geographical position of the explosion can be calculated. This signaling technique has been used to locate aviators who have been forced down in the open ocean, and explains the origin of the word sofar, which stands for sound fixing and ranging.

The refraction which makes possible this long-range propagation occurs because the velocity of sound has a minimum value at some depth (about 3600 ft in the North Atlantic). Because of refraction, a large number of rays can propagate without striking either the ocean bottom or surface. Since the sound is confined to a channel, the intensity of the sound carried by these rays decays with the inverse first power of the distance instead of as the inverse square power as it would in the absence of refraction. See UNDERWATER SOUND.

The signal received from an explosion through the sofar channel has a very special character. Because only certain rays which leave the signal source are intercepted by a receiver at a distant point, a signal from an explosion consists of a succession of pulses, each one having followed a different ray. These become closer together and more intense as time goes on, finally ceasing abruptly. The first received signal follows the ray that makes full excursions to the top or bottom, the last pulse being the one that goes along the channel axis. Although the pulse that goes along the channel axis has gone the shortest distance, it arrives last because it has followed a route in which the velocity of sound is smallest. [ROBERT W. MORSE]

Solar constant

The rate at which energy is received from the Sun just outside Earth's atmosphere. At Earth's mean distance from the Sun, the solar constant is 1.36×10^6 ergs/(cm²)(sec). Depending on the Sun's distance from the zenith, up to a third of this energy may be scattered in Earth's atmosphere. From the measured solar constant, the total radiation of the Sun is computed as 3.86×10^{33} ergs/sec.

The accuracy with which the solar constant is known at present is no better than about 2% because of observational difficulties due to variable atmospheric absorption. Solar variations within this limit have been suspected, but are questionable. Observations from above the atmosphere by satellite-borne instruments will eventually both improve the accuracy of the solar constant and settle the question of variation within the improved accuracy. See SUN.

[JOHN W. EVANS]

Solar energy

The energy transmitted from the Sun. This energy is in the form of electromagnetic radiation. The Earth receives about one-half of one-billionth of the total solar energy output. In 1971, based on radiation measurements in space, the National Aeronautics and Space Administration proposed a new space solar constant of 1353 watts per square meter (W/m²), and a standard spectral irradiance in W/m² over a small range of wavelengths (bandwidth) centered at the wavelength (in millionths of meters, or μm) shown in Fig. 1. Accordingly, the solar radiation energy in the ultraviolet is 105.8 W/m² (7.82% of the solar constant), in the visible 640.4 W/m² (47.33%), and in the infrared 606.8 W/m² (44.85%). The solar radiation energy output is essentially constant. However, because of the ellipticity of the Earth's orbit, the solar constant varies between 1398 W/m² at the winter solstice and 1308 W/m² at the summer solstice, or 3% about the mean value. Based on its cross-sectional area, the rotating Earth receives therefore 751.10¹⁵ kWhr annually.

The following are the most frequently used metric units for the radiative input area: langley ($1\ l = 1$ cal/cm² = 0.001163 Whr/cm² = 4.186 J/cm²); cal-

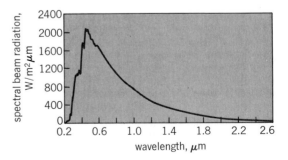

Fig. 1. NASA standard spectral irradiance at 1 astronomical unit (AU) and a solar constant of 1353 W/m². (*From Solar Electromagnetic Radiation, NASA, SP-8005, May 1971*)

orie (1 cal/cm² min = 0.0697 W/cm² = 1 l/min = 0.00418 J/cm² min); kilowatt-hour (1 kWhr/m² = 860,000 cal/m² = 86.2 l = 3.6 · 10⁶ J/m²); and joule (1 J = 0.239 cal = 2.78 · 10⁻⁴ Whr = 0.239 l·cm²).

Actually, passage through the atmosphere splits the radiation reaching the surface into a direct and a diffuse component, and reduces the total energy through selective absorption by dry air molecules, dust, water molecules, and thin cloud layers, while heavy cloud coverage eliminates all but the diffuse radiation. Figure 2 specifies the conditions for a surface perpendicular under a clear sky at mid- and low latitudes within 4 hr on either side of high noon. If these conditions were to prevail for 12 hr each day of the year (4383 hr), the energy received would lie between some 4200 and 5200 kWhr/m² yr. Actually, the number of sunshine hours even in high-insolation areas ranges from 78 to 89% of the possible, resulting in a reduction in radiative energy ("solar crude") received to the values shown in Fig. 2. Since atmospheric absorption and scattering increase strongly at low solar elevation, the average solar crude received in the most favorable areas by a horizontal surface is about 2550 kWhr/m² yr or 2.55 terawatt-hours per square kilometer per year (TWhr/km² yr). Figure 3 shows the global distribution of the average solar radiation energy incident on a horizontal surface. By following the Sun's diurnal and seasonal motion, thus facing it from near-sunrise to sunset, an instrument such as a heliostat can attain values between 3 and 3.5 TWhr/km² yr.

On a global basis, about 50% of the total incident radiation of 751.10¹⁵ kWhr/yr is reflected back into space by clouds, 15% by the surface,

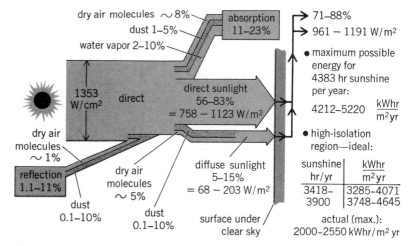

Fig. 2. Sunlight penetration of atmosphere (clear sky).

and about 5.3% is absorbed by bare soil. Of the remaining 29.7%, only about 1.7% (3.79 · 10¹⁵ kWhr/yr) is absorbed by marine vegetation and 0.2% (4.46 · 10¹⁴ kWhr/yr) by land vegetation. By far the largest portion is used to evaporate water and lift it into the atmosphere. The evaporation energy is radiated into space by vapor condensation to clouds. The solar energy spent to lift the water can be partly recovered in the form of water-power (hydraulic energy). Solar energy can be utilized in the form of heat, organic chemical energy through photosynthesis, and wind power, and also in the form of photovoltaic power (generating electricity by means of solar cells). The two greatest

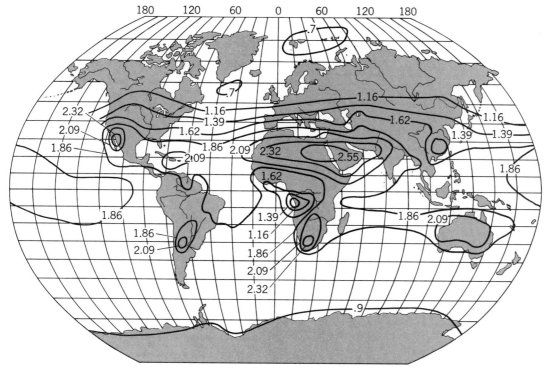

Fig. 3. Global distribution of the average annual solar radiation energy incident on a horizontal surface at the ground. The units are terawatt-hours per square kilometer year (TWhr/km² yr).

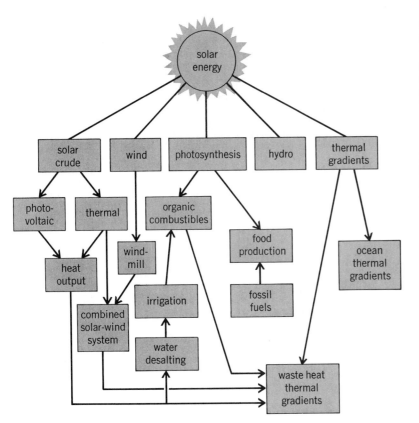

Fig. 4. Basic energy systems in solar power generation.

problems in utilizing solar energy are its low concentration and its irregular availability due to the diurnal cycle and to seasonal and climatic variations. Improved technology and investment capital are also needed.

In 1974, the U.S. Congress established the Energy Research and Development Agency (ERDA) and charged it with the development of energy conservation techniques and new technologies for extracting less readily accessible fossil energy re-

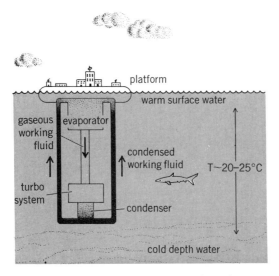

Fig. 5. Closed Rankine cycle, ocean thermal energy conversion system.

serves (such as shale oil) and for broadening the use of coal (coal gasification and liquefaction); with the advancement of nuclear (fission) technology; and with spearheading new energy technologies (geothermal, fusion, and all forms of solar energy). The basic energy systems associated with solar power generation are surveyed in Fig. 4.

Waterpower. Of the world power consumption, only about 2% is derived from waterpower, while contributions by wind power are negligible. In Europe, about 23% of available waterpower is utilized, in North America about 22%. This contrasts sharply with the lower utilization level of waterpower in other parts of the globe.

Waterpower is continuously resupplied by the Sun. Its use does not diminish a given reservoir as in the case of fossil fuels (oil, coal, gas). Waterpower is concentrated solar power and is more regularly available. It can be readily regulated and stored in the form of reservoirs. Today hydroelectric conversion efficiency from waterpower to electric power approaches 80%, compared to about 33% conversion efficiency from coal or oil. Utilization of waterpower avoids air pollution.

These are compelling reasons for increasing utilization of the world's waterpower. It is entirely within man's grasp to raise the hydroelectric power supply by a factor of 10, to some 6,400,000,000 kWhr, by the year 2000, benefiting primarily the developing areas. The main obstacle is the availability of investment capital to build the large installations required and, in some cases, the transmission facilities to distant load centers.

The development of waterpower is being advanced particularly in the Soviet Union and the People's Republic of China. Large unused waterpower resources exist still in New Guinea, Africa, South America, and Greenland.

Another form of waterpower generation is through utilization of the thermal gradient in oceans, which serve as the storage system for a vast amount of solar energy. The temperature difference between surface and bottom waters is a potentially very large source of electric power. Tropical regions, particularly between 10°N and 10°S latitude, are especially suitable, because relatively high surface temperatures provide a larger temperature gradient. Temperature differences of at least 20–23°C are available between surface and depth throughout the year, wind speeds are moderate (25 knots or less; no hurricanes), and currents are below 1 knot at all depths. One method (Fig. 5) to extract energy is to heat a suitable working fluid (for example, propane or ammonia) which is evaporated in the warm surface water and ducted to an underwater turbine, where it is allowed to expand, driving a turbogenerator system, and to condense in cool depth-waters. From there it is returned to the surface, and the process is repeated (closed Rankine cycle).

Wind power. Wind is the next largest solar-derivative power, after solar crude itself and the energy contained in the oceans. Its utilization is environmentally even more benign than fresh waterpower, because no dams and land floodings are involved.

The bulk of wind power, which lies in the upper

troposphere and the lower stratosphere, is not accessible to present-day technological potential. However, there are large areas with moderate to strong surface winds in the United States, particularly along the Aleutian chain, through the Great Plains, and along portions of the East and West Coasts. This is shown in Fig. 6 for the contiguous 48 states. A study conducted at Oklahoma State University showed that the average wind energy in the Oklahoma City area is about 0.2 kW/m² (18.5 W/ft²) of area perpendicular to the wind direction. This is roughly equivalent to the solar energy received by the same area, averaging the sunlight over 24 hr per day, all seasons, and all weather conditions. However, in contrast to solar energy, the wind energy could be converted at an efficiency of some 40%.

Agricultural utilization. The basis of the biological utilization of solar energy is the process called photosynthesis, in which solar energy provides the power within plants to convert carbon dioxide (CO_2) and water (H_2O) into sugars (carbohydrates) and oxygen. The prime conversion mechanism is the chlorophyll molecule. Organic-chemical solar energy conversion operates at very low efficiency of 0.1–0.2%; that is, for every light quantum used, 1000–500 quanta are reflected by the vegetation. However, research on algae, especially the alga *Chlorella*, has shown that higher efficiencies can be achieved. On the basis of extensive experimentation, the practically achievable yield of "chlorella farms," using sunlight as the energy source, has been estimated to be at least of the order of 35 tons of dry algae per acre. This corresponds to about 0.6% efficiency and compares very favorably even with the highest agricultural yields (10–15 tons per acre), let alone the much lower yields in less developed countries (2–2.5

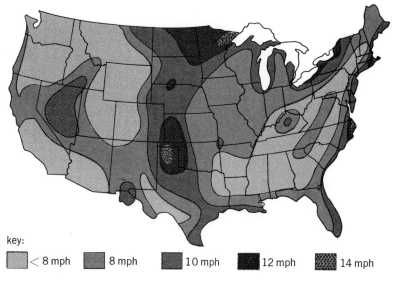

key:

| | < 8 mph | | 8 mph | | 10 mph | | 12 mph | | 14 mph |

Fig. 6. Average surface wind velocities in high-wind regions of the 48 contiguous states.

tons per acre). By building large algae farms on nonarable ground, a growing portion of the solar energy presently absorbed by bare soil, that is, about 5.3% or 400×10^{14} kWhr per annum, could be utilized for the production of organic matter for food and for conversion into synthetic liquid fuels as a complement to the world's oil supply. Again, capital and local or regional requirements determine the feasibility and worthwhileness of such endeavors.

Industrial utilization. In Japan, the United States, the southern Soviet Union, and other countries, solar energy is utilized for drying of

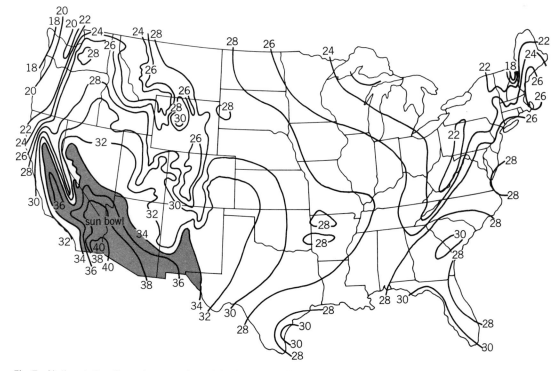

Fig. 7. National climatic center annual sunshine hours with "sun bowl" accented; values in hundreds.

Comparison of energy sources in terms of electric energy output at the bus-bar

Energy source	at electric conversion efficiency of	yields the following electric energy in kilowatt-years*
1000 metric tons of oil	35%	490 = 1.0
1000 tons of coal†	35%	312 = 0.637
1000 tons of enriched uranium in light-water reactor	31.7%	28,500,000 = 58,160
1000 tons of plutonium in liquid-metal fast-breeder reactor	40.4%	36,000,000 = 73,470
1000 km² solar absorber area‡	20%	50,300,000§ = 102,700

*The numbers at the right are indices, based on the yield of 1000 tons of oil as unity (1.0). For example, 1000 km² solar absorber area under specified conditions yields 102,700 times as much electric energy as 1000 tons of oil, and 1000 tons of coal yields 63.7% of the energy from 1000 tons of oil.

†Based on mean heating value of 26×10^6 Btu/ton.

‡Since 1000 km² solar crude yields the equivalent of 102.7×10^6 tons of oil and since a ton of oil corresponds to 7.14 British barrels, the yield of the 1000 km² corresponds to $102.7 \times 10^6 \times 7.14 = 734 \times 10^6$ bbl. Coal yield equivalent per 1000 km² solar follows from 102,700/0.637 = 161,000 per 1000 tons of coal, or the yield of 1000 km² corresponds to the yield of 161×10^6 tons of coal.

§United States production in 1969 $= 177.1 \times 10^6$ kW-yr.

fruits and vegetables by means of solar-heated air. Water evaporation in solar stills is an attractive method only where fresh water is extremely expensive, as in isolated arid regions with ready access to brackish or sea water, or under special conditions such as emergency provisions for downed flyers or astronauts, or for shipwrecked sailors. Chile, Greece, Australia, and Israel are operating and developing solar stills. Outputs may run as high as 0.15 gal of fresh water per square foot per day. Generally, it is of the order of 0.1 to 0.12 gal per square foot per day.

The most important direct use of solar radiation can be divided into two basic categories: solar heating and cooling for residential and commercial buildings, and electric power generation. As of 1975–1976, the four principal systems for buildings—water heating, space heating, space cooling, and combined system—were at widely different stages of development. Only water-heating systems had reached commercial readiness.

For solar-electric power plants, the annual sunshine hours should be as large as possible, and the humidity, which causes absorption and scattering, should be low. With its "sun bowl" (Fig. 7), the United States is the only major industrial country with high-insolation (2–2.5 TWhr/km² yr) territory within its borders; the highest values lie around 2.2 TWhr/km² yr (Fig. 3). The table shows that if in such high-insolation areas the incident solar radiation over 1000 km² is converted to electric energy at only 20% efficiency, the output is equivalent to the annual consumption of 734×10^6 bbl of crude oil or 161×10^6 metric tons of coal.

At a received solar energy level of 2.2×10^9 kWhr/km² yr and 20% conversion efficiency to electricity, a 100-GWe-yr plant requires 20,000 km² net collector area. With an arrangement of the linear parabolic reflectors such that they can follow the Sun from 30° elevation when rising to 30° elevation when setting, the total occupied area is larger. Adding 20% reserve, the collector area is 24,000 km², or about 28% of the overall occupied area of 86,000 km². Such a system would be highly modularized and would require an area that is small (Fig. 8). At 90% thermal efficiency and 40% electric efficiency, that is, a total efficiency of 36%, the required collector area is reduced from 24,000 to 13,300 km² and the overall occupied area is about 48,000 km². The desert area west of Phoenix, AZ, alone covers some 70,000 km², only a fraction of the overall available territory in the Southwest.

The energy budget of a solar station is compared with that of average desert ground in Fig. 9. Of the solar irradiation, the desert surface reflects about 40% and retains 60%, whereas the absorber area retains about 90%. Thus, 30% more of the incident solar energy is trapped. Of this, about two-thirds appears as electric energy and one-third as true, that is, extrinsic, heat production at 20% overall conversion. At 30% overall conversion from solar crude to electricity, no extrinsic heat would be

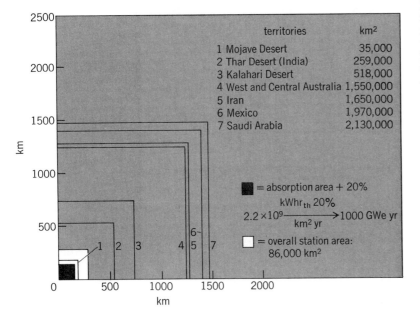

Fig. 8. Comparison of a 1000 GW-yr medium-temperature solar power complex with high-insolation territories (1 GW-yr = 8766 ×10⁶ kWhr).

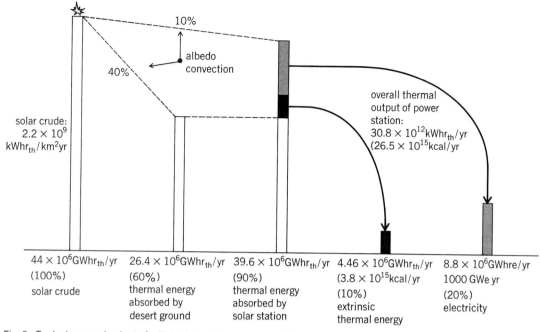

solar crude:
2.2×10^9
$kWhr_{th}/km^2yr$

10%

albedo
convection

40%

overall thermal
output of power
station:
$30.8 \times 10^{12}kWhr_{th}/yr$
$(26.5 \times 10^{15}kcal/yr$

$44 \times 10^6 GWhr_{th}/yr$
(100%)
solar crude

$26.4 \times 10^6 GWhr_{th}/yr$
(60%)
thermal energy
absorbed by
desert ground

$39.6 \times 10^6 GWhr_{th}/yr$
(90%)
thermal energy
absorbed by
solar station

$4.46 \times 10^6 GWhr_{th}/yr$
$(3.8 \times 10^{15}kcal/yr$
(10%)
extrinsic
thermal energy

$8.8 \times 10^6 GWhre/yr$
1000 GWe yr
(20%)
electricity

Fig. 9. Typical energy budget of a 1000 GWe-yr solar power station.

generated. At still higher conversion efficiency, the area occupied by the power station would be cooled, compared to the natural environment, rather than slightly heated as with the 20% system.

The overall unused thermal energy is concentrated in the power station's thermal storage system and ultimately in its electric conversion system. If the electric conversion system uses cooling water rather than air cooling, advantage can be taken of the large amount of thermal energy concentrated in the water. For this, two alternatives are available: desalination, and power generation by temperature-gradient utilization.

Three methods can be used to generate solar-electric power: (1) the solar-thermal distributed receiver system (STDS); (2) the solar-thermal central receiver system (STCRS); and (3) the photovoltaic system (PVS). In each case, the overall system may serve as backup, that is, operating only when the sun shines, equipped only with a minor energy storage capacity (for example, for 1 hr full output) to bridge temporary cloud coverage; alternatively, the overall system may include a conventional fuel system to replace solar energy at night or during cloudy days; and finally, the independent solar-electric system includes a storage system to ensure continuous power-generating capacity based on solar energy only.

In the STDS, solar radiation is absorbed over a large area covered with flat-plate (nonconcentrating) collectors or parabolic-trough concentrators, which focus the sunlight on a heat pipe carrying the working fluid (Fig. 10). The heat pipes are covered with selective coating characterized by high absorptivity to solar radiation and low emissivity at the temperature of the heat pipe. The flat-plate collector operates at turbine inlet temperatures of 250–500°F (121–260°C). With the parabolic trough, temperatures between 550–1000°F (288–538°C) at turbine inlet are attainable. The

higher the temperature, the higher the efficiency and the smaller the land area needed for a given power level, but the more expensive is the system, especially the piping and the coating.

In the STCRS, sunlight is concentrated on a receiver by a large number of mirrors designed to follow the Sun (heliostats). The receiver is a heater located atop a tower (Fig. 11) which is served by a certain collector area. The power output of such a module depends on the collector array area which, in turn, determines the tower height. The horizontal distance of the farthest mirror from the foot of

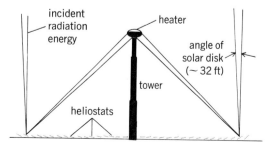

Fig. 10. Solar-thermal distributed power station system.

incident
radiation
energy

heater

angle of
solar disk
(~ 32 ft)

tower

heliostats

Fig. 11. Solar-thermal central-receiver system.

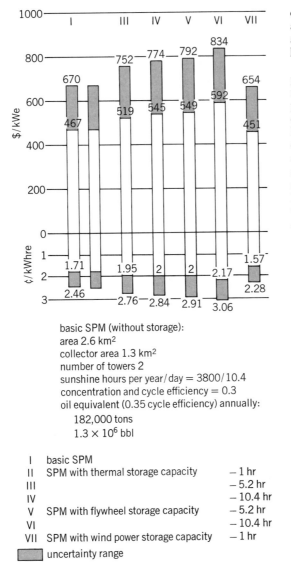

basic SPM (without storage):
area 2.6 km²
collector area 1.3 km²
number of towers 2
sunshine hours per year/day = 3800/10.4
concentration and cycle efficiency = 0.3
oil equivalent (0.35 cycle efficiency) annually:
    182,000 tons
    1.3 × 10⁶ bbl

I	basic SPM	
II	SPM with thermal storage capacity	— 1 hr
III		— 5.2 hr
IV		— 10.4 hr
V	SPM with flywheel storage capacity	— 5.2 hr
VI		— 10.4 hr
VII	SPM with wind power storage capacity	— 1 hr

⬛ uncertainty range

Fig. 12. Solar-power module (SPM), performance data, and cost summary for different energy storage types and capacities. (K. A. Ehricke).

the tower is about twice the tower height, which may range from 250 m to 500 m. A given solar power plant may consist of an arbitrary number of these modules.

Both the STDS and STCRS require large areas, due to the nature of the energy source. It is possible, of course, to subdivide the STDS into small modules, each with a small standardized electric power station, and to collect the electric current in an overall power-conditioning station. In the STCRS, the working fluid can be heated to higher temperatures, since the collector area acts as a giant parabolic mirror consisting of individual facets (heliostats). The higher temperatures attainable with the STCRS yield higher overall efficiencies (25–35%), or 150,000–200,000 kWe per square kilometer of collector area (not total area covered), compared to about 1.5 km² for the same power output in an STDS. The working fluid can be water (steam), sodium, or helium (in the order of increasing system temperature). In the simplest

case, superheated steam is generated in a heater atop the tower. The steam is ducted to the ground and used in a high-pressure and low-pressure turbogenerator system.

To provide power during off-radiation hours, part of the energy generated during sunshine hours must be stored. This means that part of a solar power module's (SPM) power output, or the output of entire modules, is not available during sunshine hours. Several options are available for storage: the energy can be stored as heat, as mechanical energy (pumping water to elevated storage basins), or as chemical energy (for example, electrolytic decomposition of water to hydrogen and oxygen). It may be possible to design the central receiver towers so that wind generators can be attached, causing the towers to generate power in the absence of sunshine as well. Figure 12 compares the cost effects of several storage modes in terms of magnitude of storage capacity (hours of full power output) and type of storage.

In the photovoltaic system (PVS), solar cells are used to produce the necessary electric energy. They are spread over a large area, as in a distributed system, to collect sunlight for the desired electric output which is in direct current. This is advantageous if long-distance transmission is involved, since losses are lower than in transmission of alternating current. For most uses, dc must be converted into ac, and if output is fed into existing distribution grids, dc-to-ac conversion equipment must be available at the plant.

The principal advantage of the PVS over the thermal systems is the absence of moving parts and fluids at high temperatures. No cooling is needed and probably no Sun-tracking. Its major disadvantages are its comparatively low efficiency (probably below 20% even after extensive development) and the fact that thermal storage cannot be used. Considerable development is required before the system can be economically competitive with the thermal systems.

Solar energy and space. Extensive use of solar energy can be anticipated in spacecraft and by extraterrestrial communities in orbit or on the Moon. Solar cells are the primary power supply for unmanned spacecraft. They are the most attractive source of electric power for manned space stations. In the 1970s and 1980s, solar cells will power tiny electric thrust units, propelling unmanned probes into the asteroid belt and on other far-flung missions in the solar system. On the Moon, undiminished solar energy can be soaked up by solar cells and solar concentrators during 14 days for immediate use, for storage in large banks of rechargeable batteries, and in fuel cells to be used during the long lunar night.

Another important relation between space and solar energy is provided by weather satellites. With improving long-term weather prediction, the practical aspects of utilizing solar energy on Earth will improve also. Advancements in the industrial utilization of space beyond the application satellites will make it possible to intercept large amounts of solar radiation in space for use on Earth or in extraterrestrial production facilities. For terrestrial applications, sunlight can be transmitted optically by reflectors (Space Light), or it can be converted to electric energy in space and transmitted to Earth via microwave beam (power generation satellites, PGS). Moreover, electric energy generated at one point on Earth can be transmitted by microwave beam to a distant load center via power relay satellite (PRS).

The general objective of Space Light is to transmit sunlight to selected areas of the Earth's night side. The key objectives can be divided into three categories: low light level for night illumination (Lunetta satellite); strong light (up to half the Sun's brightness) for stimulating food growth by enhancing photosynthetic production (photosynthetic production enhancement, or PSPE, Soletta); and light up to about 70% of the Sun's brightness as a "night sun" to provide around-the-clock solar energy for industrial purposes in a selected area of about 90,000 km².

Lunetta is designed to illuminate with the brightness of 10 to 100 full moons on a clear night (about ⅙ of that on a cloudy night), where one full moon nominally corresponds to 1/400,000 of the Sun's brightness on a clear day. In agriculture Lunetta provides the necessary brightness for sowing and harvesting at night. Lunetta's light can also accelerate large construction projects in remote areas or during long polar nights. The Lunetta can be economically placed in a 4-hr inclined orbit (about 1 earth-radius altitude) and still illuminate a fixed area (minimum of 2800 km²) if the reflector is rotated appropriately. To ensure an 8-hr illuminative period per night for a given area, six Lunettas, totaling 4.2 km² in area, are required.

The PSPE Soletta provides daylight extension (dusk and dawn illumination), especially in northern regions, and brief nighttime illumination. In a 4-hr orbit, a PSPE Soletta of a nominal 40% of full solar brightness would consist of a "swarm" of reflectors of a total area of 1040 km², all trained on the same focal area.

The night sun Soletta is in a fixed position with respect to the focal area, which preferably is located within 30° latitude on either side of the Equator. This means that the Soletta stays near the zenith of the irradiated area and does not rise or set like the Sun. Consequently, even at about 60% solar brightness the overall energy delivered per night by the Soletta is about equal to the energy delivered by the Sun per day (at equal sky conditions). The overall reflecting area of the Soletta swarm, therefore, is about 54,000 km². A solar power station in this area could operate around the clock, as if in space.

PRS and PGS are part of an overall system in which the primary electric-power station is located either on the ground or in space. The electricity is converted to microwave energy which is shaped to a controlled beam by a transmitter array antenna, which in turn is trained on a large receiver array where the microwave energy is reconverted to dc electricity by rectifiers. In the absence of optical line-of-sight connection between transmitter and receiver, a relay is required to redirect the beam.

Many costly technological problems must be solved before either optical- or microwave-beam systems can be realized. In addition, research on possible environmental effects of large numbers of microwave beams, each carrying millions of kilowatts, must be completed and evaluated for added technological requirements. *See* METEOROLOGICAL SATELLITES. [KRAFFT A. EHRICKE]

Bibliography: D. Behrman, *Solar Energy: The Awakening Science*, 1976; F. Daniels, *Direct Use of the Sun's Energy*, 1974; J. A. Duffie and W. A. Beckman, *Solar Energy Processes*, 1974; K. A. Ehricke, *Space Industrial Productivity: New Options for the Future*, Future Space Programs, Hearings of House Subcommittee on Space Science and Applications, 1975; K. A. Ehricke, *The Power Relay Satellite*, Rockwell International Rep. E74-3-1, 1974; J. A. Eibling, R. E. Thomas, and B. A. Landry, *An Investigation of Multiple-Effect Evaporation of Saline Waters by Steam for Solar Radiation*, Report from Battelle Memorial Institute to U.S. Department of Interior, December 1953; F. C. Fuglister, *Atlantic Ocean Atlas of Temperature and Salinity Profiles and Data from the International Geophysical Year of 1957–58*, Woods Hole Oceanographic Institute, 1960; P. E. Glaser, The satellite solar power station, *Proc. IEEE: International Microwave Symposium*, 1973; D. S. Halacy, Jr., *The Coming Age of Solar Energy*, 1964; H. C. Landa, *Solar Energy Handbook*, 5th ed., 1977; National Aeronautics and Space Administration, *Feasibility Study of a Satellite Solar Power Station System*, NASA CR-2357, 1974; National Aeronautics and Space Administration, *Solar Electromagnetic Radiation*, NASA SP-8005, May 1971; N. Robinson, *Solar Radiation*, 1966; U.S. Department of Health, Education and Welfare, *Proceedings of a Symposium on Biological Effects and Health Implications of Microwave Radiation*, Pub. Health Serv. Rep. BRH/DBE 70, 1969; C. Zener, Solar sea power, *Phys. Today*, 1973.

Sonar

A term that refers both to the application of underwater sound to the detection and location of objects in the sea and to the apparatus used in such applications. The word is derived from "sound navigation and ranging"; the British use the word asdic. Sonar methods are used widely in naval warfare, especially in the detection of submerged submarines. Since electromagnetic radiations, such as visible light or radar, do not penetrate the sea significantly, sonar is the most successful method of underwater detection. Except for underwater telephones, sonar is also the only means by which a fully submerged submarine can apprise itself of what is happening around it. Thus sonar is used by submarines to locate surface vessels and other submarines and to navigate through a mine field or under the ice in the Arctic Ocean.

There are two general types of sonar methods, active and passive. In an active sonar system a pulse of sound is generated by the searcher and projected into the water. This sound is reflected back from the target and detected by the searcher as an echo. Since the speed of sound in sea water is known, the range and also the bearing of the target can be determined. This method is also called echo-ranging. In a passive system the searcher detects the noises emitted by the target. Unless more than one listening station is involved, passive sonar provides information only as to the existence of a noise source and its bearing from the searcher. Because the noise radiated from a ship is often characteristic of the type of ship, passive sonar, in contrast to active sonar, may help in classifying the target.

The particular sonar system used in a given military situation depends upon operational or tactical requirements. Since active sonar systems involve the deliberate emission of sound, which discloses the searcher to easy detection by other vessels, they are used sparingly by submarines, which rely on concealment for their safety. The active systems are therefore used mainly by surface antisubmarine vessels, such as destroyers, since they generally operate in an environment which is too noisy for successful passive detection of a quiet submarine. This interfering noise may be generated by the antisubmarine vessel itself moving at high speed or by other ships in the vicinity which it may be escorting. For a discussion of ship noise and other related information *see* UNDERWATER SOUND.

There is a wide variety of sonar systems other than those carried by ships. Sonar may also be used by aircraft for submarine detection by dropping buoys (called sonobuoys) in the water, by mine sweepers in locating mines, or by acoustic torpedoes. There are also nonmilitary applications. Sonar may be used for detection of schools of fish or for the location of submerged wrecks. An echo sounder, or depth indicator, is essentially an active sonar system which sends sound pulses to the sea bottom. *See* ECHO SOUNDER; SONOBUOY.

Active sonar. Active sonar systems consist of the following: one or more transducers to send and receive sound, electrical and electronic equipment for the generation and detection of the electrical impulses to and from the transducer, and a display or recorder system for the observation of the received signals (see illustration).

The transducer is located within a water-flooded, streamlined sonar dome in order to reduce noise generated by the motion of the transducer through the water. On a surface ship the sonar dome is usually mounted on the keel in a location that puts it as deep in the water as possible. To increase this depth, a towed sonar is sometimes used. Towed sonar also can be used from blimps or helicopters. The receiving transducer of an active sonar is directional in order to resolve the bearing of the target.

Sound pulsing. A narrow bandwidth pulse of electrical energy is applied to the transducer, which in turn generates a similar pulse of acoustic energy in the water. Targets in the beam reflect part of this energy back either to the same transducer or to another receiving transducer, which converts this to electrical energy. This signal is amplified and then displayed as a function of time after the original emission, thus indicating reflecting targets in the beam of the transducer. The sonar frequencies employed are generally in the range of 5000–50,000 hertz. The pulse is repeated periodically at a rate which depends upon the maximum range obtainable.

Display systems. Various types of display or recording systems are used. Often both a visual and an aural presentation is made to the sonar operator. In an aural presentation the outgoing pulse (ping) is heard followed by the returning echoes which come back from along the sonar beam. (The frequencies which are presented are scaled down from the original ones.)

One type of visual indicator is a range recorder. In this a stylus moves across a chemically treated paper, which is darkened electrically by a received signal. The paper moves at a constant rate and the stylus is phased so that it starts each sweep as a sound pulse is emitted. The distance along the trace at which an echo appears is then a measure of the range to the target. Since successive echo pulses are displayed adjacent to one another, the motion of the target relative to the searching vessel may be determined.

An oscilloscope presentation is used with a scanning sonar. Here the emitted sound pulse is sent in all directions, and a directional receiving beam is rotated electrically in a horizontal plane. The location of the ship is at the center of the oscilloscope pattern, and the spot of the oscilloscope moves outward in a direction corresponding to the axis of the receiving transducer. The intensity of the illuminated spot varies with the received signal. This display, which gives both the range and direction of the various targets, is also used in radar and is called a plan-position indicator (PPI).

A cathode-ray tube may be operated in still another way so that the spot, initiated with the outgoing pulse, moves from left to right, the received signal deflecting the spot in the vertical direction. This is a variable-displacement indicator and shows targets as a function of range for a fixed bearing. In radar such a display is called an A-scan.

Performance factors. Many factors influence the performance of an active sonar system. Chief among these, particularly since it can be so variable, is the propagation loss in the water between the sending transducer and the target. This loss is determined by the frequency of the sound and by the thermal gradients in the ocean which refract the sound waves.

Another determining factor is the reflecting power of the target, or target strength. For an echo

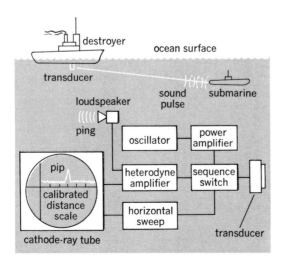

Active sonar system. (*From H. F. Olson, Acoustical Engineering, 3d ed., Van Nostrand, 1957*)

to be detected it must be of greater strength than other signals which may be received. These interfering sounds are generally due either to sonar self-noise or to reverberation.

Sonar self-noise is the noise which is generated by the motion of the ship carrying the sonar. Most of this sound is generated by cavitation and turbulence from regions very close to the transducer. Self-noise increases rapidly with the speed of the ship, especially above the speed where cavitation sets in.

Reverberation refers to the combination of all the echoes returned to an active sonar system from the ocean itself. This includes the surface and bottom, bubbles, suspended marine organisms, and inhomogeneities in the sea. A sonar operator hears reverberation as a quavering ring which sets in as soon as the outgoing sound pulse has been emitted. Since reverberation is the resultant of a large number of very weak scattered echoes, it is statistical in nature so that the individual echoes are not resolved. There are three kinds of reverberation: volume, surface, and bottom. Volume reverberation, which arises in the water itself, is evident as soon as the outgoing pulse leaves and decays fairly rapidly thereafter. Surface reverberation appears as soon as the outgoing pulse encounters the surface of the sea. This type of reverberation increases markedly with increased wave action at the sea surface. Bottom reverberation is due to irregularities in the ocean bottom and may be the most significant reverberation in shallow water.

Target strength is a term used with active sonar to give a quantitative measure of the echo returned from an object back along the direction from which the incident sound came. Since reflected or scattered sound intensities I_s are generally proportional to the incident intensity I_i, $I_s = kI_i/r^2$, where r is the distance from the target and k is a constant which measures the scattering or reflecting strength of the target. For most objects, k depends upon its orientation. The target strength T is defined as k measured in units of decibels (dB); that is, $T = 10 \log k$. For a perfectly reflecting sphere of radius a (where a is larger than the wavelength), $T = 20 \log (a/2)$. Generally the yard is chosen as the unit of length in the definition of target strength. Since $\log 1 = 0$, a sphere of 2-yd radius has a target strength $T = 0$. Therefore, an object's target strength T is a comparison of its reflecting strength with that of a sphere of 2-yd radius. For example, an object with $T = -6$ dB has the equivalent reflecting power of a perfectly reflecting sphere 1 yd in radius.

The sonar Doppler effect is the change in frequency of a sonar echo due to target motion. As in all wave motion, there is a frequency shift in a sound wave reflected from a moving target. Since the sound scatterers in the sea which are responsible for reverberation are relatively fixed, an echo from a moving target can often be detected in the presence of reverberation by having a different frequency.

Passive sonar. This is an underwater acoustic system which gains detection by merely listening for noise radiated by a possible target. Passive systems have the distinct advantage of being undetectable themselves but have the disadvantage of

having to take in all the sea and ship noise. Modern submarines carry passive sonar apparatus which is believed to be capable of detecting ships as far away as 100 mi under favorable conditions. A passive system consists of a highly directional and trainable transducer or array of transducers, electronic amplifiers, and a display system. The display is usually aural; consequently the frequencies employed generally lie in the audible region. These are also the most favorable frequencies in terms of the spectrum of noise radiated by ships. The noise emitted by a ship which is detected by passive sonar comes from the operation of machinery, primarily the propulsion machinery, and impulsive beats emanating from the rotation of the ship's propellers.

The effectiveness of passive sonar depends upon the magnitude of the radiated noise, the propagation loss between the radiating ship and the sonar, and the background noise observed by the sonar. This limiting background noise may be self-noise or ambient sea noise. The former is noise generated either by machinery on the listening ship or by its motion through the water. If the ship carrying the sonar is in a quiet condition and is moving slowly, the self-noise may be reduced sufficiently so that ambient sea noise becomes the limiting factor. This is the general continuous spectrum of noise which is found naturally in the sea. Since ambient sea noise comes from all directions, its effect relative to a target may be decreased by improving the directivity of the sonar.

[ROBERT W. MORSE]

Bibliography: A. W. Cox, *Sonar and Underwater Sound*, 1975; W. E. Kock, *Radar, Sonar, Holography: An Introduction*, 1973; R. J. Urick, *Principles of Underwater Sound for Engineers*, 1967.

Sonobuoy

A miniature broadcasting station placed in the sea to receive underwater sounds and transmit them to an aircraft or other remote point. Its original and still most common application is to provide an "ear in the water" to permit an aircraft to detect typical sounds made by submarines. The basic concept was proposed in 1941 by P. M. S. Blackett of England, but original developments were carried on by the U.S. National Defense and Research Committee because British research facilities were overtaxed with other priority tasks. Substantial numbers of sonobuoys were used in World War II, and many improvements have been made since. Sonobuoys are used by antisubmarine services of many countries.

Aircraft have unique advantages in antisubmarine warfare (ASW) because of their mobility and relative invulnerability to submarine counterattack. Without the ability to launch sonobuoys, however, the aircraft's capability against submerged submarines would be very limited, since underwater acoustics is probably the most effective means of detection. *See* UNDERWATER SOUND.

After an ASW aircraft launches a sonobuoy into the airstream, the rotochute opens, retarding and controlling the flight trajectory so that the buoy strikes the water in a nearly vertical attitude (see

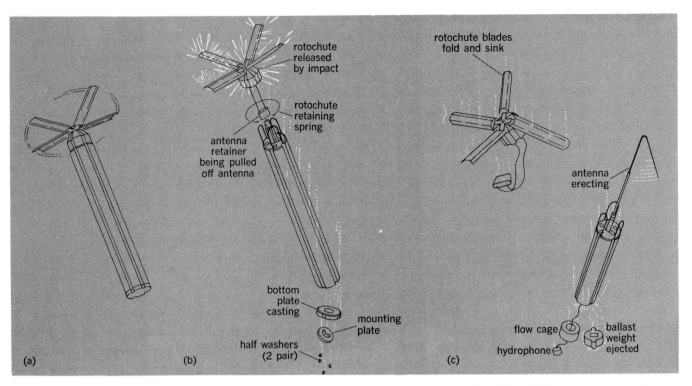

Launch of sonobuoy from an antisubmarine warfare (ASW) aircraft. (a) In flight. (b) Buoy strikes water in nearly vertical attitude. (c) Antenna and hydrophone are positioned automatically.

illustration). The impact is used to trigger the sequence illustrated: release of the rotochute; erection of the antenna; and release of the hydrophone, which sinks to its operating depth. Sea water flows into a water-activated battery which provides power for all the electronic circuitry. Pressure changes (sounds) at the hydrophone produce electrical voltages which are amplified and used to modulate the transmitter, which is pretuned to one of several channels. The modulated radio-frequency signals are received and demodulated in the aircraft, where they are listened to and analyzed for evidence of a submarine. Of course, there are many background sounds in the sea caused by wave action, marine life, ships, and other sources, making submarine detection very difficult. Because the aircraft cannot conveniently pick up the sonobuoy, at the end of its useful life a scuttling device allows water to enter the buoyancy chamber, causing the sonobuoy to sink. This expendable feature requires that sonobuoys not only be built in large quantities, but also be very carefully designed for high reliability at low cost.

While the most common type of sonobuoy merely listens for underwater sounds arriving from all directions, there are designs which even operate as miniature sonars. Such active sonobuoys are needed if the submarine is masked by natural sea background. See SONAR.

Sonobuoys have been used as geophysical instruments, providing a convenient means of picking up underwater acoustic returns in ocean bottom profiling.

[I. H. GATZKE]

Bibliography: R. E. Houtz, J. Ewing, and X. Le Pichon, Velocity of deep-sea sediments from son-

obuoy data, *J. Geophys. Res.*, vol. 73, no. 8, 1968; X. Le Pichon, J. Ewing, and R. E. Houtz, Deep-sea sediment velocity determination made while reflection profiling, *J. Geophys. Res.*, vol. 73, no. 8, 1968.

Southeast Asian waters

All the seas between Asia and Australia and the Pacific and the Indian oceans. They form a geographical and oceanographical unit because of

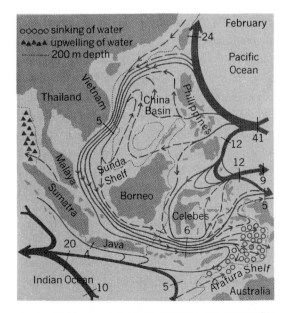

Fig. 1. Water circulation in South China and Indonesian seas, February. Transports in 10⁶ m³/sec.

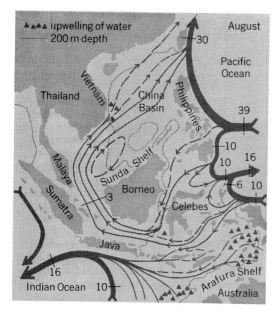

Fig. 2. Water circulation in South China and Indonesian seas, August. Transports in 10⁶ m³/sec.

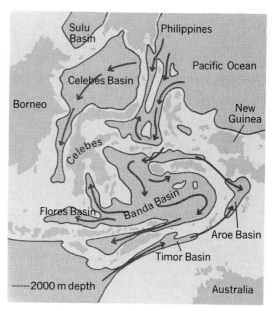

Fig. 3. Flow of bottom waters in the eastern part of the Indonesian Archipelago.

their special structure and position, and make up an area of 8,940,000 km², or about 2.5% of the surface of all oceans.

Ocean floor. The ocean floor consists of two large shelves, and a number of deep-sea depressions. The Sunda Shelf is covered with littoral sediments, and the submerged valleys of two large river systems are found on it. The Arafura Shelf connects New Guinea with Australia. The China Basin, whose maximal depth is 5016 meters (m), is connected with the Pacific Ocean over a sill about 2000 m in depth. The Sulu Basin (maximal depth 5580 m) has the highest sill (420 m). The Celebes Basin (maximal depth 6220 m) is connected with the Pacific Ocean over three sills of about 1400 m each. The Banda Basin is subdivided into several smaller basins, including the Weber Deep of 7440 m; its sill depth is 1880 m. In the Celebes, Banda, and Flores basins volcanic ashes are found.

Surface circulation. The surface circulation is completely reversed twice a year (Figs. 1 and 2) by the changing monsoon winds.

Monsoon current. During the north monsoon, which is fully developed in February, the monsoon current is formed in the northern China Sea, flows along the coast of Vietnam into the Java Sea, and into the Flores Sea. Parts of its water masses return into the Pacific Ocean; parts sink in the Banda Sea or flow into the Indian Ocean.

During the south monsoon in August, the monsoon current is formed by water flowing into the Java Sea between the Molucca Islands and through the Macassar Strait from the Pacific Ocean. It flows through the Java and the South China seas and returns north of the Philippines into the Pacific Ocean. The water masses originating from the upwelling region in the Banda Sea flow chiefly into the Indian Ocean.

The transports of the monsoon current are 3,000,000 m³/sec in August and 5,000,000 m³/sec in February, but are small compared with those in

the adjoining parts of the Pacific and Indian oceans (Figs. 1 and 2).

Salinity and temperature. The monsoons cause a pronounced rainy and dry season over most parts of the region and consequently strong annual variations of the surface salinity, which normally increases during the dry season (north monsoon), when evaporation prevails, and decreases during the rainy season (south monsoon). Regions of permanent low salinities are in the Malacca Straits, the Gulf of Thailand, and the waters between Borneo and Sumatra, because of the discharge of the large rivers. The surface temperature is always high, between 26 and 29°C, and drops only in the northern parts of the South China Sea and in the Arafura Sea to about 24°C during the dry season.

Subsurface circulation. The subsurface circulation carries chiefly the outrunners of the intermediate waters of the Pacific Ocean into these seas. These waters are identified by salinity minima at depths of 300 and 1000 m. The general subsurface flow is from the Pacific Ocean to the Indian Ocean. The deep basins are supplied with Pacific Ocean deep water that enters over the various sills (Fig. 3). Only the Timor and Aroe basins are filled with Indian Ocean deep water.

Tides. The tides are mostly of the mixed type. Diurnal tides are found in the Java Sea, in the Gulf of Tonkin, and in the Gulf of Thailand. Semidiurnal tides with high amplitudes occur in the Malacca Straits. Strong tidal currents occur in many of the small shallow passages. *See* INDIAN OCEAN; PACIFIC OCEAN.   [KLAUS WYRTKI]

Bibliography: Oceanographic and meteorological observations in the China Seas and in the Western Part of the North Pacific Ocean, Kon. Ned. Meteorol. Inst. no. 115, 1935; G. Schott, *Geographie des Indischem und Stillen Ozeans*, 1935; *The Snellius Expedition 1929–30*, 6 vols., 1933–1938; K. Wyrtki, *Physical Oceanography of the Southeast Asian Waters*, 1960.

Squall

A strong wind with sudden onset and more gradual decline, lasting for several minutes. Wind speeds in squalls commonly reach 30–60 mph, with a succession of brief gusts of 80–100 mph in the more violent squalls. Squalls may be local in nature, as with isolated thunderstorms, or may occur over a wide area in the vicinity of a well-developed cyclone, where the squalls locally reinforce already strong winds. Because of their sudden violent onset, and the heavy rain, snow, or hail showers which often accompany them, squalls cause heavy damage to structures and crops and present severe hazards to transportation.

The most common type of squall is the thundersquall or rainsquall associated with heavy convective clouds, frequently of the cumulonimbus type. Such a squall usually sets in shortly before onset of the thunderstorm rain, blowing outward from the storm and generally lasting for only a short time. It is formed when cold air, descending in the core of the thunderstorm rain area, reaches the Earth's surface and spreads out. Particularly in desert areas, the thunderstorm rain may largely or wholly evaporate before reaching the ground, and the squall may be dry, often associated with dust storms. See DUST STORM; SQUALL LINE; THUNDERSTORM.

Squalls of a different type result from cold air drainage down steep slopes. The force of the squall is derived from gravity and depends on the descending air which is colder and more dense than the air it replaces. So-called fall winds of this kind are common on mountainous coasts of high latitudes, where cold air forms on elevated plateaus and drains down fiords or deep valleys. Channeling of the air through narrow valleys increases their force. Squall winds (bora) of 100 mph or more are observed in winter, when arctic air masses from the Soviet Union intermittently spill over the mountains of Yugoslavia into the relatively warm coastal regions of the Adriatic. Violent squalls also characterize the warm foehn winds of the Alps and the similar chinook winds on the eastern slopes of the Rocky Mountains. See CHINOOK; WIND.   [CHESTER W. NEWTON]

Squall line

A line of thunderstorms, near whose advancing edge squalls occur along an extensive front. The thundery region, 20–50 km wide and a few hundred to 2000 km long, moves at a typical speed of 15 m/sec (30 knots) for 6–12 hr or more and sweeps a broad area. In the United States, severe squall lines are most common in spring and early summer when northward incursions of maritime tropical air east of the Rockies interact with polar front cyclones. Ranking next to hurricanes in casualties and damage caused, squall lines also supply most of the beneficial rainfall in some regions. See SQUALL.

A squall line may appear as a continuous wall of cloud, with forerunning sheets of dense cirrus, but severe weather is concentrated in swaths traversed by the numerous active thunderstorms. Their passage is marked by strong gusty winds,

usually veering at onset, rapid temperature drop, heavy rain, thunder and lightning, and often hail and tornadoes. Turbulent convective clouds, 10–15 km high, present a severe hazard to aircraft, but may be circumnavigated with use of radar.

Formation requires an unstable air mass rich in water vapor in the lowest 1–3 km, such that air rising from this layer, with release of heat of condensation, will become appreciably warmer than the surroundings at upper levels. Broad-scale flow patterns vary; Fig. 1 typifies the most intense outbreaks. In low levels, warm moist air is carried northward from a source such as the Gulf of Mexico. This process, often combined with high-level cooling on approach of a cold upper trough, can rapidly generate an unstable air mass. See THUNDERSTORM.

The instability of this air mass can be released by a variety of mechanisms. In the region downstream from an upper-level trough, especially near the jet stream, there is broad-scale gentle ascent which, acting over a period of hours, may suffice; in other cases frontal lifting may set off the convection. Surface heating by insolation is an important contributory mechanism; there is a marked preference for formation in midafternoon although some squall lines form at night. By combined thermodynamical and mechanical processes, they often persist while sweeping through a tongue of unstable air, as shown in Fig. 1. Squall lines forming in midafternoon over the Plains States often arrive over the midwestern United States at night. See STORM.

Figure 2 shows, in a vertical section, the simplified circulation normal to the squall line. Slanting of the drafts is a result of vertical wind shear in the storm environment. Partially conserving its horizontal momentum, rising air lags the

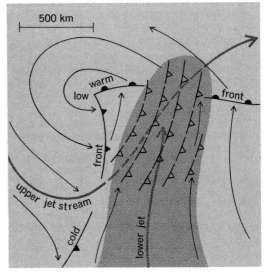

Key: ▨ moist air in warm sector of a cyclone

Fig. 1. Successive locations of squall line moving eastward through unstable northern portion of tongue of moist air in warm sector of a cyclone. Thin arrows show general flow in low levels; thick arrows, axes of strongest wind at about 1 km aboveground and at 10–12 km.

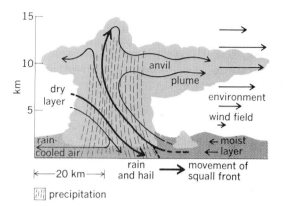

Fig. 2. Section through squall-line-type thunderstorm.

foot of the updraft on the advancing side. In the downdraft, air entering from middle levels has high forward momentum and undercuts the low-level moist layer, continuously regenerating the updraft. Buoyancy due to release of condensation heat drives the updraft, in whose core vertical speeds of 30–60 m/sec are common near tropopause level. Rain falling from the updraft partially evaporates into the downdraft branch, which enters the storm from middle levels where the air is dry, and the evaporatively chilled air sinks, to nourish an expanding layer of dense air in the lower 1–2 km that accounts for the region of higher pressure found beneath and behind squall lines. In a single squall-line thunderstorm about 20 km in diameter, 5–10 kilotons/sec of water vapor may be condensed, half being reevaporated within the storm and the remainder reaching the ground as rain or hail.   [CHESTER W. NEWTON]

Bibliography: L. J. Battan, The Thunderstorm, 1964; C. W. Newton, Severe convective storms, in Advances in Geophysics, vol. 12, 1967.

Storm

An atmospheric disturbance involving perturbations of the prevailing pressure and wind fields on scales ranging from tornadoes (1 km across) to extratropical cyclones (2–3000 km across); also, the associated weather (rain storm, blizzard, and the like). Storms influence man's activities in such matters as agriculture, transportation, building construction, water impoundment and flood control, and the generation, transmission, and consumption of electric energy.

The form assumed by a storm depends on the nature of its environment, especially the large-scale flow patterns and the horizontal and vertical variation of temperature; thus the storms most characteristic of a given region vary according to latitude, physiographic features, and season. This article is mainly concerned with extratropical cyclones and anticyclones, the chief disturbances over roughly half the Earth's surface. Their circulations control the embedded smaller-scale storms. Large-scale disturbances of the tropics differ fundamentally from those of extratropical latitudes. See HURRICANE; SQUALL LINE; THUNDERSTORM; TORNADO; TROPICAL METEOROLOGY.

Extratropical cyclones mainly occur poleward of 30° latitude, with peak frequencies in latitudes 55–65°. Those of appreciable intensity form on or near fronts between warm and cold air masses. They tend to evolve in a regular manner, from small, wavelike perturbations (as seen on a sea-level weather map) to deep waves to occluded cyclones. See FRONT; METEOROLOGY.

Dynamical processes. The atmosphere is characterized by regions of horizontal convergence and divergence in which there is a net horizontal inflow or outflow of air in a given layer. Regions of appreciable convergence in the lower troposphere are always overlaid by regions of divergence in the upper troposphere. As a requirement of mass conservation and the relative incompressibility of the air, low-level convergence is associated with rising motions in the middle troposphere, and low-level divergence is associated with descending motions (subsidence).

Fields of marked divergence are associated with the wave patterns seen on an upper-level chart (for example, at the 300-millibar level), in the manner shown in Fig. 1. The flow curvature indicates a maximum of vorticity (or cyclonic rotation about a vertical axis) in the troughs and minimum vorticity in the ridges. An air parcel moving from trough to ridge would undergo a decrease of vorticity, which by the principle of conservation of angular momentum implies horizontal divergence. In the upper troposphere, where the wind exceeds the speed of movement of the wave pattern, the divergence field is as shown in Fig. 1. To sustain a cyclone, a relative arrangement of upper and lower flow patterns as shown in Fig. 2 is ideal. This places upper divergence over the region of low-level convergence that occupies the central part and forward side of the cyclone, and upper convergence over the central and forward parts of the anticyclone where there is lower-level divergence.

The upper- and lower-level systems, although broadly linked, move relative to one another. Cyclogenesis commonly occurs when an upper-level trough advances relative to a slow-moving surface

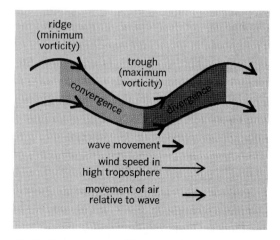

Fig. 1. Convergence and divergence in a wave pattern at a level in the upper troposphere or lower stratosphere. The lengths of arrows indicate the speed and relative motion of wind and wave.

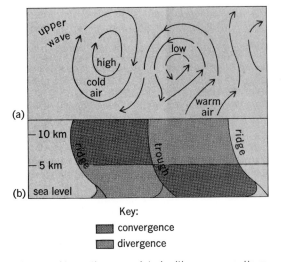

Fig. 2. Air motion associated with a wave pattern. (a) Circulation pattern in low levels in relation to upper wave. (b) West-east vertical section showing simplified regions of convergence and simplified regions of divergence. (*Adapted from J. Bjerknes*)

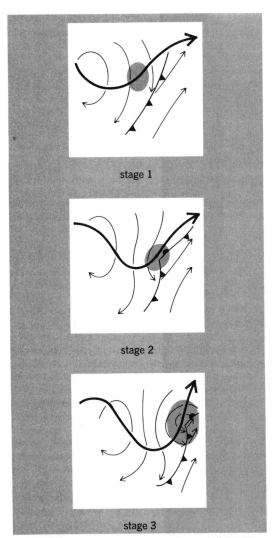

Key: ▓ region of strongest upper-level divergence

Fig. 3. Upper-level trough advancing relative to surface front and initiating cyclone. (*After S. Petterssen*)

front (stages 1 and 2 in Fig. 3). The region of divergence in advance of the trough becomes superposed over the front, inducing a cyclone that develops (stage 3 in Fig. 3) in proportion to the strength of the upper divergence. In a pattern such as Fig. 1, the divergence is strongest if the waves have short lengths and large amplitudes and if the upper-tropospheric winds are strong. For the latter reason, cyclones form mainly in close proximity to the jet stream, that is, in strongly baroclinic regions where there is a large increase of wind with height. *See* JET STREAM.

Frontal storms and weather. Weather patterns in cyclones are highly variable, depending on moisture content and thermodynamic stability of air masses drawn into their circulations. Warm and occluded fronts, east of and extending into the cyclone center, are regions of gradual upgliding motions, with widespread cloud and precipitation but usually no pronounced concentration of stormy conditions. Extensive cloudiness also is often present in the warm sector.

Passage of the cold front is marked by a sudden wind shift, often with the onset of gusty conditions, with a pronounced tendency for clearing because of general subsidence behind the front. Showers may be present in the cold air if it is moist and unstable because of heating from the surface. Thunderstorms, with accompanying squalls and heavy rain, are often set off by sudden lifting of warm, moist air at or near the cold front, and these frequently move eastward into the warm sector. *See* WEATHER.

Middle-latitude highs or anticyclones. Extratropical cyclones alternate with high-pressure systems or anticyclones, whose circulation is generally opposite to that of the cyclone. The circulations of highs are not so intense as in well-developed cyclones, and winds are weak near their centers. In low levels the air spirals outward from a high; descent in upper levels results in warming and drying aloft.

Anticyclones fall into two main categories, the warm "subtropical" and the cold "polar" highs. The large and deep subtropical highs, centered over the oceans in latitudes 25–40° and separating the easterly trade winds from the westerlies of middle latitudes, are highly persistent.

Cold anticyclones, forming in the source regions of polar and arctic air masses, decrease in intensity with height. Such highs may remain over the region of formation for long periods, with spurts of cold air and minor highs splitting off the main mass, behind each cyclone passing by to the south. Following passage of an intense cyclone in middle latitudes, the main body of the polar high may move southward in a major cold outbreak.

Blizzards are characterized by cold temperatures and blowing snow picked up from the ground by high winds. Blizzards are normally found in the region of a strong pressure gradient between a well-developed arctic high and an intense cyclone. True blizzards are common only in the central plains of North America and Siberia and in Antarctica.

Principal cyclone tracks. Principal tracks for all cyclones of the Northern Hemisphere are shown in Fig. 4. In middle latitudes cyclones form

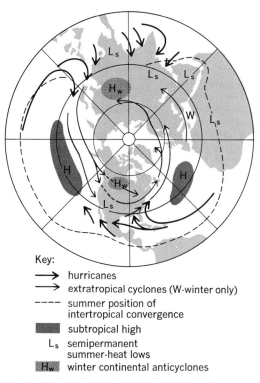

Key:

→ hurricanes

⇒ extratropical cyclones (W-winter only)

---- summer position of
intertropical convergence

▓ subtropical high

L_s semipermanent
summer-heat lows

▓ winter continental anticyclones

Fig. 4. Principal tracks of extratropical cyclones and hurricanes with significantly associated features in the Northern Hemisphere.

most frequently off the continental east coasts and east of the Rocky Mountains.

Movements of cyclones, both extratropical and tropical, are governed by the large-scale hemispheric wave patterns in the upper troposphere. The character of these waves is reflected in part by large circulation systems such as the subtropical highs, and greatest anomalies from the principal cyclone tracks occur when these highs are displaced from the mean positions shown in Fig. 4. Warm highs occasionally extend into high latitudes, blocking the eastward progression suggested by the average tracks and causing cyclones to move from north or south around the warm highs.

Over the Mediterranean, cyclones form frequently in winter but rarely in summer, when this area is occupied by an extension of the Atlantic subtropical high. Both the subtropical highs and the cyclone tracks in middle latitudes shift northward during the warmer months; on west coasts cyclones are infrequent or absent in summer south of latitudes 40–45°.

Role in terrestrial energy balance. Air moving poleward on the east side of a cyclone is warmer than air moving equatorward on the west side. Also, the poleward-moving air is usually richest in water vapor. Thus both sensible- and latent-heat transfer by disturbances contribute to balancing the net radiative loss in higher latitudes and the net radiative gain in the tropics. Air that rises in a disturbance is generally warmer than the air that sinks (Fig. 2), and latent heat is released in the ascending branches by condensation and precipitation. Hence the disturbance also transfers heat upward, as is required to balance the net radiative

loss in the upper part of the atmosphere. *See* AT-MOSPHERIC GENERAL CIRCULATION.

[CHESTER W. NEWTON]

Bibliography: E. Palmén and C. W. Newton, *Atmospheric Circulation Systems*, 1969; S. Petterssen, *Introduction to Meteorology*, 3d ed., 1969; H. Riehl, *Introduction to the Atmosphere*, 3d ed., 1978.

Storm detection

Microbarographs, radars, satellite-borne instruments, and sferics detectors (radio receivers) are used to detect storms and to assess their potential for destruction.

Microbarographs. Certain severe thunderstorms emit an identifiable kind of infrasound, or ultralow-frequency acoustic wave. Traveling at the speed of sound and ducted between the ground and high-temperature layers in the upper atmosphere, these waves are often so powerful that they can be detected by sensitive pressure detectors, called microbarographs, more than 1500 km from the emitting storm. At such distances, the pressure fluctuations of the waves are only about one-millionth of average atmospheric pressure. It takes 10–60 s for one wave cycle to pass, but the oscillations can last for hours (Fig. 1).

Similar storm-related waves travel into the ionospheric F region, 200 km above the earth. These waves were discovered by looking with a high-frequency radar for a certain kind of oscillation in the radar reflection height. Though there appears to be a causal connection between tornadic storms and ionospheric infrasound, there is no statistical evidence of the warning value of the ionospheric waves. The mechanism causing the emissions is not known, but observations of both ground-level and ionospheric waves have established that only a small fraction of all storms emit detectable infrasound, and that most of the emissions appear to come from tornadic storm systems.

In the United States, storm detection exercises have been carried out using direction-finding arrays of microbarographs for ground-level infrasound. Within a 14-state test area, 65% of the tornadic storms that occurred during the 1973 storm season were considered "detected" by three infrasound observatories, and the storm emissions had enough distinctive characteristics that false-alarm rates were considered acceptable. However, triangulation, when possible, is inaccurate, and even though the waves are often emitted prior to the observed tornadoes, their relatively slow sonic travel speed (1000 km/hr) diminishes their warning value.

Acoustical detection is thus not presently used for storm warning, and its main value lies in storm research.

[T. M. GEORGES]

Radar. Radars emit pulses of electromagnetic radiation in a wavelength region that penetrates storm clouds to provide a three-dimensional, inside view of the storm with angular resolution of about 1°. Advanced weather radars (Fig. 2) provide accurate images of both precipitation intensity inside a storm's shield of clouds and precipitation velocity. Velocity is provided by Doppler radar, using the same principle that produces the familiar

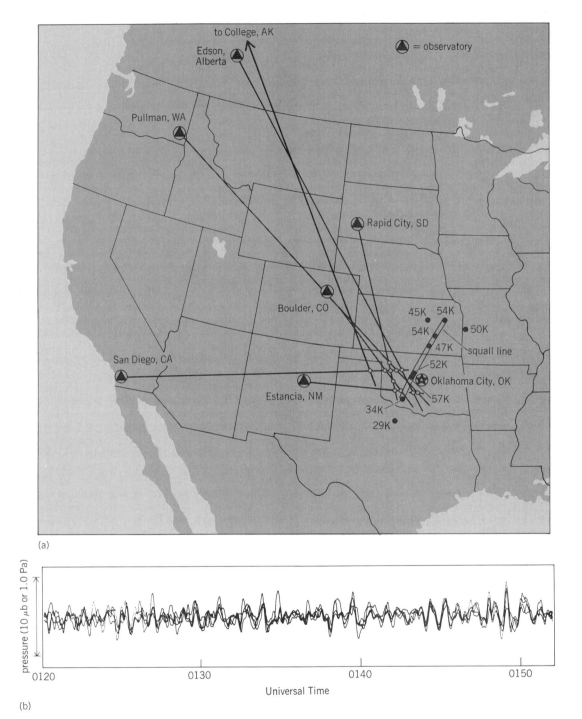

Fig. 1. Infrasound and microbarograph data during a severe-storm outbreak in Oklahoma. (a) Map of the western United States showing the intersecting infrasound bearings measured at seven observatories (the seventh being in College, AK). The numbered dots show radar-indicated storm cells whose tops reach the heights shown (K = 10³ ft or 305 m). The most violent storms often occur at the southern end of a squall line. (b) Superimposed pressure records from four microbarographs at Boulder, CO, during this event. (*Wave Propagation Laboratory, NOAA*)

change in pitch of a horn on a passing car or train. Precipitation reflects the radar's transmitted pulse and produces change in the microwave pitch proportional to the radial (Doppler) velocity toward or away from the radar. Precipitation reflectivity is proportional to the intensity of rain, snow, or hail.

Contoured reflectivity maps are routinely dis-

played by radars of the U.S. National Weather Service on widely used plan position indicator (PPI) scopes giving range and azimuth to precipitation targets whose reflectivity (intensity) is indicated by a stepped brightness scale (Fig. 3a). While a storm's reflectivity image is valuable for rainfall assessment and severe weather warnings, it is not

Fig. 2. Radar facilities at the National Severe Storms Laboratory in Norman, OK. The large hemispherical radome houses a Doppler weather radar. A conventional non-Doppler radar housed in the radome on the tall tower is the type (WSR-57) used by the National Weather Service. (*National Severe Storms Laboratory, NOAA*)

a highly reliable tornado indicator. Highest reflectivity areas often signify hail. Radar warnings are primarily based on reflectivity values, on storm top heights, and sometimes on circulatory features of hook echoes seen in the patterns of reflectivity (Fig. 3*b* and *d*).

One of the earliest images of a storm's Doppler velocity field was obtained when a pulse-Doppler radar was mated to the PPI. Significant dynamic meteorological events such as tornado cyclones, whose visual sightings are often blocked by rain showers and nightfall, produce telltale signatures in the Doppler velocity field. Such a swirling vortex signature is composed of constant Doppler velocity contours (isodops) forming a symmetric couplet of closed contours of opposite sign, that is, velocity either toward or away from the radar (Fig. 3*c* and *d*). This pattern portends tornadoes, damaging winds, and hail.

The tornadic storm depicted in Fig. 3 produced a particularly large cyclone, and its reflectivity and isodop signatures are clearly seen. A signature pattern for circularly symmetric convergence of air is similar to the cyclone pattern but is rotated clockwise by 90°. The reflectivity spiral suggests some convergence, as well as rotation, a conclusion supported by the clockwise angular displacement of isodop maxima about the cyclone center with positive maximum somewhat closer to the radar (Fig. 3*c*).

Color displays of reflectivity and Doppler velocity allow easier quantitative evaluation and better resolution of signatures. Color-coded velocities in shades of blue (toward) and red (away) of increasing brightness to signify Doppler speed provide quick and reliable detection and assessment of cyclones. Doppler radar can sort from many seemingly severe storm cells the ones that have circulation and hence potential for tornado development.

Programs have been conducted to assess the improvement in severe storm advisories when Doppler velocity is given to forecasters. The probability of detection (POD) of tornadoes with Doppler radar was found to be 0.69 and the false-alarm rate (FAR) 0.25. A weather forecast office, covering the same area with then standard techniques, showed a POD of 0.47 and an FAR of 0.4. More significant is Doppler detection of tornadoes 20 min before their touch-down on the ground, whereas the warning system dependent on visual sightings generally shows a negative lead time. These findings support plans for a national network of Doppler weather radars, to replace present non-Doppler facilities by the late 1980s.

Hurricanes have a spiraling inflow of air like that shown in Fig. 3 for the severe thunderstorm. However, the scale size of a hurricane (Fig. 4) is an order of magnitude larger than a thunderstorm. Thunderstorms are most common in hurricane outskirts and again near the storm center or "eye," which is clearly revealed and tracked by radar when the storm is within about 200 mi (320 km) of stations which line the Gulf and Atlantic coasts of the United States.   [R. J. DOVIAK]

Satellites. On Apr. 1, 1960, the first United States weather satellite *(TIROS 1)* was placed in orbit and began to send televised photographs of cloud systems back to Earth. Since then, satellite

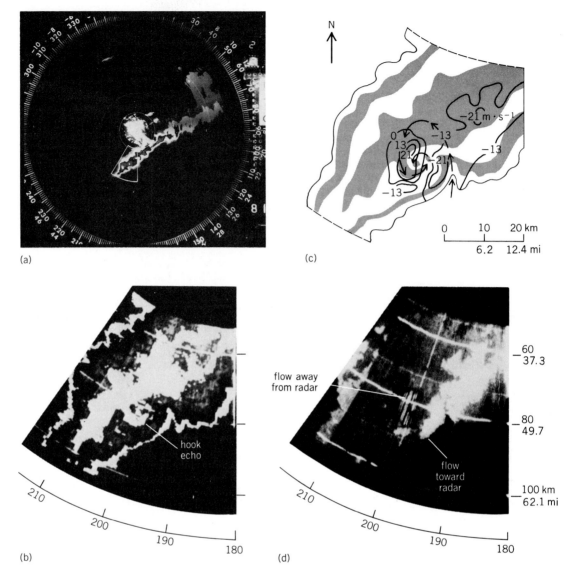

Fig. 3. Radar images of a storm. (a) WSR-57 PPI display of squall line showing its reflectivity (rainfall rate) categories as constant brightness. Dim brightness area surrounding the squall line corresponds to rainfall rate between .05 and .30 mm/hr. Then in sequence to the storm interior: bright, dark, dim, and bright areas are rainfall rates 0.3– 1.7, 1.7–9.1, 9.1–48.8, and 48.8– 262 mm/hr. Circles are range marks spaced 40 km apart. (b, c) Magnifications of area outlined in a to show Doppler radar reflectivity and Doppler velocity signatures of a storm cell. In b, reflectivity pattern from the Doppler radar shows a hook or spiral convergence (see arrows in d) that signifies the presence of a cyclone. In c, isodop pattern shows cyclone signature. Brightness levels are velocity categories: dim (<13 m · s⁻¹), bright (13– 21), and brightest (>21). Positive (away) radial velocities are angularly strobed. The Doppler signature of a tornado cyclone is 193– 203° and 75– 90 km. (d) Schematic overlay of reflectivity and isodop patterns of b and c, showing the coincidence of the reflectivity spiral and cyclone signature. (National Severe Storms Laboratory, NOAA)

data have become more comprehensive and timely. Photographs of nearly an entire hemisphere are now disseminated to users throughout the United States within 30 min of scan time. Scans are made routinely every 30 min. Intensity and development of weather systems are easily monitored by combining sequential pictures into motion picture loops. In addition, infrared data allow display of cloud systems at night, and indications of both temperature and water vapor by both night and day.

Satellite photos reveal components of storm systems heretofore inaccessible. For example, frontal boundaries can be located precisely in most cases, since sharp cloud lines usually accompany fronts. The location and configuration of jet streams can often be identified through analysis of cirrus cloud streaks. (Much of large-scale air motion which initiates convection and precipitation is associated with the jet stream.) Low-level moisture can be tracked via low clouds moving from the Gulf of Mexico into the central plains of the United States in advance of severe thunderstorm and tornado outbreaks. Developing low-pressure systems are identifiable as soon as they generate even the smallest of cloud bands. Small but intense weather disturbances of a scale less than the resolution of the widely spaced (about 150 km) surface observing stations often produce telltale cloud patterns amenable to analysis by satellite meteorologists.

Two of the most destructive weather phenomena in terms of life and property are the hurricane and the severe thunderstorm. Both events are much better monitored as a consequence of satellite technology.

Hurricanes usually form from westward-moving tropical disturbances called easterly waves. It is not known exactly why this occurs, and before satellite technology a tropical disturbance, or even a hurricane, could exist for days before discovery. With weather satellites, the hurricane is spotted and tracked routinely when it is still far at sea. The configuration of the deadly storm is unmistakable (Fig. 5). What is perhaps more significant is that the storm is becoming more recognizable in its embryonic stage (the easterly wave). Meteorologists watch easterly waves carefully as the waves evolve from mild disturbances, to tropical storms, and finally into hurricanes.

Since thunderstorms occur on a smaller time and space scale, the thunderstorm forecast problem has always involved obtaining data detailed enough to specify small zones of high storm potential. It has been known for some time, for example,

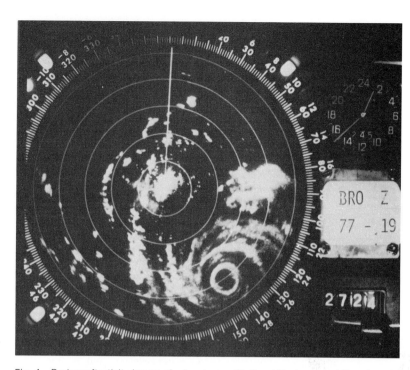

Fig. 4. Radar reflectivity image of a hurricane. (*National Hurricane and Experimental Meteorology Laboratory, NOAA*)

Fig. 5. Hurricane Belle, Aug. 8, 1976, off Florida. (*National Environmental Satellite Service, NOAA*)

that storms form along boundaries (such as cold fronts) where the convergence of two different air masses may force air to rise rapidly. But where precisely is the front at any given instant? Exactly how strong is its lifting capacity? These and other such questions are receiving at least partial answers through use of satellite photography.

Figure 6 illustrates a developing tornadic situation. Notice the distinct north-south-oriented line of thunderstorms in northern Texas. Conventional surface data indicated the existence of a nearby cold front, but it can be located only with accuracy of about 50 mi (80 km) without the supplementary satellite data. Notice also the large "comma"-shaped cloud encompassing most of the print. This identifies an intense upper-air disturbance, having the capacity to trigger thunderstorms and intensify normal storms into the severe or tornadic category. By observing these patterns, invaluable lead time is gained in gearing up public warning systems. Three hours after this photo was made, a large tornado, accompanied by 3-in.-diameter hail, struck Dallas, TX.   [JOHN WEAVER]

Sferic detectors. Lightning discharges produce a wide spectrum of electromagnetic signals that are detected by radio receivers and provide effective means for locating and tracking thunderstorms. The lower frequency portions (lower than 1 MHz) of the radio spectrum are generally selected to detect cloud-to-ground and major channels of intracloud discharges, while the higher frequencies (higher than 10 MHz) are used to sense small branches and fine structure of the discharge paths.

Both the electric and the magnetic components of lightning-produced radio signals have been used to locate and track thunderstorms through triangulation methods from the directions of arrival at two or more spaced stations. Directions in both azimuth and elevation are generally obtained through the directional responses of particular antennas or by the use of spaced antennas to measure either the simultaneous phase differences of waves or time differences separating impulse arrivals. Other methods have been devised to couple the direction of arrival with measurements of radio signal waveform or spectrum to give an estimate of range to the lightning so that it can be located from a single station.

While radar and satellite technologies provide the principal bases nowadays for storm warnings around the world, simple directional radio receivers are an inexpensive and useful alternate for thunderstorm monitoring where the more advanced facilities are unavailable. Advanced facilities for sferics detection and analysis are essential, of course, for research on storm electrical processes. *See* HURRICANE; LIGHTNING; STORM; THUNDERSTORM.   [WILLIAM L. TAYLOR]

Bibliography: American Meteorological Society, *Proceedings of the 5th Conference on Severe Storms*, 1967; D. Atlas (ed.), *Severe Local Storms*, Amer. Meteorol. Soc. Monogr., vol. 5, no. 27, 1964; L. J. Battan, *Radar Observation of the Atmosphere*, 1973; W. J. Kotsch and E. T. Harding, *Heavy Weather Guide*, 1965; J. S. Marshall (ed.), *Proceedings of the 13th Conference on Radar Meteorology*, American Meteorological Society, 1968.

Storm surge

A transient, localized disturbance at sea level, resulting from the action of a tropical cyclone, an extratropical cyclone, or a squall over the sea. Storm surges, or storm tides, are not to be confused with tsunamis, or tidal waves, which result from seismic or molar disturbances of the Earth. In the Northern Hemisphere those coastal regions which are particularly vulnerable to storm surges include the periphery of the Gulf of Mexico, the Atlantic Coast of the United States, the Gulf of Bengal, Japan and other islands of the western Pacific which lie in the typhoon belt, and the coastal regions of the North Sea. The surges occurring in the North Sea originate from the actions of large-scale extratropical storms, particularly winter storms. On the eastern coast of the United States, hurricane-induced surges, as well as surges originating from intense winter storms, occur. In the Great Lakes and the Gulf of Mexico, surges resulting from squalls are known to occur; however, hurricane-induced surges pose a more serious threat to the low-lying coastal areas of the Gulf. *See* TSUNAMI.

The time history of the surge at a given location at shore is represented by the surge hydrograph. This is a time sequence of the difference between the measured tide and the predicted periodic tide (see illustration). Maximum surge elevations of 15 ft above predicted tide are not uncommon. In the case of hurricane-induced surges, the peak water level seems to depend primarily upon the atmo-

Fig. 6. Satellite photo of a line of thunderstorms (arrow locates a line of individual thunderstorms extending northward into a mass of cloud cover with embedded storms) along a cold front. The much larger, "comma"-shaped cloud band which encompasses most of the photo is associated with the large low-pressure system which helped trigger these storms. (*National Environmental Satellite Service, NOAA*)

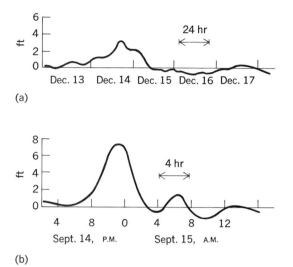

(a)

(b)

Surge hydrographs. (a) Winter storm of 1954 at Sandy Hook, N.J. (b) Hurricane of 1944 at Newport, R.I.

spheric pressure at the hurricane center. However, the horizontal scale, the direction and speed of propagation of the hurricane, and the coastal geometry and bottom topography are important influencing factors in the storm surge behavior. When a hurricane crosses the coast from the sea, the greatest surge alongshore usually occurs to the right of the hurricane path.

A storm surge is essentially a forced inertio-gravitational wave of great wave length. This implies that the duration or speed of the storm determines the dynamic augmentation of the water level at shore above that which would occur if the storm were stationary. Also, the inertial character of surges can explain quasi-periodic resurgences that often follows the primary forced surge. *See* SEA-LEVEL FLUCTUATIONS.   [ROBERT O. REID]

Bibliography: J. C. Freeman, Jr., L. Baer, and G. H. Jung, The bathystrophic storm tide, *J. Mar. Res.*, 16(1):12–22, 1957; D. L. Harris, The hurricane surge, *Proceedings of the 6th Conference on Coastal Engineering*, Council on Wave Research, pp. 96–114, 1958; U.S. Department of Commerce, *Characteristics of Hurricane Storm Surge*, 1963; P. Welander, *Numerical Prediction of Storm Surges*, in H. E. Landsberg and J. Van Mieghem (eds.), *Advances in Geophysics*, vol. 8, 1961.

Stratosphere

The stable layer of the atmosphere extending from the top of the troposphere (10–16 km) to the temperature maximum at 50–55 km. In the lower stratosphere below about 30 km, temperature slowly decreases (in winter high latitudes), remains constant (in mid-latitudes), or increases with height (in low latitudes). The minimum temperature occurs over equatorial regions, and the atmospheric motions are extensions of tropospheric systems which become progressively more damped with height. In the upper stratosphere ($\approx$ 30–50 km) there is a general increase of temperature with height to about 290°K over the summer pole and 250°K over the winter pole. Winds in summer

are steady moderate easterlies, but large-scale systems with a westerly "polar night" jet stream form in winter. In late winter and spring these systems also affect the lower stratosphere, causing large temperature changes called "sudden warmings" and also large increases in the tracer (ozone and radioactivity) content. A well-marked alternation of wind direction from west to east at low latitudes with a roughly 26-month period has also been found but has not been explained.

In composition the stratosphere is notable for its photochemically formed ozone layer (maximum mixing ratio of about 10^{-5} g/g of air at 35 km) and its low water vapor content of about 2×10^{-6} g/g of dry air. Its thermal structure can be broadly explained in terms of the radiation balance due to its minor constituents of ozone, carbon dioxide, and water vapor. *See* ATMOSPHERE; TROPOPAUSE; TROPOSPHERE.   [R. J. MURGATROYD]

Submarine canyon

A relatively narrow, deep depression in the sea floor with steep slopes, the bottom of which grades continuously downward. The sea floor has many puzzling features, but none has aroused so much controversy as the great submarine canyons, cut into the continental slopes off most coasts of the world. Many of these have rocky walls thousands of feet high. They have narrow inner gorges, winding courses, numerous tributaries and are, in fact, quite comparable to the great canyons of the land (see illustration). Some of the canyons are direct continuations of land canyons, and others occur off large rivers which flow through broad flat-floored valleys on land. The submarine canyons extend outward down the slope virtually to the deep ocean floor. The outer portions of these sea valleys are of modest dimensions and extend across broad, gently sloping fans, comparable to the piedmont fans along the fronts of mountain ranges in arid regions. *See* CONTINENTAL SHELF AND SLOPE.

The deep floors of the canyons contain sediments alternating between sands, which resemble shallow water deposits, and normal deep-sea mud deposits. It seems probable that landslides, occurring at the canyon heads, stir sediment into the

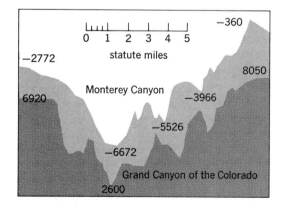

Transverse profile of the submarine canyon in Monterey Bay compared to a profile of the Grand Canyon of the Colorado River in Arizona. Elevations are in feet and the vertical is magnified five times.

water and produce a heavy suspension which sets up a current, called a turbidity current. This moves along the canyon floors, leaving behind the sand deposits when the current loses velocity. These slides occur at rather frequent intervals, changing the depths and often breaking cables laid across the canyons. Thus cable companies avoid laying cables across submarine canyons where possible.

The cause of submarine canyons is much disputed. Their close resemblance to river-cut canyons on land has convinced some geologists that they are due to river cutting followed by submergence of the valleys. The widespread distribution of submarine canyons has caused other geologists to object to this idea. Turbidity currents have been cited as an alternative cause. An unknown factor is the speed of turbidity currents, which may at times be very great although the evidence is not clear. If the canyons are caused by turbidity currents, it is difficult to understand why they should so closely resemble river-cut canyons. Furthermore, the existence of submarine canyons with hard rock walls, such as granite, has caused some dissatisfaction with the turbidity current hypothesis. Most geologists now agree that, however formed, submarine sliding of material and turbidity currents at least prevent the filling of the canyons of the sea floor. *See* MARINE GEOLOGY; TURBIDITY CURRENT.

[FRANCIS P. SHEPARD]

Sun

The star around which the Earth revolves, and the planet's source of light and heat (Fig. 1). The Sun is a globe of gas, 1.4×10^6 km in diameter, held together by its own gravity. Because of the weight of the outer layers, the density and temperature increase inward, until a central temperature of over 15,000,000°K and density more than 90 times that of water is reached. At these great temperatures and densities, thermonuclear reactions converting hydrogen into helium take place, releasing the energy which streams outward.

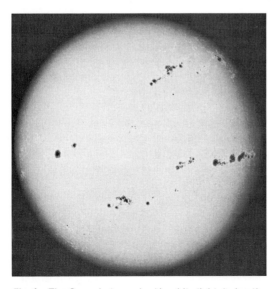

Fig. 1. The Sun, photographed in white light during the 1957 maximum of the sunspot cycle. (*Hale Observatories*)

Table 1. Principal physical characteristics of Sun

Characteristic	Value
Mean distance from Earth (the astronomical unit)	$(1.4960 \pm .0003) \times 10^8$ km
Radius	$(6.960 \pm .001) \times 10^5$ km
Mass	$(1.991 \pm .002) \times 10^{33}$ g
Mean density	$1.410 \pm .002$ g/cm^3
Surface gravity	$(2.738 \pm .003) \times 10^4$ cm/sec$^2 =$ 28 × terrestrial gravity
Total energy output	$(3.86 \pm .03) \times 10^{33}$ erg/sec
Energy flux at surface	$(6.34 \pm .07) \times 10^{10}$ erg/(cm^2)(sec)
Effective surface temperature	$5780° \pm 50°$K
Stellar magnitude (photovisual)	$-26.73 \pm .03$
Absolute magnitude (photovisual)	$+4.84 \pm .03$
Inclination of axis of rotation to ecliptic	7°
Period of rotation	About 27 days. The Sun does not rotate as a solid body; it exhibits a systematic increase in period from 25 days at the equator to 31 days at the poles.

The surface temperature of the Sun is about 6000°K; since solids and liquids do not exist at these temperatures, the Sun is entirely gaseous. Almost all the gas is in atomic form, although a few molecules exist in the coolest regions at the surface.

The Sun is a typical member of a numerous class of stars, the spectral type dG2 (d indicates dwarf). Other characteristics are given in Table 1.

Besides its great importance to human life, the Sun is of interest to all astronomers because it is the only star near enough for detailed study of its surface structure. Various surface and atmospheric phenomena, such as sunspot activity, and other behavior may be studied, and astronomers try to extrapolate these to the other stars which may be observed only as points of light.

The light and heat of the Sun make the Earth habitable. The Sun is, in fact, the ultimate source of nearly all the energy utilized by industrial civilizations in the form of water, power, fuels, and wind. Only atomic energy, radioactivity, and the lunar tides are examples of nonsolar energy.

SOLAR STRUCTURE

The interior of the Sun can be studied only by inference from the observed properties of the entire star. The mass, radius, surface temperature, and luminosity are known. Using the known properties of gases, it is possible to calculate that structure of the Sun which will produce the observed parameters at the surface (Fig. 2). The solution is complicated by uncertainties in the behavior of matter and radiation under the high temperature and density that are present in the solar interior. This is particularly true of the nature of the nuclear reactions. However, the general properties of the solution are quite reliable. A number of theoretical models using different assumptions have led to more or less similar results. A central density of near 90 g/cm³ has been found, decreasing to 10⁻⁷ g/cm³ at the surface. The central temperature is about 15,000,000°K, decreasing to 5000°K at the surface. Since this takes place over 700,000 km, the temperature gradient is only 20° per kilometer. The radiation produced at the center by nuclear interactions flows outward rapidly.

Although the material at the center of the Sun is so dense that a few millimeters are opaque, the photons created by nuclear reactions are continually absorbed and reemitted and thus make their way to the surface. The atoms in the center of the Sun are entirely stripped of their electrons by the high temperatures, and most of the absorption is by continuous processes, such as scattering of light by electrons.

In the outer regions of the solar interior, the temperature is low enough for ions and even neutral atoms to form and, as a result, atomic absorption becomes very important. The high opacity makes it very difficult for the radiation to continue outward; steep temperature gradients are established which result in convective currents. Most of the outer envelope of the Sun is in such convective equilibrium. These large-scale mass motions produce many interesting phenomena at the surface, including sunspots and solar activity.

Radiation. Electromagnetic energy is produced by the Sun in essentially all wavelengths. Important radiation has been measured from long radio waves of 300 m down to x-rays of less than 1 A (from rockets). In addition, considerable energy is emitted in the form of high-energy particles (cosmic rays). However, more than 95% of the energy is concentrated in the relatively narrow band between 2900 and 25,000 A and is accessible to routine observation from ground stations on Earth. The maximum radiation is in the green region, and the eyes of human beings have naturally evolved to be sensitive to this range of the spectrum. The total radiation and its distribution in the spectrum are parameters of fundamental significance, because they measure the total energy output of the Sun and its effective surface temperature. The total radiation received from the Sun is termed the solar constant and has, in fact, been found within the limits of observation to be constant to about ±1%. The presently accepted value is 1.97 cal/(cm²)(min). This is equivalent to 1.374×10^6 erg/(cm²)(sec).

The measurement of the solar constant has been greatly complicated by the absorption of solar radiation in all wavelengths by the Earth's atmosphere. Determinations have been made by observations at stations atop high mountains where the atmospheric perturbation is minimized, and by observing the variation during the day as the radiation passes through successively smaller distances in the terrestrial atmosphere. A great bulk of the studies has been carried out by the Smithsonian Institution in a continuous program from the 1880s until 1955.

Atmosphere. Because the Sun is visible as a two-dimensional surface rather than as a point, the study of its atmosphere and surface phenomena is most interesting.

Although the Sun is gaseous, it is seen as a discrete surface from which practically all the heat and light are radiated. One sees through the gas to the point at which the density is so high that the material becomes opaque. This layer, that is, the visible surface of the Sun, is termed the photosphere. Light from farther down reaches the Earth by repeated absorption and emission by the atoms, but the deepest layers cannot be seen directly. The surface is not actually sharp; the density drops off by about a factor of 3 each 150 km. However, the Sun is so far away that the smallest distance that can be resolved with the best telescope is about 400 km, and thus the edge appears sharp.

Looking at the Sun in isolated wavelengths absorbed by its atmospheric gases, one can no longer see down to the photosphere, but instead one sees the higher levels of the atmosphere, known as the chromosphere. This name results from the rosy color seen in this region at a solar eclipse. The chromosphere is a rapidly fluctuating region of jets and waves coming up from the surface. When all the convected energy coming up from below reaches the surface, it is concentrated in the thin material and produces considerable activty. Naturally, this outer region is considerably hotter than the photosphere, with temperatures up to 30,000°K.

When the Moon obscures the Sun at a total solar eclipse, the vast extended atmosphere of the Sun

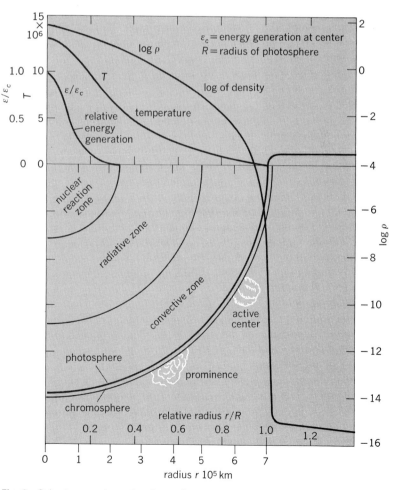

Fig. 2. Solar temperature, density, and energy generation in the interior and atmosphere of Sun. Curves in upper section show these as functions of radial distance from the center. Lower section shows the principal zones in the interior of Sun.

Fig. 3. Two sections of the Fraunhofer spectrum, showing bright continuum and dark absorption lines. The wavelength range covered by each strip is approximately 85 A. (*Sacramento Peak Observatory, operated by the Association of Universities for Research in Astronomy, Inc.*)

called the corona can be seen. The corona is transparent and is visible only when seen against the dark sky of an eclipse. Its density is low but its temperature is high (more than 1,000,000°K). The hot gas evaporating out from the corona flows steadily to the Earth and farther in what is called the solar wind.

Within the solar atmosphere many transient phenomena occur which may all be grouped under the heading of solar activity. They include sunspots and faculae in the photosphere, flares and plages in the chromosphere, and prominences and a variety of changing coronal structures in the corona. The existence and behavior of all these phenomena are connected with magnetic fields, and their frequency waxes and wanes in a great 22-year cycle called the sunspot cycle.

The sunspots and flares are sources of x-rays, cosmic rays, and radio emission which often have profound influence on interplanetary space and the upper atmosphere of the Earth.

The photospheric features of the Sun (sunspots and faculae) are readily observed with a small telescope by projection of the solar image through the eyepiece onto a shaded white card. This is the only safe method for observing the Sun without a specially designed solar eyepiece. Observations of chromospheric phenomena require special filters to exclude all light except that emitted by the atmospheric gases. Observations of the corona require either a total eclipse or a coronagraph.

Solar physics. Man's understanding of the Sun is derived from observations of the morphology of various phenomena and from physical analysis with the spectrograph. The former permits determination of what is happening, and the latter determination of the detailed physical parameters of the gas under observation.

When the light of the Sun is broken up into the different frequencies by the spectrograph, a bright continuum interrupted by thousands of dark absorption lines known as the Fraunhofer lines can be seen. Each line represents some discrete transition in a particular atom which occurs when a photon of light is absorbed and one of the atomic electrons jumps from one level to another. Of the 92 natural elements, 64 are represented in the Fraunhofer spectrum (Fig. 3), first observed by the German physicist J. Fraunhofer. The remaining atoms are undoubtedly present but remain undetected because they are rare or their lines are produced in spectral regions not accessible to present spectrographs. The relative abundances of the most numerous atoms have been estimated from the line intensities (Table 2). Many of these abundances have been confirmed by measurement of the relative abundances of different elements in the streams of particles coming from the Sun at the time of solar flares. The Sun can be described as a globe of chemically (but not spectroscopically) pure hydrogen and helium with traces of the other elements.

The nature of the formation of the spectrum lines can be understood in terms of a few simple rules which are an elaboration of those first put forth by G. R. Kirchhoff. A small amount of a hot gas will be transparent in all wavelengths except those characteristic of the atoms of the gas. In those wavelengths radiation is emitted by the jumping electrons, and an emission line spectrum is produced. If the volume of the gas is large enough and its density high enough, it will be opaque at all frequencies because of the existence of continuous processes which will absorb any wavelength of light. In this case the gas emits a continuous spectrum at all frequencies. However, at the outer edges of the gas absorb light coming from one direction (the inside) and reemit it in all directions, with a consequent reduction of intensity. This results in dark lines superposed against the bright background. Although the ultimate source of all energy is at the center of the Sun, at any one point the radiation is a result of either a scattering of light coming from within or the emission of radiation by an atom which has been excited by collisions with the electrons of the gas.

By analyzing the nature of the spectrum lines observed, it can be determined which atoms are present, as well as their stages of ionization. This is so because, as electrons are successively removed from an atom by ionization (as a result of photoelectric effect or collisions with electrons), the spectrum changes completely. Obviously, the hotter the gas, the higher the stage of ionization. Thus when lines of iron ionized 14 times are observed in the solar corona, it may be assumed that the gas is very hot. On the other hand, by observing the spectra of molecules in the upper photosphere, it may be concluded that the temperature is near the lowest possible in the Sun.

In addition, the shape of the spectrum lines is a key to the physical conditions under which they are radiated. If the temperature is high, the atoms have a high velocity and the lines are broadened by the Doppler effect. If there are strong magnetic fields, the spectrum lines are split by the Zeeman effect. If there are strong electric fields, the lines are broadened by the Stark effect. It is the task of

Table 2. Relative numbers of the most abundant atoms in Sun

Element	Number	Element	Number
Hydrogen, H	1,000,000	Silicon, Si	20
Helium, He	50,000–200,000	Sulfur, S	8
Oxygen, O	500	Aluminum, Al	2
Nitrogen, N	400	Sodium, Na	2
Carbon, C	200	Calcium, Ca	1.5
Magnesium, Mg	33	Iron, Fe	15

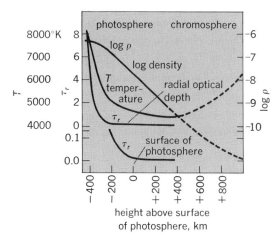

Fig. 4. The variation of temperature, density, and radial optical depth with height in the photosphere and (less certainly) in the chromosphere. Lower curve of optical depth shows variation with an expanded ordinate scale.

the spectroscopist to disentangle these various effects. He is aided by the fact that the spectrum lines can be observed at both the center of the Sun and near the limb, as well as from atoms which behave differently. For example, hydrogen atoms are very light and move extremely fast; hence the hydrogen lines show a strong broadening due to the Doppler effect. If the lines emitted by heavy atoms in the same region do not show this broadening, it may be concluded that this is purely a temperature effect, because the heavy atoms would not be expected to move so fast.

Helium, on the other hand, is inert and radiates only at high temperatures. Thus, when helium lines are seen, the gas is hot. In these ways the astrophysicist uses the spectrum lines to interpret the physical conditions in the atmosphere of the Sun.

Photosphere. As mentioned above, the photosphere is the visible surface of the Sun. In the visible wavelengths, its brightness decreases smoothly from the center of the solar disk to the limb. This limb darkening results from the fact that the line of sight to the observer passes through the atmosphere at an increasing angle to the normal as the point of observation approaches the limb. Hence the line of sight penetrates to a lower depth at the center than at the limb, where the path through the overlying material is much longer. Thus what one sees at the limb is a higher point in the atmosphere. The fact that the photosphere is darker at the limb indicates that the termperature decreases outward. Once the range of temperature is known, the variation of limb darkening with wavelength can be used to establish the absorbing properties of the materials at different wavelengths.

By combination of all the information on limb darkening at different wavelengths, it has been possible to construct reasonably reliable models of the temperature variation through the photosphere (Fig. 4).

Granulation. Except for sunspots and accompanying activity, the photosphere is quite uniform over the Sun. The only structure visible is the granulation, an irregular distribution having the shape of bright corn kernels with dark lanes in between (Fig. 5). The grains are quite small, of the order of 1000 km in diameter (about 1.3 seconds of arc as seen from the Earth), and have a life-span of about 8 min. The dark lanes between the granules are about 200 km across. Since the highest optical resolving power employed is of the order of 200–300 km, there may be many more fine lanes crossing the granules and dividing them into even smaller elements.

The granulation is visible evidence of convective activity below the surface. The bright grains are presumably the tops of hot, rising columns which bring energy up from the interior, while the dark intergranular may be the cool downward-moving material. High-resolution spectrograms (Fig. 6) show that each bright granule is marked by a violet displacement of the spectrum lines, indicating an upward velocity. The measured difference in brightness between a granule and an intergranule area is about 15%, indicating an effective temperature difference of the order of 200°K. Photographs of the granulation often show what appear to be distinct chains. However, no one has proved that these are anything other than a random association by the eye of the observer.

In Fig. 6 the brightness variations in the continuous spectrum are presumably the white-light granulation. The complex structure of the absorption lines is easy to see; there are some elements where the absorption lines are entirely missing, and others where tilted features indicate rotation. The brightest granules in the continuum correspond to arrow-shaped shifts to the blue.

Velocity fields. In recent years high-resolution Doppler effect measurements have shown the existence of important velocity fields in the photosphere.

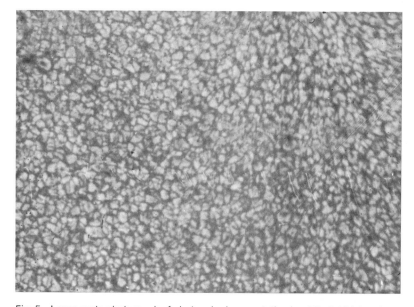

Fig. 5. Large-scale photograph of photospheric granulation in white light taken from an altitude of 80,000 ft above sea level. The length of this section is about 55,000 km on the Sun. (*Princeton University Observatory*)

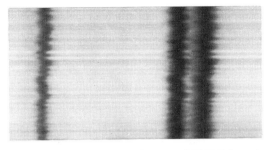

Fig. 6. High-resolution spectrogram at 5188.5 A made by J. W. Evans at Sacramento Peak Observatory, showing lines (left to right): Fe I 5187.922, Ti I 5188.700, and Ca I 5188.848. The total slit length is 120,000 km on the Sun. (*Sacramento Peak Observatory, operated by the Association of Universities for Research in Astronomy, Inc.*)

These consist of (1) periodic oscillation with an amplitude of about 0.4 km/sec and a period of about 250 sec; and (2) a cellular flow in supergranules about 30,000 km across, in which material flows outward to the edge of the cells. Superposed on these velocity fields are substantially larger velocity fields in the overlying material of the chromosphere.

The 250-sec oscillation has been shown to increase upward in amplitude and velocity, while the period decreases. It is obviously a resonant oscillation of the atmospheric material under the impetus of the convective motions below.

The cellular pattern plays an important role in the structure of the upper atmosphere. The systematic outward flow of material in each cell tends to concentrate magnetic fields near the edges. The resulting chromospheric network shows strong magnetic fields distributed around the outside of the cells, resulting in enhanced activity at those points.

Chromosphere. The chromosphere was first detected and named by early solar-eclipse observers. They saw it as a beautiful rosy arc which remained visible for a few seconds above the limb of the Moon when the photosphere had been covered. The red color is due to the dominating brightness of the H-α line of hydrogen at 6562.8 A in the chromospheric spectrum, which, except for a bare trace of continuum, is a pure emission spectrum of bright lines (Fig. 7). Because eclipses are rare and the chromosphere can be seen only edge-on at that time, astronomers have developed a special method of photographing the chromosphere in its characteristic lines, particularly H-α. In these spectrum lines the chromosphere is no longer transparent, and one looks at it instead of looking through it to the photosphere. The two principal devices for this observation are the birefringent filter, which utilizes combinations of calcite and quartz plates, and polarizers, which isolate all but a narrow band as small as 0.5 A; and the spectroheliograph, which builds up a picture by moving the spectrograph across the image of the Sun. With these devices the overpowering background of photospheric light is removed, and only the chromosphere is seen.

The study of the morphology of the chromosphere is best carried out by time-lapse motion pictures. Photographs are made every 10 sec or so, and then run through a projector at the normal rate of 16 frames per second. This gives a remarkable picture of the dynamic variations in the chromosphere. Such films of the center and limb of the Sun have provided remarkable pictures of this very complex zone just above the surface of the photosphere.

While the magnetic-field energy in the photosphere is considerably less than the energy of the material, the rapid falloff in density with height results in a situation where the magnetic field dominates the material motions, and the gas is ordered into large-scale patterns. This is not strikingly evident in the low chromosphere (0–1500 km), which seems an irregular extension of the underlying photosphere. In fact, the temperature continues to drop above the surface of the photosphere (as evidenced by limb darkening) to an apparent minimum of about 4000°K at a height of 1500 km. Even in the low chromosphere, however, the effect of the chromospheric network in the small bright regions or faculae that mark the edge of each network cell is seen. These faculae may be observed in white light near the limb of the Sun. In them the chromosphere shows a very fine structure of bright points, as though one were looking down at each line of force in the magnetic field (Fig. 8a).

Looking near the limb on chromospheric pictures or looking at a picture of the extreme limb with the disk of the Sun occulted, one can see the spicule forest as an irregular distribution of jets shooting up from the low chromosphere. At the extreme limb there are very many of these along the line of sight, resulting in the forestlike appearance. In fact, when pictures of the chromosphere on the disk are examined along with motion pictures, the spicules are seen grouped in bushes and rosettes at the edges of the chromospheric network. Because the spicules have high velocity, their spectrum lines are broad, and they are most easily seen in a chromospheric picture just off H-α (Fig. 8b). In this case they very clearly mark the chromospheric network.

Visibility through the Earth's atmosphere and instrumental inadaquecies limit one's ability to resolve the structure of the spicules (Fig. 9). They shoot up to a height of about 6000 km above the photosphere with a velocity of about 20 or 30 km/sec and then fade out. They may be very narrow, certainly less than 1000 km across. At the top of their trajectory some spicules fade out and others drop back into the chromosphere. It is believed that the spicule jets occur as a result of the focusing of mass motions in the low chromosphere by the strong magnetic fields at the edges of the

Fig. 7. The flash spectrum of the ultraviolet light from the chromosphere, photographed during an eclipse. This is a negative to enhance details. The strong H and K lines are at the right, the convergence of the Balmer series of hydrogen is near the center, and at the left the lines merge into the Balmer continuum toward shorter wavelengths. (*High Altitude Observatory, National Center for Atmospheric Research, Boulder*)

chromospheric network. They form a channel by which energy travels from below into the solar corona.

Inside the chromospheric network cells a confusing and interesting mass of small elements exists; the elements oscillate back and forth, almost like water in a bathtub, with velocities approaching 10 km/sec and a period of about 180 sec. This horizontal oscillation is very easy to see in motion pictures of the chromosphere. It is presumably coupled with the 250-sec oscillation in the photosphere, and plays an important role in the transport of energy upward.

The chromosphere represents a dynamic transition zone between the photosphere and the corona above. Since the coronal temperature is 1,-000,000°K and the temperature in the low chromosphere is about 4000°K, there must be a rapid increase of temperature through the chromosphere. This temperature increase is not steady but represents the thermal history of a spicule jet or an oscillating element as it flies through the chromosphere. Considerable effort has been devoted to the analysis and interpretation of the spectrum lines from the chromosphere in order to understand this thermal transition. The problem is confused by the fact that, in the visible region, only the spectrum of the chromosphere at the limb can be studied, where it can be separated from the tremendous radiation of the solar surface. At the limb one must look through many different elements at the same time, and the spectrum is a mixture of contributions from hot and cold regions. However, gross features can be distinguished from spectra of the chromosphere at the limb obtained at eclipse. The spectrum of the low chromosphere is essentially the same as that of the photosphere, dominated by lines of atoms and ions of low excitation. Thus the low chromosphere is essentially an upward extension of the photosphere. Above 1500 km the low chromosphere diminishes and an abrupt change takes place to a spectrum of high excitation which is dominated by lines of neutral and ionized helium that can be excited only at temperatures of 20,-000°K and higher. For a number of reasons it may be concluded that these lines are produced in the spicule jets shooting up from the low chromosphere. The high-excitation lines are also seen in the low chromosphere because spicule jets in front of the limb are seen in projection. Since the spectrum of the upper chromosphere is dominated by the spicule jets, there is little information as to what happens in the upper chromosphere inside the cells.

An important tool for the understanding of the transition region is the extreme ultraviolet spectrum as recorded by rockets and satellites (Fig. 10). Although the chromosphere is invisible against the disk in the visible region, in the ultraviolet region its high temperature makes it the dominant contributor, eventually superseded by the still hotter corona at even shorter wavelengths. In the extreme ultraviolet (XUV) spectrum, the lines of such ions as O II, O III, O IV, O V, O VI, C II, C III, C IV, and so forth, are seen. The roman numeral gives the number of electrons that have been removed, less one. From the relative intensity of these lines one can get some idea of the

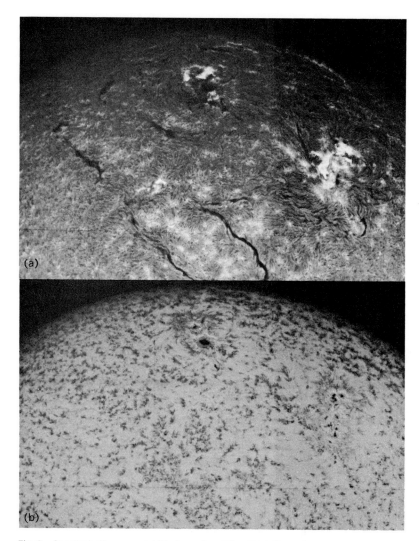

Fig. 8. Spectroheliograms. (a) Photocenter of H-α. Dark filaments form the boundary between regions of bright flocculi of different magnetic polarities. The flocculi themselves outline the cells of the chromospheric network and are brighter where the magnetic field is stronger. The fine dark features coming from each bright flocculus are spicules. At the bottom center two filaments converge in a region of weak magnetic activity. Two active regions around sunspots appear, surrounded by bright plages and "whirlpool structure." The dark filaments that curve into the sunspot regions also mark the boundary between fields of opposite polarity. These fields run horizontally along the filaments, as can be judged from the disk structure. The filament near the limb shows a typical bright rim structure underlying it. (b) At +0.7 A showing structures with wide profiles. Only the spicules at the edges of the cells remain. These features are the spicules seen at the limb, and are seen to move with appropriate velocity on disk pictures. They must stand above the surface because they appear invariably as dark areas on the limbward side of flocculi near the limb. Spicule profiles at the limb were found by R. G. Athay and by R. Michard to be broader than the photospheric H-α, so that they must be prominent in pictures in the wings. Although light from deeper layers in the Sun may reach the Earth off band (because the absorption coefficient is lower), the structures one sees are the highest because the highest have the broadest profiles. The sunspot and penumbra appear clearly, because one sees through overlying material. A bright rim is evident around the edges of the penumbrae. (R. B. Leighton and coworkers, Hale Observatories)

conditions at the successively greater heights in the atmosphere at which they are radiated. Furthermore, from the distribution of these lines on the disk it is possible to determine the temperature distribution.

Corona. The corona lies above the chromosphere (Fig. 11). The light seen from the corona originates in three distinct processes which distin-

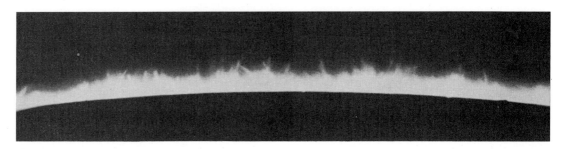

Fig. 9. Large-scale photograph of the chromosphere in H-α light, with the disk of the Sun artificially eclipsed and the hairy spicules projecting above the continuous chromosphere. The length of this section is about 140,000 km. (*Photograph by R. B. Dunn, through 15-in. telescope, Sacramento Peak Observatory, operated by the Association of Universities for Research in Astronomy, Inc.*)

guish the F (Fraunhofer), K (Kontinuierlich-continuous), and E (emission) components.

The F corona is not directly associated with the Sun. It is a halo produced by the scattering of sunlight by interplanetary dust between the Sun and the Earth, and is properly regarded as the inner zone of the zodiacal light. Because the scattering is by solid particles, the Fraunhofer spectrum is seen in this light, and hence the name. The F component is negligible compared with the K component in the inner corona, but because of the rapid falloff of density in the corona, it dominates beyond 2.5 solar radii.

The K corona is photospheric light scattered by the free electrons in the solar corona. These electrons are moving so fast that the Fraunhofer spectrum lines are blurred and a continuous spectrum is seen. The light is also polarized by the scattering process, and this polarization is used to detect the K corona against the sky background. Because the

K corona is produced by the simple process of electron scattering, it is possible to measure directly the electron density (and hence the total density) of the material in the corona. This is done by careful photometric measurements of the coronal brightness at eclipse. Since the material is predominately hydrogen and is entirely ionized, there is essentially one proton for each electron measured and the density is equal to the proton mass times the number of electrons observed. The density at the base of the corona is found to be about 4×10^8 atoms/cm^3, falling off exponentially with a scale height of 50,000 km. The structure of the K corona varies markedly with the sunspot cycle. At sunspot maximum it presents a fairly symmetrical globular appearance with many domes and streamers in all directions. At sunspot minimum the globular shell shrinks toward the solar surface and individual dominant streamers are seen. At the poles there are brushes of small,

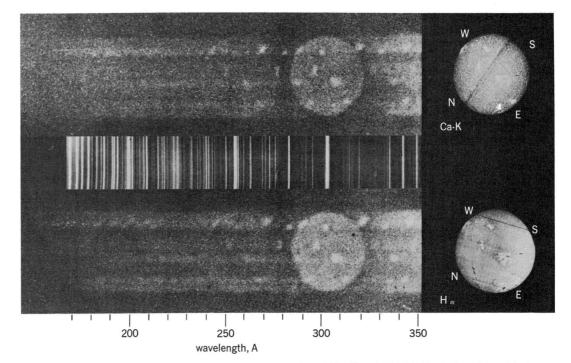

wavelength, A

Fig. 10. Pictures of the Sun in the ultraviolet made by the U.S. Naval Research Laboratory with an Aerobee rocket. A picture of the Sun in each emission line is produced. The line at 304 A is He II; that at 283 A is Fe XV. Comparison pictures taken from the ground are at right.

symmetrically diverging streamers reminiscent of an iron-filings pattern. It is this appearance which suggested the presence of a general solar magnetic field.

The electron and temperature distribution in the corona may also be studied with radio telescopes. The propagation of radio waves through the corona is such that at each frequency one can see only down to a point characterized by the plasma frequency, which increases with density. At successively higher frequencies one sees deeper and deeper into the corona. Therefore the brightness at each frequency indicates the temperature at that height. It is possible, therefore, to map the temperature as a function of height. Further information may be obtained from the distribution of brightness across the solar disk in radio radiation. At high frequencies it is possible to see completely through the corona to the chromosphere, which appears as a uniform disk of about 6000°K. At longer wavelengths the corona is no longer transparent in the slanting direction of the limb, and therefore a peak of brightness near the edge of the Sun is seen. At the lower frequencies where the radiation does not penetrate the corona, the Sun appears much larger than it does in visual wavelengths.

The E, or emission-line, corona (Fig. 12) is the true emission of the ions in the corona. It is therefore the best source of information on the physical state of the corona. In the visible region the emission lines are relatively weak compared with the photospheric radiation scattered by the K corona. But at wavelengths of 200 A, the emission lines are stronger than the radiation from the disk, which cannot contribute much at such high energies. In spite of rather formidable technical difficulties, the coronagraph makes it possible to record the spectrum of the E corona in the visible, and even to photograph the structure directly through a birefringent filter whenever the sky is sufficiently clear. Since observations of the K corona outside of eclipse are limited to rather low resolution, the emission-line corona is an important tool in the observation of the day-to-day variation of the corona. The spectrograph dilutes the sky light by spreading it into a long, continuous spectrum. The coronal emission lines, on the other hand, are merely separated by the dispersion without dilution and stand out conspicuously against the sky continuum. Figure 13 shows a short length of spectrum with a green 5303 A line of the corona and lines of a bright prominence superposed on the Fraunhofer spectrum of a sky scatter. The K corona is so weak that it cannot be seen. The emission lines are proportional to the square of the density and fall off sharply with increasing height.

Observations are continuously made at a number of observatories. Current knowledge of the E corona rests primarily on such observations of the green line of Fe XIV at 5303 A, the red line of Fe X at 6374 A, and the calcium high-temperature yellow line of Ca XV at 5694 A. Of course, hydrogen and helium are completely ionized, so that only their ultraviolet lines appear in the corona.

The history of the identification of the coronal emission lines is a remarkable chapter in astronomy. The strongest, the Fe XIV green line, was in-

Fig. 11.   The solar corona observed about 2 sec after second contact during the eclipse of Nov. 12, 1966, at Pulacayo, Bolivia (altitude 13,000 ft). This photograph, made with a radially symmetric, neutral density filter in the focal plane to compensate for the steep decline of coronal radiance with increasing distance, allows structural features to be traced from the chromosphere out to $4.5R_0$. The overexposed image of Venus appears in the NE quadrant. Typical "helmet" streamers overlie prominences at the SE and SW limbs, while another streamer at high latitude in the NW develops into a narrow ray at large distance. Arches in the corona and the absence of coronal material in domes immediately above prominences are particularly striking at the bases of the NW and SW streamers. A coronal condensation appears on the NW limb at a latitude of about 25°. (*G. A. Newkirk, High Altitude Observatory, National Center for Atmospheric Research, Boulder*)

dependently discovered by W. Harkness and C. A. Young during the 1869 eclipse. This, and weaker coronal emission lines subsequently discovered, corresponded to no known spectrum lines, and were collectively attributed to the unknown element, coronium. More than 70 years later, they

Fig. 12.   Typical stable form of structure of the E corona in the light of the green line of Fe XIV photographed without an eclipse through a birefringent filter. (*Sacramento Peak Observatory, operated by the Association of Universities for Research in Astronomy, Inc.*)

Fig. 13. Green line of Fe XIV in E corona (long arc at left) with bright metallic lines of a short prominence, and yellow chromospheric helium line D₃ (arc at right), photographed through a small coronograph. (*Sacramento Peak Observatory, operated by the Association of Universities for Research in Astronomy, Inc.*)

were identified by W. Grotrian and, more extensively, by B. Edlen as forbidden lines from highly ionized atoms which could be excited only under conditions of high temperature, low density, and enormous volumes quite beyond any conceivable laboratory resources. Most of the known lines originate from ions of Fe, Ni, Ca, or Ar from which 9 to 14 electrons have been ripped by the fierce bombardment of neighboring particles. The mere presence of these ions with ionization potentials from 233 to 814 ev is unequivocal evidence of the high kinetic temperature of the corona. The temperature calculated from the ionization which is determined from the observed absolute and relative line intensities is about 1,500,000°K. Observations of the permitted lines in the extreme ultraviolet which could be duplicated in the laboratory confirmed these identifications in recent years. In addition a host of other lines of highly ionized Mg, Si, C, O, and other atoms have been observed.

The temperature of the corona can also be measured by the Doppler broadening of the forbidden emission lines. These are very wide because of the rapid atomic motions, corresponding to temperatures well over 1,500,000°K. Thus a number of different observations confirm the high temperature of the corona.

One conclusion from consistent observation is

that the E corona in particular is highly variable. Whereas the great K-corona streamers and plumes seen at great distances from the Sun at eclipse appear to be related to large and long-lived magnetic regions on the Sun, the emission lines in the low corona vary sharply with solar activity. The Ca XV yellow line, which is produced at temperatures of 4,000,000°K, is seen only during intense eruptions accompanying the limb passage of sunspot groups. There is very little E corona at the poles, where sunspots do not occur, and not much in between the active sunspot groups. Thus the E corona is almost entirely the product of solar activity.

The high temperature of the corona is a fascinating problem. Obviously the temperature of a star is highest at the center and decreases outward to the edge. It is apparent that the steep increase in temperature in the corona is due to boundary effects connected with the steeply decreasing density at the edge of the Sun and the convective currents beneath it. These, together with magnetic fields, give a "crack of the whip" effect in which the same energy is concentrated in progressively smaller numbers of atoms. The result is the production of the high temperature of the corona. However, the mechanism is very imperfectly understood.

Solar wind. One of the most remarkable surprises of the age of rockets and satellites has been the discovery of the existence of the solar wind. This continual outflow of matter from the Sun was predicted by Eugene Parker on the basis that the high temperature in the corona must lead to a rapid outflow at great distances from the Sun. He predicted a velocity near 1000 km/sec. Satellites such as *Mariner 2* and the IMP series have detected a continual flow of plasma from the Sun with a velocity ranging from 300 to 500 km/sec and density about 1 atom/cm³ near the Earth. The flow varies with the level of solar activity. The particles flow along a spiral path dictated by magnetic fields from the Sun carried out into the interplanetary medium. The rotation of the Sun produces the spiral pattern. The magnetic field near the Earth is measured to be rather uniform over large sectors of the Sun, corresponding to one dominant polarity or another. The solar-wind flow has a continual effect on the upper atmosphere of the Earth.

SOLAR ACTIVITY

The foregoing section has been concerned with the structure of the normal steady state of the undisturbed Sun. There are, in addition, a number of transient phenomena known collectively as solar activity. These are all connected with sunspots or their remnants, and wax and wane in a remarkable cycle of activity. The sunspot cycle consists of variations in the sizes, numbers, and positions of the sunspots, expressed quantitatively as the sunspot number (Fig. 14). The number of sunspots peaks soon after the beginning of each cycle and decays to a minimum in 11 years. The magnetic polarity of the sunspot groups reverses in each successive cycle so that the complete cycle lasts 22 years. The first spots of a cycle always occur at higher latitudes, between 20 and 35°, and as the spots increase in size they approach closer to the equator.

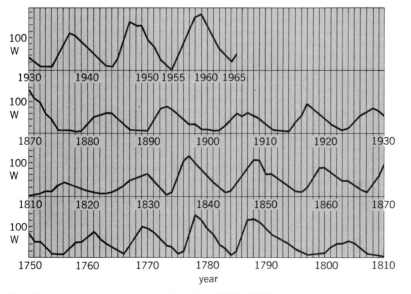

Fig. 14. Annual mean sunspot number from 1750 to 1965.

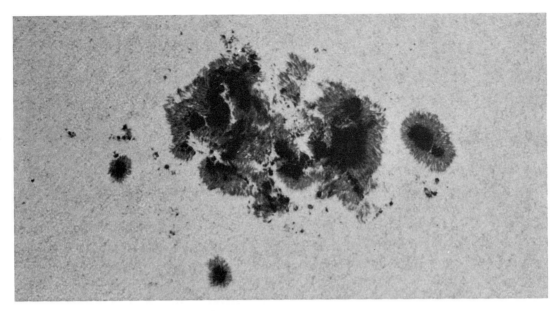

Fig. 15. Large sunspot group of May 17, 1951, photographed in white light showing the filamentary structure of the penumbra and the granulation of the surrounding surface. (*Hale Observatories*)

Almost no spots are observed outside the latitude range of 5–35°. The average duration of the cycle from the first appearance of high-latitude spots to the disappearance of the last low-latitude spots is nearly 14 years. The difference between these two figures is the result of overlap of consecutive cycles. Thus there is hardly ever a time when there are no spots on the Sun at all. In the last two cycles there has been a great dominance of sunspots in the northern hemisphere. For periods of time there were few or no spots in the southern hemisphere. In addition, the two cycles with maximum activity in 1946 and 1957 were the two largest in history. Whether these facts are connected is not known. However, there have been long periods in the past when activity was dominant in a single hemisphere.

Sunspots. Sunspots are the most conspicuous features of solar activity and are easily seen through a small telescope. They are dark areas on the Sun produced by the most intense magnetic fields. The magnetic fields produce a cooling of the surface, most likely by suppression of the normal convection which transports energy from the lower

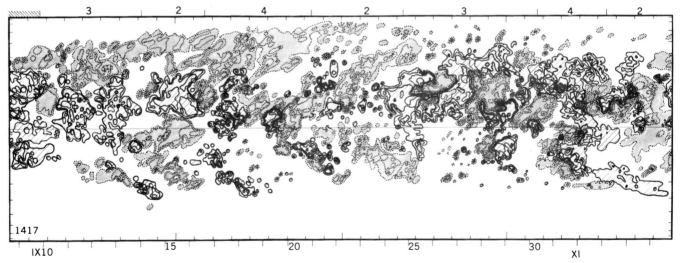

Fig. 16. Synoptic chart for magnetic fields for rotations 1417 (August, 1959), compiled by V. Bumba and R. Howard at Mount Wilson Observatory. Solid lines and hatching represent positive polarity and dotted lines and shading, negative. The quality of the observation is indicated, on a scale of 5, by the number above. Isogauss lines are drawn for 2, 6, 10, 15, and 25 gauss longitudinal field strength. The longitude is shown in 5° intervals below the map. The large unipolar regions drifting eastward and poleward from centers of activity are easily seen. Bumba and Howard found that complexes of activity start out with large spots followed by successively smaller ones, and the magnetic field spreads out into empty regions with a velocity around 100 m/sec (10,000 km/day). (*Hale Observatories*)

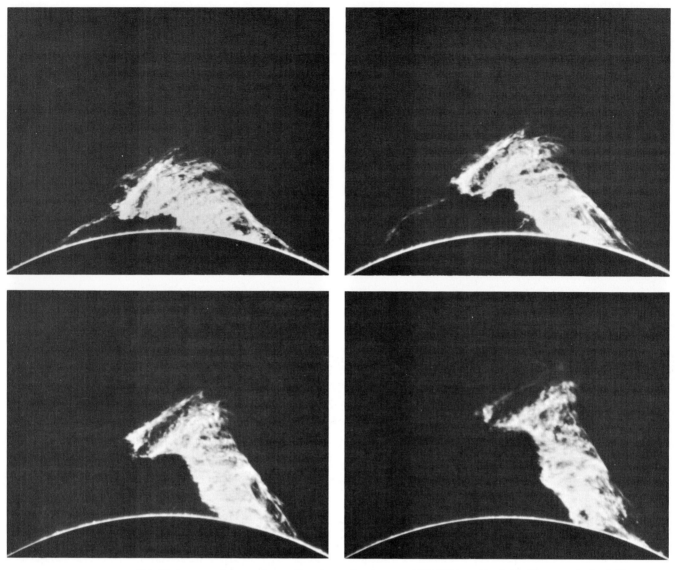

Fig. 17. Large prominence in eruptive phase after days or weeks of static inactivity. Four frames cover interval of 29 min. Horizontal width of single frame is about 670,000 km. H-α light. (*Sacramento Peak Observatory, operated by the Association of Universities for Research in Astronomy, Inc.*)

levels. Deprived of this supply of heat, the area cools by radiation and becomes a dark sunspot.

Curiously enough, the region directly above the cool sunspot is the scene of the hottest and most intense activity because of the great magnetic energy present in the sunspot.

Sunspots appear singly or in groups; the largest and most complex sunspots usually exhibit the most intense variation and activity, but often great activity is associated with ragged groups of small spots. The most important characteristics in this regard are complexity and rapid growth. Old sunspots are very often large, circular, and stable. The spot may occur with a single magnetic polarity, called a unipolar spot; more likely is the bipolar spot group made up of two main spots, the preceding one having opposite polarity from the following. A surrounding retinue of small spots is often present, with various polarities usually determined by their closeness to the leader or follower. In a given cycle, the polarities of the preceding spots of bipolar groups are all the same in one hemisphere, and opposite in the other hemisphere. In the succeeding 11-year peak these polarities will be reversed. The largest sunspot groups are much more complex, with tangled variations of magnetic field from point to point; however, there usually is a dominance of one polarity in one region of the group and the opposite polarity in the other.

Surprisingly, there are no complete observations of the detailed life cycles of sunspots. Only daily pictures of some groups which are not quite sufficient to show the details of their group are all that is available. Figure 15 is an excellent example from a very special sunspot group. Typically, an active spot group begins as an outbreak of many small spots over a fairly large area. In the growing phase it shows extremely bright plages and much flare activity. Sometimes this will last for a few days and the group will dissipate. Large spots begin to appear and the group may evolve in one of many ways. In the late stages the complexities

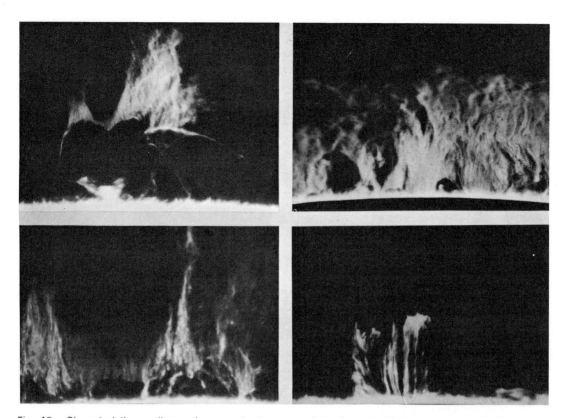

Fig. 18.   Characteristic small prominence, showing fibrous structure. The generally vertical filaments are paths of downward moving material. Horizontal width of a single frame is 125,000 km. H-α light. (*Sacramento Peak Observatory, operated by the Association of Universities for Research in Astronomy, Inc.*)

resolve, and what is left is either a simple bipolar group or an even more simple, large, round spot which then fades away. Occasionally, such a round spot will split up and show a resurgence of activity. By far the largest number of sunspots appear as small pores, grow for a few days, and then fade away. There is a strong tendency for sunspot activity to break out all over the Sun at the same time. This enhances the idea that there is a general mechanism for their growth. There also may be preferred meridians for their occurrence.

By means of the Babcock magnetograph, which permits the measurement of weak magnetic fields on the Sun, it has been possible to study the connection between the intense magnetic fields in the sunspot groups and the weak magnetic fields distributed over the surface (Fig. 16). The weak fields are concentrated on the edges of the chromospheric network, which consists of cells 30,000 km across. The fields vary considerably in strength from cell to cell. The origin of these magnetic fields appears to be the big sunspot groups. As the spots decay or are torn apart by motions in the solar atmosphere, their magnetic fields spread out and drift toward the poles. Because the poles rotate more slowly than the equator, the fields lag behind each sunspot group in a large region of one magnetic polarity shaped like the wing of a butterfly. Sometimes these regions may extend 90–120° in longitude. Their polarity is normally that of the following sunspot polarity. As time goes by, the weak fields reach the pole and establish there a dominant magnetic polarity which, because it is the same as the following spots, is op-

posite to the polarity of the preceding sunspots which dominate most spot groups.

Thus, a few years after the outbreak of the sunspot cycle, the fields generated by the broken-up

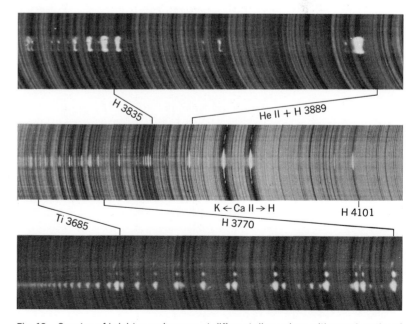

Fig. 19.   Spectra of bright prominences at different dispersions with wavelengths of corresponding lines. From top to bottom the wavelength range covered by each spectrum is 88, 545, and 129 A. The converging Balmer series merges into the Balmer continuum in bottom spectrum. (*Sacramento Peak Observatory, operated by the Association of Universities for Research in Astronomy, Inc.*)

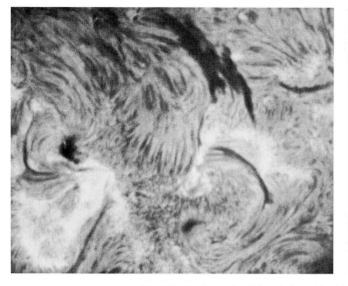

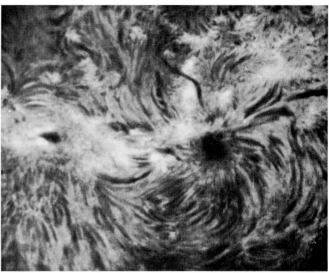

Fig. 20. Centers of activity photographed in H-α light. The length of each frame is about 280,000 km on Sun. Note the extensive area of whirlpool structure about each sunspot. (*Sacramento Peak Observatory, operated by the Association of Universities for Research in Astronomy, Inc.*)

spots reach the poles, producing at each pole magnetic fields of polarity opposite to those of the preceding sunspots in each hemisphere. Some models of solar activity propose that, as the sunspots of the parent cycle die away, this large-scale dipole field is amplified by the differential rotation to produce a new cycle of sunspots, which naturally will

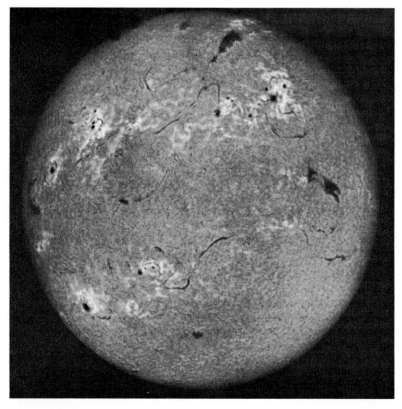

Fig. 21. The disk of the Sun photographed in H-α light, showing bright plages in centers of activity and dark filaments (prominences). (*Sacramento Peak Observatory, operated by the Association of Universities for Research in Astronomy, Inc.*)

be opposite in polarity to the preceding cycle.

The large unipolar magnetic regions spreading out from centers of activity are easily recognized on H-α or calcium monochromatic pictures of the Sun. This is because the boundaries between these regions, where the magnetic field becomes horizontal in order to change direction, are usually marked by large prominences, accumulations of material in the atmosphere supported by the horizontal field. These are seen as dark filaments, such as in Fig. 8. Because the unipolar regions are areas of enhanced magnetic field, they also correspond to enhanced emission in the chromospheric network, and thus can be distinguished in calcium spectroheliograms.

There is a grand scheme for the evolution of solar magnetic fields and, in turn, the fields produce the sunspots and all their effects. However, the explanations mostly rest on differential rotation, since there is very little understanding of why the equator of the Sun should rotate so much more rapidly than the poles.

Prominences. Although prominences appear dark against the disk, they appear bright against the dark sky. They occur only in regions of horizontal magnetic fields, because these fields support them against the solar gravity. Thus filaments on the disk are good markers of the transition from one magnetic polarity to the opposite. As can be seen from Figs. 17 and 18, prominences are among the most beautiful of solar phenomena. The spectra of prominences show a number of bright emission lines of various elements, mostly singly ionized (Fig. 19). Analysis of these lines shows that long-lived, stable prominences have a temperature of about 10,000°K, while transient prominences connected with flares show many fewer lines and are over 30,000°K.

Plages. As can be seen on H-α photographs (Fig. 20), a large sunspot is surrounded by an extensive area of whirlpool structure, which suggests magnetic control over the chromospheric features, and

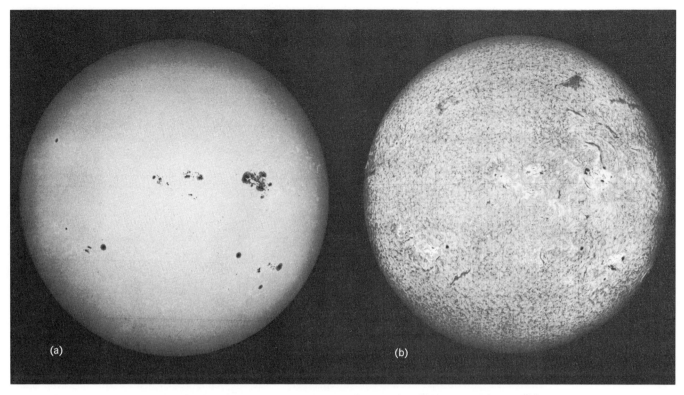

Fig. 22.   The Sun photographed nearly simultaneously (a) in white light and (b) in red H-α light, showing the appearance of plages in different wavelengths. (*Hale Observatories*)

by bright regions called plages. The plages are most easily seen in monochromatic hydrogen or calcium light with the spectroheliograph or bire-fringent filter; however, their brightest features may be seen in integrated white light near the edge of the disk, where the photosphere is somewhat darker and the relative contrast is enhanced (Fig. 21). Magnetic measurements show that most plages are the loci of enhanced magnetic fields, intermediate in strength between the sunspots (3000–4000 gauss) and the bright faculae of the chromospheric network (100 gauss). The magnetic field in the plages is also predominantly vertical. In a bipolar sunspot group, the plages will sometimes appear only on the side dominated by one magnet-ic polarity.

The appearance of plages in different wave-lengths (Fig. 22) shows clearly the growing domi-nance of the magnetic fields with increasing height. If the energy of the magnetic field is com-pared with the kinetic energy of the gas in the low photosphere, only the magnetic fields inside sun-spots are strong enough to dominate material mo-tions. But in the chromophere, the density drops so rapidly that the magnetic fields spreading out from the sunspot easily dominate. As a result little or no whirlpool structure is seen in white-light photo-spheric pictures, but it is easily seen in H-α, which reveals the chromospheric structure higher up. In white light the plages are visible only near the limb, where one sees higher in the photosphere. They are easily seen in H-α, but are even more widespread in the calcium spectroheliograms in which one looks still higher in the atmosphere.

With rockets, spectroheliograms can be made in the high-ionization lines showing emission mainly from the active regions and plages. These are the only regions in the atmosphere hot enough to pro-duce considerable ultraviolet radiation.

Although sunspots are always accompanied by plages, the reverse is not true. Occasional fields of small plages develop in the sunspot zone and fade

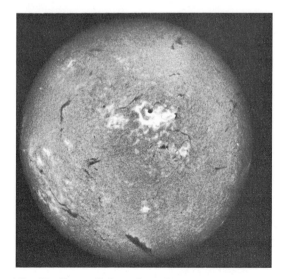

Fig. 23.   Unusually large flare of Sept. 18, 1958, photo-graphed in H-α light and enlarged from a 16-mm image. (*Sacramento Peak Observatory, operated by the Associa-tion of Universities for Research in Astronomy, Inc.*)

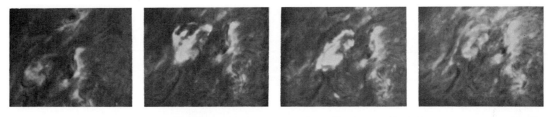

Fig. 24. Development of a large, coarsely filamentary flare in H-α light. Intervals between exposures are 4 min, 19 min, and 13 min. Each frame shows 280,000 km on *the Sun. (Sacramento Peak Observatory, operated by the Association of Universities for Research in Astronomy, Inc.)*

away without the appearance of any spots. In these cases the magnetic fields never reach sufficient strength to generate sunspots. Similarly, the plages usually remain for a few weeks to mark the location where a sunspot has died. Eventually they probably break up and spread out into the chromospheric network, but this has not been observed directly.

Flares. The most spectacular activity associated with sunspots is the solar flare (Figs. 23 and 24). A flare is defined as an abrupt increase in the H-α emission from the sunspot region. The brightness of the flare may be five times that of the associated plage; the rise time is seldom longer than a few minutes, sometimes only 10 sec. Of course, the hydrogen brightening is only a symptom; observations by other means show that there is a tremendous energy release, of as much as 10^{33} ergs in a

large flare. An active sunspot group will show many small flares the size of a large sunspot. Once or twice in the lifetime of a large group, a great flare will occur covering the entire region. The energy released in the three or four great flares each year equals that in all the small flares put together. The flares are ranked by area in three classes, with class 3 the greatest; there is also a class of subflares for smaller brightenings which are only marginally flares.

Solar flares are most numerous when a sunspot region is developing rapidly and magnetic fields are complex. The occurrence of great class 3 flares may be marked by a substantial change of the general sunspot structure. Subflares and class 1 flares will frequently appear over and over at the same point. Favorite locations for such flares are notches in the umbrae of large spots produced by the

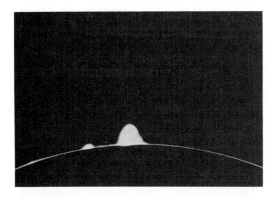

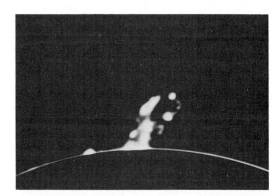

Fig. 25. Flare prominence, or spray, of Feb. 10, 1956. Universal time for successive frames: 211600, 211845, 212115, 213345 cover a total interval of 17 min 45 sec. Some fragments exceeded a velocity of 1100 km/sec. Each frame shows 670,000 km on the Sun. *(Sacramento Peak Observatory, operated by the Association of Universities for Research in Astronomy, Inc.)*

approach of a strong magnetic field of the opposite sign; small sunspots in between large ones, again indicating complex fields; and, in general, intrusion of any region of magnetic field of one sign into the fields dominantly of the opposite polarity. In any flare the brightening is rapid; at any given point, it takes but a few seconds. The brightness then fades over a much longer period. Flares are often accompanied by the ejection of large quantities of material called sprays or surges (Figs. 25 and 26). The sprays are fast expulsions of gas in all directions, with velocities as high as 1500 km/sec. They normally occur in the first moments of a big flare and appear to be part of a tremendous explosive release of energy. The spray may be preceded by the rising of a large prominence above the sunspot group, which then is expelled in the first moments of the flare. Such rising prominences are the only known precursors of solar flares. By contrast, surges normally accompany smaller flares and are an expulsion of material in one predominant direction. This must happen because the energy is insufficient to penetrate the surrounding magnetic field, and the material flows in a more organized way.

Flares are very rarely seen in white light because they are an atmospheric phenomenon and their density is so low that they are transparent. On the other hand, their temperature is so high that, in the ultraviolet region, they may equal the intensity of entire Sun. However, because the flares are most easily observed in H-α and because the H-α brightening is an extremely accurate indicator, they are mostly studied in this way.

In recent years it has become possible to observe the flare phenomenon over a great range of wavelengths, in which many diverse phenomena have been observed. In x-rays a sharp pulse of radiation is observed, with photon energies of 100,-000 volts and higher. At lower x-ray energies, the time behavior of the pulse is not so steep. The x-ray pulse always occurs during the first sharp brightening. Simultaneously, a broad-band burst of radio waves is emitted in the microwave region, peaking around 3000 MHz. This appears to be produced by the same energetic electrons which cause the x-ray burst. At longer wavelengths, of 1000–1050 MHz, more complex phenomena (Fig. 27) are observed associated with the outward movement of disturbances from the flare through the corona. The corona at each point radiates a certain frequency proportional to the density; thus as disturbances pass upward, radiation is excited at successively lower frequencies. The most intense of these is the type II burst, with a downward drift in frequency during a period of 2 to 5 min; this corresponds to an outward velocity of about 1000 km/sec and is seen only in the largest flares. The type III bursts are much more frequent and show a very rapid frequency drift corresponding to disturbances propagating in the corona at 100,000 km/sec and more. Some type III bursts are observed to turn downward, as though reflected by magnetic fields in the atmosphere. These are called U bursts. Following a large type II burst, one occasionally observes a type IV radio burst or storm which is a long-lasting wide-band emission

from a great cloud in the corona. By analogy, the intense broad-band microwave burst accompanying the x-ray emission is called microwave type IV. The outward-moving type III bursts also produce an afteremission called type V emission.

Any large sunspot group is accompanied by a radio noise storm, and the detailed bursts of this storm are normally called type I bursts. They are strongly polarized by the coronal magnetic fields.

In addition to the x-ray and radio emissions from flares, which are all electromagnetic radiation, great numbers of energetic particles, both nuclei and electrons, are observed. Since energetic particles must follow the magnetic fields, the corpuscular radiation is not observed from all big flares, but only from those favorably situated in the western hemisphere of the Sun. Because of the solar rotation, the lines of force from the western side of the Sun (as seen from the Earth) lead back to the Earth and guide the flare particles to the Earth. The nuclei are mostly protons, because hydrogen is the dominant constituent of the Sun.

The flare-produced cosmic rays are most numerous at 3–5 Mev, but range upward in energy to hundreds of millions of electron volts. Almost all

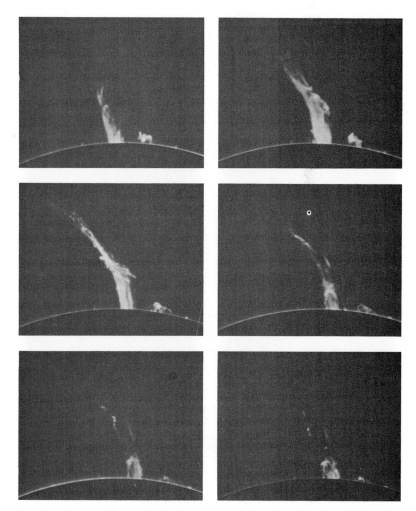

Fig. 26.   Photographs show large surge in H-α light during total time interval of 45 min. Frame spans 500,000 km on the Sun. (*Sacramento Peak Observatory, operated by the Association of Universities for Research in Astronomy, Inc.*)

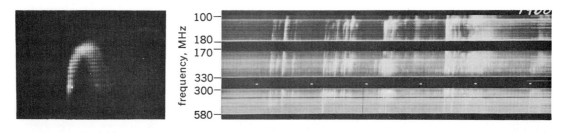

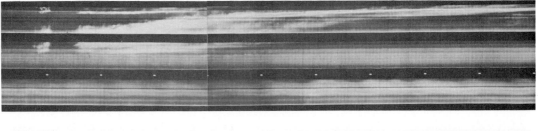

Fig. 27. Time variations in the solar radio spectrum. Frequency range from 100 to 580 MHz (the vertical coordinate) is divided into three overlapping bands with frequencies indicated in top row. Variations with time are shown in horizontal direction. Interval between the white pips between the middle and lower frequency bands represents 1 min. Top left, a highly magnified U burst in the 140-MHz band. The interval between successive vertical scans is 0.3 sec. Top right, a series of type III bursts. The middle and lower rows are continuous, showing the development of a type II burst followed by a strong continuum in the 450-MHz band. (*Recordings from radio spectrometer, Fort Davis Station, Harvard College Observatory*)

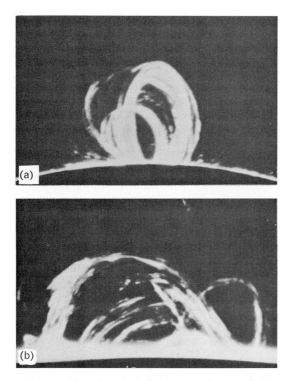

Fig. 28. Two characteristic loop prominences in H-α light. Width on Sun is (a) 340,000 km and (b) 125,000 km. (*Sacramento Peak Observatory, operated by the Association of Universities for Research in Astronomy, Inc.*)

flares produce cosmic rays, but the cosmic-ray production from the biggest flare of any year will probably equal that of all the other flares put together. The high-energy cosmic rays appear moments after the beginning of the H-α flare, but because of time of flight and the guiding by the interplanetary magnetic field, the lower-energy cosmic rays appear much later. A cosmic ray storm may last for days. The flux of low-energy particles in big flares is so intense that it endangers the lives of astronauts outside the terrestrial magnetic field. *See* COSMIC RAYS.

Relativistic electrons with energies of millions of electron volts are also observed shortly after the beginning of optical flares. They are presumably accelerated by the same phenomenon that produced the cosmic rays.

A day or two after a very large flare, the pulse of arrival of a new group of particles of low energy (1–5 Mev) but of very large numbers may be observed. These particles, which produce what is called a geomagnetic storm, are a cloud ejected directly from the flare with a velocity of 1500 km/sec or more.

It is known that any large sunspot group is continually producing large numbers of low-energy cosmic rays, presumably as the result of many small flares. The energy spectrum is very steep, about the inverse fifth power; hence, if the magnetic changes in a large flare increase the energy of all cosmic rays by a factor of 10, the number in some high-energy range will be increased by 10^5.

Current studies of solar flares are devoted to understanding why and how they occur, and how the Sun can accelerate particles to such high energies. While it is known that the flares are connected with rapid changes in the magnetic field, the exact mechanism for storing the energy for a while, to be released very suddenly (in a matter of minutes), is not clear. Current theories include magnetohydrodynamic instabilities and electric discharges.

For many years astronomers have tried to study flares by their spectra. They have been confused by the existence of many different components in the typical flare. There is the relatively low-temperature gas that radiates in H-α (were the gas fully ionized, it could not radiate in neutral hydrogen); there is the 4,000,000°K coronal cloud which produces the soft x-radiation; there is the 10^{9}°K (100,000 volt) cloud of electrons producing the hard x-rays; and there are the particles with energies of millions of electron volts. Each spectrum line shows only the material radiating that particular emission. In this way intense lines of hydrogen, helium, and metals corresponding to the main body of the flare at 20,000–30,000°K have been observed; intense corona lines corresponding to temperatures of 4,000,000°K have been observed in the cloud above a flare, leading to the conclusion that a great cloud of material at that temperature is being produced. In the extreme ultraviolet at 1.6 A, emission lines of Fe XVII–XXV have been observed, testifying to high temperatures sufficient to produce those ions. The understanding of the detailed nature of this dynamic phenomenon awaits a synthesis of all these observations.

Flare and corona. The high-energy phenomena connected with flares all occur in the atmosphere above the sunspot; the temperature in the photosphere is only 6000–7000°K and, as mentioned before, with a few exceptions flares cannot be seen in white light. When an active sunspot group is seen at the edge of the Sun, one may see the beautiful high-energy phenomena projected against the sky. These include sprays, surges, and the elegant loop prominences (Fig. 28). Even when large flares are not in progress, surges and small flares are continually throwing material into the atmosphere. In the case of a big flare, this process is tremendously accelerated. All the material injected into the atmosphere heats up very rapidly as it dissipates its original kinetic energy. Thus spectroscopic observations show that almost immediately after a big flare a cloud of hot coronal material at temperatures as high as 4,000,000°K appears. It is not possible to see the material hotter than 4,000,000°K because no lines are produced in the visible. However, rocket observations are beginning to reveal such clouds. Of course, the normal processes of heating the corona cannot sustain all the material injected by the flare, which therefore cools down by radiation. When this material cools and condenses, it rains down on the surface. Because of the great strength of magnetic fields overlying the sunspots, the gas follows the magnetic lines of force in curved trajectories, forming loop prominences. The cooler the material gets, the lower its degree of ionization and the faster it radiates and cools. In

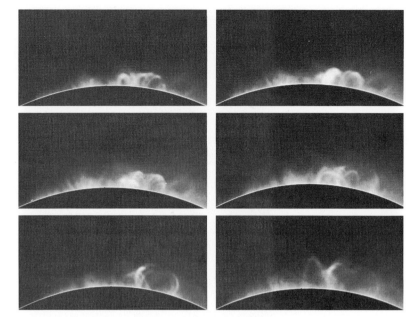

Fig. 29. Six exposures of coronal activity over an interval of 4 hr taken in the green line (5303). Concentric arch structure in first frame is a characteristic structure. (*Sacramento Peak Observatory, operated by the Association of Universities for Research in Astronomy, Inc.*)

fact, a large fraction of the H-α radiation in flares is thought to be due to the falling material from the great loop prominences coming from the coronal cloud. Since the flare may last only minutes and the loop prominences will go on for many hours, they are often seen as a signal that a great center of activity is at the limb.

The birefringent filter enables one to see the appearance of the corona by isolating the coronal emission lines. The coronal material, as well as the prominence material, follows the magnetic lines of force, and many curved arches are seen. However, these do not change as rapidly as the condensing material in the prominences (Fig. 29). Often, series of five or six concentric arches are seen over an active center; they expand slowly and occasionally have been observed to break open at the top and whip into vertical streamers with apparent velocities of up to 600 km/sec. This activity coincides with the eruption of material through the magnetic field at the time of a flare. The coronal cloud produced by the flare is responsible for low-energy x-rays in the 2–12 A range, as well as for gradual increases of radiation in the microwave region. However, the hard x-rays and impulsive microwave bursts are associated with more transient groups of hard electrons accelerated by flares.

Solar terrestrial effects. Aside from heat, light, and the solar tides, the direct influences of the Sun on the Earth are generally too delicate for direct detection by the human senses. One exception is the aurora borealis, or northern lights, which may be seen on occasions in temperate latitudes and more often in Arctic regions. The agents responsible for other solar influences on the Earth are ultraviolet and x-radiation and streams of charged particles emitted by the Sun. These are strongly variable with the level of solar activity, and the effects on the Earth are very large indeed when

Fig. 30. The 150-ft solar-tower telescope of the Mount Wilson Observatory has a vertical underground spectrograph.

observed with sensitive radio equipment and magnetic compasses, as well as with detectors on artificial Earth satellites.

The terrestrial ionosphere is a result of the steady flux of solar ultraviolet and x-radiation. Even the quiet corona and low-level chromospheric activity is sufficient to make possible an ionosphere with a low electron density. As the solar activity increases, the electron density in the ionosphere increases and higher frequences are reflected. Most of the ionization is caused by solar ultraviolet rays between 200 and 1000 A, which ionized oxygen and nitrogen in the upper atmosphere. Only a small fraction of the molecules above 100 km are ionized. *See* IONOSPHERE.

The burst of hard x-rays emitted at the onset of a solar flare causes ionization much deeper in the atmosphere, around 60 km. At this level the density of the air is sufficiently high so that the electrons cannot oscillate freely but give up their energy by collisions with neutral atoms; thus, when they are excited by radio waves (which normally would penetrate and be reflected from the higher ionospheric layers), they absorb the radio waves and produce a shortwave fade-out (SWF). Long-distance radio communication, which depends on reflection from the ionosphere, deteriorates or is blacked out altogether for a few hours. The effect is almost instantaneous with the beginning of the flare, and serves as an indicator of its occurrence. It may also be observed by recording the radio emission from galactic sources which normally penetrates the ionosphere but is suddenly absorbed during an SWF. This phenomenon is called sudden cosmic noise absorption (SCNA). Temporary ionospheric currents in the beginning of the fade-out produce changes in the geomagnetic field strength, and if they are severe they may induce currents in long land lines sufficient to stop telephone communications. SWF is always associated with a flare, although occasionally flares will not produce SWF.

The ionizing radiation in SWF is hard x-rays in the 1–2 A region. By a coincidence, there is also a small amount of ionization in this region (called the D layer) produced by the photoionization of nitric oxide by the Lyman-α line of hydrogen. However, this line does not change much during flares and does not produce a fade-out.

A whole series of other phenomena is associated with the creation of the D layer at the bottom of the ionosphere and its changing height. Among these phenomena are the sudden phase anomalies (SPA), produced by the lowering of the effective reflecting layer; and the sudden enhancement of atmospherics (SEA), produced by increased reflection of distant radio noise.

The particle effects on the Earth are extensive. The cosmic-ray storms produced by big flares do not reach the equatorial regions easily because of the Earth's magnetic field, but they spiral into the polar caps and produce what is called polar-cap absorption (PCA). This is the direct ionization of the ionosphere above the poles. Its effect is a polar blackout of radio communications across the polar regions. The very large number of particles associated with the low-energy pulse manages to penetrate the geomagnetic field to produce the geomagnetic storm. This is most intense near the poles, but may reach down into temperate and even tropic latitudes. The aurora is but one trace of the energetic particles precipitating in the upper atmosphere. The currents induced by these great numbers of particles produce sharp changes in the Earth's magnetic field, and the ionization produces considerable changes in the radio propagation.

The magnetic field changes have also been known to produce severe effects in long power lines, creating surges and pulses which trip circuit breakers and produce power outages. *See* AURORA; GEOMAGNETIC STORM.

The relation between geomagnetic storms and flares is not completely determinate. Large flares on the western side of the Sun usually produce the most geomagnetic disturbances, but great magnetic storms often occur without a flare. These appear to be connected with the emission of particles from the boundaries of unipolar regions. The effect is just beginning to be understood. The flares not associated with storms show a strong recurrence and, since for a long time their origin could not be identified, their source was termed the M region. It was only with the measurement of extensive magnetic fields on the surface that the M regions were identified. Measurements of the plasma velocity in the solar wind show that there is a strong correlation between the velocity of the plasma and the geomagnetic activity. The passage of the M region and the beginning of the geomagnetic storm are usually connected with a shock wave in the interplanetary magnetic field which impinges on the Earth's field and produces a sharp lowering of the field, known as a sudden commencement, the signal for the beginning of a geomagnetic storm. The particles which closely follow produce the storm.

There has been considerable interest in other effects of the Sun on the Earth, such as long-range weather changes, the stock market, and the birth rate. No correlations have been established. However, there are many who believe that there is some connection between solar activity and long-range changes in the weather.

SOLAR INSTRUMENTS

Although the observational instruments of the solar astronomer are the same in principle as those used by other astronomers, there are two points which determine the differences. First, the Sun is extremely bright, so that high concentration of light is not so important. Second, because of the disturbance of the Earth's atmosphere by solar heating during the day, the "seeing" is considerably worse in the daytime and solar telescopes cannot hope to utilize as sharp an image as stellar telescopes. Furthermore, the telescopes must be designed so that the great heating produced by the Sun does not distort the images. This is usually done by liberal use of white paint and careful con-

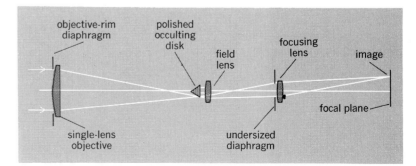

Fig. 31. Optical system of a coronagraph.

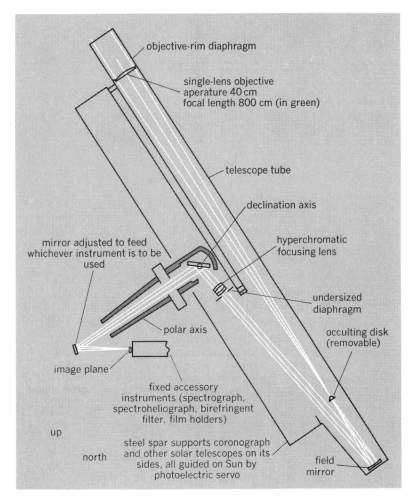

Fig. 32. Large 40-cm solar telescope of the Sacramento Peak Observatory.

objective-rim diaphragm

single-lens objective
aperature 40 cm
focal length 800 cm (in green)

telescope tube

declination axis

hyperchromatic
focusing lens

mirror adjusted to feed
whichever instrument is to be
used

undersized
diaphragm

polar axis

occulting disk
(removable)

image plane

fixed accessory
instruments (spectrograph,
spectroheliograph, birefringent
filter, film holders)

up

north

steel spar supports coronograph
and other solar telescopes on its
sides, all guided on Sun by
photoelectric servo

field
mirror

trol of the atmospheric conditions inside the telescope.

Solar telescopes fall into two general classes: those designed for observations of the brilliant solar disk, and coronagraphs designed for the study of the much fainter prominences and the still fainter corona through the relatively bright, scattered light of the sky.

Disk telescope. Excellent definition in a solar image of reasonable size is usually the first re-

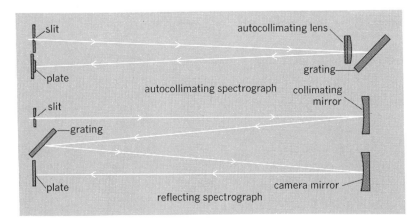

Fig. 33. Two forms of solar spectrograph. Horizontal scale greatly compressed.

slit

autocollimating lens

plate

grating

autocollimating spectrograph

collimating
mirror

slit

grating

plate

camera mirror

reflecting spectrograph

quirement in a disk telescope. Definition is limited by the quality of the optical system, by the fundamental diffraction limit of resolution inherent in a given aperture, and by the quality of the seeing. Seeing is the term used to describe the blurring effect of the changing density gradients in the terrestrial atmosphere due to thermal convection. Seeing is never perfect, and there are very few locations where excellent definition is possible during perhaps 100 hr in a year. These places are usually in lowland and maritime regions where stabilizing temperature inversions exist; unfortunately the transparency is usually poor in such locations.

The most convenient type of disk telescope is the solar tower (Fig. 30). Two flat mirrors at the top of a tower, one of which is equatorially mounted to follow the diurnal motion of the Sun, reflect sunlight into a long-focus fixed vertical telescope. The telescope may be either a refractor or a compound reflector of 30–50 cm aperture. It produces a large image of the Sun near ground level, where the light can be conveniently reflected into any one of a number of large fixed accessory instruments.

A less expensive variant of the tower telescope is the horizontal fixed telescope, with the flat mirrors at ground level. This arrangement, when carefully designed, works well for small apertures, but convective disturbances in the long horizontal air path degrade the definition of large instruments.

For some purposes an equatorially mounted telescope of standard length is preferable to the fixed types. The elimination of a pair of flat mirrors improves the definition and considerably reduces scattered light. The telescope must be fitted with a secondary optical system to magnify the solar image. Some of the finest photographs of photospheric detail have been taken with such an arrangement on telescopes of 12–20 cm aperture. It is also effective for motion-picture observations of prominences and the chromosphere through a birefringent filter, because the filter requires a secondary optical system.

Coronagraph. The coronagraph, invented by B. Lyot in 1931, is designed with one overriding consideration in mind: the elimination of instrumental scattered light. Its most delicate task is the observation of the corona immediately adjacent to the disk of the Sun, which is about a million times brighter. In an ordinary telescope, a little dust on the objective, diffraction at the edge of the aperture, and otherwise insignificant defects in the glass of the objective are all sources of diffuse scattered light which is usually some hundreds of times brighter than the corona.

The coronagraph is deceptively simple (Fig. 31). The critical component is the single-lens objective; it must be made of flawless glass, which is polished and cleaned to a perfection far beyond average standards. The rest of the system is more ordinary. The disk of the Sun is eclipsed by a polished conical disk, and photospheric light diffracted by the rim of the objective is intercepted in its image formed by the field lens, on an undersized diaphragm. The final lens then images the corona on the focal plane. The spectrum of the corona can be observed by placing the slit of the spectrograph at the focal plane, or direct photographs in the green

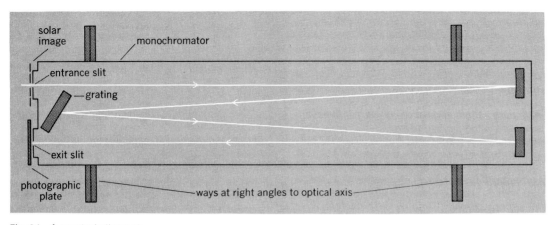

Fig. 34. A spectroheliograph.

line may be taken by inserting the appropriate birefringent filter behind the focusing lens and letting the image fall on a photographic film or plate.

The usual coronagraph has an aperture of 15 cm or less. The largest is a 40-cm instrument (Fig. 32). An internal mirror system reflects the light through the polar axis into an observing laboratory, where the image can be directed to large analyzing instruments, as in a tower telescope. The focusing lens corrects the color aberration of the objective and the telescope can be used equally well for disk and limb observations. Figures 18 and 24 were taken with it, in combination with a birefringent filter.

Spectrograph. Modern solar spectrographs for use with disk telescopes utilize the great brightness of the solar image to achieve dispersion of the order of 10 mm/A and spectroscopic resolution in excess of 500,000. These desirable characteristics require long focal lengths (10–25 m) and superlative diffraction gratings of the largest sizes that remain consistent with accuracy of ruling.

Large solar spectrographs are nearly always either of the Littrow autocollimating type or the reflecting type in which the collimator and camera element are long-focus concave mirrors (Fig. 33). They may be mounted either vertically in a well or horizontally on solid concrete piers.

The optical path inside a large spectrograph is long, particularly in the reflecting type, and internal poor seeing produced by convection in the light path is always a problem. One bold solution is complete elimination of the air by placing the spectrograph in a tube which can be evacuated. The McMath-Hulbert Observatory has been very successful with such a vacuum spectrograph.

The spectrographs used with coronagraphs for observations of solar limb phenomena are generally much smaller and simpler, with medium dispersion of 0.5 mm/A or less. They are usually simple Littrow grating spectrographs of about 200 cm focal length. The most important requirements are high light efficiency and stigmatic image. Most of them are small enough to be carried on the same mounting with the coronagraphs that feed them.

Spectroheliograph. The spectroheliograph and birefringent filter present an extended area of the solar image at one sharply defined wavelength. The wavelength chosen is usually at the center of the H-α line of hydrogen or at the H or K line of ionized calcium. The chromosphere is opaque at these particular wavelengths, and the picture obtained is, therefore, that of the chromosphere. It is evident from Figs. 8, 20, 21, and 22 that the structure is quite different from that of the photosphere, shown in Figs. 1, 5, and 15.

The spectroheliograph is a stigmatic scanning monochromator made by inserting a slit in the focal plane of a spectrograph, accurately centered on the H-α line (or any other wavelength desired), so that only the light of this line is transmitted. Variations in image intensity along the first slit are faithfully reproduced in the exit slit. Several different scanning arrangements have been successfully used; one uses fixed optics (Fig. 34). The monochromator is mounted on ways which permit smooth motion in the direction of dispersion. The solar image from a fixed telescope falls on the entrance slit of the monochromator. The light of the H-α line emerging from the second slit falls on a stationary photographic plate. As the monochromator moves along its ways, the first slit scans across the solar image, and the exit slit correspondingly scans the photographic plate. Thus an image of the Sun in the light of the H-α line is built up on the photographic plate continuously. Varying the second slit changes the output wavelength.

Birefringent filter. The birefringent filter consists of a multiple sandwich of alternate layers of polarizing films and plates cut from a birefringent crystal (usually quartz or calcite). The assembly transmits the light in a series of sharp, widely spaced wavelength bands. One or another of the polarizers absorbs the light of all intervening wavelengths. A glass filter is usually sufficient to isolate the desired band and exclude the others. Filters made for observations on the disk of the Sun generally have transmission bands 0.25–0.75 A wide, centered on the H-α line. For observations at the limb, bandwidths up to 10 A are used for prominences in H-α, and bandwidths of about 2 A for the green and red coronal lines. The birefringent filter is compact enough to be used with a conventional small telescope but is far less flexible than the bulkier spectroheliograph in choice of bandwidth and wavelength. In recent years Fabry-Perot interferometers have also been utilized.

[HAROLD ZIRIN]

Bibliography: G. Abetti, *The Sun*, rev. ed., 1957;

D. E. Billings, *A Guide to the Solar Corona*, 1966; J. C. Brandt, *The Physics and Astronomy of the Sun and Stars*, 1966; M. A. Ellison, *The Sun and Its Influence*, 1956; G. P. Kuiper (ed.), *Solar System*, vol. 1, 1953; D. H. Menzel, *Our Sun*, rev. ed., 1959; E. Tandberg-Hanssen, *Solar Activity*, 1967; H. Zirin, *The Solar Atmosphere*, 1966.

Sundog

One of two bright spots (or parhelia) which are seen on both sides of the Sun or Moon, usually with red coloring on the part closer to the Sun. They are produced by prismatic refraction on the alternate sides of the hexagonal ice crystals floating in the air with vertical axes of symmetry. Their angular distances from the Sun increase from 22° when the Sun is on the horizon to 45° when the Sun is at an elevation of 60°. For the Sun at the horizon they are situated in the inner halo; with higher Sun they approach the outer halo. The parhelia are the second most frequently observed halo phenomena. *See* HALO.   [ZDENEK SEKERA]

Sunshine-duration transmitter

A device for transmitting recordable information concerning the number of hours the Sun is visible each day compared with the number of hours of possible sunshine. These instruments must be sensitive to direct sunshine, but not to sky brightness, even though a cloudy sky near noon might be brighter than direct sunlight near sunrise or sunset. The Campbell-Stokes instrument uses a clear glass sphere to focus an image of the Sun on a paper card held in a bowl concentric to the glass sphere. The Sun's image burns a trace in the paper card to provide a record. Near sunrise and sunset the image may not be bright enough to burn the paper. The Maring-Marvin recorder is a clear-bulb and black-bulb differential gas thermometer encased in a clear glass vacuum jacket to isolate it from ambient air (see illustration). In direct sunlight the warmer black bulb moves a mercury col-

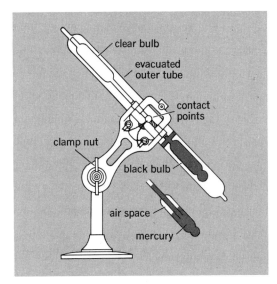

A sunshine-duration transmitter of the clear-bulb and black-bulb differential gas thermometer type. (*From F. A. Berry, Jr., E. Bollay, and N. R. Beers, Handbook of Meteorology, McGraw-Hill, 1945*)

umn in a straight connecting tube between the bulbs toward contacts embedded in the tube. The contact closure is used to record the presence of sunshine. Very bright sky near noon on an overcast day may cause this device to indicate a visible Sun in error. The above-mentioned defects are overcome in the photoelectric sunshine recorder adopted by the U.S. Weather Bureau. In this, two self-generating photoelectric cells are connected in opposition to control a relay. One of the cells is protected from direct sunlight by a shade ring; the other is exposed to both sunlight and skylight. When direct sunlight is present, the output of the shaded cell is much less and the imbalance closes a relay to actuate the recorder. Sunshine recorders usually have a provision to adjust for changes in Sun angle as the year progresses. *See* METEOROLOGICAL INSTRUMENTATION.

[VERNER E. SUOMI]

Surface water

A term commonly used to designate the water flowing in stream channels. The term is sometimes used in a broader sense as opposed to "subsurface water." In this sense, surface water includes water in lakes, marshes, glaciers, and reservoirs as well as that flowing in streams. In the broadest sense, surface water is all the water on the surface of the Earth and thus includes the water of the oceans. Subsurface water includes water in the root zone of the soil and groundwater flowing or stored in the rock mantle of the Earth. Subsurface water differs from surface water in the mechanics of its movement as well as in its location.

Surface and subsurface water are two stages of the movement of the Earth's water through the hydrologic cycle. The world's ocean and atmospheric moisture are two other main stages of the grand water cycle of the Earth. At any time there is a certain quantity of water in the various stages of this cycle. For example, the 11.5×10^9 acre-ft of water in the atmosphere is much greater than the 0.25×10^9 acre-ft of water in the stream channels. *See* HYDROLOGY.

Considerably more may be outlined concerning water's rate of movement through the several stages of the hydrologic cycle. In some stages, glaciers for example, water is locked up for long periods of time; but water in the atmosphere or in the streams is transient. A numerical value for the time of transit of water is its detention period in years—specifically, the ratio between the bulk or volume of water in a given stage of the hydrologic cycle and its mean rate of flow through that stage. For example, the Earth's supply of water as a whole amounts to about $1,100,000 \times 10^9$ acre-ft. The mean rate of flow through the hydrologic cycle is of the order of 320×10^9 acre-ft per year. Thus the detention period is about 3000 years; that is, on the average, each particle of the Earth's supply of water partakes in movement through that cycle once in 3000 years. This is an average—some particles may move more than once, some part way, and some not at all in this period of time.

Table 1 gives estimates of the amounts of water in various parts of the hydrologic cycle and their detention periods.

It may be noted that surface water on the continents is but a small part of the world's water and that the bulk of that is in fresh-water lakes. However, the detention period is also short. This means that the surface-water part, and especially the water in the streams, is rapidly discharged and replenished. That is why surface water, as well as the shallower groundwater, is called a renewable resource. Water that has a detention period of more than a generation is not renewed within sufficient time to be so considered. *See* GROUNDWATER; RIVER; TERRESTRIAL FROZEN WATER.

Source of water in streams. Precipitation that reaches the Earth is subdivided by processes of evaporation and infiltration into various routes of subsequent travel. Evaporation from wet land surfaces and from vegetation returns some of the water to the atmosphere immediately. Precipitation that falls at rates less than the local rate of infiltration enters the soil. Some of the infiltrated water is retained in the soil, sustaining plant life, and some reaches the groundwater.

Because of the slope of the land surface, the precipitation that exceeds the capacity of the soil to absorb water flows overland in the direction of the steepest slope and concentrates in rills and minor channels. During storms most of the water in surface streams is derived from that portion of the precipitation which fails to infiltrate the soil. In some forested areas of high relief and probably in some other areas, stormflow in stream channels is composed, in large part, of water which was infiltrated into the surface soil but which moved rapidly through the surficial mantle of litter and humus to the channels.

When streams are low, on the other hand, the bulk of the water in channels is the contribution of groundwater derived from precipitation that infiltrated during storms. The flow of surface streams during rainless periods represents the gradual draining of water stored temporarily in the ground. Dry-weather streamflow is the overflow of a groundwater reservoir.

The distinction between surface and subsurface water, though useful, should not obscure the fact that water on the surface and water underground is physically connected through pores, cracks, and joints in rock and soil material. In many areas, particularly in humid regions, surface water in stream channels is the visible part of a reservoir, which is partly underground; the water surface of a river is the visible extension of the surface of the groundwater.

Disposition of precipitation. Streamflow represents only a small percentage, on the average, of the water that falls as precipitation. The flow in streams under natural conditions is called runoff. The ratio of average annual runoff to average annual precipitation in the United States ranges from 20 to 40% in humid parts of the United States and from 2 to 4% in semiarid areas. On the average, the annual budget of water over the United States is roughly as follows:

Average precipitation	30 in.
Runoff by rivers to sea	9 in.
Evapotranspiration from plants and soil	21 in.
Transport of atmospheric moisture from oceans to continental area	9 in.

The 30 in. of water contributed by precipitation must be balanced by a return of water to the atmosphere. There are 21 in. returned to the atmosphere by evapotranspiration from the continental area and 9 in. flow to the oceans. Thus, to balance the atmospheric budget, the 9 in. of water that is transported as vapor from the oceans to the continents must be included.

Average runoff. There are great geographic variations within this average balance as can be visualized by a map showing annual runoff in the United States. The total runoff is greatest in areas of highest precipitation and lowest losses. In the mountains of the Northwest, where precipitation is as much as 150 in. annually, the runoff in surface streams is more than 50 in. annually over a considerable part of the high mountain country.

Much of the semiarid parts of New Mexico, Arizona, and parts of California and Nevada yield an annual runoff of only a few tenths of an inch from a precipitation of 4–8 in. In the mountainous parts of the same areas, precipitation reaches 25–30 in. and, locally, the runoff may average as much as 10 in. annually from small areas. Much of the United States west of the 100th meridian has an average annual runoff of less than 3 in.

Discharge in nearly all streams has a marked annual cycle. Spring and early summer are usually periods of high flow resulting from snowmelt and rain during a period of relatively low water loss by transpiration. Late summer is often the period of lowest flow owing to the infrequency of precipitation and to the maximum use of water by leafy vegetation. The distribution of runoff throughout the year is not the same for the whole country, however, because it varies regionally depending on the seasonal distribution of precipitation and depending on the importance of snowmelt as a source of runoff.

Individual streams and rivers in the United States have been measured over varying periods of time at about 12,000 sites. Daily discharge at most of them is published by the U.S. Geological Survey

Table 1. Distribution of the world's supply of water

Location	Volume of water, 10^9 acre-ft	Percentage of total	Detention period, years
World's oceans	1,060,000	97.39	5,000
Surface water on the continents			
Glaciers and polar ice caps	20,000	1.83	2,000
Fresh-water lakes	100	0.0093	100
Saline lakes and inland seas	68	0.0063	50
Average in stream channels	0.25	0.00002	0.05
Total surface water	20,200		700 av
Subsurface water on the continents			
Root zone of the soil	10	0.00094	0.25
Groundwater above 2500 ft	3,700	0.339	5
Groundwater below 2500 ft	4,600	0.425	100
Total subsurface water	8,300		
Atmospheric water	115	0.0011	0.03
Total world water (rounded)	1,088,000	100	3,000

Table 2. Representative values of extreme peak flows

River	Date	Drainage area, mi²	Peak discharge cfs	cfs/mi²
Big Branch near Waynesville, N.C.	Aug. 30, 1940	0.4	4,500	11,000
Big Creek near Waynesville, N.C.	Aug. 30, 1940	1.32	12,900	9,800
Laurel Creek above White Pine, W.Va.	August, 1943	2.42	7,400	3,060
Cameron Creek near Tehachapi, Calif.	Sept. 30, 1932	3.59	13,500	3,760
Unnamed Creek near York, Nebr.	July 9, 1950	6.93	23,000	3,320
Meyers Canyon near Mitchell, Ore.	July 13, 1956	12.7	54,500	4,290
Alazan Creek, below Martinez Creek, Tex.	Sept. 9, 1921	17.1	25,900	1,510
Salem Creek below Woodstown, N.J.	Sept. 1, 1940	17.5	26,100	1,490
Morgan Creek near Chapel Hill, N.C.	Aug. 4, 1924	29.1	30,000	1,030
Pine Tree Canyon, 12 mi north of Mohave, Calif.	Aug. 12, 1931	35.0	59,500	1,700
Elkhorn Creek, Keystone, W.Va.	June, 1901	44	60,000	1,360
Little Nemaha River at Syracuse, Nebr.	May 9, 1950	218	225,000	1,030
Guadalupe River near Ingram, Tex.	July 1, 1932	336	206,000	613
W. Nueces River near Brackettville, Tex.	June 14, 1935	402	580,000	1,440
W. Nueces River near Cline, Tex.	June 14, 1935	880	536,000	609
Eel River at Scotia, Calif.	Dec. 22, 1955	3,113	541,000	174
Devils River near Del Rio, Tex.	Sept. 1, 1932	4,060	597,000	147
Neosho River near Parsons, Kans.	July 14, 1951	4,817	410,000	85.1
Little River at Cameron, Tex.	Sept. 10, 1921	7,000	647,000	92.4
Ohio River at Sewickley, Pa.	Mar. 18, 1936	19,500	574,000	29.4
Susquehanna River at Marietta, Pa.	Mar. 19, 1936	25,900	787,000	30.4
Ohio River at Evansville, Ky.	Jan. 29, 1937	107,000	1,410,000	13.2
Ohio River at Metropolis, Ill.	Feb. 1, 1937	203,000	1,850,000	9.1
Columbia River at The Dalles, Ore.	June 6, 1894	237,000	1,240,000	5.2

(USGS) in the series of Water-Supply Papers entitled *Surface Water Supply of the United States.* The data are tabulated for each measuring station and are grouped in volumes by river basins.

If a stream goes dry occasionally or does not have enough flow to satisfy a desired use, storage reservoirs can be built to conserve high flows for release during low-flow periods. The design of such reservoirs requires a knowledge of how low a flow is likely to be experienced, how long it may last, and how frequently it can be expected to recur. Streamflow records can be analyzed to answer these questions.

Extremes of runoff. The amount of streamflow available during periods of extremely low flow can be shown by flow-duration curves and other types of low-flow frequency analysis. Flow-duration curves, which express the per cent of time the flow of a particular stream has been equal to or greater than any given quantity, have long been used in water-power studies, and curves showing the frequency and severity of annual lows are now being used in water-supply studies and stream-sanitation studies. The preparation of such curves is facilitated by the use of electronic computers. Many such computations have been made and published, but there is no single source of systematic publication of such material. Summaries of published data on low flows for some rivers in the United States may be obtained from the USGS.

Much more information is organized and published on floods than on low flows. Representative values of peak discharges showing the magnitude of extreme flows experienced in drainage basins of various sizes are given in Table 2.

Data similar to those included in Table 2 are published in a systematic manner under the headings of "extremes" in the tabulated flow data for individual gaging stations in the Water-Supply Papers of the USGS. Flood expectancy, even for ungaged places, may be estimated from curves the USGS is publishing in a series of papers dealing with the frequency and magnitude of floods by individual states or areas.

Relation of runoff to drainage area. Average water yield or annual runoff in a physically homogeneous area increases in direct proportion to the size of the drainage basin, but this is not true of flood potentiality. Small drainage basins produce larger peak flows per unit of drainage area than do large basins. The relation between magnitude of flood peak and the contributing drainage area may be expressed by Eq. (1), where Q is discharge

$$Q = aA^c \tag{1}$$

in cubic feet per second (cfs), A is drainage area in square miles (mi²), a and c are coefficients. If Q represents the average annual water yield, then the value of c is approximately unity for most basins in humid areas. If Q represents peak flood discharge of a given frequency or recurrence interval, such as a 10- or 50-year average recurrence interval, then past observations indicate that the value of c is between 0.7 and 0.8 for a large variety of drainage basins.

Some reasons for these differences in exponents are as follows. Each square mile of a homogeneous drainage basin contributes an equal quantity of water over the course of many years. Therefore, average annual runoff from a drainage basin is the summation of the contribution of each unit area.

Peak rates of runoff during floods, on the other hand, may be viewed roughly as the contributions

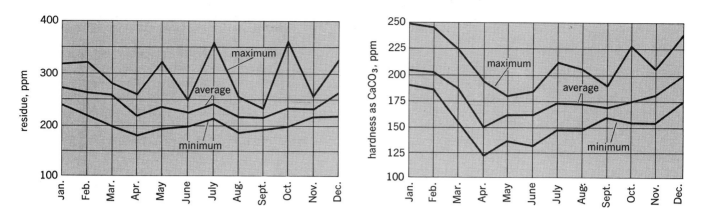

Mississippi River at Minneapolis, Minn.

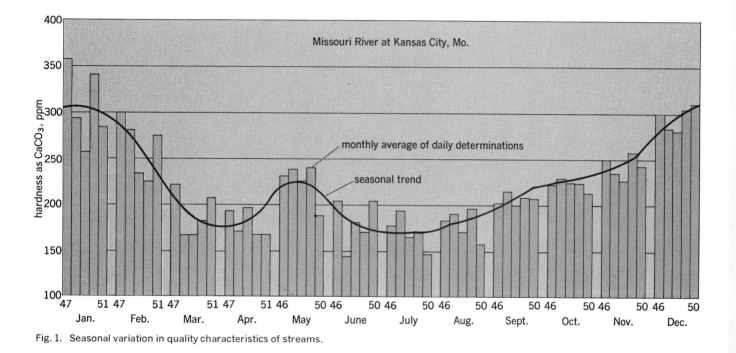

Fig. 1. Seasonal variation in quality characteristics of streams.

from the parts of the drainage basin at varying distances away. The different distances mean that the peak flood contributions reach the several points downstream at different times, and the farther downstream one goes, the greater the floodwave is spread out. The flattening action is increased by the temporary storage of water in the stream channels and in the bordering floodplains. As it moves downstream, the flood crest is increased by the contributions of tributaries, but because these contributions are not synchronized, the peak contributions are not simply additive. Owing to the delays due to channel storage, flood-peak discharges increase with drainage area, but to a power less than unity; peak flows per unit of drainage area decrease with drainage area.

Quality of water. The usefulness of available water is often limited by its quality. Good quality often is considerably more important than unlimited quantity, particularly in industry.

It is characteristic of river waters to vary in chemical and physical quality almost continuously. The chemical quality of the water in lakes, particularly large ones, remains relatively constant throughout the year. Differences between streams are caused by several factors. These include (1) the nature of soils and rocks over and through which the water flows, (2) the length of time the water is in contact with various rock types, (3) the water quality of tributary streams, (4) proportion of flow due to groundwater discharge, (5) flood and drought conditions, and, of course, (6) man-made pollution. Figure 1 illustrates seasonal variations in amounts of dissolved substance in two large streams.

Figure 2 illustrates variation in dissolved solids with streamflow. In general, streams flowing into the Atlantic Ocean, eastern Gulf of Mexico, and the northern Pacific Ocean are of good to, excellent quality for general purposes. Dissolved matter and hardness are usually below 100 parts per million (ppm) (milligrams of dissolved substance in a liter

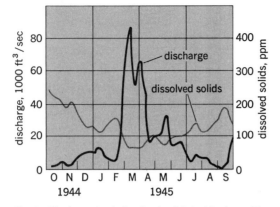

Fig. 2. Discharge and dissolved solids in Allegheny River at Kitanning, Pa.

of water) and often below 50 ppm. This applies to streams in their natural state and does not take into account the effects of pollution.

Midcontinent and southwestern streams generally have high concentrations of dissolved matter, some excessively so, and must be treated extensively for a large variety of general uses. For example, the water of the Missouri River at Kansas City has about 220–500 ppm of dissolved matter and hardness of about 190–300 ppm during a typical year. A generalized picture of the variations in surface-water quality throughout the United States is given in Fig. 3.

The temperature of streams varies continuously throughout the year, ranging from a minimum of freezing (about 32°F) in winter in northern lati-

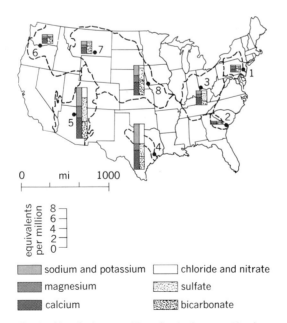

Fig. 3. Chemical composition of water in several basins. Data are for (1) Delaware River at Trenton, N.J., 1951; (2) Savannah River near Clyo, Ga., 1940; (3) Ohio River at Cincinnati, Ohio, 1951; (4) Brazos River at Richmond, Tex., 1951; (5) Colorado River near Grand Canyon, Ariz., 1951; (6) Columbia River near Rufus, Ore., 1951; (7) Yellowstone River at Billings, Mont., 1951; and (8) Missouri River at Nebraska City, Nebr., 1951.

tudes to a maximum of about 90°F in summer in southern latitudes. The monthly average water temperature generally follows rather closely the monthly average air temperatures, except in areas where the flow is made up largely of melting snow or ice, or of groundwater. The water temperature of streams fed by snowmelt is lower than the air temperature for some distances downstream from the snow fields. The water temperature of streams whose flow comes from groundwater tends to be more uniform and in summer months is usually colder than the average air temperature.

Sediment also affects surface-water quality. Nearly all streams are turbid during flood periods, some carrying tremendous quantities of sediment which must be removed before the water is suitable for industrial and most other uses. In eastern United States the amount of suspended matter in typical streams seldom exceeds 0.3% (3000 ppm) and generally averages a few hundred parts per million. In many midcontinent and western streams sediment concentrations are much greater, a maximum of 10% not being uncommon in some streams; frequently the maximum is considerably higher.

The sediment-carrying characteristics of western streams range from relatively clear-flowing mountain streams to near mud flows in the intermittent streams of arid regions. The sediment concentration of the Colorado River at Grand Canyon, Ariz., averaged about 0.6% (6000 ppm) for a period of nearly 30 years.

Data on the chemical and sediment loads of streams in the United States are published systematically in the Water-Supply Papers of the USGS.

Characteristics of river channels. Rivers and streams form channels which are the routes of transport for water and debris-load delivered to them by the basin during the slow process of landscape degradation. Channels have certain characteristics that are amazingly universal regardless of the location of the river basin. The basic mechanics which lead to these common characteristics are only imperfectly understood.

Water only partly fills the channel during periods of low flow. Generally, on more than half the days in a year, only the lowest one-fifth of the channel depth is filled with water. The channel flows bankful about once a year on the average. Though this varies somewhat from reach to reach and from one river basin to another, the generalization that the channel is constructed by the river so that it overflows once every year or every 2 years is one of the most interesting and potentially useful items of information about natural channels. Because a flood is, by definition, a flow which exceeds channel capacity, the above generalization emphasizes the fact that flooding is a natural characteristic of rivers.

Natural channels tend to be roughly trapezoidal rather than elliptical or semicircular. The width of the water surface along the channel at any given frequency of flow (high flow or low flow) generally increases as the square root of the discharge, as discharge increases downstream with the addition of tributaries. The shape of the channel cross section is asymmetric at bends, the deepest part being near the concave bank.

The depth increases downstream but not as rapidly as the width. The width-depth ratio increases downstream as about the 0.1 power of the discharge as given by expression (2), where w and d

$$\frac{w}{d} \propto Q^{0.1} \qquad (2)$$

are respectively mean width and mean depth, and Q is discharge of a given frequency.

Channel slope or gradient decreases downstream generally following an exponential or logarithmic law. Also, size of debris composing the bed tends to diminish downstream. Despite the decreasing slope in the downstream direction, mean water velocity increases slightly along the length of a river or maintains a roughly constant value.

[LUNA B. LEOPOLD]

Bibliography: V. T. Chow (ed.), *Advances in Hydroscience*, vol. 1, 1964, vol. 2, 1965, vol. 11, 1977; V. T. Chow (ed.), *Handbook of Applied Hydrology*, 1964; L. B. Leopold, *Water: A Primer*, 1974; R. L. Nace, *Water Management, Agriculture, and Ground-Water Supplies*, Annu. Meet. AAAS, 1958; A. A. Sokolow et al., *Floodflow Computation*, UNESCO, 1977; UNESCO, *World Catalogue of Very Large Floods*, 1977.

Temperature-humidity index

An index developed by the U.S. Weather Bureau to give a single numerical value, in the general range of 70–80, which would reflect the outdoor atmospheric conditions of temperature and humidity as a measure of comfort (or discomfort) during the warm season of the year. The temperature-humidity index was formerly called the discomfort index. Temperature-humidity index I_{TH} is defined by the equation below.

$$I_{TH} = 0.4 \text{ (dry-bulb temp °F} + \text{wet-bulb temp °F)} + 15$$

When the index is 70 practically all people feel comfortable; when it is 80 no one is comfortable; and when it is 75 about half the population is satisfied. The index is useful in the prediction and allocation of power-system loads resulting from the impact of air-conditioning equipment operation. *See* DEGREE-DAY; PSYCHROMETRICS.

[THEODORE BAUMEISTER]

Bibliography: American Society of Heating, Refrigeration, and Air-Conditioning Engineers, *Guide and Data Book*, 1964, 1965.

Temperature inversion

The increase of air temperature with height; an atmospheric layer in which the upper portion is warmer than the lower. Such an increase is opposite, or inverse, to the usual decrease of temperature with height, or lapse rate, in the troposphere of about 6.5°C/km or 3.3°F/1000 ft, and somewhat less on mountain slopes. However, above the tropopause, temperature increases with height throughout the stratosphere, decreases in the mesosphere, and increases again in the thermosphere. Thus inversion conditions prevail throughout much of the atmosphere much or all of the time, and are not unusual or abnormal. *See* AIR TEMPERATURE; ATMOSPHERE.

Inversions are created by radiative cooling of a lower layer, by subsidence heating of an upper layer, or by advection of warm air over cooler air or of cool air under warmer air. Outgoing radiation, especially at night, cools the Earth's surface, which in turn cools the lowermost air layers, creating a nocturnal surface inversion a few centimeters to several hundred meters thick. Over polar snowfields, inversions may be a kilometer or more thick, with differences of 30°C or more. Solar warming of a dust layer can create an inversion below it, and radiative cooling of a dust layer or cloud top can create an inversion above it. Sinking air warms at the dry adiabatic lapse of 10°C/km, and can create a layer warmer than that below the subsiding air. Air blown from over cool water onto warmer land or from snow-covered land onto warmer water can cause a pronounced inversion that persists as long as the flow continues. Warm air advected above a colder layer, especially one trapped in a valley, may create an intense and persistent inversion.

Inversions effectively suppress vertical air movement, so that smokes and other atmospheric contaminants connot rise out of the lower layer of air. California smog is trapped under an extensive subsidence inversion; surface radiation inversions, intensified by warm air advection aloft, can create serious pollution problems in valleys throughout the world; radiation and subsidence inversions, when horizontal air motion is sluggish, create widespread pollution potential, especially in autumn over North America and Europe. *See* ATMOSPHERIC POLLUTION; SMOG.

[ARNOLD COURT]

Temperature measurement

Scientifically, the measurement of thermal potential; in nontechnical terms, the measurement of that property of a body that determines the sensation of warmth or coldness received from it. If heat flows from one body to another, the first is at a higher temperature than the second.

The basic scale used for temperature measurements is the International Thermodynamic, or Kelvin, scale. On this scale the triple point of water (comprising ice, liquid, and vapor) is defined as 273.16°K and the ice point as 273.15°K. On the Celsius scale (formerly known as Centigrade), at a pressure of 1 standard atmosphere, the ice point is 0°C and the boiling point of water is 100°C. On the commonly used Fahrenheit scale the ice point is at 32°F and the boiling point of water is at 212°F.

Temperature-measuring instruments. These may be grouped under the general headings of thermometers, resistance thermometers, thermocouples, and pyrometers. Here the term thermometer includes those instruments that give a direct indication, or reading, of temperature due to expansion and contraction with heat of a substance, for example, a liquid. Resistance thermometers measure the change in the electrical resistance of a wire with heating, and this change of resistance is converted into temperature units. An elementary form of thermocouple consists of two wires of dissimilar metals, joined at each end. If the two junctions are at different temperatures, a voltage difference is set up; if a suitable potentiometer is connected into the circuit, the voltage difference can be measured and converted into temperature units. Pyrometers are of two types, optical and radiation. With optical pyrometers the brightness of an object is visually compared with a standard of brightness over a narrow wavelength interval. Radiation pyrometers measure the rate of energy emission per unit area over a broad range of wavelengths.

Temperature indicators. These indicate that a particular temperature has been reached or exceeded but do not give exact temperature values. Pyrometric cones soften and bend under the effect of increasing temperature, the effect being a combination of temperature and time. The amount of bending is a rough indication of the temperature reached.

Temperature-sensitive crayons, paints, and pellets may serve as temperature indicators by a change of color or by melting. Crayons or paints are applied to the object before being heated; the change of color with heat may be reversible or irreversible. If the change is reversible, the object must be observed while undergoing the temperature cycle; if the change is irreversible, it only indicates that the desired temperature has been reached or exceeded. Pellets are wrapped with, or inserted into, the article at the point where the temperature information is required. Their melting indicates that the desired temperature is reached.

The useful range of indicators is from about 100 to 2000°F.

Thermography. This is a method of measuring surface temperature using luminescent materials. In contact thermography a thin layer of luminescent material is spread on the surface of an object and is excited by ultraviolet radiation in a darkened room. The brightness of the coating indicates the surface temperature. In projection thermography the thermal radiation from a surface is imaged by an optical system on a thin screen of luminescent material. The pattern formed corresponds to the heat radiation of the surface.

Calibration and application. Temperature-measuring instruments are calibrated by comparing their readings under carefully controlled conditions with standard instruments, which in turn have been calibrated directly or indirectly against primary standards at the U.S. National Bureau of Standards or a similar organization. Temperature instruments are accurate only when they are used under calibration conditions. Since this is rarely possible when using temperature-measuring instruments, some or all of the following points should be considered.

1. The sensitive element or bulb and stem should be immersed to the same extent as used in the calibration or as may be indicated on the instrument.

2. The thermometer should be held or supported in a manner that will minimize or prevent heat conduction.

3. Because of thermal inertia, the thermal element may not respond to actual temperature changes quickly, the time lag being influenced by the mass and type of the element and method of support.

4. The thermal element should be shielded from radiation effects, especially in gas temperature measurements where radiation from or to the surrounding walls may alter the indication.

5. In moving streams the conversion of kinetic energy to potential energy of the stream filaments stopped by the thermal element will cause the reading to be higher than if the element were moving with the stream.

6. The possibility of temperature gradients in a body should be considered in selecting the location for a thermometer, especially in a body of fluid.

[HOWARD S. BEAN]

Bibliography: D. M. Considine (ed.), *Process Instruments and Controls Handbook*, 1974; E. F. Fiock and J. W. Murdock, Measurement of temperature in high velocity steam, *Trans. Amer. Soc. Mech. Eng.*, vol. 72, 1950; U.S. Government Printing Office, *U.S. Standard Atmosphere*, 1962.

Terrestrial atmospheric heat balance

The (heat) energy exchange between the planetary Earth-plus-atmosphere system and space, consisting of the reception, distribution, and transformation of heat received from the Sun as well as the emission, distribution, and transformation of thermal (infrared) radiation arising from the Earth's surface and atmosphere. An area of 1 m² at the Earth's mean distance from the Sun and perpendicular to the Sun's rays would receive about 1390 watts (or joules per second) if there were no intervening atmosphere. This quantity is the solar "constant" that depends on the energy output of the Sun and the Earth's distance from it. Scientists are planning to measure this quantity from satellites and space platforms to see if it is indeed as constant as is assumed.

Since the area of a sphere is four times that of the circle it presents to parallel radiation, the average meter-squared area outside the Earth's atmosphere receives about 350 W of solar energy. Thirty percent of this incident solar energy is reflected and scattered back to space, mostly by clouds in the atmosphere; it is not available for use on Earth. This quantity, the percent reflected and scattered, is called the planetary albedo. Mars and the Moon have lower albedos than Earth, and Venus has a

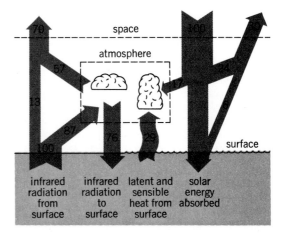

Fig. 1. Average global distribution of incident solar energy and Earth infrared energy throughout the heat balance.

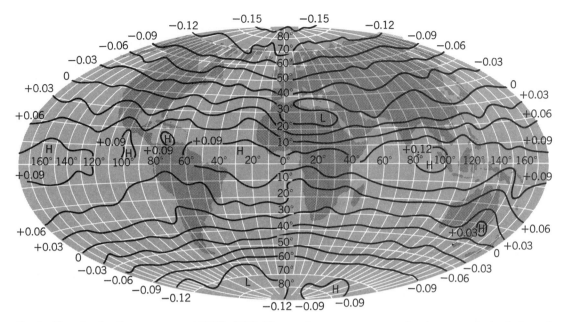

Fig. 2. Map showing the mean annual (1962–1970, 17 seasons) net radiation on a global scale; units are cal- ories per square centimeter per minute (cal · cm⁻² · min⁻¹). Highs (H) and lows (L) are indicated.

higher albedo. The 70% of the solar energy that enters the planetary system is absorbed, mostly at the Earth's surface. The oceans, which are dark, absorb solar energy very well, and cover 70% of the planet's surface. This energy is partly transformed by plants, used to evaporate water, converted to thermal (infrared) radiation from the surface, and moved from the surface into the atmosphere by conduction and convection. The clouds and gases in the atmosphere also absorb some of the solar energy, thus evaporating clouds and heating the air. *See* METEOROLOGICAL OPTICS; SOLAR ENERGY.

In order to balance this energy budget, the energy received from the Sun (from space) is compensated by infrared radiation emitted by the Earth and atmosphere and escaping to space. Since the Earth is not observed to be heating or cooling at a rapid rate, it is assumed that each average meter-squared of the Earth-atmosphere system must lose energy at the rate of $0.70 \times 350 = 245$ W. Both the albedo and the infrared radiation to space have been measured from meteorological satellites using special radiometers. These measurements continue to monitor if the heat budget is in balance or is gaining or losing energy over the entire planet or over any specific region. If the entire planet gains or loses energy, the climate would be expected to change. If certain areas have changes in heat balance, the energy-transport action of the winds and ocean currents in distributing the energy might cause regional climates to change. *See* OCEAN-ATMOSPHERE RELATIONS; TERRESTRIAL RADIATION.

The best present-day estimate of the average global disposition of the incident solar energy (350 W/m² taken as 100 units) throughout the heat balance is shown in Fig. 1. Of the 100 units, about 24 are reflected and scattered back to space by clouds and gases in the atmosphere and about 6 by the Earth's surface (land and ocean). About 17 units are absorbed by clouds and the atmosphere, and the remaining 53 units are absorbed at the surface.

Of 100 units of infrared radiation emitted from the surface of the Earth, a major portion, 87 units, is absorbed by clouds and gases in the atmosphere, while 13 units pass directly to space. The atmosphere then reradiates energy both to space (57 units) and back to the surface (76 units). Overall the atmosphere loses energy by radiation processes, the surface gains energy, and the net exchange with space is zero. (This may be verified by summing the gains and losses shown for each region.) In order to prevent excessive cooling by the atmosphere or warming by the surface, 29 units of energy are transferred from the surface to the atmosphere by sensible and latent heat processes (conduction, convection, and evaporation and condensation).

Because the Earth-atmosphere system varies from place to place, from the Poles to the Equator and from the deserts to the oceans, the average radiation balance discussed above also varies. Figure 2 demonstrates this by showing a map of only the space portion of the balance as measured by experiments on unmanned meteorological satellites. This map of "net" radiation in calories per square centimeter per minute (1 cal · cm⁻¹ · min⁻¹ equals about 698 W/m⁻²) shows that the polar regions lose more energy to space than they gain each year (negative values of the net radiation). The tropical regions gain energy, and when the atmosphere (winds) and the ocean (currents) act to move heat energy poleward toward a balanced condition, there results a major cause of the weather experienced on Earth. *See* GREENHOUSE EFFECT.

[THOMAS H. VONDER HAAR]

Bibliography: H. G. Houghton, On the annual heat balance of the Northern Hemisphere, *J. Meteorol.*, 11(1):1–9, 1954; T. H. Vonder Haar, Natural variation of the radiation budget of the earth-atmosphere system as measured from satellites, *Conference on Atmospheric Radiation of the American Meteorological Society, Boston*, pp. 211–221, 1972; T. H. Vonder Haar and V. E. Suomi, Satellite observations of the Earth's radiation budget, *Science*, 163; 667–669, 1969.

Terrestrial frozen water

Seasonally or perennially frozen waters of the Earth, exclusive of the atmosphere. Water in the frozen or solid state is the hexagonally crystallized, birefringent mineral known as ice. Terrestrial ice occurs in the form of temporary seasonal accretions during the cold months and in the perennial ice cover represented by glaciers, land-fast ice (ice shelves), and subsurface ground ice in permafrost regions.

Glaciers, past and present. Under present climatic conditions the semipermanent terrestrial ice cover is essentially glacial. Glaciers cover approximately 10% of the world's land area (14,972,138 km², estimated by R. F. Flint, 1957). Of this ice, 96% lies in Greenland (1,726,400 km²) and Antarctica (12,650,000 km²), leaving only 4% of the world's glaciers in mountains and subpolar regions. This is to be compared with approximately 32% of the world's land covered by glaciers during their maximum extension in the Pleistocene Ice Age. A close estimate of the total volume of terrestrial ice is not possible because the thickness of the Antarctic ice sheet is not yet adequately known. Flint, in 1957, tentatively estimated the volume of glacier ice existing today as about 24,-000,000 km³, "equivalent to a layer of water having the area of the present oceans and approximately 59 m thick." No estimate has been attempted of the volume of subsurface terrestrial frozen water existing in permafrost regions.

Properties of terrestrial ice. Ice is one of the more abundant minerals on the surface of the Earth. It is usually observed as colorless and transparent, but in large, dense masses it shows a vivid light blue color. The other physical properties of terrestrial frozen water vary considerably under changed conditions of internal temperature, load, crystalline orientation, and mass density. For

Table 1. Temperature and pressure relationships for forms of ice and related liquid water*

Forms	Temperature, °C	Pressure, atm
Water; ice I, III	−22.0	2,047
Ice I, II, III	−34.7	2,100
Water; ice III, V	−17.0	3,417
Ice II, III, V	−24.3	3,397
Water; ice V, VI	+ 0.16	6,175
Water; ice VI, VII	+81.6	21,700

*Based on data reported in 1940 by N. E. Dorsey.

example, ice has an indentation hardness which varies with the mass temperature and a scratch microhardness which differs with the orientation of the crystal plane. On the microhardness or Mohs scale at 0°C it has a hardness of about 2 and at −44°C a hardness of about 4. Coincident with this property is its variable plasticity or ability to deform under stress. The deformation rate is considerably reduced under colder temperatures. At a temperature of −1°C, the deformation (flow) rate of polycrystalline ice may be expected to be slowed to approximately one-fifth of that at 0°C. For example, ice chilled to −10°C has been observed by J. W. Glen to have 25 times as much ability to resist deformation as ice at 0°C.

Special hydrothermal relationships. No significant influence on the deformation rate appears in ice through changes in hydrostatic pressure, the relationship being as negligible as in liquids; but the melting point decreases as the pressure increases. This amounts to a reduction in temperature of 0.0075°C for each atmosphere of increase in pressure. As a result of this property, the melting temperature of an ice mass at any depth is fundamentally conditioned by the weight of the overlying mass. The term used to refer to this is pressure melting temperature. It is because of this characteristic that ice skating is possible. At the edge of a skate blade sufficient pressure is exerted to cause formation of a thin film of water, which serves as a lubricant. Similarly, when a wire is pressed into a block of ice, or even when pieces

Table 2. Thermal constants of various forms of frozen water compared with liquid water and other substances*

Water forms and other substances	Conductivity, cal/(°C)(cm)(sec)	Specific heat, cal/(°C)(g)	Density, g/cm³	Thermal diffusivity, cm²/sec	Relative diffusivity to ice (approx. ratio)
Frozen water					
New snow	0.0003	0.5	0.20	0.0030	0.27
Old snow	0.0006	0.5	0.30	0.0040	0.36
Average firn	0.0019	0.5	0.55	0.0070	0.64
Firn ice	0.0038	0.5	0.75	0.0100	0.91
Ice	0.0050	0.5	0.92	0.0110	1
Comparative materials					
Water (0°C)	0.0014	1.0	1.00	0.0014	0.13
Rubber	0.0005	0.40	0.92	0.0014	0.13
Steel (mild)	0.1100	0.12	7.85	0.12	11
Aluminum	0.4800	0.21	2.70	0.86	78
Copper	0.9300	0.09	8.94	1.14	104

*Based on data from U.S. Army Corps of Engineers, *Review of the Properties of Snow and Ice*, 1951, and other sources as reported in M. M. Miller, *Glaciothermal Studies on the Taku Glacier, Alaska*, 1954.

of ice are pressed together, melting occurs at the contact surface. When the pressure is released, the water refreezes, uniting the ice into a continuous mass. This is the process of regelation.

Classification of ice forms. Although there is only one ordinary form or phase of terrestrial frozen water, ice I, five other stable phases of ice exist in addition to one unstable phase. These, however, are only the product of great pressure and are not found in normal conditions outside of the experimental laboratory. The stable, extraordinary forms are known as ice II, III, V, VI, and VII; IV is the unstable category. Each of these reverts to normal terrestrial ice, water, or both when the pressure is released. Table 1 shows pressure and temperature parameters (triple points) under which these forms of ice and related liquid water exist.

Massive ice constants. Because massive ice is an aggregate of individual ice crystals, it may also be considered a monocrystalline rock. In the form of snow or firn (granular summer equivalent), it has been genetically likened to a sedimentary rock. (Similarly, pond, river, sea, or refrigerator ice may be likened to an igneous rock, and glacier ice to a metamorphic rock.) Any of these forms when dry are poor conductors of electricity. Cold dry ice, for example, has a resistivity of 10^9 ohm-cm for direct current and low-frequency alternating current and 3×10^6 ohm-cm at 60 kHz. The dielectric constant for low frequencies at 0°C is 74.6, and for high frequencies and very low temperatures it drops to 3.0. Ice is a poor conductor of heat.

A list of the thermal constants of various forms of frozen terrestrial water under normal conditions is in Table 2 (as compiled from various sources for the Juneau Icefield Research Program by M. M. Miller, 1954). The general conductivities for snow and average firn noted in this table are based on data from the U.S. Army Corps of Engineers (*Snow, Ice and Permafrost Research Establishment*, 1951). The density of firn ice is chosen arbitrarily between given values for average firn and ice. For comparison, the thermal properties listed in the International Critical Tables for rubber, steel, aluminum, and copper are also noted. The diffusivity figures in Table 2 are rounded off for convenient reference. *See* GLACIOLOGY; GROUNDWATER; HYDROLOGY; SURFACE WATER.

[MAYNARD M. MILLER]

Bibliography: V. V. Bogorodskii, *Physics of Ice*, 1971; N. E. Dorsey, *Properties of Ordinary Water-Substance*, ACS Monogr. no. 81, reprint, 1954; R. F. Flint, *Glacial and Quaternary Geology*, 1971; J. W. Glen, The creep of polycrystalline ice, *Proc. Roy. Soc. London Ser. A*, 228:528–529, 1955; M. M. Miller, *Glaciothermal Studies on the Taku Glacier, Alaska*, Ass. Int. Hydrol. Publ. no. 39, 1954; A. Post and E. R. LaChapelle, *Glacier Ice*, 1971; *Review of the Properties of Snow and Ice*, Snow, Ice, and Permafrost Research Establishment, U.S. Army Corps of Engineers Rep. no. 4, 1951.

Terrestrial radiation

Electromagnetic radiation originating from the Earth and its atmosphere at wavelengths determined by their temperature. It is sometimes called thermal radiation. The units are energy per unit area per unit time, such as calories per square centimeter per second, $cal/(cm^2)(sec)$. The atmosphere emits, absorbs, and transmits radiation, and the net flux of radiation at any point depends upon the distribution with height of temperature and water vapor. Heating and cooling due to the vertical divergence of terrestrial radiation provide a major part of the potential energy changes necessary to drive the atmospheric wind system. Terrestrial radiation is also responsible for maintaining the air temperature near the ground within limits necessary for comfortable living. *See* ATMOSPHERIC GENERAL CIRCULATION; GREENHOUSE EFFECT; TERRESTRIAL ATMOSPHERIC HEAT BALANCE.

Spectrum. At terrestrial temperature practically all emission of radiation is at wavelengths greater than 4μ $(1 \mu = 10^{-4} cm)$, that is, within the infrared part of the spectrum. The Earth's surface and all but the thinnest clouds emit, at all wavelengths, radiation of intensity only slightly less than that of blackbody radiation corresponding to their temperatures and absorb almost all the infrared radiation that reaches them. The absorption and emission by the atmosphere apart from clouds is, on the other hand, selective and depends on the spectral position and intensity of the rotation-vibration and pure rotation bands of its polyatomic molecules. The minor gases of the atmosphere have numerous absorption bands in the infrared—water vapor at 5–8 and beyond 15 μ, carbon dioxide at 13–17 μ, and ozone at 9–10 and 14 μ. Within these bands the absorption coefficient, or absorption per unit mass of absorber, has many maxima, each maximum marking the position of a spectral line. These lines have finite widths and overlap one another. The line character of atmospheric bands is of considerable importance for radiative heat transfer. Each line is the result of the transition of the molecule from one quantum state to another by absorption or emission of a photon of frequency corresponding to the line frequency. The line shape, or variation of absorption coefficient with frequency, is determined by molecular collisions in the lower atmosphere and by random thermal motions in the upper atmosphere. The absorption coefficient is also a function of temperature and pressure, and because of the marked spatial variation of these parameters it has a considerable variation, especially in the vertical.

[LEWIS D. KAPLAN]

Radiation-measuring devices. These instruments are used to obtain the radiant energy exchange, solar and terrestrial, between the Sun, space, and the Earth. The energy covers a broad range of wavelengths from about 0.17 to 100 μ. Solar radiation ranges from 0.17 to 4 μ with the maximum at 0.49 μ, whereas terrestrial (Earth) radiation ranges from about 3 to 100 μ with the maximum typically at 10 μ. It is important that the sensitivities of these instruments be independent of the wavelength of the incident radiation. Because of this requirement, practically all radiation-measuring instruments in meteorological use first convert the radiant energy into heat by absorption on a blackened target. The actual measurement is made on the resulting heat flow.

Energy in the solar beam. This is measured with a normal-incidence pyrheliometer. A number of instruments of this type are in use; however, two are considered secondary standards. The Abbot silver-disk pyrheliometer consists of a thermally insulated blackened silver disk mounted at the end of an open-end tube. The tube has a shutter and baffles to collimate the beam. During a measurement the disk is alternately exposed to and shaded from the Sun's radiation for equal periods of 2 min. A mercury thermometer embedded in the silver disk indicates the temperature change during the exposed and shaded periods. This, together with the heat capacity of the disk and other constants, is used to calculate the radiation. The Angstrom compensation pyrheliometer has two nearly identical targets equipped with electric heaters at the bottom of a similar tube. A sunshade is arranged so one target is exposed to the Sun while the other is shaded. Electric current is supplied to the shaded target until its temperature, determined by thermocouples, is equal to the sunlit one. The power required is a measure of the solar radiation for the target area.

Total Sun and sky. The total solar energy from Sun and sky striking a horizontal surface is measured with a hemispherical pyrheliometer. The Kimball-Epply and Moll-Garczynski instruments are of this type. They consist of a thermally insulated horizontal target within a partially evacuated transparent sphere or hemisphere to protect it from convective and conductive heat losses. Part of the target is a ring coated with dull black to absorb as much sunlight as possible; the remainder, a disk and a ring, is coated with white to reflect as much sunlight as possible. A thermopile measures the temperature difference between the white and black portions of the target and is the output of the instrument. Special glass or fused quartz, transparent to most of the Sun's radiation, is used for the envelope.

Net radiation measurement. An instrument which measures the net effect of all the upward and downward solar and terrestrial radiation currents must be equally sensitive to the whole range of wavelengths mentioned earlier. No window material completely transparent over this wide range exists; however, some film plastics approach complete transparency. In better instruments the detector, a horizontal plate with provision to measure the difference in temperature between upper and lower surfaces, is exposed without a wind screen. In the Albrecht net radiometer, the effect of the wind is measured by adding a known quantity of heat electrically to one of two identical plates. In the Gier and Dunkle instrument, the wind loss is held constant by a jet of air from a blower.

[VERNER E. SUOMI]

Thermocline

A layer of sea water in which the temperature decrease with depth is greater than that of the overlying and underlying water. Such layers are semipermanent features of the oceanic temperature structure, and their depth and thickness show marked variation with season, latitude and longitude, and local environmental conditions. Since the three-dimensional temperature structure has a great effect on many oceanic properties, such as the transmission of sound, the study of the nature and behavior of the thermocline is of extreme importance to many oceanographic interests, both economic and military. In general, two major types of thermocline may be identified: the permanent thermocline and the seasonal thermocline. In addition to these types, shallow thermoclines or similar stable layers often occur, owing to diurnal heating of the surface waters. *See* UNDERWATER SOUND.

Permanent thermocline. This feature is so named because its character is virtually unchanged seasonally. In Arctic and Antarctic regions, the water is cold from top to bottom. As this dense water flows south and north, respectively, it sinks beneath warmer water which moves outward from the Equator. This gives rise to the temperature discontinuity known as the permanent thermocline. The cold water flowing slowly through the deep ocean basins exhibits conservative properties throughout all the oceans; however, on top of this dense layer lie a number of shallow layers whose character varies from ocean to ocean. The top of the permanent thermocline is quite shallow at the Equator, reaches maximum depth at mid-latitudes, and becomes shallow again at about 50° latitude. The thermocline disappears between 55 to 60°N or S. In general, as the permanent thermocline deepens, it becomes thicker and the temperature gradient within it decreases. Figure 1 indicates schematically variations with latitude in the characteristics of the permanent thermocline. *See* SEA WATER.

Seasonal thermocline. This feature is a summer phenomenon found at shallower water depth than the permanent thermocline in all the world's oceans except those perennially ice-infested. As air temperatures rise above ocean temperatures in the spring season and the sea surface receives more heat than it loses by radiation and convection, the surface water begins to warm so that a negative temperature gradient develops in the first few feet (Fig. 2a). (Numbers 1–8 show sequence in development and disappearance of thermocline;

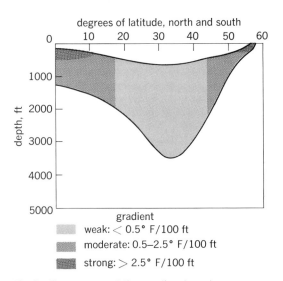

Fig. 1. The permanent thermocline, based on averages for depth, thickness, and gradient within thermocline.

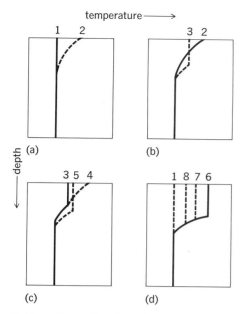

temperature ⟶

depth

(a)   (b)

(c)   (d)

Fig. 2. Formation and breakup of seasonal thermocline. (a) Formation, 1st stage. (b) Formation, 2d stage. (c) Formation, 3d stage. (d) Breakup.

profiles show temperature structure.) The surface waters are then mixed by transfer of energy from the wind. Although this mixing serves to lower the surface temperature, the net effect is a downward transport of heat and formation of an isothermal layer whose temperature is warmer than the underlying water (Fig. 2b). A strong temperature gradient, or seasonal thermocline, is thus formed between the isothermal surface layer and water beneath. This process repeats itself (Fig. 2c) until the gradient in the seasonal thermocline becomes so strong that summer winds cannot impart sufficient energy to drive the isothermal layer deeper. From July through September such a surface layer of mixed water underlain by a strong negative temperature gradient is found in most of the ocean. As air temperatures fall in autumn, the water loses heat to the atmosphere by convective and radiative processes, and the surface layer is cooled to the temperature of the water below. The seasonal thermocline breaks up (Fig. 2d), to form again the following spring. Seasonal thermoclines may be affected locally by vertical wind mixing, currents, and heat exchange across the interface between ocean and atmosphere. Further distortions may occur because the density discontinuity associated with thermoclines provides a favorable environment for internal waves. Practically all physical processes occurring in the sea have an effect on thermocline characteristics. See HALO-CLINE; INTERNAL WAVE.   [JOHN J. SCHULE, JR.]

Thunder

An acoustic wave caused by lightning. When lightning occurs, the surge of current in the return stroke suddenly heats the air in the lightning channel. The resulting expansion of the gaseous medium is similar to an explosion and produces a shock wave. The shock wave can be thought of as an expanding cylinder surrounding the original discharge channel. When the atmospheric pressure along the cylinder axis falls to the levels it had before the lightning flash, the shock wave becomes an acoustic wave and we call it thunder.

In the late 1960s investigators at Rice University and the New Mexico Institute of Mining and Technology presented new theories to explain thunder. By considering the lightning channel to consist of a series of spheres over which the electrical energy is dissipated, A. A. Few arrived at an equation for calculating the frequency spectrum of thunder.

It was estimated that the energy in a return stroke may amount to 10^6 joules/m of lightning path. Assuming that all the energy goes into the production of the shock wave, the theory indicates that the dominant frequency of thunder would be about 70 Hz and would vary inversely with the square root of the energy per unit length dissipated in the discharge channel. Earlier investigators had reported that maximum thunder power was at frequencies of about 1 Hz, a figure far smaller than that given by the new theory. Observations of thunder in Texas and New Mexico by means of modern technique revealed a broad power distribution ranging from below 10 to 150 Hz. These results are in good agreement with the theory.

In general, thunder cannot be heard more than 5–6 mi from the source because of the dispersion caused by the atmosphere. However, when conditions between the thunderstorm and the observer are stable, thunder may be heard at a distance of perhaps 10–15 mi.

Since lightning is seen almost instantaneously (speed of light = 186,000 mi/sec), while sound travels relatively slowly, one can estimate the distance to the lightning path by multiplying the number of seconds elapsed by 1000 ft/sec, an approximate value of the speed of sound.

The rumbling of thunder results mainly because the twisting nature of lightning paths causes variations in the distance to the sound sources and because multiple strokes in the same path produce a number of compression waves which interfere with one another.

Just before the loud report caused by a nearby lightning strike, a short click is sometimes heard. This sound has been ascribed to the stepped leader. See LIGHTNING.   [LOUIS J. BATTAN]

Bibliography: G. A. Dawson et al., Acoustic output of a long spark, J. Geophys. Res., 73:815–816, 1968; A. A. Few, Thunder spectrum, J. Geophys. Res., 74:6926–6934, 1969; A. A. Few et al., A dominant 200-Hertz peak in the acoustic spectrum of thunder, J. Geophys. Res., 72:6149–6154, 1967.

Thunderstorm

A convective storm accompanied by lightning and thunder and a variety of weather such as locally heavy rainshowers, hail, high winds, sudden temperature changes, and occasionally tornadoes. The characteristic cloud is the cumulonimbus or "thunderhead," a towering cloud, often with an anvil-shaped top. A host of accessory clouds, some attached and some detached from the main cloud, are often observed in conjunction with cumulonimbus. The height of a cumulonimbus base above the ground ranges from 1000 to over 10,000 ft (300–3000 m), depending on the relative humidity of air near the Earth's surface. Tops usually reach 30,000–60,000 ft (9000–18,000 m), with the taller storms occurring in the tropics or during summer in mid-latitudes. Thunderstorms travel at speeds from near zero to 60 miles per hour (mph; 27 m/s). In many tropical and temperate regions, thunderstorms furnish much of the annual rainfall.

Development. Thunderstorms are manifestations of convective overturning of deep layers in the atmosphere and occur in environments in which the decrease of temperature with height (lapse rate) is sufficiently large to be conditionally unstable and the air at low levels is moist. In such an atmosphere, a rising air parcel, given sufficient lift, becomes saturated and cools less rapidly than it would if it remained unsaturated because the released latent heat of condensation partly counteracts the expansional cooling. The rising parcel reaches levels where it is warmer (by perhaps as much as 10°C) and less dense than its surroundings, and buoyancy forces accelerate the parcel upward. The convection may be initiated by surface heating or by air flowing over rising terrain or by converging airflow in atmospheric disturbances such as fronts. The rising parcel is decelerated and its vertical ascent arrested at altitudes where the lapse rate is stable, and the parcel becomes denser than its environment. The forecasting of thunderstorms thus hinges on the identification of regions where the lapse rate is unstable, low-level air parcels contain adequate moisture, and surface heating or uplift of the air is expected to be sufficient to initiate convection.

Occurrence. Thunderstorms are most frequent in the tropics, and rare poleward of 60° latitude. In the United States, the Florida peninsula has the maximum activity with 60 thunderstorm days (days on which thunder is heard at a given observation station) per year. Thunderstorms occur at all hours of day and night, but are most common during late afternoon because of the influence of diurnal surface heating. The weak nighttime maximum of thunderstorms in the Mississippi Valley of the central United States is still a topic of debate.

Structure. Radar is used to detect thunderstorms at ranges up to 250 mi (400 km) from the observing site. Much of present-day knowledge of thunderstorm structure has been deduced from radar studies, supplemented by visual observations from the ground and satellites, and in-place measurements from aircraft, surface observing stations, and weather balloons. Thunderstorms occur in isolation, in chaotic patterns over wide areas, in the walls and spiral bands of hurricanes, in clusters within large-scale weather systems, and in squall lines perhaps several hundred miles long. An individual thunderstorm typically covers a surface area of 200–1000 mi² (500–2500 km²) and consists of one or more distinct cells, each of which is several miles across, lasts about an hour, and undergoes a characteristic life cycle. In the cumulus or growing stage, a cell consists primarily of updrafts (vertical speeds of 10–40 m/s or 20–90 mph) with precipitation suspended aloft; in the mature stage, updrafts and downdrafts coexist and heavy rain falls to the ground; in the dissipating stage, a cell contains weakly subsiding air and only light precipitation. During the mature stage, downdrafts may reach 35 mph or 15 m/s. The downdraft air is denser than its surroundings due to evaporational cooling, which occurs as clear air is entrained into the cloud from outside and forced downward by gravitational pull and by the drag of falling precipitation. The downflowing air spreads outward in all directions as it nears the surface, and forms a cold, gusty wind which is directed away from the precipitation area. This advancing cold air may provide the necessary lift in neighboring warm moist air for the formation of new updraft cells.

In an environment where the winds increase and veer with height, and mid-level air is dry enough to provide the potential for strong downdrafts, a thunderstorm may become organized so as to maintain a steady state for hours. In such strong vertical shear of the horizontal wind, the updraft is tilted so that precipitation falls out of the updraft instead of through it, and updraft and downdraft can coexist for several hours in the configuration shown in Fig. 1. A long-lived storm in a sheared environment may consist of a single intense cell (supercell) or of many cells with an organized growth of new cells on one side of the storm (generally, the southwest in the Northern Hemisphere) and decay of old cells on the opposite flank.

Severe storms. Thunderstorms are considered severe when they produce winds greater than 58 mph (26 m/s or 50 knots), hail larger than 3/4 in. (19mm) in diameter, or tornadoes. While thunderstorms are generally beneficial because of their needed rains (except for occasional flash floods), severe storms are the atmospheric equivalent of a rogue elephant with the capacity of inflicting utter devastation over narrow swaths of the countryside. Severe storms are most frequent in the Great Plains region of the United States, but even there only about 1% of the thunderstorms are severe. Severe storms are most frequently supercell storms which form in environments with high convective instability and moderate to large vertical wind shears.

Since severe storms constitute a hazard to air-

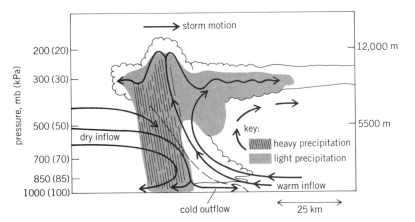

Fig. 1. Cloud boundaries and simplified circulation (arrows denote flow) of a typical mature thunderstorm in winds which blow from left to right and increase with height. Vertical scale has been exaggerated fivefold compared with the horizontal scale.

craft, their internal dynamics has been deduced largely from radar measurements. Doppler radar is specialized to measure the velocity of radar targets parallel to the radar beam, in addition to the intensity of precipitation. Doppler radar studies and analysis of surface pressure falls have shown that large hail, high winds, and tornadoes often develop from a rotating thunderstorm cell known as a mesocyclone (or tornado cyclone, if it spawns at least one tornado). Large hail, high winds, and weak tornadoes may form from nonrotating (on broad scale) multicellular storms, but are less likely. Maximum tangential winds around the typical mesocyclone are roughly 50 mph (20 m/s) and are located in a circular band which is 1–6 mi (2–10 km) in radius. A surface pressure deficit of several millibars exists at the mesocyclone center. Identification of a mesocyclone "signature" on radar has been used experimentally to issue severe weather warnings. On conventional radar displays, hook-shaped appendages to echoes are also good indications of mesocyclones, but unfortunately a large percentage of tornadic storms never exhibit such a hook. A mesocyclone sometimes is recognizable visually by rotation of a wall cloud, a discrete and distinct lowering of the cumulonimbus base (Fig. 2). The wall cloud is often seen visually to be rotating as an entity. The wall cloud is frequently the seat of intense vertical motions at low levels. The mesocyclone's rotation is apparently a combination of two effects. First, since the initial air-mass condition is one of rotation (at least the Earth's), inward motion of air parcels at low levels toward the base of an updraft is associated with increasing spin in rough accord with the conservation of angular momentum. Second, differential vertical air motion in an environment where the wind varies with height causes a tilting of vortex tubes from the horizontal into the vertical and thus also amplifies vertical vorticity (that is, vertical spin).

Attempts have been made to modify thunderstorms to increase areal rainfall and suppress hail. The results of such experiments have been inconclusive. *See* HAIL; SQUALL; STORM DETECTION; TORNADO; VORTEX.   [ROBERT DAVIES-JONES]

Storm electricity. A thunderstorm produces lightning and is thus highly electrified. A thunderstorm can be considered as a current generator that separates charge, resulting in strong electric fields and electricity for lightning and various conduction currents within, below, and above the storm. Electric currents from thunderstorms to the earth and ionosphere maintain the fair weather state of the earth and atmosphere. Typical maximum electric fields are 6 to 15 kV/m at the earth below the storm and 100 to 400 kV/m or more inside thunderstorms. Electric fields at the ground are less intense because of corona discharge from pointed objects. Lightning is the most obvious and deadly manifestation of storm electrification. Flashing rates are extremely variable. A small isolated storm produces about 3 flashes per minute, but rates as high as 100 have been observed. Lightning in squall lines and severe storms can appear nearly continuous. Lightning can propagate long distances, with at least one flash 170 km long having been observed with radar.

Initial and subsequent cloud electrification

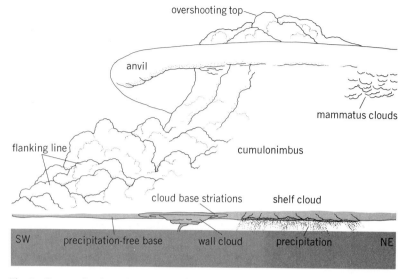

Fig. 2. Composite view of a typical tornado producing cumulonimbus as seen from a southeasterly direction. Horizontal scale is compressed, and all the features shown could not be seen from a single location. (*NOAA picture by C. Doswell and B. Dirham*)

mechanisms are unknown, but charged hydrometeors and air motion are key elements. The gross charge distribution within a thunderstorm is an upper positive charge, which spreads into the anvil, and a lower negative charge. Magnitudes of these charges are unknown; generally accepted estimates range from a few tens of coulombs to a few hundred coulombs. Most of the charge is thought to be on cloud and not on precipitation particles. Research with balloons and aircraft penetrating thunderstorms shows complex charge distributions on a smaller scale, and that the dominant charge polarity can change over distances as small as a kilometer. Knowledge of thunderstorm electrification processes awaits complete measurements. Modern research is characterized by simultaneous measurements of electrical parameters, microphysical processes, and storm dynamics in efforts to understand the complex interrelationships that result in the thunderstorm. *See* ATMOSPHERIC ELECTRICITY; LIGHTNING; THUNDER.

[W. DAVID RUST]

Tidal bore

A part of a tidal rise in a river which is so rapid that water advances as a wall often several feet high. The phenomenon is favored by a substantial tidal range and a channel which shoals and narrows rap-

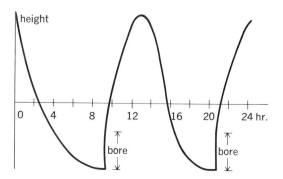

Fig. 1. Tidal curve of a river with a tidal bore.

Fig. 2. Tidal bore of the Petitcodiac River, Bay of Fundy, New Brunswick, Canada. Rise of water is about 4 ft. (*New Brunswick Travel Bureau*)

idly upstream, but the conditions are so critical that it is not common. A shoaling channel steepens the tidal curve. If the curve becomes vertical or nearly so, a bore results (Fig. 1). A narrowing channel increases the tidal range. Since the tidal range is greatest at spring tides, some rivers exhibit bores only then. Although the bore is a very striking feature, Fig. 1 shows that the tide continues to rise after the passage of the bore and that this subsequent rise may be greater. Bores may be eliminated by changing channel depth or shape. *See* RIVER TIDES; TIDE.

In North America three bores have been observed: at the head of the Bay of Fundy (Fig. 2), at the head of the Gulf of California, and at the head of Cook Inlet, Alaska. The largest known bore occurs in the Tsientang Kiang, China. At spring tides this bore is a wall of water 15 ft high moving upstream at 25 ft/sec.

[BLAIR KINSMAN]

Tide

Stresses exerted in a body by the gravitational action of another, and related phenomena resulting from these stresses. Every body in the universe raises tides, to some extent, on every other. This article deals only with tides on the Earth, since these are fundamentally the same as tides on all bodies. Sometimes variations of sea level, whatever their origin, are referred to as tides. *See* SEA-LEVEL FLUCTUATIONS.

Introduction. The tide-generating forces arise from the gravitational action of Sun and Moon, the effect of the Moon being about twice as effective as that of the Sun in producing tides. The tidal effects of all other bodies on the Earth are negligible. The tidal forces act to generate stresses in all parts of the Earth and give rise to relative movements of the matter of the solid Earth, ocean, and atmosphere. The Earth's rotation gives these movements an alternating character having principal periodicities of 12.42 and 12.00 hr, corresponding to half the mean lunar and solar day, respectively.

In the ocean the tidal forces act to generate alternating tidal currents and displacements of the sea surface. These phenomena are important to shipping and have been studied extensively. The main object of tidal studies has been to predict the tidal elevation or current at a given seaport or other place in the ocean at any given time.

The prediction problem may be attacked in two ways. Since the relative motions of Earth, Moon, and Sun are known precisely, it is possible to specify the tidal forces over the Earth at any past or future time with great precision. It should be possible to relate tidal elevations and currents at any point in the oceans to these forces, making use of classical mechanics and hydrodynamics. Such a theoretical approach to tidal prediction has not yet yielded any great success, owing in great part to the complicated shape of the ocean basins. However, use of numerical-hydrodynamical models (such as the work of K. T. Bogdanov, N. Grijalva, W. Hansen, M. C. Henderschott, and C. L. Pekeris) has yielded some satisfactory results and undoubtedly will have practical importance.

The other approach, which consists of making use of past observations of the tide at a certain place to predict the tide for the same place, has yielded practical results. The method cannot be used for a location where there have been no previous observations. In the harmonic method the frequencies of the many tidal constituents are derived from knowledge of the movements of Earth, Moon, and Sun. The amplitude and epoch of each constituent are determined from the tidal observations. The actual tide can then by synthesized by summing up an adequate number of harmonic constituents. The method might loosely be thought of as extrapolation.

A "convolution" method of tidal analysis and prediction has been proposed by W. H. Munk and D. E. Cartwright. In this method past observations at a place are used to determine a numerical operator which, when applied to the known tide-producing forces, will calculate the resulting tide.

In the following discussion only the lunar effect is considered, and it is understood that analogous statements apply to the solar effect.

Tide-generating force. If the Moon attracted every point within the Earth with equal force, there would be no tide. It is the small difference in direction and magnitude of the lunar attractive force, from one point of the Earth's mass to another, which gives rise to the tidal stresses.

According to Newton's laws, the Moon attracts every particle of the Earth with a force directed toward the center of the Moon, with magnitude proportional to the inverse square of the distance between the Moon's center and the particle. At point A in Fig. 1, the Moon is in the zenith and at point B the Moon is at nadir. It is evident that the upward force of the Moon's attraction at A is greater than the downward force at B because of its closer proximity to the Moon. Such differential forces are responsible for stresses in all parts of the Earth. The Moon's gravitational pull on the

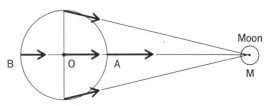

Fig. 1. Schematic diagram of the lunar gravitational force on different points in the Earth.

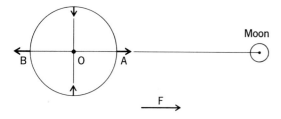

Fig. 2. Schematic diagram of the tide-generating force on different points in the Earth. The vector sum of this tide-generating force and the constant force F (which does not vary from point to point) produce the force field indicated in Fig. 1. Force F is compensated by the centrifugal force of the Earth in its orbital motion.

Earth can be expressed as the vector sum of a constant force, equal to the Moon's attraction on the Earth's center, and a small deviation which varies from point to point in the Earth (Fig. 2). This small deviation is referred to as the tide-generating force. The larger constant force is balanced completely by acceleration (centrifugal force) of the Earth in its orbital motion around the center of mass of the Earth-Moon system, and plays no part in tidal phenomena.

The tide-generating force is proportional to the mass of the disturbing body (Moon) and to the inverse cube of its distance. This inverse cube law accounts for the fact that the Moon is 2.17 times as important, insofar as tides are concerned, as the Sun, although the latter's direct gravitational pull on the Earth, which is governed by an inverse-square law, is about 180 times the Moon's pull.

The tide-generating force, as illustrated in Fig. 2, can be expressed as the gradient of the tide-generating potential, Eq. (1), where λ is the zenith

$$\psi = \frac{3}{2} \frac{\gamma M r^2}{c^3} (1/3 - \cos^2 \lambda) \qquad (1)$$

distance of the Moon, r is distance from the Earth's center, c is distance between the centers of Earth and Moon, γ is the gravitational constant, and M is the mass of the Moon. In this expression, terms containing higher powers of the smaller number r/c have been neglected. As ψ depends only on the space variables r and λ, it is symmetrical about the Earth-Moon axis.

It helps one visualize the form of the tide-generating potential to consider how a hypothetical "inertialess" ocean covering the whole Earth would respond to the tidal forces. In order to be in equilibrium with the tidal forces, the surface must assume the shape of an equipotential surface as determined by both the Earth's own gravity and the tide-generating force. The elevation of the surface is given approximately by Eq. (2), where ψ is evalu-

$$\bar{\zeta} = -\frac{\psi}{g} + \text{const} \qquad (2)$$

ated at the Earth's surface and g is the acceleration of the Earth's gravity. The elevation $\bar{\zeta}$ of this hypothetical ocean is known as the equilibrium tide. Knowledge of the equilibrium tide over the entire Earth determines completely the tide-generating potential (and hence the tidal forces) at all points within the Earth as well as on its surface.

Therefore, when the equilibrium tide is mentioned, it shall be understood that reference to the tide-generating force is also being made.

Harmonic development of the tide. The equilibrium tide as determined from relations (1) and (2) has the form of a prolate spheroid (football-shaped) whose major axis coincides with the Earth-Moon axis. The Earth rotates relative to this equilibrium tidal form so that the nature of the (equilibrium) tidal variation with time at a particular point on the Earth's surface is not immediately obvious. To analyze the character of this variation, it is convenient to express the zenith angle of the Moon in terms of the geographical coordinates θ, ϕ of a point on the Earth's surface (θ is colatitude, ϕ is east longitude) and the declination D and west hour angle reckoned from Greenwich α of the Moon. When this is done, the equilibrium tide can be expressed as the sum of the three terms in Eq. (3), where a is the Earth's radius.

$$\bar{\zeta} = \frac{3}{4} \frac{\gamma M}{g} \frac{a^2}{c^3} [(3 \sin^2 D - 1)(\cos^2 \theta - 1/3)$$
$$+ \sin 2D \sin 2\theta \cos(\alpha + \phi)$$
$$+ \cos^2 D \sin^2 \theta \cos 2(\alpha + \phi)] \qquad (3)$$

The first term represents a partial tide which is symmetrical about the Earth's axis, as it is independent of longitude. The only time variation results from the slowly varying lunar declination and distance from Earth. This tide is called the long-period tide. Its actual geographical shape is that of a spheroid whose axis coincides with the Earth's axis and whose oblateness slowly but continuously varies.

The second term of Eq. (3) represents a partial tide having, at any instant, maximum elevations at 45°N and 45°S on opposite sides of the Earth, and two minimum elevations lying at similar, alternate positions on the same great circle passing through the poles. Because of the factor $\cos(\alpha + \phi)$ the tide rotates in a westerly direction relative to the Earth, and any geographical position experiences a complete oscillation in a lunar day, the time taken for α to increase by the amount 2π. Consequently, this partial tide is called the diurnal tide. Because of the factor $\sin 2D$, the diurnal equilibrium tide is zero at the instant the moon crosses the Equator; because of the factor $\sin 2\theta$, there is no diurnal equilibrium tidal fluctuation at the Equator or at the poles.

Darwin's constituents

Constituent	Speed, deg/hr	Coefficient
Long-period		
Mf, lunar fortnightly	1.098	0.157
Ssa, solar semiannual	0.082	0.073
Diurnal		
K_1, lunisolar	15.041	0.530
O_1, larger lunar	13.943	0.377
P_1, larger solar	14.959	0.176
Semidiurnal		
M_2, principal lunar	28.984	0.908
S_2, principal solar	30.000	0.423
N_2, larger lunar elliptic	28.440	0.176
K_2, lunisolar	30.082	0.115

The third term of Eq. (3) is a partial tide having, at any instant, two maximum elevations on the Equator at opposite ends of the Earth, separated alternately by two minima also on the Equator. This whole form also rotates westward relative to the Earth, making a complete revolution in a lunar day. But any geographic position on the Earth will experience two cycles during this time because of the factor $\cos 2(\alpha + \phi)$. Consequently, this tide is called the semidiurnal tide. Because of the factor $\sin^2 \theta$, there is no semidiurnal equilibrium tidal fluctuation at the poles, while the fluctuation is strongest at the Equator.

It has been found very convenient to consider the equilibrium tide as the sum of a number of terms, called constituents, which have a simple geographical shape and vary harmonically in time. This is the basis of the harmonic development of the tide. A great number of tidal phenomena can be adequately described by a linear law; that is, the effect of each harmonic constituent can be superimposed on the effects of the others. Herein is the great advantage of the harmonic method in dealing with tidal problems. The three terms of Eq. (3) do not vary with time in a purely harmonic manner. The parameters c and D themselves vary, and the rapidly increasing α does not do so at a constant rate owing to ellipticity and other irregularities of the Moon's orbit. Actually, each of the three partial tides can be separated into an entire species of harmonic constituents. The constituents of any one of the three species have the same geographical shape, but different periods, amplitudes, and epochs.

The solar tide is developed in the same way. As before, the three species of constituents arise: long-period, diurnal, and semidiurnal. The equilibrium tide at any place is the sum of both the lunar and solar tides. When the Sun and Moon are nearly in the same apparent position in the sky (new Moon) or are nearly at opposite positions (full Moon), the lunar and solar effects reinforce each other. This condition is called the spring tide. During the spring tide the principal lunar and solar constituents are in phase. At quadrature the solar effect cancels to some extent the lunar effect, the principal lunar and solar constituents being out of phase. This condition is known as the neap tide.

The entire equilibrium tide can now be expressed by Eq. (4), where $H = 3\gamma M a^2/g\bar{c}^3 = 54$ cm,

$$\bar{\zeta} = H[\tfrac{1}{2}(1 - 3\cos^2 \theta) \sum_L f_i C_i \cos A_i$$

$$+ \sin 2\theta \sum_D f_i C_i \cos (A_i + \phi)$$

$$+ \sin^2 \theta \sum_S f_i C_i \cos (A_i + 2\phi)] \quad (4)$$

and $1/\bar{c}$ represents the mean (in time) value of $1/c$. Each term in the above series represents a constituent. Terms of higher powers of the Moon's parallax (a/c) are not included in Eq. (4) because of their different latitude dependence, but they are of relatively small importance. The subscripts L, D, and S indicate summation over the long-period, diurnal, and simidiurnal constituents, respectively. The Cs are the constituent coefficients and are constant for each constituent. They account for

the relative strength of all lunar and solar constituents. In a purely harmonic development, such as carried out by A. T. Doodson in 1921, the A parts of the arguments increase linearly with time, and the node factors f are all unity. In George Darwin's "almost harmonic" development of 1882, the constituents undergo a slow change in amplitude and epoch with the 19-year nodal cycle of the Moon. The node factors f take this slow variation into account. The As increase almost linearly with time. Tables in U.S. Coast and Geodetic Survey Spec. Publ. no. 98 enable one to compute the phase of the argument of any of Darwin's constituents at any time, and values of the node factors for each year are given.

In spite of the many advantages of the purely harmonic development, Darwin's method is still used by most agencies engaged in tidal work. In Darwin's classification, each constituent is represented by a symbol with a numerical subscript, 0, 1, or 2, which designates whether the constituent is long-period, diurnal, or semidiurnal. Some of the most important of Darwin's constituents are listed in the table.

The periods of all the semidiurnal constituents are grouped about 12 hr, and the diurnal periods about 24 hr. This results from the fact that the Earth rotates much faster than the revolution of the Moon about the Earth or of the Earth about the Sun. The principal lunar semidiurnal constituent M_2 beats against the others giving rise to a modulated semidiurnal waveform whose amplitude varies with the Moon's phase (the spring-neap effect), distance, and so on. Similarly, the amplitude of the modulated diurnal wave varies with the varying lunar declination, solar declination, and lunar phase. For example, the spring tide at full Moon or new Moon is manifested by constituents M_2 and S_2 being in phase, thus reinforcing each other. During the neap tide when the Moon is at quadrature, the constituents M_2 and S_2 are out of phase, and tend to cancel each other. The other variations in the intensity of the tide are similarly reflected in the "beating" of other groups of constituents.

Tides in the ocean. The tide in the ocean deviates markedly from the equilibrium tide, which is not surprising if one recalls that the equilibrium tide is based on neglect of the inertial forces. These forces are appreciable unless the periods of all free oscillations in the ocean are small compared with those of the tidal forces. Actually, there are free oscillations in the ocean (ordinary gravity seiches) having periods of the order of a large fraction of a day, and there may be others (planetary modes) having periods of the order of several days. For the long-period constituents the observed tide should behave like the equilibrium tide, but this is difficult to show because of their small amplitude in the presence of relatively large meteorological effects.

At most places in the ocean and along the coasts, sea level rises and falls in a regular manner. The highest level usually occurs twice in any lunar day, the times bearing a constant relationship with the Moon's meridional passage. The time between the Moon's meridional passage and the next high tide is called the lunitidal interval. The difference in level between successive high and low tides, called the range of the tide, is generally greatest near the time of full or new Moon, and smallest near the times of quadrature. This results from the spring-neap variation in the equilibrium tide. The range of the tide usually exhibits a secondary variation, being greater near the time of perigee (when the Moon is closest to the Earth) and smaller at apogee (when the Moon is farthest away).

The above situation is observed at places where the tide is predominantly semidiurnal. At many other places, it is observed that one of the two maxima in any lunar day is higher than the other. This effect is known as the diurnal inequality and represents the presence of an appreciable diurnal variation. At these places, the tide is said to be of the "mixed" type. At a few places, the diurnal tide actually predominates, there generally being only one high and low tide during the lunar day.

Both observation and theory indicate that the ocean tide can generally be considered linear. As a result of this fact, the effect in the ocean of each constituent of series in Eq. (4) can be considered by itself. Each equilibrium constituent causes a reaction in the ocean. The tide in the ocean is the sum total of all the reactions of the individual constituents. Furthermore, each constituent of the ocean tide is harmonic (sinusoidal) in time. If the amplitude of an equilibrium constituent varies with the nodal cycle of the Moon, the amplitude of the oceanic constituent varies proportionately.

As a consequence of the above, the tidal elevation in the ocean can be expressed by Eq. (5),

$$\zeta = \Sigma f_i h_i \cos (A_i - G_i) \qquad (5)$$

where $h_i(\theta,\phi)$ is called the amplitude and $G_i(\theta,\phi)$ the Greenwich epoch of each constituent. The summation in Eq. (5) extends over all constituents of all species. The fs and the As have the same meaning as in Eq. (4) for the equilibrium tide and are determined from astronomic data.

To specify completely the tidal elevation over the entire surface of the ocean for all time, one would need ocean-wide charts of $h(\theta,\phi)$, called corange charts, and of $G(\theta,\phi)$, called cotidal charts, for each important constituent. Construction of these charts would solve the ultimate problem in tidal prediction. Many attempts have been made to construct cotidal charts, the most notable those of W. Whewell, 1833; R. A. Harris, 1904; R. Sterneck, 1920; and G. Dietrich, 1944. These attempts have been based on a little theory and far too few observations.

Figures 3 and 4 show Dietrich's cotidal chart for M_2. Each curve passes through points having high water at the same time, time being indicated as phase of the M_2 equilibrium argument. A characteristic feature of cotidal charts is the occurrence of points through which all cotidal curves pass. These are called amphidromic points. Here the amplitude of the constituent under consideration must be zero. The existence of such amphidromic points has been borne out by theoretical studies of tides in ocean basins of simple geometric shape. The mechanism which gives rise to amphidromic points is intimately related to the rotation of the Earth and the Coriolis force.

The amplitude of a constituent, $h(\theta,\phi)$, is generally high in some large regions of the oceans and

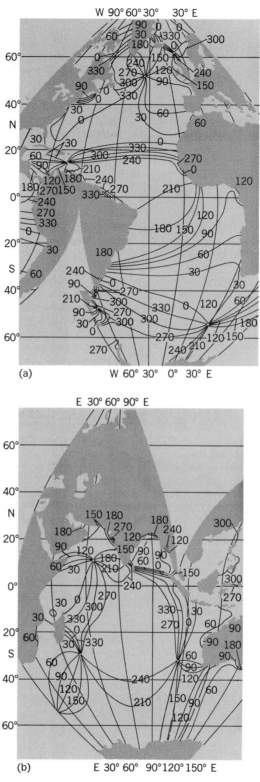

(a)

(b)

Fig. 3. Cotidal chart for M_2. (a) Atlantic Ocean. (b) Indian Ocean. (G. Dietrich, Veroeff. Inst. Meeresk., n.s. A, Geogr.-naturwiss. Reihe, no. 41, 1944)

low in others, but in addition there are small-scale erratic variations, at least along the coastline. Perhaps this is partly an illusion caused by the placement of some tide gages near the open coast and the placement of others up rivers and estuaries. It

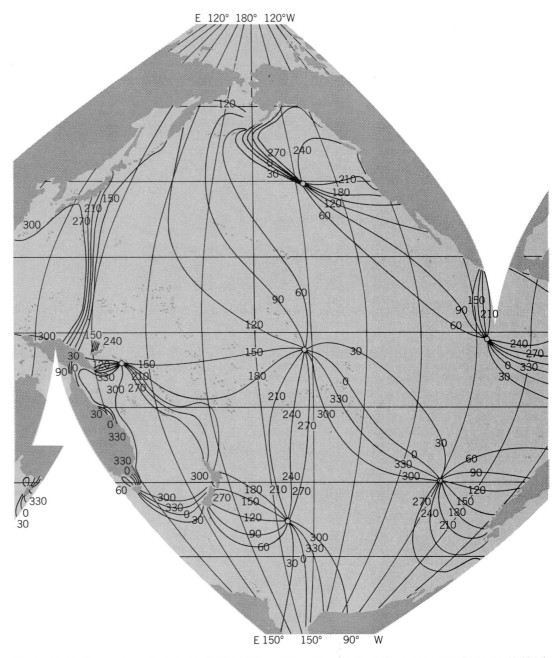

Fig. 4.   Pacific Ocean cotidal chart for M_2. (*G. Dietrich, Veroeff. Inst. Meeresk., n.s. A, Geogr.-naturwiss. Reihe, no. 41, 1944*)

is well known that the phase and amplitude of the tide change rapidly as the tidal wave progresses up a river. *See* RIVER TIDES.

The range of the ocean tide varies between wide limits. The highest range is encountered in the Bay of Fundy, where values exceeding 50 ft have been observed. In some places in the Mediterranean, South Pacific, and Arctic, the tidal range never exceeds 2 ft.

The tide may be considerably different in small adjacent seas than in the nearby ocean, and here resonance phenomena frequently occur. The periods of free oscillation of a body of water are determined by their boundary and depth configurations. If one of these free periods is near that of a large tidal constituent, the latter may be amplified con-

siderably in the small sea. The large tidal range in the Bay of Fundy is an example of this effect. Here the resonance period is nearly 12 hr, and it is the semidiurnal constituents that are large. The diurnal constituents are not extremely greater in the Bay of Fundy than in the nearby ocean. *See* SEICHE.

In lakes and other completely enclosed bodies of water the periods of free oscillation are usually much smaller than those of the tidal constituents. Therefore the tide in these places obeys the principles of statics. Since there is no tidal variation in the total volume of water in lakes the mean surface elevation does not change with the tide. The surface slope is determined by the slope of the equilibrium tide, and the related changes in elevation are usually very small, of the order of a fraction of a millimeter for small lakes.

Tidal currents. The south and east components of the tidal current can be developed in the same way as the tidal elevation since they also depend linearly on the tidal forces. Consequently, the same analysis and prediction methods can be used. Expressions similar to Eq. (5) represent the current components, each constituent having its own amplitude and phase at each geographic point. It should be emphasized that the current speed or direction cannot be developed in this way since these are not linearly related to the tidal forces.

Only in special cases are the two tidal current components exactly in or out of phase, and so the tidal current in the ocean is generally rotatory. A drogue or other floating object describes a trajectory similar in form to a Lissajous figure. In a narrow channel only the component along its axis is of interest. Where shipping is important through such a channel or port entrance, current predictions, as well as tidal height predictions, are sometimes prepared.

Owing to the rotation of the Earth, there is a gyroscopic, or Coriolis, force acting perpendicularly to the motion of any water particle in motion. In the Northern Hemisphere this force is to the right of the current vector. The horizontal, or tractive, component of the tidal force generally rotates in the clockwise sense in the Northern Hemisphere. As a result of both these influences the tidal currents in the open ocean generally rotate in the clockwise sense in the Northern Hemisphere, and in the counterclockwise sense in the Southern Hemisphere. There are exceptions, however, and the complete dynamics should be taken into account. *See* CORIOLIS ACCELERATION AND FORCE.

The variation of the tidal current with depth is not well known. It is generally agreed that the current would be constant from top to bottom were it not for stratification of the water and bottom friction. The variation of velocity with depth due to the stratification of the water is associated with internal wave motion. Serial observations made from anchored or drifting ships have disclosed prominent tidal periodicities in the vertical thermal structure of the water. *See* INTERNAL WAVE.

Dynamics of ocean tide. The theoretical methods for studying tidal dynamics in the oceans were put forth by Laplace in the 18th century. The following assumptions are introduced: (1) The water is homogeneous; (2) vertical displacements and velocities of the water particles are small in comparison to the horizontal displacements and velocities; (3) the water pressure at any point in the water is given adequately by the hydrostatic law, that is, is equal to the head of water above the given point; (4) all dissipative forces are neglected; (5) the ocean basins are assumed rigid (as if there were no bodily tide), and the gravitational potential of the tidally displaced masses is neglected; and (6) the tidal elevation is small compared with the water depth.

If assumptions (1) and (3) are valid, it can readily be shown that the tidal currents are uniform with depth. This is a conclusion which is not in complete harmony with observations, and there are internal wave modes thus left out of Laplace's theory. Nevertheless the main features of the tide are probably contained in the equations.

The water motion in the oceans is, in theory, determined by knowledge of the shape of the ocean basins and the tide-generating force (or equilibrium tide) at every point in the oceans for all time. The theory makes use of two relations: (1) the equation of continuity, which states that the rate of change of water mass in any vertical column in the ocean is equal to the rate at which water is flowing into the column; and (2) the equations of motion, which state that the total acceleration of a water "particle" (relative to an inertial system, thus taking into account the rotation of the Earth) is equal to the total force per unit mass acting on that particle. Under the above assumptions, the equation of continuity takes the form of Eq. (6), where $d(\theta,\phi)$ is

$$\frac{\partial \zeta}{\partial t} = -\frac{1}{a \sin \theta}\left[\frac{\partial}{\partial \theta}(ud \sin \theta) + \frac{\partial}{\partial \phi}(vd)\right] \quad (6)$$

the water depth. The equations of motion in the southward and eastward directions, respectively, are given by Eq. (7), where ω designates the angu-

$$\begin{aligned}\frac{\partial u}{\partial t} - 2\omega v \cos \theta &= -\frac{g}{a}\frac{\partial}{\partial \theta}(\zeta - \bar{\zeta}) \\ \frac{\partial v}{\partial t} + 2\omega u \cos \theta &= -\frac{g}{a}\csc \theta \frac{\partial}{\partial \phi}(\zeta - \bar{\zeta})\end{aligned} \quad (7)$$

lar rate of rotation of the Earth, and u and v the south and east components of the tidal current. All other quantities are as previously defined.

It is probable that exact mathematical solutions to Eqs. (6) and (7), taking even approximately into account the complicated shape of the ocean basins, will never be obtained. However, the equations have certain features which serve to give us some insight into the nature of ocean tides. For instance it is evident that if many equilibrium tides are acting simultaneously on the ocean, then the ocean tide will be the sum of the individual reactions. This linearity results directly from above assumption (6). In certain shallow regions of the ocean the tides are noticeably distorted, as would be expected if assumption (6) were violated. This distortion is usually considered as resulting from the presence of so-called shallow-water constituents having frequencies equal to harmonics and to beat frequencies of the equilibrium constituents. These must be considered, at some places, or there will be large discrepancies between prediction and observation. Certain mathematical solutions to Eqs. (6) and (7) have been obtained for hypothetical ocean basins of simple geometric shape. Laplace solved them for an ocean of constant depth covering the entire Earth. Several solutions have been obtained for an ocean of constant depth bounded by two meridians. The result of one of the solutions obtained by J. Proudman and A. Doodson is shown in Fig. 5, which represents a cotidal chart of the K_2 tide in an ocean of depth 14,520 ft bounded by meridians 70° apart. The K_2 tide was calculated because of mathematical simplifications, but the M_2 tide should be quite similar. Comparison of Fig. 5 with the Atlantic

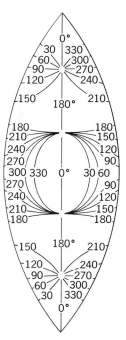

TIDE

Fig. 5. Cotidal chart for K_2 in a hypothetical ocean of constant depth bounded by meridians 70° apart. (*J. Proudman and A. T. Doodson, from A. T. Doodson and H. D. Warburg, Admiralty Manual of Tides, London, 1941*)

Ocean in Fig. 3 discloses no striking similarities except for the general occurrence of amphidromic systems.

The bodily tide. The solid part of the Earth suffers periodic deformation resulting from the tide-generating forces just as the oceans do. *See* EARTH TIDES.

The gravest known modes of free oscillation of the solid Earth have periods of the order of an hour, much shorter than those of the principal tidal constituents. Therefore, the principles of statics can be used to describe the bodily tide, in contrast to tides in the oceans and atmosphere, where the inertial effect is important.

Associated with the bodily tide are periodic changes in gravity, manifesting themselves as (1) a variation of the vertical, or plumb line, with respect to any solid structure embedded in the Earth's crust; and (2) a variation in the magnitude of the acceleration of gravity at any point. These effects arise from the gravitational attraction of the tidally displaced matter of the Earth (solid, ocean, and atmosphere) as well as directly from the tide-generating forces. The magnitude of the former factor is of the order of several tens of microgals (1 gal = 1 cm/sec^2).

Atmospheric tides. Since air, as other matter, is subject to gravitational influence, there are tides in the atmosphere possessing many features of similarity with those in the ocean. One of the characteristics of these tides is a small oscillatory variation in the atmospheric pressure at any place. This fluctuation of pressure, as in the case of the ocean tide, may be considered as the sum of the usual tidal constituents, and standard tidal analysis and prediction methods may be used. The principal lunar semidiurnal constituent M_2 of the pressure variation has been determined for a number of places, and found to have an amplitude of the order of 0.03 millibars. The dynamical theory of these tides has been the subject of considerable study. The equations which have been considered have the same general form as those for ocean tides. The S_2 constituent shows a much larger oscillation with an amplitude of the order of 1 millibar, but here diurnal heating dominates the gravitational effects. If diurnal heating were the whole story one would expect an even larger S_1 effect, and the fact that S_2 is larger is attributed to an atmospheric resonance near 12 hr. *See* UPPER-ATMOSPHERE DYNAMICS.

Tidal analysis and prediction. The distribution in space and time of the tidal forces within the Earth is precisely known from astronomic data. The effects of these forces on the oceans cannot, by present methods, be described in detail on a worldwide basis because of the difficult nature of the dynamical relationships and the complicated shape of the ocean basins. Practical prediction methods make use of past observations at the place under consideration.

The procedure is the same for prediction of any tidal variable—such as the atmospheric pressure, component displacements of the solid Earth, components of the tidal current, and so on—which depends linearly on the tidal forces. In the harmonic method the frequencies, or periods, of the tidal constituents are determined by the astronomic data, and the harmonic constants (amplitudes and epochs) are obtained from the observations. Equation (5) then represents the tide at all past and future times for the place under consideration, where the values of h are the amplitudes of whatever tidal variable is being predicted. In this discussion the sea-level elevation will be used as an example, since it is the variable for which predictions are most commonly made. The procedure is basically the same for each constituent, but is most easily described for the series of constituents, $S_1, S_2, S_3, \ldots$, whose periods are submultiples of 24 hr. Suppose that the tidal elevation at 1:00 A.M. is averaged for all the days of the tide record, and similarly for 2:00 A.M., 3:00 A.M., and for each hour of the day. The 24 values thus obtained represent the average diurnal variation during the entire record. Any constituent whose period is not a submultiple of 24 hr will contribute very little to the average of all the 1:00 A.M. values since its phase will be different from one day to the next, and its average value at 1:00 A.M. will be very close to zero for a long record. The same is true for each hour of the day, and so its average diurnal variation is small. The longer the record the freer will be the average diurnal oscillation from the effects of the other constituents. The diurnal oscillation is then analyzed by the well-known methods of harmonic analysis to determine the amplitudes and phases of all the harmonics of the 24-hr oscillation.

The same procedure is used for each other constituent; that is, the tide record is divided into consecutive constituent days, each equal to the period (or double the period in the case of the semidiurnal constituents) of the constituent. If the tide record is tabulated each solar hour, there is a slight complication due to the fact that the constituent hours do not coincide with the solar hours. This difficulty is overcome by substituting the tabulated value nearest the required time and later compensating the consistent error introduced by an augmenting factor.

Since the record length is always finite, the harmonic constants of a constituent determined by this method are somewhat contaminated by the effects of other constituents. A first-order correction of these effects can be made by an elimination procedure. In general it is more efficient to take the record length equal to the synodic (beat) period of two or more of the principal constituents. Of course, the longer the record the better. Standard analyses consist of 29 days, 58 days, 369 days, and so on.

It is not practical to determine the harmonic constants of the lesser constituents in this way if errors or uncertainties of the data are of the same order of magnitude as their amplitudes. If tidal oscillations in the oceans were far from resonance then the amplitude H of each constituent should be expected to be approximately proportional to its theoretical coefficient C, and the local epochs G all to be near the same value. In other words, for the semidiurnal constituent X, Eq. (8) should hold.

$$\frac{H(X)}{C(X)} = \frac{H(M_2)}{C(M_2)} \qquad G(X) = G(M_2) \qquad (8)$$

Here X is referred to M_2 for the reason that the latter is one of the principal constituents whose harmonic constants can be determined with best accuracy. Any other important constituent could be used. Inferring the harmonic constants of the lesser constituents by means of Eq. (8) is sometimes preferable to direct means. It should be borne in mind that a constituent of one species cannot be inferred from one of another species because their equilibrium counterparts have different geographic shapes and no general relationship such as Eq. (8) exists.

Once the harmonic constants are determined, the tide is synthesized according to Eq. (5), usually with the help of a special "tide-predicting machine," although any means of computation could be used. Usually only the times and heights of high and low water are published in the predictions.

Tidal friction. The dissipation of energy by the tide is important in the study of planetary motion because it is a mechanism whereby angular momentum can be transferred from one type of motion to another. An appreciable amount of tidal dissipation takes place in the ocean, and possibly also in the solid Earth. In 1952 Sir Harold Jeffreys estimated that about half the tidal energy present in the ocean at any time is dissipated each day. A large part of this dissipation takes place by friction of tidal currents along the bottom of shallow seas and shelves and along the coasts. The rate of dissipation is so large that there should be a noticeable effect on the tide in the oceans.

If the planet's speed of rotation is greater than its satellite's speed of revolution about it, as is the case in the Earth-Moon system, then tidal dissipation always tends to decelerate the planet's rotation, with the satellite's speed of revolution changing to conserve angular momentum of the entire system. The Moon's attraction on the irregularly shaped tidal bulge on the Earth exerts on it a decelerating torque. Thus tidal friction tends to increase the length of day, to increase the distance between Earth and Moon, and to increase the lunar month, but these increases are infinitesimal. The day may have lengthened by 1 sec during the last 120,000 years because of tidal friction and other factors.

[GORDON W. GROVES]

Bibliography: A. Defant, *Ebb and Flow: The Tides of Earth, Air, and Water*, 1958; J. J. Dronkers, *Tidal Computations and Coastal Waters*, 1964; H. Lamb, *Hydrodynamics*, 6th ed., 1945; P. Schureman, *A Manual of Harmonic Analysis and Prediction of Tides*, U.S. Coast and Geodetic Survey Spec. Publ. no. 98, 1941.

Tornado

An intense rotary storm of small diameter, the most violent of weather phenomena. Tornadoes always extend downward from the base of a convective-type cloud, generally in the vicinity of a severe thunderstorm.

Appearance ranges from a broad funnel with smallest diameter at the ground, to a narrow rope-like vortex which may not reach the ground or may intermittently lift and dip. An ill-defined cloud of dust or debris often surrounds the true tornado cloud near the ground (Fig. 2). In surface layers, air spirals inward toward the vortex, generally

rotating in a counterclockwise sense, rising rapidly in the funnel.

The visible funnel consists of cloud droplets condensed because of expansional cooling resulting from markedly lower (probably by 100–200 millibars) pressure in the vortex than in the surroundings. Height of the visible funnel depends upon the cloud base and may be 1000–10,000 ft; however, the tornado vortex probably extends a considerable distance upward within the accompanying cloud.

The path of destruction varies from a few yards to over a mile in width, and from very short to about 300 mi in length; in about 80% of cases the path is less than 50 yd wide and 3 mi long. Movement may be from any direction, most commonly from the southwest, but many tornadoes move from the northwest. Speed of movement averages 35 mph but is variable. Tornadoes may occur at any time of day, but are most frequent from mid-afternoon to early evening.

In the United States an annual average of over 700 tornadoes have been reported in recent years, with monthly percentages ranging from 20 in May to 3 in January. Maximum frequency is observed in the "tornado alley" belt, from the Texas Panhandle across Oklahoma and Kansas, thence across the Midwestern states (Fig. 1). Greatest activity is over the Southern states in winter and early spring, migrating to the tornado alley region in May, and to the northern Plains and the Midwest states in late summer. Other regions with fairly high tornado frequency include a belt from southern England across northern Europe; also Japan, the Ganges valley, southernmost Africa, southwest and southeast Australia, New Zealand, and the la Plata basin of South America.

According to T. Fujita, tornadoes in the United States east of the Rocky Mountains typically have broader and longer tracks than those in other parts of the world, and are also more intense. Based on estimates from damage to structures, 1 in 1000 may have winds exceeding 260 mph, although

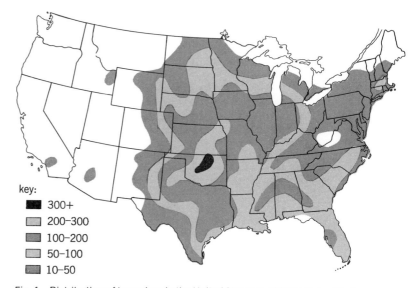

key:
■ 300+
▨ 200–300
▨ 100–200
▨ 50–100
▨ 10–50

Fig. 1. Distribution of tornadoes in the United States by 2° latitude-longitude squares (totals for 1955–1967).

Fig. 2. Tornado, at Fargo, ND. (*Fargo Forum photograph by C. Gerbert, Grand Prize Winner, 11th Annual Graflex Contest*)

more than half have winds less than 100 mph. Deaths and property damage mainly result from the 1–2% of the total number of tornadoes that are the largest and most intense and touch down in built-up areas. Damage results both from the force of the wind and from outward collapse of building walls when the atmospheric pressure is suddenly reduced.

Requisite conditions for tornado formation are pronounced thermodynamic instability combined with sufficient amounts of water vapor to produce thunderstorms, along with the presence of strong winds in the upper troposphere. These are the conditions that favor development of a squall line; at times 10–30 tornadoes occur as parasites to such a disturbance. Tornado probability increases with the thermal instability, and it is greatest with thunderstorms that have an intense radar reflectivity (large precipitation content) and high tops penetrating into the stratosphere. However, some tornadoes are not connected to the thunderstorm itself, but to vigorously developing cumulus clouds adjacent to the thunderstorm. *See* SQUALL LINE.

Both laboratory simulations and observations of natural vortexes suggest that rising motions (up to 150 mph observed) take place mainly in the sheath of the funnel, and that there is likely to be descending motion in or near its core. The intense winds are accounted for by inward-moving rings of air increasing in rotary motion under conservation of angular momentum, and speeds are consistent with movement toward the inner region of very low pressure, allowing for some frictional loss. The cause of the low pressure is inadequately understood.

Tornadoes in the United States are mostly found on the south sides of the parent thunderstorms. Heavy rain and hail often follow (but sometimes precede) passage of a tornado. The heaviest rain is likely to fall a few miles north of the tornado track; sometimes no rain falls along the track itself. Widespread thundersqualls are often observed outside the actual tornado path.

Over the past 20 years, property damage from tornadoes has averaged $75,000,000, with 113 deaths, half the previous 40-year average. Decreased fatalities (despite increased population density) are partly due to improved prediction and communications. Detection of severe thunderstorms by radar has been important. Doppler radar, which measures cloud circulations that may later develop into tornadoes, may come into widespread use.

[CHESTER W. NEWTON]

Bibliography: R. Davies-Jones and E. Kessler, Tornadoes, in W. N. Hess, (ed.), *Weather and Climate Modification*, 1974; S. D. Flora, *Tornadoes of the United States*, 1973; T. T. Fujita, Tornadoes around the world, *Weatherwise*, 26: 56–62, 78–83, 1973; G. Grigsby, *Tornado Watch*, 1977.

Transient geomagnetic variations

Transient changes of the geomagnetic field components by as much as a few percent from the quiet level due to processes in the ionosphere and magnetosphere. A month's average spectral display of the variations with periods less than several days' duration shows discrete peaks at 24, 12, 8, 6, and 3 hr, followed by a generally smooth decrease of amplitude with decreasing period T to 0.3 s (as a function of about T^{-1} to T^{-2}). The steeper changes are associated with the shorter periods, increasing disturbance level, and proximity of the observation to an auroral zone.

The 24-hr (diurnal) variations are on the order of 40 gammas (1 gamma $= 10^{-9}$ tesla) in magnetic north component H at middle latitudes, and about three times that value at the magnetic dip equator. These are the S_q variations observed clearly on the magnetograms during the quiet solar times. S_q's are produced principally by the electric currents in the E region of the ionosphere near 100-km altitude when winds there drive the charged particles through the Earth's main field (dynamo action). Because the ionization arises from direct solar radiation, the currents are mainly a daytime, subsolar, ionospheric phenomenon. A dependence upon the wind direction with respect to the local main field direction produces unique current enhancements near the magnetic dip equator. A fractional contribution to the diurnal component of the surface field arises as the Earth rotates under the dayside-to-night-side distortion of the magnetosphere caused by the solar wind. Spectral analyses of the S_q variations yield a large peak at 24 hr, with a much smaller, yet discernible, harmonic content to about the 3-hr component.

The semidiurnal lunar tidal oscillations of the atmosphere drag the E region through the Earth's main field, producing another dynamo current which is also dependent upon the day-side ionization and main field direction. Special analysis of this geomagnetic lunar variation L makes use of the fact that the lunar tidal "day" is about 50½ min longer than the solar day. L is almost a factor of 20 smaller than S_q field variations.

Temporary conductivity increases in ionization due to direct x-ray radiation from solar flares, or decreases in ionization due to solar eclipses, can modify the S_q and L dynamo currents. This variation appears as a simple half-cycle change which lasts from several minutes to about an hour, with the maximum amplitude rarely larger than 10 gammas.

Activity associated with geomagnetic storms causes rapid field variations throughout the spectra to periods as short as 0.3 s. Usually both the irregular pulsations, Pi, and the smoothly oscillating (continuous) pulsations, Pc, are observed on the magnetograms. The latter are recognized as having special periods of several minutes (Pc5, Pc4), of about 30–15 s (Pc3, Pc2), and of about 1 s (Pc1), for which amplitudes near 10, 0.3, and 0.05 gammas respectively are often reported. These oscillations arise as hydromagnetic waves whose periods and amplitudes are governed by the charged-particle population and main field configuration within the magnetosphere and the transmission of energy through the ionosphere. The amplitude variation with latitude is generally similar to that of geomagnetic storms. All these rapid variations except Pc1 are closely associated with auroral luminosity fluctuations. Averaged over a number of days, the spectrum from 3 hr to 0.3 s is generally without peaks because of the changes in Pc resonant frequencies and the dominance of Pi events. *See* ATMOSPHERIC TIDES; AURORA; GEOMAGNETIC STORM; GEOMAGNETISM; IONOSPHERE; MAGNETOSPHERE.

[WALLACE H. CAMPBELL]

Bibliography: S. Matsushita and W. H. Campbell, *Physics of Geomagnetic Phenomena*, 1967.

Tropical meteorology

The study of atmospheric structure and behavior in those parts of the world that lie astride the Equator—roughly between 25 or 30° north and south latitude. Interest centers mainly on the trade winds and other broad wind systems, on rainfall, on tropical storms, and on tropical regions as a major heat source for other parts of the atmosphere. Like all meteorology, tropical meteorology involves analysis of the disturbance patterns of the air near the ground and their relation to high-altitude features of the atmosphere. Such studies form a basis for increasingly useful predictions of precipitation, storms, and other weather conditions of concern in tropical zones. Directly or indirectly, many of these features are an influence on the weather, climate, and related human activities of extratropical areas.

Temperature. In the tropics, where there is no winter and the temperature is high at all times, only the regular diurnal and small-range seasonal cycles bring significant change. Relief from the monotonous heat is most readily obtained by going to the highlands, where air temperatures are cooler.

World heat source. The year-round directness of the Sun's rays produces temperature and heat flows of much greater than tropical significance. It is well known that the principal heat source for the atmosphere is situated in the tropics and that heat is exported by ocean and atmospheric currents from there to higher latitudes. Up to late 1940s this heat source was assumed to be steady—as if a well-controlled flame had been turned on in a physical experiment. The big changes in middle-latitude weather were ascribed to variations in the polar regions of the world. The increasing numbers of balloon observations since World War II show, however, that the heat export from the tropics is not constant but is subject to large fluctuations within weeks or months. This discovery suggests that weather on a weekly or monthly basis in higher latitudes may be affected by the fluctuations in the tropics. The computation of transfer of heat, energy, and momentum of the atmosphere from the tropical zone poleward on a short-term basis is a task of tropical meteorology, as is the assessment of importance of variations in this export upon weather patterns of the colder belts. *See* ATMOSPHERIC GENERAL CIRCULATION; TERRESTRIAL ATMOSPHERIC HEAT BALANCE.

Rainfall and water supply. Though rainfall is high in many parts of the tropics (70–80 in./year and up), abundant water supply is relatively rare. The high rate of evaporation caused by the heat greatly reduces the water supply. In addition, the distribution of rainfall over the year is uneven in many areas, especially the monsoon countries, where almost all of the annual rainfall is concentrated in one period of 2–4 months. Hence a delay of some weeks in the onset of the rainy season—a frequent occurrence—will bring serious problems, as will large variation from the normal precipitation of the rainy season—also common. *See* MONSOON.

For the most part, rainfall yielding copious amounts over wide areas occurs in disturbances of the prevailing trade or monsoon winds in the lower 5 km of the atmosphere. These disturbances have a diameter of about 1000 mi; they take the form of traveling waves, vortexes, or lines of sharp cyclonic shear. Their frequency and intensity vary greatly from place to place, and from year to year. Their relationships to high-altitude wind patterns are studied by means of balloons that measure pressure, temperature, and wind from 8 to 15 km levels above the ground. In addition, various meteorological approaches are used to estimate flood rains in mountain areas and thereby aid engineers to develop reservoirs for flood control, irrigation, and water budgeting. *See* PRECIPITATION.

[HERBERT RIEHL]

Tropical disturbances. Over most of the tropics heating and evaporation from the surface (80% oceanic) and radiative cooling at high levels maintain a conditionally unstable atmosphere. Hence convective clouds dominate, and are abundant even where the low-level convergence is weak. Weather resulting from organized disturbances is strongly modified by land-sea distribution, orography, and diurnal heating.

Intertropical convergence zone (ITCZ). A seasonally shifting zone of low pressure and variable winds, located between the northeast trade winds of the northern tropical areas and the southeast trades of the Southern Hemisphere, girdles the Earth in low latitudes. Around half the globe, the mean position of the zone is north of the Equator both winter and summer. Between eastern Africa and the central Pacific, it migrates between the summer hemispheres in association with the Asiatic-Australian monsoons. *See* METEOROLOGY.

Horizontal convergence with uplifting of moist air masses in this zone causes it to be one of the great rain belts, with frequent showers, thunderstorms, and squalls. The weather is highly variable, and at any one locale most of the seasonal rainfall is furnished by a few days of heavy rain.

Tropical lows. With concentrated bad weather on and near the ITCZ, shallow or weakly developed lows are generally found moving westward. These do not attain appreciable pressure falls while near the Equator, and only develop complete cyclonic circulations when removed poleward more than 5°. A minority of these develop into severe tropical hurricanes. *See* HURRICANE.

Easterly waves. These are north-south-oriented low-pressure troughs in the general easterly current of the trade winds. Most such waves appear near latitude 20° and, in Northern Hemisphere terms, almost entirely between late spring and fall. This is when the subtropical highs are farthest north and the easterly currents on their south sides are deep and extensive.

On approach of an easterly wave, the wind backs to a northeasterly direction; on passage of the trough the wind veers to southeast and then gradually returns to easterly. Good weather prevails west of the trough, with frequent moderate to heavy showers and thunderstorms on its east side. The waves, moving westward with speeds around 15 mph, pass at intervals of 3–4 days, and showery weather may last for 2 days after trough passage.

Severe thundersqualls. In some regions violent thundersqualls occur, similar in nature to those in middle latitudes. Notable among these are the nor'wester of northeast Indopakistan, occurring prior to onset of the summer monsoon (during which thunderstorms are more frequent but less severe); and the westward-moving "disturbance line" of central and western Africa, dominant in summer. Both occur in seasons when there is very moist air in low levels, dry air aloft, and strong winds in the upper troposphere. *See* SQUALL LINE.

[CHESTER W. NEWTON]

Bibliography: H. Riehl, *Tropical Meteorology*, 1954; World Meteorological Organization. *The Atlantic Tropical Experiment GATE*, 1974.

Tropopause

The boundary between the troposphere and the stratosphere in the atmosphere. The tropopause is broadly defined as the lowest level above which the lapse rate (decrease) of temperature with height becomes less than 2°C km^{-1}. In low latitudes the tropical tropopause is at a height of 15–17 km ($\sim$180°K), and the polar tropopause between tropics and poles is at about 10 km ($\sim$220°K). There is a well-marked "tropopause gap" or break where the tropical and polar tropopauses overlap at 30–40° latitude. The break is in the region of the subtropical jet stream and is of major importance for the transfer of air and tracers (humidity, ozone, radioactivity) between stratosphere and troposphere. Tropopause breaks also occur in the neighborhood of polar jet streams. The height of the tropopause varies seasonally and also daily with the weather systems, being higher and colder over anticyclones than over depressions. The detailed vertical temperature structure is often complex, showing multiple or laminated tropopauses, and it is often difficult to decide on the precise height of the tropopause, particularly in winter at high latitudes. *See* AIR TEMPERATURE; ATMOSPHERE; STRATOSPHERE; TROPOSPHERE.

[R. J. MURGATROYD]

Troposphere

The lowest major layer of the atmosphere. The troposphere extends from the Earth's surface to a height of 10–16 km, the base of the stratosphere. It contains about four-fifths of the mass of the whole atmosphere. *See* ATMOSPHERE.

On the average, the temperature decreases steadily with height throughout this layer, with a lapse rate of about 6.5°C km^{-1}, although shallow inversions (temperature increases with height) and greater lapse rates occur, particularly in the boundary layer near the Earth's surface. Appreciable water-vapor contents and clouds are almost entirely confined to the troposphere. Hence it is the seat of all important weather processes and the region where interchange by evaporation and precipitation (rain, snow, and so forth) of water substance between the surface and the atmosphere takes place. *See* ATMOSPHERIC GENERAL CIRCULATION; CLIMATOLOGY; CLOUD PHYSICS; METEOROLOGY; WEATHER. [R. J. MURGATROYD]

Tsunami

A radially spreading, long-period gravity-wave system caused by any large-scale impulsive sea-surface disturbance. Being only weakly dissipative in deep water, major tsunamis can produce anomalous destructive wave effects at transoceanic distances. Historically, they rank high on the scale of natural disasters, having been responsible for losses approaching 100,000 lives and uncounted damage to coastal structures and habitations. Because of their uncertainty of origin, infrequency (10 per century), and sporadicity, tsunami forecasting is impossible, but the progressive implementation, since 1946, of the effective International Tsunami Warning System has greatly reduced human casualties. Present efforts, aided by advances in the fields of geomorphology, seismicity, and hydrodynamics, are directed toward an improved understanding of tsunami source mechanisms and the quantitative aspects of transocean propagation and terminal uprush along remote coastlines. These efforts are abetted by ever-increasing coastal utilization for nuclear power plants, oil transfer facilities, and commercial ports, not to mention public recreation.

Tsunami generation. While minor tsunamis are occasionally produced by volcanic eruptions or submarine landslides, major events are now recognized to be generated by the sudden quasi-unitized dislocations of large fault blocks associated with the crumpling of slowly moving sea-floor crustal plates, where they abut normally against the continental plates. Such sources are predominately

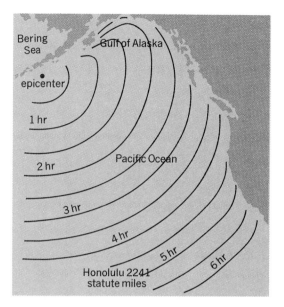

Fig. 1. Advance of tidal wave of Apr. 1, 1946, caused by an earthquake with epicenter southeast of Unimak Island. (*From L. D. Leet and S. Judson, Physical Geology, 2d ed., Prentice-Hall, © 1958; reprinted by permission of Prentice-Hall, Inc., Englewood Cliffs, NJ*)

confined to tectonically active ocean margins, the majority of which currently ring the Pacific from Chile to Japan. As opposed to secular creep, tsunami-producing block dislocations are invariably associated with major shallow-focus (< 30 km) earthquakes of intensity greater than 7 on the Gutenberg-Richter scale, followed by swarms of aftershocks of lesser intensity that decay in frequency over a week or two. Independent lines of evidence indicate that the aftershock perimeter defines the dislocated area, which is usually elongate, with its major axis parallel to the major fault trend. That only about 20% of large earthquakes fitting the above description produce major tsunamis raises the additional requirement that the dislocations have net vertical displacements, a view supported by seismic fault-plane analyses and confirmed by direct observations in the case of the 1964 Alaskan earthquake.

Because the horizontal block dimensions are characteristically hundreds of kilometers, any such vertical dislocation (a few meters) immediately and similarly deforms the water surface. The resulting tsunami disperses radially in all directions as a train of waves whose energy is concentrated at wavelengths corresponding to the block dimensions and whose initial heights are determined by the local extent of vertical dislocation.

Deep-sea propagation. Having principal wavelengths much longer than the greatest ocean depths, tsunamis are hydrodynamically categorized as shallow-water waves; to good approximation, their propagation speeds are proportional to the square root of the local water depth (600–800 km/hr in the Pacific). Thus, after determination of the source location by early seismic triangulation,

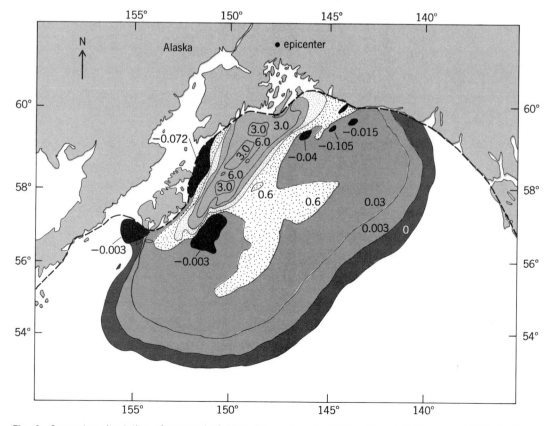

Fig. 2. Computer simulation of tsunami of Mar. 24, 1964, in Gulf of Alaska, 18 minutes after earthquake. Surface contours in centimeters; wave heights in meters; $t = 1100$ s. (*From L-S. Hwang and D. Divoky, Tsunamis, Underwater J., pp. 207–219, October 1971*)

real-time warnings of wave arrival times can be compiled from precalculated travel-time charts (Fig. 1).

Because the initial wave energy imparted by the source dislocation is spread ever thinner as the wave pattern expands across the ocean, the average wave height everywhere diminishes with travel distance, amounting to only a few centimeters (virtually undetectable in deep water) halfway around the globe. Beyond this point, energy converges again toward the antipole of the source, and wave heights increase significantly. This convergence accounts, in part, for the severity of coastal effects in Japan from Chilean tsunamis, and conversely. Additionally, azimuthal variations in local wave height are caused by source orientation and eccentricity because, as with a radio antenna, the energy is radiated more efficiently normal to the longer axis. Lastly, further variations of wave height arise from refractive effects associated with regional differences in average water depth. *See* OCEAN WAVES.

Coastal effects. After the waves cross the continental margins, the average wave height is greatly enhanced, partly by energy concentration in shallow water, and partly by strong refraction which, like a waveguide, tends to trap and further concentrate energy against the coastline. Ultimately, the shore arrests further progress. Here, the tsunami is characterized by swift currents in bays and harbors, by inundation of low-lying areas as recurrent breaking bores, and by uprush against steep cliffs, where watermarks as high as 20 m above sea level have been observed. *See* NEARSHORE PROCESSES.

Prediction methods. While most of the above behavior has been predicted qualitatively in theory, the increased accuracy required for engineering design and hazard evaluation has led to the development of numerical modeling on large computers. Taking as input a representative time- and space-dependent source dislocation, as inferred from seismic or observational evidence, and the oceanwide distribution of water depths from bathymetric charts, the computer generates the initial surface disturbance (Fig. 2), propagates it across the sea surface, and yields the time history of water motion at any desired location. Where the source motion is known (Alaska, 1964), computer simulations agree quite accurately with observations at places where the linearized equations of motion can be expected to apply. However, localized nonlinear effects, such as breaking bores, are more realistically modeled hydrodynamically by using the computer-generated wave field offshore as input.

[WILLIAM G. VAN DORN]

Bibliography: L-S. Hwang and D. Divoky, *Tsunamis, Underwater J.*, pp. 207–219, October 1971; National Academy of Sciences, *The great Alaskan earthquake of 1964, Oceanogr. Coastal Eng.*, 1972.

Turbidity current

A submarine flow of sediment or sediment-laden water which occurs when an unstable mass of sediment at the top of a relatively steep slope is jarred loose and slides downslope. As the slide or slump travels downslope, it becomes more fluid because of the loss of internal cohesive strength and because of inmixing of the superadjacent water.

Turbidity currents occur at the edge of the continental slope, in the vicinity of river mouths, and off prominent capes. They are triggered by earthquakes, hurricanes, floods, or simply by the bedload transport of rivers debouching at the edge of continental slopes.

Turbidity currents were first proposed as a hypothesis to account for the erosion of submarine and sublacustrine canyons. Supporting evidence was found in submarine telegraph cable breakage following the Grand Banks earthquake of 1929 and the Orleansville, Algeria, earthquake of 1954. In both cases submarine cables were broken consecutively in the order of increasing distance downslope. All of the cables were broken in at least two widely separated places. The sections between breaks were swept away or buried beneath sediment deposited by the current (Fig. 1).

Sediment cores taken in the suspected area of deposition revealed a recently deposited, graded bed of silt and sand on the abyssal plain. Turbidity currents deposit the heaviest grains first and the finest last, and thus their deposits are graded in size from coarse at the base to fine at the top. Graded sands containing shallow-water fossils are found in the beds of submarine canyons, at their mouths, and over the vast abyssal plains of the ocean floor. The turbidity current is now generally considered the agent responsible for the erosion of submarine canyons, the building of mid-ocean canyons, and the smoothing of abyssal plains.

The telegraph cable breaks which occurred after the Grand Banks and Orleansville earthquakes indicated by the time sequence of the breaks that the current attained velocities of at least 50 knots on a slope of 1:50, but slowed to 12 knots on slopes

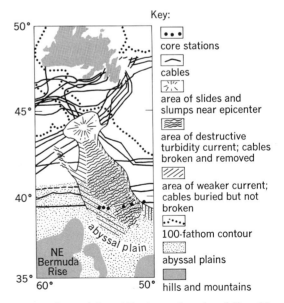

Fig. 1. Area of Grand Banks earthquake of Nov. 18, 1929. Sketch shows areas of cable breakage and cable burial, and relative position of sediment cores (from *Atlantis* cruise A180) and turbidity current.

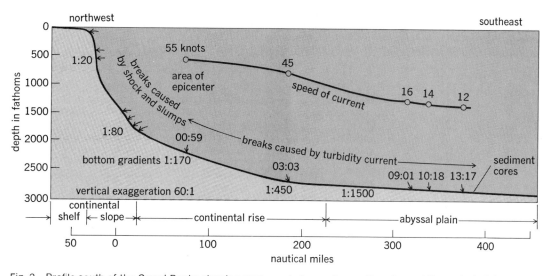

northwest / southeast

depth in fathoms

55 knots
1:20
breaks caused by shock and slumps
area of epicenter
45
speed of current
16 14 12
1:80
00:59
breaks caused by turbidity current
bottom gradients 1:170
03:03
09:01 10:18 13:17
sediment cores
vertical exaggeration 60:1
1:450
1:1500

continental shelf | slope | continental rise | abyssal plain

50 0 100 200 300 400

nautical miles

Fig. 2. Profile south of the Grand Banks showing progress of the turbidity current. Positions of 12 submarine cables are indicated by arrows. For each of the last five, the figure above the arrow indicates the time interval between the earthquake and the arrival of the current that broke the cable. Superimposed graph indicates calculated speed of the current as it passed.

of 1:1500 (Fig. 2). Turbidity current breakage of submarine cables off the mouths of the Congo River in Africa and Magdalena River in South America also indicates that turbidity currents occur off major rivers, generally during times of flood. This implies that the high bed-load transport during flood stage can generate turbidity currents.

The lowered sea level of the glacial stages must have forced most rivers of the world to empty at or near the top of the continental slope. Under those conditions turbidity currents must have been much more important than at present, and they undoubtedly transported much of the sediment deposited in ancient geosynclines. *See* CONTINENTAL SHELF AND SLOPE; MARINE GEOLOGY; MARINE SEDIMENTS; SUBMARINE CANYON.

[BRUCE C. HEEZEN]

Bibliography: H. Charnock, Turbidity currents, *Nature*, 183(4662):657–659, 1959; R. A. Daly, Origin of submarine canyons, *Amer. J. Sci.*, ser. 5, 31: 401–420, 1936; B. C. Heezen, The origin of submarine canyons, *Sci. Amer.*, 195(2):36–41, 1956; B. C. Heezen and M. Ewing, Orleansville earthquake and turbidity currents, *Bull. Amer. Ass. Petrol. Geol.*, 39(12):2505–2514, 1955; B. C. Heezen and M. Ewing, Turbidity currents and submarine slumps and the 1929 Grand Banks earthquake, *Amer. J. Sci.*, 250:849–873, 1952; P. H. Kuenen, Estimated size of the Grand Banks turbidity current, *Amer. J. Sci.*, 250:874–884, 1952; P. H. Kuenen and C. I. Migliorini, Turbidity currents as a cause of graded bedding, *J. Geol.*, 58:91–126, 1950; F. P. Shepard, *Submarine Geology*, 2d ed., 1963.

Underwater photography

The technique of using photography equipment underwater. Photography is as useful a recording system underwater as it is in air. All forms of photography are in use: single pictures, motion pictures, stereo, shadow, elapsed-time motion pictures, and so forth.

Shallow underwater photography. In shallow underwater photography, or hand-held underwater photography as it is more commonly called, the photographer, wearing some type of diving equipment, is submerged with his camera and takes photographs directly (Figs. 1 and 2). This type of underwater photography did not become a technical or scientific tool until the development of self-contained underwater breathing apparatus, or scuba. *See* DIVING.

Underwater photography is controlled by many different factors, some of which are available light, clarity of water, differential absorption of light in various wavelengths, film capability, and construction of underwater housing for cameras.

Fig. 1. Photographer using underwater motion-picture camera to record data on the bottom. This picture was taken at a depth of 120 ft. High-speed Tri-X film was used. Camera settings: 1/100 sec and *f*/4.0.

Fig. 2. Recording the action of a pressure regulator by a hand-held underwater still camera (Rolli-marine). Direct observation and photographic recording have become extremely important tools underwater.

The underwater photographer can never expect to take clear pictures at distances greater than 50 m. In most instances, he must depend on natural light for the exposure of his film. Flashbulbs and electronic flash units are often used underwater as supplemental light sources. However, care must be taken when using artificially supplied light or natural light at depths greater than a few feet if true color balance is desired, because of the differential absorption of light by water. Red light energy and other warm light is absorbed at a much higher rate than the colder light waves. Light in the blue wavelengths penetrates most deeply into the sea. Infrared and ultraviolet wavelengths are rapidly attenuated as they travel through the water medium and are not the good sources of penetrating light energy that they are in air (Fig. 3). Filters are useful, but because of the generally low light level and actual lack of light in certain wave-

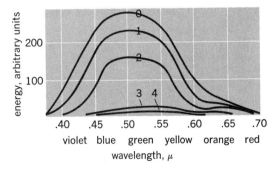

Fig. 3. Energy spectra at a depth of 10 m in different types of water. The curves marked 0, 1, 2, 3, and 4 represent energy spectra in pure, clear oceanic, average oceanic, average coastal, and turbid coastal sea water, respectively. (*From H. U. Sverdrup et al., The Oceans, Prentice-Hall, 1942*)

lengths of the light energy spectrum underwater, they have the tendency to do more harm than good.

The transparency of the water is most often the limiting factor. It is dependent on the organisms and sediment content of the water and is quite variable. Contrary to popular belief, it cannot be improved by artificial light. The small particles in the water scatter the light energy, and therefore more light in turbid water results in poorer quality pictures.

The water with the highest visibility is located away from the continental land masses in deep oceanic water or in areas where very little sediment is added to the sea water, such as off rocky or arid coasts. The highest visibility to be expected in clear oceanic waters is about 100 m. Occasionally, these clear waters lap up onto the margins of the continents and thus temporarily improve water visibility in coastal areas. However, in most instances, water visibility is less than 30 m. The average underwater visibility off southern California is about 5 m. For a discussion of water color and transparency *see* SEA WATER.

The short distances at which underwater photographs must ordinarily be taken necessitates the use of an extremely wide lens. Wide-angle lenses are also required because of the magnification (approximately 25%) of objects underwater, owing to the difference between the index of refraction of the water and air in the underwater camera housing. Special lenses have been developed to correct for this magnification.

Underwater photography equipment is available for work at almost any depth. The photographer, not the equipment, is usually the limiting factor. He can remain underwater only as long as it is safe to do so and must continually be aware of his physiological limitations. In general, if he is experienced and properly trained, a diver can safely descend to 50 m, but even so, he can remain there only a relatively short time to find a subject, set up lights if needed, and take pictures. Photographs in deeper water have been taken. Both French and United States photographic teams have taken pictures of the sunken liner *Andrea Doria* at depths as great as 77 m. This is about the maximum depth limit for hand-held underwater photography.

[ROBERT F. DILL]

Deep-sea underwater photography. Photography in the deep sea involves the design and use of camera and lighting equipment for which several requirements must be fulfilled, including the following.

1. Suitable watertight cases are required, for both the camera and the light source, to withstand the pressure of the sea. For each 10 m. of depth, approximately one additional atmosphere of pressure is exerted. The deepest ocean, about 12,000 m, requires a case to withstand 1200 kg/cm² (17,600 lb/in.²). The windows for the lens and the electrical seals must likewise be designed for such pressures.

2. Auxiliary lighting is required, especially for deep photography, since daylight is rapidly absorbed with depth, becoming almost useless even at shallow depths of a few hundred meters.

Fig. 4. Benthos-Edgerton (North Falmouth, MA) camera (left) and 200w.s. strobe (right) mounted on a diving submersible. Two observers in the plastic sphere can operate the camera when a scene of interest is in view. (*Courtesy of Harbor Branch*)

3. Corrected optics for the camera are desirable to compensate for distortions caused by the lens window – water interface when an air lens is used. However, many underwater cameras have lenses designed for use in the air even if some information is lost.

4. The camera must be positioned and triggered to get the desired photograph. Operation from a cable (with sonar sensing equipment) or from deep-diving underwater vehicles is employed (Fig. 4). Bottom-sensing switches are often used to operate the deep cameras when bottom photographs are desired.

Camera and lamp housings. Cases of cylindrical shape are the most common since they are stronger than boxlike cases and can be made from commercial pipe and cylindrical materials. A pipe with an inner diameter of 10 cm is especially useful since it can accept the standard 100-ft (31-m) motion picture spool of 35-mm film, which holds 800 photographs. Thin-base cronar film permits more exposures. Larger cameras are available that use 400 ft (122 m) of film and permit 3200 photographs to be taken on one lowering. This may be important for an underwater photographic survey, since much time is consumed in the raising and lowering of the camera gear in the sea.

Fig. 5. A 16-mm elapsed-time camera (bottom), used in study of sedimentation, and a strobe lamp (above) ready for installation by a diver. The float holds the equipment off the bottom until the site is reached.

Fig. 6. Stereo camera, designed to operate at maximum depth of sea. An electronic flash lamp is at the bottom and a compass with current-operated vane hangs from below, in the camera's view. A sonar pinger indicates the height of the camera above the ocean floor.

Fig. 7. Deep-sea photograph with digital data. The data show the time and other pertinent information at the moment the photo was taken. (*Courtesy of Robert Ballard, Woods Hole Oceanographic Institution*)

Light sources. The electronic flash lamp (xenon) has become almost universally used for deep-sea photography since it is efficient and since many photographs can be taken with it. A small battery has the ability to expose an entire roll of 35-mm film (800 photographs). The color temperature of the xenon flash is suitable for daylight color films if the camera and lamp are close to the subject so that the selective absorption of the light does not change the color balance.

Camera positioning. One of the most difficult tasks of underwater photography is positioning the camera and triggering it.

Cameras to photograph the ocean bottom are of several types: (1) the type which triggers a flash lamp when a weight suspended below the camera touches the bottom; (2) the free-floating camera which sinks to the bottom, exposes a picture, and then rises to the surface when a weight is released; (3) the type which takes many photographs at regular intervals (Fig. 5), such as every 10 sec, and depends upon positioning with sonar or other methods of adjustment; (4) the camera on a submersible which can be triggered manually by an operator in the sub: (5) the television camera, adjacent to the photographic camera, which can present the image to the surface, where a viewer can then expose the camera at the right moment, using the TV screen

as a remote viewfinder; also important is the use of a controlled ship, capable of excellent navigation, such as the *ALCOA Seaprobe*, to lower the camera/strobe/TV on a drill pipe.

The use of an active sonar transmitter (pinger; Fig. 6) of short pulse length and accurate timing mounted on the camera has proved to be a most useful method of remotely controlling a deep-sea camera from the surface without the need of electrical wires. The person controlling the operation observes the time presentation of the pulses from the active sonar. He sees first the "direct" arrival from the sonar and then the reflected arrival of the bottom-directed sonar output after reflection from the sea bottom. The distance between the camera and the ocean bottom is proportional to the time interval between these two pulses, and can be calculated since the velocity of sound in water is about 5000 ft/s (1500 m/s).

Oceanographers using corers, dredges, and numerous other cable-supported devices were quick to employ the sonar ping system of instru-

Fig. 8. An unusual picture of a bathypteroid fish photographed from the Westinghouse diving vehicle *Deepstar* at 4000 ft (1220 m) depth in the Gulf of Mexico. Tungsten lighting was used. (*Photograph by R. Church*)

Fig. 9.   Photograph taken (August 1956) in Romanche Trench, 0°10′S, 18°21′W, at depth of 25,000 ft (7625 m). A nylon cable was used from the deck of the *Calypso*. (*National Geographic Magazine*, March 1958)

ment observation, after it had been developed for use with cameras.

Many deep-sea cameras have an internal data chamber, which is photographed at the same time the deep-sea exposure is made. The most modern deep-sea cameras use photographic digital data on the film (Fig. 7). The 16 digits shown can record such information as the exact second, minute, and hour that the photo was taken as well as the compass orientation of the photo and the camera-to-subject distance. The camera-to-subject distance is fed to the camera from a sonar mounted on the camera. Because the camera-to-subject distance is shown on each photo, it is possible to calculate the field of view for each photo. Thus the size of objects in the photos can be measured. Thus the data and the film are inseparable, and chances of error are thereby greatly reduced.

The steel cable used to lower a camera to great depths has several disadvantages. First, the weight of the cable is considerable; second, the winch operator cannot tell when the camera touches the bottom, since the cable is so heavy; and third, if the camera does rest on the bottom, the cable may kink, and a break may occur when it is pulled up again or as it goes over the pulley. Many cameras and much other oceanographic equipment have been lost in this manner.

A nylon cable has some advantages for lowering cameras to great depths since the material is almost weightless in water and the winch operator can therefore tell when the camera touches the bottom. However, the winch problem is serious owing to the elastic properties of the rope. The nylon rope, after being pulled up, should be allowed to shrink before being wound on the take-up drum. Other materials, such as Kelvar, show promise due to strength and elastic properties.

When a person descends as an observer to great depths in a diving vehicle, a camera can help him remember and record what he saw. Furthermore, he can then explain to others what he saw in a visual form. Interesting examples of underwater sights are the bathypteroid fish (Fig. 8) photographed at 4000 ft (1220 m) in the Gulf of Mexico,

and a view of the Romanche Trench (Fig. 9).

A television link from the bottom to the surface is often mentioned as a substitute for underwater photography since the picture is seen immediately and the television screen can be photographed or taped at the surface. However, the quality of the television photograph or tape is not as high as that taken by a camera below because of the limitations of the television link and the lower resolution intrinsic in TV. An ideal approach is to use the television system as a remote viewfinder to aid the operator so that he can expose the photograph at the best instant. A camera-supporting cable with electrical wires to the surface and a special winch to terminate the wires are required. These details complicate the application of underwater television. *See* UNDERWATER TELEVISION.

Elapsed-time photography with a motion-picture camera in the sea is of importance in the study of sedimentation deposits caused by the tides, currents, and storms. Likewise, the speeded-up observation of biological activity reveals processes not ordinarily observed.

[HAROLD E. EDGERTON; SAMUEL O. RAYMOND]

Bibliography: H. E. Edgerton, *Electronic Flash Strobe*, 1979; H. E. Edgerton and L. D. Hoadley, Cameras and lights for underwater use, *J. Soc. Motion Pict. Telev. Eng.*, 64:345–350, 1955; J. B. Hersey (ed.), *Deep Sea Photography*, 1968; B. Kendall, *Photographs Underwater*, 1976; D. Rebikoff and P. Cherney, *Underwater Photography*, 2d ed., 1975; E. M. Thorndike, Deep-sea cameras of the Lamont Observatory, *Deep-Sea Res.*, 5:234–237, 1959.

Underwater sound

The production, transmission, and reception of sounds in the ocean are an important part of modern acoustics. Since the ocean is highly absorptive to light and other electromagnetic radiation but not to sound waves, acoustics has found many applications in probing the sea. Acoustical methods of detection, for example, are useful in locating submarines and other submerged objects, and therefore have many applications in naval science. Underwater sound is also important in scientific research: Oceanographers and geophysicists employ sound waves to determine the physical structure of the ocean and its bottom; sound is also used by marine biologists to study life in the ocean. Then, too, there are many present and potential commercial uses of underwater sound, such as seismic exploration for oil and gas beneath the ocean floor, use of sonar techniques for fish location, navigation and positioning, and communication. *See* SEA WATER; SONAR.

Controlled sound by humans. To obtain information from the sea, humans must sometimes put sound into it, maintaining control over the inception, duration, frequency, intensity, and directivity of the signal. In general, devices which transform one type of energy into another are known as transducers; when the end product is underwater sound, such devices are known as sound projectors. Most projectors use either the magnetostrictive or piezoelectric effect in which electrical energy is transformed into acoustical energy. The piezoelectric ceramics, barium titanate and lead zirconate titanate, are widely used because they can easily be molded to any desired shape. However, by no means are all underwater projectors of these types. For example, oceanographers make extensive use of explosives to study oceanographic features. Other projectors are based on spark discharge, hydraulic flow, mechanical impact, and pneumatic and other effects. High-powered, low-frequency projectors may be several feet in diameter and weigh several tons. To achieve high directivity, a number of these units may be mounted in an array. Improved directivity of small-dimension monopole sources at low frequency can also be obtained with the so-called parametric array. In this concept the source generates two high-frequency acoustic beams of nearly the same frequency which are coaxially superimposed. Nonlinear interaction in the mixing region of the medium produces a dominant difference frequency (plus additional products of nonlinearity) which effectively constitutes an end-fire source with high directivity at the difference frequency. (The same principle is used to achieve directivity at low frequency of small-size monopole receivers.)

The maximum acoustic power that can be put into the water is usually limited by the medium itself and not by the physical equipment. When the acoustic pressure reaches a certain level, called the cavitation threshold, the negative pressure cycle of the wave literally tears the liquid apart, and voids are formed which interfere with sound production and transmission. The inception of cavitation can be retarded by raising the frequency, decreasing the pulse duration, or raising the hydrostatic pressure (increasing the depth).

Ship-radiated noise. The noise radiated by a body moving through the water is an important source of underwater sound, particularly to naval scientists and operating forces. Although only an infinitesimal portion of the total energy developed by the vessel is radiated as sound energy, this may, however, be enough to cause serious concern under combat conditions. Ship noise propagated to a distance (radiated noise) might betray the presence of the vessel to hostile listening devices, while sound fields near the detecting apparatus (self-noise) interfere with the listening capability of the ship's sonar. The sources of both radiated noise and self-noise are the propulsion and auxiliary machinery, the propellers, and hydrodynamic effects. Machinery noise has a spectrum composed largely of a series of line components (single frequencies) that are related to various machinery characteristics (vane frequencies and the like). Quieting the machinery and increasing the vibration isolation between machinery and hull help to reduce this effect. Propeller noise is caused mainly by cavitation, which takes place when bubbles form on the low-pressure side of the blade. The collapse of the bubbles gives a continuous spectrum (broadband) of noise which, when it occurs, dominates the higher-frequency region of ship noise. Hydrodynamic (or flow) noise originates in the irregularly fluctuating flow of water past the vessel. The fluctuating pressure radiates sound directly but, more significantly, it excites portions of the hull into vibration which then contributes to the radiated noise.

Radiated noise consists mainly of machinery and propeller noise. The relative importance depends upon frequency, speed, and depth but, in general, the low-frequency spectrum consists largely of machinery noise, and the higher-frequency spectrum of propeller noise. The flow effects become an important component of self-noise at higher frequencies and speeds.

Ambient noise. When background noise has no clearly identifiable source, it is referred to as ambient sea noise. It is caused by marine life, seismic disturbances, distant shipping, wave action, cracking ice, rain on the surface, and so on. Ambient noise has been studied over a frequency range from about 0.1 hertz to over 100 kilohertz. It was found that surface-wave-generated and sporadic earthquakelike noises are mainly responsible for ambient noise below about 10 Hz, distant shipping traffic is an important contributor between 1 Hz and several hundred hertz, and wind-induced cavitation is thought to be the predominant influence between 30 Hz and 50 kHz. The thermal noise of the molecules sets the limit on listening capability above 50 kHz.

An important source of ambient noise is that due to marine animals. The "silent" deep frequently does not live up to its reputation, and there are grunts, growls, yelps, chirps, whistles, and so on to shatter the calm and confuse the sonar technician. A common soniferous creature is the snapping shrimp. These small crustaceans snap their claws together and, when acting in concert, produce a "frying" noise which interferes seriously with sonar listening capability. Various fish also contribute to the noise level. The toadfish, for example, produces its gruntlike noise by exciting muscles attached to its swim bladder. Great popular and professional interest has been centered on the noise made by marine mammals such as whales, dolphins, and porpoises. They produce a well-defined signal with which they locate food and communicate with one another. The porpoise, in particular, seems to have a remarkably efficient sonar system.

Speed of sound. The speed of sound in the ocean varies roughly from 4750 to 5100 feet per second (1448 to 1554 meters per second; about 3% higher than in fresh water), depending on pressure, temperature, and salinity. (The scalar quantity "speed" is preferred over the vector "velocity" in this context and has come into general use in the scientific literature.) The speed of sound in the ocean is about 4755 fps (1449 mps) for "standard" conditions of 1 atmosphere of pressure (101.325 kilopascals), a temperature of 32°F (0°C), and a salt content of 35 parts per thousand by weight. Pre-

Variation of sound speed with temperature, salinity, and depth (pressure)

Parameter	Coefficient*
Temperature, °C (ambient about 20°C)	$\dfrac{\Delta c}{\Delta T} = 2.7$ mps/°C
Salinity, parts per thousand (ppt)	$\dfrac{\Delta c}{\Delta S} = 1.2$ mps/ppt
Depth, m	$\dfrac{\Delta c}{\Delta D} = 0.017$ mps/m

*c = speed of sound in meters per second (mps).

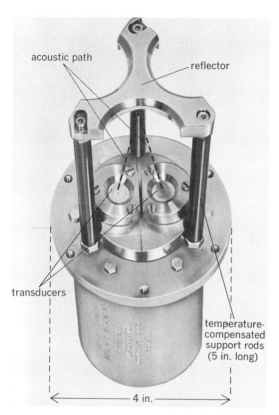

Fig. 1. A typical commercial velocimeter with protecting cage removed; 1 in.= 25.4 mm. (*NUS Corp.*)

curs during the alternate compression and rarefaction accompanying the sound wave. Absorption is greater in sea water than in fresh water over a certain frequency range because a dissociation-reassociation process occurs with the ions of magnesium sulfate, $MgSO_4$. Although $MgSO_4$ is only 4.7% by weight of the dissolved salts in sea water, it is the most important cause of the excess absorption in the sonar frequency range. Investigations have shown that there is a low-frequency relaxation in the 1-kHz region which is believed to be due to a very small amount of boric acid in sea water. The separate contributions to the absorption by water, magnesium sulfate, and boric acid are shown in Fig. 2, which gives the attenuation of a plane wave as a function of frequency in decibels per kilometer. Absorption depends in a complicated fashion on temperature and pressure.

Transmission loss. In addition to the irreversible losses in the medium, a reduction in signal intensity due to the spreading of the wave also occurs. If there is a simple, omnidirectional source in a homogeneous, unbound medium the power generated by the source is radiated in all directions so as to be equally distributed over the surface of a sphere. This is called inverse-square spreading because the intensity varies as the reciprocal of the range squared. When the medium has plane-parallel upper and lower boundaries, the spreading is cylindrical and varies as the reciprocal of the range. The latter type of spreading, with its lower attenuation, occurs when sound is trapped in a

cise experimental measurements of speed have been carried out, but there still exist small variations (less than 0.1%) among the various measurement techniques. Extreme precision is important in order to trace sound rays with satisfactory accuracy and to use instruments, such as the echo sounder or fathometer, which relate travel time to distance. *See* ECHO SOUNDER.

The relationship between sound speed and the parameters affecting it is very complicated, and several empirical and theoretical versions exist. The approximate effects of the various parameters are summarized in the table.

In the field, sound speed is usually determined with a bathythermograph, which measures water temperature as a function of depth. Speed can be computed with an appropriate equation from the pressure and temperature and from the independently determined salinity. A later instrument, called the velocimeter, directly measures the speed of sound in the field. A photograph of one of several commercial forms of this instrument is shown in Fig. 1. It consists of an electroacoustic system in which an acoustic pulse is sent over a fixed-distance path length to a receiver, which immediately initiates another acoustic pulse and so on. The instrument measures how many pulses occur in a unit time, from which the speed can be calculated, and transmits this number up a cable to the monitoring ship. This instrument is one major advance in underwater sound technology.

Absorption of sound. Sound is absorbed in sea water when an irreversible production of heat oc-

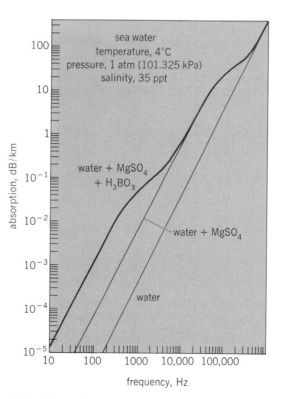

Fig. 2. Sound absorption in sea water as a function of frequency showing contributions to absorption of various components of the sea water. (*From F. H. Fisher and V. P. Simmons, Sound absorption in sea water, J. Acoust. Soc. Amer., 62:558–564, 1977*)

"channel." This is discussed in the following section.

Refraction of sound. Because of variations of sound speed with depth, sound rays are bent as they propagate from point to point. This effect becomes very important when horizontal propagation distances exceed a few hundred feet. The profile of speed versus depth depends on the geographical location, the season of the year, and even the time of day. An idealized profile is shown in Fig. 3. Just below the surface is the surface layer in which sound speed is susceptible to daily and local changes due to heating, cooling, and wind action. Below the surface layer lies the seasonal thermocline, where the temperature decreases with depth, giving a negative gradient. This region is close enough to the surface to be affected by seasonal changes. Below this is the main thermocline, which is little affected by the season, where the speed decreases rapidly with depth. Between 3000 and 4000 ft or 0.9 and 1.2 km (at mid-latitudes) a point is reached at which the temperature attains a constant value and the speed a minimum; from there to the bottom the speed increases because of the pressure effect. The region below the axis is called the deep isothermal layer. In the Arctic regions this layer may lie at or near the ice-covered surface. *See* THERMOCLINE.

In order to trace the path of a ray, the speed profile is divided into layers of constant speed and paths are then computed by Snell's law for rays leaving the source at different angles. In earlier work various types of analog computers were used, but digital computers are now favored. Four examples of refraction effects of particular interest are given below.

Mixed-layer sound channel. In cloudy, windy ocean areas of the world, a layer of isothermal water is formed just below the sea surface by turbulent wind mixing. Since the temperature is constant, the speed has a slight positive gradient because speed increases with depth. Sound rays emitted in this layer, except for those directed steeply downward, are refracted upward, as shown in the upper part of Fig. 4a, bounce off the surface, and describe a semicircle. The wave travels a considerable distance in curved arcs.

Shadow zones. When sound enters or originates in a region with a negative sound speed gradient the situation is that shown in the lower part of Fig. 4a, where a shadow zone is produced in which there is low acoustic intensity. The importance of shadow zones to submarines trying to avoid detection is obvious.

Deep sound channel. The depth at which the speed reaches a minimum defines the axis of the deep sound (or sofar) channel. A sound wave generated in this channel remains trapped in it and travels (by cylindrical spreading) for long distances (Fig. 4b). An explosion in this channel will be heard as a drawn-out signal that ceases abruptly with the passing of the slowest ray, the one which traveled directly along the axis. From this clearly defined time marker the exact location of a deep channel explosion can be determined by triangulating with two monitoring stations. This method was used during World War II by downed aviators, who dropped sofar bombs to give their location. It is also used in determining missile impact locations. (An interesting experiment was carried out in 1960 in which explosive charges detonated near Australia traveled through the "sound channel" to be detected in Bermuda, 13,000 mi or 21,000 km away, in 3¾ hr.) *See* SOFAR.

Convergence zones. When a sound ray penetrates the deep isothermal layer, it is eventually (if it does not strike the bottom first) refracted upward and strikes the surface, where it is reflected and follows similar paths to a second convergent zone, and then to subsequent zones (Fig. 4c). At intervals of 30–35 mi (48–56 km), narrow zones of high acoustic intensity occur near the surface. Shallow receivers in these zones measure signals 10–15 dB greater than those that would be caused by spherical spreading. The widths of the zones are 5–10% of the range.

Reflection of sound. Reflection occurs when sound strikes the sea surface, the sea bottom, or large obstacles in the water. If the reflecting surface is flat over distances that are large compared with the wavelength of sound, a fraction of the incident sound energy is reflected at an angle with respect to the normal which is equal to the angle of incidence. The amount of energy reflected depends on the acoustic properties of the materials involved. In the case of two fluids the fraction of incident energy reflected is given by Eq. (1), where

$$r = \left[\frac{\rho_1 c_1 - \rho_0 c_0 B}{\rho_1 c_1 + \rho_0 c_0 B}\right]^2 \tag{1}$$

the ray is incident from the medium with sound speed c_1 and density ρ_1 upon a medium having corresponding values c_0 and ρ_0. The product ρc is called the specific acoustic impedance, and B is

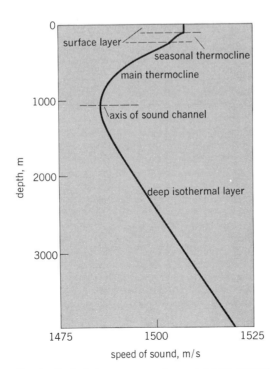

Fig. 3.   Typical deep-sea sound speed profile divided into layers. (*After R. J. Urick, Principles of Underwater Sound for Engineers, McGraw-Hill, 1967*)

given by Eq. (2), where θ is the angle of incidence (measured to the perpendicular).

$$B = [1 - (1 - c_1^2/c_0^2) \tan^2\theta]^{1/2} \qquad (2)$$

Surface reflection. The reflection from the surface of the ocean is essentially perfect because $\rho_w c_w \approx 3310\, \rho_a c_a$, where subscripts w and a refer to water and air, respectively. However, if the wavelength is small compared to the dimensions of the surface roughness, a well-defined reflected ray does not exist and the energy is scattered. A criterion (due to Lord Rayleigh) for determining conditions at which the surface becomes effectively smooth is that $\lambda > 8h \sin \theta$, where h is the distance between the top and bottom of the wave and θ is the grazing angle.

Bottom reflection. The reflection of sound from the sea bottom is similar to that from the surface but more complicated. The composition of the bottom ranges from soft mud through sand to hard rock at different locations, causing considerable variation in sound-reflecting and sound-scattering properties. Since underwater technology makes extensive use of "bottom bounce" sound paths, charts have been prepared to predict the reflective properties at a given location. Also, the ocean floor usually contains strata of material of different composition, and acoustic reflections occur at the boundaries. By using an intense source, usually an explosive, and studying the complicated echo pattern, the geophysicist can get an accurate map of the subsurface. *See* MARINE SEDIMENTS.

Shallow-water propagation. When the surface and bottom of a body of water are close together (usually less than 100 fathoms or 183 m), as in coastal waters, the sound is trapped and the result is a difficult problem in sound propagation. The theoretical treatment of shallow-water propagation is very complex and has received extensive treatment in the literature.

Reflection from large objects. In addition to reflection from the surface and the bottom, the study of underwater sound is concerned with echoes from man-made objects such as submarines, surface vessels, torpedoes, and mines and from marine animals.

It is common practice in the field of underwater sound to refer to the reflecting properties of an object as its target strength (TS), which is expressed in decibels and is defined by Eq. (3), where

$$\text{TS} = 10 \log_{10} \frac{I_0}{I_i} \qquad (3)$$

I_0 is the intensity of the reflected wave at a standard distance (usually 1 yd or 1 m) from the "acoustical center" of the reflecting object in a selected direction, and I_i is the intensity of the incident sound on the target. The value of the strength will depend on the unit of distance used, so, therefore, consistent units must be used in the "sonar equation." The target strength of shapes such as spheres, cylinders, plates, and ellipsoids can be calculated exactly if the wavelength is sufficiently small compared with the size of the object, but complicated shapes, such as that of a submarine, must be measured experimentally. The assumption that these objects are perfect reflectors is frequently not justified, and the interaction of the

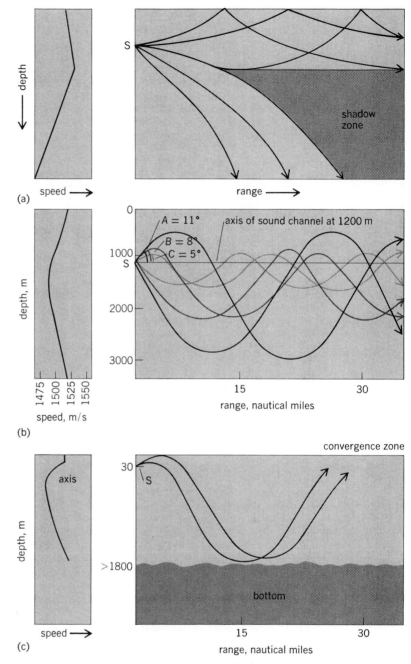

Fig. 4. Ray paths for typical sound speed profiles and different source depths (S). (*a*) Propagation in mixed-layer surface channel with shadow zone below; (*b*) propagation in deep sound channel; rays represent different initial angles of inclination to axis (*from R. J. Urick Principles of Underwater Sound for Engineers, McGraw-Hill, 1967*). (*c*) Formation of a convergence zone by a shallow water source (*after F. E. Hale, Long-range sound propagation in the deep ocean, J. Acoust. Soc. Amer., 33:956, 1961*).

wave with the object must be considered. The target strengths of different kinds of fish have been measured by a number of workers who have shown that the target strength is largely that of the swim bladder. The characteristic acoustic impedance of most of the body is close to that of water, so that little energy is reflected except from the air-filled bladder.

Scattering of sound. Scattering is the process by which sound energy is redirected from objects (or parts of objects) which have dimensions considerably smaller than a wavelength. The scattering from the surface and bottom has already been mentioned, but scattering from objects suspended in the water is also important. Contributions to such scattering are made by marine organisms, free air bubbles, suspended particles, the thermal microstructure in the sea, and the turbulences of natural or man-made origin. An important source of scattering is the so-called deep scattering layer. This is a concentration of marine animals, varying in size from a few millimeters in length to about 1 ft (30 cm), which exist in layers varying from the surface to about 0.6 mi (1 km) in depth, depending on geographical location, season, and the time of day. *See* SCATTERING LAYER.

Reception of sound. To obtain the information sought, the acoustic signal must be picked up and transformed into an electrical signal by a receiving transducer called a hydrophone. Current practice favors the use of piezoelectric ceramics for these instruments, but work on piezoelectric polymers and optical techniques looks promising. Transducer elements are usually arranged in arrays to achieve greater sensitivity, better directionality, and reduction of noise. Once the acoustical signal has been transformed into an electrical signal, the problem of extracting the information from the background noise becomes a problem in signal processing. The basic mathematical tools and electronic circuitry are the same as those used in radar, radio astronomy, geophysics, optics, and so forth. In spite of its general applications, signal processing is considered an essential part of the science of underwater acoustics. Using sophisticated electronic equipment—particularly miniaturized electronic circuitry and digital processing—and mathematical techniques, it is possible to detect the signal of interest amidst a background of noise. This signal may be an echo resulting from a transmitted signal over which one has some control (active sonar), or it may have its origin in some external source (passive sonar). In addition to detection, information is sought on target range, bearing, motion, and identification. *See* SONOBUOY.   [WILLIAM S. CRAMER]

Bibliography: V. M. Albers (ed.), *Underwater Sound*, 1972; R. J. Bobber, *Underwater Electroacoustic Measurements*, 1970; C. S. Clay and H. Medwin, *Acoustical Oceanography: Principles and Applications*, 1977; M. C. Junger and D. Feit, *Sound Structures and Their Interaction*, 1972; D. Ross, *Mechanics of Underwater Noise*, 1977; R. J. Urick, *Principles of Underwater Sound*, 2d ed., 1975.

Underwater television

Underwater television picture quality is limited by clarity of the water, available light, and the sensitivity of the camera tube. Television color cameras require white artificial light at depth to achieve the color image. Monochrome cameras can be used with a light source color-matched to the peak sensitivity of the camera tube. For example, red light is suitable for a silicon diode tube and green for the RCA 4804 silicon intensifier target (SIT) tube. A green-sensitive SIT tube is particularly suited to use with the natural light underwater. It gives a dynamic performance through a wide range of light levels. The SIT tube is sensitive to low light levels with negligible smear of moving objects, has good definition, and gives an enhanced-contrast image from the low-contrast underwater subject.

Television equipment used in underwater observation is compact and reliable and can be tailored for specific applications. Sensitive camera tubes can be protected from exposure to accidental high light levels. Underwater housings can be made to withstand any depth, and signals can be transmitted along any length of cable by appropriate circuitry. The major advantages of underwater television over conventional photography are that the observer can remain at the surface while viewing a continuously monitored scene in real time. Observations can be made and recorded for long periods on video tape either in real time or by using time-lapse video recorders. Additional information such as numerical data or a time clock can be superimposed on the picture screen, and other information can be recorded on the video tape in the form of sound tracks. Free-swimming divers can operate a television camera in much the same way as a movie camera. The portable television cameras do not yet equal the geometry and resolution standard of 16-mm film, but television gives usable images at much lower light levels. Television can be used as a monitor for taking flash photographs with underwater photographic cameras. Portable systems are available that allow the use of television recording equipment and cameras at isolated sites or in small boats. As with movie film, techniques such as time lapse, single-frame selection, and editing can be used. Stop-motion can be improved if the television camera is used with a synchronized strobed light source or with a mechanical rotating shutter system driven by a synchronized motor.

Underwater television has a growing number of users. Fishery researchers use it to study behavior of fish in the wild, the performance of fishing gear, and the interaction between fish and gear. People in the oil industry use it for inspection of oil pipes, well heads, and platform structures and for seabed surveys. Television cameras are often operated by divers in various applications. Where conditions become too hazardous for such operation, arrangements can be made for remote operation of television equipment. Remotely controlled vehicles to carry television cameras exist in great variety and are continuing to be adapted and developed for special purposes. New applications continue to arise as the television equipment is improved.

[ROBERT PRIESTLEY]

Bibliography: R. E. Craig and R. Priestley, Undersea photography in marine research, *Mar. Res.*, vol. 1, 1963; C. C. Hemmings, Underwater photography in fisheries research, *J. Cons. Int. Explor. Mer*, vol. 34, no. 3, 1972; *Oceanic Abstracts*, vols. 9–13, 1972–1976; C. L. Strickland, Underwater television, its development and future, *Underwater J.*, 5(6):244–249, 1973; A. J. Woolgar and C. J. Bennet, Silicon diode tubes and targets, *Roy. Television Soc. J.*, vol. 13, no. 3, 1970.

Upper-atmosphere dynamics

The motion of the atmosphere above 50 km. The predominant dynamical phenomena of the upper atmosphere are quite different from those encountered in the lower atmosphere. Among those encountered in the lower atmosphere are cyclones, anticyclones, tropical hurricanes, thunderstorms and shower clouds, tornadoes, dust devils, and similar phenomena. Even the largest of these phenomena do not penetrate far into the upper atmosphere. Above an altitude of about 50 km, the predominant dynamical phenomena are internal gravity waves, tides, sound waves (including infrasonic), turbulence, and large-scale circulation.

Internal gravity waves. Except under meteorological conditions characterized by convection, the atmosphere is stable against small vertical displacements of small air parcels; this results from buoyancy forces that tend to restore displaced air parcels to their original levels. An air parcel therefore tends to oscillate around its undisturbed position at a frequency known as the Brunt-Vaisala frequency; the square of this frequency is given by

$$\omega_B{}^2 = -g(\rho'_0/\rho_0 + g/C^2)$$

where ρ_0 is the undisturbed density, ρ'_0 is the vertical gradient of the density, g is the acceleration of gravity, and $C = (\gamma p_0/\rho_0)^{1/2}$ is the speed of sound; γ is the ratio of the specific heats, and p_0 is the undisturbed pressure. Typical periods for the Brunt-Vaisala oscillation are in the vicinity of 5 min.

If pressure waves are generated in the atmosphere with frequencies much greater than ω_B, they propagate as sound waves. For frequencies much less than ω_B, the waves propagate as internal gravity waves; in this case, the restoring forces for the wave motion are provided primarily by buoyancy (that is, gravity) rather than by compression. These waves have peculiar properties. The wave motion is nearly horizontal, but the inclination from the horizontal is sufficient so that gravity provides the restoring force for the displaced air parcels. The phase velocity of the waves is downward, but the group velocity is upward. Such waves are presumably excited near the Earth's surface, probably in wind systems, and the energy flow is upward, reaching well into the ionosphere. Owing to the decrease in density with altitude, the amplitudes of the waves must increase with altitude in order to preserve continuity of energy flow; above 60 km, the amplitude is so large that it dominates the observed wind profiles. An example is given in the illustration, showing a wavelike structure with a vertical wavelength of about 20 km. In the lower atmosphere, gravity waves are presumably also present, but with amplitudes so small that they cannot be detected.

Above 100 km, dissipative mechanisms come into play for internal gravity waves; these include viscosity, thermal conductivity, and eddy transport of heat and momentum. The dissipation of the waves, judged from observed amplitudes below 100 km, provides an energy source apparently large enough to be of at least minor significance in the heat budget of the upper atmosphere.

Internal gravity waves should not be confused

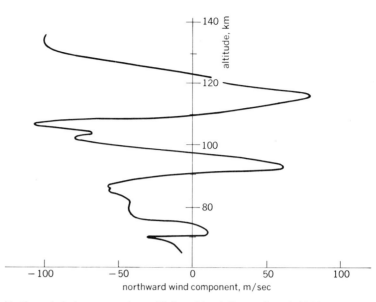

Northward wind component over Wallops Island, Va., on June 6, 1962.

with surface waves, which also occur within the atmosphere. The surface waves occur at discontinuity surfaces, or in regions of strong gradients where the gradient is strong enough to guide the propagation of the wave in a direction perpendicular to the gradient. Such waves are frequently visible on cloud sheets. *See* INTERNAL WAVE.

Tides. Tides are internal gravity waves of particular frequencies. The term tidal usually implies that the exciting force is gravitational attraction by the Moon or Sun. However, it is conventional in atmospheric tides to include also those waves excited by solar heating. One is thus concerned with three separate excitation functions—lunar gravitation, solar gravitation, and solar heating.

Examination of barometric pressures in tropical latitudes normally discloses two maxima and two minima per day, the maxima occurring near 1000–2200 hr local time; at higher latitudes the variations are smaller and the noise due to passing weather systems is much larger, and statistical analyses are needed to find the tidal oscillations. The gravitational tidal force of the Moon is 2.2 times larger than that of the Sun, and its period is 12 hr 26 min. The fact that only the 12-hr oscillation is readily seen on a barograph record therefore indicates that the oscillation is of thermal origin, or that the oscillation is so sharply resonant that the 12-hr tidal force of the Sun excites a much stronger oscillation than the larger 12-hr 26-min tidal force of the Moon. The possibility of a thermally excited oscillation of period 12 hr arises because the solar heating curve has harmonics, that is, it is not a simple sinusoidally oscillating function. Although the possibility of a resonant excitation of the 12-hr oscillation by solar gravitational force has been seriously considered and the resonant period of the atmosphere is about 12 hr, the resonance is now thought to be much too broad to discriminate between the solar and lunar gravitational tidal periods. Both are amplified several times by the resonance, but the thermal excitation must be responsible for the major observed effect.

Although the lunar tide of period 12 hr 26 min is not discernible on a barograph trace, it can be found from a statistical analysis, and it is about 15 times smaller than the 12-hr oscillation.

Although tidal oscillations are small at the Earth's surface, they are relatively large at high atmospheric levels. Only a few days' observations of the winds near the altitude of 100 km is required to disclose the diurnal pattern of behavior in which the largest components have 24- and 12-hr periods; the amplitudes of these tidal winds are about 100 times greater than those occurring at the Earth's surface.

Tidal wind patterns in the upper atmosphere also disclose their presence in another way, by generating electrical currents in the ionosphere through a dynamo action. These in turn give rise to diurnal variations in the geomagnetic field that can be observed at the Earth's surface. The lunar tidal oscillations also increase with altitude, but they are smaller than the solar tidal oscillations (about 2 m/sec as compared to 20 m/sec), so that they have not been recognized in the limited amount of wind data available. However, they are recognizable in terms of the geomagnetic variations that they produce, although a statistical analysis is needed to find them. *See* IONOSPHERE; TRANSIENT GEOMAGNETIC VARIATION.

Sound waves and infrasonics. Sound waves generated in the lower atmosphere may propagate upward; to maintain continuity of energy flow, one might expect the waves to grow in relative amplitude as they move into the more rarefied upper atmosphere. However, higher temperatures in the upper atmosphere refract most of the energy back toward the Earth's surface, giving rise to the phenomenon known as anomalous propagation. This involves the redirection of upward-moving sound waves back to the surface beyond the point where the source (usually a large explosion) can be heard by waves propagating along the surface. Thus there is normally a zone of silence, beyond which the source can again be heard because of anomalous propagation.

Sound waves or infrasonic waves (frequencies too low to be audible) generated by thunderstorms have been considered possible sources of energy input in the upper atmosphere as the waves become shock waves on entering more rarefied regions of the atmosphere. The refraction effect mentioned above makes this unlikely, but the possibility has remained under consideration. Recently the energy spectrum of sound waves generated by thunderstorms has been measured. A broad peak near 200 Hz was found, with no energy increase in the infrasonic region.

Infrasonic waves with periods from 20 to 80 sec have been observed occasionally with detectors at the Earth's surface in connection with auroral activity, the average rate of observed occurrence being about once a day. The occurrence seems to be associated with rapidly moving auroral forms where the rate of movement is supersonic. Although this rapid motion does not represent a supersonic motion of atmospheric gas, but rather only a movement of the energetic particle source causing the aurora, nevertheless, a shock wave should be produced by the supersonic motion of the generating mechanism, the aurora. The observed direction of arrival agrees with this concept of an auroral source, the preferred direction of propagation being the direction of motion of the auroral forms. The observed amplitudes range up to 10 dynes/cm².

Turbulence. There is clear visual evidence of turbulence in the upper atmosphere; this evidence is obtained by examination of vapor trails released from rockets or of long persisting meteor trails. The trails spread above about 110 km by an amount that corresponds to molecular diffusion, but at lower altitudes the spreading is far in excess of that which can be attributed to this mechanism, and turbulence is the only known means of explaining the additional spreading. Further, in the region below 110 km, the trails adopt a turbulent appearance, whereas at higher altitudes, where molecular diffusion predominates, the trails are smooth, as should be expected.

The source of the turbulence is not clear. The atmosphere is thermodynamically stable against vertical displacements throughout the region above the troposphere, and work has to be done against buoyancy forces in order to produce and maintain turbulence. The only apparent source of energy is internal gravity waves, either tidal or of random period. One concept is that wind shear associated with such waves may become great enough to overcome the stability and produce turbulence; the relevant criterion for this is that the Richardson number be smaller than some arbitrary value. The Richardson number essentially expresses the ratio of the stability (the work that must be done against buoyancy forces if turbulence is to be maintained) to the destabilizing forces involved in wind shear. It appears probable that the criterion is satisfied occasionally, but the trails do not give the impression of being turbulent only in the regions of strong shear. Another concept is that gravity waves may at times be of sufficiently large amplitude so that potentially colder air is brought over potentially warmer air; that is, the temperature distribution may be distorted in such a way that the adiabatic lapse rate is exceeded in some regions so that thermal convection can proceed. Again, the trails do not have the appearance of being convectively unstable in narrow regions as the concept requires.

The presence of turbulence and its associated eddy transport cannot be doubted, even though there are difficulties in understanding the energy source and the physical generation mechanism. In addition to the visual evidence provided by vapor and meteor trials, the effect of the eddy transport on atmospheric structure is such that much can be deduced about the average rate of eddy transport from a study of atmospheric structure. On this basis it is recognized that the average eddy diffusion coefficient varies from about 10^4 cm²/sec just above the tropopause up to about 10^7 cm²/sec near 110 km, above which it is exceeded by the molecular diffusion coefficient (above 110 km, the eddy coefficient probably decreases because of the severe damping of eddies by viscosity and heat conduction).

The use of an eddy diffusion coefficient permits an evaluation of the transport properties of eddies without specifying their physical properties. The generally accepted concept is that the larger eddies break up into progressively smaller eddies, which finally dissipate their energy against viscous forces. The largest eddies are responsible for most of the transport. Examination of vapor trails suggests that the largest eddy size is of the order of 100 m; with eddies of this size, eddy velocities near 10 m/sec would be required to produce the observed mixing rates, and the lifetimes of the largest eddies would be a few tens of seconds. *See* METEOROLOGICAL ROCKETS.

Large-scale circulation. Meteorological circulations and disturbances extend up to altitudes of about 50 km. However, at higher altitudes, internal gravity waves and tides become more prominent. Above 100 km, the circulation is not very well understood, but several interesting effects are apparent.

The atmosphere over the winter polar region remains about as warm at high altitudes as does the atmosphere at low latitudes or over the summer hemisphere. This is indicative of a poleward circulation at high levels over the winter hemisphere, with downward motion and compressional heating to compensate for the lack of solar heating; the required downward velocities are not very great—about 10^{-1} cm/sec at 50 km, 10 cm/sec at 100 km, and 10^2 cm/sec above 150 km. An interesting side effect of this circulation is that, because of diffusive separation of atmospheric constituents above 100 km, it brings an excess of lighter constituents to the winter polar region. This causes an excess concentration of helium to build up over the winter polar region, the magnitude of the effect being about a factor of 3 between summer and winter poles.

The biggest nonuniformity that has been observed in the upper atmosphere is the afternoon density bulge. This arises from heating of the upper atmosphere by solar radiation. The warmest conditions occur near the subsolar latitude and near 1400 hr local time, and the atmosphere is most distended there against the confining effects of gravity. Likely wind patterns have been calculated on the basis of the atmospheric density distribution, taking into account Coriolis force, viscosity, and ion drag; the last results from collisions of neutral atmospheric particles with ionospheric ions which are more or less anchored in place by the magnetic field. The ion drag tends to simplify the wind system, since it largely avoids the development of asymmetries that the Coriolis force would otherwise bring about. The wind system is rather simple, being a nearly symmetrical outflow from the density bulge and inflow into the region of the nighttime density minimum. This density or temperature minimum would normally be expected to occur over the winter polar region, but it is apparently displaced to low latitudes in the nighttime hemisphere by the heat transport associated with the poleward wind system. *See* ATMOSPHERIC GENERAL CIRCULATION.

[FRANCIS S. JOHNSON]

Bibliography: S. Chapman and R. S. Lindzen, *Atmospheric Tides, Thermal and Gravitational,* 1970; B. H. Haurwitz, Atmospheric tides, *Science,* 144:1415–1422, 1964; C. O. Hines, Internal atmospheric gravity waves at ionospheric heights, *Can. J. Phys.,* 38:1441–1481, 1960; F. S. Johnson, Developments in upper atmospheric science during the ISQY, *Proc. Nat. Acad. Sci.,* 58:2162–2174, 1967; J. E. Midgley and H. B. Liemohn, Gravity waves in a realistic atmosphere, *J. Geophys. Res.,* 71:3748–3749, 1966; F. Pasquill, *Atmospheric Diffusion,* 1975.

Upper synoptic air waves

Wavelike oscillations in the pattern of wind flow aloft, usually with reference to the stronger portion of the westerly current. The flow is anticyclonically curved in the vicinity of a ridge line in the wave pattern, and is cyclonically curved in the vicinity of a trough line.

Any given hemispheric upper flow pattern may be represented by the superposition of sinusoidal waves of various lengths in the general westerly flow. Analysis of a typical pattern discloses the presence of prominent long waves, of which there are three or four around the hemisphere, and of distinctly evident short waves, of about half the length of the long waves.

Typically, each short-wave trough and ridge is associated with a particular cyclone and anticyclone, respectively, in the lower troposphere. The development and intensification of one of these circulations depends in a specific instance upon the details of this association, such as the relative positions and intensities of the upper trough and the low-level cyclone. These circulations produce the rapid day-to-day weather changes which are characteristic of the climate of the middle latitudes.

The long waves aloft do not generally correspond to a single feature of the circulation pattern at low levels. They are relatively stable, slowly moving features which tend to guide the more rapid motion of the individual short waves and of their concomitant low-level cyclones and anticyclones. Thus, the long waves by virtue of their position and amplitude, can exert an indirect influence on the character of the weather over a given region for a period of the order of weeks.

It has been found that the motion of long waves and short waves can be predicted with considerable success by application of the principle of conservation of absolute vorticity to the flow pattern at middle-tropospheric elevations. *See* ATMOSPHERE; JET STREAM; STORM; VORTEX; WEATHER FORECASTING AND PREDICTION; WIND.

[FREDERICK SANDERS]

Upwelling

A process in the sea whereby subsurface water is displaced toward the surface. The upward motion may occur anywhere, but it is most conspicuous along the western coasts of the continents. In Northern Hemispheric regions where persistent winds blow nearly parallel to the coast (the coast being to the right of an observer facing the wind), the wind-driven surface water is deflected offshore (Fig. 1). This causes the subsurface water to upwell, the amount depending on the wind characteristics. Any variation in wind speed or direction causes a corresponding variation in the amount of upwelling and in the currents produced. On one occasion, off the California coast, the rate of verti-

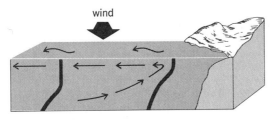

Fig. 1. Schematic diagram showing upwelling of cold, deep water off a Northern Hemispheric coast when wind blows parallel to it. Heavy lines represent temperature-depth structure. Water temperatures shown by each curve increase from left to right along horizontal scale. Note upward displacement of thermocline (layer of greatest decrease in temperature) nearshore.

cal motion in upwelling was found to be 20 m per month. In the intense upwelling off the western coast of South America the vertical displacement has been computed from wind stress to be as much as 100 m per month. Thus in one season the water reaching the surface comes from a relatively shallow depth, usually less than 200 m. Upwelling is also caused by a meeting of deep, diverging currents and by large and small cyclonic gyrals. *See* OCEAN CURRENTS; WIND STRESS OVER SEA.

Occurrence. Pronounced coastal upwellings are found off California, Peru, Morocco, and southern Africa. Some coastal upwelling is created by monsoon winds, as in Southeast Asia, where the winds blow from the southwest during the summer, then reverse and blow from the northeast during the winter. The persistence of the monsoon, especially from the southwest, and the orientation of the coasts cause upwelling along several stretches of the locale. The most intensive upwelling of the southwest monsoon season, however, is off the coast of Somaliland and Arabia. Less pronounced upwelling has been observed off the east-

ern coasts of India, Thailand, and South Vietnam. Ascending circulation has been noted around the Antarctic continent, along the Equator, and at the northern boundary of the equatorial countercurrent (Fig. 2). Localized upwelling develops in the lees of islands; lees of land promontories projecting into a current; around shoals; in counterclockwise eddies in the Northern Hemisphere, such as the Alaskan gyral; at water-mass boundaries; and in thermal domes or ridges in the open sea.

Since the water in an upwelling zone is brought up from greater depths, it is colder and denser than adjacent surface water. When intense upwelling occurs in early summer, surface temperature can be colder than in winter. The vertical temperature gradients are weaker in the region of upwelling. During this process a subsurface isothermal layer has been observed to develop at a level of the shoreward circulation.

In coastal regions, upwelling creates a distribution of mass with denser water nearshore. This, together with wind stress on the surface water, creates a relative current alongshore. Thus water movement patterns are due partly to wind currents and partly to geostrophic currents.

In the Northern Hemisphere, cyclonic winds transport surface water away from the cyclonic center and cause the heavier subsurface water to rise. *See* SEA WATER; THERMOCLINE.

Effects. Upwelled water exerts widespread influence on the meteorological conditions off a coast. If the upwelled subsurface water is colder than the air, a situation conducive to fog is developed. The cooled air also affects regional climate.

Upwelled water can introduce large quantities of plant nutrients (phosphates, nitrates, and so on) to the euphotic zone; therefore, an upwelling region is normally conducive to high photosynthetic activity with the production of organic matter. Large kelp beds thus occur on the western coast

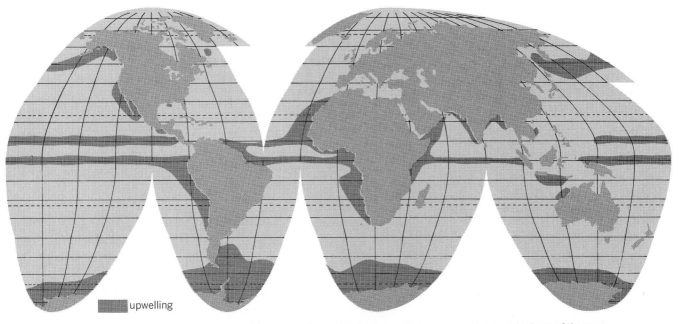

upwelling

Fig. 2. General geographic areas of the world where upwelling occurs during certain times of the year.

of Africa north of Cape Town, and extensive fishing areas are found along the California coast. In addition to the flourishing anchovy fisheries, a considerable bird population, whose cumulative guano deposits are of economic importance, exists off Peru. At the Antarctic Convergence, particularly in the Atlantic, diatoms and flagellates have an abundance of nutrients. These plankton ultimately support krill, the main food of the whale. Upwelling regions also influence the benthonic forms and even the organic composition of sea-floor sediments. If large quantities of organic matter accumulate on the sea floor and are covered with sediment, they may eventually be converted to oil. Thus sea economy is directly in balance with the important phenomenon of upwelling through its influence on the thermal, chemical, and biological properties of the sea. *See* CLIMATIC CHANGE; SEA-WATER FERTILITY.   [EUGENE C. LA FOND]

Van Allen radiation

The high-energy, charged particles which are trapped by the geomagnetic field and form belts of intense radiation in space about the Earth. The belts consist primarily of electrons and protons and extend from a few hundred kilometers above the Earth to a distance of about 7 earth radii (radius of Earth is 6371 km). The belts of trapped particles were discovered by James Van Allen and coworkers in 1958 through radiation detectors carried on satellites *Explorer 1* and *Explorer 3*. Many additional experiments have shown that the radiation has a complicated, time-dependent structure. While the precise nature of the mechanisms responsible for the origin and maintenance of the belts is uncertain, it is felt that the same mechanisms are involved in the production of aurora, magnetic storms, and other geophysical phenomena.

Motion of charged particles. A charged particle under the influence of the geomagnetic field follows a trajectory which can be conveniently described as a superposition of three separate motions. The first motion is a rapid spiral about magnetic field lines that is produced by the magnetic force acting at right angles to both the particle velocity and to the magnetic field. As the spiraling particle moves along the field line toward either the North or South Pole, the increase in magnetic field strength causes the particle to be reflected so that it bounces between Northern and Southern hemispheres. Superimposed on the spiral and bounce motions is a slow east-west drift, electrons drifting eastward and protons drifting westward. Thus individual trapped particles move completely around the Earth in a complicated pattern, their motion being constrained to lie on magnetic shells. Figure 1 illustrates the motion of a trapped proton during several north-south reflections. Subsequent bounces will carry the proton completely around the Earth. Trapped electrons behave in a similar manner except that the direction of drift is eastward. In the absence of perturbing forces such as additional electric and magnetic fields or collisions with other particles, this motion will continue for an indefinite time.

The fact that an individual particle moves on a magnetic shell whose elements consist of magnetic field lines makes it convenient to describe a location in the radiation belts by specifying the magnetic shell passing through that location. In a dipole field the magnetic shell can be labeled by a number equal to the distance from the center of the dipole to the point where the magnetic shell intersects the equatorial plane. Because the Earth's field is not a pure dipole, some adjustments must be made in converting the distance in the real geomagnetic field to a value for the magnetic shell parameter. However, with this correction one can label a magnetic shell by a length (usually given in earth radii) approximately equal to the distance from the center of the Earth to the equatorial crossing of the shell. This distance is denoted by the symbol L. Thus $L=4$ denotes the magnetic shell composed of field lines whose equatorial crossings are approximately 25,500 km from the center of the Earth. Fluxes of trapped particles are normally given as a function of L rather than as a function of geographical location.

Trapped particle populations. The distribution of charged particles in the Earth's radiation belts is illustrated schematically in Fig. 2, where electron fluxes are depicted in the cross section on the left and proton fluxes are shown on the right. The darker shading indicates more intense fluxes of trapped particles. Generally speaking the spatial structure of trapped radiation shows two maxima: an inner radiation belt centered at about $L=1.5$ and an outer belt centered at $L=4.5$. In the inner radiation belt the most penetrating particles are protons with energies extending to several hundred million electron volts (Mev). The intensity of the high-energy protons reaches 10^5 cm^{-2} sec^{-1} for energies above 15 Mev. However, the flux of high-energy protons decreases rapidly with increasing L and becomes insignificant at $L > 3.0$. Low-energy protons with energies up to a few Mev occur throughout the stable trapping region which extends between $1.15 \gtrsim L \gtrsim 8$. Electrons are also found throughout the trapping region with local maxima occurring in the inner and outer belts. The electron energies extend to several Mev, and in the outer radiation belt electrons are the most penetrating component.

Experiments have also shown that helium nuclei (alpha particles) with energies of a few Mev are also trapped in the geomagnetic field, the maximum flux of a few hundred particles cm^{-2} sec^{-1} (energy > 2 Mev) occurring at $L=3.1$. The flux of

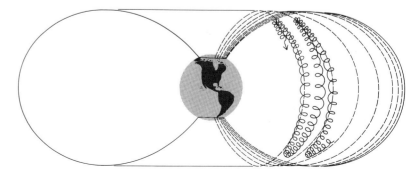

Fig. 1. Motion of a proton trapped in the geomagnetic field.

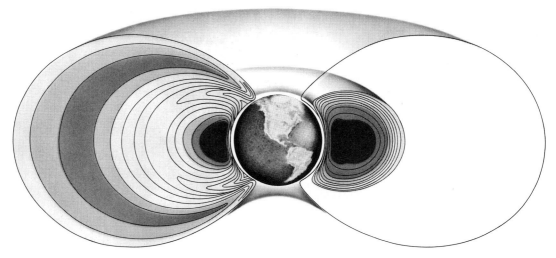

Fig. 2. Structure of Earth's radiation belts. Protons ($E >$ 15 Mev) are shown on the right, and electrons ($E > 0.5$ Mev) are given on the left. Darker shading indicates more intense fluxes of trapped particles.

alpha particles is therefore only about 2×10^{-4} that of the protons; however, the presence of helium has important implications for the origin of the radiation belts.

Sample values of the fluxes of trapped electrons and protons are given as a function of the magnetic shell parameter L in Table 1. These values are averages, and the time variations are substantial, particularly for electrons in the outer belt. Also given in Table 1 are values of the characteristic energy E_0 defined by the equation below. Here

$$N(E) = N_1 \exp \left[(E - E_1)/E_0 \right]$$

$N(E)$ is the differential flux (cm^{-2} sec^{-1} Mev^{-1}) and N_1 is the value of the differential flux at E_1. In Table 1 note the systematic decrease in E_0 with increasing L for protons. A similar trend occurs for electrons except for the region between the inner and outer belt where the electron flux exhibits a minimum and the energy spectrum is richer in low-energy electrons.

Time variations. The intensity, energy spectrum, and spatial distribution of particles within the radiation belts vary with time. The most dramatic variations are associated with magnetic storms although smaller changes accompany most magnetic fluctuations. The changes are most pro-

nounced for particles in the outer belt $L \gtrsim 3.0$ where the magnetic variations are most violent. During a magnetic storm, the electron flux in the outer belt may increase an order of magnitude or more, although the response differs greatly among individual storms.

Figure 3 illustrates the time variations in the flux of electrons above 0.5 Mev on the Equator at $L = 4$. The upper part of the figure gives the K_p index, which is a measure of the magnetic activity; the higher the value of K_p the larger are the irregular fluctuations in the geomagnetic field. It is apparent that the electron flux increases noticeably following magnetic fluctuations and decays slowly while the Earth's field is steady. The fluxes of low-energy protons in the outer belt also increase during magnetic storms although the changes in proton fluxes are generally smaller than those of electrons.

In the inner belt the high-energy proton flux is relatively constant with time, a very strong magnetic storm being required to alter the intensity appreciably. The most pronounced effect observed during storms is a reduction in the trapped proton flux, the reduction being greater with increasing distance from the Earth.

At the outer edges of the radiation belt, large time variations are common and a systematic diur-

Table 1. Omnidirectional fluxes and characteristic energies of trapped particles on the Equator

Magnetic shell parameter (L)	Low-energy protons ($E > 0.4$ Mev)		High-energy protons ($E > 15$ Mev)		Electrons ($E > 0.5$ Mev)	
	Flux, cm^{-2} sec^{-1}	E_0, Mev	Flux, cm^{-2} sec^{-1}	E_0, Mev	Flux, cm^{-2} sec^{-1}	E_0, Mev
1.25	4×10^4	2.2	1×10^4	20	4×10^7	0.9
1.5	1×10^6	2.2	1×10^5	13	6×10^7	0.4
2.0	1.6×10^7	2.1	7×10^4	8	3×10^6	0.13
2.5	1×10^8	0.6	8×10^3	8	2×10^5	0.33
3.0	2×10^8	0.4	2×10	7	6.7×10^4	0.42
3.5	1×10^8	0.34			3×10^5	0.5
4.0	3×10^7	0.3			10^6	0.4
5.0	4×10^6	0.13			2.5×10^6	0.35
6.0	9×10^5	0.11			9×10^5	0.25

nal variation exists that is caused by the warping of the overall geomagnetic field by the pressure of the solar wind, a stream of ionized plasma emitted by the Sun. The resulting asymmetry in the field produces a diurnal variation in the trapped radiation observed at a given point above the Earth. Variations in solar-wind pressure alter the degree of warping of the magnetic field and also produce variations in the flux of charged particles at high L values. *See* MAGNETOSPHERE; TRANSIENT GEOMAGNETIC VARIATION.

Artificial radiation belts. On at least nine occasions high-energy electrons were injected into the geomagnetic field and trapped as a result of high-altitude nuclear detonations. The dates and names of these explosions and the maximum flux values resulting from the detonations are shown in Table 2. Except for the Starfish event in 1962, all injections were short-lived, and the artificial belts decayed to background levels within a few days for Teak and Orange and within a few weeks for the others. Some electrons from the Starfish belt were still present in 1968, although the intensity was decreasing by a factor of about two each year.

Origin and loss of particles. Immediately after the discovery of the radiation belts it was thought that the trapped electrons and protons resulted from the spontaneous decay of neutrons produced by cosmic-ray collisions with the atmosphere. Early estimates of the numbers of neutrons available and the time required for the resulting electrons and protons to escape the magnetic confinement seemed to support this theory. However, later information showed that the number of particles produced in this manner is not adequate to maintain the belts. The high-energy protons ($E > 20$ Mev) in the inner belt are a possible exception to this conclusion, and they may indeed arise from the decay of high-energy neutrons. Trapped alpha particles, of course, cannot originate by this mechanism.

Later theories of the sources of radiation-belt particles call on fluctuating electric and magnetic fields to accelerate the trapped particles. The geomagnetic field is continually agitated by variations in solar-wind pressure, and the changing magnetic field can accelerate charged particles by means of the induced electric fields. The observed increase in trapped electron fluxes following magnetic activity as illustrated in Fig. 3 strongly supports this theory. In addition to accelerating charged particles, the fluctuating fields also cause the particles to migrate across magnetic shells and thus make it possible for a particle which is initially injected at the outer boundary of the geomagnetic field to be transported to the inner belt. A diffusion mechanism of this type does explain the widespread distribution of trapped particles, but the complex structure of inner and outer belts is not understood. One consequence of the radial diffusion of particles is that the characteristic energies of particles trapped on magnetic shells close to the Earth should be higher than the energies of particles at the outer boundaries. Such an energy dependence is observed and is given in Table 1. Detailed calculations of the acceleration and diffusion processes resulting from magnetic varia-

Table 2. Artificial radiation belts

Designation of event	Date	Maximum omnidirectional trapped flux, cm^{-2} sec^{-1}
Teak	Aug. 1, 1958	$\sim 10^6$
Orange	Aug. 12, 1958	$\sim 10^3$
Argus I	Aug. 27, 1958	$\sim 10^5$
Argus II	Aug. 30, 1958	$\sim 10^5$
Argus III	Sept. 6, 1958	$\sim 10^6$
Starfish	July 9, 1962	10^9
USSR I	Oct. 22, 1962	7×10^7
USSR II	Oct. 28, 1962	2×10^8
USSR III	Nov. 1, 1962	10^8

tions are not yet possible since there is insufficient knowledge about the electric and magnetic fields in space. As knowledge improves, it will be possible to provide more exacting tests of the theories.

The physical nature of the processes by which trapped particles are ultimately lost from the belts is also in question. From measurements of the rates at which electrons from the artificial radiation belts are lost, it is possible to estimate the average trapping time for electrons with energies of about 1 Mev. Also, the observed decay of electron fluxes during magnetically quiet periods (Fig. 3) allows a measurement of loss rates for these electrons. The average lifetime for electrons varies from about a year in the center of the inner radiation belt to a few days in the outer belt. Within a few thousand kilometers from the Earth, that is, at $L \gtrsim 1.3$, the particles are removed by collisions with atoms and molecules of the Earth's atmosphere. The cumulative effect of these collisions is that the particle eventually loses so much energy that it becomes part of the ionosphere. At higher altitudes the loss mechanisms are not known, but several possibilities have been suggested, the most promising being the interaction of the trapped particles with electromagnetic waves. The waves may

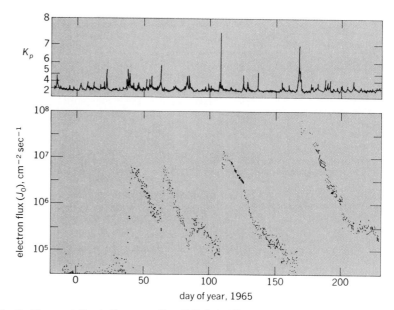

Fig. 3. Time variation in the magnetic activity index K_p and the associated changes in the flux of trapped electrons ($E > 0.5$ Mev) at $L = 4.0$.

be produced by groups of the trapped particles or may be the result of other processes occurring in the magnetosphere. In either case the electromagnetic fields of the waves perturb the motion of the trapped particles so that eventually the particles are deflected into the dense lower atmosphere where they are removed by collisions. Theoretical estimates of the loss rates produced by these effects are not conclusive because of the severe approximations which must be made in carrying out the calculations.

Relation to other phenomena. The radiation belts are directly or indirectly related to many other geophysical phenomena. Some theories for the polar aurora assume that the electrons and protons, whose interaction with the upper atmosphere produces the aurora, are temporarily stored in geomagnetically trapped orbits. However, efforts to explain the aurora as a direct result of the loss of particles from the radiation belts have not been successful because of energy considerations. The energy stored in the trapped particles is not sufficient to maintain an aurora for more than a few minutes. However, it is felt that the same processes which accelerate particles for the radiation belts may also be responsible for accelerating auroral particles. *See* AURORA.

Trapped radiation also plays an important role in magnetic storms. One of the most important features of a magnetic storm is the main phase, during which time the horizontal component of the Earth's field measured at the Equator is decreased. This decrease has been attributed to currents of trapped protons and electrons drifting in magnetic shells about the Earth. *See* GEOMAGNETIC STORM.

Particles are almost continually leaking out of the radiation belts into the atmosphere. This process produces ionization and thus contributes to the electron and ion population in the ionosphere. Also, some of the energy necessary to produce the airglow may be derived from the precipitation of charged particles into the atmosphere. *See* AIRGLOW; IONOSPHERE.

Effects on space vehicles. The fluxes of trapped electrons and protons can injure both personnel and equipment in spacecraft if the vehicle is exposed to the radiation a sufficient length of time. Because of the spatial structure of the belts the damage rate is strongly dependent on the position in space and hence on the vehicle orbit. In the inner belt region, high-energy protons penetrate material several g-cm^{-2} thick and injure components in the interior of the spacecraft. In most other regions of the radiation belts the trapped particles are less penetrating, and damage is confined to exposed equipment such as solar cells. Following the high-altitude nuclear explosion on July 9, 1962, several satellites ceased to transmit data because of loss of electrical power resulting from solar-cell damage.

[MARTIN WALT]

Bibliography: W. N. Hess, G. D. Mead, and M. P. Nakada, Diffusion of protons in the outer radiation belt, *Rev. Geophys.*, 3:4777–4791, 1965; F. S. Johnson, *Satellite Environment Handbook*, 2d ed., 1965; B. M. McCormac (ed.), *Magnetospheric Particles and Fields*, 1976; B. M. McCormac and A. Omholt, *Atmospheric Emissions*, 1969; C. E. McIlwain, Radiation belts natural and artificial, *Science*, 142:355–361, 1963; B. J. O'Brien, Review of studies of trapped radiation with satellite-borne apparatus, *Space Sci. Rev.*, 1:415–484, 1962; J. G. Roederer, *Dynamics of Geomagnetically Trapped Radiation*, 1970; R. S. White, The Earth's radiation belt, *Phys. Today*, 19:25–38, 1966.

Veering wind

A wind that changes direction in a clockwise sense; for example, a change from a southerly to a westerly direction. The wind may veer gradually, over a period of hours, or abruptly, in a few minutes or seconds on passage of a wind-shift line. The wind veers when a cyclone passes eastward on a path north of the observer. Veering with height is usually found on the east side of a cyclone, or west of an anticyclone, with warm air to the south. In the Southern Hemisphere the meaning in terms of cardinal directions is reversed. *See* WIND-SHIFT LINE.

[CHESTER W. NEWTON]

Vortex

A line vortex in two-dimensional fluid flow produces a flow or circulation around the line.

Free vortex. Consider the effect of rotating a right-circular cylinder of radius r_0 about its axis with a peripheral velocity v_0 in a fluid otherwise at rest. The fluid in contact with the surface of the cylinder rotates with the cylinder. Fluid at greater radius is also set in motion in concentric circles with velocity diminishing as the radius increases. This type of fluid motion, in which the velocity varies inversely as the radius, is referred to as a free vortex.

If the cylinder is reduced to zero radius in such a manner that $v_0 r_0$ remains constant in the limit as r_0 approaches zero, a line vortex results. The velocity at the line is infinite, so the line itself must be considered as a singular line, to be excluded from the actual fluid.

Examples of vortices occur frequently in nature. The tornado is an example of a free vortex, with high velocities near its center, and correspondingly low pressure intensities. The waterspout is its counterpart over water.

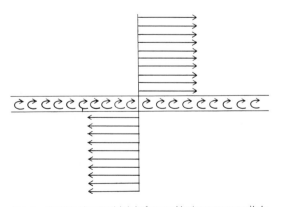

Fig. 1. Vortex sheet which is formed between oppositely directed streams.

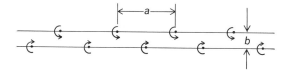

Fig. 2. Kármán vortex street consists of alternate vortices of opposite rotational sense.

The fluid motion in the case of a line vortex in an ideal (frictionless and incompressible) fluid is irrotational; that is, its motion may be described in terms of a velocity potential.

Vortex tube. If a small spherical particle of a frictionless fluid could be considered as suddenly solidified, its resulting rotation could be expressed by a vector parallel to the axis of rotation, with its length proportional to the angular velocity, and with its direction indicating the sense of rotation by the right-hand rule. When the rotation vector is everywhere zero throughout a region of fluid, the motion of that fluid is irrotational. When some finite fluid regions have nonzero values of the rotation vector, then this fluid has vorticity. A vortex line is a line drawn through the fluid such that it is everywhere tangent to the rotation vector. A collection of vortex lines through a small closed curve defines a vortex tube, which has certain special properties.

1. The circulation about a vortex tube is everywhere the same along its length. Circulation is defined as the line integral of the velocity vector around a closed path.

2. The vortex tube cannot end in the fluid. It must either extend to a boundary or close upon itself.

3. Vortex lines move with the fluid. Vorticity of a fluid is a property of the fluid itself and not the space it occupies.

A smoke ring is a practical example of a closed vortex tube. A circular vortex tube in otherwise still fluid will translate perpendicular to the plane of the ring without change in size.

Vortex arrays. A discontinuity in fluid velocity along a surface, such as slippage of one layer of fluid over another, may be handled as a vortex sheet in an otherwise continuous flow. In this case all vortex lines are in the surface (Fig. 1). A practical case of a vortex sheet is the flow downstream from an airfoil when the velocity leaving the upper surface is higher than the velocity leaving the under surface.

When real fluid flows around a body, such as wind blowing across a cable, the fluid rotation in the boundary layer causes vortices to form along the downstream side of the body. For certain conditions they remain near the body and are referred to as bound vortices.

For higher velocities the vortices form, grow, and are then systematically shed from the downstream side of the body, forming a vortex street (Fig. 2). The unsymmetrical case shown is called the Kármán vortex street after T. von Kármán, who first identified them and showed that the motion is stable when $b/a = 0.281$.

[VICTOR L. STREETER]

Water table

The upper surface of the zone of saturation in permeable rocks not confined by impermeable rocks. It may also be defined as the surface underground at which the water is at atmospheric pressure. Saturated rock may extend a little above this level, but the water in it is held up above the water table by capillarity and is under less than atmospheric pressure; therefore, it is the lower part of the capillary fringe and is not free to flow into a well by gravity. Below the water table, water is free to move under the influence of gravity. The position of the water table is shown by the level at which water stands in wells penetrating an unconfined water-bearing formation.

Where a well penetrates only impermeable material, there is no water table and the well is dry. But if the well passes through impermeable rock into water-bearing material whose hydrostatic head is higher than the level of the bottom of the impermeable rock, water will rise approximately to the level it would have assumed if the whole column of rock penetrated had been permeable. This is called artesian water, and the surface to which it rises is called the piezometric surface.

The water table is not a level surface but has irregularities that are commonly related to, though less pronounced than, those of the land surface. Also, it is not stationary but fluctuates with the seasons and from year to year. It generally declines during the summer months, when vegetation uses most of the water that falls as precipitation, and rises during the late winter and spring, when the demands of vegetation are low. The water table usually reaches its lowest point after the end of the growing season and its highest point just before the beginning of the growing season. Superimposed on the annual fluctuations are fluctuations of longer period which are controlled by climatic variations. The water table is also affected by withdrawals, as by pumping from wells. *See* GROUND-WATER.

[ALBERT N. SAYRE/RAY K. LINSLEY]
Bibliography: S. N. Davis and R. J. M. DeWiest, *Hydrogeology,* 1966; L. Huisman, *Groundwater Recovery,* 1972.

Waterspout

An intensely whirling, funnel-shaped vortex, extending from a cumulus-type cloud down several hundred to several thousand feet to the water surface. The visible funnel consists mostly of atmospheric water vapor condensed because of lower pressure in the vortex; water or salt spray drawn from the underlying surface also contributes to the structure and visibility of the funnel (see illustration). Diameters range from a few feet to several hundred, with winds occasionally strong enough to overturn small vessels.

Most waterspouts occur in moist tropical air under unstable conditions favorable for thunderstorms, from which they frequently hang. Nearly all dissipate on passing inland. Although most common in the tropical oceans, waterspouts are observed in higher latitudes, particularly in summer; they also occur over inland lakes. Many of the

WATERSPOUT

Waterspout. (*Official U.S. Navy photograph*)

larger and more violent spouts are true tornadoes, formed in association with thunderstorms over land and passing out to sea. More common are "fair weather spouts," thin vortex threads up to a few tens of feet across, associated with scattered cumulus clouds whose tops may not extend above 20,000 ft. In contrast to tornadoes most spouts form in generally undisturbed conditions, with weak winds in the surroundings. Some small spouts are thought to be akin to whirlwinds or dust devils, which form as convective whirls in extremely unstable air heated from below. *See* TORNADO.   [CHESTER W. NEWTON]

Bibliography: J. H. Golden, Waterspouts at Lower Matecumbe Key, Florida, 2 September 1967, *Weather*, 23:103–114, 1968.

Weather

The state of the atmosphere, as determined by the simultaneous occurrence of several meteorological phenomena at a geographical locality or over broad areas of the Earth. When such a collection of weather elements is part of an interrelated physical structure of the atmosphere, it is termed a weather system, and includes phenomena at all elevations above the ground. More popularly, weather refers to a certain state of the atmosphere as it affects man's activities on the Earth's surface. In this sense, it is often taken to include such related phenomena as waves at sea and floods on land.

An orderly association of weather elements accompanying a typical weather system of the Northern Hemisphere may be illustrated by a large anticyclone, or high-pressure region. In such a "high," extending over an area of many thousands of square miles, the usually gentle winds circulate clockwise around the high-pressure center. This system often brings fair weather locally, which implies a bright sunny day with few clouds. The temperature may vary widely depending on season and time of day. However, a cyclone or low-pressure region is frequently associated with a dark cloudy sky with driving rain (or snow) and strong winds which circulate counterclockwise about a low-pressure center of the Northern Hemisphere.

A weather element is any individual physical feature of the atmosphere. At a given locality, at least seven such elements may be observed at any one time. These are clouds, precipitation, temperature, humidity, wind, pressure, and visibility. Each of these principal elements is divided into many subtypes. For a discussion of a characteristic local combination of several elements, as they might be observed at a U.S. Weather Bureau station, *see* WEATHER MAP.

The various forms of precipitation are included by international agreement among the hydrometeors, which comprise all the visible features in the atmosphere, besides clouds, that are due to water in its various forms. For convenience in processing weather data and information, this definition is made to include some phenomena not due to water, such as dust and smoke. Some of the more common hydrometeors include rain, snow, fog, hail, dew, and frost.

Both a physical (or genetic) and a descriptive classification of clouds and hydrometeors have been devised. The World Meteorological Organiza-

tion, which among many other activities coordinates the taking of weather observations among the nations of the world, recognizes at least 36 cloud types and 100 classes of hydrometeors.

Certain optical and electrical phenomena have long been observed among the weather elements. These include lightning, aurora, solar or lunar corona, and halo. *See* AIR MASS; ATMOSPHERE; ATMOSPHERIC GENERAL CIRCULATION; CLOUD; FRONT; METEOROLOGY; PRECIPITATION; STORM; WEATHER STATION; WIND.

[PHILIP F. CLAPP]

Bibliography: L. J. Battan, *Weather*, 1974; H. Dickson, *Climate and Weather*, 1976; S. Petterssen, *Weather Analysis and Forecasting*, vol. 2, 2d ed., 1956; G. Sutton, *The Weather*, 1975.

Weather forecasting and prediction

Procedures for extrapolation of the future character of weather on the basis of present and past conditions. Accurate weather prediction requires a knowledge of the past state of the atmosphere, an understanding of the physical laws governing atmospheric behavior, and the availability of necessary technological aids for the rapid dissemination of meteorological information and the preparation of the forecast. The historical development of methods for forecasting the weather can be traced to innovations in these three areas. For introductory discussions of atmospheric science *see* ATMOSPHERE; METEOROLOGY.

This article is divided into five sections. The first part emphasizes the current status of the whole field of weather forecasting and prediction, and concentrates on short-range prediction, up to 48 hr in advance. The second portion deals with long-range prediction. The third section considers statistical forecasting procedures. A fourth section summarizes the bases for the developing techniques of numerical prediction. The last portion briefly characterizes the weather offices and centers through which are funneled the data of observation for analysis and processing into forecast and prediction. For résumés of weather observation and data gathering *see* METEOROLOGICAL INSTRUMENTATION; WEATHER STATION.

DEVELOPMENT

Information on the state of the atmosphere has been greatly expanded since 1939, when the demands of aircraft in World War II led to the installation of radio-sounding stations and the development of weather reconnaissance systems. These sources have since been supplemented by radar aids, rockets, and satellites. When these data are recorded on charts, the meteorologist has a three-dimensional picture of atmospheric structure. A series of such charts at 6-, 12-, or 24-hr intervals shows the development of weather systems in terms of the changes in wind, pressure, temperature, humidity, cloudiness, precipitation, and visibility. These maps permit an analysis of the dominant long-wave patterns (discussed under long-range forecasting) with linear dimensions of 10^3–10^4 mi, the migratory cyclones and anticyclones with dimensions of 10^2–10^3 mi, and the weather conditions averaged over an area on the order of 10^4 mi². The resolution of weather detail is less exact over oceanic regions, in the polar areas, and in most tropical areas. Small-scale weather systems such as land-sea breezes, mountain-valley winds, and convective showers cannot be depicted on the conventional weather map and can be studied only by means of a dense network of observing stations. Radar information gives precipitation detail within the 10^4-mi² area.

During the 1960s satellites provided the most significant developments in data acquisition. Besides cloud pictures, satellite technology will yield much needed information on upper-air winds and temperatures on a global scale. *See* METEOROLOGI-CAL SATELLITES.

Data to forecasting. The step from data to forecast is achieved by a variety of methods which may be classified as follows.

Semiempirical techniques. The forecaster first synthesizes the raw information into dynamically meaningful models of the atmosphere (as best exemplified by the polar-front model of J. Bjerknes and H. Solberg). Then, by a combination of methods including the extrapolation of past trends, expected changes based upon qualitative physical reasoning, and recollection of the behavior of similar situations in the past, he arrives at an estimate of the position of the prominent features on tomorrow's weather charts. These features include the location and intensity of cyclones, anticyclones, fronts, upper-level pressure ridges and troughs, and the like. Certain formulas can be applied to this phase of the forecast but, because of simplifying assumptions involved in their use, they yield questionable results. The details of weather such as types of clouds, rainfall, and temperature are deduced from this prediction by again referring to models and by considering such effects as advection, radiation, topographical influences, and expected stability changes.

The success of these techniques is limited by their semiempirical character and by the ability of an individual to handle the tremendous mass of significant information. Data are now so voluminous that this method of prediction utilizes only a fraction of the available information, and it becomes somewhat a matter of personal choice as to the selection process.

Numerical methods. The advent of an expanded network of weather observations and of high-speed computers has greatly stimulated meteorological research so that forecasting is passing from the pre-1945 qualitative phase to a quantitative era in which predictions are based upon computation, guided by physical principles. These methods are discussed in the sections on the numerical prediction of weather and the statistical prediction of weather.

The large-scale aspects of the atmospheric flow pattern are most accurately predicted by numerical methods. In addition, numerical methods are superseding the qualitative techniques of the past in the prediction of cloudiness, precipitation, wind, and temperature.

Evaluating forecasting. To analyze the accuracy of weather forecasts, it is necessary to recognize the statistical nature of the element being predicted. Precipitation, even when it covers wide areas, is rarely uniform in intensity. Small convective showers, which develop and dissipate rapidly, are often embedded in the general rain or snow and cause significant variations over distances of the order of 10 mi and over time periods of less than 1 hr. The shower or thunderstorm cells which are associated with air-mass showers or with cold fronts exhibit a maximum variability in time and space. This turbulent behavior presents a major forecasting problem. Consequently, the prediction must give wide range to the estimates of precipitation intensity in order to cover the probable variability over an area on the order of 10^4 mi². Only in those areas where the local variations can be attributed to orographical influences is it possible to present a more definitive estimate.

In contrast to the variability of precipitation, the turbulent fluctuations of wind and temperature are so rapid that they are not of general interest to the forecaster or the public. The specialized problem of predicting atmospheric pollution is an exception. Local variations in temperature and wind within the 1-hr, 10-mi scale can be attributed to such well-understood influences as the nature of the underlying surface, proximity to a water area, and a valley or mountain effect. Hence, detailed forecasts of these elements can be made with considerable reliability.

The final consideration with all types of forecast is the length of the time step. The larger the scale of the atmospheric system, the more persistent the phenomenon. For example, an individual summer shower has a life-span of approximately 1 hr; a cyclone is ordinarily identifiable for at least 3 days, and a particular long-wave pattern may persist for weeks. The major problem in weather prediction is the forecasting of a new development; it may be as difficult to pinpoint a summer shower a few hours in advance as to predict, a week in advance, the broad-scale features of the weather associated with a long-wave pattern. For similar reasons, the accuracy with which the weather may be predicted in detail decreases rapidly with the time elapsed since the observations were made.

Probable future developments. A major handicap in the past has been the paucity of dynamically significant information on the atmosphere over the ocean areas and over the less-populated regions of the globe. The developments in satellite technology offer real promise for filling these data gaps. With the globe's atmosphere well charted and with developments in numerical prediction it is anticipated that quantitative forecasts, via the computer, will replace completely the older qualitative techniques. The major improvement can be expected in the longer-range aspects of weather prediction. The isolated summer afternoon rain shower, which plagues many a picnic, will probably remain an unsolved forecast problem for many years to come.   [JAMES M. AUSTIN]

LONG-RANGE FORECASTING

Long-range weather forecasts are of two types. Medium- or extended-range forecasts cover periods of from 48 hr to a week in advance. Forecasts for longer periods generally extend over periods of a month, a season, or possibly two or more seasons in advance. Although meteorologists have been working on long-range prediction problems for more than 100 years, the degree of accuracy is small for all predictions exceeding a week in advance. This seems particularly true when the predictions are examined under rigid statistical controls, such as climatological probability. There is little or no evidence to indicate any sustained success for forecasts embracing periods of more than a season in advance. The reason for such limited ability lies in the utter complexity of the atmosphere's behavior—the vicissitudes of a compressible fluid responding to changing external stimuli, such as the Sun, and changing characteristics of the Earth's surface, both in space and time.

Medium or extended ranges. Scientific methods of extended-range prediction take for granted that the further out in time the forecast is projected, the more general must be the nature of the prediction. Short-range forecasts for 48 hr or less in advance specify the detail of weather in space and in time. Medium-range forecasts for a week in advance cover average conditions and trends within the week in intervals of a couple of days. Forecasts for a month or more, however, can indicate only the broad-scale (for example, areas of several hundred thousand square miles) features of average or prevailing weather. Such broad-scale aspects are usually expressed in terms of departures from seasonal norms for elements such as temperature and precipitation.

Medium-range forecast methods are apt to use one or a combination of dynamic, statistical, and synoptic techniques. Dynamic methods capitalize upon the best physical knowledge of meteorological phenomena. Statistical methods employ empirically derived equations as substitutes for physical knowledge. In the synoptic technique, various hemispheric wind and weather charts are surveyed and interpreted by an experienced meteorologist. An important part of modern dynamic methods is the principle that the vertical component of absolute vorticity remains fairly constant as air columns of the middle troposphere move from one area to another (discussed below under numerical weather prediction). When instantaneous wind and pressure charts for midtropospheric levels are averaged in time or space, certain small-scale perturbations, including short waves or vortexes in the horizontal, are suppressed. What remains are smooth, long, or planetary waves which in effect constitute a special class of motions; they are not only of larger scale (often being composed of a family of cyclones or anticyclones), but they also evolve more slowly than the individual wind charts from which they are constructed. The planetary waves which these time-averaged charts reveal are responsible for variations in the position and intensity of the well-known sea-level centers of action (like the Bermuda high, one of the subtropical oceanic highs) which largely determine prevailing weather abnormalities. For purposes of extended forecasting, the averaging process is performed on past (observed) data as well as on numerically predicted charts to 4 days in advance. Various methods of comparison of such time-averaged charts enable the synoptician to assay the continuity and trend of large-scale systems and to extrapolate them into the future for some reasona-

ble period. Dynamic methods of prediction may also be used with some success on time-averaged charts, but the physical reasons for this are not clear.

Procedures for extended forecasting vary around the world largely in accordance with facilities and availability of scientific manpower. Many countries do not have available the high-speed computing equipment necessary to prepare the dynamic component of the forecast, nor have they even the statistical components. In these cases extended forecasts are either not prepared at all or are made by educated synoptic guesswork.

After predictions of average planetary wind flows at upper levels have been made, it is possible to infer the accompanying types of weather in different areas as well as to estimate the general regions for breeding and movement of storms and air masses. Here again the statistics of the motions and weather are more predictable than is the day-to-day detail. In fact, the translation of average wind circulation into average weather is amenable to statistical stratification procedures and is fairly objective. These methods employ as input numerically predicted charts for the 700-millibar (mb) level and also surface temperature predictions to 48 hr made by field forecasters throughout the United States. From these data multiple-regression formulas make possible high-speed computation of temperature forecasts for the weather forecast centers in the United States. A similar procedure gives precipitation estimates. Other weather elements are inferred from the predicted charts for the period.

The boundary between the domain of short- and extended-range forecasting techniques has shifted: Dynamic predictions performed by computer now form the basis for 72-hr prognoses. It is possible that before 1980 dynamic predictions will form a new base for daily forecasts up to a week in advance and that time-averaging methods will no longer be necessary in that range.

Longer-period forecasting. For periods more than a week in advance, methods of forecasting rely more upon statistics and less upon physical reasoning. Concentrated efforts to explore the physics of long-range weather phenomena began about 1955 as a result of increased availability of computing facilities and hemisphere-wide data coverage, particularly from the upper air. However, attempts are being made to prepare forecasts a month ahead by employing dynamic principles in conjunction with statistical and synoptic techniques. For these purposes, another class of mean motions is defined by construction of mean maps for 30-day periods. Although real understanding of these methods is remote, experiments indicate that such objective, machine-produced prognoses are helpful and contain a large part of the accuracy of 30-day forecasts. These prognoses take rough account of the net effect of changes in insolation associated with the change of season, the tendency of certain branches of the general circulation to persist, and the compatibility of the positions of certain large-scale features like the Bermuda high and Icelandic low. Once the circulation pattern for the Northern Hemisphere is prognosticated, the average temperature and precipitation anomalies are computed with the help of elaborate statistical specification equations. The numerical results are then adjusted subjectively by attempting to consider factors not in the equations, like snow cover and wet or dry soil.

Another less expensive and less time-consuming method of longer-range prediction involves the use of statistical analogs. The historical files of weather maps for past periods (mean maps may also be used) are searched, and a wind and weather pattern is sought which is as similar as possible to the one which has been operative, on the assumption that what transpired in the earlier case will repeat. For best results the analogy should be good for large areas of the hemisphere, should hold for upper levels as well as for sea level, and should stand up for a sequence of periods preceding the forecast. The logic of this method is appealing; similar patterns under the same stimuli (such as season of the year) tend to repeat. However, the relatively short span of time for which meteorological records have been kept makes it difficult to find good analogs.

For periods beyond a month, statistical techniques seem to be the only ones sufficiently accurate; even here, there is some question as to whether the samples of data (length of record) are adequate to assure the stability of discovered relationships. However, experiments in seasonal forecasting indicate some reliability in predicting departures from normal of temperature for the contiguous United States.

Another avenue of approach to the long-range forecast problem which shows promise involves large-scale interactions between the ocean and atmosphere. J. Bjerknes has produced work which indicates that seasonally variable ocean temperatures near the Equator affect the rainfall there, and that the resulting variable release of heat of condensation controls the Hadley circulation, which in turn determines the strength of the subtropical anticyclones and the prevailing westerlies. Complementary work by J. Namias suggests that the longitudinal positioning and intensities of the long waves of the planetary circulation are also determined by air-sea interactions. Although these ideas are not developed to the point of utilizing them objectively in long-range forecasting, attempts are being made to consider them subjectively. *See* ATMOSPHERIC GENERAL CIRCULATION.

Finally, there is hope that numerical forecasts iterated day by day for weeks in advance may yield economically valuable statistics on the weather of the forthcoming month or season. A truly adequate global network of observations is necessary before this will be possible.

[JEROME NAMIAS]

STATISTICAL WEATHER FORECASTING

Statistical weather forecasting is the prediction of weather by rules based upon the statistics of weather behavior. A prediction may state the expected value of a specific weather element, such as a wind speed, or the probability of occurrence of a specific weather event, such as a thunderstorm. In the former case the prediction is understood to contain an error whose probable value may or may not be stated. The choice of the form of prediction may depend upon the intended audience; for example, the statement that there are 2 chances in 10 that tonight's temperature will fall below 32°F might aid a fruit grower more than the statement that tonight's expected minimum temperature will be 36°F.

Basic premises. Statistical forecasting is based upon the premise that the future worldwide state of the atmosphere is determined, at least approximately, by the present state, together with the intervening influences of the Sun and the underlying ocean and land, according to immutable physical laws. In theory, forecasting is equivalent to solving the equations representing these laws, but the equations are rather intractable, and because there are vast gaps between observing stations, the present weather is only partially known. It is sometimes more feasible to ascertain how future weather must evolve from present weather by studying how the observed portions of the atmosphere have previously behaved.

Prediction rules established from such study often relate future weather to present and past weather, instead of present weather alone, since past knowledge may partially compensate for incomplete present knowledge. A rule is commonly expressed as a mathematical formula. Sometimes the same information is more conveniently presented as a graph or a table.

A rule established for one location does not generally apply at another location. For example, a table established for San Francisco, showing the probability of occurrence of nighttime fog following various combinations of midafternoon temperature and relative humidity, would not be valid for predicting fog in New York. A new rule is usually needed for each new weather prediction in any particular area.

Statistical procedures. A general kind of procedure is used to establish a formula. The meteorologist chooses a set of weather elements for "predictors" and selects, commonly from past records, a set of data consisting of corresponding observed values of the predictors and the predictand. As the next step, he chooses a mathematical form with a limited number of degrees of freedom, ordinarily appearing as undetermined constants, and restricts the formula to this form. He specifies a process, ordinarily the minimization of the sum of squares of the prediction errors, by which the chosen data shall determine the constants. Evaluating the constants is then an objective and usually routine mathematical task.

The meteorologist may modify this procedure and classify combinations of values of the chosen predictors into categories. He may then construct a table by choosing, for each category, the average observed value of the predictand as the expected value or, alternatively, by choosing the observed frequency of occurrence as the probability.

The preparation of a forecast once the rule is established is objective and usually simple. The forecaster evaluates the formula after introducing the appropriate numerical values of the predictors, or reads the forecast from the appropriate location in the graph or table.

The data selected for establishing a formula constitute a finite sample of the total history of the weather. The formula is likely to succeed, when applied to future weather, only if the number of degrees of freedom is small compared to the number of values of each predictor in the sample, since virtually any finite set of numbers will fit a sufficiently complicated formula.

Ideally the sample should be made very large. When this is not feasible, because of the excessive labor involved or the absence of extensive past records, the degrees of freedom must be restricted. This is accomplished by limiting the number of predictors or restricting the formula to a more highly specialized form. Meteorological experience or physical reasoning should be used as a guide, since a blind choice of predictors, or of a mathematical form, is unlikely to yield a successful formula.

Statistical linear regression. The simplest mathematical form, and the one whose theory is most highly developed, is the linear formula. When many predictors have been chosen, the number may be reduced either by factor analysis, which selects a few linear combinations of the predictors in order of their ability to represent all the predictors, or by a procedure which selects a few predictors in order of their independent contribution to the prediction. Widespread investigation of linear formulas has followed the advent of high-speed electronic computing machines.

Appropriate nonlinear formulas are theoretically superior to linear formulas, but since they usually involve many degrees of freedom, they are more difficult to discover.

Statistical methods are highly suitable for predicting special local phenomena such as the occurrence of fog. For preparing prognostic weather maps one or two days in advance, statistical formulas are useful, but are frequently inferior to conventional subjective forecasts. For forecasting several days in advance, linear statistical formulas show a slight positive utility and compare favorably with other methods.

[EDWARD N. LORENZ]

NUMERICAL WEATHER PREDICTION

Numerical weather prediction is the prediction of weather phenomena by the numerical solution of the equations governing the motion and changes of condition of the atmosphere. More generally, the term applies to any numerical solution or analysis of the atmospheric equations of motion.

The laws of motion of the atmosphere may be expressed as a set of partial differential equations relating the instantaneous rates of change of the meteorological variables to their instantaneous distribution in space. These are developed in dynamic meteorology. A prediction for a finite time interval is obtained by summing the succession of infinitesimal time changes of the meteorological variables, each of which is determined by their distribution at the preceding instant of time. Although this process of integration may be carried out in principle, the nonlinearity of the equations and the complexity and multiplicity of the data make it impossible in practice. Instead, one must resort to finite-difference approximation techniques in which successive changes in the variables are cal-

culated for small, but finite, time intervals at a finite grid of points spanning part or all of the atmosphere. Even so, the amount of computation is vast, and numerical weather prediction remained only a dream until the advent of the modern high-speed electronic computing machine. These machines are capable of performing the millions of arithmetic operations involved with a minimum of human labor and in an economically feasible time span. Numerical methods are gradually replacing the earlier, more subjective methods of weather prediction in many United States government weather services. This is particularly true in the preparation of prognoses for large areas. The detailed prediction of local weather phenomena has not yet benefited greatly from the use of numerico-dynamic methods, as indicated above in the general section on weather forecasting.

Short-range numerical prediction. By the nature of numerical weather prediction, its accuracy depends on (1) an understanding of the laws of atmospheric behavior, (2) the ability to measure the instantaneous state of the atmosphere, and (3) the accuracy with which the solutions of the continuous equations of motion are approximated by finite-difference means. The greatest success has been achieved in predicting the motion of the large-scale (>1000 mi) pressure systems in the atmosphere for relatively short periods of time (1–3 days). For such space and time scales, the poorly understood energy sources and frictional dissipative forces may be largely ignored, and rather coarse space grids may be used.

The large-scale motions are characterized by their properties of being quasi-static, quasi-geostrophic, and horizontally quasi-nondivergent, as dicussed in another article. *See* METEOROLOGY.

These properties may be used to simplify the equations of motion by filtering out the motions which have little meteorological importance, such as sound and gravity waves. The resulting equations then become, in some cases, more amenable to numerical treatment.

A simple illustration of the methods employed for numerical weather prediction is given by the following example. Consider a homogeneous, incompressible, frictionless fluid moving over a rotating, gravitating plane in such a manner that the horizontal velocity does not vary with height. For quasi-static flow the equations of motion are Eqs. (1), and the equation of mass conservation is Eq.

$$\frac{\partial u}{\partial t} + u\frac{\partial u}{\partial x} + v\frac{\partial u}{\partial y} = -g\frac{\partial h}{\partial x} + 2\omega v$$
$$\frac{\partial v}{\partial t} + u\frac{\partial v}{\partial x} + v\frac{\partial v}{\partial y} = -g\frac{\partial h}{\partial y} - 2\omega u \tag{1}$$

(2), where u and v are the velocity components in

$$\frac{\partial h}{\partial t} + u\frac{\partial h}{\partial x} + v\frac{\partial h}{\partial y} = -h\left(\frac{\partial u}{\partial x} + \frac{\partial v}{\partial y}\right) \tag{2}$$

the directions of the horizontal rectangular coordinates x and y, t is the time, g is the acceleration of gravity, ω is the angular speed of rotation, and h is the height of the free surface of the fluid. Let the variables u, v, and h be defined at the points $x = i\,\Delta x$, $y = j\,\Delta x$ ($i = 0, 1, 2, \ldots, I$; $j = 0, 1, 2, \ldots, J$)

and at the times $t = k\,\Delta t\,(k=0, 1, 2, \ldots, K)$, and denote quantities at these points and times by the subscripts i, j, and k. Derivatives such as $\partial u/\partial t$ and $\partial u/\partial x$ may be approximated by the central difference quotients given by Eqs. (3). In this way

$$\frac{\Delta_k u_{i,j}}{2\,\Delta t} \equiv \frac{u_{i,j,k+1} - u_{i,j,k-1}}{2\,\Delta t}$$

$$\frac{\Delta_i u_{j,k}}{2\,\Delta x} \equiv \frac{u_{i+1,j,k} - u_{i-1,j,k}}{2\,\Delta x} \qquad (3)$$

Eqs. (4), the finite-difference analogs of the continuous equations, are obtained.

$$u_{i,j,k+1} = u_{i,j,k-1} - \frac{\Delta t}{\Delta x}\,(u_{i,j,k}\,\Delta_i u_{j,k} + v_{i,j,k}\,\Delta_j u_{i,k}$$
$$+ g\,\Delta_i h_{j,k}) + 4\omega v_{i,j,k}\,\Delta t$$

$$v_{i,j,k+1} = v_{i,j,k-1} - \frac{\Delta t}{\Delta x}\,(u_{i,j,k}\,\Delta_i v_{j,k} + v_{i,j,k}\,\Delta_j v_{i,k}$$
$$+ g\,\Delta_j h_{i,k}) - 4\omega u_{i,j,k}\,\Delta t \qquad (4)$$

$$h_{i,j,k+1} = h_{i,j,k-1} - \frac{\Delta t}{\Delta x}\,[u_{i,j,k}\,\Delta_i h_{j,k} + v_{i,j,k}\,\Delta_j h_{i,k}$$
$$+ h_{i,j,k}\,(\Delta_i u_{j,k} + \Delta_j v_{i,k})]$$

Equations (4) give u, v, and h at the time $(k+1)\,\Delta t$ in terms of u, v, and h at the times $k\,\Delta t$ and $(k-1)\,\Delta t$. It is then possible to calculate u, v, and h at any time by iterative application of the above equations.

It may be shown, however, that the solution of the finite-difference equations will not converge to the solution of the continuous equations unless the criterion $\Delta s/\Delta t > c\sqrt{2}$ is satisfied, where c is the maximum value of the speed of long gravity waves $\sqrt{gh}$. Under circumstances comparable to those in the atmosphere, Δt is found to be so small that a 24-hr prediction requires some 200 time steps and approximately 10,000,000 multiplications for an area the size of the Earth's surface. The computing time on a machine with a multiplication speed of 100 μsec, an addition speed of 10 μsec, and a memory access time of 10 μsec would be about 30 min. The magnitude of the computational task may be comprehended from the fact that the more accurate atmospheric models now envisaged will require some 100–1000 times this amount of computation.

A saving of time is accomplished by utilizing the quasi-nondivergent property of the large-scale atmospheric motions. If, in the above example, the horizontal divergence $\partial u/\partial x + \partial v/\partial y$ is set equal to zero, the motion is found to be completely described by the equation for the conservation of the vertical component of absolute vorticity, as developed in another article. *See* METEOROLOGY.

The solution of this equation may be obtained in far fewer time steps since gravity wave motions are filtered out by this constraint and the velocity c in the Courant-Friedrichs-Lewy criterion becomes merely the maximum particle velocity instead of the much greater gravity wave speed.

Cloud and precipitation prediction. If, to the standard dynamic variables u, v, w, p, and ρ, a sixth variable, the density of water vapor, is added, it becomes possible to predict clouds and precipi-

tation as well as the air motion. When a parcel of air containing a fixed quantity of water vapor ascends, it expands adiabatically and cools until it becomes saturated. Continued ascent produces clouds and precipitation.

To incorporate these effects into a numerical prediction schema one adds Eq. (5), which governs

$$\frac{Dr}{Dt} \equiv \frac{\partial r}{\partial t} + u\,\frac{\partial r}{\partial x} + v\,\frac{\partial r}{\partial y} + w\,\frac{\partial r}{\partial z} = S \qquad (5)$$

the rate of change of specific humidity r. Here S represents a source or sink of moisture. Then it is necessary also to include as a heat source in the thermodynamic energy equation a term which represents the time rate of release of the latent heat of condensation of water vapor. The most successful predictions made by this method are obtained in regions of strong rising motion, whether induced by forced orographic ascent or by horizontal convergence in well-developed depressions. The physics and mechanics of the convective cloud-formation process make the prediction of convective cloud and showery precipitation more difficult.

Large-scale numerical weather prediction. In 1955 the first operational numerical weather prediction model was introduced at the National Meteorological Center (NMC). This simplified barotropic model consisted of only one layer and therefore could model only the temporal variation of the mean vertical structure of the atmosphere. By the late 1960s, the speed of computers had increased sufficiently to permit the development of multilevel (usually about 6–10) models which could resolve, at least in part, the vertical variation of the wind, temperature, and moisture. These multilevel models predicted the fundamental meteorological variables mentioned above for large scales of motion. The characteristic grid size was about 400 km on a side, and the model's domain covered most of the Northern Hemisphere. Because the boundary of the domain was located in the tropics where the horizontal gradients of the atmospheric properties were weak compared with those in middle latitudes, it was possible to treat the model variables on the boundary in rather simple ways (such as holding the variables temporally constant during the forecast).

Numerical calculation of climate. While hemispheric models were being implemented for operational weather prediction 1–3 days in advance, similar research models were being developed which covered the entire Earth. These global models (also called general circulation models, or GCMs) could, in principle, be used to simulate the long-term variation of weather, that is, the climate.

The extension of numerical predictions to long time intervals requires a more accurate knowledge than now exists of the energy transfer and turbulent dissipative processes within the atmosphere and at the air-earth boundary, as well as greatly augmented computing-machine speeds and capacities. However, predictions of mean conditions over large areas may well become possible before such developments have taken place, for it is now possible to incorporate into the prediction equations estimates of the energy sources and sinks—estimates which may be inaccurate in detail but correct in the mean. Several mathematical experiments involving such simplified energy sources have yielded predictions of mean circulations that strongly resemble those of the atmosphere.

The above-mentioned experiments lead to a hope that it will be possible to explain the principal features of the Earth's climate, that is, the average state of the weather, well before it becomes possible to predict the daily fluctuations of weather for extended periods. Should these hopes be realized it would then become possible to undertake a rational analysis of paleoclimatic variation and changes induced by artificial means. If the existing climate could be understood from a knowledge of the existing energy sources, atmospheric constituents, and Earth surface characteristics, it might also be possible to predict the effects on the climate of natural or artificial modifications in one or more of these elements.

Specialized prediction models. Although the coarse grids in the hemispheric and global models are necessary for economical reasons, they are sources of two major types of forecast error. First, the truncation errors introduced when the continuous differential equations are replaced with finite difference approximations cause erroneous behavior of the scales of motion that are resolved by the models. Second, the neglect of scales of motion too small to be resolved by the mesh (for example, thunderstorms) may cause errors in the larger scales of motion. In an effort to simultaneously reduce both of these errors, models with considerably finer meshes have been tested. However, the price of reducing the mesh has been the necessity of covering smaller domains in order to keep the total computational effort within current computer capability. Thus the limited-area fine-mesh model (LFM) run at NMC has a mesh length of approximately 120 km on a side, but covers a region only slightly larger than North America. Because the side boundaries of this model lie in meteorologically active regions, the variables on the boundaries must be updated during the forecast. A typical procedure is to interpolate these required future values on the boundary from a coarse-mesh model which is run first. Although simple in concept, there are mathematical problems associated with this method, including overspecification of some variables on the fine mesh. Nevertheless, limited-area models have made significant improvements in the accuracy of short-range numerical forecasts over the United States.

Even the small mesh sizes of the LFM are far too coarse to resolve the detailed structure of many important atmospheric phenomena, including hurricanes, thunderstorms, sea- and land-breeze circulations, mountain waves, and a variety of air-pollution phenomena. In recent years considerable effort has gone into developing specialized research models with appropriate mesh sizes to study these and other small-scale systems. Thus, fully three-dimensional hurricane models with mesh sizes of 20 km simulate many of the features of real hurricanes. On even smaller scales, models with horizontal resolutions of several kilometers reproduce many of the observed features in the life cycle of thunderstorms and squall lines. It would be entirely misleading, however, to imply that

models of these phenomena differ from the large-scale models only in their resolution. In fact, physical processes which are negligible on large scales become important for some of the phenomena on smaller scales. For example, the drag of precipitation on the surrounding air is important in simulating thunderstorms, but not for modeling large scales of motion. Thus the details of precipitation processes, condensation, evaporation, freezing, and melting are incorporated into sophisticated cloud models.

In another class of special models, chemical reactions between trace gases are considered. For example, in models of urban photochemical smog, predictive equations for the concentration of oxides of nitrogen, oxygen, ozone, and reactive hydrocarbons are written. These equations contain transport and diffusion effects by the wind as well as reactions with solar radiation and other gases. Such air-chemistry models become far more complex than atmospheric models as the number of constituent gases and permitted reactions increases.   [RICHARD A. ANTHES]

CENTERS AND OFFICES OF FORECASTING

Weather forecasts for all parts of the United States are prepared by the National Weather Service (NWS), a part of the National Oceanic and Atmospheric Administration (NOAA). Various phases of the forecast work are performed at five working levels: the National Meteorological Center (NMC); National Severe Storms Forecast Center (NSSFC); National Hurricane Center (NHC); River Forecast Centers (RFCs); and Weather Service Forecast Offices (WSFOs).

National Meteorological Center. The NMC, located in Camp Springs, MD, collects weather observations, prepares meteorological analyses, and provides forecast guidance for use by NWS field offices. The aerial coverage of NMC products includes the entire globe, with most products covering the Northern Hemisphere and the tropical regions of the Southern Hemisphere. The World Meteorological Organization (WMO) has designated the NMC as the analysis and forecast arm of the Washington World Meteorological Center, which requires global responsibilities as part of the international efforts and cooperation known as the World Weather Watch. To carry out these responsibilities, all of the NMC divisions are extending coverage of their products to the entire globe.

Divisions. The NMC has four divisions as follows:

The Forecast Division applies a combination of numerical and manual techniques to produce analyses and prognoses up to 240 hr into the future, emphasizing the advance time period of 2 to 72 hr. The meteorologists manually adjust computer output to reduce errors, and interpret areas of weather and cloudiness for the guidance of NWS field offices (Fig. 1).

The Automation Division operates the NMC's computers and their interface to various NWS (Fig. 2) and Federal Aviation Administration (FAA) communication links.

The Development Division adapts research and development results in numerical weather predic-

Fig. 1.   Meteorologists in NMC Forecast Division manually adjust computer surface prognostic charts to reduce errors, and interpret areas of weather and cloudiness for guidance of NWS field offices. (*NOAA*)

tion to the NMC output and conducts research into improved numerical weather prediction, stratospheric research, and the use of satellite-derived data.

The Climate Analysis Center analyzes current and short-range climate fluctuations and prepares short-term forecasts and long-term outlooks. The forecasts and projections are used by a wide variety of government and private users.

Operations. The centralized preparation of data, analyses, and forecasts is designed to eliminate most requirements for hand-charting and independent meteorological analysis in the NWS offices as well as user groups such as airline and private meteorologists. The NMC, through the use of a large computer facility and together with numerical forecast methods, provides the NWS and other government agencies with daily forecasts and outlooks out to 10 days in advance. In the course of a day, the NMC receives 25,000 hourly aviation reports, 14,000 synoptic land station reports, 2500 ship reports, 3500 aircraft reports, and 2500 atmospheric soundings. All available cloud and temperature data derived from

Fig. 2.   Cathode-ray tube display system used throughout NWS for forecast preparation and weather watch. (*NOAA*)

Fig. 3. Master "storm room" at NHC. (*NOAA*)

weather satellites are integrated into the analyses. The NMC makes 785 facsimile and 819 teletypewriter transmissions daily to the field offices.

National Severe Storms Forecast Center. The NSSFC, located in Kansas City, MO, is responsible for preparing and releasing forecasts of areas of expected severe local storms, including tornadoes. The long-term outlook, prepared shortly after midnight, gives the geographic areas most likely to have severe thunderstorms or tornadoes during the next 24 hr. The convective forecast is used for planning purposes by preparedness groups and as input to the local forecast program by NWS offices throughout the contiguous United States. As the possibility of severe thunderstorms increases, the NSSFC severe storms meteorologist prepares tornado or severe thunderstorm watches, typically for an area of 25,000 mi^2

Fig. 4. Meteorological technician monitoring weather on NWS long-range radar. (*NOAA*)

(65,000 km^2) and for time periods 1–7 hr in advance. A second group of NSSFC meteorologists and technicians monitor weather surveillance radar reports and satellite photos to locate areas of potential hazard for aircraft inflight. The resulting warnings, called convective SIGMETs, are relayed immediately to various FAA facilities and to the airline operations offices. In a typical year, the NSSFC logs 700–900 tornadoes and 1500–2000 severe thunderstorms. Four hundred tornado or severe thunderstorm watches are issued along with 10,000–15,000 convective SIGMETs. *See* STORM DETECTION.

National Hurricane Center. The NHC, located in Miami, FL, prepares hurricane and tropical storm watches and advisories for most of the tropical Atlantic Ocean, the Caribbean, and the Gulf of Mexico. The NHC delegates parts of its public warning responsibilities to three Hurricane Watch Offices (HWO) when the United States coastal areas are threatened (Fig. 3). The HWOs are WSFOs at Boston, Washington, and San Juan. WSFOs at San Francisco and Honolulu prepare hurricane and tropical storm advisories for the eastern Pacific and central Pacific, respectively.

River Forecast Centers. There are 13 RFCs, which prepare river and flood forecasts and warnings for approximately 2500 communities. This includes forecasts for height of the flood crest as well as times when the river is expected to overflow its banks and when it will recede into its banks. At many points along larger streams or rivers, daily forecasts of river stage or discharge are routinely prepared for activities such as navigation and water management. Forecasts of seasonal snowmelt or water-year runoff are prepared for major rivers such as the Columbia, Missouri, and Mississippi and their tributaries.

Weather Service Offices. There are an additional 238 WSOs, which represent the third echelon of the system. They issue local forecasts which are adaptations of the zone forecasts. They are important in the warning and observation program and are generally located in smaller cities.

Weather Service Forecast Offices. There are 52 WSFOs within the NWS. There are no WSFOs in Rhode Island, Connecticut, Vermont, and New Hampshire, while California and Pennsylvania have two and Alaska, Texas, and New York three. Each WSFO provides whatever weather service is required for the geographical areas it serves, utilizing output from the NMC, NSSFC, NHC, and a RFC where appropriate. Each WSFO provides routine issuances of state and zone forecasts for the public plus terminal and route forecasts for aviation. Other specialized services are handled where there is a need and where resources and workload permit. These include a Disaster Preparedness Program designed to save lives and mitigate the social and economic impacts of natural disasters; Severe Local Storms Warning Program, where the weather radar is the principal method for monitoring the storms (Fig. 4); Winter Weather Warning Service; Coastal Flood Warning Program; Agricultural Weather Service; warnings of low temperatures for winter and spring crops; forecasts for flights from the United States and its possessions to other countries; meteorological support to control and combat air pollution; specialized forecasts and warnings to fire control agencies; flash flood watches and warnings;

tsunami watches and warnings; warnings for high seas, coastal waters, and inland waterways and offshore marine forecasts for recreational boating and fishing. Forecasts and warnings are transmitted via the NOAA Weather Wire, a dedicated teletype circuit, and over the NOAA Weather Radio, a continuous weather and river information broadcast on one of three high-band frequencies — 162.40, 162.475, or 162.55 MHz.

[ALLEN PEARSON]

Bibliography: J. Bjerknes and H. Solberg, Life cycle of cyclones and polar front theory of atmospheric circulation, *Geofys. Publ.*, no. 3, 1922; G. I. Marchuk, *Numerical Methods in Weather Prediction*, 1973; A. S. Monin, *Weather Forecasting as a Problem in Physics*, 1972; J. Namias, Long range weather forecasting: History, current status, and outlook, *Bull. Amer. Meteorol. Soc.*, 49(5), 1968; S. Petterssen, *Weather Analysis and Forecasting*, 2 vols., 2d ed., 1956.

Weather map

A chart portraying the state of the atmospheric circulation and weather at a particular time over a wide area. It is derived from a careful analysis of simultaneous weather observations made at many observing points in the area. Such a chart gives the weather forecaster an integrated picture of the location, structure, and, when several successive charts are available, the motion and development of the various weather systems. From this study he

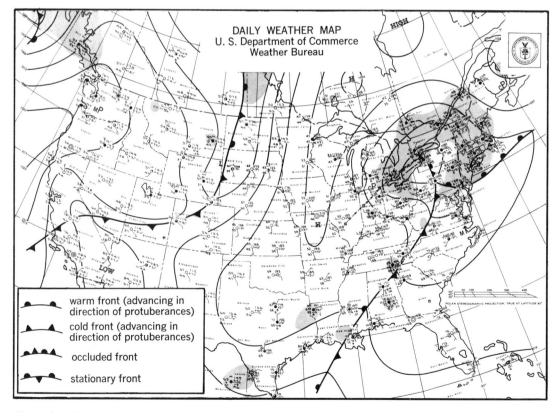

Fig. 1. A surface weather map at 0600 Greenwich Civil Time. Sea-level isobars (thin lines) are drawn for every 4 millibars and labeled in whole millibars; fronts, or transition zones separating air masses, are indicated by heavy lines; *mT* indicates tropical maritime air; *mP* indicates polar maritime air. Areas where precipitation was falling at 0600 are shaded. Previous 6-hourly positions of low-pressure center in eastern Great Lakes are indicated by crosses connected by arrows.

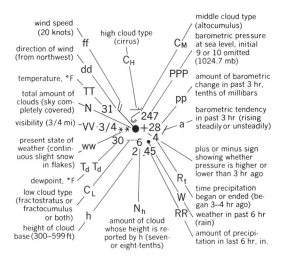

Fig. 2. Abbreviated code and station model for plotting weather elements at an observation station on the Earth's surface. Letters are symbols for the meteorological elements. In coded reports only numbers and cloud and weather plotting symbols are used. The use of symbols and abbreviations increases the density of information which can be put on the map.

may construct a prognostic chart, which portrays various weather features for selected times in the future.

Many kinds of weather maps are used, depending on the weather elements of immediate interest and their elevation above the ground. At a typical large weather analysis central, as many as 35 different charts are constructed for a given time. Among the more common of these is the surface map, which portrays the weather at the Earth's surface (Fig. 1). All the mapped weather elements except pressure are those directly observed at the weather station (Fig. 2). Except over the oceans the so-called sea-level pressure is a fictitious quantity obtained by reducing the surface pressure to sea level by a special formula so that continuous isolines (called isobars) may be drawn through regions having the same pressure value. These isobars are important in portraying the winds, the physical structure of weather systems, and the location of fronts and air masses. *See* AIR MASS; FRONT.

The surface map, at least over ocean regions, represents a section through the atmosphere along an approximately horizontal surface. This procedure may also be used in constructing upper-level maps, or charts showing the distribution of weather at fixed elevations above sea level. It is more common, however, to portray the weather at high elevations on constant-pressure surfaces (Fig. 3). The elements plotted at each upper-air station usually include the height of the constant-pressure surface, temperature, and some measure of humidity, usually the dew point. Lines of constant elevation (called isohypses or contours) then are drawn to portray the topography of the selected pressure surface. The contours bear a relation to the winds similar to that of the isobars of a constant-level chart. Thus the winds tend to blow along the contours with low heights to the left

when facing downstream. Their speed is roughly inversely proportional to the sine of the latitude and the distance between contours when these are drawn at constant-height intervals. Usually isolines of temperature, or isotherms, are also drawn, as in Fig. 3.

Other important two-dimensional or one-dimensional atmospheric sections cannot be termed weather maps, but rather meteorological charts and diagrams. These include vertical cross sections through the atmosphere, and thermodynamic diagrams similar to those used in studies of heat engines.

Weather map analysis. This branch of synoptic meteorology had its beginning at about the time the telegraph was invented, when for the first time weather observations covering large areas could be sent rapidly to a central location. Its steady development was greatly accelerated following World War I, when the techniques of air-mass analysis were developed by the Scandinavian school headed by V. Bjerknes. Upper-air charts were not commonly constructed until the 1930s, when sufficient high-level data became available.

The preparation of a weather chart at a large analysis central can be described as follows. First the encoded data at the surface and upper levels are transmitted to the central from collection centers by means of teletypes. If a map covering the Northern Hemisphere is to be prepared, data from approximately 850 surface and 400 upper-air reporting stations must be processed. This mass of data is subject to errors of observation, encoding, and transmission. Furthermore, large areas, particularly in the tropic and arctic latitudes, contain no observation stations and, hence, no data at all. The detection and correction of the errors, and the interpolation of weather in the intervals between reporting stations demand the greatest skill on the part of the chartmen who plot the data and the analysts who must interpret the data in terms of consistent physical structures of the atmosphere. After the data are corrected and plotted, the analyst then locates and draws various features such as fronts, air masses, and isobars. He must be guided by known physical principles regarding the horizontal, vertical, and temporal continuity of the atmosphere; that is, his analysis must be internally consistent. When finished, the chart is ready for the forecaster.

Data from orbiting weather satellites in the form of video cloud pictures are used on a daily basis in weather map analysis, particularly in areas of sparse surface-based observations.

Beginning in May 1969, the vertical atmospheric temperature structure over a large part of the entire Earth has been transmitted from satellites (such as *Nimbus 3* and *4* in 1969–1970) carrying satellite infrared spectrometer (SIRS) instruments. These data are used in automatic weather map analyses to supplement conventional surface-based systems. In 1970 this was done routinely over the Northern Hemisphere and experimentally over the Southern. *See* METEOROLOGICAL SATELLITES.

Automatic weather map analysis. With the introduction of high-speed electronic computers into

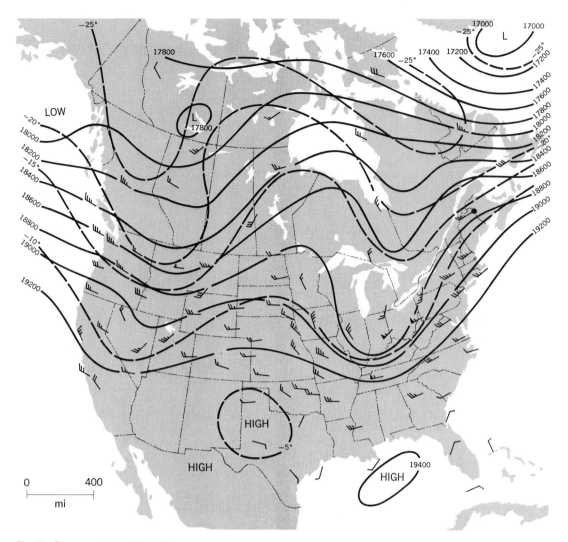

Fig. 3.   Representative 500-millibar constant-pressure chart. Isolines of topography of constant-pressure surface show height in feet above sea level. Isotherms are in °C at 5° intervals. Arrows with barbs fly with wind and show wind speed in knots. Each long barb = 10 knots; triangular barb = 50 knots. (*U.S. Weather Bureau*)

meteorology in 1951, it was realized that the slow manual methods of data collection and processing, described above, could be greatly speeded up so as to keep up with advances in automatic weather forecasting and with the large volume of data being received from all parts of the world. For these reasons experiments in automatic analysis were begun in a number of countries around 1953, and in 1957 the Joint Numerical Weather Prediction unit at Suitland, Md. (now the National Meteorological Center), began the routine production of machine-analyzed weather maps.

The objectives and final products are essentially the same as described above under weather map analysis. After the analysis has been completed by the main computer, it is stored on magnetic tapes and then displayed visually in a number of optional forms, such as machine print-outs, mechanically drawn maps using curve plotters, and photographs of pictures projected on video tubes. *See* METEOROLOGY; WEATHER FORECASTING AND PREDICTION.

[PHILIP F. CLAPP]

Weather modification

The changing of natural weather phenomena by humans. Weather is the product of the interaction of atmospheric processes on many scales, reaching from the planetary circulation to the microphysical processes in the evolution of cloud droplets and ice crystals. So far, only on the microscale of condensation and freezing nuclei have humans begun to exert modifying influences on weather. These influences may expand to a scale of several hundreds or thousands of square kilometers, that is, to what is called in meteorology the meso scale of meteorological phenomena.

There are actually four techniques by which humans may affect the natural course of weather: (1) by injection of large amounts of heat into the atmosphere in order to burn off warm fog; (2) by utilizing metastable states in the atmosphere as they happen—for instance, if clouds are being cooled below the freezing point, seeding can be effective; (3) by altering the surface condition, as by deforestation or urbanization; (4) by influencing

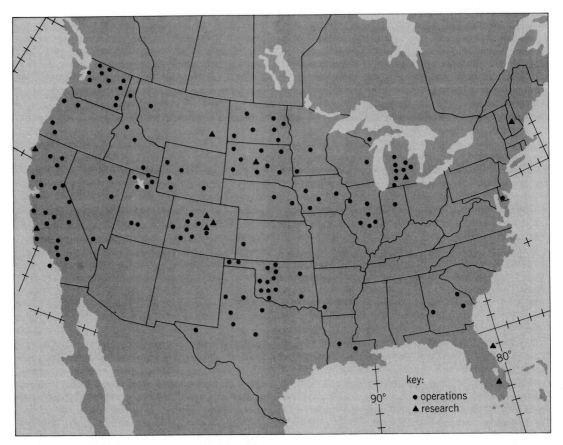

Fig. 1. Locations of research and operational weather projects in the United States, 1973–77. Two operational projects were in south-central Alaska, and no projects in Hawaii. (*NOAA Weather Modification Reporting Program, Rockville, MD*)

the weather inadvertently by pollution—for example, the exhaust gases from metal smelters contain many freezing nuclei, or smog affects the atmospheric radiation balance and, consequently, also cloud formation, which is also affected by the "heat island" over cities. Most frequently used nowadays is the second technique, the seeding of clouds with certain types of nuclei. Figure 1 shows the locations of research and operational weather modification projects in the United States during 1973–77. In 1977 about 7% of the area of the United States was covered by cloud seeding. There are 131 operational projects and 11 research projects.

Experimental principles. The numerous physical and meteorological processes which are involved in the experimental approach to weather modification make it complex and difficult. In many cases it is virtually impossible to design a classical physical experiment for determining a cause-effect relationship, and it is necessary to adopt a statistically designed experiment. Here one need not know all physical processes, feedback mechanisms, and interactions in order to derive the influence of the one artificially modified parameter, but one is required to conduct a great number of identical experiments. This calls for experimental periods which have to be counted in decades. In the meantime the environmental conditions of the experiment may change, so the appli-

cability of the statistical approach is limited. In view of these difficulties, Fig. 2 reflects the approach which should be taken during research projects by combining the classical and the statistical experimental principles.

This approach can be further sharpened by predicting the experimental result by numerical simulation of the experiment. Strangely enough, this is not the approach science has taken: it has relied more and more on statistics and given less and less weight to the physics involved in the experiment. Design, execution, and analysis were determined by statistical principles with disregard of the grave uncertainties to which the statistical approach is liable. The statistical approach postulates that one has a population of identical experimental situations which one can subdivide into control events and target events. The target events are treated with the real seeding agent, the control events with a placebo. The scientist in charge has to make the decision whether a certain cloud situation meets the experimental criteria for seeding, but in order to avoid bias in the scientist's observations the seeding decision is made from a set of random numbers and the decision is not communicated to the scientist. In rainmaking experiments, one finds the ratio of target rainfall to control rainfall and determines the significance of the results. Usually, a 5% significance is accepted as proof that the effect is not caused accidentally.

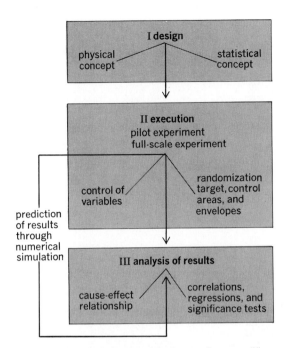

Fig. 2. Scheme for approach to weather modification experiments. Experiments must be in agreement with physical and statistical principles of experimentation.

clouds were dissipated by seeding each of three parallel 16-km-long lines 5 km apart with 100 lb of dry ice. In experiments over populated regions the weather has been modified conspicuously on the mesoscale of meteorological events: Sunshine has been made for several thousand people on a dull overcast winter day.

A contrasting case is illustrated in Fig. 4. Here, after seeding for dissipation, a miniature squall line developed over Lake Michigan consisting of heavy snow showers which lasted for about 1 hr. Similar results have been repeatedly obtained during seeding experiments over Lake Erie.

Figure 5 shows another spectacular result of cloud seeding. A cumulus cloud reaching to the $-8°C$ level was overseeded with silver iodide. Overseeding caused glaciation and hence heat of fusion was released by the glaciating cloud water. This heat increased the buoyancy of the cloud, and the cloud "exploded," growing in height and also in width. This experiment has a clear physical concept which tested a cause-effect relationship; however, randomization with seeded and control clouds contributed materially to sharpening the experimental result. Of 22 test clouds 14 were seeded and 8 were not; only 1 of the clouds not seeded grew comparatively. Of the 14 seeded clouds, 4 exploded and 10 increased in height with no increase in width.

Developments in experimental meteorology have caused an air of optimism in the approach to weather modification. In addition, radar, pulse Doppler radar, aircraft with sophisticated instrumentation serving as observing platforms, and satellites are providing unprecedented observation and analysis facilities.

Cloud seeding. Clouds being cooled below the freezing point remain liquid down to very low temperatures and only those droplets will freeze which contain a so-called freezing nucleus. It was the great discovery in the 1940s of Nobel prize winner I. Langmuir and his two assistants, V. Schaefer and B. Vonnegut, that clouds could be dissipated by seeding either with dry ice or with certain chemicals (such as silver iodide) and that, artificially, the cloud could be changed from a water cloud into an ice cloud. Judicious seeding would permit generation of only a certain number of artificial freezing nuclei and thus would influence only certain cloud conditions, such as the precipitation process or cloud dynamics.

Fog dispersal. The method of seeding is now applied operationally for the dispersal of supercooled fogs from airports all over the world. Systems have been developed for seeding with liquid propane in France (Orly Airport) and for seeding with dry ice, silver iodide, and lead iodide in the United States and the Soviet Union.

Dispersal of warm fog (temperature warmer than frost point) is accomplished by injection of heat into the atmosphere, as it is done alongside the runways of Orly Airport. *See* FOG.

Cloud modification. Some striking results of artificial weather modification have been obtained in cloud modification. Figure 3 shows an area of about 100 km² in which supercooled stratocumulus

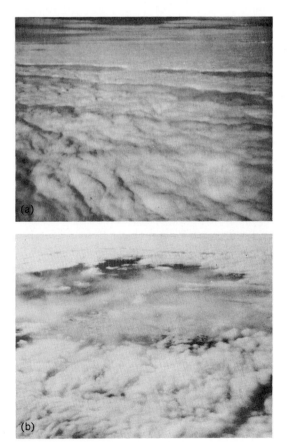

Fig. 3. Cloud dissipation. (a) Three lines in stratocumulus cloud layer 15 min after seeding. (b) Opening in stratocumulus layer 70 min after seeding. (*U.S. Army ECOM, Fort Monmouth, NJ*)

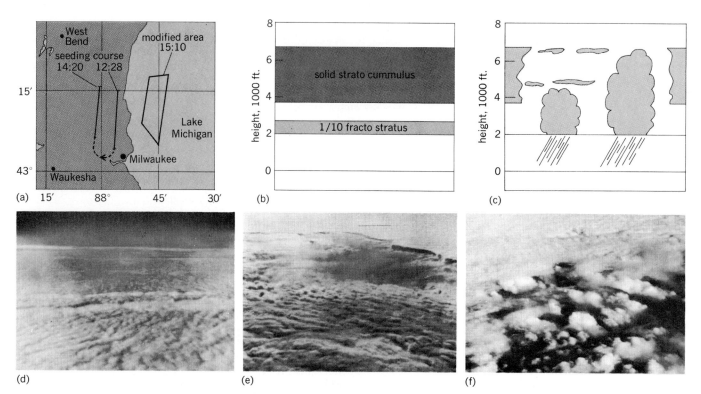

Fig. 4. Development of heavy snow showers after a seeding experiment for cloud dissipation over Milwaukee, WI, on Nov. 24, 1953. (*a*) Location of experiment. (*b*) Vertical section before seeding and (*c*) 63 min after seeding. (*d*) Cloud appearance 15 min after seeding, (*e*) 31 min after seeding, and (*f*) 63 min after seeding. (*U.S. Army ECOM, Fort Monmouth, NJ*)

The great complexity of cloud response to feedback mechanisms between microphysics and cloud dynamics was observed when an intensely growing cumulus was seeded in the updraft with 1500 g of AgI. Fifteen pyrotechnic devices were released at −20°C into the updraft at about 6000 m (mean sea level), and seeding occurred over a depth of 3000 m. At this temperature, snow crystals formed which very efficiently converted the cloud water to precipitation, so that the cloud snowed out and did not grow.

These experiments indicate that it is possible to affect materially the life history of clouds through seeding. *See* CLOUD PHYSICS.

Rainmaking. Water is one of the most abundant but also one of the most wanted substances on Earth. The interest in artificial rainmaking is therefore understandable. A panel of the U.S. National Academy of Sciences expressed restrained optimism by stating: "There is increasing but still somewhat ambiguous statistical evidence that precipitation from some types of cloud and storm systems can be moderately increased or redistributed by seeding techniques." However, the panel also recommended the early establishment of several carefully designed, randomized, seeding experiments, planned in such a way as to permit assessment of the seedability of a variety of storm types. These recommendations reflect that much corroborative evidence is still missing.

Indeed, the evidence is often controversial, particularly where the experiments can only be based on statistical design. In all such experiments it is tacitly assumed that the seeding agent gets to the right location in the cloud system and acts as desired. Seeding with ground-based generators has uncertainties because of the unknown diffusion properties of the seeding agent in the atmospheric boundary layer. Seeding from aircraft has yielded positive results in Israel, inconclusive results in Missouri, and negative results in Arizona.

The analysis of seeding data often presents great difficulties because it must be based on point measurements. Take, for instance, rainfall data. Rainfall can be measured only on discrete points whose number is limited because a balance has to be struck between the desire to have a network as dense as possible and to have one that can economically be analyzed and maintained. It has been attempted to measure rainfall by means of radar, but the relationship between the radar return and the rate of rainfall is very unreliable and changes from cloud to cloud. The variabilities which are introduced by these analysis problems may be greater than the seeding effect which should be measured.

The most noteworthy progress in this area comes from increased theoretical understanding of the precipitation process, particularly for convective clouds. Well-designed seeding experiments on 33 pairs of cumulus clouds in Australia indicated in the statistical analysis that seeding at cloud base with 20 g of silver iodide produced significant rain increases, while seeding with 0.2 g was ineffective. A seedability index can be derived from these data: lifetime of cloud in hours times depth of cloud in kilometers. Figure 6 illustrates that, for an index below 3, clouds do not rain naturally or after seeding. However, the cases which statistically determined the success through seeding were all connected with high seedability indexes.

The aim to develop an operational rainmaking technology is the purpose of the great effort which is mounted by the Bureau of Reclamation, U.S. Department of Interior. This agency's Project HIPLEX has as its goal the establishment of "a verified working technology and operational management framework capable of producing additional rain from cumulus clouds in the semi-arid Plains States." The project is planned as a 5–7-year field research project at an estimated cost of $23,000,000. Preliminary studies have indicated that the direct benefit of the added precipitation would amount to $500,000,000 annually. The plan calls for three major sites: Miles City, MT, Colby-Goodland, KS, and Big Spring-Snyder, TX. Primary research is carried out in a 120-mi-diameter (193-km) circle around each site. The project is to be developed in several phases. In the first phase, a single convective cell is taken as the experimental unit, to be succeeded in later phases by groups of cells and more complex units, up through mesoscale systems. Empirical and computational models, based on physical understanding of dynamic and microphysical processes leading to precipitation, will be used to develop and evaluate seeding hypotheses. After verification, the models will generate expected values of rainfall amounts, durations, and other characteristics for both natural and seeded clouds. These model-generated expectations will be compared with observations. Potential predictor variables and covariates available from surface data and rawinsonde data are being investigated.

The project is supported by an impressive array of modern instrumentation, including radar, raingages, nuclei counters, and weather observations, as well as radiosondes, aircraft, and satellites. *See* PRECIPITATION.

Downwind effects. Sporadic statistical analyses indicate that downwind effects occur as far away as 300 mi. Observations have been reported from the United States and Australia, with precipitation increases in most cases. In view of the rate of decay of silver iodide in daylight, the mechanism of such effects is dubious, and a strong case can be made that, downwind from any location, positive anomalies can be found simply because of the natural variability of the rate of rainfall.

Hail suppression. Scientists of the Soviet Union reported the development of an operational hail suppression project which reduces hail damage by 80–90%. It involved the detection with radar of the hail-spawning cloud regions and the delivery into them of seeding material by means of grenades. However, Soviet scientists have admitted to the difficulty of suppressing hail in very severe storms. In the United States a physical concept was developed which differed somewhat from the Soviet concept, and delivery of the seeding agent from aircraft was used rather than from the ground. The aircraft traversed the storm at the base level and discharged the seeding material, by means of rockets, vertically into the updraft. The project had to be terminated prematurely. The

Fig. 5. "Explosion" of cumulus cloud following release of heat of fusion caused by seeding with silver iodide. (*a*) Time of seeding. (*b*) Views at 9 min, (*c*) at 19 min, and (*d*) at 38 min after the seeding. (*Courtesy of J. Simpson, ESSA*)

analysis of 3 years of field experiments gave the result that an average of 60% more hail mass was measured on days when seeding was carried out (16 seeded, 16 unseeded days) than on days when seeding was not attempted. However, the 90% confidence limits for the ratio of hail mass on seeded to unseeded days range from −48 to 531% for a log normal fit of the data distribution. These results, therefore, permit the exclusion of a suppression effect in excess of 50% at the 5% confidence limit.

There are mainly two physical properties of hailstorms which make the failure to suppress hail fall (or even to increase it) plausible. First, hail does not fall in uninterrupted long swathes, but in long swathes which consist of individual hail streaks. These streaks are roughly between 10 and 20 km long and a few kilometers wide. The fine structure of a hail swath, which may be 50 km wide, is made up of numerous hail streaks whereby each streak may originate in a separate storm cell. Unless all potential hail cells are seeded at the correct time of hail formation, seeding may be ineffective. Second, as discovered by French scientists, a proportionality exists between the number of hailstones falling per square meter of surface area and their size: the more hailstones fall, the bigger they are. The reason for this relationship is unknown, but one will not go much wrong with the assumption that it is due to a feedback mechanism between the cloud dynamics and the cloud microphysics. Since the tolerance limits of the cloud are not known, it is quite possible that the artificial generation of hail embryos will allow them to grow to hailstones as large as the natural ones, and, consequently, increase the hail mass.

A project in France conducted surface seeding on a large scale, discharging tons of silver iodide into the air during the hail season. The analysis of the nonrandomized experiments indicated success. Other projects have been conducted in Germany (Bavaria), Italy, and Argentina. While the theory of hail formation suggests that hail may be suppressed by a comparatively small seeding effort, there is great difficulty in designing a field experiment which is in agreement with the scheme presented in Fig. 2. The sporadic nature of hailstorms and their large natural variability practically exclude effective randomization of the experiments. For the analysis phase it is therefore necessary to have available a store of excellent historical data on the experimental area, such as exists in the Canadian Hail Research Project in Alberta. *See* HAIL.

Lightning suppression. Two concepts have been tested in the United States. The first, developed by the Department of Agriculture, makes use of overseeding of thunderstorms with silver iodide. While physical relationships in the suppression mechanism are not fully developed, it appears that the method decreases ground strokes and increases intracloud strokes. Another approach was developed by the U.S. Army jointly with the Environmental Science Services Administration (ESSA). Discharge of the charge centers of a storm by corona discharges initiated through the introduction of metallic needles, so-called chaff. It can be shown that for thunderstorm fields, $10^6 - 10^7$ chaff particles (weighing 5 to 50 lb) can discharge several amperes, a result in agreement with the magnitude of the thunderstorm-charging current.

The result of field experiments is shown in Fig. 7. The figure is based on aircraft measurements of the number of lightning discharges in 5-min intervals of seeded and nonseeded storms. The chaff fibers were made of aluminized nylon 10 cm long and 25 μm in diameter. They initiated corona in an electric field of about 35,000 V/m (lightning requires about 300,000 V/m), thus draining away the charge that had been built up by the storm. The figure shows that the number of lightning discharges in seeded storms goes quickly to zero, while in unseeded storms lightning proceeds uninterrupted. *See* LIGHTNING.

Modification of severe storms. Modification of severe storms such as tornadoes or hurricanes is in its infancy, essentially because of incomplete understanding of the dynamic structure of these storms. ESSA and the U.S. Navy developed under Project Stormfury a concept of hurricane modification based upon overseeding of the wall clouds. Release of the heat of fusion is believed to alter the pressure gradient in such a way as to diminish the destructive winds near the storm center. As wind damage is proportional to the square of the wind velocity, a relatively small reduction of the velocity may mean a large reduction of damage.

Seeding experiments were conducted on hurricanes Esther in 1961 and Beulah in 1963. A 10%

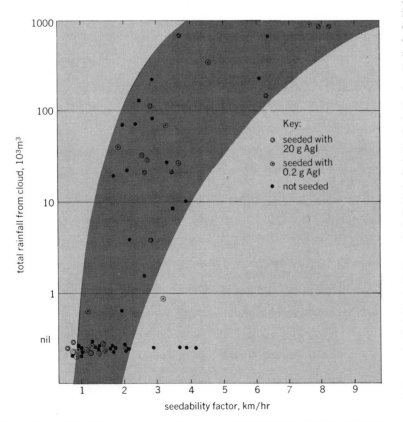

Fig. 6. Seedability factor for rainmaking from convective clouds. Seedability factor is product of cloud depth in kilometers and cloud lifetime in hours.

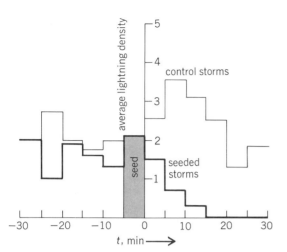

Fig. 7. Average lightning density for each time interval is for 10 seeded and 18 control storms during 1972 and 1973. (*From H. W. Kasemir et al., Lightning suppression by chaff seeding at base of thunderstorms, J. Geophys. Res., 81:1965–1970, 1976*)

reduction of wind speed was observed; however, the reduction is well within the natural wind variability of the storm. *See* HURRICANE.

Inadvertent weather modification. Modern civilization affects weather in many ways. The great industrial centers of the world are plagued with haze and smog due to effluents from factories and combustion of hydrocarbons. Haze and smog affect the heat radiation budget of the atmosphere as well as precipitation processes. Aerosols discharged by metal foundries act as freezing nuclei, and strong evidence now appears for such inadvertent weather modification. H. K. Weickmann has repeatedly observed in the Great Lakes region the formation of snow showers downwind of steelmills after measuring large numbers of freezing nuclei in the exhaust plume. It is likely that this effect is more widespread than is presently known, but systematic investigations are lacking.

The most famous case, the La Porte anomaly, is somewhat controversial. La Porte, IN, is located near the southeastern shore of Lake Michigan, 30 mi downwind of the Gary industrial complex. The rainfall measured at this site and its environment indicates that the rainfall at this station has increased with the increasing industrialization (smoke-haze days in Chicago), whereas neighboring stations remained unaffected. The difference developed particularly after about 1930, or after the last observer change for this station had taken place. The observers voluntarily measure the rate of rainfall. It appears that a systematic error in these observations caused the apparent singularity of rainfall at this location, since after this observer left in 1963, the La Porte record agrees again with the neighboring stations. For this reason, the National Oceanic and Atmospheric Administration claims observer error for the anomaly.

While the combustion of hydrocarbons causes formation of smog in large urban areas, a more subtle climate modification takes place on a global scale. The combustion process not only liberates aerosols but also carbon dioxide gas, an important constituent of the air that contributes to the greenhouse effect of the atmosphere. The gas is transparent for the Sun's visible radiation but traps the Earth's thermal radiation; thus the more carbon dioxide released, the warmer the climate becomes. Computer calculations project an average temperature increase of $1-2°C$ within two to three generations from this effect. Though this change may appear small, it would have profound influences on weather and climate. Glaciers and ice caps would melt, and coastal cities might become inundated. It is possible that other changes in the atmosphere will counteract the effect of carbon dioxide. *See* ATMOSPHERIC POLLUTION; CLIMATE MODIFICATION.

Effect on ecology and society. While influences on ecological systems appear to be often exaggerated in view of the great natural variability of weather, socioeconomic considerations are important. Fog dispersal methods at airports enable many passengers to travel, effective hail suppression would save millions of dollars, and rainmaking would be the least expensive answer to the water dilemma of modern civilization. But as interests of individuals differ, rainmaking may be a blessing to one and blight for another; socioeconomic considerations are therefore important at a time when weather modification is still in its infancy.

[HELMUT K. WEICKMANN]

Weather station

A place and facility for the observation, measurement, and recording and transmission of data of the variable elements of weather. Although there are hundreds of private observers and forecasters in the United States, the most effective observations and forecasts are those made with the aid of the great network of the U.S. Weather Bureau. This article deals with the observation and data-gathering stations of this nationwide agency.

Types. Most weather stations are classified as first-order or second-order largely on the basis of their primary or supplementary functions.

First-order stations. Major weather stations located at airports measure temperature, dew point, wind, ceiling, and visibility over grassy plots near runways.

Ceiling, the lowest height of clouds covering over half the sky, is measured by ceilometers. A beam of modulated light is projected on the cloud base. At a known distance, the spot of light is electronically scanned day and night and the cloud height computed trigonometrically. At small airports, the ceiling is measured during the day by timing the ascent to disappearance of a balloon rising at a known rate. At night, a nonmodulated light beam is used, and cloud heights are computed trigonometrically.

Prevailing visibility is determined from markers at known distances during daylight or from lights at known distances at night. At jet terminals, transmissometers parallel to the instrumented runway are calibrated in units of runway visibility. At airports equipped with high-intensity runway lights, runway visual range, the distance a pilot can see high-intensity lights along the runway as he approaches for a landing, is reported. This is determined by an electronic computer, using transmissivity, background illumination, and the intensity setting of the runway lights.

All observational data are remotely recorded in weather offices near the field. Rainfall intensity is remotely recorded, using tipping bucket, weighing, or float-type gages. Snowfall rates present numerous problems in telemetering. Pressure, pressure change, and altimeter settings are measured by mercurial or precision aneroid barometers and barographs. Clouds, current weather conditions, and obstructions to vision are determined by visual observations. Remote automatic stations transmit to a weather office by teletypewriter all observations of elements for which sensors are available.

Numerous airport and other first-order stations in larger cities measure solar radiation, ozone content of the air, and gradients of temperature and moisture at short intervals above and below the surface.

Many first-order stations are also equipped to make upper-air measurements. Precise theodolite observations of the course of a pilot balloon rising at a known rate provide data on the horizontal velocity of upper-level winds. A smaller number of stations measure pressure, temperature, and humidity by means of a radiosonde, carried by a larger balloon. Many stations use radio direction-finding or radar equipment to track radiosondes.

Second-order stations. Supplementary or second-order stations furnish detailed surface weather data from key points at specified hours. Climatological stations manned by volunteers record the extremes of temperature and the amount and type of precipitation each day. Many of these stations also serve hydrologic purposes and in addition report snow density, river stage, rates of stream discharge, evaporation, and, in some cases, wind.

Over ocean and lake areas, reports are received from moving ships, specially equipped fixed ships, and a few automatic stations on anchored buoys.

Radiosonde data over oceans are limited to the fixed weather ships and a few moving ships. Only the fixed ships are equipped for upper-wind measurements. Important upper-air data from over ocean areas are obtained by reconnaissance planes flying regular patterns but using alternate courses during storms. Limited data from over oceans can be obtained from transosondes. These large balloons, carrying radio-transmitting equipment, fly at a constant altitude and provide temperature and wind data. In spite of the difficulties involved, both rockets and satellites must be used to obtain data from levels not otherwise probed.

Over land, at sea, and in the air, radar surveys all weather within a radius of 100–200 mi. Hydrologists can use radar photographs to interpolate between rainfall measurements and thus draw more complete isohyetal maps. For air safety, radar locates the active precipitation areas and areas of turbulence and, even when skies are clear, may identify important wind-shear lines.

Frequency of observations. Most first-order stations report to national networks of weather offices and centers at hourly intervals and to international networks at 6-hr intervals. Second-order stations

report at 3- or 6-hr intervals. Upper winds are observed at 6- or 12-hr intervals, and radiosondes are released every 12 hr. Moving ships report every 6 hr and fixed ships every 3. Hydrologic stations report daily during storms. Climatological stations mail reports monthly. For hydrologic purposes, storage precipitation gages in mountainous areas are visited seasonally.

Density of observational networks. Spacing of observation points is strongly influenced by population density, with many observations near coasts, in valleys, and along main transportation routes. The required density of stations varies for each meteorological user, and there is no universally acceptable and economically realistic plan. However, meteorologists agree on broad objectives. Roughly, these are surface stations not more than 100 mi apart, upper-air stations not more than 300 mi apart, climatological stations 30 mi apart, and radar stations 200 mi apart. Most meteorologists would accept an ocean spacing one-fifth as dense as that over the continents. *See* METEOROLOGICAL INSTRUMENTATION; WEATHER FORECASTING AND PREDICTION.

[ALBERT K. SHOWALTER]

Wind

The motion of air relative to the Earth's surface. The term usually refers to horizontal air motion, as distinguished from vertical motion, and to air motion averaged over a chosen period of 1–3 min. Micrometeorological circulations (air motion over periods of the order of a few seconds) and others small enough in extent to be obscured by this averaging are thereby eliminated. The choice of the 1- to 3-min interval has proven suitable for the study of (1) the hour-to-hour and day-to-day changes in the atmospheric circulation pattern; and (2) the larger-scale aspects of the atmospheric general circulation.

The direct effects of wind near the surface of the Earth are manifested by soil erosion, the character of vegetation, damage to structures, and the production of waves on water surfaces. At higher levels wind directly affects aircraft, missile and rocket operations, and dispersion of industrial pollutants, radioactive products of nuclear explosions, dust, volcanic debris, and other material. Directly or indirectly, wind is responsible for the production and transport of clouds and precipitation and for the transport of cold and warm air masses from one region to another. *See* ATMOSPHERIC GENERAL CIRCULATION; WIND MEASUREMENT.

Cyclonic and anticyclonic circulation. Each is a portion of the pattern of airflow within which the streamlines (which indicate the pattern of wind direction at any instant) are curved so as to indicate rotation of air about some central point of the cyclone or anticyclone. The rotation is considered cyclonic if it is in the same sense as the rotation of the surface of the Earth about the local vertical, and is considered anticyclonic if in the opposite sense. Thus, in a cyclonic circulation, the streamlines indicate counterclockwise (clockwise for anticyclonic) rotation of air about a central point on the Northern Hemisphere or clockwise (counterclockwise for anticyclonic) rotation about a point on the Southern Hemisphere. When the streamlines close completely about the central point, the pattern is denoted respectively a cyclone or an anticyclone. Since the gradient wind represents a good approximation to the actual wind, the center of a cyclone tends strongly to be a point of minimum atmospheric pressure on a horizontal surface. Thus the terms cyclone, low-pressure area, or "low" are often used to denote essentially the same phenomenon. In accord with the requirements of the gradient wind relationship, the center of an anticyclone tends to coincide with a point of maximum pressure on a horizontal surface, and the terms anticyclone, high-pressure area, or "high" are often used interchangeably.

Cyclones and anticyclones are numerous in the lower troposphere at all latitudes. At higher levels the occurrence of cyclones and anticyclones tends to be restricted to subpolar and subtropical latitudes, respectively. In middle latitudes the flow aloft is mainly westerly, but the streamlines exhibit wavelike oscillations connecting adjacent regions of anticyclonic circulation (ridges) and of cyclonic circulation (troughs).

Although the atmosphere is never in a completely undisturbed state, it is customary to refer to cyclonic and anticyclonic circulations specifically as atmospheric disturbances. Cyclones, anticyclones, ridges, and troughs are intimately associated with the production and transport of clouds and precipitation, and hence convey a connotation of disturbed meteorological conditions.

A more rigorous definition of circulation is often employed, in which the circulation over an arbitrary area bounded by the closed curve S is given by Eq. (1), where the integration is taken complete-

$$C = \oint v_t \, dS \qquad (1)$$

ly around the boundary of the area. Here v refers to the wind at a point on the boundary, the subscript t denotes the component of this wind parallel to the boundary, and dS is a line element of the boundary. The component v_t is considered positive or negative according to whether it represents cyclonic or anticyclonic circulation along the boundary S. In this context, the circulation may be positive (cyclonic) or negative (anticyclonic) even when the streamlines within the area are straight, since the distribution of wind speed affects the value of C. See ATMOSPHERE; CLOUD; GEOSTROPHIC WIND; GRADIENT WIND; PRECIPITATION; STORM.

Convergent or divergent patterns. These are said to occur in areas in which the (horizontal) wind flow and distribution of air density is such as to produce a net accumulation or depletion, respectively, of mass of air. Rigorously, the mean horizontal mass divergence over an arbitrary area A bounded by the closed curve S is given by Eq. (2),

$$D = \frac{1}{A} \oint \rho v_n \, dS \qquad (2)$$

where the integration is taken completely around the boundary of the area. Here ρ is the density of air, v refers to the wind at a point on the boundary, the subscript n denotes the component of this wind perpendicular to the boundary, and dS is an element of the boundary. The component v_n is taken positive when it is directed outward across the boundary and negative when it is directed inward. Convergence is thus synonymous with negative divergence. If spatial variations of density are neglected, the analogous concept of velocity divergence and convergence applies.

The horizontal mass divergence or convergence is intimately related to the vertical component of motion. For example, since local temporal rates of change of air density are relatively small, there must be a net vertical export of mass from a volume in which horizontal mass convergence is taking place. Only thus can the total mass of air within the volume remain approximately constant. In particular, if the lower surface of this volume coincides with a level ground surface, upward motion must occur across the upper surface of this volume. Similarly, there must be downward motion immediately above such a region of horizontal mass divergence.

The horizontal mass divergence or convergence is closely related to the circulation. In a convergent wind pattern the circulation of the air tends to become more cyclonic; in a divergent wind pattern the circulation of the air tends to become more anticyclonic.

Regions which lie in the path of an approaching cyclone are characterized by a convergent wind pattern in the lower troposphere and by upward vertical motion throughout most of the troposphere. Since the upward motion tends to produce condensation of water vapor in the rising air current, abundant cloudiness and precipitation typically occur in this region. Conversely, the area in advance of an anticyclone is characterized by a divergent wind pattern in the lower troposphere and by downward vertical motion throughout most of the troposphere. In such a region, clouds and precipitation tend to be scarce or entirely lacking.

A convergent surface wind field is typical of fronts. As the warm and cold currents impinge at the front, the warm air tends to rise over the cold air, producing the typical frontal band of cloudiness and precipitation. See FRONT.

Zonal surface winds. Such patterns result from a longitudinal averaging of the surface circulation. This averaging typically reveals a zone of weak variable winds near the Equator (the doldrums) flanked by northeasterly trade winds in the Northern Hemisphere and southeasterly trade winds in the Southern Hemisphere, extending poleward in each instance to about latitude 30°. The doldrum belt, particularly at places and times at which it is so narrow that the trade winds from the two hemispheres impinge upon it quite sharply, is designated the intertropical convergence zone, or ITC. The resulting convergent wind field is associated with abundant cloudiness and locally heavy rainfall. A westerly average of zonal surface winds prevails poleward of the trade wind belts and dominates the middle latitudes of both hemispheres. The westerlies are separated from the trade winds by the subtropical high-pressure belt, which occurs between latitudes 30 and 35° (the horse latitudes), and are bounded on the poleward side in each hemisphere between latitudes 55 and 60° by the subpolar trough of low pressure. Numerous cyclones and anticyclones progress eastward in the zone of prevailing westerlies, producing the abrupt day-to-day changes of wind, temperature, and weather which typify these regions. Poleward of the subpolar low-pressure troughs, polar easterlies are observed.

The position and intensity of the zonal surface wind systems vary systematically from season to season and irregularly from week to week. In general the systems are most intense and are displaced toward the Equator in a given hemisphere during winter. In this season the subtropical easterlies and prevailing westerlies attain mean speeds of about 15 knots, while the polar easterlies are somewhat weaker. In summer the systems are displaced toward the pole by 5 to 10° of latitude and weaken to about one-half their winter strength.

When the pattern of wind circulation is averaged with respect to time instead of longitude, striking differences between the Northern and Southern Hemispheres are found. On the Southern Hemisphere, variations from longitude to longitude are relatively small and the averaged pattern is described quite well in terms of the zonal surface wind belts. On the Northern Hemisphere there are large differences from longitude to longitude. In winter, for example, the subpolar trough is mainly manifested in two prominent low centers, the Icelandic low and the Aleutian low. The subtropical ridge line is drawn northward in effect over the continents and is seen as a powerful and extensive high-pressure area over Asia and as a relatively weak area of high pressure over North America. In summer the Aleutian and Icelandic lows are weak or entirely absent, while extensive areas of low pressure over the southern portions of Asia and western North America interrupt the subtropical high-pressure belt. See CLIMATOLOGY; MONSOON.

Upper air circulation. Longitudinal averaging indicates a predominance of westerly winds. These westerlies typically increase with elevation and culminate in the average jet stream, which is found in lower middle latitudes near the tropopause at elevations between 35,000 and 40,000 ft. The subtropical ridge line aloft is found equatorward of its surface counterpart and easterlies occur at upper levels over the equatorward portions of the trade wind belts. In high latitudes, weak westerlies aloft are found over the surface polar easterlies. Seasonal and irregular fluctuations of the circulation aloft are similar to those which characterize the surface winds. See JET STREAM.

Minor terrestrial winds. In this category are circulations of relatively small scale, attributable indirectly to the character of the Earth's surface. One example, the land and sea breeze, is a circulation driven by pronounced heating or cooling of a given area in comparison with little heating or cooling in a horizontally adjacent area. During the day, air rises over the strongly heated land and is replaced by a horizontal breeze from the relatively cool sea. At night, air sinks over the cool land and spreads out over the now relatively warm sea.

Another example is formed by the mountain and valley winds. These result from cooling and heating, respectively, of the mountain slopes relative to the horizontally adjacent free air above the valley floor. During the day, air flows up from the valley along the strongly heated mountain slopes, but at night, air flows down the relatively cold mountain slopes toward the valley bottom. A similar type of descending current of cooled air is often observed along the sloping surface of a glacier. This nighttime air drainage, under proper topographical circumstances, can lead to the accumulation of a pool of extremely cold air in nearby valley bottoms.

Local winds. These commonly represent modifications by local topography of a circulation of large scale. They are often capricious and violent in nature and are sometimes characterized by extremely low relative humidity. Examples are the mistral which blows down the Rhone Valley in the south of France, the bora which blows down the gorges leading to the coast of the Adriatic Sea, the foehn winds which blow down the Alpine valleys, the williwaws which are characteristic of the fiords of the Alaskan coast and the Aleutian Islands, and the chinook which is observed on the eastern slopes of the Rocky Mountains. Local names are also given in some instances to currents of somewhat larger scale which are less directly related to topography.

Some examples of this type of wind are the norther, which represents the rapid flow of cold air from Canada down the plains east of the Rockies and along the east coast of Mexico into Central America; the nor'easter of New England, which is part of the wind circulation about intense cyclones centered offshore along the Middle Atlantic coastal states; and the sirocco, a southerly wind current from the Sahara which is common on the coast of North Africa and sometimes crosses the Mediterranean Sea. See CHINOOK; SIROCCO.

[FREDERICK SANDERS]

Wind measurement

The determination of three parameters: the size of an air sample, its speed, and its direction of motion.

Size of sample. The size of the air sample is highly dependent on how the measurement is made. When the wind measurement is taken with small, sensitive, and rapid-response instruments, or by the drift of small suspended particles, the air sample can have a scale of millimeters or less. If the wind measurement is made from the pressure gradient, as measured on a weather map, the scale can be hundreds of kilometers. Wind measurements also depend on the system used. When the wind is measured at a point fixed with respect to the Earth's surface with air moving by, the measurement is called Eulerian. In this, one is continuously measuring different air samples. Continuing measure of the speed and direction of the same air sample is called a Lagrangian measurement. This is obtained by measuring the drift of balloons, clouds, or smoke. The two systems only give the same value when the wind is perfectly steady. Ordinarily when a wind speed or direction is given, some sort of averaging, usually a time average, is implied. *See* GEOSTROPHIC WIND; GRADIENT WIND.

Wind direction. This is designated as the direction from which the wind is blowing, given in terms of 8, 16, or 32 points of the compass and in degrees or tens of degrees from north, measured clockwise. A calm wind is reported at 00; a north wind is 36 (representing 360°). A wind vane is used to measure the direction of the surface wind. Basically, this is a grossly unsymmetrical body mounted near its center of gravity and free to rotate about a vertical axis. This simple and old meteorological instrument now usually includes an electrical device for remote indication or recording of the wind direction. A wind vane free to rotate about a vertical and a horizontal axis (called a bivane) also indicates the vertical wind component.

Wind speed. In terms of its force on common objects such as leaves, smoke, or waves, wind speed can be estimated by the Beaufort scale. The scale of Beaufort numbers ranges from 0 for winds calm with velocity under 1 mph (smoke rises vertically) to 12 for hurricane-force winds with velocity over 75 mph. Meteorologists often measure the wind speed in terms of the cause of the wind, that is, the atmospheric pressure gradient, which is inversely proportional to the spacing of the isobars on a weather map. For information concerning instruments which measure surface winds *see* ANEMOMETER. For further information on this and on measurement of upper-air wind speed *see* METEOROLOGICAL INSTRUMENTATION.

[VERNER E. SUOMI]

Bibliography: W. E. K. Middleton, *Invention of the Meteorological Instruments*, 1969; W. E. K. Middleton and A. E. Spilhaus, *Meteorological Instruments*, 3d ed., 1953.

Wind rose

A diagram in which statistical information concerning the direction and speed of the wind at a particular location may be conveniently summarized. In the standard wind rose a line segment is drawn in each of perhaps eight compass directions from a common origin (see illustration). The length of a particular segment is proportional to the frequency with which winds blow from that direction. Parts of a given segment are given various thicknesses, indicating frequencies of occurrence of various classes of wind speed from the given direction. *See* WIND MEASUREMENT.

[FREDERICK SANDERS]

Wind-shift line

A line along which the wind veers abruptly. In the Northern Hemisphere a shift from easterly or southeasterly winds to southerly or southwesterly is normally observed on passage of a warm front, while a shift from southerly or southwesterly to westerly or northerly winds is usually found on passage of a cold front, or a squall line. Passage of wind-shift lines is commonly accompanied by marked changes in general weather conditions. Lowest barometric pressure is generally found on passage of a cold-front type of wind-shift line. *See* FRONT; SQUALL LINE.

[CHESTER W. NEWTON]

Wind stress

The drag or tangential force per unit area exerted on the surface of the Earth by the adjacent layer of moving air. Erosion of ground surfaces and the production of waves on water surfaces are manifestations of wind stress. Surface wind stress determines the exchange of momentum between the Earth and the atmosphere and, together with internal atmospheric viscous stresses, exerts a strong influence on the typical variation of wind through the lowest few thousand feet of the atmosphere. Estimated values of the surface wind stress range up to several dynes/cm², depending upon the nature of the surface and upon the character of the adjacent airflow. For a discussion of the relation of wind stress to ocean currents and waves *see* OCEAN CURRENTS; OCEAN WAVES; WIND STRESS OVER SEA.

Internal horizontal stresses. Significant stresses arise within the lower atmosphere because of the strong shear of the wind between the slowly moving air near the ground and the more rapidly moving air a few thousand feet above and because of the turbulent nature of the airflow in this region. The turbulent eddies referred to here have characteristic dimensions ranging up to a few hundreds of meters.

The effectiveness of this turbulent viscosity in the transferring of momentum from level to level may, in favorable circumstances, be 1,000,000 times as great as the effectiveness of purely molecular viscosity. These turbulent viscous stresses exert an indirect effect on the surface wind stress by effecting the transfer of large amounts of momentum from higher levels down to levels adjacent to the surface of the Earth.

The torque exerted by the surface wind stress averaged over the entire Earth and over a sufficiently long period of time must be equal to zero. Otherwise the net torque acting between the Earth and the atmosphere would tend to alter the rate of rotation of the Earth about its axis and to alter the mean wind circulation of the atmosphere. In the atmosphere the angular momentum supplied, in effect, by the Earth in regions of easterly surface flow balances the angular momentum drained by the Earth in regions of westerly wind flow, so that the average torque exerted by the surface wind stress is indeed zero, or very nearly so. The average magnitude of the surface wind stress is, of course, not equal to zero. It determines the average rate at which kinetic energy of the winds is dissipated by surface friction.

Wind pressure. This is the force exerted by the wind per unit area of solid surface exposed normal to the wind direction. In contrast to shearing stresses, the wind pressure arises from the difference in pressure between the windward and lee sides of the exposed surface. Wind pressure thus represents a substantial force when the wind speed is high. *See* METEOROLOGY; WIND.

[FREDERICK SANDERS]

Bibliography: S. L. Hess, *Introduction to Theoretical Meteorology*, 1959; S. Petterssen, *Weather Analysis and Forecasting*, 2d ed., vol. 1, 1956; P. Sachs, *Wind Forces in Engineering*, 1977.

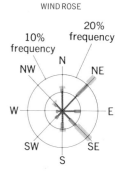

WIND ROSE

20% frequency

10% frequency

N

NW   NE

W   E

SW   SE

S

▨ strong winds
▮ moderate winds
▮ light winds

Standard wind rose.

Wind stress over sea

The drag or tangential force of the wind on the sea is expressed in units of dynes per square centimeter or micronewtons per square meter but is normally taken to represent the mean drag over an undefined area, perhaps several kilometers square, containing many waves. It is usually related to an appropriate time and space average of the wind near the sea surface (at 10 m above the mean level, for example).

Over solid surfaces overland, in the absence of vertical density gradients, the drag varies as the square of the wind speed and is expressed, following aerodynamic usage, by the equation shown below, where τ is the shearing stress, ρ is the air

$$\tau = C_D \rho v^2$$

density, and v is the wind speed at a specified height, say 10 m. For a given (uniform) surface geometry, or roughness, C_D is a constant known as the drag coefficient. See WIND STRESS.

Over a water surface a drag coefficient is defined in the same way, but since the surface geometry varies with the wind speed, C_D is to be regarded as a variable. The drag coefficient over the sea is an important variable in both meteorology and oceanography since it relates the wind speed to the drag which generates ocean waves, drives the ocean currents, and sets the scale of the atmospheric turbulence that transfers water vapor and heat from the ocean to the atmosphere to provide the energy for clouds and weather systems. *See* ATMOSPHERIC GENERAL CIRCULATION; OCEAN CURRENTS; OCEAN WAVES; OCEAN-ATMOSPHERE RELATIONS.

Methods of measurement. Many methods have been used to measure the drag of the wind on the sea, but they are of four main types. The first technique attempts to get information from the wind tilt of enclosed bodies of water. When wind blows over a lake, the surface becomes tilted upward toward the lee end. Measurements of the mean slope, which typically would be of the order of 1 cm in 10 km, can provide estimates of the wind drag. There are theoretical difficulties associated with end effects, but the major difficulties are practical: Steady conditions are required, and although the surface waves can be filtered out to permit observations of the necessary accuracy, it is often hard to eliminate the long-period oscillations of the whole body of water, which are known as seiches.

The second method involves the measurement of the vertical distribution of the mean wind in the lowest few meters of air over the sea. When air and water are at the same temperature, wind speed increases as the logarithm of the height, and its rate of increase provides a measure of the wind drag. This method is empirical, but it is well established; the difficulties are of engineering rather than of science. It is difficult to get instruments of the necessary precision (the wind speed has to be measured to about 1 cm/sec) and even harder to expose them over water in such a way that their supports do not interfere with the wind or the waves or the currents. For this reason many of the observations have been made in lakes and in shal-

low water, although some workers have developed techniques for making wind measurements accurately enough from small ships or buoys.

The third technique again involves measuring the vertical distribution of the mean wind, but in this case to greater heights. In steady conditions the drag of the wind on the sea can be inferred from the component of the wind across the isobars; therefore, for this method accurate measurements of the distribution of atmospheric pressure, as well as accurate wind observations, are necessary. This method does not work well in middle latitudes because in those areas the wind blows at only a small angle to the isobars in the lowest kilometer or so. It works better in the trade winds, where attempts to use it have met with moderate success.

The fourth technique again requires detailed wind measurements, in this case detailed in time by using an instrument of rapid response, reacting, for example, in about one-tenth of a second. Measurements are made of both the horizontal and the vertical components of the wind, and their mean product provides the drag. There are again some theoretical difficulties in the interpretation of the results, but they can probably be overcome. The major difficulties are again of getting suitable instruments suitably exposed, and of the large amount of numerical analysis that has to be done to get the results. On the other hand, the method provides detailed information on the structure of the turbulent flow, and it is being increasingly used. A further attraction of this method is that estimates of evaporation and direct heat transfer can be made by measuring fluctuations of humidity and temperature as well as those of the vertical velocity component. The drag can also be inferred from observations of the fluctuations at high frequencies by deducing the energy dissipation, with the assumption that the small-scale turbulence is isotropic.

Measured values. It might be thought that with all these methods available there would be substantial agreement about the drag on the sea produced by a given wind. This is not the case; there are many discrepancies in the published results which are difficult to reconcile. There is some agreement that the drag is small relative to that of a fixed solid surface with the same geometry, and that the drag increases with wind speed at a rate somewhat more rapid than the square of the wind speed. It has also been suggested that naturally occurring thin films of surface-active agents have a greater effect on the hydrodynamics and aerodynamics of the sea surface than has generally been supposed.

A summary of the results obtained was given by H. U. Roll in 1965. The general value of C_D at moderate wind speeds is about 1.5×10^{-3}, and there is some evidence for an increase in C_D with increasing wind speed, from about 1.0×10^{-3} in light winds to 2.0×10^{-3} in strong winds. The values in strong winds and at long fetches, which are of the greatest importance, are the most difficult to estimate. Here the values are uncertain and new methods are being devised.   [H. CHARNOCK]

Bibliography: H. U. Roll, *Physics of the Marine Atmosphere*, 1965.

McGRAW-HILL
ENCYCLOPEDIA OF OCEAN AND ATMOSPHERIC SCIENCES

List of Contributors

List of Contributors

A

Aagard, Knut. *Department of Oceanography, University of Washington, Seattle.* ARCTIC OCEAN.

Ackerman, Dr. Edward A. *Deceased; formerly, Carnegie Institution of Washington.* RIVER—coauthored.

Allee, Dr. Paul A. *Atmospheric Physics and Chemistry Laboratory, Environmental Science Services Administration, Boulder.* BROCKEN SPECTER.

Allen, Dr. L. H. *University of Florida.* CROP MICROMETEOROLOGY.

Anthes, Richard A. *Department of Meteorology, Pennsylvania State University.* WEATHER FORECASTING AND PREDICTION—in part.

Armstrong, Francis A. J. *Freshwater Institute, Winnipeg.* SEA-WATER FERTILITY—in part.

Austin, Prof. James M. *Professor of Meteorology, Massachusetts Institute of Technology.* WEATHER FORECASTING AND PREDICTION—in part.

Axford, Dr. W. I. *Professor of Physics and of Applied Physics and Information Science, University of California, San Diego.* MAGNETOSPHERE.

B

Battan, Louis J. *Director, Institute of Atmospheric Physics, University of Arizona.* LIGHTNING—in part; RADAR METEOROLOGY; THUNDER.

Baumeister, Prof. Theodore. *Consulting Engineer; Stevens Professor of Mechanical Engineering, Emeritus, Columbia University; Editor in Chief, "Standard Handbook for Mechanical Engineers."* TEMPERATURE-HUMIDITY INDEX.

Bean, Howard S. *Deceased; formerly, Consultant on Fluid Metering, Liquids and Gases, Sedona, AZ.* HYGROMETER; TEMPERATURE MEASUREMENT.

Beaumariage, Dr. Donald C. *Director, Manned Undersea Science and Technology, National Oceanic and Atmospheric Administration, Rockville, MD.* OCEANOGRAPHIC SUBMERSIBLES—in part.

Beaumont, Robert T. *Director of Meteorological Operations, E.G.G., Inc., Boulder.* SNOW GAGE; SNOW SURVEYING.

Benfer, Neil A. *BOMAP Scientific Editor, Barbados Oceanographic and Meteorological Analysis Project.* SEA STATE; SEA WATER—in part.

Benson, Dr. Bruce B. *Department of Physics, Amherst College.* SEA WATER—in part.

Bergeron, Prof. Tor. *Deceased; formerly, Institute of Meteorology, Uppsala.* METEOROLOGY—in part.

Biscaye, Dr. Pierre E. *Lamont-Doherty Geological Observatory, Palisades, NY.* MARINE SEDIMENTS—in part.

Bjerknes, Dr. Jacob. *Deceased; formerly, Professor Emeritus, Department of Meteorology, University of California, Los Angeles.* CLIMATIC CHANGE—in part.

Blad, Dr. Blaine L. *Institute of Agriculture and Natural Resources, University of Nebraska.* EVAPOTRANSPIRATION.

Blanchard, Dr. Duncan C. *State University of New York, Albany.* SEA WATER—in part.

Blumenstock, Dr. David I. *Professor of Geography, University of California, Berkeley.* CLIMATIC PREDICTION.

Brinton, Dr. Edward. *Scripps Institution of Oceanography, La Jolla.* SEA-WATER FERTILITY—in part.

Buettner, Dr. Konrad J. K. *Decreased; formerly, Pierce Foundation, Yale University.* BIOCLIMATOLOGY.

C

Cain, Dr. Joseph C. *U.S. Geological Survey, Denver.* GEOMAGNETISM—in part.

Calvert, Dr. Stephen E. *Grant Institute of Geology, University of Edinburgh.* GULF OF CALIFORNIA.

Campbell, Dr. Wallace. *U.S. Geological Survey, Denver.* GEOMAGNETIC STORM; TRANSIENT GEOMAGNETIC VARIATIONS.

Carritt, Dr. Dayton E. *American Dynamics International, Inc., Fort Lauderdale.* SEA WATER—in part.

Chamberlain, Dr. Joseph W. *Department of Space Physics and Astronomy, Rice University.* AURORA.

Chapman, Dr. Sydney. *High Altitude Observatory, University of Colorado.* GEOMAGNETISM—in part.

Chapman, Dr. Valentine. *University of Auckland.* SALT-MARSH.

Charnock, Henry. *Department of Oceanography, University of Southampton.* WIND STRESS OVER SEA.

Clapp, Philip F. *Weather Bureau, U.S. Department of Commerce.* WEATHER; WEATHER MAP.

Clarke, Dr. George L. *Professor of Biology, Harvard University, and Marine Biologist, Woods Hole Oceanographic Institution.* SEA WATER—in part.

Cloud, Dr. Preston E., Jr. *Department of Geology, University of California, Santa Barbara.* REEF.

Coachman, Prof. Lawrence K. *Department of Oceanography, University of Washington, Seattle.* BERING SEA.

Cochrane, Dr. John D. *Department of Oceanography, Texas A&M University.* SEA WATER—in part.

Coroniti, S. C. *Climatic Impact Assessment Program, Office of the Secretary of Transportation, Washington, DC.* ATMOSPHERIC POLLUTION—in part.

Court, Dr. Arnold. *Professor of Climatology, Department of Geography, San Fernando Valley State College.* FROST; ISOTHERMAL CHART; SENSIBLE TEMPERATURE; TEMPERATURE INVERSION.

Cox, Dr. Charles S. *Department of Oceanography, University of California, La Jolla.* CAPILLARY WAVE; INTERNAL WAVE.

Craig, Dr. Harmon. *Professor of Geochemistry, Scripps Institution of Oceanography, La Jolla.* SEA WATER—in part.

Cramer, Dr. William S. *Ship Acoustics Department, Naval Ship Research and Development Center, Washington, DC.* UNDERWATER SOUND.

Crittenden, Dr. Charles V. *Geographer, Economic Development Administration, U.S. Department of Commerce.* AIR; CYCLONE.

Crow, Loren W. *Certified Consulting Meteorologist, Denver.* INDUSTRIAL METEOROLOGY.

Cruickshank, Michael J. *Geological Survey, Department of the Interior, Reston, VA.* MARINE MINING.

Cuckler, Lee E. *Interdivisional Coordinator, Aeronautical and Instrument Division, Robertshaw Controls Company, Anaheim, CA.* MOISTURE-CONTENT MEASUREMENT.

Currie, Ronald I. *Secretary, Scottish Marine Biological Association, and Director, Dunstaffnage Marine Research Laboratory, Oban, Scotland.* SEA-WATER FERTILITY—in part.

Cushing, Dr. D. H. *Fisheries Laboratory, Ministry of Agriculture, Fisheries and Food, Lowestoft, England.* SEA-WATER FERTILITY—in part.

D

Davies-Jones, Robert. *National Severe Storms Laboratory, National Oceanic and Atmospheric Administration, Norman, OK.* THUNDERSTORM—in part.

Deland, Dr. Raymond J. *Department of Meteorology and Oceanography, New York University.* AIR PRESSURE; AIR TEMPERATURE.

Dietrich, Prof. Gunter O. *Deceased; formerly, Institute of Oceanography, University of Kiel.* ATLANTIC OCEAN.

Dietz, Dr. Robert S. *Atlantic Oceanographic and Meteorological Laboratories, ESSA, U.S. Department of Commerce, Miami.* CONTINENTAL SHELF AND SLOPE.

Dill, Dr. Robert F. *Ocean Mining Administration, U.S. Department of the Interior.* DIVING; UNDERWATER PHOTOGRAPHY—in part.

Dinsmore, Robertson P. *International Ice Patrol, Woods Hole Oceanographic Institution.* ICEBERG.

Donn, Prof. William L. *Department of Geology, City College of New York.* SEICHE.

Doty, Dr. Robert. *Intermountain Forest and Range Experiment Station, Ogden, UT.* HYDROLOGY—in part.

Doviak, R. J. *National Severe Storms Laboratory, National Oceanic and Atmospheric Administration, Norman, OK.* STORM DETECTION—in part.

Duntley, Dr. Seibert Q. *Director, Visibility Laboratory, Scripps Institution of Oceanography, La Jolla.* SEA WATER—in part.

Dyer, Dr. K. R. *Institute of Oceanographic Sciences, Somerset, England.* ESTUARINE OCEANOGRAPHY.

E

Edgerton, Dr. Harold E. *Institute Professor, Emeritus, Massachusetts Institute of Technology.* UNDERWATER PHOTOGRAPHY—in part.

Ehricke, Dr. Krafft A. *Autonetics Division, North American Rockwell Corporation, Anaheim, CA.* SOLAR ENERGY.

Eliassen, Prof. Arnt. *Institute of Geophysics, University of Oslo.* METEOROLOGY—in part.

Emiliani, Dr. Cesare. *Rosenstiel School of Marine and Atmospheric Sciences.* MARINE SEDIMENTS—in part.

Ericson, David B. *Lamont-Doherty Geological Observatory, Palisades, NY.* MARINE SEDIMENTS—in part.

Evans, Dr. John W. *Director, Sacramento Peak Observatory, Air Force Cambridge Research Laboratories, Sunspot, NM.* SOLAR CONSTANT.

Evenson, Dr. Paul. *Fermi Institute, University of Chicago.* COSMIC RAYS.

Everetts, Dr. John, Jr. *Professor of Architectural Engineering, Pennsylvania State University.* PSYCHROMETRICS.

Ewing, Dr. Gifford C. *Woods Hole Oceanographic Institution.* SEA WATER—in part.

Ewing, John. *Senior Research Associate, Lamont-Doherty Geological Observatory, Palisades, NY.* MARINE GEOLOGY—in part.

Ewing, Dr. Maurice. *Deceased; formerly, Lamont-Doherty Geological Observatory, Palisades, NY.* MARINE GEOLOGY—in part.

F

Fisher, Dr. Frederick H. *Scripps Institution of Oceanography, La Jolla.* SEA WATER—in part.

Fox, Prof. Herbert. *Chairperson, Department of Mechanics and Aeronautics, New York Institute of Technology.* AIR-VELOCITY MEASUREMENT; ANEMOMETER.

Fritz, Dr. Sigmund. *Chief Space Scientist, National Environmental Satellite Center, Environmental Science Services Administration.* INSOLATION.

Fulks, J. R. *National Weather Service, Chicago.* DEW; DEW POINT; FOG; HAIL; HUMIDITY; PRECIPITATION; RAIN SHADOW; SNOW.

G

Gatzke, Irvin H. *Surveillance Technology Administrator, Naval Air Systems Command, Washington, DC.* SONOBUOY.

Georges, T. M. *Wave Propagation Laboratory, National Oceanic and Atmospheric Administration, Boulder.* STORM DETECTION—in part.

Godson, Dr. Warren L. *Meteorological Service of Canada, Toronto.* GEOPHYSICS—in part.

Goldberg, Dr. Edward D. *Scripps Institution of Oceanography, La Jolla.* HYDROSPHERE; HYDROSPHERIC GEOCHEMISTRY; SEA WATER—in part.

Goodell, Prof. H. G. *Department of Geology and Oceanography, Florida State University.* MARINE SEDIMENTS—in part.

Gordon, Dr. Arnold. *Lamont-Doherty Geological Observatory, Palisades, NY.* ANTARCTIC OCEAN.

Grobecker, Dr. Alan J. *Project Manager, Climatic Impact Assessment Program, Office of the Secretary of Transportation, Washington, DC.* ATMOSPHERIC POLLUTION—in part.

Groves, Dr. Gordon W. *Instituto de Geofisica, Torre de Ciencias, Ciudad Universitaria, Mexico.* TIDE.

H

Hainsworth, Bruce D. *Assistant to the President, Foxboro Company, Foxboro, MA.* BAROMETER.

Haltiner, Dr. George J. *U.S. Navy Post Graduate School, Monterey, CA.* NAVAL METEOROLOGY.

Hamilton, Dr. Edwin L. *U.S. Navy Electronics Laboratory, San Diego.* OCEANIC ISLANDS; SEAMOUNT AND GUYOT—coauthored.

Harley, Dr. John H. *Director, Health and Safety Laboratory, U.S. Atomic Energy Commission.* RADIOACTIVE FALLOUT.

Harrison, Henry T. *Consulting Meteorologist, Weaverville, NC.* AERONAUTICAL METEOROLOGY—in part.

Harrison, Louis P. *(Retired) Techniques Development Laboratory, Systems Development Office, Weather Bureau, ESSA, U.S. Department of Commerce.* ISOBAR.

Haurwitz, Dr. Bernhard. *Professor of Atmospheric Science, Colorado State University.* ATMOSPHERIC TIDES.

Heezen, Dr. Bruce C. *Lamont-Doherty Geological Observatory, Palisades, NY.* MARINE GEOLOGY—in part; TURBIDITY CURRENT.

Hela, Dr. Ilmo. *Institute of Marine Research, Helsinki.* BALTIC SEA.

Hersey, Dr. John B. *Office of Naval Research, U.S. Department of the Navy.* SCATTERING LAYER.

Hewson, Dr. E. W. *Chairperson, Department of Atmospheric Sciences, Oregon State University.* ATMOSPHERIC POLLUTION—in part; SMOG.

Holland, Dr. Heinrich D. *Department of Geological and Geophysical Sciences, Harvard University.* ATMOSPHERE, EVOLUTION OF.

Howell, Dr. Benjamin F., Jr. *Professor of Geophysics, Department of Geology and Geophysics, and Assistant Dean, Graduate School, Pennsylvania State University.* GEOPHYSICS—in part.

Hromadik, J. J. *Director, Amphibious and Harbor Division, U.S. Naval Civil Engineering Laboratory, Port Hueneme, CA.* OCEANOGRAPHIC PLATFORMS.

Hubert, Lester. *National Environmental Satellite Center, Environmental Science Services Administration, Suitland, MD.* METEOROLOGICAL SATELLITES.

Hull, Dr. McAllister H., Jr. *Department of Physics and Astronomy, State University of New York, Buffalo.* RADIATION.

Hunkins, Dr. Kenneth L. *Lamont-Doherty Geological Observatory, Palisades, NY.* ICE ISLAND.

Hunten, Dr. Donald M. *Kitt Peak National Observatory, Tucson.* AIRGLOW; ALKALI EMISSIONS.

I

Inman, Dr. Douglas L. *Professor of Oceanography, Scripps Institution of Oceanography, La Jolla.* NEAR-SHORE PROCESSES.

J

Jacobs, Dr. Woodrow C. *Senior Scientist, Ocean Data Systems, Inc., Rockville, MD.* CONTINENTALITY; MARINE INFLUENCE ON WEATHER AND CLIMATE; OCEAN-ATMOSPHERE RELATIONS.

Johnson, Dr. Francis S. *Acting President, University of Texas.* IONOSPHERE; UPPER-ATMOSPHERE DYNAMICS.

K

Kaplan, Prof. Lewis D. *Professor of Meteorology, University of Chicago.* GREENHOUSE EFFECT; TERRESTRIAL RADIATION—in part.

Ketchum, Dr. Bostwick H. *Woods Hole Oceanographic Institution.* SEA WATER—in part.

Kinsman, Blair. *College of Marine Studies, University of Delaware.* RIVER TIDES; TIDAL BORE.

Kutzbach, Prof. John E. *Department of Meteorology, University of Wisconsin.* CLIMATIC CHANGE—in part.

L

Ladd, Dr. Harry S. *National Museum, Washington, DC.* ATOLL.

LaFond, Dr. Eugene C. *Senior Scientist and Consultant for Oceanography, Naval Undersea Center, San Diego.* SEA WATER—in part; UPWELLING.

Landsberg, Dr. H. E. *Professor and Director, Institute for Fluid Dynamics and Applied Mathematics, University of Maryland.* Validator of BIOCLIMATOLOGY; CLIMATE MODIFICATION.

Leiby, Jonathan. *Woods Hole Oceanographic Institution.* OCEANOGRAPHIC VESSELS.

Leopold, Dr. Luna B. *U.S. Geological Survey.* SURFACE WATER.

Lettau, Prof. Heinz H. *Department of Meteorology, University of Wisconsin.* MICROMETEOROLOGY.

Linsley, Prof. Ray K. *Professor Emeritus of Civil Engineering, Stanford University.* ATMOSPHERIC WATER VAPOR; GROUNDWATER; HYDROLOGY—in part; WATER TABLE—in part.

Lorenz, Dr. Edward N. *Department of Meteorology, Massachusetts Institute of Technology.* WEATHER FORECASTING AND PREDICTION—in part.

Ludlam, Prof. Frank. *Deceased; formerly, Department of Meteorology, Imperial College, London.* CLOUD.

Lyman, Dr. John. *Professor of Oceanography, University of North Carolina, Chapel Hill.* MEDITERRANEAN SEA; OCEANS AND SEAS; SEA; SEA WATER—in part.

Lyon, Dr. Waldo K. *Arctic Submarine Research Laboratory, Naval Undersea Warfare Center, San Diego.* SEA ICE.

M

McHugh, Dr. J. L. *Professor of Marine Resources, Marine Sciences Research Center, State University of New York, Stony Brook.* MARINE FISHERIES.

McLellan, Dr. Hugh J. *National Science Foundation.* GULF OF MEXICO.

Marshall, Dr. Norman Bertram, Sr. *Senior Principal Scientific Officer, British Museum (Natural History), London.* DEEP-SEA FAUNA; MARINE BIOLOGICAL SAMPLING.

Mason, Dr. Basil J. *Director General, Meteorological Office, Bracknell, England.* CLOUD PHYSICS.

Menard, Dr. Henry W., Jr. *Scripps Institution of Oceanography, La Jolla.* OCEANIC ISLANDS; SEAMOUNT AND GUYOT—coauthored.

Metcalf, William G. *Woods Hole Oceanographic Institution.* CARIBBEAN SEA.

Miller, Elizabeth L. *Lamont-Doherty Geological Observatory, Palisades, NY.* OCEANS—coauthored.

Miller, Prof. Maynard M. *Professor of Geology, Michigan State University, and Director, Foundation for Glacial and Environmental Research, Seattle.* GLACIOLOGY; SNOW LINE; SNOWFIELD AND NÉVÉ; TERRESTRIAL FROZEN WATER.

Mitchell, Dr. J. Murray, Jr. *Senior Research Climatologist, National Oceanic and Atmospheric Administration, McLean, VA.* CLIMATOLOGY.

Morse, Dr. Robert W. *Associate Director and Dean of Oceanographic Studies, Woods Hole Oceanographic Institution.* ECHO SOUNDER; SOFAR; SONAR.

Murgatroyd, Dr. R. J. *Meteorological Office, Bracknell, England.* ATMOSPHERE; STRATOSPHERE; TROPOPAUSE; TROPOSPHERE.

N

Nafe, Dr. John E. *Lamont-Doherty Geological Observatory, Palisades, NY.* MARINE SEDIMENTS—in part.

Namais, Dr. Jerome. *Scripps Institution of Oceanography, La Jolla.* DROUGHT; WEATHER FORECASTING AND PREDICTION—in part.

Neumann, Dr. A. Conrad. *School of Marine and Atmospheric Sciences, University of Miami.* RED SEA—coauthored.

Neumann, Dr. P. Gerhard. *Professor of Oceanography, New York University.* OCEAN WAVES—in part.

Newcombe, Dr. Curtis L. *Director, San Francisco Bay Marine Research Center, San Francisco State College.* MARINE ECOSYSTEM.

Newton, Dr. Chester Whittier. *National Center for Atmospheric Research, Boulder.* ATMOSPHERIC GENERAL CIRCULATION; BACKING WIND; DUST STORM; SQUALL; SQUALL LINE; STORM; TORNADO; TROPICAL METEOROLOGY—in part; VEERING WIND; WATERSPOUT; WIND-SHIFT LINE.

Nierenberg, Prof. William A. *Director, Scripps Institution of Oceanography, La Jolla.* OCEANOGRAPHY—in part.

P

Patton, Dr. Donald. *Carnegie Institution of Washington.* RIVER—coauthored.

Pattullo, Dr. June G. *Department of Oceanography, Oregon State University.* SEA-LEVEL DATUM PLANES; SEA-LEVEL FLUCTUATIONS; SEA WATER—in part.

Payne, Harold G. *Manager of Municipal Electric Department, North Attleboro, MA.* BAROMETER.

Pearson, Allen. *Director, National Severe Storms Forecast Center, National Oceanic and Atmospheric Administration, Kansas City, MO.* WEATHER FORECASTING AND PREDICTION—in part.

Pierson, Dr. Willard J., Jr. *Department of Meteorology and Oceanography, New York University.* OCEAN WAVES—in part.

Priestley, Dr. Robert. *Department of Agriculture and Fisheries for Scotland, Marine Laboratory, Aberdeen.* UNDERWATER TELEVISION.

Prinn, Prof. Ronald G. *Department of Meteorology, Massachusetts Institute of Technology.* ATMOSPHERIC CHEMISTRY; ATMOSPHERIC OZONE.

Q

Quiroz, Dr. Roderick S. *Research Meteorologist, Upper Air Branch, National Oceanic and Atmospheric Administration, U.S. Department of Commerce.* MESOSPHERE.

R

Rasmusson, Dr. Eugene M. *Geophysical Fluid Dynamics Laboratory, Environmental Science Services Administration, Princeton.* HYDROMETEOROLOGY.

Raymond, Samuel O. *Department of Electrical Engineering and Computer Science, Massachusetts Institute of Technology.* UNDERWATER PHOTOGRAPHY — in part.

Reed, Prof. Richard J. *Department of Atmospheric Sciences, University of Washington, Seattle.* FRONT — in part.

Reid, Joseph L. *Scripps Institution of Oceanography, La Jolla.* PACIFIC OCEAN; SEA WATER — in part.

Reid, Prof. Robert O. *Department of Oceanography, Texas A&M University.* STORM SURGE.

Reiter, Prof. Elmar R. *Department of Atmospheric Science, Colorado State University.* AERONAUTICAL METEOROLOGY — in part; CLEAR-AIR TURBULENCE.

Richards, Dr. Francis A. *Department of Oceanography, University of Washington, Seattle.* BLACK SEA.

Riehl, Dr. Herbert. *Department of Meteorology, University of Chicago.* TROPICAL METEOROLOGY — in part.

Rochford, David J. *Division of Fisheries and Oceanography, Commonwealth Scientific and Industrial Research Organization, Australia.* BASS STRAIT.

Rosenberg, Norman J. *Department of Agricultural Meteorology, University of Nebraska.* AGRICULTURAL METEOROLOGY.

Ross, Dr. David A. *Woods Hole Oceanographic Institution.* RED SEA — coauthored.

Rudnick, Dr. Philip. *Scripps Institution of Oceanography, La Jolla.* SEA WATER — in part.

Rust, W. David. *National Severe Storms Laboratory, National Oceanic and Atmospheric Administration, Norman, OK.* THUNDERSTORM — in part.

Ryan, William B. F. *Lamont-Doherty Geological Observatory, Palisades, NY.* OCEANS — coauthored.

Ryther, Dr. John H. *Woods Hole Oceanographic Institution.* SEA-WATER FERTILITY — in part.

S

Salanave, Dr. Leon E. *Department of Physics and Astronomy, San Francisco State University.* LIGHTNING — in part.

Sanders, Dr. Frederick. *Department of Meteorology, Massachusetts Institute of Technology.* BAROCLINIC FIELD; BAROTROPIC FIELD; CHINOOK; CORIOLIS ACCELERATION AND FORCE; GEOSTROPHIC WIND; GRADIENT WIND; ISENTROPIC SURFACES; ISOPYCNIC; ISOTACH; JET STREAM; METEOROLOGICAL SOLENOID; MONSOON; SIROCCO; UPPER SYNOPTIC AIR WAVES; WIND; WIND ROSE; WIND STRESS.

Sayre, Dr. Albert N. *Deceased; formerly, Consulting Groundwater Geologist, Behre Dolbear and Company.* HYDROLOGY; WATER TABLE — both in part.

Schaefer, Dr. Karl E. *Chief, Physiological Sciences Division, U.S. Naval Submarine Center, Groton, CT.* DIVING PHYSIOLOGY.

Schmitt, Dr. Walter R. *Scripps Institution of Oceanography, La Jolla.* MARINE RESOURCES.

Schotland, Richard M. *Institute of Atmospheric Physics, University of Arizona.* PSYCHROMETER.

Schule, John Joseph, Jr. *Acting Director, Department of Marine Science, U.S. Naval Oceanographic Office.* HALOCLINE; THERMOCLINE.

Segeler, C. George. *Director, Technical Service, Dave Sage, Inc., New York.* DEGREE-DAY.

Sekera, Prof. Zdenek. *Deceased; formerly, Department of Meteorology, Institute or Geophysics and Planetary Physics, University of California, Los Angeles.* HALO; METEOROLOGICAL OPTICS; MIRAGE; RAINBOW; SUNDOG.

Shepard, Prof. Francis P. *Scripps Institution of Oceanography, La Jolla.* SUBMARINE CANYON.

Sheppard, Prof. P. A. *Deceased; formerly, Department of Meteorology, Imperial College, London.* METEOROLOGY — in part.

Showalter, Albert K. *Senior Scientist, Office of Hydrology, Environmental Science Services Administration, Weather Bureau.* WEATHER STATION.

Shrock, Prof. Robert R. *Department of Earth and Planetary Sciences, Massachusetts Institute of Technology.* EARTH SCIENCES.

Silberman, Dr. Edward. *Civil Engineering Department, University of Minnesota.* HYDROLOGY — in part.

Simpson, Dr. Joanne. *Department of Environmental Sciences, University of Virginia.* HURRICANE — in part.

Simpson, Prof. Robert H. *Research Professor, Department of Environmental Sciences, University of Virginia.* HURRICANE — in part.

Slichter, Prof. Louis B. *Deceased; formerly, Institute of Geophysics, University of California, Los Angeles.* EARTH TIDES.

Sloss, Dr. L. L. *Department of Geological Sciences, Northwestern University.* SALINE EVAPORITE.

Snodgrass, Frank E. *Institute of Geophysics and Planetary Physics, University of California, La Jolla.* OCEAN WAVES — in part.

Soberman, Dr. Robert K. *Manager, Meteor and Planetary Physics, General Electric Company, Valley Forge Space Center, Philadelphia.* NOCTILUCENT CLOUDS.

Stergis, Dr. Christos G. *Department of the Air Force, Los Angeles.* SFERICS.

Stommel, Dr. Henry. *Deceased; formerly, Woods Hole Oceanographic Institution.* GULF STREAM.

Streeter, Prof. Victor L. *Department of Civil Engineering, University of Michigan.* VORTEX.

Sturges, Dr. Wilton. *Department of Oceanography, Florida State University.* OCEAN CURRENTS.

Suomi, Prof. Verner E. *Department of Meteorology, University of Wisconsin.* METEOROLOGICAL INSTRUMENTATION; MILLIBAR; PRECIPITATION GAGES; SUNSHINE-DURATION TRANSMITTER; TERRESTRIAL RADIATION; WIND MEASUREMENT.

T

Tait, Dr. John B. *Deceased; formerly, Oceanographer, Marine Laboratory, Aberdeen, Scotland.* NORTH SEA.

Taylor, William L. *National Severe Storms Laboratory, National Oceanic and Atmospheric Administration, Norman, OK.* STORM DETECTION — in part.

Tully, Dr. John P. *Pacific Oceanographic Group, Fisheries Research Board of Canada, Nanaimo, British Columbia.* NORTHWEST PASSAGE.

Tyler, John E. *Scripps Institution of Oceanography, La Jolla.* SEA WATER — in part.

U

Untersteiner, Prof. Norbert. *Arctic Ice Dynamics Joint Experiment, Division of Marine Resources, University of Washington, Seattle.* POLAR METEOROLOGY.

V

Vadus, Joseph R. *National Oceanic and Atmospheric Administration, Rockville, MD.* OCEANOGRAPHIC SUBMERSIBLES — in part.

Van Andel, Dr. Tjeerd H. *Department of Oceanography, Oregon State University.* DELTA; MARINE SEDIMENTS — in part.

Van Dorn, William G. *Scripps Institution of Oceanography, La Jolla.* TSUNAMI.

Vaucouleurs, Prof. Gerard de. *Department of Astronomy, University of Texas.* ALBEDO.

Veronis, Dr. George. *Chairperson, Department of Geology, Yale University.* OCEANOGRAPHY — in part.

Vine, Allyn C. *Oceanographer, Woods Hole Oceanographic Institution.* SEA WATER — in part.

Vonder Haar, Prof. Thomas H. *Department of Atmospheric Sciences, Colorado State Universty.* TERRESTRIAL ATMOSPHERIC HEAT BALANCE.

Vonnegut, Dr. Bernard. *Department of Atmospheric Science, State University of New York, Albany.* ATMOSPHERIC ELECTRICITY.

W

Walt, Dr. Martin. *Research Laboratories, Lockheed Missiles and Space Company, Palo Alto, CA.* VAN ALLEN RADIATION.

Weaver, John. *National Severe Storms Laboratory, National Oceanic and Atmospheric Administration, Norman, OK.* STORM DETECTION — in part.

Webb, Willis L. *Chief, Meteorological Satellite TA, Atmospheric Sciences Laboratory, U.S. Army Electronic Command, White Sands, NM.* METEOROLOGICAL ROCKETS.

Weickmann, Dr. Helmut K. *Director, Atmospheric Physics and Chemistry Laboratory, Environmental Science Services Administration, Boulder.* WEATHER MODIFICATION.

Willett, Prof. Hurd C. *Department of Meteorology, Massachusetts Institute of Technology.* AIR MASS.

Williams, Dr. Roger T. *Department of Meteorology, Naval Postgraduate School, Monterey.* FRONT — in part.

Winchester, Prof. John W. *Chairperson, Department of Oceanography, Florida State University.* HALOGEN ATMOSPHERIC CHEMISTRY.

Wood, Dr. Edward J. F. *Institute of Marine Science, University of Miami.* MARINE MICROBIOLOGY.

Woodcock, Dr. Alfred H. *Institute of Geophysics, University of Hawaii.* SEA WATER — in part.

Wooster, Dr. Warren S. *Scripps Institution of Oceanography, La Jolla.* SEA WATER — in part.

Worthington, L. Valentine. *Woods Hole Oceanographic Institution.* SARGASSO SEA.

Wyrtki, Dr. Klaus. *Professor of Oceanography, University of Hawaii.* INDIAN OCEAN; SOUTHEAST ASIA WATERS.

Z

Zirin, Dr. Harold. *Hale Observatories.* SUN.

McGRAW-HILL
ENCYCLOPEDIA OF OCEAN AND ATMOSPHERIC SCIENCES

Asterisks indicate page references to article titles.